# Organometallic Chemistry of the Transition Elements

# MODERN INORGANIC CHEMISTRY

Series Editor: John P. Fackler, Jr.
*Texas A&M University*

CARBON-FUNCTIONAL ORGANOSILICON COMPOUNDS
Edited by Václav Chvalovský and Jon M. Bellama

GAS PHASE INORGANIC CHEMISTRY
Edited by David H. Russell

HOMOGENEOUS CATALYSIS
WITH METAL PHOSPHINE COMPLEXES
Edited by Louis H. Pignolet

THE JAHN-TELLER EFFECT AND
VIBRONIC INTERACTIONS IN MODERN CHEMISTRY
I. B. Bersuker

METAL INTERACTIONS WITH BORON CLUSTERS
Edited by Russell N. Grimes

MÖSSBAUER SPECTROSCOPY
APPLIED TO INORGANIC CHEMISTRY
*Volumes 1 and 2* • Edited by Gary J. Long
*Volume 3* • Edited by Gary J. Long and Fernande Grandjean

ORGANOMETALLIC CHEMISTRY OF THE
TRANSITION ELEMENTS
Florian P. Pruchnik
Translated from Polish by Stan A. Duraj

A Continuation Order Plan is available for this series. A continuation order will bring delivery of each new volume immediately upon publication. Volumes are billed only upon actual shipment. For further information please contact the publisher.

# Organometallic Chemistry of the Transition Elements

**Florian P. Pruchnik**
*University of Wroclaw*
*Wroclaw, Poland*

*Translated from Polish by*
**Stan A. Duraj**
*Cleveland State University*
*Cleveland, Ohio*

*PLENUM PRESS* • *NEW YORK AND LONDON*

Library of Congress Cataloging in Publication Data

Pruchnik, Florian P.
    Organometallic chemistry of the transition elements / Florian P. Pruchnik: trans-
lated from Polish by Stan A. Duraj.
        p.      cm. — (Modern inorganic chemistry.)
    Includes bibliographical references.
    ISBN 0-306-43192-0
    1. Organometallic chemistry. 2. Transition metal compounds. I. Title. II. Series.
QD4.11. — P78   1990                                                    89-71115
547′.056 — dc20                                                             CIP

© 1990 Plenum Press, New York
A Division of Plenum Publishing Corporation
233 Spring Street, New York, N.Y. 10013

Printed in the United States of America

# *Preface*

Organometallic chemistry belongs to the most rapidly developing area of chemistry today. This is due to the fact that research dealing with the structure of compounds and chemical bonding has been greatly intensified in recent years. Additionally, organometallic compounds have been widely utilized in catalysis, organic synthesis, electronics, etc.

This book is based on my lectures concerning basic organometallic chemistry for fourth and fifth year chemistry students and on my lectures concerning advanced organometallic chemistry and homogeneous catalysis for Ph.D. graduate students.

Many recent developments in the area of organometallic chemistry as well as homogeneous catalysis are presented. Essential research results dealing with a given class of organometallic compounds are discussed briefly. Results of physicochemical research methods of various organometallic compounds as well as their synthesis, properties, structures, reactivities, and applications are discussed more thoroughly. The selection of tabulated data is arbitrary because, often, it has been impossible to avoid omissions. Nevertheless, these data can be very helpful in understanding properties of organometallic compounds and their reactivities.

All physical data are given in SI units; the interatomic distances are given in pm units in figures and tables.

I am indebted to Professor S. A. Duraj for translating and editing this book. His remarks, discussions, and suggestions are greatly appreciated. I also express gratitude to Virginia E. Duraj for editing and proofreading.

I sincerely thank Professor A. Wojcicki for his comments, discussions, and editorial help. I also extend my thanks to Professors J. P. Fackler, Jr., F. A. Cotton, and R. A. Walton for their suggestions and remarks.

My special thanks go to the chemistry faculty of Cleveland State University, especially to Professors A. H. Andrist and R. L. R. Towns for their support throughout the translation of the book. I also sincerely thank Richelle P. Emery for her excellent word processing skills in connection with the preparation of the English manuscript.

The first Polish edition of this book appeared in 1984, and the second, expanded edition is now in press.

I should also like to acknowledge the comments and discussions of Professors S. Pasynkiewicz, A. Ratajczak, and W. Wojciechowski during the preparation of the Polish editions of this book.

Finally, I sincerely thank my wife, Halina, for typewriting, proofreading, and for all her help during the writing and preparation of the book.

F. Pruchnik

# Contents

*Chapter 1.  Introduction to Organometallic Chemistry*
1. Number of Valence Electrons. Characteristics of Ligands ................ 1
2. The Main Group Elements ............................................. 7
3. The Transition Metals ............................................... 11
4. Elementary Reactions of Organometallic Compounds .................... 15
5. Nomenclature of Organometallic Compounds .......................... 15
References ........................................................... 20


*Chapter 2.  Metal Carbonyls*
1. Preparation of Metal Carbonyls ..................................... 23
2. Bonding in Metal Carbonyls ........................................ 24
3. IR Spectra of Metal Carbonyls ..................................... 32
4. Electronic Spectra ................................................. 41
5. $^{13}$C NMR Spectra ................................................ 45
6. Photoelectron Spectroscopy ........................................ 49
7. Mass Spectrometry ................................................. 53
8. Properties of Metal Carbonyls ..................................... 55
   a. Group 3 Metal Carbonyls ....................................... 60
   b. Group 4 Metal Carbonyls ....................................... 63
   c. Group 5 Metal Carbonyls ....................................... 64
   d. Group 6 Metal Carbonyls ....................................... 69
   e. Group 7 Metal Carbonyls ....................................... 75
   f. Group 8 Metal Carbonyls ....................................... 79
   g. Group 9 Metal Carbonyls ....................................... 85
      Cluster Compounds of Co, Rh, and Ir ........................... 89
   h. Group 10 Metal Carbonyls ...................................... 92
      Metal Carbonyl Halides of Group 10 Metals ..................... 94
   i. Group 11 Metal Carbonyls ...................................... 95
9. Metal Carbonyl Hydrides ........................................... 96
   a. Bonding ....................................................... 96
   b. Methods of Identification ..................................... 97
   c. Preparation of Hydrido Complexes ............................. 104

  d. Acidic Properties of Metal Carbonyl Hydrides . . . . . . . . . . . . . . . . . . . . . . . . 107
  e. Ligand Field Strength and trans Effect of the Hydrido Ligand . . . . . . . . . . . 107
  f. Structures of Hydrido Complexes . . . . . . . . . . . . . . . . . . . . . . . . . . . . . . . . . 107
  g. Properties of Metal Carbonyl Hydrides . . . . . . . . . . . . . . . . . . . . . . . . . . . . 108
10. Metal Carbonyls Containing Bridging CO Groups Bonded Through C and O  110
11. Thiocarbonyl and Selenocarbonyl Complexes . . . . . . . . . . . . . . . . . . . . . . . . . 112
  a. Structure, Bonding, and Spectroscopy . . . . . . . . . . . . . . . . . . . . . . . . . . . . . 112
  b. Preparation of Thiocarbonyl and Selenocarbonyl Complexes . . . . . . . . . . . . 115
  c. Reactions of Thiocarbonyl Complexes . . . . . . . . . . . . . . . . . . . . . . . . . . . . . 116
12. Applications of Metal Carbonyl Complexes . . . . . . . . . . . . . . . . . . . . . . . . . . . 119
13. Applications of Organometallic Compounds for Preparation of Metals,
    Oxides, Carbides, and Nitrides, and for Deposition of Metal Layers and Their
    Compounds . . . . . . . . . . . . . . . . . . . . . . . . . . . . . . . . . . . . . . . . . . . . . . . . . . . 119
14. Metal Carbonyl Complexes Possessing Column Structures . . . . . . . . . . . . . . . . 121
References . . . . . . . . . . . . . . . . . . . . . . . . . . . . . . . . . . . . . . . . . . . . . . . . . . . . . . . . 122


*Chapter 3.   Metal–Metal Bonds and Clusters*
  1. Bonding and Structure . . . . . . . . . . . . . . . . . . . . . . . . . . . . . . . . . . . . . . . . . . . 129
    a. Theory of Skeletal Electron Pairs . . . . . . . . . . . . . . . . . . . . . . . . . . . . . . . . . 133
    b. The Isolobal Analogy of Fragments . . . . . . . . . . . . . . . . . . . . . . . . . . . . . . . 141
    c. Topological Calculations of CVMO . . . . . . . . . . . . . . . . . . . . . . . . . . . . . . . 147
    d. Reactions between Clusters . . . . . . . . . . . . . . . . . . . . . . . . . . . . . . . . . . . . . . 149
    e. Molecular Orbital Theory . . . . . . . . . . . . . . . . . . . . . . . . . . . . . . . . . . . . . . 150
  2. Clusters as Models of Metallic Lattices of Catalysts . . . . . . . . . . . . . . . . . . . . 155
  3. Interstitial (Encapsulated) Clusters . . . . . . . . . . . . . . . . . . . . . . . . . . . . . . . . . 157
  4. "Electron Poor" Transition Metal Clusters . . . . . . . . . . . . . . . . . . . . . . . . . . . 160
  5. NMR Spectra of Clusters . . . . . . . . . . . . . . . . . . . . . . . . . . . . . . . . . . . . . . . . . 163
  6. IR Spectra . . . . . . . . . . . . . . . . . . . . . . . . . . . . . . . . . . . . . . . . . . . . . . . . . . . . 165
  7. The M – M Bond Energy in Clusters . . . . . . . . . . . . . . . . . . . . . . . . . . . . . . . . 166
  8. Synthesis of Metal Clusters . . . . . . . . . . . . . . . . . . . . . . . . . . . . . . . . . . . . . . . 166
  9. Properties of Clusters . . . . . . . . . . . . . . . . . . . . . . . . . . . . . . . . . . . . . . . . . . . . 169
    a. Group 4 . . . . . . . . . . . . . . . . . . . . . . . . . . . . . . . . . . . . . . . . . . . . . . . . . . . . . 169
    b. Group 5 . . . . . . . . . . . . . . . . . . . . . . . . . . . . . . . . . . . . . . . . . . . . . . . . . . . . . 170
    c. Group 6 . . . . . . . . . . . . . . . . . . . . . . . . . . . . . . . . . . . . . . . . . . . . . . . . . . . . . 170
    d. Group 7 . . . . . . . . . . . . . . . . . . . . . . . . . . . . . . . . . . . . . . . . . . . . . . . . . . . . . 171
    e. Group 8 . . . . . . . . . . . . . . . . . . . . . . . . . . . . . . . . . . . . . . . . . . . . . . . . . . . . . 172
    f. Group 9 . . . . . . . . . . . . . . . . . . . . . . . . . . . . . . . . . . . . . . . . . . . . . . . . . . . . . 174
    g. Group 10 . . . . . . . . . . . . . . . . . . . . . . . . . . . . . . . . . . . . . . . . . . . . . . . . . . . . 175
    h. Group 11 . . . . . . . . . . . . . . . . . . . . . . . . . . . . . . . . . . . . . . . . . . . . . . . . . . . . 177
10. Reactions of Clusters . . . . . . . . . . . . . . . . . . . . . . . . . . . . . . . . . . . . . . . . . . . . 179
    a. Reactions Leading to Structural Changes in Clusters . . . . . . . . . . . . . . . . . . 180
    b. Migration of Ligands Within Clusters . . . . . . . . . . . . . . . . . . . . . . . . . . . . . 183
    c. Acid–Base Reactions . . . . . . . . . . . . . . . . . . . . . . . . . . . . . . . . . . . . . . . . . . 187
    d. Ligand Substitution . . . . . . . . . . . . . . . . . . . . . . . . . . . . . . . . . . . . . . . . . . . 190
    e. Oxidative Addition . . . . . . . . . . . . . . . . . . . . . . . . . . . . . . . . . . . . . . . . . . . . 190
References . . . . . . . . . . . . . . . . . . . . . . . . . . . . . . . . . . . . . . . . . . . . . . . . . . . . . . . . 193

*Chapter 4.   Compounds Containing 1e Carbon–Donor Ligands*
 1. Bonding in Compounds with Metal–Carbon σ Bonds..................... 200
 2. Stability of Compounds Containing 1e Ligands......................... 201
 3. Structure of Compounds Containing Metal–Carbon σ Bonds.............. 208
 4. IR and Raman Spectra................................................ 212
 5. NMR Spectra ....................................................... 213
 6. ESR Spectra......................................................... 215
 7. Photoelectron Spectra .............................................. 215
 8. Magnetic Properties................................................. 217
 9. Electronic Spectra................................................... 218
10. Preparation of Complexes Containing σ Bonded Carbon Ligands ........... 219
    a. Reactions of Transition Metal Compounds with Alkylating or Arylating
       Reagents ....................................................... 219
    b. Reactions Involving Oxidative Addition of Organic Halides............. 220
    c. Migration (Insertion) Reactions ................................... 222
    d. Reactions of Anionic Complexes with Organic Halides ................. 224
    e. Elimination of Neutral Group Connecting Carbon with Metal
       (Deinsertion) ................................................... 224
    f. Preparation of Acetylide Complexes................................ 225
    g. Photochemical Synthesis............................................ 225
    h. Electrochemical Synthesis.......................................... 226
11. Properties of Complexes Containing Metal–Carbon σ Bonds .............. 227
    a. Complexes of Group 3 Metals...................................... 227
    b. Complexes of Group 4 Metals...................................... 229
    c. Complexes of Group 5 Metals...................................... 232
    d. Complexes of Group 6 Metals...................................... 232
    e. Complexes of Group 7 Metals...................................... 234
    f. Complexes of Group 8 Metals...................................... 238
    g. Complexes of Group 9 Metals...................................... 242
    h. Complexes of Group 10 Metals..................................... 244
    i. Complexes of Group 11 Metals..................................... 249
12. Reactions of Complexes Containing Metal–Carbon σ Bonds............... 253
    a. The Breaking of the Metal–Carbon Bond ........................... 253
    b. Migration (Insertion) Reactions ................................... 255
    c. Oxidative Addition Reactions ..................................... 260
    d. The σ–π Rearrangement of Complexes Containing σ Metal–Carbon Bonds 261
13. Activation of Hydrocarbons ........................................... 262
References ............................................................... 269

*Chapter 5.   Carbene and Carbyne Complexes*
 1. Bonding.............................................................. 277
 2. Structures of Carbene and Carbyne Complexes ......................... 281
 3. IR Spectra .......................................................... 285
 4. NMR Spectra ....................................................... 285
 5. Photoelectron Spectra ............................................... 287
 6. Electronic Spectra.................................................... 290
 7. ESR and Magnetic Properties......................................... 291

8. Methods of Preparation of Carbene Complexes.......................... 292
   a. From Metal Carbonyls.............................................. 292
   b. Nucleophilic Addition to Isocyanide Ligands ......................... 292
   c. Hydrogen Atom Abstraction from $\sigma$ Bonded Carbon Ligands ............ 294
   d. Alkylation of Complexes with Some $M-C \sigma$ Bonds .................. 295
   e. Electrophilic Addition to Coordinated Ligands........................ 295
   f. Reactions of Complexes with Mixtures of Acetylenes and Alcohols ........ 295
   g. Synthesis from Ionic Complexes .................................... 295
   h. Oxidative Addition Reactions ...................................... 296
   i. Reactions of Complexes with Neutral Carbene Precursors .............. 297
   j. Nucleophilic Addition to Carbyne Complexes......................... 298
   k. Oxidation Reactions .............................................. 298
   l. Addition to Carbonyl Ligands...................................... 298
   m. Synthesis of $\mu$-Methylene Complexes ............................. 299
9. Methods of Preparation of Carbyne Complexes......................... 300
   a. Removal of Alkoxy, Alkylthio, and Amino Groups from the Carbene
      Ligand ......................................................... 300
   b. Hydrogen Elimination from the Carbene Ligand ..................... 300
   c. Cleavage of $C \equiv C$ Bonds in Acetylenes and $C \equiv N$ Bonds in Cyanides...... 300
   d. Reactions of Carbyne Complexes with Anionic Transition Metal
      Complexes ...................................................... 301
   e. Spontaneous Carbene Ligand Rearrangement......................... 301
   f. Reactions of Acyl Complexes with Dibromotriphenyl Phosphorus(V) ..... 301
   g. Oxygen Abstraction from Acyl Ligands ............................. 306
   h. Protonation of Isocyanide Ligands................................. 306
   i. Miscellaneous Methods of Synthesis of Carbyne Complexes............. 306
10. Properties of Carbene and Carbyne Complexes ......................... 307
11. Reactions of Carbene Complexes ..................................... 308
    a. Nucleophilic Substitution Reactions ................................ 308
    b. Rearrangement Reactions Caused by Addition ....................... 311
    c. Hydrogen Substitution at $\alpha-C$................................. 312
    d. Reduction of Carbene Complexes .................................. 312
    e. Oxidation Reactions .............................................. 313
    f. Reactions with Electrophiles ...................................... 314
    g. Substitution Reactions of Noncarbene Ligands....................... 314
    h. Release of Carbene Ligands ....................................... 315
    i. Migration Reactions .............................................. 317
12. Reaction of Carbyne Complexes....................................... 317
    a. Reactions with Nucleophiles ...................................... 317
    b. Reactions Leading to Bridging Carbyne Complexes................... 318
    c. Reactions of Carbyne Complexes with CO and $CO_2$ ................. 318
    d. Removal of Carbyne Ligands....................................... 319
    e. Exchange Reactions Involving Noncarbyne Ligands .................. 320
    f. Reactions with Acids.............................................. 320
    g. Miscellaneous Reactions.......................................... 320
13. Reactions of Complexes Containing Methylene Bridges .................. 320
14. Applications of Carbene and Carbyne Complexes ...................... 322
References ............................................................. 324

*Chapter 6. Compounds Containing Two-Electron π-Ligands*

A. Olefin Complexes

1. Bonding and Structure.............................................. 327
2. IR Spectra of Olefin Complexes ..................................... 333
3. ¹H NMR Spectra .................................................... 335
4. ¹³C NMR Spectra.................................................... 337
5. Photoelectron Spectra .............................................. 339
6. Electronic Spectra.................................................. 341
7. Stability of Olefin Complexes ....................................... 343
8. Methods of Synthesis of Olefin Compounds........................... 347
   a. Olefin Addition to the Transition Metal Salts....................... 347
   b. Addition of Olefins to Coordination Compounds of the Transition Metals 347
   c. Substitution of Coordinated Ligands with Olefins .................... 347
   d. Reduction in the Presence of Olefins............................... 349
   e. Preparation of Olefin Complexes from Hydrocarbon Ligands Coordinated
      to the Metal .................................................... 350
   f. Exchange Reactions.............................................. 350
   g. Dehydration of Alcohols.......................................... 351
   h. Electrochemical Syntheses ........................................ 351
9. Properties of Transition Metal Olefin Complexes ...................... 351
   a. Group 4 Metal Complexes......................................... 352
   b. Group 5 Metal Complexes......................................... 352
   c. Group 6 Metal Complexes......................................... 353
   d. Group 7 Metal Complexes......................................... 356
   e. Group 8 Metal Complexes......................................... 358
   f. Group 9 Metal Complexes......................................... 361
   g. Group 10 Metal Complexes........................................ 368
   h. Group 11 Metal Complexes........................................ 377
10. Complexes Containing Heteroolefins................................. 379
11. Applications of π-Olefin Complexes.................................. 381
    a. Reactions of Olefin Complexes with Nucleophiles .................. 381
    b. Reactions of Olefin Complexes with Electrophiles .................. 384
    c. Amination of Olefins............................................. 386
    d. Olefin Exchange Reactions........................................ 386
    e. Substitution of Olefin Ligands.................................... 386
    f. Transformation of Olefin Complexes into Vinyl Compounds .......... 386
    g. Hydrogen/Deuterium Exchange in Olefin Compounds ................ 387
    h. Photochemical Reactions ......................................... 388

B. Acetylene Complexes

1. Bonding and Structure.............................................. 389
2. IR Spectra ........................................................ 392
3. NMR Spectra ...................................................... 394
4. Methods of Preparation of Acetylene Complexes....................... 396
   a. Alkyne Addition to the Transition Metal Salts and Complexes .......... 396
   b. Ligand Substitution in Coordination Compounds .................... 397
5. Properties of Acetylene Complexes................................... 398
6. Reactions of Acetylene Complexes ................................... 404
7. Metallacyclization Reactions........................................ 409

8. Complexes Containing $R - C \equiv Y$ and $R - Y \equiv C$ Ligands . . . . . . . . . . . . . . . . . . 412

C. Allene Complexes

1. Bonding and Structure . . . . . . . . . . . . . . . . . . . . . . . . . . . . . . . . . . . . . . . . . . . . . . 415
2. IR and NMR Spectra. Reactions of Allene Complexes . . . . . . . . . . . . . . . . . . . . 417

References . . . . . . . . . . . . . . . . . . . . . . . . . . . . . . . . . . . . . . . . . . . . . . . . . . . . . . . . . . . . 421

*Chapter 7. Complexes Containing Three-Electron π-Ligands*

1. The Metal–Allyl Bond . . . . . . . . . . . . . . . . . . . . . . . . . . . . . . . . . . . . . . . . . . . . . . 427
2. IR Spectra of Allyl Complexes . . . . . . . . . . . . . . . . . . . . : . . . . . . . . . . . . . . . . . . . 431
3. NMR Spectra . . . . . . . . . . . . . . . . . . . . . . . . . . . . . . . . . . . . . . . . . . . . . . . . . . . . . 432
4. Electronic Spectra . . . . . . . . . . . . . . . . . . . . . . . . . . . . . . . . . . . . . . . . . . . . . . . . . 439
5. Structural Properties . . . . . . . . . . . . . . . . . . . . . . . . . . . . . . . . . . . . . . . . . . . . . . . 441
6. Photoelectron Spectra . . . . . . . . . . . . . . . . . . . . . . . . . . . . . . . . . . . . . . . . . . . . . . 441
7. Methods of Preparation of π-Allyl Complexes . . . . . . . . . . . . . . . . . . . . . . . . . . . 444
   a. Reactions of Transition Metal Complexes with Allyl Compounds of Main Group Metals . . . . . . . . . . . . . . . . . . . . . . . . . . . . . . . . . . . . . . . . . . . . . . . . . 444
   b. Reactions of Allyl Halides with Transition Metal Complexes in the Presence of Reducing Agents . . . . . . . . . . . . . . . . . . . . . . . . . . . . . . . . . . . . . . 444
   c. Oxidative Addition of Allyl Halides . . . . . . . . . . . . . . . . . . . . . . . . . . . . . . . . 445
   d. Reactions of Olefins with Metal Complexes . . . . . . . . . . . . . . . . . . . . . . . . . . 445
   e. Reactions of 1,3-Dienes with Metal Compounds . . . . . . . . . . . . . . . . . . . . . . . 446
   f. Rearrangements of σ-Allyl Compounds to π-Allyl Compounds . . . . . . . . . . . 447
   g. Synthesis of Metal Complexes Containing Chelated Allyl Ligands . . . . . . . . . 447
   h. Miscellaneous Methods . . . . . . . . . . . . . . . . . . . . . . . . . . . . . . . . . . . . . . . . . . 448
8. Properties of π-Allyl Metal Compounds . . . . . . . . . . . . . . . . . . . . . . . . . . . . . . . . 450
   a. Complexes of Group 4 Metals . . . . . . . . . . . . . . . . . . . . . . . . . . . . . . . . . . . . . 450
   b. Complexes of Group 5 Metals . . . . . . . . . . . . . . . . . . . . . . . . . . . . . . . . . . . . . 451
   c. Complexes of Group 6 Metals . . . . . . . . . . . . . . . . . . . . . . . . . . . . . . . . . . . . . 451
   d. Complexes of Group 7 Metals . . . . . . . . . . . . . . . . . . . . . . . . . . . . . . . . . . . . . 453
   e. Complexes of Group 8 Metals . . . . . . . . . . . . . . . . . . . . . . . . . . . . . . . . . . . . . 454
   f. Complexes of Group 9 Metals . . . . . . . . . . . . . . . . . . . . . . . . . . . . . . . . . . . . . 458
   g. Complexes of Group 10 Metals . . . . . . . . . . . . . . . . . . . . . . . . . . . . . . . . . . . . 459
   h. Complexes of Group 3 Metals . . . . . . . . . . . . . . . . . . . . . . . . . . . . . . . . . . . . . 460
9. Cyclopropenyl Complexes . . . . . . . . . . . . . . . . . . . . . . . . . . . . . . . . . . . . . . . . . . . 461
10. Heteroallyl Complexes . . . . . . . . . . . . . . . . . . . . . . . . . . . . . . . . . . . . . . . . . . . . . . 462
11. Reactions of Allyl Complexes . . . . . . . . . . . . . . . . . . . . . . . . . . . . . . . . . . . . . . . . 463
    a. Thermal Decomposition . . . . . . . . . . . . . . . . . . . . . . . . . . . . . . . . . . . . . . . . . 463
    b. Interactions with Nucleophilic Reagents; Oxidative Hydrolysis . . . . . . . . . . . 464
    c. Electrophilic Attack on Allyl Groups . . . . . . . . . . . . . . . . . . . . . . . . . . . . . . . 466
    d. Formation of σ-Allyl Complexes . . . . . . . . . . . . . . . . . . . . . . . . . . . . . . . . . . . 466
    e. Reduction . . . . . . . . . . . . . . . . . . . . . . . . . . . . . . . . . . . . . . . . . . . . . . . . . . . . . 466
    f. Reductive Elimination . . . . . . . . . . . . . . . . . . . . . . . . . . . . . . . . . . . . . . . . . . . 467
    g. Migration (Insertion) Reactions of Ligands . . . . . . . . . . . . . . . . . . . . . . . . . . . 467

References . . . . . . . . . . . . . . . . . . . . . . . . . . . . . . . . . . . . . . . . . . . . . . . . . . . . . . . . . . . . 468

*Chapter 8. Compounds Containing Four-Electron π-Ligands*

1. The Bonding . . . . . . . . . . . . . . . . . . . . . . . . . . . . . . . . . . . . . . . . . . . . . . . . . . . . . 471
2. Structure of Compounds with 4e Ligands . . . . . . . . . . . . . . . . . . . . . . . . . . . . . . 476

3. NMR Spectra ........................................................... 480
4. Photoelectron Spectra .................................................. 485
5. IR Spectra ............................................................. 485
6. Methods of Preparation of 1,3-Diene Complexes ......................... 485
   a. Reaction of Coordination Compounds with Dienes .................... 485
   b. Reactions Involving Nucleophilic Attack on Diene Complexes ........ 488
7. Methods of Preparation of Cyclobutadiene Complexes .................... 489
8. Methods of Preparation of Trimethylenemethane Complexes .............. 496
9. Reactions of 1,3-Diene Complexes ...................................... 498
10. Reactions of Cyclobutadiene Complexes ................................ 500
    a. Thermal Decomposition ............................................. 500
    b. Ligand Substitution Reactions ..................................... 500
    c. Reactions Involving the Cyclobutadiene Ligand .................... 501
       Nucleophilic Attack ............................................... 501
       Electrophilic Attack .............................................. 502
    d. Oxidation Reactions ............................................... 502
    e. Reduction Reactions ............................................... 503
11. Reactions of Trimethylenemethane Complexes ........................... 504
12. Properties of Diene Complexes ........................................ 504
References .............................................................. 505

*Chapter 9.   Compounds Containing Five-Electron π-Ligands*
1. The Metal–Cyclopentadienyl Bond ...................................... 510
2. Structure of Cyclopentadienyl Complexes .............................. 518
3. NMR Spectra ........................................................... 520
4. IR Spectra ............................................................. 522
5. Magnetic Properties ................................................... 524
6. EPR Spectra ........................................................... 525
7. Electronic Spectra .................................................... 525
8. Photoelectron Spectra ................................................. 527
9. Thermochemistry of Cyclopentadienyl Complexes ........................ 528
10. Mass Spectra ......................................................... 529
11. Methods of Preparation of Cyclopentadienyl Complexes ................. 530
    a. Reactions Utilizing Cyclopentadiene in the Presence of Bases or
       Cyclopentadienyl Compounds of the Main Group Elements ........... 530
    b. Syntheses Utilizing Fulvenes ..................................... 531
    c. Reactions of Transition Metal Compounds with Unsaturated
       Hydrocarbons ...................................................... 532
    d. Syntheses Utilizing Cyclopentadiene .............................. 533
    e. Reactions of Dienes with Metal Vapors ............................ 533
    f. Preparation of Triple-Decker Cyclopentadienyl Complexes .......... 534
    g. Preparation of Compounds Containing Bridging Cyclopentadienyl Ligands 536
    h. Synthesis of Trinuclear $M-P(O)(OR)_2$–Cyclopentadienyl Complexes ..... 536
    i. Electrochemical Synthesis ........................................ 537
    j. Synthesis of Dienyl Complexes .................................... 537
12. Properties of Cyclopentadienyl Complexes ............................ 538
    a. Group 4 Metal Complexes .......................................... 538
    b. Group 5 Metal Complexes .......................................... 541

c. Group 6 Metal Complexes . . . . . . . . . . . . . . . . . . . . . . . . . . . . . . . . . . . . . . . . . . 543
d. Group 7 Metal Complexes . . . . . . . . . . . . . . . . . . . . . . . . . . . . . . . . . . . . . . . . . . 545
e. Group 8 Metal Complexes . . . . . . . . . . . . . . . . . . . . . . . . . . . . . . . . . . . . . . . . . . 545
f. Group 9 Metal Complexes . . . . . . . . . . . . . . . . . . . . . . . . . . . . . . . . . . . . . . . . . . 548
g. Group 10 Metal Complexes . . . . . . . . . . . . . . . . . . . . . . . . . . . . . . . . . . . . . . . . . 550
h. Group 11 Metal Complexes . . . . . . . . . . . . . . . . . . . . . . . . . . . . . . . . . . . . . . . . . 551
i. Group 3 Metal Complexes Including Compounds of the Lanthanides and
   the Actinides. . . . . . . . . . . . . . . . . . . . . . . . . . . . . . . . . . . . . . . . . . . . . . . . . . . . . 551
13. Polynuclear and Cluster Cyclopentadienyl Complexes . . . . . . . . . . . . . . . . . . . . 554
14. Pentadienyl Complexes . . . . . . . . . . . . . . . . . . . . . . . . . . . . . . . . . . . . . . . . . . . . . 554
15. Pyrrole Complexes . . . . . . . . . . . . . . . . . . . . . . . . . . . . . . . . . . . . . . . . . . . . . . . . 555
16. Multidecker Complexes . . . . . . . . . . . . . . . . . . . . . . . . . . . . . . . . . . . . . . . . . . . . 556
17. Reactions of Cyclopentadienyl Compounds . . . . . . . . . . . . . . . . . . . . . . . . . . . . 557
    a. Oxidation of Metallocenes and Electron Transfer Reactions. . . . . . . . . . . . . 557
    b. Reduction of Cyclopentadienyl Complexes . . . . . . . . . . . . . . . . . . . . . . . . . . 558
    c. Protonation . . . . . . . . . . . . . . . . . . . . . . . . . . . . . . . . . . . . . . . . . . . . . . . . . . . 559
    d. The M — cp Bond Rupture . . . . . . . . . . . . . . . . . . . . . . . . . . . . . . . . . . . . . . . 559
    e. Metallation Reactions. . . . . . . . . . . . . . . . . . . . . . . . . . . . . . . . . . . . . . . . . . . 561
    f. Aromatic Substitution Reactions. . . . . . . . . . . . . . . . . . . . . . . . . . . . . . . . . . . 563
    g. Preparation of Polymetallocenes . . . . . . . . . . . . . . . . . . . . . . . . . . . . . . . . . . 564
    h. Ligand Substitution Reactions. . . . . . . . . . . . . . . . . . . . . . . . . . . . . . . . . . . . 564
    i. Reactions with Substituents Containing Various Functional Groups. . . . . . . 564
    j. Activation of C — H Bonds . . . . . . . . . . . . . . . . . . . . . . . . . . . . . . . . . . . . . . 565
    k. Photochemistry of Cyclopentadienyl Complexes. . . . . . . . . . . . . . . . . . . . . . 565
18. Application of Cyclopentadienyl Complexes. . . . . . . . . . . . . . . . . . . . . . . . . . . . . 565
References . . . . . . . . . . . . . . . . . . . . . . . . . . . . . . . . . . . . . . . . . . . . . . . . . . . . . . . . . . . 569

*Chapter 10. Complexes Containing Six-Electron π-Ligands*
1. The Bonding . . . . . . . . . . . . . . . . . . . . . . . . . . . . . . . . . . . . . . . . . . . . . . . . . . . . . . 575
2. Electronic Spectra of Arene Complexes . . . . . . . . . . . . . . . . . . . . . . . . . . . . . . . . 578
3. Magnetic Properties . . . . . . . . . . . . . . . . . . . . . . . . . . . . . . . . . . . . . . . . . . . . . . . . 578
4. IR Spectra . . . . . . . . . . . . . . . . . . . . . . . . . . . . . . . . . . . . . . . . . . . . . . . . . . . . . . . . 579
5. NMR Spectra . . . . . . . . . . . . . . . . . . . . . . . . . . . . . . . . . . . . . . . . . . . . . . . . . . . . . 579
6. Photoelectron Spectra . . . . . . . . . . . . . . . . . . . . . . . . . . . . . . . . . . . . . . . . . . . . . . 580
7. ESR Spectra. . . . . . . . . . . . . . . . . . . . . . . . . . . . . . . . . . . . . . . . . . . . . . . . . . . . . . . 582
8. Structure of Arene Complexes. . . . . . . . . . . . . . . . . . . . . . . . . . . . . . . . . . . . . . . . 583
9. Thermochemistry of Arene Complexes. . . . . . . . . . . . . . . . . . . . . . . . . . . . . . . . . 584
10. Mass Spectra . . . . . . . . . . . . . . . . . . . . . . . . . . . . . . . . . . . . . . . . . . . . . . . . . . . . . 585
11. Properties of Arene Compounds. . . . . . . . . . . . . . . . . . . . . . . . . . . . . . . . . . . . . . 586
12. Asymmetric Arene Complexes. . . . . . . . . . . . . . . . . . . . . . . . . . . . . . . . . . . . . . . . 588
13. Triple Layer and Heteroarene Complexes. . . . . . . . . . . . . . . . . . . . . . . . . . . . . . . 588
14. Methods of Preparation of Arene Complexes . . . . . . . . . . . . . . . . . . . . . . . . . . . . 590
    a. Reactions of Transition Metal Compounds with Aluminum and Aluminum
       Chloride . . . . . . . . . . . . . . . . . . . . . . . . . . . . . . . . . . . . . . . . . . . . . . . . . . . . . . 590
    b. Reactions of Salts of Metals with Grignard Compounds . . . . . . . . . . . . . . . . 590
    c. Cyclotrimerization of Acetylenes . . . . . . . . . . . . . . . . . . . . . . . . . . . . . . . . . . 591
    d. Dehydrogenation of 1,3-Cyclohexadiene . . . . . . . . . . . . . . . . . . . . . . . . . . . . 592
    e. Substitution of Ligands. . . . . . . . . . . . . . . . . . . . . . . . . . . . . . . . . . . . . . . . . . 592

    f.  Reactions Involving Rearrangement of Ligands ....................... 593
    g.  Condensation of Metal Vapors with Ligands. ....................... 594
    h.  Miscellaneous Reactions. ......................................... 594
  15.  Reactions of Arene Complexes ...................................... 594
    a.  Redox Reactions ................................................. 594
    b.  Electrophilic Substitution Reactions. ............................. 596
    c.  Nucleophilic Substitution and Addition Reactions ................... 596
    d.  Displacement of Ligands. ......................................... 598
  16.  Application of Arene Complexes. .................................... 599
References ............................................................. 600

*Chapter 11.  Complexes Containing Seven- and Eight-Electron π-Ligands*
  1.  Bonding, Structure, Spectroscopy, and Magnetic Properties .............. 603
  2.  Properties of Cycloheptatrienyl and Cyclooctatetraene
     Complexes ....................................................... 607
  3.  Methods of Preparation of Cycloheptatrienyl and Cyclooctatetraene
     Complexes ....................................................... 609
  4.  Reactions of Cycloheptatrienyl and Cyclooctatetraene Complexes .......... 611
    a.  Reactions with Nucleophiles ...................................... 611
    b.  Oxidation and Reduction Reactions ................................ 613
    c.  Ligand Exchange Reactions in Complexes Containing CHT. ............. 614
    d.  The $M-C_nH_n$ Bond Breaking. .................................... 614
    e.  Metallation Reactions. ........................................... 615
References ............................................................. 616

*Chapter 12.  Isocyanide Complexes*
  1.  Bonding and Structure. ............................................ 617
  2.  Electronic Spectra. ............................................... 620
  3.  NMR Spectra ..................................................... 621
  4.  IR Spectra ....................................................... 622
  5.  ESR Spectra and Magnetic Properties. ............................... 624
  6.  Ultraviolet and X-Ray Photoelectron Spectra ........................ 624
  7.  Properties of Isocyanide Complexes. ................................ 624
    a.  Group 3 Metal Complexes. ......................................... 624
    b.  Group 4 Metal Complexes. ......................................... 625
    c.  Group 5 Metal Complexes. ......................................... 626
    d.  Group 6 Metal Complexes. ......................................... 626
    e.  Group 7 Metal Complexes. ......................................... 626
    f.  Group 8 Metal Complexes. ......................................... 627
    g.  Group 9 Metal Complexes. ......................................... 627
    h.  Group 10 Metal Complexes. ........................................ 628
    i.  Group 11 Metal Complexes. ........................................ 629
  8.  Methods of Synthesis of Isocyanide Complexes ....................... 629
    a.  Preparations Involving Metal Cyanides and Cyanide Metal Complexes.... 629
    b.  Reactions of Isocyanides with Transition Metal Compounds ............ 631
    c.  Oxidation and Reduction Reactions. ............................... 632
  9.  Reactions of Isocyanide Complexes ................................. 632
    a.  Cleavage of the $M-CNR$ Bond ..................................... 632

    b. Oxidation–Reduction Reactions . . . . . . . . . . . . . . . . . . . . . . . . . . . . . . . . .  633
    c. Oxidative Addition Reactions . . . . . . . . . . . . . . . . . . . . . . . . . . . . . . . . . . .  634
    d. Nucleophilic Addition. . . . . . . . . . . . . . . . . . . . . . . . . . . . . . . . . . . . . . . . .  634
    e. Formation of Metal Carbyne Complexes. . . . . . . . . . . . . . . . . . . . . . . . . .  635
    f. Insertion Reactions . . . . . . . . . . . . . . . . . . . . . . . . . . . . . . . . . . . . . . . . . . .  636
    g. Dimerization of Isocyanides. . . . . . . . . . . . . . . . . . . . . . . . . . . . . . . . . . . .  639
10. Applications of Isocyanide Complexes. . . . . . . . . . . . . . . . . . . . . . . . . . . . . .  640
References . . . . . . . . . . . . . . . . . . . . . . . . . . . . . . . . . . . . . . . . . . . . . . . . . . . . . . . .  641

*Chapter 13.*   *Application of Organometallic Compounds in Homogeneous Catalysis*
  1. Basic Principles. . . . . . . . . . . . . . . . . . . . . . . . . . . . . . . . . . . . . . . . . . . . . . . . .  647
    a. Migration Reactions . . . . . . . . . . . . . . . . . . . . . . . . . . . . . . . . . . . . . . . . . .  652
    b. Oxidative Coupling Reactions . . . . . . . . . . . . . . . . . . . . . . . . . . . . . . . . . .  654
    c. Oxidative Addition Reactions . . . . . . . . . . . . . . . . . . . . . . . . . . . . . . . . . .  654
  2. Hydrogenation . . . . . . . . . . . . . . . . . . . . . . . . . . . . . . . . . . . . . . . . . . . . . . . . .  655
    a. Olefin Hydrogenation . . . . . . . . . . . . . . . . . . . . . . . . . . . . . . . . . . . . . . . .  655
    b. Hydrogenation of Aromatic Compounds. . . . . . . . . . . . . . . . . . . . . . . . .  662
    c. Asymmetric Hydrogenation . . . . . . . . . . . . . . . . . . . . . . . . . . . . . . . . . . .  662
    d. Hydrogenation of the $C=O$, $C=N$, and $C\equiv N$ Bonds. . . . . . . . . . . . . . . . . .  667
  3. Isomerization of Olefins. . . . . . . . . . . . . . . . . . . . . . . . . . . . . . . . . . . . . . . . . .  670
    a. Migration of the Double Bond . . . . . . . . . . . . . . . . . . . . . . . . . . . . . . . . .  670
    b. *Cis–Trans* Isomerization . . . . . . . . . . . . . . . . . . . . . . . . . . . . . . . . . . . . . .  673
  4. Oligomerization and Polymerization of Olefins and Acetylenes. . . . . . . . . . . . .  674
    a. Polymerization . . . . . . . . . . . . . . . . . . . . . . . . . . . . . . . . . . . . . . . . . . . . .  674
    b. Oligomerization . . . . . . . . . . . . . . . . . . . . . . . . . . . . . . . . . . . . . . . . . . . .  680
  5. Carbonylation Reactions . . . . . . . . . . . . . . . . . . . . . . . . . . . . . . . . . . . . . . . . .  690
    a. Hydroformylation Reactions . . . . . . . . . . . . . . . . . . . . . . . . . . . . . . . . . . .  690
    b. Carbonylation Reactions . . . . . . . . . . . . . . . . . . . . . . . . . . . . . . . . . . . . . .  694
    c. The Stability of Phosphine-Containing Catalysts for Hydroformylation . . . .  701
  6. Hydrocyanation . . . . . . . . . . . . . . . . . . . . . . . . . . . . . . . . . . . . . . . . . . . . . . . .  702
  7. Hydrosilation . . . . . . . . . . . . . . . . . . . . . . . . . . . . . . . . . . . . . . . . . . . . . . . . . .  703
  8. Metathesis . . . . . . . . . . . . . . . . . . . . . . . . . . . . . . . . . . . . . . . . . . . . . . . . . . . .  704
    a. Olefin Metathesis. . . . . . . . . . . . . . . . . . . . . . . . . . . . . . . . . . . . . . . . . . . .  704
    b. Mechanism of the Metathesis Reaction . . . . . . . . . . . . . . . . . . . . . . . . . . .  707
    c. Metathesis of Functionalized Olefins . . . . . . . . . . . . . . . . . . . . . . . . . . . .  713
    d. Metathesis of Alkynes. . . . . . . . . . . . . . . . . . . . . . . . . . . . . . . . . . . . . . . .  713
  9. Disproportionation Reactions. . . . . . . . . . . . . . . . . . . . . . . . . . . . . . . . . . . . . .  713
10. Reduction of Carbon Monoxide; the Fischer–Tropsch Synthesis. . . . . . . . . . . .  714
    a. Mechanism of the Fischer–Tropsch Reaction . . . . . . . . . . . . . . . . . . . . . .  716
11. Reactions of Carbon Dioxide . . . . . . . . . . . . . . . . . . . . . . . . . . . . . . . . . . . . . .  725
12. Water–Gas Shift Reaction . . . . . . . . . . . . . . . . . . . . . . . . . . . . . . . . . . . . . . . . .  729
13. Amination and Amoxidation Reactions. . . . . . . . . . . . . . . . . . . . . . . . . . . . . . .  731
14. Heterogenized Homogeneous Catalysts. . . . . . . . . . . . . . . . . . . . . . . . . . . . . . .  732
References . . . . . . . . . . . . . . . . . . . . . . . . . . . . . . . . . . . . . . . . . . . . . . . . . . . . . . . .  736

Abbreviations . . . . . . . . . . . . . . . . . . . . . . . . . . . . . . . . . . . . . . . . . . . . . . . . . . . . .  745

Index . . . . . . . . . . . . . . . . . . . . . . . . . . . . . . . . . . . . . . . . . . . . . . . . . . . . . . . . . . . .  749

*Organometallic Chemistry of the Transition Elements*

# Chapter 1

# Introduction to Organometallic Chemistry

## 1. NUMBER OF VALENCE ELECTRONS. CHARACTERISTICS OF LIGANDS

Organometallic compounds are those in which there is a metal–carbon bond. According to this definition, in the case of transition metals, this group of compounds includes not only metal carbonyls, olefin complexes, cyclopentadienyl, and other $\pi$-complexes, but also cyanide and fulminate compounds. Certain difficulties arise in defining the metal of the main group elements. Usually, organometallic compounds are comprised not only of compounds of typical metals, but also of metalloids such as boron, silicon, phosphorus, arsenic, selenium, etc. In compounds of metals as well as in those of metalloids, the bond is generally polarized as follows: $M^{\delta+} - C^{\delta-}$. Consequently, the metal or metalloid atom will be susceptible to nucleophilic attack while the carbon atom will be susceptible to electrophilic attack. In all other compounds (e.g., with oxygen, nitrogen, fluorine, chlorine, bromine, etc.), the polarity of the element–carbon bond is opposite.

Therefore, organometallic compounds contain carbon atoms bonded to elements which are more electropositive than carbon itself.

The development of organometallic chemistry occurred unusually rapidly in recent years. This development could be attributed to the relation between general theory of structural chemistry and organometallic chemistry, which contributed to generalization and broadening of certain concepts such as multicenter bonding and cluster compounds. The continuing development of structural chemistry (including structural organometallic chemistry) in turn stimulates the synthesis and structural investigations of new organometallic compounds. Another important factor influencing the intensification of research in this field is application of organometallic compounds in organic synthesis, catalysis, and technology, for example, preparation of metals and their new compounds.

*1*

In most organometallic compounds the $M-C$ bond has, to a significant degree, covalent character. Definite ionic character of this bond is generally manifested only in compounds of the alkali and alkaline earth metals. The ionic or covalent contribution to the bond depends on ionization potential of the metal, the size of a resulting ion, the ratio of the ionic charge to its radius, and $\sigma$-donor, $\sigma$-acceptor, $\pi$-donor, and $\pi$-acceptor properties of ligands and their structure.

The main group elements have four valence orbitals, $ns$ and $np$, and transition metals have nine valence orbitals, $(n-1)d$, $ns$, and $np$. Because of these orbitals, compounds of the main group elements which contain bonds of considerable covalent character obey the octet rule, that is, they form $8e$ compounds, while transition metals can form $18e$ complexes since all valence orbitals are utilized to create molecular orbitals.

The $nd$ orbital energy of the far-right main group elements decreases as a result of an increase in the nuclear charge to such a level that these orbitals may be utilized to form bonds. Therefore, compounds with more than 8 valence electrons can be obtained. The $f$-electron elements with valence orbitals $(n-2)f$, $ns$, and $np$ may form $22e$ complexes or, if $(n-1)d$ orbitals are also used, $32e$ compounds may be formed. The structures, properties, and valence electron counting are conveniently described assuming that a complex $MX_n$ is formed from an atom M and $n$ ligands X, which represent atoms or groups of atoms. Table 1.1 gives a classification of ligands according to the number of contributed electrons for the $M-X$ bond.[1-6, 12, 14-17] Also given are qualitatively estimated properties of ligands, such as tendency to form stable ions, number of possible coordination sites, and $\sigma$- and $\pi$-donor and acceptor properties.

Examples of valence electron count are given in Table 1.2.

In most cases, the total number of electrons which are donated by a given ligand can be determined unequivocally. However, it is possible to interpret differently the electron donor properties of tetrafluoroethylene and some acetylene ligands. Ethylene in $[PtCl_3C_2H_4]^-$ is a $2e$ ligand; however, $C_2F_4$ in $[(Ph_3P)_2PtC_2F_4]$ should be regarded rather as a $2 \times 1e$ ligand because carbon atoms lie in the plane in which phosphorus and platinum atoms are also located. Therefore, the bonding of carbon in the $C_2F_4$ ligand corresponds more to $sp^3$ than to $sp^2$ hybridization, and the complex may be regarded as an analog of $cis-[(Ph_3P)_2Pt(CF_3)_2]$. Similarly, $HOOCC \equiv CCOOH$ may be regarded as a $2 \times 1e$ ligand in $[Pt(PPh_3)_2C_2(COOH)_2]$. Acetylenes may also act as $4e$ ligands, mainly in compounds in which they play a bridging role (Figure 1.1) and less frequently in nonbridging complexes, such as $[ML(C_2R_2)_3]$ (M = Mo, W) (see Figure 6.23).

Dienes, trienes, and similar molecules with isolated double bonds are $n \times 2e$ ligands. Conjugated dienes are $4e$ ligands (not $2 \times 2e$ ligands); in compounds of these organic molecules, there is delocalization of $\pi$ electrons over all carbon atoms of $sp^2$ hybridization.

All chelating ligands may be divided into two groups. The first group contains ligands in which coordinating atoms are separated; therefore these are equivalent to $n$ analogous unidentate ligands. These are, among others, $R_2N(CH_2)_n NR_2$, $R_2P(CH_2)_n PR_2$, or $RS(CH_2)_n SR$. The second group comprises ligands for which bond formation with metal cannot be represented simply as a sum of interactions of particular segments of the coordinating molecule, but rather as a result of interaction of the entire ligand itself, for example, 1,3-dienes, 1,10-phenanthroline, 2,2'-dipyridine, acetylacetonate ligand ($3e$).

*Table 1.1. Properties of Ligands*[a]

| | | Ligand properties | | | | Tendency to form stable ion | | Stability, properties |
|---|---|---|---|---|---|---|---|---|
| Ligand | $ne$ | $\sigma$-donor | $\sigma$-acceptor | $\pi$-donor | $\pi$-acceptor | | | |
| 1 | 2 | 3 | 4 | 5 | 6 | 7 | | 8 |
| $BF_3$ | 0 | — | s | w | w | — | | Inert |
| F | 1 | vw | vs | w | — | $F^-$ | vs | Strongly electronegative |
| Cl | 1 | vw | s | wm | vw | $Cl^-$ | vs | Strongly electronegative, as a bridge is $(1+2)e$ ligand |
| Br | 1 | vw | ms | wm | vw | $Br^-$ | vs | Electronegative, as a bridge is $(1+2)e$ ligand |
| I | 1 | w | m | wm | w | $I^-$ | s | |
| OH | 1 | ms | w | m | w | $OH^-$ | m | Very reactive, easily forms bridges $(1+2)e$, easily gives off $H^+$ |
| SH | 1 | m | w | w | — | $SH^-$ | m | |
| OMe | 1 | ms | w | m | w | $OMe^-$ | m | Inert, easily forms bridges $(1+2)e$ |
| SMe | 1 | m | w | w | w | $SMe^-$ | m | |
| H | 1 | m | m | — | — | $H^+$ $H^-$ | — w | Stereochemically labile, polarizability of the M—H bond changes in wide ranges |
| $CH_3$ | 1 | ms | w | — | — | $CH_3^-$ | vw | $CH_3^-$ very reactive, inert as a ligand, very unstable radical |
| $C_2H_5$ | 1 | ms | w | — | — | $C_2H_5^-$ | vw | $C_2H_5^-$ very reactive, shows tendency to eliminate $H^-$ from $\beta$ carbon |
| $C_3H_7$ | 1 | ms | w | — | — | $C_3H_7^-$ | vw | |
| $CH_2CH=CH_2$ | 1 | ms | w | — | — | $C_3H_5^-$ | vw | Undergoes 1,3-migration, stereochemically labile |
| $C_6H_5$ | 1 | m | w | vw | w | $C_6H_5^-$ | vw | Shows tendency to give off *ortho* H |
| COMe | 1 | m | w | vw | wm | $COMe^+$ | vw | |
| CN | 1 | m | w | wm | m | $CN^-$ | s | Stable radical, the M—CN bond is susceptible to nucleophilic attack |
| −SCN | 1 | mw | vw | wm | m | $SCN^-$ | s | Inert, may form $(1+2)e$ bridges |
| −NCS | 1 | m | vw | w | wm | $SCN^-$ | s | M—SCN—M |
| $NH_2$ $(sp^3)$ | 1 | m | m | — | — | $NH_2^-$ | w | Inert, forms $(1+2)e$ bridges |

[a] m—moderate, s—strong, w—weak, v—very.

*(Table continued)*

Table 1.1. (continued)

| Ligand | ne | σ-donor | σ-acceptor | π-donor | π-acceptor | Tendency to form stable ion | Stability, properties |
|---|---|---|---|---|---|---|---|
| | 1 | 3 | 4 | 5 | 6 | 7 | 8 |
| $NMe_2$ (sp³) | 1 | | m | — | — | $NMe_2^-$  w | Inert, forms bridges, strong π-acceptor as sp² ligand |
| $PMe_2$ | 1 | m | w | w | m | $PMe_2^-$ | Inert, forms $(1+2)e$ bridges |
| $SiCl_3$ | 1 | wm | wm | vw | m | — | Very reactive |
| $GeCl_3$ | 1 | wm | wm | — | m | — | Very reactive |
| $SnCl_3$ | 1 | wm | wm | w | m | $SnCl_3^-$  m | Very reactive, forms bridges |
| HgCl | 1 | m | wm | w | w | | Quite inert |
| $Mn(CO)_5$ | 1 | m | w | s | w | $Mn(CO)_5^-$  m | Quite inert, π-basic |
| $Co(CO)_4$ | 1 | m | w | s | wm | $Co(CO)_4^-$  m | |
| $V(CO)_6$ | 1 | m | w | s | w | $V(CO)_6^-$  m | Inert, π-basic |
| $BH_4$ | 1 | wm | wm | | — | $BH_4^-$  m | Reactive |
| O | 2×1 | wm | ms | ms | w | $O^{2-}$  m | Shows strong tendency to react with electrophiles |
| N | 3×1 | wm | m | sm | w | $N^{3-}$  mw | |
| $C_2F_4$ | 2×1 | vw | m | vw | wm | — | Inert, bidentate |
| $NH_3$ | 2 | s | vw | — | — | $NH_3^-$  vw | Moderately reactive |
| $NMe_3$ | 2 | s | vw | — | — | — | Inert |
| $NR_3$ | 2 | s | vw | — | — | — | Inert |
| $PH_3$ | 2 | wm | w | vw | wm | — | Reactive |
| $PF_3$ | 2 | w | w | w | s | — | Quite inert |
| $PMe_3$ | 2 | m | vw | vw | m | — | Inert |
| $PR_3$ | 2 | wm | vw | vw | m | — | Inert |
| $PPh_3$ | 2 | w | vw | w | ms | — | Inert, shows tendency to give off ortho H |
| $H_2O$ | 2 | m | w | — | — | — | Has labile H atoms |
| $R_2O$ | 2 | m | w | vw | — | — | |
| $R_2S$ | 2 | m | vw | w | w | — | Inert |

| Ligand | n | | | | | | Ion | Remarks |
|---|---|---|---|---|---|---|---|---|
| CO | 2 | w | vw | vw | ms | | — | Inert, the carbon atom is susceptible to nucleophilic attack |
| CSe | 2 | wm | vw | w | s | | — | Unstable in the free state |
| CS | 2 | wm | vw | w | s | | — | Unstable in the free state |
| N$_2$ | 2 | w | vw | vw | wm | | — | Inert |
| RNC | 2 | wm | w | w | ms | | — | Inert, susceptible to nucleophilic attack |
| RCN | 2 | m | w | w | wm | | — | Inert |
| Pyridine | 2 | ms | w | w | w | | — | Inert |
| CH$_2$=CHR | 2 | wm | vw | m | m | vw | — | Inert, undergoes migration reactions |
| R$_2$C=CR$_2$ | 2 | wm | vw | ms | ms | | — | Inert, stereochemically labile, undergoes migration (insertion) reactions |
| C$_2$H$_4$ | 2 | wm | vw | ms | ms | | — | Inert, stereochemically labile, undergoes migration (insertion) reactions |
| NO | 1 + 2 | vw | vw | m | ms | s / vw | NO$^+$ / NO$^-$ | Inert |
| $\pi$-C$_3$H$_5$ | 3 | wm | w | m | m | | C$_3$H$_5$ | pseudobidentate, stereochemically labile (shows dynamic properties) |
| $\pi$-C$_3$R$_x$H$_{5-x}$ | 3 | w | w | m | m | | C$_3$R$_x$H$_{5-x}^-$ | Inert, forms bridges |
| CH$_3$COO | 3 | ms | w | m | wm | | CH$_3$COO$^-$ | Bidentate, shows delocalization of $\pi$-electrons all over the molecule |
| Acetylacetonyl | 3 | ms | w | w | m | | C$_5$H$_7$O$_2^-$ | Bidentate, shows delocalization of $\pi$-electrons all over the molecule |
| Dimethylglyoximyl | 3 | ms | w | w | m | | Hdmg$^-$ | Bidentate, shows delocalization of $\pi$-electrons all over the molecule |
| RCSS | 3 | ms | w | w | m | wm | RCSS$^-$ | Inert, forms bridges |
| R$_2$NCSS | 3 | m | w | w | m | wm | R$_2$NCSS$^-$ | Inert, forms bridges |
| Butadiene | 4 | wm | w | m | ms | | — | Pseudobidentate |
| Cyclobutadiene | 4 | wm | w | m | ms | | — | Pseudobidentate, unstable in the free state |
| *o*-Phenanthroline | 4 | m | w | vw | wm | | anion | Bidentate, rigid geometry |
| 2,2'-Bipyridyl | 4 | m | vw | vw | wm | | anion | Bidentate, quite rigid geometry |
| $\pi$-C$_5$H$_5$ | 5 | m | w | m | m | | anion | Pseudotridentate |
| Benzene | 6 | wm | w | vw | m | | anion | Pseudotridentate |
| $\pi$-Cycloheptatrienyl | 7 | mw | m | m | m | | cation | Pseudotridentate, unstable in the free state |
| Cyclooctatetraene | 8 | m | w | w | m | | C$_8$H$_8^{2-}$ | Pseudotetradentate |

*Table 1.2. Valence Electron Counts*

| Compound | Metal and ligands | Number of electrons |
|---|---|---|
| 1) $[V(CO)_6]$ | V | 5 |
| | 6CO | $\dfrac{12}{17}$ |
| 2) $[Mn_2(CO)_{10}]$ | Mn | 7 |
| | 5CO | 10 |
| | Mn $-$ Mn bond | $\dfrac{1}{18}$ |
| | or | |
| | Mn | 7 |
| | 5CO | 10 |
| | $Mn(CO)_5$ | $\dfrac{1}{18}$ |

Each $Mn(CO)_5$ group may be treated as a 1*e* ligand.

| | | |
|---|---|---|
| 3) $[(OC)_3Fe\{(CH_2)_3C\}]$ | Fe | 8 |
| | 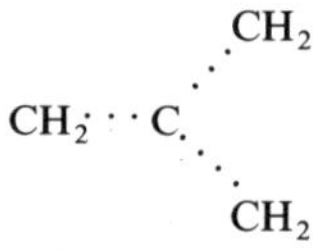 $CH_2 \cdots C$ (with $CH_2$, $CH_2$, $CH_2$) | 4 |
| | 3CO | $\dfrac{6}{18}$ |
| 4) $[Ti(C_5H_5)_2(CO)_2]$ | Ti | 4 |
| | $2C_5H_5$ | 10 |
| | 2CO | $\dfrac{4}{18}$ |
| 5) $[U(C_8H_8)_2]$ | U | 6 |
| | $2C_8H_8$ | $\dfrac{16}{22}$ |
| 6) 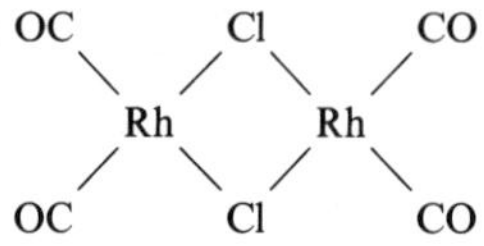 (see structure) | Rh | 9 |
| | Cl | $1+2$ |
| | 2CO | $\dfrac{4}{16}$ |
| 7) $[Mo(C_5H_5)_3NO]$ | Mo | 6 |
| | $\eta^5$-$C_5H_5$ | 5 |
| | $\eta^3$-$C_5H_5$ | 3 |
| | $\sigma$-$C_3H_5$ | 1 |
| | NO | $\dfrac{1+2}{18}$ |

Structure for 3):

$$CH_2 \cdots C \begin{matrix} CH_2 \\ \\ CH_2 \end{matrix}$$

Structure for 6):

```
OC      Cl      CO
  \   /   \   /
   Rh     Rh
  /   \   /   \
OC      Cl      CO
```

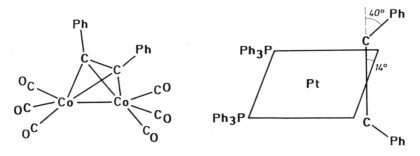

Figure 1.1   The structures of $Co_2(C_2Ph_2)$ $(CO)_6$ and $Pt(C_2Ph_2)$ $(PPh_3)_2$.

The number of $d$ or $f$ electrons in a complex may be calculated from the following formula: $x = m - l - p$, where $m$ is the number of valence electrons of the metal, $l$ is the number of ligands possessing unpaired electrons, and $p$ is the charge of the ion. For example:

| | | |
|---|---|---|
| $[FeF_6]^{3-}$ | $8 - 6 + 3 = 5$ | $d^5$ |
| $[Fe(H_2O)_6]^{2+}$ | $8 - 0 - 2 = 6$ | $d^6$ |
| $[HMn(CO)_5]$ | $7 - 1 = 6$ | $d^6$ |
| $[H_2Fe(CO)_4]$ | $8 - 2 = 6$ | $d^6$ |
| $[HRh(PPh_3)_4]$ | $9 - 1 = 8$ | $d^8$ |
| $[CrO_4]^{2-}$ | $6 - 4(2 \times 1) + 2 = 0$ | $d^0$ |
| $[PrCl_3]$ | $5 - 3 = 2$ | $f^2$ |
| $[MoNCl_5]^{2-}$ | $6 - 3 \times 1 - 5 + 2 = 0$ | $d^0$ |
| $[Fe_3O(CH_3COO)_6 (H_2O)_3]^+$ | $\frac{1}{3}(24 - 2 \times 1 - 6 - 1) = 5$ | $d^5$ |

According to the given examples, oxygen and nitrogen should be classified as $2 \times 1e$ and $3 \times 1e$ ligands, respectively. Such classification of ligands helps to determine properties and structures of compounds of the main group elements and transition metals. The main group elements are known for their tendency to form bonds utilizing all valence $ns$ and $np$ orbitals which leads to the formation of $8e$ compounds.

## 2. THE MAIN GROUP ELEMENTS

With a few exceptions, the ionization potential of an atom increases within a given period with increase in atomic number. Because of this phenomenon, the alkali and alkaline earth metals form ionic compounds with electronegative ligands such as halogens, while the metalloids and nonmetals form compounds with considerable covalent character. Exceptions to the $8e$ rule are encountered most frequently in ionic compounds as well as in compounds of metalloids and nonmetals. In the latter case, compounds with more than eight valence electrons may be formed. These deviations are a function of properties of the central atom and the ligands.

The elements of the second period have considerably higher ionization potentials and smaller atomic and ionic radii than their homologs in respective groups. Thus, the chemical properties of the first element of a given group are substantially different from the properties of the remaining elements. These differences are considerably greater in going from the first element to the second than between any other elements within the group.

The difference between energies of *ns* and *np* orbitals plays an important role. This difference is small for the elements of the second period, and it increases within the group, with increase in atomic number, achieving maximum value for the elements of the last period (Hg, Tl, Pb, Bi), which consequently may form low valent compounds (including organometallic ones). The increased stabilization of *ns* orbitals as compared to *np* orbitals with increase of atomic number within a group is caused by stronger penetration of *ns* orbitals into the inner electron shells and greater screening effects of *np* orbitals.

The most electropositive elements, the alkali and the alkaline earth metals—with the exception of beryllium—easily form positive ions due to low ionization potentials. This tendency increases as atomic number increases within a group because heavier elements have lower ionization energies. However, this tendency is weakened because the lattice and solvation energies are lowered as ion size becomes greater. Compounds of these metals with strongly electronegative ligands showing a strong tendency to form negative ions such as halogens, oxygen, and hydroxy groups will therefore have predominantly ionic character.

In labile complexes possessing donor ligands ($H_2O$, $NH_3$) a decisive role is played by the ion–dipole interactions. Coordination number, which changes in a continuous manner, is therefore determined by the size of the ion rather than the number of valence orbitals. For less electronegative ligands such as alkyls, aryls, $PhCH_2$, and $C_5H_5$, the ionic character of the metal–ligand bond is not so clearly pronounced. The $M-C$ bond is more ionic if stabilization of the negative charge occurs due to its delocalization over several carbon atoms. An example of such a ligand is cyclopentadienyl. Cyclopentadienyl complexes with group 1 and 2 metals* have decisively ionic character. Similarly, alkali metal compounds with aromatic hydrocarbons in ether solutions are ionic, for example, $Na^+(THF)C_{10}H_8^-$.

An important factor is the ratio of the charge of an ion to its radius. In ionic compounds possessing both high values of this ratio and easily polarizable ligands, disturbance of the electron cloud of the anion may occur, leading to increased covalent character of the metal–ligand bond, and even to the decomposition of the compound. $BeCO_3$, as well as lithium, beryllium, and magnesium peroxides exemplify such instability. The alkali metals and, even more so, elements of groups 2 and 13 form, with less electronegative ligands such as alkyls, aryls, and H, compounds in which covalent character is more pronounced, leading to formation of dimers, oligomers, and even polymeric compounds. Consequently, these ligands form bridges connecting two or three metal atoms. Bonding in such compounds is best described by employing multi-center molecular orbitals rather than utilizing the ionic model.

The four-center bonding exists in methyllithium (Figure 1.2) and methylsodium,

---

* According to the rules of the Nomenclature Commission, IUPAC, the groups of the Periodic Chart are designated by Arabic numerals 1–18. Block *s* elements are numbered 1 and 2, block *d* as 3–12, block *f* as 3, and block *p* as 13–18. This system avoids letters A and B, which were not consequently utilized. The new system is unequivocal, and precisely differentiates the transition metals. In block *p*, the second cipher designates the previously used number of the group. Previous groups IIIA and IIIB are now designated as 3 and 13, etc.

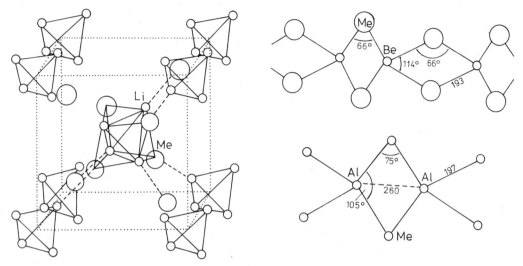

Figure 1.2.   The structures of $LiCH_3$, $Be(CH_3)_2$, and $Al_2(CH_3)_6$.

which form tetramers in the solid state. The tetramers are connected with each other, and therefore methyllithium has a high melting point and is not soluble in nonpolar organic solvents.

Other compounds of lithium also form tetramers or hexamers. Such compounds are stable in solutions containing even small amounts of Lewis bases, such as amines, because each lithium atom has one available orbital. The presence of amines, particularly chelating ones and in larger quantities, for example, tetramethylethylene-diamine, TMED, leads to the formation of monomeric compounds involving the coordinated amine.

Beryllium compounds have polymeric structures due to the formation of three-center bonds, $Be-C-Be$ (Figure 1.2). $Be(CH_3)_2$ is a *4e* compound. As a result of coordinating Lewis bases, it achieves the octet structure.

The formation of coordination compounds easily takes place. The beryllium chloride is also polymeric, structurally analogous to dimethylberyllium. However, this is not an electron-deficient compound, because each bridging chlorine atom is a $(1+2)e$ ligand. Also, magnesium compounds have polymeric structures. These polymers are formed quite easily despite more ionic character of the bonds, because the magnesium atom is larger than the beryllium atom and thus the ratio of the size of the magnesium atom to the alkyl group is more suitable for the formation of stable compounds.

Steric factors often influence the structures of the compounds.[1, 8, 14, 15] Very small hydrogen atoms form bridges between small boron atoms in diborane, $B_2H_6$, or in other boranes. However, boron compounds with ligands such as alkyls, aryls, F, Cl, and Br, which have greater sizes, are monomeric. Aluminum forms dimeric compounds, $Al_2R_6$, because its atomic radius is larger. Steric factors in organometallic compounds are probably the most favorable for magnesium and aluminum. The structural changes in those compounds occur relatively easily because of the rather low energy of dimer–monomer dissociation for compounds of group 2 and 13 elements. It is equal only *ca.* 42 kJ/bridging bond for $[Me_2AlMe_2AlMe_2]$. Bridges formed by carbon atoms of $sp^2$ hybridization are slightly more stable than bridges formed by $sp^3$-hybridized carbon atoms. Phenyl and vinyl are better bridging ligands than methyl groups in aluminum and gallium compounds.

The weakening of the tendency to form dimers in heavier group 13 elements is also due to the increasing difference between $np$ and $ns$ orbital energies. Most notably, this is seen for thallium, which may adopt a $+1$ oxidation state in organometallic compounds. Increase in energy difference between $ns$ and $np$ orbitals for group 13–14 elements which have filled $(n-1)d$ and $4f$ orbitals as compared to the elements immediately preceding the transition metals (groups 1 and 2) is caused by an increase of the nuclear charge resulting from filled $d$ and $f$ orbitals. Increase in the nuclear charge causes greater stabilization of $ns$ orbitals than of $np$ orbitals. Because of the increase in nuclear charge the difference in the spatial arrangement of $p$ orbitals as compared to $s$ orbitals is also greater.

These effects are particularly marked for Hg, Tl, Pb, and Bi, which are after $4f$- and $5d$-electron elements in the periodic Chart, and thus have their atomic numbers greater by 32 as compared to respective elements of the preceding period (Cd, In, Sn, ...). Therefore, the main group elements, and particularly those of groups 14–16, show, as the atomic number increases, greater tendency to form compounds in which the oxidation state is $m-2$ ($m=$ the number of valence electrons) especially with ligands showing greater ability to form ions, for example, Cl, O, S, and $C_5H_5$. Most of these ligands also form compounds in which the central atom has the highest oxidation state of the given group.

The most stable compounds in lower oxidation states are formed by period 6 elements. Examples of these compounds are TlCl, TlOH, $Tl(C_5H_5)$, $SnCl_2$, SnO, PbO, $Sn(C_5H_5)_2$, $PbCl_2$, PbS, and $BiCl_3$. Most of them are $8e$ compounds. Such structures are obtained owing to the formation of bridges and metal-to-metal bonds:

In compounds with ligands that do not show the ability to form stable ions, the main group elements adopt the highest oxidation state. According to Pauling's electroneutrality rule, these compounds are covalent; the charge of the central ion ranges from $+0.5$ to $-0.5$ and is independent of the formal oxidation state of the central ion itself. Compounds of metals in $n-2$ oxidation states possess free electron pairs and may form complexes, preferably with transition metals, such as $[Rh_2Cl_2(SnCl_3)_4]^{4-}$. Stronger abilities for complex formation are observed in compounds of elements possessing higher ionization potentials such as group 15 elements, for example, $PR_3$, $AsR_3$, and $SbR_3$.

As the atomic number increases in the periods as a result of the nuclear charge increase, $nd$ orbital energy becomes lower, to such a degree that the $nd$ orbitals may participate in bond formation. Consequently, groups 13–18 elements may form compounds which have more than eight valence electrons. Such compounds form ligands possessing strong $\sigma$-acceptor properties, which increase the positive charge of the central atom and simultaneously decrease $nd$ orbital energy. The following compounds may serve as examples: $PF_5$, $SF_4$, $ClF_3$, $XeF_2$ ($10e$ compounds), $SiF_6^{2-}$, $PF_6^-$, $SF_6$, $IF_5$ and $XeF_4$ ($12e$ compounds). An *ab initio* calculation points out that nonconformity to the octet rule is caused by the formation of three-center four-electron bonds ($3c$–$4e$) and other electron-rich bonds, which are a combination of $p$ orbitals of atoms composing the molecule. The participation of $d$ orbitals in bond formation is very important, although their population is generally considerably lower than would be predicted from the hybridization theory, e.g., $sp^3 d^n$. The $d$ orbitals participate mainly in the formation of multicenter bonds and particularly in back-donation bonds.[15] The tendency to increase the number of valence electrons in compounds of elements of a given group becomes greater as the atomic number increases.

The elements of periods 5 and 6 may also form such complexes with less electronegative ligands, for example, $SnCl_6^{2-}$, $SnBr_6^{2-}$, $TeCl_6^{2-}$, $TeBr_6^{2-}$. This situation is caused by a greater decrease in valence orbital energies of $4d$, $5d$, or $6d$ as compared to $3d$, because the former orbitals more strongly penetrate the inner electron shells.

For the transition metals this is different. Compounds with electronegative ligands possessing considerable $\sigma$-acceptor abilities generally do not form $18e$ complexes, because an increase in the positive charge on the central atom causes too great a decrease in $(n-1)d$, $ns$, or $np$ orbital energies, some of which become the core orbitals.

The ability of ligands to form compounds in the highest oxidation states involving different central atoms decreases in a series: $F > O > OR > NR_2 > CR_3 > Cl > H > Br$.

## 3. THE TRANSITION METALS

The energy of the valence orbitals of the transition metals decreases in a given period as the atomic number increases. Therefore, $d$ orbital energy of the far-right transition metals is very low. Thus, the $d$ electrons become the inner electrons (Figure 1.3).[1–4, 7, 9–14]

The decrease in the energy of the $s$ electron to the level of core energy is seen as an inert pair effect for $p$-block metals, while the lowering of energy of $p$ orbitals is observed as chemical inertness of the noble gases. The $d$ orbital energy decreases faster for the first row of transition metals as compared to the second and the third. However, the $s$ orbitals lower their energy faster for the third row of transition metals. This situation

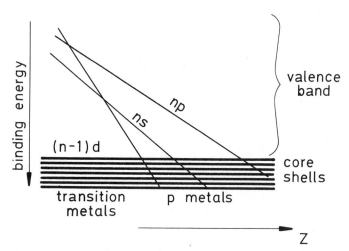

Figure 1.3.  The dependence of valence orbital energy on atomic number.

happens as a result of increasing nuclear charge as $4f$ orbitals become filled for the lanthanides.

The sum of ionization potentials of valence electrons is smaller for elements of the second and third rows of the transition metals than for $3d$ electron metals. This phenomenon is caused by a faster increase in ionization energy of $3d$ electrons as compared to that of $4d$ and $5d$ electrons due to increasing charge of the metal ion. Because $3d$ electrons, in contrast to $4d$ and $5d$ electrons, do not penetrate inner electron shells, their energy is therefore more dependent on the ionic charge.

In the case of the first row of transition elements, the ratios of successive ionization potentials are approximately equal to the ratios of the squares of ionic charges, because the outer $3d$ orbitals of ions possessing $3d^n$ configurations feel the effective charge as approximately equal to the ionic charge, $I_1:I_2:\cdots I_n = z_1^2:z_2^2:\cdots z_n^2$, where $I_i$ is the ionization potential and $z_i$ is the charge of the ion.

The important factor influencing the $d$ electron energy is the energy required to cause pairing of two spins. This energy is higher for $3d$ orbitals than for $4d$ and $5d$ orbitals. Because of this situation, the bonding energy for $d^{6-10}$ electron configurations is lower for $3d$ orbitals than for $4d$ orbitals. The opposite is true for $s$ electrons, which explains the easier formation of Cu(II) compounds than Ag(II) compounds. These considerations show that the second- and third-row elements should form more stable compounds in high as well as in low oxidation states as compared to the first row.

In groups, the stability of higher oxidation states increases with increasing atomic number. Therefore, heavier elements achieve the highest oxidation states within a given group. However, in series, owing to the lowering of $d$ electron energy, the far-right metals will not achieve the highest oxidation state. Most notably, these properties are observed for compounds with ligands possessing strong $\sigma$-acceptor properties (F, Cl, O, etc.), for example, $TiO_2$, $V_2O_5$, $CrO_3$, $Mn_2O_7$, $CoO(OH)$, $Ni_2O_3 \times 2H_2O$, $TiCl_4$, $VCl_4$, $CrCl_3$, $MnCl_3$, $FeO_4^{2-}$, $RuO_4$, $OsO_4$, $CrCl_3$, $MoCl_6$, and $WCl_6$.

The stability of compounds is determined not only by atomic orbital energy of elements which form bonds, but also by the effectiveness of orbital overlap. The $d$ orbital span for the second and third series of metals is greater than that for the first series.

Therefore 4*d* and 5*d* electron metals interact with ligands in a more effective manner and thus form more covalent compounds. Because of valence orbital energy and orbital sizes, compounds of these elements in their lower oxidation states, particularly organometallic ones, are more stable than analogous complexes of 3*d* electron metals. The increased stability of olefin and acetylene compounds with increasing atomic number in a given group may serve as an example. Olefin complexes of cobalt are few and very unstable, while rhodium and iridium olefin compounds are quite common and usually air-stable.

The transition metals form 18*e* complexes most commonly with ligands showing donor–acceptor properties. Such compounds may also be prepared when some coordinated groups are donor ligands and the remainder are acceptor ligands. In these cases the bonds are mainly covalent.

Coordination compounds containing exclusively $\sigma$-donor or $\sigma$-acceptor ligands have predominantly ionic character of the metal–ligand bond. In such cases the formation of 18*e* compounds is accidental. These complexes can most satisfactorily be described by the crystal field theory. The number of valence electrons in complexes of this kind is usually different from 18.

Ligands exhibiting strong $\sigma$-acceptor properties cause an increase of positive charge on the central atom. Therefore, they often form anionic coordination compounds, e.g., $[CrCl_6]^{3-}$, $[CoBr_4]^{2-}$, and $[FeF_6]^{3-}$. Ligands showing strong $\sigma$-donor properties increase negative charge on the central ion. According to Pauling's electroneutrality rule, complexes of this kind are generally cationic, for examples $[Fe(NH_3)_6]^{3+}$ and $[V(NH_3)_6]^{3+}$. Therefore, the formation of anionic or cationic complexes reduces the negative or positive charge on the central atom. Thus, complexes analogous to the metal carbonyls with ligands possessing strong $\sigma$-donor properties such as $Cr(NH_3)_6$, $Fe(NH_3)_5$, and $Ni(NH_3)_4$ are unknown.

The dependence of the valence electron number in a complex and its structure on properties of ligands is given in Table 1.3. Complexes with ligands exhibiting strong $\pi$-acceptor properties are most commonly 18*e* compounds. Often, complexes possessing different ligands also confirm to the rule of utilizing all valence orbitals. However, some of them must have $\pi$-acceptor properties. Exceptions to the 18-electron rule are common for the elements with a smaller number of *d* electrons or for those of the far-right transition metals. For those elements with few *d* electrons, this situation is most commonly caused by steric factors and relatively lower ionization potentials which, with more electronegative ligands, leads to the formation of complexes with stronger ionic character.

The titanium group metals do not form 18*e* complexes with 1*e* or even 2*e* ligands. This would require coordination of 9 or 7 ligands, respectively, which on steric grounds is not possible or, in the case of 2*e* ligands, energetically unfavorable. With strongly electronegative ligands, 8*e* complexes are usually formed, in which orbitals *s* and *p* are used for bond formation, e.g., $TiCl_4$ and $TiR_4$ (R = Me, Et). Owing to a relatively small number of *d* electrons, which make it impossible to form strong $d_\pi$–$\pi$ bonds between the metal and ligand, stable carbonyls of the titanium group metals are unknown. Titanium atoms in CO matrices at only several absolute degrees form, among others, hexacarbonyltitanium, which is a 16*e* compound. $Ti(CO)_7$, the 18*e* compound, is not formed; in this case steric factors are also undoubtedly responsible.

However, 18*e* carbonyl complexes of group 4*d* metals are known for ligands creating relatively small steric crowding and those which strongly increase electron

Table 1.3. The Dependence of the Valence Electron Number in a Complex and its Structure on Properties of Ligands

| σ-Acceptor ligands | | | σ- and π-Donor-acceptor ligands | | σ-Donor ligands | |
|---|---|---|---|---|---|---|
| 8 | 9 | 10 | 17 | 18 | 13 | 14 |
| $[TiCl_4]$ | $[VCl_4]$ | $[VOCl_3]$ | $[V(CO)_6]$ | $[V(CO)_6]^-$ | $[Ti(NH_3)_6]^{3+}$ | $[V(NH_3)_6]^{3+}$ |
| 11 | 12 | 13 | $[V(dipy)_3]$ | $[Mo(CO)_6]$ | 15 | 16 |
| $[VCl_5]^-$ | $[TiF_6]^{2-}$ | $[VO(acac)_2]$ | $[Mo(C_6H_6)_2]^+$ | $[Co(cp)_2]^+$ | $[Cr(NH_3)_6]^{3+}$ | $[Mn(NH_3)_6]^{3+}$ |
| 14 | 15 | 16 | $[Co(CN)_5]^{3-}$ | $[H_2Fe(CO)_4]$ | 17 | 18 |
| $[TaF_7]^{2-}$ | $[CrCl_6]^{3-}$ | $[NiCl_4]^{2-}$ | | $[Cr(CO)_5NH_3]$ | $[Fe(NH_3)_6]^{3+}$ | $[Co(NH_3)_6]^{3+}$ |
| 17 | 18 | | | | 19 | 20 |
| $[FeF_6]^{3-}$ | $[PtCl_6]^{2-}$ | | | | $[Co(NH_3)_6]^{2+}$ | $[Ni(NH_3)_6]^{2+}$ |
| | $[CoF_6]^{3-}$ | | | | | |

density on the central atom, for example, $M(cp)_2(CO)_2$, $M = Ti$, Zr, Hf. Elements possessing a greater number of $d$ electrons may, with smaller and less electronegative ligands, form coordination compounds exhibiting large and rare coordination numbers: $[ReH_9]^{2-}$ (18$e$), $[ReH_7(PR_3)_2]$ (18$e$), $[WH_6(PR_3)_2]$ (16$e$), $[WH_6(PMe_2Ph)_3]$ (18$e$), $[MoH_4(PR_3)_4]$ (18$e$), etc.

As the atomic number increases in transition metal series, the difference between $ns$ and $np$ orbital energy strongly increases. Therefore, the nickel group metals show a strong tendency to form 16$e$ planar complexes in which the $p_z$ orbital does not participate in the bond formation. This tendency is exceptionally marked in complexes possessing $d^8$ electron configurations such as Au(III), Pd(II), Pt(II), Rh(I), and Ir(I). The copper group metals, for which the difference between $ns$ and $np$ orbitals is even greater, form generally linear complexes with coordination numbers 2 and $+1$ oxidation states. Square-planar complexes with labile ligands easily change their coordination numbers and oxidation states. Therefore, these complexes are often active catalysts for many reactions.

## 4. ELEMENTARY REACTIONS OF ORGANOMETALLIC COMPOUNDS

The transition metals may form many compounds unknown for the main group elements due to partly filled $d$ or $f$ orbitals. The transition metal complexes may be inert or labile. The inert complexes react slowly, especially with regard to ligand substitution reactions. Thus, these complexes are kinetically stable. However, this situation is not equivalent to thermodynamic stability. A thermodynamically unstable compound may be inert because, for certain electron configurations, the $S_N1$ dissociative mechanism of ligand substitution as well as the $S_N2$ associative mechanism require high activation energy. High activation energies, according to the ligand field theory, are observed for octahedral complexes comprising central atoms which have $d^3$ and low-spin $d^4$–$d^6$ electron configurations.[12, 13, 18–20] The reactivity of complexes, including organometallic ones, mainly depends upon their electron configuration, symmetry, coordination number, metal-to-ligand bond energy, ligand properties, intermolecular interactions, etc.[18–22] Reactions of organometallic compounds are usually complex. However, they can often best be described with the aid of elementary organometallic reaction steps (Table 1.4).[23] These reactions are significant in catalytic processes. During the course of one chemical reaction, several elementary steps usually take place, for example, nucleophilic ligand substitution may occur through dissociation of a Lewis base followed by association of another Lewis base.

## 5. NOMENCLATURE OF ORGANOMETALLIC COMPOUNDS

According to the IUPAC rules, organometallic compounds are those in which the carbon atoms are bonded to any other element with the exception of hydrogen, carbon, nitrogen, oxygen, fluorine, chlorine, bromine, iodine, and astatine. In the process of naming organometallic compounds, simple and consequent nomenclature used for coordination complexes should be applied. The name of a given organometallic compound is formed in such a way that the ligands are first listed in alphabetical order and then—in the case of neutral compounds and complex cations—followed by the name of the metal. The name of a complex anion is formed by adding the termination -ate. The number of the same ligands in the molecule is indicated by multiplying affixes: di, tri,

Table 1.4. Elementary Reactions of Organometallic Compounds

| Reaction | $\Delta$NVE[a] | $\Delta$OS[b] | $\Delta$CN[c] | Example | Reverse reaction | $\Delta$NVE | $\Delta$OS | $\Delta$CN |
|---|---|---|---|---|---|---|---|---|
| 1 Lewis acid dissociation | 0 | −2 | −1 | $[FeH(CO)_5]^+ \rightleftarrows [Fe(CO)_5] + H^+$ | Lewis acid association | 0 | 2 | 1 |
| | 0 | −2 | −1 | $[CoH(CO)_4] \rightleftarrows H^+ + [Co(CO)_4]^-$ | | 0 | 2 | 1 |
| | 0 | 0 | −1 | $[IrCl(BF_3)(CO)(PPh_3)_2] \rightleftarrows [IrCl(CO)(PPh_3)_2] + BF_3$ | | 0 | 0 | 1 |
| 2 Lewis base dissociation | −2 | 0 | −1 | $[CoH(CO)_4] \rightleftarrows [CoH(CO)_3] + CO$ | Lewis base association | 2 | 0 | 1 |
| | −2 | 0 | −1 | $[RhH(CO)(PPh_3)_3] \rightleftarrows [RhH(CO)(PPh_3)_2] + PPh_3$ | | 2 | 0 | 1 |
| 3 Reductive elimination | −2 | −2 | −2 | $[IrCl(H)_2(CO)(PPh_3)_2] \rightleftarrows [IrCl(CO)(PPh_3)_2] + H_2$ | Oxidative addition | 2 | 2 | 2 |
| 4 Migration (insertion) | −2 | 0 | −1 | $[Mn(Me)(CO)_5] \rightleftarrows [Mn(COMe)(CO)_4]$ | Remigration (deinsertion) | 2 | 0 | 1 |
| | −2 | 0 | −1 | $[CoH(C_2H_4)(CO)_3] \rightleftarrows [Co(C_2H_5)(CO)_3]$ | | 2 | 0 | 1 |
| 5 Oxidative coupling (cycloaddition) | −2 | 2 | 0 | 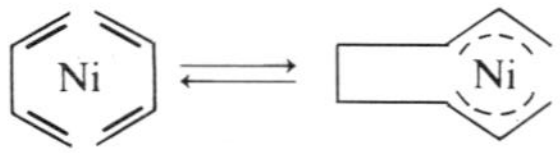 $\overset{\displaystyle \bigcirc}{Ni} \rightleftarrows \overset{\displaystyle \text{(Ni)}}{\phantom{x}}$ | Reductive decoupling (cycloelimination) | 2 | −2 | 0 |
| | −2 | 2 | 0 | $[Fe(C_2F_4)_2(CO)_3] \rightleftarrows \begin{matrix} CF_2-CF_2 \\ | \quad\quad\; Fe(CO)_3 \\ CF_2-CF_2 \end{matrix}$ | | 2 | −2 | 0 |

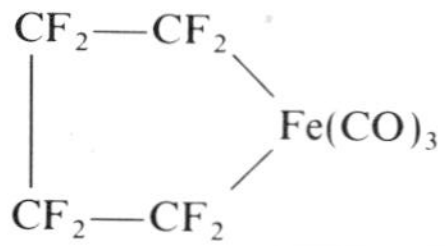

[a] Change in the number of valence electrons.
[b] Change in the oxidation state.
[c] Change in the coordination number.

*Table 1.5. Names of Ligands*

| Formula | Ion | Ligand |
|---|---|---|
| $O^{2-}$ | Oxide | Oxo |
| $H^-$ | Hydride | Hydrido |
| $OH^-$ | Hydroxide | Hydroxo |
| $O_2^{2-}$ | Peroxide | Peroxo |
| $HO_2^-$ | Hydrogenperoxide | Hydrogenperoxo |
| $S^{2-}$ | Sulfide | Thio |
| $S_2^{2-}$ | Disulfide | Disulfido |
| $HS^-$ | Hydrogensulfide | Mercapto |
| $CN^-$ | Cyanide | Cyano |
| $CH_3O^-$ | Methoxide or methanolate | Methoxo or methanolato |
| $CH_3S^-$ | Methanethiolate | Methanethiolato or methylthio |
| $N^{3-}$ | Nitride | Nitrido |

tetra, hexa, nona, etc. If the name of the ligand is complex, multiplicative affixes are used, such as bis, tris, tetrakis, pentakis, hexakis, etc., the ligand name being in parentheses. The names of anionic, organic, or inorganic ligands end in -o. In general, if the anion name ends in -ide, -ite, or -ate, the final -e is replaced by -o, giving -ido, -ito, and -ato, respectively. For some anionic ligands, the names are formed in a modified way by changing the termination or the name itself, for example, fluoride–fluoro, chloride–chloro. The names of selected anions and their corresponding ligands are given in Table 1.5.

The names of organic radicals are given with the ending -yl: phenyl, methyl, etc., and neutral ligands are named without changes. Ligands $H_2O$, $NH_3$, CO, and NO are named aqua, ammine, carbonyl, and nitrosyl, respectively. The ending -ato is given to ligands derived from organic compounds by the loss of protons. The oxidation state of the metal is given in Roman numerals or the cipher 0 in parentheses after the compound name. The charge of the ion with the exception of 0 (zero) may also be given in parentheses. Formulas of coordination compounds are placed in square brackets. The symbol for the central atom is given first, followed by the ionic and subsequently the neutral ligands. The sequence of ligands in given classes is listed in alphabetical order according to the symbols of atoms which are actually connected to the central atom. Ligands forming a bridge between two central atoms are preceded by the letter $\mu$. If the ligand constitues a bridge for three, four, or more atoms, designations are denoted by a subscript immediately after the letter $\mu$, for example, $\mu_3$, $\mu_4$, etc.

| | |
|---|---|
| $[RhCl(CO)(PEtMePh)_2]$ | carbonylchlorobis(ethylmethylphenylphosphine)rhodium(I) |
| $[Rh_2Cl_2(CO)_4]$ | tetracarbonyldi-$\mu$-chloro-dirhodium |
| | or di-$\mu$-chloro-bis(dicarbonylrhodium) |
| $[\{PtI(CH_3)_3\}_4]$ | tetra-$\mu_3$-iodo-tetrakis(trimethylplatinum) |
| | tetra-$\mu_3$-iodo-tetrakis[trimethylplatinum(IV)] |
| | tetra-$\mu_3$-iodo-dodecamethyltetraplatinum(IV) |
| $[Re(CO)_6][BF_4]$ | hexacarbonylrhenium(I) |
| | tetrafluoroborate |
| | hexacarbonylrhenium(1+)tetrafluoroborate(1−) |
| $K[IrCl_4(CO)]$ | potassium carbonyltetrachloroiridate(III) |
| | potassium carbonyltetrachloroiridate(1−) |
| $[PtBrCl(NH_3)_2]$ | diamminebromochloroplatinum |
| $[Ni(C_4H_7N_2O_2)_2]$ | bis(2,3-butanedione dioximato)nickel(II) |

For compounds in which the metal atom is bonded to two or more neighboring carbon atoms, a general name π-complexes has been utilized. However, this name is not used for individual compounds. If all carbon atoms in a chain or ring, or all carbon–carbon double bonds, are bound to the metal atom, the name of the ligand is preceded by the Greek letter $\eta$ which may be read as eta or hapto (from the Greek haptein, to fasten).[24] In some journals (printed via the photoreproduction technique) the letter h rather than $\eta$ is utilized; however, IUPAC does not recommend this. If only part of the chain or ring is bound to the metal, designators of atoms forming bonds with the metal are placed before the letter $\eta$. When successive atoms form bonds, designators of the first and last atoms are given. Currently, in the literature, the number of carbon atoms connected to the metal appears as a superscript immediately following the letter $\eta$, e.g., $Fe(\eta^5 - C_5H_5)_2$.

The names of dinuclear and polynuclear chain compounds containing metal–metal bonds are formed analogously to the mononuclear complexes, treating the metal from the farther group in the Periodic Chart with its attached groups as a ligand, the name of which terminates in -io.

In compounds containing two atoms of the same group of the Periodic Chart, the ligand is the metal and its ligands possessing smaller atomic number. If the metal atoms are connected with a bridge, and, moreover, there is a metal–metal bond between them, the name is formed as is that for bridging compounds, and the number of bonds between metal atoms is given in italics after the name. The geometrical shape of clusters is specified by *triangulo*, *quadro*, *tetrahedro*, *octahedro*, *triprismo*, *dodecahedro*, *antiprismo*, and *icosahedro*. Metal atoms are numbered if necessary.

$[(CO)_4Co - Mn(CO)_5]$ — pentacarbonyl(tetracarbonylcobaltio)manganese

$[(CO)_5Mn - Re(CO)_5]$ — pentacarbonyl(pentacarbonylmanganio)rhenium

$[Fe(CO)_4(AuPPh_3)_2]$ — tetracarbonylbis(triphenylphosphineaurio)iron

$[(OC)_3 Co(CO)_2 Co(CO)_3]$ — di-μ-carbonyl-bis(tricarbonylcobalt)(*Co–Co*)

$[(\eta\text{-}C_5H_5)NiPhC \equiv CPhNi(\eta\text{-}C_5H_5)]$ — μ-diphenylacetylene-bis(η-cyclopentadienylnickel)(*Ni–Ni*)

$[(OC)_3 Fe(EtS)_2 Fe(CO)_3]$ — bis-μ-ethylthio-bis(tricarbonyliron)(*Fe–Fe*)

$[(Ni\text{-}\eta\text{-}C_5H_5)_3 (CO)_2]$ — di-μ₃-carbonyl-*cyclo*-tris(η-cyclopentadienylnickel)(3 *Ni–Ni*)

$[Os_3(CO)_{12}]$ — *cyclo*-tris(tetracarbonylosmium)(3 *Os–Os*)

dodecacarbonyl-*triangulo*-triosmium

$[Mo_6Cl_8Cl_3\{(C_6H_5)_2$
$PCH_2CH_2P(C_6H_5)_2\}py]Cl$ — octa-μ₃-chloro-trichloro[ethylenebis(diphenylphosphine)](pyridine)-*octahedro*-hexamolybdenum(1 + ) chloride

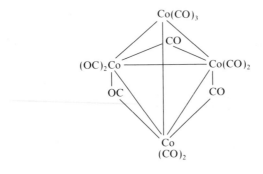

2,3;3,4;4,2-tri-μ-carbonyl-1,1,1,2,2,3,3,4,4,-nonacarbonyl-*tetrahedro*-tetracobalt

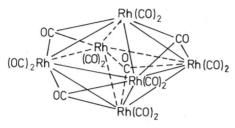

1,2,3;1,4,5;2,5,6;3,4,6-tetra-$\mu_3$-carbonyl-dodecacarbonyl-*octahedro*-hexarhodium

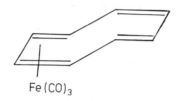

tricarbonyl(1-4-η-cyclooctatetraene)iron

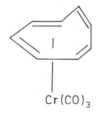

tricarbonyl(1-6-η-cyclooctatetraene)chromium

ReH(C₅H₅)₂    bis(η-cyclopentadienyl)hydridorhenium
Cr(C₆H₆)₂    bis(η-benzene)chromium

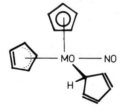

(η-cyclopentadienyl)(1-3-η-cyclopentadienyl)-(σ-cyclopentadienyl)nitrosylmolybdenum

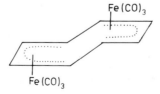

trans-μ-(1-4-η:5-8-η-cyclooctatetraene)-bis(tricarbonyliron)

Biscyclopentadienyl compounds of the transition metals are generally named metallocenes; however, this name should not be used for individual compounds, such as nickelocene, ruthenocene, osmocene, etc., with the exception of the iron complex, which is commonly named ferrocene.

Ferrocene derivatives are named according to organic nomenclature:

$Fe(C_5H_5)_2$          bis($\eta^5$-cyclopentadienyl)iron(II)
                        ferrocene

$[Fe(C_5H_5)_2][BF_4]$     bis($\eta$-cyclopentadienyl)iron(1 +) tetrafluoroborate
                        bis($\eta$-cyclopentadienyl)iron(III) tetrafluoroborate
                        ferrocenium tetrafluoroborate

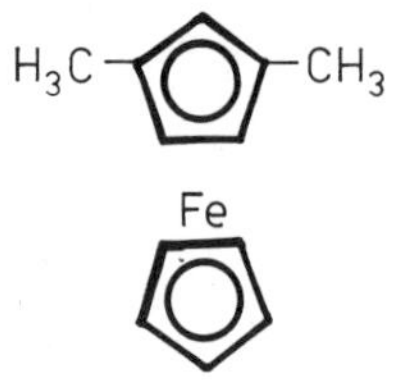

1,3-dimethylferrocene

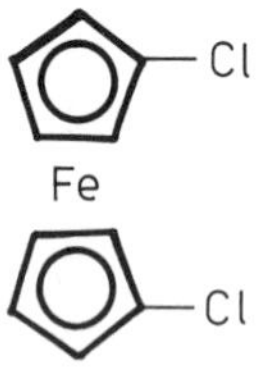

1,1'-dichloroferrocene

The names of compounds of trivalent phosphorus, arsenic, antimony, and bismuth as well as those of tetravalent silicon, germanium, tin, and lead may be formed either according to the nomenclature of coordination compounds or according to the nomenclature of substituted respective hydrides: phosphine ($PH_3$), arsine ($AsH_3$), stibine ($SbH_3$), bismuthine ($BiH_3$), silane ($SiH_4$), germane ($GeH_4$), stannane ($SnH_4$), and plumbane ($PbH_4$). The following names exemplify both rules:

$PPh_3$         triphenylphosphine
            triphenylphosphorus(III)

$PHPh_2$       diphenylphosphine
            hydridodiphenylphosphorus(III)

$GeHMe_3$      trimethylgermane
            hydridotrimethylgermanium(IV)

$PbEt_4$        tetraethylplumbane
            tetraethyllead(IV)

$SnMe_4$       tetramethylstannane
            tetramethyltin(IV)

$SiMe_4$        tetramethylsilane
            tetramethylsilicon(IV)

## REFERENCES

1. M. L. H. Green and P. Powell, *Comprehensive Inorganic Chemistry* (J. C. Bailar, Jr., ed.), Vol. 1, p. 1295, Pergamon Press, Oxford (1973).

2. R. B. King, *Transition Metal Organometallic Chemistry*, Academic Press, New York (1969).
3. G. E. Coates, M. L. H. Green, and K. Wade, *Organometallic Compounds*, Vols. 1 and 2, Methuen, London (1968).
4. D. A. Boczwar, *et al.*, *Metody Elementoorganicheskoy Khimii*, Vols. 1 and 2, Izd. Nauka, Moscow (1975).
5. A. N. Niesmiejanow, *et al.*, *Metody Elementoorganicheskoy Khimii*, Vols. 1 and 2, Izd. Nauka, Moscow (1974).
6. S. P. Gubin, *et al.*, *Metody Elementoorganicheskoy Khimii*, Izd. Nauka, Moscow (1976).
7. C. S. G. Phillips and R. J. P. Williams, *Inorganic Chemistry*, Oxford University Press, Oxford (1966).
8. J. P. Oliver, *Adv. Organomet. Chem.*, **15**, 235 (1977).
9. D. M. P. Mingos, *Adv. Organomet. Chem.*, **15**, 1 (1977).
10. E. M. Szustorowicz, *Khimicheskaya Svyaz*, Izd. Nauka, Moscow (1973).
11. J. B. Bersuker, *Elektronnoye Stroyenye i Svoystva Koordinacyonnykh Soyedinyeniy*, Khimia, Leningradskoye Otdelenie (1976).
12. F. A. Cotton and G. Wilkinson, *Basic Inorganic Chemistry*, Wiley, New York (1976); F. A. Cotton and G. Wilkinson, *Advanced Inorganic Chemistry*, 4th ed., Wiley, New York (1980).
13. L. Pajdowski, *Chemia ogolna*, PWN, Warsaw (1981).
14. M. E. O'Neil and K. Wade, *Comprehensive Organometallic Chemistry* (G. Wilkinson, F. G. A. Stone, and E. W. Abel, eds.), Vol. 1, Chapter 1, Pergamon, Oxford (1982).
15. W. Kutzelnigg, *Angew. Chem., Int. Ed. Engl.*, **23**, 272 (1984).
16. C. M. Lukehart, *Fundamental Transition Metal Organometallic Chemistry*, Brooks/Cole Publishing Company, Monterey (1985).
17. S. P. Gubin and G. B. Szulpin, *Khimia Kompleksov so Svazyami Metall-uglerod*, Nauka, Sibirskoye Otdelenie, Novosibirsk (1984).
18. F. Basolo and R. G. Pearson, *Mechanisms of Inorganic Reactions*, Wiley, New York (1967).
19. M. L. Tobe, *Inorganic Reaction Mechanisms*, Thomas Nelson & Sons Ltd., London (1972).
20. A. Bielanski, *Chemia nieorganiczna*, PWN, Warsaw (1982).
21. V. Gutmann, *Coord. Chem. Rev.*, **18**, 225 (1976).
22. E. Cesarotti, R. Ugo, and L. Kaplan, *Coord. Chem. Rev.*, **43**, 275 (1982).
23. C. A. Tolman, *Chem. Soc. Rev.*, **1**, 337 (1972).
24. F. A. Cotton, *J. Am. Chem. Soc.*, **90**, 6230 (1968).

# Chapter 2

# Metal Carbonyls

Metal carbonyls are compounds of transition metals with carbon monoxide ligands. They find applications in many catalytic processes and in organic synthesis. Thus, research on preparations, structures, and applications of metal carbonyls and their derivatives has been intensive for several years. The first metal carbonyl, $Ni(CO)_4$, was prepared by A. Mond, G. Langer, and F. Quincke[1] in 1890 by the reaction of metallic nickel with carbon monoxide. It found application immediately; it has been used for industrial preparation of pure nickel. Many other carbonyls were synthesized shortly thereafter.[2]

## 1. PREPARATION OF METAL CARBONYLS

Only $Ni(CO)_4$ and $Fe(CO)_5$ can be prepared by direct reaction of the metal with carbon monoxide. Small amounts of $Fe(CO)_5$ are formed, for example, in steel cylinders in which carbon monoxide is stored. Other metals do not react with carbon monoxide at all or react with it with great resistance, so that only small amounts of the carbonyls can be obtained by the reaction of metals with carbon monoxide.

Most commonly, metal carbonyls are prepared by reduction of transition metal compounds. As reducing agents, the following are employed: excess of carbon monoxide, carbon monoxide and hydrogen mixture, metals of groups 1, 2, and 13 and their organometallic compounds, such as RMgX, $Al_2R_6$, and LiR. The synthesis of the carbonyls may also be carried out via electrochemical reduction[3] of transition metal compounds in organic solvents by employing as an anode aluminum or a metal of a carbonyl being synthesized. Stainless steel, copper, etc., cathodes are used. Another recent, commonly utilized method for metal carbonyl preparation, especially for unstable ones, is direct condensation of metal atoms and carbon monoxide in a low-temperature matrix (usually $< 10$ K).[4] Many organometallic and inorganic complexes are prepared by similar reactions involving low-temperature cocondensation of metal vapors and other substrates (Chapters 8–10).

Commonly utilized starting transition metal materials are oxides, halides, salts of fatty acids, complexes containing sulfur, etc. The role of the reducing agents is to lower the oxidation state of a metal and not to reduce the compound to an active metallic phase because this would, in most cases, lead to the inhibition of the reaction as shown by thermodynamic calculations. This point is also indicated by the formation of carbonyl derivatives possessing CO and other coordinating ligands as intermediates.

The role of some promoters (i.e., ligands containing sulfur) is to lower the oxidation–reduction potential of a metal due to a complex formation with the promoter. The synthesis is carried out with or without solvent. Usually, high pressures (up to 30–40 MPa) and elevated temperatures (up to 600 K) are employed.

The essential effect of a solvent is not limited to its coordinating ability; no less important is its interaction with a reducing agent. Group 1 and 2 metals, in an ether solvent such as tetrahydrofuran, are slightly soluble and easier reduction of the transition metal occurs as a result of an electron transfer from the reducing agent to the reduced compound via the solvent molecule. When substances that are able to form stable radical anions (for example, pyridine, naphthalene, benzophenone) are added to solvents, the $d$-electron metal is reduced by the radical anions. The syntheses carried out with or without solvents may be, in both cases, homogeneous or heterogeneous. The preparation of $Os(CO)_5$, $Re_2(CO)_{10}$, and $Tc_2(CO)_{10}$ from $OsO_4$, $Re_2O_7$, and $Tc_2O_7$, respectively, at elevated temperatures without a solvent is carried out in a homogeneous medium. However, in some syntheses involving solvents, the reagents are in different phases.

## 2. *BONDING IN METAL CARBONYLS*

Carbon monoxide (Figure 2.1) has the following electron configuration[5–7]: $1\sigma^2 2\sigma^2 3\sigma^2 4\sigma^2 1\pi_x^2 1\pi_y^2 5\sigma^2$. The molecular orbital $5\sigma$ is located mainly on the carbon atom and corresponds to the lone-pair electrons in the valence-bond theory. Its character is slightly antibonding.[5–7]

Carbon monoxide is a $2e$ ligand; it is a moderately weak $\sigma$-donor and quite a strong $\pi$-acceptor (Table 1.1). In carbonyls, CO is bonded to metals as terminal or bridging ligands. Bridging usually involves bonds to two or three metal atoms (Figure 2.2).

Almost always CO is bonded to metal atoms through carbon.[8–13] In exceptional cases, CO may act as a bridging $4e$ ligand bonded simultaneously to metals through carbon and oxygen, for example, $Mn_2(CO)_5 (Ph_2PCH_2PPh_2)_2$[14] and $[Fe_4(CO)_{13}H]^-$,[15] and even as a $6e$ ligand in $Nb_3cp_3(CO)_7$[16, 17] (Figure 2.3).

In $Mn_2(CO)_5 (Ph_2PCH_2PPh_2)_2$, the bridging carbonyl group is bonded to one manganese atom only through carbon, while with the other manganese atom it forms a bond through carbon and oxygen. The latter bond is a result of an overlap between $1\pi$ and $2\pi$ CO orbitals and $d\sigma$ and $d\pi$ manganese orbitals; this bonding is therefore analogous to $\pi$-olefin bonding in alkene complexes such as $[PtCl_3C_2H_4]^-$.

In $Nb_3Cp_3(CO)_7$ the bridging carbon monoxide is bonded to one niobium atom through carbon, and with the other two Nb atoms it forms a $\pi$-olefin bond (a six-electron ligand).

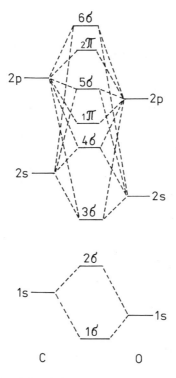

Figure 2.1.  MO scheme for carbon monoxide molecule.

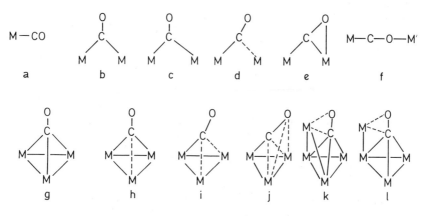

Figure 2.2.  Possible modes of bonding of CO to metal atoms: (a) terminal ligand, (b) symmetrical $\mu$-CO group, (c) asymmetrical $\mu$-CO group, the CO vector is perpendicular to M—M, (d) asymmetrical $\mu$-CO group, semibridging, the CO vector is not perpendicular to M—M, (e) asymmetrical $\mu$-$\eta^1$:$\eta^2$-CO group, 4e ligand, (f) $\mu$-CO($C, O$) group, M-soft acid, M′-hard acid, (g) symmetrical $\mu_3$-CO group, (h, i) asymmetrical $\mu_3$-CO groups usually encountered in heterometallic complexes, (j) $\mu_3$-$\eta^1$:$\eta^2$-CO group, 6e ligand, (k, l) $\mu_4$-$\eta^1$:$\eta^2$-CO groups, 4e ligands.

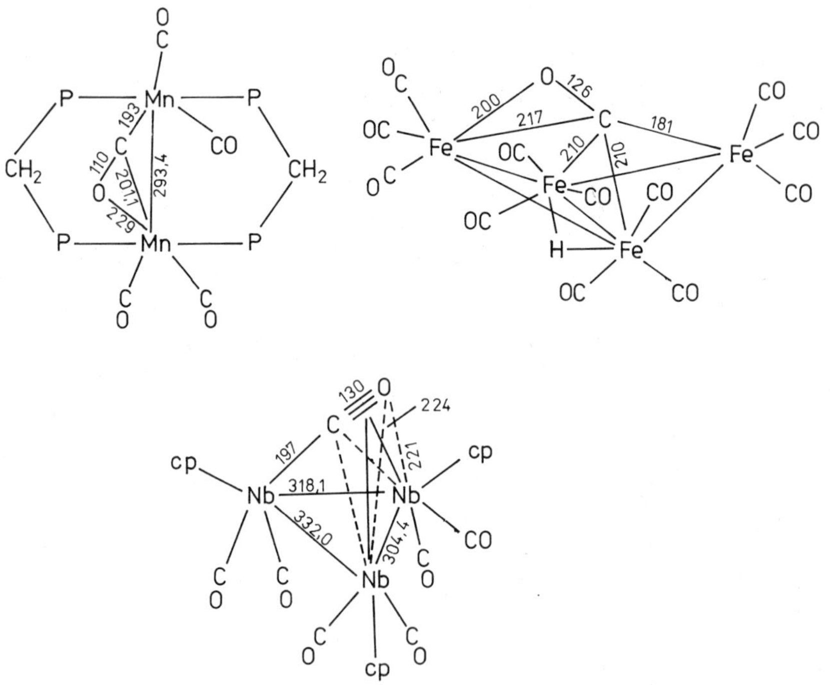

Figure 2.3. Structures of $[Mn_2(CO)_5][(Ph_2PCH_2PPh_2)_2]$, $[Fe_4H(CO)_{13}]^-$, and $[Nb_3Cp_3(CO)_7]$. The distances are in pm units.

Semibridging $\mu_2$-CO ligands (Figure 2.2d) or asymmetrical $\mu_3$-CO groups bonded to central atoms only through carbon are most common in heterometallic complexes. The differences between distances MC and M′C are sometimes very substantial. In $(OC)_2Co(\mu\text{-}CO)_2RhCO(PEt_3)_2$, these differences are 78 pm and 70 pm.[18]

Often, bridging carbonyl groups exist in pairs:

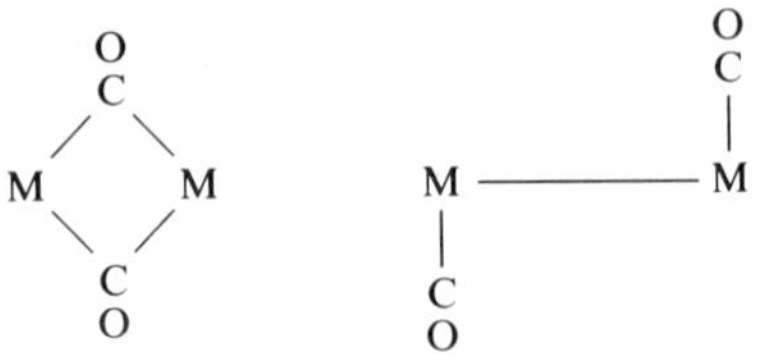

In such cases, compounds may exist in equilibrium involving bridging structures (a) as well as structures possessing terminal CO groups (b). In intermediate cases, asymmetrical bridges with great differences in M–C distances are formed. Examples are $Fe_3(CO)_{12}$ and $Fe_3(CO)_8(SC_4H_8)_2$ (Figure 2.4).

The tendency to form carbonyls with bridging CO groups decreases with increase in atomic size of the metal. This trend is observed in groups as well as in series of the

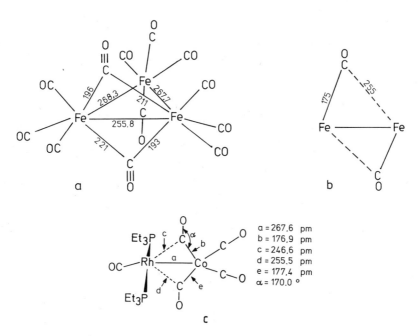

Figure 2.4.  Structures of (a) $Fe_3(CO)_{12}$, (c) $(OC)_2Co(\mu\text{-}CO)_2Rh(CO)PEt_3)_2$, and (b) the distances of bridging bonds in $Fe_3(CO)_8(SC_4H_8)_2$. The distances are in pm units.

Periodic Table. In $Fe_3(CO)_{12}$, carbon monoxide forms bridges, while in $Ru_3(CO)_{12}$ and $Os_3(CO)_{12}$, all CO groups are terminal because atomic radii of Ru and Os are larger than the radius of Fe. Similarly, manganese, having a relatively larger atomic radius, forms a metal carbonyl, $(CO)_5Mn-Mn(CO)_5$, with terminal CO groups. However, in the case of cobalt, which has a smaller atomic radius, there is equilibrium between bridged and nonbridged forms of $Co_2(CO)_8$.

In complexes with increased electron density on the central atom, the oxygen of the CO group shows somewhat basic properties, and with Lewis acids such as $AlCl_3$, $AlR_3$, $Ln(Cp)_3$ (where Ln is the lanthanide metal), etc., this oxygen may form compounds in which the bridging CO group is bonded to two different metal atoms through carbon and oxygen (Figure 2.2), for example, $Fe_2(Cp)_2(CO)_4 \cdot 2AlEt_3$.[170, 171]

The metal–carbonyl group bond may be described by the following resonance structures:

$$M^- - C^+ \equiv O : \longleftrightarrow M = C = O :$$

However, a more accurate description of this bonding comes from molecular orbital theory. Thus, the metal–CO bond (Figure 2.5) is formed as a result of an overlap between a weakly antibonding $\sigma$-orbital, which is mainly localized on the carbon atom (lone pair on carbon), and empty hybridized metal orbitals, as well as a result of the formation of back-bonding due to an interaction between filled $d\pi$ metal orbitals and antibonding $2\pi$ orbitals of CO.[7–10, 12, 19–24]

The formation of $\sigma$ M$-$C bonds should lead to concentration of electron density on the metal atom. The $\pi$ back-bonding prevents this concentration, since a reverse process takes place—there is donation of electron density back into the antibonding $2\pi$ orbitals of CO. Thus, the $\pi$-bond enhances the $\sigma$-acceptor properties of the metal, as well as the $\sigma$-donor properties of CO. This situation leads to the synergic effect: the increase in stability of $\sigma$ bond strengthens the $\pi$ bonding, and *vice versa*. Because the $5\sigma$ orbital is of antibonding character, greater stability of the $\sigma$ M$-$C bond leads to increased multiplicity of the C$-$O bond. However, in the case of $\pi$-bonding, this effect is reversed; the electron density increases on antibonding orbitals $2\pi_x$ and $2\pi_y$.

A more precise description of electronic structures of metal carbonyls is achieved by employing molecular orbital theory for specific cases. For octahedral metal carbonyls, the molecular orbital scheme is represented in Figure 2.6.[8, 10, 12, 19, 22–24]

In the $O_h$ group, the $\pi$ bonding and $\pi^*$ antibonding orbitals of six CO ligands form four triply degenerate linear combinations belonging to the representations $t_{1g}$, $t_{1u}$, $t_{2g}$, and $t_{2u}$. The $t_{2g}(\pi)$ and $t_{2g}(\pi^*)$ orbitals may overlap with the $t_{2g}$ metal orbitals. Less important is the interaction of metal orbitals with the filled $\pi$ orbitals of the CO group,

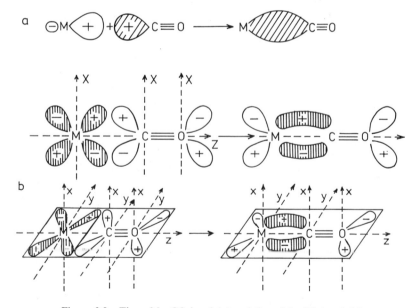

Figure 2.5.  The $\sigma$ M$-$CO bond (a) and the $\pi$ M$-$CO bond (b).

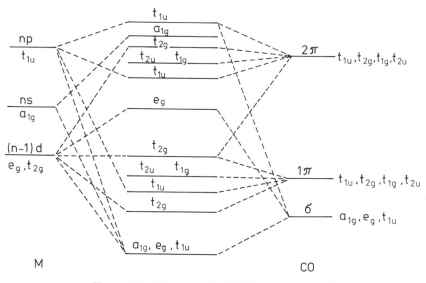

Figure 2.6.  MO scheme for $M(CO)_6$ molecules ($O_h$).

since they have considerably lower energy than the metal $d$ orbitals (Figure 2.1). The overlap of $t_{2g}(\pi^*)$ orbitals with $d_{xy}$, $d_{xz}$, and $d_{yz}$ orbitals causes an increase of the splitting parameter $\varDelta$ and a lowering of the multiplicity of the carbon–oxygen bond. The bond order, assuming that the $t_{2g}$ orbital energy of the metal and the $\pi^*$ orbital energy of carbon monoxide is the same, may be as low as 2.75. However, this bond order is quite arbitrary and unattainable, because the $d$-orbital energy of a free metal atom is lower than the $\pi^*$-orbital energy of CO, and therefore the interaction of such orbitals is less effective. Although the interaction between $2\pi$ antibonding orbitals with $d\pi$ orbitals is far more important than the interaction between $d\pi$ and $1\pi$ orbitals, the latter interaction cannot be neglected.[24, 25] The relative importance of the $\sigma$-donor and $\pi$-redonor factors is a controversial matter. Calculations show that, while the contribution of the $\sigma$ donor factor is 80%, that of the $\pi$ acceptor is 20%.[26] However, other authors give more balanced contributions of these two factors: 55% $\sigma$-donor, 45% $\pi$-redonor.[27]

In the case of tetrahedral complexes, the situation becomes more complicated because linear combinations of both $\sigma$ orbitals as well as $\pi$ and $\pi^*$ orbitals possess functions belonging to $t_2$ representation (Figure 2.7). This leads to mixing of $\sigma$, $\pi$, and $\pi^*$ orbitals of the CO group. In such tetrahedral complexes, the interaction of metal $d$ orbitals with $2\pi$ antibonding orbitals of the CO molecule is more important than the interaction of metal $d$ orbitals with $1\pi$ orbitals, since the latter have relatively low energy. As in octahedral complexes, the formation of back-bonding due to an overlap between $t_2$ and $e$ metal orbitals and $2\pi$ orbitals of the CO group possessing the same symmetry plays an important role. Cu(I) confirms this scheme, showing only a weak tendency to coordinate CO, and Zn(II) does not form carbonyls at all, because the energy of the $d$ orbitals is considerably lower than that of $2\pi$ orbitals, which therefore makes interaction of these orbitals ineffective. Similarly, the vibrational frequency $\nu(CO)$ decreases, causing an increase of electron density on $2\pi$ orbitals when going from $Ni(CO)_4$

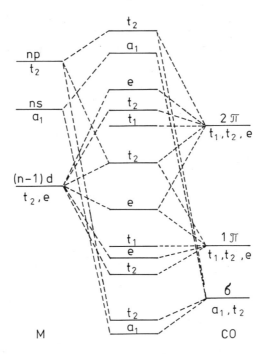

Figure 2.7.   MO scheme for $M(CO)_4$ molecules $(T_d)$.

through $[Co(CO)_4]^-$ to $[Fe(CO)_4]^{2-}$ because, as the atomic number decreases and the negative charge on the metal increases, the $d$-orbital energy becomes greater. From the presented molecular orbital schemes, it may be concluded that ligands which show strong $\pi$-acceptor properties form covalent bonds with the metal.

Molecular orbitals in complexes may be divided into three groups. The first group contains orbitals which are bonding and localized mainly on ligands. The second group comprises nonbonding orbitals, and to the third group belong antibonding orbitals. The first two groups of orbitals are filled. Therefore, in covalent coordination compounds, all valence metal orbitals can be utilized: $(n-1)\,d$, $ns$, and $np$. Thus, these are $18e$ compounds which obey Sidgwick's rule.[28] This rule states that the sum of valence electrons possessed by the metal plus electrons contributed by the ligands is equal to the atomic number of the nearest, heavier, noble gas atom. Metal carbonyls show such a great tendency to form $18e$ complexes that, for the metals possessing odd atomic numbers such as Mn, Tc, Re, Co, Rh, and Ir, only polynuclear stable carbonyls are known.

As a result of the $M-M$ bond formation, each central atom achieves $18e$ configuration. In $Re_2(CO)_{10}$, the structure of which is shown in Figure 2.8, each rhenium atom has 18 valence electrons:

$$
\begin{array}{ll}
5CO & 10 \\
Re & 7 \\
Re(CO)_5 & \underline{\phantom{0}1\phantom{0}} \\
 & 18e
\end{array}
$$

The $Re(CO)_5$ group is a $1e$ ligand, since a single $Re-Re$ bond is formed.

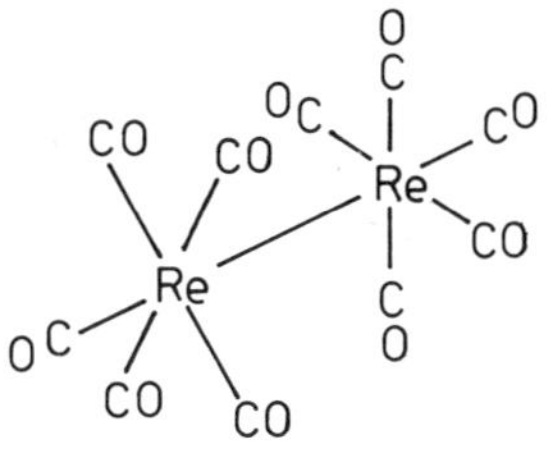

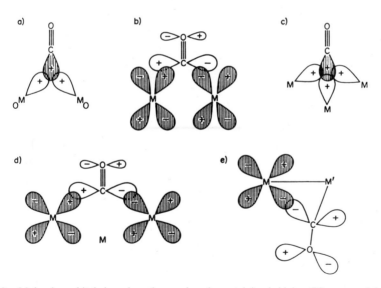

Figure 2.8. The structure of $[Re_2(CO)_{10}]$.

In polynuclear complexes, CO groups may constitute bridging ligands bonding two, three, or four metal atoms. Figure 2.9 represents the overlap between $5\sigma$ and $2\pi$ orbitals of CO with the corresponding metal orbitals. Such bonds should be treated as involving multicenter, delocalized orbitals. Structural studies seem to confirm this type of bonding; the $M-C-M$ angles are in the range of 80–85°, and are similar to three-center bonds in $Al_2(CH_3)_6$ and $[Be(CH_3)_2]_n$.

The formation of polynuclear carbonyl clusters containing an $M-M$ bond is only possible when the metal–metal bond energy is quite substantial.[29] As a measure of this energy, the enthalpy of metal vaporization is usually used. The number of clusters, as well as their stability, ordinarily increases when the atomic number of the metal increases within a given group of the Periodic Table, since the enthalpy of vaporization increases in the same direction. The values of heats of vaporization of some metals in kJ $mol^{-1}$ are as follows: Mn—289, Tc—648, Re—778, Fe—418, Ru—648, Os—678, Co—427, Rh—556, Ir—669, Ni—427, Pd—354, Pt—565, Cu—339, Ag—285, and Au—364.

Figure 2.9. Molecular orbitals in polynuclear carbonyls containing bridging CO groups: (a), (c) the $\sigma$ donor bond, (b) the $\pi$ redonor bond, (d) the back-bonding to three metal atoms (one of the two degenerate orbitals), (e) the $d\pi-p\pi$ redonor bond for the semibridging CO ligand.

Palladium and silver, which have very low metal–metal bond energy values, form considerably fewer clusters than their homologs, and their stable polynuclear carbonyls are not known.

## 3. IR SPECTRA OF METAL CARBONYLS

Infrared spectra of metal carbonyls constitute a valuable source of information concerning both structure and bonding.[20, 8, 9, 24, 32] The stretching frequency $v(CO)$ in carbonyls is lower than the stretching frequency of the CO molecules due to back-bonding. Because of the weakly antibonding character of the $5\sigma$ orbital on carbon monoxide, the formation of $\sigma M - CO$ should strengthen the CO bond, and thus increase the frequency $v(CO)$. The range in which the $v(CO)$ bands are found depends on the nature of the carbonyl groups present in the complex. For terminal CO groups, $v(CO)$ has a higher value than it does for bridging groups. From positions of $v(CO)$ bands in the spectrum, the electronic structure of the $M - C \equiv O$ fragment may be inferred, and the structure of the complex may be determined from the number of bands and their intensities.

The bands due to $v(CO)$, $v(MC)$, $\delta(MCO)$, and $\delta(CMC)$ vibrations are presented in Table 2.1.

Table 2.1. Bands due to $v(CO)$, $v(MC)$, $\delta(MCO)$, and $\delta(CMC)$ Vibration

| Vibration | Structure | $v$ (cm$^{-1}$) |
|---|---|---|
| $v(CO)$ | $M - CO$ | 1900–2200, for anions 1600–2000 |
| | 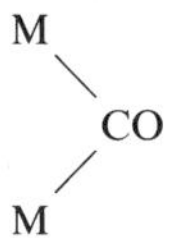 M\\CO/M | 1800–1900 |
| | 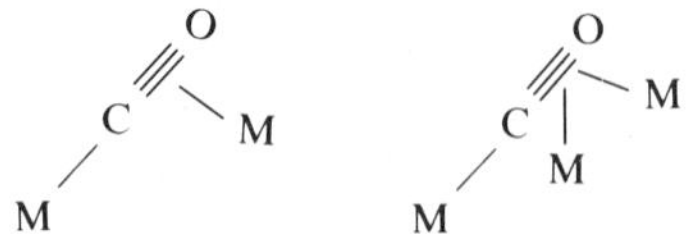 C(=O)(M) with M; and C(=O)–M with M/M | 1330–1650 |
| | 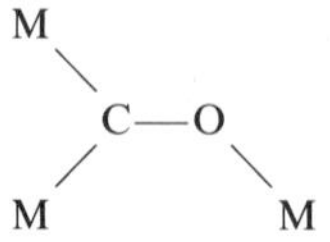 M\\C—O/M \\M | 1270–1700 |
| | 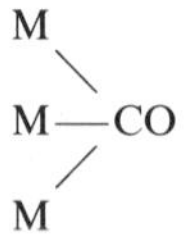 M\\M—CO/M | 1700–1800 |
| | $M - C \equiv O - M'$ | 1650 |
| $v(MC)$ and $\delta(MCO)$ | | 300–700 |
| $\delta(CMC)$ | | 50–120 |

The intensity of $v(CO)$ bands is usually strong. This intensity depends on the square of the size of the oscillating dipole. The relatively large value of the oscillating dipole is caused by the polarity of the bond, or as a result of changes in the electron structure of the bond. In the case of carbon monoxide and metal carbonyls, polarity of the $C-O$ bond is weak. Thus, the relatively strong intensity of $v(CO)$ bands in metal carbonyls is caused by the character of a metal–carbonyl bond, which leads to considerable changes in electronic structure with changes in distances between carbon and oxygen. The amplitude of $C-O$ vibrations is 6 pm in the first excited vibrational state. This lowers the $2\pi$-orbital energy by 20,000 cm$^{-1}$, and strengthens metal–carbonyl back-bonding, thus changing charge distribution in the $M-C\equiv O$ group. This explains the significant value of the oscillating dipole and strong intensity of $v(CO)$ bands in metal carbonyls. The intensity may be small, and even the $v(CO)$ band could be forbidden if the oscillating dipoles of particular CO groups are oppositely directed, leading to compensation.

The bands caused by $M-C$ vibrations are less intense than $v(CO)$ bands, but in general stronger than $\delta(MCO)$. The changes in the $M-C$ bond distances have therefore less influence on the character of the $M-C$ bond than the changes in the $C-O$ bond distances. Consequently, the analysis of IR spectra in the $v(CO)$ vibrational region furnishes valuable information concerning the electronic structure and symmetry of the carbonyls.

*Table 2.2. The Number of $v(CO)$ Bands in Carbonyl Complexes*[a]

| Compound, group | Symmetry | Vibration symmetry | Activity and intensity of bands |
|---|---|---|---|
| $M(CO)_2$: CO groups equivalent, $<CMC=\theta$ | $D_{\infty v}$ | $\Sigma_g^+$ | $R$(pol) |
| | | $\Sigma_g^-$ | IR, intensity ($I$) large |
| | $C_{2v}$ | $A_1$ | $R$(pol), IR (for $\theta\to180°$, $I\to0$) |
| | | $B_1$ | IR, $I$ large, $R$(dp), $I(A_1)/I(B_1)=\mathrm{ctg}^2(\theta/2)$ |
| | $C_s$ | $A'$ | IR, $R$(pol) |
| | | $A''$ | IR, large, $R$(dp) |
| | $D_{3h}$ | $A_1'$ | $R$(pol) |
| | | $A_2''$ | IR, $I$ large |
| $M(CO)_3$: CO groups equivalent, $<CMC=\theta$ | $C_3$ | $A$ | IR, $R$(pol) |
| | | $E$ | IR, $R$(dp) |
| | $C_{3v}$ | $A_1$ | IR, $I$ medium, $R$(pol) |
| | | $E$ | IR, $I$ large, $R$(dp), $I(A_1)/I(E)=\frac{3}{4}\mathrm{ctg}(\theta/2)-\frac{1}{4}$ |
| $M(CO)_3$: CO groups nonequivalent | $C_s$ | $2A'$ | IR, $R$ |
| | | $A''$ | IR, $R$ |
| | $C_{2v}$ mar- $M(CO)_3L_3$, $M(CO)_3L$ (square) $M(CO)_3L_2$ trigonal bipyramidal with equatorial L | $2A_1$ | IR, $R$(pol) |
| | | $B_2$ | IR, $R$(dp) |

[a] The number of bands are given for solution spectra.

*(Table continued)*

*Chapter 2*

*Table 2.2. (continued)*

| Compound, group | Symmetry | Vibration symmetry | Activity and intensity of bands |
|---|---|---|---|
| $M(CO)_4$: CO groups equivalent | $T_d$, $Ni(CO)_4$ | $A_1$ | $R$(pol) |
|  |  | $T_2$ | IR, $R$(dp) |
|  | $D_{4h}$ | $A_{1g}$ | $R$(pol) |
|  |  | $B_{2g}$ | $R$(dp) |
|  |  | $E_u$ | IR, $I$ large |
| $M(CO)_4$: CO groups nonequivalent | $C_{3v}$ | $2A_1$ | IR, $R$(pol) |
|  | $M(CO)_4^- (M')^+$ | $E$ | IR, $R$(dp) |
|  | $(OC)_3MCO-M'$ |  |  |
|  | $C_{2v}$ | $A_1$ | IR, $I$ small, $R$ |
|  | cis-$M(CO)_4L_2$, incorporation | $A_1$ | IR, $I$ medium, $R$ |
|  | of an ionic tetrahedron | $B_1$ | IR, $I$ large, $R$ |
|  | into an ion pair of | $B_2$ | IR, $I$ large, $R$ |
|  | $C_{2v}$ symmetry |  |  |
|  | $C_{4v}$ | $2A_1$ | IR, $R$(pol) |
|  | $[M(CO)_4L]$ | $E$ | IR, $R$(dp) |
|  | $D_{2d}$ | $A_1$ | $R$ |
|  | $M(CO)_4$ intermediate | $B_2$ | IR, $R$ |
|  | structure between | $E$ | IR, $R$ |
|  | square and tetrahedron |  | $I(B_2)/I(E) = \mathrm{tg}^2\,\omega$ |

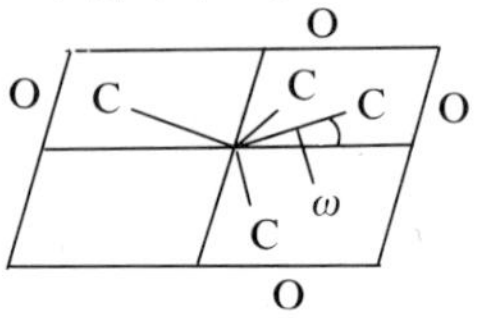

| $M(CO)_5$: CO groups nonequivalent | $D_{3h}$ | $2A_1'$ | $R$ |
|---|---|---|---|
|  |  | $A_2''$ | IR |
|  | $Fe(CO)_5$ | $E'$ | IR, $R$ |
|  | $C_{4v}$ | $2A_1$ | IR, $I$ medium ($A_1^1$), $I$ small ($A_1^2$), $R$ |
|  | $[M(CO)_5L]$ | $B_2$ | $R$ |
|  |  | $E$ | IR, $I$ large, $R$ |
| $M(CO)_6$: CO groups equivalent | $O_h$ | $A_{1g}$ | $R$ |
|  |  | $E_g$ | $R$ |
|  | $[Cr(CO)_6]$ | $T_{1u}$ | IR, $I$ large |
| $[M(CO)_2]_2$ | $D_2$ | $A$ | $R$ |
|  | 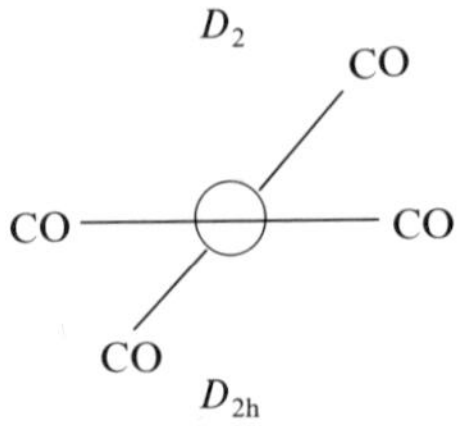  | $B_1$ | IR, $R$ |
|  |  | $B_2$ | IR, $R$ |
|  |  | $B_3$ | IR, $R$ |

|  | $D_{2h}$ | $A_g$ | $R$ |
|---|---|---|---|
|  | 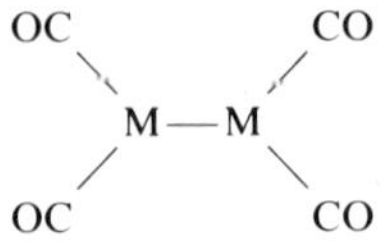  | $B_{1u}$ | IR |
|  |  | $B_{3u}$ | IR |
|  |  | $B_{2g}$ | $R$ |

*(Table continued)*

Carbonyls* 35

*Table 2.2. (continued)*

| Compound, group | Symmetry | Vibration symmetry | Activity and intensity of bands |
|---|---|---|---|
| $[M(CO)_2]_2$ | $D_{2d}$ 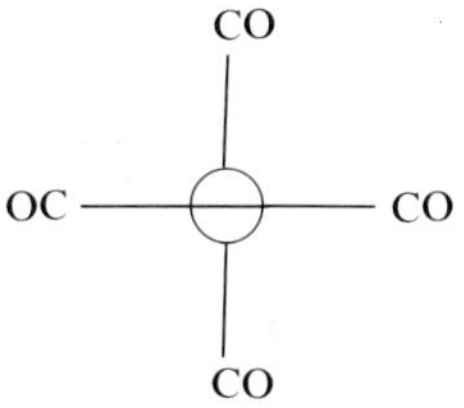 | $A_1$ | $R$ |
| | | $B_2$ | IR |
| | | $E$ | IR, $R$ |
| | $C_{2v}$ 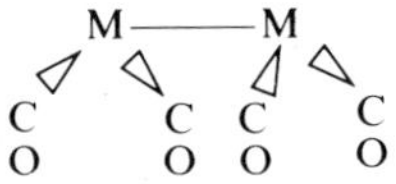 | $A_1$ | IR, $R$ |
| | | $B_1$ | IR, $R$ |
| | | $B_2$ | IR, $R$ |
| | | $A_2$ | $R$ |
| | $C_{2h}$ 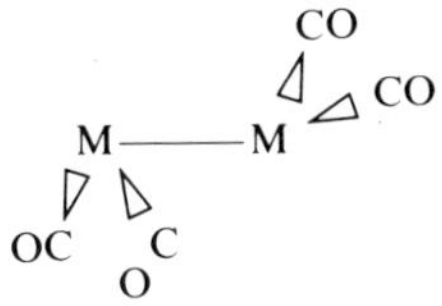 | $A_g$ | $R$ |
| | | $B_u$ | IR |
| | | $A_u$ | IR |
| | | $B_g$ | $R$ |
| $[M(CO)_3]_2$ | $D_{3d}$ 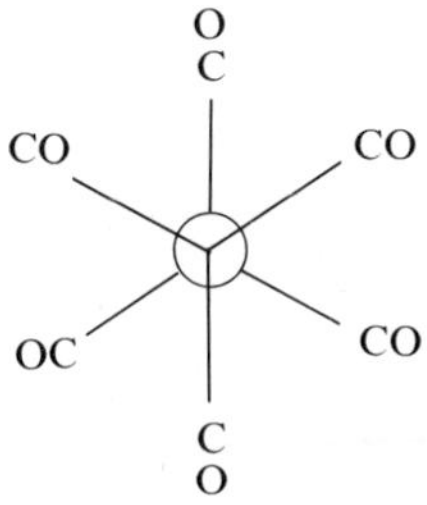 | $A_{1g}$ | $R$ |
| | | $A_{2u}$ | IR |
| | | $E_g$ | $R$ |
| | | $E_u$ | IR |
| | $D_{3h}$ 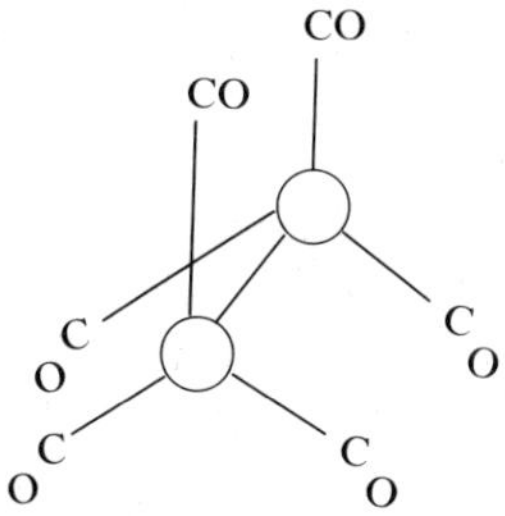 | $A_1'$ | $R$ |
| | | $A_2''$ | IR |
| | | $E'$ | IR, $R$ |
| | | $E''$ | $R$ |

*(Table continued)*

*Table 2.2.* *(continued)*

| | | Vibration symmetry | Activity and intensity of bands |
|---|---|---|---|
| $[M(CO)_3]_4$ | $T_d$ | $A_1$ | $R$ |
| | $[Ir_4(CO)_{12}]$ | $E$ | $R$ |
| | | $T_1$ | inactive |
| | | $2T_2$ | IR, $R$ |
| $[M(CO)_4]_2$ | $D_{2h}$ 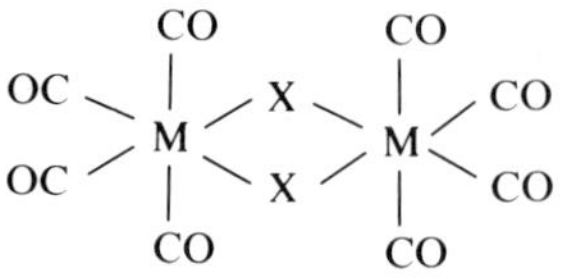 | $2A_g$ | $R$ |
| | | $2B_{1u}$ | IR |
| | | $B_{2u}$ | IR |
| | | $B_{3u}$ | IR |
| | | $B_{2g}$ | $R$ |
| | | $B_{3g}$ | $R$ |
| $[M(CO)_4]_3$ | $D_{3h}$ | $2A_1'$ | $R$ |
| | $Os_3(CO)_{12}$ | $A_2'$ | |
| | | $A_2''$ | IR |
| | | $3E'$ | IR, $R$ |
| | | $E''$ | $R$ |
| $[M(CO)_4]_2$ | $D_{4d}$ 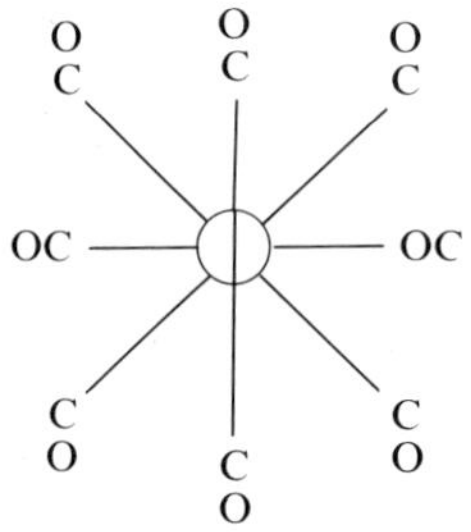 | $A_1$ | $R$ |
| | | $B_2$ | IR |
| | | $E_1$ | IR |
| | | $E_2$ | $R$ |
| | | $E_3$ | $R$ |
| | $D_{4h}$ 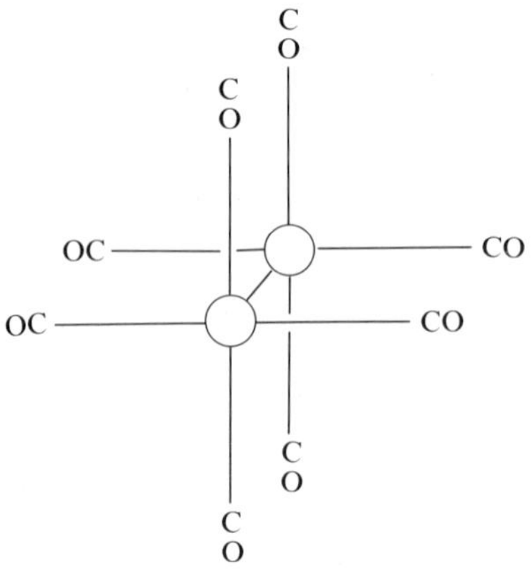 | $A_{1g}$ | $R$ |
| | | $A_{2u}$ | IR |
| | | $B_{2g}$ | $R$ |
| | | $B_{1u}$ | inactive |
| | | $E_g$ | $R$ |
| | | $E_u$ | IR |
| $[M(CO)_5]_2$ | $D_{4d}$ | $2A_1$ | $R$ |
| | $[Mn_2(CO)_{10}]$ | $2B_2$ | IR |
| | | $E_1$ | IR |
| | | $E_2$ | $R$ |
| | | $E_3$ | $R$ |

Metal carbonyls may be classified into the following two groups:

1. Carbonyls in which all CO groups are equivalent,
2. Carbonyl complexes possessing at least two different types of CO groups.

The first group consists of carbonyls in which the following groups are present: $M(CO)_3$ ($D_{3h}, C_{3v}$), $M(CO)_4$ (tetrahedron $T_d$, square $D_{4h}$, distorted square $D_{2d}$) as well as $M(CO)_6$ ($O_h, D_{3d}, D_{3h}$). The second category includes: $M(CO)_5$ ($D_{3h}, C_{4v}$), cis-$L_2M(CO)_4$ ($C_{2v}$), and mer-$L_3M(CO)_3$ ($C_{2v}$). The number of $v(CO)$ bands and their intensities as well as their symmetries are given in Table 2.2. The number of $v(CO)$ bands given in this table refers to the spectra in solution. For solid carbonyls, this number is usually greater. This results not only due to a different symmetry of a compound in the solid state and symmetry of the crystal, but also due to intermolecular interactions between CO groups.

The position of $v(CO)$ bands depends on the electronic configuration of the metal, its $d$ valence electron energy, and the interaction of filled $d$ orbitals with $2\pi$ antibonding orbitals of the CO molecules. All of this determines the multiplicity of the carbon–oxygen bond. Thus, an increase in the $M-C$ bond strength causes a decrease in the $C-O$ bond stability. As the effective metal nuclear charge is lowered and the negative charge on the metal atom increases, the $M-CO$ interaction also increases. This is illustrated by the lowering of $v(CO)$ frequencies and thus also of the $k(CO)$ force constant, and by increasing the $k(M-C)$ force constant for the following isoelectronic series of complexes: $[M(CO)_4]^{n-}$ and $[M(CO)_6]^{m\pm}$ (Table 2.3). These effects are so strong for these compounds that the $v(CO)$ bands occur in the region typical for the bridging CO group. As a result of mutual compensation of kinematic and dynamic interactions of $v(MC)$ and $v(CO)$ vibrations, the $v(CO)$ band energy may be directly used as a measure of metal–carbonyl bond strength. For this purpose, the average value of energy of all $v(CO)$ bands is used, calculated from the formula $\langle v(CO) \rangle = \sum g_i v_i / \sum g_i$, where $g_i$ represents degeneracy of the $i$th vibration of frequency $v_i$.

The $\pi$ back-bonding causes a shift of the charge from the metal atom to the CO group which lowers the $1s$ oxygen electron energy, and therefore increases the contribution of the structure (**b**) in the resonance of the $M-C\equiv O$ bond,

$$M^- - C^+ \equiv O : \longleftrightarrow M = C = O :$$

$$\quad\quad\text{a} \quad\quad\quad\quad\quad \text{b}$$

It has been found that a very good linear relationship exists between the average band energies $\langle v(CO) \rangle$ or corresponding degeneracy-weighted stretching force constants $\langle k(CO) \rangle$ and the $1s$ oxygen or carbon atom electron energy for many different metal carbonyls[30] (Figure 2.10, Table 2.7 below).

To calculate the $k(CO)$ force constants, the Cotton–Kraihanzel model[31] is often applied, which assumes that CO force constants can be calculated from the CO stretching frequencies alone because these are at much higher frequencies ($>1800$ cm$^{-1}$) than all other vibrations ($<700$ cm$^{-1}$) in simple metal carbonyls and in many substituted carbonyls. Furthermore, this model assumes that the vibrations are truly harmonic and that the interaction constants between cis- and trans-CO groups are related in a simple manner. It is assumed that the interaction of two trans-CO groups is twice as large as

Table 2.3. Dependence of $\nu(CO)$ $(cm^{-1})$ on the Charge of $M(CO)_m^{n\pm}$

| Frequency | $[Ni(CO)_4]$ | $[Co(CO)_4]^-$ | $[Fe(CO)_4]^{2-}$ | $[Mn(CO)_4]^{3-}$ | $[Re(CO)_4]^{3-}$ | $[Cr(CO)_4]^{4-}$ [a] |
|---|---|---|---|---|---|---|
| $\nu(CO)$, $A_1$, $R$ | 2128 | 1918 | 1788 | 1670[b] | 1690[b] | 1462[79] |
| $\nu(CO)$, $T_2$, IR, $R$ | 2037 | 1883 | 1788 | | | |
| $\nu(MC)$ | 380 | 439 | 464 | | | |
| $k(CO)$ (N m$^{-1}$) | 1760 | 1320 | 1140 | | | |
| $k(MC)$ (N m$^{-1}$) | 260 | 360 | 410 | | | |
| | $[Fe(CO)_5]$ | $[Mn(CO)_5]^-$ | $[Cr(CO)_5]^{2-}$ | $[Mo(CO)_5]^{2-}$ [c] | $[V(CO)_6]^-$ | |
| $\nu(CO)$, $A''$, IR | 2003 | 1898 | 1760 | 1769 | | |
| $\nu(CO)$, $E'$, IR, $R$ | 1982 | 1863 | 1722 | 1723 | | |
| | $[Re(CO)_6]^+$ | $[W(CO)_6]$ | $[Cr(CO)_6]$ | $[V(CO)_6]$ [d] ($D_{4b}$) | | |
| $\nu(CO)$, $T_{1u}$, IR | 2085 | 1986 | 1987 | 1980 ($A_{2u}$) 1972 ($E_u$) | 1843 | |

[a] Unknown structure.
[b] J. E. Ellis and R. A. Faltynek, *J. Chem. Soc., Chem. Commun.*, 966 (1975).
[c] J. E. Ellis, S. G. Hentges, D. G. Kalina, and G. P. Hagen, *J. Organomet. Chem.*, **97**, 79 (1975).
[d] T. C. DeVor and H. F. Franzen, *Inorg. Chem.*, **15**, 1318 (1976).

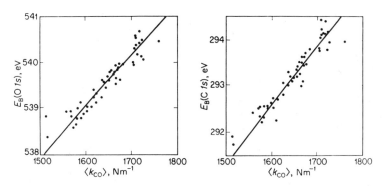

Figure 2.10. The dependence of 1s orbital energy of carbon and oxygen on the average values of $\langle k(\text{CO}) \rangle$ and $\langle v(\text{CO}) \rangle$ for carbonyl complexes[30] (see Table 2.7).

the interaction of *cis*-CO groups, because the *trans*-CO groups interact with the same two $d$ orbitals while the *cis*-CO groups share only one $d$ orbital. These assumptions are not valid rigorously. Therefore, it is not possible to obtain absolute force constants but, for a series of similar compounds, the obtained values are relatively correct. In the Cotton–Kraihanzel method, it is assumed that the interaction constant for stretching vibration of two CO groups on the same metal atom is positive. This was confirmed by experiment. Therefore, symmetric CO group vibrations occur at higher frequencies than asymmetric vibrations, facilitating the interpretation of the IR spectra of metal carbonyls. The MO analysis by Fenske–Hall method[7, 24] shows that, in metal carbonyls, the populations of CO orbitals are constant, with the exception of the donor orbital of the ligand, and the lowest, unoccupied orbitals, i.e., $5\sigma$ and $2\pi$. Therefore, the following relationships are obeyed:

$$k(\text{CO}) = 3744 - 888a - 1645b \quad \text{for neutral carbonyls} \qquad (2.1a)$$

$$k(\text{CO}) = 3932 - 978a - 1892b \quad \text{for cationic carbonyl} \qquad (2.1b)$$

$$k(\text{CO}) = 3642 - 993a - 1176b \quad \text{for } M(\text{CO})_5X \text{ and } M(\text{CO})_4X_2 \text{ complexes} \qquad (2.1c)$$

where $k$ is the force constant calculated by the Cotton–Kraihanzel method in $[\text{N m}^{-1}]$ while $a$ and $b$ are populations of $5\sigma$ and $2\pi$ orbitals in the CO group. Both coefficients $a$ and $b$ are negative, as could be expected from the antibonding character of $2\pi$ and $5\sigma$ orbitals. It was found that the relationship[33] is obeyed precisely between Cotton–Kraihanzel force constants $k(\text{CO})$, for $M(\text{CO})_n$-type carbonyls and the force constant $k_d$, for a monocarbonyl $M(\text{CO})$ containing the same number of $d$ electrons; for

Table 2.4. Force Constant $k_d$ $(N\,m^{-1})$ for Monocarbonyl Fragments

| M − CO | $k_5$ | $k_6$ | $k_7$ | $k_8$ | $k_9$ | $k_{10}$ |
|---|---|---|---|---|---|---|
| First series of the transition metals | 1373 | 1387 | 1444 | 1498 | 1554 | 1610 |
| Second series of the transition metals | — | 1381 | — | 1506 | — | 1636 |
| Third series of the transition metals | 1353 | 1381 | 1445 | 1498 | — | 1613 |

example, $k_d = k_6$ for $Cr(CO)$ and $Mn(CO)^+$, as well as the sum of $E_L^\theta$ constants, which encompass the influence of all ligands L forming angle $\theta$ with the considered CO group:

$$k_{CO} = k_d + \sum_L E_L^\theta \qquad (2.2)$$

The constant $k_d$ depends on electron configuration of the metal atom and on the position of the element in the Periodic Table. It is slightly different for each series of the transition metals (Table 2.4). Equation (2.2) can be generalized for ionic carbonyl complexes of the type $[M(CO)_n Y_m]^q$:

$$k_{CO} = k_d + \sum_L E_L^\theta + q E_c \qquad (2.3)$$

where $q$ is the charge of the complex and $E_c$ is a constant equal to $197 \pm 10 \ N m^{-1}$. The constants $k_d$ and $E_L^\theta$ were calculated from force constants determined for $M(CO)_n$ by

Table 2.5. *Timney's Constants $E_L^\theta$ ($N m^{-1}$) for Ligands*

| Ligand | $E_L^{cis\ a}$ | $E_L^{trans}$ | $E_L^{td}$ | $E_L^{ax,eq}$ | $E^{eq,eq}$ |
|---|---|---|---|---|---|
| CO | 33.5 | 126.1 | 37.3 | 25.5 | 51.4 |
| CS | 56 | 160 | 65 | | |
| NO | 42 | 232 | 30.0 | 45 | 22 |
| $N_2$ | 14.0 | 52.0 | 6 | | |
| Cl | 143 | 106 | 145 | | |
| Br | 134 | 101 | 141 | | |
| I | 112 | 104 | 125 | | |
| H | 75 | 129 | — | 70 | — |
| Me | 71 | 92 | 71 | | |
| $PF_3$ | 33.2 | 141.6 | 44.9 | 16.0 | 44.6 |
| $PCl_3$ | 30.6 | 109.3 | 35.3 | 21 | |
| $PCl_2Ph$ | 14 | 82 | 13 | | |
| $PClPh_2$ | −5 | 55 | −11 | | |
| $PPh_3$ | −21 | 29 | −31.7 | −52 | |
| $PMe_3$ | −27.7 | 29.8 | −38.7 | −61 | |
| $P(OMe)_3$ | −15.2 | 66.3 | −11.2 | −30 | |
| $P(OPh)_3$ | 1.3 | 94 | −0.3 | | |
| $PCy_3$ | −35 | 28 | −51 | | |
| $PE_3$ | −32 | 26 | | | |
| MeNC | −14 | 30 | −21.9 | −18 | |
| EtNC | −9 | 30 | −22.9 | −22 | |
| BuNC | −9 | 31 | −23.1 | −22 | |
| $C_2H_4$ | 6 | 87 | 4 | | |
| $\eta^5$-$C_5H_5$ | | 159 | 99 | | |
| en | −60 | −54 | | | |
| py | −29 | −43 | | | |
| bipy | −24 | −62 | | | |
| EtOH | −41 | −26 | | | |
| $Pr_2^nO$ | −35 | −32 | | | |
| $Et_2O$ | −38 | −5 | | | |
| DMF | −57 | −94 | | | |

[a]The $E_L^{cis}$ and $E_L^{trans}$ refer to octahedral $E_L^{td}$ to tetrahedral and $E_L^{ax,eq}$ and $E_L^{eq,eq}$ to trigonal bipyramidal complexes, for which $E_L^{ax,ax}$ is practically equal to $E_L^{trans}$.

*Table 2.6. Dependence of Constants A and B on Angle*
*CMC*

| Angle CMC | $A$ (N m$^{-1}$) | $B$ |
|---|---|---|
| 90° | 180 | 0.0929 |
| 109° | 203 | 0.0983 |
| 120° | 221 | 0.1115 |
| 180° | 241 | 0.1134 |

other precise methods (Table 2.5). The frequencies $v(CO)$ may be calculated from the above formulas if an interaction constant between two CO groups is taken into consideration,

$$k_{12} = A - B\frac{k_1 + k_2}{2}$$

where $k_1$ and $k_2$ are force constants of interacting CO groups while constants $A$ and $B$ depend only on the angle CMC as seen in Table 2.6.

The IR spectra of carbonyls are given in Tables 2.7–2.11.

## 4. ELECTRONIC SPECTRA

In electronic spectra of metal carbonyls[8, 20, 24] there are bands due to *d–d* transitions between metal orbitals, transitions within the ligand, charge-transfer transitions M → L and L → M, and also, in the case of polynuclear carbonyls, transitions between orbitals of metal–metal bonds. Transitions within the CO group probably occur in the vacuum ultraviolet region.

The lowest spin allowed transitions $\pi \to \pi^*$ and $n \to \pi$ for the free CO molecule

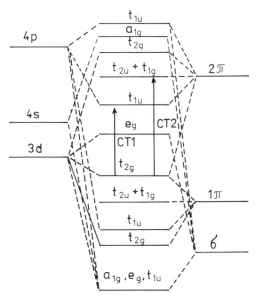

Figure 2.11. Charge-transfer transitions for M(CO)$_6$.

occur at 87,000 and 63,000 cm$^{-1}$. Their energy should increase after a complex has been formed. Octahedral $M(CO)_6^{n\pm}$ complexes should exhibit, in addition to $d$–$d$ transitions, bands due to charge transfer from metal to ligand (MLCT) as well as from ligand to metal (CT). The latter occur for $18e$ compounds in the region of high energy, at the end of quartz ultraviolet or at higher energies. This is due to a very large value of the splitting parameter $10Dq$, and thus due to high energy of the $e_g$ orbitals. The ligand-to-metal charge-transfer bands may have lower energy in the case of compounds in which the valence electron number is less than 18, such as $Cr(CO)_6^+$, $V(CO)_6$, $Ti(CO)_6$, etc. In octahedral metal hexacarbonyls, two intense bands are found due to MLCT transitions from $t_{2g}$ metal orbital to the $t_{1u}$ (the band has lower energy) and $t_{2u}$ orbitals (Figure 2.11).

*Table 2.7. Carbon 1s and Oxygen 1s Binding Energies, Degeneracy-Weighted Average C—O Stretching Frequencies, and Number-Weighted Average C—O Force Constants*[30]

| Carbonyl compound | $E_B(C\ 1s)$ (eV) | $E_B(O\ 1s)$ (eV) | $\langle v(CO)\rangle$ (cm$^{-1}$) | $\langle k(CO)\rangle$ (N m$^{-1}$) |
|---|---|---|---|---|
| $C_6H_6Cr(CO)_3$ | 291.74 | 538.37 | 1938 | 1518 |
| $C_7H_7V(CO)_3$ | 291.9 | 538.8 | 1935 | 1514 |
| $Mo(CO)_5P(NMe_2)_3$ | 292.2 | 538.62 | 1977 | 1580 |
| $C_5H_5Cr(CO)_2NS$ | 292.24 | 539.12 | 1996 | 1610 |
| $C_5H_5Mn(CO)_3$ | 292.30 | 538.90 | 1974 | 1575 |
| $Mo(CO)_5PMe_3$ | 292.30 | 538.75 | 1978 | 1582 |
| $C_5H_5V(CO)_4$ | 292.4 | 538.8 | 1963 | 1558 |
| $C_5H_5(Me)Mo(CO)_3$ | 292.5 | 538.8 | 1971 | 1570 |
| $Me(OMe)CCr(CO)_5$ | 292.5 | 539.1 | 1984 | 1591 |
| $C_5H_5Cr(CO)_2NO$ | 292.51 | 538.91 | 1986 | 1594 |
| $C_5H_5Mo(CO)_2NO$ | 292.52 | 538.77 | 1984 | 1591 |
| $C_5H_5W(CO)_2NO$ | 292.52 | 538.57 | 1974 | 1575 |
| $C_5H_5Mn(CO)_2CS$ | 292.53 | 538.97 | 1980 | 1585 |
| $Mo(CO)_5P(OEt)_3$ | 292.6 | 538.85 | 1989 | 1599 |
| $C_5H_5Co(CO)_2$ | 292.68 | 539.01 | 1998 | 1614 |
| $Mo(CO)_5P(OMe)_3$ | 292.7 | 539.0 | 1992 | 1604 |
| $C_7H_8Fe(CO)_3$ | 292.79 | 539.09 | 2004 | 1623 |
| $W(CO)_5CS$ | 292.97 | 539.50 | 2030 | 1666 |
| $C_3H_5Fe(CO)_2NO$ | 292.98 | 539.22 | 2012 | 1636 |
| $C_4H_6Fe(CO)_3$ | 293.03 | 539.29 | 2005 | 1625 |
| $Mo(CO)_5PCl_3$ | 293.06 | 539.41 | 2017 | 1645 |
| $Cr(CO)_5CS$ | 293.07 | 539.60 | 2034 | 1672 |
| $Cr(CO)_6$ | 293.16 | 539.66 | 2026 | 1659 |
| $W(CO)_6$ | 293.17 | 539.54 | 2024 | 1656 |
| $C(CH_2)_3Fe(CO)_3$ | 293.17 | 539.17 | 2016 | 1643 |
| $Mo(CO)_6$ | 293.22 | 539.58 | 2028 | 1663 |
| $HCCo_3(CO)_9$ | 293.26 | 539.63 | | |
| $BrCCo_3(CO)_9$ | 293.28 | 539.66 | | |
| $Mn_2(CO)_{10}$ | 293.28 | 539.57 | 2017 | 1645 |
| $MeCCo_3(CO)_9$ | 293.32 | 539.56 | | |
| $ClCCo_3(CO)_9$ | 293.34 | 539.67 | | |
| $Mo(CO)_5PF_3$ | 293.34 | 539.77 | 2024 | 1656 |
| $Cr(CO)_5PF_3$ | 293.37 | 539.72 | | |

*(Table continued)*

*Table 2.7. (continued)*

| Carbonyl compound | $E_B(C\,1s)$ (eV) | $E_B(O\,1s)$ (eV) | $\langle v(CO)\rangle$ (cm$^{-1}$) | $\langle k(CO)\rangle$ (N m$^{-1}$) |
|---|---|---|---|---|
| $Co_2(CO)_8$ | 293.40 | 539.78 | 2025 | 1658 |
| $MeCOMn(CO)_5$ | 293.44 | 539.86 | 2035 | 1674 |
| $C_3H_5Mn(CO)_5$ | 293.47 | 539.82 | 2036 | 1676 |
| cis-$Mo(CO)_3(PF_3)_3$ | 293.49 | 539.73 | 2030 | 1666 |
| $Mn(CO)_4NO$ | 293.49 | 539.77 | 2026 | 1659 |
| $MeRe(CO)_5$ | 293.62 | 539.64 | 2033 | 1671 |
| $MeMn(CO)_5$ | 293.62 | 539.91 | 2032 | 1669 |
| $Fe(CO)_5$ | 293.71 | 539.96 | 2036 | 1676 |
| $Ni(CO)_4$ | 293.78 | 540.11 | 2066 | 1725 |
| $Fe(CO)_2(NO)_2$ | 293.79 | 539.91 | 2059 | 1714 |
| $HMn(CO)_5$ | 293.80 | 539.95 | 2039 | 1681 |
| $Co(CO)_3NO$ | 293.81 | 539.99 | 2056 | 1709 |
| $HCo(CO)_4$ | 293.94 | 540.06 | 2068 | 1729 |
| $Mn(NO)_3CO$ | 293.98 | 540.47 | 2088 | 1762 |
| $IMn(CO)_5$ | 294.05 | 540.30 | 2057 | 1710 |
| $SiCl_3Mn(CO)_5$ | 294.12 | 540.37 | 2058 | 1712 |
| $BrMn(CO)_5$ | 294.14 | 540.43 | 2063 | 1720 |
| $H_2Fe(CO)_4$ | 294.15 | 540.17 | 2060 | 1715 |
| $ClMn(CO)_5$ | 294.16 | 540.49 | 2066 | 1725 |
| $CF_3COMn(CO)_5$ | 294.22 | 540.54 | | 1711 |
| $CF_3Mn(CO)_5$ | 294.40 | 540.67 | 2062 | 1719 |
| $SiF_3Mn(CO)_5$ | 294.43 | 540.59 | 2056 | 1709 |
| CO | 296.24 | 542.57 | 2143 | 1856 |

*Table 2.8. The R, IR (cm$^{-1}$) Spectra of Octahedral Carbonyls[a]*

| Band designation | Compound | | | | |
|---|---|---|---|---|---|
| | $[V(CO)_6]^-$ in $CH_3CN$ | $[Cr(CO)_6]$ in $CCl_4$ | $[Mo(CO)_6]$ in $CCl_4$ | $[W(CO)_6]$ in $CS_2$ | $[Re(CO)_6]^+$ in $CH_3CN$ |
| $A_{1g}$ $v_1(CO)$ | 2020 | 2112 | 2117 | 2119 | 2197 |
| $v_2(MC)$ | 374 | 381 | 402 | 427 | 441 |
| $E_g$ $v_3(CO)$ | 1894 | 2018 | 2019 | 2010 | 2122 |
| $v_4(M-C)$ | 393 | 394 | 392 | 412 | 426 |
| $T_{1g}$ $v_5(\delta MCO)$ | 356 | | | | 354 |
| $T_{1u}$ $v_6(CO)$ | 1858 | 1984 | 1986 | 1977 | 2085 |
| $v_7(\delta MCO)$ | 650 | 665 | 593 | 583 | 584 |
| $v_8(MC)$ | 460 | 444 | 367 | 374 | 356 |
| $v_9(\delta CMC)$ | 92 | 103 | 91 | 92 | 82 |
| $T_{2g}$ $v_{10}(\delta MCO)$ | 517 | | | | 486 |
| $v_{11}(\delta CMC)$ | 84 | 101 | 91 | 92 | 82 |
| $T_{2u}$ $v_{12}(\delta MCO)$ | 506 | | | | 522 |
| $k(CO)$ (N m$^{-1}$) | | 1704 | 1715 | 1702 | |
| $k(MC)$ (N m$^{-1}$) | | 210 | 200 | 232 | |

[a] According to Jones *et al.*[172] and Abel *et al.*[173]

*Table 2.9. The IR Spectra ($cm^{-1}$) of Some Metal Pentacarbonyls*

| Compound | Solvent | $v_1(A_1')$ | $v_2(A_1')$ | $v_6(A_2'')$ | $v_{10}(E')$ |
|---|---|---|---|---|---|
| $[Fe(CO)_5]^a$ | solid | 2115 | 2033 | 2003 | 1982 |
| $[Ru(CO)_5]^b$ | $C_7H_{16}$ | | | 2022 | 2000 |
| $[Os(CO)_5]^b$ | $C_7H_{16}$ | | | 2035 | 1999 |
| $[Mn(CO)_5]^{-c}$ | THF | | | 1898 | 1863 |

[a] P. Cataliotti, A. Foffani, and L. Marchetti, *Inorg. Chem.*, **10**, 1594 (1971).
[b] F. Calderazzo and P. L'Eplattenier, *Inorg. Chem.*, **6**, 1220 (1967).
[c] W. F. Edgel, J. Huff, J. Thomas, H. Lehman, C. Angell, and G. Asato, *J. Am. Chem. Soc.*, **82**, 1254 (1960).

*Table 2.10. The R and IR Spectra ($cm^{-1}$) of Tetrahedral Carbonyls*

| Band designation | Compound, Solvent | | | | |
|---|---|---|---|---|---|
| | $[Fe(CO)_4]^{2-}$ $H_2O^a$ | $[Co(CO)_4]^-$ $DMF^b$ | $[Ni(CO)_4]$ $CCl_4^c$ | $[Pd(CO)_4]$ $CO^d$ | $[Pt(CO)_4]$ $CO^d$ |
| $A_1$   $v_1(CO)$ | 1788 | 2002 | 2125 | 2122 | 2119 |
| $v_2(MC)$ | | 431 | 380 | | |
| $E$   $v_3(\delta MCO)$ | | | 380 | | |
| $v_4(\delta CMC)$ | | | 78 | | |
| $T_2$   $v_5(CO)$ | 1786 | 1888 | 2044 | 2066 | 2049 |
| $v_6(\delta MCO)$ | 556 | 523 | 455 | | |
| $v_7(MC)$ | 646 | 556 | 422 | | |

[a] H. Stammereich, K. Kawai, Y. Tawares, P. Krumholz, J. Behmoiras, and S. Brill, *J. Chem. Phys.*, **32**, 1482 (1960).
[b] W. F. Edgell and J. Lyforg, IV, *J. Chem. Phys.*, **52**, 4329 (1970).
[c] L. H. Jones, R. S. McDowell, and M. Goldblatt, *J. Chem. Phys.*, **48**, 2663 (1968).
[d] E. P. Kundig, M. Moskovits, and G. A. Ozin, *J. Mol. Struct.*, **14**, 137 (1972).

*Table 2.11. The IR Spectra of Trinuclear Carbonyls$^a$*

| Compound | Solvent | $v(CO)$ ($cm^{-1}$) |
|---|---|---|
| $[Fe_3(CO)_{12}]$ | n-Hexane | 2046s, 2023m, 2013sh, 1867vw, 1835w |
| $[RuFe_2(CO)_{12}]$ | n-Hexane | 2115w, 2052s, 2040s, 2021sh, 2000m, 1989sh, 1860vw, 1828w |
| $[OsFe_2(CO)_{12}]$ | n-Hexane | 2117w, 2055s, 2041s, 2036m, 2013m, 2001m, 1990sh, 1860vw, 1827vw |
| $[Et_4N][MnFe(CO)_{12}]$ | THF | 2063w, 1999s, 1990s, 1972s, 1944m, 1903w, 1827w, 1785w |
| $[Et_4N][TcFe_2(CO)_{12}]$ | THF | 2077w, 2008m, 1987s, 1943m, 1903w, 1814w, 1783w |
| $[Et_4N][ReFe_2(CO)_{12}]$ | THF | 2075w, 2006m, 1991s, 1946m, 1903w, 1814w, 1785w |
| $[Os_3(CO)_{12}]$ | Cyclo- | 2068s, 2035s, 2014m, 2002m |
| $[Ru_3(CO)_{12}]$ | hexane | 2061,2s, 2031,3s, 2018,7w, 2012,5m |

[a] According to References 146–149.

*Table 2.12. Splitting and Racah's Parameters for Carbonyls*

| Compound | $10Dq$ $(cm^{-1})$ | $B$ $(cm^{-1})$ | $C$ $(cm^{-1})$ | $B_0$ $(cm^{-1})$ | $\beta = \dfrac{B}{B_0}$ |
|---|---|---|---|---|---|
| $[Ti(CO)_6]^a$ | 28255 | 450 | | 560 | 0.80 |
| $[V(CO)_6]^b$ | 29300 (10Dq) | | | | |
| | 1215 (Ds) | 450 | 1800 | 578 | 0.78 |
| | 365 (Dt) | | | | |
| $[V(CO)_6]^{-c,d}$ | 25500 | 430 | 1300 | 790 | 0.54 |
| $[Cr(CO)_6]^c$ | 32200 | 520 | 1700 | 790 | 0.66 |
| $[Mn(CO)_6]^{+c}$ | 41650 | | 2600 | 870 | |
| $[Mo(CO)_6]^c$ | 32150 | 380 | 1100 | 460 | 0.83 |
| $[W(CO)_6]^c$ | 32200 | 390 | 1300 | 370 | 1.1 |
| $[Re(CO)_6]^{+c}$ | 41000 | 470 | 1375 | 470 | 1.0 |

[a] A. B. Lever and G. A. Ozin, *Inorg. Chem.*, **16**, 2012 (1977).
[b] T. C. DeVore and H. F. Franzen, *Inorg. Chem.*, **15**, 1318 (1976).
[c] N. A. Beach and H. B. Gray, *J. Am. Chem. Soc.*, **90**, 5713 (1968).
[d] G. T. W. Wittmann, G. N. Krynauw, S. Lotz, and W. Ludwig, *J. Organomet. Chem.*, **293**, C33 (1985).

On the side of the high energy of the second band, there is a shoulder which is most probably a result of the transition $t_{2g}(d) \to t_{2g}(2\pi)$. The band is not intense because this transition is forbidden. The energy difference between the shoulder and the main band decreases with the increase in the positive charge on the metal, as is expected. The $d$-orbital energy becomes lower, leading to a decrease in the interaction between $d$ orbitals and $2\pi$ orbitals. Simultaneously, the energy difference between $t_{2g}(2\pi)$ and $t_{2u}$ orbitals becomes smaller. Despite the large value of the splitting parameter $10Dq$, the $d$–$d$ energy transition is lower than that of charge-transfer transitions. This is a result of interelectron repulsion. The energy differences between terms are not equal to energy differences between orbitals.

The $d$–$d$ bands may be divided into two groups: (1) spin allowed, caused by transitions between terms of the same spin multiplicity, and (2) spin forbidden. The following transitions serve as respective examples: $^1A_{1g}(t_{2g}^6) \to {}^1T_{1g}(t_{2g}^5 e_g)$ and $^1A_{1g}(t_{2g}^6) \to {}^3T_{1g}(t_{2g}^5 e_g)$ for $Cr(CO)_6$.

Carbonyl groups, due to their $\sigma$-donor and $\pi$-acceptor properties, create a very strong ligand field. Consequently, the values of the splitting parameter for carbonyl complexes are very high, even higher than those of cyanide complexes (Table 2.12). Therefore, high-spin carbonyls are unknown.

## 5. $^{13}C$ NMR SPECTRA

The application of $^{13}C$ NMR for investigation of metal carbonyls is generally difficult owing to low abundance of $^{13}C$ in natural compounds (1.1%) and long spin–lattice relaxation time. The development of NMR spectrophotometers, as well as the use of paramagnetic complexes which accelerate relaxation, caused the recent, widely adopted utilization of the $^{13}C$ NMR method.[34, 35, 20] As a result of the addition of 0.2 M Cr(acac)$_3$ to Fe(CO)$_5$, a fortyfold increase in the intensity of the $^{13}C$ signal occurred without causing a shift. However, it is still difficult to register spectra of carbonyls which contain 100% or nearly 100% of metal isotopes possessing spins of $I > \frac{1}{2}$.

The interaction of $^{13}C$ nuclei with $^{55}Mn(I = \frac{5}{2})$ and $^{59}Co(I = \frac{7}{2})$ causes splitting of the $^{13}C$ bands into 6 (for $^{55}Mn$) and 8 (for $^{59}Co$) signals of the same intensity. The shifts given in Table 2.13 are calculated with tetramethylsilane as an internal $^{13}C$ reference. The positive values of the shifts correspond to higher frequencies (lower fields) relative to $Si(CH_3)_4$: $\delta(TMS) = \delta(CS_2$ internal$) + 192.8$; $\delta(TMS) = \delta(C_6H_6) + 128.5$; $\delta(TMS) = \delta(CH_3{}^{13}COOH) + 178.3$.

The $^{13}C$ chemical shifts for terminal carbonyl groups occur in the 180–240 ppm range, and for bridging groups in the 230–280 ppm range (Table 2.13). For bridging carbonyl groups, these shifts are markedly different from characteristic shifts for organic compounds possessing CO groups, such as acetone, 204.1 ppm. However, these bridging CO shifts have values which are close to those of metal carbene complexes.

*Table 2.13. The $^{13}C$ and $^{17}O$ Spectra of Some Carbonyls*

| Compound | $\delta(CO)$ (ppm) | Coupling constants (Hz) |
|---|---|---|
| $[Ni(CO)_4]^a$ | 192.5 | |
| $[Fe(CO)_5]^a$ | 211.9 | $^1J(^{57}Fe-C) = 23.4$ |
| $[Cr(CO)_6]^b$ | 212.3 | |
| $[Mo(CO)_6]^b$ | 204.1 | $^1J(^{95}Mo-C) = 68$ |
| $[W(CO)_6]^b$ | 191.9 | $^1J(W-C) = 126$ |
| $[W_2(OPr^i)_6(\mu_3\text{-}CO)]_2{}^{(82)}$ | 305.5 | $^1J(^{183}W-{}^{13}C) = 188.9$ |
| $[W(CO)_3(\text{durene})]^b$ | 213.7 | |
| $[Mo(CO)_3(\text{durene})]^c$ | 224.4 | |
| $[Cr(CO)_3(\text{durene})]^c$ | 235.5 | |
| $Na_3[V(^{13}CO)_5]^{(73)}$ | 290.1 | |
| $Na_2(VH(^{13}CO)_5]^{(73)}$ | 250.3 | |
| $Na[V(^{13}CO)_6]^{(73)}$ | 224.2 | |
| $cis\text{-}[RuCl_2(CO)_2(PEt_3)_2]^d$ | 195.4 | $^2J(P-Ru-C) = 10.6$ |
| $cis\text{-}[OsCl_2(CO)_2(PPr_2^n Bu^t)_2]^d$ | 177.6 | $^2J(P-Os-C) = 10.3$ |
| $[Co_3(CO)_9CBr]^e$ | 186.2 | |
| $[Rh_2(CO)_3cp_2]^f$ | 191.8$^{\text{ter.}}$ | |
| | 231.8$^{\text{br.}}$ (193 K) | |
| $[Fe(CO)_2cp]_2{}^g$ | 210.9; 285.1 | |
| $[Rh_4(CO)_{12}]^h$ | 189.5 (quintet) | |
| | (323 K) | $J(^{103}Rh-C) = 17.1$ |
| $[Rh_4(CO)_{12}]^h$ | 228.8$^{\text{br.}}$ | $J(^{103}Rh-C^{\text{br.}}) = 35;$ |
| | 183.4; 181.8; | $J(^{103}R-C^{\text{ter.}}) = 75; 64; 62$ |
| | 175.5 (208 K) | |
| $[Rh_6(CO)_{16}]$ | 180.1$^{\text{ter.}}$, 231.5$^{\text{br.}}$ | |
| $[Rh_7(CO)_{16}I]^{2-}$ | 218.5 (octet) | |

$^a$ P. Lauterbur and R. B. King, *J. Am. Chem. Soc.*, **87**, 3266 (1965).
$^b$ B. E. Mann, *J. Chem. Soc., Dalton Trans.*, 2012 (1973).
$^c$ B. E. Mann, *Chem. Commun.*, 976 (1971).
$^d$ D. F. Gill, B. E. Mann, and B. L. Shaw, *J. Chem. Soc., Dalton Trans.*, 311 (1973).
$^e$ L. F. Farnell, E. W. Randall, and E. Rosenberg, *Chem. Commun.*, 1078 (1971).
$^f$ J. Evans, B. F. G. Johnson, J. Lewis, and J. R. Norton, *J. Chem. Soc., Chem. Commun.*, 79 (1973).
$^g$ O. A. Gansow, A. R. Burke, and W. D. Vernon, *J. Am. Chem. Soc.*, **94**, 2550 (1972).
$^h$ J. Evans, B. F. G. Johnson, J. Lewis, J. R. Norton, and F. A. Cotton, *J. Chem. Soc., Chem. Commun.*, 807 (1973).

*(Table continued)*

*Table 2.13. (continued)*

| Compound | $\delta(CO)$ (ppm) | Coupling constants (Hz) |
|---|---|---|
| | The $^{17}O$ Spectra[i] | |
| $[W(CO)_6]$ | 356.8 | 124.5 $^1J(^{183}W - ^{13}C)$ |
| $[W(CO)_5CPh_2]$ | 364.7 (c) | 127.0 (c), 102.5 (t) |
| | 452.6 (t) | $^1J(^{183}W - ^{13}C)$ |
| $[W(CO)_5C(OMe)Ph]$ | 357.1 (c) | 127.0 (c), 115.2 (t) |
| | 388.9 (t) | $^1J(^{183}W - ^{13}C)$ |
| $[W(CO)_5C(NH_2)Ph]$ | 352 (c) | 126.9 (c), 124.8 (t) |
| | 372.5 (c) | $^1J(^{183}W - ^{13}C)$ |
| $[W(CO)_5 \overline{CCBu^t = C}Bu^t]$ | 350.69 (c) | 125.7 (c), 133.0 (t) |
| | 367.6 (t) | $^1J(^{183}W - ^{13}C)$ |
| $[W(CO)_5P(OMe)_3]$ | 353.7 (c) | 125.1 (c), 139.1 (t) |
| | 359.0 (t) | $^1J(^{183}W - ^{13}C)$ |
| $[W(CO)_5PPh_3]$ | 353.6 (c) | |
| | $\geqslant 353.6$, sh (t) | |
| $[W(CO)_5PBu_3^n]$ | 354.1 (c) | |
| | 354.1 (t) | 124.4 (c), 142.1 (t) |
| | | $^1J(^{183}W - ^{13}C)$ |
| $[W(CO)_4(Ph_2PC_2H_4PPh_2)]$ | 349.4 (c) | |
| | 358.1 (t) | |
| $[Bu_4N]^+[W(CO)_5I]^-$ | 349.0 (c) | |
| | 349.0 (t) | 127.0 (c), 175.8 (t) |
| | | $^1J(^{183}W - ^{13}C)$ |

$c = cis$, $t = trans$, ter. = terminal, br. = bridging

[i] Y. Kawada, T. Sugawara, and H. Iwamura, *J. Chem. Soc., Chem. Commun.*, 291 (1979).

With good approximation, the $^{13}C$ chemical shift is influenced by three factors:

1. Paramagnetic shielding factor, $\delta_p$.
2. Diamagnetic shielding factor, $\delta_d$.
3. Anisotropic shielding factor.

The first factor is the most important and results from the mixing of the excited states with the ground state:

$$\delta_p = \frac{K}{\Delta E} \langle r^{-3} \rangle_{2p} (Q_{AA} + Q_{AB}) \tag{2.4}$$

where $\Delta E$ is an average excitation energy, $K$ is a constant, $\langle r^{-3} \rangle_{2p}$ is the average expectation value of $r^{-3}$ for the $2p$ orbitals, $Q_{AA}$ depends on the electron density near the nucleus, and $Q_{AB}$ depends on the influence of the neighboring atom B, and on the multiplicity of the A $-$ B bond. It is therefore easy to see how the chemical shift is affected by $\Delta E$, by inductive effects on $\langle r^{-3} \rangle$, and by the formation of metal–carbon $\pi$-bonding. The relative importance of these effects is not clear.

There is a relationship between the multiplicity of the metal–carbon bond and the carbonyl carbon shielding. This relationship is confirmed by the following: there is a linear dependence of $^{13}C$ chemical shift on force constants $k(CO)$ for $W(CO)_{6-n}L_n$

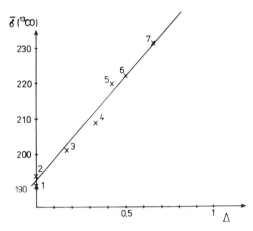

Figure 2.12.   The dependence of the average value of the chemical shift $\delta$ ($^{13}CO$) on the negative charge per atom of the rhodium atom ($\Delta$) [36]: (1) $[Rh_6(CO)_{16}]$, (2) $[Rh_4(CO)_{12}]$, (3) $[Rh_{12}(CO)_{30}]^{2-}$, (4) $[Rh_6(CO)_{15}]^{2-}$, (5) $[Rh_7(CO)_{16}]^{3-}$, (6) $[Rh_4(CO)_{11}]^{2-}$, (7) $[Rh_6(CO)_{14}]^{4-}$.

complexes,[34] and there is a dependence of $\delta(^{13}C)$ on the magnitude of the negative charge on the rhodium atom in polynuclear rhodium carbonyls[36] (Figure 2.12). The magnitude of the charge on the metal atom influences the bond order of the metal–carbon bond (see above) and thus causes the multiplicity of the carbon–oxygen bond to become lower. For compounds in which the metal adopts lower oxidation states, $\delta(^3CO)$ assumes exceptionally high values which are typical for metal carbene complexes, e.g., for $Na_3V(CO)_5$ $\delta(^{13}CO)$ is 290.1 ppm.[73] For the tungsten carbonyls $[W_2(OPr^i)_6(\mu_3\text{-}CO)]_2$ and $[W_2(OPr^i)_6 py(\mu_3\text{-}CO)]_2$ containing the bridging ligand $\mu_3$-CO connected through the carbon atom with two metal atoms and through the oxygen atom with the third central metal atom, as well as for other dinuclear carbonyls of the type $[M_2(OR)_6(\mu\text{-}CO)]$ (Figure 2.17 below), very high chemical shifts $\delta(^{13}CO)$ are also observed: 291–341 ppm.[82]

Many metal carbonyls show dynamic properties. $Fe(CO)_5$, as well as its derivatives $Fe(CO)_4 L$, give only one $^{13}C$ NMR signal at $-170\ °C$. Iron pentacarbonyl, which has a trigonal–bipyramidal structure, should exhibit two resonances of the carbonyl groups. From the measurements of the coupling constant $^1J(^{57}Fe - ^{13}C)$, it is determined that one $^{13}C$ NMR signal occurs due to intramolecular and not intermolecular exchange of CO groups. Dynamic properties are also shown by $[Co_2(CO)_8]$, $[Rh_4(CO)_{12}]$, $[Rh_6(CO)_{15}]^{2-}$ and $[Rh_7(CO)_{16} I]^{2-}$ (Table 2.13).

$$\text{I} \quad\rightleftharpoons\quad \text{II} \quad\rightleftharpoons\quad \text{III} \tag{2.5}$$

I                                    II                                    III

In the case of dicobalt octacarbonyl, the equilibria in equation (2.5) exist in solution. These equilibria shift either to the bridged or nonbridged forms, depending upon conditions.[37] Form (III) exists in considerable quantities at higher temperatures and high pressures of CO.

## 6. PHOTOELECTRON SPECTROSCOPY

Photoelectron spectroscopy investigates the intensity and energy of electrons which are removed from the valence orbital (low-energy photoelectron spectroscopy) or inner orbitals (high-energy photoelectron spectroscopy, also known as Electron Spectroscopy for Chemical Analysis, ESCA), by photons.[38-44]

For low-energy photoelectron spectroscopy, helium is used as a source of radiation. It generates photons possessing energy of 21.22 eV or 40.81 eV by the process $He(1s2p) \rightarrow He(1s^2)$ or $He^+(2p) \rightarrow He^+(1s)$, respectively. The source of high-energy photoelectron spectroscopy is an X-ray tube, which emits soft X-ray radiation, usually AlK$\alpha$ (1486.6 eV). The kinetic energy of the ejected electron ($E$) is equal to the energy of the photon ($hv$) minus the ionization potential ($I$) and the changes in the rotational and vibrational energies $\Delta E(\text{rot})$ and $\Delta E(\text{vib})$:

$$E = hv - \Delta E(\text{rot}) - \Delta E(\text{vib}) - I \qquad (2.6)$$

The energy required to eject the electron is only an approximate measure of this electron's binding energy, because the ejection of the electron is accompanied by such a change in the remaining electrons, that the resulting ion's energy is minimal. Therefore, the observed values of the ionization potential ($I$) are smaller than those resulting from $1e$ ionization model.

In almost all carbonyls, ionization corresponds to the removal of an electron from the completely filled shells, which leads to the appearance of only one band for each level. Intensity of the bands is proportional to the number of electrons occupying a given shell. $V(CO)_6$ is an exception having the electron structure $t_{2g}^5$ and, as a result of ionization, may give one of the $t_{2g}^4$ configurations described by terms $^3T_{1g}$, $^1T_{2g}$, $^1E_g$, and $^1A_{1g}$.

The low-energy photoelectron spectra are obtained in the vapor phase in order to avoid the influence of the electron band structure of the solid as well as that of the static charges. ESCA spectra may be collected for solids and for vapors, because the interaction between inner electrons of different molecules is negligible, and there is no broadening of the peaks resulting from the electron band structures of solids. The bands of the low-energy photoelectron spectra are relatively broad (halfwidth *ca* 0.2 eV) as a result of the Franck–Condon principle.

The photoelectron spectrum of free CO exhibits three bands due to removal to electrons from $5\sigma$ orbitals (14.01 eV), from $1\pi$ orbitals (a series of peaks with maximum at 16.91 eV), and from $4\sigma$ orbitals (19.72 eV). These bands are also found, approximately in the same range, in spectra of the carbonyls. However, due to interactions between CO groups, these bands usually are diffused and it is difficult to identify particular transitions. The ejection of $d$ electrons requires 8–10 eV of energy.

In the spectrum of vanadium hexacarbonyl, the presence of the following bands was confirmed: the band at 7.52 eV was assigned as $^2T_{2g}(t_{2g}^5) \rightarrow {}^3T_{1g}(t_{2g}^4)$ and the band

Table 2.14. The Photoelectron Spectra of Carbonyls

| Compound | Electron configuration | d electrons | Ionization energy (eV) CO | Others |
|---|---|---|---|---|
| [V(CO)$_6$]$^a$ | $d^5$ | 7.52; 7.88 | 12.95, 13.64, 17.3 | |
| [Cr(CO)$_6$]$^{a,b,c}$ | $d^6$ | 8.42 | 13.38, 14.21, 14.40, 15.12, 15.60, 16.2, 17.82 | 10.00 ($\sigma$M−L) |
| [Cr(CO)$_5$PMe$_3$]$^d$ | $d^6$ | 7.58 (e), 7.72 ($b_2$) | 11.2–15.5, 16.9 | |
| [Mo(CO)$_6$]$^{b,c}$ | $d^6$ | 8.50 | 13.32, 14.18, 14.41, 14.66, 15.2, 15.6, 16.2, 17.71 | |
| [W(CO)$_6$]$^{a,b}$ | $d^6$ | 8.56 | 13.27, 14.20, 14.42, 14.88, 15.2, 15.54, 17.84 | |
| [Mn(CO)$_5$H]$^{e,f,g}$ | $d^6$ | 8.85 (e), 9.14 ($b_2$) | 13.0–17.0 | 10.55 ($\sigma$Mn−H) |
| [Mn(CO)$_5$Cl]$^{g,h,i}$ | $d^6$ | 10.56 (e), 11.18 ($b_2$) | 14.0, 14.4–15.2, 17.0 | 8.94 ($p_\pi$Cl) 9.56 ($p_\sigma$Cl) |
| [Mn(CO)$_5$I]$^{g,h,i}$ | $d^6$ | 9.69 (e), 10.44 ($b_2$) | 13.0–16.6, 17.7 | (8.44, 8.74) ($p_\pi$I) |
| [Mn$_2$(CO)$_{10}$]$^{h,i}$ | | 8.02 ($a_1$), (8.35, 9.03) (e) | | 9.0 ($p_\delta$I) |
| [Re(CO)$_5$H]$^{f,i,j}$ | $d^6$ | (8.86, 9.15) (e), 9.53 ($b_2$) | 13.5–17.0 | 10.5 ($\delta$Re−H) |
| [Re$_2$(CO)$_{10}$]$^{f,i}$ | | 8.07 ($a_1$), (8.57, 8.86, 9.27, 9.58) (e) | 13.0–17.3 | |
| [Fe(CO)$_5$]$^b$ | $d^8$ | 8.60 (e'), 9.86 (e'') | >13.5 | |

| Compound | | | | |
|---|---|---|---|---|
| [Fe(CO)$_4$H$_2$]$^e$ | $d^6$ | 9.65 | 14–16 | (10.95, 11.30) ($\delta$Fe–H) |
| [Fe(CO)$_3$(COD)]$^l$ | $d^8$ | 7.45, 8.27 | 10.87 ($\delta$ otien + CO) | 8.87 $\pi_1 a_2$ C=C–C=C) |
| | | | | 10.44 ($\pi_1 b_1$ C=C–C=C) |
| | | | | 11.5 ($\delta$CO–H) |
| [Co(CO)$_4$H]$^j$ | $d^8$ | 8.90 ($e'$), 9.80 ($e''$) | 13.8–17 | |
| [Ni(CO)$_4$]$^{b,k}$ | $d^{10}$ | 8.90 ($t_2$), 9.77 ($e$) | 14.12, 14.95, 15.7, 18.25 | |
| [Co(CO)$_3$(NO)]$^{b,k}$ | $d^{10}$ | 8.90 ($t_2$), 9.82 ($e$) | 14.53, 14.92, 18.17 | |
| [Fe(CO)$_2$(NO)$_2$]$^{b,k}$ | $d^{10}$ | (8.56, 8.97) ($t_2$), 9.74 ($e$) | 14.71, 15.43, 16.27, 18.20 | |
| [Ru$_3$(CO)$_{12}$]$^m$ | $d^8$ | 7.7, 8.0, 9.8, 10.1 | | |
| [Os$_3$(CO)$_{12}$]$^m$ | $d^8$ | 7.83, 8.28, 8.50, 9.24, 9.60, 10.44 | | |
| [Os$_6$(CO)$_{18}$]$^m$ | $d^8$ | 7.50, 8.09, 9.47; 10.44 | | |

[a] S. Evans, J. C. Green, A. F. Orchard, T. Saito, and D. W. Turner, *Chem. Phys. Lett.*, **4**, 361 (1969).

[b] D. R. Lloyd and E. W. Schlag, *Inorg. Chem.*, **8**, 2544 (1969).

[c] B. R. Higginson, D. R. Lloyd, P. Burroughs, D. M. Gibson, and A. F. Orchard, *J. Chem. Soc., Faraday Trans. 2*, **69**, 1659 (1973).

[d] B. R. Higginson, D. R. Lloyd, J. A. Connor, and I. H. Hillier, *J. Chem. Soc., Faraday Trans. 2*, **70**, 1418 (1974).

[e] M. F. Guest, B. R. Higginson, D. R. Lloyd, and I. H. Hillier, *J. Chem. Soc., Faraday Trans. 2*, **71**, 902 (1975).

[f] M. B. Hall, *J. Am. Chem. Soc.*, **97**, 2057 (1975).

[g] M. F. Guest, M. F. Hall, and J. H. Hillier, *Mol. Phys.*, **25**, 629 (1973).

[h] S. Evans, J. C. Green, A. F. Orchard, and D. W. Turner, *Discuss. Faraday Soc.*, **47**, 112 (1969).

[i] B. R. Higginson, D. R. Lloyd, S. Evans, and A. F. Orchard, *J. Chem. Soc., Faraday Trans. 2*, **71**, 1913 (1975).

[j] S. Gradock, E. A. Ebsworth, and A. Robertson, *J. Chem. Soc., Dalton Trans.* 22 (1973).

[k] J. H. Hillier, M. F. Guest, B. R. Higginson, and D. R. Lloyd, *Mol. Phys.*, **27**, 215 (1974).

[l] J. C. Green, P. Powell, and J. van Tilborg, *J. Chem. Soc., Dalton Trans.*, 1974 (1976).

[m] J. C. Green, E. A. Seddond, and D. M. P. Mingos, *J. Chem. Soc., Chem. Commun*, 94 (1979).

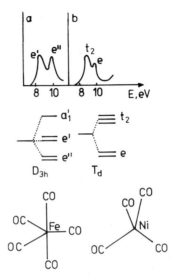

Figure 2.13. The photoelectron spectra of Fe(CO)₅ and Ni(CO)₄.

*Table 2.15. ESCA Spectra of Carbonyl Compounds*

| Compound | C(1s) | O(1s) | M($^2p_{3/2}$) |
|---|---|---|---|
| | | Binding energy (eV) | |
| CO | 295.9[a]; 296.24[b]; | 542.1[a]; 542.57[b]; 542.4[c] | |
| [Cr(CO)₆] | 293.11 | 539.96[a] | |
| [Mo(CO)₆] | 293.06 | 539.91[a] | |
| [W(CO)₆] | 292.98 | 539.87[a] | |
| [MeCMn(CO)₅] (O‖) | 293.36[c] (CO) 292.0[c] (COMe) 290.36[c] (Me) | 539.75 537.01 | |
| [Fe₂Cp₂(CO)₄][c] | | 538.55[ter.]; 537.36[br.] | |
| [Co₄(CO)₁₂][c] | | 539.62[ter.]; 538.37[br.] | |
| CO[d] | 290.6 | 535.8 | |
| [Cr(CO)₆][d] | 288.1 | 534.1 | 576.5 |
| [Cr(CO)₅NH₃][d] | 287.3 | 533.4 | 575.7 |
| [Cr(CO)₅PH₃][d] | 287.1 | 533.2 | 575.5 |
| [Cr(CO)₅PMe₃][d] | 286.8 | 533.1 | 575.4 |
| [Fe(CO)₅][d] | 288.2 | 534.2 | 709.8 |
| [Fe(CO)₂(NO)₂][d] | 288.4 | 534.0 (NO-; 534.4) | 709.7 |
| [Co(CO)₃NO][d] | 288.4 | 534.0 (NO-; 534.6) | 780.9 |
| [Ni(CO)₄][d] | 288.4 | 534.4 | 855.0 |

ter. = terminal, br. = bridging

[a] W. L. Jolly, *Coord. Chem. Rev.*, **13**, 47 (1974).
[b] T. D. Thomas, *J. Electron Spectrosc.*, **8**, 45 (1976).
[c] S. C. Avanzino and W. L. Jolly, *J. Am. Chem. Soc.*, **98**, 6505 (1976).
[d] M. Barber, J. A. Connor, M. F. Hall, J. H. Hillier, and W. N. E. Meredith, *Faraday Discuss. Chem. Soc.*, **54**, 219 (1972).

at 7.88 eV was assigned as $^2T_{2g}(t_{2g}^5) \rightarrow {}^1T_{2g}, {}^1E_g(t_{2g}^4)$. The $^2T_{2g}(t_{2g}^5) \rightarrow {}^1A_{1g}(t_{2g}^4)$ transition is probably hidden beneath this second band.

The spectrum of hexacarbonylchromium consists of one band at 8.42 eV, which occurs due to the removal of an electron from the $t_{2g}^6$ shell. Pentacarbonyliron gives, according to expectation for a trigonal bipyramid, two bands of equal intensity at 8.60 eV and 9.86 eV [electron configuration $d^8$, $(e'')^4 (e')^4$]. For Ni(CO)$_4$, calculations predicted[44] that the $t_2$ orbital enery is lower than that of the $e$ orbitals. However, it is observed in the photoelectron spectrum that this ordering is reversed and in accordance with the predictions of ligand field theory. The spectrum of Ni(CO)$_4$ shows two bands at 8.90 eV and 9.77 eV (configuration $d^{10}$, $e^4 t_2^6$) with intensities in the ratio $3:2$ respectively (Table 2.14, Figure 2.13).

The high-energy photoelectron spectroscopy of carbonyls focuses primarily on $1s$ orbital energy of the carbon and oxygen atoms. In complexes, this energy is less than that of carbon $1s$ in free CO, because the electron density on the coordinated carbonyl group increases. Therefore the binding energy of $1s$ electrons of the carbon atom in carbonyls is lowered by 2.0–4.5 eV; $1s$ energy of the oxygen atom is reduced by 1.5–5 eV.[44, 45] (Tables 2.7 and 2.15). As a result of the positive charge increase on the metal atom, the binding energy of its inner electrons also increases. The binding energy of $3p$ electrons in Cr(CO)$_6$ is equal to 45.8 eV and is close to the energy of these orbitals in Cr(III) compounds such as CrCl$_3 \cdot 6$H$_2$O (45.4 eV) and K$_3$[Cr(CN)$_6$] (44.8 eV).

Similar binding energies of $3p$ orbitals were also found for [Fe$_2$(CO)$_9$] (54.6 eV) and K$_4$[Fe(CN)$_6$] (54.0 eV).[44] There is a linear relationship between $1s$ electron energy of the oxygen and carbon atoms and (see IR spectra) the average value of the $\langle v(CO) \rangle$[30] (Table 2.7, Figure 2.10).

The $1s$ electron energy of the oxygen atom of bridging CO groups is even lower than that of terminal COs, which argues for stronger influence of the back-bonding in the case of the bridging carbonyl groups.


## 7. MASS SPECTROMETRY

The investigations of mass spectra of metal carbonyls were started by Aston in the 1920s and 1930s in order to determine the isotopic distribution of metals. Later, metal carbonyls were used for electromagnetic separation and preparation of larger quantities of particular metal isotopes.

Most metal carbonyls give parent ions in the mass spectrum. From the investigations of clastograms,[46–49] that is, the dependency of intensity of a given band on the energy of electrons which are used to ionize the compound, it appears that particular ions are formed as a result of the progressive loss of CO groups:

$$M(CO)_x^+ \xrightarrow{-CO} M(CO)_{x-1}^+ \xrightarrow{-CO} \cdots \xrightarrow{-CO} M^+ \qquad (2.7)$$

This is also confirmed by the fact that the higher the appearance potentials of the ions, the lower the number of carbonyl groups bonded to the metal.[46, 47] For some carbonyls, such as Fe(CO)$_5$, such a fragmentation pattern was confirmed by the observation of metastable ions.

Usually, mass spectra of the carbonyls also show ions possessing $+2$ charge, for example, [Fe(CO)$_x$]$^{2+}$ ($x = 1$–4). Their intensities are sometimes considerable; for

$[W(CO)_2]^{2+}$ the intensity equals 23% of the main band. The $C-O$ bond may also be broken as seen by the appearance of ions of the type $[Fe(CO)_x C]^{n+}$, $FeC^{n+}$, and $FeO^{n+}$ ($n = 1$ or 2). From the appearance potentials of the ions, it is possible to calculate the average $M-CO$ bond energy. However, this value is quite incorrect, because the appearance potentials are a complex function, dependent on the spectrometer. For a given instrument, these values may be different, even for a series of similar compounds (e.g., mononuclear carbonyls), by at least 0.3 eV as seen from one compound to another. Additionally, the resulting ion $M^+$ is in an excited state. After taking this fact into consideration, it is possible to obtain quite good agreement between results which are obtained from mass spectra and those obtained by calorimetric measurements concerning the value of the $M-CO$ bond energy.

In a manner similar to that of mononuclear complexes, polynuclear metal carbonyls undergo a fragmentation in which a stepwise loss of carbonyl groups occurs. Generally, the presence of cluster ions of the type $M_n^+$ is observed, for example, $M_3^+$ is formed from $[M_3(CO)_{12}]$ where $M = Fe$, Ru, Os, or $Os_8^+$ is formed from $Os_8(CO)_{23}$. Spectra of dinuclear carbonyls, such as $[Mn_2(CO)_{10}]$, $[Re_2(CO)_{10}]$, $[MnRe(CO)_{10}]$,

Table 2.16. Thermodynamic Data for Metal Carbonyls[a]

| Compound | $\Delta H$ of formation (kJ mol$^{-1}$) | $\Delta H$ of dissociation (kJ mol$^{-1}$) | Average energy of dissociation of $M-M$ bond (kJ mol$^{-1}$) | Average energy of dissociation of $M-C$ bond (kJ mol$^{-1}$) | Vapor pressure equation $p$ (mm Hg) |
|---|---|---|---|---|---|
| $[Cr(CO)_6]$ | $-908.3 \pm 2.0$ | 646 | | 108 | lg $p = 11.832 - 3755.2/T$ (323–413 K) |
| $[Mo(CO)_6]$ | $-916.2 \pm 1.8$ | 910 | | 152 | lg $p = 11.174 - 3561.3/T$ (328–418 K) |
| $[W(CO)_6]$ | $-884.5 \pm 2.5$ | 1069 | | 178 | lg $p = 11.523 - 3872.0/T$ (328–418 K) |
| $[Mn(CO)_5]$ | $-767.8 \pm 6$ | 496 | | 99 | |
| $[Mn_2(CO)_{10}]$ | $-1597.5 \pm 5.3$ | 1068 | 67 | 100 | lg $p = 9.225 - 3262.6/T$ (375–419 K) |
| $[Re(CO)_5]$ | $-686.2 \pm 6$ | 908 | | 182 | |
| $[Re_2(CO)_{10}]$ | $-1559.4 \pm 21$ | 2029 | 128 | 187 | lg $p = 10.68 - 4152/T$ (351–409 K) |
| $[Fe(CO)_5]$ | $-1335 \pm 25$ | 585 | | 117 | lg $p = 8.3098 - 2050.7/T$ (238–418 K) |
| $[Ni(CO)_4]$ | $-600.4 \pm 4.0$ | 588 | | 147 | lg $p = 7.690 - 1519/T$ (238–318 K) |
| $[Fe_2(CO)_9]$ | $-1335 \pm 25$ | 1173 | 82 | 117$^t$, 64$^m$ | |
| $[Fe_3(CO)_{12}]$ | $-1753 \pm 28$ | 1676 | 82 | 117$^t$, 64$^m$ | |
| $[Ru_3(CO)_{12}]$ | $-1820 \pm 28$ | 2414 | 117 | 172 | |
| $[Os_3(CO)_{12}]$ | $-1644 \pm 28$ | 2666 | 130 | 190 | |
| $[Co_2(CO)_8]$ | $-1172 \pm 10$ | 1160 | 83 | 136$^t$, 68$^m$ | lg $p = 17,600 - 5420.6/T$ |
| $[Co_4(CO)_{12}]$ | $-1749 \pm 28$ | 2130 | 83 | 136$^t$, 68$^m$ | |
| $[Rh_4(CO)_{12}]$ | $-1749 \pm 28$ | 2649 | 114 | 166$^t$, 83$^m$ | |
| $[Rh_6(CO)_{16}]$ | $-2299 \pm 28$ | 3496 | 114 | 166$^t$, 83$^m$ | |
| $[Ir_4(CO)_{12}]$ | $-1715 \pm 26$ | 3051 | 130 | 190 | |

[a] According to References 2, 8, 50–52, 56.

$[Tc_2(CO)_{10}]$, $[CoRe(CO)_9]$, $[Fe_2(CO)_9]$, and $[Co_2(CO)_8]$, show ions of the type $[M_2(CO)_x]^+$ as well as $[M(CO)_y]^+$. The percent of $[Fe_2(CO)_x]^+$ ions is very small compared to that of mononuclear ions (unlike for other dinuclear carbonyls). The energy of dissociation of $M-M$ bonds for carbonyls of group 7 metals was determined.[46, 47] Energy values are as follows: $103\ kJ\ mol^{-1}$ for $[Mn_2(CO)_{10}]$, $177\ kJ\ mol^{-1}$ for $[Tc_2(CO)_{10}]$, $187\ kJ\ mol^{-1}$ for $[Re_2(CO)_{10}]$, and $217\ kJ\ mol^{-1}$ for $[MnRe(CO)_{10}]$. These values are higher than those obtained by calorimetry[50] (Table 2.16).

In the case of $[Ni(CO)_4]$, $[Fe(CO)_5]$, and $[M(CO)_6]$ where $M = Cr, Mo, W$, spectra of negative ions were also investigated. For $M(CO)_n$-type carbonyls, the following ions were observed: $M(CO)_{n-1}^-, M(CO)_{n-2}^- \cdots M(CO)^-, M^-$. The parent ions are not stable, because the extra electron must occupy an antibonding molecular orbital. An exception is $V(CO)_6$, a $17e$ complex, in which the electron occupies a non-bonding orbital. $[V(CO)_6]^-$ is easily formed in solution; it was also found in the mass spectrum of vanadium hexacarbonyl.

The investigations of clastograms and metastable ions reveal that negative ions are also formed as a result of progressive loss of CO groups. If the intensity of the positive ion peak $M(CO)_x^+$ is significant, the content of the corresponding negative ion $M(CO)_x^-$ is usually small, and *vice versa*. The number of miscellaneous negative ions in the spectra is generally small. An exception is $[W(CO)_5(PPh_3)]$, which gives more peaks in the negative ion spectrum than in the positive ion spectrum.

# 8. *PROPERTIES OF METAL CARBONYLS*

Metal carbonyls are extremely toxic, particularly those which have high vapor pressure at room temperature. Some of them are even more toxic than carbon monoxide itself. Because of high volatility, tetracarbonylnickel and pentacarbonyliron are exceptionally harmful.

Most metal carbonyls easily undergo oxidation reactions while exposed to air, and some of them even inflame in air. Therefore, reactions involving metal carbonyls should be carried out in the atmosphere of inert gas under a well-ventilated hood. Owing to their covalent bonds, metal carbonyls are volatile and generally well soluble in nonpolar organic solvents (Table 2.16).

The average $M-C$ bond energies for $[Ni(CO)_4]$, $[Fe(CO)_5]$, and $[Cr(CO)_6]$ are 147, 117, and $108\ kJ\ mol^{-1}$, respectively, and are not in agreement with the observed reactivity of these compounds. Basolo and Wojcicki[54] found that the rate of CO exchange is large for $Ni(CO)_4$ and small for $Fe(CO)_5$ and $Cr(CO)_6$. The exchange is first order with respect to the complex and zero order with respect to the ligand. Therefore, the exchange proceeds by dissociative mechanism $S_N1$, in which the rate of the reaction is determined by $M-C$ bond breaking.[51, 53, 54]

$$[M(CO)_n] \longrightarrow [M(CO)_{n-1}] + CO \qquad \text{slow} \qquad (2.8)$$

$$[M(CO)_{n-1}] + CO \longrightarrow [M(CO)_{n-1}(CO)] \qquad \text{fast} \qquad (2.9)$$

The values of bond energies are not in agreement with the observed direction of reactivity of these carbonyls. The energy of the breaking of the first $M-CO$ bond determines the reactivity of carbonyls if a dissociative mechanism occurs. This energy is considerably different from the average energy in compounds of the first transition metal series. For $[Fe(CO)_5]$, $[Cr(CO)_6]$, $[Mo(CO)_6]$, and $[W(CO)_6]$, the dissociation energies of the first $M-CO$ bond obtained by the laser pyrolysis method are as follows: 173.5, 154, 169, and 192 kJ mol$^{-1}$, respectively.[52] The differences between the average dissociation energy of $M-CO$ bonds and the dissociation energy of the first $M-CO$ bond are smaller for the second and third transition metal series.[52]

The activation energy of the CO exchange reaction for $Ni(CO)_4$ is 100.3 kJ mol$^{-1}$; this value is considerably lower than the average bond energy. From the properties of $Ni(CO)_4$, it is known, however, that the loss or substitution of the first CO molecule is considerably easy, and thus the energy required for the breaking of the first $M-CO$ bond is much lower than the average energy. The rates of CO exchange in the dark for some carbonyls are given in Table 2.17.

The reactions of carbonyl complexes may be divided into the following five groups:

1.  Reactions involving oxidation state changes.
2.  Substitution reactions.
3.  Reactions involving the breaking of metal–metal bonds in polynuclear carbonyls.
4.  Ligand migration reactions and insertions.
5.  Reactions of coordinated CO groups.

The oxidation reactions of carbonyls by acids proceed according to the equation

$$M(CO)_m + nH^+ \longrightarrow M^{n+} + \frac{n}{2}H_2 + mCO \qquad (2.10)$$

Mineral and organic acids react in this manner, as do other proton donors, such as 1,3-diketones. Therefore, carbonyls may be utilized for the preparation of many com-

*Table 2.17. Rates for Exchange Reaction of Carbon Monoxide with Metal Carbonyls*[a]

| Compound | $T(K)$ | Solvent | Rate constant $(s^{-1}$ or $t_{1/2})$ | $\Delta H^{\neq}$ (kJ mol$^{-1}$) | $\Delta S^{\neq}$ (kJ mol$^{-1}$) |
|---|---|---|---|---|---|
| $[Ni(CO)_4]$ | 303 | n-Hexane | $2.1 \times 10^{-2}$ | 100.3 | 55 |
| $[Fe(CO)_5]$ | 298 | Benzene | (4 years) | | |
| $[Cr(CO)_6]$ | 390 | Gas | $2 \times 10^{-5}$ | 163 | |
| $[Mo(CO)_6]$ | 389 | Gas | $7.5 \times 10^{-5}$ | 130 | |
| $[W(CO)_6]$ | 415 | Gas | $2.6 \times 10^{-6}$ | 168 | |
| $[V(CO)_6]$ | 283 | Heptane | (7 hours) | | |
| $[Mn_2(CO)_{10}]$ | 298 | Benzene | (10 years) | | |
| $[Fe_3(CO)_{12}]$ | 298 | Benzene | (350 hours) | | |
| $[Co_2(CO)_8]$ | 273 | Toluene | $1.5 \times 10^{-3}$ | 96 | −54 |
| $[Co_3(CO)_9CH]$ | 313 | Toluene | $9.5 \times 10^{-7}$ | | |
| $[Co_3(CO)_9CF]$ | 313 | Toluene | $1.1 \times 10^{-5}$ | 63 | |
| $[Co_3(CO)_9CCl]$ | 313 | Toluene | $6.3 \times 10^{-6}$ | 84 | |
| $[Co_3(CO)_9CBr]$ | 313 | Toluene | $3.5 \times 10^{-6}$ | 96 | |
| $Hg[Co(CO)_4]_2$ | 273 | Toluene | $3.5 \times 10^{-3}$ | 54 | −100 |
| $Cd[Co(CO)_4]_2$ | 252 | Toluene | $1.9 \times 10^{-3}$ | 46 | −117 |

[a] According to References 51, 53, and 55.

plexes because they are currently easily accessible and relatively free of impurities involving other metals. Two such examples are: $M(acac)_n$, $M = Fe(II)$, $Fe(III)$, $Cr(III)$, $Mo(III)$, etc., and $M_2(O_2CR)_4$, $M = Cr$, $Mo$, $W$.

The alkali or alkali earth metals are often used for reduction of metal carbonyls in a variety of solvents.[2, 9, 10, 57]

$$[Co_2(CO)_8] + 2Na \xrightarrow[200\ K]{NH_3} 2Na[Co(CO)_4] \qquad (2.11)$$

The first group of reactions includes also disproportionation reactions, which occur in the presence of Lewis bases.[2, 9, 10, 58]

$$3[Co_2(CO)_8] + 12py \longrightarrow 2[Co(py)_6][Co(CO)_4]_2 + 8CO \qquad (2.12)$$

This reaction takes place with the formation of intermediates; in the first step, cationic carbonyl is formed:

$$[Co_2(CO)_8] + B \longrightarrow [Co(CO)_4 B][Co(CO)_4] \qquad (2.13)$$

where B represents a Lewis base.

Some intermediates are quite stable and may be isolated from the reaction mixture. $[Co_2(CO)_8]$ and $[V(CO)_6]$ very readily react with Lewis bases. With Lewis bases B, $[V(CO)_6]$ forms compounds of the formula $[VB_n][V(CO)_6]_2$ containing $V^{2+}$ and $V^{1-}$; an intermediate complex $[V(CO)_5 B][V(CO)_6]$ is also formed.

The oxidation–reduction reactions also proceed in the presence of the alkali metal hydroxides:

$$[Fe(CO)_5] + 4OH^- \longrightarrow [Fe(CO)_4]^{2-} + CO_3^{2-} + 2H_2O \qquad (2.14)$$

The mechanism of this reaction is complicated and, depending on conditions, polynuclear compounds are also formed:

$$2[HFe(CO)_4]^- \longrightarrow H_2 + [Fe_2(CO)_8]^{2-} \qquad (2.15)$$

$$2[Mo(CO)_6] + 9OH^- \longrightarrow [Mo_2(OH)_3 (CO)_6]^{3-} + 6HCOO^- \qquad (2.16)$$

In basic medium, in the first step, nucleophilic attack of $OH^-$ on the coordinated CO molecule occurs and, as a result, a hydroxycarbonyl complex (metallocarboxylic acid) is formed. Such compounds are formed during the conversion of CO (water–gas shift reaction, wgsr).[59–61]

$$Fe(CO)_5 + OH^- \longrightarrow \left[ Fe(CO)_4 \left( C \overset{\displaystyle O}{\underset{\displaystyle OH}{\big\Vert}} \right) \right] \longrightarrow [FeH(CO)_4]^- + CO_2 \qquad (2.17)$$

$$[FeH(CO)_4]^- + H_2O \longrightarrow [FeH_2(CO)_4] \xrightarrow[-H_2]{+CO} Fe(CO)_5 \qquad (2.18)$$

Metal carbonyls may be oxidized by a variety of oxidizing agents, such as $O_2$, $X_2$

$(X = Cl, Br, I)$, $Fe(CN)_6^{3-}$, $Fe^{3+}$, $Ce^{4+}$, $Cu^{2+}$, $Rh^{3+}$, etc., as well as oxidized electrochemically.[3, 8–10, 62, 63]

$$[M(CO)_6] - e \longrightarrow [M(CO)_6]^+ \longrightarrow [M(CO)_6] + [M(CO)_6]^{2+}$$

$$\downarrow \qquad\qquad (2.19)$$

$$M = Cr, Mo, W \qquad\qquad\qquad M^{2+} + 6CO$$

Metal carbonyls also undergo electrochemical reduction:

$$[M(CO)_6] + e \rightarrow [M(CO)_6]^- \xrightarrow[-CO]{} [M(CO)_5]^- \rightarrow \tfrac{1}{2}[M_2(CO)_{10}]^{2-} \qquad (2.20)$$

$$Fe(CO)_5 + e \rightarrow [Fe(CO)_5]^- \xrightarrow[-CO]{} [Fe(CO)_4]^- \rightarrow \tfrac{1}{2}[Fe_2(CO)_8]^{2-} \qquad (2.21)$$

$$[M(CO)_6 + 2e \xrightarrow[-CO]{} [M(CO)_5]^{2-} \qquad (2.22)$$

Substitution reactions may proceed with displacement of one or more carbonyl groups. Bridging as well as terminal carbonyl groups may be displaced by other ligands, such as olefins, acetylenes, aromatic hydrocarbons, polyenes, and Lewis bases (oxygen, nitrogen, phosphine ligands, etc.). This class of reactions also comprises described isotope exchange reactions. Substitution reactions proceed most commonly by a dissociative mechanism. In 18$e$ complexes, the substitution reactions occur by associative mechanism when, in the transition state, no 20$e$ complex is formed, i.e., an $n$-electron ligand may rearrange itself into a $(n-2)$-electron ligand, for example: $\eta^6$-arene $\rightarrow \eta^4$-arene, $\eta^5$-$C_5H_5 \rightarrow \eta^3$-$C_5H_5$, $\eta^3$-$C_3H_5 \rightarrow \sigma$-$C_3H_5$, linear nitrosyl $(3e) \rightarrow$ bent nitrosyl $(1e)$.[65]

The CO substitution in $M(CO)_n(NO)$ leading to the $M(CO)_{n-1}L(NO)$ product proceeds according to an $S_N2$ mechanism possibly because, in the transition state, an 18$e$ complex is formed, $M(CO)_nL(NO)$, in which NO acts as a 1$e$ ligand.

The substitution reactions occur considerably faster under high-energy irradiation of carbonyls, which leads to dissociation of CO and formation of compounds possessing free coordination sites.[63–67, 58]

$$Cr(CO)_6 \xrightarrow{h\nu} Cr(CO)_5 + CO \qquad\qquad (2.22a)$$

$$Cr(CO)_5 + L \longrightarrow Cr(CO)_5 L \qquad\qquad (2.22b)$$

$$Mn_2(CO)_{10} \xrightarrow{h\nu} Mn_2(CO)_9 + CO \qquad\qquad (2.22c)$$

In the case of thermal and photolytic reactions of dinuclear carbonyls, the breaking of the $M-M$ bond occurs, leading to many products which result from chain radical reactions as well as disproportionation reactions.[52, 58, 63–67, 70] The $Co-Co$ bond breaking in $[Co_2(CO)_8]$ takes place with particular ease.

The fourth class of reactions comprises processes in which a new ligand is formed from two ligands of the coordination sphere. It has been shown that the majority of such reactions proceed via migration of one ligand to another.[51, 67, 68]

$$\text{migration: } L_n M - Y \rightleftharpoons L_n M - Y - X$$
$$| \quad \nearrow$$
$$X \quad (2.23)$$

where   $X = H^-, R^-, OR^-, OH^-, H_2O, NR_2^-, NR_3$; $Y = CO$,
alkenes, acetylenes, RCHO, RCN, $SO_2$, $O_2$.

The migration occurs for ligands which are in a *cis* position to one another. The free coordination site is rapidly attacked by a solvent molecule. This explains the dependence of the rate of migration on the solvent. The migration of alkyl groups to CO is also accelerated by some cations, aluminum halides, aluminum oxides, etc., most likely due to the coordination of the oxygen atom of CO by the electrophile [cf. reaction (4.178)].

For pentacarbonylmethylmanganese, the migration reaction is first order with respect to the complex as well as the ligand. The proposed reaction mechanism is as follows:

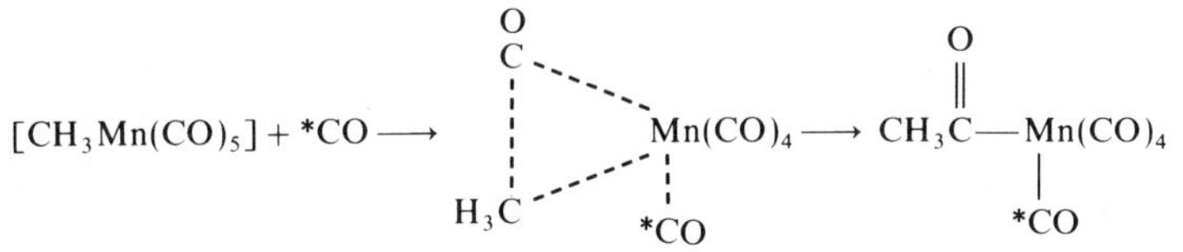

$$[CH_3Mn(CO)_5] + {}^*CO \longrightarrow \quad \cdots \cdots Mn(CO)_4 \longrightarrow CH_3C\!\!-\!\!Mn(CO)_4 \qquad (2.24)$$

However, not all reactions of ligands in the coordination sphere proceed by migration; some reactions proceed by a different mechanism involving insertion of ligand Y into the M−X bond[68, 69]:

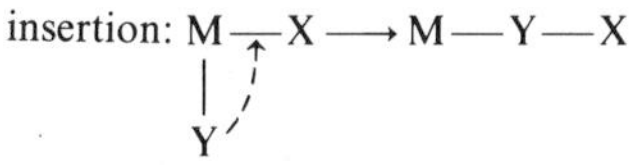

$$\text{insertion: } M\!\!-\!\!X \longrightarrow M\!\!-\!\!Y\!\!-\!\!X \qquad (2.25)$$

Brunner and Vogt reported that in the case of chiral iron(II) complexes, migration of the $CH_3$ group to a coordinated CO occurs because the reaction proceeds with change of the configuration on the iron atom:

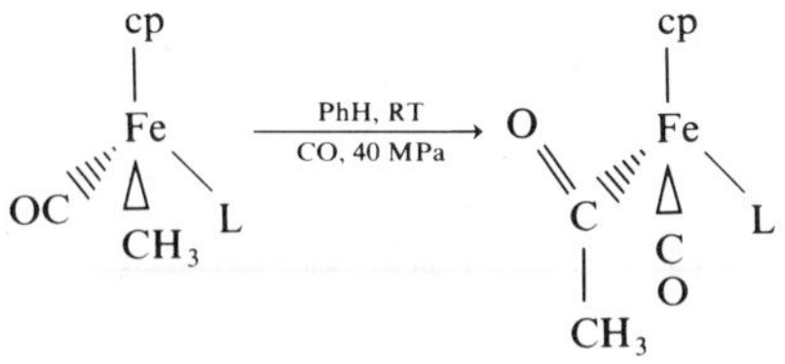

$$(2.26)$$

$$L = PPh_2NMe\text{-}\underline{S}\text{-}CHMePh$$

The stereospecificity of this reaction is very high ($>90\%$). Flood and Campbell[69] investigated reactions of the formation of acyl derivatives from chiral iron complexes of the type $Fe(cp)(Et)(CO)PR_3$, where $PR_3 = PPh_3$ and $P(OCH_2)_3CMe$, and reported that ethyl group migration or CO insertion is solvent dependent. These reactions proceed with ethyl migration in the atmosphere of CO in nitroethane, nitromethane, and acetonitrile. The optical yields are high for the first two solvents. In HMPA, DMSO, DMF, or PDC, the optical purity of the product is moderate or low, and the configuration of the complex formally corresponds to CO migration into the $Fe-C_2H_5$ bond, that is, insertion of CO.

*Chapter 2*

Many catalytic processes such as oxo reactions, for example, hydroformulation, proceed via migration and insertion mechanisms.

Similar to the preceding class are reactions during which attack on CO takes place by an uncoordinated external reagent. Attack of external nucleophile occurs on the carbon atom. Alkylation of resulting acyl complexes leads to the formation of metal carbene complexes (see Chapter 5).

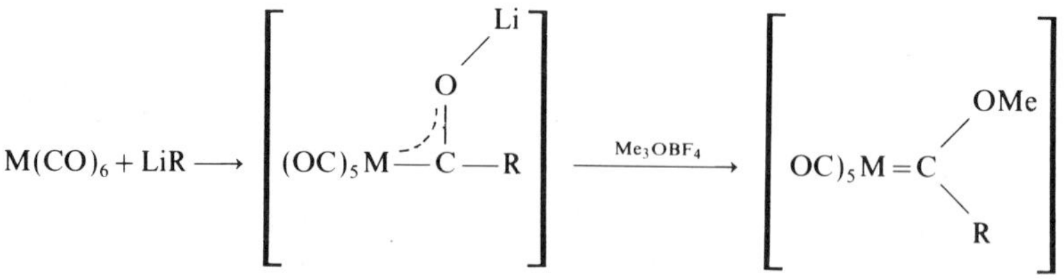

$$M(CO)_6 + LiR \longrightarrow \left[ (OC)_5 M \underset{}{\overset{\overset{\displaystyle O}{\|}}{-}} C - R \right] \xrightarrow{Me_3OBF_4} \left[ OC)_5 M = C \overset{\displaystyle OMe}{\underset{\displaystyle R}{<}} \right] \quad (2.27)$$

R = alkyl, aryl, PR_2, NR_2

Ions such as $OH^-$ and $OR^-$ also react with carbonyl ligands to give metalloacids and metalloesters, respectively [see equation (2.17)].

Carbonyls of almost all transition metals and even some of the main group metals are presently known (Tables 2.18–2.20). Carbonyls of the main group metals, some carbonyls of the transition metals possessing small numbers of $d$ electrons, as well as platinum, palladium, copper, and silver carbonyls are unstable, and were obtained in carbon monoxide or argon matrices at low temperatures (4–10 K).

## a. Group 3 Metal Carbonyls

Elements of this group do not form stable carbonyls. $M(CO)_y$ compounds, where M = Pr, Nd, Gd, Ho, Eu, Er, Yb, and U, may be prepared by condensation of metal atoms with carbon monoxide at low temperatures.[4] Some low-valent lanthanide

*Table 2.18. Some Stable Homonuclear and Heteronuclear Transition Metal Carbonyls*

| 5 | 6 | 7 | 8 | 9 | 10 |
|---|---|---|---|---|---|
| $[V(CO)_6]$ | $[M(CO)_6]$ (M = Cr, Mo, W) | $[M_2(CO)_{10}]$ (M = Mn, Tc, Re) | $[M(CO)_5]$ $[M_3(CO)_{12}]$ (M = Fe, Ru, Os) $[Fe_2(CO)_9]$ $[Os_2(CO)_9]$ $[Os_5(CO)_{16}]$ $[Os_6(CO)_{18}]$ $[Os_7(CO)_{21}]$ $[Os_8(CO)_{23}]$ | $[M_2(CO)_8]$ $[M_4(CO)_{12}]$ $[M_6(CO)_{16}]$ (M = Co, Rh, Ir) | $[Ni(CO)_4$ |

$[MnRe(CO)_{10}]$, $[MnTc(CO)_{10}]$, $[TeRe(CO)_{10}]$, $[MnCo(CO)_9]$, $[TeCo(CO)_9]$, $[ReCo(CO)_9]$, $[Mn_2Fe(CO)_{14}]$, $[Re_2Fe(CO)_{14}]$, $[MnReFe(CO)_{14}]$, $[ReFe_2(CO)_{12}]_2$, $[Fe_2Ru(CO)_{12}]$, $[FeRu_2(CO)_{12}]$, $[Fe_2Os(CO)_{12}]$, $[FeOs_2(CO)_{12}]$, $[Ru_2Os(CO)_{12}]$, $[RuOs_2(CO)_{12}]$, $[Co_2Rh_2(CO)_{12}]$, $[Co_2Rh_4(CO)_{16}]$, $[Co_2Os(CO)_{11}]$, $[Co_3Rh(CO)_{12}]$, $[Co_2Ir_2(CO)_{12}]$, $[Rh_3Ir(CO)_{12}]$, $[Rh_2Ir_2(CO)_{12}]$.

*Table 2.19. Unstable Homonuclear Metal Carbonyls*

| 3,13 | 4,14 | 5 | 6 | 7 | 8 | 9 | 10 | 11 |
|---|---|---|---|---|---|---|---|---|
| $[La(CO)_n]$ ($n=1$–$6$) | $[Ge(CO)_n]$ | $[V(CO)_n]$ ($n=1$–$5$) | $[M(CO)_n]$ ($n=1$–$5$) $M=Cr, Mo, W$ | $[M(CO)_n]$ $M=Mn$ ($n=1$–$5$) | $[Fe(CO)_n]$ ($n=1$–$4$) | $[M(CO)_n]$ $M=Co,$ **Rh, Ir** ($n=1$–$4$) | $[Ni(CO)_n]$ $[Pd(CO)_m]$ | $[Cu(CO)_m]$ $[Ag(CO)_m]$ |
| $[Al_x(CO)_n]$ | $[Sn(CO)_n]$ | | $[Cr_2(CO)_{10}]$ | | $[Os_4(CO)_{13}]$ $[Fe_2(CO)_8]$ | | $[Pt(CO)_m]$ ($n=1$–$3, m=1$–$4$) | $[Au(CO)_n]$ ($m=1$–$3, n=1, 2$) |
| $[U(CO)_m]$ ($m=1$–$6$) | $[Ti(CO)_6]$ $[Ti_2(CO)_n]$ | $[Ta(CO)_m]$ ($m=1$–$6$) $[V_2(CO)_{12}]$ | | | | | $[Pt_x(CO)_{2x}]$ | $[Cu_2(CO)]$ $[M_2(CO)_6]$ $M=Cu, Ag$ |

Table 2.20. Ionic Metal Carbonyls

| 5 | 6 | 7 | 8 | 9 | 10 |
|---|---|---|---|---|---|
| $[M(CO)_6]^-$ | $[M(CO)_5]^{2-}$ | $[M(CO)_5]^-$ | $[M(CO)_4]^{2-}$ | $[M(CO)_4]^-$ | $[Ni_2(CO)_6]^{2-}$ |
| $[M(CO)_5]^{3-}$ | $[M(CO)_4]^{4-}$ | $[M(CO)_6]^+$ | $[Fe_2(CO)_8]^{2-}$ | $[Co_3(CO)_{10}]^-$ | $[Ni_3(CO)_8]^{2-}$ |
| (M = V, Nb, Ta) | $[M_2(CO)_{10}]^{2-}$ | $[M_2(CO)_9]^{2-}$ | $[Fe_3(CO)_{11}]^{2-}$ | $[M(CO)_3]^{3-}$ | $[Ni_5(CO)_{12}]^{2-}$ |
| | $[M_3(CO)_{14}]^{2-}$ | (M = Mn, Tc, Re) | $[Fe_4(CO)_{13}]^{2-}$ | $[M_6(CO)_{15}]^{2-}$ | $[Ni_6(CO)_{12}]^{2-}$ |
| | (M = Cr, Mo, W) | $[Re_4(CO)_{16}]^{2-}$ | $[Ru_6(CO)_{18}]^{2-}$ | $[M_6(CO)_{14}]^{4-}$ | $[Ni_9(CO)_{18}]^{2-}$ |
| | | $[Mn(CO)_4]^{3-}$ | $[M(CO)_6]^{2+}$ | (M = Co, Rh, Ir) | $[Ni_8H_2(CO)_{14}]^{2-}$ |
| | | $[Re(CO)_4]^{3-}$ | (M = Fe, Ru, Os) | $[Rh_{12}(CO)_{30}]^{2-}$ | $[Ni_{11}H_2(CO)_{20}]^{2-}$ |
| | | | | $[Ir_4(CO)_{10}H_2]^-$ | $[Pt_3(CO)_6]_n^{2-}$ |
| | | | | $[Rh_7(CO)_{16}]^{3-}$ | (n = 1–10) |
| | | | | $[Rh_4(CO)_{11}]^{2-}$ | $[Pt_{26}(CO)_{32}]^{2-}$ |
| | | | | $[Ir_8(CO)_{20}]^{2-}$ | $[Pt_{38}(CO)_{44}H_x]^{2-}$ |
| | | | | $[Rh_{22}(CO)_{37}]^{4-}$ | $[Ni_{38}Pt_6(CO)_{48}H_{6-n}]^{n-}$ |
| | | | | $[Rh_{17}(CO)_{32}(S)_2]^{3-}$ | (n = 5, 4) |

compounds react with carbon monoxide. The cyclopentadienyl samarium(II) complex, $Sm(C_5Me_5)_2 (THF)_2$, reacts with carbon monoxide under atmospheric pressure to give $[(C_5Me_5)_2 Sm(\mu_3\text{-}O_2CCCO) Sm(C_5Me_5)_2 (THF)]_2$,[187] which has the following structure:

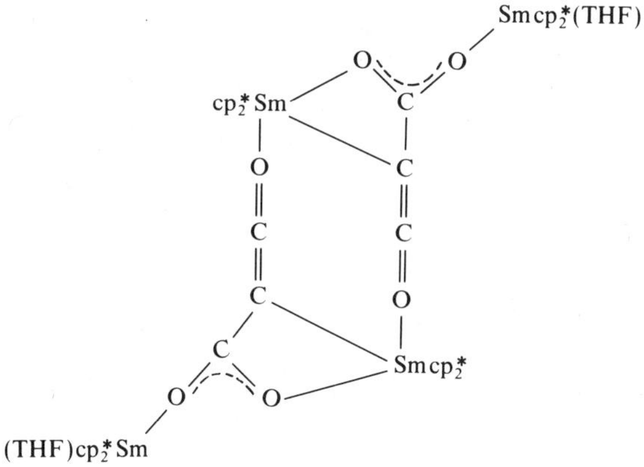

Sm cp$_2^*$(THF)

cp$_2^*$ Sm

(THF)cp$_2^*$ Sm

Sm cp$_2^*$

cp* = C$_5$Me$_5$

Similar complexes such as $[Sm(C_5Me_5)_2]$ and $[Sm(\mu\text{-}I)(C_5Me_5)(THF)_2]_2$ also react with carbon monoxide. Therefore, homologation reactions involving one electron reduction of CO may be important in Fischer–Tropsch processes.

### b. Group 4 Metal Carbonyls

Carbonyls of these metals which do not contain any other ligands in the coordination sphere are not stable (Table 2.21). Presently known are $Ti(CO)_n$ ($n = 1$–$6$) carbonyls, which were prepared by reaction of atomic titanium with carbon monoxide at low temperatures. However, the following compounds are stable: $[Ti(CO)_2cp_2]$, $[Zr(CO)_2cp_2]$, and $[Hf(CO)_2cp_2]$. Titanium and zirconium compounds are obtained by the following reactions:

$$[TiCl_2cp_2] \xrightarrow[\text{2, CO, 11 MPa}]{\text{1, 2Nacp, PhH}} [Ti(CO)_2cp_2] \qquad (18\%) \qquad (2.28)$$

$$[TiCl_4] \xrightarrow[\text{2, CO}]{\text{1, >4 Nacp}} [Ti(CO)_2cp_2] \qquad\qquad\qquad (2.29)$$

$$[Ti(BH_4)cp_2] \xrightarrow[\text{CO, 4 days}]{\text{Et}_3\text{N}} [Ti(CO)_2cp_2] \qquad (80\%) \qquad (2.30)$$

$$[Zr(BH_4)_2cp_2] \xrightarrow[\text{CO, 10 MPa}]{\text{Et}_3\text{N, 323 K}} [Zr(CO)_2cp_2] \qquad (15\%) \qquad (2.31)$$

The reaction of titanium atoms with CO at low temperatures (10–15 K) gives $[Ti(CO)_6]$ and $[Ti_2(CO)_y]$. The $[Ti(CO)_7]$ compound is not formed. Therefore, the existence of heptacarbonyltitanium described earlier in the patent literature (1 MPa CO, 570–1070 K) seems doubtful. The dinuclear complex $[Ti_2(CO)_y]$ does not contain any

*Table 2.21. Properties of Group 4 Metal Carbonyls*

| Compound | Color | Melting point (K) | $\nu(CO)$ (cm$^{-1}$) | Remarks |
|---|---|---|---|---|
| $[Ti(CO)_6]^a$ | | | 1985, 1953, 1945 | 10–15 K, CO, CO/Kr = 1/10 |
| $[Ti_2(CO)_n]^a$ | | | | |
| $[Ti(CO)_2cp_2]^b$ | Red-brown | 363 dec | 1965, 1883 (in THF) | Pyrophoric |
| $[Ti(CO)_2(C_5Me_5)_2]^c$ | Yellow-brown | 353–8 | 1930, 1850 | |
| $[Zr(CO)_2cp_2]^d$ | Violet | | 1887, 1976 (in $C_6H_{14}$) | |
| $[Hf(CO)_2cp_2]^e$ | Purple | | | d (HfC) 216 pm |

[a] R. Busby, Klotzbücher, and G. A. Ozin, *Inorg. Chem.*, **16**, 822 (1977).
[b] J. G. Murray, *J. Am. Chem. Soc.*, **83**, 1281 (1961).
[c] F. Calderazzo, J. J. Salzmann, and P. Mosinan, *Inorg. Chim. Acta*, **1**, 65 (1967).
[d] G. Fachinetti, G. Fochi, and C. Floriani, *J. Chem. Soc., Chem. Commun.*, 230 (1976).
[e] D. J. Sikora, M. D. Rausch, R. D. Rogers, and J. L. Atwood, *J. Am. Chem. Soc.*, **101**, 5079 (1979).

bridging carbonyl groups. The cyclopentadienyl complexes have tetrahedral structures assuming that the $C_5H_5$ ligand occupies only one coordination site. The $[Ti(CO)_6]$ carbonyl is octahedral.

## c. Group 5 Metal Carbonyls

The preparation of $[V(CO)_6]^-$ is accomplished most commonly by the reaction of $VCl_3$ or $VCl_4$ with the main group metals such as sodium, magnesium and zinc, or aluminum in the atmosphere of CO:

$$VCl_3 + 4\,Na + 6\,CO \xrightarrow[\text{30 MPa, 430 K}]{\text{diglyme}} [Na\{(MeOC_2H_4)_2O\}_2][V(CO)_6] + 3\,NaCl \quad (2.32)$$

$$2\,VCl_4 + 5\,Mg/Zn + 12\,CO \xrightarrow{\text{pyridine}} [Mg(py)_n][V(CO)_6]_2 + 4\,MgCl_2 \quad (2.33)$$

$Na[M(CO)_6]$, where M = V, Nb, and Ta, may also be prepared by the reduction of corresponding chlorides of the group 5 metals with sodium simultaneously utilizing high-energy ultrasounds.[74]

Hexacarbonylvanadate $(1-)$ does not undergo thermal substitution with triphenylphosphine, although one of the CO groups is easily given off in photochemical reactions with concomitant substitution by $PR_3$, $AsPh_3$, $SbPh_3$, and THF.[85] The THF molecule in $[V(CO)_5(THF)]^-$ may easily be substituted by different ligands.[85]

On heating, $[V(diphos)_3][V(CO)_6]$ gives $[V(CO)_4(diphos)]$. The reaction of $[V(CO)_6]^-$ with chloride complexes of certain metals affords compounds possessing metal–metal bonds:

$$Na[V(CO)_6] \begin{cases} \xrightarrow{Ph_3PAuCl} [Ph_3PAuV(CO)_6] \\ \\ \xrightarrow{Ph_3SnCl} [Ph_3SnV(CO)_6] + NaCl \end{cases} \quad (2.34)$$

In the solid state, the $[V(THF)_4][V(CO)_6]_2$ complex, which may be prepared by the reaction of $[V(CO)_6]$ with THF, is trinuclear, containing the $[V(THF)_4]^{2+}$ cation bonded to the two $[V(CO)_6]^-$ anions through oxygen of carbonyl groups[72] (Table 2.22, Figure 2.15 below).

Similar complexes of $[VL_4][V(CO)_6]_2$, where L = acetone, ether, and pyridine, are also known. In solution, these complexes have analogous structures because their IR spectra exhibit an intensive band in the region $1657-1686 \text{ cm}^{-1}$. Recently, anionic complexes of the type $[M(CO)_5]^{3-}$ were prepared by the reduction of $[M(CO)_6]^-$ with sodium in liquid ammonia at 195 K.[73]

The reaction between $[V(CO)_6]^-$ and hydrochloric or phosphoric acids forms $HV(CO)_6$, which is stable only in solution. This carbonyl hydride decomposes after evaporation of the solvent according to the equation:

$$[HV(CO)_6] \longrightarrow \tfrac{1}{2}H_2 + [V(CO)_6] \tag{2.35}$$

Hexacarbonylvanadium is the only carbonyl which does not contain any other ligands, is relatively stable, and does not obey Sidgwick's rule. $[V(CO)_6]$ is paramagnetic in the solid state and in solution (Table 2.22). With strong Lewis bases and with ligand coordinating through nitrogen and oxygen, $[V(CO)_6]$ undergoes disproportionation reactions:

$$3[V(CO)_6] + nL \longrightarrow [VL_n][V(CO)_6]_2 + 6CO \tag{2.36}$$

With soft Lewis bases $[V(CO)_6]$ gives CO group substitution products:

$$[[V(CO)_6] + 2\,PR_3 \longrightarrow \textit{trans-}[V(CO)_4\,(PR_3)_2] + 2CO \tag{2.37}$$

$$R = Ph, Et, Pr^n$$

$$[V(CO)_6] + 2P(C_6H_{11})_3 \longrightarrow [V(CO)_4\,\{P(C_6H_{11})_3\}_2]_2 \tag{2.38}$$

The vanadium–vanadium bond in $[V(CO)_4\,\{P(C_6H_{11})_3\}_2]_2$ is weak because in solution this compound is partially dissociated. $[V(CO)_6]$ is easily reduced to $[V(CO)_6]^-$ in the presence of $[Co(CO)_4]^-$, $[Mn(CO)_5]^-$, $[M_2(CO)_{10}]^{2-}$, $[Mcp(CO)_3]^-$ (M = Cr, Mo, W), and $Cocp_2$. However, $Mncp_2$ and $Nicp_2$ readily oxidize hexacarbonylvanadium to $Vcp(CO)_4$. Two-electron oxidation of $[Nb(CO)_6]^-$ by halides and acetylacetonates of $Hg^{II}$, $Cu^{II}$, $Ag^I$, $Fe^{III}$ affords $[Nb_2X_3(CO)_8]^-$.[91] The 18e complex $[V(CO)_5\,(NO)]$ which results from the reaction of $[V(CO)_6]$ with NO is very unstable; it cannot be isolated from the solution. More stable is the following NO compound containing triphenylphosphine molecule:

$$[V(CO)_4\,(PPh_3)_2] \xrightarrow{\text{NO}} [V(CO)_4\,(NO)(PPh_3)] + PPh_3 \tag{2.39}$$

Treatment of hexacarbonylvanadium with arenes results in the formation of $[V(CO)_4\,(Ar)]$ which is oxidized by the excess of $[V(CO)_6]$ during the reaction, and an ionic complex of the composition $[V(CO)_4\,(Ar)][V(CO)_6]$ is obtained. The reaction of $[V(CO)_6]$ with (2-allylphenyl)diphenylphosphine or (2-propenyl-phenyl)diphenylphosphine affords compounds containing olefins $\pi$-bonded to the metal[71] (Figure 2.14).

Table 2.22. Group 5 Metal Carbonyls

| Compound | Color | Melting point (K) | $\nu$(CO) (cm$^{-1}$) | Solubility | Remarks |
|---|---|---|---|---|---|
| Na$_3$[V(CO)$_5$][73] | | | | | $\delta$ $^{13}$C = 290, 1 ppm, $J(^{51}\text{V} - {}^{13}\text{C}) = 139$ Hz |
| [Na(diglim)$_2$][V(CO)$_6$][a] | Yellow | | 1858 | H$_2$O, THF | Diamagnetic |
| [V(THF)$_4$][V(CO)$_6$]$_2$[b] | Yellow | | 1859 | THF, CH$_2$Cl$_2$ | |
| [V(Et$_2$O)$_4$][V(CO)$_6$]$_2$ (72) | Green | | 2038w, 1961m, 1942sh, 1878vs, 1686s | | |
| [Et$_4$N][V(CO)$_5$PPh$_3$][c] | Yellow | 453–455 | 1953, 1842, 1805, 1767 | A, E | Stable in air |
| [V(CO)$_4$(PPh$_3$)$_2$][d] | Orange | 415 | 1845 | THF, C$_6$H$_6$ CHCl$_3$ | $\mu_{\text{eff}}$ = 1.78 BM, Diamagnetic at low temperatures |
| [V(CO)$_6$][e] | Green-blue | 343 dec | 1980 | Hydrocarbons | $\mu_{\text{eff}}$ = 1.81, |
| [V(CO)$_6$]$_{1-5}$[f] | | | 2046, 2021, 2014, | | CO matrix, 15 K |
| [V$_2$(CO)$_{12}$][g] | | | 1987, 1852 | | CO matrix |
| [Vcp(CO)$_4$][h] | Red | 412–413 | 2031, 1931 | Organic solvents | Stable in air |
| [V(CO)$_5$] (75) | | | | | Matrix Kr, 20 K, BPT (deformed), ground state $^2B_2$ |
| [V(CO)$_4$] (75) | | | | | Matrix Kr, 20 K, $T_{\text{d}}$, ground state $^6A_1$ |

| Compound | Color | M.p. (dec) | IR bands | Solubility | Stable in air |
|---|---|---|---|---|---|
| $[Vcp(CO)_3PPh_3]$[h] | Red | 433–434 | 1962, 1884, 1865 | Organic solvents | |
| $[V_2cp_2(CO)_5]$[h,i] | Dark-green | 372 dec | 2006, 1953, 1893, 1869, 1826 | Organic solvents | Stable in air |
| $Cs_3[Nb(CO)_5]$[73] | Dark-red | | 1810w, 1566vs | | |
| $[Na(diglim)_2][Nb(CO)_6]$[j] | Yellow | 418 dec | 1830 | Diglyme | |
| $[Nbcp(CO)_4]$[k,l] | Red | 417–418 | 2036, 1928 | Organic solvents | |
| $[Na(diglim)_2][Ta(CO)_6]$[j] | Yellow | | 1850 | | |
| $[Tacp(CO)_4]$[j,m] | Red | 444–446 | 2020, 1900 | Organic solvents | |
| $[Ta(CO)_{1-6}]$[n] | | | | | |
| $Cs_3[Ta(CO)_5]$[73] | Dark brown-red | | 1813w, 1562vs | | |
| $Cs_2[Tacp(CO)_3]^{2-}$[76] | Light-orange | 523 dec | 1736s, 1570s | | |

[a] E. W. Abel, R. A. N. McLean, S. P. Tyfield, P. S. Braterman, and A. P. Walker, *J. Mol. Spectrosc.*, **30**, 29 (1969).
[b] M. Schneider and E. Weiss, *J. Organomet. Chem.*, **121**, 365 (1976).
[c] H. Behrens, H. Brandl, and K. Lutz, *Z. Naturforsch.*, **22B**, 99 (1967).
[d] R. P. M. Werner, *Z. Naturforsch.*, **16B**, 477 (1961).
[e] H. Haas and K. Sheline, *J. Am. Chem. Soc.*, **88**, 3219 (1966); R. Ercoli, F. Calderazzo, and A. Alberoda, *J. Am. Chem. Soc.*, **82**, 2955 (1960); L. V. Interrante and G. V. Nelson, *J. Organomet. Chem.*, **25**, 153 (1970).
[f] L. Hanlan, H. Huber, and G. A. Ozin, *Inorg. Chem.*, **15**, 2592 (1976).
[g] T. A. Ford, H. Huber, W. Klotzbücher, M. Moskovits, and G. A. Ozin, *Inorg. Chem.*, **15**, 1666 (1976); T. C. De Vore and H. F. Franzen, *Inorg. Chem.*, **15**, 1318 (1976).
[h] E. O. Fischer and R. J. J. Schneider, *Chem. Ber.*, **103**, 3584 (1970).
[i] E. O. Fischer and R. J. J. Schneider, *Angew. Chem., Int. Ed. Engl.*, **6**, 569 (1967).
[j] R. P. M. Werner, A. H. Fibey, and S. M. Manastyrskyj, *Inorg. Chem.*, **3**, 298 (1964).
[k] K. N. Anisimow, N. E. Kolobowa, and A. A. Pasynskij, *Izv. AN SSSR, Ser. Chim.*, 2238 (1969).
[l] A. A. Pasynskij, in: *Metody Elementoorganiczeskoj Chimii*, Vol. 1, p. 434, Izd. Nauka, Moscow (1974).
[m] A. A. Pasynskij, in: *Metody Elementoorganiczeskoj Chimii*, Vol. 1, p. 453, Izd. Nauka, Moscow (1974).
[n] R. L. DeKock, *Inorg. Chem.*, **10**, 1205 (1971).

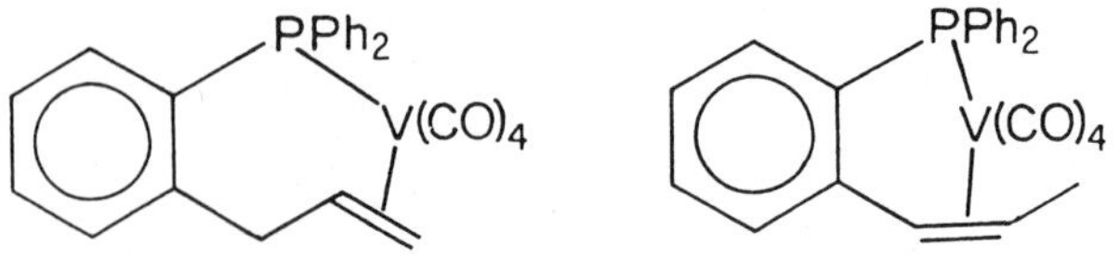

Figure 2.14. Structures of vanadium(0) complexes of the type $LV(CO)_4$ where $L = $ (2-allylphenyl) diphenylphosphine and (2-*cis*-propenylphenyl)diphenylphosphine.

$[Vcp(CO)_4]$ represents one of the most stable organometallic compounds of vanadium. It may be prepared by the following reactions:

$$VCl_3 + Nacp + 2Na \xrightarrow[\text{(CO) 31MPa}]{\text{THF, 403K, Fe(CO)}_5} [Vcp(CO)_4] \quad (50\%) \tag{2.40}$$

$$Vcp_2 + 4CO \xrightarrow[\text{(CO) 25MPa}]{413K,\ H_2\ 7MPa} [Vcp(CO)_4] \tag{2.41}$$

No solvent is needed for the second reaction. $[Vcp(CO)_4]$ may be reduced with sodium to $[Vcp(CO)_3]^{2-}$:

$$[Vcp(CO)_4] + 2Na \longrightarrow Na_2[Vcp(CO)_3] + CO \tag{2.42}$$

After acidification, this product does not form a hydride, but undergoes oxidation with $H_2$ evolution:

$$2Na_2[Vcp(CO)_3] + 4HCl \longrightarrow [V_2cp_2(CO)_5] + CO + 2H_2 + 4NaCl \tag{2.43}$$

$[Nb(CO)_6]^-$, hexacarbonylniobate$(1-)$, is prepared by reduction of $NbCl_5$ in the atmosphere of CO:

$$NbCl_5 + 6Na + 6CO \xrightarrow[\text{378 K, 30 MPa}]{\text{diglyme}} [Na(diglyme)_2][Nb(CO)_6] \quad (26\%) \tag{2.44}$$

Acidification of the solution of this compound results in formation of $[HNb(CO)_6]$. This hydride easily undergoes substitution with $PPh_3$ to give $[HNb(CO)_5(PPh_3)]$. The sodium salt of $[Nb(CO)_6]^-$ reacts with chlorotriphenyltin to afford a seven-coordinate metal–metal bonded complex, which may further react with $PPh_3$:

$$Ph_3SnCl + [Nb(CO)_6]^- \longrightarrow Ph_3Sn-Nb(CO)_6 + Cl^-$$

$$\xrightarrow{PPh_3} Ph_3Sn-Nb(CO)_5PPh_3 \tag{2.45}$$

The hexacarbonyl compound is unstable; on the other hand, the phosphine complex does not undergo decomposition for several months if stored under inert atmosphere.

$[Nbcp(CO)_4]$ is prepared by reduction of $NbCl_5$ in the presence of Nacp:

$$NbCl_5 + 2Nacp + 3Na \xrightarrow[\text{CO, 31 MPa, 353 K}]{\text{THF, Fe(CO)}_5} [Nbcp(CO)_4] \tag{2.46}$$

$[Ta(CO)_6]^-$ is obtained in a manner analogous to that of niobium and vanadium hexacarbonyl anions:

$$TaCl_5 + 6Na + 6CO \xrightarrow{\text{diglyme}} [Na(diglyme)_2][Ta(CO)_6] \tag{2.47}$$

Hexacarbonyltantalate($1-$) is more resistant to oxidation than $Na[Nb(CO)_6]$; however, it is less resistant than $Na[V(CO)_6]$. The carbonyl groups in $[Ta(CO)_6]^-$ in basic medium are not substituted by $PPh_3$. After acidification of the solution in the presence of $PPh_3$, a hydride, $[HTa(CO)_5PPh_3]$, is formed which, after subsequent reduction, furnishes $[Ph_4As][Ta(CO)_5PPh_3]$.

Compounds possessing metal–tantalum bonds may be prepared in the following way:

$$RHgCl + Na[Ta(CO)_6] \longrightarrow [RHgTa(CO)_6] + NaCl \tag{2.48}$$

$$(R = Et, Me, Ph, allyl, cp),$$

$$Ph_3SnCl + Na[Ta(CO)_6] \longrightarrow Ph_3SnTa(CO)_6 + NaCl \tag{2.49}$$

$[Tacp(CO)_4]$ is prepared by reduction of $TaCl_5$ with sodium in the presence of Nacp:

$$TaCl_5 + 2Nacp + 3Na \xrightarrow[\text{35 MPa}]{\text{THF, Fe(CO)}_5,\ 403\ K} [Tacp(CO)_4] \quad (8\%) \tag{2.50}$$

In photochemical reactions the carbonyl groups of $Tacp(CO)_4$ are substituted by $PPh_3$ and acetylenes.

Six-coordinate compounds of group 5 elements are octahedral. In the case of $[V(CO)_6]$, there is a slight distortion of the octahedron because of the Jahn–Teller effect (electronic configuration $t_{2g}^5$). This distortion is manifested by the broadening of the $v(CO)$ band in the IR spectrum. The V(0) compounds are monomeric. The effective magnetic moment is *ca* 1.8 BM. For $[V(CO)_6]$ at temperatures below 60 K, the magnetic moment decreases because of the formation of $[V_2(CO)_{12}]$, and at 4 K the paramagnetism of $[V(CO)_6]$ corresponds to only 1% of its monomeric form. The IR spectra of $[V_2(CO)_{12}]$ obtained in CO matrices or in mixtures containing CO and noble gases showed that $[V_2(CO)_{12}]$ possesses bridging CO groups: $[(CO)_5V(\mu\text{-}CO)_2V(CO)_5]$.[77] Compounds of the type $Mcp(CO)_4$ adopt structures of a square pyramid (Figure 2.15).

### d. Group 6 Metal Carbonyls

Hexacarbonylchromium is prepared by reduction of Cr(III) or Cr(II) compounds such as halides, acetylacetonates, acetates, etc., in the presence of CO (pressure 0.1–30 MPa) utilizing RMgX, $Al_2R_6$, group 1 and 2 metals, Al, Zn, Fe, etc.[8–10] Most common solvents are ether, pyridine, benzene, and saturated hydrocarbons.[3, 10, 78]

$$CrCl_2 + PhMgBr + 6CO \xrightarrow[\text{Et}_2\text{O, 277 K}]{\text{CO 0.1 MPa}} [Cr(CO)_6] \quad (14\%) \tag{2.51}$$

$$CrCl_3 + 6CO + Mg \xrightarrow[\text{Et}_2\text{O, 300 K}]{\text{CO 7 MPa}} [Cr(CO)_6] \quad (87\%) \tag{2.52}$$

Hexacarbonylchromium is a white compound which is air stable as a solid and in

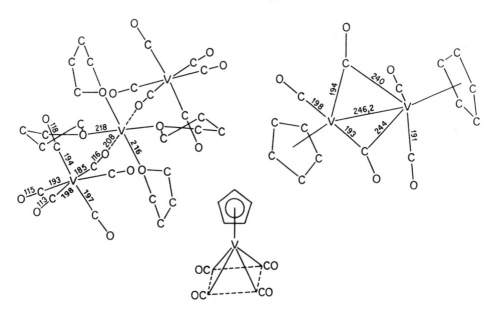

Figure 2.15.  The structures of $[V(C_4H_8O)_4][V(CO)_6]_2$, $V_2cp_2(CO)_5$, and $Vcp(CO)_4$. The distances are in pm units.

solution. At low temperatures, it does not react with hydrochloric acid, sulfuric acid, $Br_2$, and $I_2$. However, it is readily oxidized by $HNO_3$ and chlorine. With $\beta$-diketones, $Cr(CO)_6$ forms diketo Cr(III) complexes, for example:

$$Cr(CO)_6 + 3Hacac \longrightarrow Cr(acac)_3 + \tfrac{3}{2}H_2 + 6CO \tag{2.53}$$

Hexacarbonylchromium can be reduced by group 1 and 2 metals in liquid ammonia or by NaBH$_4$ in tetrahydrofuran to form anionic complexes: $[Cr(CO)_5]^{2-}$, $[CrH(CO)_5]^-$, $[Cr_2H(CO)_{10}]^-$, $[Cr_3(CO)_{14}]^{2-}$.

$$2[Cr(CO)_6] + 2NaBH_4 + 6NH_3 \xrightarrow[\text{330 K}]{\text{liquid NH}_3} Na_2[Cr_2(CO)_{10}] + 2B(NH_2)_3 + 2CO + 7H_2 \tag{2.54}$$

$$2[M(CO)_6] + 2NaBH_4 \xrightarrow[\text{reflux}]{\text{THF}} Na_2[M_3(CO)_{14}] + 4CO + H_2 + B_2H_6$$

$$(M = Cr, Mo) \tag{2.55}$$

Hexacarbonylmolybdenum was synthesized in low yields by L. Mond by the reaction of pyrophoric molybdenum with carbon monoxide. Presently, it is prepared by carbonylation of $MoCl_5$, molybdic acid, molybdenum sulfide, and other molybdenum compounds. The following reducing agents may be employed: Mg, Al, Na, Zn, Fe, Grignard reagents, aluminum organometallic compounds, etc. Sometimes such syntheses involve "dry methods" whereby no solvent is used. However, most commonly,

$Mo(CO)_6$ is prepared in solutions utilizing such solvents as $Et_2O$, THF, $Et_2O$ and benzene, benzene, dichloroethane and ether, etc.

$$2MoCl_5 + 5Zn + 12CO \xrightarrow[C_2H_4Cl_2 + Et_2O]{5\ MPa\ CO,\ Zn,\ >338\ K} 2[Mo(CO)_6] + 5ZnCl_2 \qquad (2.56)$$
$$(91\%)$$

$$MoCl_5 + [Fe(CO_5)] \xrightarrow{CO + H_2\ 10\ MPa} [Mo(CO)_6] \qquad (2.57)$$

$$2MoCl_5 + 12CO + 5Mg \xrightarrow[CO,\ 0.1\ MPa]{Mg,\ THF} 2[Mo(CO)_6] + 5MgCl_2 \qquad (2.58)$$

Hexacarbonylmolybdenum is an air stable solid which is slowly oxidized in solution when exposed to air. It does not react with hydrochloric acid; it is slowly oxidized by concentrated sulfuric acid and, in the presence of concentrated nitric acid, it rapidly undergoes decomposition. Fluorine oxidizes hexacarbonylmolybdenum to fluorides, while chlorine, bromine, and iodine form carbonyl halides, $[Mo(CO)_4X_2]$, at low temperatures.

Carboxylic acids and $\beta$-diketones react with $[Mo(CO)_6]$ at high temperatures (433 K and 393 K, respectively) to afford appropriate complexes:

$$[Mo(CO)_6] + 3Hacac \longrightarrow Mo(acac)_3 + \tfrac{3}{2}H_2 + 6CO \qquad (2.59)$$

Hexacarbonylmolybdenum is readily reduced by the alkali metals and $NaBH_4$ to the following anionic complexes: $[Mo(CO)_5]^{2-}$, $[Mo_2(CO)_{10}]^{2-}$, and $[Mo_3(CO)_{14}]^{2-}$. $[Mo_2(OH)_3(CO)_6]^{3-}$ is formed by reactions with bases:

$$2[Mo(CO)_6] + 9OH^- \longrightarrow [Mo_2(OH)_3(CO)_6]^{3-} + 6HCOO^- \qquad (2.60)$$

In this case no hydrogen evolution was observed and no formation of carbonates occurred; therefore, this is a substitution reaction which occurs in hot alcohols containing KOH.

Hexacarbonyltungsten is formed in a way similar to $Mo(CO)_6$ with the application of analogous reducing agents, solvents, and reaction conditions. In the case of synthesis utilizing triethylaluminum as a reducing agent, $[W(CO)_6]$ is best prepared in diethyl ether. Most commonly, $WCl_6$ is employed because it can be more readily carbonylated compared to other salts of group 6 and 7 metals:

$$WCl_6 > MoCl_5 > Mn(OAc)_3 > CrCl_3 > MnCl_2$$

$$WCl_6 + 3Zn + 6CO \xrightarrow[338\ K]{CO,\ 5\ MPa,\ Et_2O} [W(CO)_6] + 3ZnCl_2 \qquad (2.61)$$
$$(85\%)$$

Nesmeyanov developed an additional synthetic method for $[W(CO)_6]$ preparation utilizing $Fe(CO)_5$ as starting reagent; he also developed a similar method for $[Mo(CO)_6]$ preparation.

Chemical properties of $[W(CO)_6]$ are very similar to those of hexacarbonylmolybdenum. In solutions, $[W(CO)_6]$ is slowly oxidized by air; in the solid state, it is resistant to air oxidation. $[W(CO)_6]$ does not react with hydrochloric acid, concentrated sulfuric acid, and diluted nitric acid. At lower temperatures (195 K) chlorine, bromine, and iodine oxidize hexacarbonyltungsten to $[W(CO)_4X_2]$. Carboxylic acids and

*Table 2.23. Group 6 Metal Carbonyls*

| Compound | Color | Melting point (K) | $\nu(CO)$, IR($R$) (cm$^{-1}$) | Solubility | Remarks |
|---|---|---|---|---|---|
| $[Cr(CO)_6]^a$ | White | 427-8 | 1984 (CCl$_4$) (2107, 2007) | CCl$_4$, THF CS$_2$, C$_n$H$_{2n+2}$ | Cr−C: 190.9 C−O: 113.7 |
| $[Et_4N]_2[Cr_2(CO)_{10}]^b$ | Yellow | | 1983vw, 1917, 1890,1793sh | MeCN, THF | |
| $Na_4[Cr(CO)_4]^{(79)}$ | Yellow | | 1657w, 1462s | | |
| $[Mo(CO)_6]^a$ | White | 421 | 1986 (CCl$_4$) (2112, 2003) | CCl$_4$, THF C$_n$H$_{2n+2}$, CS$_2$ | Mo−C: 206.3 C−O: 114.5 |
| $Na_4[Mo(CO)_4]^{(79)}$ | Yellow-orange | | 1680w, 1471s | | |
| $(Et_4N)_2[Mo_2(CO)_{10}]^b$ | Yellow | | 1983vw, 1937, 1895, 1795 | MeCN, THF | |
| $[W(CO)_6]^a$ | White | 442 | 1977 (CS$_2$) (2115, 1999) | CS$_2$, THF C$_n$H$_{2n+2}$ | W−C: 205.8 C−O: 114.8 |
| $(Et_4N)_2[W_2(CO)_{10}]^b$ | Yellow | | 1978vw, 1943, 1894, 1795 | THF, MeCN | |
| $Na_4[W(CO)_4]^{(79)}$ | Light-orange | | 1679w, 1529s, 1478s | | |

$^a$ L. M. Jones, R. S. McDowell, and M. Goldblatt, *Inorg. Chem.*, **8**, 2349 (1969).
$^b$ E. Lindner, H. Behrens, and S. Birkle, *J. Organomet. Chem.*, **15**, 165 (1968).

$\beta$-diketones react with $[W(CO)_6]$ to give corresponding salts. Sodium, NaBH$_4$, and similar strong reducing agents react with hexacarbonyltungsten in liquid ammonia or THF to afford carbonyltungstates: $[W(CO)_5]^{2-}$ and $[W_2(CO)_{10}]^{2-}$.

In KOH methanol solutions, hexacarbonyltungsten undergoes reduction to a hydroxy complex:

$$2[W(CO)_6] + 12KOH + 2MeOH$$
$$\longrightarrow K_4[W_2(OH)_2(CO)_6(MeOH)_2] + 2K_2CO_3 + 4KOOCH + 2H_2O + H_2 \quad (2.62)$$

Group 6 metal carbonyls form methoxy or methanol complexes in basic methanol solutions; however, they form oxo complexes in ethanol. In solutions of $[M(CO)_6]$ which have been irradiated by UV, unstable five-coordinate $[M(CO)_5]$ complexes are formed. These complexes have tetragonal–pyramidal structures at 93 K as shown by X-ray, UV, and IR studies. These compounds are diamagnetic, monomeric, and readily react with Lewis bases to give substitution products $[M(CO)_5L]$. From the latter, photochemical reactions afford further substitution products: $[M(CO)_4L_2]$, $[M(CO)_3L_3]$, etc. In a similar way, complexes with olefins, diolefins, arenes, and olefin-phosphines, etc., are also prepared. For the synthesis of $[M(CO)_{6-x}L_x]$ ($x > 1$) derivatives, $[M(CO)_5(THF)]$ is commonly utilized because the ether molecule readily undergoes substitution. The following complexes are obtained in this fashion: $[M(CO)_4DAB]$ (DAB = RN=C−C=NR, diazabutadiene) and $[M(CO)_{4-x}(DAB)(PR_3)_x]$.[89] For $[Cr(CO)_5L]$ where L denotes CO, halogen, or a ligand containing P or N, and for fac-$[Mo(CO)_3(Pr_3)_3]$, there is a linear relationship between redox potentials and $\delta(^{13}CO)$, which shows that $E°$ and $\delta(^{13}C)$ depend only on electronic factors.[90]

Group 6 hexacarbonyls also afford carbene complexes:

$$[M(CO)_6] + LiR \longrightarrow (CO)_5 M = C \begin{smallmatrix} O-Li^+ \\ \\ R \end{smallmatrix} \xrightarrow{MeI} (CO)_5 M = C \begin{smallmatrix} O-Me \\ \\ R \end{smallmatrix} \qquad (2.63)$$

$[M(CO)_6]$ and $[M(CO)_{6-x}L_x]$ compounds are octahedral; the central atoms in $[M_2(OH)_3(CO)_6]^{3-}$ also have octahedral coordination with bridging $OH^-$ ligands. Structurally, $[M_2(CO)_{10}]^{2-}$ ions are analogous to $[Mn_2(CO)_{10}]$; these compounds have metal–metal bonds[87] (Figure 2.16). Like elements of other groups, group 6 metals may form carbonyls in which the metal atom adopts low as well as high oxidation states. The reduction of $[M(CO)_4(TMED)]$ by sodium in liquid ammonia[79] gives $Na_4[M(CO)_4]$, where M = Cr, Mo, and W (Table 2.23). In these compounds, the central atoms formally have $-4$ oxidation states. The structures of these compounds are not known. Low values of $v(CO)$ are explained by the following formula: $[Cr(CONa)_4]$. However, the formation of acetylenediolate ion cannot be ruled out:

$$M \begin{smallmatrix} CO \\ \\ CO \end{smallmatrix} \underset{-2e}{\overset{2e}{\rightleftharpoons}} M - \begin{smallmatrix} C \\ \| \| \\ C \end{smallmatrix} \begin{smallmatrix} O^- \\ \\ O^- \end{smallmatrix} \longleftrightarrow M^{2-} \begin{smallmatrix} C \\ \\ C \end{smallmatrix} \begin{smallmatrix} O \\ \\ O \end{smallmatrix} \qquad \text{or cyclo-}C_4O_4^{4-} \text{ ion}$$

$Na_4[M(CO)_4]$ reacts with $NH_4Cl$ ($M/NH_4Cl = 1/2.5$) in ammonia to give $[M_2H_2(CO)_8]^{2-}$. $[NEt_4]_2[W_2H_2(CO)_8]$ interacts with $K[B(sec\text{-}Bu)_3H]$ to form $K_2[WH_2(CO)_4]$. Also known are tetrahedral clusters of the type $[NR_4]_4[\{MH(CO)_4\}_4]$, in which the hydrogen atoms probably form $\mu_3$ bridges, capping the four faces of the tetrahedron.[80] $[M(CO)_5]^{2-}$ complexes have trigonal–bipyramidal structures.[86] In gas phase, reactions of $[Cr(CO)_5]^-$ and $[Mo(CO)_5]^-$ ions [prepared from $M(CO)_6$ by ionization with electrons] with oxygen

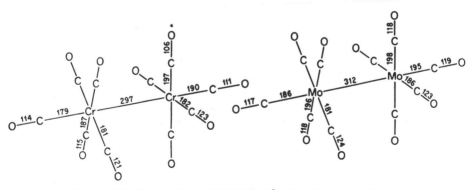

Figure 2.16. The structure of $[M_2(CO)_{10}]^{2-}$. The distances are in pm.

form[81] $[CrO_2(CO)_3]^-$ and $[MoO_2(CO)_3]^-$. With CO hydrocarbon solutions of dimeric alkoxy complexes of Mo(III) and W(III) furnish carbonyls[82]:

$$2Mo_2(OR)_6 + 6CO \xrightarrow[\text{295 K}]{\text{0.1 MPa CO}} Mo(CO)_6 + 3MO(OR)_4 \qquad (2.64)$$

In the first stage, a carbonyl Mo(III) complex, $Mo_2(OR)_6$ ($\mu$-CO), is formed reversibly. In the presence of pyridine, the following compound is obtained: $Mo_2(OR)_6$ (py)$_2$ ($\mu$-CO) (Figure 2.17). $W_2(OR)_6$ reacts with carbon monoxide in a similar way to give $W(CO)_6$. However, the splitting of the $W \equiv W$ bond leads to a disproportionation reaction in which W(0) and W(VI) are formed. In the presence of a stoichiometric quantity of carbon monoxide the following complexes may be prepared: $[\{W_2(OPr^i)_6$ $(\mu_3\text{-}CO)\}_2]$ and $[W_2(OBu^i)_6 (\mu\text{-}CO)]$[82] (Figure 2.17).

Complexes containing coordinated molecules with donor atoms, which may coordinate to other metal atoms, play an important role in Mo and W carbonyl chemistry. Examples of such ligands are complexes of cobalt and rhodium possessing the following general formula:

$$\left[ \begin{array}{c} \text{cp} \\ | \\ \text{M} \\ \diagup \quad | \quad \diagdown \\ R_2P \quad R_2P \quad R_2P \\ \| \quad \| \quad \| \\ O \quad O \quad O \end{array} \right]^- \qquad \begin{array}{l} R = \text{alkyl,} \\ \text{aryl,} \\ O\text{-alkyl} \end{array}$$

Because these ligands coordinate through oxygen atoms, they stabilize high as well as low oxidation states.[83] Two representative Mo(II) and W(II) dinuclear complexes are given in equation (2.65). The 18$e$ osmium complexes may serve as 2$e$ ligands, which may coordinate to other metals, such as $(Me_3P)(OC)_4 OsW(CO)_5$. There is a donor–acceptor Os$-$W bond in this compound. In the solid state

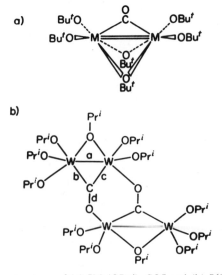

Figure 2.17.   The structures of (a) $[M_2(OBu^i)_6 CO]$ and (b) $[\{W_2(OPr^i)_6 CO\}_2]$.

$(Me_3P)(CO)_4OsW(CO)_5$ does not show bridging CO groups. The compound has a staggered structure with the $PMe_3$ ligand in *trans* position to tungsten. The Os–W distance is 307.56(5) pm. In solutions, the compound exhibits dynamic properties. The *trans* form exists in equilibrium with the compound in which the phosphine ligand is in *cis* position to tungsten. The carbonyl ligands undergo exchange between osmium and tungsten atoms; the intermediate form contains bridging CO groups.[84]

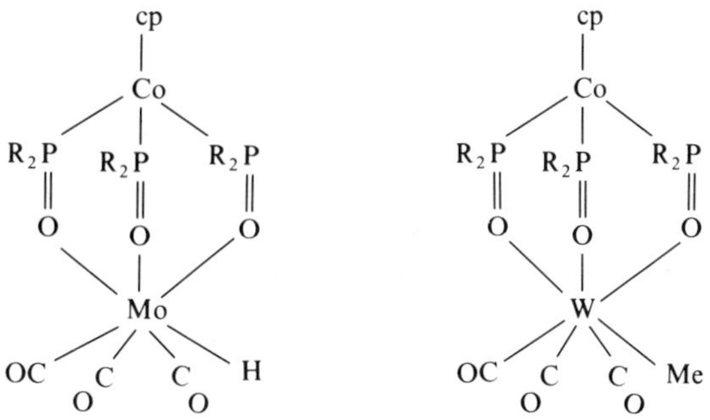

(2.65)

The $M(CO)_6$ carbonyls react with $Ph_2PPPh_2$ to give dimeric complexes $[(OC)_4M(\mu\text{-}PPh_2)_2M(CO)_4]$. These compounds may be obtained with better yields from $[M(CO)_5(PPh_2H)]$.[88] These dimeric complexes possess metal–metal bonds. The tungsten–tungsten bond in $[(OC)_4W(\mu\text{-}PR_2)_2W(CO)_4]$ is slightly more than 300 pm (303–305 pm); both central atoms have octahedral coordination. The organolithium compounds (LiR) attack the equatorial CO group of $[(OC)_4Mo(\mu\text{-}PPh_2)_2Mo(CO)_4]$ to afford the acyl complex $[(OC)_4Mo(\mu\text{-}PPh_2)_2Mo(COR)(CO)_3]^-$ which subsequently reacts with $[Me_3O][BF_4]$, forming the carbene complex $[(OC)_4Mo(\mu\text{-}PPh_2)_2Mo\{CR(OMe)\}(CO)_3]$.

### e. Group 7 Metal Carbonyls

Most commonly, $[Mn_2(CO)_{10}]$ is obtained by the reduction of Mn(II) salts in ethers such as $Et_2O$ and $(Pr^i)_2O$ as well as in DMF, pyridine, benzene, and xylenes. Reducing agents employed are Grignard reagents, organometallic compounds of aluminum, Mg, Na, $NaPh_2CO$, and $NaC_{10}H_8$. Decarbonyldimanganese is also formed by electrolytic reduction of $Mn(acac)_2$ in water-free pyridine at CO pressures of 20 MPa:

$$2Mn(OAc)_2 + 10CO \xrightarrow[(i\text{-}Pr)_2O,\ 313\ K,\ 20\ h]{Al_2(i\text{-}Pr)_6,\ 12.5\ MPa} [Mn_2(CO)_{10}] \quad (50\%) \tag{2.66}$$

$$2MnCl_2 + 10CO + 4Ph_2CONa \xrightarrow{20\ MPa,\ 473\ K} [Mn_2(CO)_{10}] + 4Ph_2CO + 4NaCl \tag{2.67}$$

The product is isolated from the reaction mixture by sublimation or steam distillation. $[Mn_2(CO)_{10}]$ is a yellow crystalline compound, which is soluble in organic solvents and in $[Ni(CO)_4]$, but is insoluble in water. In the solid state, in air, $[Mn_2(CO)_{10}]$ is stable for a long time, but in solutions it oxidizes with carbon monoxide evolution. The halogens easily oxidize $[Mn_2(CO)_{10}]$, breaking its Mn–Mn bond and forming the halides $[MnX(CO)_5]$ (Table 2.24).

*Table 2.24. Group 7 Metal Carbonyls*

| Compound | Color | Melting point (K) | $\nu(CO)$, IR($R$) (cm$^{-1}$) | Solubility |
|---|---|---|---|---|
| [Mn$_2$(CO)$_{10}$]$^a$ | Yellow | 427–8 | 2046, 2015, 1984 (2116, 2018, 1976, 1947) | Organic, nonsoluble in H$_2$O |
| (Mn(CO)$_5$Cl]$^b$ | Light-yellow | | 2139, 2055, 1998 (cyclohexane) | |
| Na[Mn(CO)$_5$]$^c$ | | | 1898, 1863 | THF, diglyme |
| [Tc$_2$(CO)$_{10}$]$^d$ | White | 432–3 | 2065, 2017, 1984 | Organic, nonsoluble in H$_2$O |
| [Tc(CO)$_5$Cl]$^e$ | White | | 2153, 2057, 2029, 1991 (CCl$_4$) | |
| [Re$_2$(CO)$_{10}$]$^d$ | White | 450 | 2074, 2013, 1974 (2126, 2027, 1989, 1934) | Organic, nonsoluble in H$_2$O |
| [Re(CO)$_5$Cl]$^b$ | White | | 2155, 2046, 1983 (CCl$_4$) | THF, diglyme |

$^a$ G. Bor, *Chem. Commun.*, 641 (1969); D. M. Adams, M. A. Hooper, and A. Squire, *J. Chem. Soc. A*, 71 (1971).
$^b$ H. D. Kaesz, R. Bau, D. Hendrickson, and J. M. Smith, *J. Am. Chem. Soc.*, **89**, 2844 (1967).
$^c$ W. F. Edgell, J. Huff, J. Thomas, H. Lehman, C. Angell, and G. Asato, *J. Am. Chem. Soc.*, **82**, 1254 (1960).
$^d$ N. Flitcroft, D. K. Huggins, and H. D. Kaesz, *Inorg. Chem.*, **3**, 1123 (1964).
$^e$ J. C. Hileman, D. K. Huggins, and H. D. Kaesz, *Inorg. Chem.*, **1**, 933 (1962).

The manganese carbonyl may be converted to pentacarbonylhydridomanganese by hydrogen in the presence of CO, by magnesium in the mixture of H$_2$SO$_4$, water, and methanol, or by LiAlH$_4$. [MnH(CO)$_5$] is a colorless, weakly acidic compound, well soluble in water p$K$ = 7.1) and in organic solvents. In air this hydride rapidly decomposes to [Mn$_2$(CO)$_{10}$] and H$_2$.

[Mn(CO)$_5$]$^-$ is generally prepared by the reaction of [Mn$_2$(CO)$_{10}$] with KOH in water or ethanol solution or by reduction with metallic sodium.

$$[Mn_2(CO)_{10}] + 2Na \xrightarrow{\text{THF}} 2Na[Mn(CO)_5] \qquad (2.68)$$

$$13[Mn_2(CO)_{10}] + 44KOH \longrightarrow 24K[Mn(CO)_5] + 2Mn(OH)_2 + 10K_2CO_3 + 20H_2O \qquad (2.69)$$

Na[Mn(CO)$_5$] is almost colorless and easily undergoes oxidation reactions. The potential of the reaction [Mn(CO)$_5$]$^- \rightleftarrows \frac{1}{2}$[Mn$_2$(CO)$_{10}$] is 0.68 V. [Mn$_2$(CO)$_9$]$^{2-}$ is isoelectronic with [Fe$_2$(CO)$_9$]. [Mn$_2$(CO)$_9$]$^{2-}$ may be obtained by the reaction of [Mn$_2$(CO)$_{10}$] with NaBH$_4$ in tetrahydrofuran. [Mn(CO)$_4$(NO)] and [Mn(CO)$_3$(NO)(PPh$_3$)] are reduced with sodium amalagam or K[B(Bu$^t$)$_3$H] to Na$_2$[Mn(CO)$_3$(NO)] or K$_2$[Mn(CO)$_3$(NO)],$^{(97)}$ which represent rare examples of anionic carbonyl–nitrosyl complexes.

[Mn(CO)$_6$]$^+$ is prepared from [MnX(CO)$_5$] under carbon monoxide in the presence of AlCl$_3$ or other halogen acceptors:

$$[MnCl(CO)_5] + AlCl_3 + CO \longrightarrow [Mn(CO)_6]^+ [AlCl_4]^- \qquad (2.70)$$

[Mn(CO)$_6$]$^+$ [AlCl$_4$]$^-$ is colorless, resistant to heating, and reactive toward water:

$$[Mn(CO)_6]^+ + H_2O \longrightarrow [HMn(CO)_5] + H^+ + CO_2 \qquad (2.71)$$

There are many compounds of $[Mn(CO)_6]^+$ with anionic complexes, including anionic carbonyls, for example, $[Mn(CO)_6][Co(CO)_4]$ and $[Mn(CO)_6][V(CO)_6]$.

Manganese(I) and rhenium(I) form with $P_3O_9^{3-}$ hexacoordinated compounds corresponding to the formula $[NBu_4]_2[M(P_3O_9)(CO)_3]$, in which the metal is bonded to three oxygen atoms of the metaphosphato ligand and to three CO groups ($C_{3v}$ symmetry).[94]

$[Tc_2(CO)_{10}]$ is formed by the reduction of $Tc_2O_7$ or $TcO_2$ with carbon monoxide:

$$2TcO_2 + 14CO \xrightarrow[\text{2 MPa}]{\text{523 K}} [Tc_2(CO)_{10}] + 4CO_2 \quad (50\%) \tag{2.72}$$

$[Tc_2(CO)_{10}]$ is a colorless compound which easily reacts with halogens to give $[TcX(CO)_5]$.

$[Re_2(CO)_{10}]$ is obtained from $Re_2O_7$, $ReO_3$, $ReO_2$, $Re_2S_7$, or $KReO_4$ by the action of CO without solvent, or from $ReCl_5$ or $ReCl_3$ in THF utilizing Na as a reducing agent:

$$Re_2O_7 + 17CO \xrightarrow[\text{35 MPa}]{\text{523 K, 16 h}} [Re_2(CO)_{10}] + 7CO_2 \quad (100\%) \tag{2.73}$$

$[M_2(CO)_{10}]$ compounds are colorless and react with halogens to form $[MX(CO)_5]$. In the solid state, $[M_2(CO)_{10}]$ compounds are resistant to air oxidation; however, in solutions, they slowly undergo oxidation. Diluted and even concentrated acids which do not possess oxidizing properties do not react with $[Re_2(CO)_{10}]$. $ReF_6$ breaks the carbon–oxygen bond to give $ReOF_4$. Analogously, hexafluororhenium reacts with molybdenum and tungsten carbonyls. In reactions of $Cl_2$, $Br_2$, and $I_2$ with $M_2(CO)_{10}$ where $M_2 = Mn_2$, $Re_2$, and MnRe, $[MX(CO)_5]$ and $[M(CO)_5(MeCN)]^+$ are formed in MeCN; in nonpolar solvents, however, $I_2$ forms only $[MI(CO)_5]$. The rate of the reaction decreases in the order $Re_2(CO)_{10} > MnRe(CO)_{10} > Mn_2(CO)_{10}$, and thus is the reverse of that anticipated from the $M-M$ bond energy. $MnBr(CO)_5$ reacts with $R_2NP(=NR)_2$ to give stable 16e complexes [Equation (2.74)]. At low temperatures (195 K) the first compound may react with CO to give an 18e hexacoordinated complex $[Mn\{(RN)_2PBr(NR_2)\}(CO)_4]$, which at room temperatures is transformed back into the starting complex.[192]

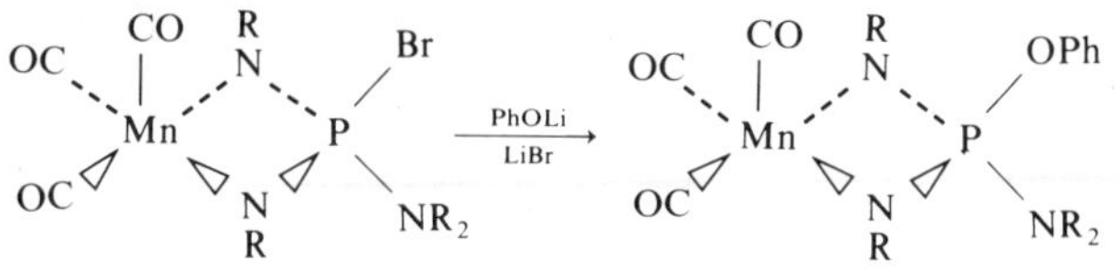

$$\tag{2.74}$$

From the standpoint of CO activation and reduction, $(CO)_5M-\overset{\overset{\displaystyle O}{\|}}{C}-\overset{\overset{\displaystyle O}{\|}}{C}-M(CO)_5$ compounds, where M = Mn or Re, are of particular interest. They are prepared from $Na[M(CO)_5]$ and oxalyl chloride[96]:

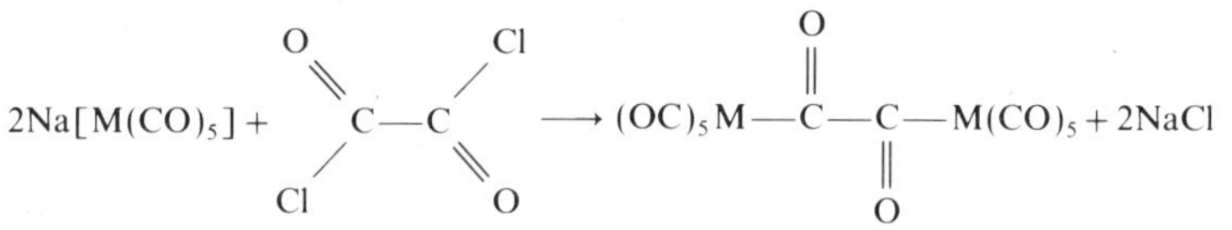

$$2Na[M(CO)_5] + \quad \longrightarrow (OC)_5M-C-C-M(CO)_5 + 2NaCl \tag{2.75}$$

In the solid state as well as in solutions, these compounds decompose to $[M_2(CO)_{10}]$. $[(OC)_5M-CO]_2$ compounds are stable only at low temperatures. As postulated (in Chapter 13), the formation of such compounds by oxidative coupling is an alternative way for preparation of organic compounds from carbon monoxide. However, these compounds have not yet been directly prepared from coordinated CO molecules. It is possible that such ligands are formed in strongly reduced complexes of chromium group metals of the type $Na_4[M(CO)_4]$. Compounds possessing $O=C=C-C{<}^{O}_{O}$ ligands are formed by the reaction of carbon monoxide with $Sm(pmcp)_2 (THF)_2$, which represents a very strong reducing agent.[187]

[$Re_2(CO)_{10}$] is more stable than analogous compounds of technetium and manganese. $[M_2(CO)_{10}]$ carbonyls, where $M = Re$ or $Tc$, undergo reduction reactions with sodium amalgam or dispersed alkali metals to give $[M(CO)_5]^-$, which upon treatment with nonoxidizing acids form carbonyl hydrides, $[HM(CO)_5]$:

$$Na[Re(CO)_5] + H_3PO_4 \longrightarrow [ReH(CO)_5] + NaH_2PO_4 \qquad (2.76)$$

Hydrides are volatile liquids which are easily transformed into $[M_2(CO)_{10}]$ in the presence of oxygen. The reduction of $[Re_2(CO)_{10}]$ with sodium borohydride in THF leads to the formation of $[Re_4(CO)_{16}]^{2-}$, which is isolated as a $Bu_4N^+$ salt.

Decacarbonylrhenium reacts with bases to give $[Re_2(CO)_8O_2H]^-$ (a different product is obtained from $[Mn_2(CO)_{10}]$):

$$[Re_2(CO)_{10}] + 3KOH + H_2O \longrightarrow K[Re_2(CO)_8O_2H] + K_2CO_3 + CO + 2H_2 \qquad (2.77)$$

$K[Re_2(CO)_8O_2H]$ is a dinuclear compound with oxo and hydroxo bridges. $[M(CO)_6]^+$ compounds, where $M = Tc$ or $Re$, are obtained similarly to the corresponding manganese complex. The salts of these Tc or Re cations are more stable than those of $[Mn(CO)_6]^+$. $[Re(CO)_6]^+$ compound, in contrast to the manganese complex, does not react with water.

As mentioned above, the halogens react with $[M_2(CO)_{10}]$ where $M = Mn$, Tc, and Re, causing splitting of the $M-M$ bond. At higher temperatures, dinuclear compounds with halogen bridges are formed.

$$[M_2(CO)_{10}] + Br_2 \longrightarrow 2[M(CO)_5Br] \longrightarrow \left[ (CO)_4M {<}^{\overset{Br}{\diagup\diagdown}}_{\underset{Br}{\diagdown\diagup}} M(CO)_4 \right] + 2CO \qquad (2.78)$$

Manganese also forms anionic halogen complexes: $[MnX_2(CO)_4]^-$ and $[Mn_2X_2(CO)_8]^{2-}$.

Under irradiation, $[Mn_2(CO)_{10}]$ and $[Re_2(CO)_{10}]$ undergo splitting of the $M-M$ bond as well as of the $M-CO$ bond. These processes lead to the formation of radicals $Mn(CO)_5$ and $Mn_2(CO)_x$, where $x = 8$ or 9.[58, 64, 92, 93] By photolytic reactions, many carbonyls of the type $M_2(CO)_{10-x}L_x$ have been prepared. If $[Mn_2(CO)_{10}]$ interacts with ligands which force certain types of coordination because of either electronic or steric effects, complexes of different composition may be formed. An example of such a ligand is $Ph_2PCH_2PPh_2$, which when reacted with $[Mn_2(CO)_{10}]$ gives $[Mn_2(CO)_6(\mu\text{-}Ph_2PCH_2PPh_2)_2]$. Because of steric repulsion, this compound loses one CO molecule, furnishing $[Mn_2(\mu\text{-}CO)(CO)_4(\mu\text{-}Ph_2PCH_2PPh_2)_2]$, in which the bridging CO group is a $4e$ ligand (Figure 2.3).[14]

$[MnBr(CO)_3(Bu'N=CHCH=NBu')]$ reacts with $[Fecp(CO)_2]^-$ to form homodinuclear compounds rather than heterodinuclear compounds. Products of the reaction are $Fe_2cp_2(CO)_4$, $(OC)_3Mn(\mu\text{-}\eta^4\text{-}Bu'N=CHCH=NBu')Mn(CO)_3$ and $(OC)_3Mn(\mu\text{-}CO)\{\mu\text{-}Bu'N=CHCH(NBu')CH(NBu')CH=NBu'\}Mn(CO)_2]$.

In the first manganese compound, $Bu'N=CHCH=NBu'$ is an $8e$ ligand; in the second Mn compound, a new ligand exists which forms via oligomerization of diazo-butadiene with the $C-C$ bond formation.[95]

$[M_2(CO)_{10}]$ (M = Mn, Tc, and Re) carbonyls react with Lewis bases giving substitution products (for the rhenium carbonyl) or, dependent upon the ligand basicity, substitution or disproportionation products (for the manganese carbonyl).

With phosphines, $P(OR)_3$, aniline, or $o$-phenylenediamine, $[Mn_2(CO)_{10}]$ forms complexes which are products of the substitution reactions; it gives disproportionation products with stronger, nitrogen-containing bases (Z) such as pyridine, morpholine, ethylenediamine, etc.:

$$3[Mn_2(CO)_{10}] + 2nZ \longrightarrow 2[Mn(Z)_n][Mn(CO)_5]_2 + 10CO \qquad (2.79)$$

Olefins and acetylenes do not furnish stable substitution products of the composition $[M_2(CO)_{10-x}L_x]$ where L = alkene or alkyne, although it is assumed that these products are formed in olefin oligomerization and polymerization reactions.

With phosphines ($PR_3$) and phosphine oxides ($OPR_3$), dependent upon R, paramagnetic monomeric complexes $Mn(CO)_4PR_3$ (R = Ph, Et) or diamagnetic dimers $[Mn(CO)_4PR_3]_2$ (R = Cy, OPh) are formed. Also known are ionic substitution products $[M(CO)_3C_6H_6]^+$, $[Mn(CO)_4(CN)_2]^-$, $[Tc(CO)_4(PPh_3)_2]^+$, and $[Re(CO)_4(C_2H_4)_2]^+$.

Compounds possessing a $\sigma$ M$-$C bond are usually prepared by the following reactions:

$$Na[M(CO)_5] + RX \longrightarrow [RM(CO)_5] + NaX \qquad (2.80)$$

$$[M(CO)_5X] + MR \longrightarrow [RM(CO)_5] + MX \qquad (M = Li, MgX) \qquad (2.81)$$

Group 7 elements form many ionic cluster carbonyls as well as cluster hydrides, for example, $[M_3H_3(CO)_{12}]$, $[Mn_3(CO)_{14}]^-$, $[Re_3H_4(CO)_{10}]^-$, $[Re_4(CO)_{16}]^{2-}$.

Metal atoms in $[M_2(CO)_{10}]$ compounds have coordination number 6 (Figure 2.18). There is a direct bond between the two metal atoms and the equatorial CO groups occupy staggered positions. The $[M(CO)_5]^-$ anions have trigonal–bipyramidal structures. The cationic complexes $[M(CO)_6]^+$ are octahedral, as are isoelectronic group 6 carbonyls.

## f. Group 8 Metal Carbonyls

$[Fe(CO)_5]$ may be obtained directly by the reaction of carbon monoxide under atmospheric pressure with metallic iron, which is usually prepared by reduction of its oxides or other salts (Table 2.25). However, higher pressures and elevated temperatures are often used in order to achieve better yields. $[Fe(CO)_5]$ is also formed from iron

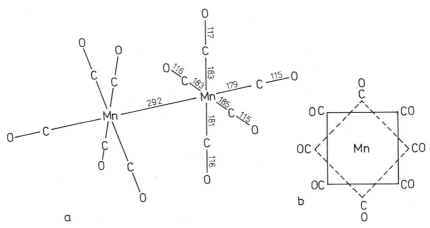

Figure 2.18. The structure of $Mn_2(CO)_{10}$: (a) side view, (b) view along the Mn–Mn axis. The distances are in pm.

halides in the presence of copper and silver. The yields of $[Fe(CO)_5]$ are lower if it is prepared from $FeSO_4$ in $NH_3(aq)$ solutions.

$$Fe + 5CO \xrightarrow{\text{5–20 MPa, 473 K}} [Fe(CO)_5] \qquad (94.5\%) \qquad (2.82)$$

$$FeI_2 + 5CO \xrightarrow[\text{Cu/Ag}]{\text{20 MPa, 473 K}} [Fe(CO)_5] \qquad (98\%) \qquad (2.83)$$

*Table 2.25. Group 8 Metal Carbonyls*

| Compound | Color | Melting point (K) | $\nu(CO)$, IR($R$) (cm$^{-1}$) | Solubility |
|---|---|---|---|---|
| $[Fe(CO)_5]^a$ | Yellow | 252.5 | 2002, 1979 (2116, 2030) | Organic solvents |
| $[Fe_2(CO)_9]^b$ | Orange | 373 dec | 2066, 2038, 1855, 1851 | Nonsoluble in organic solvents |
| $[Fe_3(CO)_{12}]^c$ | Dark-green | 423 dec | 2046, 2023, 2013, 1867, 1835 | Organic solvents |
| $[Ru(CO)_5]^d$ | White | 251 | 2035, 1999 | Organic solvents |
| $[Ru_3(CO)_{12}]^e$ | Orange | 427 dec | 2062, 2026, 2002, 1989 (2127, 2034, 2004, 1994) | Organic solvents |
| $[Os(CO)_5]^d$ | White | 258 | 2034, 1991 | Organic solvents |
| $[Os_2(CO)_9]^f$ | Yellow-orange | 337–340 dec | 2080s, 2038vs, 2024m 2000w, 1778m | Organic solvents |
| $[Os_3(CO)_{12}]$ | Yellow | 497 | 2070, 2019, 1998, 1986 (2130, 2028, 2006, 2000, 1989, 1979) | Organic solvents |

[a] M. Bigorgne, *J. Organomet. Chem.*, **24**, 211 (1970).
[b] M. Poliakoff and J. J. Turner, *J. Chem. Soc. A*, 2403 (1971).
[c] J. Knight and M. J. Mays, *Chem. Commun.*, 1006 (1970); M. Poliakoff and M. Turner, *Chem. Commun.*, 1008 (1970).
[d] F. Calderazzo and P. L'Eplattenier, *Inorg. Chem.*, **6**, 1220 (1967).
[e] C. O. Quicksall and T. G. Spiro, *Inorg. Chem.*, **7**, 2365 (1968).
[f] J. R. Moss and W. A. G. Graham, *Chem. Commun.*, 835 (1970); J. R. Moss and W. A. Graham, *J. Chem. Soc., Dalton Trans.*, 95 (1977).

Pentacarbonyliron, [Fe(CO)$_5$], is easily available and therefore is utilized for preparations of many carbonyls of other metals as well as for polynuclear carbonyls of iron itself. [Fe(CO)$_5$] is a liquid, strongly toxic, relatively stable in air; however, it should be handled under inert atmosphere.

[Fe$_2$(CO)$_9$] is formed during UV irradiation of [Fe(CO)$_5$] solution in glacial acetic or acetic anhydride. [Fe$_2$(CO)$_9$] is an orange compound which is insoluble in organic media despite its dimeric structure in the solid state as confirmed by X-ray studies.

[Fe$_3$(CO)$_{12}$] is prepared by oxidation of alkali solutions containing carbonyl-ferrates which are formed by reaction of [Fe(CO)$_5$] with aqueous alkali:

$$4[Fe_3(CO)_{11}H]^- + 6H^+ \longrightarrow \tfrac{11}{3}[Fe_3(CO)_{12}] + Fe^{2+} + 5H_2 \qquad (2.84)$$

[Fe$_3$(CO)$_{12}$] is a dark-green compound which is slightly soluble in organic solvents.

[Ru(CO)$_5$] is formed by the reaction of CO with RuI$_3$ or RuI$_2$(CO)$_2$ in the presence of an excess of dispersed copper or silver. During the reduction of RuI$_3$, [Ru$_3$(CO)$_{12}$] is obtained simultaneously. [Os(CO)$_5$] and [Os$_3$(CO)$_{12}$] are formed at high pressures (20–30 MPa) and temperatures (420–570 K) by the action of carbon monoxide on OsCl$_3$ or Os$_2$Br$_9$ mixed with pulverized copper. Presently, there are many low-pressure synthetic methods known for preparation of polynuclear carbonyls of group 8, 9, and 10 metals.[98, 99]

[Os(CO)$_5$] is prepared quantitatively by the reaction of CO with OsO$_4$:

$$OsO_4 + 9CO \xrightarrow{\text{30 MPa, 573 K}} [Os(CO)_5] + 4CO_2 \quad (100\%) \qquad (2.85)$$

Pentacarbonyliron adopts a trigonal–bipyramidal structure in the solid, liquid, and gaseous states. Physicochemical studies show that analogous structures are also adopted by [Ru(CO)$_5$] and [Os(CO)$_5$] (Figure 2.19).

For [Fe$_3$(CO)$_{12}$], considerable asymmetry of the bridging Fe$-$CO$-$Fe bonds is characteristic. In solutions, this trinuclear iron compound exhibits dynamic properties. The IR spectra show that the number of isomers with bridging carbonyl groups

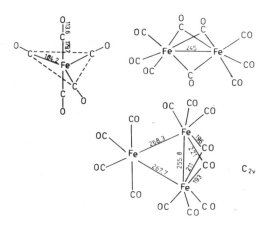

Figure 2.19. The structures of Fe(CO)$_5$, Fe$_2$(CO)$_9$, and Fe$_3$(CO)$_{12}$. The distances are in pm.

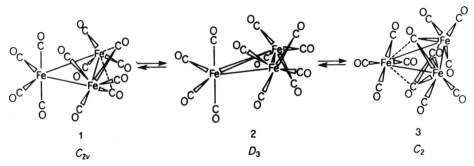

Figure 2.20.   The structure of $Fe_3(CO)_{12}$ in solutions.

possessing $C_{2v}$ or $C_2$ symmetry is small. In the $^{13}C$ NMR spectrum, there is a sharp singlet up to 123 K. This means that the energy barrier for the process of averaging carbonyl groups is very low. Therefore, it is now thought that the equivalency of the CO groups does not result from their reorientation which would require higher activation energy, but rather from the rotation of iron atoms about the $C_2$ axis passing through the Fe(1) atom.

The dynamic properties of $[Fe_3(CO)_{12}]$ may be represented by the following equilibrium: $1 \rightleftarrows 2 \rightleftarrows 3^{(189, 190)}$ (Figure 2.20).

In $[Ru_3(CO)_{12}]$ and $[Os_3(CO)_{12}]$, there are only terminal carbonyl groups; each metal atom is bonded to four CO groups and two other metal atoms. Therefore, each metal atom assumes distorted octahedral coordination (Figure 2.21). The CO groups are located at vertices of a truncated hexagonal bipyramid. The formation of compounds with terminal CO groups occurs most commonly in clusters possessing relatively longer M−M distances, and thus for $5d$ electron metals, rarely for $4d$ electron metals. The Os–Os distance in $[Os_3(CO)_{12}]$ is 288 pm and the Ru–Ru distance in $[Ru_3(CO)_{12}]$ is 285 ppm. The smaller Fe–Fe distance is one of the reasons that in $[Fe_3(CO)_{12}]$ there are bridging CO groups, and that carbonyl groups are located at vertices of the icosahedron.

In air, $[Fe(CO)_5]$ undergoes oxidation less readily than analogous carbonyls of ruthenium and osmium, because with the increase in atomic number the tendency toward formation of hexacoordinated compounds increases. When pentacarbonyliron is oxidized by a halogen in an uncontrollable manner, Fe(II) salts are formed; if careful oxidation occurs, $FeX_2(CO)_4$ compounds are obtained. The stability of the $FeX_2(CO)_4$

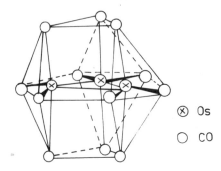

⊗  Os

○  CO

Figure 2.21.   The structures of $[Os_3(CO)_{12}]$ (truncated hexagonal bipyramid).

compound increases from chlorine to iodine; the iodo derivative, in contrast to the chloro and bromo derivatives, does not undergo hydrolysis in cold water.

$[RuX_2(CO)_4]$ and $[OsX_2(CO)_4]$, where X = Br or I, are usually prepared by the reaction of Ru(III) or Os(III) halides with carbon monoxide. The reactions of polynuclear carbonyls with halogens give mononuclear compounds; polynuclear halogen complexes may also be formed. $[Fe_3(CO)_{12}]$ reacts with iodine to give $[FeI_2(CO)_4]$ and $[Fe_2I_2(CO)_8]$. At room temperature, in the reaction of $[Ru_3(CO)_{12}]$ with halogens, $[Ru_2X_4(CO)_6]$ and $[Ru_3X_6(CO)_{12}]$ are formed. In the dimer, the two ruthenium atoms are connected by halogen bridges. At higher temperatures, the polymeric complex $[RuX_2(CO)_2]_n$ is formed. $[M_3(CO)_{12}]$, where M = Ru or Os, furnishes linear compounds of the composition $M_3X_2(CO)_{12}$ when treated with halogens. Also known are compounds of the formula $[MX_2(CO)_2]_n$.

The reduction of Fe, Ru, or Os carbonyls with sodium in alkali solutions or in liquid ammonia leads to the formation of carbonylate anions:

$$[Fe(CO)_5] + 4OH^- \longrightarrow [Fe(CO)_4]^{2-} + CO_3^{2-} + 2H_2O \tag{2.86}$$

$$[Fe(CO)_5] + 2Na \xrightarrow{\text{NH}_3} Na_2[Fe(CO)_4] + CO \tag{2.87}$$

$$[Fe_2(CO)_9] + 4Na \xrightarrow{\text{NH}_3} 2Na_2[Fe(CO)_4] + CO \tag{2.88}$$

From $Fe(CO)_5$ as well as from polynuclear carbonyls, dependent upon the basicity of the solutions, the following anions may be obtained: $[Fe(CO)_4]^{2-}$, $[Fe_2(CO)_8]^{2-}$, $[Fe_3(CO)_{11}]^{2-}$, $[Fe_4(CO)_{13}]^{2-}$, etc. Corresponding carbonyl hydrides are also known: $[FeH(CO)_4]^-$, $[Fe_2H(CO)_8]^-$, $[Fe_3H(CO)_{12}]^-$, $[Fe_4H(CO)_{13}]^-$, etc. $[Fe_2(CO)_8]^{2-}$ and $[Os_2(CO)_8]^{2-}$ have similar structures.[100, 101] Each metal atom is bonded to four CO groups which occupy vertices of the trigonal bipyramid, and to the second metal atom located at the fifth axial vertex. Equatorial CO groups are staggered. $[Ru_2(CO)_8]^{2-}$ has a different structure; one ruthenium atom has a trigonal–bipyramidal coordination while the second metal atom adopts a square-pyramidal coordination. Four CO ligands bonded to the first Ru atom lie at the vertices of a trigonal bipyramid; the fifth equatorial coordination site is occupied by the second ruthenium atom. In the second fragment of the $[Ru_2(CO)_8]^{2-}$ molecule, the four vertices of the base of a square pyramid are occupied by the three CO groups and the ruthenium atom, and the remaining vertex is occupied by the fourth CO molecule.

In the ionic compounds $K_2[Fe(CO)_4]$[104] and $Na_2[Fe(CO)_4] \cdot 1.5C_4H_8O_2$[105], cations interact with anions. In the sodium compound some $Na^+$ cations are connected to oxygen atoms of CO groups, while the other $Na^+$ cations are quite close to iron atoms (the $Na^+$–Fe distance equals 309 pm) and therefore interact with the OC–Fe–CO grouping. The deformation of the anion is considerable because one of the C–Fe–C angles is increased to 130°. In the potassium compound, the $K^+$ ions interact with oxygen (the smallest $K^+$–O distance equals 271 pm) and carbon atoms (the smallest $K^+$–C distance equals 328 pm). As a result of these interactions, the C–Fe–C angle equals 121.0°. Therefore, in the bonding between cations and anions, there is considerable transfer of electron density from the electron-rich Fe center to the $M^+$ acceptor. In compounds which have $Na^+$ coordinated through different ligands, for example, $[Na\{N(CH_2CH_2OCH_2CH_2OCH_2CH_2)_3N\}]_2[Fe(CO)_4]$, the tetrahedral structure of the iron is essentially undistorted (the C–Fe–C angle equals 109.5°).[104]

In contrast to $[Fe_4H_2(CO)_{13}]$, the analogous hydrides of ruthenium or osmium are stable. Besides tri- and tetranuclear clusters, also known are clusters containing five or six iron or ruthenium atoms and from five to eight osmium atoms as well as carbonyl carbide clusters in which metal atoms are bonded to the carbon atom[36, 99] (Figure 2.22): $[Fe_5(CO)_{15}C]$, $[Fe_5(CO)_{14}C]^{2-}$, $[Fe_6(CO)_{16}C]^{2-}$, $[Ru_5(CO)_{15}C]$, $[Ru_6(CO)_{17}C]$, $[Ru_6H_2(CO)_{18}]$, $[Os_5(CO)_{16}]$, $[Os_5H_2(CO)_{15}]$, $[Os_6(CO)_{18}]$, $[Os_6H_2(CO)_{18}]$, $[Os_7(CO)_{21}]$, $[Os_8(CO)_{23}]$, $[Os_8(CO)_{21}C]$. Such compounds are most commonly prepared by the reduction of $[Fe(CO)_5]$ or $[Ru_3(CO)_{12}]$ with $[Co(CO)_4]^-$, $[Fe(CO)_4]^{2-}$, $[Mn(CO)_5]^-$, $[V(CO)_6]^-$, or $[Fecp(CO)_2]^-$ as well as by pyrolysis of $[Os_3(CO)_{12}]$. Many compounds containing different metal atoms may also be obtained, such as $[FeM_3(CO)_{13}H_2]$ (M = Ru, Os) or $[MOs_3(CO)_{13}H_3]$ (M = Mn, Re).

Iron carbonyls undergo disproportionation reactions in the presence of Lewis bases:

$$5[Fe(CO)_5] + 6py \longrightarrow [Fe(py)_6][Fe_4(CO)_{13}] + 12CO \qquad (2.89)$$

The iron, ruthenium, and osmium carbonyls react with unsaturated hydrocarbons, phosphines, arsines, stibines, and ligands which coordinate through nitrogen and sulfur atoms to give the following substitution products: $[M(CO)_{5-x}L_x]$, $[M_2(CO)_{9-x}L_x]$ and $[M_3(CO)_{12-x}L_x]$.

Of particular interest are phosphido-bridged derivatives of iron carbonyls, such as $(OC)_3Fe(\mu-PR_2)_2Fe(CO)_3$.[88, 102, 103] Upon reduction, these compounds give anions of the type $[(OC)_3Fe(\mu-PR_2)_2Fe(CO)_3]^{2-}$, in which the Fe−Fe bond is cleaved. The reduction with $[BEt_3H]^-$ leads to the formation of formyl complexes and the reaction with LiR′ furnishes acyl coordination compounds.[102] In the reaction of

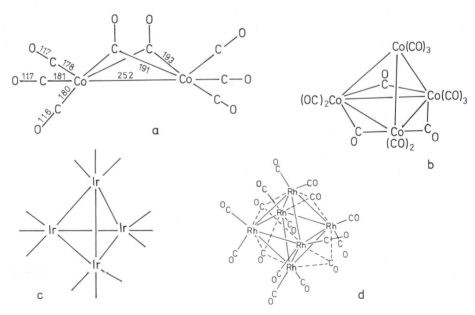

Figure 2.22.  The structures of (a) $Co_2(CO)_8$, (b) $Co_4(CO)_{12}$, (c) $Ir_4(CO)_{12}$, and (d) $Rh_6(CO)_{16}$. The distances are in pm.

$[\{\mu\text{-}1,2\text{-}C_6H_4(CH_2PPh)_2\}\{Fe(CO)_3\}_2]$ with LiBu$^n$, first deprotonation of the $CH_2$ group occurs followed by nucleophilic attack of the resulting carbanion on the iron atom with the formation of the $Fe-C$ bond.[103]

### g. Group 9 Metal Carbonyls

$[Co_2(CO)_8]$ is prepared by the reaction of carbon monoxide with dispersed metal or with its compounds (halides, carboxylates, etc.) in the presence of copper. At atomospheric pressure in aqueous solutions containing $CN^-$ ions, Co(II) salts from $[Co(CO)_4]^-$ up to 100% yield if the $CN^-/Co(II)$ ratio is from 1 to 2. Oxidation of $[Co(CO)_4]^-$ gives $[Co_2(CO)_8]$.

$$2Co + 8CO \xrightarrow[423\ K]{3\text{-}4\ MPa} [Co_2(CO)_8] \quad (90\%) \tag{2.90}$$

$$2Co(CN)_2 + 11CO + 12OH^- \longrightarrow 2[Co(CO)_4]^- + 4CN^- + 3CO_3^{2-} + 6H_2O \tag{2.91}$$

$$2[Co(CO)_4]^- + 2Ag^+ \longrightarrow 2Ag + [Co_2(CO)_8] \tag{2.92}$$

$[Co_4(CO)_{12}]$ is formed during decomposition of $[Co_2(CO)_8]$ when the latter is heated up to 333 K under inert atmosphere. $[Co_4(CO)_{12}]$ may also be obtained by reflux of solutions of $[Co_2(CO)_8]$ in hydrocarbons. Dodecacarbonyltetracobalt is purified by crystallization from benzene.

$$2[Co_2(CO)_8] \longrightarrow [Co_4(CO)_{12}] + 4CO \tag{2.93}$$

$[Co_6(CO)_{16}]$ is synthesized by the reduction of $[Co_4(CO)_{12}]$ with sodium or by heating ethanol solution of $[Co_2(CO)_8]$. $[Co_6(CO)_{15}]^{2-}$ is formed first, and subsequently is oxidized by $Hg^{++}$:

$$3[Co_2(CO)_8] + 2xEtOH \longrightarrow 2[Co(EtOH)_x][Co(CO)_4]_2 + 8CO \tag{2.94}$$

$$7[Co(EtOH)_x][Co(CO)_4]_2 \xrightarrow[vacuum]{333\ K,\ EtOH} 3[Co(EtOH)_x][Co_6(CO)_{15}] + 11CO \tag{2.95}$$

$$[Co_6(CO)_{15}]^{2-} \xrightarrow[H_2O]{Hg^{++},\ 298\ K} [Co_6(CO)_{16}] + [Co_4(CO)_{12}] + Co^{++} \quad (10\%) \tag{2.96}$$

$$9[Co_4(CO)_{12}] + 20Na \xrightarrow[THF]{273\text{-}298\ K} 4Na_2[Co_6(CO)_{15}] + 12Na[Co(CO)_4] \tag{2.97}$$

The existence of $[Rh_2(CO)_8]$ under normal conditions has not been confirmed, although its synthesis from dispersed metal under pressure (28 MPa) and at 473 K had been reported. However, it was shown that $[Rh_2(CO)_8]$ exists in solutions only at high pressures and at low temperatures.[106] $[Rh_2(CO)_8]$ and $[Ir_2(CO)_8]$ were also prepared by reactions of atomic Rh and Ir with carbon monoxide in CO matrices. During condensation at 10 K, $[Rh(CO)_4]$ or $[Ir(CO)_4]$ are formed first, and subsequently at 50 K react to give $[M_2(CO)_8]$ compounds. At 223 K, the dinuclear carbonyls are transformed into $[M_4(CO)_{12}]$. This means that with respect to tetranuclear compounds, the dinuclear complexes are thermodynamically unstable. $[Rh_2(CO)_8]$ in solutions as well as in the matrix and $[Ir_2(CO)_8]$ in the matrix have bridging carbonyl groups (Table 2.26).

*Table 2.26. Group 9 Metal Carbonyls*

| Compound | Color | Melting point (K) | $\nu$(CO), IR($R$) (cm$^{-1}$) | Remarks |
|---|---|---|---|---|
| $[Co_2(CO)_8]^a$ | Orange | 324 dec | 1) 2112w, 2071s, 2042s, 2001w, 1863w, 1853s | 1) Spectrum in solution |
| | | | 1') 2106w, 2069s, 2031s, 2022s, 1991m | 1') Nonbridging form |
| $[Co_4(CO)_{12}]^b$ | Black | 373 dec | 2104vw, 2063vs, 2055vs, 2048sh, 2038w, 2027w, 1898vw, 1867m | |
| $[Co_6(CO)_{16}]^b$ | Black | 378 dec | 2103w, 2061s, 2057sh, 2026w, 2020mw, 2018mw, 1806w, 1772s | |
| $[Rh_2(CO)_8]^c$ | | | 2060s, 2040s, 1852w, sh 1830mw | CO/Rh matrix, 50 K |
| $[Rh_2(CO)_8]^d$ | | | 2086s, 2061s, 1860mv, 1845s | stable in solution only at high pressures of CO$^d$ |
| $[Rh_4(CO)_{12}]^{a,b}$ | Red | 423 dec | 2105w, 2077vs, 2055w, 2028s, 1848s | |
| $[Rh_6(CO)_{16}]^a$ | Black | 493 dec | 2077s, 2074s, 2041m, 2016m, 1770s | |
| $[Ir_2(CO)_8]^c$ | | | 2095ms, 2068s, br, 1848w, sh, 1822m | IR/CO matrix, 50 K |
| $[Ir_4(CO)_{12}]^e$ | Yellow | 483 dec | 2072s, 2032s | |
| $[Ir_4(CO)_{12}]^c$ | | | 2068s, 2035ms, 1869m | IR/CO matrix, 50 K |
| $[Co(CO)_3]^{3-f}$ | Red-orange | | 1744m, 1600vs | |
| $[Rh(CO)_3]^{3-f}$ | Brown | | 1735m, 1598vs | |
| $[Ir(CO)_3]^{3-f}$ | Brown | | 1744m, 1615vs | |
| $[Ir_6(\mu_3\text{-CO})_4(CO)_{12}]^{(109)}$ | Red | | | $d(\text{Ir}-\text{Ir}) = 277.9(1)$ pm |
| $[Ir_6(\mu\text{-CO})_4(CO)_{12}]^{(109)}$ | Black | | | $d(\text{Ir}-\text{Ir}) = 277.8(b)$ pm |

$^a$ W. P. Griffith and A. J. Wickham, *J. Chem. Soc. A*, 834 (1969).
$^b$ G. Bor, *Spectrochim. Acta*, **19**, 1209 (1963); L. Malatesta, G. Caglio, and M. Angoletta, *Chem. Commun.*, 532 (1970); G. Bor, G. Sbrignadello, and K. Noack, *Helv. Chim. Acta*, **58**, 815 (1975).
$^c$ L. A. Hanlan and G. A. Ozin, *J. Am. Chem. Soc.*, **96**, 6324 (1974).
$^d$ R. Whyman, *J. Chem. Soc., Dalton Trans.*, 1375 (1972).
$^e$ R. Whyman, *J. Organomet. Chem.*, **24**, C35 (1970); R. Whyman, *J. Chem. Soc., Dalton Trans.*, 2294 (1972).
$^f$ J. E. Ellis, P. T. Barger, and M. L. Winzenburg, *J. Chem. Soc., Chem. Commun.*, 686 (1977).

$[Rh_4(CO)_{12}]$ and $[Rh_6(CO)_{16}]$ are formed by the reaction of CO with $RhCl_3$ [equation (2.98)]. $[Rh_4(CO)_{12}]$ may also be prepared by the reduction of $[Rh_2Cl_2(CO)_4]$ with CO in the presence of sodium carbonate. Various low-pressure,

$$4RhCl_3 + 12CO \xrightarrow[\text{20 MPa, 323-353 K}]{\text{Cu(Ag, Zn)}} [Rh_4(CO)_{12}]$$

$$\xrightarrow[\text{20 MPa, 353-503 K}]{\text{Cu}} [Rh_6(CO)_{16}]$$

$$(2.98)$$

high-yield synthetic methods are also available for preparation of $[Rh_4(CO)_{12}]$ and $[Rh_6(CO)_{16}]^{(36, 98, 10)}$:

$$[Rh_2Cl_2(CO)_4] + CO \xrightarrow[\text{0.1 MPa}]{\text{LiOAc, MeOH}} [Rh_6(CO)_{16}] \qquad (77\%) \qquad\qquad (2.99)$$

$$K_3[RhCl_6] + CO \xrightarrow[\text{0.1 MPa, 80–90\%}]{\text{Cu, OH}^-,\, H_2O} [Rh_4(CO)_{12}] \xrightarrow[\text{333 K}]{\text{n-heptane}} [Rh_6(CO)_{16}] \qquad (2.100)$$

$[Ir_4(CO)_{12}]$ is synthesized from $IrX_3$, $M_3[IrX_6]$, or $M_2[IrX_6]$ where $X = Cl$, Br, or I and $M = Na$, K, or $NH_4$, and carbon monoxide in the presence of an excess of copper or silver[2] or from $[IrCl(CO)_2\,(p\text{-toluidine})]$[107]:

$$IrX_3 + CO \xrightarrow[\text{373–413 K}]{\text{35 MPa, Cu}} [Ir_4(CO)_{12}]$$

$$[IrCl(CO)_2\,(p\text{-toluidine})] + CO \xrightarrow{\text{0.4 MPa}} [Ir_4(CO)_{12}] \qquad (2.101)$$

The reduction of $[Ir_4(CO)_{12}]$ in THF with sodium leads to the formation of $[Ir_6(CO)_{15}]^{2-}$, which after oxidation with an acid gives $[Ir_6(CO)_{16}]$:

$$[Ir_6(CO)_{15}]^{2-} + CO + 2H^+ \xrightarrow[\text{HOAc}]{\text{CO}} [Ir_6(CO)_{16}] + H_2 \qquad (2.102)$$

$[Ir_4(CO)_{12}]$ may be prepared quantitatively by heating the solution of $IrCl_3$ or $IrCl_4$ in anhydrous formic acid at 373 K in the autoclave.[108]

$[Co_2(CO)_8]$ and $[Co_4(CO)_{12}]$ are air-sensitive; however, large crystals of these compounds do not undergo noticeable oxidation in a short period of time. Solutions of these carbonyls rapidly undergo decomposition. The halogens quantitatively oxidize $[Co_2(CO)_8]$ and $[Co_4(CO)_{12}]$ to Co(II) salts. Oxidizing acids oxidize $[Co_2(CO)_8]$ to Co(II) compounds and nonoxidizing acids react with this carbonyl only slowly and partially.

Rhodium and irridium carbonyls of the type $[M_4(CO)_{12}]$ are considerably more stable and in the solid state in air do not undergo decomposition. Strong acids oxidize $[Rh_4(CO)_{12}]$ in acetonitrile to $[Rh(MeCN)_2\,(CO)_2]^+$. However, $[Ir_4(CO)_{12}]$ undergoes protonation to $[Ir_4(CO)_{12}H_2]^{2+}$ with strong acids. Therefore, the stability of tetranuclear clusters increases when the $M-M$ bond energy becomes greater (Table 2.16). All carbonyls of the group 9 metals undergo reduction with the alkali metals to mono- or polynuclear anionic complexes. In liquid ammonia $[Co_2(CO)_8]$ and $[Co_4(CO)_{12}]$, like $[M_4(CO)_{12}]$ (M = Rh, Ir) in THF, are reduced in the atmosphere of CO to $[M(CO)_4]^-$. During reduction of $Na[Co(CO)_4]$ by sodium in liquid ammonia, $[Co(CO)_3]^{3-}$ is formed; analogous compounds $[M(CO)_3]^{3-}$ of rhodium and iridium are obtained by reduction of $[M_4(CO)_{12}]$ by sodium in liquid $NH_3$, or in hexamethylphosphoramide. The $v(CO)$ bands of these compounds occur at very low frequencies (1750–1600 cm$^{-1}$) (Table 2.26).

The Co, Rh, and Ir carbonyls also react with hydroxides to give metal carbonylate complexes:

$$11[Co_2(CO)_8] + 32OH^- \longrightarrow 2Co^{2+} + 20[Co(CO)_4]^- + 8CO_3^{2-} + 16H_2O \qquad (2.103)$$

$$[Rh_6(CO)_{16}] + 4OH^- \xrightarrow{\text{N}_2,\text{ MeOH}} [Rh_6(CO)_{15}]^{2-} + CO_3^{2-} + 2H_2O \qquad (2.104)$$

In the presence of Lewis bases, $[Co_2(CO)_8]$ and $[Co_4(CO)_{12}]$ easily undergo disproportionation reactions:

$$3[Co_2(CO)_8] + 2nZ \longrightarrow 2[Co(Z)_n][Co(CO)_4]_2 + 8CO \qquad (2.105)$$

$$3[Co_4(CO)_{12}] + 24py \longrightarrow 4[Co(py)_6][Co(CO)_4]_2 + 4CO \qquad (2.106)$$

Tetracarbonylcobaltate$(1-)$ forms ionic complexes with group 1 elements. However, compounds of the type $M[Co(CO)_4]_n$, where $n \geqslant 2$ and M = zinc, cadmium, mercury, indium, etc., are covalent, possessing M–Co bonds in which the main group metal has normal coordination number. These compounds are monomeric in the solid state. $Ag[Co(CO)_4]$ and $Cu[Co(CO)_4]$ are tetrameric clusters in which the metal atoms form planar, eight-membered rings. Each of the distorted $[Co(CO)_4]^-$ tetrahedrons is bonded to two atoms of silver or copper.[111]

All group 9 carbonyls easily form CO ligand substitution products of the type $[M_2(CO)_{8-x}L_x]$, $[M_4(CO)_{12-x}L_x]$, and $[M_6(CO)_{16-x}L_x]$, where L = olefin, acetylene, phosphine, etc.

In octacarbonyldicobalt two CO groups are bridging, and there is a direct Co–Co bond (Figure 2.22). The bridging CO groups exist also in $[Rh_2(CO)_8]$.[106] In solutions, the cobalt carbonyl exists in equilibrium in three of its forms, two of which do not contain bridging CO groups [equation (2.5)]. In the solid state, $[Co_4(CO)_{12}]$ and $[Rh_4(CO)_{12}]$ possess $C_{3v}$ symmetry. The metal atoms lie at the vertices of the tetrahedron, and three carbonyl groups form bridges at its base. The $[Ir_4(CO)_{12}]$ molecule has $T_d$ symmetry; the iridium atoms lie at the vertices of the tetrahedron and all carbonyl groups are terminal. Rhodium atoms in hexadecacarbonylhexarhodium form an octahedron. Two terminal CO groups are bonded to each metal atom. Each of the four bridging CO groups forms bonds with three rhodium atoms and is located on one of the four threefold axes which pass alternately through the faces of the octahedron. In this manner each rhodium atom is bonded to two bridging CO groups. $[Co_6(CO)_{16}]$ is isomeric with the rhodium compound. The IR spectra of hexanuclear cobalt, rhodium, and iridium compounds are analogous; therefore, the structure of the cobalt complex is probably the same as that of $[Rh_6(CO)_{16}]$ (Figure 2.22). $[Ir_6(CO)_{16}]$ exists as two isomers. The red compound is structurally analogous to $[Rh_6(CO)_{16}]$ and, in

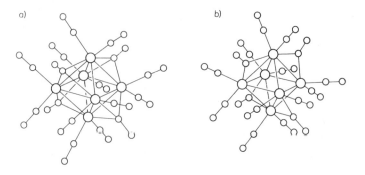

Figure 2.23.   The structures of $Ir_6(CO)_{16}$: (a) red isomer, (b) black isomer.

the black isomer, each of the six iridium atoms lies at the vertices of an octahedron; additionally, each iridium atom is connected with two terminal and two bridging carbonyl groups ($\mu$-CO) (Figure 2.23).[109]

*Cluster Compounds of Co, Rh, and Ir.* In addition to the above-mentioned neutral cluster compounds, there is a large number of anionic carbonyl clusters and metal carbonyl carbides. Carbonyl carbides are formed when the interstice inside the metal cluster is sufficiently large to accommodate the carbon atom. Carbonyl carbides possessing at least four metal atoms are known. The most thoroughly investigated carbides are those of rhodium because they are very stable and resist air oxidation. Carbonyl clusters of group 9 elements containing even more than 20 metal atoms are now known: $[M_6(CO)_{15}]^{2-}$ (M = Co, Rh, Ir), $[M_6(CO)_{14}]^{4-}$, $[M_6(CO)_{15}C]^{2-}$ (M = Co, Rh), $[Co_8(CO)_{18}C]^{2-}$, $[Rh_7(CO)_{16}]^{3-}$, $[Rh_8(CO)_{19}C]$, $[Ir_8(CO)_{22}]^{2-}$, $[Rh_{12}(CO)_{30}]^{2-}$, $[Rh_{12}(CO)_{25}(C_2)]$, $[Rh_{13}(CO)_{24}H_{5-n}]^{n-}$, $[Rh_{14}(CO)_{25}]^{4-}$, $[Rh_{15}(CO)_{27}]^{3-}$, $[Rh_{15}(CO)_{28}(C)_2]^-$, $[Rh_{17}(CO)_{32}(S)_2]^{3-}$, $[Rh_{14}(CO)_{33}(C)_2]^{2-}$, $[Rh_{22}(CO)_{37}]^{4-}$.

The cobalt carbonyls are prepared by the disproportionation reaction of $[Co_2(CO)_8]$ in the presence of Lewis bases or by the reduction of cluster cobalt carbonyls with the alkali metals. The iridium compounds are obtained during reduction of $[Ir_4(CO)_{12}]$ with sodium in ether solution. The rhodium carbonyls are usually synthesized by reduction of $[Rh_2Cl_2(CO)_4]$ or $[RhCl_6]^{3-}$ with carbon monoxide in basic medium or by nucleophilic attack of bases on the carbonyl group of carbonyl clusters (see preparation of $[M_4(CO)_{12}]$ and $[M_6(CO)_{16}]$).

$$[Rh_6(CO)_{16}] + OMe^- \longrightarrow [Rh_6(CO)_{15}(COOMe)]^- \tag{2.107}$$

$$[Rh_6(CO)_{16}] + 2RNH_2 \longrightarrow [RNH_3]^+ [Rh_6(CO)_{15}(CONHR)]^- \tag{2.108}$$

$$[Rh_6(CO)_{15}(COOMe)]^- + 2OH^- \longrightarrow [Rh_6(CO)_{15}]^{2-} + HCO_3^- + MeOH \tag{2.109}$$

$$6[RhCl_6]^{3-} + 23OH^- + 26CO + CHCl_3 \xrightarrow{\text{MeOH}} [Rh_6(CO)_{15}C]^{2-} + 39Cl^- + 11CO_2 + 12H_2O \tag{2.110}$$

The reaction between $[Rh_4(CO)_{12}]$ and $[Co_2(CO)_8]$ affords $[CoRh(CO)_7]$ from which other different heterometallic carbonyls may be prepared.[110] $[CoRh(CO)_7]$ is also obtained by the reaction of $Na[Co(CO)_4]$ with $Rh_2Cl_2(CO)_4$.

$$Rh_4(CO)_{12} + 2Co_2(CO)_8 \rightleftharpoons 4CoRh(CO)_7 \rightleftharpoons 2Co_2Rh_2(CO)_{12} + 4CO \tag{2.111}$$

$$Na[Co(CO)_4] + Rh_2Cl_2(CO)_4 \xrightarrow[\text{CO}]{195\,\text{K}} CoRh(CO)_7 \tag{2.112}$$

Cobalt exhibits a characteristic tendency to form clusters of the formula $RCCo_3(CO)_9$, which have trigonal–pyramidal structures. The cobalt atoms occupy the three positions at the base of the pyramid, and the carbon group lies at its top vertex. These cobalt–carbon clusters are prepared by reactions of $RCX_3$ with $Co_2(CO)_8$ or $NaCo(CO)_4$ in the presence of Lewis bases[123]:

$$3Na[Co(CO)_4] + MX_3 \longrightarrow 3NaX + [MCo_3(CO)_9] + 3CO$$

$$(M = RC, P, As, \text{etc.}) \tag{2.113}$$

Cobalt carbonyl halides, in contrast to rhodium and iridium carbonyl halides, are very unstable, and only few have been prepared and investigated; $[CoX(CO)_4]$ is

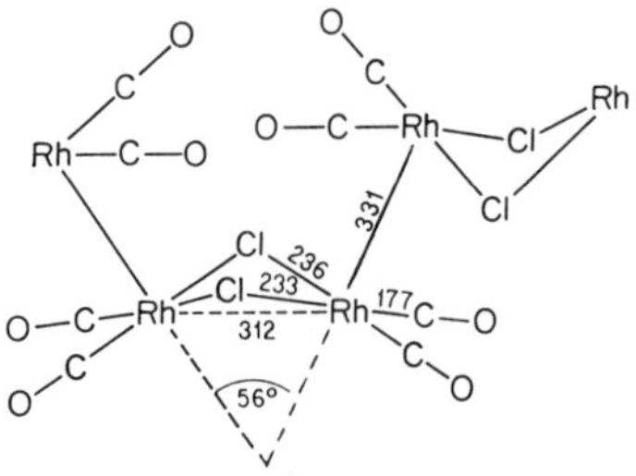

Figure 2.24.   The structure of $Rh_2Cl_2(CO)_4$. The distances are in pm.

formed by the reaction of halogens with $SnCl_2[Co(CO)_4]_2$. Also known are $[CoI(CO)_3PPh_3]$ and $[CoI(CO)_2(PPh_3)_2]$.

Rhodium and iridium carbonyl halides are easily formed from $[MX_6]^{3-}$ or $MX_3$ by the action of CO, most commonly in solutions, although in some cases these compounds may be obtained in the solid state under atmospheric pressure of CO:

$$RhCl_3(H_2O)_3 + CO \xrightarrow[\text{0.1 MPa}]{\text{373 K}} [Rh_2Cl_2(CO)_4] \qquad (2.114)$$

Rhodium and iridium carbonyl halides are also prepared from organic compounds, for example, anhydrous HCOOH and DMF, which decompose with carbon monoxide evolution. The following rhodium carbonyl halides are known: $[Rh_2X_2(CO)_4]$, $[RhX_2(CO)_2]^-$ (X = Cl, Br, I), $[Rh_2X_4(CO)_2]^{2-}$ (X = Br, I), $[RhX_3(CO)]$, $[RhX_4(CO)_2]^-$, and $[RhX_5(CO)]^{2-}$ (X = Cl, Br, I). However, the most numerous are iridium carbonyl halides, e.g., $[IrX_5(CO)]^{2-}$ (X = Cl, Br), $[IrX_4(CO)_2]^-$, $[IrI_4(CO)]^-$, $[IrI_3(CO)_3]$, $[Ir_2I_6(CO)_4]$, $[Ir_2I_6(CO)_3]$, $[IrX_2(CO)_2]^-$, $[IrX(CO)_3]$. Some of these compounds are easily prepared, for example, $[Rh_2X_2(CO)_4]$ or $[IrX(CO)_3]$, and therefore serve as common starting material for preparation of other rhodium or iridium complexes. $[Rh_2X_2(CO)_4]$ complexes are dimeric, containing bridging halogen atoms. The $[Rh_2Cl_2(CO)_4$ molecule is bent; there is a weak interaction between rhodium atoms of neighboring molecules (the intermolecular Rh–Rh distance is 331 pm) (Figure 2.24). The bent structure is also present in solution (the dipole moment is 1.64 $D$), which is attributed to another weak interaction occurring between rhodium atoms of the same molecule (the intramolecular rhodium distance is 312 pm). This interaction is considerably weaker than that of a normal 2e bond.[112] The rhodium–rhodium interaction also occurs in dimers of the type $[Rh_2(RCOO)_2(CO)_4]$:

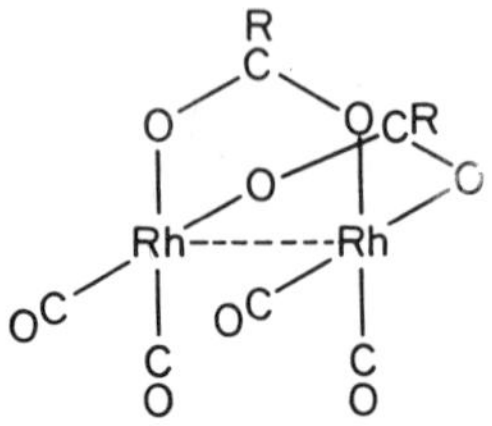

In polar solvents such as alcohols, the carboxylate bridges are broken, and the following monomeric complexes are formed: $[Rh(RCOO)(CO)_2(ROH)]$. In methanol–benzene solutions, an equilibrium exists between dimers $Rh_2(RCOO)_2(CO)_4$ and monomers $Rh(RCOO)(CO)_2(MeOH)$, where R = H or Me.[122]

Rhodium and iridium also give rise to complexes of the following composition: $[RhX(CO)L_2]$ and $[IrX(CO)L_2]$ where $X = Cl$, Br, or I and $L$ = phosphines, arsines, or stibines. In the case of monodentate ligands, these compounds adopt *trans* square-planar coordinations. These complexes, like $[RhH(CO)L_3]$ and $[IrH(CO)L_3]$ hydrides, are active homogeneous catalysts for hydrogenation, oxidation, and hydro-formylation reactions, and therefore have been studied intensively. Rhodium also forms carbonyl complexes with tris(2-pyridyl)phosphine,[114, 115] e.g., $[RhH(CO)(PPh_3)_2$ $\{P(2\text{-pyridyl})_3\}]$ and *trans*-$[RhCl(CO)\{P(2\text{-pyridyl})_3\}_2]$. The rhodium carbonyl halide is five-coordinate in solution, because one of the pyridylphosphine ligands is coordinated to the rhodium atom through the phosphorus and nitrogen atoms.[115] The $[RhX(CO)(PR_3)_2]$ compounds ($X = Cl$, Br, or I) react with CO to give five-coordinate complexes of the type $RhX(CO)_2 (PR_3)_2$, possessing a trigonal–bipyramidal structure in which the CO ligands occupy axial positions.[132]

Rhodium gives characteristic carbonyl complexes with diphosphines $R_2P(CH_2)_nPR_2$. Monomeric complexes with *cis*-arrangement are obtained only for $n = 2$. In other cases, the structures (2.115) are formed.[113, 116] The dinuclear complex

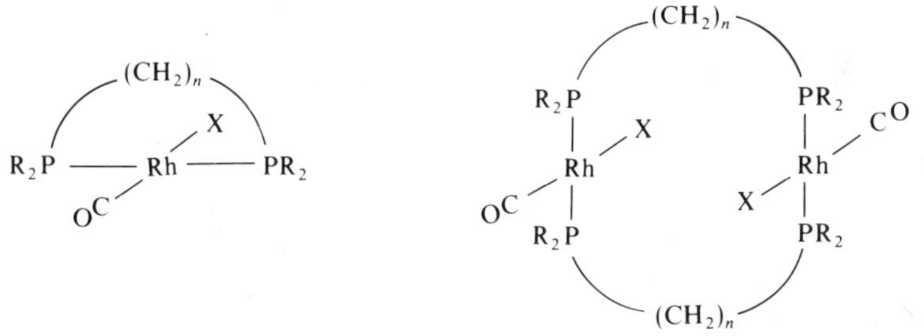

(2.115)

$(OC) ClRh(\mu\text{-}Ph_2PCH_2PPh_2)_2 RhCl(CO)$[116–118] reacts with nucleophiles, and as a result of reduction or thermal CO evolution forms interesting "A-frame" compounds (2.116).[119–121] Similar compounds are formed by $Ph_2P(2\text{-pyridyl})$.[120, 121] Phosphido

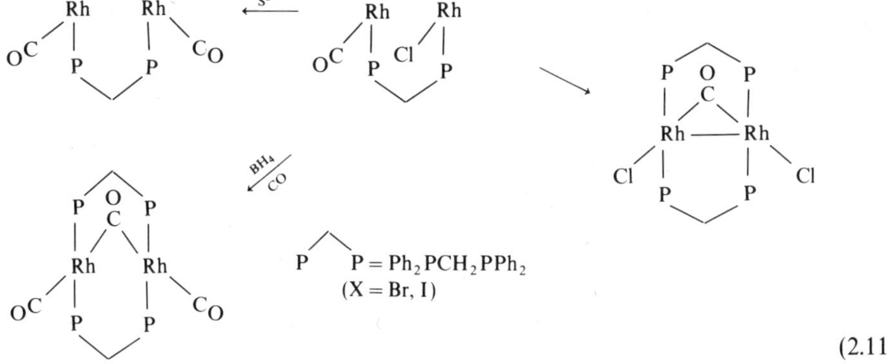

(2.116)

complexes, $Rh_2(PR_2)_2(CO)_{4-x}L_x$ $(L = PR_3)$, exist as isomers in which the rhodium atoms may have square or tetrahedral local coordination; one rhodium atom may adopt square coordination and the other atom tetrahedral coordination.[137] The relative energies of these isomers are close.

There are a few known examples of Rh(II) and Ir(II) carbonyl complexes.[133-136] For rhodium, these are dinuclear with strong $Rh-Rh$ bonds,[133, 134] for example, $[Rh_2(OOCMe)_4(CO)_2]$. Iridium readily forms mononuclear complexes, although dinuclear iridium carbonyls such as $[Ir_2(tcbi)_2(CO)_2(NCMe)_2\{P(OEt)_3\}_2]\cdot MeCN$,[135] $[Ir(H)(\mu\text{-}SBu')(CO)(PR_3)]_2$, and $[Ir_2I_2(\mu\text{-}SBu')_2(CO)_4]$[135] are known.

## h. Group 10 Metal Carbonyls

Metallic nickel prepared by reduction of its compounds readily reacts with carbon monoxide to give $Ni(CO)_4$ (Table 2.27). Many other metals (Fe, Co, Rh, and Re) may also react with CO; however, the rate of formation of the carbonyl is slow, and the process must be carried out under forcing conditions.

$$Ni + 4CO \xrightarrow[\text{0.1 MPa}]{\text{298 K}} [Ni(CO)_4] \tag{2.117}$$

Tetracarbonylnickel may also be prepared in high yields from Ni(II) complexes in aqueous solutions. Particularly high yields of $Ni(CO)_4$ were obtained from compounds containing sulfur ligands such as nickel sulfide or dithiocarboxylates, or from other complexes, e.g., $NiCl_2$ in the presence of reducing agents:

$$NiS + 5CO + 4OH^- \xrightarrow[\text{0.1 MPa}]{\text{1}\,n\,\text{NaOH}} [Ni(CO)_4] + CO_3^{2-} + S^{2-} + 2H_2O \tag{2.118}$$

Under normal conditions $Pd(CO)_4$ and $Pt(CO)_4$ are not stable; however, these carbonyls were prepared and investigated in CO matrices. The fact that $Pd(CO)_4$ and

*Table 2.27. Group 10 Metal Carbonyls*

| Compound | Color | Melting point (K) | $v(CO)$ (cm$^{-1}$) | Solubility, Remarks |
|---|---|---|---|---|
| $[Ni(CO)_4]$ | White | 254 | 2044 (2125$R$) | Organic solvents |
| $[Ni(CO)_{1-3}]^a$ | | | 1996-Ni(CO) | Stable up to 26 K |
| | | | 1967-Ni(CO)$_2$ | CO/Ni matrix |
| $[Pd(CO)_4]^b$ | | | 2070 | Tetrahedral, CO/Pd matrix |
| $[Pd(CO)_{1-3}]^b$ | | | | CO/Pd matrix |
| $[Pt(CO)_4]^c$ | | | 2053 | Tetrahedral, CO/Pt matrix |
| $[Pt(CO)_{1-3}]^c$ | | | 2052-PtCO, | |
| | | | 2057-Pt(CO)$_2$ | CO/Pt matrix |
| | | | 2049-Pt(CO)$_3$ | |
| | | | ($D_{3h}$) | |

[a] R. L. DeKock, *Inorg. Chem.*, **10**, 1205 (1971).
[b] E. P. Kündig, M. Moskovits, and G. A. Ozin, *Can. J. Chem.*, **50**, 3587 (1972); J. H. Darling and J. S. Ogden, *Inorg. Chem.*, **11** 666 (1972).
[c] D. McIntosh, E. P. Kündig, M. Moskovits, and G. A. Ozin, *J. Am. Chem. Soc.*, **95**, 7234 (1973); E. P. Kündig, M. Moskovits, and G. A. Ozin, *J. Mol. Struct.*, **14**, 137 (1972).

$Pt(CO)_4$ are unstable is explained by an increased energy difference between $ns$ and $np$ orbitals. The energy of $p$ orbitals is too great and the energy of $d$ orbitals is too small for the former orbitals to be effective $\sigma$-acceptor orbitals, and the latter orbitals to be effective $\pi$-donor orbitals. Therefore, the group 10 metals often form 16$e$ complexes in which the $p_z$ orbital does not participate in the formation of the metal–ligand bond. Similarly for group 11 metals, all $p$ orbitals are not always utilized.

Ni(II), in contrast to Pd(II) and Pt(II), does not form stable carbonyl halides. Thus the ability to form $\pi$ bonds is greater for $Ni^0$ than for $Pd^0$ and $Pt^0$. The reverse situation happens for these metals in $+2$ oxidation states; Pd(II) and Pt(II) form $\pi$ back-bonding more readily than Ni(II) does. These differences result from the various electronic configurations of these metals, i.e., Ni-$d^8s^2$, Pd-$d^{10}$, Pt-$d^9s^1$, and therefore an excitation energy required for formation of complexes in which central atoms have zero oxidation numbers is different than for formation of complexes possessing $+2$ oxidation states. These facts are confirmed by the existence of stable carbonyl complexes of $Pd^0$ and $Pt^0$ containing relatively weaker $\pi$-acceptor ligands and stronger $\sigma$-donor groups, for example, $[Pd_3(CO)_3(PPh_3)]$ and $[Pt_3(CO)_3(PPh_3)_4]$. Such a combination of ligands increases electron density on the metal and therefore increases its $\pi$-donor properties. Because of such reasons, anionic carbonyl complexes of platinum are also more stable, e.g., $[Pt_3(CO)_6]_n^{2-}$.

$[Pt(CO)_4]$, $[Pd(CO)_4]$, as well as $[Ni(CO)_4]$ have tetrahedral structures.[124] $[Ni(CO)_4]$ is one of the most toxic compounds; it is a stronger poison than HCN. Certain amounts of $[Ni(CO)_4]$ are formed during cigarette smoking. Cobalt and iron carbonyls are also present in the cigarette smoke.[129] At 293–375 K in the atmosphere of CO, it is possible to remove up to 90% of Co, 80% of Ni, and 60% of Fe present in tobacco. The content of nickel in one liter of tobacco smoke may reach more than 100 nanomoles. The total content of Fe, Ni, and Co may reach one micromole per one liter of tobacco smoke. As a result of such concentrations, some countries introduced recommendations that the content of $Ni(CO)_4$ in tobacco smoke be lowered to 1 ppb (part per billion).

Tetracarbonylnickel is readily oxidized by air to a Ni(II) salt and in the gas phase it reacts with $CS_2$ according to the following equation:

$$2[Ni(CO)_4] + CS_2 \longrightarrow 2NiS + C + 8CO \tag{2.119}$$

The reduction reactions of $[Ni(CO)_4]$ by sodium in liquid ammonia or by the alkali metal amalgams in THF afford polynuclear anions for which the following formulas are assigned[2, 8–10]: $[Ni_2(CO)_6]^{2-}$, $[Ni_3(CO)_8]^{2-}$, $[Ni_4(CO)_9]^{2-}$, $[Ni_5(CO)_9]^{2-}$. However, the existence of these anions is not certain.[36] The reaction of $[Ni(CO)_4]$ with amalgams in THF or with the alkali metal hydroxides in MeOH gives $[Ni_5(CO)_{12}]^{2-}$ and $[Ni_6(CO)_{12}]^{2-}$. In solution these anions exist in equilibrium:

$$[Ni_5(CO)_{12}]^{2-} + [Ni(CO)_4] \longrightarrow [Ni_6(CO)_{12}]^{2-} + 4CO \tag{2.120}$$

The pentanuclear cluster has an elongated trigonal–bipyramidal structure; in $[Ni_6(CO)_{12}]^{2-}$ the nickel atoms lie at the vertices of a trigonal antiprism. From the latter compound, the following clusters may be obtained: $[Ni_9(CO)_{18}]^{2-}$ and $[Ni_{12}(CO)_{21}H_n]^{(4-n)-}$ where $n = 2$–4.[36, 126, 127]

$[Pt(CO)_2]_n$ is prepared by hydrolysis of $[PtCl_2(CO)_2]$ in benzene as well as by the carbonylation reaction of $Na_2[PtCl_4]$ in ethanol.[36, 125] $[Pt(CO)_2]_5$ is formed by the reaction of $[PtCl_4]^{2-}$ with CO in HCl solution.[128] Recently, the following anionic clusters were investigated: $[Pt_3(CO)_6]_n^{2-}$ ($n = 1$–6, 10).[36, 130] Structurally, they consist of planar $[Pt_3(\mu_2\text{-}CO)_3(CO)_3]$ units stacked in a column along the threefold axis (Figure 3.26); when $n \geqslant 3$ the platinum atoms of the base and the top of the column are twisted with respect to each other (by 26° for $n = 3$ and by 64° for $n = 5$). $[Ni_9(CO)_{18}]^{2-}$ has a similar structure. Pt–Pt distances in triangles are 266 pm and distances between any two neighboring planes are 303–309 pm. In solution, $[Pt_9(CO)_{18}]^{2-}$ shows dynamic properties; the other triangles rotate about the threefold axis with respect to the middle triangle.[131] The $[Pt(CO)_2]_n$ is probably also anionic of the type $[Pt_3(CO)_6]^{2-}_{1-10}$ because it always contains cations. The ratio $Pt/M^+$ is 15. Similarly $[Pt(CO)_2]_5$ may contain $H^+$ ions because it was obtained in acidic medium.

There are many carbonyl complexes of nickel, palladium, and platinum containing phosphines (L). Nickel compounds of the type $[Ni(CO)_{4-x}L_x]$ are readily formed in substitution reactions of $[Ni(CO)_4]$. Palladium and platinum phosphine carbonyls are prepared by reactions of compounds of these metals with carbon monoxide in the presence of phosphines. The following complexes are known: $[M(CO)L_3]$, $[M_3(CO)_3L_3]$, $[M_3(CO)_3L_4]$, $[Pt(CO)_2L_2]$ and $[M_4(CO)_5L_4]$ (M = Pd, Pt). Trinuclear platinum compounds resist oxidation.

*Metal Carbonyl Halides of Group 10 Metals.*   Group 10 metal carbonyl halides are known, but for nickel such compounds are very unstable. The only examples are $M[NiX(CO)_3]$ (X = Cl, Br, I), which are formed from $[Ni(CO)_4]$ and the alkali metal halides.

$[PdCl_2(CO)_2]$ is obtained by reaction of $PdCl_2$ suspended in MeOH or $SOCl_2$ with carbon monoxide.[138] $[Pd_2Cl_4(CO)_2]^{2-}$ is prepared from $[PdCl_4]^{2-}$ and CO in concentrated HCl[139, 140] (Figure 2.25). Various polynuclear cluster carbonyls of palladium are also known, such as $[Pd_4(CO)_2(phen)_4(RCOO)_4]$, $[Pd_4(OAc)_4(CO)_4]$, and $[Pd_2X_2(PR_3)_2(CO)_2]$ (X = Cl, Br, I). Palladium also forms complexes of the following type: $[PdCl_3(CO)]^-$, $[PdX(CO)]_n$ (X = Cl, Br, I), and $[Pd_2Cl_2(\mu\text{-}CO)(PEt_2Ph)_2]$.[188]

The most stable carbonyl halides of group 10 metals are those of platinum. $[PtX_2(CO)_2]$ and $[PtX_2(CO)]_2$ are prepared by reaction of $PtX_2$ (X = Cl, Br, I) with carbon monoxide under pressure. These compounds have the following structures:

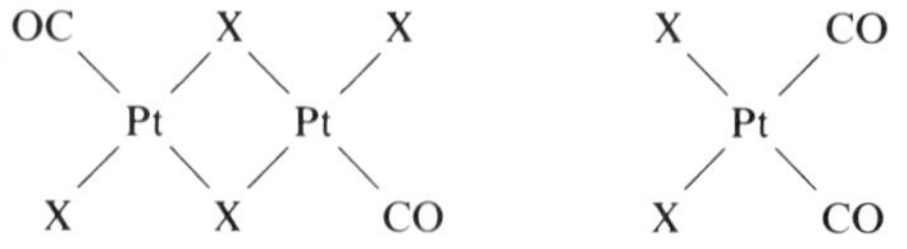

$$\text{(2.121)}$$

The thermal stability of these compounds increases from iodo carbonyls to chloro compounds, and the resistance to hydrolysis increases in the opposite direction. Reactions of these compounds with HX acids or MX halides give anions of the type $[PtX_3(CO)]^-$. Such anions are also readily formed by reactions of formic acid with platinum halides or platinum halide complexes. For palladium and platinum, "A-frame" carbonyls are characteristic (Figure 2.25).[121, 142]

The reaction of $[Pt(CO)_2]_5$ with $Fe^{3+}$ in the presence of halides in aqueous solutions affords Pt(IV) carbonyls such as $[PtCl_2H_2(CO)]$ and $[PtBr_3H_2(CO)]^-$.[157] Recently, a synthesis of nickel(II) carbonyl $[Ni(SiCl_3)_2(CO)_3]$ possessing a

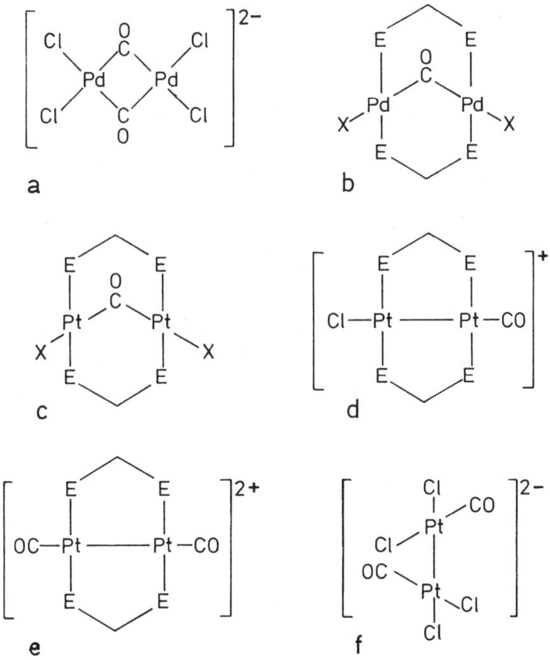

Figure 2.25. Structures of carbonyl halides of Pd and Pt; $E\frown E = R_2ECH_2ER_2$ ($E = P, As$) ($X = Cl, Br, I$):
  (a) $[Pd_2Cl_4(CO)_2]^{2-}$,[140] (b) $[Pd_2X_2(CO)(R_2ECH_2ER_2)_2]$,[121,142]
  (c) $[Pt_2X_2(CO)(R_2ECH_2ER_2)_2]$,[121,142] (d) $[Pt_2Cl(CO)(R_2ECH_2ER_2)_2]^{+}$,[121,142]
  (e) $[Pt_2(CO)_2(R_2ECH_2ER_2)_2]^{2+}$,[121,142] and (f) $[Pt_2Cl_4(CO)_2]^{2-}$.[156]

trigonal–bipyramidal structure in which the $SiCl_3$ ligands occupy the axial positions was reported.[143]

## i. Group 11 Metal Carbonyls

Stable, dinuclear carbonyls of this group are not known. The reason for this is the too low energy of $(n-1)d$ orbitals and too high energy of $np$ orbitals (as in the case of Pd and Pt). Nevertheless, many metal carbonyls of group 11 metals were prepared by condensation of metal atoms with carbon monoxide at temperatures *ca* 15 K: $[M(CO)]$, $[M(CO)_2]$ where $M = Cu, Ag,$ and $Au$, $[M(CO)_3]$, and $[M_2(CO)_6]$ where $M = Cu$ and $Ag$. These compounds have $C_{\infty v}$, $D_{\infty h}$, $D_{3h}$, and $D_3$ symmetries, respectively. In dimers, there are direct metal–metal bonds.[4, 124, 144] Based on the IR spectrum of $Au(CO)_2$, it is assumed that $Au(CO)_2$ has an isocarbonyl ligand, in which the bonding of one carbonyl group to the gold atom occurs through the oxygen atom: $OC-Au-O\equiv C$.[145] It is probably a situation forced by the crystal structure of the matrix of solid carbon monoxide and not by the tendency to form a $Au-O$ bond. The observed IR spectrum may also be explained by minimal nonequivalency of the carbonyl groups assuming the difference between the force constants for the two carbonyl groups to be of the order $10\ Nm^{-1}$.[124]

Copper and gold carbonyl halides such as $[CuCl(CO)]_n$, $[CuCl(CO)]\cdot 2H_2O$,

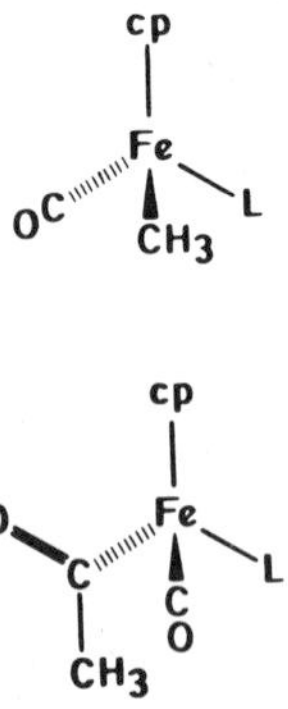

Figure 2.26. The structure of $Cu_2L^{2+}$.

and [AuCl(CO)] are known. The anhydrous copper compound is formed by reaction of copper(I) chloride with CO and the hydrate by the reaction involving the same reagents but carried out in hydrochloric acid. Cu(I) compounds are utilized for removal of CO traces because they bind CO. The reaction of gold(I) chloride suspended in benzene with CO furnishes [AuCl(CO)]. It is a white compound which readily undergoes hydrolysis in water. It is a linear monomer.

1,4-bis(1-oxa-4,10-dithia-7-azacyclododecan-7-ylmethyl)benzenedicopper(I) tetrafluoroborate in nitromethane reversibly binds carbon monoxide (Figure 2.26).[150]

Similar properties are exhibited by the following two compounds: [Cu(histamine)(CO)]$^+$, $v(CO) = 2091$ cm$^{-1}$ and [(histamine)Cu($\mu$-histamine) Cu(histamine)(CO)$_2$]$^{2+}$, $\mu(CO) = 2066$ and $2055$ cm$^{-1}$. In solutions these complexes reversibly bind CO. In the solid state these compounds are very stable.[151]

## 9. METAL CARBONYL HYDRIDES

### a. Bonding

Hydrogen may act as a ligand possessing moderate $\sigma$-donor properties or relatively strong $\sigma$-acceptor properties. Therefore the properties of hydrogen vary depending on the complex. The transition metal hydrides including carbonyl hydrides in solution may be acidic, neutral, or basic, because the polarity of the metal–hydrogen bond changes from $M^{\delta+} - H^{\delta-}$ to $M^{\delta-} - M^{\delta+}$.[8, 29, 152–155, 161–163]

The other ligands in a hydrido compound also influence the polarity of the $M - H$ bond. The hydrogen atom may be bound as a terminal, bridging, or $\mu_3$-type ligand:

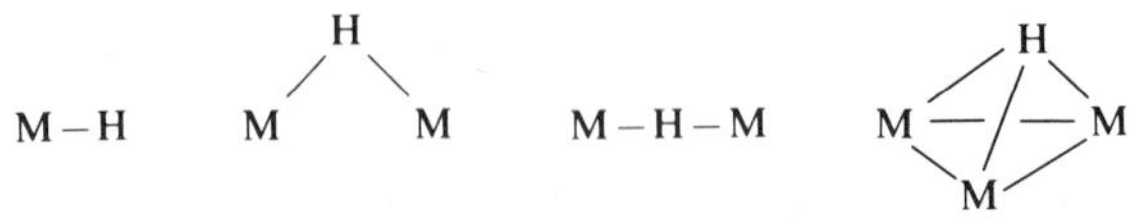

The bridge $M - H - M$ may be bent or linear. In the case of bridging structures, the M H M bond may be described by multicenter molecular orbitals (Figure 2.27). Recently, complexes containing coordinated hydrogen molecules were obtained,[166] for instance, [M(CO)$_3$ (PR$_3$)$_2$ (H$_2$)] where M = Mo, W and R = Cy, Pr$^i$ (Figure 2.28).

In larger cluster compounds, hydrogen atoms occupy positions inside the cluster in a cavity between metal atoms, and therefore represent interstitial atoms (Figure 2.29).

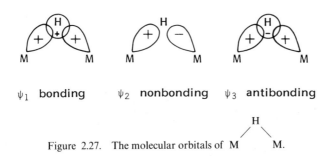

$\psi_1$ bonding  $\psi_2$ nonbonding  $\psi_3$ antibonding

Figure 2.27. The molecular orbitals of M $\overset{\text{H}}{\diagup\diagdown}$ M.

Compounds of this type include $[Co_6H(CO)_{15}]^-$ ($\tau = -13.2$), $[Ni_{12}H(CO)_{21}]^{3-}$ ($\tau = 28.0$),[161] $[Ni_{12}H_2(CO)_{21}]^{2-}$ ($\tau = 34.0$), $[Rh_{13}H_2(CO)_{24}]^{3-}$ ($\tau = 36.7$), $[Rh_{13}H_3(CO)_{24}]^{2-}$ ($\tau = 39.3$),[158] and $[Ru_6H(CO)_{18}]^-$ ($\tau = -6.4$).[159, 160] The hydrogen atoms may be located in the following positions: (1) in ruthenium and cobalt complexes in the octahedral interstices, (2) in rhodium compounds on the surface of the compound in the middle of an uncompleted octahedron and thus in the base of a square pyramid which is formed by $AB_2C_2$ atom (Figure 2.30), and (3) in nickel complexes inside, on the face of an octahedral hole closer to the central layer of nickel atoms[126, 161]; these hydrogen atoms may therefore be considered as $\mu_3$-H species. In the first case, the $^1$H NMR signals occur at very low fields; in the second and third cases, the $^1$H NMR absorption range is typical for the transition metal hydrides. Such studies are important for understanding the hydrogen migration in metals.

### b. Methods of Identification

Metal carbonyl hydrides are generally less stable than the carbonyls themselves. A majority of the transition metal hydrides are thermodynamically unstable because of the oxidation reaction; they are also kinetically unstable in this reaction. Many mononuclear metal carbonyl hydrides undergo decomposition at room temperature; only rarely do they exist above 450 K. The stability of metal carbonyl hydrides usually becomes greater when atomic number increases in periods as well as in groups of the Periodic Chart. $[H_2Os(CO)_4]$ is stable in contrast to $[H_2Ru(CO)_4]$ and $[H_2Fe(CO)_4]$. Group 4 metal carbonyl hydrides are not known, and group 5 metal carbonyl hydrides are unstable. However, many carbonyl hydrides of Fe, Co, and Ni have been prepared and characterized. Often the stability of metal carbonyl hydrides

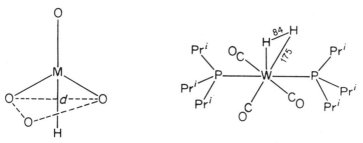

Figure 2.28. The general structure of pentacoordinated hydrido complexes and structure of $W(CO)_3(H_2)(PPr_3^i)_2$.

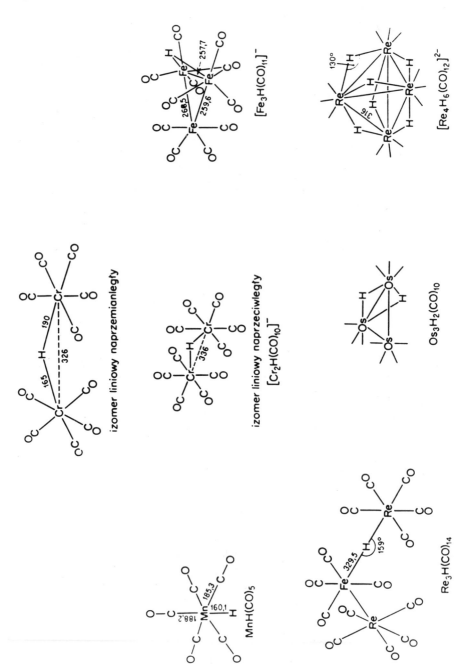

$[Fe_3H(CO)_{11}]^-$

$[Re_4H_6(CO)_{12}]^{2-}$

izomer liniowy naprzemianległy

izomer liniowy naprzeciwległy
$[Cr_2H(CO)_{10}]^-$

$Os_3H_2(CO)_{10}$

$MnH(CO)_5$

$Re_3H(CO)_{14}$

Figure 2.29.   Structures of some hydrido carbonyls. The distances are in pm.

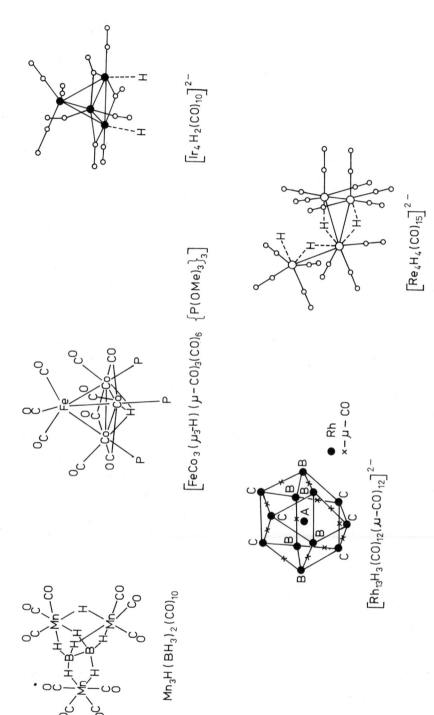

Figure 2.30.   Structures of hydrido complexes.

*Table 2.28. Dissociation Constants for Hydridocarbonyls*

| Compound | $K$ | Solvent | Method |
|---|---|---|---|
| $[HCo(CO)_4]$ | $<2$ | $H_2O$ | Titration of the solution by base |
| $[HCo(CO)(PF_3)_3]$ | $\sim 1$ | $H_2O$ | Titration of the solution by base |
| $[HCo(CO)_4]$ | Close to HCl | MeOH | |
| $[HCo(CO)_3PPh_3]$ | $1.09 \times 10^{-7}$ | $H_2O$ | Titration of the anion by acid |
| $[HCo(CO)_3P(OPh)_3]$ | $1.13 \times 10^{-5}$ | $H_2O$ | Titration of the anion by acid |
| $[HCo(CO)_2L_2]$ | Very weak acid | $H_2O$ | |
| $\quad$ L = PPh$_3$ 1ub P(OPh)$_3$ | | | |
| $[H_2Fe(CO)_4]$ | $K_1\ 3.6 \times 10^{-5}$ | | |
| | $K_2\ 1 \times 10^{-14}$ | $H_2O$ | Titration of the anion by acid |
| $[HMn(CO)_5]$ | $0.8 \times 10^{-7}$ | $H_2O$ | Titration of the anion by acid |
| $[HV(CO)_6]$ | Strong acid | $H_2O$ | |
| $[HV(CO)_5PPh_3]$ | $1.5 \times 10^{-7}$ | $H_2O$ | |

increases if some carbonyl groups are replaced by phosphines. In order to identify the presence of hydrogen coordinated to the metal the following methods are employed: pyrolysis, reactions with aqueous solutions of acids leading to hydrogen evolution, reactions with halogens (especially bromination and iodination) to give hydrogen halides or hydrogen, reactions with $CCl_4$ in which chloroform and chloroolefins are formed, and reduction of $HgCl_2$ to give mercury. However, none of the above methods gives reliable quantitative or even qualitative results.

The best methods for investigation of hydride complexes are IR, NMR, and mass spectroscopy. The $v(M-H)$ bands for terminal hydrogen atoms occur in the range 1700–2300 cm$^{-1}$. The bridging H atoms connected to two or more metal atoms have their frequencies at 800–1550 cm$^{-1}$. However, the identification of metal hydrides based on IR spectra is not reliable because the intensity of the $v(M-H)$ bands changes considerably and other bands such as $v(CO)$, $v(CN)$, $v(C\equiv C)$, and $v(N\equiv N)$ are present in the $M-H$ range. In some metal hydrides, the $v(M-H)$ band was not even observed.

Distinct evidence of existence of the $M-H$ bond is provided by the $v(M-H)$ shift in deuterated complexes to such frequencies that the ratio $v(M-H)/v(M-D)$ is equal to 1.3–1.4. The bending vibrations $\delta(M-H)$ occur in the 700–950 cm$^{-1}$ range, and because their intensity is relatively low, they constitute a source of supplemental information. The IR spectra do not reveal how many hydrogen atoms are bonded to the metal. This information cannot be obtained from the number of $v(M-H)$ bands either, because the higher number of $v(M-H)$ bands may be caused by the presence of different isomers as well as by various crystal effects.

The best method for the transition metal hydride identification is $^1H$ NMR. The signals of protons occur at high fields ($\tau > 11$), most commonly in the 15–30$\tau$ range, although for some complexes the proton resonances may reach $\tau > 50$; for example, in $[IrHCl_2(PBu_2^tMe)_2]$, the $^1H$ NMR signal of the hydrido ligand occurs at 60.5$\tau$ (see Table 2.29). Therefore, the range of $^1H$ NMR signal for the transition metal hydrides is unique because no other hydrogen compounds absorb in this range, including the main group metal hydrides. Almost all organic compounds give $^1H$ NMR signals at lower fields ($\tau < 10$) than tetramethylsilane. The diamagnetic and paramagnetic factors are the most important in shielding the hydrido ligand. It is believed that the paramagnetic factor is crucial in the overall shielding for compounds with partly filled

Table 2.29. *Properties of Selected Hydridocarbonyls*

| Compound | Preparation | Color | Melting point (K) | $^1$H NMR τ (ppm) multiplicity | J (Hz) | M–H (M–D) |
|---|---|---|---|---|---|---|
| [NEt$_4$]$_2$[VH(CO)$_5$][73] | [V(CO)$_5$]$^{3-}$ + EtOH | Yellow | | 14.78 | 27.7 (VH) / 10.8 (CH) | |
| [VH(CO)$_3$(C$_6$H$_3$Me$_3$)][a] | | | 15.8 | | | |
| [VH(CO)$_6$][b] | [V(CO)$_6$]$^-$ + H$^+$ | | | | | |
| [VH(CO)$_5$(PPh$_3$)][b] | [V(CO)$_6$]$^-$ + PPh$_3$ + H$^+$ | | | | | |
| [NbH(CO)$_5$PPh$_3$][c] | [Nb(CO)$_6$]$^-$ + PPh$_3$ + H$^+$ | | | | | |
| [NEt$_4$]$_2$[Cr$_2$(H)$_2$(CO)$_8$][79] | [Cr(CO)$_4$]$^{4-}$ + NH$_4$Cl | | | 24.6 (1) | | |
| [Cr$_2$H(CO)$_{10}$]$^{-,d,e}$ | [Cr(CO)$_6$]$^-$ + NaBH$_4$ | Yellow | | 29.47 (1) | | |
| [CrMoH(CO)$_{10}$]$^{-,e}$ | [Cr(CO)$_6$]$^-$ + Mo(CO)$_6$ + NaBH$_4$ | | | 25.31 (1) | | |
| [CrWH(CO)$_{10}$]$^{-,e}$ | [Cr(CO)$_6$]$^-$ + W(CO)$_6$ + NaBH$_4$ | | | 25.43 | 41.9 (WH) | |
| [Mo$_2$H(CO)$_{10}$]$^{-,e}$ | [Mo(CO)$_6$]$^-$ + NaBH$_4$ | Yellow | | 22.15 (1) | | |
| [W$_2$H(CO)$_{10}$]$^{-,e}$ | [W(CO)$_6$]$^-$ + NaBH$_4$ | | | 22.52 | 42.1 (WH) | |
| K$_2$[W(H)$_2$(CO)$_4$][79] | | Orange-yellow | | 22.37 | 33.6 (WH) | |
| [MoWH(CO)$_{10}$]$^-$ | [Mo(CO)$_6$] + W(CO)$_6$ + NaBH$_4$ | Yellow | | 13.4 (3) | 42.3 (WH) | |
| [W(CO)$_3$(PPr$^i_3$)$_2$($\eta^2$-H$_2$)][166] | [W(CO)$_3$(PPr$^i_3$)$_2$] + H$_2$ | Violet | | 14.21 (1) | | |
| [NBu$_4$]$_2$[{WH(CO)$_3$}$_4$][80] | | | | 15.5 (3) | 44.5 (WH) | |
| [MnH(CO)$_5$][j] | [Mn(CO)$_5$]$^-$ + H$^+$ | White | 248.4 | 17.5 | | 1780 |
| cis-[MnH(CO)$_4$PPh$_3$][g] | [MnH(CO)$_5$] + PPh$_3$ | | 410 | 16.94 (2) | 34 (PH) | |
| t-[MnH(CO)$_3$(PPh$_3$)$_2$][h] | [MnH(CO)$_5$] + PPh$_3$ | Yellow | 483 dec | 17.4 (3) | 29 (PH) | |
| [Mn$_3$H$_3$(CO)$_{12}$][i] | ([Mn$_2$(CO)$_{10}$] + KOH) + H$^+$ | Red | 333 | 34.0 (1) | | |
| [ReH(CO)$_5$][j] | [Re(CO)$_5$]$^-$ + H$^+$ | White | 285.5 | 15.66 | | 1832 |

[a] A. Davison and D. L. Reger, *J. Organomet. Chem.*, **23**, 491 (1970).
[b] W. Hieber, E. Winter, and E. Schubert, *Chem. Ber.*, **95**, 3070 (1962).
[c] A. Davison and J. E. Ellis, *J. Organomet. Chem.*, **23**, C1 (1970).
[d] J. Roziere, P. Teulon, and M. D. Grillone, *Inorg. Chem.* **22**, 557 (1983); L. B. Handy, P. M. Treichel, L. F. Dahl, and R. G. Hayter, *J. Am. Chem. Soc.*, **88**, 366 (1966).
[e] R. G. Hayter, *J. Am. Chem. Soc.*, **88**, 4376 (1966); D. W. Hart, R. Bau, and T. F. Koetzle, *Organometallics*, **4**, 1590 (1985).
[f] W. F. Edgell and W. N. Riesen, Jr., *J. Am. Chem. Soc.*, **88**, 5451 (1966); W. F. Edgell, J. W. Fischer, G. Asato, and W. M. Riesen, Jr., *Inorg. Chem.*, **8**, 1103 (1969).
[g] W. Hieber and H. Duchatsch, *Chem. Ber.*, **98**, 2933 (1965).
[h] R. Ugo and F. Bonati, *J. Organomet. Chem.*, **8**, 189 (1967).
[i] B. F. G. Johnson, R. D. Johnston, J. Lewis, and B. H. Robinson, *J. Organomet. Chem.*, **10**, 105 (1967).
[j] A. Davison, J. A. McCleverty, and G. Wilkinson, *J. Chem. Soc.*, 1133 (1963).

(*Table continued*)

*Table 2.29.* (*continued*)

| Compound | Preparation | Color | Melting point (K) | $^1$H NMR $\tau$ (ppm) multiplicity | $J$ (Hz) | M–H (M–D) |
|---|---|---|---|---|---|---|
| cis-[ReH(CO)$_4$PPh$_3$]$^k$ | [ReH(CO)$_5$] + PPh$_3$ | White | 373 | 14.33 (2) | 22 (PH) | 1828 |
| t-[ReH(CO)$_3$(PPh$_3$)$_2$]$^{k,l}$ | [Re$_3$H$_3$(CO)$_{12}$] + PPh$_3$ | White | 480 dec | 14.48 (3) | 18 (PH) | 1785 |
| [Re$_3$H$_2$(CO)$_{12}$]$^{-m}$ | [Re$_2$(CO)$_{10}$] + NaBH$_4$ | Yellow | | 27.2 (1) | | 1100R |
| [Re$_3$H$_3$(CO)$_{12}$]$^n$ | [Re$_2$(CO)$_{10}$] + NaBH$_4$ | White | 333 subl. | 27.1 (1) | | 1100R |
| [Re$_4$H$_4$(CO)$_{12}$]$^p$ | [Re$_3$H$_3$(CO)$_{12}$] + $\Delta$ | Red | | 15.08 (1) | | |
| [Re$_4$H$_4$(CO)$_{15}$]$^{2-o}$ | [Re$_2$(CO)$_{10}$] (1) KOH, MeOH} (2) H$_2$O → [Re$_4$H$_4$(CO)$_{15}$]$^{2-}$ | Yellow | | 20.5 / 26.9 | | |
| cis-[FeH$_2$(CO)$_4$]$^{q(152)}$ | [Fe(CO)$_4$]$^{2-}$ + 2H$^+$ | Yellow | 203 | 21.1 | | 1887 |
| [(OC)$_4$Fe($\mu$-H)W(CO)$_5$]$^{-q}$ | | | | 21.8 | 15.0 (WH) | |
| [FeH(CO)$_5$]$^{+r}$ | [Fe(CO)$_5$] + H$^+$ | Green | | −6.41 (1) | | 1900 |
| [Ru$_6$H(CO)$_{18}$]$^{-s}$ | [Ru$_3$(CO)$_{12}$] + Mn(CO)$_5^-$ + H$^+$ | Dark-red | | | | |
| c-[RuH$_2$(CO)$_4$]$^t$ | [Ru(CO)$_4$]$^{2-}$ + 2H$^+$ | White | 210 | 17.62 (1) | | 1980 |
| RuH(CO)$_5$]$^{+u}$ | [Ru(CO)$_5$] + H$^+$ | | | 18.0 (1) | | |
| c-[RuH$_2$(CO)(PPh$_3$)$_3$]$^v$ | [RuCl$_2$L$_3$] + NaBH$_4$ + EtOH | White | 420 | 16.69 / 18.67 | 30; 16 (PH) / 29; 74; 6 (PH) | 1960 / 1900 |
| $\alpha$-[Ru$_4$H$_4$(CO)$_{12}$]$^w$ | [Ru$_3$(CO)$_{12}$] + H$_2$ | Yellow | | 28.0 | | 1248 (902) |
| [OsH$_2$(CO)$_4$]$^x$ | OsO$_4$ + H$_2$ + CO | White | | 18.73 (1) | | 1845 (1427) |
| [OsH(CO)$_5$]$^{+y}$ | [Os$_3$(CO)$_{12}$] + H$_2$SO$_4$ + NH$_4$PF$_6$ H$_2$ + CO + OsO$_4$ | White | | 18.2 | | |
| [Os$_3$H$_2$(CO)$_{12}$]$^z$ | | White | 368 | 19.85 (1) | | |
| [FeRuOs$_2$H$_2$(CO)$_{13}$]$^{aa}$ | [RuOs$_2$(CO)$_{12}$] + Na$_2$[Fe(CO)$_4$] + H$^+$ | Orange | 385 | 29.7 | | |
| [FeRu$_3$H$_2$(CO)$_{13}$]$^{bb}$ | [Fe(CO)$_5$] + [Ru$_3$(CO)$_{12}$] | Orange | 247 | 28.7 (1) | | |
| [CoH(CO)$_4$]$^{cc}$ | [Co$_2$(CO)$_8$] + H$_2$; Co(CO)$_4^-$ + H$^+$ | Yellow | 293 dec | 20.7 | | 1934 |
| [CoH(CO)$_3$(PPh$_3$]$^g$ | [Na/Hg + (Co$_2$(CO)$_6$L$_2$]) + H$^+$ | Yellow | | 20.7 | 51 (PH) | 1960 |
| [CoH(CO)(PPh$_3$)$_3$]$^{dd}$ | [CoH(N$_2$)(PPh$_3$)$_3$] + CO | Orange | | 22.0 (4) | 48 (PH) | 1960 |
| [FeCo$_3$H(CO)$_9${P(OMe)$_3$}$_3$]$^{ee}$ | [FeCo$_3$H(CO)$_{12}$] + P(OMe)$_3$ | | | | | |
| [RhH(CO)(PPh$_3$)$_3$]$^{ff}$ | [RhCl(CO)(L$_2$)] + L + NaBH$_4$ | Yellow | | 19.7 | 14 (PH) | 2000 |

| Complex | Preparation | Color | | | | $\nu(CO)$ |
|---|---|---|---|---|---|---|
| $[RhCl_2CO(bipy)]$[gg] | $[RhCl_2(CO)_2]^- + L$ | Brown | | 39.3 | 23.1; 5.5; 4.2 (Rh–H) | 2115 |
| $[Rh_{13}H_3(CO)_{24}]^{2-}$[hh] | $[Rh_{12}(CO)_{30}]^{2-} + H_2$ | | | | | |
| $[IrH(CO)(PPh_3)_3]$[ii] | $[IrH(CO)(PPh_3)_2] + PPh_3$ | Yellow | 418 | 20.9 | 24 (PH) | 2070 |
| $[IrHCl_2(PBu^t_2Me)_2]$[ll] | $IrCl_3 \cdot 3H_2O + PBu^tMe$ | | | 60.5 | 10.7 (PH) | |
| $[Ir_4H(CO)_{11}]^-$[jj] | $[Ir_4(CO)_{12}] + K_2CO_3$ | Yellow | | 5.5 (1) | | |
| $[Ir_4H_2(CO)_{11}]$[jj] | $[Ir_4H(CO)_{11}]^- + H^+$ | Yellow | | 5.53 (1) | | 2130 |
| t-PtHCO(PEt$_3$)$_2$]$^+$[kk] | t-[PtHCl$_2$]+CO | Black | | 14.76 (3) | 13.5 (PH) | 2167 |
| t-PtHCO(AsEt$_3$)$_2$]$^+$[kk] | t-[PtHCl$_2$]+CO | Black | | 15.65 (1) | | 2106 |

k N. Flitcroft, J. M. Leach, and F. J. Hopton, *J. Inorg. Nucl. Chem.*, **32**, 137 (1970).

l M. Freni, D. Giusto, and V. Valenti, *J. Inorg. Nucl. Chem.*, **27**, 755 (1965).

m M. R. Churchill, P. H. Bird, H. D. Kaesz, R. Bau, and B. Fontal, *J. Am. Chem. Soc.*, **90**, 7135 (1968).

n J. M. Smith, W. Fellmann, and L. H. Jones, *Inorg. Chem.*, **4**, 1361 (1965).

o V. G. Albano, G. Ciani, M. Freni, and P. Romiti, *J. Organomet. Chem.*, **96**, 259 (1975).

p R. Saillant, G. Barcelo, and H. D. Kaesz, *J. Am. Chem. Soc.*, **92**, 5739 (1970).

q K. Farmery and M. Kilner, *J. Chem. Soc. A*, 634 (1970); L. Arndt, T. Delord, and M. Y. Darensbourg, *J. Am. Chem. Soc.*, **105**, 456 (1984).

r Z. Iqbal and T. C. Waddington, *J. Chem. Soc. A*, 2958 (1968).

s C. R. Eady, B. F. G. Johnson, J. Lewis, M. C. Malatesta, P. Machin, and M. McPartlin, *J. Chem. Soc. Chem. Commun*, 945 (1976).

t I. D. Cotton, M. I. Bruce, and F. G. A. Stone, *J. Chem. Soc. A*, 2162 (1968).

u L. Vaska and D. L. Catone, *J. Am. Chem. Soc.*, **88**, 5324 (1966).

v P. S. Hallman, B. R. McGarvey, and G. Wilkinson, *J. Chem. Soc. A*, 3143 (1968).

w H. D. Kaesz, A. R. Knox, J. W. Koepke, and R. B. Saillant, *J. Chem. Soc. Chem. Commun.*, 477 (1971).

x F. L'Eplattenier and F. Calderazzo, *Inorg. Chem.*, **6**, 2092 (1967).

y A. I. Deeming, D. F. G. Johnson, and J. Lewis, *J. Chem. Soc. A*, 2967 (1970).

z J. R. Moss and W. A. G. Graham, *J. Organomet. Chem.*, **23**, C47 (1970).

aa G. L. Geoffroy and W. L. Gladfelter, *J. Am. Chem. Soc.*, **99**, 304 (1977).

bb D. B. W. Yawney and F. G. A. Stone, *J. Chem. Soc. A*, 502 (1969).

cc W. F. Edgell and R. Summitt, *J. Am. Chem. Soc.*, **83**, 1772 (1961).

dd A. Misono, Y. Uchida, M. Hidai, and T. Kuse, *Chem. Commun.*, 981 (1968); A. Misono, Y. Uchida, M. Hidai, and T. Kuse, *Chem. Commun.*, 208 (1969); S. Otsuka and M. Rossi, *J. Chem. Soc. A*, 497 (1969).

ee B. T. Huie, B. Knobler, and H. D. Kaesz, *J. Chem. Soc. Chem. Commun.*, 684 (1975).

ff D. Evans, G. Yagupsky, and G. Wilkinson, *J. Chem. Soc. A*, 2660 (1968).

gg J. V. Kingston and G. R. Scollary, *Chem. Commun.*, 670 (1970).

hh S. Martinengo, B. T. Heaton, R. J. Goodfellow, and P. Chini, *J. Chem. Soc., Chem. Commun.*, 39 (1977).

ii J. F. Harrod, D. F. R. Gilson, and R. Charles, *Can. J. Chem.*, **47**, 1431 (1969).

jj L. Malatesta, G. Caglio, and M. Angoletta, *Chem. Commun.*, 532 (1970).

kk M. J. Church and M. J. Mays, *J. Chem. Soc. A*, 3074 (1968); M. J. Church and M. J. Mays, *J. Chem. Soc. A*, 1938 (1970); C. Master, B. L. Shaw, and R. E. Stainbank, *J. Chem. Soc. Dalton Trans*, 2069 (1976).

ll H. D. Empsall, E. M. Hyde, E. Mentzer, B. L. Shaw, and M. F. Utley, *J. Chem. Soc., Dalton Trans.*, 664 (1972).

*d* orbitals when the energy difference, $\Delta E$, between the ground state and the excited state is small. It is less important for carbonyl and cyclopentadienylcarbonyl hydrides, which have smaller chemical shifts than any other hydrido complexes.

The M–H distances are approximately equal to the sum of covalent radii as shown by neutron and X-ray studies. In order to find out the precise structure of metal hydrides, both neutron and X-ray studies are needed. The distances obtained from $^1$H NMR studies in the solid state based on the second moment calculated from the line shape are too small. This results from the inaccuracy of the Van Vleck equation. The Mn–H distance in $[MnH(CO)_5]$ calculated from the second moment of the Van Vleck equation is 128 pm. The longer distance (144 pm) was calculated from NMR data based on the modified Van Vleck equation.[164] A similar distance was calculated from electron diffraction studies in the gas phase. All these distances are lower than the sum of covalent radii which equals 157 pm. The Mn–H distance in $[MnH(CO)_5]$ obtained from neutron diffraction studies equals 160.1 pm. Similar distances were found for other hydride complexes.

Hydrogen is a normal ligand occupying one coordination site. Neighboring ligands are distorted toward hydrogen. Mass spectra establish the composition of hydrido complexes because, in most cases, parent ions are observed. Also, it is possible to distinguish between complexes possessing terminal or bridging hydrido ligands, because the former lose hydrogen from the parent ion first, while the latter usually split off other ligands first.[165] Therefore, it is easier to find the bonding mode of hydrogen from mass spectra than from IR spectra, where there are many bands in the $v(M-H)$ range, and from $^1$H NMR spectra because the chemical shifts of terminal and bridging hydrogens are similar.

### c. Preparation of Hydrido Complexes

There are several methods for the preparation of metal carbonyl hydrides:

1.  Reactions with hydrogen.
(a)  Direct hydrogenation

$$[Os_3(CO)_{12}] + H_2 \xrightarrow[6\,h]{8\,MPa,\,373\,K} [H_2Os(CO)_4] \tag{2.122}$$

$$OsO_4 + H_2 + CO \xrightarrow[\text{heptane, 6 h}]{18\,MPa,\,433\,K} [H_2Os(CO)_4] \tag{2.123}$$

$$[Co_2(CO)_8] + H_2 \xrightarrow{373\,K,\,25\,MPa} 2[HCo(CO)_4] \tag{2.124}$$

In this way, from $[Re_2(CO)_{10}]$, $[Ru_3(CO)_{12}]$, and $[Os_3(CO)_{12}]$, it is possible to prepare the following complexes under atmospheric pressure: $[H_3Re_3(CO)_{12}]$, $[H_4Re_4(CO)_{12}]$, $[H_4Ru_4(CO)_{12}]$, $[H_2Os_3(CO)_{10}]$, and $[H_4Os_4(CO)_{12}]$.

(b)  Ligand displacement by $H_2$

$$[Os(CO)_4PPh_3] + H_2 \xrightarrow[\text{THF}]{12\,MPa,\,403\,K} [H_2Os(CO)_3PPh_3] \tag{2.125}$$

$$[Os(CO)_3(PPh_3)_2] + H_2 \xrightarrow[\text{THF}]{12\,MPa,\,403\,K} [H_2Os(CO)_2(PPh_3)_2] \tag{2.126}$$

(c)   Oxidative addition of $H_2$

$$[IrCl(CO)(PPh_3)_2] + H_2 \xrightarrow{\text{0.1 MPa}} [H_2IrCl(CO)(PPh_3)_2] \qquad (2.127)$$

2.   Reactions of complexes with reducing agents and metals.

(a)   In these syntheses, the most common are $NaBH_4$ and $LiAlH_4$. $NaBH_4$ is utilized in ethers and alcohols, while $LiAlH_4$ is used in ethers. Sometimes $Na[BH(CH_3)_3]$ may be used, since this reagent is soluble in aromatic hydrocarbons and the resulting trialkylboron does not react with the transition metal hydride. Also employed are other compounds of the type $Na[BH_{4-x}R_x]$ where R = alkyl.

$$NaBH_4 + [M(CO)_6] \xrightarrow[\text{diglyme}]{\text{330–420 K}} Na[HM_2(CO)_{10}] \qquad (2.128)$$

$$(M = Mo, W, Cr)$$

In a similar manner, $NaBH_4$ reacts with almost all transition metal carbonyls to give anionic, polynuclear hydrides. Analogous reactions involving carbonyl complexes possessing other ligands are represented by the following:

$$[Mocp(CO)_4]^+ \xrightarrow{\text{NaBH}_4} [HMocp(CO)_3] + CO \qquad (2.129)$$

(b)   The alkali metals, especially sodium, reduce metal carbonyls to anionic mono- and polynuclear complexes in ethers or liquid ammonia. The hydrides are obtained by acidification of anionic complexes:

$$[Tc_2(CO)_{10}] \xrightarrow[\text{2. H}^+]{\text{1. Na/Hg, THF}} [HTc(CO)_5] + [H_3Tc_3(CO)_{12}] \qquad (2.130)$$
$$\text{(traces)}$$

$$[Re_2(CO)_{10}] \xrightarrow[\text{2. H}^+]{\text{1. Na/Hg, THF}} [HRe_3(CO)_{14}] + [H_3Re_3(CO)_{12}] \qquad (2.131)$$

(c)   The alkyl derivatives of lithium, magnesium, and aluminum react with complexes in the presence of carbon monoxide to furnish alkyl carbonyl complexes, which subsequently form hydrides by elimination.

(d)   Some metal carbonyl halides are reduced by hydrazine to afford hydrido complexes:

$$[MCl(CO)(PPh_3)_2] + PPh_3 \xrightarrow{\text{N}_2\text{H}_4} [MH(CO)(PPh_3)_3] \qquad (2.132)$$

$$M = Rh, Ir$$

Various other reducing agents may also be employed, such as $H_3PO_2$, $NaH_2PO_2$, $Na_2S_2O_4$, formic acid, etc.

3.   Reactions involving a hydrogen atom transfer from a solvent or a coordinated group.

(a)   In the presence of a ligand or a ligand in the basic medium, (KOH, NaOH) alcohols react with complexes to give hydrides. Possible mechanisms are as follows:

$$
\begin{array}{c}
\text{H} \\
| \\
CH_3\!-\!\overset{\displaystyle}{\underset{\displaystyle |}{C}}\!-\!O\!-\!ML_x \\
| \\
\text{H}
\end{array}
\longrightarrow
\begin{array}{c}
\text{H} \\
| \\
CH_3\!-\!\overset{\displaystyle}{\underset{\displaystyle |}{C}}\!=\!O\!-\!ML_y \\
| \\
\text{H}
\end{array}
\longrightarrow
\begin{array}{c}
CH_4 + C\!\equiv\!O\!-\!ML_y \\
| \\
\text{H}
\end{array}
\longrightarrow
\begin{array}{c}
O\!\equiv\!C\!-\!ML_y \\
| \\
\text{H}
\end{array}
\qquad (2.133)
$$

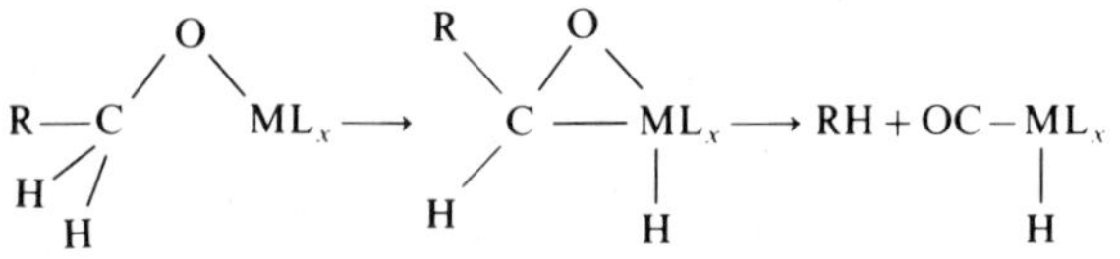

$$\underset{\underset{H}{\overset{H}{\diagdown}}}{\overset{\overset{O}{\diagup}}{R-C}}ML_x \longrightarrow \underset{\underset{H}{\diagup}}{\overset{R}{\diagdown}}\underset{\underset{H}{\diagup}}{\overset{O}{\diagup}}{C-ML_x} \longrightarrow RH + \underset{H}{\overset{}{OC-ML_x}} \qquad (2.134)$$

$$[RhX_3L_3] \xrightarrow{RO^-} [Rh(OR)X_2L_3] \longrightarrow [RhHX_x(CO)L_2] + R'H \qquad (2.135)$$

(b)   Some carbonyl clusters are able to oxidize relatively unreactive solvents, such as alkanes:

$$4[Ru_3(CO)_{12}] + 6C_nH_{2n+2} \longrightarrow 3[H_4Ru_4(CO)_{12}] + 6C_nH_{2n} \qquad (2.136)$$

(c)   Often coordinated aromatic phosphines transfer the *ortho* H atom to afford hydrido complexes:

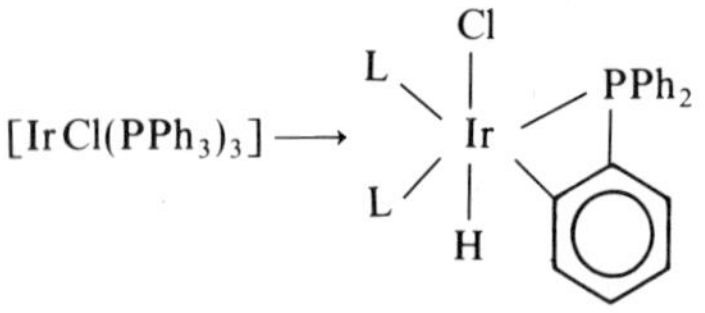

$$[IrCl(PPh_3)_3] \longrightarrow \qquad (2.137)$$

Similar reactions are also known for other coordinated molecules. The majority of isomerization reactions involving unsaturated hydrocarbons are catalyzed by *d*-electron metal complexes. In the transition state, hydrido complexes are usually formed:

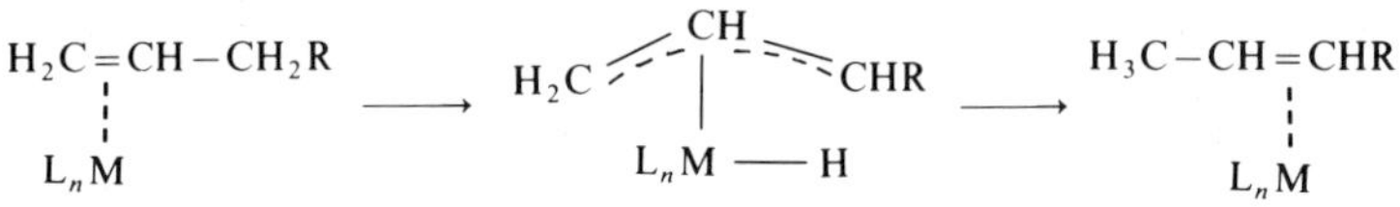

$$\underset{L_nM}{\overset{}{H_2C=CH-CH_2R}} \longrightarrow \underset{L_nM-H}{\overset{CH}{H_2C \diagup \quad \diagdown CHR}} \longrightarrow \underset{L_nM}{\overset{}{H_3C-CH=CHR}} \qquad (2.138)$$

4.   Metal carbonyl disproportionation reactions. The following example illustrates this method:

$$5[Fe(CO)_5 + 6py \longrightarrow [Fe(py)_6][Fe_4(CO)_{13}] \xrightarrow{H^+} [Fe_4(CO)_{13}]^- + [H_2Fe_4(CO)_{13}] \qquad (2.139)$$

Many other examples were discussed earlier.

5.   Oxidative addition reactions involving HX:

$$[Fe_3(CO)_{12}] + RSH \longrightarrow [Fe_3H(CO)_9SR] \qquad (2.140)$$

$$[IrCl(CO)(PPh_3)_2] + HBr \longrightarrow [IrH(Cl)Br(CO)(PPh_3)_2] \qquad (2.141)$$

$$[IrCl(CO)(PPh_3)_2] + R_3MH \longrightarrow [IrH(Cl)(CO)MR_3(PPh_3)_2] \qquad (2.142)$$
$$(M = Si, Ge, Sn)$$

In nonpolar solvents that do not contain any traces of water, the addition of HX to IrCl(CO)(PPh_3)_2 leads to the formation of *cis*-complexes. However, in moist solvents, a mixture of *cis*- and *trans*-complexes is obtained.[29, 152–155, 161–163]

6.   Protonation of metal carbonyls.
Protonation of neutral transition metal complexes such as carbonyls leads to cationic hydrides:

$$[M_3(CO)_{12}] + H^+ \xrightarrow{\text{concentrated } H_2SO_4} [HM_3(CO)_{12}]^+ \qquad (2.143)$$

$$(M = Ru, Os)$$

$$[Ir_4(CO)_{12}] + 2H^+ \xrightarrow{\text{concentrated } H_2SO_4} [H_2Ir_4(CO)_{12}]^{2+} \qquad (2.144)$$

$$[Fe(CO)_5] + H^+ \xrightarrow[CF_3COOH]{BF_3H_2O} [HFe(CO)_5]^+ \qquad (2.145)$$

## d. Acidic Properties of Metal Carbonyl Hydrides

Some metal carbonyl hydrides dissolved in polar solvents such as water, alcohols, nitriles, ketones, etc., are acids. The acidic character of these complexes is due not only to the polarity of the metal–hydrogen bond, but to solvation of the proton and the carbonylate anion. Acidic metal carbonyl hydrides dissolved in nonpolar solvents do not dissociate, and their chemical properties are like those of nonacidic metal hydrides. $[HCo(CO)_4]$ and $[HV(CO)_6]$ are strong acids (Table 2.28).

## e. Ligand Field Strength and trans Effect of the Hydrido Ligand

The majority of spectroscopic studies of complexes involving $\pi$-acceptor ligands indicate that hydrogen causes a strong field. However, Wilkinson's studies on $[RhHX(en)_2]$ show that H creates a relatively weak field, and that its place in the spectrochemical series is between oxygen and nitrogen ligands: $OH^- < H_2O < H^- < NCS^- < NH_3$.[191]

As a ligand, hydrogen causes a very strong *trans* effect comparable to those of CO and phosphines. This phenomenon was confirmed by kinetic studies as well as by structural findings, which showed that the metal–ligand distance for ligands with *trans* orientation to hydrogen is longer than the metal–ligand distance for some other ligands such as halogens in an analogous complex.

## f. Structures of Hydrido Complexes

$[PtH(CO)L_2]^+$ complexes have planar structures with ligands occupying vertices of the square; phosphines or arsines occupy *trans* positions. All five-coordinate complexes adopt the structure of a distorted trigonal bipyramid with the hydrogen atom lying at the axial position. The equatorial ligands are shifted toward the hydrogen atom (Figure 2.28).

The distance $d$ between the metal atom and the plane containing equatorial ligands depends on the sizes of all ligands. This distance increases when sizes of ligands become greater. For small ligands, the distance $d$ is small, as for $[MnH(CO)_5]$; however, for $[RhHCO(PPh_3)_3]$, $d = 36$ pm and for $[RhH(PPh_3)_4]$, $d = 70$ pm.

Structural deformations are also observed for hexacoordinated complexes. Compounds $[MH_2L_4]$ have structures which are intermediate between octahedral and tetrahedral arrangements with hydrogen atoms located in the two planes of the tetrahedron.

The deviation from ideal structure for a given coordination number increases when sizes of ligands become greater and the ligand bond which is *trans* to hydrogen becomes stronger. Complexes with borohydrides possess $M-H-B$ bridges and often $B-B$ bonds. The location of hydrogen atoms in hydrido complexes based on X-ray studies is generally impossible. Therefore, their position is indirectly determined on the basis of an incomplete coordination sphere and, for bridging hydrogen atoms, on the basis of the metal–metal distances and angles. Usually, carbonyl bridges cause shortening of the metal–metal distances, while hydrogen bridges cause lengthening of these distances.

## g. Properties of Metal Carbonyl Hydrides

The majority of metal carbonyl hydrides show dynamic properties, that is, stereochemical lability. The positions of atoms change with a noticeable rate (stereo-isomerization). These properties are most often investigated by NMR. The nuclear magnetic resonance method may be used only if the stereoisomerization occurs fast enough to influence the line shape and simultaneously allow the interaction between nuclear spins.

The activation energy for stereoisomerization of metal carbonyl hydrides varies considerably. Complexes possessing the coordination numbers 6 and 4 are generally inert, while compounds which have coordination numbers 5 or higher than 6 are usually stereochemically labile. Because of crucial importance in catalytic processes, the isotopic H–D exchange between hydrido complexes and hydroxyl solvents, deuterium, and olefins is of particular interest. Often, exchange reactions with solvents such as $D_2O$ or EtOD are catalyzed by acids and bases.

The exchange of coordinated hydrido ligands with deuterium readily takes place if oxidative addition is possible:

$$MH + D_2 \longrightarrow MHD_2 \longrightarrow MD + HD \tag{2.146}$$

The exchange reaction between deuterium of $[MDL_n]$ and the *ortho* hydrogen in triarylphosphines also occurs through oxidative addition. The exchange between $[MHL_n]$ and $C_2D_4$ takes place when the ligand migration reaction is possible:

$$MHL_n + C_2D_4 \rightleftarrows HM(C_2D_4)L_n \rightleftarrows M(C_2D_4H)L_n \rightleftarrows DM(C_2D_3H)L_n \rightleftarrows MDL_n + C_2D_3H \tag{2.147}$$

Therefore, the hydrido complexes are catalysts or are formed as intermediates in many reactions catalyzed by transition metal complexes, as in the isomerization of unsaturated hydrocarbons:

$$ML_n + RCH_2CH = CH_2 \rightleftarrows L_nM - \begin{matrix} CH_2 \\ \| \\ CH \\ | \\ CH_2R \end{matrix} \rightleftarrows L_nM - \begin{matrix} H \quad CH_2 \\ | \quad \diagdown \\ {}^iCH \\ \diagup\!\!\diagup \\ CHR \end{matrix} \rightleftarrows L_nM - \begin{matrix} CH_3 \\ | \\ CH \\ \| \\ CHR \end{matrix} \tag{2.148}$$

Lewis bases may replace hydrido ligands from many complexes. Such reactions occur according to a mechanism of reductive elimination:

$$[Ru^{II}H(OAc)(PPh_3)_3] + 3CO \longrightarrow [Ru^o(CO)_3(PPH_3)_2] + HOAc + PPh_3 \tag{2.149}$$

$$[FeH_2(CO)_4] + PPh_3 \longrightarrow [Fe(CO)_4PPh_3] + H_2 \tag{2.150}$$

$$[FeH_2(CO)_4] + 2PPh_3 \longrightarrow [Fe(CO)_3(PPh_3)_2] + H_2 + CO \tag{2.151}$$

With acids, metal carbonyl hydrides react with hydrogen evolution. However, at times, some M−H bonds remain unbroken, especially in compounds possessing coordinated phosphines or in polynuclear hydrido carbonyls. Thus, for some cases, it is not possible to quantitatively determine the hydrido ligands by this method.

$$[FeH_2(CO)_4] + 2HA \longrightarrow [FeA_2(CO)_4] + 2H_2 \qquad (2.152)$$

Sometimes in the presence of an acid, instead of M−H bond breaking there is formation of a new M−H bond:

$$[IrH(CO)(PPh_3)_3] + HClO_4 \longrightarrow [IrH_2(CO)(PPh_3)_3]^+ ClO_4^- \qquad (2.153)$$

Hydrides react with the alkali metals and bases to give anionic complexes. Metal carbonyl hydrides possessing strong acidic properties react in this manner most readily:

$$[VH(CO)_3(mesitylene)] + Na \longrightarrow Na[V(CO)_3(mesitylene)] + \tfrac{1}{2}H_2 \qquad (2.154)$$

$$[Re_3H_3(CO)_{12}] + KOH \longrightarrow K[Re_3H_2(CO)_{12}] + H_2O \qquad (2.155)$$

Carbonyl hydrides react with some Lewis bases to afford addition or substitution products. Addition reactions occur for 16$e$ hydrido compounds:

$$[IrH(CO)(PPh_3)_2] + L \longrightarrow [IrH(CO)(PPh_3)_2L] \qquad (2.156)$$

$$[ReH(CO)_5] + L \longrightarrow [ReH(CO)_4L] + CO \qquad (2.157)$$

$$[CoH(CO)_4] + L \longrightarrow [CoH(CO)_3L] + CO \qquad (2.158)$$

$$[RuClH(CO)L_3] + L' \longrightarrow [RuClH(CO)L_2L'] + L \qquad (2.159)$$

Owing to high reactivity, hydrido complexes are utilized in the synthesis of compounds which either cannot be prepared or are prepared with difficulty in a different way. For example, tetrafluoroethylene reacts with hydrido carbonyls to give $CF_2CF_2H$ complexes:

$$[MnH(CO)_5] + CF_2=CF_2 \longrightarrow [Mn(CF_2CF_2H)(CO)_5] \qquad (2.160)$$

Other unsaturated compounds may react analogously.

$N$-methyl-nitroso-$p$-toluenesulfonamide, a nitrosation reagent, is utilized in the synthesis of nitroso complexes. It reacts with $[MnH(CO)_5]$ to furnish $[Mn(CO)_4NO]$. Disulfides and analogous selenium compounds react with many metal carbonyl hydrides to give thiolate and selenolate complexes:

$$[ReH(CO)_5] + R_2S_2 \longrightarrow [(OC)_4 Re(SR)_2 Re(CO)_4] \qquad (2.161)$$

$$[MoHcp(CO)_3] + R_2S_2 \longrightarrow [Mo(SR) cp(CO)_2]_2 \qquad (2.162)$$

$$[MnH(CO)_5] + R_2Se_2 \longrightarrow [Mn(SeR)(CO)_4]_2 \qquad (2.163)$$

Metal carbonyl hydrides may also be used for preparation of complexes possessing heteronuclear metal–metal bonds involving the main group metals and the transition metals:

$$[OsH_2(CO)_4] + SnCl_4 \longrightarrow \textit{cis-}[OsH(SnCl_3)(CO)_4] + SnCl_3H \qquad (2.164)$$

$$[RuH_2(CO)_4] + 2Sn(CH_3)_3H \longrightarrow [Ru(SnMe_3)_2(CO)_4] + 2H_2 \qquad (2.165)$$

$$2[MnH(CO)_5] + CdR_2 \longrightarrow [(CO)_5MnCdMn(CO)_5] + 2RH \qquad (2.166)$$

Reactions between hydrido complexes and unsaturated hydrocarbons generally proceed in two steps. The first step involves coordination of the hydrocarbon to form a $\pi$-complex followed by hydrido ligand migration and the formation of the product possessing $\sigma$ metal–carbon bond. Dependent upon the hydrocarbon alkyl, alkenyl, allyl, or vinyl complexes may be formed:

$$[CoH(CO)_4] + C_2H_4 \longrightarrow [CoH(C_2H_4)(CO)_3] + CO \longrightarrow [Co(C_2H_5)(CO)_4] \qquad (2.167)$$

$$[MnH(CO)_5] + CH_2{=}CH{-}CH{=}CH_2$$
$$\longrightarrow [Mn(CH_2CH{=}CHCH_3)(CO)_5] \xrightarrow{hv} [Mn(\eta^3\text{-}C_4H_7)(CO)_4] + CO \qquad (2.168)$$

$$[MnH(CO)_5] + MeOOCC{\equiv}CCOOMe \longrightarrow [(OC)_5MnC(COOMe){=}CHCOOMe] \qquad (2.169)$$

$$[ReH(CO)_5] + F_3CC{\equiv}CH \longrightarrow \textit{cis-}[(OC)_5ReCH{=}C(CF_3)H] \qquad (2.170)$$

Some metal carbonyl hydrides react with $CO_2$ to give formate complexes. Reactions involving $CS_2$ give thiocarbonyl and dithioformate complexes[169]:

$$[RuClH(CO)(PPh_3)_3] + CS_2 \longrightarrow \quad + PPh_3 \qquad (2.171)$$

## 10. METAL CARBONYLS CONTAINING BRIDGING CO GROUPS BONDED THROUGH C AND O

In all previously discussed metal carbonyls, the CO group is bonded to the transition metal through the carbon atom which is in accordance with the theory of hard and soft acids and bases. Transition metals in their low oxidation states are soft acids (they have large sizes and contain electrons possessing low energies of excitation) and therefore form more stable compounds with soft bases (donor atoms are readily polarizable, have low electronegativities, and are easily oxidized). Oxygen ligands are hard bases, and thus there is a tendency to form $M-C$ bonds in metal carbonyls.

Calculations utilizing the valence bond method show that structure **I** is more stable for the $HCO^+$ molecule and structure **II** is more stable for the $COH^+$ ion:

$$:C{\equiv}O: \longleftrightarrow :C{=}O:$$
$$\textbf{(I)} \qquad \textbf{(II)} \qquad\qquad (2.172)$$

Therefore, it may be concluded that the coordinated molecule CO will form a bond with another metal possessing hard acid character when the multiplicity of the carbon–oxygen bond becomes considerably lower as a result of an increase in the strength of the $\pi$ $M-CO$ back-bonding.[170, 171] Thus, compounds containing the $M-CO-M'$ (M' represents hard acid) bond should form bridging carbonyl groups for which exceptionally strong lowering of the $v(CO)$ frequency is observed. A first-prepared compound of this kind is $[cpFe(CO)_2]_2 \cdot 2AlEt_3$ possessing the structure (2.173).

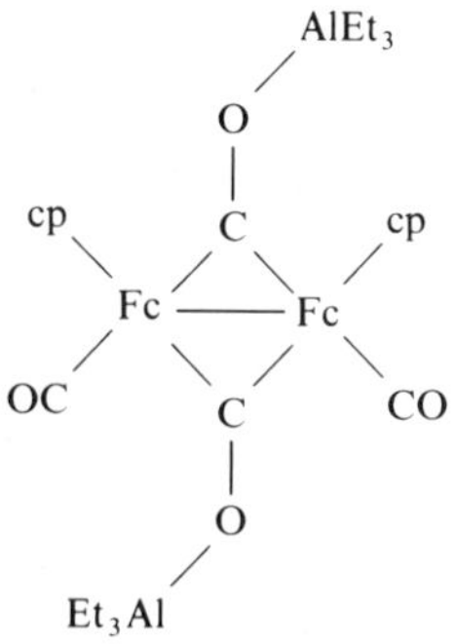

(2.173)

Thus, the bridging carbonyl groups are more basic than the terminal ones. Therefore, the presence of hard acids may induce the formation of $M-C-M$ bridges [equation (2.174)].

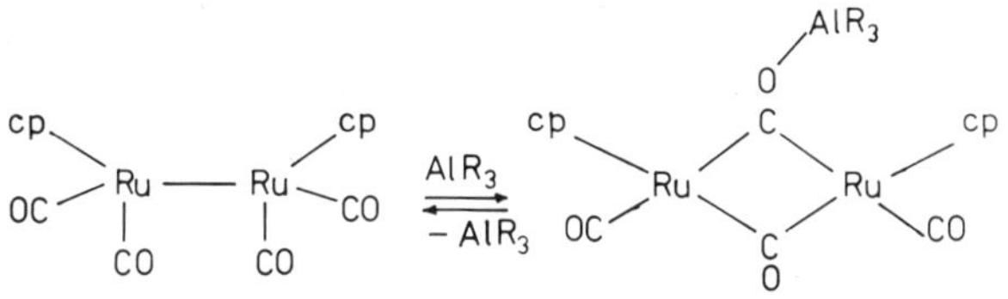

(2.174)

Compounds containing the $M-CO-M'$ grouping should even more readily be formed from carbonyls possessing $\mu_3$-CO in which the bridge is formed between three metal atoms. Stronger basicity of $\mu_3$-CO than that of $\mu$-CO is observed for $[Ni_3cp_3(\mu_3\text{-}CO)_2]$ and $[Nicp(\mu\text{-}CO)]_2$. The former compound forms 1:1 and 1:2 adducts with $AlR_3$, while the latter complex forms only 1:1 adducts.

In the presence of hexaalkyldialuminum, $[Co_2(CO)_8]$ and $[Co_4(CO)_{12}]$ undergo decomposition; however, with $AlBr_3$, adducts may be formed and probably have structures (2.175).

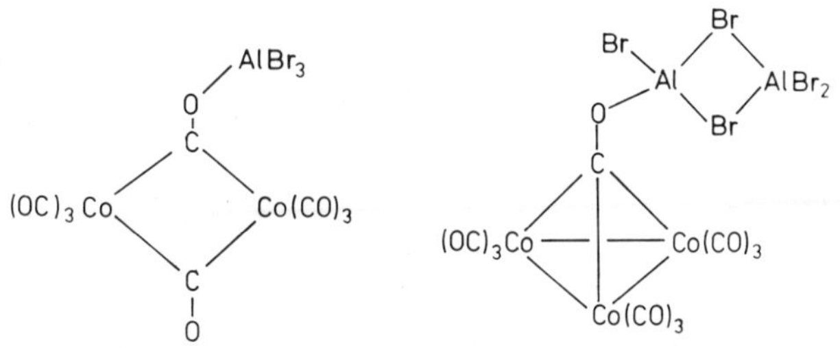

(2.175)

Similar adducts may be formed by cyclopentadienyl lanthanide compounds, such as $Lncp_3$, which are relatively strong Lewis acids.

The terminal carbonyl groups may also play the role of a base if there is considerable concentration of electron density on them. Examples are $[Co(CO)_4]^-$, $[Fe(CO)_4]^{2-}$, $[Mn(CO)_5]^-$, $[Mocp(CO)_3]^-$, and $[Fecp(CO)_2]^-$ which, with $Mg^{2+}$, $Al^{3+}$, and $Mn^{2+}$, form the $M-CO-M'$ bonds. In $[\{Mocp(CO)_3\}_2 Mg(py)_4]$ the magnesium atom is six-coordinate; it is bonded to four pyridine molecules as well as to two oxygen atoms of carbonyl groups (in *trans* orientation). A similar structure is adopted by $[V(THF)_4][V(CO)_6]_2$. Four molecules of tetrahydrofuran are coordinated to the V(II) atom as well as two carbonyl groups of two $[V(CO)_6]^-$ anions through oxygen atoms each (Table 2.22, Figure 2.15). The frequencies of vibrations $v(CO)$ for bridging carbonyl groups bonded through the oxygen and carbon atoms, $M-C-O-M'$, are very low; they occur in the $1270-1650 \text{ cm}^{-1}$ range.[82]

$[Mo(phen)(PPh_3)_2(CO)_2]$ forms the adduct with $AlEt_3$ which has the following structure:

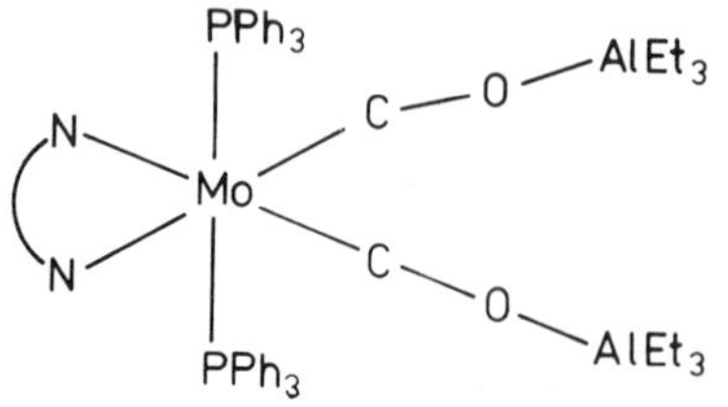

In the electron spectrum of the molybdenum complex, there is an intense metal–ligand charge transfer band (MLCT) from the $d$ orbital on the $\pi^*$ orbital of phenanthroline. After the formation of the adduct with triethylaluminum, the MLCT band becomes shifted to higher energies. This constitutes evidence for stronger back-bonding in the adduct than in the $[Mo(CO)_2(phen)(PPh_3)_2]$ complex.

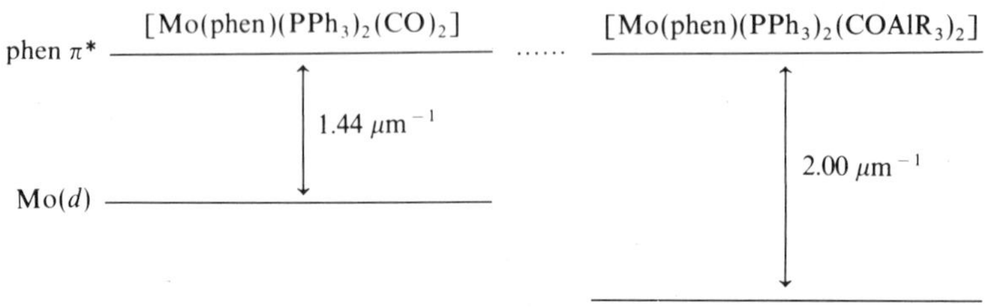

## 11. THIOCARBONYL AND SELENOCARBONYL COMPLEXES

### a. Structure, Bonding, and Spectroscopy

In contrast to the CO molecule, the CS and CSe molecules are not stable. Carbon monosulfide polymerizes, sometimes explosively, above 113 K. However, both CS and CSe form complexes in which they show themselves as stable ligands (Table 2.30).

The thiocarbonyl group may be bonded to the metal through the carbon atom as a terminal or bridging 2e ligand (a, b, c),[169, 174–176, 186, 11] bridging 4e ligand (d),[176, 11] or bridging 6e ligand (e).[186]

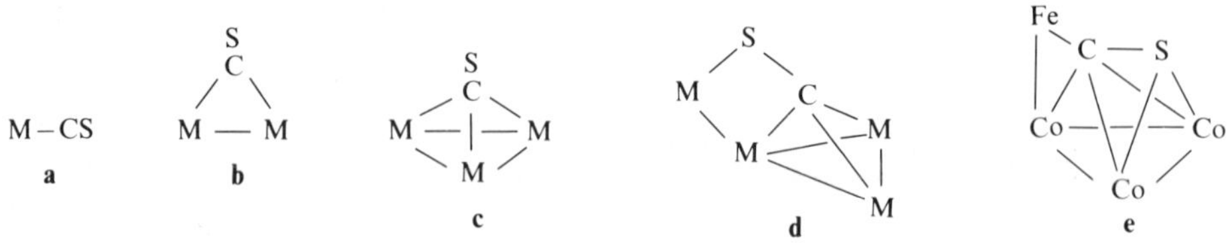

It is assumed that in $(diphos)_2(CO) WCSW(CO)_5$ there is a thiocarbonyl group bonded to tungsten atoms through the carbon and sulfur atoms: $W-C-S-W$.

*Table 2.30. Thiocarbonyl and Selenocarbonyl Compounds*

| Compound | Color | Melting point (K) | $\nu$(CS) (cm$^{-1}$) | (Solvent) |
|---|---|---|---|---|
| [Cr(CO)$_5$(CS)]*[a,b] | Yellow | | 1253 | (Hexane) |
| [Cr(CO)$_3$(CS)(diphos)][b] | Yellow-orange | | 1209 | (CS$_2$) |
| [Mo(CO)$_5$(CS)][a,b] | | | 1247 | (Hexane) |
| [W(CO)$_5$(CS)]**[a,b] | Yellow | | 1258 | (Hexane) |
| mer-[W(CO)$_3$(CS)(diphos)][b] | Yellow | | 1215 | (CS$_2$) |
| [W(CO)(CS)(diphos)$_2$][b] | Light-yellow | | 1161 | (CH$_2$Cl$_2$) |
| [Mncp(CO)$_2$(CS)][d] | Yellow | 325–6 | 1266 | (CS$_2$) |
| [Mncp(CO)(CS)PPh$_3$][c] | Orange | 453 dec | 1231 | (CS$_2$) |
| [Mncp(CO)(CS)$_2$][d] | | 344–5 | 1305, 1238 | (CS$_2$) |
| [Mncp(CS)$_3$][d] | | | 1338, 1240 | (CS$_2$) |
| [Recp(CO)$_2$(CS)][177] | | | 1277 | (CS$_2$) |
| [Fe$_4$($\mu_3$-S)($\mu_4$-CS)(CO)$_{12}$][176] | Black | 393 dec | 963 | |
| [Fecp(CO)(CS)(PPh$_3$)]PF$_6$[e] | Yellow | | 1320 | (Nujol) |
| c-[Fecp(CO)($\mu$-CS)]$_2$[f] | Black | | 1124 | (CS$_2$) |
| [RuCl$_2$(CO)(CS)(PPh$_3$)$_2$][g] | White | | 1302 | (Nujol) |
| [RuCl$_2$(CS)(PPh$_3$)$_2$]$_2$[h] | Red | 438–9 | 1290 | (Nujol) |
| [OsClH(CS)(PPh$_3$)$_3$][179] | | | 1280 | (Nujol) |
| [OsClH(CO)(CS)(PPh$_3$)$_2$][179] | | | 1295 | (Nujol) |
| [Os$_4$($\mu_3$-S)($\mu_4$-CS)(CO)$_{12}$][176] | | | | |
| [OsCl$_2$(CO)(CS)(PPh$_3$)$_2$][g] | White | | 1315 | (Nujol) |
| [Co$_3$cp$_3$($\mu_3$-S)($\mu_3$-CS)][175] | Black | 562 dec | | |
| [cpFeCo$_3$(CO)$_9$($\mu_4$-CS)][186] | | | | |
| [RhCl$_3$(CS)(PPh$_3$)$_2$][i] | Red-brown | 414 dec | 1364 | (Nujol) |
| [IrCl$_3$(CS)(PPh$_3$)$_2$][j] | Light-yellow | 466 dec | 1368 | (Nujol) |
| [IrCl$_2$(CO)(CS)(PPh$_3$)$_2$]PF$_6$[i] | White | 408 dec | 1409 | (Nujol) |
| [Ni(CS)$_4$][j] | | | 1305 | |
| [PtCl(CS)(PPh$_3$)$_2$]$^+$[k] | | | 1400 | |
| [Mncp(CO)$_2$(CSe)][177] | Gold | 337–8 | 1107 | (CS$_2$) |
| [Mn($\eta$-C$_5$H$_4$Me)(CO)$_2$(CSe)][177] | | | 1106 | (CS$_2$) |
| [Recp(CO)$_2$(CSe)][177] | Light-yellow | 370–3 | 1124 | (CS$_2$) |
| [Cr($\eta$-C$_6$H$_5$CO$_2$Me)(CO)$_2$(CSe)][177] | Orange-red | 458 | 1063 | (CS$_2$) |
| [RuCl$_2$(CO)(CSe)(PPh$_3$)$_2$][l] | White | | | |

* = $\delta$($^{13}$CS) 331.1 ppm.
** = $\delta$($^{13}$CS) 298.7 ppm.
[a] B. D. Dombek and R. J. Angelici, *J. Am. Chem. Soc.*, **95**, 7516 (1973).
[b] B. D. Dombek and R. J. Angelici, *Inorg. Chem.*, **15**, 1089 (1976).
[c] N. J. Coville and I. S. Butler, *J. Organomet. Chem.*, **64**, 101 (1974).
[d] A. E. Fenster and I. S. Butler, *Inorg. Chem.*, **13**, 915 (1974).
[e] L. Bussetto and A. Palazzi, *Inorg. Chim. Acta*, **19**, 233 (1976).
[f] J. W. Dunker, J. S. Finer, J. Clardy, and R. J. Angelici, *J. Organomet. Chem.*, **114**, C49 (1974).
[g] K. R. Grundy, R. O. Harris, and W. R. Roper, *J. Organomet. Chem.*, **90**, C34 (1975).
[h] P. W. Armit, T. A. Stephenson, and E. S. Switkes, *J. Chem. Soc., Dalton Trans.*, 1134 (1974).
[i] C. J. Curtis and M. Kubota, *Inorg. Chem.*, **13**, 2277 (1974).
[j] G. V. Calder, J. G. Verkade, and L. W. Yarbrough, II, *Chem. Commun.*, 705 (1973).
[k] J. M. Lisey, E. D. Dobrzynski, R. J. Angelici, and J. Clardy, *J. Am. Chem. Soc.*, **97**, 656 (1975).
[l] G. R. Clark, K. R. Grundy, R. O. Harris, S. M. James, and W. R. Roper, *J. Organomet. Chem.*, **90**, C37 (1975).

In contrast to the carbonyl group, this is possible because ligands containing sulfur as a donor atom are soft bases and may form quite stable bonds with soft acids, in this case with the tungsten atom. Such bridges also occur in $[(diphos)_2(CO) WCSAgSCW(CO)(diphos)_2] BF_4$.

Complexes containing bridging selenocarbonyl groups are as yet unknown. Most probably, in all complexes the CSe group is bonded to the metal atom via carbon as in $[RuCl_2(CO)(CSe)(PPh_3)_2]$, for which the X-ray structure has been determined. The M—C bond strength is greater for CS and CSe groups than for the CO group since it was determined that the M—C distance for the M—C(O) bond is longer than the distance for M—C(S) and M—C(Se) bonds.[169, 174-177] Based on IR and $^1$H NMR studies, it was also postulated that the selenocarbonyl group in such complexes as $[Mcp(CO)_2(CX)]$ and $[Cr(\eta$-$C_6H_5COOMe)(CO)_2(CX)]$ where M = Mn, Re and X = S, Se is slightly more strongly bonded to the metal than the thiocarbonyl group.[177] The vibrational frequency $v(CS)$ for the free carbon monosulfide in the $CS_2$ matrix is 1274 cm$^{-1}$ at 85 K[169] and 1259 at 77 K.[177] For the coordinated CS molecule, this frequency may be larger or smaller in contrast to the CO group for which, in metal carbonyls, lowering of the vibrational frequency $v(CO)$ always occurs in comparison to the free CO molecule.

The vibrational $v(CS)$ bands in complexes are narrow and very intensive and occur in the 1150–1410 cm$^{-1}$ range for the terminal CS groups and in the 960–1150 cm$^{-1}$ range for the bridging ones. For bridging CS groups this range may actually be wider, since only few complexes with $\mu$-CS have been investigated to date. The integrated intensities of the $v(CS)$ bands are greater than those of the $v(CO)$ bands.[177] The $\delta(MCS)$ bands occur at frequencies $ca$ 60 cm$^{-1}$ lower than those of the $\delta(MCO)$ vibrations. The increase of the $v(CS)$ frequency to a value which is higher than that of the free CS results from the interaction between $v(CS)$ and $v(M$—$CS)$ frequencies, despite the $\pi$-acceptor properties of the thiocarbonyl group and despite the existence of M—C back-bonding.

The vibrational $v(CSe)$ bands occur in the 1060–1140 cm$^{-1}$ range and are very intense. The free CSe molecule has a $v(CSe)$ frequency of 1036 cm$^{-1}$, which has been obtained by extrapolation from the electronic spectrum of CSe.[177] The $^{13}$C NMR chemical shifts for metal thiocarbonyl complexes are large. The $^{13}$C NMR signals for CS groups occur in the range 310–340 ppm; $^{13}$C NMR signals for carbene complexes also occur in the same range.

The mass spectra of thiocarbonyl complexes resemble those of carbonyl complexes; the fragmentation patterns are similar.

The M—C distances for selenocarbonyl and thiocarbonyl groups are shorter by 5–12 pm than those for CO groups. The *trans* effect of CS or CSe is very strong, probably stronger than that of the CO group as illustrated by the structure of $RuCl_2(CO)(CSe)(PPh_3)_2$ (Figure 2.31). The Ru—Cl distance *trans* to the CSe group is longer than the analogous distance *trans* to the CO group.

The highest occupied molecular orbital $7\sigma$ in the CS molecule has higher energy than the corresponding $5\sigma$ orbital in the CO molecule. However, the energy of $\pi$-acceptor orbitals is lower for the thiocarbonyl group compared to the carbon monoxide molecule.[178] As a result of this phenomenon, the CS molecule represents a stronger $\sigma$-donor and a better $\pi$-acceptor than the carbonyl group. This explains higher M—CS bond strength observed in many thiocarbonyl complexes. Similar behavior is exhibited by selenocarbonyl complexes. CS and CSe are softer bases compared to CO.

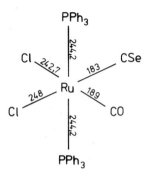

Figure 2.31. The structure of $RuCl_2(CO)(CSe)(PPh_3)_2$. The distances are in pm.

Therefore, $\pi$-acceptor versus $\sigma$-donor properties of CS and CSe change more drastically than those of CO. In thiocarbonyl complexes additional bonding contribution comes from $2\pi$ orbitals.[178]

From the considerations concerning the dependence of the stretching frequency $v(CS)$ of thiocarbonyl complexes $M(CO)_n(CS)$ on the force constant $k(CO)$ of $M(CO)_{n+1}$ compounds, it is evident that the ratio of $\pi$-acceptor properties to the sum of $(\sigma + \pi)$ donor properties for the CS group changes more considerably than for the carbonyl group.[174] The thiocarbonyl group is a stronger $\pi$-acceptor ligand compared to CO in complexes possessing high electron density on the central atom and weaker in compounds with lower electron density on the metal. The former compounds constitute anionic complexes or compounds containing ligands which exhibit quite strong $\sigma$-donor and $\pi$-acceptor properties, for example, *cis*-$[W(CO)(CS)(diphos)_2]$ and $[Mncp(CO)(CS)PCy_3]$. The latter compounds comprise cationic complexes or compounds with ligands possessing strong $\sigma$-acceptor and relatively weak $\pi$-donor properties such as $[Mncp(CO)(CS)(NO)]^+$ and $[RhCl_3(CS)(PPh_3)_2]$. Because of better acceptor properties of CS and CSe in the majority of thiocarbonyl and selenocarbonyl complexes, the force constants $k(CO)$ are greater than those in corresponding metal carbonyl compounds. The force constants $k(CO)$ for complexes $[Mncp(CO)_3]$, $[Mncp(CO)_2(CS)]$, and $[Mncp(CO)_2(CSe)]$ are 1549, 1574, and 1592 $Nm^{-1}$, respectively.[177]

## b. Preparation of Thiocarbonyl and Selenocarbonyl Complexes

Thiocarbonyl and selenocarbonyl metal complexes are generally prepared from carbon disulfide, $CS_2$; thiophosgene, $CSCl_2$; and carbon diselenide, $CSe_2$, as well as from transition metal complexes such as metal carbonyls. Commonly, triphenylphosphine is utilized to eliminate sulfur or selenium:

$$[Mcp(CO)_2L] + CS_2 + PPh_3 \longrightarrow [Mcp(CO)_2(CS)] + PPh_3S + L \qquad (2.176)$$

$$(M = Mn, Re; L = C_8H_{14}, THF)$$

$$[Mcp(CO)_2L] + CSe_2 + PPh_3 \longrightarrow [Mcp(CO)_2(CSe)] + PPh_3Se + L \qquad (2.177)$$

$$[Cr(\eta\text{-}C_6H_5CO_2Me)(CO)_2(THF)] + CSe_2 + PPh_3$$

$$\longrightarrow [Cr(\eta\text{-}C_6H_5CO_2Me)(CO)_2(CSe)] + PPh_3Se + THF \qquad (2.178)$$

Thiophosgene is used in reactions involving anionic complexes:

$$Na_2[M_2(CO)_{10}] + CSCl_2 \longrightarrow [M(CO)_5(CS)] + 2NaCl + [M(CO)_6] \cdots$$

$$(M = Cr, W) \tag{2.179}$$

With some complexes, particularly those of platinum group metals, thiophosgene may undergo oxidative addition reaction:

$$[IrCl(N_2)(PPh_3)_2] + CSCl_2 \longrightarrow [IrCl_3(CS)(PPh_3)_2] + N_2 \tag{2.180}$$

$$[RhCl(PPh_3)_3] + CSCl_2 \longrightarrow [RhCl_3(CS)(PPh_3)_2] + PPh_3 \tag{2.181}$$

## c. Reactions of Thiocarbonyl Complexes

The coordinated thiocarbonyl group more readily undergoes nucleophilic attack on the carbon atom and electrophilic attack on the sulfur atom than does the carbonyl group. Moreover, thiocarbonyl complexes may participate in reactions in which the thiocarbonyl group itself does not change, for example, substitution reactions involving ligands other than CS.

Primary amines react with $[M(CO)_5(CS)]$ to form isocyanide complexes; carbene complexes are probably formed as intermediates:

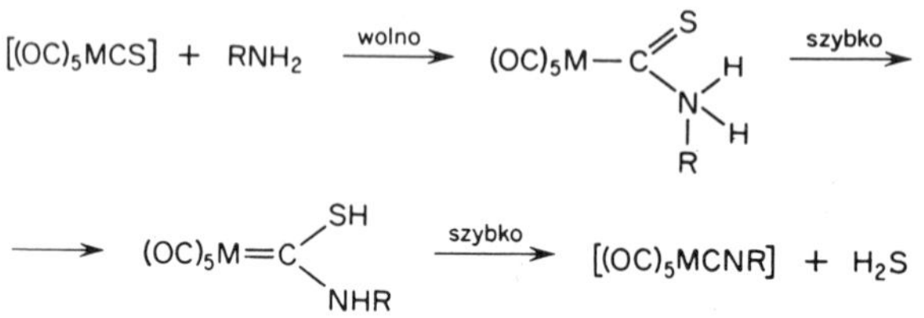

$$[(OC)_5MCS] + RNH_2 \xrightarrow{\text{wolno}} (OC)_5M - C \overset{S}{\underset{N}{\diagdown}} \cdots \xrightarrow{\text{szybko}}$$

$$\longrightarrow (OC)_5M = C \overset{SH}{\underset{NHR}{\diagup}} \xrightarrow{\text{szybko}} [(OC)_5MCNR] + H_2S \tag{2.182}$$

In the same manner, analogous tungsten complexes such as $[W(CO)_{5-x}(CS)L_x]$ containing phosphine ligands react, but more slowly, and increase electron density on the thiocarbonyl group. It has also been found that in complexes of the type $[W(CO)_4(CS)PPh_3]$, the CS group *trans* to $PPh_3$ undergoes nucleophilic attack more slowly than that which is *cis*. This observation suggests that the *trans* phosphine ligand causes a stronger increase in the negative charge on the CS group than the same *cis* ligand.

Secondary amines react with $[W(CO)_5(CS)]$ to give thioformamide complexes:

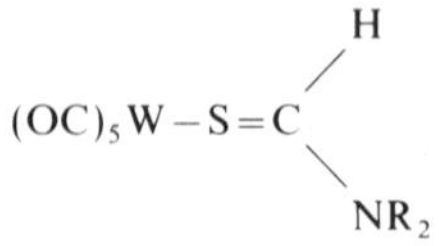

$$(OC)_5W - S = C \overset{H}{\underset{NR_2}{\diagup}}$$

The nucleophilic attack on the carbon atom of the CS group readily occurs in the case of cationic thiocarbonyl complexes. Even compounds possessing weak nucleophilic properties such as alcohols and water attack the CS group.

$$[PtCl(CS)(PPh_3)_2]^+ \xrightarrow{\text{MeOH}} [PtCl\{C(S)OMe\}(PPh_3)_2] \tag{2.183}$$

$$[PtCl(CS)(PPh_3)_2]^+ \xrightarrow{\text{H}_2\text{O}} [PtCl(CO)(PPh_3)_2]^+ \tag{2.184}$$

$$[Fecp(CO)_2CS]^+ \begin{cases} \xrightarrow{N_3^-} [Fecp(CO)_2NCS] + N_2 \\ \xrightarrow{N_2H_4} [Fecp(CO)_2NCS] + NH_4^+ \\ \xrightarrow{NCO^-} [Fecp(CO)_2CN] + COS \end{cases} \quad (2.185)$$

Such a type of reaction may further be illustrated by migration of the hydrido ligand to the CS group in hydrido complexes [equation (2.186)].[179] The thioformyl

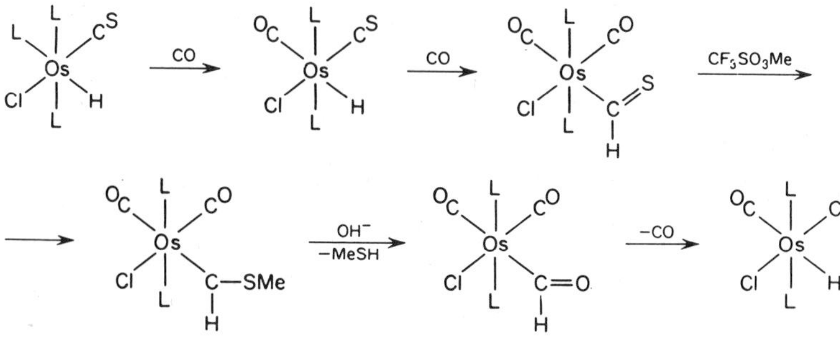

$$(2.186)$$

complex $[\tau^1H(HCS) = -6.93(t), \; {}^3J(PH) = 2.2 \; Hz]$ readily reacts with $CF_3SO_3Me$ to give a carbene cationic compound, which undergoes hydrolysis reaction to form a formyl complex. The thiocarbonyl group in the reaction of $[IrH(CS)(PPh_3)_3]$ with hydrogen undergoes reduction to give methanethiolate.[180]

$$[IrH(CS)(PPh_3)_3] + 2H_2 \longrightarrow [Ir(H)_2(SMe)(PPh_3)_3] \quad (2.187)$$

In the first step the migration of the $H^-$ ligand probably occurs to the carbon atom to form a thioformyl complex, which in subsequent hydrogen migration and oxidative addition reactions becomes reduced.

The thiocarbonyl group also undergoes electrophilic attack on the sulfur atom more easily than the CO group does on the oxygen atom. As a result of the increased negative charge on the sulfur atom and the lower electronegativity of the sulfur atom compared to the oxygen atom, Lewis acids react with the coordinated CS group more easily than they do with the coordinated CO molecule.

The formation of dinuclear complexes possessing the $M-C-S-M$ bridges represents the above-mentioned type of reactions:

$$2[W(CO)(CS)(diphos)_2] + AgBF_4 \longrightarrow [(diphos)_2(OC)WCSAgSCW(CO)(diphos)_2] \quad (2.188)$$

$$[W(CO)(CS)(diphos)_2] + HgCl_2 \longrightarrow [(diphos)_2(CO)WCSHgCl_2] \quad (2.189)$$

Thiocarbonyl complexes possessing increased electron density on the central atom readily react with organic electrophilic reagents to give S-alkyl derivatives:

$$[W(CO)(CS)(diphos)] + RSO_3F \longrightarrow [(diphos)_2(CO)W(CSR)]^+ \; SO_3F^- \quad (R = Me, Et)$$
$$(2.190)$$

This complex contains coordinated S-alkylthiocarbonyl ligand $CSR^+$ (or alkylthio-carbyne ligand $\equiv C-S-R$).

Other thiocarbonyl complexes react similarly:

$$cis\text{-}[Fecp(CO)(CS)]_2 + MeOSO_2F \longrightarrow [Fe_2cp_2(CO)_2(CS)(CSMe)]^+ SO_3F^- \qquad (2.191)$$

$$trans\text{-}[WI(CO)_4(CS)]^- + (MeCO)_2O \longrightarrow trans\text{-}\left[ WI(CO)_4 \left( CSC\!\!\begin{array}{c}\nearrow^O\\[2pt]\searrow_{Me}\end{array} \right) \right] + MeCOO^- \qquad (2.192)$$

The iron complex contains a bridging methylthiocarbonyl ligand. Only those thiocarbonyl complexes which possess a larger negative charge on the sulfur atom react with electrophilic reagents; therefore, they are the thiocarbonyls whose frequency $v(CS)$ is low. Complexes for which this frequency is higher than $1200\ cm^{-1}$ generally do not react with electrophiles.

Some thiocarbonyls undergo oxidative addition reactions:

$$[IrH(CS)(PPh_3)_3] + H_2 \longrightarrow [IrH_3(CS)(PPh_3)_2] + PPh_3 \qquad (2.193)$$

$$[W(CO)_5(CS)] + Br_2 + 2PPh_3 \longrightarrow [WBr_2(CO)_2 CS(PPh_3)_2] + 3CO \qquad (2.194)$$

Thiocarbonyl metal complexes also give products of substitution of other ligands in a given complex without any changes involving the thiocarbonyl group, because this group is bonded to the central atom more strongly than the CO group and some other ligands:

$$[M(CO)_5(CS)] + PPh_3 \longrightarrow [M(CO)_4(CS)(PPh_3)] + CO \qquad (M = Cr, W) \qquad (2.195)$$

$$[W(CO)_3(CS)(diphos)] + diphos \xrightarrow[diphos]{melt} [W(CO)(CS)(diphos)_2] \qquad (2.196)$$

$$[Mncp(CO)_2(CS)] + C_8H_{14} \xrightarrow{hv} [Mncp(CO)(CS)(C_8H_{14})] + CO$$

$$[Mncp(CO)(CS)_2] + C_8H_{14} + PPh_3S \xleftarrow{\quad\begin{array}{c}CS_2\\PPh_3\end{array}\quad} \qquad (2.197)$$

## 12. *APPLICATIONS OF METAL CARBONYL COMPLEXES*

Metal carbonyls have various applications, the most important being in catalysis, particularly in the oxo synthesis. Carbonyls may also be utilized as catalysts in many other processes such as teleomerization of unsaturated compounds by polyhalogenated alkanes, hydrosilylation, hydrogenation of carbon–carbon multiple bonds, reduction of selected organic compounds, isomerization of unsaturated compounds, oligomerization and polymerization of hydrocarbons, hydrocyanation and preparation of optically active compounds from hydrogenation and hydroformylation reactions. Metal carbonyls and their derivatives are also commonly utilized in other areas. High purity metals, their metallic layers for electronics, and other technologies may be obtained from metal carbonyls. Additionally, metal carbonyls may be used for preparation of other transition metal complexes, and may also be useful in organic synthesis. Applications of metal carbonyls in catalysis will be discussed in Chapter 13 and the properties of cluster metal carbonyls in Chapter 3.

## 13. *APPLICATIONS OF ORGANOMETALLIC COMPOUNDS FOR PREPARATION OF METALS, OXIDES, CARBIDES, AND NITRIDES, AND FOR DEPOSITION OF METAL LAYERS AND THEIR COMPOUNDS*

Organometallic compounds are often volatile and may decompose at low temperatures to give the metal or, dependent upon the conditions of the decomposition, to give such compounds as oxides, carbides, nitrides, etc. These decompositions are of practical significance. Because of their volatility and the fact that some are prepared through selective synthesis, many organometallic compounds may be obtained in a very pure state, free from other metals. Such methods may therefore be utilized for preparation of highly pure metals as well as for deposition of layers of metals and their compounds on parts of various apparatuses, which is significant in electronics, particularly for deposition of layers of conductors and insulators as well as for preparation of magnetic materials.[181]

In order to decompose organometallic compounds, the following methods may be used:

- thermodecomposition,
- photolytic decomposition under the influence of electromagnetic radiation,
- irradiation by electron beam,
- decomposition in plasma,
- electrolytic reduction.

Decomposition reactions of compounds may be carried out in the gas phase or in solution. The type of applied atmosphere or solvent depends upon products needed. The preparation of metal layers resulting from decomposition of organometallic compounds is often very profitable technologically. However, its drawback is the formation, in many cases, of byproducts which contaminate the metal. For preparation of metals and their compounds, organometallic and volatile complexes other than metal carbonyls are utilized (Table 2.31). Metals or metallic layers deposited on objects made of different metals, glasses, ceramics, etc., are relatively pure, even quite pure in some instances. For example, through pyrolysis of triisobutylaluminum, metal containing 99.999% of Al is obtained, and through electrolysis of this compound, metal containing 99.9999% of Al is prepared. As a result of pyrolysis of $Fe(CO)_5$, iron not containing traceable quantities of Cr, Mn, Co, Ni, Cu, P, and S, but possessing only $5 \times 10^{-3}\%$ of C and $3 \times 10^{-3}\%$ of O is obtained. Quite often, however, prepared metals contain considerable quantities of carbon and oxygen, especially if no additional treatment is applied. Carbon may be found in carbides, solid solutions, and mixtures.

The decomposition of mixtures of two or more organometallic compounds allows

*Table 2.31. Organometallic and Inorganic Compounds Utilized for Preparation of Metals*

| 11 | 12 | 4 |
|---|---|---|
| [Cu(HCOO)$_2$] (620–720 K)$^a$ | ZnR$_2$ (692 K) | [TicpCl$_3$] (920–970 K) |
| [Cu(acac)$_2$] (530–720 K) | CdR$_2$ (485–610 K) | [TicpBr$_3$] |
| [CuCl(CO)] |  | [Ti(C$_5$Me$_5$)(CO)$_2$] |
| [AuCl(CO)] |  | [ZrcpCl$_3$] |
| [AuCl(PR$_3$)] |  |  |

| 5 | 6 | 7 |
|---|---|---|
| [Mcp(CO)$_4$] | [M(CO)$_6$] (573–873 K) | [M$_2$(CO)$_{10}$] (570–870 K) |
| [V(CO)$_4$(PPh$_3$)$_2$] | [M(aren)$_2$] (723–900 K) | [MX(CO)$_5$] |
| [M(aren)$_2$] | [M(CO)$_3$aren] | [Mcp(CO)$_3$] |
| Mcp$_2$ | [M(aren)$_2$]X | [Mcp$_2$] |
| (M = V, Nb, Ta) | (X = Cl, Br, I; M = Cr, Mo, W) | (M = Mn, Re) |

| 8 | 9 | 10 |
|---|---|---|
| [Fe(CO)$_5$] (640–720 K) | [Co$_2$(CO)$_8$] (453–500 K) | [Ni(CO)$_4$] (453–473 K) |
| [Fe(acac)$_3$] | [Co(CO)$_3$NO] (423–700 K) | [Nicp(NO)] |
| [Ru(CO)$_5$] | [Co(acac)$_3$] (600–613 K) | [Ni(acac)$_2$] (620–720 K) |
| [RuCl$_2$(CO)$_2$] | [Cocp$_2$] | [Nicp$_2$] |
| [Rucp$_2$] | [Rh$_2$Cl$_2$(CO)$_4$] | [Pd$_2$Cl$_2$(allyl)$_2$] |
| [Ru(acac)$_3$] | [Rh$_4$(CO)$_{12}$] | [Pd(acac)(allyl)] |
| [Oscp$_2$] | [IrCl(CO)$_3$] | [Pd(acac)$_2$] |
| [OsCl$_2$(CO)$_4$] |  | [PtCl$_2$(CO)$_2$] |
|  |  | [Pt(acac)$_2$] |

| 2 | 13 | 14 |
|---|---|---|
| BeR$_2$ | BR$_3$ | SiCl$_x$R$_{4-x}$ |
| (R = Me, Et, Bu) | (R = Me, Et, Pr) | (x = 1–3; R = Me, Et, Pr, Bu) |
| MgR$_2$ | AlR$_3$ | SiR$_4$ |
| (R = Ph, Me, Et, cp) | (R = i-Bu, Me, Et, n-Pr, i-Pr) | (R = Me, Et, Pr, Bu) |
|  | MR$_3$ | GeR$_4$ |
|  | (R = Me, Et, Pr, cp; M = Ga, In, Tl) | (R = C$_5$H$_{11}$, Bu, Pr, Et) |

| 14 | 15 |
|---|---|
| SnR$_4$, Sn$_2$R$_6$ | MR$_3$ |
| SnH$_x$R$_{4-x}$ | (M = As, Sb, Bi; R = Me, Et) |
| (R = Me, Et, Pr, Bu, Ph) |  |
| PbR$_4$ |  |
| (R = Et, Pr, Me, Ph) |  |

$^a$ Optimal temperature range for decomposition.

the preparation of metal alloys and their compounds. This is exceptionally important for preparation of various magnetic materials.

These decompositions may also be used for preparation of many other compounds such as semiconductors of the $A^{III}B^V$, $A^{II}B^{VI}$ type and insulators (SiO$_2$, TiO$_2$, ZrO$_2$, Ta$_2$O$_5$, Al$_2$O$_3$) for production of condensers:

$$CdR_2 + H_2S \longrightarrow CdS + 2RH \tag{2.198}$$

$$ZnEt_2 + (NH_4)_2Se \longrightarrow ZnSe + 2C_2H_6 + 2NH_3 \tag{2.199}$$

$$GaR_3 + P \xrightarrow{600\,K} GaP \tag{2.200}$$

$$Fe(CO)_5 + Ni(CO)_4 \xrightarrow{500\,K} Fe + Ni \tag{2.201}$$

$$Fe(CO)_5 + O_2 \longrightarrow Fe_2O_3 \tag{2.202}$$

$$Cr(CO)_6 + O_2 \longrightarrow Cr_2O_3 \tag{2.203}$$

## 14. *METAL CARBONYL COMPLEXES POSSESSING COLUMN STRUCTURES*[182–184]

In the solid phase, square-planar coordination compounds may have chain (column) structures with direct interactions between metal atoms (Figure 2.32).

Typical examples of such compounds are complexes of $d^8$ electronic configuration. Such compounds contain metals of groups 8, 9, and 10. The interaction between $d_{z^2}$ orbitals of infinite number of metal atoms in the column leads to the formation of the $d_{z^2}$ band. The electrical conductivity along the chain ($\sigma_\parallel$) is generally considerably greater than the conductivity in the perpendicular direction ($\sigma_\perp$). The ratio $\sigma_\parallel/\sigma_\perp$ is usually $10^2$–$10^5$. For partially oxidized compounds in which some energy levels of the $d_{z^2}$ band were not occupied, there is a great increase in the $\sigma_\parallel$ conductivity. It may reach values comparable with the conductivity of metals. The best investigated compounds of this type are complexes of platinum such as $K_2[Pt(CN)_4X_{0.3}] \cdot nH_2O$ where $X = F$, Cl, and Br.

Column structures have also been determined for carbonyl complexes of rhodium, iridium, and platinum. For platinum complexes of the formula $[Pt_3(CO)_6]_n^{2-}$, the maximum value of $n$ probably does not reach more than 20 (Figure 3.26) and therefore these carbonyls do not show anisotropy of conductivity. Various Ir(I) and Rh(I) complexes possessing column structures are known: $[IrX(CO)_3]$ ($X = Cl$, Br, I), $[IrCl_{1.07}(CO)_{2.93}]$, $[Ir(acac(CO)_2]$, $H_{0.38}IrCl_2(CO)_2(H_2O)_{2.9}$, $K_{0.58}[IrCl_2(CO)_2]$, and $K_2[Ir_2Cl_{4.8}(CO)_4]$.

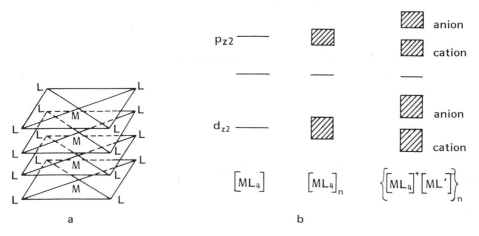

Figure 2.32. (a) Column-like structure of square complexes. (b) Molecular orbital scheme for complexes exhibiting metal–metal interactions.

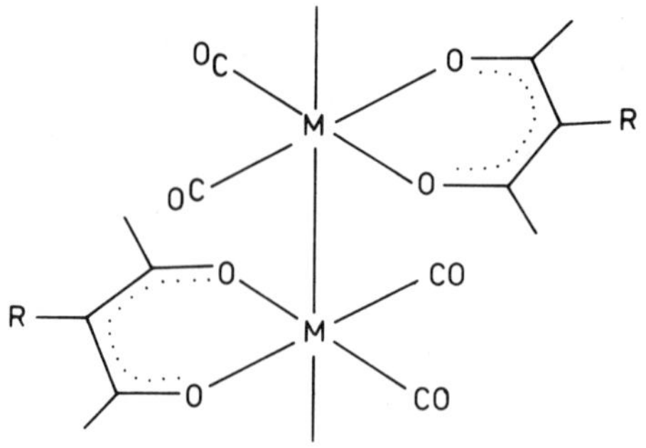

Figure 2.33. Column-like structure of $[M(acac)(CO)_2]$.

Some compounds which contain a small percent of the Ir(II) complex $[IrCl_2(CO)_2]$ show great conductivity. For $[IrCl_{1.07}(CO)_{2.93}]$, the conductivity reaches a value of $0.2\ ohm^{-1} cm^{-1}$. This compound has a column structure consisting of $[IrCl(CO)_3]$ units among which cis-$[IrCl_2(CO)_2]$ molecules are accidentally located. Similar structures are found for rhodium complexes such as $[Rh_2X_2(CO)_4]$ (Figure 2.24), $[Rh(acac)(CO)_2]$ (Figure 2.33). and $[Rh(CO)_2(L-L)][RhX_2(CO)_2]$, where L–L = 2,2′-bipyridine, 1,10-phenanthroline, 2,9-dimethyl-1,10-phenanthroline, and 4,7-diphenyl-1,10-phenanthroline, and X = Cl or Br.[185]

However, these compounds exhibit smaller conductivities compared to non-stoichiometric iridium complexes because the $d_{z^2}$ band is completely occupied. Moreover, the overlap of $d_{z^2}$ orbitals in those rhodium complexes is less effective than in iridium compounds.

## REFERENCES

1. L. Mond, C. Langer, and F. Quincke, *J. Chem. Soc.*, 749 (1980).
2. I. Wender and P. Pino (eds.), *Organic Syntheses via Metal Carbonyls*, Vol. 1, Interscience, New York (1968).
3. A. P. Tomilov, I. N. Chernykh, and Yu. M. Kargin, in: *Elektrokhimia Elementooranicheskich Soyedinenii*, Nauka, Moscow (1985).
4. M. Moskovits and G. A. Ozin (eds.), *Cryochemistry*, Wiley, New York (1976); K. J. Klabunde, *Acc. Chem. Res.*, **8**, 393 (1976); P. L. Timms, *Angew. Chem., Int. Ed. Engl.*, **14**, 273 (1975); P. S. Kell and M. J. McGlinchey, *Angew. Chem., Int. Ed. Engl.*, **14**, 195 (1975); G. A. Ozin, M. P. Andrews, L. F. Nazar, H. X. Huber, and C. G. Francis, *Coord. Chem. Rev.*, **48**, 203 (1983); J. K. Burdett, *Coord. Chem. Rev.*, **27**, 1 (1978).
5. C. J. Balhausen and H. B. Gray, *Molecular Orbital Theory*, Benjamin, New York (1965).
6. L. H. Jones, *J. Mol. Spectrosc.*, **9**, 130 (1962); *Acc. Chem. Res.*, **9**, 128 (1976).
7. R. F. Fenske, *Abstracts of Papers*, XIII ICCC Krakow-Zakopane (Sept. 1970); *Pure App. Chem.*, **27**, 61 (1971); M. B. Hall and R. F. Fenske, *Inorg. Chem.*, **11**, 1619 (1972).
8. F. A. Cotton and G. Wilkinson, *Basic Inorganic Chemistry*, Wiley, New York (1976); F. A. Cotton and G. Wilkinson, in: *Advanced Inorganic Chemistry*, Wiley, New York (1980); W. P. Griffith, *Comprehensive Inorganic Chemistry* (J. C. Bailar, Jr., ed.), Vol. 4, p. 105, Pergamon, Oxford (1973).
9. D. A. Bochvar, N. P. Gambaryan, and D. A. Bochvar, *Metody Elementoorganicheskoy Khimii*, Vol. 1s, p. 9, Nauka, Moscow (1975); R. A. Sokolik, *Metody Elementoorganicheskoy Khimii*, Vols. 1, 2, Nauka, Moscow (1974); Vols. 1, 2; S. P. Gubin and G. B. Shulpin, *Khimia Kompleksov so Svyazami Metalluglerod*, Nauka, Novosibirsk (1984).
10. W. G. Syrkin, *Karbonily Metallov*, Khimia, Moscow (1983).

11. E. Sappa, A. Tiripicchio, and P. Braunstein, *Coord. Chem. Rev.*, **65**, 219 (1975).
12. C. M. Lukehart, *Fundamental Transition Metal Organometallic Chemistry*, Brooks/Cole, Monterey (1985).
13. C. P. Horwitz and D. F. Shriver, *Adv. Organomet. Chem.*, **23**, 219 (1984).
14. R. Colton, C. J. Commons, and B. F. Hoskins, *J. Chem. Soc., Chem. Commun.*, 363 (1975); R. Colton and C. J. Commons, *Aust. J. Chem.*, **28**, 1673 (1975).
15. M. Manassero, M. Sansoni, and G. Longoni, *J. Chem. Soc., Chem. Commun.*, 919 (1976).
16. W. A. Herrmann, M. L. Ziegler, K. Weidenhammer, and H. Biersack, *Angew. Chem., Int. Ed. Engl.*, **18**, 960 (1979).
17. L. N. Lewis and K. G. Caulton, *Inorg. Chem.*, **19**, 3201 (1980).
18. D. A. Roberts, W. C. Mercer, S. M. Zaburak, G. L. Geoffrey, C. N. DeBrosse, M. E. Cass, and C. G. Pierpont, *J. Am. Chem. Soc.*, **104**, 910 (1982).
19. P. S. Braterman, *Structure and Bonding*, **10**, 57 (1972).
20. P. S. Braterman, *Metal Carbonyl Spectra*, Academic Press, London (1975).
21. G. R. Dobson, *Acc. Chem. Res.*, **9**, 300 (1976).
22. A. Gołebiewski, *Chemia Kwantowa Zwiazków Nieorganicznych*, pp. 446–461, PWN, Warsaw (1969).
23. R. F. Fenske, *Prog. Inorg. Chem.*, **21**, 179 (1976).
24. D. M. P. Mingos, *Comprehensive Organometallic Chemistry* (G. Wilkinson, F. G. A. Stone, and E. W. Abel, eds.), Vol. 3, p. 1, Pergamon, Oxford (1982).
25. R. Hoffman, M. M.-L. Chen, and D. L. Thorn, *Inorg. Chem.*, **16**, 503 (1977).
26. J. B. Johnson, and W. G. Klemperer, *J. Am. Chem. Soc.*, **99**, 7132 (1977).
27. D. E. Sherwood, Jr. and M. B. Hall, *Inorg. Chem.*, **19**, 1805 (1980).
28. N. V. Sidgwick and R. W. Bailey, *Proc. Soc.* London, Ser. A, **144**, 521 (1934).
29. H. D. Kaesz, *Chem. Br.*, **9**, 344 (1973).
30. W. L. Jolly, S. C. Avanzino, and R. R. Rietz, *Inorg. Chem.*, **16**, 964 (1977); S. C. Avanzino, A. A. Bakke, H.-W. Chen, C. J. Donahue, W. L. Jolly, T. H. Lee, and A. J. Ricco, *Inorg. Chem.*, **19**, 1931 (1980).
31. F. A. Cotton and C. S. Kraihanzel, *J. Am. Chem. Soc.*, **84**, 4432 (1962).
32. P. S. Braterman, *Structure and Bonding*, **26**, 1 (1976).
33. J. A. Timney, *Inorg. Chem.*, **18**, 2502 (1979).
34. B. E. Mann, *Adv. Organomet. Chem.*, **12**, 135 (1974).
35. M. H. Chisholm and S. Godleski, *Prog. Inorg. Chem.*, **20**, 299 (1976).
36. P. Chini, G. Longoni, and V. G. Albano, *Adv. Organomet. Chem.*, **14**, 285 (1976); P. Chini, *J. Organomet. Chem.*, **200**, 37 (1980).
37. G. Bor, U. K. Dietler, and K. Noack, *J. Chem. Soc., Chem. Commun.*, 914 (1976); G. Bor and K. Noack, *J. Organomet. Chem.*, **64**, 367 (1974).
38. A. Hamnett and A. F. Orchard, *Electronic Structure and Magnetism of Inorganic Compounds*, Vol. 1, p. 1, London Chemical Society (1972).
39. T. L. James, *J. Chem. Educ.*, **48**, 712 (1971).
40. C. Furlani and C. Cauletti, *Structure and Bonding*, **35**, 119 (1978).
41. V. I. Nefedov, *J. Electron Spectrosc.*, **12**, 459 (1977); *Rentgenoelektronnaja Spektroskopia Khimicheskih Soyedinenii*, Khimia, Moscow (1984).
42. J. C. Green, *Structure and Bonding*, **43**, 37 (1981).
43. A. H. Cowley, *Prog. Inorg. Chem.*, **26**, 46 (1979).
44. W. L. Jolly, *Coord. Chem. Rev.*, **13**, 47 (1974).
45. M. Barber, J. A. Connor, M. F. Guest, M. B. Hall, J. H. Hillier, and W. N. E. Meredith, *Faraday Discuss. Chem. Soc.*, **54**, 219 (1972).
46. M. R. Litzow and T. R. Spalding, *Mass Spectrometry of Inorganic and Organometallic Compounds*, Elsevier, Amsterdam (1973).
47. R. W. Kiser, in: *Mass Spectroscopy of Organometallic Compounds, Characterization of Organometallic Compounds* (M. Tsutsui, ed.), Part I, Interscience, New York (1969).
48. J. M. Miller and G. L. Wilson, *Adv. Inorg. Chem. Radiochem.*, **18**, 229 (1976).
49. *A Specialist Periodical Report, Mass Spectrometry 4*, The Chemical Society, London (1977).
50. J. A. Connor, *Top. Curr. Chem.*, **71**, 71 (1977); B. F. G. Johnson (ed.), *Transition Metal Clusters*, p. 345, Wiley, Chichester (1980).

51. F. Basolo and R. G. Pearson, *Mechanism of Inorganic Reactions*, p. 7, Wiley, New York (1967); F. Basolo, *Coord. Chem. Rev.*, **43**, 7 (1982); *Inorg. Chim. Acta*, **50**, 65 (1981).
52. K. E. Lewis, D. M. Golden, and G. P. Smith, *J. Am. Chem. Soc.*, **106**, 3905 (1984).
53. D. A. Brown, in: *Theory, Structure, Properties of Complex Compounds* (B. Jeżowska-Trzebiatowska, ed.), p. 37, PWN, Warsaw (1979); D. J. Darensbourg, *Adv. Organometal. Chem.*, **21**, 113 (1982); J. A. S. Howell and P. M. Burkinshaw, *Chem. Rev.*, **83**, 557 (1983).
54. F. Basolo and A. Wojcicki, *J. Am. Chem. Soc.*, **83**, 520 (1961).
55. R. Angelici, *Organomet. Chem. Rev.*, **3**, 173 (1968).
56. A. Marcomini and A. Poé, *J. Chem. Soc., Dalton Trans.*, 95 (1984).
57. H. Behrens, *Adv. Organomet. Chem.*, **18**, 1 (1980).
58. A. E. Stiegman and D. R. Tyler, *Coord. Chem. Rev.*, **63**, 217 (1985); *Acc. Chem. Res.*, **17**, 61 (1984).
59. P. C. Ford, *Acc. Chem. Res.*, **14**, 31 (1981).
60. A. D. King, R. B. King, and D. B. Yang, *J. Am. Chem. Soc.*, **102**, 1028 (1980); *J. Chem. Soc., Chem. Commun.*, 529 (1980).
61. D. M. Vandenberg, T. M. Suzuki, and P. C. Ford, *J. Organomet. Chem.*, **272**, 309 (1984).
62. F. A. Cotton, *Angew. Chem., Int. Ed. Engl.*, **24**, 274 (1985).
63. M. O. Albers and N. J. Coville, *Coord. Chem. Rev.*, **53**, 227 (1984).
64. T. Kobayashi, K. Yasufuku, J. Iwai, H. Yesaka, H. Noda, and H. Ohtani, *Coord. Chem. Rev.*, **64**, 1 (1985).
65. F. Basolo, *Inorg. Chim. Acta*, **100**, 33 (1985); D. A. Sweigart, in: *Mechanisms of Inorganic and Organometallic Reactions* (M. K. Twigg, ed.), p. 237, Plenum, New York (1984).
66. R. B. Hitam, K. A. Mahmoud, and A. J. Rest, *Coord. Chem. Rev.*, **55**, 1 (1984).
67. A. Wojcicki, *Adv. Organomet. Chem.*, **11**, 87 (1973).
68. C. K. Anderson and R. J. Cross, *Acc. Chem. Res.*, **17**, 67 (1984).
69. H. Brunner, B. Hammer, I. Bernal, and M. Draux, *Organometallics*, **2**, 1595 (1983); T. C. Flood and K. D. Campbell, *J. Am. Chem. Soc.*, **106**, 2853 (1984); T. C. Flood, K. D. Campbell, H. H. Downs, and S. Nakanishi, *Organometallics*, **2**, 1590 (1983).
70. G. Palyi, F. Ungvary, V. Galamb, and L. Marko. *Coord. Chem. Rev.*, **53**, 37 (1984).
71. L. V. Interrante and V. G. Nelson, *J. Organomet. Chem.*, **25**, 153 (1970).
72. T. G. Richmond, Q.-Z. Shi, W. G. Trogler, and F. Basolo, *J. Am. Chem. Soc.*, **106**, 76 (1984); Q.-Z. Shi, T. G. Richmond, W. C. Trogler, and F. Basolo. *J. Am. Chem. Soc.*, **106**, 71 (1984).
73. G. F. P. Warnock, S. B. Philson, and J. E. Ellis, *J. Chem. Soc., Chem. Commun.*, 893 (1984); G. F. P. Warnock, J. Sprague, K. L. Fjare, and J. E. Ellis. *J. Am. Chem. Soc.*, **105**, 672 (1983).
74. K. S. Suslick and R. E. Johnson, *J. Am. Chem. Soc.*, **106**, 6856 (1984).
75. J. R. Morton and K. F. Preston, *Organometallics*, **3**, 1386 (1984).
76. K. M. Pfahl and J. E. Ellis, *Organometallics*, **3**, 230 (1984).
77. H. Huber, T. A. Ford, M. Moskovitz, G. A. Ozin, and W. Klotzbücher, *Inorg. Chem.*, **15**, 1666 (1976); E. P. Kündig, M. Moskovitz, and G. A. Ozin, *Angew. Chem., Int. Ed. Engl.*, **14**, 292 (1975).
78. J. Grobe and H. Zimmerman, *Z. Naturforsch.*, **35B**, 533 (1980); M. Guainazzi, G. Silvestri, S. Gambino, and G. Filardo, *J. Chem. Soc., Dalton Trans.*, 927 (1972).
79. J. T. Lin, G. P. Hagen, and J. E. Ellis, *J. Am. Chem. Soc.*, **105**, 2296 (1983).
80. J. T. Lin and J. E. Ellis, *J. Am. Chem. Soc.*, **105**, 6252 (1983).
81. K. Lane, L. Sallans, and R. R. Squires, *J. Am. Chem. Soc.*, **105**, 2719 (1984).
82. M. H. Chisholm, D. M. Hoffman, and J. C. Huffman, *Organometallics*, **4**, 986 (1985); *Chem. Soc. Rev.*, **14**, 69 (1985).
83. W. Klaeui, A. Müller, and M. Scotti, *J. Organomet. Chem.*, **253**, 45 (1983); W. Klaeui, W. Eberspach, and R. Schwarz, *J. Organomet. Chem.*, **252**, 347 (1983).
84. F. W. B. Einstein, T. Jones, R. K. Pomeroy, and P. Rushman, *J. Am. Chem. Soc.*, **106**, 2707 (1984).
85. K. Ihmels and P. Rehder, *Organometallics*, **4**, 1334 (1985); *Organometallics*, **4**, 1340 (1985); *Chem. Ber.*, **118**, 895 (1985).
86. J. M. Maher, R. P. Beatty, and N. J. Cooper, *Organometallics*, **4**, 1354 (1985).
87. L. B. Handy, J. K. Ruff, and L. F. Dahl, *J. Am. Chem. Soc.*, **92**, 7312 (1970).
88. A. Wojcicki, *Inorg. Chim. Acta*, **100**, 125 (1985); S.-G. Shyu and A. Wojcicki, *Organometallics*, **3**, 809 (1984).
89. H. Bock and H. tom Dieck, *Chem. Ber.*, **100**, 228 (1967); *Angew. Chem.*, **78**, 549 (1966); H. tom

Dieck and I. W. Renk, *Chem. Ber.*, **105**, 1419 (1972); G. van Koten and K. Vrieze, *Adv. Organomet. Chem.*, **21**, 151 (1982).

90. A. M. Bond, S. W. Carr, and R. Colton, *Organometallics*, **3**, 541 (1984); A. M. Bond, S. W. Carr, R. Colton, and D. P. Kelly, *Inorg. Chem.*, **22**, 989 (1983).
91. F. Calderazzo and G. Pampaloni, *J. Chem. Soc., Chem. Commun.*, 1249 (1984).
92. D. G. Leopold and V. Vaida, *J. Am. Chem. Soc.*, **106**, 3720 (1984).
93. I. R. Dunkin, P. Harter, and C. J. Sheilds, *J. Am. Chem. Soc.*, **106**, 7248 (1984).
94. C. J. Beseker, V. W. Day, and W. G. Klemperer, *Organometallics*, **4**, 564 (1985).
95. J. Keijsper, G. van Koten, K. Vrieze, M. Zoutberg, and C. H. Stam, *Organometallics*, **4**, 1306 (1985).
96. E. J. M. de Boer, J. de With, N. Meijboom, and A. G. Orpen, *Organometallics*, **3**, 259 (1985).
97. Y.-S. Chen and J. E. Ellis, *J. Am. Chem. Soc.*, **105**, 1689 (1983).
98. *Inorg. Synth.*, **20**, 209–242 (1980).
99. Handbuch der Präparativen Anorganischen Chemie (G. Brauer, ed.), Ferdinand Enke Verlag, Stuttgart, 1981.
100. H. B. Chin, M. B. Smith, R. D. Wilson, and R. Bau, *J. Am. Chem. Soc.*, **96**, 5285 (1974).
101. L.-Y. Hsu, N. Bhattacharyya, and S. G. Shore, *Organometallics*, **4**, 1483 (1985).
102. Y.-F. Yu, J. Gallucci, and A. Wojcicki, *J. Chem. Soc., Chem. Commun.*, 653 (1984); S.-G. Shyn and A. Wojcicki, *Organometallics*, **4**, 1457 (1985).
103. D. Seyferth, T. G. Wood, J. P. Fackler, and A. M. Mazany, *Organometallics*, **3**, 1121 (1984).
104. R. G. Teller, R. G. Finke, J. P. Collman, H. B. Chin, and R. Bau, *J. Am. Chem. Soc.*, **99**, 1104 (1977).
105. H. B. Chin and R. Bau, *J. Am. Chem. Soc.*, **98**, 2434 (1976).
106. R. Whyman, *J. Chem. Soc., Dalton Trans.*, 1375 (1972); *J. Chem. Soc., Dalton Trans.*, 2294 (1972).
107. P. Chini and B. T. Heaton, *Top. Curr. Chem.*, **71**, 1 (1977).
108. F. Pruchnik and K. Wajda-Hermanowicz (in preparation).
109. L. Garlaschelli, S. Martinengo, P. L. Bellon, F. Demartin, M. Manassero, M. Y. Chiang, C.-Y. Wei, and R. J. Bau, *J. Am. Chem. Soc.*, **106**, 6664 (1984).
110. O. T. Horvath, G. Bor, and P. Pino, *Abstracts*, XIIth International Conference on Organometallic Chemistry, pp. 180, 336, Vienna (1985).
111. P. Kluefers, *Angew. Chem., Int. Ed. Engl.*, **23**, 307 (1984).
112. J. G. Norman and D. J. Gmur, *J. Am. Chem. Soc.*, **99**, 1446 (1977).
113. R. Mason, G. Scollary, B. Moyle, K. I. Hardcastle, B. L. Shaw, and C. J. Moulton, *J. Organomet. Chem.*, **113**, C49 (1976); F. C. March, R. Mason, K. M. Thomas, and B. L. Shaw, *J. Chem. Soc., Chem. Commun.*, 584 (1975).
114. K. Kurtev, D. Ribola, R. A. Jones, C. J. Cole-Hamilton, and G. Wilkinson, *J. Chem. Soc., Dalton Trans.*, 55 (1980).
115. F. Pruchnik, K. Wajda, and T. Lis, *Inorg. Chim. Acta*, **40**, 207 (1980); *Abstracts*, 9th International Conference on Coordination Chemistry, p. 57, Dijon (1979).
116. J. T. Mague and J. P. Mitchener, *Inorg. Chem.*, **8**, 119 (1969).
117. A. R. Sanger, *Chem. Commun.*, 893 (1975); *J. Chem. Soc., Dalton Trans.*, 120 (1977).
118. M. Couwie and S. K. Dwight, *Inorg. Chem.*, **19**, 2500 (1980).
119. C. P. Kubiak and R. Eisenberg, *Inorg. Chem.*, **19**, 2726 (1980); *J. Am. Chem. Soc.*, **102**, 3637 (1980).
120. J. P. Farr, M. M. Olmstead, and A. L. Balch, *J. Am. Chem. Soc.*, **102**, 6654 (1980).
121. R. J. Puddephatt, *Chem. Soc. Rev.*, **12**, 99 (1983).
122. F. Pruchnik and K. Wajda, *Inorg. Chim. Acta*, **40**, 203 (1980).
123. B. R. Penfold and B. H. Robinson, *Acc. Chem. Res.*, **6**, 73 (1973).
124. J. K. Burdett, *Coord. Chem. Rev.*, **27**, 1 (1978).
125. G. Booth, J. Chatt, and P. Chini, *Chem. Commun.*, 639 (1965).
126. P. Chini, *J. Organomet. Chem.*, **200**, 37 (1980).
127. P. W. Jolly, in: *Comprehensive Organometallic Chemistry* (G. Wilkinson, F. G. A. Stone, and E. W. Abel, eds.), Vol. 6, p. 3, Pergamon, Oxford (1982).
128. K. J. Matveev, L. N. Rachkovskaya, and N. K. Eremenko, *Izv. Sib. Otd. Akad. Nauk. SSSR*, No. 4 (139), 81 (1968).
129. E. E. Stahly and E. W. Lard, *Chem. Ind. (London)*, 85 (1977).

130. G. Longoni and P. Chini, *J. Am. Chem. Soc.*, **98**, 7225 (1976).
131. C. Brown and B. T. Heaton, *J. Chem. Soc., Chem. Commun.*, 309 (1977).
132. A. R. Sanger, *Can. J. Chem.*, **63**, 571 (1985); *Can. J. Chem.*, **62**, 2168 (1984).
133. T. R. Felthouse, *Prog. Inorg. Chem.*, **29**, 73 (1982).
134. E. B. Boyar and S. D. Robinson, *Coord. Chem. Rev.*, **50**, 109 (1983).
135. P. G. Rasmussen, J. E. Anderson, O. H. Bailey, M. Tamres, and J. C. Bayon, *J. Am. Chem. Soc.*, **107**, 279 (1985); A. Thorez, A. Maisonnat, and R. Poilblanc, *Chem. Commun.*, 518 (1977); J.-J. Bonnet, P. Kalak, and R. Poilblanc, *Angew. Chem.*, **92**, 572 (1980).
136. G. J. Leigh and R. L. Richards, in: *Comprehensive Organometallic Chemistry* (G. Wilkinson, F. G. A. Stone, and E. W. Abel, eds.), Vol. 5, p. 541, Pergamon, Oxford (1982).
137. S.-K. Kang, T. A. Albright, T. C. Wright, and R. A. Jones, *Organometallics*, **4**, 666 (1985).
138. D. Belli Dell'Amico, F. Calderazzo, and N. Zandona, *Inorg. Chem.*, **23**, 137 (1984).
139. A. D. Gelman and E. Mejlakh, *Dokl. Akad. Nauk SSSR*, **36**, 188 (1942).
140. P. L. Goggin and J. Mink, *J. Chem. Soc., Dalton Trans.*, 534 (1974); P. L. Goggin, R. J. Goodfellow, I. R. Herbert, and A. G. Orpen, *J. Chem. Soc., Chem. Commun.*, 1077 (1981).
141. O. N. Temkin and Ł. G. Bruk, *Uspiechi Chimii*, **52**, 206 (1983).
142. P. M. Maitlis, P. Espinet, and M. J. H. Russell, in: *Comprehensive Organometallic Chemistry* (G. Wilkinson, F. G.A. Stone, and E. W. Abel, eds.), Vol. 6, p. 265, Pergamon, Oxford (1982).
143. S. K. Janikowski, L.-J. Radonovich, T. J. Groshens, and K. J. Klabunde, *Organometallics*, **4**, 396 (1985).
144. H. Huber, E. P. Kündig, M. Moskovits, and G. A. Ozin, *J. Am. Chem. Soc.*, **97**, 2097 (1975).
145. D. McIntosh and G. A. Ozin, *Inorg. Chem.*, **16**, 51 (1977).
146. J. Knight and M. J. Mays, *Chem. Commun.*, 1006 (1970).
147. M. Poliakoff and J. J. Turner, *Chem. Commun.*, 1008 (1970).
148. G. A. Battiston, G. Bor, U. K. Dietler, S. F. A. Kettle, R. Rossetti, G. S. Sbrignadello, and P. L. Stanghellini, *Inorg. Chem.*, **19**, 1961 (1980); G. A. Battiston, G. A. Brignadello, and G. Bor, *Inorg. Chem.*, **19**, 1973 (1980).
149. T. R. Gilson, *J. Chem. Soc., Dalton Trans.*, 149 (1984); *J. Chem. Soc., Dalton Trans.*, 155 (1984).
150. J. E. Bulkowski, P. L. Burk, M. F. Ludman, and J. A. Osborn, *J. Chem. Soc., Chem. Commun.*, 498 (1977).
151. M. Pasquali, C. Floriani, A. Gaetani-Manfredotti, and C. Guastini, *J. Chem. Soc., Chem. Commun.*, 197 (1979).
152. M. L. H. Green and D. J. Jones, *Adv. Inorg. Chem. Radiochem.*, **7**, 115 (1965); A. P. Ginsberg, *Transition Metal Chem.*, **1**, 111 (1965).
153. E. L. Muetterties, (ed.), *Transition Metal Hydrides*, Marcel Dekker, New York (1971).
154. J. P. McCue, *Coord. Chem. Rev.*, **10**, 265 (1973).
155. H. D. Kaesz and R. B. Saillant, *Chem. Rev.*, **72**, 231 (1972).
156. A. Modinos and P. Woodward, *J. Chem. Soc., Dalton Trans.*, 1516 (1975).
157. L. N. Rachkovskaya, N. K. Eremenko, and K. I. Mayveev, *Dokl. Akad. Nauk SSSR*, 190, 1396 (1970); *Izv. Sib. Otd. Akad. Nauk SSSR, Ser. Khim. Nauk*, **5**, 78 (1970); Yu. I. Mironov, L. M. Plyasova, and V. V. Bakakin, *Sov. Phys. Crystallogr.*, **19**, 317 (1974).
158. S. Martinengo, B. R. Heaton, R. J. Goodfellow, and P. Chini, *J. Chem. Soc., Chem. Commun.*, 39 (1977).
159. C. R. Eady, B. F. G. Johnson, J. Lewis, M. C. Malatesta, P. Machin, and M. McPartlin, *J. Chem. Soc., Chem. Commun.*, 945 (1976).
160. P. J. Jackson, B. F. G. Johnson, J. Lewis, P. R. Raithby, M. McPartlin, W. J. H. Nelson, K. D. Rouse, J. Allibon, and S. A. Mason, *Chem. Commun.*, 295 (1980).
161. R. Bau, R. G. Teller, S. W. Kirtley, and T. F. Koetzle, *Acc. Chem. Res.*, **12**, 176 (1979); R. G. Teller and R. Bau, *Structure and Bonding*, **44**, 1 (1981).
162. L. M. Venanzi, *Coord. Chem. Rev.*, **43**, 251 (1982).
163. G. G. Hlathy and R. J. Crabtree, *Coord. Chem. Rev.*, **65**, 1 (1985).
164. D. L. Van der Hart, H. S. Gutowsky, and T. C. Farrer, *J. Am. Chem. Soc.*, **89**, 5056 (1967).
165. J. Lewis and B. F. G. Johnson, *Acc. Chem. Res.*, **1**, 245 (1968).
166. G. J. Kubas, R. R. Ryan, B. I. Swanson, P. J. Vergamini, and H. J. Wasserman, *J. Am. Chem. Soc.*, **106**, 451 (1984).

167. R. F. Jordan and J. R. Norton, *J. Am. Chem. Soc.*, **104**, 1255 (1982).
168. R. G. Pearson and P. C. Ford, *Comments Inorg. Chem.*, **1**, 279 (1982).
169. I. S. Butler and A. E. Fenster, *J. Organomet. Chem.*, **66**, 161 (1974); P. V. Yaneff, *Coord. Chem. Rev.*, **23**, 183 (1977).
170. D. F. Shriver, Sr. and A. Alich, *Coord. Chem. Rev.*, **8**, 15 (1972).
171. A. E. Crease and P. Legzdins, *J. Chem. Educ.*, **52**, 499 (1975).
172. L. H. Jones, R. S. McDowell, and M. Goldblatt, *Inorg. Chem.*, **8**, 2349 (1969).
173. E. W. Abel, R. A. N. McLean, S. P. Tyfield, P. S. Braterman, and A. P. Walker, *J. Mol. Spectrosc.*, **30**, 29 (1969).
174. M. A. Andrews, *Inorg. Chem.*, **16**, 496 (1977).
175. H. Werner and K. Leonhard, *Angew. Chem., Int. Ed. Engl.*, **18**, 627 (1979).
176. P. V. Broadhurst, B. F. G. Johnson, J. Lewis, and P. R. Raithby, *Chem. Commun.*, 812 (1980); *J. Chem. Soc., Dalton Trans.*, 1395 (1982).
177. I. S. Butler, D. Cozak, and S. R. Stobart, *Inorg. Chem.*, **16**, 1779 (1977); I. M. Balbich, A. M. English, and I. S. Butler, *Organometallics*, **3**, 1786 (1984).
178. D. L. Lichtenberg and R. F. Fenske, *Inorg. Chem.*, **15**, 2015 (1976).
179. T. J. Collins and W. R. Roper, *J. Chem. Soc., Chem. Commun.*, 1044 (1976).
180. W. R. Roper and K. G. Town, *J. Chem. Soc., Chem. Commun.*, 781 (1976).
181. G. A. Razuvayev, B. G. Gribov, G. A. Domrachev, and B. Salamatin, *Metallorganicheskie Soyedinenia v Elektronikie*, Izd. Nauka, Moscow (1977); B. G. Gribow, G. A. Domrachev, B. V. Zhuk, B. C. Kaverin, B. I. Kozyrkin, V. V. Melnikov, and O. N. Suvorova, *Osazhdenie Plenok i Pokrytii Razlozheniem Metalloorganicheskikh Soyedinenii*, Nauka, Moscow (1981).
182. J. S. Miller and A. J. Epstein, *Prog. Inorg. Chem.*, **20**, 1 (1976).
183. T. W. Thomas and A. E. Underhill, *Chem. Soc. Rev.*, **1**, 99 (1972).
184. J. R. Ferraro, *Coord. Chem. Rev.*, **43**, 205 (1982); *Coord. Chem. Rev.*, **36**, 357 (1981).
185. F. Pruchnik and K. Wajda, *J. Organomet. Chem.*, **164**, 71 (1979).
186. V. G. Albano, D. Braga, L. Busetto, M. Monari, and V. Zannotti, *Abstracts*, p. 445, XIIth ICOMC, Vienna (Sept. 8–13, 1985).
187. W. J. Evans, J. W. Grate, L. A. Hughes, H. Zhang, and J. L. Atwood, *J. Am. Chem. Soc.*, **107**, 3728 (1985).
188. R. D. Feltham, G. Elbaze, R. Ortega, C. Eck, and J. Dubrawski, *Inorg. Chem.*, **24**, 1503 (1985).
189. B. F. G. Johnson, *J. Chem. Soc. Chem. Commun.*, 703 (1976).
190. J. Evans, *Adv. Organomet. Chem.*, **16**, 319 (1977).
191. R. D. Gillard and G. Wilkinson, *J. Chem. Soc.*, 3594 (1963); J. A. Osborn, R. D. Gillard, and G. Wilkinson, *J. Chem. Soc.*, 3168 (1964).
192. O. J. Scherer, J. Kerth, and W. S. Sheldrick, *Angew. Chem. Int. Ed. Engl.*, **23**, 156 (1984).

# Chapter 3

# Metal–Metal Bonds and Clusters

## 1. BONDING AND STRUCTURE

Polynuclear compounds in which metal atoms are separated by relatively short distances have been known for more than 70 years. In 1913 it was demonstrated that tantalum chloride, $TaCl_2 \cdot 2H_2O$, is in reality $Ta_6Cl_{14} \cdot 7H_2O$, and in 1935 it was shown that in $[W_2Cl_9]^{3-}$, the tungsten–tungsten distance is only about 250 pm. In 1938 the X-ray structure determination of $Fe_2(CO)_9$ revealed that the two iron atoms are separated by about 250 pm, and this observation started investigations of polynuclear metal carbonyl compounds. However, the real development of the chemistry of polynuclear compounds began in the early 1960s when the existence of metal–metal bonds was clearly recognized and the term "metal atom cluster" introduced.[1a]

There is an enormous number of compounds which possess metal–metal bonds. Therefore, a metal atom cluster[1a] will be defined as a polynuclear compound in which there are substantial and direct bonds between the metal atoms. The following groups of metal atoms will constitute clusters: $M-M$, *triangulo*-$M_3$, *tetrahedro*-$M_4$, *octahedro*-$M_6$, linear groups $M_n$, where $n = 3, 4,...$, etc. The metal atoms of a cluster are named skeletal atoms. At present, clusters of almost all elements are known. Clusters may be divided into two groups[1–15,22–25]:

1. "Electron rich" clusters, containing a large number of $d$ electrons of the metal.

2. "Electron poor" clusters, possessing a relatively small number of valence electrons.

The first group comprises clusters of the late transition elements, especially those of groups 8–11; the second group contains clusters of the early transition metals (groups 4–7) as well as compounds of some main group elements. Some groups of the Periodic Chart show a greater tendency for cluster formation than do some other elements. Thus, elements of the transition metal groups 5–10 and those of the main groups 14 and 15 as well as the element boron show the greatest tendency to form clusters. Therefore, there is a correlation between the ease of cluster formation and the energy of homonuclear bond because, for the above-mentioned elements, such energy is the highest. Also, generally, heavier elements of any given group form more clusters. However, this is not the only factor, as evidenced by Hg(I) compounds.

Oxidation states of metals forming clusters are low. For "electron rich" clusters, oxidation states equal zero, e.g., in metal cardonyl clusters $[M_p(CO)_q]$, and they are even negative as in metal cardonylate anions, $[M_p(CO)_q]^{x-}$. However, in "electron poor" clusters, metal ions generally adopt oxidation states $+2$ and $+3$, although for some Mo, W and Re compounds the metal possesses $+4$–$+6$ oxidation states, for example, $[Mo_3O_4(C_2O_4)_3(H_2O)_3]^{2-}$, $[W_3O_2(OOCMe)_6(H_2O)_3]^{2+}$, $[Re_3(pmcp)_3O_6]^{2+}$.

The development of cluster chemistry has been influenced by possible, practical application of clusters in catalysis and organic synthesis. This development also comes about because of interest in metal–metal bonds and the theory of multicenter bonds, relationships between the structural theories of complexes and clusters, as well as analogies between clusters and metal surfaces. In clusters there are types of bonds which are not encountered in any other compounds, for example, bridges:

The simplest clusters are by definition dinuclear complexes possessing strong metal–metal bonds.[1a–c] In the case of compounds of the main group elements, the formation of bonds of the E–E type is achieved by utilization of the $s$ and $p$ orbitals, while in the transition metal complexes, metal–metal bonds are mainly formed as a result of interactions between metal $d$ orbitals which do not create bonds with ligands. When two metal atoms approach one another, the bonding orbitals for the $M_2$ molecule are formed as seen in Figure 3.1. Owing to the effectiveness of the overlap of the orbitals $\delta \ll \pi < \sigma$, the molecular orbital energy increases in the order $\sigma < \pi \ll \delta < \delta^* \ll \pi^* < \sigma^*$, if it is assumed that Hückel's approximation, stating that the molecular orbital energies are proportional to the overlap integrals, is obeyed. In complexes, the $M-M$ bond order depends on many factors, the most important of which are the number of $d$ electrons of the metal, the number and properties of ligands, the symmetry of the molecule, and the metal–metal distance.[1a–f] In $M_2$ molecules possessing $D_{\infty h}$ symmetry, the $\pi$ and $\delta$ orbitals are degenerate. In complexes, this degeneracy is removed, however. In compounds which have $D_{4h}$ symmetry, such as $[Re_2Cl_8]^{2-}$, the $d_{x^2-y^2}$ orbitals are directed toward ligands and mainly form $\sigma$ metal–ligand bonds. Therefore, in dinuclear complexes, the maximum $M-M$ bond order equals 4 for cases in which central atoms contain four $d$ electrons each. Consequently, coordination compounds possessing $\sigma^2\pi^4\delta^2$ electronic configuration are formed.

Molecular orbital theory explains the formation of multiple bonds both in the case of transition metal complexes as well as diatomic molecules of main group elements and compounds of these elements. There is an analogy between $E_2$ molecules of the elements of the second period of the Periodic Chart (and their compounds containing multiple bonds) and $M_2$ molecules as well as dimeric transition metal complexes which possess metal–metal bonds. The molecules $C_2$, $N_2$, and $O_2$ have $\pi^4\sigma^0$, $\pi^4\sigma^2$, and $\sigma^2\pi^4\pi^{*2}$ electronic configurations, respectively, with the corresponding bond orders 2, 3, and 2. The

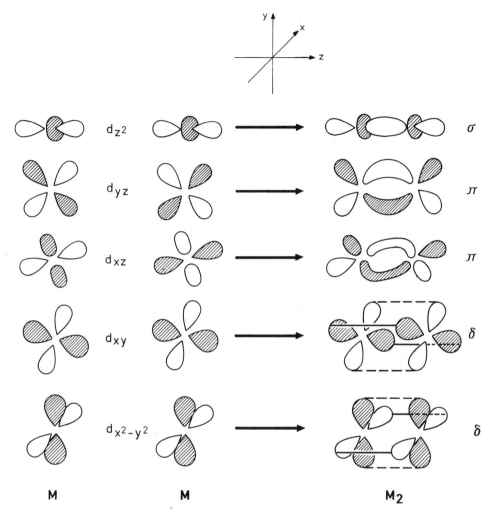

Figure 3.1. The bonding orbitals for the $M_2$ molecule that are formed when two metal atoms approach one another.

oxygen molecule is paramagnetic because it contains two unpaired electrons occupying $\pi^*$ degenerate orbitals. The heteronuclear molecule NO and the ion $O_2^+$ each have bond orders of 2.5, which is a result of the $\sigma^2\pi^4\pi^{*1}$ configuration. The ions $O_2^-$ and $O_2^{2-}$ possess the electronic configurations $\sigma^2\pi^4\pi^{*3}$ (bond order 1.5) and $\sigma^2\pi^4\pi^{*4}$ (single bond). For a series of similar compounds, it is possible to correlate the bond order with the bond distance, the energy of dissociation of the bond, the force constants, and the IR stretching vibrations.

The metal–metal bond order in transition metal $M_2$-type molecules and in transition metal complexes may be higher than in the case of the main group elements because of the participation of the $d$ orbitals. The diatomic molecule $Mo_2$ has a quintuple bond,[1b, 1d] with an electronic configuration of $\sigma^2\pi^4\delta^4\sigma(s)^2$ (the Mo–Mo distance equals 194 pm). The sixth electron pair $\sigma_g(5s)$ is effectively nonbonding as shown by spectroscopic measurements. Strong $d$–$d$ bonds are also encountered for the molecules $Cr_2$ [$\sigma^2\pi^4\delta^4\sigma(s)^2$ electronic configuration] and $V_2$ [$\sigma^2\pi^4\sigma(s)^2\delta^2$ electronic configuration] in which the M–M distance are 168 and 177 pm, respectively.

There are hundreds of compounds containing quadruple, triple, and single metal–metal bonds, as well as less numerous examples of compounds in which the multiplicity of the M$-$M bonds equals 3.5, 2.5, 2.0, 1.5, 2/3, and 0.5. The formation of complexes containing metal–metal bonds depends on the position of an element in the Periodic Chart. The interaction of the effective nuclear charge with valence electrons decreases as the atomic number increases in a given group. Metals of the first transition series which possess common oxidation states rarely form strong metal–metal bonds. This is particularly true for groups 6 and 7, where Mo(III), W(III), Tc(IV), and Re(IV) form numerous dinuclear complexes containing M$\equiv$M bonds, while analogous Cr(III) and Mn(IV) compounds are not known.[1b, 1c, 1e]

The formation of M$-$M bonds also strongly depends on the nature of ligands.[1f] Of the two $d^1$–$d^1$ dimers, $Mo_2Cl_{10}$ and $Mo_2Cl_4(OPr^i)_6$, only the second contains Mo$-$Mo single bonds. Similarly, the [$Re_2Cl_8$]$^{2-}$ compound possesses a quadrauple Re$\overset{4}{-}$Re bond, while the high-spin complex [$Re_2(\mu-Cl)_2 Cl_4(dppe)_2$] does not exhibit any metal–metal bonding between the central atoms.

Due to the energy order of the molecular orbitals, the quadruple M$\overset{4}{-}$M bond may exist in complexes possessing $d^4$ electronic configuration of the central atoms, as in the following compounds of Cr(II), Mo(II), W(II), Tc(III), Re(III), for example: [$M_2(OOCR)_4$] (M = Cr, Mo, W), [$M_2X_8$]$^{4-}$ (M = Cr, Mo, W; X = alkyl; M = Mo, W, X = Cl), [$M_2(allyl)_4$] (M = Cr, Mo), [$M_2(OOCR)_4X_2$] (M = Tc, Re), [$M_2X_8$]$^{2-}$ (M = Tc, Re, X = Cl, Br), [$M_2X_4L_4$] (M = Mo, W), [$Re_2X_6L_2$], [$Re_2R_8$]$^{2-}$, etc. These complexes have eclipsed configurations which are preferred, because for such configurations the overlap between $d_{xy}$ and $d'_{xy}$ ($\delta$) is the greatest. The overlap is precisely zero in complexes having staggered structures:

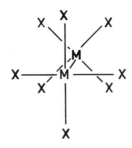

The overlap integral[1b] depends on the angle of internal rotation $\chi$ and is proportional to $\cos(2\chi)$. Therefore, even for relatively high $\chi$ values, the integral is large; for example, rotation by 30° causes the decrease of the overlap integral by half ($\chi = 45°$ for staggered conformation). Because of steric factors, deviations from eclipsed configuration are often observed.

In complexes, the quadruple M$\overset{4}{-}$M distances are considerably smaller than in the case of metals themselves. For example, in $Mo_2(OOCH)_4$ the Mo–Mo distance equals 209.1 pm, in $Cr_2(OOCCH_3)_4$ the Cr–Cr distance is 228.8 pm, and in

$[Cr_2(5\text{-Me-2-MeOC}_6H_3)_4]$ the metal-to-metal separation is 182.8 pm. The Mo–Mo and Cr–Cr distances in metals are 272.5 pm and 249.8 pm, respectively.

Compounds containing triple $M \equiv M$ bonds have $\sigma^2\pi^4$ or $\sigma^2\pi^4\delta^2\delta^{*2}$ electronic configurations. Thus, such bonds are formed by metals which possess $d^3$ or $d^5$ electronic structures. Examples of complexes containing $d^3$ central ions are compounds of Mo(III), W(III), and Re(IV), for instance, $M_2X_6$ (M = Mo, W, X = OR, $NR_2$, R). Compounds of such ions as Cr(I), Mo(I), W(I), Re(II), Ru(III), Os(III), for instance, $M_2cp_2(CO)_4$ (M = Cr, Mo, W), $Re_2Cl_4(PR_3)_4$, $Re_2(C_3H_5)_4$, and $[Os_2Cl_2(2\text{-OC}_5H_4N)_4]$, are representative of complexes possessing $d^5$ configurations.

Oxidation or reduction of complexes possessing quadruple $M \overset{4}{-} M$ bonds may lead to compounds in which the bond order equals 3.5. Such compounds may have $\sigma^2\pi^4\delta^1$ or $\sigma^2\pi^4\delta^2\delta^{*1}$ electronic configuration, for example, $[Mo_2(SO_4)_4]^{3-}$ and $[MoW(OOCBu')_4I]$ or $[Tc_2X_8]^{3-}$ and $[Re_2Cl_4(PR_3)_4]^+$, respectively. Less numerous are complexes having double $M = M$ bonds with electronic configuration $\sigma^2\pi^2$, for instance, $Mo_2(OPr^i)_8$ and $W_2Cl_4(OEt)_4(HOEt)_2$, or with $\sigma^2\pi^4\delta^2\delta^{*2}\pi^{*2}$ electronic configuration, such as $Ru_2L_2$ (L = anion dibenzotetraazaannulene, $C_{22}H_{22}N_4^{2-}$) and $Ru_2(mhp)_4$. Examples of compounds which show 2.5 bond orders are complexes of the type $Ru_2(OOCR)_4X$ in which the electronic configuration is $\sigma^2\pi^4\delta^2\delta^{*1}\pi^{*2}$. Many dimers of rhodium(II) and platinum(III) which contain $M - M$ single bonds are known, for example, $Rh_2(OOCR)_4$, $Rh_2(OOCR)_2(L\text{–}L)_2X_2$, $[Rh_2(SO_4)_4(H_2O)_2]^{4-}$, and $[Pt_2(SO_4)_4(H_2O)_2]^{2-}$. These compounds have $\sigma^2\pi^4\delta^2\delta^{*2}\pi^{*4}$ electronic configuration. Oxidation of $Rh_2^{4+}$ dimers leads to $[Rh_2X_4L_2]^+$ complexes with electronic configuration $\sigma^2\pi^4\delta^2\delta^{*2}\pi^{*3}$ in which the Rh–Rh bond order is 1.5. On the other hand, reduction of $Rh_2^{4+}$ dimers affords $Rh_2^{3+}$ complexes ($[Rh_2X_4L_2]^-$) in which the $M - M$ bond order equals 0.5 (electronic configuration $\sigma^2\pi^4\delta^2\delta^{*2}\pi^{*4}\sigma^{*1}$).

### a. *Theory of skeletal electron pairs*

Historically, the structural theory of borane and carborane clusters was developed first. This theory was later expanded to include clusters of the main group elements as well as those of the transition metals.[3,13] This approach allowed the anticipation, in many cases, of the structure and the number of valence electrons of a given cluster as well as its mode of transformation during oxidation and reduction reactions.

Boranes form three types of clusters: closo (closed), nido (nestlike), and arachno (weblike). Boranes of the general formula $B_nH_n^{2-}$ and isoelectronic carboranes such as $CB_{n-1}H_n^-$ and $C_2B_{n-2}H_n$ have closo structures; $n$ skeletal atoms lie at the vertices of a polyhedron composed of triangles (Figure 3.2).

Boranes or carboranes of nido type are compounds possessing formulas $B_nH_{n+4}$, $CB_{n-1}H_{n+3}$, $C_2B_{n-2}H_{n+2}$, etc. Their structures are analogous to closo compounds with the exception that one vertex is not occupied. In this way, a cluster containing $n$ atoms is represented by a polyhedron possessing $n+1$ vertices. For example, in $B_5H_9$, five boron atoms are located at the vertices of an octahedron; the sixth vertex is not occupied. Compounds of arachno type have the following formulas: $B_nH_{n+6}$, $CB_{n-1}H_{n+5}$, $C_2B_{n-2}H_{n+4}$, etc. The arrangement of $n$ skeletal atoms for these compounds is determined by the corners of a polyhedron having $n+2$ vertices, for instance, the boron atoms in $B_4H_{10}$ lie at the vertices of an octahedron in which two vertices are not occupied (Figure 3.3).

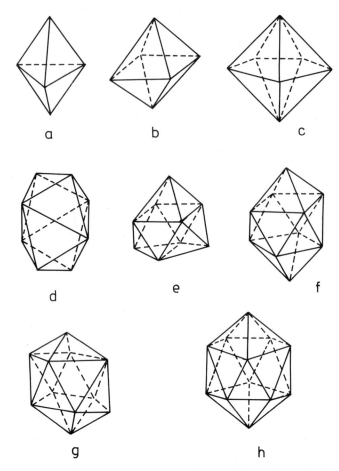

Figure 3.2.    Polyhredra of closo clusters: (a) trigonal bipyramid ($D_{3h}$), (b) octahedron ($O_h$), (c) pentagonal bipyramid ($D_{5h}$), (d) dodecahedron ($D_{2d}$), (e) tricapped trigonal prism ($D_{3h}$), (f) bicapped Archimedean antiprism ($D_{4d}$), (g) octadecahedron ($C_{2v}$), (h) icosahedron ($I_h$).

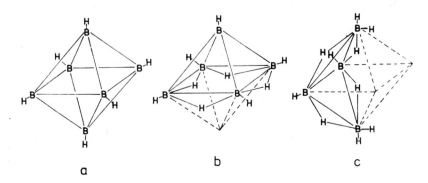

Figure 3.3.    Clusters: (a) closo $B_6H_6^{2-}$, (b) nido $B_5H_9$, (c) arachno $B_4H_{10}$.

The nido and arachno structures may formally be obtained from hypothetical $B_nH_n^{4-}$ and $B_nH_n^{6-}$ anions, respectively, by addition of protons to the terminal $B-H$ group or to the $B-H-B$ bridges, or by replacement of the $BH^-$ unit by the iso-electronic CH groups.

Each boron or carbon atom has sp hybridization. One of the hybridized orbitals of each atom forms a bond with the terminal hydrogen atom, and the remaining three orbitals (one sp and two p orbitals) give multicenter orbitals of the cluster grouping. The molecular orbital theory shows that borane polyhedra having $n$ vertices have such symmetry that $3n$ atomic orbitals of the cluster possessing $n$ skeletal atoms form $n + 1$ bonding molecular orbitals.[3]

In the case of the borane $B_6H_6^{2-}$ ($n = 6$) seven electron pairs are contributed to the bonding within the cage and the remaining twelve electrons are used to form two-center, two-electron (2c–2e) $B-H$ bonds. Similarly, the electron counting of the nido and arachno structures may be accounted for because these structures are derivatives of $B_nH_n^{4-}$ and $B_nH_n^{6-}$ hypothetical anions. Each $B-H$ group donates two electrons (and three orbitals) for the formation of the $B_n$ cage. Therefore, the $B_nH_n^{4-}$ and $B_nH_n^{6-}$ anions contain $n + 2$ and $n + 3$ electron pairs, respectively, which form bonds between the skeletal atom (Figure 3.4). Wade, Mingos, and Williams showed that, for the formation of other cages possessing an analogous structure, the same number of orbitals is needed. Thus, the same number of skeletal electrons is required regardless of the presence of the BH, CH, $M(CO)_3$, or $ML_n$ units in the cluster. In metal carbonyls, analogously to the $B-H$ unit, the metal atom uses three orbitals to form cluster bonds, and the remaining six orbitals are utilized to form bonds with the ligands or are

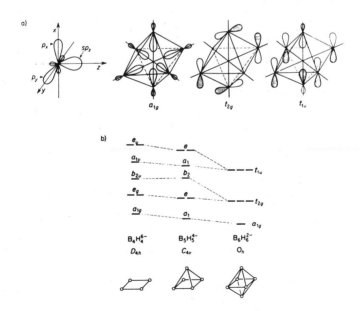

Figure 3.4. (a) Orbitals of the BH group which participate in the formation of skeletal bonds of clusters and bonding skeletal orbitals of the $B_6H_6^{2-}$ cluster. (b) Correlation diagram of molecular orbitals for $B_4H_4^{6-}$, $B_5H_5^{4-}$, and $B_6H_6^{2-}$.

occupied by nonbonding electrons. Therefore, metal carbonyl clusters of closo, nido, and arachno types have $n+1$, $n+2$, and $n+3$ ($n=$ number of metal atoms in the cluster) skeletal electron pairs. Accordingly, the number of valence electrons $N$ for clusters of the transition metals should be equal to

$$N = 12V + 2(V + x) = 14V + 2x \tag{3.1}$$

and for clusters of the main group metals

$$N = 2V + 2(V + x) = 4V + 2x \tag{3.2}$$

where $V$ denotes the number of metal atoms and $x = 1$ for closo, 2 for nido, and 3 for arachno clusters, respectively. The quantity $12V$ (or $2V$ for the main group metals) represents the sum of nonbonding and ligand–metal bonding electrons. Mononuclear covalent compounds of the transition metals generally have 18 valence electrons (Sidgwick's rule). The clusters obey the 18-electron rule if the number of valence electrons $N$ for the transition metals is given by

$$N = 18V - 2E \tag{3.3}$$

where $E$ denotes the number of the edges of the cluster (the number of the metal–metal bonds). For clusters of the main group metals, the number of valence electrons, according to Sidgwick's rule, should be as follows:

$$N = 8V - 2E \tag{3.4}$$

The atoms residing inside the cluster donate all their valence electrons for the formation of the skeletal bonds. For example, the compound $[Ru_6(CO)_{17}C]$ has 86 valence electrons and therefore represents a closo cluster. The placement of an atom of the main group elements or the transition metals in the cluster hole does not change the number of valence electrons. The value of $x$ is zero for clusters in which the group $M(CO)_n$ is added to one face of a regular polyhedron such as an octahedron, or $x$ is $-1$ if two groups are added, etc.

Table 3.1 lists clusters obeying Wade's and Sidgwick's rules.[1–4] Note that

*Table 3.1. Fulfillment of Sidgwick's and Wade's Rules in Clusters*

| Geometry of the cluster | Formula of the compound | Number of edges $E$ | Number of valence electrons $N$ | Sidgwick's rule $18V - 2E$ | Wade's rule $14V + 2X$ |
|---|---|---|---|---|---|
| Triangle (*arachno*) | $[Os_3(CO)_{12}]$ | 3 | 48 | 48 | 48 |
| Tetrahedron (*nido*) | $[Rh_4(CO)_{12}]$ | 6 | 60 | 60 | 60 |
| Trigonal bipyramid (*closo*) | $[Os_5(CO)_{16}]$ | 9 | 72 | 72 | 72 |
| Square pyramid (*nido*) | $[Fe_5(CO)_{15}C]$ | 8 | 74 | 74 | 74 |
| Trigonal prism | $[Rh_6(CO)_{15}C]^{2-}$ | 9 | 90 | 90 | 90 |
| Trigonal prism | $[Pt_6(CO)_{12}]^{2-}$ | 9 | 86 | 90 | 90 |
| Octahedron | $[Rh_6(CO)_{16}]$ | 12 | 86 | 84 | 86 |
| Capped octahedron[a] | $[Rh_7(CO)_{16}]^{3-}$ | 15 | 98 | 96 | 98 |
| Archimedean antiprism | $[Co_8(CO)_{18}C]^{2-}$ | 16 | 114 | 112 | 114 |

[a] An octahedron possessing one rhodium atom over the center of one face.

considerably more metal carbonyls obey Wade's rule than Sidgwick's rule. However, exceptions to Wade's rule occur quite commonly, and can be explained by the different geometry of carbonyl clusters compared to borane clusters, which causes a change in the number of bonding orbitals in the cluster. According to Wade's rule the change in the number of valence electrons caused by oxidation or reduction should lead to the geometry changes of the cluster:

$$\text{closo} \underset{-2e}{\overset{+2e}{\rightleftharpoons}} \text{nido} \underset{-2e}{\overset{+2e}{\rightleftharpoons}} \text{arachno} \qquad (3.5)$$

Other examples of the main group and transition metal clusters are given in Tables 3.2 and 3.3.

The analogy betweeen clusters of the transition metals and boranes is clearly seen from the molecular orbital schemes[14] for $Co_6$ and $B_6$ (Figure 3.5).

The interaction between $d$ orbitals is weak for given experimental distances between the metal atoms. Therefore, these $d$ orbitals only form a narrow band of molecular orbitals. In the case of more diffused $s$ and $p$ orbitals, the energy of molecular orbitals created from their overlap becomes more differentiated. The number of bonding orbitals resulting from $s$ and $p$ orbitals is the same for $Co_6$ and $B_6$.

*Table 3.2. Some Clusters of the Main Group Elements*[1,3,9,11–13,22–25]

| Cluster | Cluster structure |
|---|---|
| $Hg_2^{2+}$, $Cd_2^{2+}$, $Te_2^{2+}$ | |
| $P_4$ | Tetrahedron, $T_d$ |
| $Pb_4^{4-}$ | Tetrahedron, $T_d$ |
| $M_4^{2+}$ (M = S, Se, Te) | Square, $D_{4h}$ |
| $Hg_4^{6-}$ | Square, $D_{4h}$ |
| $Bi_4^{2-}$ | Square, $D_{4h}$ |
| $Bi_5^{3+}$ | Trigonal bipyramid, $D_{3h}$ |
| $Sn_5^{2-}$ | Trigonal bipyramid, $D_{3h}$ |
| $Pb_5^{2-}$ | Trigonal bipyramid, $D_{3h}$ |
| $Te_6^{4+}$ | Trigonal prism |
| $Sb_7^{3+}$ | Capped octahedron, $C_{3v}$  |
| $Pb_7^{4-}$ | Capped octahedron, $C_{3v}$ |
| $P_7^{3-}$ | $C_{3v}$ 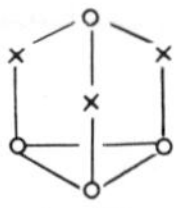 |
| $P_4S_3$ | $C_{3v}$ |
| $Bi_8^{2+}$ | Square Archimedean antiprism |
| $Bi_9^{5+}$ | Tricapped trigonal prism, $D_{3h}$ |
| $Sn_9^{4-}$ | Capped Archimedean antiprism, $C_{4v}$ |
| $Pb_9^{4-}$ | Capped Archimedean antiprism, $C_{4v}$ |
| $Ge_9^{4-}$ | Capped Archimedean antiprism, $C_{4v}$ |
| $Ge_9^{2-}$ | Tricapped trigonal prism |
| $Rb_9O_2^{5+}$ | Face sharing bioctahedron, oxygen ions are located in octahedral holes |
| $Cs_{11}O_3^{5+}$ | Three octahedra connected by faces, oxygen ions are located in octahedral holes |
| $P_{11}^{3-}$ | Trishomocubane |

A striking similarity of molecular orbital schemes of these clusters confirms Wade's rule. Orbital diagrams for other elements of the 8–10 groups are also analogous. However, for gold clusters, such diagrams are different, because the *p–p* orbital overlap of gold is very small. The analogy between clusters of the transition metals and compounds of the main group elements becomes more visible if the number of orbitals able to form skeletal bonds and the number of electrons occupying these orbitals are compared with corresponding fragments of clusters: $M(CO)_n$ and BH or CH. The groups BH and CH have three orbitals available for skeletal cluster bond formation, i.e., between atoms residing at the corners of the cluster. These three orbitals are occupied by two electrons in the BH group and by three electrons in the CH fragment. The cluster fragment, $M(CO)_n$, which possesses axial symmetry, has $6 - n$ occupied, non-bonding orbitals with definite *d*-orbital character as well as *n* orbitals which form

*Table 3.3. Examples of Transition Metal Clusters*[1–15,22–25]

| Cluster | Type of cluster, structure, type of metallic lattice |
|---|---|
| $[Cr_2(CO)_{10}]^{2-}$, $[Mn_2(CO)_{10}]$, $[Fe_2(CO)_9]$, $[Co_2(CO)_8]$, $[Pd_2(CNR)_6]^{2+}$, $[Rh_2(PR_3)_8]$, $[Pt_2(CNR)_6]^{2+}$ | Dinuclear |
| $[Mn_3H_3(CO)_{12}]$, $[Mn_3cp_3(NO)]$, $[Re_3H_2(CO)_{12}]^-$, $[M_3(CO)_{12}]$ (M = Fe, Ru, Os), $[Co_2Os(CO)_{11}]$, $[Rh_3cp_3(CO)_3]$, $[Rh_3H_3\{P(OR)_3\}_6]$, $[M_3(CNR)_6]$ (M = Pd, Pt), $[Pt_3(CO)_6]^{2-}$ | Trinuclear, triangle |
| $[M_4(CO)_{12}]$ (M = Co, Rh, Ir), $[Fe_4cp_4(CO)_4]$, $[Fe_4(CO)_{13}]^{2-}$, $[M_4H_4(CO)_{12}]$ (M = Ru, Os), $[Co_4cp_4H_4]$, $[Ni_4(CO)_6(PR_3)_4]$, $[Ni_4(CNR)_7]$, $[Ni_4cp_4H_3]$, $[Pt_4(CO)_5(PR_3)_4]$ | Tetranuclear, tetrahedron |
| $[Re_4(CO)_6]^{2-}$, $[Fe_4(CO)_{13}H]^-$, $[Co_2Pt_2(CO)_8(PPh_3)_2]$, $[Pt_4(CO)_5(PPhMe_2)_4]$, $[Os_3Au(CO)_{10}PPh_3Cl]$, $[Co_4(CO)_{10}S_2]$ | Tetranuclear, "butterfly" structure or square |
| $[Os_5(CO)_{16}]$, $[Ni_5(CO)_{12}]^{2-}$, $[Ni_3M_2(CO)_{16}]^{2-}$ (M = Mo, W), $[Pt_3Sn_2(COD)_2Cl_6]$ | Pentanuclear, trigonal bipyramid |
| $[M_6(CO)_{16}]$ (M = Co, Rh, Ir), $[M_6(CO)_{15}]^{2-}$ (M = Co, Rh, Ir), $[Ni_6(CO)_{12}]^{2-}$, $[Os_6(CO)_{18}]^{2-}$, $[M_6(CO)_{18}H]^-$ (M = Ru, Os), $[Ru_6H_2(CO)_{18}]$, $[Au_6(PR_3)_6]^{2+}$, $[Cu_6(C_6H_4NR_2)_4Br_2]$, $[Cu_6H_6(PR_3)_6]$ | Octahedron |
| $[Pt_6(CO)_{12}]^{2-}$ | Trigonal prism |
| $[Os_6(CO)_{18}]$ | Bicapped tetrahedron |
| $[Os_6H_2(CO)_{18}]$ | Capped square pyramid |
| $[Rh_7(CO)_{16}]^{3-}$, $[Rh_7(CO)_{16}I]^{2-}$, $[Os_7(CO)_{21}]$ | Capped octahedron, $C_{3v}$ |
| $[Ni_8(CO)_8(PR)_6]$ | Cube |
| $[Cu_2Rh_6(CO)_{15}(NCR)_2]$ | $D_{3h}$ |
| $[Ni_9(CO)_{18}]^{2-}$, $[Pt_9(CO)_{18}]^{2-}$ | $C_3$ |
| $[Au_9(PR_3)_8]^{3+}$ | |
| $[Fe_4Pt_6(CO)_{22}]^{2-}$ | $S_4$ |
| $[Pd_{10}(CO)_{12}(PBu_3^n)_6]$, $[Os_{10}(CO)_{24}C]^{2-}$ | Tetracapped octahedron |
| $[Au_{11}I_3(PR_3)_7]$ | $C_{3v}$ |
| $[Rh_{12}(CO)_{30}]^{2-}$ : $[(OC)_{15}Rh_6\text{-}Rh_6(CO)_{15}]^{2-}$ | $C_{2h}$ |
| $[Pt_{12}(CO)_{24}]^{2-}$ | $C_3$ |

*(Table continued)*

*Table 3.3. (continued)*

| Cluster | Type of cluster, structure, type of metallic lattice |
|---|---|
| $[Ni_{12}(CO)_{21}H_2]^{2-}$, $[Fe_6Pd_6(CO)_{24}H]^{3-}$ | $D_{3h}$ |
| $[Rh_{12}(CO)_{25}C_2]$ | $C_i$ |
| $[Rh_{12}(CO)_{24}(C)_2]^{2-}$ | $D_{2h}$ |
| $[Co_{13}(CO)_{24}(C)_2H]^{4-}$ | $C_2$ |
| $[Rh_{13}(CO)_{24}H_3]^{2-}$ | $D_{3h}$, hcp[a] |
| $[Rh_{13}(CO)_{24}H_2]^{3-}$ | $D_{3h}$, hcp |
| $[Rh_{14}(CO)_{25}]^{4-}$ | $C_{4v}$, bcc[a] |
| $[Rh_{14}(CO)_{26}]^{2-}$ | $C_s$, bcc/hcp |
| $[Rh_{15}(CO)_{27}]^{3-}$ | $C_1$, bcc/hcp |
| $[Rh_{15}(CO)_{28}(C)_2]^-$ | $C_{2v}$ |
| $[Pt_{15}(CO)_{30}]^{2-}$ | $C_3$ |
| $[Rh_{17}(CO)_{32}(S)_2]^{3-}$ | $D_{4d}$ |
| $[Pt_{18}(CO)_{36}]^{2-}$ | $C_3$ |
| $[Pt_{19}(CO)_{22}]^{4-}$ | $D_{5h}$ |
| $[Rh_{22}(CO)_{37}]^{4-}$ | $C_{3v}$, $ABAC$[a] |
| $[Pt_{26}(CO)_{32}]^{2-}$ | $D_{3h}$, hcp |
| $[Pt_{38}(CO)_{44}]^{2-}$ | $O_h$, ccp[a] |
| $[Ni_{38}Pt_6(CO)_{48}H_{6-n}]^{n-}$ | $O_h$, ccp |

[a] hcp (hexagonal close-packed) hexagonal lattice possessing the densest packing arrangement, $A_3$ (stacking arrangement of layers AB); bcc (body-centered cubic) regular lattice body-centered, $A_2$; ccp (cubic close-packed) regular lattice face centered possessing the densest packing arrangement, $A_1$ (stacking arrangements of layers ABC); ABAC lattice possessing the densest packing arrangement (stacking arrangement of layers ABAC).

metal–carbonyl bonds with $n$ CO groups. The remaining three orbitals may form skeletal bonds between metal atoms. These three orbitals are occupied by $y = q - 2(6 - n)$ electrons. For any $ML_m$ fragments, the number of electrons which are utilized for skeletal bond formation may be calculated as follows for the transition elements:

$$y = q - (12 - m \cdot z) = q + p - 12 \qquad (3.6)$$

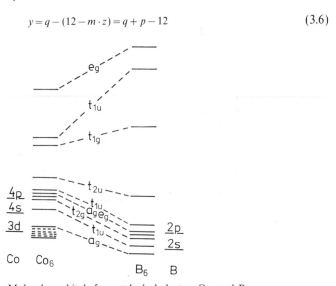

Figure 3.5. Molecular orbitals for octahedral clusters $Co_6$ and $B_6$.

For the main group elements, the following expression is used:

$$y = q - (2 - m \cdot z) = q + p - 2 \tag{3.7}$$

where $q$ is the number of valence electrons of the metal, $z$ is the number of electrons donated by the ligand to form a metal–ligand bond, and $p$ is the number of electrons donated by all ligands. The number of skeletal bonding electrons donated by $EL_m$ cluster fragments of the main group and the transition elements is given in Table 3.4.[3] The three-electron fragments $M(CO)_3$ ($M = Ir$, Co) may be replaced by the CH unit. Thus, it is easy to calculate the number of skeletal electron pairs in the clusters of the type $E_n L_m$ [clusters (3.8)]. These are nido clusters for which $n = 4$. The number of

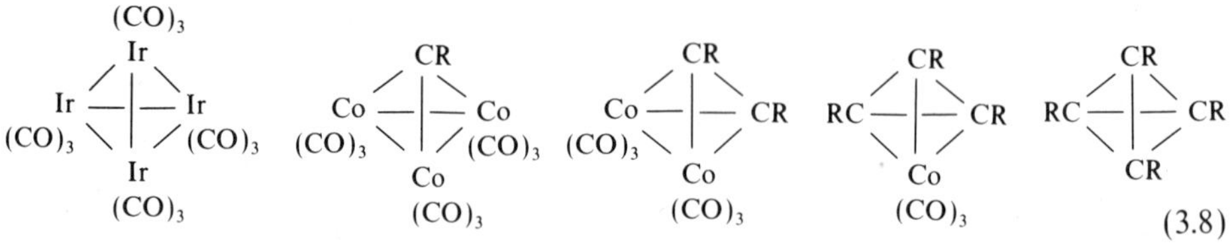

$$\tag{3.8}$$

skeletal electrons is $2(n + 2) = 12$. Because $M(CO)_3$ and CH are $3e$ fragments, the number of skeletal electrons is truly $4 \times 3 = 12$. The (EtCCEt) $Co_4(CO)_{10}$ molecule is a closo cluster in which the acetylene carbon atoms and the cobalt atoms lie at the vertices of an octahedron; the fragments $Co(CO)_{2.5}$ are $2e$ donors. Therefore, this compound has 14 skeletal electrons ($3e$ from CEt). Analogously, benzene may be treated as an arachno cluster possessing $6 + 3$ skeletal electron pairs. Six $3e$ CH fragments actually form a cluster containing 18 skeletal electrons. The remaining 12 valence electrons of benzene form $C - H$ bonds. Table 3.5 and Figure 3.6 show examples of compounds which may be considered as clusters.

Table 3.4. Number of Skeletal Electrons of $EL_m$ Fragments

| $q$ | Element | $y = q + p - 2$ | | |
|---|---|---|---|---|
| | | $E\ (p = 0)$ | $EH\ (p = 1)$ | $EH_2$, EL $(p = 2)$ |
| 1 | Li, Na | −1 | 0 | 1 |
| 2 | Be, Mg, Zn, Cd, Hg | 0 | 1 | 2 |
| 3 | B, Al, Ga, In, Tl | 1 | 2 | 3 |
| 4 | C, Si, Ge, Sn, Pb | 2 | 3 | 4 |
| 5 | N, P, As, Sb, Bi | 3 | 4 | 5 |
| 6 | O, S, Se, Te | 4 | 5 | 6 |
| 7 | F, Cl, Br, I | 5 | 6 | 7 |
| | | $y = q + p - 12$ | | |
| | | $M(CO)_2$ | Mcp | $M(CO)_3$ | $M(CO)_4$ |
| 6 | Cr, Mo, W | −2 | −1 | 0 | 2 |
| 7 | Mn, Te, Re | −1 | 0 | 1 | 3 |
| 8 | Fe, Ru, Os | 0 | 1 | 2 | 4 |
| 9 | Co, Rh, Ir | 1 | 2 | 3 | 5 |
| 10 | Ni, Pd, Pt | 2 | 3 | 4 | 6 |

*Table 3.5. Number of Skeletal Electrons of Hydrocarbons and Transition Metal Complexes*

| Number of skeletal electron pairs, nep | Basic polyhedron | Cluster | | |
|---|---|---|---|---|
| | | closo nep = V + 1 | nido nep = V + 2 | arachno nep = V + 3 |
| 6 | Trigonal bipyramid | $[(PhCCPh)Fe_3(CO)_9]$ | $[(PhCCPh)Co_2(CO)_6]$ | $[(CH_2CH_2)Pt(PPh_3)_3]$ |
| 7 | Octahedron | $[(EtCCEt)Co_4(CO)_{10}]$ | $[(C_4H_4)Fe(CO)_3]$ $[C_5H_5]^+$ | $[(\pi-C_3H_5)Co(CO)_3]$ $[C_4H_4]^{2-}$ |
| 8 | Pentagonal bipyramid | $[(PhCCPh)_2Fe_3(CO)_8]$ | $[(C_5H_5)Mn(CO)_3]$ $[(C_5H_5BeMe]$ | $[(H_2CCHCHCH_2)Fe(CO)_3]$ $C_5H_5^-$ |
| 9 | Hexagonal bipyramid | | $[(C_6H_6)_2Cr]$ | $C_6H_6$ |
| 10 | Heptagonal bipyramid | | $[(C_7H_7)V(CO)_3]$ | $C_7H_7^+$ |

## b. *The isolobal analogy of fragments*

Table 3.4 shows that there is electron equivalency of groups contributing to a compound. Thus, the BH group is equivalent to $M(CO)_3$ (M = Fe, Ru, Os), Mcp (M = Co, Rh, Ir) and $M(CO)_4$ (M = Cr, Mo, W) while the CH group is equivalent to Mcp (M = Ni, Pd, Pt), $M(CO)_3$ (M = Co, Rh, Ir), $M(CO)_4$ (M = Mn, Tc, Re), etc. Halpern first turned attention to the electron equivalency of certain organic and inorganic groups.[17] He showed the similarity in reactivity patterns between alkyl radicals $CR_3$

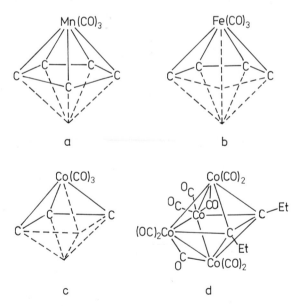

Figure 3.6. Skeletal structures of complexes showing their relationship with polyhedra:
(a) $Mn(C_5H_5)(CO)_3$, (b) $Fe(CH_2=CHCH=CH_2)(CO)_3$, (c) $Co(C_3H_5)(CO)_3$,
(d) $Co_4(EtC\equiv CEt)(CO)_{10}$.

and the $Co(CN)_5^{3-}$ complex as well as between carbenes and four-coordinate complexes possessing $d^8$ electron configuration of the central atom.

The reactivity of compounds or fragments of compounds is determined by the frontier orbitals, namely the highest occupied molecular orbital (HOMO) and the lowest unoccupied molecular orbital (LUMO). In order to define similarity between fragments of compounds or compounds themselves, the term "isolobality" was introduced.[14, 18–20] Fragments are said to be isolobal if the number, symmetry, energy, and shape of their frontier orbitals are similar, and the number of electrons occupying these orbitals is the same. The isolobal analogy is best considered by examples of fragments which are formed by removal of certain ligands from the coordination sphere of octahedral complexes and from tetrahedral carbon compounds. It can easily be seen that $ML_m$ fragments of an octahedron are analogous to $CH_n$ fragments.

$$\text{—M} \qquad \text{M} \qquad \text{M} \qquad \text{C} \qquad \text{C} \qquad \text{C} \qquad (3.9)$$

According to the theory, the formation of an octahedral complex may be viewed as follows: from nine valence orbitals of the central atom, six orbitals undergo hybridization. These hybridized orbitals interact with six two-electron ligands, creating six bonding and six antibonding orbitals. In the case of complexes containing ligands which create a strong field (such ligands show quite strongly $\pi$-acceptor properties), bonding and nonbonding orbitals are occupied; therefore, $18e$ compounds are formed (Figure 3.7a). The removal of one ligand leads to a pentacoordinate complex $ML_5$ possessing $C_{4v}$ symmetry. Consequently, the energy of one hybrid does not change, and this hybrid belongs to frontier robitals, just as "$t_{2g}$" orbitals. For $ML_4$ and $ML_3$ fragments, the frontier orbitals are those hybrids which are directed to positions previously occupied by removed ligands (Figure 3.8).

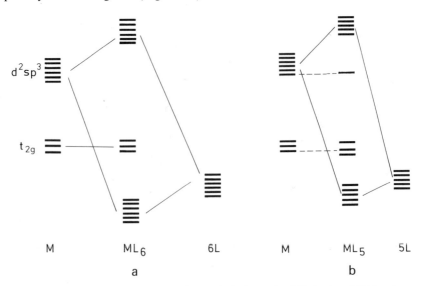

Figure 3.7. Simplified MO schemes for (a) the octahedral complex $ML_6 (O_h)$ and (b) for the complex $ML_5$ ($C_{4v}$).

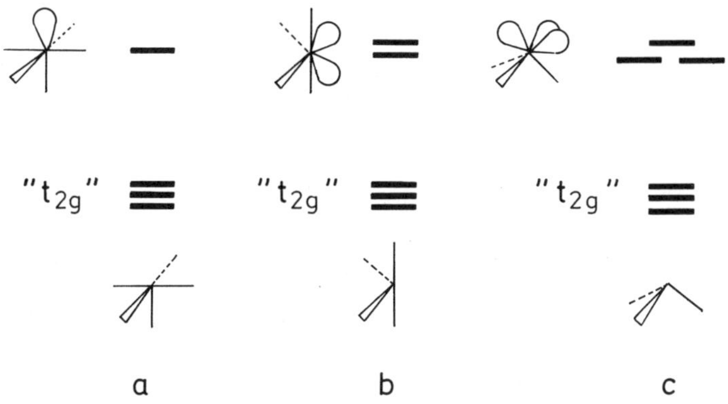

Figure 3.8. Schemes of the frontier orbitals for the fragments: (a) $ML_5$, (b) $ML_4$, (c) $ML_3$.

The CR fragment is also isolobal with the $d^5$-$ML_5$ groups which have three unpaired electrons occupying the $e$ and $a_1$ frontier orbitals (Figure 3.9). The isolobal relatioship is described by a "two-headed" arrow with half an orbital below:

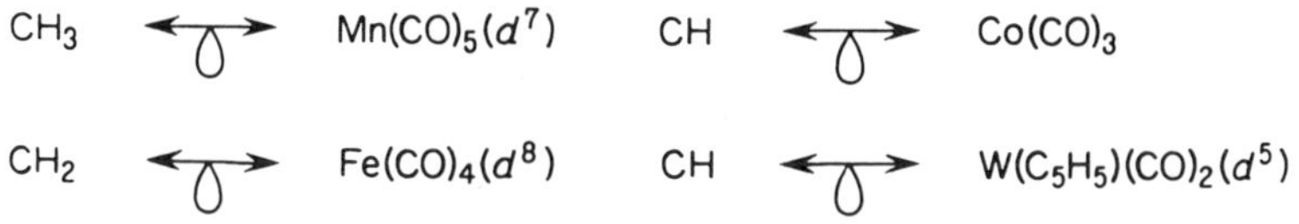

$$CH_3 \quad \longleftrightarrow \quad Mn(CO)_5(d^7) \qquad CH \quad \longleftrightarrow \quad Co(CO)_3$$

$$CH_2 \quad \longleftrightarrow \quad Fe(CO)_4(d^8) \qquad CH \quad \longleftrightarrow \quad W(C_5H_5)(CO)_2(d^5) \tag{3.10}$$

The removal of two ligands from $ML_5$ and $ML_4$ fragments which lie on one axis (for example, $z$) causes an increase in the number of nonbonding $d$ orbitals because the metal $d_{z^2}$ atomic orbital is lowered in energy. It changes from the metal–ligand $\sigma$-antibonding orbital into a nonbonding one. From an $ML_5$ complex, $ML_3$ is formed, which is T-shaped (Figure 3.10). Therefore, it is obvious that the following fragments are isolobal: $d^n$-$ML_5$ and $d^{n+2}$-$ML_3$ ($C_{2v}$, T-shaped) as well as $d^n$-$ML_4$ ($C_{2v}$) and $d^{n+2}$-$ML_2$ ($C_{2v}$).

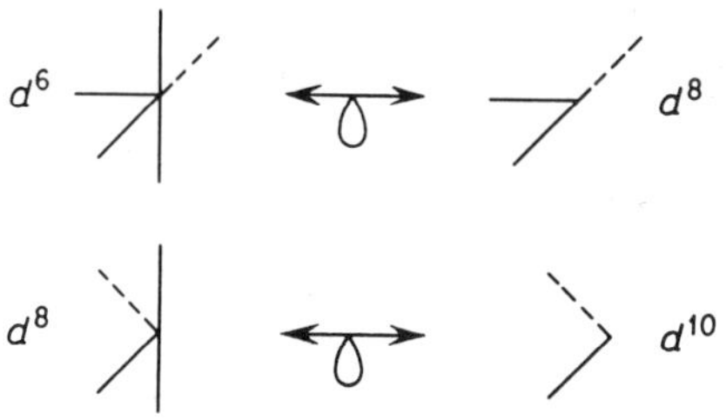

$$d^6 \quad \longleftrightarrow \quad d^8$$

$$d^8 \quad \longleftrightarrow \quad d^{10} \tag{3.11}$$

For the transition metal complexes possessing the coordination number $n$, the number of bonding and nonbonding orbitals equals nine and, in the case of the main group compounds, the number of these orbitals is four. This leads to the effective atomic number rule (Sidgwick's rule) and the octet rule, respectively. In $ML_{n-x}$ and $EL'_{m-x}$ fragments which result from the removal of $x$ $2e$ ligands from the coordination sphere of an $18e$ complex $ML_n$ or from an $8e$ compound $EL'_m$, $x$ directed hybrids are created.

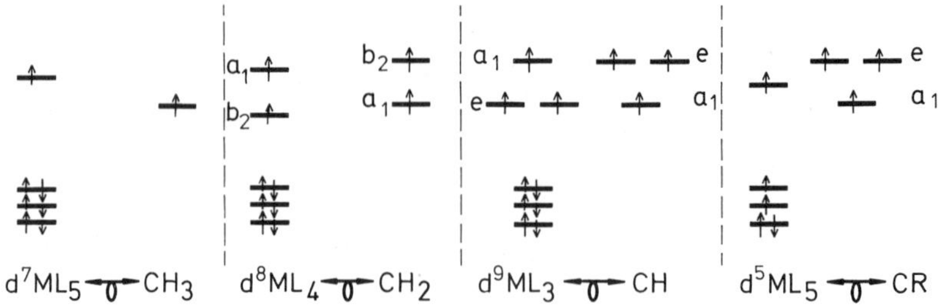

$$d^7ML_5 \longleftrightarrow CH_3 \qquad d^8ML_4 \longleftrightarrow CH_2 \qquad d^9ML_3 \longleftrightarrow CH \qquad d^5ML_5 \longleftrightarrow CR$$

Figure 3.9.   Isolobal analogies between the fragments $CH_m$ and $d^xML_n$.

Thus, $ML_{n-x}$ and $EL'_{m-x}$ fragments are isolobal, provided that they contain the appropriate number of electrons (Figure 3.11). For fragments which are obtained from $18e$ complexes $ML_7$ ($d^4$ configuration) and $ML_8$ ($d^2$ configuration), the relationships (3.12) are obeyed. Isolobal fragments obtained from various complexes are given in Table 3.6.

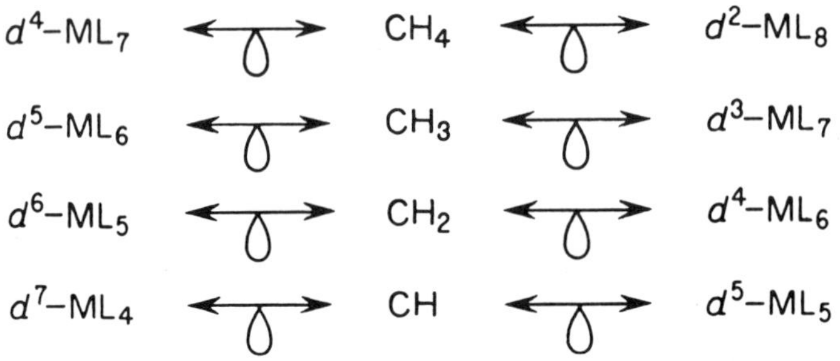

$$
\begin{array}{ccccc}
d^4\text{--}ML_7 & \longleftrightarrow & CH_4 & \longleftrightarrow & d^2\text{--}ML_8 \\
d^5\text{--}ML_6 & \longleftrightarrow & CH_3 & \longleftrightarrow & d^3\text{--}ML_7 \\
d^6\text{--}ML_5 & \longleftrightarrow & CH_2 & \longleftrightarrow & d^4\text{--}ML_6 \\
d^7\text{--}ML_4 & \longleftrightarrow & CH & \longleftrightarrow & d^5\text{--}ML_5
\end{array}
\qquad (3.12)
$$

Tables 3.4–3.7 and Figures 3.7–3.11 show that there are many analogies between organic compounds and inorganic compounds including clusters of the main group

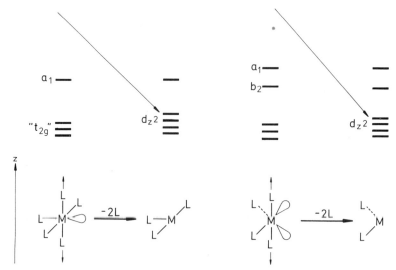

Figure 3.10.   Schemes of frontier orbitals for $T - ML_3$ and $ML_2$ ($C_{2v}$).

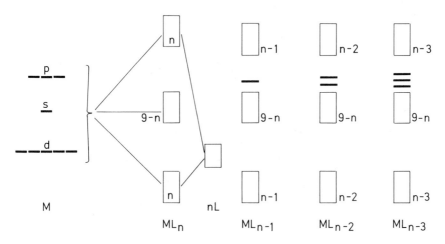

Figure 3.11. Simplified MO schemes for $ML_n$ complexes and $ML_{n-x}$ fragments.

elements and the transition metals. Cationic carbon compounds[70,71] of the type $[C_m R_m]^{n+}$ such as $[C_5 H_5]^+$ and $[C_6 Me_6]^{2+}$ are isoelectronic and isostructural with boron clusters $B_5 H_9$ and $B_6 H_{10}$ as well as with the following transition metal complexes: $Fe(C_4 H_4)(CO)_3$ abd $Fe_2(C_4 H_4)(CO)_6$.

The inorganic compound $P_4 O_6$ is analogous to adamantane $C_{10}H_{16}$ and urotropine because $CH_2 \longleftrightarrow O$ and $P \longleftrightarrow CH \longleftrightarrow N$ [see relationships (3.13) and

$$(3.13)$$

Table 3.6. The Isolobal Analogy[18]

| Organic fragment | Coordination number and electron configuration of fragments derived from transition metal complexes | | | | |
|---|---|---|---|---|---|
| $CH_3$ | $d^1$-$ML_8$ | $d^3$-$ML_7$ | $d^5$-$ML_6$ | $d^7$-$ML_5$ | $d^9$-$ML_4$ |
| $CH_2$ | $d^2$-$ML_7$ | $d^4$-$ML_6$ | $d^6$-$ML_5$ | $d^8$-$ML_4$ | $d^{10}$-$ML_3$ |
| $CH$ | $d^3$-$ML_6$ | $d^5$-$ML_5$ | $d^7$-$ML_4$ | $d^9$-$ML_3$ | |

*Chapter 3*

*Table 3.7. Isolobal Fragments*

---

$CH_3$, $V(CO)_6$, $Mocp(CO)_3$, $Mn(CO)_5$, $Fecp(CO)_2$, $Co(CO)_4$, $PtH(PR_3)_2$, $Au(PPh_3)$,

$CH_2$, $Ticp_2(CO)$, $Ta(Me)cp_2$, $Vcp(CO)_3$, $Cr(CO)_5$, $Mncp(CO)_2$, $Rhcp(CO)$, $IrCl(CO)_2$, $Ni(PR_3)_2$, $Pt(PR_3)_3$, $Cucp$,

$CH$, $Ticp(C_6H_6)$, $V(CO)_5$, $TacpCl(PMe_3)_2$, $Wcp(CO)_2$, $[Mncp(CO)_2]^+$, $Mn(CO)_4$, $ReBr_2(CO)_3$, $Fecp(CO)$, $Co(CO)_3$, $Nicp$

$CH_3^+$, $Cr(CO)_5$, $Mncp(CO)_2$, etc., $BH_3$

$CH_2^+$, $BH_2$, $Mn(CO)_4$, $Fecp(CO)$

$CH^+$, $BH$, $Fe(CO)_3$, $Cocp$

---

(3.14)]. The compounds $Ru_3(CO)_{12}$ and $Os_3(CO)_{12}$ as well as their derivatives are analogs of cyclopropane because the $M(CO)_4$ and $CH_2$ fragments are isolobal [rela-

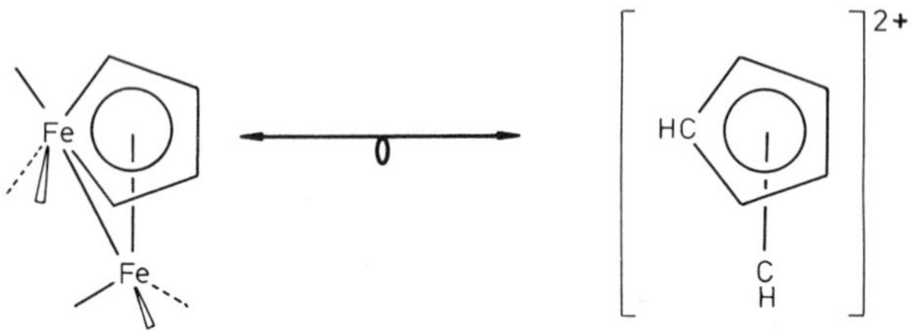

(3.14)

tionship (3.15)]. The structure of $Os_5(CO)_{19}$ is better understood if this compound is represented as an "olefin" complex, since $Os_2(CO)_8$ is isolobal with $R_2C=CR_2$ [rela-

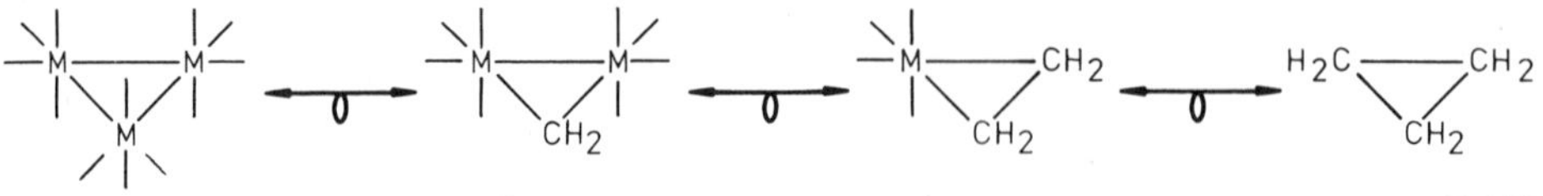

(3.15)

tionship (3.16)]. The isolobal fragments need not have similar chemical properties. The carbonyl $V(CO)_6$ does not form a dimer under normal conditions, while $CH_3$ radicals

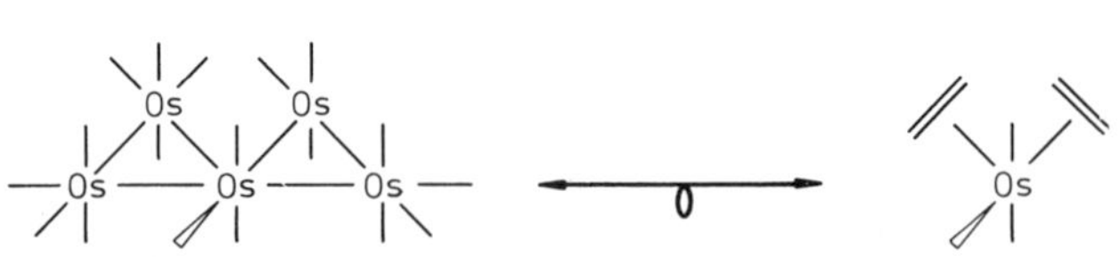

(3.16)

are very unstable, and rapidly undergo recombination to give $C_2H_6$. The properties of the isolobal fragments $Pt(PR_3)_2$ and $CH_2$ are decisively different; only $CH_2$ dimerizes to ethylene or polymerizes. However, "inorganic methylenes" readily form "cyclopropene" or "cyclopropane" derivatives:

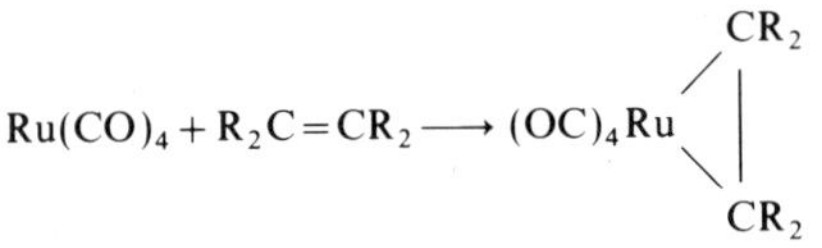

$$\text{Ru(CO)}_4 + R_2C = CR_2 \longrightarrow \text{(OC)}_4\text{Ru} \underset{CR_2}{\overset{CR_2}{\big|}} \tag{3.17}$$

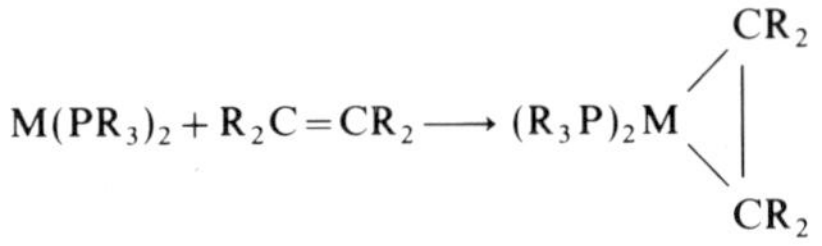

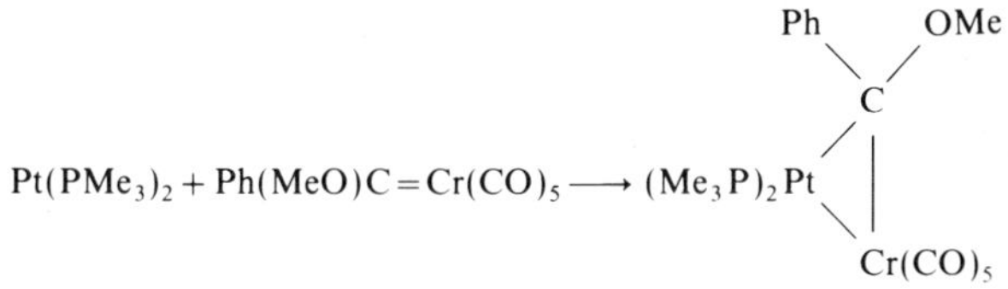

$$\text{M(PR}_3)_2 + R_2C = CR_2 \longrightarrow (R_3P)_2M \underset{CR_2}{\overset{CR_2}{\big|}} \qquad (M = \text{Ni, Pd, Pt}) \tag{3.18}$$

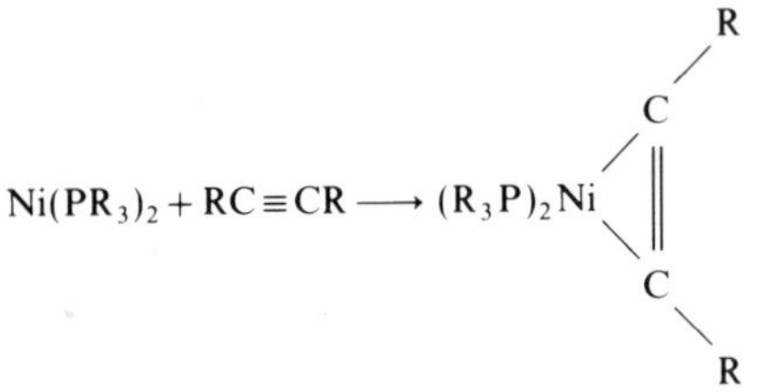

$$\text{Ni(PR}_3)_2 + RC \equiv CR \longrightarrow (R_3P)_2\text{Ni} \underset{\underset{R}{C}}{\overset{\overset{R}{C}}{\big\|}} \tag{3.19}$$

$$\text{Pt(PMe}_3)_2 + \text{Ph(MeO)C} = \text{Cr(CO)}_5 \longrightarrow (Me_3P)_2\text{Pt} \underset{\text{Cr(CO)}_5}{\overset{\overset{\text{Ph} \quad \text{OMe}}{C}}{\big|}} \tag{3.20}$$

$$\text{Pt(PMe}_3)_2 + \text{(OC)}_2\text{cpW} \equiv \text{CC}_6\text{H}_4\text{Me-p} \longrightarrow (Me_3P)_2\text{Pt} \underset{\text{Wcp(CO)}_2}{\overset{\overset{C_6H_4Me\text{-}p}{C}}{\big\|}} \tag{3.21}$$

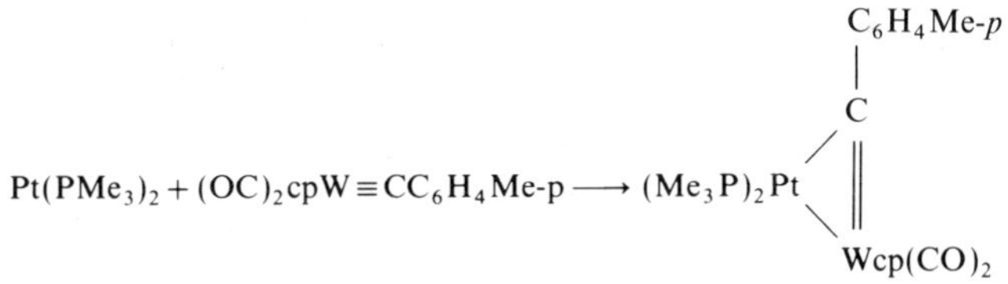

The formation of many other larger clusters may also be explained by isolobal analogy.

### c. *Topological calculations of CVMO*

The number of valence orbitals of clusters having various geometries may also be calculated utilizing Euler's theorem, which states that the number of edges $E$ of a polyhedron possessing $V$ vertices and $F$ faces is given by the formula

$$E = V + F - 2 \tag{3.22}$$

$$V = V_n + V_m \tag{3.23}$$

where $V_n$ is the number of atoms of the main group elements obeying the octet rule, and $V_m$ is the number of the transition metal atoms. Utilizing Euler's theorem and the effective atomic number rule, Teo[21] derived the expression for the number of molecular orbitals of a cluster $T$(CVMO). This number, assuming that each edge corresponds to a two-center, two-electron bond (2c–2e) should, according to Sidgwick's rule, be equal to

$$T = 3V_n + 8V_m - F + 2 \qquad (3.24)$$

because

$$T = N/2 = 4V_n + 9V_m - E \qquad (3.25)$$

and

$$N = 8V_n + 18V_m - 2E \qquad (3.26)$$

However, in delocalized systems, not all metal–metal interactions are considered to be two-center, two-electron bonds. Therefore, a constant $Y$ is introduced in order to achieve the agreement between the calculated and observed number of bonding orbitals:

$$T = 3V_n + 8V_m - F + 2 + Y \qquad (3.27)$$

Some rules for determination of the value of $Y$ are[21]:

1.  $Y = 0$ for polyhedra in which each skeletal atom is bonded to three metal atoms (3-connected polyhedra): triangular and pentagonal prisms, cubes, cuneans, and pentagonal dodecahedrons.

2.  Capping of an $n$-gonal face of a polyhedron by $ML_m$ increases $Y$ by $n - 3$; thus, capping of a triangular, tetragonal, or pentagonal face causes an increase in $Y$ of 0, 1, or 2, respectively.

3.  $Y = 0$ for all pyramids.

4.  For bipyramids, $Y$ may be determined by utilizing rule number 2 by capping a corresponding pyramid: $Y = 0$ or 2 for a trigonal bipyramid, $Y = 1$ (or 3) for a tetragonal bipyramid, and $Y = 2$ for a pentagonal bipyramid. In the case of higher $Y$ values (2 for a trigonal bipyramid and 3 for a tetragonal bipyramid) the "extra" electron pairs occupy tangential orbitals $xy$ of the bipyramid. For elongated bipyramids these orbitals become bonding orbitals.[21] Lauher[4] pointed out that in deformed bipyramids the number of bonding skeletal pairs is increased.

5.  $Y = 1$ for a trigonal antiprism, 1 or 3 for a square antiprism, as well as 3 for a pentagonal antiprism.

6.  $Y = S$ for polyhedra connected by vertices or edges; $S$ denotes the number of common vertices or edges. $Y = -H$ for polyhedra containing common faces, where $H$ is the number of hidden edges.

Both the skeletal electron pair theory of Wade, Mingos, and Williams[3,14] and Teo's topological theory of calculating electrons[21] are based on Sidgwick's effective atomic number rule. Both theories linked the number of bonding orbitals with the number of metal atoms, assuming that three orbitals of each metal are utilized to form bonds in the cluster. Both theories also assume that the metal–metal bonds are single. Therefore, both theories give very similar or identical results. Both are based on the molecular orbital theory, but Teo's topological theory links $Y$ with "missing" antibonding orbitals

($Y = E - A$, where $E$ is the number of edges and $A$ the number of antibonding orbitals), while the skeletal electron pair theory gives the number of bonding orbitals of the cluster. The number of valence electrons for analogous clusters of the main group elements and the transition metals, based on the discussed cluster theories, are given in Table 3.8.

### d. *Reactions between clusters*

In recent years, it was discovered that a considerable number of large clusters is obtained as a result of condensation of smaller fragments such as tetrahedral, octahedral, or trigonal prismatic units, etc. The condensation may proceed through a vertex, an edge, or a face. Mingos[14] developed rules for the calculation of the number of valence electrons for the condensation products. The formation of a bioctahedron linked by a vertex may be represented as condensation of an octahedral cluster and a square-pyramidal cluster (Figure 3.12). Molecular orbital calculations show that the nido fragment has frontier orbitals $a_1$ and $e$ which have donor properties; therefore, the nido cluster is a six-electron donor interacting with empty orbitals of the closo fragment. If the nido compound contains $m$ metal atoms, and the closo cluster has $n$ metal atoms, then the created bioctahedron possessing a common corner has $14(n + m)$ valence electrons [$x = 0$ in equation (3.1)] (Figure 3.12a):

$$14(n + m) = (14m + 4) + (14n + 2) - 6 \qquad (3.28)$$

If this process is represented as a condensation of two closo polyhedra with $n$ and $p = m + 1$ atoms, then

$$14(n + m) = [14(m + 1) + 2] + (14n + 2) - 18 \qquad (3.29)$$

The formation of a bioctahedron possessing a common face ($x = -1$) may be represented as a junction of closo and arachno fragments; the arachno fragment is a 10$e$ donor having $a_1$, $e_1$, and $e_2$ type frontier orbitals:

$$14(m + n) - 2 = (14m + 6) + (14n + 2) - 10 \qquad (3.30)$$

$$14(m + n) - 2 = [14(m + 2) + 2] + (14n + 2) - 34 \qquad (3.31)$$

*Table 3.8. Number of Valence Electrons in Clusters*[14]

| Polyhedron | Main group compounds | | Transition metal compounds |
|---|---|---|---|
| Closo | $4n + 2$ ($n \geqslant 5$) | $B_n H_n^{2-}$ | $14n + 2$ $Os_5(CO)_{16}$ |
| | | | $Rh_6(CO)_{16}$ |
| Nido | triangular polyhedra | $4n + 4$ ($n \geqslant 4$) $B_n H_{n+4}$ | $14n + 4$ $Ru_5C(CO)_{15}$ |
| Arachno | | $4n + 6$ ($n \geqslant 4$) $B_n H_{n+6}$ | $14n + 6$ $Fe_4H(CO)_{13}$ |
| | | | $[Rh_6C(CO)_{15}]^{2-}$ |
| 3-connected clusters (tetrahedron, prisms, pentagonal dodecahedron) | $5n$ | $C_n H_n$ | $15n$ $M_4(CO)_{12}$ (M = Co, Rh, Ir), $[Rh_6(CO)_{15}C]^{2-}$ |
| Ring compounds | $6n$ ($n \geqslant 3$) | $C_n H_{2n}$, $S_n$ | $16n$ $Os_3(CO)_{12}$, |
| Isolated vertex held together by bridging groups | $8n$ | $Ph_4Al_4^- N_4Ph_4$ | $18n$ $Cu_4I_4(AsMe_3)_4$ |

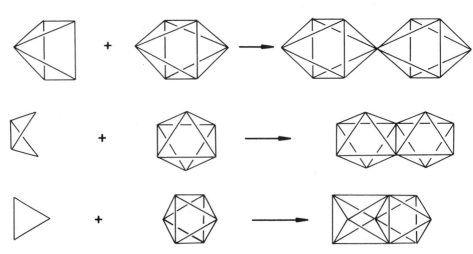

Figure 3.12.  Condensation of clusters.

Such analysis leads to the generalization that the number of electrons in the resulting joined polyhedron is equal to the sum of numbers of parent electrons of polyhedra A and B minus the number of electrons characteristic for the atom, for the pair of atoms of an edge, or for atoms of a plane:

$$N_{AB} = N_A + N_B - N_C \tag{3.32}$$

where $N_{AB}$ denotes the number of valence electrons in the joined polyhedron while $N_A$ and $N_B$ equal the number of valence electrons in the parent polyhedra A and B; $N_C = 18$, 34, or 48 for a polyhedra in which the A and B fragments are linked by one atom, two atoms (an edge), and three atoms (a face). If two polyhedra (in which one results from a triangular-faced polyhedron) undergo condensation to form a common face composed of four atoms, then $N_C = 62$. In the case of condensation of 3-connected polyhedra by a square face, $N_C = 64$. Examples of condensation of clusters are given in Figures 3.12 and 3.13.

e. *Molecular orbital theory*

The dependence of the number of bonding and antibonding orbitals, and therefore the number of valence electrons on the structure of the cluster grouping, was developed using semiempirical calculations LCAO–MO by Lauher.[4] Since the strongest interaction occurs between $s$ and $p$ orbitals of the metal atoms, the energy difference in the $d$ band of the cluster is smaller than in the $s$ and $p$ band. Because of mixing of orbitals of the same symmetry, some orbitals have, to a considerable degree, $s$ and $p$ character. Since each transition metal atom has nine valence orbitals, clusters containing three metal atoms have 27 molecular orbitals ($D_{3h}$ symmetry). Calculations show that 24 of these orbitals are cluster valence molecular orbitals (CVMOs), i.e., bonding and non-bonding, while the remaining three have energy which is considerably higher than the energy of $p$ orbitals. Therefore, the destabilization of $p$ orbitals is the most important factor determining the electronic structure of small clusters, and clusters of this type

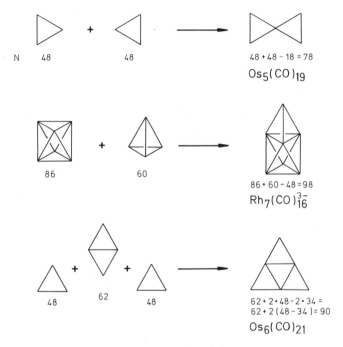

Figure 3.13. Condensation of clusters.

are 48 valence electron compounds. An analogous situation occurs for compounds possessing lower symmetry, such as $C_{2v}$. Examples of such complexes are carbonyls of the type $M_3(CO)_{12}$ (M = Fe, Ru, Os). Similar calculations were also carried out for many other clusters containing as many as several hundred atoms. Molecular orbitals of a cluster are considered to be valence orbitals if their energy is not higher than the energy of $p$ orbitals of a free metal atom. The number of valence orbitals for various clusters are given in Figure 3.14 and in Table 3.9.

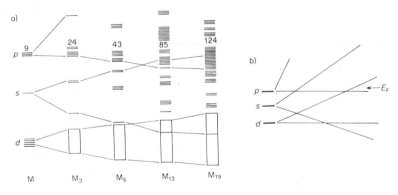

Figure 3.14. (a) Atomic orbitals of the metal and molecular orbitals of some clusters: $M_3$ (triangle), $M_6$ (octahedron), $M_{13}$ (cuboctahedron), and $M_{19}$ octahedron). (b) Qualitative scheme of the dependence of molecular orbitals energy on the number of the metal atoms in the cluster showing the final band structure of a typical metal. The arrow denotes the Fermi energy $E_F$.

Table 3.9. Number of Orbitals and Valence Electrons in Clusters[4]

| Cluster's structure | Number of metal atoms $N$ | $9N$ | CV $MO^a$ | $CVE^a$ | Examples |
|---|---|---|---|---|---|
| Triangle | 3 | 27 | 24 | 48 | $M_3(CO)_{12}$ (Fe, Ru, Os) |
| Tetrahedron, $T_d$ | 4 | 36 | 30 | 60 | $M_4(CO)_{12}$ (Co, Rh, Ir) |
| "Butterfly" structure, $C_{2v}$ | 4 | 36 | 31 | 62 | $[Fe_4(CO)_{12}(\mu_4\text{-}CO)H]^-$ |
| Square plane | 4 | 36 | 32 | 64 | $[Pt_4(OAc)_8]$, $[Co_4S_2(CO)_{10}]$ |
| Trigonal bipyramid | 5 | 45 | 36 | 72 | $[Os_5(CO)_{16}]$ |
| Elongated trigonal bipyramid | 5 | 45 | 38 | 76 | $[Ni_5(CO)_{12}]^{2-}$ |
| Square pyramid | 5 | 45 | 37 | 74 | $[Fe_5(CO)_{15}C]$, $[Fe_5(CO)_{14}C]^{2-}$ |
| Bicapped tetrahedron$^a$ | 6 | 54 | 42 | 84 | $[Os_6(CO)_{18}]$ |
| Octahedron | 6 | 54 | 43 | 86 | $[M_6(CO)_{16}]$ (Co, Rh, Ir) |
| Bitetrahedron, sharing an edge | 6 | 54 | 43 | 86 | |
| Capped square pyramid | 6 | 54 | 43 | 86 | $[Os_6H_2(CO)_{18}]$ |
| Prism | 6 | 54 | 45 | 90 | $[Rh_6(CO)_{15}C]^{2-}$ |
| Capped octahedron | 7 | 63 | 49 | 98 | $[Rh_7(CO)_{16}]^{3-}$ |
| Capped prism | 7 | 63 | 51 | 102 | |
| Bicapped octahedron | 8 | 72 | 55 | 110 | $[Os_8(CO)_{22}]^{2-}$, $[Re_8C(CO)_{24}]^{2-}$ |
| Antiprism | 8 | 72 | 57 | 114 | $[Co_8(CO)_{18}C]^{2-}$ |
| Bicapped prism | 8 | 72 | 57 | 114 | $[Cu_2Rh_6C(CO)_{15}(NCMe)_2]$ |
| Cube | 8 | 72 | 60 | 120 | $[Ni_8(PPh)_6(CO)_8]$ |
| Tricapped octahedron | 9 | 81 | 63 | 126 | |
| Tricapped prism | 9 | 81 | 64 | 128 | |
| Capped square antiprism | 9 | 81 | 65 | 130 | |
| Capped cube | 9 | 81 | 66 | 132 | |
| Bicapped cube | 10 | 90 | 71 | 142 | |
| Bicapped square antiprism | 10 | 90 | 71 | 142 | |

$^a$ Capped, bicapped, tricapped denotes clusters obtained by adding to a regular polyhedron additional $M(CO)_m$ fragments; CVMO represents cluster valence MOs; CVE designates cluster valence electrons.

Electron-rich clusters may be divided into two groups. Group I comprises compounds which utilize all cluster valence orbitals (CVMO), and Group II contains those clusters in which the number of antibonding molecular orbitals is greater than in group I compounds. Group I clusters are mainly formed by the iron and cobalt group metals, while group II compounds are obtained by elements found at the end of a given series of transition metals (Pt, Cu, Au) in which the energy difference between the $ns$ orbitals and the $np$ orbitals is considerably greater than in the remaining transition elements. Because of this, the number of valence electrons in group II clusters is smaller. Therefore, platinum, copper, and gold readily form $16e$ complexes and even compounds containing 14 valence electrons. Only group I compounds show "magic numbers" of valence electrons of the cluster. Between these clusters and clusters of the main group elements, there is an analogy; thus, Wade's theory of skeletal electron pairs can be applied to these clusters. The following clusters belong to group I: $[M_3(CO)_{12}]$ (M = Fe, Ru, Os) $(48e)$, $[M_4(CO)_{12}]$ (M = Co, Rh, Ir) $(60e)$, $[M_6(CO)_{16}]$ (Co, Rh, Ir)

$(86e)$, $[Ru_6(CO)_{18}]^{2-}$ $(86e)$. Examples of group II clusters are $[Pt_3(CO)_6]^{2-}$ $(44e)$ and $[Pt_3(CO)_3(PPh_3)_3]$ $(42e)$, in which the $p_z$ orbitals are not occupied. The ion $[Pt_{19}(CO)_{22}]^{4-}$ is also an example of a group II cluster which, according to calculations,[4] should have 123 orbitals, and therefore 246 valence electrons. However, this ion possesses only 238 valence electrons. Analysis of calculations established that the $e_2'$ and $e_2''$ orbitals are not occupied. The $e_2'$ and $e_2''$ orbitals are combinations of tangential $p$ orbitals (Figure 3.15). These orbitals do not have appropriate symmetry in order to interact with CO orbitals, and their energy is too high for electron occupancy.

Exceptions among "electron rich" clusters include compounds in which certain $d$ orbitals are empty, while the $s$ and $p$ orbitals are occupied. This is equivalent to the formation of multiple $M-M$ bonds. Only a few such compounds are presently known, for example, $[Re_4H_4(CO)_{12}]$ $(56e$ instead of $60e)$, and $[Os_3H_2(CO)_{10}]$ $(46e$ instead of $48e)$. The skeletal bonds in those compounds may be represented as follows:

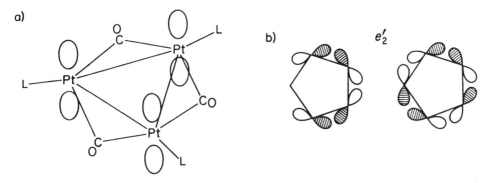

Another example is $[Os_{10}C(CO)_{24}]^{2-}$ $(134e$ instead of $140e)$ which has a tetrahedral structure with the carbon atom centrally located in the octahedral interstice. In this case, the highest degenerate $t_2$ orbitals (formally $d$ orbitals) are not occupied. The molecular orbital theory and Wade's theory predict that the change in the number of valence electrons should lead to structural changes of the skeletal atoms of the cluster. In the case of tetranuclear compounds, in which the skeletal atoms lie at the vertices of a tetrahedron, the number of valence electrons is 60, and increases when the number of $M-M$ bonds decreases (Figure 3.16). As the cluster becomes larger, the ratio of the number of valence orbitals to the number of metal atoms (CVMO/$N$) decreases. For an octahedron, this ratio equals $43/6 = 7.17$ (Tables 3.10 and 3.11). For the octahedral rhodium cluster, each metal should be connected with 2.67 two-electron ligands such as CO, because rhodium has 9 valence electrons, $(2 \times 7.17 - 9)/2 = 2.67$. Indeed, this ratio

Figure 3.15. (a) Empty $p_z$ orbitals in clusters $[Pt_3(CO)_6]^{2-}$ and $[Pt_3(CO)_3L_3]$. (b) Nonoccupied tangential $p$ orbitals possessing $e_2'$ symmetry in clusters $[Pt_{19}(CO)_{22}]^{4-}$.

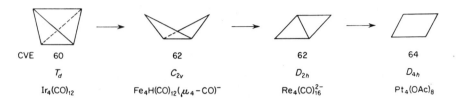

| CVE | 60 | 62 | 62 | 64 |
|-----|-----|-----|-----|-----|
| | $T_d$ | $C_{2v}$ | $D_{2h}$ | $D_{4h}$ |
| | $Ir_4(CO)_{12}$ | $Fe_4H(CO)_{12}(\mu_4-CO)^-$ | $Re_4(CO)_{16}^{2-}$ | $Pt_4(OAc)_8$ |

Figure 3.16. The dependence of the number of valence electrons on the symmetry for tetranuclear clusters.

L/Rh is observed in $Rh_6(CO)_{16}$. The average ratio, CVMO/$N$, is valid for clusters in which all skeletal atoms are equivalent. There are three groups of atoms in an octahedral cluster containing 19 metal atoms (Figure 3.17): six corner atoms, twelve atoms located on edges, and one atom located in the middle of the octahedron. These atoms have four, seven, and twelve nearest neighbors. The contribution of each group of atoms to the total number of CVMOs to the cluster is different. It is possible to calculate the number of localized CVMOs on particular atoms (Tables 3.10 and 3.11). The number of CVMOs decreases for a given atom as the number of neighbors increases. Magnetic measurements of Fe, Co, Ni, Cu and their alloys show that the number of CVMOs for metals equals 5.28 because the strongest ferromagnetic properties were found for the iron and cobalt alloy containing 28% cobalt, and therefore 8.28 electrons. The nickel and copper alloy (44% nickel), in which there are 10.56 electrons per metal atom, is diamagnetic. Table 3.10 shows that even central atoms of different clusters possessing twelve nearest neighbors have *ca* 5.7 CVMOs, which is higher by 0.4 compared to atoms in metallic lattices. This is probably due to the contribution of *p*

Table 3.10. Number of Valence Molecular Orbitals (CVMO) for Various Atoms in Clusters of Transition Metals

| Cluster | Number of M atoms | Location of atoms | Number of atoms of a given type $N_S$ | Number of nearest neighbors | CV MO | Contribution of atomic orbitals | | |
|---------|-------------------|-------------------|----------------|-----------------------------|-------|-----|-----|-----|
| | | | | | | $s$ | $p$ | $d$ |
| Octahedron | 6 | $A^a$ | 6 | 4 | 7.17 | 0.71 | 1.56 | 4.90 |
| Octahedron | 19 | A | 6 | 4 | 7.13 | 0.71 | 1.52 | 4.90 |
| | | B | 12 | 7 | 6.29 | 0.55 | 0.89 | 4.86 |
| | | C | 1 | 12 | 5.71 | 0.44 | 0.54 | 4.74 |
| Tetrahedron | 20 | D | 4 | 3 | 7.09 | 0.76 | 1.42 | 4.91 |
| | | E | 12 | 6 | 6.23 | 0.57 | 0.80 | 4.86 |
| | | F | 4 | 9 | 5.98 | 0.48 | 0.71 | 4.79 |
| Hexagon | 7 | G | 6 | 3 | 6.99 | 0.73 | 1.35 | 4.92 |
| | | H | 1 | 6 | 6.05 | 0.54 | 0.66 | 4.85 |
| Hexagon | 19 | G | 6 | 3 | 6.92 | 0.71 | 1.29 | 4.91 |
| | | I | 6 | 4 | 6.45 | 0.62 | 0.93 | 4.90 |
| | | H | 1 | 6 | 5.93 | 0.50 | 0.58 | 4.85 |
| | | H' | 6 | 6 | 5.98 | 0.52 | 0.59 | 4.86 |

[a] Designations of atoms are given in Figure 3.17.

Table 3.11. *Theoretical Ability of Clusters to Form Bonds Expressed as Numbers of 2-Electron Ligands per One Rhodium Atom*

| Octahedral cluster | $L/Rh$ | $L/Rh_s{}^a$ | Cluster's size (pm) | Hexagonal cluster | $L/Rh$ | Cluster's size (pm) |
|---|---|---|---|---|---|---|
| $Rh_1$ | 4.50 | 4.50 | 270 | $Rh_1$ | 4.50 | 270 |
| $Rh_6$ | 2.67 | 2.67 | 650 | $Rh_7$ | 2.36 | 810 |
| $Rh_{19}$ | 2.03 | 2.14 | 1030 | $Rh_{19}$ | 1.92 | 1340 |
| $Rh_{44}$ | 1.77 | 2.05 | 1410 | $Rh_{37}$ | 1.76 | 1880 |
| $Rh_{85}$ | 1.62 | 2.09 | 1790 | $Rh_{61}$ | 1.68 | 2420 |
| $Rh_{146}$ | 1.53 | 2.19 | 2170 | $Rh_{91}$ | 1.63 | 2960 |
| $Rh_{231}$ | 1.51 | 2.38 | 2550 | $Rh_{127}$ | 1.60 | 3500 |
| $Rh_{344}$ | 1.48 | 2.58 | 2930 | $Rh_{169}$ | 1.58 | 4040 |
| | | | | $Rh_{217}$ | 1.56 | 4570 |
| | | | | $Rh_{271}$ | 1.54 | 5110 |
| | | | | $Rh_{331}$ | 1.53 | 5650 |

$^a$ Number of 2-electron ligands per surface rhodium atom.

orbitals to molecular orbitals of the cluster. The contribution for internal cluster atoms possessing coordination number twelve equals 0.54 of the $p$ orbital, and for the atoms in a metallic lattice the low-lying $p$ orbitals are empty.

## 2. CLUSTERS AS MODELS OF METALLIC LATTICES OF CATALYSTS

In large clusters, metal atoms often constitute fragments of typical metallic lattices: $A_3$—hexagonal close-packed (hcp), $A_1$—cubic close-packed (ccp), and $A_2$—body-centered cubic (bcc). Fragments of hexagonal close packed structure ($A_3$) are found in the following clusters: $[Rh_{13}(CO)_{24}H_{5-n}]^{n-}$ $(n = 2, 3)$, $[Pt_{26}(CO)_{32}]^{2-}$, and $[Ni_{12}(CO)_{21}H_{4-n}]^{n-}$ $(n = 2, 3, 4)$. Fragments of $A_1$ structures are encountered in $[Pt_{38}(CO)_{44}]^{2-}$ and $[Ni_{38}Pt_6(CO)_{48}H_{6-n}]^{n-}$, while $A_2$ fragments are found in $[Rh_{14}(CO)_{25}]^{4-}$. In $[Rh_{22}(CO)_{37}]^{4-}$, the rhodium atoms form a fragment of the close-packed arrangement of the type ABAC, which is typical for certain lanthanides (La, Nd, Pr) (Table 3.3, Figure 3.18). Large clusters may serve as models of metallic lattices of catalysts deposited on supports. Metals possessing the cubic close-packed structure are most often investigated by chemisorption on the (100) and (111) planes. The base of the square pyramid is an example of the (100) plane (Figure 3.17). The atom which is located in the middle of the base has the coordination number eight. The faces of the 20-atom tetrahedron represent the (111) planes. The atom which lies in the middle of the plane has nine nearest neighbors. Microcrystallites deposited on supports may have structures similar to those of clusters.[2,9,26] Silver deposited on $A$-type zeolite forms octahedra $Ag_6$ localized in interstices of the zeolite,[27] while gold deposited on mica forms microcrystallites exhibiting a fivefold symmetry axis.[28] This type of symmetry is impossible for periodic crystallographic lattices. Later, it was explained using molecular orbital calculations that the 13-atom cluster should not have cubo-

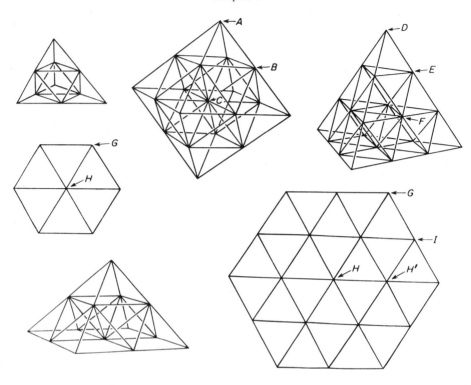

Figure 3.17. Clusters: $M_{10}$ (tetrahedron), $M_{19}$ (octahedron), $M_{20}$ (tetrahedron), $M_{14}$ (square pyramid), $M_7$ (hexagon), $M_{19}$ (hexagon). Atoms are denoted as in Table 3.10.

octahedral structure (a fragment of the cubic close-packed lattice) but the icosohedral structure, and therefore that of a polyhedron with a fivefold symmetry axis. The platinum compound, $[Pt_{19}(CO)_{22}]^{4-}$, has similar structure. Many metals deposited on $Al_2O_3$, $SiO_2$, and $TiO_2$ form planar, two-dimensional crystallites measuring from 1000 to 6000 pm (10–60 Å). Rhodium deposited on aluminum oxide forms planar clusters possessing average dimensions 1510 pm with a maximum value of 4000 pm.[29] Ruthenium and osmium form planar clusters on $SiO_2$; the diameters of these clusters are as large as 6000 pm.[30] Also, platinum deposited on $TiO_2$ and reduced at high temperatures by $H_2$ forms planar clusters. As a result of oxidation, these clusters become three-dimensional. After subsequent reduction with hydrogen, they again become planar.[31] Calculations for planar hexagonal rhodium clusters show that the number of CVMOs localized on particular atoms is smaller than for octahedral structures despite slightly lower coordination numbers.[4] Only atoms possessing coordination number 12 in octahedral clusters have a smaller number of CVMOs than inner atoms of hexagonal grouping $Rh_n$ neighboring with six metal atoms. Because of a small number of valence electrons, clusters must coordinate ligands in order to have completely occupied valence orbitals. This is difficult for larger, spatial clusters which contain many inner metal atoms. Therefore, deformed clusters often result which coordinate additional ligands. Such clusters have a lower number of CVMOs. The

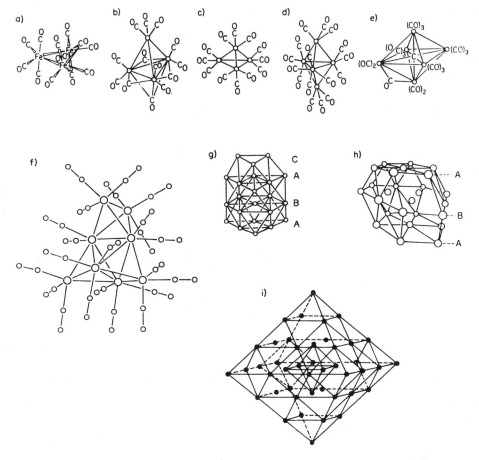

Figure 3.18. Structures of some clusters of the transition metals: (a) $[Fe_3(CO)_{12}]$, (b) $[Fe_4(CO)_{13}]^{2-}$, (c) $[Ir_4(CO)_{12}]$, (d) $[Os_5(CO)_{16}]$, (e) $[Ru_6C(CO)_{17}]$, (f) $[Os_8H_2(CO)_{22}]^-$, (g) $[Rh_{22}(CO)_{37}]^{4-}$, (h) $[Pt_{26}(CO)_{32}]^{2-}$, (i) $[Pt_6Ni_{38}(\mu_3-CO)_{18}(\mu-CO)_{12}(CO)_{18}H_2]^{4-}$.

average number of two-electron ligands which must be coordinated by each rhodium atom is given above in Table 3.11. Rhodium atoms residing on the surface of the three-dimensional cluster should form bonds with a large number of ligands. However, because of steric reasons, this is difficult.

## 3. INTERSTITIAL (ENCAPSULATED) CLUSTERS

For many clusters, the number of valence electrons may be increased as a result of formation of encapsulated compounds containing H, C, N, P, Sb, and S in the interstices between metal atoms (Table 3.12). Recently "electron poor" clusters containing encapsulated metal atoms such as K and Be and atoms of other elements of the second

Table 3.12. Interstitial Clusters of Transition Metals

| Cluster | Interstitial atom | Structure[a] | References |
|---|---|---|---|
| $[Nb_6HI_{11}]$ | H | Octahedron $Nb_6$ | 32 |
| $[Ru_6H(CO)_{18}]^-$ | H | Octahedron $Rh_6$ | 33 |
| $[Co_6H(CO)_{15}]^-$ | H | Octahedron $Co_6$ | 34 |
| $[Fe_6C(CO)_{16}]^{2-}$ | C | Octahedron $Fe_6$ | 2, 8 |
| $[Ru_6C(CO)_{17}]$ | C | Octahedron $Ru_6$ | 2, 8 |
| $[Rh_6C(CO)_{15}]^{2-}$ | C | Prism $Rh_6$ | 2, 8 |
| $[Rh_6N(CO)_{15}]^-$ | N | Prism $Rh_6$ | 35, 8 |
| $[Rh_8C(CO)_{19}]$ | C | Prism $Rh_6$ | 8 |
| $[Co_8C(CO)_{18}]^{2-}$ | C | Archimedean antiprism $Co_8$ | 2, 8 |
| $[Rh_9P(CO)_{21}]^{2-}$ | P | Capped Archimedean antiprism, the fragment $Rh_8P$ has square antiprism structure | 36 |
| $[Rh_9As(CO)_{21}]^{2-}$ | As | Capped Archimedean antiprism, $C_{4v}$ | 36 |
| $[Rh_{10}P(CO)_{22}]^{3-}$ | P | Bicapped Archimedean antiprism | 36 |
| $[Rh_{10}As(CO)_{22}]^{3-}$ | As | Bicapped Archimedean antiprism | 36 |
| $[Rh_{12}Sb(CO)_{27}]^{3-}$ | Sb | Deformed icosahedron | 37 |
| $[Ni_{12}H_2(CO)_{21}]^{2-}$ | H | $D_{3h}$, H in the octahedral holes | 38 |
| $[Ni_{12}H(CO)_{21}]^{3-}$ | H | $D_{3h}$, H in the octahedral holes | 38 |
| $[Rh_{12}(C)_2(CO)_{24}]^{2-}$ | C | $D_{2h}$, C atoms are in holes of two units having trigonal prism structure | 2 |
| $[Co_{13}(C)_2H(CO)_{24}]^{4-}$ | C | $C_2$, C atoms are in holes of two units having trigonal prism structure | 2 |
| $[Rh_{13}H_3(CO)_{24}]^{2-}$ | H | $D_{3h}$, truncated hexagonal bipyramid, hcp-type lattice, H atoms are located in holes of square pyramids | 39 |
| $[Rh_{15}(C)_2(CO)_{28}]^-$ | C | $C_{2v}$, C atoms are in octahedral holes | 8 |
| $[Rh_{17}S_2(CO)_{32}]^{3-}$ | S | S atoms are located in holes of two units having square antiprism structure which are connected by Rh atom | 40 |

[a] Archimedean antiprism = square antiprism.

period of the Periodic Chart (B, C) have been prepared,[72] for example, $[Zr_6(\mu_6\text{-}K)I_{14}]$ and $[W_4(C)(O-i\text{-}Pr)_{12}(NMe)]$ (Table 3.14 below).

Deformation and formation of encapsulated compounds do not necessarily take place if the number of metal atoms becomes so great that the aggregate of these atoms loses its molecular property to such an extent that this grouping becomes a typical bulk metal. It is difficult to assess at what size a cluster loses its molecular properties. This probably happens in the case of several hundred metal atoms. The formation of planar clusters on surfaces of supports renders all valence orbitals to be occupied by electrons, because metal atoms may form bonds with ligands at both sides of a cluster. One side bonds with the oxygen of the support, while the other side may form bonds with the adsorbed molecules. The average size of planar clusters of rhodium deposited on aluminum oxide is 1510 pm. On the basis of CO adsorption studies, it was found that the highest CO/Rh ratio equals 1.7, which agrees well with the calculated value of L/Rh for hexagonal rhodium clusters of this size (Table 3.11). These considerations show that monoatomic layers of metal atoms are often formed. This should favor deposition which renders impossible the formation of encapsulated compounds. Studies of encapsulated clusters are important because such clusters may serve as models of interstitial compounds of the transition metals,[1,2,8,9,37,73–76] and therefore compounds of these metals with hydrogen, carbon, nitrogen, etc. Information about dynamic properties of hydrido encapsulated clusters may allow a better understanding of hydrogen diffusion in metals, or hydrogen migration on surfaces of metallic catalysts (spillover). Such clusters may find application in catalysis and organic synthesis in the following processes: ammonia synthesis, Fischer–Tropsch synthesis, water–gas conversion, hydroformylation, ethyl glycol synthesis, hydrogenation of olefins, acetylenes, arenes and $\diagdown$CO bonds, activation of saturated hydrocarbons, etc.

Encapsulated clusters are also significant for the development of the theory of chemical bonds.

Theories of interstitial compounds assumed either electron shifting from the conductivity band of metals to the nonmetal atom and negative ion formation, or valence electron transfer of the encapsulated atom to the conductivity band leading to positive ion formation. In this case the encapsulated phase is considered as an alloy; therefore, metallic properties are retained. Both theories explain some experimental data, such as the decrease of conductivity of encapsulated compounds.

Recently, the formation of covalent metal–nonmental bonds has been taken into consideration. In all encapsulated compounds, nonmetal atoms have high coordination numbers. In encapsulated clusters, high coordination numbers (higher than in normal compounds) of nonmetals are also found. Most commonly, these coordination numbers are in the 5–10 range. The compounds $M_4(C)L_n$ and $M_4(N)L_n$ may be considered as interstitial compounds or as clusters possessing trigonal bipyramidal structure in which one corner is occupied by the carbon or the nitrogen atom, respectively. The coordination number of the nonmetal atom and the geometry of the cluster depend on the electron structure of the cluster and on the ratio $r_E/r_M$, where $r_E$ equals the nonmetal's radius and $r_M$ equals the metal's radius. The cluster $[Rh_6C(CO)_{15}]^{2-}$ adopts a prismatic structure while $[Ru_6C(CO)_{17}]$ assumes an octahedral arrangement despite the similarity of the Ru and Rh radii. Thus, electronic structure determines the geometry of these clusters. Larger elements, such as P, As, and S, occupy larger interstices in clusters possessing the following structures: square antiprism, capped or bicapped antiprisms (Table 3.12). In $[Rh_{12}Sb(CO)_{27}]^{3-}$.[37] the large antimony atom is located in the interstice of a deformed icosohedron and its coordination number is 12. The nonmetal–metal distances are shorter by 15–20 pm than the sum of covalent radii, excluding hydrogen and antimony compounds. The principle of close packing gives the following atomic radii in clusters: H 51–64 (37) pm, C 57–74 (77) pm, N 67 (71) pm, P 93–102 (110) pm, S 89 (104) pm, As 105 (124) pm, where the values in parentheses represent covalent radii. Unusually high values of frequencies $v(M-C)$ were also observed. This indicates that the $M-E$ bonds are very strong and have covalent character. Therefore, the encapsulated compounds are more stable than other clusters. The bonds of hydrogen atoms to the $M_n$ grouping are considerably weaker than those of other elements which utilize four orbitals for bond formation with metal atoms, while the hydrogen atom interacts with the metal atoms only though the $s$ orbital. The same $M_n$–C and $M_n$–H distances indicate that the interstices are too large for the hydrogen atom. Consequently, hydrogen atoms easily migrate in clusters. The encapsulated atoms located inside clusters (e.g., $[M_5(C)L_n]$ and $[M_6(C)L_n]$) are not reactive. However, the carbon atom in $[Fe_4(C)(CO)_{12}]^{2-}$ reacts with hydrogen and electrophiles such as $R^+$, and carbon monoxide attacks the nitrogen atom in the cluster $[Ru_3(N)(CO)_{12}]^-$ to afford $[Ru_3(NCO)(CO)_{11}]^-$. Therefore, those nonmetallic atoms which may be treated as skeletal cluster atoms capable of coordinating are reactive.

## 4. "ELECTRON POOR" TRANSITION METAL CLUSTERS

The most important "electron poor" clusters are those containing $M_3$ and $M_6$[(7)] groupings. These clusters are formed by the transition elements of groups 4–7 (Table 3.13). Less numerous in these groups are organometallic compounds, although recently their number has increased considerably. Trinuclear clusters may be divided

Table 3.13. Electron Poor Clusters

| Cluster | Distance M–M (pm) | Cluster structure | Reference |
|---|---|---|---|
| $[Ti_7Cl_{16}]$ | 295.4 | Triangle, $M_3$ | 41 |
| $[Nb_3Cl_8]$ | 281 | Triangle, $M_3$ | 42 |
| $[Nb_3Br_8]$ | 288 | Triangle, $M_3$ | 43 |
| $[Nb_3Cl_6(C_6Me_6)_3]^+$ | 333.4 | Triangle, $U_6$ | 44 |
| $[Nb_3Cl_6(C_6Me_6)_3]^{2+}$ | 333.5 | Triangle, $U_6$ | 45 |
| $[Ta_3Cl_6(C_6Me_6)_3]^+$ | | Triangle, $U_6$ | 44 |
| $[Mo_3O_4F_9]^{5-}$ | 250.2 | Triangle, $M_3$ | 46 |
| $[Mo_3S_4cp_3]^+$ | 281.2 | Triangle, $M_3$ | 47 |
| $[Mo_3O_4(C_2O_4)_3(H_2O)_3]^{2-}$ | 248.6 | Triangle, $M_3$ | 48 |
| $[Mo_3S_4(CN)_9]^{5-}$ | 276.6 | Triangle, $M_3$ | 42 |
| $[W_3(OCH_2CMe_3)O_3Cr_3(OOCCMe_3)_{12}$ | 261.0 | Triangle, $M_3$ | 49 |
| $[W_3O_3Cl(OAc)Cl_4(PBu_3)_3]$ | 260.9 | Triangle, $M_3$ | 50 |
| $[Mo_3S_{13}]^{2-}$ | 272.2 | Triangle, $M_3$ | 51 |
| $[Mo_3S_7Cl_4]$ | 274.2 | Triangle, $M_3$ | 52 |
| $[Mo_3(OAc)_6(OEt)_2(H_2O)_3]^{2+}$ | 285.4 | Triangle, $B_6$ | 53 |
| $[Mo_3(CMe)_2(OAc)_6(H_2O)_3]^{2+}$ | 280.7–282.0 | Triangle, $B_6$ | 54 |
| $[Mo_3O(CMe)(OAc)_6(H_2O)_3]^+$ | 275.2 | Triangle, $B_6$ | 55 |
| $[W_3O_2(OAc)_6(H_2O)_3]^{2+}$ | 274.7 | Triangle, $B_6$ | 56 |
| $[W_3O_2(OOCCMe_3)_6(OOCCMe_3)_2(H_2O)]$ | 276 | Triangle, $B_6$ | 56 |
| $[W_3O_2(OAc)_6(OAc)_3]^-$ | 276.9 | Triangle, $B_6$ | 56 |
| $[Re_3Cl_{12}]^{3-}$ | 248.9 | Triangle, $U_3$ | 57 |
| $[Re_3(\mu\text{-}Me)_3Me_6(PEt_2Ph)_2]$ | 243.3–247.4 | Triangle, $U_3$ | 58 |
| $[Re_3(\mu\text{-}Cl)_3(CH_2SiMe_3)_6]$ | 238.6 | Triangle, $U_3$ | 59 |
| $[Mo_4I_{11}]^{2-}$ | | Tetrahedron | 60 |
| $[Mo_5Cl_{13}]^{2-}$ | | Square pyramid | 61 |
| $[Zr_6I_{12}]$ | | Octahedron | 63 |
| $[Nb_6Cl_{12}]^{2+}$ | | Octahedron | 64 |
| $[Nb_6HI_{11}]$ | 284 | Octahedron | 32 |
| $[Nb_6I_{11}]$ | 285 | Octahedron | 62 |
| $[Ta_6(\mu\text{-}Cl)_{12}(H_2O)_6]Cl_2$ | 290 | Octahedron | 64 |
| $[Mo_6(\mu_3\text{-}Cl)_8Cl_6]^{2-}$ | 265 | Octahedron | 65 |
| $[Mo_6Br_{12}]\cdot 2H_2O$ | 263–264 | Octahedron | 66 |

into three groups[7,67] according to the number of bridging ligands $\mu_3$. Those clusters which do not possess any $\mu_3$ bridging ligands are denoted by the letter U, those containing one $\mu_3$ bridge are known as M, and those having two $\mu_3$ bridges are labeled B (Figure 3.19). The number of *d* electrons in "electron poor" clusters is equal to that required for the formation of two-center M—M bonds between neighboring metal atoms.[7,15] Therefore, trinuclear clusters contain 5–8 electrons that are localized on metal atoms, while hexanuclear clusters have 16–24 skeletal electrons.

There are also $\mu_2$ or $\mu_3$ bridging atoms in octahedral clusters, for example, in

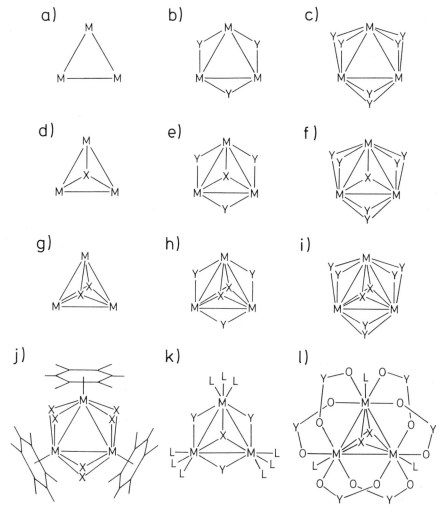

Figure 3.19. Structures of trinuclear clusters:
(a) $[M_3]-U_0$, (b) $[M_3(\mu\text{-}Y)_3]-U_3$, (c) $[M_3(\mu\text{-}Y)_6]-U_6$, (d) $[M_3(\mu_3\text{-}X)]-M_0$,
(e) $[M_3(\mu_3\text{-}X)(\mu\text{-}Y)_3]-M_3$, (f) $[M_3(\mu_3\text{-}X)(\mu\text{-}Y)_6]-M_6$, (g) $[M_3(\mu_3\text{-}X)_2]-B_0$,
(h) $[M_3(\mu_3\text{-}X)_2(\mu\text{-}Y)_3]-B_3$, (i) $[M_3(\mu_3\text{-}X)_2(\mu\text{-}Y)_6]-B_6$,
(j) $[M_3X_6C_6Me_6]^{n+}$ (M = Nb, Ta, X = Cl, Br, $n$ = 1, 2), (k) $[W_3(\mu_3\text{-}O)(\mu\text{-}O)_3F_9]^{5-}$,
(l) $[W_3(\mu_3\text{-}O)_2(\mu\text{-}OOCMe)_6(H_2O)_3]^{2+}$.

$[Mo_6Cl_8]^{4+}$ and $[Nb_6Cl_{12}]^{2+}$. In the molybdenum compounds the chlorine atoms are located above the middle of the octahedral faces (each chlorine atom is bonded to three molybdenum atoms), while in the niobium complex the chlorine atoms are localized above the edges of the octahedron (each chlorine atom is bonded to two niobium atoms). In trinuclear clusters, the coordination number of each metal atom excluding $M-M$ bonds may equal 6 (the pseudooctahedral structure), 4 (the tetrahedral structure), or 7 ($C_{2v}$ symmetry) (Figure 3.19). In compounds in which the metal atoms possess the octahedral coordination, the $t_{2g}$ orbitals of each metal form nine molecular orbitals, four of which are bonding (Figure 3.20). As a result of the small difference between $1e$ and $2a_1$ orbitals, complexes containing 6–8$d$ electrons are formed. In trinuclear compounds in which metal atoms have tetrahedral coordination, the $e_g$ orbitals of each metal atom form three bonding orbitals $1e(d_{x^2-y^2})$ and $a_1(d_{z^2})$ and three antibonding orbitals $2e(d_{z^2})$ and $a_2(d_{x^2-y^2})$. In such compounds the metal–metal bonds are single because they contain six $d$ electrons.

The number of $d$ electrons in these types of clusters is usually shown by the superscript of the symbol denoting the number of skeletal atoms: $[M_3]^6$, $[M_3]^7$, $[M_3]^8$. Clusters containing more electrons are rarely formed. The only exceptions are $[Mo_3]^9$ compounds. Rhenium(III) clusters, $Re_3X_9$, and similar compounds possess twelve $d$ electrons. In these cases the electronic structure is different; the skeletal atoms form multiple metal–metal bonds.[68] In the case of rhenium(III) compounds and for octahedral clusters, the $M-M$ bonds can be satisfactorily described considering only interactions of $d$ orbitals.[15,68,69]

In $[Re_3X_9L_3]$, $[Mo_6Cl_8]^{4+}$, and $[Nb_6Cl_{12}]^{2+}$ (Figure 3.21), bonding orbitals are

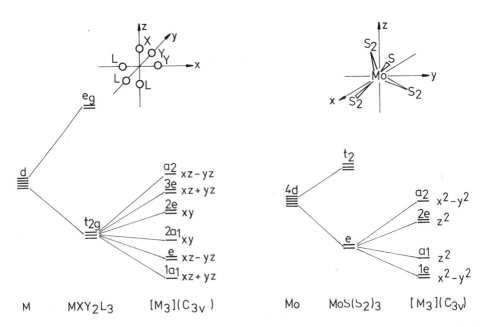

Figure 3.20. (a) Local coordinate system for the isolated $MXY_2L_3$ fragment and molecular orbital scheme for $[M_3(\mu_3\text{-}X)(\mu\text{-}Y)_3L_9]$. (b) Local coordinate system for the $MoS(S_2)_3$ fragment and the molecular orbital diagram for $[Mo_3(\mu_3\text{-}S)(\mu\text{-}S_2)_3(S_2)_3]^{2-}$.

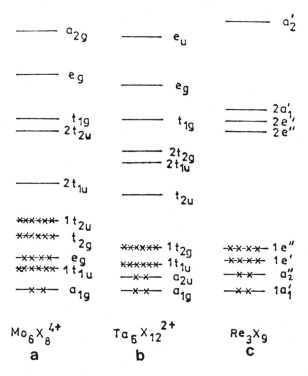

Figure 3.21. Molecular orbital scheme for the following clusters: (a) $[Mo_6X_8]^{4+}$, (b) $[Ta_6X_{12}]^{2+}$, (c) $[Re_3X_9]$.

occupied by 12, 24, and 16 electrons, respectively. Therefore, the multiplicity of the metal–metal bond is 2 for the rhenium clusters, 1 for the molybdenum cluster, and 2/3 for the niobium compound.

## 5. NMR SPECTRA OF CLUSTERS

$^{13}$C NMR spectra of carbonyls and hydridocarbonyls including carbonyl clusters were described in Chapter 2. In this chapter, only nuclear magnetic resonance spectra of encapsulated clusters (mainly carbide clusters) and compounds containing $\mu_3$ and $\mu_2$ bridging ligands coordinating through carbon atoms (clusters with alkylidene and alkylidyne ligands) will be discussed. Carbon atoms that are located inside the cluster and those in $M_4(C)L_n$ (which may be treated as one of the skeletal atoms) have their $^{13}$C NMR chemical shifts in the 250–500 ppm range. Often, these chemical shifts are even greater than for carbene and carbyne complexes. Splitting of signals due to $^{13}$C–$^{103}$Rh interactions in carbide compounds of rhodium leads to the conclusion that the carbon atom interacts equally with all metal atoms. For $\mu_3$-CR ligands, the $^{13}$C chemical shifts are in the 200–350 ppm range (Table 3.14), which is also typical for alkylidyne complexes. In the case of $M_n(\mu_3\text{-CH})L_m$ complexes, the $^1$H NMR signals occur at relatively high fields, most commonly in the 15–10 ppm range, although these signals sometimes occur at $-6$ ppm; therefore, the $^1$H NMR $\mu_3$-CH chemical shifts are sometimes like those of metal hydrido complexes.

At low fields, signals due to other inner atoms (nitrogen, phosphorus) are also observed. As in carbide complexes, in clusters of the type $Rh_n(N)L_m$ and $Rh_n(P)L_m$, there is an interaction of all metal atoms with nonmetal atoms. The ratio $J(^{13}C - {}^{103}Rh)/J(^{15}N - {}^{103}Rh)$ agrees well with the gyromagnetic ratio for $^{13}C$ and $^{15}N$.[77] The $^{15}N$ chemical shifts for $[Co_6N(CO)_{15}]^-$ and $[Rh_6N(CO)_{15}]^-$ are smaller than expected from the $\delta^{13}C$ for similar carbide complexes such as $[M_6C(CO)_{15}]^{2-}$.[78-80] For the following compounds the $^{15}N$ NMR chemical

*Table 3.14.* $^{13}C$ and $^1H$ NMR Spectra of Clusters

| Compound | $\delta(^{13}C)$ (ppm) | $\delta(^1H)$ (ppm) |
|---|---|---|
| $[W_4(C)(O-i\text{-}Pr)_{12}NMe]^a$ | 366.8 $W_4(C)$ | |
| $[PPN]_2[Fe_3CrC(CO)_{13}]^b$ | 483.4 $M_4C$, 233–214 CO | |
| $[PPN]_2[Fe_3WC(CO)_{13}]^b$ | 470.3 $(J_{C-W} = 62$ Hz) $M_4C$ | |
| | 221–212 CO | |
| $[PPN][RhFe_3(C)(CO)_{12}]^c$ | 461.1 $(J_{Rh-C} = 19.5$ Hz) $M_4C$ | |
| | 218.0; 215.9; 190.4 | |
| | (d, $J_{Rh-C} = 63.5$ Hz) CO | |
| $[PPN][Fe_3Mn(C)(CO)_{13}]^c$ | 468.4 $M_4C$ | |
| | 237.5; 224.1; 215.0; 214.0; | |
| | 208.7 CO | |
| $[PPN]_2[Ni_3Fe_3C(CO)_{13}]^c$ | 435.1 $M_6C$ | |
| | 243.4; 226.2; 197.3 CO | |
| $[PPN][CoFe_4C(CO)_{14}]^c$ | 476.9 $M_5C$ | |
| | 216.1; 214.7; 209.5; 197.2 CO | |
| $[Co_6C(CO)_{15}]^{2-e}$ | 330.5 $Co_6C$, | |
| | 252; 234.9; 224.9 CO | |
| $[Rh_6C(CO)_{15}]^{2-f}$ | 264.7 $(J_{Rh-C} = 13.7$ Hz) | |
| | 236.3 $(J_{Rh-C} = 51.8$ Hz) | |
| | 225.2 $(J_{Rh-C} = 30.8$ Hz) | |
| | 198.1 $(J_{Rh-C} = 77.1; 3.8$ Hz) | |
| $[Re_7C(CO)_{21}]^{3-g}$ | 423.6 $Re_6C$ | |
| $[PPN]_2[Ru_{10}(C)_2(CO)_{24}]^i$ | 457.1 $Ru_6C$, 241.8 $(\mu-CO)$ | |
| | 205.3; 203.0; 202.3; 200.7; | |
| | 193.8 | |
| $[Fe_3W(H)(CH)(CO)_{13}]^b$ | 339.7 $(J_{C-H} = 97$ Hz) CH | $-0.51$ CH, $-16.1$ MH |
| | 212–206 CO | |
| $[PPN][CoFe_2(CO)_9(\mu_3\text{-}CCO)]^b$ | 82.8 $(J_{C-C} = 79.4$ Hz) $M_3C$ | |
| | 172.5 CCO | |
| | 213.5 CO | |
| $[MnFe_3CH(CO)_{13}]^c$ | 335.5 $(J_{C-H} = 107.4$ Hz) | $-0.33$ |
| | 224.5; 222.4; 218.6; 210.0; | |
| | 209.6; 207.3; 206.8; 204.4 CO | |
| $Fe_4(\mu\text{-}H)(\mu_4\text{-}\eta^2\text{-}CH)(CO)_{12}^d$ | 215(d) $(J_{\mu-H-C} = 7$ Hz) CH | 1.31 CH,   27.9 (FeH) |
| $Os_3CH(H)(CO)_{10}^h$ | 219 CH, 176.7; 173.4; | 14.16(d) CH, $-18.13$(d), |
| | 171.8; 171.1; 169.0; | $J = 1.6$ Hz |
| | 168.0 CO | |

*(Table continued)*

Table 3.14. *(continued)*

| Compound | $\delta(^{13}C)$ (ppm) | $\delta(^{1}H)$ (ppm) |
|---|---|---|
| $[Rh_3(\mu_3\text{-}CH)cp_3(\mu\text{-}CO)_2]PF_6$[j] | | 4.13 cp; −6.20 ($J_{Rh-H} = 2.4$ Hz) |
| $[PPN][Fe_4H(\mu_4\text{-}CO)(CO)_{12}]$[k] | 279.7 $\mu_4$-CO; 221.5; 219.7; 217.5; 215.6; 211.4; 209.5; 209.1 CO | −24.9 |
| $Os_3(\mu\text{-}H)_2(\mu_3\text{-}\eta^1\text{-}CCO)(CO)_9$[l] | 8.6 $Os_3C$; 160.3 CCO | |
| $Os_3(\mu\text{-}CH_2)(CO)_{11}$[m] | 62.5 $CH_2$, 193–171 CO | 7.75(d), 6.47(d) ($J_{H-H} = 7.2$ Hz) −1.48 |
| $[RhFe_3(\mu_4\text{-}CH)(CO)_{12}]$[c] | 340 $M_4C$ ($J_{C-H} = 102$ Hz) | |
| $[Ru_3(\mu_3\text{-}CCO)(\mu\text{-}CO)_3 (CO)_6]^{2-}$[n] | −28.3 ($\mu_3$-C); 159.1 (CCO) 273.3 ($\mu$-CO), 204.0 and 202.3 (CO) | |

[a] H. M. Chisholm, K. Folting, J. C. Huffman, J. Leonelli, N. S. Merchant, C. A. Smith, and L. C. E. Taylor, *J. Am. Chem. Soc.*, **107**, 3722 (1985).

[b] J. W. Kolis, E. M. Holt, J. A. Hriljac, and D. F. Shriver, *Organometallics*, **3**, 496 (1984).

[c] J. A. Hriljac, P. N. Swepston, and D. F. Shriver, *Organometallics*, **4**, 158 (1985).

[d] E. M. Holt, K. H. Whitmire, and D. F. Shriver, *J. Organomet. Chem.*, **213**, 125 (1981); M. A. Beno, J. M. Williams, M. Tachikawa, and E. L. Muetterties, *J. Am. Chem. Soc.*, **103**, 1485 (1981); M. Tachikawa and E. L. Muetterties, *J. Am. Chem. Soc.*, **102**, 4541 (1980).

[e] D. Braga, K. Henrick, B. F. G. Johnson, J. Lewis, M. McPartlin, W. J. H. Nelson, and J. Puga, *Chem. Commun.*, 1083 (1982).

[f] V. G. Albano, P. Chini, S. Martinengo, D. J. A. McCaffrei, and D. Strumolo, *J. Am. Chem. Soc.*, **96**, 8106 (1974).

[g] G. Ciani, G. D'Alfonso, M. Freni, P. Romiti, and A. Sironi, *Chem. Commun.*, 339 (1982); G. Ciani, G. D'Alfonso, M. Freni, P. Romiti, and A. Sironi, *Chem. Commun.*, 705 (1982).

[h] J. R. Shapley, M. E. Cree-Uchiyama, and G. M. St. George, *J. Am. Chem. Soc.*, **105**, 140 (1983).

[i] C.-M. T. Hayward, J. R. Shapley, M. R. Churchill, C. Bueno, and A. Rheingold, *J. Am. Chem. Soc.*, **104**, 7347 (1982).

[j] P. A. Dimas, E. N. Duesler, R. J. Lawson, and J. R. Shapley, *J. Am. Chem. Soc.*, **102**, 7787 (1980).

[k] C. P. Horwitz and D. F. Shriver, *Organometallics*, **3**, 756 (1984).

[l] J. R. Shapley, D. S. Strickland, G. M. St. George, M. R. Churchill, and C. Bueno, *Organometallics*, **2**, 185 (1983).

[m] G. R. Steinmetz, E. D. Morrison, and G. L. Geoffroy, *J. Am. Chem. Soc.*, **106**, 2559 (1984).

[n] M. J. Sailor and D. F. Shriver, *Organometallics*, **4**, 1476 (1985).

shifts are very high: $[FeRu_3N(CO)_{12}]^-$, $[FeRu_4N(CO)_{14}]^-$, $[Ru_6N(CO)_{16}]^-$, $[Ru_5N(CO)_{14}]^-$, and $[Ru_4N(CO)_{12}]^-$.[78] $^{31}P$ NMR spectra were also investigated for these complexes: $[Co_6P(CO)_{16}]^-$[81] $[Rh_9P(CO)_{21}]^{2-}$,[36] and $[Rh_{10}P(CO)_{22}]^{3-}$.[83] Metal NMR spectra such as $^{195}Pt$ for $[Pt_n(CO)_{2n}]^{2-}$ ($n = 3, 6, 9$)[82] and $^{103}Rh$ for $[Rh_9P(CO)_{21}]^{2-}$ and $[Rh_{17}S_2(CO)_{32}]^{3-}$[84] were also measured. Very large clusters often exhibit paramagnetic behavior which precludes use of NMR spectroscopy for structural studies.

## 6. IR SPECTRA

IR spectra of polynuclear and cluster carbonyls were discussed in Chapter 2. The $\nu(M-C)$ frequencies in carbide clusters occur in the 500–850 cm$^{-1}$ range, and therefore are usually higher than the $\nu(M-CO)$ frequencies in carbonyls. The $k(Co-C)$ force constant for the Co−C bond in $[Co_6C(CO)_{12}S_2]$ is 155 N m$^{-1}$ [85] and is lower than for the M−CO bond in carbonyls, although almost seven times greater than the $k(Ru-H)$ force constant in $[Ru_6H(CO)_{18}]^-$.[86] The $\nu(M-C)$ bands were determined from isotopic shifts for $M_n(^{13}C)L_m$. This shift agrees well with the calculated value of the diatomic model $v(M-^{12}C)/v(M-^{13}C) = (13/12)^{1/2}$. The spectrum of the $[Os_{10}(C)(CO)_{24}]^{2-}$ anion shows only one band at 753 cm$^{-1}$, which indicates that the carbon atom is located in the center of the octahedron. However, at low temperatures, this band splits into a doublet, which indicates that the octahedron is distorted.[87] There are three $v(M-C)$ bands of the $Os_6C$ group at 773, 760, and 735 cm$^{-1}$ in the spectrum of $[Os_{10}(H)_2(C)(CO)_{24}]$. These bands are shifted to 744, 730, and 708 cm$^{-1}$ in $[Os_{10}(H)_2(^{13}C)(CO)_{24}]$. The presence of those three bands results from an interaction of a central octahedron with the hydrido ligands.[87] In the IR spectrum of $[Co_6(C)(CO)_{12}S_2]$ (which adopts a prismatic structure), there are two $v(Co-C)$ bands at 810 and 548 cm$^{-1}$. These bands are shifted to 790 and 535.5 cm$^{-1}$ for the compound containing the encapsulated $^{13}C$ atom.[85] The $v(M-C)$ frequencies for some carbide iron hydrido compounds are as follows: $[NEt_4][Fe_5RhC(CO)_{16}]$—807, 784, and 750 cm$^{-1}$; $[NEt_4][Fe_4RhC(CO)_{14}]$—784 and 760 cm$^{-1}$; $[Fe_4Rh_2(C)(CO)_{16}]$—766 and 695 cm$^{-1}$; $[NEt_4][Fe_3Rh_3C(CO)_{15}]$—780 and 750 cm$^{-1}$.[93]

## 7. THE M−M BOND ENERGY IN CLUSTERS

The metal–metal bond energies may be calculated from calorimetric, kinetic, electron bombardment data, etc.[88–92] If, in the gas phase, the molecular structure of the compound is the same as in the solid state, then its enthalpy of decomposition may be represented by the sum of enthalpies of particular bonds, neglecting the distances of these bonds. For iron carbonyls Fe(CO)$_5$, Fe$_2$(CO)$_9$, and Fe$_3$(CO)$_{12}$, as well as for cobalt complexes Co(CO)$_4$, Co$_2$(CO)$_8$, and Co$_4$(CO)$_{12}$, the following relationships are obeyed:

$$
\begin{array}{llll}
Fe(CO)_5 & 5T = 585 & Co(CO)_4 & 4T = 544 \\
Fe_2(CO)_9 & 6T+6B+M = 1173 & Co_2(CO)_8 & 6T+4B+M = 1160 \\
Fe_3(CO)_{12} & 10T+4B+3M = 1676 & Co_4(CO)_{12} & 9T+6B+6M = 2121
\end{array}
$$

where $T$, $B$, and $M$ denote enthalpies of the following bonds: terminal metal–carbon, bridging metal–carbon, and metal–metal. The numbers represent values of enthalpies of decomposition of compounds $\Delta H_D$ in kJ mol$^{-1}$. The numerical coefficients at $T$, $B$, and $M$ denote the number of bonds of a given type in the molecule. The solution of these equations for iron compounds gives the following values: $T = 117$, $B = 65$, and $M = 82$ kJ mol$^{-1}$. In the case of cobalt carbonyls, $T = 136$, $B = 65$, and $M = 84$ kJ mol$^{-1}$. These data indicate that the following empirical relationships are obeyed: $T \sim 2B$ and $M \sim 0.66\, T$, and therefore $2T \approx 4B \approx 3M$. The M − M bond energies for polynuclear carbonyls are given in Table 2.16. The bond energies for dinuclear transition metal complexes and $M_2$ molecules are given in the literature.[89,90]

## 8. SYNTHESIS OF METAL CLUSTERS

The electron-rich metal clusters are most commonly prepared from metal compounds in which the transition metal has the most stable and common oxidation state.

Reaction conditions are selected in order to synthesize either the cluster or a low-valent metal complex from which subsequently the cluster may be obtained by removal of some ligands. This type of synthetic methodology may be utilized for the preparation of most transition metal clusters.

Polynuclear anionic metal carbonyl compounds are usually prepared by reduction reactions of metal carbonyls $M(CO)_n$ with such reducing agents as the alkali metals, $NaBH_4$ in ethers, hydrocarbons, liquid ammonia, and similar solvents [see, for example, reactions (2.54), (2.55), (2.84), (2.89), (2.95)–(2.102), and (2.108)–(2.113)]. In alkali medium the metal carbonyls may be reduced by certain solvents (e.g., alcohols) or by the CO ligand itself, and in the presence of Lewis bases the carbonyls disproportionate to give anionic clusters. The mixed metal clusters containing platinum and rhodium are formed by reduction reactions of chloro complexes[94]:

$$[PtCl_6]^{2-} + RhCl_3 \xrightarrow[CO]{NaOH, MeOH} [PtCOCl_3]^- \xrightarrow{CO} [Pt_3(CO)_6]^{2-}_{6-5} + [RhCl_2(CO)_2]^- \xrightarrow{CO}$$

$$[Pt_3(CO)_6]^{2-}_{5-4} + [RhCl_2(CO)_2]^- \xrightarrow{CO} [PtRh_5(CO)_{15}]^- \xrightarrow{CO} [PtRh_4(CO)_{14}]^{2-} \quad (3.33)$$

Formally, the preparation of $[PtRh_5(CO)_{15}]^-$ may be described by the following stoichiometric equation:

$$[PtCl_6]^{2-} + 5RhCl_3 + 20CO_3^{2-} + 25CO$$

$$+ 10H_2O \xrightarrow[0.1 \text{ MPa CO}]{MeOH, 295 K} [PtRh_5(CO)_{15}]^- + 21Cl^- + 20HCO_3^- + 10CO_2 \quad (3.34)$$

The same cluster, $[PtRh_5(CO)_{15}]^-$, may also be obtained from anionic metal carbonyls of platinum and rhodium:

$$[Pt_{12}(CO)_{24}]^{2-} + 5[Rh_{12}(CO)_{30}]^{2-} \xrightarrow[0.1 \text{ MPa CO}]{THF, 295 K} 12[PtRh_5(CO)_{15}]^- \quad (3.35)$$

Similar methods may be applied to many other salts. Reaction of $RhCl_3(tripod)$ with $AgCF_3SO_3$ gives the rhodium–silver cluster[95]:

$$3RhCl_3(tripod) + 12AgCF_3SO_3 \longrightarrow [Rh_3Ag_3H_9(tripod)_3]^{3+}$$

Other interesting syntheses of group 10 and 11 metal clusters are

$$Pd(OAc)_2 + CO + PBu_3^n \longrightarrow Pd_{10}(CO)_x(PBu_3^n)_y \quad (3.36)$$

$$x = 12, \ y = 6^{[96]} \quad \text{or} \quad x = 14, \ y = 4$$

$$NiCl_2 \xrightarrow[PhP(SiMe_3)_2]{PPh_3} Ni_8(PPh)_6(PPh_3)_4 + SiMe_3Cl^{[97]} \quad (3.37)$$

$$Zncp_2 + Ni(COD)_2 \xrightarrow[348 K]{PhH} Ni_2Zn_4cp_6^{[98]} \quad (3.38)$$

Metal clusters are also formed during thermal decomposition of mononuclear or polynuclear compounds because, at higher temperatures, the dissociation or decomposition of ligands more easily takes place which, in turn, facilitates metal–metal bond formation.

$$3Rh_4(CO)_{12} \xrightarrow{\text{400–430 K}} 2Rh_6(CO)_{16} + 4CO \tag{3.39}$$

More complex metal carbonyl clusters are often obtained by reductive condensation, which involves sequential addition of nucleophilic mononuclear anionic compounds or simple anionic clusters to the main electrophilic cluster.[2]

$$[Rh_4(CO)_{12}] + Rh(CO)_4^- \xrightarrow[\text{0.1 MPa CO}]{\text{298 K, THF}} [Rh_5(CO)_{15}]^- + CO \tag{3.40}$$

Condensations of nucleophilic anionic clusters with electrophilic complexes are also effective routes to some clusters:

$$[Rh_{14}(CO)_{25}]^{4-} + [Rh(CO)_2(MeCN)_2]^+ \xrightarrow[\text{MeCN}]{\text{298 K}} [Rh_{15}(CO)_{27}]^{3-} \tag{3.41}$$

$$7[Ni_6(CO)_{12}]^{2-} + 6PtCl_2 \xrightarrow{\text{MeCN}} [Ni_{38}Pt_6(CO)_{48}H_{6-n}]^{n-}$$
$$+ 2Ni^{2+} + 2Ni(CO)_4 + 28CO + 12Cl^{-\,(99)} \tag{3.42}$$

A promising method appears to be photochemical condensation or decomposition of some complexes or coordinated ligands by complexes or clusters:

$$[Co(CO)_4]^- + Os_3(CO)_{12} \xrightarrow[\text{366 nm}]{hv} [CoOs_3(CO)_{13}]^{-\,(100)} \tag{3.43}$$

$$33\%$$

Metal carbonyl carbide cluster compounds may be formed either by the carbon–oxygen bond breaking of the CO molecule or by the decomposition of an organic compound present in the reaction mixture. Chloroform is a convenient source of the encapsulated carbon atom because in this case the synthesis may be carried out under mild conditions:

$$6RhCl_6^{3-} + 23OH^- + 26CO + CHCl_3 \xrightarrow[\text{0.1 MPa CO}]{\text{298 K, MeOH}} [Rh_6(C)(CO)_{15}]^{2-}$$
$$+ 39Cl^- + 11CO_2 + 12H_2O^{(2)} \tag{3.44}$$

The carbon atom in $[W_4(C)(NMe)(O-i\text{-}Pr)_{12}]$ probably comes from the decomposition of the $NMe_2$ ligand (Table 3.14, footnote reference *a*):

$$W_2(NMe_2)_6 + i\text{-PrOH} \xrightarrow[\text{298 K}]{\text{hexane}} W_4(C)(NMe)(O-i\text{-}Pr)_{12}$$

The carbon–oxygen bond breaking of CO generally requires application of high temperatures, often above 370 K. Under mild conditions, the formation of carbide clusters may occur in the presence of strong acids:

$$Fe(CO)_5 + RhCl_3 \cdot nH_2O \xrightarrow[388\,K,\,2h]{diglyme} [Rh_5FeC(CO)_{16}]^{-\,(93)} \qquad (3.45)$$

$$Ru_3(CO)_{12} + Na \xrightarrow[\substack{1.\ reflux,\ 12h \\ 2.\ Et_4NCl\ in\ H_2O}]{diglyme} [NEt_4]_2[Ru_6C(CO)_{16}]^{(101)} \qquad (3.46)$$

$$[PPN]_2[Fe_4(CO)_{13}]^{2-} \xrightarrow{HSO_3CF_3} [PPN][HFe_4C(CO)_{12}] \qquad (3.47)$$

(Table 3.14, footnote reference *d*)

Nitride-encapsulated clusters are obtained from compounds containing coordinated NO, $N_3^-$, or $NCO^-$ molecules or ions:

$$[PPN][Ru_4(NCO)(CO)_{13}] \xrightarrow[reflux,\,24h]{THF} [PPN][Ru_4N(CO)_{12}]^{(78)} \qquad (3.48)$$

$$[PPN]N_3 + Ru_3(CO)_{12} \xrightarrow[reflux,\,12h]{THF} [PPN][Ru_6N(CO)_{16}]^{(78)} \qquad (3.49)$$

$$[PPN][Fe(CO)_3NO] + Ru_3(CO)_{12} \longrightarrow [PPN][FeRu_3N(CO)_{12}]^{(78)} \qquad (3.50)$$

The clusters $[Rh_9P(CO)_{21}]^{2-}$, $[Rh_{10}(P, As)(CO)_{22}]^{3-}$, and $[Rh_{12}Sb(CO)_{27}]^{3-}$ were prepared by reaction of $[Rhacac(CO)_2]$ with $EPh_3$.[36,37]

Methods for preparation of mixed metal clusters such as $[M_xM'_{4-x}(CO)_n]$, $[MM'M''_2(CO)_n]$, $[M_xM_{3-x}(CO)_m]$, etc., have also been described.[102]

Clusters may also be synthesized by electrolysis of mono- or polynuclear compounds.[103–106]

$$M_2(CO)_{10} \xrightarrow{e} [M_3(CO)_{14}]^- \qquad (M = Mn, Re) \qquad (3.51)$$

$$[PtCo_2(\mu\text{-}CO)(CO)_7(PPh_3)] \xrightarrow{e} Pt_2Co_2(\mu\text{-}CO)_3(CO)_5(PPh_3)_2 \qquad (3.52)$$

## 9. PROPERTIES OF CLUSTERS

The "electron rich" clusters of the first series of the transition metals are generally more sensitive to oxidation than clusters of the second and third series. The most stable electron-rich clusters are those of groups 8, 9, and 10. Clusters of rhodium and iridium are exceptionally stable. However, some clusters of the first series of the transition metals can be stable. This is particularly true for compounds with stable electron configurations containing $n+1$, $n+2$, or $n+3$ skeletal electron pairs. For example, $Co_3(CO)_9CR$ complexes contain six skeletal electron pairs $[Co(CO)_3 \leftrightsquigarrow CR]$ and some of them are not oxidized by air in the solid state or in solution.

### a. *Group 4*

Metals of this group form triangular clusters of the general formula $[M_3Cl_6(C_6Me_6)_3]^+$, where $M = Ti$ or $Zr$ (See Chapter 10). Also known is an 86*e* cluster of Ti(III, IV) possessing the composition $Ti_6O_8cp_6$[107]; its structure is similar to other octahedral clusters. The titanium atoms lie at the vertices of the octahedron. Eight bridging oxygen atoms, each bonded to three titanium atoms, reside over the triangular faces of the octahedron. Additionally, each titanium atom is coordinated by one cyclopentadienyl ligand ($\eta^5$). The average Ti–Ti distance is 289.1 pm, which is smaller than in the titanium metal itself in spite of the fact that the formal oxidation state of Ti is $+3.5$. Thus, only three $d$ electrons are bonding the titanium atoms in $Ti_6O_8cp_6$.

Recently, octahedral compounds of the type $[Zr_6Cl_{12}]$, $[Zr_6CCl_{14}]$, $KZr_6CCl_{15}$, and $CsKZr_6BCl_{15}$ have been prepared.[72]

### b. *Group 5*

Metals of this group form di- and trinuclear complexes such as $(CO)cp_2Nb(\mu\text{-}H)Fe(CO)_4$,[115] $V_2(\mu\text{-}PMe_2)_2(CO)_8$, $V_2cp_2(CO)_5$ (Figure 2.15, Table 2.22), $[M_2cp_2(C_2Ph_2)_2(CO)_2]$ (M = Tb, Ta) (Chapter 6), $M_2Cl_4(C_6Me_6)_2$, $[M_3(\mu\text{-}Cl)_6(C_6Me_6)_3]^{n+}$ (Chapter 10), $[Nb_3(\mu\text{-}Br)_6(C_6Me_6)_3]^+$, and $Nb_3cp_3(CO)_7$ (Figure 2.3). Clusters possessing two $\mu_3$ bridges of the type $B_6$ are also known (Figure 3.19): $[Nb_3(\mu_3\text{-}O)_2(\mu\text{-}OOCR)_6(THF)_3]^+$ (R = Ph, $CMe_3$),[108] $[Nb_3(\mu_3\text{-}O)_2(\mu\text{-}SO_4)_6(H_2O)_3]^{5-}$.[109] Additionally, hexanuclear halogen clusters of Nb and Ta have been reported (Table 3.13).[1,7,11,12]

### c. *Group 6*

Metals of this group are characteristic for their ability to form inorganic and organometallic compounds that possess triple and quadruple metal–metal bonds.[1,90] Some illustrative examples are $[M_2Me_8]^{4-}$ (Chapter 4) (M = Cr, Mo, W), $M_2(C_3H_5)_4$ (M = Cr, Mo), $M_2(COT)_3$, $M_2(mhp)_4$, $M_2(chp)_4$, $M_2(dmhp)_4$, $M_2(map)_4$ (M = Cr, Mo, W), $M_2R_6$ (R = $CH_2SiMe_3$, $CH_2Ph$, $CH_2CMe_3$, $NMe_2$, OR′; M = Mo, W), $M_2(\mu\text{-}CO)(OBu')_6$ (M = Mo, W), $[W_2(\mu_3\text{-}\eta^2\text{-}CO)(OPr^i)_6]_2$ (Figure 2.17). Compounds containing single and double metal–metal bonds are also known, for example, $[M_2(\mu\text{-}PR_2)_2(CO)_8$, $M_2(\mu\text{-}CO)_2(CO)_4(NN)_2]$ (NN = bipy, phen), and $[M_2(CO)_{10}]^{2-}$. In contrast to elements of groups 4 and 5, group 6 metals form the trinuclear carbonyl complexes $Cr_3(CO)_{14}^{2-}$ and $[Cr_3(\mu_3\text{-}S)(\mu\text{-}CO)_3(CO)_9]^{2-}$.[110]

In the case of tungsten, carbonyl compounds containing the main group elements are encountered: $[E_2\{W(CO)_5\}_3]$ (E = As, Sb, Bi)[111] and $[W_2(CO)_8(\mu\text{-}\eta^2\text{-}Bi_2)(\mu\text{-}BiMe)W(CO)_5]$.[112] In the first type of compounds, only bonds between the main group elements are formed; there is no direct interaction between the transition metal atoms. The second compound contains, in addition to the bismuth–bismuth bond, also a tungsten–tungsten bond. An analogous situation occurs in $[cp_2Mo_2Fe_xTe_2(CO)_7]$ (x = 1 or 2).[113]

The elements of group 6 form many electron-rich mixed metal clusters, for example, $[Pd_2M_2cp_2(\mu_3\text{-}CO)_2(\mu\text{-}CO)_4(PEt_3)_2]$ (M = Cr, Mo, W)[114] and $[Ni_3M_2(CO)_{16}]^{2-}$ (see also Table 3.14).

Molybdenum and tungsten form trinuclear (Table 3.13) and tetranuclear clusters, such as $[W_4(\mu_4\text{-}C)(OPr^i)_{12}(NMe)]$ or $Mo_4(N)_2(OPr^i)_{12}$ (Table 3.14, footnote reference *a*).

Cyclopentadienyl clusters of the group 6 metals are numerous.[116] A few examples are $[Mo_3cp_3(PPh_2)_3(CO)_3]$, $[(OC)cpRe(\mu\text{-}CO)_2Mocp_2]$, [(OC) cpCr($\mu$-CO)$_2$Nicp(CO)], [cp$_2$Mo$_2$Fe$_2$($\mu_3$-CO)$_2$(CO)$_6$($\mu_3$-S)$_2$] (butterfly structure), [(C$_5$H$_4$Me)$_2$Mo$_2$Fe$_2$($\mu$-CO)$_2$(CO)$_6$($\mu_3$-S)$_2$] (rhombus structure) (Figue 3.22), cp$_2$Mo$_2$Pt$_2$($\mu$-CO)$_6$(PEt$_3$)$_2$, Pt$_2$Mo$_2$cp$_2$(CO)$_6$(dppe), etc.

## d. *Group 7*

In the case of this group, many dinuclear and trinuclear electron-poor clusters are known (Chapter 4),[1,90] for instance, [(Me$_3$P)(Me$_3$SiCH$_2$) Mn(CH$_2$SiMe$_3$)$_2$ Mn(CH$_2$SiMe$_3$)(PMe$_3$)],[118] Re$_2$Me$_8^{2-}$, Re$_2$Me$_2$(OAc)$_4$, Re$_2$(CH$_2$SiMe$_3$)$_4$(OAc)$_2$, Re$_2$Me$_6$L$_2$, [Re$_2$($\mu$-H)$_3$(CO)$_6$]$^-$, Re$_3$Me$_9$, and Re$_3$Me$_6$(acac)$_3$.

Also quite numerous are hexanuclear clusters: [Re$_6$Cl$_6$H(CH$_2$SiMe$_3$)$_9$], [Re$_6$Cl$_6$H$_6$(CH$_2$SiMe$_3$)$_6$], and [Re$_6$Cl$_6$(CH$_2$SiMe$_3$)$_6$]. These compounds form adducts with Lewis bases.

Metal carbonyl clusters are also numerous. In addition to dinuclear carbonyls such as M$_2$(CO)$_{10}$ (M = Mn, Tc, Re), many hydrido clusters and ionic species exist which contain three or more metal atoms: [M$_3$H$_3$(CO)$_{12}$] (M = Mn, Tc, Re), Mn$_3$(CO)$_{14}^-$ (linear compound, $D_{4h}$), and Re$_3$($\mu$-H)(CO)$_{14}$ (the Re−Re−Re angle is 90°). Rhenium forms a number of tetranuclear compounds. Some of them have a butterfly structure while others adopt a triangle of rhenium atoms with one Re attached to it: Re$_4$($\mu_3$-H)$_4$(CO)$_{12}$, [(Re$_4$($\mu$-H)$_4$(CO)$_{13}$]$^{2-}$, [Re$_4$($\mu$-H)$_6$(CO)$_{14}$]$^{2-}$, and [Re$_4$(CO)$_{16}$]$^{2-}$. Rhenium also forms mixed metal carbonyls possessing several metal atoms (Table 3.14). There are numerous clusters of the group 7 metals containing cyclopentadienyl groups,[116]

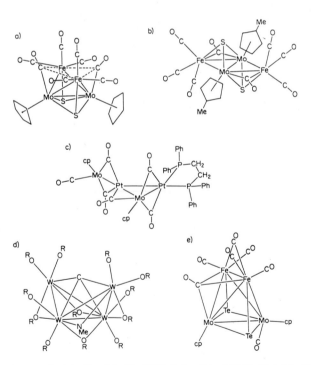

Figure 3.22. Structures of (a) [cp$_2$Mo$_2$Fe$_2$(CO)$_8$S$_2$], (b) [(C$_5$H$_4$Me)$_2$Mo$_2$Fe$_2$(CO)$_8$S$_2$], (c) [Pt$_2$Mo$_2$cp$_2$(CO)$_6$(dppe)], (d) [W$_4$C(OPr$^i$)$_{12}$(NMe)], (e) [cp$_2$Mo$_2$Fe$_2$Te$_2$(CO)$_7$].

for example, $Mn_3cp_3(NO)_4$, $(OC)cpMn(\mu\text{-}CO)_2Rhcp(CO)$, $Re_2(C_5Me_5)_2(CO)_3$, $MnFe_2cp(\mu_3\text{-}PPh)(\mu\text{-}CO)_2(CO)_6$, and $[(C_5Me_5)_3Re_3(\mu\text{-}O)_6]^{2+}(ReO_4^-)_2$.[119] Many clusters possessing phisphido ligands, $PR_2$, are also known; some of them are linear: $[(OC)_4Mn(\mu\text{-}PPh_2)PtCO(\mu\text{-}PPh_2)(Mn(CO)_4]$, $[(OC_4Mn(\mu\text{-}PPh_2)Pt(\mu\text{-}PPh_2)_2$ $Pt(\mu\text{-}PPh_2)Mn(CO)_4]$.

Among recent additions in this area are the large rhenium clusters $[Re_7C(CO)_{21}]^{3-}$ and $[NEt_4]_2[Re_8C(CO)_{24}]$,[117]

Structures of some manganese and thenium clusters are given in Figure 3.23.

### e. Group 8

Metals of this group form many stable clusters such as neutral and anionic carbonyls, encapsulated complexes, polynuclear compounds possessing cyclopentadienyl, phosphido, phosphine, and ligands (Figure 3.24).

The following carbonyls and hydridocarbonyls are known: $M_2(CO)_9$ (M = Fe, Os), $M_3(CO)_{12}$ (M = Fe, Ru, Os), $Os_5(CO)_{19}$, $Os_5(CO)_{16}$, $Os_6(CO)_{18}$, $Os_7(CO)_{21}$, $Os_8(CO)_{23}$, $M_3H_2(CO)_{11}$, $M_3H(CO)_{11}^-$, $M_4(CO)_{13}^{2-}$, $[M_4H(CO)_{13}]^-$, $M_4H_2(CO)_{13}$, $M_4H_4(CO)_{12}$, $[M_6H(CO)_{18}]^-$ (M = Fe, Ru, Os), $Os_7H_2(CO)_{20}$, $[Os_8H(CO)_{22}]^-$,[120] $[Fe_4C(CO)_{12}]^{2-}$, $[Fe_4C(CO)_{13}]$, $[Fe_3MC(CO)_{13}]^{2-}$ (M = Cr, W; Table 3.14), $[M_5C(CO)_{15}]$ (M = Fe, Ru, Os), $[Ru_6C(CO)_{17}]$, $[Ru_6C(CO)_{16}]^{2-}$, $[Fe_4MC(CO)_m]^{n-}$, $[Fe_4M_2C(CO)_m]^{n-}$, $[Fe_5MC(CO)_m]^{n-}$ (M = Cr, Mo, W, Mn, Re, Rh, Ir, Ni, Pd, Cu),[122] $[Ru_{10}(C)_2(\mu\text{-}CO)_4(CO)_{20}]^{2-}$,[123] $[Ru_6N(CO)_{16}]^-$, $[Ru_5N(CO)_{14}]^-$, and $[Ru_4N(CO)_{12}]^-$.[78]

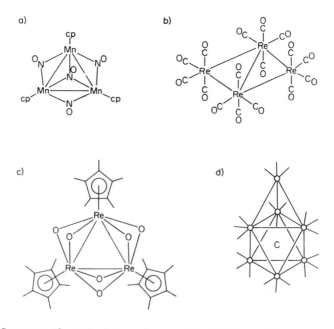

Figure 3.23. Structures of metal clusters of group 7d: (a) $Mn_3cp_3(NO)_4$, (b) $Re_4(CO)_{16}^{2-}$, (c) $[Re_3(C_5Me_5)_3(\mu\text{-}O)_6]^{2+}$, (d) $[Re_7C(CO)_{21}]^{3-}$.

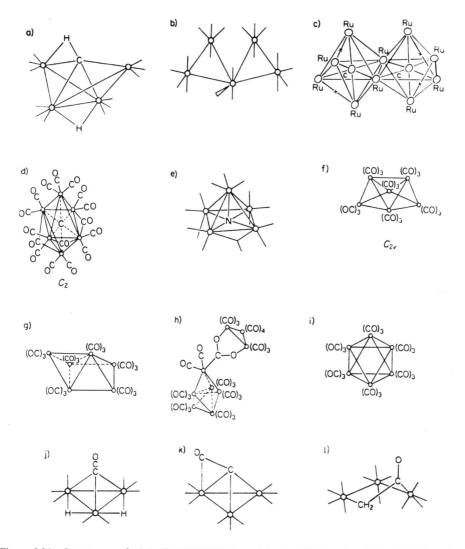

Figure 3.24. Structures of (a) $Fe_4H(CH)(CO)_{12}$, (b) $Os_5(CO)_{19}$, (c) $[Ru_{10}(C)_2(CO)_{24}]^{2-}$, (d) $[Fe_6(CO)_{13}(\mu\text{-}CO)_3C]^{2-}$, (e) $[Ru_5N(\mu\text{-}CO)(CO)_{13}]^-$, (f) $Os_6(CO)_{18}$, (g) $Os_6H_2(CO)_{18}$, (h) $[Os_3H(CO)_{10}(\mu\text{-}O_2C)Os_6(CO)_{17}]^-$, (i) $Os_6(CO)_{18}^{2-}$, (j) $Os_3(\mu_3\text{-}\eta^1\text{-}CCO)H_2(CO)_9$, (k) $Fe_3(\mu_3\text{-}\eta^2\text{-}CCO)(CO)_9^{2-}$, (l) $Os_3\{\mu\text{-}\eta^2(CC)\text{-}CH_2CO\}(CO)_{12}$ ($\times$ denotes $\mu\text{-}CO$).

Owing to their structures and reactivities, clusters containing $M_4C$, $M_3CR$, and $M_3CR_2$ skeletons are of particular interest. In addition to the previously described four metal atom compounds (Table 3.14, footnote references *b* and *c*), they comprise trinuclear $M_3$ complexes having $\mu_3\text{-}CH$, $\mu_3\text{-}CCO$, and $\mu_2\text{-}CH_2$ bridges; such clusters are intermediate products of CO reduction: $[Os_3(CO)_{11}(\mu\text{-}CH_2)]$, $[Os_3H_2(CO)_9(\mu_3\text{-}CCO)]$ (Table 3.14, footnote references *l* and *m*), and $[Fe_3(CO)_9(CCO)]^{2-}$.[121]

Mixed metal clusters containing metals of groups 8, 9, and others have also been obtained and investigated [102, 1, 2, 3, 5, 6, 8, 14, 21–25]: $[M_xM'_{3-x}(CO)_{12}]$ (M = Fe, Ru, Os, M' = Ru, Os) and $[CoM_3(CO)_{13}]^-$, etc.

Iron–sulfur clusters are important and much investigated because they serve as models for iron–sulfur proteins, which constitute a large and significant group of biologically active substances. These substances are probably fundamental for all forms of life. They are important for such processes as the uptake and evolution of hydrogen, nitrogen fixation, ATP formation, electron transfer, and the like.

Models for ferredoxins are cubic clusters of the general formula $[Fe_4(\mu_3\text{-}S)_4(SR)_4]^{n-}$ ($n = 1$–$4$).[165] Recently, the newly discovered cluster $[NEt_4]_2^+[Fe_6S_6I_6]^{2-}$ possessing a $Fe_6S_6$ core has been investigated.[174]

## f. *Group 9*

Clusters of this group are among the most stable, and therefore a great number of them has been prepared. Rhodium forms the largest number of clusters of various sizes containing from 7 to 55 rhodium atoms.[184] In general, clusters of the group 9 metals have been investigated intensively because of their exceptional activity as catalysts (cobalt and especially rhodium compounds) for hydroformulation, water–gas shift reaction, hydrogenation of unsaturated hydrocarbons, alkene isomerization, etc. These clusters are also interesting as models for encapsulated compounds.

The most important carbonyl clusters of the group 9 metals are $M_4(CO)_{12}$, $M_6(CO)_{16}$ (see Chapter 2, Figures 2.22 and 2.23, Tables 3.3, 3.9, 3.12, and 3.14) $[Rh_5(CO)_{15}]^-$, $[Rh_6C(CO)_{15}]^{2-}$, $[Co_6C(CO)_{15}]^{2-}$, $[Co_2Rh_4C(CO)_{15}]^{2-}$, $[Rh_7(CO)_{16}]^{3-}$ (capped octahedron) $[Rh_9(CO)_{19}]^{3-}$ (face-sharing bioctahedron), $[Rh_{11}(CO)_{23}]^{3-}$ (trioctahedron connected by faces),[124] $[Rh_{12}(C)_2(CO)_{24}]^{2-}$, $[Rh_{12}(C)_2(CO)_{25}]$, $[Rh_{15}(C)_2(CO)_{28}]^-$, $[Rh_{13}H_{5-n}(CO)_{24}]^{n-}$, $[Rh_{14}(CO)_{25}]^{4-}$, $[Rh_{15}(CO)_{30}]^{3-}$, $[Rh_{15}(CO)_{27}]^{3-}$, $[Rh_{22}(CO)_{37}]^{4-}$, $[Ir_8(CO)_{22}]^{2-}$, $[Co_8(C)(CO)_{18}]^{2-}$, $[M_6(CO)_{15}]^{2-}$ (M = Co, Rh, Ir), and $[Co_{13}(C)_2(H)(CO)_{24}]^{4-}$.

The rhodium compound $[Rh_6(\mu\text{-}AsBu_2^t)_2(\mu_4\text{-}AsBu^t)(\mu\text{-}CO)_2(CO)_9]$[125] is probably the only example of a $Rh_6$ cluster in which the $Rh_6$ framework is a pentagonal pyramid. Recently, the mixed metal cluster $[Co_6Ni_2(C_2)(CO)_{16}]^{2-}$ having the structure of a biprism sharing a rectangular face (the carbon–carbon distance is 149.4 pm) was prepared.[126] An important class of clusters are alkylidyne compounds of the type $[Co_3CR(CO)_9]$.

There is a large number of clusters containing cyclopentadienyl ligands: $M_3cp_3(\mu_3\text{-}CR)_2$[127] (M = Co, Rh, Ir), $M_3cp_3(CO)_3$ (M = Co, Rh),[128] $Co_4cp_4(\mu_3\text{-}H)_4$ (see Chapter 9), $Rh_4cp_4(\mu_3\text{-}CO)_2$, $[Rh_3cp_3(\mu_3\text{-}CH)(\mu\text{-}CO)_2]^+PF_6^-$ (Table 3.14), $[Pt\{Rhpmcp(CO)\}_4]$, and $[cp_2Rh_2Fe(CO)_8]$.[116]

Cobalt forms the pentanuclear compound $Co_5(CO)_{11}(PMe_2)_3$ possessing a trigonal–pyramidal structure. In this compound, the fifth cobalt atom is attached to one of the edges, and all $PMe_2$ groups form bridges.[129] Rhodium affords several clusters with phosphido ligands: $Rh_3(\mu\text{-}PPh_2)_3(CO)_3(PPh_3)_2$, $Rh_3(\mu\text{-}PPh_2)_3(CO)_5$, and $Rh_3(\mu\text{-}PBu_2^t)_3(CO)_3$. The latter compound is electron deficient. The $Rh_3P_3$ group is planar in contrast to the complexes containing $PPh_2$.[130]

Of particular interest are electron-deficient hydridophosphite compounds of Rh(I): $[Rh(\mu\text{-}H)(PR_3)_2]_x$ ($x = 3$ for $R = OMe$, $OEt$, $x = 2$ for $R = OPr^i$) and $[Rh(\mu\text{-}H)(R_2PCH_2CH_2PR_2)]_x$ ($x = 4$ for $R = OMe$, $OEt$, $x = 2$ for $R = Pr^i$).[138, 139] The

X-ray and neutron studies of some complexes containing monophosphites have been reported.[138] In $Rh_2(\mu\text{-}H)_2(PR_3)_4$ the phosphorus, hydrogen, and rhodium atoms lie in the same plane, while in the triangular trimer the cyclic fragment $Rh_3(\mu\text{-}H)_3P_6$ is not planar. The Rh–Rh distances are 265 pm in the dimer and 281 pm in the trimer. Recently, the large cluster of rhodium with chloro and phosphine ligands was prepared, $Rh_{55}Cl_6(PPh_3)_{12}$. This cluster is obtained by reduction of $RhCl(PPh_3)_3$ with $B_2H_6$.[184]

The structures of some group 9 metal clusters are presented in Figure 3.25.

## g. *Group 10*

Metals of this group form many large clusters containing numerous skeletal atoms. So far, the largest structurally investigated clusters probably are $[Ni_{38}Pt_6(\mu_3\text{-}CO)_{18}(\mu\text{-}CO)_{12}(CO)_{18}H_{6-n}]^{n-}$ ($n = 3\text{–}6$). These clusters consist of an inner octahedron constructed of six platinum atoms which are located inside an outer octahedron composed of 38 nickel atoms. Thus, the structure of such clusters is that of

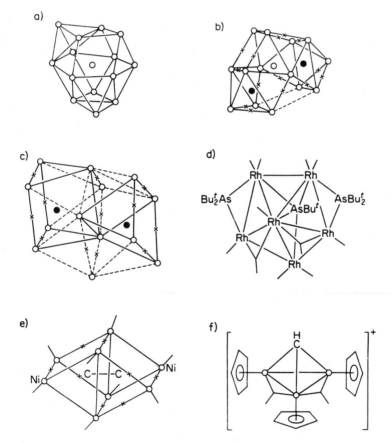

Figure 3.25. Structures of metal clusters of group 9d: (a) $[Rh_{15}(CO)_{27}]^{3-}$, (b) $[Rh_{15}(C)_2(CO)_{28}]^-$, (c) $[Co_{13}H(C)_2(CO)_{24}]^{4-}$ (continuous lines connect corners of prisms), (d) $[Rh_6(AsBu_2^t)_2(AsBu^t)(CO)_{11}]$, (e) $[Co_6Ni_2(C_2)(CO)_{16}]^{2-}$, (f) $[Rh_3cp_3(CH)(CO)_2]^+$ ($\times$ denotes $\mu$-CO).

cubic close-packed metals (ccp). An analogous structure is found for $[Pt_{38}(CO)_{44}H_x]^{2-}$ (Figure 3.18).[99] It may be possible to obtain clusters having the general formulas $[Ni_{44-x}Pt_x(CO)_{48}H_{6-n}]^{n-}$ ($x \geqslant 6$) which would be analogous to the series of dodecanuclear anionic compounds $[Ni_{12-x}Pt_x(CO)_{21}H_{4-n}]^{n-}$ ($x = 2, 3$; $n = 2$–4).[99] Nickel and platinum form many anionic carbonyls: $[Ni_5(CO)_{12}]^{2-}$, $[Ni_6(CO)_{12}]^{2-}$, $[Ni_9(CO)_{18}]^{2-}$, $[Ni_{12}(CO)_{21}H_{4-n}]^{n-}$ ($n = 2$–4),[131] $[Pt_3(CO)_6]_n^{2-}$ ($n = 1$–10),[132] $[Pt_{19}(CO)_{22}]^{4-}$,[133] $[Pt_{26}(CO)_{32}]^{2-}$,[2] and $[Pt_{38}(CO)_{44}H_x]^{2-}$.[2,99] The compound $[Pt_{19}(CO)_{22}]^{4-}$ has $D_{5h}$ symmetry which cannot occur for metal crystal lattices (Figure 3.26). Recently, carbide clusters of the type $[Ni_8C(\mu\text{-}CO)_8(CO)_8]^{2-}$ and $[Ni_9C(\mu_3\text{-}CO)_4(\mu\text{-}CO)_4(CO)_9]^{2-}$ were prepared. The former has the structure of an Archimedean antiprism and the latter has the structure of a capped Archimedean antiprism.[134]

Palladium forms fewer carbonyl clusters than nickel and platinum. However, palladium clusters containing ligands possessing strong $\sigma$ donor properties such as phosphines,[135] 2, 2′-bipyridine and 1, 10-phenanthroline[136,137] are common:

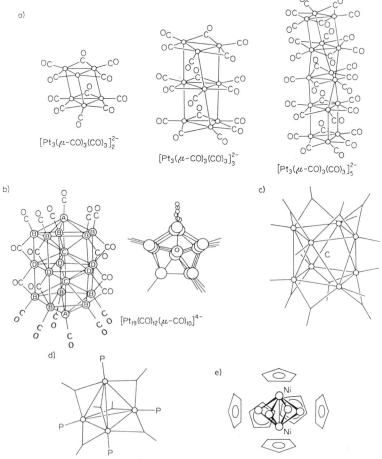

Figure 3.26. Structures of (a) $[Pt_3(CO)_6]_n^{2-}$, (b) $[Pt_{19}(CO)_{22}]^{4-}$, (c) $[Ni_8C(CO)_{16}]^{2-}$, (d) $[Pd_4(CO)_6(PBu_3)_4]$, (e) $[Ni_2Zn_4cp_6]$.

$[Pd_4(CO)_2(phen)_4](OAc)_4$, $M_3(CO)_3(PR_3)_n$ (M = Pd, Pt, $n = 3, 4$), $M_4(CO)_5(PR_3)_4$ (M = Pd, Pt), $Pt_5(CO)_6L_4$, $Pd_7(\mu_3\text{-}CO)_4(\mu\text{-}CO)_3(PMe_3)_7$, and $Pd_{10}(CO)_{12}(PR_3)_6$.

Lower stability of palladium clusters compared to nickel and platinum ones is explained by lower energy of the Pd–Pd bond. It is assumed that the energy of a M–M bond can be measured by the energy of sublimation, which for Ni, Pd, and Pt is 427, 354, and 565 kJ mol$^{-1}$, respectively. For Cu, Ag, and Au, the energy of sublimation is 339, 285, and 364 kJ mol$^{-1}$, respectively. It turns out that in the case of group 11 metals, according to predictions, the number of silver clusters is smaller than that of copper and gold clusters.

Cyclopentadienyl clusters are characteristic for nickel; they include $Ni_3cp_3(CO)_2$, $Ni_4cp_4H_3$, $Ni_3cp_3(\mu_3\text{-}NBu')$, and $Ni_3(\mu_3\text{-}S)_2cp_3$.

Also, octahedral clusters of the type $Ni_2Zn_4cp_6$[98] (Figure 3.26), $Ni_6cp_6$, and $Ni_6cp_6^+$ [140] have been described.

A method for synthesis of clusters containing PR and $PR_3$ or CO ligands, e.g., $[Ni_8(\mu_4\text{-}PPh)_6(PPh_3)_4]$, $Ni_8Cl_4(PPh)_6(PPh_3)_4$, and $Ni_8(CO)_4(PPh)_6(PPh_3)_4$,[97] has been elaborated. It involves reactions of the transition metal halides with $P(SiMe_3)_2R$ in the presence of $PR_3$. These compounds are cubic. The PPh ligands create $\mu_4$-type bridges as in $Ni_8(PR)_6(CO)_8$. $CoCl_2$ reacts in a similar way to give $Co_4(\mu_3\text{-}PPh)_4(PPh_3)_4$.[97]

Analogously, reactions involving $S(SiMe_3)_2$ afford the following clusters[142]: $Co_6(\mu_3\text{-}S)_8(PPh_3)_6^+$ (octahedron), $[Co_7Cl_2(\mu_4\text{-}S)_3(\mu_3\text{-}S)_3(PPh_3)_5][Ni_8(\mu_4\text{-}S)_6Cl_2(PPh_3)_6]$ (cube), and $[Ni_8(\mu_3\text{-}S)_3(\mu_4\text{-}S)_2(PPh_3)_7]$ (bis trigonal-bipyramid with common edge). Reactions of $Se(SiMe_3)_2$ with $[MCl_2(PPh_3)_2]$ give selenium containing clusters such as $[Co_4(\mu_3\text{-}Se)_4(PPh_3)_4]$ (tetrahedron), $[Co_6(\mu_3\text{-}Se)_8(PPh_3)_6]$ (octahedron), $[Co_9(\mu_4\text{-}Se)_3(\mu_3\text{-}Se)_8(PPh_3)_6]$, and $[Ni_{34}Se_{22}(PPh_3)_{10}]$.[183] In the complex $Pt_3Ph(\mu\text{-}PPh_2)_3(PPh_3)_2$ each platinum atom bonds with two bridging $PPh_2$ ligands and with one terminal $PPh_3$ or Ph group. There is no metal–metal bonding between the platinum atoms connected to the phosphine molecules because the Pt–Pt distance is 336 pm (the remaining Pt–Pt distances are 278.5 pm).[141]

Nickel forms compounds with $P_2$ (diphosphorus) and $R'P = PR'$ (diphosphene) which may be coordinated in a similar way to olefins: $[Ni(R'P = PR')(R_2PCH_2CH_2PR_2)]$ and $[Ni_2(P \equiv P)(R_2PCH_2CH_2PR_2)_2]$.[144] In the latter compound, the nickel and phosphorus atoms are located at the vertices of a distorted tetrahedron. The phosphene cluster $[Ni_5Cl(CO)_6\{(Me_3Si)_2CHP = PCH(SiMe_3)_2\}_2]$ [145] is known.

Yet another class of group 10 metal clusters is provided by isonitrile compounds $Ni_4(CNBu')_7$, $Ni_4(CNBu')_4(\mu_3\text{-}PhCCPh)_3$,[143] and $M_3(CNR)_6$ (M = Pd, Pt). The cluster $Pt_3(\mu\text{-}CNBu')_3(CNBu')_3$ is a 42$e$ triangular complex with three $\mu_2$ bridges and three terminal isocyanides; the compound $Pd_3(CNBu')_5(\mu\text{-}SO_2)_2$ also has a triangular structure, with palladium atoms bridged by one $SO_2$ molecule and two CNBu' ligands. In contrast to $Ni(CNBu')_4$, isocyanide clusters of nickel catalyze cyclization of acetylene to benzene, and cyclization of 1, 3-butadiene to cyclooctadienes, as well as hydrogenation of acetylenes.[143]

## h. *Group 11*

The differences between properties of clusters of copper, silver, and gold and those of group 8–10 metals are substantial. As a result of an even larger lowering of the energy of $d$ orbitals of these elements compared to that of the metals of the preceding groups, stable metal carbonyl complexes are not known since $d$ orbitals do not effectively interact with carbon monoxide $\pi^*$ orbitals. The formation of clusters should be facilitated by ligands possessing strong $\sigma$ donor properties, because the increase in electron density on the metal causes an increase in energy and size of valence orbitals which in turn leads to an increase in stability of the $M-M$ bond. Clusters of group 11 metals are stabilized by ligands which coordinate through nitrogen, phosphorus, or sulfur.

Tetranuclear copper complexes have tetrahedral, square-planar, or butterfly structures. The compound $Cu_4\{(Pr^iO)_2PS_2\}_4$ has a distorted tetrahedral skeleton of the metal atoms with one sulfur atom symmetrically bonded to two copper atoms, while the other sulfur atom forms a bond with the third copper atom. Copper and silver form many clusters containing alkyl and aryl ligands (See Chapter 4). The complex $Cu_4(CH_2SiMe_3)_4$, like $Cu_4\{N(SiMe_3)_2\}_4$, has a square-planar structure with bridging alkyl ligands. In the alkyl compound, in contrast to the amide one, there are $Cu-Cu$ bonds.

Hydrocarbonyl ligands containing other coordinating atoms such as N or O form chelate complexes with copper, for example, $Cu_4(2\text{-}Me_2NCH_2\text{-}5\text{-}MeC_6H_3)_4$. This compound has a butterfly structure; the Cu–Cu distances are 238 pm. The carbon atom of the phenyl ring is coordinated to two copper atoms and the nitrogen atom is bonded to only one copper atom.

There are many other clusters of copper and silver including organometallic ones[146]: $[Cu_6(2\text{-}Me_2NC_6H_4)_4(C\equiv CC_6H_4Me-4)_2]$, $Cu_8(2\text{-}MeOC_6H_4)_8$ (Archimedean antiprism),[147] $Cu_6(2\text{-}Me_2NC_6H_4)_4(CF_3SO_3)_2$, $Cu_4M_2(2\text{-}Me_2NC_6H_4)_4(CF_3SO_3)_2$

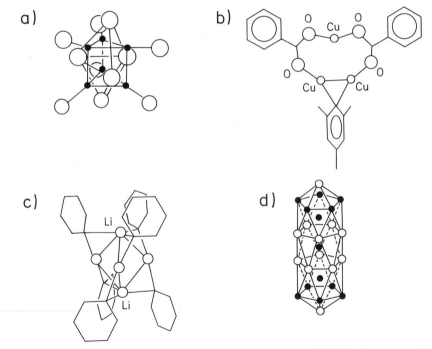

Figure 3.27. Structures of (a) $[Cu_6I_{11}]^{5-}$, (b) $[Cu_3C_6H_2Me_3)(OOCPh)_2]$, (c) $Ag_3Li_2Ph_6^-$, (d) $Au_{13}Ag_{12}Cl_6(PPh_3)_{12}^{n+}$.

$(M = Ag, Au)$,[148] $Cu_3(\mu-C_6H_2Me_3-2,4,6)(\mu-PhCOO)_2$,[149] $Cu_5(C_6H_2Me_3-2,4,6)_5$,[150] and $[Cu_6(\mu_3-I)_2(\mu_4-I)_3I_6]^{5-}$ (prism with bridging iodine atoms above the center of each face).[151]

In heteronuclear clusters such as $(CuCo(CO)_4]_4$ and $[AgCo(CO)_4]_4$,[152] the CO molecules are connected only to the cobalt atoms, and the metal atoms form rings. In the compound $[(OC)_4CoZn_2Co_2(CO)_7]$, the zinc atoms also are connected only to the cobalt atoms.[153] Similarly, the silver atoms of the cluster $[Rh_3Ag_3H_9(tripod)_3]^{3+}$ are bonded only to the rhodium atoms.[95] The cluster $[Ir_2Cu_3(\mu-H)_6(NCMe)_3(PMe_2Ph)_6]^{3+}$ has a trigonal–bipyramidal structure in which the iridium atoms occupy the axial positions and the hydrogen atoms form bridges between the iridium and copper atoms.[155]

Many heteronuclear clusters possessing the main group metals have been investigated, e.g., $[Cu_3Li_2Ph_6]^-$,[154] $[Li_6Br_4(Et_2O)_{10}]^{2+}$ $[Ag_3Li_2Ph_6]^-_2$,[156] $Cu_4MgPh_6$, and $[Li(Et_2O)_4]^+[Cu_4LiPh_6]^- \cdot 2Et_2O$.[157] They have trigonal–bipyramidal structures. There are no bonds between equatorial silver atoms or equatorial copper atoms.

Numerous clusters of gold which are prepared by reduction of $AuX(PR_3)$ with $NaBH_4$ are known. The largest known gold cluster $Au_{55}Cl_6(PPh_3)_{12}$ can be obtained by reduction of $AuCl(PPh_3)$ with $B_2H_6$ in benzene solution.[158,184] The formula of this compound was established by elemental analysis and molecular weight determination. The $[Au_{13}Cl_2(PMe_2Ph)_{10}]^{3+}$ ion adopts an icosahedral structure with an encapsulated gold atom. Structures of other related gold clusters can best be explained assuming that some metal atoms were removed from the vertices of an icosahedron. This series of compounds includes $[Au_8L_7]^{2+}$, $[Au_8L_8]^{2+}$, $[Au_9L_8]^{3+}$, and $[Au_{11}L_{10}]^{3+}$.[22] Of particular interest is the recently prepared cluster $[Au_{13}Ag_{12}Cl_6(PPh_3)_{12}]^{n+}$, which may be structurally regarded as three interpenetrating icosahedra[159] (Figure 3.27).

## 10. REACTIONS OF CLUSTERS

Sometimes reactions of clusters are considerably different from those of mononuclear compounds. For example, in some clusters, it is possible to coordinate CO as a $\eta^2$ ligand, and when this ligand is bonded to several metal atoms it can easily undergo a variety of reactions, such as carbon–oxygen bond scission. Hydrocarbons, alkyls, alkylidenes, and alkylidynes are often differently coordinated in clusters compared to mononuclear complexes. Therefore, reactions of ligands themselves in clusters may be proceeding differently (reduction, oxidation, bond breaking, oxidative addition, reductive elimination, etc.). Because of interesting reactions, clusters find application in organic synthesis, catalysis, synthesis of inorganic complexes and organometallic compounds, etc. Therefore, intensive studies concerning syntheses and structures of clusters are being carried out. In fact, the number of publications devoted to syntheses and structures of clusters is larger than that concerning reactions of clusters, although the number of publications dealing with the latter is currently increasing.[1,2,5,14,18,103,138,160–164]

Reactions of clusters may lead to changes in their structures or they may only change the clusters' coordination sphere. They may also cause changes within the ligands themselves.

The dependence between the cluster structure and the number of its valence electrons shows that oxidation, reduction, or other reactions leading to changes in the valence electron number must cause rearrangement of the clusters' skeleton, or that such reactions must either increase or decrease the multiplicity of the metal–metal bond involved. The structural changes of clusters may also be caused by association or dissociation of Lewis acid type ligands within the coordination sphere, that is, protonation or deprotonation.

### a. Reactions leading to structural changes in clusters

Such changes may be realized by the reduction or oxidation reactions of clusters. Reduction and oxidation processes involving transition metal compounds are much more common compared to main group compounds. This is particularly true for certain clusters which may be considered as multistep redox systems. Among others, the cubic iron clusters $Fe_4cp_4(\mu_3\text{-}S)_4^n$ ($n = +3, +2, +1, 0, -1$) and $Fe_4cp_4(\mu\text{-}CO)_4^n$ ($n = +2, +1, 0, -1$)[160, 103] are such systems. In these complexes the metal atoms lie at the vertices of a tetrahedron and the bridging ligands are located symmetrically above its four faces. The metal–metal distances become larger with lowering of the oxidation state of the metal.[160, 165] Ferredoxin analogs containing redox centers of the cubic structural type [4Fe−4S], that is, $[Fe_4S_4(SR)_4]^{n-}$ clusters where $R = Ph$, $CH_2Ph$, etc., undergo multistep redox processes.[103, 165]

So far, dependence of cluster sizes on the number of oxidation states has not been unequivocally established. Metal carbonyls and their simple derivatives do not form stable radical ions, while for cyclopentadienyl clusters many reversible redox processes have been observed, particularly in the case of the following compounds: $M_3cp_3(\mu_3\text{-}E)$ ($E = S$, PR, CR, etc.), $M_3cp_3(\mu_3\text{-}E)$ ($E = S$, PR, CO), and $M_4cp_4(\mu_3\text{-}E)_4$ ($E = S$, SR, NR, CO). The EPR studies showed that for trinuclear clusters, the LUMO have predominantly $d$ metal character, and therefore reduction must lead to the increase of the $M−M$ bond distance. The M–M distances in $Co_3(\mu_3\text{-}E)(CO)_9$ ($E = S$, PhS) type clusters in which one electron occupies an antibonding orbital are 10 pm longer than in the 48e cluster $FeCo_2(\mu_3\text{-}E)(CO)_9$. This effect is smaller for larger clusters, i.e., for the pair[166] $Ni_6cp_6/Ni_6cp_6^+$.

In clusters $Co_2(CO)_6(PhCCPh)$ and $Co_3(\mu_3\text{-}CY)(CO)_9$ where $Y = Ph$ or Cl, substitution of CO ligands by $PR_3$ groups is initiated by reduction of the cluster. Thermal substitution is very slow. The extra electron is placed in the antibonding molecular orbital formed by the metal–metal interaction. Such reduction causes reversible metal–metal bond breaking with subsequent fast addition of the ligand to the cluster. Oxidation of the complex may occur by electron transfer either to the reagent or to the electrode[167]:

$$(3.53)$$

In both cases the reaction is catalytic.

The metal–metal bond breaking may also take place as a result of ligand addition to the cluster, oxidative addition of XY molecule, or as a result of changes in bonding properties of the ligand itself, for instance, conversion of an $n$ electron ligand into an $n + 2$ electron one, such as $\eta^1$-CO(2e) $\to \eta^2$-CO(4e) [equations (3.54) and (3.55)].[160–164] In contrast to mononuclear complexes, oxidative addition of an $X-Y$

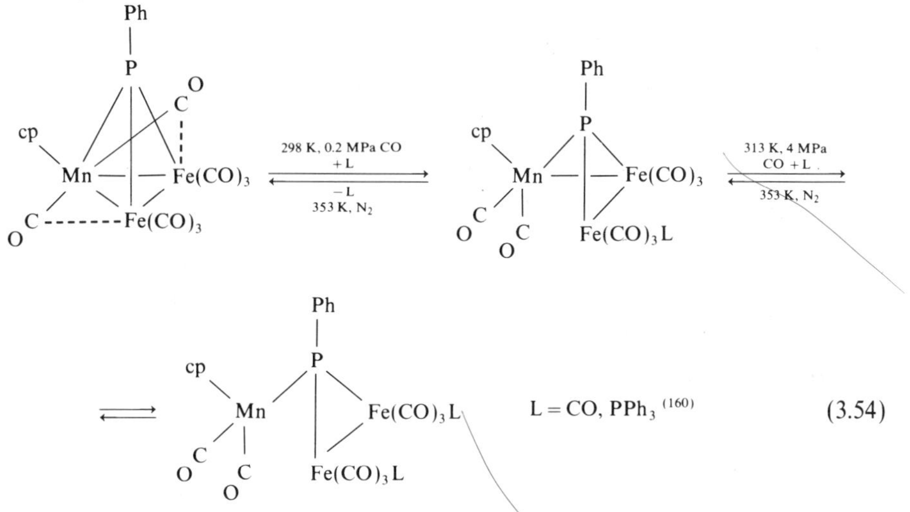

$$L = CO, PPh_3 \quad {}^{(160)} \tag{3.54}$$

$$[Re_3(\mu\text{-}H)_4(CO)_{10}]^- + 2CO \underset{373\text{ K, 10 MPa }H_2}{\overset{298\text{ K, 5 MPa CO}}{\rightleftharpoons}} Re_3(\mu\text{-}H)_2(CO)_{12}^- + H_2 \quad {}^{(168)} \tag{3.55}$$

$$46e \qquad\qquad\qquad\qquad\qquad 48e$$

molecule to the cluster may lead to a product in which X and Y are bonded to different metal atoms. Such a process may be termed a two-center oxidative addition.

$$Os_3(CO)_{12} + X_2 \longrightarrow (OC)_4 XOs\text{-}Os(CO)_4\text{-}OsX(CO)_4 \tag{3.56}$$

Often, oxidative addition is accompanied by dissociation of a Lewis base. Thus, in such a case the $M-M$ bond is not broken:

$$Os_3(CO)_{12} \xrightarrow{+H_2} Os_3(\mu\text{-}H)\,H(CO)_{11} \xrightarrow{-CO} Os_3(\mu\text{-}H)_2(CO)_{10} \tag{3.57}$$

$$48e \qquad\qquad 48e \qquad\qquad 46e$$

Owing to dissociation of the CO molecule, the 46e cluster $Os_3(\mu\text{-}H)_2(CO)_{10}$ achieves the Os=Os double bond. The trinuclear ruthenium compound $[Ru_3(CO)_8(R\text{-}DAB)]$ (R-DAB: $RN=CH-CH=NR$, $R = Pr^i$, cy, cyCH$_2$, neopentyl) undergoes oxidative addition of hydrogen to give $Ru_4(\mu\text{-}H)_4(CO)_{12}$ and the linear cluster $[(OC)_2HRu(\mu\text{-}R\text{-}DAB)\,RuCO(\mu\text{-}CO)_2RuCO(\mu\text{-}R\text{-}DAB)\,RuH(CO)_2]$.[169]

Often the skeletal change occurs after proton addition to the cluster.

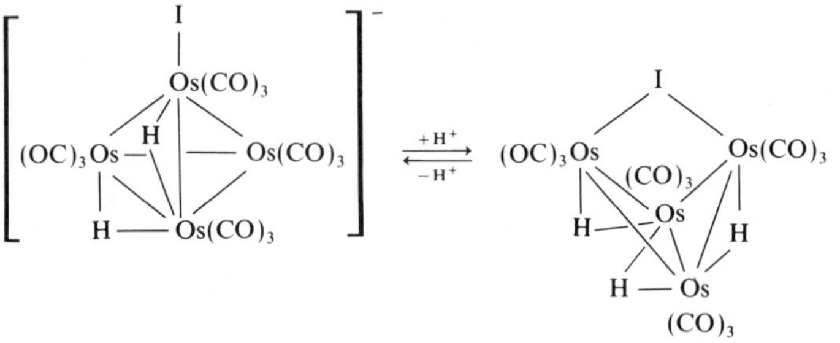

$$(3.58)$$

In this case, the $Os - Os$ bond breaking is accompanied by the changes in coordination properties of iodine, which is a $1e$ ligand before the reaction and a $3e$ bridging ligand after $H^+$ had been added to the cluster.

Protonation of $[Fe_4(CO)_{13}]^{2-}$ causes rearrangement of this tetrahedral compound into a "butterfly" structure. During this process one of the $2e$ $\eta^1$-CO groups becomes a $4e$ $\mu_4$-$\eta^2$-CO ligand (Figures 3.18 and 2.3):

$$[Fe_4(\mu_3\text{-CO})(\mu\text{-CO})_3(CO)_9]^{2-} + H^+ \longrightarrow Fe_4(\mu\text{-H})(\mu_4\text{-CO})(CO)_{12}]^- \qquad (3.59)$$

Structural changes of larger clusters caused by protonation are often more pronounced.

$$octahedro\text{-}[Os_6H(CO)_{18}]^- \xrightarrow{+H^+} [Os_6(H)_2(CO)_{18}] \text{ capped square pyramid} \qquad (3.60)$$

Some rhodium and platinum clusters exhibit dynamic properties, i.e., skeletal fluxionality. In the $^{103}$Rh NMR of $[Rh_9P(CO)_{21}]^{2-}$, which has a capped antiprism structure, three signals possessing the intensity ratio $4:4:1$ are observed at 183 K, while at 298 K there is only one signal. This experiment shows that, due to fluxionality of the skeleton, all rhodium atoms becomes equivalent.[84] The same is indicated also by the presence of a decet in the $^{31}$P NMR spectrum of this compound.[36] Similarly, $[Rh_{10}S(CO)_{22}]^{2-}$ and $[Rh_{10}E(CO)_{22}]^{3-}$ (E = P, As) are rigid at low temperatures but fluxional at high temperatures.[170]

In $[Pt_3(CO)_6]_n^{2-}$-type complexes, the $Pt_3$ triangles rotate about the threefold axis of symmetry, and $Pt_3(CO)_6$ units undergo an intermolecular exchange. In addition to those processes requiring the $Pt - Pt$ bond breaking, there is a movement of the sphere of ligands with respect to the skeleton of the cluster.[82]

The formation of $M - M$ bonds takes place during the synthesis of larger clusters from compounds possessing a smaller number of metal atoms. Nucleophilic anionic mononuclear complexes or clusters may be added to a neutral cluster which is electrophilic in character [reactions (3.40), (3.43)].

$$Ru_2Os(CO)_{12} + Fe(CO)_4^{2-} \xrightarrow[2,\ H^+]{1,\ \Delta} FeRu_2Os(H)_2(CO)_{13} \qquad (3.61)$$

$$Ru_3(CO)_{12} + Co(CO)_4^- \xrightarrow{\Delta} [CoRu_3(CO)_{13}]^- \qquad (3.62)$$

Often, reverse reactions take place which involve giving off mononuclear complexes[102]:

$$[FeRu_3H(CO)_{13}]^- + 3CO \longrightarrow [Ru_3H(CO)_{11}]^- + Fe(CO)_5 \qquad (3.63)$$

Anionic complexes are good nucleophiles and their preparations are often facile. Most mononuclear complexes, however, have electrophilic character as seen from ligand substitutions. Therefore, reactions of anionic clusters with neutral or cationic mononuclear complexes lead to an enlargement of clusters and also to the formation of M−M bonds. For example, in the reaction of $[Ru_3H(CO)_9PPh]^-$ with $Rh(CO)_3(PEt_3)_2^+$, $[Ru_3RhH(PPh)(CO)_9(PEt_3)_2]$ is formed.[171] The cluster $[Fe_5C(CO)_{14}]^{2-}$ reacts with $M(CO)_3L_3$ (M = Cr, Mo, W), $Rh_2Cl_2(CO)_4$, $Ir_2Cl_2(COD)_2$, $Ni(COD)_2$, $[Pd_2Cl_2(allyl)_2]$, etc., to give $Fe_5C(CO)_{14}M(CO)_n]^{x-}$ (n = 2, 3; x = 2, 1, 0).[122]

## b. *Migration of ligands within clusters*

In triangular clusters of the type $M_3(CO)_{12}$, $M_3(CO)_{12-x}L_x$ equatorial CO molecules undergo exchange according to the planar merry-go-round mechanism (3.64).[162] The CO groups which are approximately coplanar with the $M_3$ ring rotate

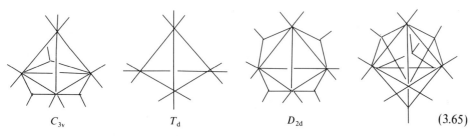

$$(3.64)$$

about metal atoms. Exchange according to this mechanism may take place in any clusters containing planar-triangular or square units. In the case of $Os_3(CO)_{10}L_2$ (L = PR₃, MeCN) the exchange according to a planar merry-do-round occurs only if L occupies axial positions at different osmium atoms (L = MeCN); the exchange does not occur if L occupies equatorial positions (L = PR₃).

Migration of CO ligands in tetranuclear clusters $M_4(CO)_{12}$ (M = Co, Rh, Ir) and their derivatives $M_4(CO)_{12-x}L_x$ may also proceed by the same merry-go-round mechanism. During this process the compound possessing $T_d$ symmetry which does not have bridging CO groups becomes transformed into the cluster containing three bridging CO ligands ($C_{3v}$ symmetry). Formation of other highly symmetrical forms in the transition state seems probable as well, for example, forms with four $\mu_2$-CO bridges ($D_{2d}$) or with four $\mu_3$-CO ligands [structures (3.65)]. Therefore, a large number of

$$C_{3v} \qquad T_d \qquad D_{2d} \qquad (3.65)$$

possible mechanisms of migration exists for $M_4(CO)_{12}$ clusters. The planar merry-go-round mechanism for the $T_d \leftrightarrows C_{3v}$ process is equivalent to rocking of metal atoms inside a densely packed array of CO ligands. This is accompanied by the ligand symmetry change from $O_h$ to $I_h$ because, in complexes possessing $T_d$ symmetry, the CO molecules define an octahedron ($O_h$), and in compounds with $C_{3v}$ symmetry the CO groups define an icosahedron ($I_h$).

Migration of CO in $Fe_3(CO)_{12}$ may also be explained by rocking of the $Fe_3$ skeleton within the sphere of ligands (Chapter 2).

The cluster $Fe_3(CO)_{12}$ exhibits dynamic properties even in the solid state. The solid state $^{13}C$ NMR spectrum of this compound shows six signals possessing the same intensity. None of them corresponds to a $\mu_2$-CO ligand although such ligands are present in the complex. This may be explained by rapid changes in orientation of $Fe_3$ units in a densely packed sphere of CO ligands adopting somewhat distorted icosahedron geometry [equation (3.66)].[175]

$$(3.66)$$

In many clusters, both small and large, there is also a possibility of exchanges proceeding according to the modified process of conical surface merry-go-round [equation (3.67)].

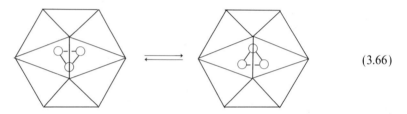

$$(3.67)$$

Migration of CO ligands in larger clusters may occur according to several mechanisms simultaneously. An important role is played by the pairwise two-center exchange mechanism which requires coplanarity or near coplanarity of migrating ligands [equation (3.68)]. Because of this requirement, migration of ligands in clusters

$$(3.68)$$

containing more than three metal atoms does not proceed by the pairwise two-center exchange mechanism.

More possible is migration occurring by concerted two-center one-for-one exchange mechanism (3.69) or two-center *cis* pairwise exchange (3.70).

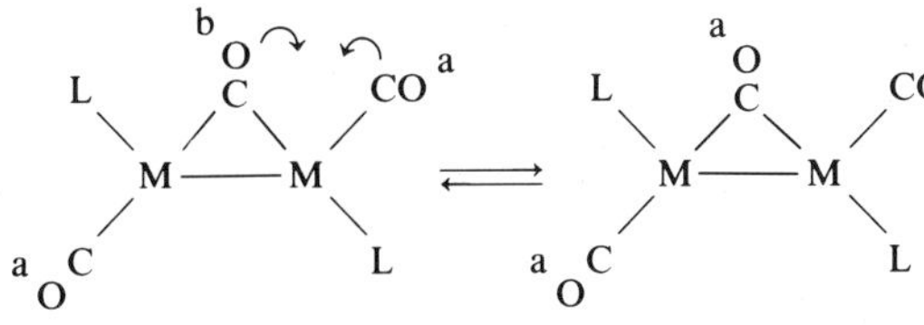

$$(3.69)$$

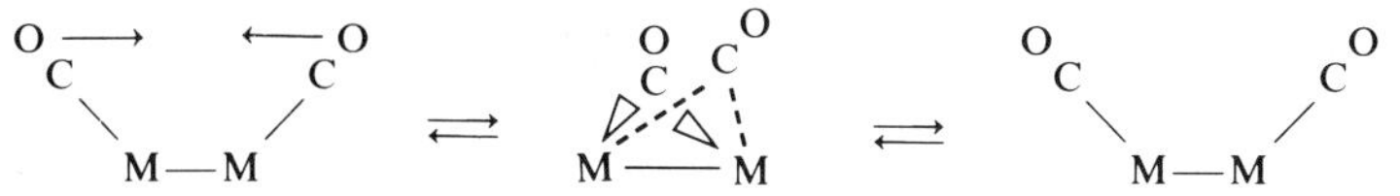

$$(3.70)$$

Migration of CO ligands depends on the cluster charge. In the case of isostructural clusters, migration happends more easily for compounds possessing a higher negative charge. The compound $[Rh_{13}H_2(CO)_{24}]^{3-}$ is stereochemically nonrigid at 193 K, while $[Rh_{13}H_3(CO)_{24}]^{2-}$ is fluxional above 298 K. Similar differences in dynamic properties are exhibited also by some other clusters.

Of particular interest are migration processes involving hydrido ligands, which are often accompanied by migration of other ligands such as CO. Mechanistic studies of these reactions are important since they may lead to better understanding of the Fischer–Tropsch process. Like in metals, in clusters the mobility of the hydrido ligands is exceptionally high. The energy of activation for hydrogen migration in clusters assumes very low values, sometimes even as low as 10–20 kJ mol$^{-1}$. For example, the $^1$H NMR spectrum of $[Rh_3H_3\{P(OMe)_3\}_6]$ in the hydrido region at 183 K exhibits a septet of quartets showing that hydrido ligands interact with all phosphorus and rhodium atoms.[138,162]

Other interesting lingand migration processes in four metal atom carbide clusters are provided by the two iron compounds $Fe_4C(CO)_{13}$ and $Fe_4C(CO)_{12}^{2-}$ as well as alkylidyne, ketenylidene, and alkylidene clusters such as $Fe_4(\mu_4\text{-CH})(\mu\text{-H})(CO)_{12}$ (Table 3.14), $Ru_3(\mu_3\text{-CH})(\mu\text{-H})(CO)_{10}$,[173] $Ru_3(\mu\text{-CH}_2)(\mu\text{-CO})(CO)_{10}$,[173] $Os_3(\mu_3\text{-CPh})(\mu\text{-H})(CO)_{10}$,[172]$Fe_3(\mu_3\text{-}\eta^2\text{-CCO})(CO)_9^{2-}$, $Os_3(\mu_3\text{-}\eta^1\text{-CCO})(\mu\text{-H})_2(CO)_9$ (Table 3.14, footnote reference *l*), and $[Rh_3(\mu_3\text{-CH})\,cp_3(\mu\text{-CO})_2]^+$ (Table 3.14, footnote reference *j*). The $[Fe_3(CO)_{11}]^{2-}$ anionic complex reacts with acetyl choride to give $[Fe_3(CO)_{10}(COOCMe)]^-$ which, after reduction, forms $[Fe_3(\mu_3\text{-}\eta^2\text{-CCO})(CO)_9]^{2-}$.[121] Protonation of the ketenylidene compound causes splitting of the CCO ligand and subsequent migration of CO to the skeleton of the cluster [equation (3.71)].

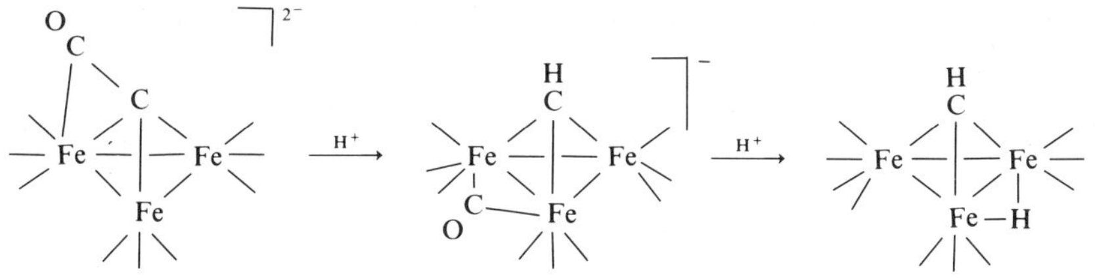

$$(3.71)$$

Transformations of ligands and their migration has also been observed in osmium clusters. For example, $Os_3(CH)H(CO)_{10}$ in reflucing hexane forms $Os_3(\mu_3\text{-CCO})(\mu\text{-H})_2(CO)_9$ (Table 3.14, footnote reference *l*). $Ru_3(CH)H(CO)_{10}$ reacts much more readily ($t_{1/2} = 9$ min at 261 K) in the same manner to give $Ru_3(\mu_3\text{-CCO})(\mu\text{-H})_2(CO)_9$ [equation (3.72)].[173] The same compound is formed by pyrolysis of $Ru_3(\mu\text{-CH}_2)(\mu\text{-CO})(CO)_{10}$.

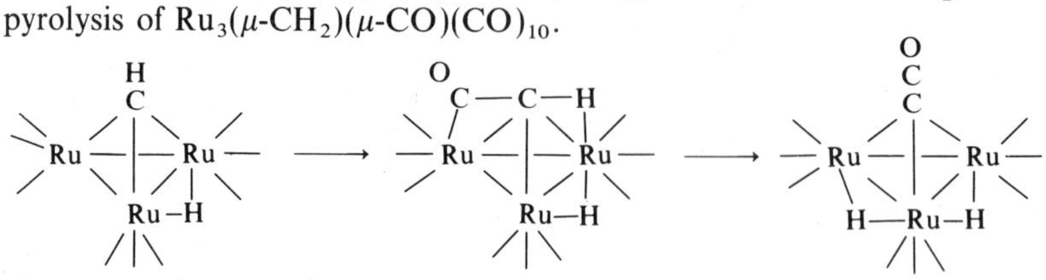

$$ (3.72) $$

Oxidation of $[Fe_4C(CO)_{12}]^{2-}$ with $Ag^+$ leads to the formation of $[Fe_4C(CO)_{12}]$, which is very reactive; with hydrogen it gives $[Fe(CH)(H)(CO)_{12}]$.[138] As seen by NMR, the methylidyne proton rapidly scrambles among iron atoms [equation (3.73)].

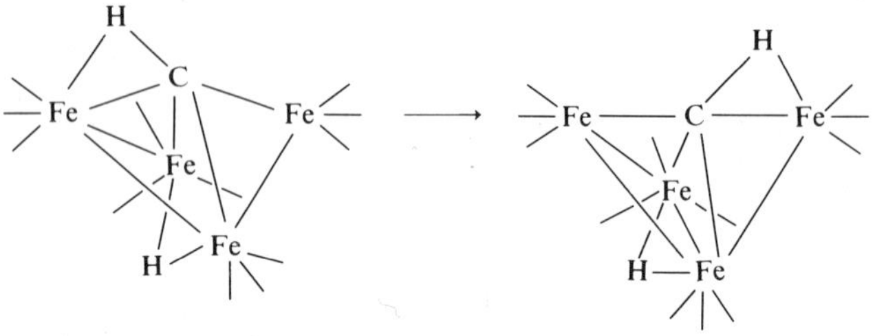

$$ (3.73) $$

Moreover, there is also a slower exchange of hydrogen atoms between the positions $C-H-Fe$ and $Fe-H-Fe$.[138] These examples suggest that the carbon atom in $M_4C$- and $M_3C$-type clusters is to be treated as one of the skeletal atoms. Ligands coordinated to such a carbon atom should exhibit properties closely related to those connected to the metal atoms.

Many other examples of migration reactions are known, especially those typical for catalytic processes, such as CO insertion, catalytic hydrogenation of unsaturated hydrocarbons, reduction of nitriles, water–gas shift reaction, etc. These processes will be discussed in Chapter 13.

Insertion of CO into the $M-CH_2-M$ bond has been investigated [equations (3.74) and (3.75)].[179]

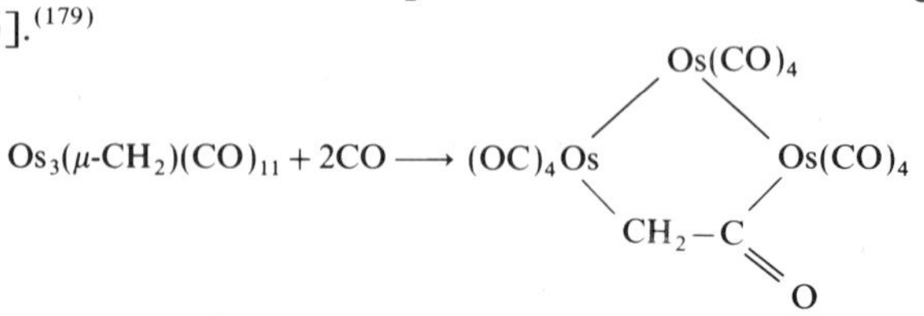

$$ Os_3(\mu\text{-CH}_2)(CO)_{11} + 2CO \longrightarrow \quad (3.74) $$

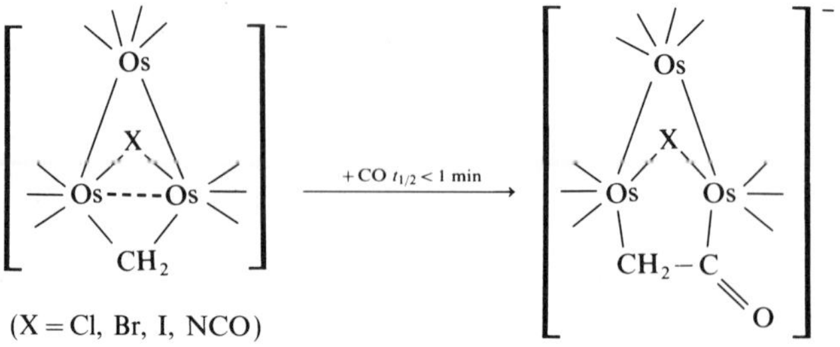

$$ (X = Cl, Br, I, NCO) \qquad (3.75) $$

Reversible CO migration occurs also in the case of encapsulated ruthenium complex $[Ru_4N(CO)_{12}]^-$ [equation (3.76)].[78]

$$[Ru_4N(CO)_{12}]^- \underset{0.1\ MPa\ CO}{\overset{\overset{21\ MPa\ CO}{359\ K}}{\rightleftarrows}} [Ru_3(NCO)(CO)_{11}]^- + [Ru_3H(CO)_{11}]^- + Ru(CO)_5 \quad (3.76)$$

### c. *Acid–base reactions*

Many neutral and all anionic clusters are Lewis bases. Dependent upon the nucleophilic character of the metal atoms and ligands, addition of a proton may occur either to the metal or to the ligand. Usually, $M-H$ bonds are formed first. There are known cases of addition of two protons to neutral electron-rich clusters, and the formation of anionic clusters possessing charges ranging from $-1$ to $-6$, for example, $[Ni_{38}Pt_6(CO)_{48}]^{6-}$.[99]

Protonation of small clusters ($M_3$, $M_4$) usually does not cause their structural changes. However, there are exceptions, among others, $[Fe_4(CO)_{13}]^{2-}$ (Figure 2.3, Section 3.10a). Similarly, $Fe_4(COMe)(CO)_{12}^-$ undergoes a structural rearrangement during protonation (Table 3.14, footnote reference *d*); see equations (3.77).

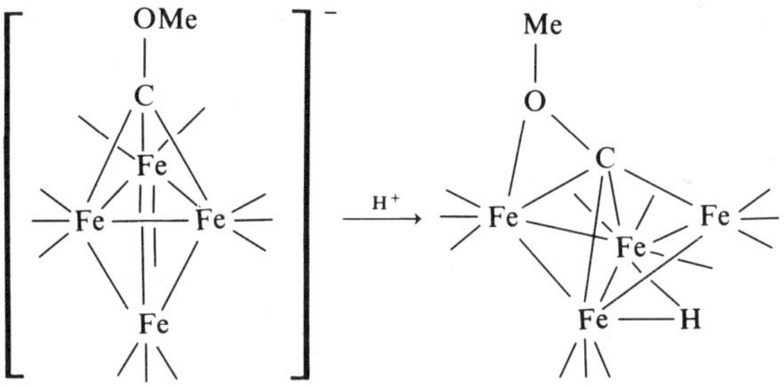

$$\qquad\qquad\qquad\qquad\qquad\qquad\qquad\qquad\qquad\qquad (3.77)$$

Reactions with acids sometimes lead to protonation of ligands. Bridging ligands such as $\mu_2$-, $\mu_3$-, and $\mu_4$-CO undergo protonation with particular ease [equations (3.78) and (3.79)].

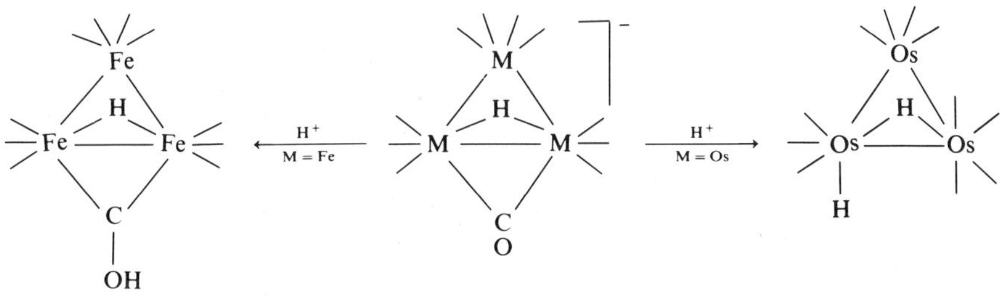

(M = Fe, Os)

$$\qquad\qquad\qquad\qquad\qquad\qquad\qquad\qquad\qquad\qquad (3.78)$$

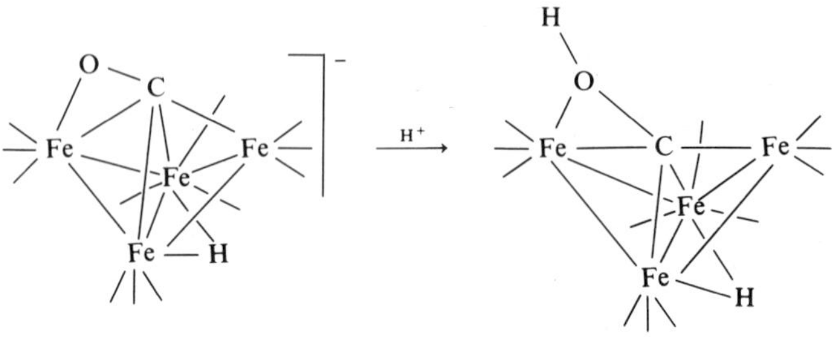

$$(3.79)$$

Other electrophiles, such as $Me^+$ or $Me-C^+=O$, do not attack metal atoms but react with ligands. Considerably larger sizes of all electrophiles compared to $H^+$ prevent their addition to metal atoms. Therefore, such electrophiles react with ligands possessing the highest electron density in the cluster. This is one reason why the following bonds are formed: the main group element electrophile-oxygen of the bridging CO ligand or electrophile-sulfur in clusters containing $\mu_3$-S bridges [equations (3.80) and (3.81)].

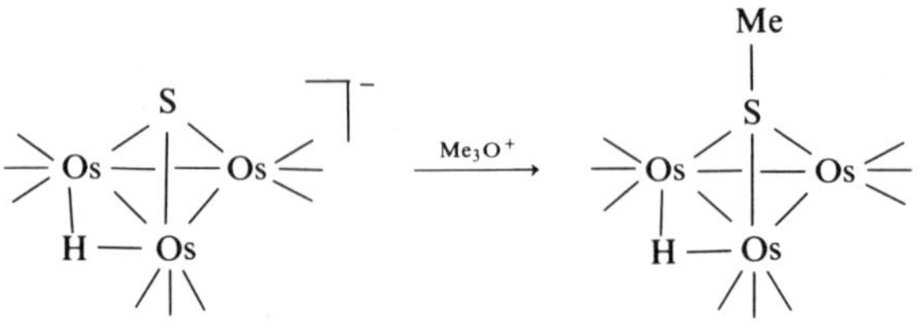

$$(3.80)$$

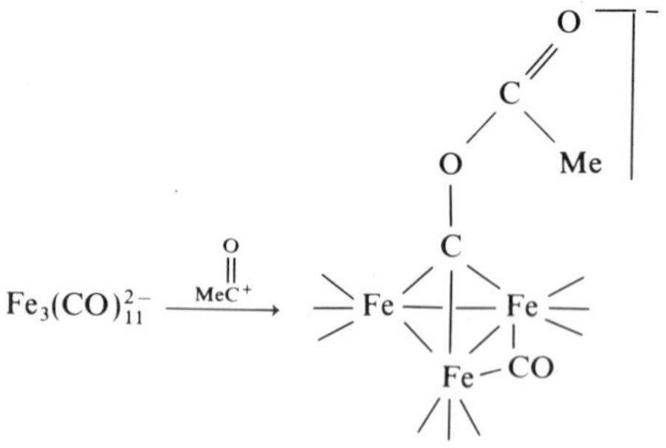

$$(3.81)$$

Diverse reactivity is shown by ketenylidene clusters which have been itensively studied because of CO activation, carbon–carbon bond formation, and synthesis of new clusters. Reactivity of the CCO ligand is drastically different, depending upon the charge of the cluster and the character of the metal forming the cluster. For example, in $[Fe_3(CCO)(CO)_9]^{2-}$ and $[Fe_2Co(CCO)(CO)_9]^-$ electrophiles $(H^+)$ attack the $\alpha$ carbon of the CCO group while in neutral or cationic ketenylidene clusters such as $M_3(CCO)(H)_2(CO)_9$ (M = Ru, Os) and $[Co_3(CCO)(CO)_9]^+$ the $\beta$ carbon is attacked by nucleophiles (Table 3.14, footnote references *b, c, l, m, n*). The reaction of $[Ru_3(CCO)(CO_9]^{2-}$ with an acid gives a product in which the proton is added to the ruthenium atoms; in the case of $[Fe_3(CCO)(CO)_9]^{2-}$ the proton is added to the $\alpha$

carbon atom of the CCO ligand [reaction (3.71), (3.72), (3.79), (3.83)]. This can be explained by stronger basicity of ruthenium than iron [see equations (3.82) and (3.83)].

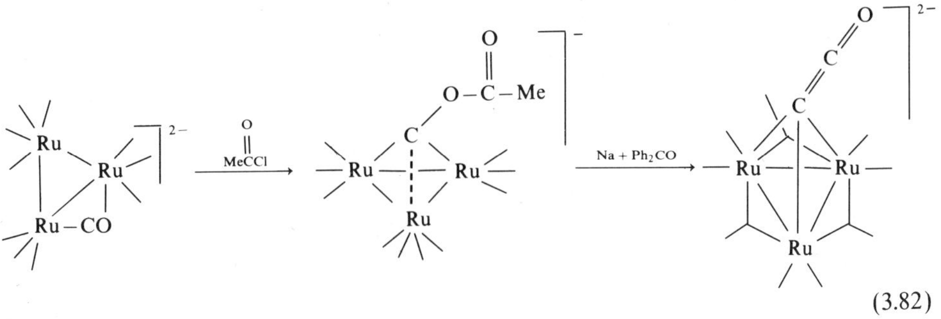

$$(3.82)$$

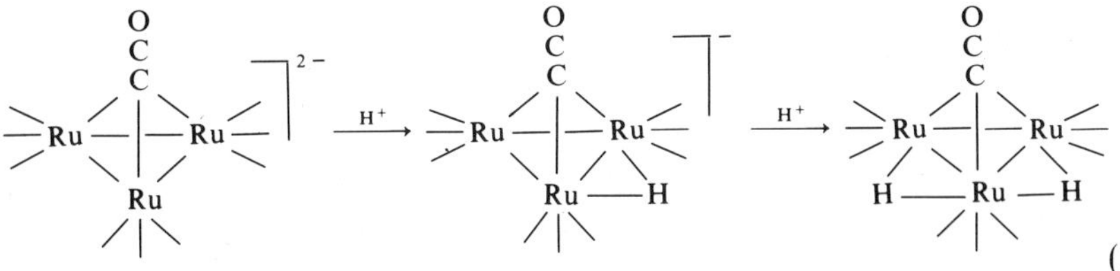

$$(3.83)$$

In carbyne osmium compounds such as $[Os_3(\mu\text{-COMe})(\mu\text{-H})(CO)_{10}]$ nucleophilic attack on the carbyne carbon atom also takes place. By carrying out sequential reactions with nucleophiles and electrophiles, it is possible to break the C—O bond (Table 3.14, footnote reference *h*). The reaction furnishes an isolable carbene complex, $[Os_3(\mu\text{-CHOMe})(\mu\text{-H})(CO)_{10}]^-$.

$$[Os_3(\mu\text{-COMe})(\mu\text{-H})(CO)_{10}] \xrightarrow[\text{2. H}^+]{\text{1. H}^-} [Os_3(\mu_3\text{-CH})(\mu\text{-H})(CO)_{10}] \qquad (3.84)$$

However, a different reaction mechanism operates for another nucleophile–electrophile pair, Ph⁻ and Me⁺ [180]; see scheme (3.85).

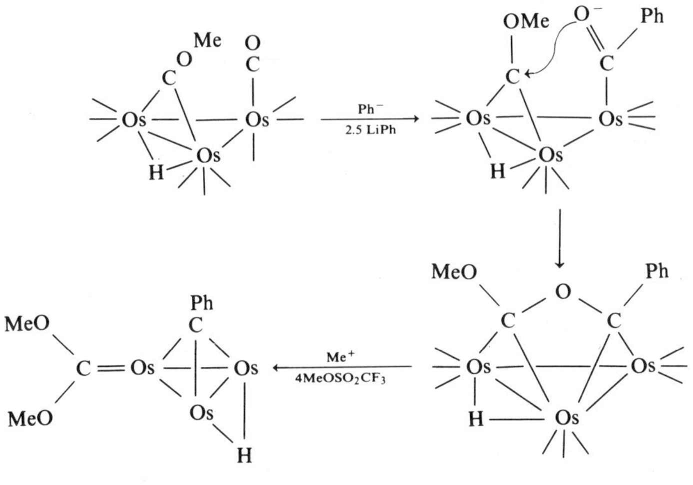

$$(3.85)$$

### d. *Ligand substitution*

Substitution of CO ligands in clusters is most commonly realized in the same way as in the case of mononuclear metal carbonyls. Substitution may be induced by one of the following, most frequently utilized methods: thermal, electrochemical, chemical (reactions with N-oxide of trimethylamine or $Bu_3^n PO$), photochemical, catalysis by radicals, catalysis by transition metal compounds, etc.[176]

In general, substitution of CO ligands in clusters proceeds by a dissociative mechanism as in the case of mononuclear metal carbonyls. Ligand displacement may also readily occur according to an associative mechanism in the case of complexes containing $M = M$ double bonds as in clusters in which the $M - M$ bond breaking takes place relatively easily (see Section 3.10a).

$$Os_3 H_2(CO)_{10} \xrightarrow[293\,K]{PPh_3} Os_3 H_2(CO)_{10}(PPh_3) \xrightarrow[343\,K]{-CO} Os_3 H_2(CO)_9(PPh_3) \qquad (3.86)$$

$$Os_3(CO)_{12} \xrightarrow{PPh_3} Os_3(CO)_{12-x}(PR_3)_x \qquad (x = 1\text{--}3) \qquad (3.87)$$

Of great importance for the induction of CO substitution is $Me_3 NO$.[176]

$$M_x(CO)_y + L + Me_3 NO \longrightarrow M_x(CO)_{y-1}L + Me_3 N + CO_2 \qquad (3.88)$$

According to this method the following compounds have been obtained: $Ru_4 H_4(CO)_{10}(dppe)$, $Os_3(CO)_{11}(C_2 H_4)$, and $Os_3(CO)_{11}(NCMe)$.

The electrochemical method involves reduction of the cluster which leads to $M - M$ bond breaking and coordination by additional ligands (see Section 3.10a). Oxidation of clusters to radical cations also facilitates substitution reaction.

Generation of cluster radicals by photochemical methods or by means of radical anions such as $Ph_2 CO^{\bar{\cdot}}$ in THF is frequently utilized in substitution reactions:

$$Rh_6(CO)_{16} + Bu'NC \xrightarrow{Ph_2 CO^{\bar{\cdot}}} Rh_6(CO)_{12}(CNBu')_4 \qquad (3.89)$$

The following transition metal compounds are used as good catalysts for CO substitution: $CoX_2$, $NiX_2$, $RhCl(PPh_3)_3$, $PdO$, $Fe_2 cp_2(CO)_4$, $Mo_2 cp_2(CO)_6$, as well as metallic palladium.

$$Ir_4(CO)_{12} + RNC \xrightarrow{Pd/C} Ir_4(CO)_{12-x}(CNR)_x \qquad (3.90)$$

Except for olefins, other unsaturated hydrocarbons may also substitute CO ligands. In the osmium complex, $Os_3(CO)_9(\mu_3\text{-}C_6 H_6)$, the benzene molecule is coordinated to all osmium atoms substituting three axial CO molecules.[177]

### e. *Oxidative addition*

In general, oxidative addition reactions of clusters are identified with dissociative chemisorption of heterogeneous solid-state catalysts. Therefore, studies of oxidative addition of clusters are useful in finding connections between these areas of chemistry. In the case of clusters, oxidative addition may be one-center if both fragments of the XY molecule add to the same metal atom or two-center if X and Y are coordinated to two metal atoms. One cannot exclude multicenter oxidative addition of polyatomic molecules, for example, hydrocarbons.

Oxidative addition of hydrogen [reactions (3.55)–(3.57)] to clusters may change the multiplicity of $M-M$ bonds, or cause their breaking. Most often, however, this process is accompanied by dissociation of a Lewis base and, in such cases, of course, the skeleton of the cluster does not change. Oxidative addition and reductive elimination of trinuclear ruthenium clusters have been investigated. Based on the kinetic equation, activation parameters, and isotope effects, the pathway for the reversible reaction

$$[Ru_3(\mu\text{-COMe}) H(CO)_{10}] + H_2 \rightleftharpoons [Ru_3(\mu_3\text{-COMe})(H)_3(CO)_9] + CO$$

has been elucidated. It was found that reductive elimination occurs via a three-center transition state while oxidative addition of $H_2$ is preceded by dissociation of CO [equation (3.91)].[178]

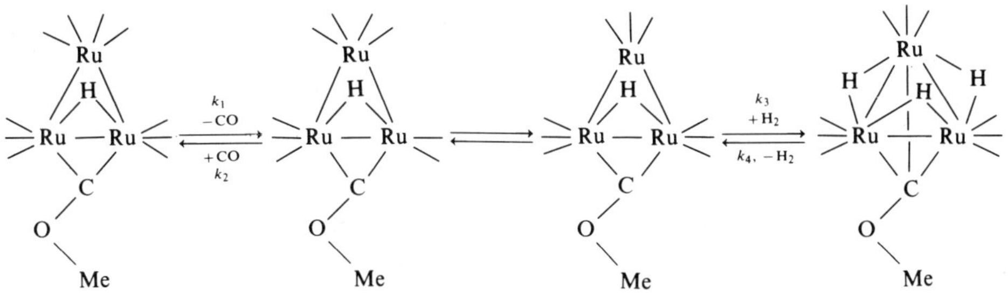

(3.91)

Bond breaking involving the vinyl $C-H$ rather than alkyl $C-H$ bond shows that the influence of several metal atoms creates unique ligand chemistry in clusters which differs from the chemistry of mononuclear complexes[138, 161, 162]; see scheme (3.92).

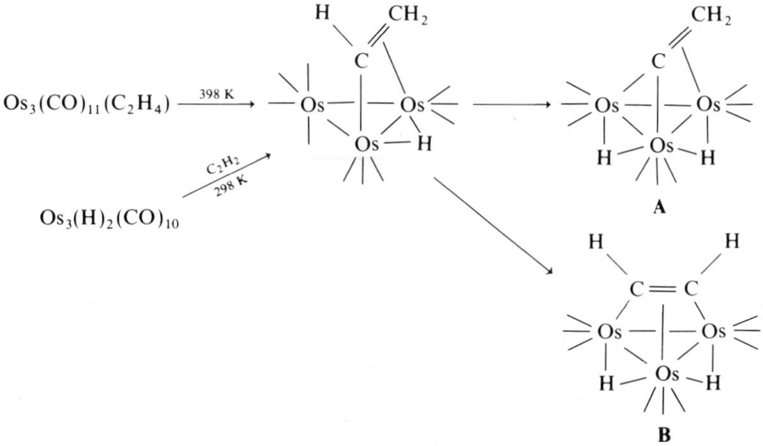

(3.92)

For cyclic olefins, compounds possessing structure (**B**) with a carbon–carbon double bond parallel to one of the $M-M$ bonds are formed. Ruthenium carbonyls in analogous reactions give both (**A**) and (**B**) forms, as does $Os_3(CO)_{12}$ in the reaction with $MeCH=CH_2$. Also, benzene, although chemically very different from cyclic olefins, gives complexes which have structure (**B**): $Os_3(\mu_3\text{-}C_6H_4)(H)_2(CO)_9$.[161]

Oxidative addition of alkynes occurs in a similar way [see structures (3.93)]

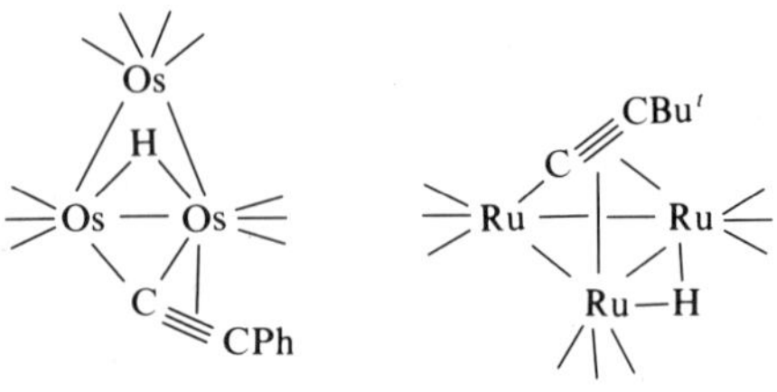

$$(3.93)$$

Also, ligands coordinated through N, O, P, etc. undergo oxidative addition to give $\alpha$, $\beta$ or $\gamma$ metallation products [see structures (3.94)].

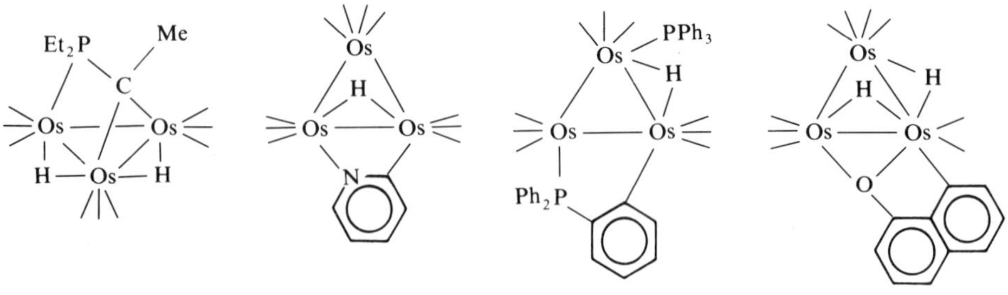

$$(3.94)$$

The methyl group in $Os_3(\mu\text{-}CH_3)(\mu\text{-}H)(CO)_{10}$ is asymmetrically bridged to osmium atoms through carbon and hydrogen and exists in equilibrium with the methylene compound.[181] This is a classical example of a facile breaking of the $C-H$ bond by metal atoms of a cluster. This cluster reaction represents a simple model of $C-H$ bond cleavage occurring on the surface of the metal [equation (3.95)].

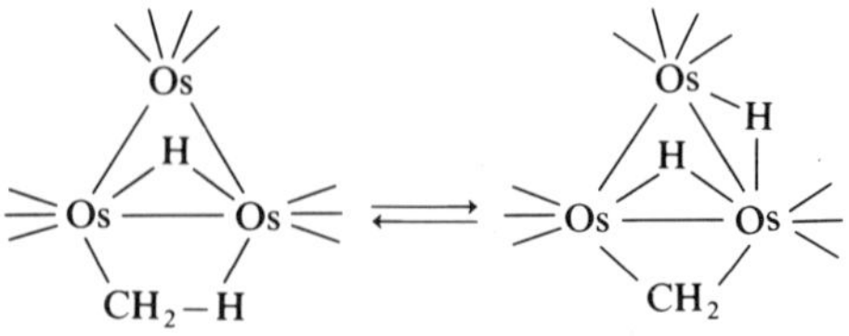

$$(3.95)$$

In some cases, typical only for clusters, oxidative addition of acetylenes involving $C-C$ bond scission takes place and leads to carbyne complexes.[161, 182, 127]

$$Os_6(CO)_{18} + PhCCPh \longrightarrow Os_6(\mu_3\text{-}CPh)(\mu_4\text{-}CPh)(CO)_{16} \qquad (3.96)$$

$$Rh_3cp_3(\mu_3\text{-}CO)(\mu_3\text{-}PhCCPh) \xrightarrow{-CO} Rh_3(\mu_3\text{-}CPh)_2cp_3 \qquad (3.97)$$

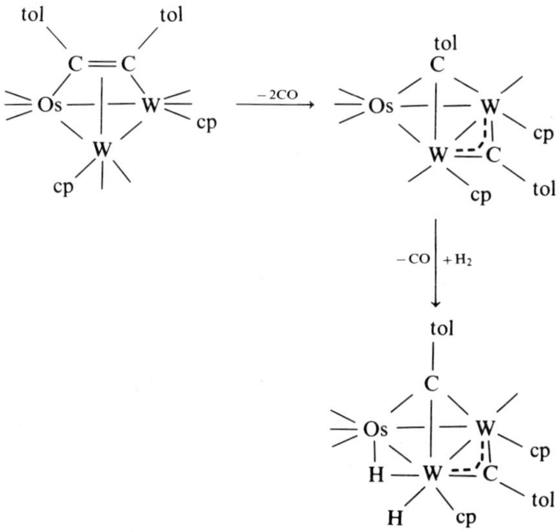

(3.98)

## REFERENCES

1. a. F. A. Cotton, *Q. Rev. Chem. Soc.*, **20**, 389 (1966); F. A. Cotton, *Acc. Chem. Res.*, **1**, 257 (1968).
b. F. A. Cotton and R. A. Walton, *Multiple Bonds Between Metal Atoms*, Wiley, New York (1982); F. A. Cotton and R. A. Walton, *Structure and Bonding*, **62**, 1 (1985). c. F. A. Cotton and G. Wilkinson, *Advanced Inorganic Chemistry*, 4th ed., Wiley, New York (1980). d. B. E. Bursten and F. A. Cotton, *Faraday Soc. Symp.*, **14**, 443 (1980). e. M. H. Chisholm and I. P. Rothwell, *Prog. Inorg. Chem.*, **29**, 1 (1982). f. S. Shaik, R. Hoffmann, C. R. Fisel and R. H. Summerville, *J. Am. Chem. Soc.*, **102**, 4555 (1980).
2. P. Chini, *J. Organomet. Chem.*, **200**, 37 (1980).
3. K. Wade, *Chem. Br.*, **11**, 177 (1975); K. Wade, *Adv. Inorg. Chem. Radiochem.*, **18**, 1 (1976); D. M. P. Mingos, *Nature (London)*, **236**, 99 (1972); R. E. Williams, *Adv. Inorg. Chem. Radiochem.*, **18**, 67 (1976); R. E. Williams, *Inorg. Chem.*, **10**, 210 (1971).
4. J. W. Lauher, *J. Am. Chem. Soc.*, **100**, 5305 (1978); J. W. Lauher, *J. Am. Chem. Soc.*, **101**, 2604 (1979); J. W. Lauher, *J. Catal.*, **66**, 237 (1980); J. W. Lauher, *J. Organomet. Chem.*, **213**, 25 (1981).
5. E. L. Muetterties, T. N. Rhodin, E. Band, C. F. Brucker, and W. R. Pretzer, *Chem. Rev.*, **79**, 91 (1979).
6. E. Band and E. L. Muetterties, *Chem. Rev.*, **78**, 639 (1978).
7. A. Müller, R. Jostes, and F. A. Cotton, *Angew. Chem., Int. Ed. Engl.*, **19**, 875 (1980).
8. P. Chini, G. Longoni, and V. G. Albani, *Adv. Organomet. Chem.*, **14**, 285 (1976).
9. E. L. Muetterties, *J. Organomet. Chem.*, **200**, 177 (1980).
10. H. D. Kaesz, *J. Organomet. Chem.*, **200**, 145 (1980).
11. A. Simon, *Angew. Chem., Int. Ed. Engl.*, **20**, 1 (1981).
12. H. G. Schnering, *Angew. Chem., Int. Ed. Engl.*, **20**, 33 (1981).
13. R. J. Gillespie, *Chem. Soc. Rev.*, **8**, 315 (1979).
14. D. M. P. Mingos, *Adv. Organomet. Chem.*, **15**, 1 (1977); D. M. P. Mingos, *Acc. Chem. Res.*, **17**, 311 (1984); D. M. P. Mingos, *Inorg. Chem.*, **24**, 114 (1985); D. M. P. Mingos, *J. Chem. Soc., Chem. Commun.*, 706 (1983); D. G. Evans and D. M. P. Mingos, *J. Organomet. Chem.*, **251**, C13 (1983).
15. D. L. Kepert and K. Vrieze, in: *Comprehensive Inorganic Chemistry* (J. Bailar, ed.), Vol. 4, p. 197, Pergamon, Oxford (1973).
16. N. N. Greenwood, *Chem. Soc. Rev.*, **13**, 353 (1984).
17. J. Halpern, *Discuss. Faraday Soc.*, **46**, 7 (1968); J. Halpern, *Adv. Chem. Ser.*, **70**, 1 (1968).

18. R. Hoffmann, *Angew. Chem., Int. Ed. Engl.*, **21**, 711 (1982).
19. F. G. A. Stone, *Angew. Chem., Int. Ed. Engl.*, **23**, 89 (1984).
20. M. Elian, M. M. L. Chen, D. M. P. Mingos, and R. Hoffmann, *Inorg. Chem.*, **15**, 1148 (1976).
21. B. K. Teo, *Inorg. Chem.*, **23**, 1251 (1984); B. K. Teo, *Inorg. Chem.*, **24**, 115 (1985); B. K. Teo, *Inorg. Chem.*, **24**, 1627 (1985); B. K. Teo, G. Longoni, and F. R. K. Chung, *Inorg. Chem.*, **23**, 1257 (1984).
22. Yu. L. Slovokhotov and Yu. T. Struchkov, *Usp. Khim.*, **54**, 556 (1985).
23. S. P. Gubin, *Usp. Khim*, **54**, 529 (1985).
24. F. Pruchnik, *Wiadomosci Chemiczne*, **37**, 391 (1983).
25. W. Romanowski, *Mikroczasteczki i Clustery Metaliczne w Katalizie*, PWN, Warsaw–Wroclaw (1983).
26. M. Moskovits, *Acc. Chem. Res.*, **12**, 229 (1979).
27. Y. Kim and K. Seff, *J. Am. Soc.*, **99**, 7055 (1977).
28. J. G. Allpress and J. V. Sanders, *Surf. Sci.*, **7**, 1 (1967).
29. D. J. C. Yates, L. L. Murell, and E. B. P. Prestridge, *J. Catal.*, **57**, 41 (1979).
30. E. B. Prestridge, G. H. Via, and J. H. Sinfelt, *J. Catal.*, **50**, 115 (1977).
31. R. T. K. Baker, E. B. Prestridge, and R. L. Garten, *J. Catal.*, **59**, 293 (1979).
32. A. Simon, *Z. Anorg. Allg. Chem.*, **335**, 311 (1967).
33. P. F. Jackson, B. F. G. Johnson, J. Lewis, P. R. Raithby, M. McPartlin, W. J. H. Nelson, K. D. Rose, J. Allibon, and S. A. Mason, *J. Chem. Soc., Chem. Commun.*, 295 (1980).
34. D. W. Hart, R. G. Teller, C.-Y. Wei, R. Bau, G. Longoni, S. Campanella, P. Chini, and T. F. Koetzle, *J. Am. Chem. Soc.*, **103**, 1458 (1981).
35. S. Martinengo, G. Ciani, A. Sironi, B. T. Heaton, and J. Mason, *J. Am. Chem. Soc.*, **101**, 7095 (1979).
36. J. L. Vidal, W. E. Walker, R. L. Pruett, and R. C. Schoening, *Inorg. Chem.*, **18**, 129 (1979); J. L. Vidal, W. E. Walker, and R. C. Schoening, *Inorg. Chem.*, **20**, 20 (1981); J. L. Vidal, *Inorg. Chem.*, **20**, 243 (1981).
37. J. L. Vidal and J. M. Troup, *J. Organomet. Chem.*, **213**, 351 (1981).
38. R. W. Broach, L. F. Dahl, G. Longoni, P. Chini, A. J. Schultz, and J. M. Williams, *Adv. Chem. Ser.*, **167**, 93 (1978).
39. V. G. Albano, A. Ceriotti, P. Chini, G. Ciani, S. Martinengo, and M. Anker, *J. Chem. Soc., Chem. Commun.*, 859 (1975); V. G. Albano, G. Ciani, S. Martinengo, and S. Sironi, *J. Chem. Soc., Dalton Trans.*, 978 (1979).
40. J. L. Vidal, R. A. Fiato, L. A. Cosby, and R. L. Pruett, *Inorg. Chem.*, **17**, 2574 (1978).
41. H. Schäfer, R. Laumanns, B. Krebs, and G. Henkel, *Angew. Chem., Int. Ed. Engl.*, **18**, 325 (1979).
42. H. G. Schnering, H. Wohrle, and H. Schäfer, *Naturwissenschaften*, **48**, 159 (1961).
43. A. Simon and H. G. Schnering, *J. Less-Common Met.*, **11**, 31 (1966).
44. R. B. King, D. M. Braitsch, and P. N. Kapoor, *J. Am. Chem. Soc.*, **97**, 60 (1975); E. O. Fischer and F. Röhrscheid, *J. Organomet. Chem.*, **6**, 53 (1966).
45. S. Z. Goldberg, B. Spivack, G. Stanley, R. Eisenberg, D. M. Braitsch, J. S. Miller, and M. Abkowitz, *J. Am. Chem. Soc.*, **99**, 110 (1977).
46. A. Müller, A. Ruck, M. Dartmann, and U. Reinsch-Vogell, *Angew. Chem., Int. Ed. Engl.*, **20**, 483 (1981).
47. P. J. Vergamini, H. Vahrenkamp, and L. F. Dahl, *J. Am. Chem. Soc.*, **93**, 6327 (1971).
48. A. Bino, F. A. Cotton, and Z. Dori, *J. Am. Chem. Soc.*, **100**, 5252 (1978).
49. V. Katovic, J. L. Templeton, and R. E. McCarley, *J. Am. Chem. Soc.*, **98**, 5705 (1976).
50. P. R. Sharp and R. R. Schrock, *J. Am. Chem. Soc.*, **102**, 1430 (1980); F. A. Cotton, T. R. Felthouse, and D. G. Lay, *J. Am. Chem. Soc.*, **102**, 1430 (1980).
51. A. Müller, R. G. Bhattacharyya, and B. Pfefferkorn, *Chem. Ber.*, **112**, 778 (1979); A. Müller, S. Pohl, M. Dartmann, J. P. Cohen, J. M. Bennett, and R. M. Kirchner, *Z. Naturforsch.*, **B34**, 434 (1979).
52. J. Marcoll, A. Rabenau, D. Mootz, and H. Wunderlich, *Rev. Chim. Miner.*, **11**, 607 (1974).
53. A. Bino, M. Ardon, J. Maor, M. Kaftory, and Z. Dori, *J. Am. Chem. Soc.*, **98**, 7093 (1976).
54. M. Ardon, A. Bino, F. A. Cotton, Z. Dori, M. Kapon, and B. W. S. Kolthammer, *Inorg. Chem.*, **20**, 4082 (1981).
55. A. Bino, F. A. Cotton, and Z. Dori, *J. Am. Chem. Soc.*, **103**, 243 (1981); A. Bino, F. A. Cotton,

Z. Dori, and B. W. S. Kolthammer, *J. Am. Chem. Soc.*, **103**, 5779 (1981); A. Bino, F. A. Cotton, Z. Dori, L. R. Falvello, and G. M. Reisner, *Inorg. Chem.*, **21**, 3750 (1982).

56. A. Bino, F. A. Cotton, Z. Dori, S. Koch, H. Küppers, M. Millar, and J. C. Sekutowski, *Inorg. Chem.*, **17**, 3245 (1978).
57. J. A. Bertrand, F. A. Cotton, and W. A. Dollase, *Inorg. Chem.*, **2**, 1166 (1963).
58. P. Edwards, K. Mertis, G. Wilkinson, M. B. Hursthouse, and K. M. A. Malik, *J. Chem. Soc., Dalton Trans.*, 334 (1980).
59. A. F. Masters, K. Mertis, J. F. Gibson, and G. Wilkinson, *Nouv. J. Chem.*, **1**, 389 (1977).
60. S. Stensvad, B. J. Helland, M. W. Babich, R. A. Jacobson, and R. E. McCarley, *J. Am. Chem. Soc.*, **100**, 6257 (1978).
61. K. Jödden, H. G. Schnering, and H. Schäfer, *Angew. Chem., Int. Ed. Engl.*, **14**, 570 (1975).
62. L. R. Bateman, J. F. Blount, and L. F. Dahl, *J. Am. Chem. Soc.*, **88**, 1082 (1966).
63. J. D. Corbett, R. L. Doake, K. R. Poeppelmeier, and D. H. Guthrie, *J. Am. Chem. Soc.*, **100**, 652 (1978).
64. P. A. Vaughan, J. H. Sturdivant, and L. Pauling, *J. Am. Chem. Soc.*, **72**, 5477 (1950); F. A. Cotton and T. E. Haas, *Inorg. Chem.*, **3**, 10 (1964).
65. P. A. Vaughan, *Proc. Natl. Acad. Sci. U.S.A.*, **36**, 461 (1950).
66. L. J. Guggenberger and A. W. Sleigh, *Inorg. Chem.*, **8**, 2041 (1969).
67. Y. Jiang, Y.-S. Kiang, A. Tang, R. Hoffmann, J. Huang, and J. Lu, *Organometallics*, **4**, 27 (1985).
68. F. A. Cotton and G. G. Stanley, *Chem. Phys. Lett.*, **58**, 450 (1978); B. E. Bursten, F. A. Cotton, J. C. Green, E. A. Seddon, and G. G. Stanley, *J. Am. Chem. Soc.*, **102**, 955 (1980).
69. F. A. Cotton and T. E. Haas, *Inorg. Chem.*, **3**, 10 (1964); B. E. Bursten, F. A. Cotton, and G. G. Stanley, *Isr. J. Chem.*, **19**, 132 (1980).
70. H. Hogeveen and P. W. Kwant, *Acc. Chem. Res.*, **8**, 413 (1975).
71. H. Schwarz, *Angew. Chem., Int. Ed. Engl.*, **20**, 991 (1981).
72. R. P. Ziebarth and J. D. Corbett, *J. Am. Chem. Soc.*, **107**, 4571 (1985); J. D. Smith and J. D. Corbett, *J. Am. Chem. Soc.*, **106**, 4618 (1984).
73. G. L. Soloveichik, B. M. Bulychev, and K. N. Semenenko, *Koord. Khim.*, **9**, 1585 (1983).
74. S. D. Wijeyesekera, R. Hoffmann, and C. N. Wilker, *Organometallics*, **3**, 962 (1984); *Organometallics*, **3**, 9 (1984).
75. E. L. Muetterties, *Prog. Inorg. Chem.*, **28**, 203 (1982).
76. B. F. G. Johnson, J. Lewis, W. J. H. Nelson, J. N. Nicholls, and M. D. Vargas, *J. Organomet. Chem.*, **249**, 252 (1983).
77. S. Martinengo, G. Ciani, A. Sironi, B. T. Heaton, and J. Mason, *J. Am. Chem. Soc.*, **101**, 7095 (1979).
78. M. L. Blohm and W. L. Gladfelter, *Organometallics*, **4**, 45 (1985); D. E. Fjare and W. L. Gladfelter, *J. Am. Chem. Soc.*, **106**, 4799 (1984).
79. J. Mason, *Chem. Rev.*, **81**, 205 (1981).
80. B. T. Heaton, L. Strona, and S. T. Martinengo, *J. Organomet. Chem.*, **215**, 415 (1981).
81. P. Chini, G. Ciani, S. Martinengo, A. Sironi, L. Longetti, and B. T. Heaton, *Chem. Commun.*, 188 (1979).
82. C. Brown, B. T. Heaton, P. Chini, A. Fumagalli, and G. Longoni, *Chem. Commun.*, 309 (1977); C. Brown, B. T. Heaton, A. D. C. Towl, G. Longoni, A. Fumagalli, and P. Chini, *J. Organomet. Chem.*, **181**, 233 (1979).
83. J. L. Vidal, W. E. Walker, and R. C. Schoening, *Inorg. Chem.*, **20**, 238 (1981).
84. O. A. Gansow, D. S. Gill, J. F. Bennis, J. R. Hutchison, J. L. Vidal, and R. C. Schoening, *J. Am. Chem. Soc.*, **102**, 2449 (1980).
85. G. Bor, U. K. Dietler, P. L. Stanghellini, G. Gervasio, R. Rosetti, G. Sbrignadello, and G. A. Battiston, *J. Organomet. Chem.*, **213**, 277 (1981); G. Bor and P. L. Stanghellini, *Chem. Commun.*, 886 (1979).
86. I. A. Oxton, S. F. A. Kettle, P. F. Jackson, B. F. G. Johnson, and J. Lewis, *Chem. Commun.*, 687 (1979).
87. I. A. Oxton, S. F. A. Kettle, P. F. Jackson, B. F. G. Johnson, and J. Lewis, *J. Mol. Struct.*, **71**, 117 (1981); P. F. Jackson, B. F. G. Johnson, J. Lewis, M. McPartlin, and W. J. H. Nelson, *Chem. Commun.*, 224 (1980).

88. J. A. Connor, *Top. Curr. Chem.*, **71**, 71 (1977).
89. J. A. Connor, in: *Transition Metal Cluster* (B. F. G. Johnson, ed.), p. 345, Wiley, Chichester (1980).
90. F. A. Cotton, *Chem. Soc. Rev.*, **12**, 35 (1983); F. A. Cotton, *Chem. Soc. Rev.*, **4**, 27 (1975); M. H. Chisholm, D. M. Hoffmann, and J. C. Huffmann, *Chem. Soc. Rev.*, **14**, 69 (1985).
91. C. E. Housecroft, M. E. O'Neil, and K. Wade, *J. Organomet. Chem.*, **213**, 35 (1981).
92. A. K. Baer, J. A. Connor, N. I. Et-Saied, and H. A. Skinner, *J. Organomet. Chem.*, **213**, 151 (1981).
93. S. P. Gubin, N. M. Mikova, M. C. Cybenov, and V. E. Lopatin, *Koord. Khim*, **10**, 625 (1984).
94. A. Fumagalli, S. Martinengo, P. Chini, D. Galli, B. T. Heaton, and R. Della Pergola, *Inorg. Chem.*, **23**, 2947 (1984).
95. F. Bachechi, J. Ott, and L. M. Venanzi, *J. Am. Chem. Soc.*, **107**, 1760 (1985).
96. E. G. Mednikov, N. K. Eremenko, V. A. Mikhailov, S. P. Gubin, Yu. L. Slovokhotov, and Yu. T. Struchkov, *Chem. Commun.*, 989 (1981).
97. D. Fenske, J. Hachgenei, and F. Rogel, *Angew. Chem., Int. Ed. Engl.*, **23**, 982 (1984); D. Fenske, R. Basoglu, J. Hachgenei, and F. Rogel, *Angew. Chem., Int. Ed. Engl.*, **23**, 160 (1984).
98. P. H. M. Budzelaar, J. Boersma, G. J. M. van der-Kerk, A. L. Spek, and A. J. M. Duisenberg, *Organometallics*, **4**, 680 (1985).
99. A. Cerrotti, F. Demartin, G. Longoni, N. Manassero, M. Marchionna, G. Piva, and M. Sansoni, *Angew. Chem., Int. Ed. Engl.*, **24**, 697 (1985).
100. E. W. Burkhardt and G. L. Goeffroy, *J. Organomet. Chem.*, **198**, 179 (1980).
101. C.-M. T. Hayward and J. R. Shapley, *Inorg. Chem.*, **21**, 3816 (1982).
102. G. Geoffroy, *Acc. Chem. Res.*, **13**, 469 (1980).
103. P. Lemoine, *Coord. Chem. Rev.*, **47**, 55 (1982).
104. H. Vahrenkamp, *Adv. Organomet. Chem.*, **22**, 169 (1983).
105. P. Lemoine, A. Givaudeau, and M. Gross, *Electrochim. Acta*, **21**, 1 (1976).
106. P. Lemoine, A. Giraudeau, M. Gross, R. Bender, and P. Braunstein, *J. Chem. Soc., Dalton Trans.*, 2059 (1981).
107. J. C. Huffmann, J. G. Stone, W. C. Krusell, and K. G. Caulton, *J. Am. Chem. Soc.*, **99**, 5830 (1977).
108. F. A. Cotton, S. A. Duraj, and W. J. Roth, *J. Am. Chem. Soc.*, **106**, 3257 (1984).
109. A. Bino, *J. Am. Chem. Soc.*, **102**, 7990 (1980); A. Bino, *Inorg. Chem.*, **21**, 1917 (1982).
110. D. J. Darensbourg and D. J. Zalewski, *Organometallics*, **3**, 1598 (1984).
111. B. Siegwarth, L. Zsolnai, H. Berke, and G. Huttner, *J. Organomet. Chem.*, **222**, C5 (1982); G. Huttner, U. Weber, B. Siegwarth, and O. Scheidsteger, *Angew. Chem., Int. Ed. Engl.*, **21**, 215 (1982); G. Huttner, U. Weber, and L. Zsolnai, *Z. Naturforsch.*, **B37**, 707 (1982).
112. A. M. Arif, A. H. Cowley, N. C. Norman, and M. Pakulski, *J. Am. Chem. Soc.*, **107**, 1062 (1985).
113. L. E. Bogan, Jr., T. B. Rauchfuss, and A. L. Rheingold, *J. Am. Chem. Soc.*, **107**, 3843 (1985).
114. R. Bender, P. Braunstein, J.-M. Jud, and Y. Dusausoy, *Inorg. Chem.*, **22**, 3394 (1983).
115. J. A. Labinger, K. S. Wong, and R. Scheidt, *J. Am. Chem. Soc.*, **100**, 3254 (1978).
116. E. Sappa, A. Tiripicchio, and P. Braunstein, *Coord. Chem. Rev.*, **65**, 219 (1985).
117. G. Ciani, G. D'Alphonso, M. Freni, P. Romiti, and A. Sironi, *Chem. Commun.*, 339 (1982); G. Ciani, G. D'Alphonso, M. Freni, P. Romiti, and A. Sironi, *Chem. Commun.*, 705 (1982).
118. J. I. Davies, C. G. Howard, A. C. Skapski, and G. Wilkinson, *Chem. Commun.*, 1077 (1982).
119. W. A. Herrmann, R. Serrano, M. L. Ziegler, H. Pfisterer, and B. Nuber, *Angew. Chem., Int. Ed. Engl.*, **24**, 50 (1985).
120. B. F. G. Johnson, J. Lewis, W. J. H. Nelson, M. D. Vargas, D. Braga, K. Henrick, and M. McPartlin, *J. Chem. Soc., Dalton Trans.*, 2151 (1984).
121. J. W. Kolis, E. M. Holt, and D. W. Shriver, *J. Am. Chem. Soc.*, **105**, 7307 (1983).
122. M. Tachikawa, R. L. Geerts, and E. L. Muetterties, *J. Organomet. Chem.*, **213**, 11 (1981).
123. C.-M. T. Hayward, J. R. Shapley, M. R. Churchill, C. Bueno, and A. L. Rheingold, *J. Am. Chem. Soc.*, **104**, 7347 (1982).
124. A. Fumagalli, S. Martinengo, G. Ciani, and A. Sironi, *J. Chem. Soc., Chem. Commun.*, 453 (1983); S. Martinengo, A. Fumagalli, R. Bonfichi, G. Ciani, and A. Sironi, *J. Chem. Soc., Chem. Commun.*, 825 (1982).
125. R. A. Jones and B. R. Whittlescy, *J. Am. Chem. Soc.*, **107**, 1078 (1985).
126. A. Arrigoni, A. Ceriotti, R. Della Pergola, G. Longoni, M. Manassero, N. Masciocchi, and M. Sansoni, *Angew. Chem., Int. Ed. Engl.*, **23**, 322 (1984).

127. A. D. Clauss, J. R. Shapley, C. N. Wilker, and R. Hoffman, *Organometallics*, **3**, 619 (1984).
128. W. I. Bailey, Jr., F. A. Cotton, J. D. Janerson, and B. W. S. Kolthammer, *Inorg. Chem.*, **21**, 3131 (1982).
129. E. Keller and H. Vahrenkamp, *Angew. Chem.*, **89**, 738 (1977).
130. J. L. Atwood, W. E. Hunter, R. A. Jones, and T. C. Wright, *Inorg. Chem.*, **22**, 993 (1983).
131. A. Ceriotti, P. Chini, R. Della Pergola, and G. Longoni, *Inorg. Chem.*, **22**, 1595 (1983).
132. G. Longini and P. Chini, *J. Am. Chem. Soc.*, **98**, 7225 (1976); J. C. Calabrese, L. F. Dahl, P. Chini, G. Longoni, and S. Martinengo, *J. Am. Chem. Soc.*, **96**, 2614 (1974).
133. D. M. Washecheck, E. J. Wucherer, L. F. Dahl, A. Ceriotti, G. Longoni, M. Manassero, M. Sansoni, and P. Chini, *J. Am. Chem. Soc.*, **101**, 6110 (1979).
134. A. Ceriotti, G. Longoni, M. Manassero, M. Perego, and M. Sansoni, *Inorg. Chem.*, **24**, 117 (1985).
135. N. K. Eremenko, E. G. Mednikov, and S. S. Kurasov, *Usp. Khim.*, **54**, 671 (1985); N. K. Eremenko, E. G. Mednikov, and S. P. Gubin, *Koord. Khim.*, **10**, 617 (1984).
136. T. A. Stromnova, M. N. Vargaftik, T. S. Khodashova, M. A. Poraj-Koszyc, and I. I. Moiseev, *Koord. Chim.*, **7**, 132 (1981); T. A. Stromnova, M. N. Vargaftik, N. S. Khodashova, M. A. Poraj-Koszic, and I. Moiseev, *Izv. Akad. Nauk SSSR, Ser. Khim.*, 1690 (1980).
137. O. N. Temkin and L. G. Bruk, *Usp. Khim.*, **52**, 206 (1983); T. A. Stromnova and M. H. Vargaftik, *Izv. Akad. Nauk SSSR, Ser. Khim.*, 1411 (1980).
138. E. L. Muetterties, *Catal. Rev. Sci. Eng.*, **23**, 69 (1981); E. L. Muetterties, *Chem. Soc. Rev.*, **11**, 283 (1982).
139. M. D. Fryzuk, *Can. J. Chem.*, **61**, 1347 (1983); M. D. Fryzuk, *Organometallics*, **1**, 408 (1982).
140. M. S. Paquette and L. F. Dahl, *J. Am. Chem. Soc.*, **102**, 6623 (1980).
141. R. Bender, P. Braunstein, A. Tiripicchio, and M. Tiripicchio Camellini, *Angew. Chem., Int. Ed. Engl.*, **24**, 861 (1985).
142. D. Fenske, J. Hachgenei, and J. Ohmer, *Angew. Chem., Int. Ed. Engl.*, **24**, 706 (1985).
143. M. G. Thomas, E. L. Muetterties, and R. O. Day, *J. Am. Chem. Soc.*, **98**, 4645 (1976); V. W. Day, R. O. Day, J. S. Kristoff, F. J. Hirsekorn, and E. L. Muetterties, *J. Am. Chem. Soc.*, **97**, 2571 (1975).
144. H. Schaefer, D. Binder, and D. Fenske, *Angew. Chem., Int. Ed. Engl.*, **24**, 522 (1985).
145. M. M. Olmstead and P. P. Power, *J. Am. Chem. Soc.*, **106**, 1495 (1984).
146. P. R. Raithby, in: *Transition Metal Clusters* (B. F. G. Johnson, ed.), p. 5, Wiley, Chichester (1980).
147. A. Camus, N. Marsich, G. Nardin, and L. Randaccio. *J. Organomet. Chem.*, **174**, 121 (1979).
148. G. van Koten, J. T. B. H. Jastrzebski, and J. G. Noltes, *Chem. Commun.*, 203 (1977).
149. H. L. Aalten, G. van Koten, K. Goubitz, and C. H. Stam, *Chem. Commun.*, 1252 (1985).
150. T. Tsuda, T. Yazawa, K. Watanabe, T. Fujii, and T. Saegusa, *J. Org. Chem.*, **46**, 192 (1981); S. Gambarotta, C. Floriani, A. ChiesiVilla, and C. Guastini, *Chem. Commun.*, 1156 (1983).
151. F. Mahdjour-Hassan-Abadi, H. Hartl, and J. Fuchs, *Angew. Chem., Int. Ed. Engl.*, **23**, 514 (1984).
152. P. Kluefers, *Angew. Chem., Int. Ed. Engl.*, **23**, 307 (1984).
153. J. M. Burlitch, S. E. Hayes, and J. T. Lemley, *Organometallics*, **4**, 167 (1985).
154. H. Hope, D. Oram, and P. Power, *J. Am. Chem. Soc.*, **106**, 1149 (1984).
155. L. F. Rhodes, J. C. Huffman, and K. G. Caulton, *J. Am. Chem. Soc.*, **107**, 1759 (1985).
156. M. Y. Chiang, E. Boehlen, and R. Bau, *J. Am. Chem. Soc.*, **107**, 1679 (1985).
157. S. I. Khan, P. G. Edwards, H. S. H. Yuan, and R. J. Bau, *Am. Chem. Soc.*, **107**, 1682 (1985).
158. G. Schmid, R. Pfeil, R. Boese, F. Bandermann, S. Meyer, G. H. M. Calis, and J. W. A. van der Felden, *Chem. Ber.*, **114**, 3634 (1981).
159. B. K. Teo and K. Keating, *J. Am. Chem. Soc.*, **106**, 2224 (1984).
160. H. Vahrenkamp, *Adv. Organomet. Chem.*, **22**, 169 (1983).
161. A. J. Deeming, in: *Transition Metal Clusters* (B. F. G. Johnson, ed.), p. 391, Wiley, Chichester (1980).
162. E. Band and E. L. Muetterties, *Chem. Rev.*, **78**, 639 (1978).
163. V. A. Maksakov and S. P. Gubin, *Koord. Khim.*, **10**, 689 (1984).
164. D. A. Sweigart, in: *Mechanisms of Inorganic and Organometallic Reactions* (M. V. Twigg, ed.), Vol. 2, p. 237, Plenum, New York (1984).
165. C. D. Garner, in: *Transition Metal Clusters* (B. F. G. Johnson, ed.), p. 265, Wiley, Chichester (1980).
166. M. S. Poquette and L. F. Dahl, *J. Am. Chem. Soc.*, **102**, 6621 (1980).

167. G. J. Bezems, P. H. Rieger, and S. Visco, *Chem. Commun.*, 265 (1981).

168. T. Beringhelli, G. Ciani, G. D'Alfonso, H. Molinari, and A. Sironi, *Inorg. Chem.*, **24**, 2666 (1985).

169. J. Kijsper, L. H. Polm, G. van Koten, K. Vrieze, E. Nielsen, and C. H. Stam, *Organometallics*, **4**, 2006 (1985).

170. L. Garlaschelli, A. Fumagalli, S. Martinengo, B. T. Heaton, D. O. Smith, and L. Strona, *J. Chem. Soc., Dalton Trans.*, 2265 (1982).

171. M. J. Mays, P. R. Raithby, P. L. Taylor, and K. Henrick, *J. Organomet. Chem.*, **224**, C45 (1982).

172. W.-Y. Yeh, J. R. Shapley, Y.-J. Li, and M. R. Churchill, *Organometallics*, **4**, 767 (1985).

173. J. S. Holmgren and J. R. Shapley, *Organometallics*, **3**, 1322 (1984); J. S. Holmgren and J. R. Shapley, *Organometallics*, **4**, 793 (1985).

174. W. Saak, G. Henkel, and S. Pohl, *Angew. Chem., Int. Ed. Engl.*, **23**, 150 (1984).

175. H. Dorn, B. E. Hanson, and E. Motell, *Inorg. Chim. Acta*, **54**, L71 (1981).

176. M. O. Albers and N. J. Coville. *Coord. Chem. Rev.*, **53**, 227 (1984).

177. M. A. Gallop, A. H. Wright, B. F. G. Johnson, and J. Lewis, *Abstracts*, p. 408, 12th ICOMC, Vienna (September 8–13, 1985).

178. L. M. Bavaro, P. Montangero, and J. B. Keister, *J. Am. Chem. Soc.*, **105**, 4977 (1983).

179. E. D. Morrison, G. R. Steinmetz, G. L. Geoffroy, W. C. Fultz, and A. L. Rheingold, *J. Am. Chem. Soc.*, **106**, 4783 (1984); E. D. Morrison and G. L. Geoffroy, *J. Am. Chem. Soc.*, **107**, 3541 (1985).

180. J. R. Shapley, W.-Y. Yeh, M. R. Churchill, and Y.-J. Li, *Organometallics*, **4**, 1898 (1985).

181. R. B. Calvert, J. R. Shapley, A. J. Schultz, J. M. Williams, S. L. Suib, and G. D. Stucky, *J. Am. Chem. Soc.*, **100**, 6240 (1978); R. B. Calvert and J. R. Shapley, *J. Am. Chem. Soc.*, **100**, 7726 (1978); **99**, 5225, 6544 (1977).

182. Y. Chi and J. R. Shapley, *Organometallics*, **4**, 1900 (1985).

183. D. Fenske, J. Ohmer, and J. Hachgenei, *Angew. Chem., Int. Ed. Engl.*, **24**, 993 (1985).

184. G. Schmid, *Structure and Bonding*, **62**, 51 (1985).

# Chapter 4

# Compounds Containing
# 1e Carbon–Donor Ligands

The following organic radicals are $1e$ ligands: alkyl, aryl, alkenyl, alkynyl, etc. Therefore, compounds with metal–carbon $\sigma$ bonds are those in which there are $M-C(sp^3)$, $M-C(sp^2)$, and $M-C(sp)$ groupings. Many compounds possessing metal to carbon $\sigma$ bonds are known both homoleptic $(MR_n)$ as well as heteroleptic $(MR_nL_m)$.[1-6] Like sandwich compounds, complexes containing metal to carbon $\sigma$ bonds very often do not obey Sidgwick's rule. The number of valence electrons in compounds with $M-C$ $\sigma$ bonds varies from 6 to 18. Among other reasons, this is due to steric factors. For example, the titanium group metals would have to coordinate several alkyl ligands in order to achieve the 18 valence electron configuration. Often, central metal atoms in complexes containing metal to carbon $\sigma$ bonds are in their high oxidation states. Alkyl and aryl ligands may be terminal or bridging. In the case of bridging ligands, a three-center, two-electron (3c–2e) bond is formed.

Compounds possessing bridging groups such as $-CH_2(CH_2)_nCH_2-$ as well as complexes with chelating ligands are known[1-6, 13, 14, 85-88, 110, 111, 123]:

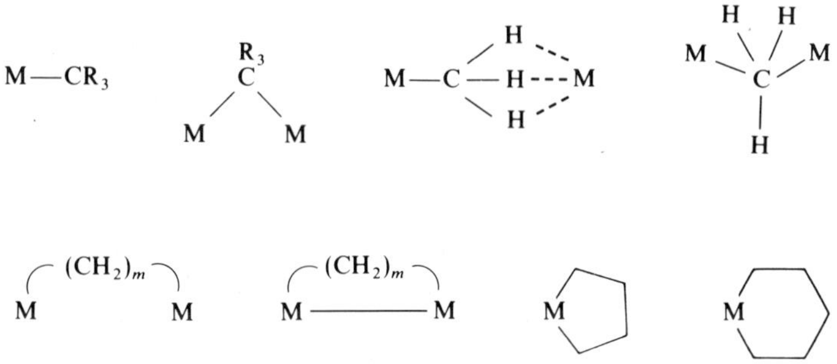

$$(4.1)$$

Although containing metal–carbon $\sigma$ bonds and possessing sp$^3$ hybridization of carbon atoms, the $\mu$-alkylidene and $\mu_3$-alkylidyne compounds

$$
\begin{array}{ccc}
& CR_2 & \\
\diagup & & \diagdown \\
M & & M
\end{array}
\qquad \text{and} \qquad
\begin{array}{ccc}
& CR & \\
\diagup & | & \diagdown \\
M & & M \\
& M &
\end{array}
$$

will be discussed in Chapter 5, which is devoted to carbene and carbyne complexes.

In the case of the main group metals, there are also compounds containing four-center, two-electron (4c–2e) bonds, for example, in $Li_4R_4$ (Chapter 1). Homoleptic organometallic compounds with 1$e$ ligands may be neutral $[MR_n]$ or anionic $[MR_m]^{x-}$. Cationic complexes containing these ligands are rare. Exceptions are ions encountered in mass spectrometry and some cationic compounds of copper(I) and gold(I).

Homoleptic compounds are those which have the $[MR_n]$ composition, while heteroleptic complexes possessing the general formula $[MR_nX_m]_y$ contain at least two different kinds of ligands in addition to central atoms. The following examples illustrate homoleptic complexes $[TiMe_4]$, $[TiMe_5]^-$, and $[Mo_2Me_8]^{4-}$ while the compounds $[TiMe_3Cl]$ and $[ZrMe_2Cl_2]$ represent heteroleptic compounds. It is more convenient to use the adjective homoleptic rather than binary, because the latter may only be used for compounds of the type $[MX_n]$, where X represents a single atom.

In contrast to homoleptic cationic complexes with metal–carbon $\sigma$ bonds, heteroleptic compounds are quite stable. The best known of such cationic complexes are amine and aqua chromium(III) complexes, for example, $[CrR_2(bipy)_2]^+$ (R = Ph, $CH_2SiMe_3$, anisyl, tolyl), $[Cr(R)(H_2O)_5]^{2+}$ (R = $CH_2Ph$, $CHCl_2$), as well as 16$e$ and 18$e$ complexes of group 8–10 metals. Cationic complexes of other metals containing C–M $\sigma$ bonds are less investigated: $[Nbcp_2\{CS(SR)\}(\sigma\text{-allyl})]^+$, $[Tacp_2(Me)(CH_2SiMe_3)]^+$, $[Tacp_2(CH_2SiMe_3)_2]^+$, $[MRL_3]^+$ (M = Pd, Pt), $[PtR_3L_3]^+$, $[NiRL_4]^+$, $[FeR(CO)_3(L-L)]^+$, etc.

## 1. BONDING IN COMPOUNDS WITH METAL–CARBON $\sigma$ BONDS

The local symmetry of the M–C bond is $C_{\infty v}$. In this symmetry group, orbitals of the metal and carbon atoms always form $\sigma$ and $\pi$ bonds[5, 13, 80] (Table 4.1).

The $\pi$-interaction only slightly influences the metal–alkyl bond because the carbon $p_x$ and $p_y$ orbitals form bonds with hydrogen (sp$^3$ carbon atoms). In the case of alkenyl,

Table 4.1. *Metal and Carbon Orbitals of the* M–C *Grouping in the* $C_{\infty v}$ *Point Group*

| Symmetry | Orbitals of the metal | Orbitals of the ligand | Type of bond |
|---|---|---|---|
| $A_1$ ($\Sigma^+$) | $s, p_z, d_{z^2}$ | $s, p_z$ | $\sigma$ |
| $E_1$ ($\pi$) | $p_x, p_y, d_{xz}, d_{yz}$ | $p_x, p_y$ | $\pi$ |
| $E_2$ ($\Delta$) | $d_{x^2-y^2}, d_{xy}$ | — | — |

aryl, and alkynyl complexes which possess metal–carbon $sp^2$ and $sp$ bonds, there is a possibility of charge transfer from the ligand to the metal because of the overlap between occupied $\pi$ orbitals and empty metal orbitals. Also, there is electron flow in the opposite direction due to the formation of back $\pi$-bonding which comes about by the interaction of occupied $d\pi$ orbitals with antibonding $\pi^*$ ligand orbitals. Often aryl complexes are more stable than alkyl compounds, which may be explained by higher electronegativity of aryl groups or by additional $\pi$-interaction occurring within the $M-C$ bond.

## 2. STABILITY OF COMPOUNDS CONTAINING 1e LIGANDS

The transition metal homoleptic complexes containing alkyl ligands are kinetically and thermally less stable than analogous compounds of the main group metals. For some time, it was believed that this is a result of the diminished strength of the carbon–transition metal bond. However, thermodynamic studies showed that the $M-C(sp^3)$ bond energy for the transition metals and the main group metals is the same (Table 4.2).

In order to calculate $D(M-L)$ values, experimentally determined energies of dissociation of alkanes are used. Instead of $D(M-L)$ values, equivalent $E(M-L)$ values are sometimes utilized (Table 4.3) which are calculated from the assumed constant value for $E(C-H)$ modified by introducing corrections for steric and electronic effects.[1, 7]

In order to determine the energies of dissociation of the metal–ligand bond, the following methods are commonly utilized: calorimetric combustion, solution calorimetry, ion cyclotron resonance, mass spectrometry, ion beam technique, equilibrium constant measurements, kinetic methods, and photochemical determination of the threshold wavelength of the photolytic dissociation of the metal–ligand bond.[7, 75–77]

The energy of the $M-C$ bond of homoleptic transition metal compounds increases as the atomic number of the element in the group becomes greater. In the case of main group complexes, this trend is opposite. The stability of the $M-C$ bond also increases

Table 4.2. *Average Bond Dissociation Enthalpies $D(M-L)$ $(kJ\,mol^{-1})$ for Homoleptic Complexes of the Type $ML_n{}^a$*

|  | M = Ti | Zr | Hf | Nb | Ta | Mo | W |
|---|---|---|---|---|---|---|---|
| L | n = 4 | 4 | 4 | 5 | 5 | 6 | 6 |
| Me | $(260)^b$ | (310) | (330) |  | 261 |  | 159 |
| $CH_2CMe_3$ | 188 | 227 | 224 |  |  |  |  |
| $CH_2Ph$ | 203 | 252 |  |  |  |  |  |
| $NMe_2$ | 307 | 350 | (370) |  | 328 |  | 220 |
| $NEt_2$ | 309 | 343 | 367 |  |  |  |  |
| OMe |  |  |  | 419 | 439 |  | (360) |
| $OPr^i$ | 444 | 518 | 534 |  |  |  |  |
| F | 586 | 648 | 665 | 573 | 603 | 449 | 508 |
| Cl | 430 | 490 | 497 | 406 | 430 | 305 | 347 |

$^a$ According to Reference 7.
$^b$ Estimated or extrapolated values are given in parentheses.

Table 4.3. Bond Energies $E$ $(M-L)$

| Bond | Compound | $E(M-L)$ (kJ mol$^{-1}$) |
|---|---|---|
| Ge$-$C | [GeM$_4$] | 247 |
| Ge$-$C | [GeEt$_4$] | 238 |
| Ge$-$N | [Me$_3$GeNMe$_2$] | 230 |
| Ge$-$O | [Me$_3$GeOEt] | 331 |
| Sn$-$C | [SnMe$_4$] | 218 |
| Sn$-$C | [SnEt$_4$] | 192 |
| Ti$-$C | [Ti(CH$_2$CMe$_3$)$_4$] | 184 |
| Ti$-$C | [Ti(CH$_2$SiMe$_3$)$_4$] | 268 |
| Ti$-$C | [Ti(CH$_2$Ph)$_4$] | 264 |
| Ti$-$N | [Ti(NMe$_2$)$_4$] | 331 |
| Ti$-$O | [Ti(OPr$^i$)$_4$] | 469 |
| Zr$-$C | [Zr(CH$_2$CMe$_3$)$_4$] | 226 |
| Zr$-$C | [Zr(CH$_2$SiMe$_3$)$_4$] | 314 |
| Zr$-$C | [Zr(CH$_2$Ph)$_4$] | 310 |
| Hf$-$C | [Hf(CH$_2$CMe$_3$)$_4$] | 243 |

[a] According to Reference 1.

with the lowering of the oxidation state of the metal (like the stability of the $M-X$ bond for halides, $MX_n$). Thus the thermal resistance of [NbMe$_5$] is lower than that of [TaMe$_5$] and, for vanadium, only a compound possessing $+4$ oxidation state is known, i.e., [VMe$_4$]. In mixed alkyl–carbonyl complexes, the $M-CH_3$ bond energy is the same or higher than that of the $M-CO$ bond.

The average metal–carbon bond energy is higher for complexes of the left-hand side of the Periodic Chart than that of the middle and right-hand side of the Periodic Chart. For hydrocarbyl complexes of the scandium, titanium, and vanadium groups, the average enthalpy of dissociation of the $M-C$ bond usually equals 160–350 kJ mol$^{-1}$, and for compounds of the remaining transition metals, this enthalpy is most commonly in the 80–170 kJ mol$^{-1}$ range. The enthalpy of dissociation of the $M-R$ bond for selected compounds is as follows[75–78]: MnR(CO)$_5$, R = Me: 155 kJ mol$^{-1}$; R = Ph: 171 kJ mol$^{-1}$; R = CH$_2$Ph: 88 kJ mol$^{-1}$; R = CF$_3$: 171 kJ mol$^{-1}$; R = COMe: 130 kJ mol$^{-1}$; [CoR(saloph)py], R = Pr$^n$: 105 kJ mol$^{-1}$; Pr$^i$: 84 kJ mol$^{-1}$; CH$_2$CMe$_3$: 75 kJ mol$^{-1}$; CH$_2$Ph: 92 kJ mol$^{-1}$. In the case of complexes of the type [CoR(dmgH)L] where R = CHMePh, and L = py, 4-aminopyridine, 4-methylpyridine, 4-cyanopyridine, and imidazole, the values of $D(Co-R)$ change in the range 75–89 kJ mol$^{-1}$. They increase linearly with the increase of p$K_a$ of the base which is coordinated in the *trans* position to the alkyl group. Greater values of the $\sigma$ $M-C$ bond are reported also for cyclopentadienyl–alkyl compounds of group 3 and 4 metals than for those of higher group metals: MoMe$_2$cp$_2$—125 kJ mol$^{-1}$, WMe$_2$cp$_2$—180 kJ mol$^{-1}$, and TiPh$_2$cp$_2$—272 kJ mol$^{-1}$. In thorium complexes such as [Th(pmcp)$_2$R$_2$] and [Th(pmcp)$_2$(R)Cl] the Th$-$C $\sigma$ bonds are stable; their enthalpies of dissociation are 285–395 kJ mol$^{-1}$.[77] In this case, there is no great difference between the energy of dissociation of the first $M-C$ bond and the second one (this second value is slightly higher). The enthalpy of dissociation decreases in the following order: Ph > Me $\approx$ CH$_2$SiMe$_3$ > Et $\gtrsim$ CH$_2$Ph $\approx$ n-Bu $\approx$ CH$_2$CMe$_3$. A similar order was found for compounds possessing the same type of ligands and other metals (Tables 4.2 and 4.3). In the case of $CH_3$, $CH_2Me$, $CHMe_2$, and $CH_2CMe_3$, steric factors have essential influence on the stability of the M–C bond. However, the electronic influence of Si on the $M-CH_2SiMe_3$ bond energy cannot be excluded.

The knowledge of enthalpies of dissociations for M–H, M–R, M–C(O)R, and Y–C(O)R (Y = H, R) bonds allows the estimation of enthalpies of various organometallic reactions (Table 4.4). This table shows that the following reactions are thermodynamically favorable: oxidative addition of hydrogen, CO insertion into the M–R bond, olefin insertion into the M–H bond, and olefin insertion into the M–R bond. Oxidative addition of aldehydes to the transition metal complexes is possible. However, due to the near-zero value of the enthalpy of this reaction, oxidative addition of aldehydes does not occur easily in the above-mentioned processes. Thermodynamically unfavorable are oxidative addition of the unstrained C–C bonds, oxidative addition of C–H bonds, and CO insertion into the M–H bonds.

The greatly diminished thermostability of homoleptic alkyl compounds of the transition metals compared to analogous complexes of the main group elements is not a result of smaller energies of metal–carbon bonds, but is due to hydrocarbon elimination processes which take place because the transition metals possess *d* orbitals.[1–3, 5–9, 79–84]

Most commonly, $\beta$ elimination occurs leading to the formation of a hydrido complex and an olefin. During decomposition reactions of platinum complexes of the type $PtRR'(PR_3'')_2$ (R, R' = Et, $Pr^n$, $Pr^i$, $Bu^n$, $Bu^i$, and $n-C_5H_{11}$), $\beta$ elimination occurs first to give an olefin followed by subsequent reductive elimination of RH.[91]

$$[Cu(CH_2CD_2Et)(PBu_3)] \longrightarrow [CuD(PBu_3)] + CH_2=CDEt \qquad (4.2)$$

$$[PtBu_2(PPh_3)_2] \xrightarrow{-PPh_3} [Pt(Bu)(H)(C_4H_8)(PPh_3)] \longrightarrow n\text{-}C_4H_8 + n\text{-}C_4H_{10} + [Pt(PPh_3)_2] \qquad (4.3)$$

*Table 4.4. Estimated Values of Enthalpies for Some Addition and Insertion Reactions[a]*

| Reaction | $\Delta H^\circ$ (kJ mol$^{-1}$) |
|---|---|
| $MR + CH_2=CH_2 \longrightarrow MCH_2CH_2R$ | $-100$ |
| $M + H_2 \longrightarrow M(H)_2$ | $-50$ to $-70$ |
| $MH + CH_2CH_2 \longrightarrow MCH_2CH_3$ | $-30$ to $-70$ |
| $MR + CO \longrightarrow MC(O)R$ | $-20$ to $-60$ |
| $M + RCHO \longrightarrow M(CR)H$ (O double bond) | $0$ |
| $M + RH \longrightarrow M(R)H$ | $+20$ to $+70$ |
| $MH + CO \longrightarrow MCH$ (O double bond) | $+30$ to $+70$ |
| $M + R_2 \longrightarrow M(R)_2$ | $+50$ to $+110$ |

[a] R = alkyl, M = first-row transition metal; according to References 7, 76, 99–101.

These reactions are thermally activated, and therefore the generated $M-H$ bonds are thermodynamically more stable than $M-C$ bonds. Low-temperature photolytic decomposition of such complexes as $M(C_5R'_5)R(CO)_3$ ($M = Mo$, $W$, $R' = H$, $Me$, $R = Et$, $Pr^n$, $Pr^i$, $Bu^n$, and $n\text{-}C_5H_{11}$) affords $16e$ compounds $M(C_5R'_5)R(CO)_2$ and $[M(C_5R'_5)H(\text{olefin})(CO)_2]$.[83, 84]

Quite often $\alpha$ elimination processes also take place in which carbenes are formed. They may remain in the coordination sphere of the metal.

$$[M(CHR_2)L_n] \longrightarrow [MHL_n] + CR_2 \tag{4.4}$$

The $\alpha$ elimination reaction is well documented for the main group metals. Nevertheless, $\alpha$ elimination also explains decomposition of many of the transition metal complexes, particularly methyl complexes:

$$[TiMe_4] \longrightarrow [Me_2TiCH_2] + CH_4 \tag{4.5}$$

The $\alpha$ elimination process occurs in the reaction of $[Wcp_2Me(C_2H_4)]^+$ with $PMe_2Ph$ and the following cation is formed: $[Wcp_2(H)(CH_2PMe_2Ph)]^+$. The formation of this product is explained by proposing the existence of an equilibrium between the $16e$ methyl complex and the $18e$ hydrido methylene compound [scheme (4.6)].

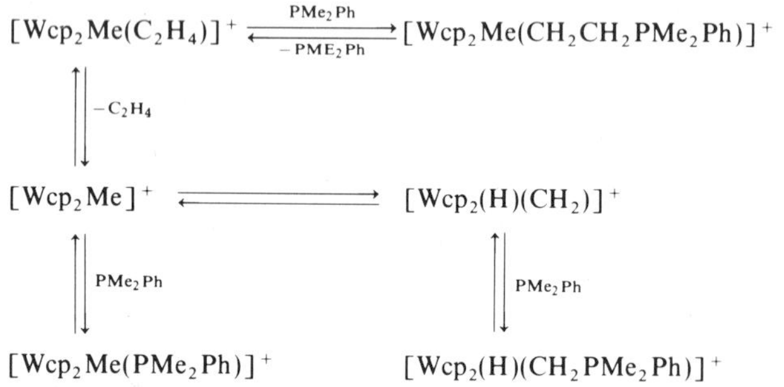

$$ \tag{4.6}$$

In the case of photolysis of $Crcp(Me)(CO)_3$ in $Ar$, $CH_4$, $N_2$, and $CO$ matrices, hydrogen abstraction occurs leading to the formation of the hydrido-carbene compound $Crcp(CH_2)(H)(CO)_2$. This reaction is reversible.[82]

$$L_nCr-CH_3 \rightleftharpoons L_nCr(CH_2)H \tag{4.7}$$

For compounds containing at least two $1e$ ligands, reductive elimination is possible (Table 1.4):

$$L_nM \overset{\diagup X}{\underset{\diagdown Y}{}} \longrightarrow ML_n + X-Y \tag{4.8}$$

Typical examples involve eliminations of $R-R$ and $R-H$ compounds for complexes possessing two alkyl groups or one alkyl group and the hydrogen ligand in the *cis* position:

$$[PtMe_3I(PMe_2Ph)_2] \longrightarrow [PtMeI(PMe_2Ph)_2] + C_2H_6 \tag{4.9}$$

$$[NiR_2(bipy)] + \text{olefin} \longrightarrow [NiR_2(\text{olefin})(bipy)] \longrightarrow [Ni(\text{olefin})(bipy)] + R_2 \tag{4.10}$$

There are two possible mechanisms for such reactions: (1) homolytic $M-R$ bond breaking and (2) concerted intramolecular *cis* elimination.

In contrast to $\alpha$ and $\beta$ elimination reactions, it is not necessary to have free coordination sites in reductive elimination reactions, although in some Pt(IV) six-coordinate complexes and Au(III) square compounds, there is dissociation of a neutral ligand prior to reductive elimination, and this step determines the rate of the entire process.[1, 2, 10] In alkyl–hydrido *cis* complexes, reductive elimination may occur by intramolecular or by intermolecular processes. In Pt(II) and Rh(III) complexes[1, 2, 10, 79, 80] such as $[cis\text{-}Pt(R)HL_2]$ and $mer\text{-}[Rh(R)Cl(H)L_3]$, reductive elimination occurs via an intramolecular path; however, in some osmium compounds,[8] such as $cis\text{-}Os(Me)H(CO)_4$, reductive elimination is intermolecular. Usually reductive elimination proceeds most easily for hydrido complexes, less easily for hydrido–alkyl compounds, and least easily for dialkyl derivatives. Methane elimination from $PtMe(H)(PPh_3)_2$ takes place rapidly at 298 K, while the complex $PtMe_2(PPh_3)_2$ is stable; it decomposes at 410 K. These observations were confirmed by calculations.[89, 90] Calculated activation energies for elimination reactions of $H_2$, RH, and $R_2$ from $PdH_2$, PdHR, and $PdR_2$, respectively, are lowest for $H_2$ and highest for $R_2$.[90]

Decomposition of metallacyclopentane compounds may occur either via reductive cycloelimination (reductive decoupling) (Table 1.4) with ethylene evolution or because of reductive elimination leading to the formation of cyclobutane.

$$cp_2Ti \overset{CH_2-CH_2}{\underset{CH_2-CH_2}{\big|}} \longrightarrow Ticp_2 + 2CH_2{=}CH_2 \tag{4.11}$$

$$cis\text{-}(Ph_3P)_2Ni \longrightarrow Ni(PPh_3)_2 + \square \tag{4.12}$$

Reductive elimination of cyclobutane is allowed by symmetry rules for *cis* square-planar complexes, while it is forbidden for *trans* compounds.[169] Complexes such as $L_3Ni$, e.g., $(Ph_3P)_3Ni$, decompose to give ethylene because dissociation of the ligand L makes it possible to form a four-coordinate complex possessing the *trans* structure:

$$L_3Ni \longrightarrow \overset{L}{\underset{L}{Ni}} \longrightarrow NiL_2 + 2C_2H_4 \tag{4.13}$$

These conclusions are in accord with experimental data for many nickel and iron complexes. The evolution of ethylene also takes place in the reaction of $WCl_6$ with $Li(CH_2)_4Li$.

Decomposition of compounds containing $1e$ ligands may also occur by homolytic cleavage of the $M-C$ bond. However, this type of decomposition is not commonly encountered because it is characterized by higher energy of activation compared to other concurrent processes. Homolytic cleavage is dominant for decomposition reactions of the following compounds: $[PtcpMe_3]$, $[NicpMe(PR_3)]$, $[PtMeI(bipy)]$, $[Pt(\sigma\text{-allyl}) Me_2Br(PMe_2Ph)_2]$, $[MnCH_2Ph(CO)_5]$, and $[Cu(CH_2CMe_2Ph)(PBu_3)]$. Relatively common are decomposition processes of alkyl compounds (or the like) occurring due to the formation of dinuclear or polynuclear (rarely) complexes. The formation of dinuclear compounds during elimination reactions involving a change in the oxidation state of the metal is assumed for decomposition of many transition metal complexes containing $1e$ ligands in such cases which exclude formation of considerable quantities of free radicals. These alkyl complexes are of the following metals: Ti(IV), Cr(III), Mn(II), Fe(III), Cu(I), Ag(I) and Au(I).

$$[AuMe(PPh_3)] \rightleftharpoons AuMe + PPh_3 \tag{4.14}$$

$$AuMe + AuMe(PPh_3) \longrightarrow [Au_2Me_2(PPh_3)] \longrightarrow C_2H_6 + 2Au + PPh_3 \tag{4.15}$$

Alkyl groups may be eliminated as dimers or as disproportionation products if they contain hydrogen atoms in the $\beta$ position. The $\beta$ elimination process affords the metal hydride, which causes hydrocarbon elimination in the protonolysis reaction of MR:

$$MCH_2CH_2R \longrightarrow MH + RCH=CH_2 \tag{4.16}$$

$$MH + MCH_2CH_2R \longrightarrow 2M + RCH_2CH_3 \tag{4.17}$$

Elimination reactions involving dinuclear intermediate compounds (without evolution of the free radicals) may also occur without changes in the oxidation state of the metal. Tetramethyltitanium decomposes according to this mechanism. During decomposition, methane is given off in the amount of about 3 moles per mole of $TiMe_4$, leaving a black diamagnetic precipitate which, after hydrolysis, gives mainly methane and traces of higher hydrocarbons. Studies involving D and $^{13}C$ labeled compounds showed that the products do not contain hydrogen and carbon atoms derived from a solvent and that the precipitate contains compounds possessing the following bonds: $TiMe$, $Ti_2CH_2$, $Ti_3CH$, and $Ti_4C$.

In general, the mechanism of the decomposition may be represented as follows:

$$LMCH_2R \longrightarrow \begin{matrix} LM-CHR-H \\ | \quad\quad | \\ LM\text{------}CH_2R \end{matrix} \longrightarrow (LM)_2CHR + RCH_3 \tag{4.18}$$

This mechanism also accounts for decomposition of other transition metal compounds.[8]

$$[OsR(H)(CO)_4] \longrightarrow \left[ Os\left(C\overset{\displaystyle O}{\underset{}{\diagup\!\!\diagup}}\!\!-R\right)(H)(CO)_3 \right] \tag{4.19}$$

$$[Os(COR)(H)(CO)_3] + [OsR(H)(CO)_4] \longrightarrow (OC)_4(H)Os-OsR(CO)_4 + RH \tag{4.20}$$

Frequently, decomposition of organometallic coordination compounds takes place via a reaction of a $\sigma$-hydrocarbyl group with another ligand; both ligands are bonded to the same metal atom. Such a reaction may also occur because of coordination of new ligands:

$$[TiMe_2(\eta\text{-}C_5Me_5)_2] \longrightarrow CH_4 + [TiMe(\eta\text{-}C_5Me_5)(C_5Me_4CH_2)] \tag{4.21}$$

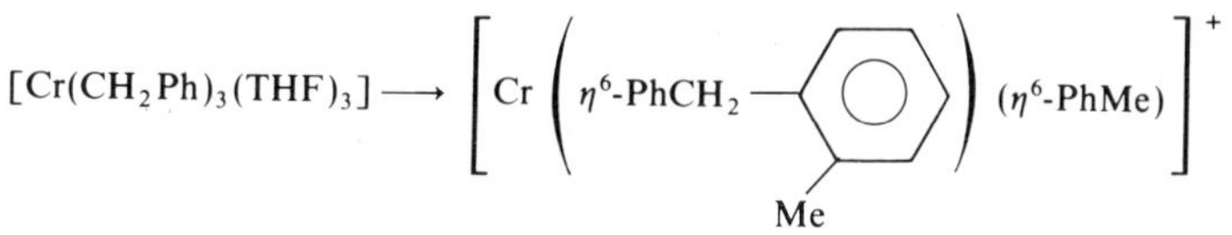

$$[Cr(CH_2Ph)_3(THF)_3] \longrightarrow \left[ Cr\left( \eta^6\text{-}PhCH_2 - \bigcirc\text{-Me} \right)(\eta^6\text{-}PhMe) \right]^+ \tag{4.22}$$

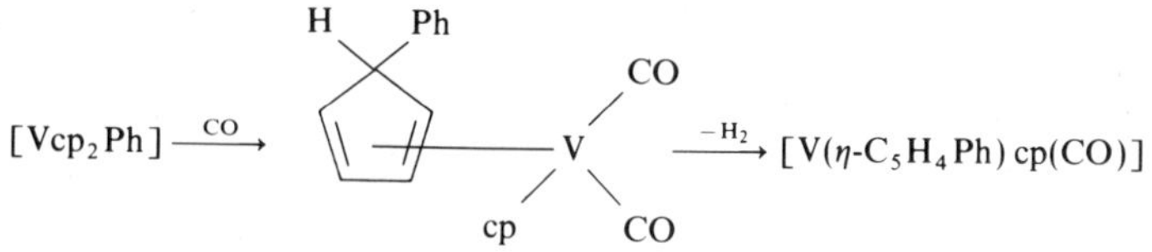

$$[Vcp_2Ph] \xrightarrow{CO} \text{(structure)} \xrightarrow{-H_2} [V(\eta\text{-}C_5H_4Ph)\,cp(CO)] \tag{4.23}$$

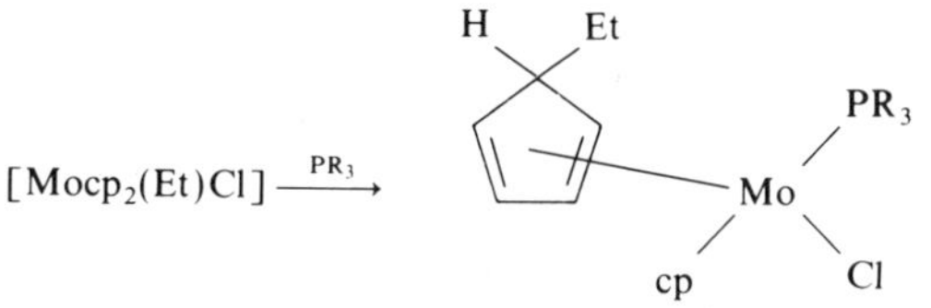

$$[Mocp_2(Et)Cl] \xrightarrow{PR_3} \text{(structure)} \tag{4.24}$$

$$3[Ticp_2(C_6D_5)_2] \begin{cases} 2[Ticp_2(C_6D_4)] + 2C_6D_6 \\ 2Ti(C_{10}H_8) + 2C_6D_4H_2 \\ \quad Ti(C_{10}H_8) + C_6D_5H \\ [Ticp(C_5H_4)(C_6D_5)] + C_6D_5H \end{cases} \tag{4.25}$$

In the decomposition reaction (4.25) the hydrogen atom is given off via a dinuclear intermediate complex [reaction (4.18)].

The above-mentioned data show that decomposition of complexes containing $\sigma$ M$-$C bonds occurs readily if such complexes have free coordination sites. The instability of compounds increases strongly if ligands possess hydrogen atoms in the $\beta$ position ($\beta$ elimination takes place readily). The stability of complexes with alkyl and aryl ligands decreases usually in the following order:

$$1\text{-norbornyl} > PhCH_2 > Me_3SiCH_2 \sim Me_3CCH_2 > Ph > Me \gg Et > sec\text{- and } t\text{-alkyls} \tag{4.26}$$

In comparison with compounds containing unidentate alkyl and aryl ligands, complexes containing fluoroalkyl or fluoroaryl groups as well as chelating ligands are more stable due to higher electronegativity of the ligands or owing to chelate effect.

$$C_6F_5 > Ph$$

$$\text{chelating ligands} > \text{nonchelating } 1e \text{ ligands} \tag{4.27}$$

The formation of anionic complexes leads to the increase of stability. However, it is not known to what degree this corresponds to the increase of the negative charge and to the blocking of the available coordination site (for example, $[TiMe_4]$ and $Li[TiMe_5]$). Because of the same reasons, coordination of the additional neutral ligands also causes increase of stability. For titanium complexes, this stability varies as follows:

$$[TiMe_4(Me_2PC_2H_4PMe_2)] > [TiMe_4(PMe_3)_2] > [TiMe_4] \qquad (4.28)$$

The electronic structure of the central metal ion also influences the stability of complexes containing $1e$ ligands. Exceptionally stable are octahedral alkyl complexes of Cr(III), Co(III), and Rh(III) containing amine and aqua ligands in which the central metal ion has $d^3$ and $d^6$ electron configuration. A change in the electron density distribution associated with the formation of the intermediate activated complex is facilitated (the activation energy is lower) if the direct product of ground-state wave functions and that of the lowest-lying excited state has the same symmetry as does the normal vibration of the molecule which is the basis for the dissociation reaction coordinate.

For octahedral complexes the energy of activation of the reaction involving homolytic breaking of the $M-C$ bond decreases depending upon the electronic structure in the following manner:

$$t_{2g}^6 > t_{2g}^3 > t_{2g}^1 \qquad (4.29)$$

The orbital symmetry of the ground and excited states shows that complexes possessing $t_{2g}^6$ and $t_{2g}^3$ electron configuration are inert toward dissociation as well as toward substitution reactions, while compounds with $t_{2g}^1$ configuration are labile in the substitution reactions.[9] Experimental data show that the most stable $\sigma$-hydrocarbyl complexes are those in which the metal atom has $d^0$, $d^3$, $d^6$ (low spin), $d^8$, and $d^{10}$ electron configuration, and therefore in such cases, where the electronic shell or subshell is completely filled, half-filled or empty. Heteroleptic compounds with $\sigma$ metal–carbon bonds are usually more stable if they contain other ligands which create strong fields. Such ligands include CO, cyclopentadienyl, and $CN^-$.

## 3. STRUCTURE OF COMPOUNDS CONTAINING METAL–CARBON σ BONDS

In general, M–C distances in transition metal complexes vary in the 190–240 pm range. Greater distances are encountered in the case of alkyl compounds of the lanthanides and the actinides. For a given metal the bond distance decreases according to the series:

$$M-C(sp^3) > M-C(sp^2) > M-C(sp) \qquad (4.30)$$

Coordination numbers change from 2 to 8. Complexes of elements of groups 4–7 possessing compositions $MR_4$, $MR_5$, and $MR_6$ have tetrahedral (often considerably distorted), trigonal–bipyramidal, and octahedral structures, respectively.

Many electron-deficient transition metal complexes such as chromium, manganese, nickel, and copper group compounds are characteristic in their ability to form polynuclear compounds containing alkyl bridges. This is also true for many compounds of other metals[85] (Figures 4.1 and 4.2, Table 4.5).

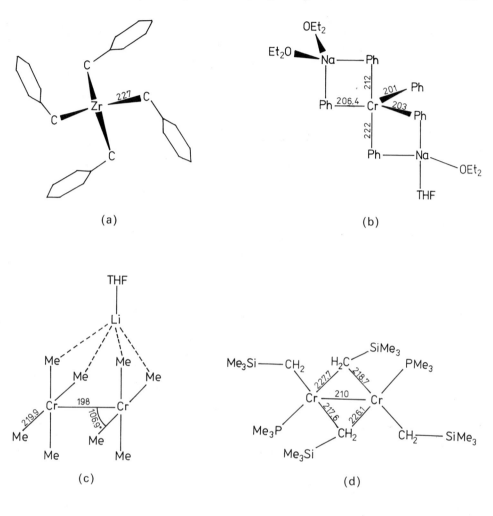

Figure 4.1. Structures of (a) $Zr(CH_2Ph)_4$, (b) $Na_2CrPh_5 \cdot 3ET_2O \cdot THF$, (c) $Li_4Cr_2Me_8 \cdot 4THF$, (d) $Cr_2(CH_2SiMe_3)_4(PMe_3)_2$, (e) $Ru_2(CH_2PMeC_2H_4PMe_2)_2H_2(dmpe)_2$ (distances are in pm units).

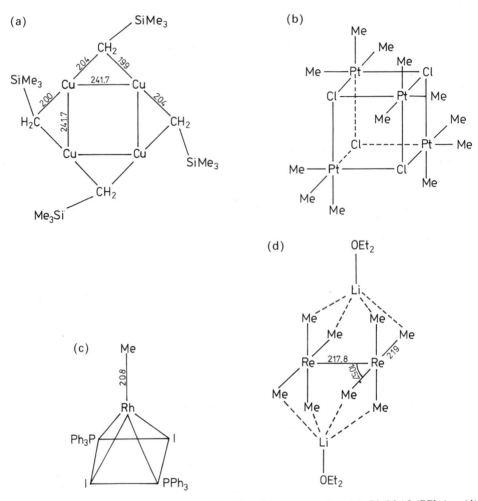

Figure 4.2. Structures of (a) $[Cu(CH_2SiMe_3)]_4$, (b) $[PtClMe_3]_4$, (c) $Rh(Me)I_2(PPh_3)_2$, (d) $Li_2Re_2Me_8 \cdot 2OEt_2$, (distances are in pm units).

Table 4.5. Structures of σ-Hydrocarbyl Complexes of Transition Metals

| | | Distance (pm) | |
|---|---|---|---|
| Compound | Coordination | M–C | M–X |
| $[Ticp(o\text{-}C_6H_4CH_2NMe_2)_2]^a$ | Square pyramid | 219.7 | 238.9 (Ti–N) |
| $cp_2Ti \begin{matrix} CPh{=}CPh^b \;\; etc. \\ | \\ CPh{=}CPh \end{matrix}$ | Tetrahedron | 217.2; 214.1 | |

(Table continued)

*Table 4.5. (continued)*

| Compound | Coordination | Distance (pm) M–C | Distance (pm) M–X |
|---|---|---|---|
| cp$_2$Hf with CPh=CPh[b] (bridging two CPh=CPh) | Tetrahedron | 222; 218 | |
| $[V(2,4,6-C_6H_2Me_3)_4]^{(33)}$ | Tetrahedron | 207.1–209.5 | |
| $Li_4\left[Cr\left(\begin{array}{c}CH_2{-}CH_2\\ \vert \\ CH_2{-}CH_2\end{array}\right)_2\right]\cdot 4Et_2O$[c] | | 223.8 | 197.5 (Cr–Cr) |
| $Li_4[Mo_2Me_8]\cdot 4THF$[d] | | 229 | 214.8 (Mo–Mo) |
| $[(OC)cpRh(\mu\text{-}CH_2)Rhcp(CO)]$[l] | | 202.9 | 183.2 (Rh–Co) 266.5 (Rh–Rh) |
| $Li[Lu(2,6-Me_2C_6H_3)_4]4THF$[e] | Tetrahedron | 245 | |
| $[Au_2Cl_2\{(CH_2)_2PEt_2\}_2]$[f] | Square | 194.2 | 235.9 (Au–Cl) |
| $[Mo_2(CH_2SiMe_3)_6]$[g] | Tetrahedron | 213.1 | 216.7 (Mo–Mo) |
| $[Re_2(\mu\text{-}CSiMe_3)_2(CH_2SiMe_3)_4]$[h] | | 193.2 (Re–$\mu$-C) 192.6 (Re–$\mu$-C) 209.2 (Re–C) 205.9 (Re–C) | 255.7 (Re–Re) |
| $[(Me_3P)_4Ru(\mu\text{-}CH_2)_2 Ru(\mu\text{-}CH_2)_2Ru(PMe_3)_4]$[i] | | 197.9 (Ru–C) 209.4 | 263.7 (Ru–Ru) 239.1–242.7 (Ru–P) |
| (structure with Me$_2$, Me$^2$—Re$^3$—Me$^3$, Me$_2$Re$^2$—Re$^1$Me$_2$, PhEt$_2$P, Me$^1$, PEt$_2$Ph)[j] | | 193.8 (Re$^1$C$^1$) 244.7 (Re$^1$C$^3$) 193.0 (Re$^2$C$^1$) 244.3 (Re$^2$C$^2$) 221.5 (Re$^3$C$^2$) 223.2 (Re$^3$C$^3$) 210.5–220.3 (Re$^{1-3}$–C$^{4-9}$) | 247.4 (Re$^1$–Re$^2$) 243.5 (Re$^1$–Re$^3$) 243.3 (Re$^2$–Re$^3$) |
| $[Tacp_2(\eta^2\text{-}C_6H_4)Me_2]$[k] | | 216.9 (Ta–Me) 218.1 (Ta–Me) | 205.9 / 209.1 Ta–C$_6$H$_4$ |

[a] L. E. Manzer, R. C. Gearharat, J. Guggenberger, and J. F. Whitney, *Chem. Commun.*, 942 (1976).
[b] J. L. Atwood, W. E. Hunter, H. Alt, and M. D. Rausch, *J. Am. Chem. Soc.*, **98**, 2454 (1976).
[c] J. Krausse and G. Schoedl, *J. Organomet. Chem.*, **27**, 59 (1971).
[d] F. A. Cotton, J. M. Troup, T. R. Webb, D. H. Williamson, and G. Wilkinson, *J. Am. Chem. Soc.*, **96**, 3824 (1974).
[e] S. A. Cotton, F. A. Hart, M. B. Hursthouse, and A. J. Welch, *Chem. Commun.*, 1225 (1972).
[f] H. Schmidbaur, J. R. Mandl, A. Frank, and G. Huttner, *Chem. Ber.*, **109**, 466 (1976).
[g] F. Huq, W. Mowat, A. Shortland, A. C. Skapski, and G. Wilkinson, *Chem. Commun.*, 1079 (1971).
[h] Bochman, G. Wilkinson, A. M. R. Galas, M. B. Hursthouse, and K. M. A. Malik, *J. Chem. Soc., Dalton Trans.*, 1797 (1980).
[i] R. A. Jones, G. Wilkinson, A. M. R. Galas, M. B. Hursthouse, and K. M. A. Malik, *J. Chem. Soc., Dalton Trans.*, 1771 (1980).
[j] P. Edwards, K. Mertis, G. Wilkinson, M. B. Hursthouse, and K. M. A. Malik, *J. Chem. Soc., Dalton Trans.*, 334 (1980).
[k] M. R. Churchill and W. J. Youngs, *Inorg. Chem.*, **18**, 1697 (1979).
[l] W. A. Herrmann, C. Kruger, R. Goddard, and I. Bernal, *Angew. Chem., Int. Ed. Engl.*, **16**, 334 (1977).

## 4. IR AND RAMAN SPECTRA[23]

The IR spectra of methyl complexes exhibit the types of bands shown in display (4.31). Other alkyl compounds show characteristic bands for both $CH_3$ and $CH_2$ groups.

$$
\begin{array}{lll}
\text{I} & 2900\text{–}3000 \text{ cm}^{-1} \text{ (w)} & \nu(CH) \\
\text{II} & \sim 1400 \text{ cm}^{-1} \quad \text{ (w)} & \delta_d(CH_3) \\
\text{III} & 1300\text{–}1100 \text{ cm}^{-1} \text{ (s)} & \delta_s(CH_3) \\
\text{IV} & 950\text{–}700 \text{ cm}^{-1} \quad \text{ (s)} & \rho_r(CH_3) \\
\text{V} & 600\text{–}300 \text{ cm}^{-1} \quad \text{ (m)} & \nu(M\text{–}C) \\
\text{VI} & 300\text{–}100 \text{ cm}^{-1} \quad \text{ (m–w)} & \delta(CMC)
\end{array}
\qquad (4.31)
$$

Table 4.6. Frequencies $\nu(M\text{–}C)$ for $\sigma$-Hydrocarbyl Complexes of Transition Metals[23]

| Compound | $\nu$ (M–C) (cm$^{-1}$) |
|---|---|
| cis-[PtMe$_2$(PMe$_3$)$_2$] | 525 (s), 508 (s) |
| cis-[PtEt$_2$(PMe$_3$)$_2$] | 511 (w), 500 (w) |
| cis-[Pt(Pr$^n$)$_2$(PMe$_3$)$_2$] | 609 (w), 601 (w) |
| cis-[PtEt$_2$(PEt$_3$)$_2$] | 516 (w), 496 (w) |
| cis-[PtMe$_2$(PEt$_3$)$_2$] | 526 (m), 506 (m) |
| trans-[PtMe(NO)(PMe$_3$)$_2$] | 567 (w) |
| trans-[PtMeCl(PMe$_3$)$_2$] | 551 (m) |
| trans-[PtMeBr(PMe$_3$)$_2$] | 546 (m) |
| trans-[PtMeI(PMe$_3$)$_2$] | 538 (m) |
| [Pt(acac)Cl{CH(COMe)$_2$}] | 568 (m) |
| [Pt(acac){CH(COMe)$_2$}$_2$] | 565 (m), 542 (m) |
| Li$_2$[PtMe$_4$]$^a$ | 511 (pol), 503 (dp) |
| Li[PtMe$_3$(PPh$_3$)]$^a$ | 538, 515 |
| Li$_2$[PtMe$_6$]$^a$ | 508 (pol), 505 (dp), 478 (vs), (IR) |
| [PtMe$_4$(PMe$_3$)$_2$]$^a$ | 529 (R), 511 (R), 525 (s), (IR), 475 (s), (IR) |
| [Pt(OH)Me$_3$]$_4$$^b$ | 590 (w), 570 (vw), (sh) |
| WMe$_6$$^c$ | 482 |
| [ReOMe$_4$][39] | 552, 529 |
| [ReMe$_6$]$^d$ | 500 |
| [ReO(CH$_2$SiMe$_3$)$_4$][39] | 500 |
| [Re$_2$O$_3$(CH$_2$SiMe$_3$)$_4$][39] | 540, 515 |
| TiMe$_4$$^e$* | 577, 489 |
| Li[AuMe$_4$]$^f$** | 530 (pol), 522 (dp), 484 (dp) |
| Li[AuMe$_2$]$^f$*** | 526, 490 |

* $\delta$ (CtiC) 180.
** $\delta_s$ (Me) 1212 (pol), 1176 (dp).
*** $\delta_s$ (Me) 1173.

$^a$ G. W. Rice and R. S. Tobias, *J. Am. Chem. Soc.*, **99**, 2141 (1977).
$^b$ G. L. Morgan, R. D. Rennick, and C. C. Soong, *Inorg. Chem.*, **5**, 372 (1966).
$^c$ A. J. Shortland and G. Wilkinson, *J. Chem. Soc., Dalton Trans.*, 872 (1973).
$^d$ K. Mertis and G. Wilkinson, *J. Chem. Soc., Dalton Trans.*, 1488 (1976).
$^e$ H. H. Eysel, H. Siebert, G. Groh, and H. J. Berthold, *Spectrochim. Acta*, **26A**, 1595 (1970).
$^f$ G. W. Rice and R. S. Tobias, *Inorg. Chem.*, **15**, 489 (1976); G. W. Rice and R. S. Tobias, *Inorg. Chem.*, **14**, 2402 (1975).

The bands in display (4.32) encountered in ethyl complexes.

| I | 3000–2900 cm$^{-1}$ (m–s) | $v$(CH) |
|---|---|---|
| II | ~1460 cm$^{-1}$ (m) | $\delta$(CH$_2$) or $\delta_d$(CH$_3$) |
| III | ~1410 cm$^{-1}$ (m) | $\delta_d$(CH$_3$) or $\delta$(CH$_2$) |
| IV | 1385–1370 cm$^{-1}$ (w) | $\delta_s$(CH$_3$) |
| V | 1300–1200 cm$^{-1}$ (m) | $\delta_t$(CH$_2$) or $\rho_w$(CH$_2$) |
| VI | 1100–950 cm$^{-1}$ (w) | $v$(CC) |
| VII | 950–600 cm$^{-1}$ (s) | $\rho_r$(CH$_3$) |
| VIII | 600–400 cm$^{-1}$ (w) | $v$(M–C) |
| IX | 300–100 cm$^{-1}$ (m–w) | $\delta$(CMC) |

(4.32)

The $v$(M–C) bands for the other $\sigma$ metal–carbon compounds are found in the same range as those for the alkyl complexes. Usually, the intensity of such bands is low with the exception of methyl complexes, for which bands of high intensities are often observed. Platinum compounds furnished the largest amount of data. For *trans*-[PtXMe(PR$_3$)$_2$] and *cis*-[PtMe$_2$(PR$_3$)$_2$], $v$(Pt–C) and $\delta$(Me) increase with increase in the electronegativity of the ligand located in the *trans* position. The $v$(Pt–C) decreases with increase in the *trans* effect of the ligand. The frequencies $v$(M–C) for $\sigma$-hydrocarbyl complexes of the transition metals are given in Table 4.6. Alkyl ligands rotate about the M–C axis. The energy barrier of this process is very low. In the case of methyl complexes it is possible to determine the C–H bond distance as well as its $D_{298}^{\circ}$ energy from IR spectra in the range of CH stretching vibrations.[93] The CH bonds in MnMe(CO)$_5$ are somewhat stronger than in ReMe(CO)$_5$. Conversely, the Re–Me bond strength is greater than the Mn–Me one.[93]

## 5. NMR SPECTRA[1, 2, 5, 13, 24, 25]

The chemical shifts ($\tau$) of methyl group protons coordinated to the transition metals are encountered in the 7–12.5 range. In ethyl complexes the $\tau$ values of the methyl groups are sometimes lower than those of the methylene units. In titanium compounds TiMeXcp$_2$ (X = Cl, Br, and I) the $\tau$(Me) values increase in the series Cl < Br < I.

A reverse sequence was found for Pt(IV) complexes of the type [PtMe$_3$X]$_4$ where X = halogen. The coupling constants $^2J$(Pt–H$_{Me}$) assume values in the 40–90 Hz range. They decrease considerably from Pt(II) to Pt(IV) compounds. This is attributed to the decrease of $s$ character of the Pt–C bonds with a change in the hybridization of Pt(II) complexes from $dsp^2$ to $d^2sp^3$ in Pt(IV) compounds. Exceptions are Li$_2$[PtMe$_4$] and Li$_2$[PtMe$_6$] for which the $^2J$(Pt–H) coupling constants are almost equal and have values of 43.5 and 40.0 Hz, respectively. For square-planar complexes such as *trans*-PtXRL$_2$, the $^2J$(Pt–H) coupling constant is smaller the stronger the *trans* effect of the ligand X. In the case of alkyl rhodium compounds, the coupling constant $^2J$(Rh–H) is considerably smaller (2–3 Hz) because of, among others, the smaller nuclear moment of $^{103}$Rh.

The $^{13}$C chemical shifts of methyl groups of the transition metal complexes vary from $-30$ to $+20$ ppm (Table 4.7). The coupling constant $^1J(Pt-C_{Me})$ for square *trans*-[PtMeXL$_2$] complexes is in the 350–700 Hz range. This coupling constant decreases for compounds containing ligands which cause a stronger *trans* effect.

For *cis*-[PtMe$_2$L$_2$] complexes, coupling constants $^1J(Pt-C)$ are higher, reaching values of 590–820 Hz. The $^{13}$C chemical shifts and coupling constants $^1J(Pt-C_{Me})$ for methyl derivatives of Pt(IV) are similar to those of Pt(II) complexes possessing *trans* structures. Exceptionally larger shifts are observed for methylene carbon atoms in some trimethylsilylmethyl complexes as well as in compounds containing the (CH$_2$)$_2$SiMe$_2$ group.

*Table 4.7.* $^{13}$C NMR Spectra for Metal σ-Carbyl Complexes

| Compound | Shift (ppm) (assignment) and $J$(Hz) |
|---|---|
| [MoMe$_2$(PhH)(PMe$_3$)$_2$][a] | $-8.6$ (Me–Mo), $J$(PC) = 24.0 |
| [MoMe$_2$(PhMe)(PMe$_3$)$_2$][a] | $-8.2$ (Me–Mo), $J$(PC) = 23.3 |
| [MoMe$_2$(p-C$_6$H$_4$Me$_2$)(PMe$_2$Ph)$_2$][a] | $-4.6$ (Me–Mo), $J$(PC) = 22.0 |
| [MoMe$_2$(PhH)(PMe$_2$Ph)$_2$][a] | $-7.0$ (Me–Mo), $J$(PC) = 22.5 |
| [MoMe$_2$(p-C$_6$H$_4$Me$_2$)(PMe$_3$)$_2$][a] | $-7.5$ (Me–Mo), $J$(PC) = 23.0 |
| [Mo$_2$Me$_4$(PMe$_3$)$_4$][b] | 2.69 (Me–Mo), $J$(CP) = 6.0 |
| [Mo$_2$(CH$_2$SiMe$_3$)$_2$(OAc)$_2$(PMe$_3$)$_2$][b] | 9.05 (CH$_2$–Mo), $J$(PC) = 29.2 |
| [(Me$_3$SiCH$_2$)$_2$Mo(μ-CH$_2$SiMe$_2$CH$_2$)Mo(PMe$_3$)$_3$][b] | 43.81 (CH$_2$–Mo), 73.62 (μ-CH$_2$–Mo) |
| [Mo$_2${(CH$_2$)$_2$SiMe$_2$}(CH$_2$SiMe$_3$)$_2${P(OMe)$_3$}$_3$][b] | 73.19 (CH$_2$–Mo), 79.87 {(CH$_2$)$_2$SiMe$_2$–Mo} |
| *cis*-[RuMe$_2$(PMe$_3$)$_4$][b] | $-3.38$ (Me–Ru), $J$(CP) = 63.8; 12.6 |
| [Ru{(CH$_2$)$_2$SiMe$_2$}(PMe$_3$)$_4$][b] | $-28.98$ {(CH$_2$)$_2$SiMe$_2$–Ru}, $J$(PC) = 46.0; 7.7 |
| *fac*-[RhMe$_3$(PMe$_3$)$_3$][b] | 7.64 (Me–Rh), $J$(PC) = 100.0 |
| *fac*-[Rh{(CH$_2$)$_2$SiMe$_2$}(CH$_2$SiMe$_3$)(PMe$_3$)$_3$][b] | $-16.12$ {(CH$_2$)$_2$SiMe$_2$ + CH$_2$SiMe$_3$Rh}, $J$(PC) = 72.7 |
| [Mo$_2$Ph$_3$(OAc)(PMe$_3$)$_3$][c] | 124.68–127.29 (Ph); 141.33–142.88 (Ph) |
| [Mo$_2$(C$_6$H$_4$F)$_3$(OAc)(PMe$_3$)$_3$][c] | 112.62–114.18 and 142.18–143.47 (C$_6$H$_4$F) |
| [Re$_2$(2-C$_6$H$_4$OMe)$_6$][c] | 108.41; 111.44; 119.83; 120.52; 127.21; 128.46; 130.56; 141.5 (C$_6$H$_4$) |
| [Re$_3$Me$_9$(PEt$_2$Ph)$_3$][d] | 9.46 (Me–Re); $-11.83$ (Re–Me–Re) |
| [Re$_2$Me$_6$(PMe$_2$Ph)$_2$][d] | 12.4 (Me–Re) |
| [MnMe(CO)$_5$] | $-9.4$ |
| [WMecp(CO)$_3$] | $-28.9$ |
| [FeMecp(CO)$_2$] | $-23.5$ |
| [IrMe$_3$(PEt$_3$)$_3$] | $-8.9$ |
| [IrMe(Cl)I(CO)(AsEt$_2$Ph)$_2$] | $-13.8$ |
| *cis*-[PtMe$_2$(COD)] | 4.7 |
| *cis*-[PtMe$_2$(AsMe$_3$)$_2$] | $-4.1$ |
| *trans*-[PtMeCl(AsMe$_3$)$_2$] | $-28.4$ |
| *fac*-[PtMe$_3$Cl(AsMe$_2$Ph)$_2$] | 8.1 (Me *trans* to As), $-9.4$ |
| *fac*-[PtMe$_3$I(AsMe$_3$)$_2$] | 2.0 (Me *trans* to As), $-0.8$ |

(25)

[a] See Ref. *o*, Table 4.15.
[b] See Ref. *k*, Table 4.15.
[c] See Ref. *r*, Table 4.15.
[d] See Ref. *p*, Table 4.16.

## 6. *ESR SPECTRA*

ESR spectra of many $\sigma$ metal–carbon bonded complexes have been investigated (Table 4.8). Numerous four-coordinate $MR_4$-type complexes have symmetry lower than tetrahedral, for example, $[Cr(CH_2SiMe_3)_4]^-$, $[V(CH_2SiMe_3)_4]$, $[Cr(CH_2SiMe_3)_4]$, etc. The compound $ReMe_6$ has a structure of a distorted octahedron because of the Jahn–Teller effect. In contrast to $ReF_6$, the distortion of $ReMe_6$ is not dynamic. This situation may be due to a considerably larger $10Dq$ value for the methyl complex compared to the fluoro derivative. Based on ESR spectra, a square-antiprism structure was proposed for $[ReMe_8]^{2-}$.

## 7. *PHOTOELECTRON SPECTRA*[26-32, 92]

Some alkyl complexes have been studied by photoelectron spectra. These include tetrahedral $MR_4$, octahedral $MR_6$, $TaMe_5$, and selected alkyl heteroleptic compounds such as $[MnMe(CO)_5]$, $[ReMe(CO)_5]$, $[MMe_2cp_2]$, $[MMecp(CO)_3]$ (M = Mo, W), $[MMecp(CO)_2]$ (M = Fe, Ru), $[Fe(CH_2CN) cp(CO)_2]$, $[ReOMe_4]$, $[ReO(CH_2SiMe_3)_4]$. The photoelectron spectra of homoleptic complexes exhibit bands arising from ionizations involving the following orbitals: (1) $d$ of the central ion, and (2) $\sigma M - C$ and $\sigma C - H$.

In the $T_d$ point group (for $MMe_4$) four $\sigma M - C$ bonds have symmetry $a_1$ and $t_2$, while twelve $\sigma C - H$ bonds have the following irreducible representation: $a_1 + e + t_1 + 2 \cdot t_2$ (Figure 4.3).

*Table 4.8. ESR Spectra of $\sigma$-Carbyl Complexes*

| Compound | ESR parameters |
|---|---|
| $[V(adme)_4]^a$ | $g = 1.985 \|A_{izo}\| \sim 5.0$ mT (298 K, in toluene) |
| $[V(adme)_2(OBu^t)_2]^a$ | $g = 1.994 \|A_{izo}\| = 5.1$ mT (298 K) |
| $[V(CH_2SiMe_3)_4]^{(37)}$ | $g = 1.968$ |
| $[Cr(adme)_4]^a$ | $g = 2.004$ (298 K), $g = 1.996; 4.056$ (120 K) |
| $[Cr(CH_2SiMe_3)_4]^{-(37)}$ | $g_{\|\|} = 2.0 \pm 0.05$; $g_\perp = 3.9 \pm 0.1$ |
| $[Cr(CH_2SiMe_3)_4]^{(37,38)}$ | $g = 1.993$ |
| $[Mn(adme)_2]^a$ | $g = 1.97$ (298 K) |
| $[ReOMe_4]^{(36)}$ | $g_x = 1.958$, $g_y = 1.936$, $g_z = 2.252$ |
| | $A_x = -0.0330$, $A_y = -0.0187$, $A_z = -0.0422$ |
| $[ReO(CH_2SiMe_3)_4]^{(36)}$ | $g_x = 1.951$, $g_y = 1.941$, $g_z = 2.218$ |
| | $A_x = -0.0355$, $A_y = -0.0191$, $A_z = -0.0416$ |
| $[ReMe_6]^{(35)}$ | $g_x = 2.122$, $g_y = 2.095$, $g_z = 2.155$ |
| | $A_x = +0.0217$, $A_y = -0.0328$, $A_z = -0.0428$ |
| $[ReMe_8]^{2-(35)}$ | $g_{\|\|} = 1.977$, $g_\perp = 2.109$ |
| | $A_{\|\|} = -0.0179$, $A_\perp = -0.0396$ |
| $[Nbcp_2Me_2]^b$ | $g = 1.9984$ |
| $[Nbcp_2Ph_2]^b$ | $g = 1.9982$ |
| $[Ticp_2Me_2]^{-b}$ | $g = 1.990$ |
| $[Tacp_2Ph_2]^b$ | $g = 1.988$ |

$^a$ See Ref. *a*, Table 4.16.
$^b$ I. H. Elson, J. K. Kochi, U. Klabunde, L. E. Manzer, G. W. Parshall, and F. N. Tebbe, *J. Am. Chem. Soc.*, **96**, 7374 (1974).

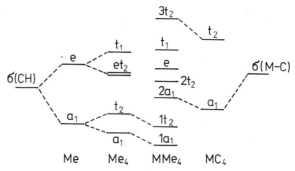

Figure 4.3. Qualitative MO scheme for MMe$_4$ complexes.

Table 4.9. *Photoelectron Spectra of Metal Alkyl Complexes, Ionization Energy (eV)*
*and Assignment of Bands*

| Compound | Orbitals | | | |
|---|---|---|---|---|
| | $d$ | $\sigma(M-C)$ | $\sigma(C-C)$ | $\sigma(C-H)$ |
| SiMe$_4$[32] | | 10.57; 15.85 | | 13.06; 14.08 |
| GeMe$_4$[32] | | 10.23; 15.9 | | 13.0 (sh), 13.85 |
| SnMe$_4$[32] | | 9.70 | | ~13.4 |
| PbMe$_4$[32] | | 8.81; 9.09; 9.86; 15.3 | | ~13.3 |
| [Ge(CH$_2$CMe$_3$)$_4$][30] | | 9.01 | | |
| [Sn(CH$_2$CMe$_3$)$_4$][30] | | 8.58 | | |
| [Sn(CH$_2$SiMe$_3$)$_4$][28] | | 8.71 | | |
| [Pb(CH$_2$SiMe$_3$)$_4$][28] | | 8.14; 8.86 | | |
| [Ti(CH$_2$CMe$_3$)$_4$][30] | | 8.33 | 11.35 | 12.59 |
| [Zr(CH$_2$CMe$_3$)$_4$][30] | | 8.33 | 11.28 | 12.50 |
| [Hf(CH$_2$CMe$_3$)$_4$][30] | | 8.51 | 11.40 | 12.54 |
| [Cr(CH$_2$CMe$_3$)$_4$][28,92] | 7.25 | 8.37 | 11.0 | 12.2 |
| [Cr(CH$_2$SiMe$_3$)$_4$][28,92] | 7.26 | 8.69 | 10.4 (Si—C) | 13.6 |
| TaMe$_5$[29,92] | | 8.83; 9.25; 10.10; 11.14 | | 12.07; 13.51 |
| WMe$_6$[29,92] | | 8.59; 9.33; 10.17 | | 11.55 (sh); 11.97; 13.14; 20.8 |
| ReMe$_6$[29,92] | 7.89 | 8.47; 9.77; 10.48 | | 12.2; 13.4 |
| ReOMe$_4$[29,92] | 8.86 | 9.5; 9.95 | | 13.0 |
| [ReO(CH$_2$SiMe$_3$)$_4$][29,92] | 8.0 | 8.46; 8.93 | 10.2 (Si—C) | 12.9 |
| [MnMe(CO)$_5$] | 8.65 (e); 9.12 ($b_2$) | 9.48 | | 12.6 |
| [ReMe(CO)$_5$] | 8.71; 8.93 (e); 9.51 ($b_2$) | 9.51 | | 12.8 |
| [MoMe$_2$cp$_2$] | 6.1 | 8.3; 9.6 | | |
| [WMe$_2$cp$_2$] | 6.0 | 8.3; 9.6 | | |
| [MoMecp(CO)$_3$] | 7.78 | 9.07 | 9.7; 10.0 (e$_1$cp) | |
| [WMecp(CO)$_3$] | 7.6; 7.77 | 9.26 | 9.92; 10.2 (e$_1$cp) | |
| [FeMecp(CO)$_2$] | 7.78; 8.53 | 9.15 | 9.90 (e$_1$cp) | |
| [Fe(CH$_2$CN)cp(CO)$_2$] | 8.29; 8.90; 9.48 | 11.14 | 10.25 (e$_1$cp) | |
| [RuMecp(CO)$_2$] | 8.13; 8.29; 8.96 | 9.48 | 9.98; 10.51 (e$_1$cp) | |

The block from [MnMe(CO)$_5$] through [RuMecp(CO)$_2$] is bracketed with reference (31,92).

The highest occupied $\sigma(M-C)$ orbitals lie in the 8–9 eV range. For tetrahedral transition metal complexes, the energy of these orbitals depends only slightly upon the transition metal itself, while for group 14 metals the ionization potential decreases as a result of the lowering of the electronegativity of the metal as the atomic number increases. For octahedral complexes, $MMe_6$, six $\sigma(M-C)$ orbitals transform as $a_{1g}+t_{1u}+e_g$. If the local symmetry for the $CH_3$ group is cylindrical, the eighteen $\sigma(C-H)$ orbitals have the following irreducible representation: $t_{1g}+t_{2g}+2t_{1u}+t_{2u}+a_{1g}+e_g$. As long as the interaction between neigboring bonding orbitals is stronger than the interaction between more distant pairs of bonding orbitals, the $\sigma(C-H)$ orbitals may be divided into two groups: (1) $\sigma(C_{2s}-H)$ orbitals having representations $a_{1g}$, $e_g$, and $t_{1u}$, and (2) $\sigma(C_{2p}-H)$ orbitals having representations $t_{1g}$, $t_{2g}$, $t_{2u}$, and $t_{1u}$. The photoelectron spectra of many alkyl and dialkylamide complexes (Table 4.9) show strong dependence of band intensities caused by ionization of the $M-C$ or $M-N$ orbitals on the character of metal orbitals. In general, the intensity of the $M-C$ and $M-N$ bands is rather weak compared to the intensity of bands arising from ionization of $C-H$ orbitals.[29] Therefore, it is not possible to assign the bands based on their intensities.

## 8. MAGNETIC PROPERTIES

Complexes containing metal–carbon $\sigma$ bonds have magnetic properties which are analogous to other compounds of $d$ or $f$ electron metals. Magnetically diluted com-

*Table 4.10. Magnetic Properties of Metal $\sigma$-Carbyl Complexes*

| Compound | $\mu_{eff}$ (BM) | Electronic structure of the metal |
|---|---|---|
| $[Ti(o\text{-}C_6H_4CH_2NMe_2)_2cp]$[68] | 1.7 | $d^1$ |
| $[Ticp_2Ph]$[69] | 1.58 | $d^1$ |
| $[Ticp_2(2,6\text{-}C_6H_3Me_2)]$[69] | 1.66 | $d^1$ |
| $[V(CH_2SiMe_3)_4]$[37] | 1.55 | $d^1$ |
| | (in benzene) | |
| $[V(2,4,6\text{-}C_6H_2Me_3)_4]$[33] | 1.65–1.80 | $d^1$ |
| $[V(norbornyl)_4]$[70] | 1.82 | $d^1$ |
| $[V(2,4,6\text{-}C_6H_2Me_3)_3]\cdot 1.25\,THF$[71] | 2.74 | $d^2$ |
| $[Cr(CH_2CMe_3)_4]$[38] | 2.7 | $d^2$ |
| $[Cr(CH_2CMe_2Ph)_4]$[38] | 2.8 | $d^2$ |
| $[Cr(CH_2SiMe_3)_4]$[38] | 2.9 | $d^2$ |
| $[Cr(norbornyl)_4]$[70,95] | 2.84 | $d^2$ |
| $[Mn(norbornyl)_4]$[70] | 3.78 | $d^3$ |
| $[Cr(CH_2(SiMe_3)_4]^-$[37] | 3.7 | $d^3$ |
| $[Co(1\text{-}norbornyl)_4]$[70] | 2.00 | $d^5$ |
| $[Fe(1\text{-}norbornyl)_4]$[70] | 0 | $d^4$ |
| $Li_2[CrPh_4]\cdot 4\,THF$[72] | 4.73 | $d^4$ |
| $Li_2[CrPh_4]\cdot 2\,THF\cdot 2\,bipy$[72] | 3.70 | $d^4$ |
| $[CrPh_2(bipy)_2]$[72] | 3.22 | $d^4$ |
| $Li_2(MnMe_4)\cdot 2\,TMED$[73] | 5.50 | $d^5$ |
| $Li_2[MnMe_4]$[73] | 2.2 | $d^5$ |
| $[Cr_2(CH_2SiMe_3)_4(PMe_3)_4]$[74] | 0.9 | $d^4$ |

pounds of $d$ electron elements have magnetic moments close to the "spin only" values, $\mu_{\text{eff}} = 2\sqrt{S(S+1)}$ BM, while complexes of the lanthanides and the actinides show magnetic moments resulting from the Russell–Saunders coupling: $\mu = g\sqrt{J(J+1)}$ BM. Dinuclear compounds of Cr(II), Mo(II), W(II), and Re(III) in which there are quadruple metal–metal bonds, are diamagnetic. Some polynuclear compounds (for example, those of Mn) have magnetic moments lowered, indicating direct or indirect exchange interaction through $\sigma$ bonded carbon ligands (Table 4.10). Other dinuclear complexes in which there are metal–metal bonds are also diamagnetic, for example, $Rh_2(Ph_2PC_6H_4)_2(OAc)_2$.

## 9. ELECTRONIC SPECTRA

In electronic spectra of complexes containing $\sigma$ bonded carbon ligands (Table 4.11), bands due to electron transfer from the ligand to the metal occur at low energies because electronegativity of carbon and ligands coordinated through carbon is low, and the charge transfer energy is proportional to the electronegativity difference between the ligand and the central metal atom: $X_L - X_M$. Therefore, for many complexes, par-

*Table 4.11. Electronic Spectra of Metal $\sigma$-Carbyl Complexes*

| Compound | Energy of the band (extinction coefficient) $(cm^{-1})$ $(\varepsilon)$ |
|---|---|
| $[Ticp(o\text{-}C_6H_4NMe_2)_2]^a$ | 14 200 (sh) (211), 17 400 (268) |
| $[V(CH_2SiMe_3)_4]^{(37)}$ | 15 625, 23 600 (310) |
| $[V(2,4,6\text{-}C_6H_2Me_3)_4]^{(33)}$ | 13 370 (254) $(^2E \to {}^2T_2)$, 18 350 (CT), 21 480 (CT), 23 870 (CT) |
| $[Cr(CH_2CMe_3)_4]^{(38)}$ | 18 500 $(^3B_1 \to {}^3A_2)$, 21 100 (1090) $(^3B_1 \to {}^3E)$ $(D_{2d}$ symmetry) |
| $[Cr(CH_2CMe_2Ph)_4]^{(38)}$ | 18 200, 20 500 (1380) |
| $[Cr(CH_2CPh_3)_4]^{(38)}$ | 17 600, 20 200 (1380) |
| $[CrMe_4]^{(38)}$ | 20 000, 22 200 $(\sim 600)$ |
| $[Cr(CH_2SiMe_3)_4]^{(38,37)}$ | 17 100, 19 400 (1060), $10Dq = 12\,500$ |
| $[Cr(adme)_4]^b$ | 20 000 (1000) $(^3B_1 \to {}^3E, {}^3A_2)$ $(D_{2d})$ |
| $[Cr(CH_2SiMe_3)_4]^{-(37)}$ | 7 800 (50), 16 250 (760), $10Dq = 8\,800$, $B = 700$ |
| $[ReOMe_4]^{c(34)}$ | 17 850, 35 700 |
| $[Co(CN)_5(CF_2CF_2H)]^{3-d}$ | 32 150 $(^1A_{1g} \to {}^1T_{1g})$ |
| $[Co(CN)_5Me]^{3-e}$ | 31 450 $(^1A_{1g} \to {}^1T_{1g})$ |
| $[(NC)_5CoCF_2CF_2Co(CN)_5]^{6-d}$ | 31 250 $(^1A_{1g} \to {}^1T_{1g})$, 36 250 $(^1A_{1g} \to {}^1T_{2g})$ |
| $[CoEt(salen)]^f$ | 15 100 (lg$\varepsilon = 3.13$), 21 600 (3.42), 24 800 (3.67), 29 200 (4.13) |
| $[RhMe(dmgH)_2(H_2O)]^g$ | 24 800 (4800) |
| $[Cr(norbornyl)_4]^{(95)}$ | 16 100 sh $(^3A_2 \to {}^3T_2)$, 18 100 sh and 18 700 sh $(^3A_2 \to {}^3E)$, 20 600 (1340) $(^3A_2 \to {}^3T_1)$, 29 900 sh CT, 34 000 sh CT, 37 700 (29 700) CT |

[a] See Ref. $a$, Table 4.5.
[b] See Ref. $a$, Table 4.16.
[c] See Refs. $k$ and $l$, Table 4.16.
[d] M. J. Mays and G. Wilkinson, *J. Chem. Soc.*, 6629 (1965).
[e] J. Halpern and J. P. Maher, *J. Am. Chem. Soc.*, **86**, 2311 (1964).
[f] See Ref. $e$, Table 4.18.
[g] See Ref. $n$, Table 4.18.

ticularly for those which are octahedral, the parts of bands due to $d$–$d$ transition are masked by charge transfer (CT) bands. The position of $\sigma$ bonded carbon ligands in the spectrochemical series is high for complexes of group 8–10 metals, while it is low for complexes of elements containing a smaller number of $d$ electrons such as titanium.

For complexes $[RuXCl(diphos)_2]$, the ligand field strength decreases in the following order: $CN > Ph > Me > H \gg Cl$. For ions $[CoX(CN)_5]^{3-}$ the field strength for $X = Me$ is almost the same as for $X = CN$. For $[CrX_6]^{3-}$ the following spectrochemical series is observed: $CN \gg NH_3 > Me > H_2O > F$, while for $[TiCl_2Xpy_3]$ it is $Cl > Ph > Me$. For tetrahedral Cr(IV) complexes containing $M-C$ $\sigma$ bonds such as $CrR_4$, the splitting parameter $10Dq$ decreases as follows: $Me \backsim$ 1-adamantylmethyl $> CH_2CMe_3 > CH_2CMe_2Ph > CH_2CPh_3 > CH_2SiMe_3$.

## 10. *PREPARATION OF COMPLEXES CONTAINING $\sigma$ BONDED CARBON LIGANDS*[1–5, 11–14, 85–88, 96–99]

### a. *Reactions of transition metal compounds with alkylating or arylating reagents*

This is the most commonly utilized method for synthesis of transition metal complexes with $M-C$ $\sigma$ bonds. The following substrates are utilized: halides, acetylacetonates, acetates, $d$-electron element alcoholates, as well as organometallic compounds of lithium, magnesium, aluminum, sodium, zinc, mercury, elements of group 14, etc. Ethers and hydrocarbons are used as solvents. In general, the synthesis should be carried out at lower temperatures and under inert atmosphere.

$$CrCl_3 \cdot 3THF + 3MgBrPh \longrightarrow CrPh_3 \cdot 3THF + 3MgBrCl \tag{4.33}$$

$$TiCl_3 + LiMe \xrightarrow{223\ K} TiMe_3 \tag{4.34}$$

$$TiCl_4 + Mg(CH_2Ph)Cl \xrightarrow{255\ K} [Ti(CH_2Ph)_4] \tag{4.35}$$

$$TiCl_4 + LiMe \xrightarrow{193\ K} TiMe_4 \tag{4.36}$$

$$VCl_4 + LiCH_2SiMe_3 \longrightarrow [V(CH_2SiMe_3)_4] \tag{4.37}$$

$$MnCl_2 + LiMe \longrightarrow MnMe_2 \tag{4.38}$$

$$NbCl_5 + ZnMe_2 \longrightarrow [NbMe_3Cl_2] + [NbMe_2Cl_3] \tag{4.39}$$

$$WCl_6 + LiMe \xrightarrow{Et_2O} WMe_6 \tag{4.40}$$

$$WCl_4 + 7LiMe \xrightarrow[Et_2O]{195\ K} Li_4[W_2Me_8] \cdot 4Et_2O \tag{4.41}$$

$$Cr_2(OAc)_4 + Mg(CH_2SiMe_3)_2 + PMe_3 \longrightarrow [Cr_2(\mu\text{-}CH_2SiMe_3)_2(CH_2SiMe_3)_2(PMe_3)_2] \tag{4.42}$$

$$TiCl_4 + Al_2Me_6 \xrightarrow[193\ K]{hexane} [TiMeCl_3] \tag{4.43}$$
(excess)

$$TiCl_4 + Al_2Me_6 \longrightarrow [TiMe_xCl_{4-x}] \tag{4.44}$$

Al/Ti < 1           $x = 1$

Al/Ti = 1           $x = 2$

Al/Ti > 1           $x = 3$

$$AgNO_3 + PbR_4 \xrightarrow{\text{EtOH}} AgR \tag{4.45}$$

$$cis\text{-}[PtCl_2(PEt_3)_2] + LiPh \longrightarrow cis\text{-}[PtPh_2(PEt_3)_2] \tag{4.46}$$

$$trans\text{-}[PtCl_2(PEt_3)_2] + LiPh \longrightarrow trans\text{-}[PtPh_2(PEt_3)_2] \tag{4.47}$$

$$[PtCl_2(CDO)] + MgMeCl \longrightarrow [PtMe_2(CDO)] \tag{4.48}$$

$$PtCl_4 + HgMe_2 \longrightarrow [PtClMe_3]_4 \tag{4.49}$$

$$n\,NaC \equiv CR + MX_n \longrightarrow [M(C \equiv CR)_n] + n\,NaX \tag{4.50}$$

## b. Reactions involving oxidative addition of organic halides

Many transition metal complexes in low oxidation states react with organic halides to give oxidative addition products. These products have coordination numbers increased by 2 and the oxidation states are also higher by 2 as compared to starting compounds. Complexes possessing $d^{10}$ and $d^8$ electron configurations of the central metal atom most commonly undergo these oxidative addition reactions.

$$[Pd(PPh_3)_4] + PhCl \longrightarrow [Pd(Ph)\,Cl(PPh_3)_2] + 2PPh_3 \tag{4.51}$$

$$2[Ni(PPh_3)_4] + 2\alpha\text{-}C_5H_5NBr \longrightarrow [Ni_2Br_2(\alpha\text{-}C_5H_4N)_2\,(PPh_3)_2] + 6PPh_3 \tag{4.52}$$

$$2[Ni(PPh_3)_4] + 2\beta\text{-}C_5H_4NCl \xrightarrow{\text{PhMe}} [Ni_2Cl_2(\beta\text{-}C_5H_4N)_2\,(PPh_3)_2] + 6PPh_3 \tag{4.53}$$

$$[Ni(PPh_3)_3] + RCl \xrightarrow[\text{PhH}]{293\,K} [Ni(R)\,Cl(PPh_3)_2] + PPh_3 \tag{4.54}$$

$$[Pt(PPh_3)_4] + PhC \equiv CBr \longrightarrow [PtBr(C \equiv CPh)(PPh_3)_2] + PPh_3 \tag{4.55}$$

The rates of reactions of aryl halides with $Ni(PPh_3)_3$ decrease in the series[20]: $p\text{-}ClC_6H_4CN > p\text{-}ClC_6H_4COPh > p\text{-}ClC_6H_4COMe > m\text{-}ClC_6H_4CN > m\text{-}ClC_6H_4COOMe > m\text{-}ClC_6H_4Cl > m\text{-}ClC_6H_4OPh > m\text{-}ClC_6H_4Me > p\text{-}ClC_6H_4Cl = ClC_6H_5 > p\text{-}ClC_6H_4Me$,

$$[Pt(PPh_3)_4] + RI \longrightarrow [Pt(R)\,I(PPh_3)_2] + 2PPh_3 \tag{4.56}$$

$$[Pt(R)\,I(PPh_3)_2] + RI \longrightarrow [PtR_2I_2(PPh_3)_2] \tag{4.57}$$

$$cis\text{-}[PtMe_2(PMe_2Ph)_2] + MeX \longrightarrow fac\text{-}[PtMe_3X(PMe_2Ph)_2] \tag{4.58}$$

Many alkyl complexes of Ir(III) and Rh(III) have been obtained by addition of alkyl halides to square-planar coordination compounds of Ir(I) and Rh(I). The addition of RX to $[IrCl(CO)(PPh_3)_2]$ gives cis products, while in the case of complexes of the type $[IrCl(CO)L_2]$ $(L = PMe_2Ph, PMePh_2, AsMe_2Ph)$, the addition yields *trans* derivatives [equations (4.59) and (4.60)].

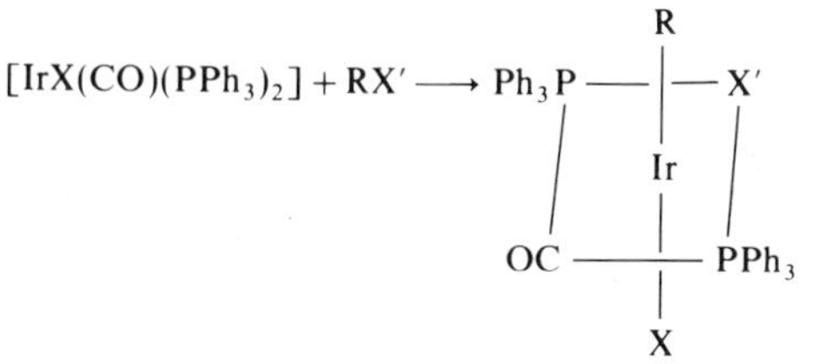

$$[IrX(CO)(PPh_3)_2] + RX' \longrightarrow \begin{array}{c} R \\ | \\ Ph_3P \!-\!\!-\!\!-\! \underset{\overset{|}{\underset{X}{Ir}}}{\overset{|}{\phantom{x}}} \!-\!\! X' \\ \phantom{Ph_3P}OC \!-\!\!-\!\!-\! PPh_3 \end{array} \qquad (4.59)$$

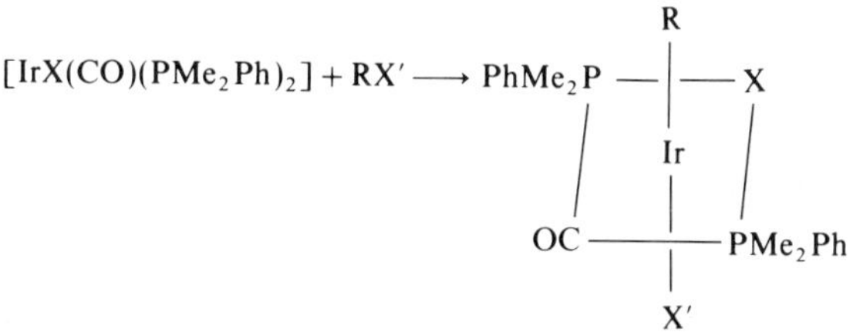

$$[IrX(CO)(PMe_2Ph)_2] + RX' \longrightarrow \begin{array}{c} R \\ | \\ PhMe_2P \!-\!\!-\! \underset{\overset{|}{\underset{X'}{Ir}}}{\overset{|}{\phantom{x}}} \!-\!\! X \\ \phantom{Ph}OC \!-\!\!-\! PMe_2Ph \end{array} \qquad (4.60)$$

The Wilkinson complex $[RhCl(PPh_3)_3]$ reacts with methyl iodide differently than with many other $16e$ complexes. In addition to the five-coordinate compound $[Rh(Me)I_2(PPh_3)_2]$, a derivative possessing coordinated MeI molecule is formed.

$$[RhCl(PPh_3)_3] + 2MeI \longrightarrow [Rh(Me)ClI(PPh_3)_2(MeI)] \qquad (4.61)$$

$$[RhCl(PPh_3)_3] + MeI \longrightarrow [Rh(Me)I_2(PPh_3)_2] \qquad (4.62)$$

In the case of some strong reducing reagents, oxidative addition reactions involve an increase by one unit of the oxidation state of the metal.

$$2ML_n + RX \longrightarrow MRL_n + MXL_n \qquad (4.63)$$

Such reactions, like the reaction of $[IrCl(CO)(PMe_3)_2]$ with alkyl halides, occur according to a free radical mechanism, although in certain cases other mechanisms may also be possible.

$$M^{II}L_n + RX \longrightarrow M^{III}XL_n + R \qquad (4.64)$$

$$M^{II}L_n + R \xrightarrow{\text{fast}} M^{III}RL_n \qquad (4.65)$$

The free radical mechanism has been observed for reactions of Co(II) and Cr(II) compounds with RX.

$$2[Co(CN)_5]^{3-} + RX \longrightarrow [Co(CN)_5R]^{3-} + [Co(CN)_5X]^{3-} \qquad (4.66)$$

$$(R = \text{alkyls, aryls, } R'CH=CHCH_2, \text{ vinyl, etc.})$$

The addition of such a complex to acetylene or tetrafluoroethylene may also take place:

$$2[Co(CN)_5]^{3-} + HCCH \longrightarrow [(NC)_5CoCHCHCo(CN)_5]^{6-} \qquad (4.67)$$

$$2Co(CN)_5^{3-} + C_2F_4 \longrightarrow [(CN)_5CoCF_2CF_2Co(CN)_5]^{6-} \qquad (4.68)$$

Also, Cr(II) compounds react with many organic halides:

$$2[Cr(H_2O)_6]^{2+} + RX \xrightarrow[H_2O]{} [CrR(H_2O)_5]^{2+} + [CrX(H_2O)_5]^{2+} \qquad (4.69)$$

The formation of hydrido complexes containing $M-C$ $\sigma$ bonds may also be achieved by oxidative addition involving $C-H$ bond cleavage, particularly if such a bond is activated by the proximity of a phosphorus or nitrogen atom coordinated to the metal (Figure 4.1e).[97, 98]

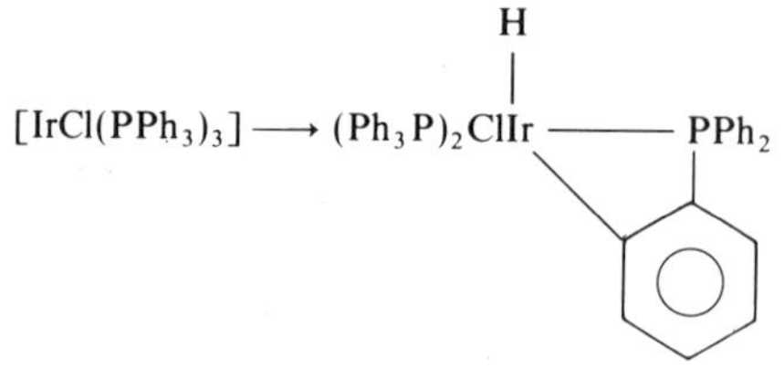

$$[IrCl(PPh_3)_3] \longrightarrow (Ph_3P)_2ClIr \qquad (4.70)$$

$$2[Ru(Me_2PC_2H_4PMe_2)_2] \longrightarrow [Ru_2\{CH_2P(Me)C_2H_4PMe_2\}_2H_2(Me_2PC_2H_4PMe_2)_2] \qquad (4.71)$$

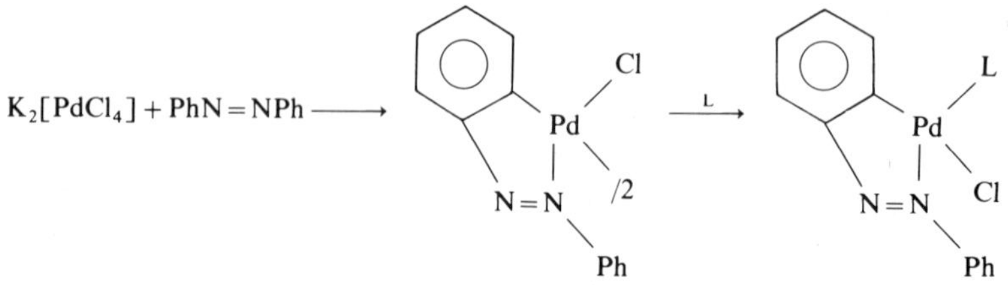

$$K_2[PdCl_4] + PhN=NPh \longrightarrow \qquad \xrightarrow{L} \qquad (4.72)$$

Acetylenes may react in a similar manner involving $C-H$ bond cleavage:

$$2RCCH + [Pt(PPh_3)_4] \longrightarrow [Pt(C\equiv CR)_2(H)_2(PPh_3)_2] + 2PPh_3 \qquad (4.73)$$

$$[IrCl(CO)(PPh_3)_2] + HCCCOOR \longrightarrow [Ir(C\equiv CCOOR)ClH(CO)(PPh_3)_2] \qquad (4.74)$$

### c. *Migration* (*insertion*) *reactions*

Migration of a coordinated nucleophile to an alkene or acetylene bonded to the central metal atom leads to the formation of alkyl or alkenyl complexes, respectively. The addition of an external nucleophile (not bonded to the metal) to coordinated alkenes occurs rarely (Chapter 13). Most commonly, the migrating ligand is H. However, many alkyl, halogen, etc., migrations are known.

$$trans\text{-}[PtClH(PEt_3)_2] \xrightarrow{C_2H_4} trans\text{-}[PtClEt(PEt_3)_2] \qquad (4.75)$$

$$[MnH(CO)_5] + BD \longrightarrow [Mn(CH_2CH=CHMe)(CO)_5] \qquad (4.76)$$

$$[RhHCl_2(PPh_3)_2] + C_2H_4 \longrightarrow [RhEtCl_2(PPh_3)_2] \qquad (4.77)$$

$$[RhHCl_2(PPh_3)_2] + HCCH \longrightarrow [Rh(CH=CH_2)Cl_2(PPh_3)_2] \qquad (4.78)$$

$$trans\text{-}[PtClH(PEt_3)_2] + CF_3CCCF_3 \longrightarrow [Pt(CCF_3=CHCF_3)Cl(PEt_3)_2] \qquad (4.79)$$

Reactions of rhodium hydrido complexes with fluoroolefins give migration products, which are analogous to compounds formed in the reactions of these complexes with olefins. However, during the reaction involving *trans*-Pt(H)XL$_2$, hydrogen fluoride is given off:

$$\text{trans-PtClH(PEt}_3)_2 + \text{C}_2\text{F}_3\text{H} \longrightarrow \text{trans-[Pt(CF=CFH)Cl(PEt}_3)_2] + \text{HF} \qquad (4.80)$$

$$[\text{RhCl}_2\text{H(PPh}_3)_2] + \text{C}_2\text{F}_4 \longrightarrow [\text{Rh(CF}_2\text{CHF}_2)\text{Cl}_2(\text{PPh}_3)_2] \qquad (4.81)$$

$$\text{trans-PtClH(PEt}_3)_2 + \text{C}_2\text{F}_4 \longrightarrow \text{trans-[Pt(CF=CF}_2)\text{Cl(PEt}_3)_2] + \text{HF} \qquad (4.82)$$

$$\text{trans-[PtMe}_2(\text{PMe}_2\text{Ph})_2] + \text{C}_2\text{F}_4 \longrightarrow \text{trans-[Pt(CF}_2\text{CHF}_2)_2(\text{PMe}_2\text{Ph})_2] \qquad (4.83)$$

$$\text{cis-[PtMe}_2(\text{PMe}_2\text{Ph})_2] + \text{C}_2\text{F}_4 \longrightarrow \text{cis-[PtMe(CF}_2\text{CHF}_2)(\text{PMe}_2\text{Ph})_2] \qquad (4.84)$$

Reactions of diazomethane with transition metal complexes also furnish derivatives possessing M$-$C $\sigma$ bonds:

$$[\text{MnH(CO)}_5] + \text{CH}_2\text{N}_2 \longrightarrow [\text{MnMe(CO)}_5] \qquad (4.85)$$

$$[\text{IrCl(CO)(PPh}_3)_2] + \text{CH}_2\text{N}_2 \longrightarrow [\text{Ir(CH}_2\text{Cl)(CO)(PPh}_3)_2] \qquad (4.86)$$

$$[\text{PdCl}_2(\text{PPh}_3)_2] + \text{NCCHN}_2 \longrightarrow [\text{Pd(CHClCN)}_2(\text{PPh}_3)_2] \qquad (4.87)$$

Dichlorocarbene, which is formed by thermal decomposition of CCl$_3$COONa, reacts with the tungsten hydrido complex to give the dichloromethyl compound:

$$[\text{Wcp}_2\text{H}_2] + \text{CCl}_3\text{COONa} \xrightarrow[-\text{CO}_2]{-\text{NaCl}} [\text{Wcp}_2(\text{CHCl}_2)\text{H}] \qquad (4.88)$$

In the case of cyclic compounds, hydrogen migration may be accompanied by ring opening:

$$[\text{CoH(CO)}_4] + \text{CH}_2\underset{\diagdown\;\;\;\diagup}{\overset{}{\text{—CH}_2}} \longrightarrow [\text{Co(CH}_2\text{CH}_2\text{OH)(CO)}_4] \qquad (4.89)$$

$$[\text{CoH(CO)}_4] + \;\diamondsuit\text{O} \longrightarrow [\text{Co(CH}_2\text{CH}_2\text{CH}_2\text{OH)(CO)}_4] \qquad (4.90)$$

Insertion of CF$_2$=CF$_2$ into the Pt$-$OMe bond has been observed[106]:

$$\text{PtMe(OMe)(dppe)} \xrightleftharpoons{\text{C}_2\text{F}_4} \text{PtMe(OMe)(CF}_2\text{=CF}_2)(\text{dppe}) \longrightarrow \text{PtMe(CF}_2\text{CF}_2\text{OMe)(dppe)} \qquad (4.91)$$

It appears interesting that the OMe ligand rather than the methyl group migrates. Many complexes are prepared by migration reactions:

$$[\text{MnMe(CO)}_5] + \text{CO} \longrightarrow [\text{Mn(COMe)(CO)}_5] \qquad (4.92)$$

$$[\text{CoR(CO)}_4] + \text{CO} \longrightarrow [\text{Co(COR)(CO)}_4] \qquad (4.93)$$

Almost all migration (insertion) reactions are utilized in catalytic processes such as hydrogenation, polymerization, oligomerization, isomerization, hydroformylation, hydrosilylation, etc. (Chapter 13).

## d. *Reactions of anionic complexes with organic halides*

Anionic complexes, particularly those possessing low oxidation states, afford compounds with $M-C$ $\sigma$ bonds via reaction with organic halides:

$$Na[Co(CO)_4] + MeCOCl \longrightarrow [Co(COMe)(CO)_4] + NaCl \qquad (4.94)$$

$$Na_2[Fe(CO)_4] + RCOX \longrightarrow Na[Fe(COR)(CO)_4] + NaX \qquad (4.95)$$

$$[Mn_2(CO)_{12}] + Na/Hg \xrightarrow{\text{THF}} Na[Mn(CO)_5] \xrightarrow{\text{MeI}} [MnMe(CO)_5] + NaI \qquad (4.96)$$

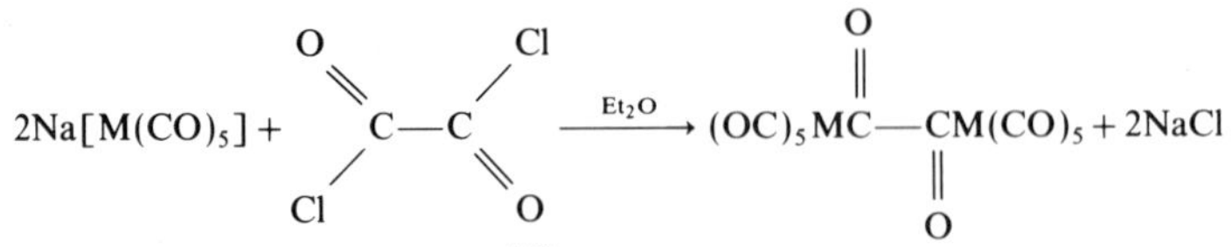

$$2Na[M(CO)_5] + \underset{\text{Cl}}{\overset{\text{O}}{\underset{\diagdown}{\overset{\diagdown}{C}}}} - \underset{\text{O}}{\overset{\text{Cl}}{\underset{\diagup}{\overset{\diagup}{C}}}} \xrightarrow{\text{Et}_2\text{O}} (OC)_5MC\overset{\text{O}}{\underset{\text{O}}{\|}}CM(CO)_5 + 2NaCl \qquad (4.97)$$

$$(M = Mn, Re)^{[170]}$$

$$Na[Fecp(CO)_2] + BrCH_2C\equiv CH \longrightarrow [Fe(CH_2C\equiv CH)\,cp(CO)_2] + NaBr \qquad (4.98)$$

$$Na[Wcp(CO)_3] + EtI \longrightarrow [WEtcp(CO)_3] + NaI \qquad (4.99)$$

These reactions involve heterolytic splitting of the carbon–halogen bond. The nucleophilicity of anionic complexes changes as follows:

$$[Fecp(CO)_2]^- > [Rucp(CO)_2]^- > [Nicp(CO)]^- > [Re(CO)_5]^- > [Wcp(CO)_3]^- > [Mn(CO)_5]^-$$

$$> [Mocp(CO)_3]^- > [Crcp(CO)_3]^- > [Co(CO)_4]^- > [CrCN(CO)_5]^-$$

$$> [MoCN(CO)_5]^- > [WCN(CO)_5]^-.$$

## e. *Elimination of neutral group connecting carbon with metal (deinsertion)*

Many complexes containing $M-C$ $\sigma$ bonds may be prepared by elimination of the neutral group which bonds carbon with the metal. The following molecules are given off: $CO_2$, $CO$, $N_2$, and $SO_2$.

The elimination may proceed as a result of photolytic or thermal decomposition:

$$[Mn(COMe)(CO)_5] \xrightarrow{\Delta} [MnMe(CO)_5] + CO \qquad (4.100)$$

$$[Fe(COPh)\,cp(CO)_2] \xrightarrow{h\nu} [FePhcp(CO)_2] + CO \qquad (4.101)$$

$$[CuOOCC_6F_5] \xrightarrow[\text{quinoline}]{333\,K} [Cu(C_6F_5)(C_9H_7N)] + CO_2 \qquad (4.102)$$

$$[Ni(OOCAr)_2L_2] \xrightarrow{\Delta} [NiAr_2L_2] + CO_2 \qquad (4.103)$$

$$[IrCl_2(SO_2C_6H_4Me)(CO)(PPh_3)_2] \longrightarrow [IrCl_2(C_6H_4Me)(CO)(PPh_3)_2] + SO_2 \qquad (4.104)$$

The protonated phenyldiazoplatinum complex, which is prepared by the reaction of diazonium tetrafluoroborate with *trans*-PtClH(PEt$_3$)$_2$, evolves nitrogen and HBF$_4$ during passage through a column containing aluminum oxide:

$$[PtCl(NNHPh)(PEt_3)_2]BF_4 \longrightarrow [PtCl(Ph)(PEt_3)_2] + N_2 + HBF_4 \qquad (4.105)$$

## f. *Preparation of acetylide complexes*[4, 11, 88]

Ethynide and ethynediide complexes with $M-C$ bonds may be prepared according to methods which were already described as well as by reactions of acetylene with transition metal salts:

$$[M(H_2O)_x]^{n+} + C_2H_2 \longrightarrow [MCCH(H_2O)_y]^{(n-1)+} + H^+ \qquad (4.106)$$

$$2[M(H_2O)_x]^+ + C_2H_2 \longrightarrow M_2C_2 + 2H^+ + xH_2O \qquad (4.107)$$

These reactions take place in water, alcohols, and other polar solvents. The type of the compound formed depends on the transition metal and the acidity of the solution. From ammonia solutions of Ag(I) and Cu(I) salts, acetylides possessing the composition $M_2C_2$ are precipitated, while $Ag_2SO_4$ and $Cu_2SO_4$ in sulfuric acid solution afford soluble ethynides:

$$[M(H_2O)_x]^+ + HCCH \longrightarrow [M(CCH)(H_2O)_y] + H^+ \qquad (4.108)$$

Acetylides are also prepared by direct reaction of metals, their oxides and hydrides with carbon at high temperatures (1800–2300 K) or by reaction of metals with hydrocarbon vapors. The latter method is utilized for preparation of acetylides of scandium, yttrium, and the lanthanides.

## g. *Photochemical synthesis*[81, 96, 100–105]

Many complexes and organometallic compounds undergo photochemical reactions under the influence of electromagnetic irradiation leading to the formation of compounds with $M-C$ $\sigma$ bonds. The metal–hydrocarbyl bond undergoes cleavage relatively readily and therefore, from polycarbyl complexes, it is possible to prepare compounds containing a smaller number of $\sigma M-C$ bonds. Alkyl complexes are also obtained as a result of photochemical decomposition of acyl compounds:

$$[WMe_2cp_2]PF_6 \xrightarrow[L]{hv} [WMecp_2L]PF_6 \qquad (4.109)$$

$$L = py, \ MeCN, \ PMe_2Ph^{(102)}$$

$$Re(CH_2R)_3(NBu')_2 \xrightarrow{hv} (Bu'N)_2Re \overset{\displaystyle CHR}{\underset{\displaystyle CH_2R}{\big\langle}} \qquad (4.110)$$

$$R = SiR_3', \ Ph^{(103)}$$

$$Mn(COMe)(CO)_5 \xrightarrow[-CO]{hv} Mn(COMe)(CO)_4 \longrightarrow MnMe(CO)_5^{(81)} \qquad (4.111)$$

Thus, substitution of some ligands (mainly CO) takes place readily and is utilized in the synthesis of compounds containing $M-C$ bonds. The reverse reaction, CO insertion into the $M-R$ bond, may also be accelerated by light.[101]

$$Th(CH_2SiMe_3)(pmcp)_2 + CO \xrightarrow{hv} (pmcp)_2Th \overset{\displaystyle O}{\overset{\displaystyle \triangle}{-\!\!-\!\!-}} CCH_2SiMe_3 \qquad (4.112)$$

In many cases, oxidative addition of $C-H$ bonds also occurs under the influence of light[100]:

$$Fe(C_2H_4)(Ph_2PCHCHPPh_2)_2 \xrightarrow[-C_2H_4]{h\nu} Fe(Ph_2PCHCHPPh_2)_2 \longrightarrow \left( \begin{array}{c} \text{structure} \end{array} \right) \qquad (4.113)$$

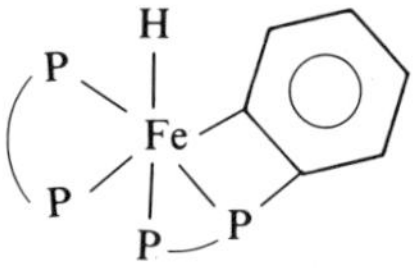

$$Rh(pmcp)\,H_2(PMe_3) \xrightarrow[213\,K]{h\nu/RH} Rh(pmcp)(PMe_3)$$

$$\underset{253\,K}{\overset{RH}{\rightleftharpoons}} Rh(pmcp)(R)(H)(PMe_3) \xrightarrow[213\,K]{CHBr_3} Rh(pmcp)(R)\,Br(PMe_3) \qquad (4.114)$$

$$R = Et,\ Pr^n,\ \text{cyclopropyl}[104]$$

$$Ta(CD_3)_3(OC_6H_3Bu_2^t\text{-}2,6)_2 \xrightarrow[-CD_4]{h\nu} Ta(CD_3)(CD_2)(OC_6H_3Bu_2^t\text{-}2,6)_2$$

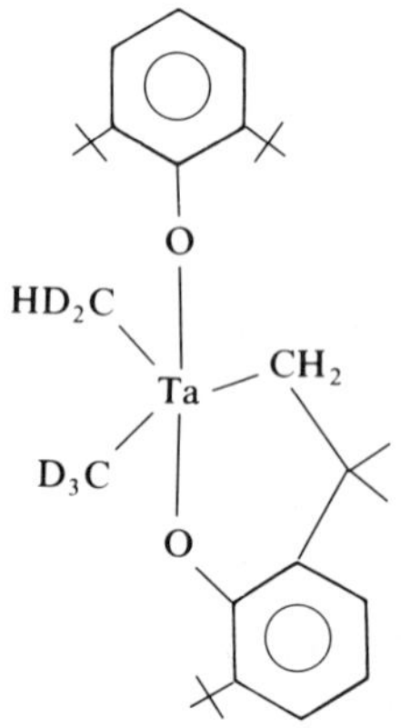

$$(4.115)$$

## h. *Electrochemical synthesis*

Processes of electrochemical reduction, oxidation, and synthesis of metalloorganic compounds have been investigated intensively. Many metalloorganic complexes and catalysts may be obtained via electrochemical methods.[107] Electrolytic redox processes may be accompanied by reductive elimination or homolytic cleavage of the $M-C$ bond and the formation of various products of radical reactions:

$$\text{trans-}[NiBrPh(PPh_3)_2] \xrightarrow[+PPh_3]{+e} NiPh(PPh_3)_3 \xrightarrow{e} [NiPh(PPh_3)_3]^- \qquad (4.116)$$

The complex $NiPh(PPh_3)_3$ decomposes, giving off biphenyl as a result of homolytic cleavage of the $Ni-C$ bond. The anionic compound $NiPh(PPh_3)_3^-$ in heterolytic reaction with $NiBrPh(PPh_3)_2$ also gives biphenyl and $Ni(PPh_3)_4$ which combines with PhBr, regenerating the starting complex.[108] Therefore, it is possible to form the $C-C$ bond by electrocatalytic reactions. Diethylbis(bipyridine)iron is oxidized electrolytically:

$$FeEt_2(bipy)_2 \xrightarrow[THF]{-e} FeEt_2(bipy)_2^+ \qquad (4.117)$$

The cation undergoes homolytic decomposition to afford typical products of solvent-cage radical reactions.[109]

## 11. *PROPERTIES OF COMPLEXES CONTAINING METAL–CARBON* σ *BONDS*

### a. *Complexes of group 3 metals*[14, 77, 101, 110–119]

Elements of group 3 (Table 4.12) form the following compounds:
$MR_3$ [R = aryl, alkyl, alkenyl, alkynyl, $CH(SiMe_3)$, $CH_2SiMe_3$, $C_6F_5$, etc.].
$MR_4^-$, $[Li(TMED)]_3[LnMe_6]$, $Li(TMED)[Mcp_2Me_2]$,[113] $[Mcp_2(\mu\text{-}R)]_2$, $[Mcp_2R(THF)]$, $Li(OEt_2)_2[Mcp_2(Me)Cl]$, $[cp_2M(\mu\text{-}R)_2AlR_2]$ (R = Me, Et, M = Sc, Y, La, Ln = lanthanide), $[Li(OR_2)_2][Mcp_2(R)X]$, $[M(CH_2C_6H_4NMe_2\text{-}4)_3]$, $[Lu(pmcp)_2\{(CH_2)_2PMe_2\}]$,[116] and

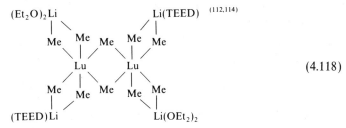

$$(4.118)$$

*Table 4.12. Properties of σ-Carbyl Complexes of 3d Group Metals*

| Compound | Color | Mp (K) | Remarks |
|---|---|---|---|
| $[Sc(C\equiv CPh)cp_2]^a$ | Yellow | 527 dec | |
| $[Sc(C\equiv CPh)_3]^a$ | Dark-brown | >523 dec | |
| $[ScPh_3]^a$ | Yellow-brown | | |
| $[Sc\{CH(SiMe_3)_2\}_3]\cdot 2\,THF^a$ | White | | $\nu(ScC)$ 415 cm$^{-1}$ |
| $YMe_3\cdot nTHF^a$ | Orange-yellow | | |
| $Li[LaPh]_4$ | Dark-brown | | |
| $Li[PrPh_4]^a$ | Dark-brown | | |
| $Li[Lu(2,6\text{-}C_6H_3Me_2)_4]\cdot 4\,THF^b$ | White | | |
| $[Tb(CH_2SiMe_3)_3(THF)_2]^c$ | White | 323–327 | $\mu_{eff} = 4.64$ BM |
| $[Y\{CH(SiMe_3)_2\}_3]^i$ | White | | $\nu(YC)$ 402 cm$^{-1}$ |
| $[Gd(CCPh)cp_2]^d$ | Yellow | | $\mu_{eff} = 7.98$ BM |
| $[ErPhcp_2]^d$ | Pink | | $\mu_{eff} = 9.53$ BM |
| $[ErMecp_2]^d$ | Pink | | $\mu_{eff} = 9.41$ BM |
| $[YbMecp_2]^d$ | Orange | | $\mu_{eff} = 4.41$ BM |
| $[Th(CH_2Ph)_4]^e$ | Yellow | | |
| $Li_2[U(CH_2SiMe_3)_6]\cdot 7(Me_2NC_2H_4NMe_2)^f$ | Green | | |
| $[Ucp_3Me]^{g(14)}$ | Yellow-brown | | |
| $[Ce\{1,1'\text{-}(CH_2)_3(\eta^5\text{-}C_5H_4)_2\}(C\equiv CPh)]^h$ | | >373 | |

$^a$ F. A. Hart, A. G. Massey, and M. S. Saran, *J. Organomet. Chem.*, **21**, 147 (1970); F. A. Hart and M. S. Saran, *Chem. Commun.*, 1614 (1968).

$^b$ See Ref. *e*, Table 4.5.

$^c$ J. L. Atwood, W. D. Hunter, R. D. Rogers, J. Holton, J. McMeeking, R. Pearce, and M. F. Lappert, *Chem. Commun.*, 140 (1978); H. Schumann and J. Mueller, *J. Organomet. Chem.*, **146**, C5 (1978); H. Schumann and J. Mueller, *J. Organomet. Chem.*, **169**, C1 (1979).

$^d$ N. Ely and M. Tsutsui, *J. Am. Chem. Soc.*, **97**, 1280 (1975); N. Ely and M. Tsutsui, *J. Am. Chem. Soc.*, **97**, 2551 (1975); N. Ely and M. Tsutsui, *J. Am. Chem. Soc.*, **96**, 4042 (1974).

$^e$ E. Kohler, W. Bruser, and K.-H. Thiele, *J. Organomet. Chem.*, **76**, 235 (1974).

$^f$ R. Andersen, E. Carmona-Guzman, K. Mertis, E. Sigurdson, and G. Wilkinson, *J. Organomet. Chem.*, **99**, C19 (1975).

$^g$ T. J. Marks, A. M. Seyan, and J. R. Kolb, *J. Am. Chem. Soc.*, **95**, 5529 (1973).

$^h$ J. John and M. Tsutsui, *J. Coord. Chem.*, **10**, 177 (1980).

$^i$ See Ref. *d*, Table 4.14.

A large group of ylid complexes of the type $[M\{(CH_2)(CH_2)PMe_2\}_3]$ is also known.[115] As shown by $^1H$, $^{13}C$, and $^{31}P$ NMR spectra, these complexes exist in equilibrium with oligomers because ylids can act as either chelating ligands or bridging ligands.

Group 3 element complexes with $M-C$ $\sigma$ bonds are extremely sensitive to oxidizing agents and water.

Homoleptic complexes $MR_3$ are very unstable, and consequently their chemical properties and structures have been little investigated. Anionic complexes of the type $MR_4^-$ and $MR_6^{3-}$ possessing $M-R-Li$ bridges have higher stabilities. In heteronuclear neutral complexes such as $cp_2M(\mu\text{-}R)_2AlR_2$, the aluminum and lanthanide atoms are also bonded by alkyl bridges. Complexes $[cp_2M(\mu\text{-}R)_2Mcp_2]$ are centrosymmetric; in the yttrium derivative, the positions of hydrogen atoms were determined and Y–H interactions were not found.

The tendency to form bridging bonds $M-R-M$ is extremely strong for lanthanides because these elements tend to achieve high coordination numbers. The complex $[Lu(pmcp)_2Me]$ does not possess an analogous structure to cyclopentadienyl compounds because pentamethylcyclopentadienyl ligands cause great steric hindrance. Therefore, there is only one, almost linear bridge in this complex in the solid state as well as in solution. The electron deficiency of lutetium atoms is so great that even hydrogen atoms of the methyl group may play the role of donors.

$$Lu-C-H \text{-----} Lu, \quad \angle LuCLu = 170(4)^\circ \text{ }^{(111)}$$

(pm)

$$Lu \xrightarrow{244.0(9)} CH_3 \xrightarrow{275.6} Lu$$

Me

234.4(12) (pm)

                                                                    (4.119)

The actinides also show a strong tendency for the formation of $M-R-M$ bridges. Monomeric, homoleptic compounds of the type $UR_4$ exist only in the case of large ligands. In the presence of an excess of LiR, the compound $UCl_4$ gives anionic complexes $UR_6^{2-}$. Uranium(V) forms octacoordinate anions of the type $UR_8^{3-}$. The benzyl compound of thorium(IV), $Th(CH_2Ph)_4$, probably has a structure which is similar to that of $M(CH_2Ph)_4$ ($M = Zr$, $Hf$). Thus, the ThCC angle should be equal to *ca* 90°. The following complexes are also known: $Mcp_2R_2$, $Mcp_3R$ ($M = U$, $Th$), $[Li(TMED)]_3[ThMe_7] \cdot TMED$,[117] $Th(pmcp)_2\{(CH_2)_2CMe_2\}$,[118] and $Ucp\{(CH_2)_2PPh_2\}_3$.[119] The bond is covalent to a significant degree in compounds of the actinides, while it is covalent to a lesser degree in complexes of the lanthanides. The dependence of the magnetic moment on temperature shows that, in compounds of lanthanides, there is a certain contribution of covalent bonding because in the strong field of ligands the orbital momentum is partly quenched. Compounds of lanthanides containing $\sigma M - C$ bonds react with Hg(II) to give HgRCl. They also react with carbon dioxide to give corresponding carboxylic acids:

$$MPh_3 \xrightarrow[\text{2. H}_2\text{O}]{\text{1. CO}_2} PhCOOH \qquad (4.120)$$

## b. *Complexes of group 4 metals*[1-5, 13, 85-88, 120-123]

All titanium group metals (Table 4.13) form the following compounds: $MR_4$ [R = Me, $Me_3SiCH_2$, $Me_3CCH_2$, norbornyl, C(Cl) = NMe, $PhCH_2$, etc.], [$MR_nX_{4-n}$] (X = Cl, Br, OR, etc.; $n = 1, 2$, rarely $n = 3$), [$TiMe_5$]$^-$, [$ZrMe_6$]$^{2-}$, [$Ti(o-C_6H_4OC_6H_4-o)_2$]$^-$, [$Mcp_2R_2$], [$Mcp_2R$]$^-$ (R = aryl, alkynyl, $CH_2Ph$), [$Mcp_2RX$], [$TicpR_3$], etc.

*Table 4.13. Properties of σ-Carbyl Complexes of Titanium Group Metals*

| Compound | Color | Mp (K) | Remarks |
|---|---|---|---|
| TiMe$_4$[a] | Bright-yellow | 195 dec | |
| [Ti(CH$_2$Ph)$_4$][b] | Red | 343–344 | |
| [TiPh(OPr$^i$)$_3$][c] | White | 361–363 | |
| TiPh$_4$[d] | Orange-yellow | 253 dec | |
| [Ti(1-adamantyl)$_4$][e] | White | 506–508 | $\tau = 8.03; 8.38$ |
| [TiPhCl$_3$][f] | Black-violet | 268 dec | |
| [Ticp($o$-C$_6$H$_4$CH$_2$NMe$_2$)$_2$][g] | Dark-green | | |
| [Ti(CH$_2$SiMe$_3$)$_4$][h] | Bright-green-yellow | 273–274 | $\nu$(TiC) 500 cm$^{-1}$ (br, m) |
| [Ti(CH$_2$CMe$_3$)$_4$][i] | Bright-yellow | 378–380 r | $\nu$(TiC) 540 (br, s); 507 cm$^{-1}$ (br, s) |
| [Ti{CH(SiMe$_3$)$_2$}$_3$][j] | Green-blue | | $g = 1.968$, $\nu$(TiC) 403; 432 cm$^{-1}$ |
| [Ticp$_2$Ph$_2$][k] | Orange | 519–521 dec | |
| ZrMe$_4$[l] | Red | 258 dec | |
| Li$_2$[ZrMe$_6$][l,m] | Bright-yellow | | |
| [Zr(CH$_2$SiMe$_3$)$_4$][h] | White | 283–284 dec | $\nu$(ZrC) 470 cm$^{-1}$ (br, m) |
| [Zr(CH$_2$CMe$_3$)$_4$][i] | White | 381–384 dec | $\nu$(ZrC) 530 (br, s), 488 (br, s) |
| [Zr(CH$_2$Ph)$_4$][b] | Yellow | 406–407 | |
| [Hf(CH$_2$Ph)$_4$][n] | Bright-yellow | 385–387 | |
| [Hf(CH$_2$SiMe$_3$)$_4$][h] | White | 281–283 | $\nu$(Hf–C) 470 cm$^{-1}$ (br, m) |
| [Hf(CH$_2$CMe$_3$)$_4$][i] | White | 288–289 | $\nu$(Hf–C) 536 (br, s), 491 (br, s) |

[a] J. J. Berhold and G. Groh, *Z. Anorg. Allg. Chem.*, **319**, 230 (1963).
[b] U. Zucchini, E. Albizzati, and U. Giannini, *J. Organomet. Chem.*, **26**, 357 (1971); W. Bruser, K.-H. Thiele, P. Zdunneck, and F. Brune, *J. Organomet. Chem.*, **32**, 335 (1971).
[c] H. Holloway, *Chem. Ind.*, 1096 (1957).
[d] G. A. Razuvaev, V. N. Latjaeva, A. V. Malysheva, and A. Kiljakova, *Dokl. Akad. Nauk SSSR*, **150**, 566 (1963); V. N. Latjaeva, G. A. Razuvaev, A. V. Malisheva, and G. A. Kiljakova, *J. Organomet. Chem.*, **2**, 388 (1964).
[e] R. M. G. Roberts, *J. Organomet. Chem.*, **63**, 159 (1973).
[f] K.-H. Thiele, P. Zdunneck, and D. Baumgart, *Z. Anorg. Allg. Chem.*, **378**, 62 (1970); K.-H. Thiele, *Usp. Khim.*, **41**, 1180 (1972).
[g] See Ref. *a*, Table 4.5.
[h] M. R. Collier, M. F. Lappert, and R. Pearce, *J. Chem. Soc., Dalton Trans.*, 445 (1973).
[i] P. J. Davidson, M. F. Lappert, and R. Pearce, *J. Organomet. Chem.*, **57**, 269 (1973).
[j] See Ref. *d*, Table 4.14.
[k] J. Summers and R. H. Uloth, *J. Am. Chem. Soc.*, **76**, 2278 (1954); J. Summers, R. H. Uloth, and A. Holmes, *J. Am. Chem. Soc.*, **77**, 3604 (1955).
[l] H. J. Berthold and G. Groh, *Angew. Chem.*, **78**, 495 (1966).
[m] H. J. Berthold and G. Groh, *Z. Anorg. Allg. Chem.*, **372**, 292 (1970).
[n] J. J. Felten and W. P. Anderson, *J. Organomet. Chem.*, **36**, 87 (1972).

Complexes of titanium with chelating diyl ligands represent a large group, for example,

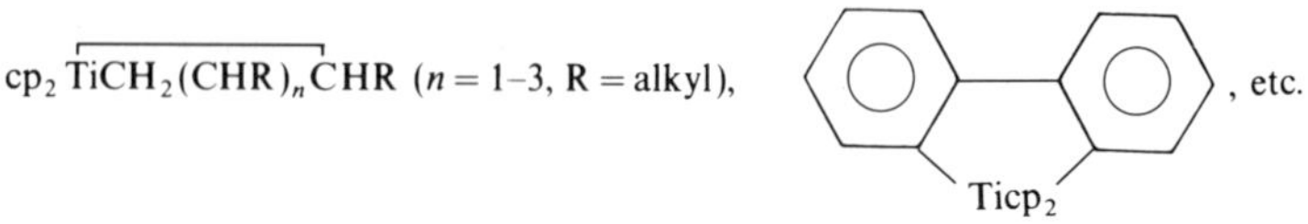

$$cp_2 \overline{TiCH_2(CHR)_nCHR} \ (n = 1\text{-}3, \ R = \text{alkyl}),$$

, etc.

The dinuclear zirconium compound containing the ethylene bridge is interesting [structure (4.121)].[122] Similarly, the zirconium compound possessing the methyl bridge is of

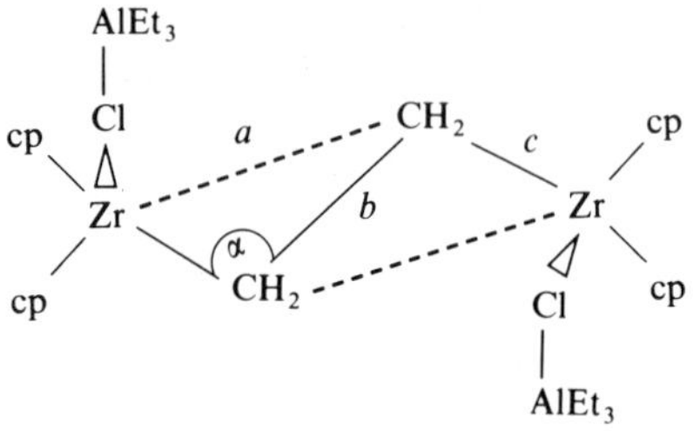

(4.121)

$\alpha = 75.9°$
$a = 249$ pm
$b = 155$ pm
$c = 236$ pm

interest because it has a trigonal–bipyramidal structure, and the carbon atom is shifted with respect to the plane of hydrogen atoms only by 8 pm [structure (4.122)].[123]

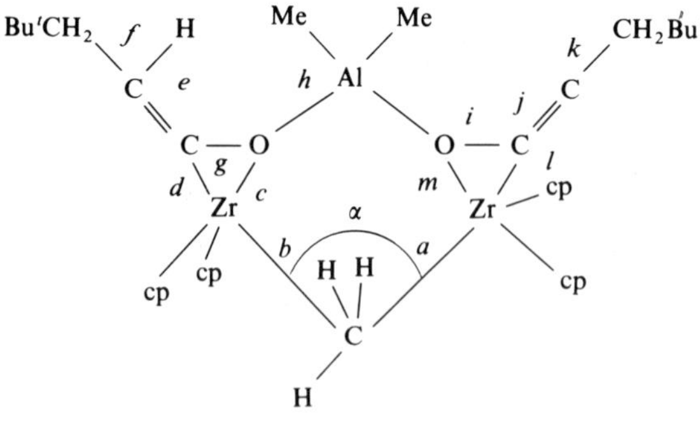

(4.122)

| | |
|---|---|
| $a = 245.6$ pm | $h = 181.9(4)$ pm |
| $b = 255.9(7)$ pm | $i = 141.3(7)$ pm |
| $c = 217.9$ pm | $j = 133.0(9)$ pm |
| $d = 218.7$ pm | $k = 149.8(10)$ pm |
| $e = 132.3$ pm | $l = 218.2(6)$ pm |
| $f = 150.1$ pm | $m = 218.3(4)$ pm |
| $g = 139.8(7)$ pm | $\alpha - 147.8(3)°$ |

In the case of titanium and zirconium, complexes possessing $+3$ and $+2$ oxidation states of the central ion are known, for example, $TiR_2$ ($R = PhCH_2$, Ph), $TiR_3$ [$R = (Me_3Si)_2CH$, $(NC)_3C$, $CCl=NMe$], [$ZrPh_2(ether)_2$], etc. Alkyl compounds of the titanium group, particularly those containing hydrogen atoms in the $\beta$ position, readily undergo thermal decomposition; moreover, they are extremely sensitive to air and moisture. However, some compounds possessing large ligands are exceptionally stable, i.e., Ti(1-adamantyl)$_4$. This compound undergoes decomposition with difficulty at 443 K when treated with the mixture $HNO_3 + HF + H_2O_2$. Anionic compounds such as $[TiMe_5]^-$ and $[ZrMe_6]^{2-}$ are formed rather easily. Also, titanium group compounds readily expand the coordination number and the number of valence electrons by the addition of Lewis bases, for example, $[TiMe_4(diamine)]$.

*Table 4.14. Properties of $\sigma$-Carbyl Complexes of Vanadium Group Metals*

| Compound | Color | Mp (K) | Remarks |
|---|---|---|---|
| $[V(CH_2Ph)_4] \cdot Et_2O^a$ | Black | | $g = 1.965$ BM |
| $[V(CH_2SiMe_3)_4]^b$ | Dark-green | 316 | |
| $Li_4(VPh_6) \cdot 3,5Et_2O^c$ | Black-violet | | $\mu_{eff} = 3.85$ BM |
| $[VO(CH_2SiMe_3)_3]^b$ | Bright-yellow | 348 | |
| $[V\{CH(SiMe_3)_2\}_3]^d$ | Green-blue | | $\nu$(VC) 400, 460 cm$^{-1}$ |
| $[V(CH_2Ph)_2Cl_2]^{e,a}$ | Dark-brown | >473 dec | |
| $[VMe_2Cl_2]^f$ | Bright-violet | 423–453 dec | |
| $[Vcp_2Ph]^g$ | Black | 365 | |
| $[Nb(2,4,6-C_6H_2Me_3)_3Cl_2]^h$ | Orange | | |
| $[NbMe_3Cl_3]^{i(2)}$ | Orange | 295 dec | $\tau$(Me) 7.10 (CH$_2$Cl$_2$) 307 K |
| $[NbMe_3Cl_2]^i$ | Yellow | 295 dec | $\tau$(Me) 7.40 (CH$_2$Cl$_2$) 307 K |
| $[NbMeCl_4]^i$ | Orange-brown | 337 dec | $\tau$(Me) 6.55 (CH$_2$Cl$_2$) 307 K |
| $[Nbcp_2Me_2]^j$ | Dark-red | | |
| $Li_4(NbPh_6) \cdot 3,5Et_2O^k$ | Black | | $\mu_{eff} = 1.81$ BM |
| $TaMe_5^{l(2)}$ | Bright-yellow | 283 | |
| $[TaMeCl_4]^i$ | Yellow | 295 dec | $\tau$(Me) 7.22 (CH$_2$Cl$_2$) |
| $[TaMe_3Cl_2]^i$ | Bright-yellow | 295 dec | $\tau$(Me) 8.23 (CH$_2$Cl$_2$) |
| $[TaMe_2(C_5H_4Me)_2]^j$ | Dark-red | | |
| $[TaMe_3(acac)_2]^m$ | Yellow | 356 dec | $\nu$(TaC) 535 cm$^{-1}$ |
| $[Ta_2(\mu-CSiMe_3)_2(CH_2SiMe_3)_4]^n$ | Orange | 443 | |
| $Li_4(TaPh_6) \cdot 3,5Et_2O^{o,k}$ | Black | | $\mu_{eff} = 1.68$ BM |

[a] G. A. Razuvaev, V. N. Latyaeva, A. N. Linyova, and W. W. Drobotenko, *Dokl. Akad. Nauk SSSR*, **208**, 876 (1973); G. A. Razuvaev, V. N. Latyaeva, L. I. Vyshinskaya, A. N. Linyova, V. V. Drobotenko, and V. K. Cherkasov, *J. Organomet. Chem.*, **93**, 113 (1975).

[b] W. Mowat, A. Shortland, G. Yagupskyy, N. J. Hill, M. Yagupsky, and G. Wilkinson, *J. Chem. Soc., Dalton Trans.*, 533 (1972); G. Yagupsky, W. Mowat, A. Shortland, and G. Wilkinson, *Chem. Commun.*, 1369 (1970).

[c] E. Kurras, *Angew. Chem.*, **72**, 635 (1960).

[d] G. K. Barker and M. F. Lappert, *J. Organomet. Chem.*, **76**, C45 (1974).

[e] A. N. Linyova, *Thesis*, University of Gorkii (1971).

[f] K.-H. Thiele and S. Wagner, *J. Organomet. Chem.*, **20**, P25 (1969).

[g] F. W. Siegert and H. J. de Liefde Meijer, *J. Organomet. Chem.*, **15**, 131 (1968); F. W. Siegert and H. J. de Liefde Meijer, *Recl. Trav. Chim. Pays-Bas*, **89**, 764 (1970).

[h] P. R. Sharp, D. Astruc, and R. R. Schrock, *J. Organomet. Chem.*, **182**, 477 (1979).

[i] G. W. A. Fowles, D. A. Rice, and J. D. Wilkins, *J. Chem. Soc., Dalton Trans.*, 961 (1973).

[j] I. H. Elson, J. K. Kochi, U. Klabunde, L. E. Manzer, G. W. Parshall, and F. N. Tebbe, *J. Am. Chem. Soc.*, **96**, 7374 (1974).

[k] B. Sarry, V. Dobrusskin, and H. Singh, *J. Organomet. Chem.*, **13**, 1 (1968).

[l] R. R. Schrock and P. Meakin, *J. Am. Chem. Soc.*, **96**, 5288 (1974).

[m] D. H. Williamson, C. Santini-Scampucci, and G. Wilkinson, *J. Organomet. Chem.*, **77**, C25 (1974); C. Santini-Scampucci and G. Wilkinson, *J. Chem. Soc., Dalton Trans.*, 807 (1976).

[n] W. Mowat and G. Wilkinson, *J. Am. Chem. Soc., Dalton Trans.*, 1120 (1973); G. Wilkinson, *Pure Appl. Chem.*, **30**, 627 (1972); F. Huq, W. Mowat, A. C. Skapski, and G. Wilkinson, *Chem. Commun.*, 1477 (1971).

[o] B. Sarry and M. Schon, *J. Organomet. Chem.*, **13**, 9 (1968); B. Sarry, *Angew. Chem., Int. Ed. Engl.*, **6**, 571 (1967).

Cyclopentadienyl complexes of the type $Mcp_2R_2$ are more stable. However, this is not a result of higher energy of the $\sigma$ bond between the metal and the carbon containing the ligand, but a result of the higher kinetic inertness of such compounds due to the presence of relatively large cyclopentadienyl rings which hinder access to the central atom.

### c. Complexes of group 5 metals[1–5, 13, 85–88, 105, 107, 124–126]

The vanadium group metals (Table 4.14) form the following complexes containing $\sigma M - C$ bonds: $VPh_2$, $[V\{CH(SiMe_3)_2\}_3]$, $[V\{C(CN)_3\}_3]$, $[V_2(dmp)_4]$, $[V_2(tmp)_4]$,[126] $Li[V(dmp)_4]$, $[V(2,4,6-C_6H_2Me_3)_3] \cdot 1.25$ THF, $VR_4$ ($R = CH_2Ph$, $CH_2SiMe_3$, norbornyl), $VR_2Cl_2$ ($R = Me$, Et, Ph, $CH_2Ph$), $VRCl_3$, $[VO(CH_2SiMe_3)_3]$, $[MPh_6]^{4-}$, $[Mcp_2R_2]$, $[Mcp_2R]$ ($M = V$, Nb, Ta), $[Vcp_2(C\equiv CPh)_2]$, $[Nb(C_6F_6)_4]$, $[MX_4R]$, $[MX_3R_2]$, $[MX_2R_3]$, $MXR_4$ ($M = Nb$, Ta; $X = Cl$, Br; $R = Me$, $CH_2SiMe_3$, $CH_2CMe_3$, etc.), $MMe_5$, $[M(CH_2CMe_3)_3(CHCMe_3)]$, $[(Me_3SiCH_2)_2M(\mu\text{-}CSiMe_3)_2M(CH_2SiMe_3)_2]$, $[MMe_5(PMe_3)_2]$, $[MMe_5(dmpe)]$, $[MMe_7]^{2-}$, $[MPh_6]^{4-}$, $[MMe(\eta\text{-}COT)(\eta^4\text{-}COT)]$, $[MR_2X_2]$ ($M = Nb$, Ta), $[Tacp_2R_3]$, $[TaPh_6]^{n-}$ ($n = 1,3$), $[NbPh_6]^{2-}$, $[Tacp_2R_2]^+$, $[Nbcp_2R_nX_{3-n}]$, $[Nbcp_2R(O)]$.

The vanadium group element homoleptic compounds with $\sigma M - C$ bonds are relatively thermally unstable and very sensitive to oxidizing agents and water. They also have the tendency to expand the coordination number as a result of $R^-$ or Lewis base (amine, phosphine, etc.) addition. Heteroleptic complexes possessing cyclopentadienyl, halogen, oxo, acetylacetonate, etc., ligands are usually more thermally stable than homoleptic complexes such as $MR_4$.

### d. Complexes of group 6 metals[1–5, 13, 85–87, 105, 107, 127–134]

The chromium group metal complexes containing hydrogen atoms in the $\beta$ position are very unstable. The lowest oxidation state which the chromium group complexes (Table 4.15) possess in their compounds containing $\sigma M - C$ bonds equals $+2$. As the

Table 4.15. *Properties of $\sigma$-Carbyl Complexes of Chromium Group Metals*

| Compound | Color | Mp (K) | Remarks |
|---|---|---|---|
| $[Cr(2,4,6-C_6H_2Me_3)_2] \cdot 3$ THF[a] | Red | 358 dec | $\mu_{eff} = 4.82$ BM |
| $[CrPh_3] \cdot 3$ THF[b] | Dark-red | 328 dec | |
| $[Cr(o\text{-}C_6H_4NMe_2)_3] \cdot 3$ THF[c] | Red | | |
| $K_3[Cr(C\equiv CH)_6]^d$ | Orange | | $\sim 448$ K explosive |
| $Li_4(Cr_2Me_8) \cdot 4$ THF[e] | Yellow | | $\nu(CrC)$ 421 vs, 422 sh, 465 sh, m |
| $Li_4\left[ Cr\left\{ \begin{array}{c} CH_2\!-\!CH_2 \\ | \\ CH_2\!-\!CH_2 \end{array} \right\}_2 \right]_2 \cdot 4Et_2O^f$ | Yellow | 403 dec | |

*(Table continued)*

*Table 4.15. (continued)*

| Compound | Color | Mp (K) | Remarks |
|---|---|---|---|
| $Li_3[CrPh_6] \cdot 2,5Et_2O$[g] | Orange | >373 dec | |
| $Na_2[CrPh_5] \cdot 3Et_2O$[h] | Blue-green | | $\mu_{eff} = 3.66$ BM |
| $Li_2[CrPh_4] \cdot 4$ THF[i] | Yellow | 363 | |
| $Li[Cr(CH_2SiMe_3)_4]$[j] | Blue-green | | |
| $[Cr\{CH(SiMe_3)_2\}_3]$[v] | Bright-green | | $g = 2.0$ and 3.9 |
| | | | $\mu_{eff} = 3.7$ BM |
| $[Cr(CH_2SiMe_3)_4]$[j] | Dark-red | 313 | |
| $[Cr_2(\mu\text{-}CH_2SiMe_3)_2(CH_2SiMe_3)_2(PMe_3)_2]$[k] | Dark-red | 353–373 dec | |
| $cis\text{-}[CrPh_2(bipy)_2]I$[l] | Orange-yellow | 503 dec | |
| $[Cr_2\{(CH_2)_2PMe_2\}_4]$[m] | Gold-yellow | 438–440 | |
| $[Mo_2\{(CH_2)_2PMe_2\}_4]$[n,m] | Orange-red | 529–548 dec | $\nu$(MoC) 464, |
| | | | 566 cm$^{-1}$ |
| $[MoMe_2(\eta\text{-}PhH)(PMe_2Ph)_2]$[o] | Dark-red | | (Mo–Me) 230 pm, |
| | | | $d$(Mo–C$_{Ph}$) 226–237 pm |
| $[Mo_2(CH_2SiMe_3)_6]$[j] | Yellow | 372 | |
| $[Mo_2Me_4(PMe_2Ph)_4]$[k] | Blue | 403 | |
| $[Mo_2(CH_2CMe_3)_6]$[p] | Yellow | 408–411 | |
| $Li_4[Mo_2Me_8] \cdot 4$ THF[q] | Bright-red | | |
| $[Mo_2(2\text{-}MeOC_6H_4)_4]$[r] | Pink | | |
| $[W_2(CH_2SiMe_3)_6]$[j] | Orange-brown | 383 | |
| $WMe_6$[s,t] | Red | 303 | pyrophoric $\tau = 8.2$ |
| | | | $J(^{183}W - H) = 3$Hz |
| | | | $(CCl_3F)$ |
| $Li_2[WMe_8] \cdot 2C_4H_8O_2$[s,t] | Orange-red | | |
| $Li_4(W_2Me_8) \cdot 4$ THF[u] | Bright-red | | $\tau$(Me) 9.63 (PhH) 7.23; |
| | | | 8.48 (THF) |
| $Li_4[W_2Me_{4.8}Cl_{3.2}] \cdot 4$ THF[u] | Red | | |

[a] M. Tsutsui and H. H. Zeiss, *J. Am. Chem. Soc.*, **82**, 6255 (1960).

[b] W. Herwig and H. H. Zeiss, *J. Am. Chem. Soc.*, **81**, 4798 (1959).

[c] F. Hein and D. Tille, *Z. Anorg. Allg. Chem.*, **329**, 72 (1964).

[d] R. Nast and E. Sirtl, *Chem. Ber.*, **88**, 1723 (1955).

[e] E. Kurras and J. Otto, *J. Organomet. Chem.*, **4**, 114 (1965); E. Kurras and K. Zimmerman, *J. Organomet. Chem.*, **7**, 348 (1967); J. Krausse, G. Marx, and G. Schoedl, *J. Organomet. Chem.*, **21**, 159 (1970).

[f] J. Krausse and J. Schoedl, *J. Organomet. Chem.*, **27**, 59 (1971).

[g] F. Hein and R. Weiss, *Z. Anorg. Allg. Chem.*, **295**, 145 (1958).

[h] F. Hein and K. Schmiedeknecht, *J. Organomet. Chem.*, **5**, 454 (1966).

[i] W. Seidel, K. Fischer, and K. Schmiedeknecht, *Z. Anorg. Allg. Chem.*, **390**, 273 (1972).

[j] See Ref. b, Table 4.14.

[k] R. A. Andersen, R. A. Jones, G. Wilkinson, M. B. Hoursthouse, and K. M. A. Malik, *Chem. Commun.*, 283 (1977); R. A. Andersen, R. A. Jones, and G. Wilkinson, *J. Chem. Soc., Dalton Trans.*, 446 (1978).

[l] H. Mueller, *Z. Chem.*, **9**, 311 (1969); J. J. Daly, F. Sanz, R. P. A. Sneeden, and H. H. Zeiss, *J. Chem. Soc., Dalton Trans.*, **73**, 1497 (1973); R. P. A. Sneeden and H. H. Zeiss, *J. Organomet. Chem.*, **47**, 125 (1973).

[m] E. Kurras, U. Rosenthal, H. Mennega, and G. Oehme, *Z. Chem.*, **14**, 160 (1974).

[n] E. Kurras, H. Mennega, G. Oehme, U. Rosenthal, and G. Engelhardt, *J. Organomet. Chem.*, **84**, C13 (1975).

[o] E. Carmona-Guzman and G. Wilkinson, *J. Chem. Soc., Dalton Trans.*, 1139 (1978); E. Carmona-Guzman and G. Wilkinson, *J. Chem. Soc., Dalton Trans.*, 1519 (1979).

[p] W. Mowat and G. Wilkinson, *J. Chem. Soc., Dalton Trans.*, 1120 (1973).

[q] F. A. Cotton, J. M. Troup, T. R. Webb, D. H. Williamson, and G. Wilkinson, *J. Am. Chem. Soc.*, **96**, 3824 (1974).

[r] R. A. Jones and G. Wilkinson, *J. Chem. Soc., Dalton Trans.*, 472 (1979).

[s] A. Shortland and G. Wilkinson, *Chem. Commun.*, 318 (1972); A. Shortland and G. Wilkinson, *J. Chem. Soc., Dalton Trans.*, 872 (1973); L. Galyer and G. Wilkinson, *J. Chem. Soc., Dalton Trans.*, 2235 (1976).

[t] L. Galyer, K. Mertis, and G. Wilkinson, *J. Organomet. Chem.*, **85**, C37 (1975).

[u] F. A. Cotton, S. Koch, K. Mertis, M. Millar, and G. Wilkinson, *J. Am. Chem. Soc.*, **99**, 4989 (1977).

[v] See Ref. d, Table 4.14.

atomic number increases, the stability of compounds with metals in higher oxidation states becomes greater. In the case of tungsten, the compound $WMe_6$ is known. The remaining elements do not form analogous compounds. Characteristic for metals of this triad are polynuclear compounds possessing direct M(II)-to-M(II) and M(III)-to-M(III) metal–metal bonds. Chromium and molybdenum form many relatively stable octahedral complexes in which the metal has a $+3$ oxidation state possessing $d^3$ electron configuration of the central ion. Therefore, these complexes are inert toward reactions that break the $M-C$ bond and ligand substitution (Section 4.2). The chromium(II) aqua complexes react with organic halides to give many compounds of the type $[CrR(H_2O)_5]^{2+}$. Such complexes constitute relatively rare examples of cationic compounds containing $\sigma M-C$ bonds. Additional examples are *cis*-$[CrR_2(L–L)_2]^+$ (R = Ph, $CH_2SiMe_3$, anisyl, tolyl; L–L = bipy, phen).

The following compounds of the chromium triad which have $\sigma M-C$ bonds are known: $[Cr_2(CH_2SiMe_3)_4(PMe_3)_2]$, $CrMe_2$, $[Cr(o-C_6H_4OMe)_2]_2$, $[Cr(2,4,6-C_6H_2Me_3)_2]$, $[M_2R_4]$ [M = Cr, Mo; R = $Me_2Ph(CH_2)_2$], $[Cr\{CH(SiMe_3)_2\}_3]$, $[M_2Me_8]^{4-}$, $M_2(chel)_4$ (chel = dmp, tmp, mhp, map, dmhp) (M = Cr, Mo, W),[134] $[CrPh_5]^{2-}$, $[CrR_6]^{3-}$ (R = Me, $C\equiv CH$, Ph),

$$\left[ Cr \left\langle \begin{array}{c} CH_2{-}CH_2 \\ | \\ CH_2{-}CH_2 \end{array} \right\rangle_3 \right]^{3-} , \quad \left[ Cr \left\langle \begin{array}{c} CH_2{-}CH_2 \\ \\ CH_2{-}CH_2 \end{array} \right\rangle CH_2 \right)_3 \right]^{3-} ,$$

$$\left[ Cr \left( \begin{array}{c} CH_2{-}CH_2 \\ \\ CH_2{-}CH_2 \end{array} \right\rangle CH_2 \right)_3 \right]^{2-} ,$$

$[Cr_2Ph_6]^{2-}$, $[CrR_4]^{2-}$, $[Cr(mes)_3]^-$, $[Cr_2(\mu-CH_2SiMe_3)_2(CH_2SiMe_3)_2(PMe_3)_2]$, $[Mo_2\{\mu-(CH_2)_2SiMe_2\}(CH_2SiMe_3)_2(PR_3)_3]$ (R = Ph, Me), $[CrR_4]^-$ (R = Ph, $CH_2SiMe_3$, $CH_2CMe_3$), $CrR_4$ (R = $CH_2SiMe_3$, $CH_2CMe_3$, $Bu^t$), $[MoPh_6]^{3-}$, $[M_2R_6]$ (M = Mo, W, R = $CH_2SiMe_3$, $CH_2CMe_3$), $[W(CH_2Ph)_4]$, $[W(C_6F_5)_5]$, $WMe_6$, $[W(C_6F_5)_5]^-$, $[WPh_6]^{2-}$, $[WMe_8]^{2-}$, $[CrRCl_2]$, $[MRcp(CO)_3]$ (M = Cr, Mo, W; R = alkyl, Ph, $CH_2Ph$, $p-C_6H_4Me$), $[CrPh_2Cl(MeOC_2H_4OMe)]$, $[WRCl_5]$ (R = Me, Et, Bu, Ph), $[WPhCl_3]$, $[W_2R_{8-x}Cl_x]^{4-}$, $[(O) np_3 W(\mu-O) Wnp_3(O)]$,[128] $[W_2(\mu-CSiMe_3)(\mu-CPhCPhCSiMe_3)(CH_2SiMe_3)_4]$,[129] $[\{W(pmcp)Me_3\}_2N_2]$,[130] $K_3[M(C\equiv CR)_3(CO)_3]$ (R = H, Me, Ph; M = Cr, Mo, W), $[W_2(CH_2R)_2(\mu-OOCEt)_4]$ (R = Ph, np),[131] $[Wcp(CH_2CH_2Ph)(CO)_3]$,[132] *cis*-$[WMe(CO)_4 PMe_3]^-$,[133] $[(OC)_3cpM(\mu-CH_2CH_2) M'cp(CO)_3]$ (M = Mo, W; M' = Mo, W).[87]

## e. *Complexes of group 7 metals*[1–5, 13, 85, 88, 97, 134–137]

Manganese forms compounds (Table 4.16) containing $\sigma Mn-C$ bonds in its $+2$ oxidation state. These compounds are of the following composition: $MnR_2$ [R = $C(CN)_3$, $C(CN)_2CPh=C(CN)_2$, $CH_2SiMe_3$, $2,4,6-C_6H_2Me_3$, $CH_2CMe_3$], $[MnR_3]^-$ (R = Me, Et, n-Bu), $[MnR_4]^{2-}$ (R = Me, $C\equiv CH$, $C\equiv CPh$, $C\equiv CMe$),

*Table 4.16. σ-Carbyl Complexes of Manganese Group Metals*

| Compound | Color | Mp (K) | Remarks |
|---|---|---|---|
| $[Mn(1\text{-adamantylmetyl})_2]^a$ | Amber | 403 dec | Reacts with oxygen very easily |
| $[MnPh_2] \cdot THF^{b,c}$ | Green | 338 dec | |
| $[MnMe_2]^b$ | Bright-yellow | 353 | |
| $[Mn(CH_2SiMe_3)_2]^d$ | Bright-orange | 423 | |
| $[Mn(CH_2CMe_3)_2]^d$ | Red-brown | 381 | |
| $[Mn(norbornyl)_4]^e$ | Green | | $\mu_{eff} = 3.78$ BM |
| $K_2[Mn(C\equiv CH)_4]^f$ | Pink | | $\mu_{eff} = 5.89$ BM |
| $[K_3(Mn(C\equiv CH)_6]^f$ | Brown-black | | Explosive |
| $Li_2[MnMe_4] \cdot Me_2NC_2H_4NMe_2{}^d$ | Red | 383 | |
| $Li_2[MnMe_4]^d$ | Brown-yellow | | |
| $Li_2[Mn(o\text{-}C_6H_4OC_6H_4\text{-}o)_2] \cdot 6\,THF^h$ | Yellow-orange | | $\mu_{eff} = 5.84$ BM |
| $Li_2[Re_2Me_8] \cdot 2Et_2O^i$ | Bright-red | | $\tau(Me)$ 9.21, $\tau(Et_2O)$ 6.58 (q), 8.88 (t) (PhH) |
| $Li_2[Re_2Me_8] \cdot phen^i$ | Purple | | |
| $[RePh_3(PEt_2Ph)_2]^j$ | Blue | 396–398 dec | $d(Re\text{–}Ph)$ 202.7 pm |
| $[RePh_2(N)(PPh_3)_2]^j$ | Orange | 451–453 dec | |
| $[ReOMe_4]^{k,l}$ | Red-purple | 317 | |
| $ReMe_6{}^{l,g}$ | Green | 283–285 dec | |
| $Li_2[ReMe_8]^l$ | Red | | |
| $cis\text{-}[ReO_2Me_3]^g$ | Yellow | 283 | |
| $[Re_3(CH_2SiMe_3)_6(\mu\text{-}Cl)_3]^m$ | Dark-blue | | $d(Re\text{–}Re)$ 238.6 pm $d(Re\text{–}C)$ 209.9 pm |
| $[Re_2Me_2(OAc)_4]^n$ | Black | 473–523 dec | |
| $[Re_2(CH_2SiMe_3)_4(OAc)_2]^n$ | Black | 408–410 | |
| $[Re_6Me_{12}(OAc)_6]^o$ | Orange | 383 dec | |
| $[Re_3Me_6(acac)_3]^o$ | Red | 443 | |
| $[Re_3Me_9(PEt_2Ph)_2]^p$ | Violet-pink | | |
| $[Re_2Me_6(PEt_2Ph)_2]^p$ | Green | | |

[a] M. Bochmann, G. Wilkinson, and G. B. Joung, *J. Chem. Soc., Dalton Trans.*, 1879 (1980).
[b] C. Beermann and K. Clauss, *Angew. Chem.*, **71**, 627 (1959); M. Tamura and J. Kochi, *J. Organomet. Chem.*, **29**, 111 (1971); H. Gilman and J. G. Bailie, *J. Org. Chem.*, **2**, 84 (1937); H. Gilman and R. H. Kirby, *J. Am. Chem. Soc.*, **63**, 2046 (1941).
[c] R. Riemschneider, H.-G. Kassahn, and W. Schneider, *Z. Naturforsch.*, **15b**, 547 (1960).
[d] R. A. Andersen, E. Carmona-Guzman, K. Mertis, E. Sigurdson, and G. Wilkinson, *J. Organomet. Chem.*, **99**, C19 (1975); R. A. Andersen, E. Carmona-Guzman, J. F. Gibson, and G. Wilkinson, *J. Chem. Soc., Dalton Trans.*, 2204 (1976).
[e] B. K. Bower and H. G. Tennet, *J. Am. Chem. Soc.*, **94**, 2512 (1972).
[f] R. Nast and H. Griesshammer, *Chem. Ber.*, 1315 (1957).
[g] See Ref. *t*, Table 4.15.
[h] H. Drevs, *Z. Chem.*, **15**, 451 (1975).
[i] F. A. Cotton, L. D. Gage, K. Mertis, L. W. Shive, and G. Wilkinson, *J. Am. Chem. Soc.*, **98**, 6922 (1976).
[j] J. Chatt, J. D. Garforth, and G. A. Rowe, *J. Chem. Soc. (A)*, 1834 (1966); W. E. Carroll and R. Bau, *Chem. Commun.*, 825 (1978).
[k] K. Mertis, J. F. Gibson, and G. Wilkinson, *Chem. Commun.*, 93 (1974).
[l] K. Mertis and G. Wilkinson, *J. Chem. Soc., Dalton Trans.*, 1488, 1492 (1976).
[m] A. F. Masters, K. Mertis, J. F. Gibson, and G. Wilkinson, *Nouv. J. Chem.*, **1**, 389 (1977); M. B. Hursthouse and K. M. A. Malik, *J. Chem. Soc., Dalton Trans.*, 1334 (1978).
[n] R. A. Jones and G. Wilkinson, *J. Chem. Soc., Dalton Trans.*, 1063 (1978).
[o] P. G. Edwards, F. Felix, K. Mertis, and G. Wilkinson, *J. Chem. Soc., Dalton Trans.*, 361 (1979).
[p] P. Edwards, K. Mertis, G. Wilkinson, M. B. Hursthouse, and K. M. A. Malik, *J. Chem. Soc., Dalton Trans.*, 334 (1980).

$[Mn(C \equiv CH)_2] \cdot 4NH_3$, and $[Mn(o\text{-}C_6H_4OC_6H_4\text{-}o)_2]^{2-}$. The following complexes of Mn(III), Mn(IV), and Mn(I) are also known: $[Mn(o\text{-}C_6H_4OC_6H_4\text{-}o)_2]^-$, $[Mn(norbornyl)_4]$, and $[MnR(CO)_5]$.

The compound $[Mn(CH_2SiMe_3)_2]$ is a polymer containing bridging alkyl ligands; the manganese atoms have tetrahedral coordination. The structure of this compound is therefore analogous to that of dimethylberyllium (Figure 1.2). The complex $Mn(CH_2CMe_3)_2$ is a linear tetramer possessing alkyl bridges and two terminal ligands connected to the terminal manganese atoms which have coordination number 3. The two manganese atoms which are located in the middle of the chain have tetrahedral coordination. Like rhenium, manganese forms the following typical complexes: $[MnR(CO)_5]$, $[MnR(CO)_{5-x}L_x]$ (see Chapter 2), $[(OC)_5 (Mn(\mu\text{-}CH_2CH_2CH_2) Mn(CO)_5]$. Also, the following $\alpha$-hydroxyalkyl manganese(I) complex has been prepared[137]:

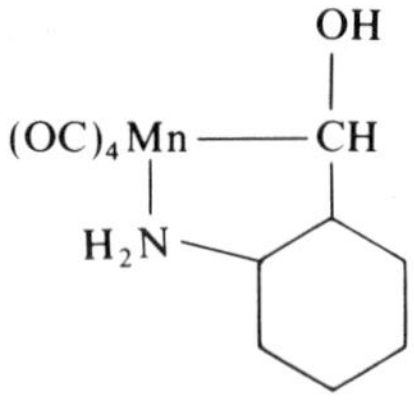

$$
(OC)_4 Mn \underset{H_2N}{\overset{\overset{\displaystyle OH}{|}}{\underset{\diagdown}{—CH}}}
\quad\quad (4.123)
$$

The following compounds are characteristic for rhenium in higher oxidation states such as $+3$ to $+7$: $Li_2[Re_2Me_8] \cdot 2Et_2O$, $[ReOMe_4]$, $[ReMe_6]$, $[Re_2Me_2(OAc)_4]$, $[Re_2R_4(OAc)_2]$, $[Re_3Cl_3R_6]$ ($R = Me$, $CH_2CMe_3$, $CH_2SiMe_3$),

Table 4.17. σ-Carbyl Complexes of Iron Group Metals

| Compound | Color | Mp (K) | Remarks |
|---|---|---|---|
| $Li_2[FeMe_4]^a$ | Bright-yellow | | |
| $[FeEr_2(bipy)]^b$ | Dark | | 1050 nm ($\varepsilon = 2760$), 610 (4580), 395 (1030) |
| $[Fe(o\text{-}C_6H_4CH_2NMeC_2H_4NMeC_6H_4\text{-}o)]^c$ | Dark-red | | |
| $[Fe(CH_2Ph)(salen)]^d$ | Red | | $\mu_{eff} = 5.1\text{--}5.9$ BM (90–297 K) |
| trans-$[Fe(C_6Cl_5)_2(PEt_2Ph)_2]^e$ | Gold-yellow | 408–415 dec | $\mu_{eff} = 3.6$ BM |
| $[FePhcp(CO)_2]^f$ | Yellow | 308–309 | |
| $[Fe(norbornyl)_4]^g$ | Purple | | Diamagnetic |
| $K_4[Fe(CCH)_6]^h$ | Yellow | | Diamagnetic |
| $K_3[Fe(CCH)_6]^h$ | Brown-violet | | Explosive |
| 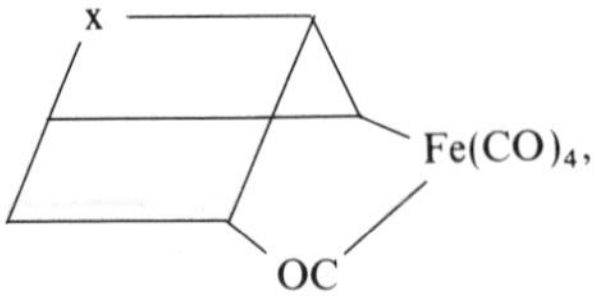 (structure) $Fe(CO)_4$, $x = C_6H_4{}^i$ | White | 388 | |

(Table continued)

*Table 4.17.* *(continued)*

| Compound | Color | Mp (K) | Remarks |
|---|---|---|---|
| cis-[RuMe$_2$(DPPE)$_2$][j] | Yellow | 523–529 dec | |
| cis-[RuMe$_2$(PMe$_3$)$_4$][k] | White | 473–483 dec | |
| [RuPhH(Me$_2$PC$_2$H$_4$PMe$_2$)$_2$][l] | White | 409–411 | |
| [Ru(CH$_2$COCH$_3$)(H)(DMPE)$_2$][m] | | | $\nu$(RuH) 1795, $\tau$(HRu) = 26.74 (*trans*) and 18.89 (*cis*) |
| [Ru{o-C$_6$H$_4$OP(OPh)$_2$}Cl(CO){P(OPh)$_3$}$_2$][n] | White | 436–438 | |
| [Ru(2-CH$_2$OC$_6$H$_4$)(PMe$_3$)$_4$][o] | White | 513 dec | |
| [(Me$_3$P)$_4$Ru($\mu$-CH$_2$)$_2$Ru($\mu$-CH$_2$)$_2$Ru(PMe$_3$)$_4$][r] | Dark-red | 563–583 dec | |
| [RuPh$_2$(PMe$_3$)$_4$][o] | White | 403–413 dec | |
| cis-[OsMeCl(DPPE)$_2$] · $\frac{1}{2}$PhH[j] | Bright-yellow | 568–569 | |
| [Os$_3$(C$_6$H$_4$(CO)$_7$(PMe$_2$)$_2$][16] | Orange-red | | |
| [Os$_3$(C$_6$H$_4$)H(CO)$_9$(PMe$_2$)][16] | Yellow | | |
| [Os$_3$(C$_6$H$_4$)(CO)$_7$(PPh$_2$)$_2$][15] | Red | 511 | |
| [OsMe$_2$(CO)$_4$][p,q(8)] | White | 338–339 | $\nu$(CO) 2126 w, 2042 vs, 2009 s, 1976 w cm$^{-1}$ |
| [Os(Me)I(CO)$_4$][q] | White | 378–379 dec | |
| [Os(Me)H(CO)$_4$][q(8)] | White | 298 | $\nu$(CO) 2130 w, 2060 m, 2045 vs, 2028 s $\nu$(OsH) 1950 w |
| [(OC)$_4$HOs–OsMe(CO)$_4$][(8)s] | White | | $\nu$(CO) 2103 s, 2065 s, 2047 m, 2034 m, 2020, 2013 m, 2002 w |
| [(OC)$_4$MeOs–Os(CO)$_4$OsMe(CO)$_4$][(8)] | White | | $\nu$(CO) 2134, 2097 s, 2068 w, 2054 m, 2039 s, 2031 s, 2009 m, 1992 m |
| [Os$_3$($\mu$-H)$_2$($\mu$-CH$_2$)(CO)$_{10}$][r] | Yellow | | |

[a] H. J. Spiegel, G. Groh, and H. J. Berhold, *Z. Anorg. Allg. Chem.*, **398**, 225 (1973).
[b] A. Yamamoto, K. Morifuji, S. Ikeda, T. Saito, Y. Uchida, and A. Misono, *J. Am. Chem. Soc.*, **87**, 4652 (1965); A. Yamamoto, K. Morifuji, S. Ikeda, T. Saito, Y. Uchida, and A. Misona, *J. Am. Chem. Soc.*, **90**, 1878 (1968); T. Saito, Y. Uchida, A. Misono, A. Yamamoto, K. Morifuji, and S. Ikeda, *J. Am. Chem. Soc.*, **88**, 5198 (1966).
[c] F.-W. Kuepper, *J. Organomet. Chem.*, **13**, 219 (1968).
[d] Floriani and F. Calderazzo, *J. Chem. Soc. (A)*, 3665 (1971); C. Floriano and F. Calderazzo, *Chem. Commun.*, 417 (1968).
[e] J. Chatt and B. L. Shaw, *J. Chem. Soc.*, 285 (1961).
[f] R. B. King and M. B. Bisnette, *J. Organomet. Chem.*, **2**, 15 (1964); T. S. Piper and G. Wilkinson, *J. Inorg. Nucl. Chem.*, **2**, 136 (1956).
[g] See Ref. *e*, Table 4.16.
[h] R. Nast and F. Urban, *Z. Anorg. Allg. Chem.*, **287**, 17 (1956).
[i] R. Aumann and H. Averbeck, *Angew. Chem., Int. Ed. Engl.*, **15**, 610 (1976).
[j] J. Chatt and R. G. Hayter, *J. Chem. Soc.*, 6017 (1963).
[k] See Ref. *k*, Table 4.15.
[l] J. Chatt and J. M. Davidson, *J. Chem. Soc.*, 843 (1965).
[m] S. D. Ittel, C. A. Tolman, A. D. English, and J. P. Jesson, *J. Am. Chem. Soc.*, **100**, 7577 (1978).
[n] M. Preece, S. D. Robinson, and J. N. Wingfield, *J. Chem. Soc., Dalton Trans.*, 613 (1976).
[o] R. A. Jones and G. Wilkinson, *J. Chem. Soc., Dalton Trans.*, 472 (1979).
[p] J. Evans. S. J. Okrasinski, A. J. Pribula, and J. R. Norton, *J. Am. Chem. Soc.*, **99**, 5835 (1977).
[q] F. L'Eplattenier, *Inorg. Chem.*, **8**, 965 (1969); F. L'Eplattenier, *Chimia*, **23**, 144 (1969); F. L'Eplattenier and C. Pelichet, *Helv. Chim. Acta*, **53**, 1091 (1970); R. D. George, S. A. R. Knox, and F. G. A. Stone, *J. Chem. Soc., Dalton Trans.*, 972 (1973).
[r] R. B. Calvert and J. R. Shapley, *J. Am. Chem. Soc.*, **99**, 5225 (1977); R. B. Calvert, J. R. Shapley, A. J. Schultz, J. M. Williams, S. L. Suib, and G. D. Stucky, *J. Am. Chem. Soc.*, **100**, 6240 (1979).
[s] J. Evans, S. J. Okrasinski, A. J. Pribula, and J. R. Norton, *J. Am. Chem. Soc.*, **98**, 4000 (1976).

[RePh$_3$(PPhEt$_2$)$_2$], [Re$_6$Me$_{12}$(OAc)$_6$], [Re$_3$R$_6$Cl$_3$L$_6$], [Re$_6$(CH$_2$SiMe$_3$)$_6$Cl$_6$H$_6$], [Re$_6$(CH$_2$SiMe$_3$)$_9$Cl$_6$H], Re$_3$R$_{12}$, [Re$_3$Me$_6$(acac)$_3$], [Re$_3$Me$_9$(PR$_3$)$_x$] ($x = 0, 1, 2$), [Re$_3$Cl$_3$R$_3$(R'COO)$_3$], [ReR$_3$(PEt$_2$Ph)$_2$] (R = Ph, $p$-tolyl), $cis$-[ReMe$_3$(O)$_2$], [ReMe(O)$_3$], [O{Re(CH$_2$SiMe$_3$)$_3$(O)}$_2$], [N$_2${Re(CH$_2$SiMe$_3$)$_4$}$_2$], [ReR$_2$(N)(PPh$_3$)$_2$] (R = Ph, $p$-tolyl, $o$-tolyl). However, many water and air resistant rhenium(I) compounds such as [ReR(CO)$_5$] are known, as well as the rhenium(II) compound [RePh$_2$(PPh$_3$)$_2$], [Re$_2$(CH$_2$SiMe$_3$)$_2$Cl$_2$L$_2$] (L = CO, PPh$_3$), and [Re$_3$(CH$_2$SiMe$_3$)$_3$Cl$_3$py$_3$]. Rhenium complexes containing polymethylene bridging ligands, such as [(OC)$_5$ Re($\mu$-CH$_2$CH$_2$) Re(CO)$_5$], are also known.[87]

Homoleptic manganese(II) compounds are very unstable, pyrophoric, and very sensitive to water. Homoleptic rhenium complexes also readily react with oxygen, water, and alcohols. The compound Li$_2$[Re$_2$Me$_8$] is also pyrophoric. Heteroleptic compounds, like compounds of elements of other groups, are less reactive.

### f. *Complexes of group 8 metals*[1–5, 13, 85, 87, 88, 97, 98, 107, 138–155]

For this triad, only iron (Table 4.17) gives homoleptic complexes containing $\sigma$M−C bonds. The remaining elements form heteroleptic compounds. The stability of phosphine complexes of the type M(aryl)$_2$ (PR$_3$)$_2$ decreases considerably according to the series Ni > Co > Fe. Ruthenium and osmium complexes characteristically activate the C−H bond in coordinated ligands and form compounds with $\sigma$M−C bonds as a result of oxidative addition. The following reactions serve as examples:

$$[\text{Ru(arene)(Me}_2\text{PC}_2\text{H}_4\text{PMe}_2)_2] \longrightarrow [\text{Ru(aryl)(H)(Me}_2\text{PC}_2\text{H}_4\text{PMe}_2)_2] \tag{4.124}$$

$$2[\text{Ru(Me}_2\text{PC}_2\text{H}_4\text{PMe}_2)_2] \longrightarrow [\text{Ru}_2(\mu\text{-CH}_2\text{P(Me)C}_2\text{H}_4\text{PMe}_2)_2 \text{ (Me}_2\text{PC}_2\text{H}_4\text{Me}_2)_2 \text{ (H)}_2] \tag{4.125}$$

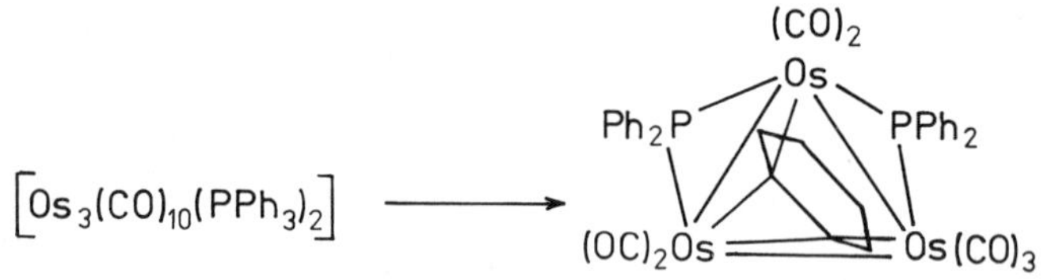

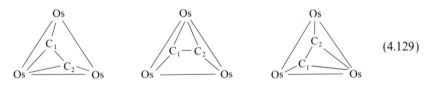

$$\left[\text{Os}_3(\text{CO})_{10}(\text{PPh}_3)_2\right] \longrightarrow \tag{4.126}$$

Similar reactions take place for complexes with PMe$_2$Ph which give products possessing analogous structures[15, 16]:

$$[\text{Os}_3(\text{CO})_{10} (\text{PMe}_2\text{Ph})_2] \longrightarrow [\text{Os}_3(\text{C}_6\text{H}_4)(\text{CO})_7 (\text{PMe}_2)_2] \tag{4.127}$$

$$[\text{Os}_3(\text{CO})_{11} (\text{PMe}_2\text{Ph})] \longrightarrow [\text{Os}_3(\text{C}_6\text{H}_4) \text{H(CO)}_9 (\text{PMe}_2)] \tag{4.128}$$

In this case, a C$_6$H$_4$ ligand is formed owing to breaking of a C−H bond as well as a P−C linkage. The C$_6$H$_4$ ligands in the two latter complexes show fluxional behavior; they undergo rapid rotation with respect to the osmium atoms (C$_1$ and C$_2$ denote atoms of C$_6$H$_4$ bonded to the metal atoms):

$$\tag{4.129}$$

Oxidative addition of benzene to $Os_3(CO)_{12}$ readily takes place, leading to the formation of $Os_3(\mu_3\text{-}C_6H_4)H_2(CO)_9$. Olefins react with $M_3(CO)_{12}$ (M = Ru, Os) in a similar way (see Chapter 3).

The reaction of the polynuclear complex $Ru_2(OAc)_4Cl$ with bis(trimethylsilylmethyl)magnesium, $Mg(CH_2SiMe_3)_2$, leads to the abstraction of the $\delta$ hydrogen atom and formation of the chelating compound[17]:

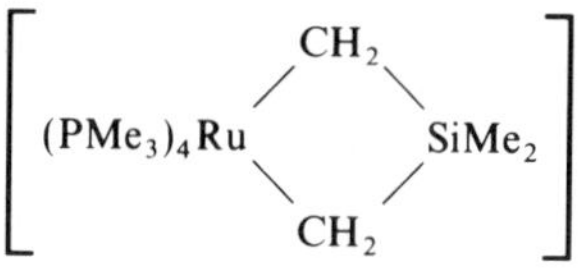

$$\left[ (PMe_3)_4Ru \underset{CH_2}{\overset{CH_2}{<\quad>}} SiMe_2 \right]$$

The iron group metals form the following representative compounds containing $\sigma M - C$ bonds:

$Li_2[FeMe_4]$, $[FeEt_2(bipy)]$, $[Fe(o\text{-}C_6H_4CH_2NMeCH_2CH_2NMeCH_2C_6H_4\text{-}o)]$, $[Fe(C_6H_4CH_2NMe_2)_2]$, $[FePh(salen)]$, $[FeR_2(PR_3')_2]$, $[Fe(o\text{-}C_6H_4OC_6H_4\text{-}o)_2]^{2-}$, $[Fe(o\text{-}C_6H_4CH_2OMe)_2]$, $[FeI(COMe)(CO)_2(PMe_3)_2]$, $[FeRcp(CO)_2]$, $[(OC)_2cpFe\{\mu\text{-}(CH_2)_n\}Fecp(CO)_2]$ ($n = 3\text{--}12$),

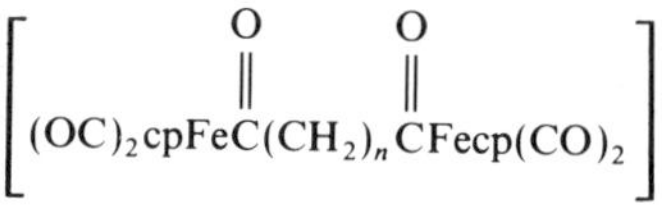

$$\left[ (OC)_2cpFe\overset{O}{\overset{\|}{C}}(CH_2)_n\overset{O}{\overset{\|}{C}}Fecp(CO)_2 \right] \quad (n = 3, 4, 7)$$

$[FeR(CO)_4]^-$, $K_3[Fe(CCR)_6]$, $K_4[Fe(CCR)_6]$, $[Fe(norbornyl)_4]$, $[Ru(aryl)H(diphos)_2]$, $[MXR(diphosphine)_2]$ (M = Ru, Os; R = alkyl, aryl), $[RuMe_2(PMe_3)_4]$,[17] $[Ru\{o\text{-}C_6H_4OP(OPh)_2\}X(CO)\{P(OPh)_3\}_2]$ (X = H, Cl),

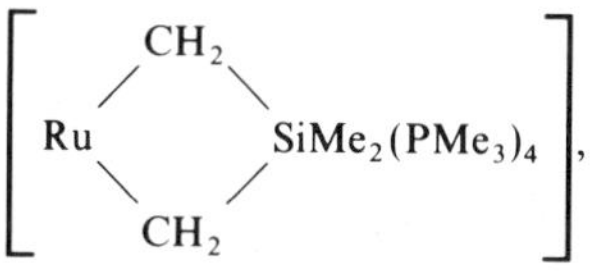

$$\left[ Ru \underset{CH_2}{\overset{CH_2}{<\quad>}} SiMe_2(PMe_3)_4 \right], \quad [Ru_2(CH_2PMeC_2H_4PMe_2)_2(H)_2(Me_2PC_2H_4PMe_2)_2]$$

$[(OC)_2cp(Ru)\{\mu\text{-}(CH_2)_n\}Rucp(CO)_2]$, $[Os_3(C_6H_4)(CO)_7(PR_2)_2]$, $[OsR_2(CO)_4]$ (R = alkyl), $[Os(Me)H(CO)_4]$, $[Os(Me)I(CO)_4]$, $[OsMe(CO)_4]^-$, $[(OC)_4MeOsOs(CO)_4OsMe(CO)_4]$, $[(OC)_4HOsOsMe(CO)_4]$, and $[(OC)_4Os\{\mu\text{-}(CH_2)_n\}Os(CO)_4]^+$ ($n = 2,3$).

Processes of the C—H bond activation as well as migratory insertion have been investigated intensively. Therefore, many complexes of group 8 metals with $\sigma M - C$ bonds have been studied. The reaction of fac-$[FeMe(CO)_3(PP)]^+$ with CO or $P(OMe)_3$ leads to the formation of a corresponding acyl compound. However, in the case of $[FeMe(I)(CO)_2(PP)]$ the reaction involving the same Lewis base (L) furnishes products of apparent insertion (L = CO) or substitution $\{L = P(OMe_3)\}$: fac-$[FeAc(CO)_2L(PP)]^+$ [L = CO, $P(OMe)_3$], $[\{FeMe(I)(CO)_2LP\}P]$ $\{L = P(OMe)_3$, PP = $o\text{-}C_6H_4(PMe_2)_2\}$.[139] The direction of the reaction depends on steric factors. In the case of optically active compounds of the type $[Fe(R)cp(CO)L]$, the reaction with CO gives products of R or CO migration[141] depending upon the solvent [reaction (2.26)]. For synthesis of oxygen $C_2$ compounds from the *syn* gas, CO + $H_2$, it is important to study the insertion of CO into the $M-C$ bond in complexes containing oxygen hydrocarbyl ligands. For example, the complex $[Fe(CH_2OR)\,cp(CO)(L)]$ gives the following compounds:

$$\left[\underset{Fe(CCH_2OR)cp(CO)L}{\overset{O}{\overset{\|}{}}}\right], \qquad [Fe(CH_2COOR)cp(CO)L] \quad [L=CO, P(OMe_3), PPh_3]^{(144)}$$

The CO insertion was also studied for iron hexacoordinate methyl complexes. Thus the complexes $FeMe(X)(CO)_2(PMe_3)_2$ ($X=Cl$, Br, NCS, CNO, and CN) give $[Fe(COMe)\,X(CO)_2(PMe_3)_2]$.[148]

The reactions of Fe, Ru, Os, and Rh complexes containing $PMe_3$ may lead to the $C-H$ bond cleavage during its oxidative addition to the metal[140, 143]:

$$M(PMe_3)_4 \rightleftharpoons [\overline{M(CH_2}PMe_2)\,H(PMe_3)_2] \quad (M=Fe, Ru, Os) \qquad (4.130)$$

Other similar compounds are known: $[\overline{Ru(CH_2}PMe_2)\,Cl(PMe_3)_3]$ and $[\overline{Ru(CH_2}PMe_2)_2(PMe_3)_2]$. The complex $[Os(2,3,5\text{-}\eta\text{-}C_{10}H_{13})(\eta\text{-}C_6H_3Me_3\text{-}1,3,5)][PF_6]$ possesses a three-center, two-electron (3c–2e) $C-H-Os$ bond.[138]

Recently, further orthometallation reactions of phosphines in dinuclear carboxylate complexes were carried out to give compounds of the type $[Os_2(\mu\text{-}OOCR)_2\{\mu\text{-}P(C_6H_4)Ph_2\}_2Cl_2]$.[146] The complex $[Ru(\mu\text{-}PhCONH)_4Cl]$ reacts with $PPh_3$ to cause breaking of the phosphorus–phenyl bond with the formation of the phenyl compound $[Ru_2Ph_2(PhCONH)_2\{Ph_2POC(Ph)N\}_2]$.[146]

Complexes of iron and other metals which are prepared from $\sigma$ allyl or propargyl organometallic compounds and $(NC)_2C=C(CN)_2$, $SO_2$, or related unsaturated electrophilic molecules via cycloaddition reactions are important because such complexes may find application in organic synthesis[147]:

$$MCH_2CR^1=CR^2R^3 + (NC)_2C=C(CN)_2 \longrightarrow \qquad (4.131)$$

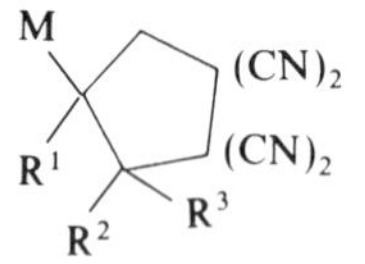

$$MCH_2C\equiv R + SO_2 \longrightarrow \qquad (4.132)$$

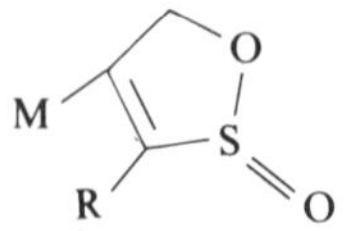

Also, many new aryl porphyrine high- and low-spin iron complexes were obtained.[142] The aryl group may migrate from the iron metal to the nitrogen atom.

Very reactive iron complexes containing bonded chelating ligands are interesting. They are obtained by the following reaction[150]:

$$[Fe(CO)_3L]^{2-} + [F_3CSO_3CH_2CH_2]_2 \xrightarrow{-2F_3CSO_3^-} \left[L(OC)_3Fe\right] \qquad (4.133)$$

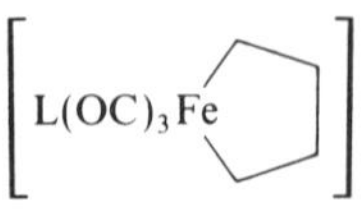

Also known are

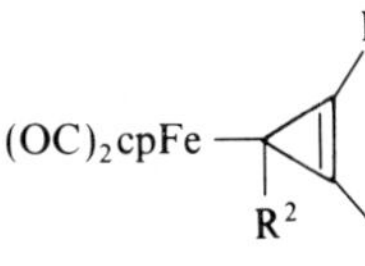

$(OC)_2cpFe$ ![cyclopropenyl structure with R$^1$, R$^2$, R$^1$ substituents] $(R^1 = R^2 = Ph; R^1 = Ph, R^2 = H; R^1 = Bu^t, R^2 = Me)^{(149)}$

as well as osmium(VI) compounds[145] such as $[Osnp_4'(N)]^-$, $[Osnp_2'(Cl_2)(N)]^-$ $(np' = CH_2SiMe_3)$, $[OsMe_3(N)(THF)_2]$, $[Os(CH_2Ph)_4N]^-$, $[Osnp_4(N)]^-$, $[OsMe_4(N)]^-$.

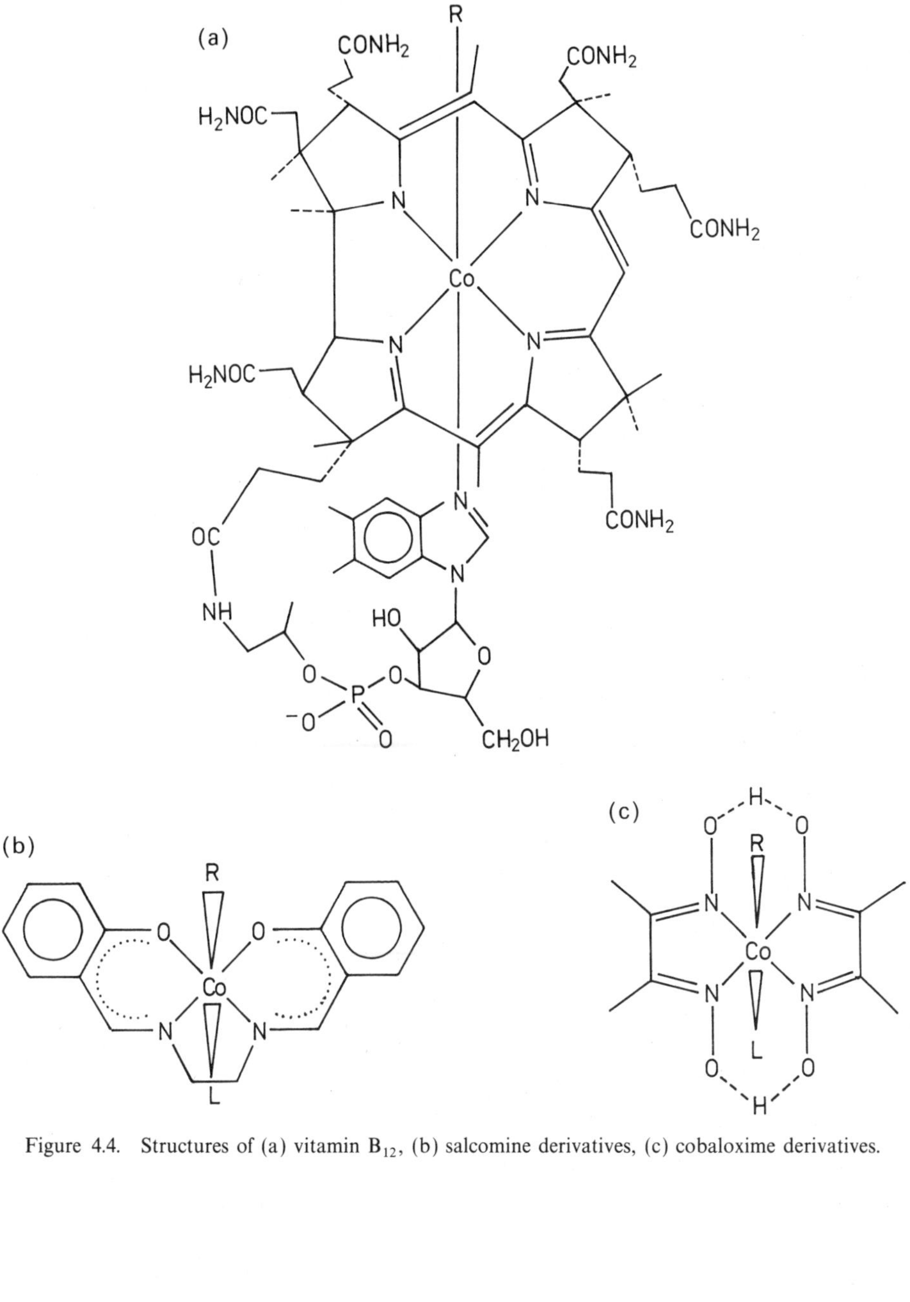

Figure 4.4. Structures of (a) vitamin $B_{12}$, (b) salcomine derivatives, (c) cobaloxime derivatives.

## g. Complexes of group 9 metals[1-5, 13, 85, 87, 88, 97, 98, 156-168]

Cobalt forms its most stable compounds containing $\sigma M - C$ bonds in $+2$ and $+3$ oxidation states, while rhodium and iridium form such compounds in $+1$ and $+3$ oxidation states (Table 4.18). Cobalt(I) complexes, such as $CoR(CO)_4$, are labile and very unstable in the case of alkyl ligands. Characteristic for cobalt are compounds of the type $[Co(CN)_5R]^{3-}$ which are stable in neutral aqueous solutions. Also stable are the following cobalt compounds: vitamin $B_{12}$, i.e., cyanocobalamin, and its models such as cobaloximes (Figure 4.4c) as well as alkyl derivatives of salcomine (Figure 4.4b), etc. The vitamin $B_{12}$ is a hexacoordinate Co(III) compound possessing a ring system of the corrin type. Along the axis which is perpendicular to the corrin ring, the following ligands are coordinated: 5,6-dimethylbenzimidazole and R (cyanide, $H_2O$, nitrile, alkyl, etc.) (Figure 4.4a).

Aquacobalamin, $R = H_2O$ (Figure 4.4a), undergoes reduction to the Co(II) compound, vitamin $B_{12r}$, under the influence of a weak reducing agent. However, if strong reducing agents are used, the Co(I) compound, vitamin $B_{12s}$, is obtained, which may react with alkylating agents, acetylenes, or olefins to form more stable Co(III) complexes containing a Co−C bond. Prepared in this manner, alkyl derivatives do not decompose under the influence of acids. However, they are sensitive to light; under its influence, they form vitamin $B_{12r}$, and as a result of the cobalt–alkyl bond splitting, various organic compounds are obtained.

In model compounds (Figure 4.4b, c),[161] there is a possibility of an exchange of both axial and equatorial ligands which allows formation of compounds possessing Co−C bonds of various degrees of stability. In cobaloximes of the type $[CoR(dmgH)_2L]$ the Co−N distances change very little, while the differences between the Co−C distances are relatively great, and they depend on steric hindrance of R and the type of the ligand occupying the *trans* position.[161] Also, the Co−C bond energy depends on steric and electronic factors of L and R.[76] The energy of the cobalt–carbon bond $D(Co-C)$ in the coenzyme $B_{12}$ (adenosine cobalamine) equals $108.7 \pm 8.4$ kJ mol$^{-1}$.[76] Octahedral cobalt complexes may lose the axial ligand and form pentacoordinate compounds possessing square-pyramidal structures. Similar derivatives are also formed by other elements of group 8–10 metals, for example: $[FeR(salen)]$, $[RhR(L)(salen)]$, $[IrRL(salen)]$ $(L = PR_3, py)$, and $[Ru_2R_2(salen)_2]$.

Table 4.18. σ-Carbyl Complexes of Cobalt Group Metals

| Compound | Color | Mp (K) | Remarks |
|---|---|---|---|
| $[Co(CH_2Ph)(CO)_3PPh_3]^a$ | Yellow | 408 | |
| $[CoEt_2(dipy)]^b$ | | | |
| $[Co(COCH_2Ph)(CO)_3(PPh_3)]^a$ | Yellow | 396 | |
| *trans*-$[Co(2,4,6-C_6H_2Me_3)_2(PEt_2Ph)_2]^c$ | Yellow | | |
| $[Li(TMED)]_2[CoMe_4]^d$ | Blue | | |
| $[CoMe(salen)]^e$ | Dark-red | | |
| $[CoMe(dmgH)_2]^f$ | Dark-red | | |
| $[CoCF_3(CO)_4]^g$ | Bright-amber | 283.5–284 | |
| $K_4[Co(CCH)_6]^h$ | Black-green | | Strongly explosive |
| $K_3[Co(CCH)_6]^h$ | Green | | Explosive |

*(Table continued)*

*Table 4.18. (continued)*

| Compound | Color | Mp (K) | Remarks |
|---|---|---|---|
| [Co(norbornyl)$_4$]$^i$ | Brown | | $\mu_{eff} = 2.9$ BM |
| [RhPh(PPh$_3$)$_3$]$^k$ | Yellow | 433–443 dec | |
| [RhPh(PMe$_3$)$_3$]$^j$ | Orange-yellow | 358–363 dec | |
| [RhMe(PPh$_3$)$_3$]$^k$ | Orange-yellow | 393–413 dec | |
| [RhBr(1-C$_{10}$H$_7$)$_2$(PEt$_2$Ph)$_2$]$^l$ | Bright pink | 417–419 | |
| [RhMe(py)(dmgH)$_2$]$^{m,n}$ | | 435 dec | |
| [Rh(COEt)(CO)$_2$(PPh$_3$)$_2$]$^o$ | Bright-yellow | | $\nu$(CO), 1975 s, 1940 s, 1628 s |
| *trans*-[Rh$_2$($\mu$-CH$_2$)cp$_2$(CO)$_2$]$^{(19)}$ | Bright-red | 367 | $\tau$(CH$_2$) 2.88, $\tau$(cp) 4.50, $\nu$(CO) 1950 cm$^{-1}$ |
| [Rh$_2$($\mu$-CHMe)cp$_2$(CO)$_2$]$^{(19)}$ | Red-brown | 384–385 dec | $\tau$(H) 1.32, $\tau$(cp) 4.57, $\tau$(Me) 7.47 {(CD$_3$)$_2$CO}, $\nu$(CO) 1948 cm$^{-1}$ (KBr) |

$$\left[\begin{array}{c} \mathrm{Rh} \underset{\mathrm{CH_2-CH_2}}{\overset{\mathrm{CH_2-CH_2}}{\diagup \diagdown}} \mathrm{CH_2(C_5Me_5)PPh_3} \end{array}\right]^p$$

| Compound | Color | Mp (K) | Remarks |
|---|---|---|---|
| (above) | Orange-red | 433 dec | |
| [Rh{$o$-C$_6$H$_4$OP(OPh)$_2$}{P(OPh)$_3$}$_3$]$^q$ | Yellow | | |
| *fac*-[RhMe$_3$(PMe$_3$)$_3$]$^t$ | Yellow | 379–381 | |
| [Ir(COEt)(CO)$_2$(PPh$_3$)$_2$]$^o$ | White | 389 dec | $\nu$(CO) 1978 s, 1920 s, 1628 s |
| [Ir$_2$Cl$_4$Et$_2$(CO)$_4$]$^r$ | White | >409 dec | |
| [IrCl$_2$Et(CO)$_2$py]$^r$ | White | 409–410 dec | |
| *fac*-[IrMe$_3$(PMe$_2$Ph)$_3$]$^s$ | White | 464–468 dec | $\nu$(Ir$-$C) 510 m |

$^a$ Z. Nagy-Magos, G. Bor, and L. Marko, *J. Organomet. Chem.*, **14**, 205 (1968).
$^b$ T. Saito, Y. Uchido, A. Misono, A. Yamamoto, K. Morifuji, and S. Ikeda, *J. Organomet. Chem.*, **6**, 572 (1966).
$^c$ P. G. Owston and J. M. Rowe, *J. Chem. Soc.*, 3411 (1963).
$^d$ See Ref. *f*, Table 4.12.
$^e$ G. Costa, G. Mestroni, and L. Stefani, *J. Organomet. Chem.*, **7**, 493 (1967); G. Costa, G. Mestroni, and E. deSavagnani, *Inorg. Chim. Acta*, **3**, 323 (1969); G. Costa and G. Mestroni, *Tetrahedron Lett.*, **41**, 4005 (1967).
$^f$ G. N. Schrauzer and R. J. Windgassen, *J. Am. Chem. Soc.*, **88**, 3738 (1966); G. N. Schrauzer and R. J. Windgassen, *J. Am. Chem. Soc.*, **89**, 1999 (1967).
$^g$ W. R. McClellan, *J. Am. Chem. Soc.*, **83**, 1598 (1961).
$^h$ R. Nast and H. Lewinsky, *Z. Anorg. Allg. Chem.*, **282**, 210 (1955).
$^i$ See Ref. *e*, Table 4.16.
$^j$ See Ref. *o*, Table 4.17.
$^k$ W. Keim, *J. Organomet. Chem.*, **14**, 179 (1968); W. Keim, *J. Organomet. Chem.*, **8**, P25 (1967).
$^l$ J. Chatt and A. E. Underhill, *J. Chem. Soc.*, 2088 (1963).
$^m$ V. B. Panov, M. L. Khidekel, and S. A. Shchepilov, *Izv. Akad. Nauk SSSR, Ser. Khim.*, 2397 (1968); B. F. Rogachev and M. L. Khidekel, *Izv. Akad. Nauk SSSR, Ser. Khim.*, 141 (1969).
$^n$ J. H. Weber and G. N. Schrauzer, *J. Am. Chem. Soc.*, **92**, 726 (1970).
$^o$ G. Yagupsky, C. K. Brown, and G. Wilkinson, *Chem. Commun.*, 1244 (1969); G. Yagupsky, C. K. Brown, and G. Wilkinson, *J. Chem. Soc. (A)*, 1392 (1970).
$^p$ P. Diversi, G. Ingrosso, and A. Lucherini, *Chem. Commun.*, 52 (1977).
$^q$ See Ref. *n*, Table 4.17.
$^r$ B. L. Shaw and E. Singleton, *J. Chem. Soc. (A)*, 1683 (1967).
$^s$ J. Chatt and B. L. Shaw, *J. Chem. Soc. (A)*, 1836 (1966); B. L. Shaw and A. C. Smithies, *J. Chem. Soc. (A)*, 1047 (1967).
$^t$ See Ref. *k*, Table 4.15.

Cobalt(IV) complexes are also known. Complexes of the type $[MR(pmcp) H(PR_3)]$ (M = Rh, Ir) are intensively studied because they activate C−H bonds.[104, 162–164] Cobalt complexes of the type $[cpCo\{\mu\text{-}(CH_2)_n\}(\mu\text{-CO})_2Cocp]$ ($n = 2$–4) with bridging $(CH_2)_n$ ligands possess the Co−Co bond.[165]

Dinuclear cobalt compounds such as $cp(R)Co(\mu\text{-CO})_2 Co(R)cp$ decompose to give ketones and other products.[166]

Rhodium(II) complexes of the type $[Rh_2(\mu\text{-Ph}_2PC_6H_4)(\mu\text{-OAc})_2]\cdot 2L$ (L = HOAc, py) may be obtained by orthometallation of triphenylphosphine during a reaction with $Rh_2(OAc)_4$.[167] Dicarbonyl-1,3-diphenyl-1,3-propanedionatorhodium(I) splits the Sb−C bond in SbPh$_3$ leading to the formation of the phenyl rhodium(III) complex $[RhPh_2(PhCOCHCOPh)(SbPh_3)_2]$.[168]

The cobalt group metals form the following representative compounds: $[Co(CH_2Ph)(CO)_3(PPh_3)]$, $[Co_3(CR)(CO)_9]$, $[CoR(CO)_4]$ (R = alkyl, aryl, acyl), $[CoEt(dipy)_2]$, $[Co(o\text{-}C_6H_4N=NPh)(CO)_2(PMePh_2)]$, $[Co(o\text{-}C_6H_4N=NPh)(CO)_3]$, $[CoR(X)(PR_3)_2]$, $[CoR_2L_2]$, $[CoR_2(dipy)]$, $[Co(o\text{-}C_6H_4OC_6H_4\text{-}o)_2]^{2-}$, $[CoR_4]^{2-}$ (R = Me, CH$_2$SiMe$_3$), $[Co(o\text{-}C_6H_4OC_6H_4\text{-}o)_3]^{3-}$, $[Co(C\equiv CR)_6]^{4-}$, $[Co(CCR)_6]^{3-}$, $[Co(CN)_5R]^{3-}$, $[CoR_2cpL]$, $[CoR_3(PR_3)_3]$, $[CoR_2X(PR_3)_3]$, $[CoR_2(dipy)_2]^+$, $[Co(norbornyl)_4]$, $[CoR(X)cpL]$, $[Co(o\text{-}C_6H_4CH_2NR_2)_3]$, $[CoR(salen)]$, $[CoR(dmgH)_2]$, $[(OC)cpRh(\mu\text{-CHR})Rhcp(CO)]$ (R = H, Me),[19] $[MR(PPh_3)_3]$ (M = Co, Rh, Ir), $[MR(CO)(PPh_3)_2]$ (M = Rh, Ir), $[MR_3L_3]$ (M = Rh, Ir),[17] $[RhX_2RL_2]$, $[RhX_2R(CO)L_2]$, $[RhXR_2L_2]$, $K_2[Rh(CN)_4 Me(H_2O)]$, $[Rh(o\text{-}C_6H_4N=NPh)_2(OAc)]$, $[Rh\{(CH_2)_2SiMe_2\}(CH_2SiMe_3)(PMe_3)_3]$,[17] $[RhR(porphyrine)]$,[18] $[Rhcp(Me)\{P(OMe)_3\}_2]^+ X^-$, $[Rhcp(Me)\{P(OMe)_3\}\{P(O)(OMe)_2\}]$,

$$\left[ Rh \begin{array}{c} CH_2 \\ \diagdown \\ \diagup \\ CH_2 \end{array} (CH_2)_n(\eta\text{-}C_5Me_5)(PPh_3) \right] \quad (n = 2, 3, 4)$$

$[Rh\{o\text{-}C_6H_4OP(OPh)_2\}\{P(OPh)_3\}_3]$, $[IrX_2R(CO)L_2]$, $[IrX_2RL_3]$, and $[IrXR_2L_3]$.

## h. *Complexes of group 10 metals*[1–5, 11, 13, 63, 85, 87, 88, 90, 91, 97, 98, 106, 171–184]

Typical for these metals are square-planar complexes such as $[MR_4]^{2-}$, $[MR(X)L_2]$, and $MR_2L_2$ (Table 4.19). The stability of square compounds of the type

*Table 4.19. σ-Carbyl Complexes of Nickel Group Metals*

| Compound | Color | Mp (K) | Remarks |
|---|---|---|---|
| $[Ni(CMe_3)_2]^a$ | Violet | 393 dec | |
| $[Ni(o\text{-}CH_2C_6H_4PPh_2)_2]^b$ | Yellow | | |
| $[NiEt_2(bipy)]^c$ | Dark-green | | Diamagnetic 720 nm ($\varepsilon = 4900$), 430 nm (4500) |
| $[NiCl(2\text{-}C_5H_4N)(PPh_3)_2]^d$ | Brown-yellow | 460 | |

*(Table continued)*

*Table 4.19. (continued)*

| Compound | Color | Mp (K) | Remarks |
|---|---|---|---|
| $[Ni(1\text{-adme})_2(bipy)]^e$ | Brown-green | 451 dec | |
| $[NiCl(3\text{-}C_5H_4N(PPh_3)_2]^d$ | Brown-yellow | 397 | |
| $[Ni_2\{(CH_2)_2PMe_2\}_4]^{f,g}$ | Yellow | | |
| $[NiMe_2(diphos)]^h$ | Yellow | 403 dec | |
| $Li_2[NiMe_4]\cdot 2\,THF^i$ | Gold-yellow | 402–403 dec | |
| $Li_2[NiPh_4]\cdot 4\,THF^i$ | Yellow | 375–376 dec | |
| $K_4[Ni(CCH)_4]^j$ | Orange | | Pyrophoric, does not explode |
| $[Ni(C_6F_5)_2(\eta\text{-PhMe})^k$ | Red-brown | 412–413 | |
| $K_2[Ni(CCH)_4]^{l,j}$ | Yellow | | |
| $[Ni(2,4,6\text{-}C_6H_2Me_3)_2(PEt_2Ph)_2]^w$ | Orange-yellow | | |
| $[Pd(1\text{-adme})_2(bipy)]^e$ | Red | 433 dec | |
| $[Pd(o\text{-}CH_2C_6H_4PPh_2)_2]^b$ | White | 520 dec | |
| $cis\text{-}[Pd(o\text{-}C_6H_4CH_2NMe_2)_2]^m$ | White | 453 | |
| $[PdMe_2(PPh_3)_2]^n$ | | 470–471 | |
| $[Pd(o\text{-}C_6H_4N=NPh)(acac)]^o$ | Orange-brown | 453.5–454.5 | |
| $trans\text{-}[Pd(C_6H_5)_2(SC_4H_8)_2]^{p,q}$ | | 423 | |
| $[Bu_4^nN][Pd(C_6F_5)_3(C_4H_8S)]^{p,q}$ | | 375 | |
| $[Bu_4^nN]_2[Pd(C_6F_5)_4]^{p,q}$ | | 431 | |
| $[PtMe_2(PPh_3)_2]^r$ | White | | |
| $[PtMe_2(COD)]^s$ | White | | |
| $[Pt(1\text{-adme})_2(COD)]^e$ | Yellow-white | 434 | |
| $[Me_2Pt(\mu\text{-COT})PtMe_2]^t$ | Yellow | 441–444 dec | |
| $[Bu_4^nN][Pt(C_6F_5)_3PPh_3]^{p,q}$ | | 445 | |
| $Li_2[PtMe_4]^u$ | White | | |
| $[PtClMe_3]_4^v$ | White | | |
| $[PtI(Me_3)]_4^v$ | Orange | 468–473 dec | |
| $Li_2[PtMe_6]^u$ | White | | |
| $[PtMe_4(PMe_3)_2]^u$ | White | 429–430 | |
| $[PtMe_4(PMePh_2)_2]^u$ | White | 393 | |

[a] G. Wilke and H. Schott, *Angew. Chem.*, **78**, 592 (1966).

[b] G. Longoni, P. Chini, F. Canziani, and P. Fantucci, *Chem. Commun.*, 470 (1971).

[c] T. Saito, Y. Uchida, A. Misono, A. Yamamoto, K. Morifuji, and S. Ikeda, *J. Am. Chem. Soc.*, **88**, 5198 (1966).

[d] K. Isobe, Y. Nakamura, and S. Kawaguchi, *Chem. Lett.*, 1383 (1977).

[e] See Ref. *a*, Table 4.16.

[f] H. H. Karsch and H. Schmidbaur, *Angew. Chem., Int. Ed. Engl.*, **12**, 853 (1973); H. H. Karsch and H. Schmidbaur, *Chem. Ber.*, **107**, 3684 (1974).

[g] D. J. Brauer, C. Krueger, P. J. Roberts, and Y.-H. Tsay, *Chem. Ber.*, **107**, 3706 (1974).

[h] M. L. H. Green and M. H. Smith, *J. Chem. Soc. (A)*, 639 (1971).

[i] R. Taube and G. Honymus, *Angew. Chem., Int. Ed. Engl.*, **14**, 261 (1975).

[j] R. Nast and K. Vester, *Z. Anorg. Allg. Chem.*, **279**, 146 (1955).

[k] S.-T. Lin, R. N. Narske, and K. Klabunde, *Organometallics*, **4**, 571 (1985).

[l] R. Nast and H. Kasperl, *Z. Anorg. Allg. Chem.*, **295**, 227 (1958).

[m] G. Longoni, P. Fantucci, P. Chini, and F. Canziani, *J. Organomet. Chem.*, **39**, 413 (1972).

[n] G. Calvin and G. E. Coates, *J. Chem. Soc.*, 2008 (1960).

[o] R. F. Heck, *J. Am. Chem. Soc.*, **90**, 313 (1968).

[p] R. Uson, J. Fornies, P. Espinet, R. Navarro, F. Martinez, and M. Tomas, *Chem. Commun.*, 789 (1977).

[q] R. Uson, J. Fornies, F. Martinez, and M. Tomas, *J. Chem. Soc., Dalton Trans.*, 888 (1980).

[r] J. Chatt and B. L. Shaw, *J. Chem. Soc.*, 705 (1959).

[s] H. C. Clark and L. E. Manzer, *J. Organomet. Chem.*, **59**, 411 (1973).

[t] J. Mueller and P. Goeser, *Angew. Chem., Int. Ed. Engl.*, **6**, 363 (1967); C. R. Kistner, *Inorg. Chem.*, **2**, 1255 (1963).

[u] G. W. Rice and R. S. Tobias, *J. Am. Chem. Soc.*, **99**, 2141 (1977).

[v] A. Robson and M. R. Truter, *J. Chem. Soc.*, 630 (1965); J. J. Chatt and L. A. Duncanson, *J. Chem. Soc.*, 2939 (1953); H. Gilman, M. Lichtenwalter, and R. A. Benkeser, *J. Am. Chem. Soc.*, **75**, 2063 (1953).

[w] J. Chatt and B. L. Shaw, *J. Chem. Soc.*, 1718 (1960).

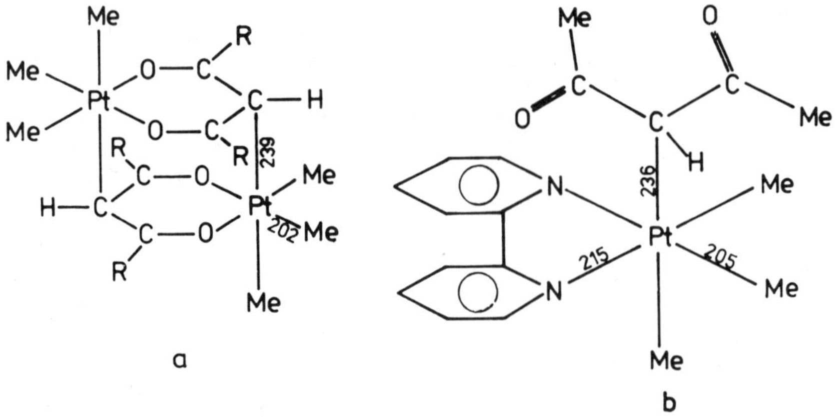

Figure 4.5. Structures of (a) $Pt_2(C_3H_7COCHCOC_3H_7)_2Me_6$ and (b) $PtMe_3(acac)(bipy)$ (distances are in pm units).

$MR_2L_2$ as a function of M, R, and L varies as follows: $Pt > Pd \sim Ni > Co > Fe$; $o$-tolyl $> m$-tolyl $> p$-tolyl $> Ph > $ alkyl; $PR_3 > AsR_3 > SbR_3$.

Nickel complexes are yellow or brown, while platinum and palladium derivatives are colorless or light yellow. The thermal stability and resistance to oxidation of aryl compounds of the type $[Ni(Ar)X(PR_3)_2]$ and $[Ni(Ar)_2(PR_3)_2]$ (Ar represents an aryl group substituted in *ortho* position) may be explained by formation of the $d\pi$–$p\pi$ bond between the metal and the aryl. Alkyl complexes are stable if they contain chelating ligands or bulky ligands; both increase the stability or inertness of such compounds. Nickel(0) forms characteristic anionic complexes which are obtained by reactions of olefin nickel compounds with LiR or $NaR^{(63)}$: $[Li(TMEDA)_2][NiR(C_2H_4)_2]$ (R = Me, Et, n-Bu, Ph), $[Li(THF)_4][NiPh(CDT)]$, $[(NaPh)_2 (ether)_x NiC_2H_4]_2$ (ether = THF, $Et_2O$), and $[(LiPh)_2 (THF)_4 NiC_{14}H_{10}]$, while platinum gives, unlike palladium and nickel, octahedral Pt(IV) compounds. These compounds may be mononuclear or polynuclear. Exceptions are platinum(III) compounds such as $[PtR_2(OOCR')L]_2^{(173)}$ which possess platinum–platinum bonds. Coordination compounds of platinum with $\beta$-diketones often exhibit bonding between platinum and the central carbon atom of the diketone. Examples of such compounds are $[Pt_2(diketone)_2Me_6]$ and $[PtMe_3(acac)(bipy)]$ (Figure 4.5). Platinum and palladium give alkyl[bis(diphenyl-phosphino)methane] complexes which, in some cases, are "A-frame" compounds [structures (4.134) and (4.135)].$^{(177)}$ Migratory insertion reactions into $M-C$ bonds

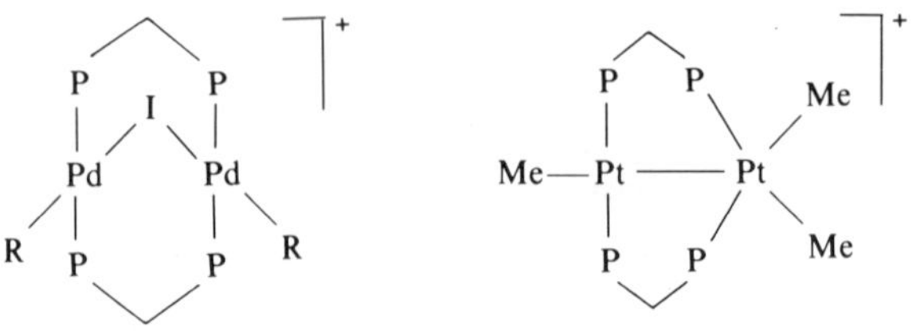

(4.134)

(R = Me, MeCO)

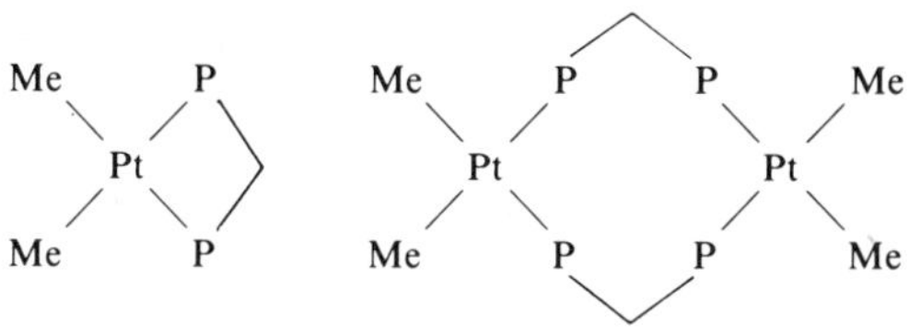

$$
\text{(4.135)}
$$

$$
(P \frown P = \text{dppm, dmpm, depm})
$$

continue to be studied intensively. The hydride complex *trans*-$[\text{PtClH}\{\text{P(OMe)}_3\}_2]$ reacts with methyl acrylate to give *cis*-$[\text{Pt}\{\text{CH(Me)COOMe}\}\text{Cl}\{\text{P(OMe}_3)_2\}]$.[178] Analogous compounds containing $\text{PEt}_3$ do not react with $\text{CH}_2=\text{CHCOOMe}$. Intramolecular insertion of acetylene in *trans*-$[\text{L}_2\text{ClPdCO}_2(\text{CH}_2)_n\text{C}\equiv\text{CMe}]$ proceeds via tetracoordinate rather than pentacoordinate compounds in which the $\text{C}\equiv\text{C}$ bond occupies a site of one molecule of phosphine [equation (4.136)].[179]

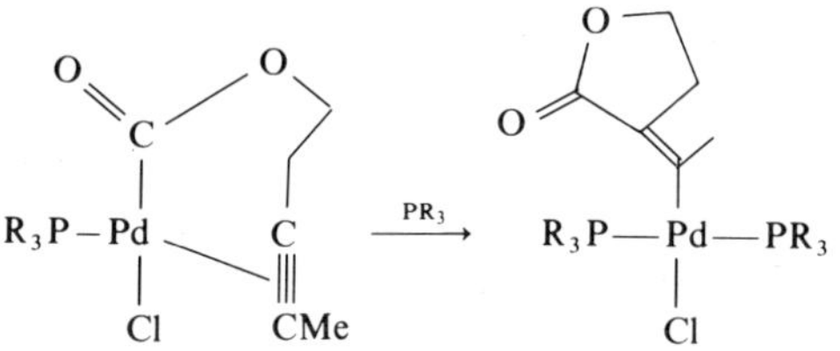

$$
\text{(4.136)}
$$

In the case of the platinum compound with the chelating ligand 2,2-dimethyl-4-pentene-1-yl, reversible insertion takes place.[180] This reaction represents one of few examples of such reversible reactions [equation (4.137)]. Carbonyl ligand migration may also occur to the nitrogen atom [equation (4.138)].[181] Contrary to the behavior of the linear bis(trimethylsilyl)amides, in this case *N,N*-disilylamide is formed.

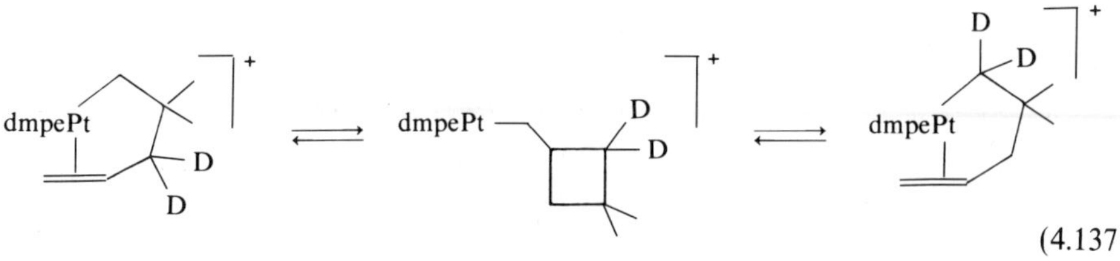

$$
\text{(4.137)}
$$

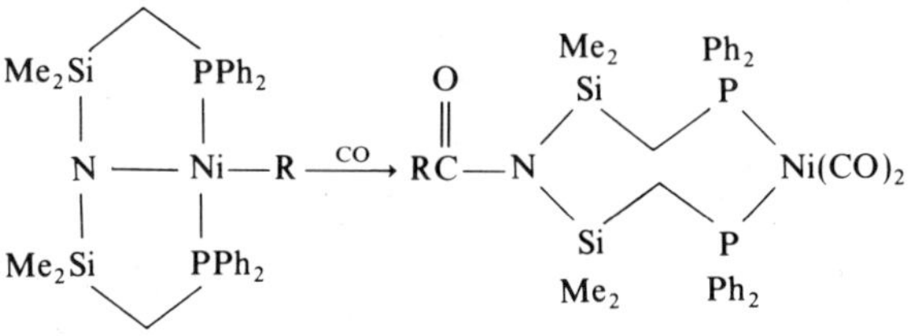

$$
\text{(4.138)}
$$

$$
(R = \text{Me, CH}_2\text{CH}=\text{CH}_2, \text{CH}=\text{CH}_2, \text{Ph})
$$

The complexes $[Pt\{C_6H_3(CH_2NMe_2)_2\text{-}2,6\}X]$ react with MeI and $CF_3SO_3Me$ to give the complex $[Pt\{MeC_6H_3(CH_2NMe_2)_2\text{-}2,6\}X]^+ CF_3SO_3^-$, in which the methyl group is bonded to the carbon atom attached to platinum.(182) In the first step of the reaction, oxidative addition of MeI takes place followed by methyl migration to the phenyl ring. Oxidative addition of the C–H bond in tetramethylthiourea (tmtu) leads to the formation of the platinum(IV) carbyl complex (4.139).(183) Platinum(II) $\sigma$-allyl complexes react with tetracyanoethylene to give cycloaddition products(184):

NMe2
C
S
NMe
Br
CH2
Pt
CH2
Br
S
NMe
C
NMe2

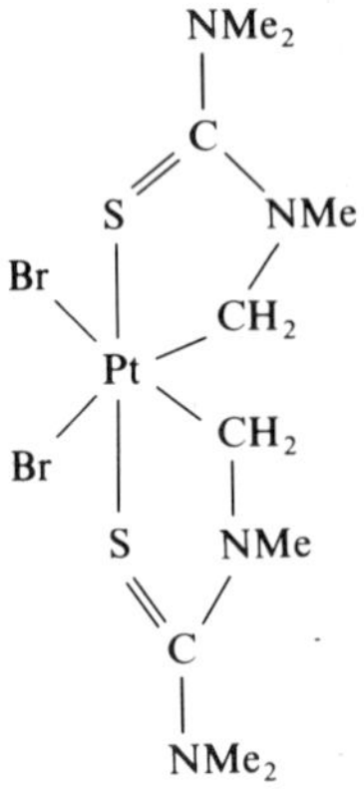

(4.139)

PtClL2 + (NC)2C=C(CN)2
CN
NC
PtClL2
NC
CN

(4.140)

The elements of the Ni, Pd, and Pt triad form the following representative compounds: $NiR_2$ ($R = CPh_3$, $o\text{-}CH_2C_6H_4PPh_2$, $2,4,6\text{-}C_6H_2Me_3$), $[MXRL_2]$, $[MR_2L_2]$ (M = Ni, Pd, Pt), $[M_2X_2R_2L_2]$ (M = Pd, Pt), $[Ni_2\{CH_2)_2PMe_2\}_4]$, $MR_2$ ($R = o\text{-}CH_2C_6H_4PPh_2$, $o\text{-}C_6H_4CH_2NR'_2$; M = Pd, Pt), $[Ni(o\text{-}C_6H_4OC_6H_4\text{-}o)_2]^{2-}$ $[NiR_4]^{2-}$ (R = Me, Ph, $C_6F_5$, $C_6Cl_5$), $[NiPh_3]^-$, $[Pd(C_6F_5)_4]^{2-}$, $[Pd(C_6F_5)_3(C_4H_8S)]^-$, $[Pt(C_6F_5)_3L]^-$, $[PtMe_4]^{2-}$, $[MR_2(CDO)]$ (M = Pd, Pt), $[PtXMe_3]_4$ (X = Cl, Br, OH), $[PtMe_3cp]$, $[Pt(acac)Me_3]_2$, $[Pt(acac)Me_3(bipy)]$, $K[Pt(acac)_3]$, $K[PtCl(acac)_2]$, $K[PtCl_2(acac)]$, $[PtR_4L_2]$, $[PtMe_5L]^-$, $[PtMe_5Et]^{2-}$, $[PtMe_6]^{2-}$, $K_4[Ni(CCR)_4]$, $K_2[M(CCR)_4]$ (M = Ni, Pd, Pt), $[Ni(CCR)_2]$, $K_2[M(CCR)_2]$ (M = Pd, Pt), $[M(CCR)_2L_2]$ (M = Ni, Pd, Pt), $[NiRL_4]^+$ ($L = PR_3$, py, dipy, RNC, etc.),

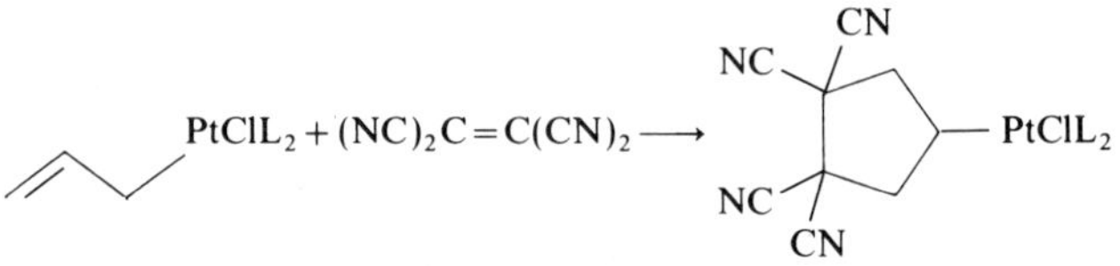

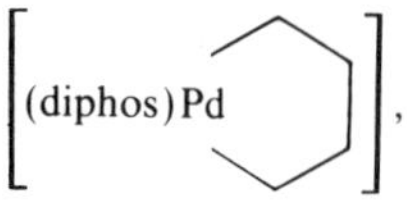

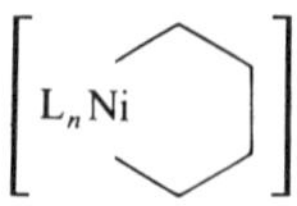

(diphos)Pd

, 
LnNi

($n = 2, 3$), 
L3Ni

($L = PR_3$, dipy, etc.)

$[MRcpL]$ (M = Ni, Pd, Pt), 
Cl2Pt
R
2−

, 
L2X2Pt
(CH2)n

($n = 1, 2$)

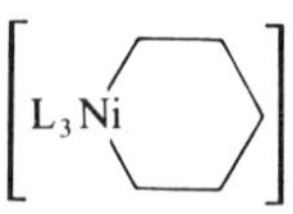

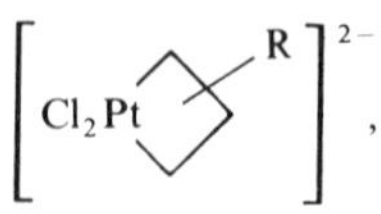

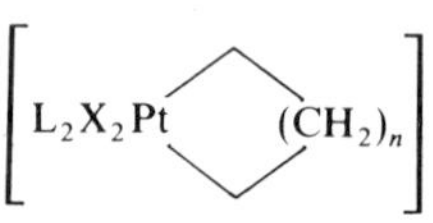

$[Pt_2Me_4(\mu\text{-OOCR})_2L_2]$, $[PtR_3L_3]^+$, $[PtR_2(OOCR')L]_2$, $[PtRL_3]^+$, $[Ni_2\{(CH_2)$
$(CH_2)PMe_2\}_4]$, $[\{Ni_2Ph_4(N_2)Li_4(Et_2O)_2Li_2Ph_2\}_2]$, $[Ni_2(\mu\text{-Me})_2(MeCHCHCHMe)_2]$,
$[phen(I)Me_2Pt\{\mu\text{-}(CH_2)_n\}PtMe_2(I)phen]$ $(n = 2\text{--}5)$.[176]

### i. Complexes of group 11 metals[1–5, 11, 13, 88, 186–198]

Copper and silver form compounds containing $\sigma M-C$ bonds only in the $+1$ oxidation states, while gold gives compounds in $+1$, $+2$, and $+3$ oxidation states (Table 4.20). Copper compounds such as CuR are polynuclear or polymeric. Often, tetramers are formed; however, dimers, trimers, hexamers, and octamers are also known. The compound $[Cu(CH_2SiMe_3)]_4$ is a tetramer possessing alkyl bridges and direct $Cu-Cu$ bonds (Figure 4.2a). Also known are compounds of the types $[LiCuR_2]$ and $[Li_2CuR_3]$. Cationic complexes $[MR_2]^+$ $(M = Cu, Ag, Au)$ containing phosphine $\sigma$-hydrocarbyl ligands may also be obtained. Pentanuclear complexes possessing the following compositions are interesting: $[Li(THF)_4]^+$ $[Cu_5Ph_6]^-$, $[Li_4Cl_2(Et_2O)_{10}]^{2+}$ $[Cu_3Li_2Ph_6]_2^-$, $[Li(Et_2O)_4]^+$ $[Cu_4LiPh_6]^-$, $[Cu_4MgPh_6]$, $[Li_6Br_4(Et_2O)_{10}]^{2+}$ $[Ag_3Li_2Ph_6]_2^-$. Their structures are depicted in (4.141).[185–188]

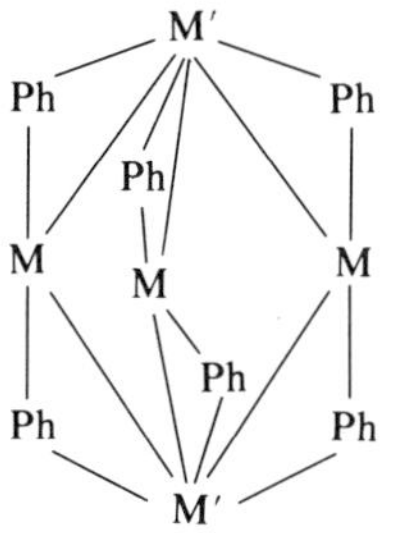

for M = Cu; M′ = Cu, Li, Mg
for M = Ag; M′ = Li
(4.141)

These complexes are obtained by the action of LiPh or MgPh$_2$ on CuI or CuBr. The type of compound formed depends on the Cu:Li(Mg) ratio and on the solvent. In the case of methyl compounds, depending on the ratio of CuX(CuMe) to LiMe, various complexes are formed which remain in equilibrium. Gilman's reagent, LiCuMe$_2$, the best known and most commonly utilized organometallic compound of copper, is considerably more complex than can be judged by its stoichiometric composition.[21, 189]

Based on the $^1$H and $^7$Li NMR studies,[21, 189] it was found that for the ratio LiMe:CuMe = 1:2, the compound LiCu$_2$Me$_3$ is formed which, at higher values of Li/Cu, remains in equilibrium with Li$_2$Cu$_2$Me$_4$. At ratios of LiMe:CuMe changing from 2:3 to 1:1, the compound Li$_2$Cu$_3$Me$_5$ is obtained which, however, is probably a mixture of various compounds of copper and lithium. Among others, this mixture contains Cu$_2$Li$_2$Me$_4$. For the Li/Cu ratio of one, the compounds LiCuMe$_2$ and Li$_2$CuMe$_3$ are obtained and were assigned the following structures based on the $^1$H NMR studies:

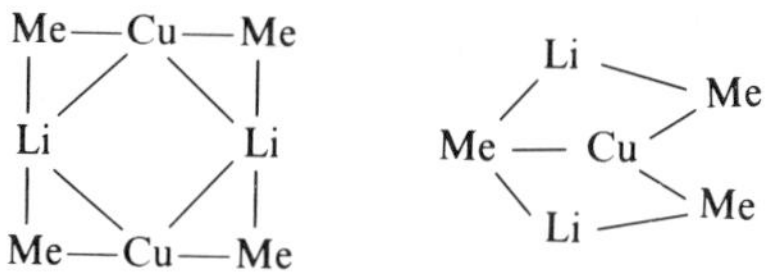

(4.142)

*Table 4.20. σ-Carbyl Complexes of Copper Group Metals*

| Compound | Color | Mp (K) | Remarks |
|---|---|---|---|
| $(CuMe)_n{}^a$ | Bright-yellow | | Explosive |
| $(CuPh)_n{}^b$ | White | | |
| $[Cu_4(CH_2SiMe_3)_4]^c$ | White | 351–352 dec | |
| $[Cu_2\{(CH_2)_2PMe_2\}_2]^d$ | White | 409–411 | |
| $[Cu_6(o\text{-}C_6H_4NMe_2)_4(CF_3SO_3)_2]^{(22)}$ | | 391–393 dec | |
| $[Cu_4Au_2(o\text{-}C_6H_4NMe_2)_4(CF_3SO_3)_2]^{(22)}$ | | 423 dec | |
| $[Cu_4Ag_2(o\text{-}C_6H_4NMe_2)_4(CF_3SO_3)_2]^{(22)}$ | | 298 dec | |
| $AgPh^{e,f}$ | White | | |
| $5AgPh \cdot 2AgNO_3{}^{e,f}$ | Bright-yellow | | |
| $[Ag_2\{(CH_2)_2PMe_2\}_2]^d$ | White | 427–428 | |
| $Li[Ag(o\text{-}C_6H_4CH_2NMe_2)_2]^g$ | Cream | 383 dec | |
| $[Ag(CH_2PPh_3)_2]Cl^h$ | White | 465–469 | |
| $[Ag(C_6H_4Me\text{-}p)]^r$ | White | | |
| $Li[AgPh_2] \cdot Et_2O^r$ | White | | |
| $K_2[Cu(CCH)_3]^i$ | White | | |
| $K[Cu(CCMe)_2]^i$ | White | | |
| $K[Ag(CCH)_2]^j$ | White | | |
| $AgCCH^j$ | Yellow | | Explodes under pressure at 193 K |
| $K[Au(CCH)_2]^k$ | White | | |
| $Au(CCH)^k$ | Bright-yellow | | |
| $[Au(CH=CH_2)PPh_3]^l$ | White | 396–398 dec | |
| $[Au_2Cl_2(Et)_4]^m$ | White | 321 dec | |
| $[Au_2Cl_4Ph_2]^n$ | Yellow | 346–348 dec | |
| $[Li(PMDT)][AuMe_2]^o$ | White | 393–396 dec | |
| $[Li(PMDT)][AuMe_4]^o$ | White | 359–361 | |
| $[Au(C_6F_5)(SC_4H_8)]^p$ | White | 382 | |
| $[Bu_4N][Au(C_6F_5)_4]^p$ | | 513 | |
| $[Au_2\{(CH_2)_2PEt_2\}_2]^q$ | White | 408–410 | |
| $[Au_2\{(CH_2)_2PEt_2\}_2Cl_2]^q$ | Dark-yellow | 429 | |

[a] H. Gilman, R. G. Jones, and L. A. Woods, *J. Org. Chem.*, **17**, 1630 (1952).

[b] G. Costa, A. Camus, L. Gatti, and N. Marsich, *J. Organomet. Chem.*, **5**, 568 (1966).

[c] J. A. J. Jarvis, B. T. Kilbourn, R. Pearce, and M. F. Lappert, *Chem. Commun.*, 475 (1973); M. F. Lappert and R. Pearce, *Chem. Commun.*, 24 (1973).

[d] H. Schmidbaur, J. Adlkofer, and W. Buchner, *Angew. Chem., Int. Ed. Engl.*, **12**, 415 (1973); H. Schmidbaur, J. Adlkofer, and M. Heimann, *Chem. Ber.*, **107**, 3697 (1974).

[e] H. Gilman and J. M. Straley, *Recl. Trav. Chim. Pays-Bas*, **55**, 821 (1936).

[f] C. D. M. Beverwijk and G. J. M. van der Kerk, *J. Organomet. Chem.*, **43**, C11 (1972); A. Costa and M. M. Camus, *Gazz. Chim. Ital.*, **86**, 77 (1956).

[g] A. J. Leusink, G. van Koten, J. W. Marsman, and J. G. Noltes, *J. Organomet. Chem.*, **55**, 419 (1973).

[h] Y. Yamamoto and H. Schmidbaur, *J. Organomet. Chem.*, **96**, 133 (1975).

[i] R. Nast and W. Pfab, *Chem. Ber.*, **89**, 415 (1956).

[j] R. Nast and H. Schindel, *Z. Anorg. Allg. Chem.*, **326**, 201 (1963).

[k] R. Nast and U. Kirner, *Chem. Ber.*, **97**, 207 (1964); R. Nast and U. Kirner, *Z. Anorg. Allg. Chem.*, **330**, 311 (1964).

[l] A. N. Nesmeyanov, E. G. Perevalova, E. W. Krivykh, A. N. Kosina, K. I. Grandberg, and E. I. Smyslova, *Izv. Akad. Nauk SSSR, Ser. Khim.*, 653 (1972).

[m] M. S. Kharasch and H. S. Isbell, *J. Am. Chem. Soc.*, **53**, 2701 (1931); M. S. Kharasch and H. S. Isbell, *J. Am. Chem. Soc.*, **52**, 2923 (1930).

[n] M. S. Kharasch and H. S. Isbell, *J. Am. Chem. Soc.*, **53**, 3053 (1931).

[o] G. W. Rice and R. S. Tobias, *Inorg. Chem.*, **15**, 489 (1976); G. W. Rice and R. S. Tobias, *Inorg. Chem.*, **14**, 2402 (1975).

[p] R. Uson, A. Laguna, and J. Vicente, *Chem. Commun.*, 353 (1976).

[q] H. Schmidbaur, J. R. Mandl, A. Frank, and G. Hutner, *Chem. Ber.*, **109**, 466 (1976).

[r] H. K. Hofstee, J. Boersma, and J. M. van der Kerk, *J. Organomet. Chem.*, **168**, 241 (1979); J. Blenkers, H. K. Hofstee, J. Boersma, and G. J. M. van der Kerk, *J. Organomet. Chem.*, **168**, 251 (1979).

The type of compounds obtained and their structures depend both on the solvent and the presence of halides. The use of very large ligands such as $-C(SiMe_3)_3$[234] or $-C_6H_2Me_3$-2,4,6[235] allows the isolation of the monomeric linear organocuprates of the formula $[R-Cu-R]^-$. Linear monomers $[CuXR]^-$ with small hydrocarbyl groups crystallize as their lithium crown ether salts[233]: $[Li(12\text{-crown-4})_2][CuMe_2]$, $[Li(12\text{-crown-4})_2][CuPh_2]\cdot THF$, $[Li(12\text{-crown-4})_2][Cu(Br)\,CH(SiMe_3)_2]\cdot PhMe$. Structural studies of phenyl cuprates show that cations may be both mono- and polynuclear lithium compounds: $Li(THF)_4^+$, $Li(Et_2O)_4^+$, $[Li_4Cl_2(Et_2O)_{10}]^{2+}$, $[Li_6Br_4(Et_2O)_{10}]^{2+}$.[185–188] This proves that unequivocal determination of structures of metalloorganic copper–lithium compounds based on the $^1H$ and $^7Li$ NMR spectra is not always possible and requires further studies.

More stable in solution are copper complexes containing chelating ligands such as $C_6H_4CH_2NR_2$-2 and $C_6H_4NR_2$-2. The complex $Cu_4(C_6H_3CH_2NMe_2$-2-Me-5$)_4$ is tetranuclear both in the solid state as well as in solutions.[191] Similarly, the structure of the hexanuclear complex $[Cu_6Br_2(C_6H_4NMe_2$-2$)_4]$ remains unchanged in solution. The association numbers of other copper complexes vary from 2 to 6 depending upon the solvent, temperature, and concentration. For example, in $[Cu_n\{C_6H_2(OMe)_3\}_n]$, $n$ is between 5 and 7, while $Cu(CH_2SiMe_3)$ is a tetramer in benzene and a hexamer in cyclohexane.

In the case of silver, complexes containing nitrogen chelating ligands are labile. For example, $Ag(C_6H_4CH_2NMe_2$-2$)$ exists as a mixture of tetranuclear and hexanuclear complexes in solution. For compounds of the type $[M_xLi_yR_{x+y}]$ ($M = Cu$, Ag; $R = Me$, $C_6H_4CH_2NMe_2$-2, $C_6H_4Me$-4), the most stable are tetranuclear complexes. For chelating ligands coordinating via carbon and nitrogen, the structural units are as follows (Figure 4.6)[191, 192]:

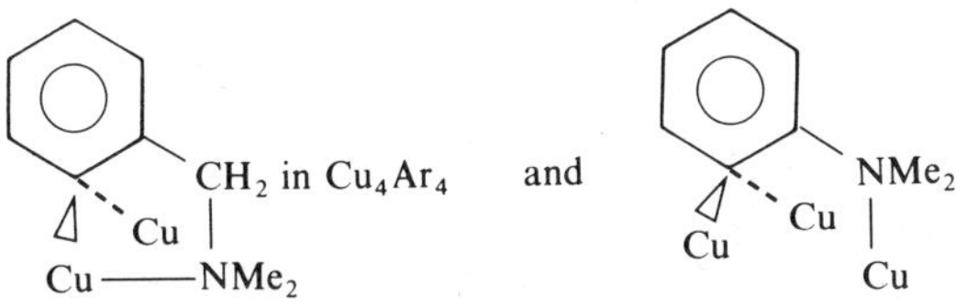

$$(4.143)$$

Some organometallic copper compounds are polymeric, such as phenylcopper, which possesses $-Cu-Ph-Cu-$ chains, some acetylene complexes, etc.

The ylid-type ligands such as $Me_3PCH_2$ furnish dimeric complexes containing bridges:

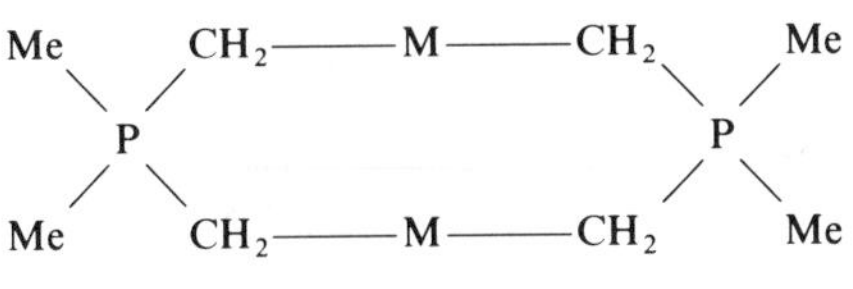

$$(4.144)$$

$$(M = Cu, Ag, Au)$$

The following gold compound undergoes oxidation to afford the Au(II) and Au(III) derivatives [equation (4.145)].

a)

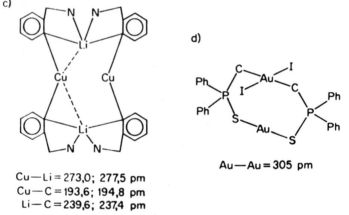

Cu—Cu = 242; 247 pm

b)

x = 247,2(5) pm
y = 272,6(5) pm

c)

Cu—Li = 273,0; 277,5 pm
Cu—C = 193,6; 194,8 pm
Li—C = 239,6; 237,4 pm

d)

Au—Au = 305 pm

e)

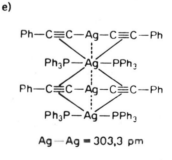

Ph—C≡C—Ag—C≡C—Ph

Ph₃P—Ag—PPh₃

Ph—C≡C—Ag—C≡C—Ph

Ph₃P—Ag—PPh₃

Ag—Ag = 303,3 pm

Figure 4.6. Structures of carbyl complexes of copper group metals: (a) [CuC≡CPh]$_n$, (b) Cu$_8$(C$_6$H$_4$OMe)$_8$, (c) Cu$_2$Li$_2$(C$_6$H$_4$CH$_2$NMe$_2$−2)$_4$,[190] (d) [Au$^I$(SPPh$_2$CH$_2$)$_2$Au$^{III}$I$_2$],[193] (e) [Ag(PPh$_3$)$_2$][Ag(CCPh)$_2$] (distances are in pm units).

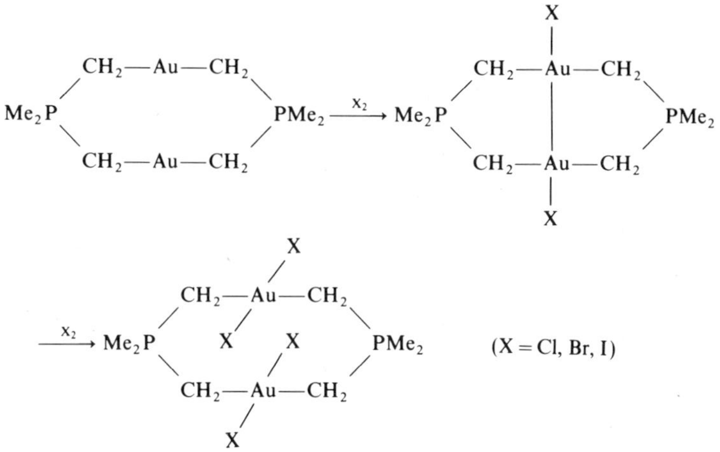

$$\begin{array}{c} CH_2-Au-CH_2 \\ Me_2P \qquad\qquad PMe_2 \xrightarrow{X_2} \\ CH_2-Au-CH_2 \end{array}$$

$$Me_2P \qquad\qquad\qquad PMe_2$$

$$\xrightarrow{X_2} Me_2P \qquad\qquad PMe_2 \qquad (X = Cl, Br, I) \qquad (4.145)$$

These complexes represent the few examples of such Au(II) complexes. The Au–Au distance in $[Au_2\{(CH_2)_2PEt_2\}_2Cl_2]$ is exceptionally small (259.7 pm, Table 4.5). The majority of copper and silver compounds are unstable, while gold derivatives are more stable.

Some compounds of the copper group metals are as follows:
$[CuR]_n$, $[CuLiR_2]$, $[CuLi_2R_3]$, $[Cu_2LiR_3]$, $[Cu_2MgR_4]$, $[Cu_2Li_2R_4]$, $[Cu_3Li_2R_6]^-$, $[Cu_4LiR_6]^-$, $[Cu_4MgR_6]$, $[Cu_4LiR_5]$, $[CuR_2]^+$ $[R = CH_2PMe_3$, $CH(SiMe_3)PMe_3$, $CH_2PPh_3$, $MeCHPPh_3$, $Pr^iCHPPh_3]$, $[CuLiRR']$ $(R = C \equiv CR''$, $R' = Me$, Et, Pr, Bu, Ph; $R = C_6H_4F$, $R' = C_6Cl_5)$, $[Cu_4R_4]$, $[Cu_8R_8]$, $[CuRL_2]$, $[Cu_2R_2(dppe)]$, $[CuR \cdot CuX]_n$, $[Cu_6Ar_4X_2]$ $(Ar = o\text{-}C_6H_4NMe_2$, $o\text{-}C_6H_4CH_2NMe_2$, X = halogen), $[Cu_6Ar_4L_2]$, $[Cu_4Ag_2Ar_4L_2]$, $[Cu_4Au_2Ar_4L_2]$, $[Ag_4Au_2Ar_4L_2]$ $(Ar = o\text{-}C_6H_4NMe_2$, $L = CF_3SO_3)$,[22] $[AgR]_n$, $[AgRL]$, $[Ag_2R_2]$, $[(AgR)_n \cdot (AgNO_3)_m]$, $[AgLiR_2]$ (R = aryl, alkenyl), $[Ag_2Li_2R_4]$, $[Ag_2BrR]_n$, $[Ag_3BrR_2]$, $[AgR_2]^+$ $(R = Ph_3PCHR'$, $R' = H$, Me, $Pr^i)$, $[Au_2Li_2(C_6H_4CH_2NMe_2\text{-}2)_4]$, $[Au_2Zn_2Ph_6]$, $[AuR]_n$ [for $R = R_2'P(CH_2)_2$, $n = 2$; for $R = C \equiv CPh$ and 2-pyridyl, $n = 3$], $[AuR_2]^-$, $[AuRL]$, $[AuRX_2]_2$, $[AuR_2X]_2$, $[Au(CN)R_2]_4$, $[AuR_3L]$, $[AuR_2XL]$, $[AuR_4]^{2-}$,

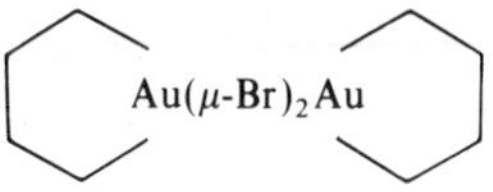

$$Au(\mu\text{-Br})_2Au \quad , \qquad [Au_2\{(CH_2)_2PMe_2\}_2]$$

$[Au_2\{(CH_2)_2PR_2\}_2X_2]$, $[Au_2\{(CH_2)_2PR_2\}_2X_4]$, $[AuR_2L_2]^+$, $[AuR_2X_2]^-$.[192, 194, 195]

## 12. REACTIONS OF COMPLEXES CONTAINING METAL–CARBON $\sigma$ BONDS[1–6, 12, 13, 40]

### a. The breaking of the metal–carbon bond

The $\sigma M - C$ bond breaking may take place as a result of reactions with compounds containing active hydrogen atoms, with halogens, alkyl halides, hydrogen, metal halides, LiAlH$_4$, or as a result of homolytic splitting or reductive elimination (Section 4.2), etc.

All compounds containing $\sigma$M$-$C bonds react with acids with varying ease, and often react with water, alcohols, and other proton compounds.

$$[\text{MRL}]_n + \text{HX} \longrightarrow [\text{MXL}_n] + \text{RH} \tag{4.146}$$

Complexes of platinum(IV) and gold(III) are most stable in such reactions because they do not decompose under the influence of water and alcohols. Also resistant to water are some compounds of the type MR$_n$ of the remaining metals containing bulky ligands that do not possess hydrogen atoms in $\beta$ position with respect to the central atom. The complex $[\text{V}(2,4,6\text{-C}_6\text{H}_2\text{Me}_3)_4]$ is an example. Alkyl and aryl complexes of transition metals usually show low reactivity with respect to anions because the transition metals generally have low electropositive character. The electronegativity of carbon is considerably lower than that of halogen, oxygen, nitrogen, etc., ligands. Therefore, the M$-$C bond is covalent to a significant degree, and some compounds containing M$-$C $\sigma$ bonds may be recrystallized from alcohols. Also, such compounds do not decompose under the influence of dilute acids, although water-free hydrogen halides usually cause rapid splitting of the M$-$C bond.

$$cis\text{-}[\text{PtPh}_2(\text{PEt}_3)_2] \xrightarrow[-\text{C}_6\text{H}_6]{+1\ \text{HCl}} cis\text{-}[\text{PtClPh}(\text{PEt}_3)_2] \xrightarrow{\text{HCl}} cis\text{-}[\text{PtCl}_2(\text{PEt}_3)_2] \tag{4.147}$$

It is assumed that this reaction proceeds through an unstable hydrido Pt(IV) complex.

The M$-$C bond breaking by means of reagents containing active hydrogen also occurs in other reactions:

$$[\text{PdMe}_2(\text{PEt}_3)_2] + 2\text{PhSH} \longrightarrow [\text{Pd}(\text{SPh})_2(\text{PEt}_3)_2] + 2\text{CH}_4 \tag{4.148}$$

$$[\text{NiMe}(\text{cp})\text{PPh}_3] + \text{HBr} \longrightarrow [\text{NiBr}(\text{cp})\text{PPh}_3] + \text{CH}_4 \tag{4.149}$$

$$[\text{M}(\text{CH}_2\text{Ph})_4] + \text{HX} \longrightarrow [\text{M}(\text{CH}_2\text{Ph})_3\text{X}] + \text{PhMe} \quad (\text{M} = \text{Ti, Zr}) \tag{4.150}$$

$$[\text{WMe}_8]^{2-} + 8\text{H}^+ \longrightarrow \text{W(VI)} + 8\text{CH}_4 \tag{4.151}$$

$$[\text{Wcp}_2\text{H}(\text{Ph})] + \text{HCl} \longrightarrow [\text{Wcp}_2\text{Cl}_2] + \text{PhH} \tag{4.152}$$

$$[\text{Ti}(\text{CH}_2\text{Ph})_4] + \text{MeOD} \longrightarrow [\text{Ti}(\text{CH}_2\text{Ph})_3(\text{OMe})] + \text{PhCH}_2\text{D} \tag{4.153}$$

$$[\text{Cu}_2\text{C}_2] + 2\text{H}^+ \longrightarrow 2\text{Cu}^+ + \text{HCCH} \tag{4.154}$$

$$[\text{NiR}_2(\text{bipy})] + \text{HCl} \longrightarrow [\text{NiCl}_2\text{bipy}] + 2\text{RH} \tag{4.155}$$

$$[\text{Ni}(\text{C}_2\text{R})_4]^{2-} + 4\text{CN}^- + 4\text{H}_2\text{O} \longrightarrow [\text{Ni}(\text{CN})_4]^{2-} + 4\text{RCCH} + 4\text{OH}^- \tag{4.156}$$

$$[\text{ReMe}(\text{CO})_5] + \text{HX} \longrightarrow [\text{ReX}(\text{CO})_5] + \text{CH}_4 \tag{4.157}$$

Reactions with nucleophiles may also lead to the $\sigma$M$-$C bond splitting.

$$[\text{Co}(\text{COR})(\text{CO})_4] + \text{R}'\text{OH} \longrightarrow [\text{CoH}(\text{CO})_4] + \text{RCOOR}' \tag{4.158}$$

$$[\text{Co}(\text{COR})(\text{CO})_4] + 2\text{R}'\text{NH}_2 \longrightarrow [\text{R}'\text{NH}_3]^+ [\text{Co}(\text{CO})_4]^- + \text{RCONHR}' \tag{4.159}$$

$$[\text{MR}(\text{CO})_n] + [\text{AlH}_4]^- \longrightarrow [\text{MH}(\text{CO})_n] + \text{RH} \tag{4.160}$$

$$[\text{Co}(\text{dmgH})_2\text{CH}_2\text{CH}_2\text{CN}] + \text{OH}^- \longrightarrow [\text{Co}(\text{dmgH})_2] + \text{CH}_2{=}\text{CHCN} + \text{H}_2\text{O} \tag{4.161}$$

Reactions with $LiAlD_4$ are utilized to determine which carbon atom is bonded to the metal.

The decomposition of complexes containing $\sigma M - C$ bonds is also caused by oxidizing agents ($O_2$, $X_2$, etc.) and inorganic metallic compounds ($HgCl_2$, $SnCl_4$, $BiCl_3$, $TiCl_3$):

$$[ReMe_6] + O_2 \longrightarrow [ReOMe_4] \tag{4.162}$$

$$[PtPh_2(PEt_3)_2] + I_2 \longrightarrow [PtIPh(PEt_3)_2] + PhI \tag{4.163}$$

$$[NiMecp(PPh_3)_2] + I_2 \longrightarrow [NiIcp(PPh_3)_2] + MeI + nC_2H_6 \tag{4.164}$$

$$[Au_2Br_2Et_4] + 2Br_2 \longrightarrow [Au_2Br_4Et_2] + 2EtBr \tag{4.165}$$

$$[Pt(2\text{-}C_6H_4Me)_2py_2] + PhI \longrightarrow [PtI(2\text{-}C_6H_4Me)py_2] + MeC_6H_4Ph \tag{4.166}$$

$$VPh_4 + 2HgCl_2 \xrightarrow{203K} 2HgClPh + Ph - Ph + VCl_2 \tag{4.167}$$

$$4[TiMeCl_3] + SnCl_4 \longrightarrow SnMe_4 + 4TiCl_4 \tag{4.168}$$

$$[CoMe(CN)_5]^{3-} + HgCl_2 \longrightarrow HgMeCl + [Co(CN)_5Cl]^{3-} \tag{4.169}$$

$$[Cr(CH_2C_5H_4N)(H_2O)_5]^{2+} + HgCl_2 \longrightarrow Hg(CH_2C_5H_4N)Cl + [Cr(H_2O)_5Cl]^{2+} \tag{4.170}$$

$$[Co(CH_2C_5H_4N)(CN)_5]^{3-} + TlCl_3 \longrightarrow Tl(CH_2C_5H_4N)Cl_2 + [Co(CN)_5Cl]^{3-} \tag{4.171}$$

Reactions with mercury chloride are often utilized for identification of the $M - C$ bonds.

## b. *Migration (insertion) reactions* (see Section 4.10.c)

Migration reactions of $\sigma$ carbyl ligands are fundamental for many catalytic processes such as oligomerization and polymerization of olefins, polyolefins, and acetylenes, isomerization of hydrocarbons, hydroformylation, etc.

The migration of the hydrocarbyl ligand to the coordinated Lewis base such as CO, RNC, etc., or CO insertion into the metal–hydrocarbyl bond, usually takes place in the case of 18$e$ complexes via 16$e$ compounds or solvated 18$e$ compounds[199–203] (see Chapter 2):

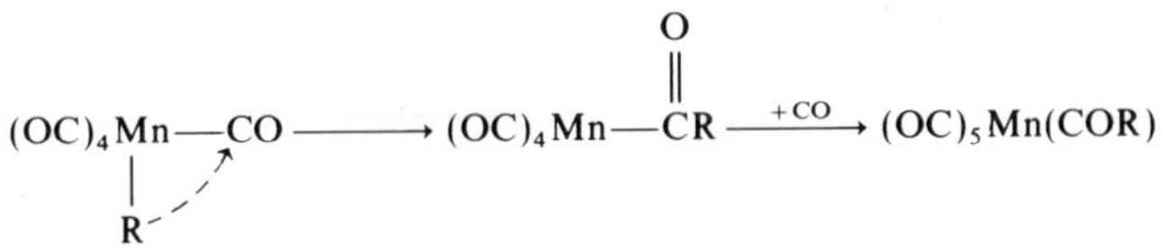

$$\tag{4.172}$$

The rate of migration of the ligand R in $MnR(CO)_5$ decreases in the series n-Pr > Et > Ph > Me > PhCH$_2$ $\gtrsim$ CF$_3$. During this reaction, the optically active R

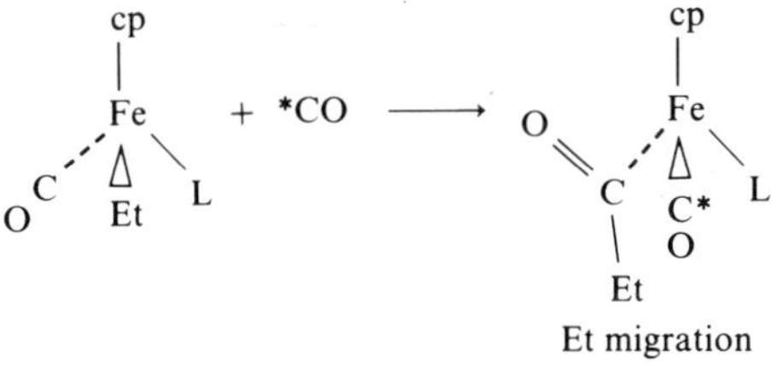

$$\tag{4.173}$$

Et migration

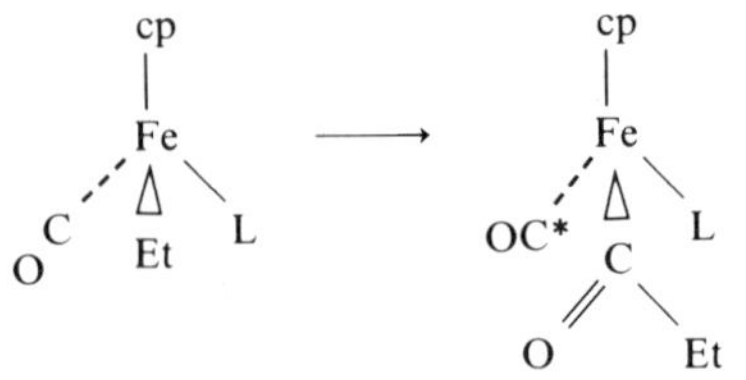

$$\text{(4.174)}$$

CO migration

groups retain their configurations. Also, the stereochemistry corresponding to insertion of CO into the $M-R$ bond was observed.[141, 202] In the case of $[Fe(Et)\,cp(CO)L]$ $(L = PPh_3)$ in weakly coordinating or noncoordinating $(EtNO_2, MeNO_2, MeCN)$ solvents, ethyl migration takes place, while in strongly coordinating (and relatively hard) solvents (HMPA, DMSO, DMF), CO insertion apparently occurs.[141]

Migration reactions also take place in $16e$ square-planar complexes of platinum(II), such as $[PtR(X)(CO)L]$.[203]

The mechanism of this reaction is shown in equation (4.175). In the transition

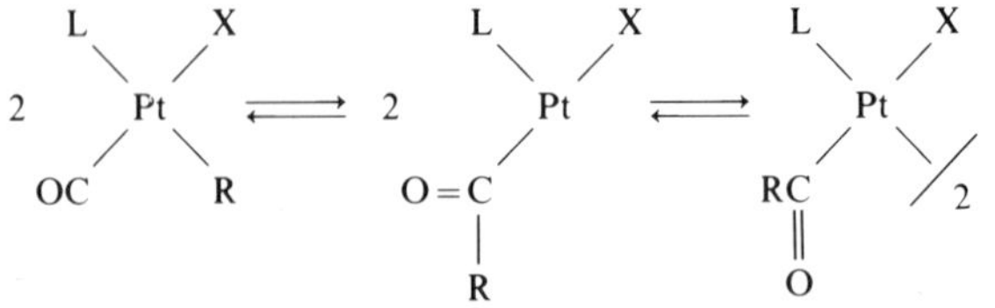

$$\text{(4.175)}$$

state, or intermediate, $14e$ complexes are formed. The tendency for R migration in square complexes decreases according to the series: $R = Et > Ph > Me > CH_2Ph$, $C \equiv CMe$, $C \equiv CPh$, $CCl = CCl_2$, and $C_5H_5$. The formation of dimers was not demonstrated for some hydrocarbyl groups. The migration of R occurs only if there is a phosphine in the *trans* position with respect to R; then the *trans* effect is sufficiently strong to cause relatively easy $Pt-R$ bond splitting. The group R migration proceeds more easily if the carbon atom bonded to the metal possesses a negative charge and if the carbon atom of CO or CNR has a partial positive charge.

The migration of alkyls to coordinated alkene or alkyne represents a very important process:

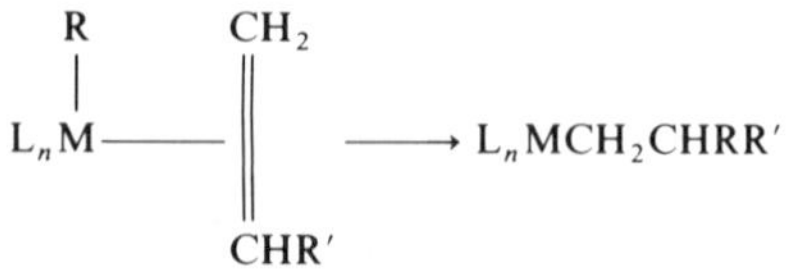

$$\text{(4.176)}$$

Intramolecular insertion of acetylene in *trans*-$[L_2ClPdCOO(CH_2)_nC \equiv CMe]$ also takes place via a four-coordinate complex[179] [equation (4.136)]. The migration reaction is reversible in some cases[180] [equation (4.137)].

The carbyl transfer to one of the CO groups in $[RFe(CO)_4]^-$ and $[RMn(CO)_5]$ is catalyzed by Lewis acids via ion pair formation.[42, 43, 204] For reaction (4.177), the

$$[RFe(CO)_4]^- + PPh_3 \xrightarrow{\text{THF}} [R\overset{\overset{\textstyle O}{\|}}{C}-Fe(CO)_3(PPh_3)]^- \qquad \text{(4.177)}$$

rate increases 170-fold if the sodium ion complexed by the crown ether dicyclohexyl-18-crown-6 is replaced by the $Na^+$ ion solvated by THF. The rate is also increased if $Na^+$ is exchanged for $Li^+$. Mechanism (4.178) is proposed for this reaction.

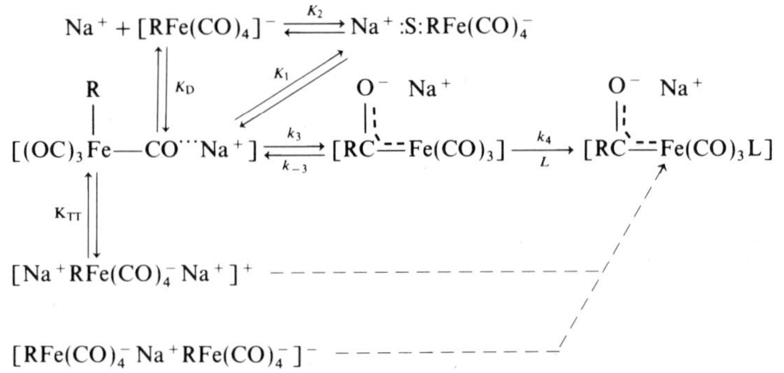

$$Na^+ + [RFe(CO)_4]^- \underset{}{\overset{K_2}{\rightleftharpoons}} Na^+ :S:RFe(CO)_4^-$$

(4.178)

The migration (insertion) reaction may also occur in the case of many other unsaturated molecules such as $SO_2$, $CO_2$, $N_2$, NO, nitriles, isonitriles, and ketones. The insertion of sulfur dioxide may proceed very rapidly. The NMR and IR measurements show that the reaction proceeds through O-sulfinates which subsequently undergo rearrangement to give stable S-sulfinates:

$$[FeRcp(CO)_2] + SO_2 \longrightarrow [(CO)_2 cpFe\text{-}SO_2 R]$$

In certain cases, the reaction is sterospecific at metal [see equation (4.179)].

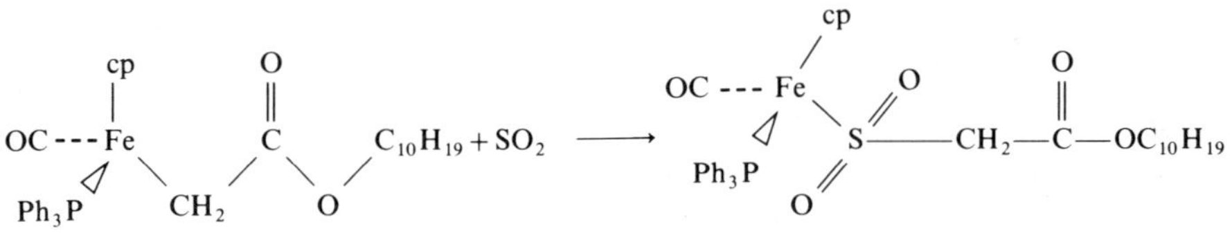

(4.179)

The reaction of sulfur dioxide with ethylene, catalyzed by palladium complexes,[41] affords *trans*-2-butenyl ethyl sulphone [equation (4.180)].

$$CH_2{=}CH_2 + H-Pd-X \longrightarrow EtPdX \xrightarrow{SO_2} EtSO_2PdX \xrightarrow{2C_2H_4} EtSO_2C_4H_8PdX$$

$$\longrightarrow EtSO_2CH_2CH_2CH{=}CH_2 + HPdX \qquad (4.180)$$

$$EtSO_2CH_2CH{=}CHMe \longleftarrow\rfloor$$

Many examples of reactions of complexes containing $\sigma M-C$ bonds with carbon dioxide are known.[44-46, 205] The $\sigma$-hydrocarbyl ligand may migrate to the carbon atom as well as to the oxygen atom of the carbon dioxide molecule:

$$M^{+\delta} - R^{-\delta} + CO_2 \longrightarrow M - O\overset{\overset{\displaystyle O}{\parallel}}{C} - R \qquad (4.181)$$

$$M^{-\delta} - R^{+\delta} + CO_2 \longrightarrow M - \underset{\underset{\displaystyle O}{\parallel}}{C} - O - R \qquad (4.182)$$

The reaction of $CO_2$ with $Ticp_2Me_2$ and $Ticp_2Ph_2$ forms the carbon–carbon bond while the reaction of $CO_2$ with $CoEt(CO)(PPh_3)_2$ illustrates both types of migration:

$$[Ticp_2Me_2] + 2CO_2 \longrightarrow [cp_2Ti(OOCMe)_2] \qquad (4.183)$$

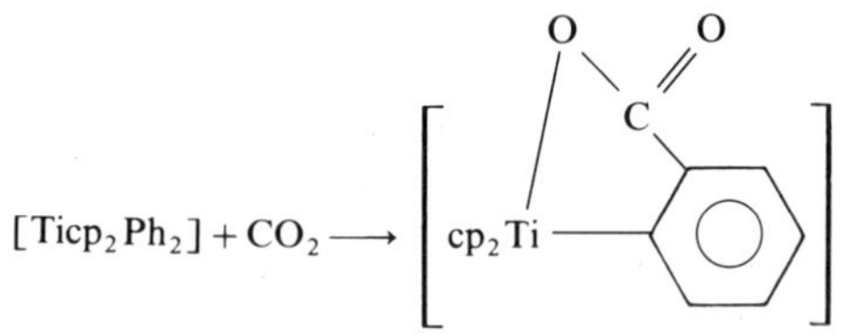

$$[Ticp_2Ph_2] + CO_2 \longrightarrow \left[ cp_2Ti \underset{}{\overset{}{\diagdown}} \right] \qquad (4.184)$$

$$[NiEt_2(bipy)] \xrightarrow[PhH]{CO_2} \left[ (bipy)Ni \underset{O-C-Et}{\overset{Et}{\diagup}} \right] \qquad$$

$$NiCO_3(bipy) + Et_2CO \longleftarrow \overset{|CO_2}{} \qquad \overset{|CO_2}{}$$

$$EtCOOH \longleftarrow [Ni(OOCEt)_2] \qquad (4.185)$$

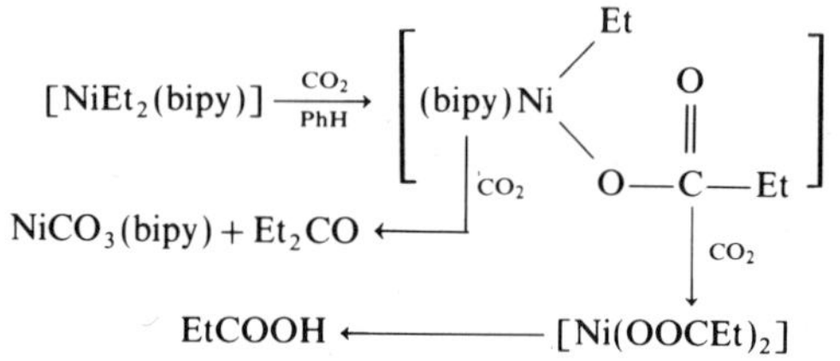

$$CoEt(CO)(PPh_3)_2 + CO_2 \xrightarrow{MeI} EtCOOMe + MeCOOEt \qquad (4.186)$$

The dithiocarboxylato complexes are formed by the reaction of $CS_2$ with $MR_xL_y$.[47]

$$[MR(CO)_5] + CS_2 \longrightarrow [M - S\overset{\overset{\displaystyle S}{\parallel}}{C}R(CO)_4] + CO \qquad (4.187)$$
$$(M = Mn, Re)$$

The oxidation of complexes containing $\sigma M - C$ bonds leads to the formation of a variety of products. In the transition state, peroxide complexes are sometimes formed.[51] However, the mechanism of this reaction has not been established definitely. Based on the products of the reaction and its dependence on the solvent, a radical mechanism was proposed for the oxidation processes of many complexes[51]:

$$ROO^* + [RMX_n] \longrightarrow [ROOMX_n] + R^* \qquad (4.188)$$

Alcohols, phenols, aldehydes, and hydrocarbons are often products of the oxidation. The peroxides of transition metals are formed during oxidation of the following complexes: $M(CH_2SiMe_3)_4$ $(M = Ti, Zr)$, $[Mo_2(CH_2SiMe_3)_6]$, $[Mo_2(CH_2CMe_3)_6]$, $[W_2(CH_2CMe_3)_6]$, $[Zr(CH_2CMe_3)_4]$, $[CoR(dmgH)_2L]$, and $[RTiCl_3]$.

$$[Co(dmgH)_2(py)R] + O_2 \longrightarrow [Co(dmgH)_2(py)(OOR)] \qquad (4.189)$$

The oxidation of $[M(CH_2Ph)_4]$ $(M = Ti, Zr)$ proceeds with the consumption of two moles of oxygen. After hydrolysis of the reaction mixture, benzyl alcohol is obtained. The complex $[Cr(CH_2Ph)(H_2O)_5]^{2+}$ is oxidized to benzoic aldehyde, while the oxidation products of $[TiPh(OPr)_3]$ and $[TiPh(OBu)_3]$ are phenoxytitanium complexes, benzene, and diphenyl.

The formation of amines by reactions of complexes possessing $\sigma M - C$ bonds with dinitrogen, $N_2$, in the presence of strong reducing agents proceeds probably via hydrocarbyl ligand migration.[52-55, 206, 207] Nucleophilic attack of $R^-$ at ligated $N_2$ is also known.

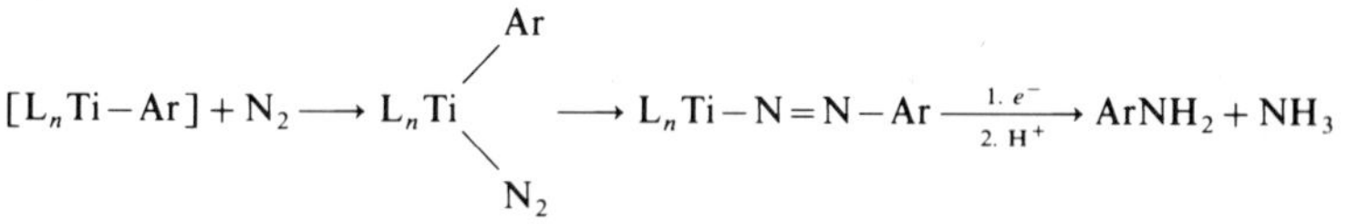

$$[L_nTi-Ar] + N_2 \longrightarrow L_nTi\underset{N_2}{\overset{Ar}{<}} \longrightarrow L_nTi-N=N-Ar \xrightarrow[2.\ H^+]{1.\ e^-} ArNH_2 + NH_3 \qquad (4.190)$$

$$[Mncp(CO)_2(N_2)] \xrightarrow[THF]{R^-} [Mncp(CO)_2(NR=N)]^- \xrightarrow[THF]{R^+} [Mncp(CO)_2(NR=NR)]$$

$$(R = Me, Ph) \qquad (4.191)$$

The migration of alkyl ligands to nitrogen oxide enables preparation of compounds containing the $N-C$ bond.[48-50]

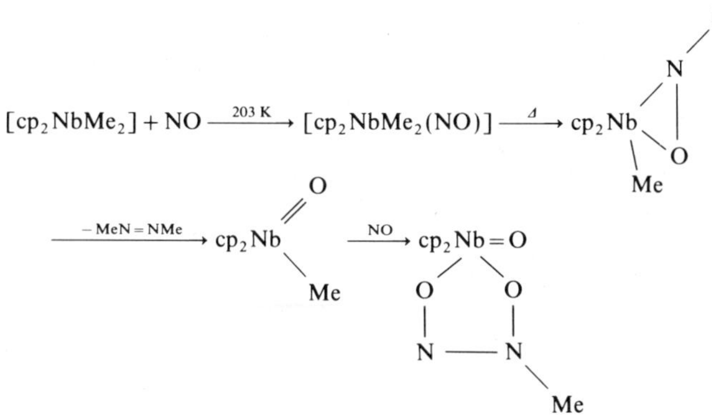

$$[cp_2NbMe_2] + NO \xrightarrow{203\ K} [cp_2NbMe_2(NO)] \xrightarrow{\Delta} cp_2Nb\begin{array}{c}N(Me)\\ |\\ O\end{array}Me$$

$$\xrightarrow{-MeN=NMe} cp_2Nb\begin{array}{c}O\\ \|\\ Me\end{array} \xrightarrow{NO} cp_2Nb=O \quad \begin{array}{c}O\quad O\\ |\quad\ |\\ N---N\\ \quad\ \ Me\end{array} \qquad (4.192)$$

$$NbMe_5 + NO \longrightarrow Me_2Nb\left(\begin{array}{c}O-N\\ |\\ O-N-Me\end{array}\right)_3 \qquad (4.193)$$

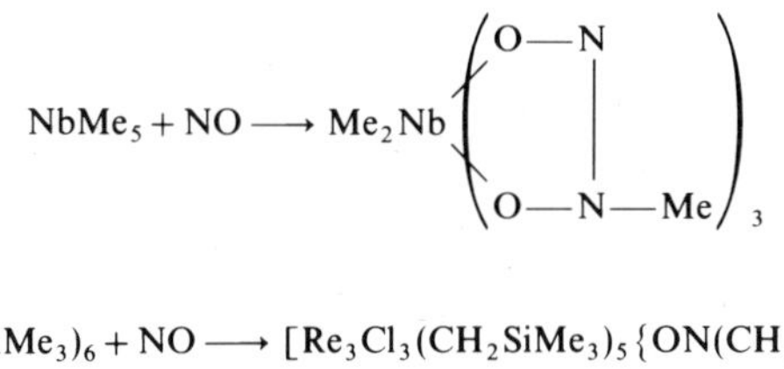

$$Re_3Cl_3(CH_2SiMe_3)_6 + NO \longrightarrow [Re_3Cl_3(CH_2SiMe_3)_5\{ON(CH_2SiMe_3)NO\}] \qquad (4.194)$$

The ion $[CoX(CO)_2NO]^-$ $(X = Br, I)$ interacts stoichiometrically with benzyl halides to give benzaldoxime and phenylacetic acid after acidification. It is assumed that an oxidative addition product is formed as an intermediate.[50]

$$PhCH_2X + [CoX(CO)_2NO]^- \longrightarrow [CoX_2(CO)_2(PhCH_2)NO]^-$$

$$\xrightarrow{H_2O} \begin{cases} PhCH_2NO \longrightarrow PhCH=NOH \\ \\ PhCH_2COOH \end{cases}$$

(4.195)

Nitrosyl complexes of ruthenium react in a similar manner.[50] In this case, supposedly, nucleophilic attack of the coordinated NO molecule on the benzyl halide takes place, or electrophilic addition of $PhCH_2^+$ to the coordinated NO group occurs with concomitant addition of X to the metal:

$$[Ru(NO)_2(PPh_3)_2] + PhCH_2Br \longrightarrow [RuBr(NO)(\overset{\overset{\displaystyle O}{\parallel}}{N}-CH_2Ph)(PPh_3)_2]$$

$$\longrightarrow [RuBr(NO)(\overset{\overset{\displaystyle OH}{\mid}}{N}=CHPh)(PPh_3)_2] \xrightarrow{-H_2O} [RuBr(NO)(NCPh)(PPh_3)_2]$$

$$\xrightarrow{CO} [RuBr(CO)(NO)(PPh_3)_2] + PhCN$$

(4.196)

The complexes $[CoR(CN)_5]^{3-}$ form nitriles in the acidic media [equation (4.197)].

$$[CoR(CN)_5]^{3-} \xrightarrow{H^+} [Co(\overset{\overset{\displaystyle CNH}{\mid}}{\underset{\displaystyle R}{}})(CN)_4]^{2-} \longrightarrow [Co-\overset{\overset{\displaystyle C=NH}{\mid}}{\underset{\displaystyle R}{}}(CN)_4]^{2-}$$

$$\xrightarrow{CN^-} [Co(CN)_6]^{3-} + RCN + H^+$$

(4.197)

The methyl nickel complex $[NiMecp(PPh_3)]$ reacts with cyclohexyl isonitrile to give the insertion product:

$$[Ni(\overset{\overset{\displaystyle C=NC_6H_{11}}{\mid}}{\underset{\displaystyle Me}{}})(CNC_6H_{11})cp]$$

(4.198)

### c. Oxidative addition reactions

Alkyl ligands have low electronegativity and therefore they enhance reactivity of the central atom in oxidative addition reactions[1–5, 13, 99] (see Section 4.10.b).

$$cis\text{-}[PtMe_2(PMe_2Ph)_2] + MeX \longrightarrow [PtMe_3X(PMe_2Ph)_2] \tag{4.199}$$

$$cis\text{-}[PtMe_2(PR_3)_2] + X_2 \longrightarrow [PtMe_2X_2(PR_3)_2] \tag{4.200}$$

$$trans\text{-}[Pt(Me)XL_2] + MeX \longrightarrow [PtMe_2X_2L_2] \tag{4.201}$$

$$[RhMe(PPh_3)_3] + PhI \longrightarrow [RhPh(Me)I(PPh_3)_3] \longrightarrow PhMe + [RhI(PPh_3)_3] \tag{4.202}$$

The oxidative addition of RX to $[Au(CH_2)_2PPh_2]_2$ is reversible.[196]

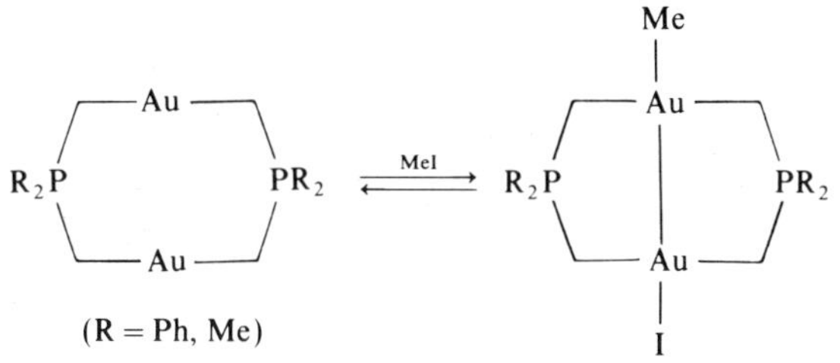

$$(4.203)$$

$$(R = Ph, Me)$$

This is an intermolecular, two-center process (the methyl group and iodine add to different central atoms). The complex $[Au(CH_2)_2PPh_2]_2$ catalyzes the halogen exchange between MeBr and MeI as well as between $[Au(CH_2)_2PPh_2]_2Br_2$ and MeI in CDCl$_3$ solution. The halogen exchange also takes place between complexes (4.204).

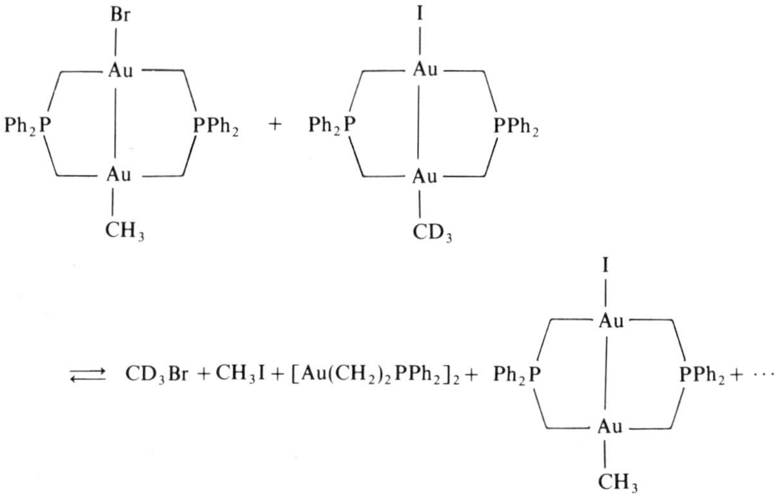

$$(4.204)$$

## d. The σ–π rearrangement of complexes containing σ metal–carbon bonds[56]

The $\sigma M - C$ bond in complexes with 1e ligands is formed as a result of the overlap between the $d$ orbital (or the $d^x sp^y$ orbital) of the metal and the sp$^n$ hybridized orbital of the carbon atom. This bond has little, if any, $\pi$ bonding character. In $\pi$ complexes, the bonding consists of two components, $\sigma$ and $\pi$. The $\sigma$ component is created by interaction between the $d_\sigma$ orbital of the metal and the $\pi$ orbital of the ligand. The $\pi$ bond is formed by the overlap between the $d_\pi$ orbital or the $p_\pi d_\pi$ hybrid of the metal and the $\pi^*$ orbital of the ligand. The stability of these two kinds of bonds and therefore the equilibrium between the $\sigma$ and $\pi$ complex depends to a certain degree on the electron density on the central atom. The increase of this density will favor the $\pi$ complex formation because the $\pi$ bond component will be increased. Therefore, ligands showing strong $\pi$ accepting properties, particularly if they are located in the *trans* position with respect to the olefin, will weaken the metal–alkene $\pi$ bonding. The complex *trans*-$[PtCl_2(C_2H_4)_2]$

readily undergoes rearrangement to the *cis* isomer because, in the *trans* complex, both ethylene molecules form the bond with the same platinum *d* orbital. The decrease of the electron density on the metal as a result of the presence of $\pi$ accepting ligands will favor the formation of $\sigma$ complexes and *vice versa*. The rearrangement reactions of complexes containing $\sigma M - C$ bonds into $\pi$ complexes occurs in numerous catalytic processes such as polymerization of unsaturated hydrocarbons and their isomerization, hydrogenation, hydroformylation, and also oxidation (Chapter 13). Alkyl complexes may rearrange into $\pi$ olefin metal derivatives as a result of interaction with $(CPh_3)^+ X^-$ or, in the case of the alkenyl complex $[Fecp(CO)_2(CH_2CH=CHR)]$, with hydrogen chloride (see Section 6.8.e). In the reaction of the crotyl complex $[(OC)_2 cpFe(CH_2CH=CHMe)]$ with $D^+$, the attack takes place on the third carbon atom:

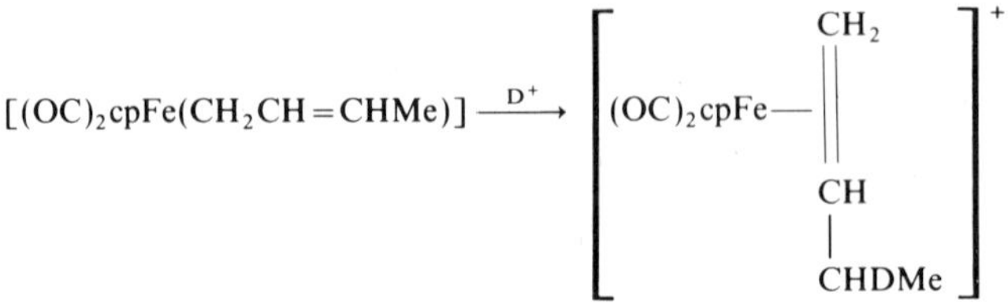

$$[(OC)_2cpFe(CH_2CH=CHMe)] \xrightarrow{\;D^+\;} \left[ (OC)_2cpFe - \begin{array}{c} CH_2 \\ \| \\ CH \\ | \\ CHDMe \end{array} \right]^+ \qquad (4.205)$$

The phenyl compound of $Cr(III)CrPh_3 \cdot 3THF$ undergoes rearrangement to an arene coordination compound (see Section 10.14.b).


## 13. *ACTIVATION OF HYDROCARBONS*

The formation of compounds containing $\sigma M - C$ bonds was observed during activation of the $C-H$ bonds by means of transition metal complexes.[57-61, 76, 89, 104, 105, 162-164, 167, 208-229]

Alkanes may be activated by superacids[62, 220] or complexes of the transition metals. The superacid reactions lead to the formation of cations possessing a five-coordinate carbon atom:

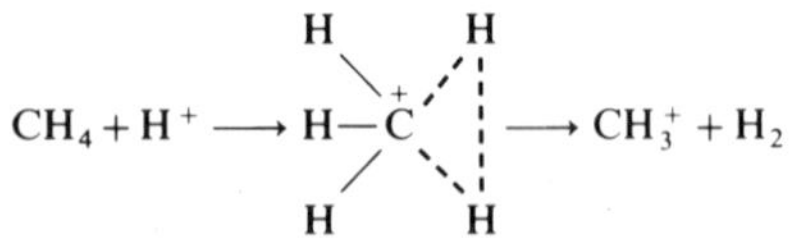

$$CH_4 + H^+ \longrightarrow H-\overset{+}{C}\underset{H}{\overset{H}{\diagdown}}\underset{H}{\overset{H}{\diagup}} \longrightarrow CH_3^+ + H_2 \qquad (4.206)$$

Many other carbonium cations such as $(CH)_n^{x+}$ ($n = 5, 6, 7; x = 1, 2, 3$) possessing cluster structures are also known (Chapter 3). Reactions of hydrocarbons with superacids lead to the H–D exchange.

The $C-H$ bond activation may occur according to various mechanisms.[57-62, 66, 67, 76, 89, 104, 208-231] One of them is a radical mechanism. In the presence of transition metal complexes having hard acid character, the $C-H$ bond splitting takes place. Complexes of the following metals belong to this group: Cr, Mn, Fe, Co, and Ni.[230] Reactions of organic compounds RH with complexes of metals possessing high oxidation–reduction potentials may proceed via electron transfer,[230]

$$RH + [M^{n+}L_m] \longrightarrow RH^{+*} + [M^{(n-1)+}L_m] \qquad (4.207)$$

$$RH^{+*} \longrightarrow R^* + H^+ \qquad (4.208)$$

electrophilic substitution,

$$RH + [M^{n+}L_m] \longrightarrow [M^{n+}RL_m]^- + H^+ \tag{4.209}$$

$$[M^{n+}RL_m]^- \longrightarrow [M^{(n-1)+}L_m]^- + R^* \tag{4.210}$$

or homolytic splitting,

$$RH + M^{n+}L_m \longrightarrow R^* + [MHL_m] \tag{4.211}$$

$$[MHL_m] \longrightarrow [M^{(n-1)+}L_m]^- + H^+ \tag{4.212}$$

As a result of the instability of metal compounds, it is very often impossible to prove which mechanism of the $C-H$ bond activation operates, since $R^*$ radicals are formed in all cases.

Electron transfer from the hydrocarbon molecule to the cobalt(III) or manganese(III) occurs during direct oxidation of arenes:

$$ArH + Co^{III} \longrightarrow ArH^{+*} + Co^{II} \tag{4.213}$$

Electrophilic substitution was confirmed in the case of arene oxidation by means of $Pb(OOCCF_3)_4$:

$$ArH + (OOCCF_3)_4 \longrightarrow PbAr(OOCCF_3)_3 + CF_3COOH \tag{4.214}$$

Strong reducing agents behave as electron donors. Bis(hexamethylbenzene)iron, the $20e$ neutral complex possessing strong reducing properties, rearranges to an $18e$ compound in the presence of oxygen.[213]

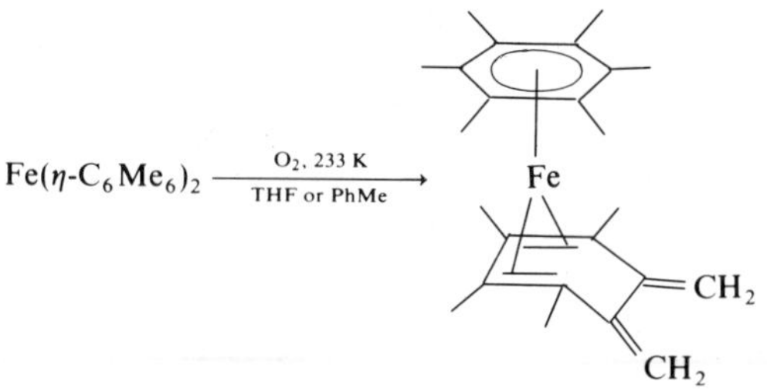

$$ \tag{4.215}$$

Thus, the $C-H$ bonds are activated during this reaction.

Transition metal complexes possessing the character of soft acids may activate $C-H$ bonds according to the following mechanisms:

(a)  electrophilic substitution

$$[M^{n+}L_m] + RH \longrightarrow [M^{n+}RL_m]^- + H^+ \tag{4.216}$$

(b)  one-center oxidative addition

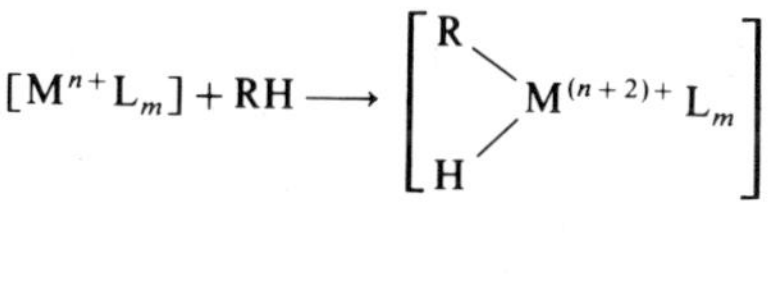

$$ \tag{4.217}$$

(c)   two-center oxidative addition

$$2[M^{n+}L_m] + RH \longrightarrow [M^{(n+1)+}RL_m] + [M^{(n+1)+}HL_m] \qquad (4.218)$$

The activation of hydrogen proceeds according to mechanisms (4.216)–(4.218); therefore, the activation of hydrocarbons probably occurs in a similar way. It was confirmed experimentally[76, 230] that the activation takes place via mechanisms (a), (b), and (c). The activation of alkenes by Pt(II) complexes probably occurs by electrophilic substitution in the first stage:

$$RH + [Pt^{II}Cl_n] \longrightarrow H^+ + [PtRCl_n]^- \qquad (4.219)$$

Oxidative addition of RH to transition metal complexes is thermodynamically unfavorable (see Table 4.4). The enthalpy of the reaction (4.217) usually falls in the $\Delta H = 40$–$60 \text{ kJ mol}^{-1}$ range. Therefore, the oxidative addition may take place in the case of hydrocarbons which contain large internal strains. The $C-H$ bond activation is best achieved by complexes which do not have large steric hindrance, because the metal–carbon bond energy is higher the smaller the repulsion between the ligands and the created $\sigma$ bonded group. The influence of steric factors is observed if the $M-C$ bond energies in $MRL_n$ complexes and those in MR, $MR^+$, etc., occurring in mass spectra, are compared. For the former, this energy is usually $ca$ $120 \text{ kJ mol}^{-1}$, and for the latter, in the range of $200$–$240 \text{ kJ mol}^{-1}$.

Therefore, for the $C-H$ bond activation, $16e$ or even $14e$ complexes containing compact ligands which do not cause large steric hindrance should be selected. All presently known complexes which activate the $C-H$ and $C-C$ bonds obey these conditions.

These conclusions are also confirmed by calculations concerning activation of the $H_2$ and $CH_4$ molecules.[89] The calculations show that the hydrogen molecule interacts with $d^6$ $ML_5$ and $d^8$ $ML_4$ complexes in the most effective way if the approach is "parallel," i.e., if both hydrogen atoms are at equal distances from the central atom. The activation of the $H_2$ molecule, which perpendicularly approaches the complex, is considerably less effective. Because of steric factors, the best interaction of methane with the coordination compound takes place through perpendicular orientation of the $H-CH_3$ bond, and therefore in the case of linear $M-H-C$ grouping. Also, stabilization is achieved for the $M-H-C$ angles higher than $130°$.[89] The calculations reveal that $d^{10}$ $ML_3$ and $d^{10}$ $ML_2$ systems should also well activate the $C-H$ bonds.[89] Under certain conditions, even saturated hydrocarbons may react with transition metal complexes with low energies of activation.[76, 89, 208–228] For example, the complex [Ir(pmcp)(CO)] reacts with $CH_4$ even at 12 K.

Many transition metal complexes form agostic (from Greek, to clasp) $M-H-C$ bonds which may be linear or angular.[89, 138, 211, 218, 222] The term agostic is used in the case of the hydrogen atom which is covalently bonded simultaneously to the metal and carbon atoms. The presence of the agostic hydrogen atom is detected by low values of $J(^{13}C-H)$ ($70$–$105 \text{ Hz}$) and $v(C-H)$ ($2350$–$2850 \text{ cm}^{-1}$). For organic compounds which have saturated $C(sp^3)-H$ bonds, the $J(^{13}C-H)$ constant assumes values in the range of $120$–$130 \text{ Hz}$. The $^1H$ NMR signals for agostic protons usually occur at strong fields; the chemical shift $\delta$ changes from $-30 \text{ ppm}$ to $+6 \text{ ppm}$. The $C-H$ distances are

5–10% longer than normal nonbridging CH bonds, and the M–H distances are 15–20% longer than in usual hydrido complexes. The agostic $M-H-C$ bonds occur in mono- as well as in polynuclear hydrocarbyl and carbene complexes (cf. Chapter 3).

The following iridium complexes are effective activators of the $C-H$ bonds: $[IrH_2(Me_2CO)L_2]A$ $[L = PPh_3,$ $P(C_6H_4F\text{-}4)_3;$ $A = BF_4^-,$ $SbF_6^-]$, $[IrH_2(8\text{-methyl-}$ quinoline$)L_2]BF_4$ (the methyl group of quinoline forms agostic bond with iridium.[211] Based on the structural data of complexes containing $M-H-C$ bonds, the mechanism for oxidative addition of the $C-H$ bond was proposed. The preference for inter- molecular $C-H$ bond splitting in the external reactant over intramolecular cyclo- metallation depends mainly on conformation and steric factors. Usually, despite the fact that numerous examples of cyclometallations are known, in the case of complexes which do not cause large steric hindrance, the $C-H$ bond breaking takes place in the hydrocarbon molecule which is not coordinated to the central atom before the reaction.[76, 89, 211] Cyclopentadienyl complexes of rhodium(III) and iridium(III), such as $Rh(pmcp)H_2(PMe_3)$,[104, 162, 164] $[Ir(pmcp)H_2(PMe_3)]$, $[Ir(pmcp)(PMe_3)]$, and $[Ir(pmcp)R(H)(PMe_3)]$, have been thoroughly investigated.[163, 209] After irradiation with UV light the rhodium complex evolves hydrogen and reacts with saturated hydrocarbons and arenes to give $[Rh(pmcp)R(H)(PMe_3)]$. The complex $[Ir(pmcp)H_2(PMe_3)]$ behaves similarly in photolytic and thermal reactions.[163, 209] The vinyl complex $[Ir(pmcp)(CH=CH_2)(H)(PMe_3)]$ is formed by the reaction of ethylene with $[Ir(pmcp)(cyclohexyl)(H)(PMe_3)]$. The $C(sp^2)-H$ bonds, despite higher energies than those of the $C-H$ bonds ($sp^3$ hybridization), are more readily activated. The activation of the CH bond in ethylene by pentamethylcyclopentadienyl iridium complex does not proceed via $\pi$ olefin complex. Therefore mechanism (4.220) of the reaction was proposed.[209]

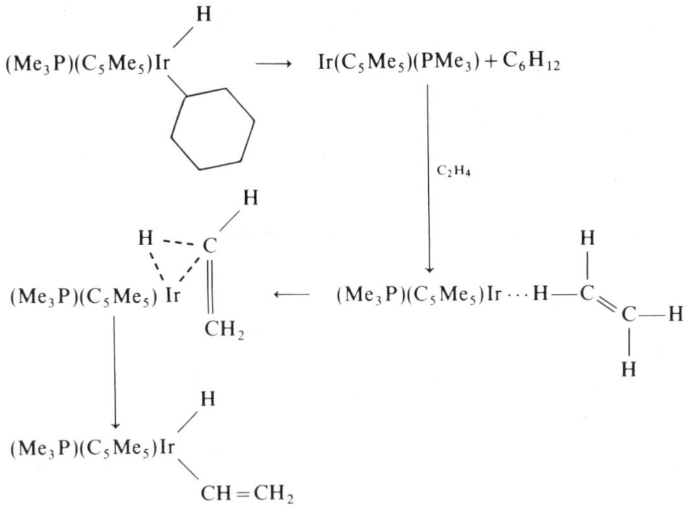

$$(4.220)$$

Allyl rhodium complexes deposited on silica gel activate methane. The products of the reaction are allyl and hydrido rhodium complexes, as well as propylene and small amounts of butene and butane.[210]

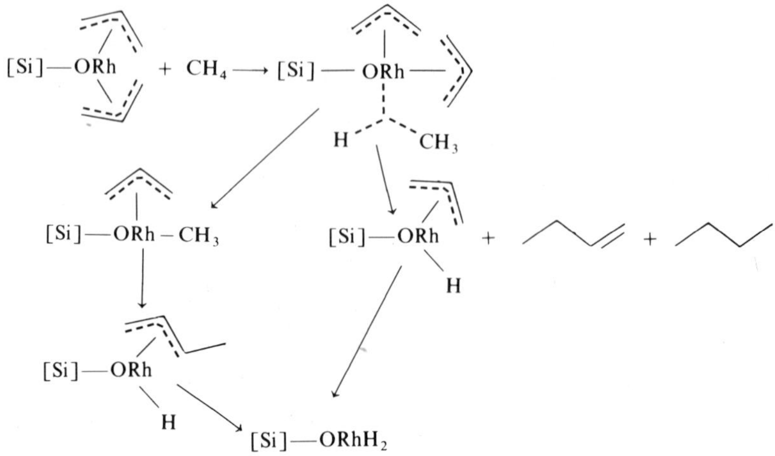

$$[Si]—ORh + CH_4 \longrightarrow [Si]—ORh— \quad \text{with} \quad H \cdots CH_3$$

(diagram of reaction scheme)

$$[Si]—ORh—CH_3$$

$$[Si]—ORh \quad + \quad \text{(alkene)} \quad + \quad \text{(alkene)}$$

with H

$$[Si]—ORh \text{ with } H$$

$$[Si]—ORhH_2 \tag{4.221}$$

Polyhydrido complexes are often utilized for the CH bond activation, for example, $ReH_7(PCy_3)_2$.[214] During heating of this complex in $C_6D_6$ (333–353 K) reductive elimination of hydrogen takes place and the 16$e$ hydride $[ReH_5(PCy_3)_2]$ is formed which exchanges hydrogen for deuterium not only in the case of hydrido ligands, but also one hydrogen atom at C2 and C3 each in all cyclohexyl rings.

Dicyclopentadienylhydridomethyltungsten(IV), $[Wcp_2(Me)H]$, undergoes reductive elimination in solution at low concentrations, while the intermolecular hydrogen exchange between the methyl and hydrido ligands proceeds faster than methane evolution at higher concentrations[215]:

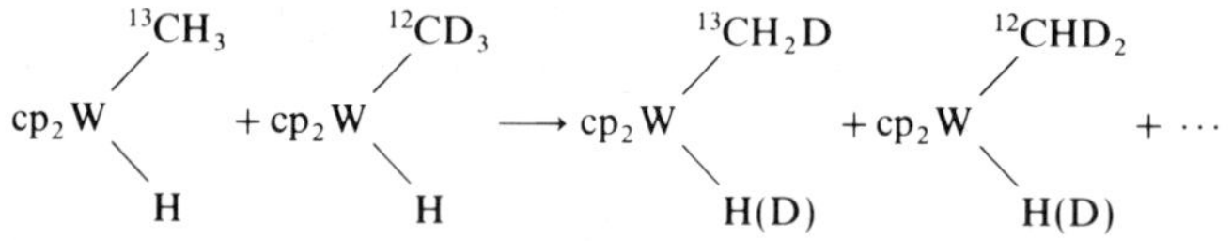

$$cp_2W \overset{^{13}CH_3}{\underset{H}{}} + cp_2W \overset{^{12}CD_3}{\underset{H}{}} \longrightarrow cp_2W \overset{^{13}CH_2D}{\underset{H(D)}{}} + cp_2W \overset{^{12}CHD_2}{\underset{H(D)}{}} + \cdots \tag{4.222}$$

The thorium complex

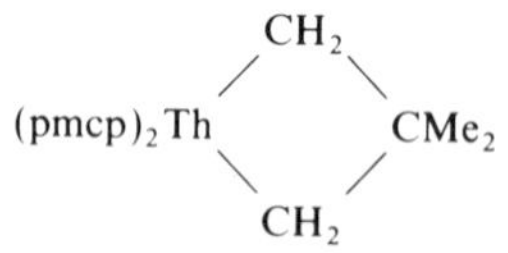

$$(pmcp)_2Th \overset{CH_2}{\underset{CH_2}{}} CMe_2$$

which contains a strained, four-membered metallacyclobutane ring,[118] reacts with tetramethylsilane and methane to give

$$[(pmcp)_2Th(CH_2SiMe_3)np] \quad \text{and} \quad (pmcp)_2Th \overset{CH_2}{\underset{CH_2}{}} SiMe_2$$

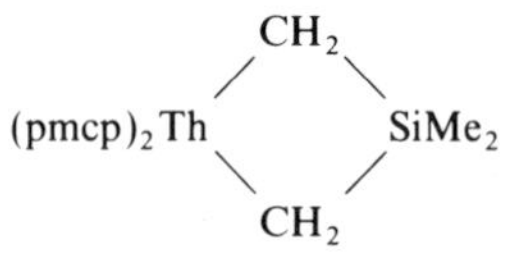

as well as

$$(pmcp)_2Th \overset{Me}{\underset{np}{}}$$

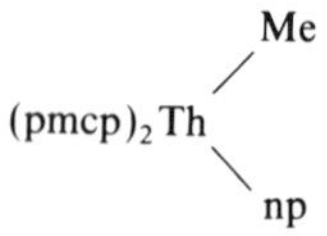

The activation of alkanes by means of $[M(pmcp)_2 Me]$ $(M = Y, Lu)$ probably depends on electrophilicity of the central atoms.[217] In the first stage of the reaction, the methane molecule probably forms an agostic $M-H-C$ bond with the metal atom.

The following $C-H$ bond activators containing compounds of the first and middle groups of the transition metals should be mentioned: $MX_n + AlMe_2Cl$ $[MX_n = TiCl_4,$ $Ticp_2Cl_2$, $V(acac)_3$, $VCl_3]$ and $Vcp_2$, $Mncp_2$. The Ziegler–Natta catalysts and cyclopentadienyl complexes are the most noteworthy.[66, 208] Cyclopentadienyl complexes catalyze the H–D exchange in the following mixture: $CD_4 + C_2H_4 + C_6H_6$. The exchange occurs between ethylene and benzene, and simultaneously between methane and ethylene, while this exchange is not observed between methane and benzene. Therefore, ethylene plays a role of the deuterium "carrier."

The activation of methane in the presence of ethylene effects methane addition to the olefin with the formation of propane. This reaction is catalyzed by the system $Ti(OBu)_4 + AlEt_3$[208]:

$$CD_4 + CH_2 = CH_2 \xrightarrow[\text{0.1 MPa}]{\text{293 K, PhH}} C_3H_4D_4 \tag{4.223}$$

The reaction of methane with acetylene proceeds in a similar manner.[208]

$$CH_4 + CH \equiv CH \xrightarrow[\text{PhH}]{\text{Fe(acac)}_3 + \text{AlEt}_3} CH_3CH = CH_2 \tag{4.224}$$

The $C-H$ and $C-C$ bonds are effectively activated by the transition metal ions as well as by polyatomic ions such as $M_2^+$, $MM_2'^+$, $MCH_2^+$. In contrast to metal complexes, the oxidative addition of the $C-H$ bonds to the $M^+$ ions is thermodynamically favorable because the $M^+ - CH_3$ and $M^+ - H$ bond energies most commonly equal $200-240$ kJ mol$^{-1}$ and $160-240$ kJ mol$^{-1}$, respectively. Thus, their sum is higher than the $C-H$ bond energies. The ions Fe$^+$ and Cu$^+$ dehydrogenate linear alkanes as a result of 1,4 and 1,2 elimination, while the Ni$^+$ ion dehydrogenates these alkanes via 1,4 elimination.[223] During reactions of alkanes[223] and cycloalkanes[229] with Fe$^+$, Co$^+$, and Ni$^+$, oxidative addition takes place because of the $C-C$ bond splitting.

The ions Ru$^+$, Rh$^+$, and Pd$^+$ react similarly, causing the $C-C$ bond splitting.[224]

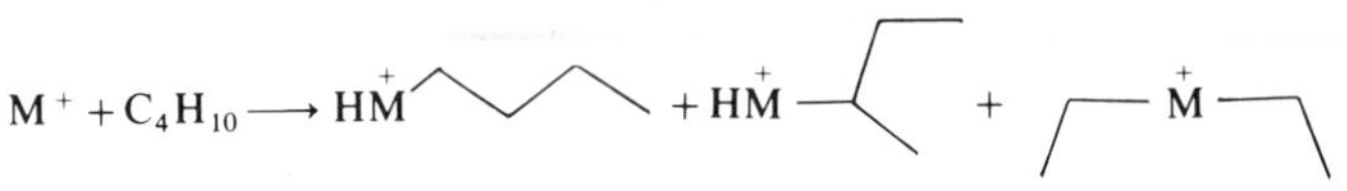

$$\tag{4.225}$$

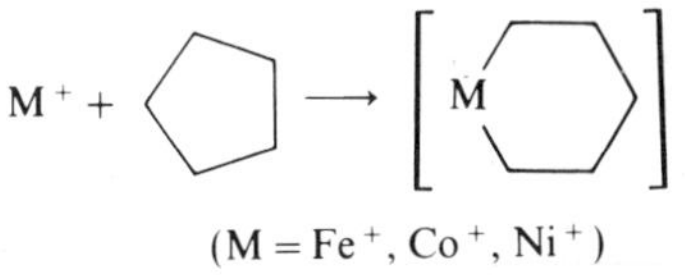

$$(M = Fe^+, Co^+, Ni^+) \tag{4.226}$$

$$M^+ + C_2D_6 \longrightarrow MCD_3^+ + CD_3 \tag{4.227}$$
$$(M = Ru^+, Rh^+, Pd^+)$$

The mechanism of the reaction of the ion Sc$^+$ with alkanes is probably different than that of ions of group 8–10 metals. In this case, 1,3 dehydrogenation is observed.

It is proposed that, first, insertion of $Sc^+$ into the $C-H$ bond occurs, followed by the addition of the $\gamma$ $C-H$ alkyl bond with $H_2$ elimination.[225]

The ion $Co_2^+$ does not react with alkanes. However, CO addition to this ion leads to the formation of $Co_2CO^+$, which effectively attacks the $C-H$ bond in alkanes. The rate constant of the reaction of activation of n-butane is at least two orders of magnitude higher for $Co_2CO^+$ than for $Co_2^+$.[226] In contrast to dimeric cations $Co_2^+$ and $FeCo^+$, the trinuclear ion $FeCo_2^+$ reacts with alkanes to give dehydrogenation products.[227] The mononuclear ions $M^+$ generally split the $C-C$ bond.

Carbene cations such as $FeCH_2^+$ and $CoCH_2^+$ activate alkanes and cycloalkanes, and in the case of olefins both their methathesis as well as activation takes place as inferred from the hydrogen evolution.[228]

Numerous examples of oxidative addition of $C-H$ bonds of ligand molecules leading to cyclometallation are known [reactions (4.70)–(4.72), (4.113), (4.115), (4.139), Figures 4.1e and 4.5].[97, 98, 100, 105, 167] The $C-H$ bond breaking most easily takes place in phosphorus and nitrogen donor ligands.

Methane undergoes activation with the greatest difficulty, followed by other alkanes, cycloalkanes, aromatic hydrocarbons, and hydrocarbon derivatives. Methane is activated by means of platinum(II) complexes in acetic acid solution.

$$CH_4 + D_2O \xrightarrow[\text{HOAc, 373 K}]{K_2[PtCl_4]} CH_3D + HDO \tag{4.228}$$

The exchange reaction products contain all possible compounds $CD_xH_{4-x}$ ($x = 1-4$).

The presence of dominant amounts of asymmetrically substituted higher alkanes, like the H–D exchange in methane, proves that not only compounds with $\sigma M-C$ bonds but also carbene complexes, and perhaps even carbyne compounds, are formed as intermediates:

$$
\begin{array}{ccccccccc}
CH_4 & & CH_3D & & CH_2D_2 & & CHD_3 & & CD_4 \\
\downarrow \quad \nearrow & & \nearrow & & \nearrow & & \nearrow & & \nearrow \\
PtCH_3 & & PtCH_2D & & PtCHD_2 & & PtCD_3 \\
\downarrow \quad \nearrow \quad \searrow & & \nearrow \quad \searrow & & \nearrow \quad \searrow & & \nearrow \\
Pt=CH_2 & & Pt=CHD & & Pt=CD_2
\end{array} \tag{4.229}
$$

The catalytic activity of platinum group complexes in the H–D exchange reaction in alkanes decreases in the series $Pt > Ir > Rh > Pd \sim Ru$.

Platinum group compounds, particularly those of platinum itself, and also copper complexes, catalyze oxidation reactions of alkanes to alkyl halides as well as to oxygen derivatives. The mechanism of oxidation is in many cases probably analogous to the H–D exchange reaction because the rate constants and the energies of activation of these reactions only differ slightly. The energy of activation of the H–D exchange reaction for alkanes and their derivatives occurring in the presence of Pt(II) complexes assumes values of $75-100$ kJ mol$^{-1}$.

The $C-C$ bonds in polycyclic hydrocarbons which exhibit strain are easily broken. Examples of such compounds are shown in (4.230).

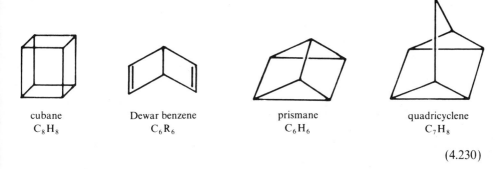

| cubane | Dewar benzene | prismane | quadricyclene |
|--------|---------------|----------|---------------|
| $C_8H_8$ | $C_6R_6$ | $C_6H_6$ | $C_7H_8$ |

$$(4.230)$$

Transition metal complexes, particularly those of rhodium, ruthenium, iridium, palladium, platinum, and silver, catalyze rearrangements of such hydrocarbons. Compounds containing $M-C$ bonds are formed as intermediates.[64, 208] The rearrangement of quadricyclene into norbornadiene is exothermic. The reverse reaction takes place under the influence of sunlight. Quadricyclene is stable in the absence of catalysts which rearrange it into NBD. Therefore, the system of these two compounds may be applied for solar energy storage.[65, 232] In this way, *ca* 1000 J cm$^{-3}$ may be accumulated and stored.

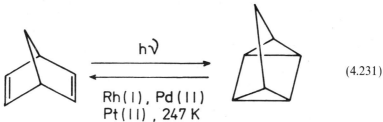

$$(4.231)$$

# REFERENCES

1. P. J. Davidson, M. F. Lappert, and R. Pearce, *Chem. Rev.*, **76**, 219 (1976).
2. R. R. Schrock and G. W. Parshall, *Chem. Rev.*, **76**, 243 (1976); R. Nast, *Coord. Chem. Rev.*, **47**, 89 (1982).
3. L. G. Makarova, in: *Metody Elementoorganicheskoi Khimii* (A. N. Nesmeyanov and K. A. Kocheshkov, eds.), p. 229, Izd. Nauka, Moscow (1976).
4. L. P. Yureva, *Metody Elementoorganicheskoi Khimii* (A. N. Nesmeyanov and K. A. Kocheshkov, eds.), p. 350, Izd. Nauka, Moscow (1976).
5. M. L. H. Green, *Organometallic Compounds. The Transition Elements*, Methuen, London (1968).
6. P. S. Braterman and R. J. Cross, *Chem. Soc. Rev.*, **2**, 271 (1973).
7. J. A. Connor, *Top. Curr. Chem.*, **71**, 71 (1977); H. A. Skinner and J. A. Connor, *Pure Appl. Chem.*, **57**, 79 (1985); G. Pilcher and H. A. Skinner, in: *The Chemistry of the Metal–Carbon Bond* (F. R. Hartley and S. Patai, eds.), Chapter 2, Wiley, Chichester, 1982.
8. J. R. Norton, *Acc. Chem. Res.*, **12**, 139 (1979).
9. D. M. Mingos, *Chem. Commun.*, 165 (1972).
10. S. Komiya, T. A. Albright, R. Hoffmann, and J. K. Kochi, *J. Am. Chem. Soc.*, **98**, 7255 (1976).
11. O. N. Temkin and P. M. Flid, *Kataliticheskie Prevrashchenia Acetilenovykh Soedimenii v Rastvorakh Kompleksov Metallov*, Izd. Nauka, Moscow (1968).
12. J. Tsuji, *Organic Synthesis by Means of Transition Metal Complexes*, Springer-Verlag, Berlin (1975).

13. B. L. Shaw and N. I. Tucker, *Comprehensive Inorganic Chemistry* (J. C. Bailar, Jr., ed.), Vol. 4, p. 781, Pergamon Press, Oxford (1973).

14. M. Tsutsui, N. Ely, and R. Dubois, *Acc. Chem. Res.*, **9**, 217 (1976).

15. C. W. Bradford, R. S. Nyholm, G. J. Gainsford, J. M. Guss, P. R. Ireland, and R. Mason, *Chem. Commun.*, 87 (1972).

16. A. J. Deeming, R. S. Nyholm, and M. Underhill, *Chem. Commun.*, 224 (1972).

17. R. A. Andersen, R. A. Jones, G. Wilkinson, M. B. Hursthouse, and K. M. A. Malik, *Chem. Commun.*, 283 (1977).

18. A. M. Abeysekera, R. Grigg, J. Trocha-Grimshaw, and V. Viswanathaa, *Chem. Commun.*, **227** (1976).

19. W. A. Herrmann, C. Krueger, R. Goddard, and I. Bernal, *Angew. Chem., Int. Ed. Engl.*, **16**, 334 (1977).

20. M. Foa and L. Cassar, *J. Chem. Soc., Dalton Trans.*, 2572 (1975).

21. E. C. Ashby and J. J. Watkins, *Chem. Commun.*, 784 (1976).

22. G. van Koten, J. T. B. H. Jastrzebski, and J. G. Noltes, *Chem. Commun.*, 203 (1977).

23. K. Nakamoto, in: *Characterization of Organometallic Compounds* (M. Tsutsui, ed.), Part I, p. 73, Interscience, New York (1969); E. Maslowsky, Jr., *Chem. Soc. Rev.*, **9**, 25 (1980).

24. M. H. Chisholm and S. Godleski, *Prog. Inorg. Chem.*, **20**, 299 (1976).

25. B. E. Mann, *Adv. Organomet. Chem.*, **12**, 135 (1974).

26. A. H. Cowley, *Prog. Inorg. Chem.*, **26**, 45 (1979).

27. C. Furlani and C. Cauletti, *Structure Bonding*, **35**, 119 (1978).

28. S. Evans, J. C. Green, and S. E. Jackson, *J. Chem. Soc., Faraday Trans. 2*, **69**, 191 (1973).

29. J. C. Green, D. R. Lloyd, L. Galyer, K. Mertis, and G. Wilkinson, *J. Chem. Soc., Dalton Trans.*, 1403 (1978).

30. M. F. Lappert, J. B. Pedley, and G. Sharp, *J. Organomet. Chem.*, **66**, 271 (1974).

31. J. C. Green and S. E. Jackson, *J. Chem. Soc., Dalton Trans.*, 1698 (1976).

32. S. Evans, J. C. Green, P. J. Joachim, A. F. Orchard, D. D. Turner, and J. P. Maier, *J. Chem. Soc., Faraday Trans. 2*, 905 (1972).

33. T. Głowiak, R. Grobelny, B. Jeżowska-Trzebiatowska, G. Kreisel, W. Seidel, and E. Uhlig, *J. Organomet. Chem.*, **155**, 39 (1978).

34. K. Mertis and G. Wilkinson, *J. Chem. Soc., Dalton Trans.*, 1488 (1976).

35. J. F. Gibson, G. M. Lack, K. Mertis, and G. Wilkinson, *J. Chem. Soc., Dalton Trans.*, 1492 (1976).

36. J. F. Gibson, K. Mertis, and G. Wilkinson, *J. Chem. Soc., Dalton Trans.*, 1093 (1975).

37. W. Mowat, A. Shortland, G. Yagupsky, N. J. Hill, M. Yagupsky, and G. Wilkinson, *J. Chem. Soc., Dalton Trans.*, 533 (1972).

38. W. Mowat, A. J. Shortland, N. J. Hill, and G. Wilkinson, *J. Chem. Soc., Dalton Trans.*, 770 (1973).

39. K. Mertis, D. H. Williamson, and G. Wilkinson, *J. Chem. Soc., Dalton Trans.*, 607 (1975).

40. F. Basolo and R. G. Pearson, *Mechanisms of Inorganic Reactions*, Wiley, New York (1967).

41. H. S. Klein, *Chem. Commun.*, 377 (1968); S. O'Brien, *J. Chem. Soc. (A)*, 9 (1970).

42. J. P. Collman, R. G. Finke, J. N. Cawse, and J. I. Brauman, *J. Am. Chem. Soc.*, **100**, 4766 (1978).

43. S. B. Butts, E. M. Holt, S. H. Strauss, N. W. Alcock, R. E. Stimson, and D. F. Shriver, *J. Am. Chem. Soc.*, **101**, 5864 (1979).

44. M. E. Volpin, *Pure Appl. Chem.*, **30**, 607 (1972).

45. M. E. Volpin and I. S. Kolomnikow, *Organomet. React.*, **5**, 313 (1975); P. Sobota, and B. Jeżowska-Trzebiatowska, *Coord. Chem. Rev.*, **26**, 71 (1978).

46. T. Ito and A. Yamamoto, *J. Soc. Org. Synth. Chem. Tokyo*, **34**, 308 (1976); T. Yamamoto and A. Yamamoto, *Chem. Lett.*, 615 (1978).

47. E. Lindner, R. Grimmer, and H. Weber, *Angew. Chem.*, **82**, 639 (1970).

48. A. R. Middleton and G. Wilkinson, *J. Chem. Soc., Dalton Trans.*, 1888 (1980).

49. P. Edwards, K. Mertis, G. Wilkinson, M. B. Hursthouse, and K. M. A. Malik, *J. Chem. Soc., Dalton Trans.*, 344 (1980).

50. J. A. McCleverty, *Chem. Rev.*, **79**, 53 (1979).

51. Yu. A. Aleksandrov, *Zhydkofaznoe Avtookislenie Elementoorganicheskikh Soedinenii*, Izd. Nauka, Moscow (1978).

52. M. E. Volpin, *J. Organomet. Chem.*, **200**, 319 (1980).

53. A. E. Shilov, *Usp. Khim.*, **43**, 863 (1974).

54. M. E. Volpin and V. B. Shur, *Organomet. React.*, **1**, 55 (1970).
55. D. Sellmann and W. Weiss, *Angew. Chem., Int. Ed. Engl.*, **17**, 269 (1978); D. Sellmann and W. Weiss, *Angew. Chem., Int. Ed. Engl.*, **16**, 880 (1977).
56. M. Tsutsui and A. Courtney, *Adv. Organomet. Chem.*, **16**, 241 (1978).
57. A. E. Shilov and A. A. Shteinman, *Coord. Chem. Rev.*, **24**, 97 (1977); A. E. Shilov and A. A. Shteinman, *Kinetika i Kataliz.*, **18**, 1129 (1977).
58. N. F. Goldshleger, W. M. Nikipelov, A. T. Nikitaev, K. I. Zamaraev, A. E. Shilov, and A. A. Shteinman, *Kinetika i Kataliz*, **20**, 538 (1979).
59. D. E. Webster, *Adv. Organomet. Chem.*, **15**, 147 (1977); G. W. Parshall, *Acc. Chem. Res.*, **8**, 113 (1975).
60. J. J. Ziółkowski, in: *Metal Complex Catalysis* (J. J. Ziółkowski, ed.), Wrocław University Press, p. 71, Wrocław (1979).
61. J. L. Garnett, *Catal. Rev.*, **5**, 229 (1972).
62. H. Hogeveen and P. W. Kwant, *Acc. Chem. Res.*, **8**, 413 (1975).
63. K. Jonas and C. Krueger, *Angew. Chem., Int. Ed. Engl.*, **19**, 520 (1980); K. Jonas, *Adv. Organomet. Chem.*, **19**, 97 (1981).
64. K. C. Bishop, III, *Chem. Rev.*, **76**, 461 (1976).
65. N. Kamiya and T. Ohta, in: *Solar-Hydrogen Energy Systems* (T. Ohta, ed.), Pergamon, Oxford (1979).
66. E. A. Grigoryan, F. S. Dyachkovskii, and I. R. Mullagalev, *Dokl. Akad. Nauk SSSR*, **224**, 859 (1975).
67. G. B. Shulpin, L. P. Rosenberg, R. P. Shibaeva, and A. E. Shilov, *Kinetika i Kataliz*, **20**, 1570 (1979).
68. L. E. Manzer, R. C. Gearhart, L. Guggenberger, and J. F. Whitney, *Chem. Commun.*, 942 (1976).
69. J. H. Teuben and H. J. de Liefde Meijer, *J. Organomet. Chem.*, **46**, 313 (1972).
70. B. K. Bower and H. G. Tennet, *J. Am. Chem. Soc.*, **94**, 2512 (1972).
71. W. Seidel and G. Kreisel, *Z. Chem.*, **14**, 25 (1974).
72. W. Seidel, K. Fischer, and K. Schmiedeknecht, *Z. Inorg. Allg. Chem.*, **390**, 273 (1972).
73. R. A. Andersen, E. Carmona-Guzman, K. Mertis, E. Sigurdson, and G. Wilkinson, *J. Organomet. Chem.*, **99**, C19 (1975); R. A. Andersen, E. Carmona-Guzman, J. E. Gibson, and G. Wilkinson, *J. Chem. Soc., Dalton Trans.*, 2204 (1976).
74. R. A. Andersen, R. A. Jones, G. Wilkinson, M. B. Hoursthouse, and K. M. A. Malik, *Chem. Commun.*, 283 (1977); R. Andersen, R. A. Jones, and G. Wilkinson, *J. Chem. Soc., Dalton Trans.*, 446 (1978).
75. G. Pilcher and H. A. Skinner, in: *The Chemistry of the Metal–Carbon Bond* (F. R. Hartley and S. Patai, eds.), p. 43, Wiley, New York (1982).
76. J. Halpern, *Acc. Chem. Res.*, **15**, 238 (1982); J. Halpern, *Inorg. Chim. Acta*, **100**, 41 (1985); J. Halpern, S.-H. Kim, and T. W. Leung, *J. Am. Chem. Soc.*, **106**, 8317 (1984).
77. J. W. Bruno, T. J. Marks, and L. R. Morss, *J. Am. Chem. Soc.*, **105**, 6824 (1983).
78. J. A. Connor, M. T. Zafarani-Mottar, J. Bickerton, N. I. Saied, S. Suradi, R. Carson, G. A. Takhin, and H. A. Skinner, *Organometallics*, 1982, **1**, 1166 (1982).
79. D. Milstein, *Acc. Chem. Res.*, **17**, 221 (1984).
80. D. M. P. Mingos, in: *Comprehensive Organometallic Chemistry* (G. Wilkinson, F. G. A. Stone, and E. W. Abel, eds.), Vol. 3, p. 1, Pergamon Press, Oxford (1982).
81. R. B. Hitam, K. A. Mahmoud, and A. J. Rest, *Coord. Chem. Rev.*, **55**, 1 (1984).
82. K. A. Mahmoud, A. J. Rest, and H. G. Alt, *Chem. Commun.*, 1011 (1983).
83. R. J. Kazlauskas and M. S. Wrighton, *J. Am. Chem. Soc.*, **102**, 1727 (1980); R. J. Kazlauskas and M. S. Wrighton, *J. Am. Chem. Soc.*, **104**, 6005 (1982).
84. K. A. Mahmoud, A. J. Rest, H. G. Alt, M. E. Eichner, and B. M. Jansen, *J. Chem. Soc., Dalton Trans.*, 175 (1984).
85. J. Holton, M. F. Lappert, R. Pearce, and P. I. W. Yarrow, *Chem. Rev.*, **83**, 135 (1983).
86. G. A. Razuvaev and L. I. Wyshinskaya, *Usp. Khim.*, **52**, 1648 (1983.)
87. J. R. Moss and L. G. Scott, *Coord. Chem. Rev.*, **60**, 171 (1984).
88. R. Nast, *Coord. Chem. Rev.*, **47**, 89 (1982).
89. J. Saillard and R. Hoffmann, *J. Am. Chem. Soc.*, **106**, 2006 (1984).
90. J. J. Low and W. A. Goddard, III, *J. Am. Chem. Soc.*, **106**, 8321 (1984).

91. S. Komiya, Y. Morimoto, A. Yamamato, and T. Yamamoto, *Organometallics*, **1**, 1528 (1982).
92. J. C. Green, *Structure and Bonding*, **43**, 37 (1981).
93. C. Long, A. R. Morrison, D. C. McVean, and G. P. McQuillan, *J. Am. Chem. Soc.*, **106**, 7418 (1984).
94. K. L. Brandenburg and H. B. Abrahamson, *J. Organomet. Chem.*, **293**, 371 (1985).
95. H. B. Abrahamson, K. L. Brandenburg, B. Lucero, M. E. Martin, and E. Dennis, *Organometallics*, **3**, 1379 (1984).
96. H. G. Alt, *Angew. Chem., Int. Ed. Engl.*, **23**, 766 (1984).
97. I. Omae, *Coord. Chem. Rev.*, **32**, 235 (1980); I. Omae, *Angew. Chem., Int. Ed. Engl.*, **21**, 889 (1982).
98. M. I. Bruce, *Angew. Chem., Int. Ed. Engl.*, **16**, 73 (1977).
99. J. U. Mondal and D. M. Blake, *Coord. Chem. Rev.*, **47**, 205 (1982); G. K. Anderson and R. J. Cross, *Acc. Chem. Res.*, **17**, 67 (1984).
100. G. L. Geoffroy and M. S. Wrighton, *Organometallic Photochemistry*, Academic Press, New York (1979).
101. D. C. Sonnenberger, E. A. Mintz, and T. J. Marks, *J. Am. Chem. Soc.*, **106**, 3484 (1984).
102. S. M. B. Costa, A. R. Dias, and F. J. S. Pina, *J. Chem. Soc., Dalton Trans.*, 314 (1981).
103. D. S. Edwards and R. R. Schrock, *J. Am. Chem. Soc.*, **104**, 6806 (1982); D. S. Edwards, L. V. Biondi, J. W. Ziller, M. R. Churchill, and R. R. Schrock, *Organometallics*, **2**, 1505 (1983).
104. R. A. Periana and R. G. Bergman, *Organometallics*, **3**, 508 (1984); R. A. Periana and R. G. Bergman, *J. Am. Chem. Soc.*, **106**, 7272 (1984).
105. L. R. Chamberlain, A. P. Rothwell, and I. P. Rothwell, *J. Am. Chem. Soc.*, **106**, 1847 (1984).
106. H. E. Bryndza, *Organometallics*, **4**, 406 (1985).
107. N. G. Connelly and W. E. Geiger, *Adv. Organomet. Chem.*, **23**, 1 (1984).
108. G. Schiavon, G. Bontempelli, M. De Nobili, and B. Corain, *Inorg. Chim. Acta*, **42**, 211 (1980).
109. W. Lau, J. C. Huffman, and J. K. Kochi, *Organometallics*, **1**, 155 (1982).
110. T. J. Marks and R. D. Ernst, in: *Comprehensive Organometallic Chemistry* (G. Wilkinson, F. G. A. Stone, and E. W. Abel, eds.), Vol. 3, Chapter 21, Pergamon Press, Oxford (1982).
111. P. L. Watson and G. E. Parshall, *Acc. Chem. Res.*, **18**, 51 (1985); P. L. Watson, *J. Am. Chem. Soc.*, **105**, 6491 (1983).
112. H. Schumann, *Angew. Chem., Int. Ed. Engl.*, **23**, 474 (1984).
113. H. Schumann, H. Lauke, E. Hahn, M. J. Heeg, and D. van der Helm, *Organometallics*, **4**, 321 (1985).
114. H. Schumann, H. Lauke, E. Hahn, and J. Pickardt, *J. Organomet. Chem.*, **263**, 29 (1984).
115. H. Schumann and S. Hohmann, *Chem.-Ztg.*, **100**, 336 (1976); H. Schumann and F. W. Reier, *J. Organomet. Chem.*, **235**, 287 (1982).
116. H. Schumann, I. Albrecht, F.-W. Reier, and E. Hahn, *Angew. Chem., Int. Ed. Engl.*, **23**, 522 (1984).
117. H. Lauke, P. J. Swepston, and T. J. Marks, *J. Am. Chem. Soc.*, **106**, 6841 (1984).
118. C. M. Fendrick and T. J. Marks, *J. Am. Chem. Soc.*, **106**, 2214 (1984).
119. R. E. Cramer, A. L. Mori, R. B. Maynard, J. W. Gilje, K. Tatsumi, and A. Nakamura, *J. Am. Chem. Soc.*, **106**, 5920 (1984).
120. M. Bottrill, P. D. Gavens, J. W. Kelland, and J. McMeeking, in: *Comprehensive Organometallic Chemistry* (G. Wilkinson, F. G. A. Stone, and E. W. Abel, eds.), Vol. 3, Chapters 22.1, 22.3, 22.4, Pergamon Press, Oxford (1982); M. Bottrill, P. D. Gavens, and J. McMeeking, in: *Comprehensive Organometallic Chemistry* (G. Wilkinson, F. G. A. Stone, and E. W. Abel, eds.), Vol. 3, Chapter 22.2, Pergamon Press, Oxford (1982).
121. D. J. Cardin, M. F. Lappert, C. L. Raston, and P. I. Riley, in: *Comprehensive Organometallic Chemistry* (G. Wilkinson, F. G. A. Stone, and E. W. Abel, eds.), Vol. 3, Chapters 23.1–23.3, Pergamon Press, Oxford (1982).
122. W. Kaminsky, J. Kopf, H. Sinn, and H.-J. Vollmer, *Angew. Chem., Int. Ed. Engl.*, **15**, 629 (1976).
123. R. M. Waymouth, B. D. Santarsiero, and R. H. Grubbs, *J. Am. Chem. Soc.*, **106**, 4050 (1984).
124. N. G. Connelly, in: *Comprehensive Organometallic Chemistry* (G. Wilkinson, F. G. A. Stone, and E. W. Abel, eds.), Vol. 3, Chapter 24, Pergamon Press, Oxford (1982).
125. J. A. Labinger, in: *Comprehensive Organometallic Chemistry* (G. Wilkinson, F. G. A. Stone, and E. W. Abel, eds.), Vol. 3, Chapter 25, Pergamon Press, Oxford (1982).
126. F. A. Cotton, G. E. Lewis, and G. N. Mott, *Inorg. Chem.*, **22**, 560 (1983).
127. S. W. Kirtley, in: *Comprehensive Organometallic Chemistry* (G. Wilkinson, F. G. A. Stone, and E. W. Abel, eds.), Vol. 3, Chapters 26.1, 27.1, 28.1, Pergamon Press, Oxford (1982).

128. I. Feinstein-Jaffe, D. Gibson, S. J. Lippard, R. R. Schrock, and A. Spool, *J. Am. Chem. Soc.*, **106**, 6305 (1986).
129. M. H. Chisholm, J. A. Heppert, and J. C. Huffman, *J. Am. Chem. Soc.*, **106**, 1151 (1984).
130. R. C. Murray and R. R. Schrock, *J. Am. Chem. Soc.*, **107**, 4557 (1985).
131. M. D. Braydich, B. E. Bursten, M. H. Chisholm, and D. L. Clark, *J. Am. Chem. Soc.*, **107** 4459 (1985); M. H. Chisholm, D. M. Hoffman, J. C. Huffman, W. G. van der Sluys, and S. Russo, *J. Am. Chem. Soc.*, **106**, 5386 (1984).
132. S.-C. H. Su and A. Wojcicki, *Organometallics*, **2**, 1296 (1983).
133. D. J. Darensbourg and R. Kudaroski, *J. Am. Chem. Soc.*, **106**, 3672 (1984).
134. F. A. Cotton and R. A. Walton, *Multiple Bonds Between Metal Atoms*, Wiley, New York (1982).
135. P. M. Treichel, in: *Comprehensive Organometallic Chemistry* (G. Wilkinson, F. G. A. Stone, and E. W. Abel, eds.), Vol. 4, Chapter 29, Pergamon Press, Oxford (1982).
136. N. M. Boag and H. D. Kaesz, in: *Comprehensive Organometallic Chemistry* (G. Wilkinson, F. G. A. Stone, and E. W. Abel, eds.), Vol. 4, Chapter 30, Pergamon Press, Oxford (1982).
137. G. D. Vaughn and J. A. Gladysz, *Organometallics*, **3**, 1596 (1984).
138. M. A. Bennett, I. J. McMahon, S. Pelling, G. B. Robertson, and W. A. Wickramasinghe, *Organometallics*, **4**, 754 (1985).
139. C. R. Jablonski and Y.-P. Wang, *Organometallics*, **4**, 465 (1985).
140. V. V. Mainz and R. A. Andersen, *Organometallics*, **3**, 675 (1984).
141. T. C. Flood and K. D. Campbell, *J. Am. Chem. Soc.*, **105**, 2853 (1984); T. C. Flood, K. D. Campbell, H. H. Downs, and S. Nakamishi, *Organometallics*, **2**, 1590 (1983); H. Brunner, J. Bernal, and M. Draux, *Organometallics*, **2**, 1595 (1983).
142. D. Lancon, P. Cocolios, R. Guilard, and K. M. Kadish, *J. Am. Chem. Soc.*, **106**, 4472 (1984); R. Guilard, B. Boisselier-Cocolios, A. Tabard, P. Cocolios, B. Simonet, and K. M. Kadish, *Inorg. Chem.*, **24**, 2509 (1985).
143. J. Gotzig, A. L. Rheingold, and H. Werner, *Angew. Chem., Int. Ed. Engl.*, **23**, 814 (1984); H. Werner and J. Gotzig, *Organometallics*, **3**, 547 (1984); H. Werner and R. Werner, *J. Organomet. Chem.*, **209**, C60 (1981).
144. E. J. Crawford, C. Lambert, K. P. Menard, and A. R. Cutler, *J. Am. Chem. Soc.*, **107**, 3130 (1985).
145. P. A. Belmonte and Z.-Y. Own, *J. Am. Chem. Soc.*, **106**, 7493 (1984).
146. A. R. Chakravarty, F. A. Cotton, and D. A. Tocher, *J. Am. Chem. Soc.*, **106**, 6409 (1984); A. R. Chakravarty, F. A. Cotton, and D. A. Tocher, *Inorg. Chem.*, **23**, 4697 (1984); A. R. Chakravarty, F. A. Cotton, and D. A. Tocher, *Chem. Commun.*, 501 (1984).
147. A. Wojcicki, in: *Fundamental Research in Organometallic Chemistry* (M. Tsutsui, Y. Ishii, and Y. Huang, eds.), p. 563, Van Nostrand–Rheinhold, New York (1982).
148. G. Reichenbach, G. Cardaci, and G. Bellachioma, *J. Chem. Soc., Dalton Trans.*, 847 (1982).
149. R. Gompper and E. Bartmann, *Angew. Chem., Int. Ed., Engl.*, **24**, 209 (1985).
150. E. Lindner, E. Schauss, W. Hiller, and R. Fawzi, *Angew. Chem., Int. Ed. Engl.*, **23**, 711 (1984).
151. T. W. Nikitina, *Metody Elementoorganicheskoi Khimii. Metalloorganicheskie Soedinenia Zheleza*, Nauka, Moscow (1985).
152. A. A. Koridze, in: *Metody Elementoorganicheskoi Khimii. Kobalt. Nikiel. Platinovyje Metally* (A. N. Nesmeyanov and K. A. Kocheshkov, eds.), Chapters 3 and 4, Nauka, Moscow (1978).
153. M. D. Johnson, in: *Comprehensive Organometallic Chemistry* (G. Wilkinson, F. G. A. Stone, and E. W. Abel, eds.), Vol. 4, Chapter 31.2, Pergamon Press, Oxford (1982).
154. M. A. Bennett, in: *Comprehensive Organometallic Chemistry* (G. Wilkinson, F. G. A. Stone, and E. W. Abel, eds.), Vol. 4, Chapter 32.3, Pergamon Press, Oxford (1982).
155. R. D. Adams, in: *Comprehensive Organometallic Chemistry* (G. Wilkinson, F. G. A. Stone, and E. W. Abel, eds.), Vol. 4, Chapter 33, Pergamon Press, Oxford (1982).
156. R. D. W. Kemmitt and D. R. Russell, *Comprehensive Organometallic Chemistry* (G. Wilkinson, F. G. A. Stone, and E. W. Abel, eds.), Vol. 5, Chapter 35, Pergamon Press, Oxford (1982).
157. R. P. Hughes, in: *Comprehensive Organometallic Chemistry* (G. Wilkinson, F. G. A. Stone, and E. W. Abel, eds.), Vol. 5, Chapter 35, Pergamon Press, Oxford (1982).
158. G. J. Leigh and R. L. Richards, *Comprehensive Organometallic Chemistry* (G. Wilkinson, F. G. A. Stone, and E. W. Abel, eds.), Vol. 5, Chapter 36, Pergamon Press, Oxford (1982).
159. 159. E. W. Leonova and W. Ch. Syundyukova, in: *Metody Elementoorganicheskoi Khimii. Kobalt.*

*Nikiel, Platinovye Metally* (A. N. Nesmeyanov and K. A. Kocheshkov, eds.), Chapter 1, Nauka, Moscow, (1978).

160. W. A. Khankarova, in: *Metody Elementoorganicheskoi Khimii. Kobalt. Nikiel, Platinovye Metally* (A. N. Nesmeyanov and K. A. Kocheshkov, eds.), Chapters 5 and 6, Nauka, Moscow (1978).

161. N. Bresciani-Pahor, M. Forcolin, L. G. Marzilli, L. Randaccio, M. F. Summers, and P. J. Toscano, *Coord. Chem. Rev.* **63**, 1 (1985).; N. Bresciani-Pahor, L. Randaccio, and E. Zangrando, 12th ICOMC.

162. W. D. Jones and F. J. Feher, *J. Am. Chem. Soc.*, **106**, 1650 (1984).

163. M. J. Wax, J. M. Stryker, J. M. Buchanan, C. A. Kovac, and R. G. Bergman, *J. Am. Chem. Soc.*, **106**, 1121 (1984).

164. W. D. Jones and F. J. Feher, *J. Am. Chem. Soc.*, **107**, 620 (1985).

165. K. H. Theopold and R. G. Bergman, *Organometallics*, **1**, 1571 (1982).

166. N. E. Schore, C. S. Ilenda, M. A. White, H. E. Bryndza, M. G. Matturro, and R. G. Bergman, *J. Am. Chem. Soc.*, **106**, 7451 (1984).

167. A. R. Chakravarty, F. A. Cotton, D. A. Tocher, and J. H. Tocher, *Organometallics*, **4**, 8 (1985).

168. G. J. Lamprecht, J. G. Leipoldt, and C. P. van Biljon, *Inorg. Chim. Acta*, **88**, 55 (1984).

169. A. Stockis and R. Hoffmann, *J. Am. Chem. Soc.*, **102**, 2952 (1980); R. J. McKinney, D. L. Thorn, R. Hoffmann, and A. Stockis, *J. Am. Chem. Soc.*, **103**, 2595 (1981).

170. E. J. M. de Boer, J. de With, N. Meijboom, and A. G. Orpen, *Organometallics*, **4**, 259 (1985).

171. P. W. Jolly, in: *Comprehensive Organometallic Chemistry* (G. Wilkinson, F. G. A. Stone, and E. W. Abel, eds.), Vol. 6, Chapter 37.4, Pergamon Press, Oxford (1982).

172. P. M. Maitlis, P. Espinet, and M. J. H. Russell, in: *Comprehensive Organometallic Chemistry* (G. Wilkinson, F. G. A. Stone, and E. W. Abel, eds.), Vol. 6, Chapter 38.4, Pergamon Press, Oxford (1982).

173. F. R. Hartley, in: *Comprehensive Organometallic Chemistry* (G. Wilkinson, F. G. A. Stone, and E. W. Abel, eds.), Vol. 6,. Chapter 39, Pergamon Press, Oxford (1982).

174. F. S. Denisov, in: *Metody Elementoorganicheskoi Khimii. Kobalt. Nikiel. Platinovye Metally* (A. N. Nesmeyanov and K. A. Kocheshkov, eds.), Chapter 2, p. 128, Nauka, Moscow, (1985).

175. A. Z. Rubeszhov, in: *Metody Elementoorganicheskoi Khimii. Kobalt. Nikiel. Platinovye Metally* (A. N. Nesmeyanov and K. A. Kocheshkov, eds.), Chapter 7, p. 566, Nauka, Moscow (1985); A. Z. Rubeshov, in: *Metody Elementoorganicheskoi Khimii. Kobalt. Nikiel. Platinovye Metally* (A. N. Nesmeyanov and K. A. Kocheshkov, eds.), Chapter 8, p. 627, Nauka, Moscow (1985).

176. P. K. Monaghan and R. J. Puddephatt, *Inorg. Chim. Acta*, **76**, L237 (1983).

177. L. Manojlovic-Muir, K. W. Muir, A. A. Frew, S. S. M. Ling, M. A. Thomson, and R. J. Puddephatt, *Organometallics*, **3**, 1637 (1984); R. J. Puddephatt, *Chem. Soc. Rev.*, **12**, 99 (1983).

178. W. R. Meyer and L. M. Venanzi, *Angew. Chem., Int. Ed. Engl.*, **23**, 529 (1984).

179. E. G. Samsel and J. R. Norton, *J. Am. Chem. Soc.*, **106**, 5505 (1984).

180. T. C. Flood and S. P. Bitler, *J. Am. Chem. Soc.*, **106**, 6076 (1984).

181. M. D. Fryzuk and P. A. MacNeil, *J. Am. Chem. Soc.*, **106**, 6993 (1984).

182. J. Terheijden, G. van Koten, I. C. Vinke, and A. L. Spek, *J. Am. Chem. Soc.*, **107**, 2891 (1985).

183. P. Castan, J. Jaud, N. P. Johnson, and R. Soules, *J. Am. Chem. Soc.*, **107**, 5011 (1985).

184. M. Calligaris, G. Carturan, G. Nardin, A. Scrivanti, and A. Wojcicki, *Organometallics*, **2**, 865 (1983).

185. P. G. Edwards, R. W. Geller, M. W. Marks, and R. Bau, *J. Am. Chem. Soc.*, **104**, 2072 (1982).

186. H. Hope, D. Oram, and P. P. Power, *J. Am. Chem. Soc.*, **106**, 1149 (1984).

187. S. I. Khan, P. G. Edwards, H. S. H. Yuan, and R. Bau, *J. Am. Chem. Soc.*, **107**, 1682 (1985).

188. M. Y. Chiang, E. Bohlen, and R. Bau, *J. Am. Chem. Soc.*, **107**, 1679 (1985).

189. B. H. Lipshutz, J. A. Kozlowski, and C. M. Breneman, *J. Am. Chem. Soc.* **107**, 3197 (1985).

190. G. van Koten and J. T. B. H. Jastrzebski, *J. Am. Chem. Soc.* **107**, 697 (1985).

191. G. van Koten and J. G. Noltes, *Comprehensive Organometallic Chemistry* (G. Wilkinson, F. G. A. Stone, and E. W. Abel, eds.), Vol. 2, Chapter 14, Pergamon Press, Oxford (1982).

192. R. J. Puddephatt, in: *Comprehensive Organometallic Chemistry* (G. Wilkinson, F. G. A. Stone, and E. W. Abel, eds.), Vol. 2, Chapter 15, Pergamon Press, Oxford (1982).

193. A. M. Mazany and J. P. Fackler, Jr., *J. Am. Chem. Soc.*, **106**, 801 (1984).

194. G. K. Anderson, *Adv. Organomet. Chem.*, **20**, 32 (1982).

195. H. Schmidbaur and K. C. Dash, *Adv. Inorg. Radiochem.*, **25**, 239 (1982).
196. J. P. Fackler, Jr. and H. H. Murray, *Organometallics*, **3**, 821 (1984).
197. A. J. Markwell, *J. Organomet. Chem.*, **293**, 257 (1985).
198. G. van Koten, J. T. B. H. Jastrzebski, C. H. Stam, and N. C. Niemann, *J. Am. Chem. Soc.*, **106**, 1880 (1984).
199. C. Masters, *Homogeneous Transition Metal Catalysis—A Gentle Art*, Chapman and Hall, London (1981).
200. B. Cornills, in: *New Syntheses with Carbon Monoxide* (J. Falbe, ed.), Chapter 1, Springer-Verlag, Berlin (1980).
201. B. K. Nefedov, *Syntezy Organicheskikh Soedinenii na Osnowie Ugleroda*, Nauka, Moscow (1978).
202. T. G. Attig and A. Wojcicki, *J. Organometal. Chem.*, **82**, 397 (1974); P. Reich-Rohrwig and A. Wojcicki, *Inorg. Chem.*, **13**, 2457 (1974); T. G. Attig, R. G. Teller, S.-M. Wu, R. Bau, and A. Wojcicki, *J. Am. Chem. Soc.*, **101**, 619 (1979).
203. G. K. Anderson and R. J. Cross, *Acc. Chem. Res.*, **17**, 67 (1984).
204. E. Cesarotti, R. Ugo, and L. Kaplan, *Coord. Chem. Rev.*, **43**, 275 (1982).
205. D. A. Palmer and R. van Eldik, *Chem. Rev.*, **83**, 651 (1983).
206. M. E. Volpin and V. B. Shur, in: *New Trends in the Chemistry of Nitrogen Fixation* (J. Chatt, ed.), Chapter 3, Academic Press, London (1980).
207. G. J. Leigh, in: *New Trends in the Chemistry of Nitrogen Fixation* (J. Chatt, ed.), Chapter 8, Academic Press, London (1980).
208. R. H. Crabtree, *Chem. Rev.*, **85**, 245 (1885); E. A. Grigorian, *Uspiechi Chimii*, **53**, 347 (1984).
209. P. E. Stoutland and R. G. Bergman, *J. Am. Chem. Soc.*, **107**, 4581 (1985).
210. N. Kitajima and J. Schwartz, *J. Am. Chem. Soc.*, **106**, 2220 (1984).
211. R. H. Crabtree, E. M. Holt, M. Lavin, and S. M. Morehouse, *Inorg. Chem.*, **24**, 1986 (1985).
212. S. Sprouse, K. A. King, P. J. Spellane, and R. J. Watts, *J. Am. Chem. Soc.*, **106**, 6647 (1984).
213. A. M. Madonik and D. Astruc, *J. Am. Chem. Soc.*, **106**, 2437 (1984).
214. E. H. K. Zeiher, D. G. De Wit, and K. G. Caulton, *J. Am. Chem. Soc.*, **106**, 7006 (1984).
215. R. M. Bullock, C. E. L. Headford, S. E. Kegley, and J. R. Norton, *J. Am. Chem. Soc.*, **107**, 727 (1985).
216. M. D. Curtis, L. G. Bell, and W. M. Butler, *Organometallics*, **4**, 701 (1985).
217. P. L. Watson, *J. Am. Chem. Soc.*, **105**, 6491 (1983).
218. N. Koga, S. Obara, and K. Morokuma, *J. Am. Chem. Soc.*, **106**, 4625 (1984).
219. M. J. Burk, R. H. Crabtree, C. P. Parnell, and R. J. Uriarte, *Organometallics*, **3**, 816 (1984).
220. H. Schwarz, *Angew. Chem., Int. Ed. Engl.*, **20**, 991 (1981).
221. E. L. Muetterties, *Catal. Rev. Sci. Eng.*, **23**, 69 (1981); E. L. Muetterties, *Chem. Soc. Rev.*, **11**, 283 (1982).
222. M. Brookhart and M. L. H. Green, *J. Organomet. Chem.*, **250**, 395 (1983).
223. R. Houriet, L. F. Halle, and J. L. Beauchamp, *Organometallics*, **2**, 1818 (1983).
224. M. L. Mandich, L. F. Halle, and J. L. Beauchamp, *J. Am. Chem. Soc.*, **106**, 4403 (1984).
225. M. A. Tolbert and J. L. Beauchamp, *J. Am. Chem. Soc.*, **106**, 8117 (1984).
226. R. B. Freas and D. P. Ridge, *J. Am. Chem. Soc.*, **106**, 825 (1984).
227. D. B. Jacobson and B. S. Freiser, *J. Am. Chem. Soc.*, **106**, 5351 (1984).
228. D. B. Jacobson and B. S. Freiser, *J. Am. Chem. Soc.*, **107**, 4373 (1985); D. B. Jacobson and B. S. Freiser, *J. Am. Chem. Soc.*, **107**, 67 (1975); D. B. Jacobson and B. S. Freiser, *J. Am. Chem. Soc.*, **107**, 2605 (1985).
229. D. B. Jacobson and B. S. Freiser, *Organometallics*, **3**, 513 (1984).
230. J. K. Kochi, *Organometallic Mechanisms and Catalysis*, Academic Press, New York (1978).
231. A. E. Shilov, *Activation of Saturated Hydrocarbons by Transition Metals Complexes*, Reidel, Dordrecht (1984).
232. K. I. Zamarajew and W. N. Parmon, in: *Fotokataliticzeskoje Preobrazowanie Solnecznoj Energii* (K. I. Zamarajew, ed.), Izd. Nauka, Sibirskoje Otdelenie, Nowosibirsk (1985).
233. M. Hope, M. M. Olmstead, P. P. Power, J. Sandell, and X. Xu, *J. Am. Chem. Soc.*, **107**, 4337 (1985).
234. C. Eaborn, P. B. Hitchcock, J. D. Smith, and A. C. Sullivan, *J. Organomet. Chem.*, **263**, C 23 (1984).
235. P. Leoni, M. Pasquali, and C. A. Ghilardi, *J. Chem. Soc., Chem. Commun.*, 240 (1983).

# Chapter 5

# Carbene and Carbyne Complexes

## 1. BONDING

Carbenes, $R^1R^2C:$, and carbynes, $RC\vdots$, are molecules which have short lives in the free state. However, they may form stable complexes with the transition metals. Carbenes are $2e$ ligands while carbynes are $3e$ ligands. In carbenes, the carbon atom assumes $sp^2$ hybridization; in carbynes it has sp hybridization.

The following types of carbene (alkylidene) and carbyne (alkylidyne) ligands are known[1-7]:

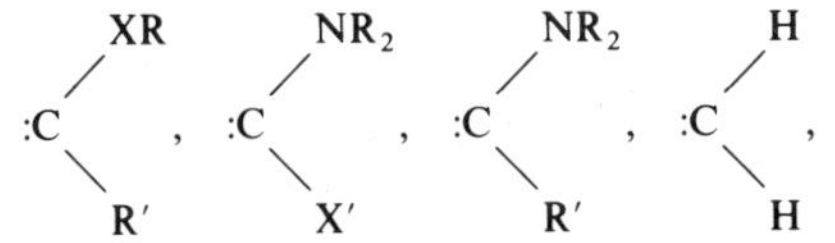

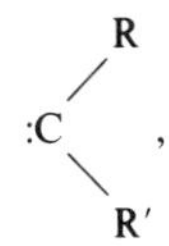

$$
:C\!\!\begin{array}{c}{}^{XR}\\{}^{R'}\end{array}\;,\;\;:C\!\!\begin{array}{c}{}^{NR_2}\\{}^{X'}\end{array}\;,\;\;:C\!\!\begin{array}{c}{}^{NR_2}\\{}^{R'}\end{array}\;,\;\;:C\!\!\begin{array}{c}{}^{H}\\{}^{H}\end{array}\;,
$$

$$
:C\!\!\begin{array}{c}{}^{R}\\{}^{R'}\end{array}\;,\;\;:C-R,\;CNR_2,\;\text{etc.,}\;(R,\,R'\text{-alkyl, aryl, hydrogen}; X = O,\,S\cdots; X' = \text{halogen}) \tag{5.1}
$$

Also, carbene complexes encompass dinuclear compounds containing such bridging units as

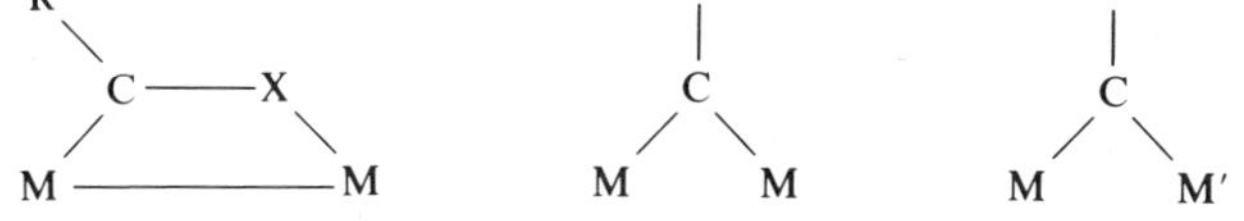

$$
\begin{array}{c}
R\\
\diagdown\\
C\!-\!\!-X\\
\diagup\quad\diagdown\\
M\!-\!\!-\!\!-\!\!-\!\!-M
\end{array}
\qquad
\begin{array}{c}
SiMe_3\\
|\\
C\\
\diagup\;\diagdown\\
M\qquad M
\end{array}
\qquad
\begin{array}{c}
R\\
|\\
C\\
\diagup\;\diagdown\\
M\qquad M'
\end{array}
\tag{5.2}
$$

$$(R = OMe,\ p\text{-}C_6H_4Me,\ \text{etc.})$$

Germylene complexes, analogs of carbene compounds, have also been prepared[8]:

$$(R = 2,4,6\text{-}C_6H_2Me_3, Me)$$

This chapter also deals with compounds possessing the following bridges:

$$\begin{array}{ccc} X & X & \\ & \diagdown\diagup & \\ & C & \quad\text{and}\quad \\ & \diagup\diagdown & \\ M & M & \end{array} \qquad \begin{array}{c} Z \\ | \\ C \\ \diagup | \diagdown \\ M\ M\ M \end{array}$$

Although the bridging carbon atom in these types of compounds has sp$^3$ hybridization resulting from its tetrahedral coordination, i.e., in $[(CO)_5W\{\mu\text{-}C(OMe)Ph\}Pt(PMe_3)_2]^{(9)}$ and $[Co_3(CZ)(CO)_9]$ (Chapter 2, Z = halogen, H, alkyl, aryl, SiMe$_3$, OMe, etc.), in many cases such compounds react like carbene complexes. Some compounds containing $M-CXY-M$ bridges were discussed in Chapter 4.

Owing to their physicochemical properties and reactivities, carbene complexes may conveniently be divided into (1) Fischer-type coordination compounds[1-5,34-41] and (2) Schrock-type complexes.[6,10,34-41] Fischer-type complexes encompass 18$e$ compounds of metals in their low oxidation states having carbene ligands containing heteroatoms with free electron pairs or aryl groups. Thus, Fischer carbene complexes possess ligands in which the substituents may form a $\pi$ bond with the carbene carbon atom. Generally, compounds of group 6–10 metals most commonly containing carbonyl and cyclopentadienyl ligands belong to the class of Fischer carbene complexes. Schrock-type carbene complexes are most often electron-deficient compounds of metals in their high oxidation states containing ligands that create a weak or moderate field. The most typical complexes of this class are those of nobium, tantalum, and tungsten with ligands such as alkyls, halogens, cyclopentadienyls, and phosphines. In Fischer complexes, the carbene atom is electrophilic while in Schrock compounds it is nucleophilic. It seems unexpected that in Fischer complexes in which metals in their low oxidation states have a large number of $d$ electrons, the carbene is electrophilic, and in Schrock compounds which have their central metal atoms in high oxidation states and therefore a small number of $d$ electrons, the carbene ligand has increased electron density and exhibits nucleophilic properties. The shortening of the $M=C$ bond distance in Fischer complexes with respect to the $M-C$ bond distance in $\sigma$ bonded alkyl compounds reaches 20 pm, and in Schrock complexes this shortening may even attain values from 20 to 40 pm. However, Schrock compounds containing arylcarbene ligands are known, for example, $[cp_2(PhCH_2)Ta=CHPh]$ with the Ta=C distance of 207 pm which is greater than in other complexes of this type. Also, compounds possessing ligands which are characteristic for Fischer complexes but a carbene ligand characteristic for Schrock complexes are known: $[(OC)_2cpFe=CHMe]^+$ and $[(OC)_2cpMn=CMe_2]$. The latter two compounds have electrophilic carbene ligands.

The $\sigma$ and $\pi$ metal–carbene bond is formed as a result of the interaction between the $\sigma$ $d$sp orbital of the metal and the $\sigma$ (sp$^2$) orbital of the carbene as well as the interaction between $d_\pi$ metal orbital and p$_\pi$ carbene orbital.

In the case of Fischer complexes the main role is played by the $\sigma$ metal–carbene bond, while $\pi$ backbonding is less important because there is an interaction between the empty $p\pi$ orbital of the carbene carbon atom and the occupied $p\pi$ orbitals of the substituents.

Structures of carbene complexes may be represented by formulas (5.4). In Schrock

$$
\begin{array}{ccc}
L_nM=C\overset{R}{\underset{R'}{\diagup\diagdown}} &
L_nM^-\!-\!C^+\overset{R}{\underset{R'}{\diagup\diagdown}} &
L_nM\!-\!C\overset{R}{\underset{R'}{\diagup\diagdown}} \\[18pt]
L_nM\!-\!C\overset{R}{\underset{R'}{\diagup\diagdown}} &
L_nM\!\overset{\mathbf{R}}{\underset{R'}{\diagup\diagdown}}C &
L_nM\text{---}C\overset{R}{\underset{R'}{\diagup\diagdown}} \\[18pt]
L_nM\text{---}C\overset{R}{\underset{R}{\diagup\diagdown}} &
L_nM\!-\!C\overset{R}{\underset{R'}{\diagup\diagdown}} &
L_nM\text{---}C\overset{R}{\underset{R'}{\diagup\diagdown}}
\end{array} \qquad (5.4)
$$

complexes, an important role is played by the donor $\sigma$ [$\sigma$ $dsp$-$\sigma(sp^2)$] bond as well as by the $\pi$ backbonding which is formed as a result of interaction of the $d\pi$ metal orbital with the empty $p\pi$ carbon atom orbital. However, the structure of these complexes is more complicated than may be judged from this simple model. In the fragment $M-CHR$, the MCR angle is considerably greater than $120°$, reaching up to $170°$, and the MCH angle is very small ($\sim 80°$). Because substituents R in such complexes are bulky ($CMe_3$), an essential role in the deformation of the complex may be played by steric factors, although the influence of electronic factors cannot be neglected.[39]

Calculations show[41] that complexes of the type $L_2Ti=CHR$ and $L_2Ti=CR_2$ ($L = H$, Cl, cp, $R = H$, OH, SH) which have their central ion configurations $d^2$ in addition to isomers I and II may form a stable compound III possessing a pyramidal structure. In the case of the latter isomer, coordination of an olefin or acetylene may readily occur and facilitate reactions of these compounds involving the carbene ligand [scheme (5.5)]. Optimum $\beta$ values are $34°$, $57.5°$, and $75°$ for $L = $ cp, Cl, and H, respectively. The formation of the pyramidal isomer is more difficult in complexes possessing

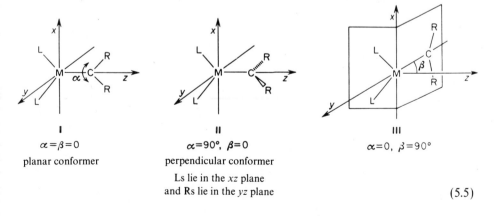

$$
\begin{array}{ccc}
\mathbf{I} & \mathbf{II} & \mathbf{III} \\
\alpha=\beta=0 & \alpha=90°,\ \beta=0 & \alpha=0,\ \beta=90° \\
\text{planar conformer} & \text{perpendicular conformer} & \\
& \text{Ls lie in the } xz \text{ plane} & \\
& \text{and Rs lie in the } yz \text{ plane} & \qquad (5.5)
\end{array}
$$

$d^3$ and $d^4$ electronic configurations. Thus, these complexes should have classical planar structure I.

The energy of activation of the rotation about the metal–carbon bond is considerably greater for Schrock complexes $M=CR_2$ (R = H, alkyl, aryl) than for Fischer compounds M-C$\overset{\displaystyle XR}{\underset{\displaystyle XR}{\Big\backslash}}$ for which this energy is basically higher than zero only if there is steric hindrance. For $cp_2(Me)Ta=CH_2$ the barrier of rotation is higher than

$88\ kJ\ mol^{-1}$,[10] for $cis$-$\left[(Ph_3P)(OC)_4Cr=C\overset{NMe}{\underset{NMe}{\diagdown}}\right]$ it equals zero, while for $cis$-

$\left[(OC)_4Cr\!\!=\!\!\left(C\overset{NMe}{\underset{NMe}{\diagdown}}\right)_2\right]$, in which steric hindrance plays an essential role, the

barrier of rotation is $ca\ 37.5\ kJ\ mol^{-1}$.[11]

In contrast to compounds with carbene ligands containing two heteroatoms, in complexes such as $[(OC)_5Cr=C(NMe_2)Me]$ and $[(OC)_5Cr=C(OMe)Me]$, the activation energy of rotation is considerable: for the former compound it is higher than $104\ kJ\ mol^{-1}$ and for the latter, $50\ kJ\ mol^{-1}$.[12,13] Usually, $\Delta G^+$ of rotation of the carbene about the $M-C$ axis is $50$–$100\ kJ\ mol^{-1}$ for Schrock complexes and up to $60\ kJ\ mol^{-1}$ for Fischer carbene compounds.[43] Also, energies of dissociation of metal–carbene bonds and frequencies $v(M=C)$ for Schrock complexes are higher compared to Fischer compounds. This was confirmed by calculations performed for $[H_2(Me)Nb=CH_2]$, $[(OC)_5Cr=CHOH]$, and $[(OC)_4Fe=CHOH]$.[40]

Carbynes, $(RC\colon)$, are $3e$ ligands and their carbon atoms have sp hybridization (two $p$ orbitals are not hybridized). Therefore, in carbyne complexes $L_nM-CR$, two molecular orbitals $d_\pi$–$p_\pi$ are formed. Thus, formally, the metal–carbon bond is triple. Calculations of electronic structures of $[(OC)_5Cr\equiv CH]^+$ and $[(OC)_4ClCr\equiv CH]$[40] show that the carbyne carbon charge is negative. Because in cationic carbyne complexes the carbon atom reacts with nucleophiles, this carbon's reactivity is not controlled by the charge, contrary to these calculations. The contradictions between calculations and experiments may be explained[40] by means of the frontier orbital theory (reactions controlled by frontier orbitals). Nucleophiles react with the atom of a molecule on which the LUMO is located, while electrophiles react with the atom of a molecule on which the HOMO is mainly localized. In the case of $[(OC)_5Cr\equiv CH]^+$, the maximal coefficient in LUMO is at the carbyne carbon atom, and therefore it reacts with the nucleophile despite the fact that it is negative.

The frontier orbital theory also agrees with the observed experimental reactivity of Fischer and Schrock carbene complexes.[40] Calculations show that the linear $Cr\equiv C-R$ group is more stable than the bent one.[40] Thus, the deviations of $M\equiv C-R$ fragments from linearity probably result from intermolecular interactions within the crystal lattice.

In complexes containing bridging ligand $CH_2$, the bonds within $M\overset{\displaystyle CH_2}{\underset{}{\diagup\diagdown}}M$ have the same character as in cyclopropane. The formation of $L_n\overline{M(\mu\text{-}CR_2)M}L_n$ may be represented as an addition of a carbene $CR_2$ to an "inorganic olefin" $L_nM=ML_n$, such as $[(OC)\,cpRh=Rhcp(CO)]^{(44,45)}$ (cf. Section 3.1.b). Therefore, a $\mu$-alkylidene compound in which there is a metal–metal bond may be considered as dimetallacyclopropane, analogously to a corresponding olefin complex, i.e., metallacyclopropane.

The empty carbene $CH_2$ orbital $p_\pi$ has energy somewhat lower than that of HOMO in the dinuclear compound $(OC)_2cpMn=Mncp(CO)_2$. Thus, the formation of $[(OC)_2cp\overline{Mn(\mu\text{-}CH_2M}ncp(CO)_2]$ leads to a charge transfer from HOMO to the methylene group$^{(45)}$; the calculated electron density is $0.526e$.

## 2. STRUCTURES OF CARBENE AND CARBYNE COMPLEXES

The $\pi$-backbonding in alkylidene and alkylidyne complexes is evidenced by considerably shorter $M-C$ distances compared to the metal–carbon distances in compounds containing $\sigma\,M-C$ bonds. The metal–carbon distances in compounds containing $M-C$ single bonds are usually *ca* 20 pm longer than the metal–carbene carbon distances and even up to 40 pm longer compared to the metal–carbyne carbon distances.

*Table 5.1. Structures of Carbene and Carbyne Complexes*

| Compound | Distance M–CXY (pm) | Distance M–L (pm) |
|---|---|---|
| $[Tacp_2(CH_2)(Me)]^{(6)}$ | 202.6 | 224.6 (Ta–Me) |
| $[Tacp_2(CHPh)(CH_2Ph)]^{(6)}$ | 207 | 230 (Ta–CH$_2$Ph) |
| $[Tacp_2\{CH(CMe_3)\}Cl]^{(6)}$ | 203 | |
| $[Ta_2(\mu\text{-}Cl)_2Cl_4\{CH(CMe_3)\}(PMe_3)_2]^{(6)}$ | 189.8 | 281.5, 244.8 (Ta–$\mu$-Cl) |
| $[Ta(\eta\text{-}C_5Me_5)Cl(CPh)(PMe_3)_2]^a$ | 184.9 (Ta$\equiv$C) | |
| $[CrC(OEt)\{\overline{C(OH)C}S(CH_2)_3S\}(CO)_4]^b$ | 199 | 238 (Cr–S) 181 (Cr–CO) |
| $[Ru\{\overline{CN(C_6H_4Me\text{-}4)(CH_2)_2N}(C_6H_3Me\text{-}4)\}Cl(PEt_3)_2]^c$ | 190.8 | 235.2 (Ru–P) 199.4 (Ru–Csp$^2$) 245.2 (Ru–Cl) |
| $[\overline{Ir\{Bu_2^tP(CH_2)_2C(CH_2)_2PBu_2^t}\}Cl]^d$ | 200.6 | 241.4 (Ir–Cl) 230.3 (Ir–P) |
| *trans*-$[Pt\{\overline{CN(Ph)CH_2CH_2NPh}\}Cl_2(PEt_3)]^{(4)}$ | 202.0 | 231.1 (Pt–Cl) 229.1 (Pt–P) |
| *cis*-$[Pt\{\overline{CN(Ph)CH_2CH_2NPh}\}Cl_2(PEt_3)]^{(4)}$ | 200.9 | 238.1 (Pt–Cl *trans* to carbene) 236.2 (Pt–Cl *trans* to P) 223.4 (Pt–P) |

$^a$ S. J. McLain, C. D. Wood, L. W. Messerle, R. R. Schrock, F. J. Hollander, W. J. Youngs, and M. R. Churchill, *J. Am. Chem. Soc.*, **100**, 5962 (1978).

$^b$ H. G. Raubenheimer, S. Lotz, and J. Coetzer, *Chem. Commun.*, 732 (1976).

$^c$ P. B. Hitchcock, M. F. Lappert, and P. L. Pye, *Chem. Commun.*, 196 (1977).

$^d$ H. D. Empsall, E. M. Hyde, R. Markham, W. S. McDonald, M. C. Norton, B. L. Shaw, and B. Weeks, *Chem. Commun.*, 589 (1976).

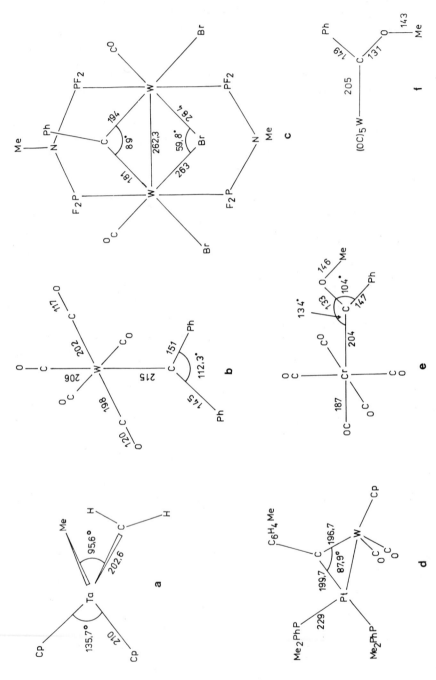

Figure 5.1. Structures of carbene complexes: (a) [Tacp₂Me(CH₂)]; (b) [W(CO)₅(CPh₂)], *J. Am. Chem. Soc.*, **99**, 2127 (1977); (c) [W₂(μ-Br) Br₂(μ-F₂PNMePF₂)₂(CO)₂, *J. Organomet. Chem.*, **199**, C24 (1980); (d) [(Me₂PhP)₂Pt(μ-CC₆H₄Me) Wcp(CO)₂], *J. Chem. Soc., Dalton Trans.*, 1609 (1980); (e) [CrC(OMe) Ph(CO)₅]; (f) [WC(OMe) Ph(CO)₅].

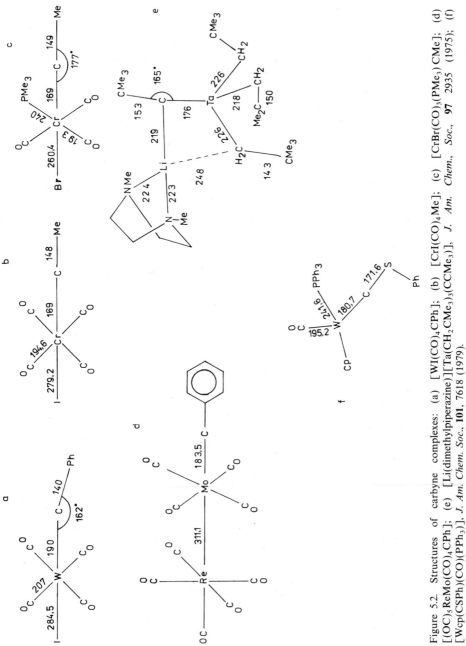

Figure 5.2. Structures of carbyne complexes: (a) [WI(CO)₄CPh]; (b) [CrI(CO)₄Me]; (c) [CrBr(CO)₃(PMe₃) CMe]; (d) [(OC)₅ReMo(CO)₄CPh]; (e) [Li(dimethylpiperazine)][Ta(CH₂CMe₃)₃(CCMe₃)], *J. Am. Chem. Soc.,* **97** 2935 (1975); (f) [Wcp(CSPh)(CO)(PPh₃)], *J. Am. Chem. Soc.,* **101,** 7618 (1979).

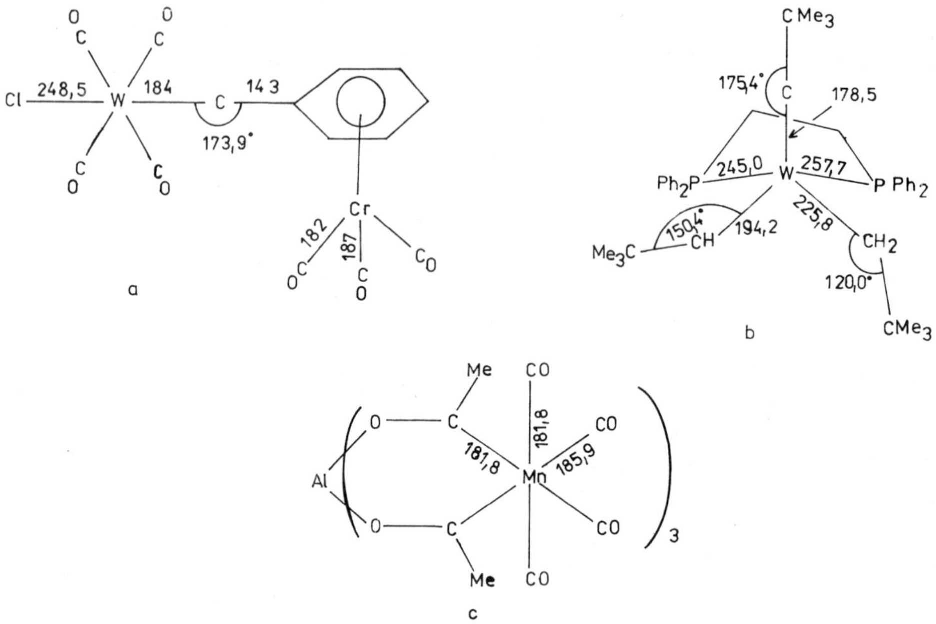

Figure 5.3. Structures of (a) $[(OC)_5ClWC(Ph-\eta)Cr(CO)_3]$, *Chem. Ber.*, **113** 1010 (1980); (b) $[W(CH_2CMe_3)(CHCMe_3)(CCMe_3)(Ph_2PC_2H_4PPh_2)]$, *Inorg. Chem.*, **18**, 2454 (1979); (c) $Al[(OCMe)_2Mn(CO)_4]_3$, *J. Am. Chem. Soc.*, **97**, 6903 (1975).

Therefore, in alkylidyne complexes, there is a triple bond $M\equiv CR$, and in alkylidene complexes, the multiplicity of the $M-C$ bond considerably exceeds one. The $M$–CPh distance in *trans*-(phenylcarbyne)tetracarbonyl(pentacarbonylrhenio)molybdenum is 183.5 pm. Thus, within the limits of experimental error, it equals the sum of the radii of the triple bonds $Mo\equiv Mo$ and $C\equiv C$ $(122.4 + 60.5 = 182.9$ pm). In carbene complexes containing oxygen as the heteroatom, the multiplicity of the carbon–oxygen bond is so great that isomers exist:

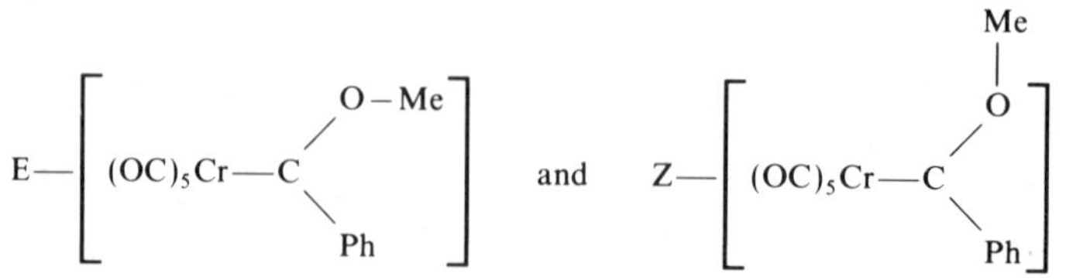

$$E-\left[(OC)_5Cr-C\underset{Ph}{\overset{O-Me}{<}}\right] \quad \text{and} \quad Z-\left[(OC)_5Cr-C\underset{Ph}{\overset{O}{<}}\overset{|}{\underset{}{Me}}\right]$$

There is no $\pi$ bond between the phenyl and carbene carbon atom. In these, as in other carbene complexes, phenyl rings do not lie in the $M-C\overset{X}{<}$ plane. The $M=C\overset{X}{<}_Y$ group is planar, while $M\equiv C-R$ in carbyne complexes is linear. Small deviations from linearity are explained by the existence of steric hindrance (Figures 5.1–5.3 and Table 5.1).

The structures *trans*- or *cis*-Pt⎯C$\left(\begin{smallmatrix} NPh \\ \\ NPh \end{smallmatrix}\right)$Cl$_2$(PEt$_3$) show that the *trans* effect of the carbene is somewhat greater than that of the phosphine, because of distance between chlorine and platinum in the *trans* arrangement with respect to carbene is greater than the analogous distance for the *cis*-chlorine arrangement. In $\mu$-methylene complexes, there is generally a metal–metal bond. Usually, methylene and alkylidene compounds are symmetrical, and the MCM angles are very acute, most commonly in the range of 76–81°.[42]

## 3. IR SPECTRA

In carbene carbonyl complexes [M(CO)$_n$CXY] bands caused by $v$(CO) vibrations occur at lower energies than in corresponding carbonyl complexes M(CO)$_{n+1}$. This proves that carbene ligands possess weaker $\pi$-acceptor properties compared to the carbonyl group. Conversely, in the case of carbyne complexes, $v$(CO) bands occur at higher energies than in corresponding carbonyl and carbene compounds. Therefore, the formation of a strong M≡CR bond because of the presence of two $\pi$ bonds causes considerable decrease in the electron density on the metal. Based on the force constants $k$(CO), it was found that CPh$_2$ is a stronger $\pi$ acceptor than a similar carbene ligand containing oxygen such as CPh(OMe) (Table 5.2). The $v$(CH) band in alkylidene complexes containing less than 18 valence electrons occurs at unusually low frequencies. In the spectrum of [Tacp(CHCMe$_3$)Cl$_2$], the energy of the $v$(CH) is 2510 cm$^{-1}$. The 18$e$ compounds, such as Mcp$_2$(CHR)X, show the $v$(CH) band in normal ranges at higher energies. The lower value of the $v$(CH) band for electron-deficient complexes ($<18e$) also corresponds to a lower value of the coupling constant $^1$J(CH) which equals 75–100 Hz. For 18$e$ compounds of tantalum, this coupling constant is 105–130 Hz.

The $v$(M≡C) frequencies for carbyne complexes of chromium and tungsten [(OC)$_4$XM≡CR] (X = Cl, Br, I; R = Me, Ph) occur in the 1250–1380 cm$^{-1}$ range, and they strongly interact with $v$(C—Ph) vibrations and ring deformation oscillations or $\delta s$(CH$_3$) and $v$(C—CH$_3$). The force constants $k$(M≡C) change for these compounds in the 720–740 N m$^{-1}$ range, and show that the W≡C bond is stronger than the Cr≡C bond. The spectra of carbene complexes containing C(OR)R′ ligand show the $v$(C$_{carb}$—O) band at *ca* 1300 cm$^{-1}$. Based on the analysis of infrared spectra of clusters containing the following ligands: $\mu$-CMe, $\mu_3$-CMe, and $\mu$-CHMe, as well as $\mu$-C$_2$H$_n$ and $\mu_3$-C$_2$H$_n$ ($n = 1$–4), Evans and McNulty came to the conclusion that it is possible to use IR spectroscopy for determination of the bonding modes of these ligands.[47]

## 4. NMR SPECTRA[14,15]

The $^{13}$C NMR spectra were investigated for many carbene complexes and a considerable number of carbyne complexes. The signals for carbene carbon atoms occur at low fields, in many cases lower than for carbon atoms of sp$^2$ hybridization in tri-

Table 5.2. IR Spectra of Carbene and Carbyne Complexes

| Compound | $v(CO)$ (cm$^{-1}$) | $v(M-L)$ and other bands (cm$^{-1}$) |
|---|---|---|
| $[Cr(CO)_5\{C(OMe)Me\}]^a$ | 2070 s, 1992 s, 1953 vs | |
| $[Cr(CO)_5\{CCl(NMe_2)\}]^b$ | | 779 $v(CCl)$, 1520 $v(CN)$ |
| $[Rh(CO)Cl_3\{CCl(NMe_2)\}]_2{}^b$ | | 860 $v(CCl)$, 1578 $v(CN)$ |
| $[RhCl_3(PEt_3)_2\{CCl(NMe_2\}]^b$ | | 807 $v(CCl)$, 1533 $v(CN)$ |
| $[Mn(CO)_5\{CCl(NMe_2)\}]ClO_4{}^b$ | | 826 $v(CCl)$, 1566 $v(CN)$ |
| $[(OC)_5ReMo(CO)_4(CPh)]^c$ | 2104, 2049, 2019, 1975, 1968 | |
| $[W(CO)_5(CPh_2)]^d$ | 2070, 1971, 1963 | $k(CO)$ 1587, 1608 Nm$^{-1}$ |
| $[W(CO)_5CPh(OMe)]^d$ | 2070 m, 1984 w, 1958 s, 1945 s | $k(CO)$ 1571, 1590 Nm$^{-1}$ |
| $[Cr(CO)_5CF(NEt_2)]^e$ | 2064 w, 1930 vs | |
| $[Cr(CO)_5CNEt_2][BF_4]^e$ | 2132 w, 2024 vs | |
| $[Cr(CO)_5(CNMe_2)][BF_4]^f$ | 2158 m, 2110 w, 2033 s | |
| $[Cr(CO)_5C(NMe_2)_2]^f$ | 2052 w, 1920 vs | |
| $[(OC)_4BrW\equiv C(\eta\text{-}C_5H_4Fecp)]^g$ | 2112 w, 2020 s | |
| $[Fe\{CN(Me)CH_2CH_2NMe\}_2(CO)_2PPh_3][BF_4]^h$ | 1973 vs, 1904 s | |
| $[PtCl_4\{C(NHMe)_2\}(PEt_3)]^i$ | | 3310 $v(NH)$, 1605, 1565 $v(NC)$; 342, 297, 263 $v(PtCl)$ |
| $[Wcp(CSPh)(CO)(PPh_3)]^j$ | 1886 s | 1225 $v(CS)$ |
| $[Tacp(CHCMe_3)Cl_2]^k$ | | 2510, 2490 sh $v(C_{carb}-H)$ |
| $[Tacp(CHCMe_3)Me_2]^k$ | | 2475, 2510 sh $v(C_{carb}-H)$ |
| $[Nbcp(CHCMe_3)Cl_2]^k$ | | 2520 $v(C_{carb}-H)$ |
| $[Mo(CO)_5\{C(OEt)CH(SiMe_3)_2\}]^{(19)}$ | 2070 m $(A_1)$, 1977 w $(B_1)$, 1949 s $(A_1)$, 1946 s $(E)$ | $B_1$ IR forbidden |
| $[Mo(CO)_5\{C(OEt)CH_2SiMe_3\}]^{(19)}$ | 2068 s $(A_1)$, 1939 s $(A_1)$, 1954 s, 1951 s $(E)$ | $B_1$ IR forbidden |
| $trans\text{-}[(OC)_4ClWCMe]^{(12b,c)}$ | 2132 $(R)$, 2065 $(R)$, 2040 vs (IR) | $v(W\equiv C)$ 1355 m |
| $trans\text{-}[(OC)_4BrWCMe]^{(12b,c)}$ | 2130 $(R)$, 2054 $(R)$, 2020 s (IR) | $v(W\equiv C)$ 1352 s |
| $trans\text{-}[(OC)_4BrWCCD_3]^{(12b,c)}$ | 2130 $(R)$, 2100 $(R)$, 2065 $(R)$, 2000 (IR) | $v(W\equiv C)$ 1310 vs |

[a] E. O. Fischer and A. Maasboel, *Chem. Ber.*, **100**, 2445 (1967); C. G. Kreiter and E. O. Fischer, *Chem. Ber.*, **103**, 1561 (1970).
[b] A. J. Hartshorn, M. F. Lappert, and K. Turner, *Chem. Commun.*, 929 (1975).
[c] E. O. Fischer, G. Huttner, T. L. Lindner, A. Frank, and F. R. Kreissl, *Angew. Chem., Int. Ed. Engl.*, **15**, 157 (1976).
[d] C. P. Casey, T. Burkhardt, C. A. Bunnell, and J. C. Calabrese, *J. Am. Chem. Soc.*, **99**, 2127 (1977).
[e] E. O. Fischer, W. Kleine, and F. R. Kreissl, *Angew. Chem., Int. Ed. Engl.*, **15**, 616 (1976).
[f] A. J. Harstshorn and M. F. Lappert, *Chem. Commun.*, 761 (1976).
[g] E. O. Fischer, M. Schluge, and J. O. Besenhard, *Angew. Chem., Int. Ed. Engl.*, **15**, 683 (1976).
[h] M. F. Lappert, J. J. MacQuitty, and P. L. Pye, *Chem. Commun.*, 411 (1977).
[i] J. Chatt, R. L. Richards, and G. H. D. Royston, *J. Chem. Soc., Dalton Trans.*, 599 (1976).
[j] W. W. Greaves, R. J. Angelici, B. J. Helland, R. Klima, and R. A. Jacobson, *J. Am. Chem. Soc.*, **101**, 7618 (1979).
[k] C. D. Wood, S. J. McLain, and R. R. Schrock, *J. Am. Chem. Soc.*, **101**, 3210 (1979).

alkylcarbenium cations $CR_3^+$. The chemical shifts $\delta(^{13}C)$ of carbene carbon atoms fluctuate in a very broad range: 140–400 ppm. They depend mainly on the X and R groups of the carbene CXR as well as, to a smaller degree, on the metal, although even in this case, on going from six-coordinate complexes to five- and particularly to four-coordinate compounds, considerable differences in their values are observed. In the case of alkylidene and alkylidyne bridging groups, the chemical shifts $\delta(^{13}C)$ assume the following values: 20–210 ppm and 110–400 ppm, respectively[42,47] (Chapter 3, Table 3.14). It is assumed that the highest contribution to the chemical shift of the carbene carbon is that of the paramagnetic screening factor [equation (2.4)]. It seems improbable for the carbene carbon atoms to be more electron deficient than trialkylcarbenium ions despite the fact that these carbene atoms are sometimes more deshielded than $CR_3^+$. The shifts toward lower fields also result from lower $\Delta E$ values and higher $Q_{AB}$ values. The increase of $Q_{AB}$ is caused by the multiple character of the $M-CXR$ bond owing to the interaction of $d$ orbitals of the metal and the $p_z$ orbital of the carbene. By changing an alkoxy group OR for an amino group, there is an increase in screening of the carbene carbon atom, most commonly by 70–80 ppm. This is probably due to stronger $\pi$-donor properties of the $NR_2$ group compared to OR substituents. The chemical shift of carbenes as a function of R decreases for complexes of chromium and tungsten as follows: Me > Ph > 1-ferrocenyl > 2-thienyl > 2-furyl. This decrease is therefore in accordance with the lowering of donor properties of these substituents. For complexes of the third series of the transition metals, the chemical shifts are lower compared to compounds of lighter metals of a given group; in general, this lowering does not exceed 15 ppm. The chemical shifts $\delta(^1H)$ for terminal carbene complexes of the type $M=CHR$ are in the 6–13 ppm range, and for bridging compounds such as $\overline{M(\mu\text{-}CHR)M}$ these shifts are 5–11 ppm. Thus, neither $^{13}C$ NMR spectra nor $^1H$ NMR spectra can serve for unequivocal determination of these two groups of compounds[42,43] (cf. Chapter 3).

In general, screening of carbyne carbon atoms is smaller compared to carbene carbon atoms. The $\delta(^{13}C)$ NMR shifts for alkylidyne complexes are in the 250–350 ppm range, and therefore are higher than shifts for sp hybridized carbon atoms in acetylenes by about 200 ppm or more. The presence of only one signal for carbonyl groups in complexes of the type $MX(CR)(CO)_4$ (M = Cr, Mo, W) proves that they have *trans* geometry. For $WX(CR)(CO)_4$, $\delta C(CO)$ is lower than for carbene compounds such as $W(CO)_5(CXR)$, which indicates (like IR spectra) that the W-CO $\pi$ back bond is weaker in alkylidyne compounds compared to alkylidene complexes (Table 5.3).

## 5. PHOTOELECTRON SPECTRA[16,17,18a,19,45,46,71,72]

Carbene complexes of the formula $M(CO)_5(CXR)$ have $C_s$ symmetry (Table 5.4). Consequently, the degeneracies of the metal $d$ orbitals and the molecular orbitals are lifted. The $d$ orbitals are split into sublevels $2a' + a''$ while the five $\delta M-CO$ orbitals gain $3a' + 2a''$ representations. The $\sigma M$–carbene orbital transforms as $a'$. The qualitative molecular orbital diagram for $M(CO)_5(CXR)$ is given in Figure 5.4.

The first bands in photoelectron spectra of carbene complexes occur at low energies (7–8 eV), and they are due to ejections of electrons from $d$ metal orbitals possessing $a'$ and $a''$ symmetry. The subsequent bands in the 9.5–10.5 eV range result from ionization of electrons in bonding $\sigma M-CXR$ orbital.

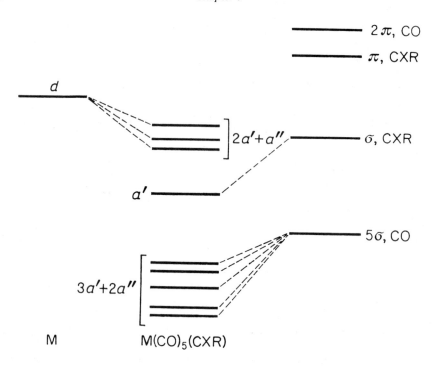

Figure 5.4. The molecular orbital scheme for $[M(CO)_5(CXR)]$.

*Table 5.3. $^{13}C$ NMR Chemical Shifts for Carbene and Carbyne Complexes*

| Compound | $\delta(C_{\text{carbene}})$ (ppm) | Coupling constant (Hz), chemical shifts of other atoms C (ppm) | |
|---|---|---|---|
| $[Cr(CO)_5CMe(OMe)]^{(14,15)}$ | 362.3 | 217.6 | 223.6 |
| $[Cr(CO)_5CMe(OEt)]^{(14,15)}$ | 357.0 | 217.0 | 226.0 |
| $[Cr(CO)_5CPh(OMe)]^{(14,15)}$ | 354.5 | 218.4 | 226.0 |
| $[Cr(CO)_5CMe(NHMe)]^{(14,15)}$ | 284.8 | 219.2 | 224.4 |
| $[Cr(CO)_5CPh(NHMe)]^{(14,15)}$ | 289.0 | 218.0 (CO-*cis*) | 224.0 (CO-*trans*) |
| $[W(CO)_5CMe(OMe)]^{(14,15)}$ | 332.9 | 197.6 | 203.6 |
| $[W(CO)_5CMe(SMe)]^{(14,15)}$ | 332.5 | 198.1 | 207.4 |
| $[W(CO)_5CPh(OMe)]^{(14,15)}$ | 321.9 | 198.6 | 204.6 |
| $[W(CO)_5CMe(SPh)]^{(14,15)}$ | 332.0 | 197.1 | 207.5 |
| $[Cr(CO)_5CF(NEt_2)]^a$ | 245.7 | 216.5 | 221.5 |
| $[CrC(OEt)\{C(OH)\overline{CS(CH_2)_3S}\}(CO)_4]^b$ | 313.3 | | |
| $[Nb_2(CSiMe_3)_2(CH_2SiMe_3)_4]^c$ | 406 | | |
| $[Ta_2(CSiMe_3)_2(CH_2SiMe_3)_4]^c$ | 406 | | |
| $[W_2(CSiMe_3)_2(CH_2SiMe_3)_4]^c$ | 354 | $J(^{183}W-^{13}C)$ 74 Hz | |
| $[W(CPh_2)(CO)_5]^d$ | 358.3 | | |

*(Table continued)*

*Table 5.3. (continued)*

| Compound | $\delta(C_{\text{carbene}})$ (ppm) | Coupling constant (Hz), chemical shifts of other atoms $C$ (ppm) |
|---|---|---|
| $[Cr(CPh_2)(CO)_5]$[1] | 399.4 | |
| $[Tacp_2(CH_2Ph)(CHPh)]$[m] | 246 | $J(CH)$ 127 Hz, $\tau^1H(C_{\text{carb}}\text{-}H) = -0.86$ |
| $[Tacp_2(CHCMe_3)Cl]$[m] | 274 | $J(CH)$ 121 Hz, $\tau^1H(C_{\text{carb}}\text{-}H) = -0.1$ |
| $[Nbcp_2(CHCMe_3)Cl]$[m] | 299 | $J(CH)$ 131 Hz, $\tau^1H(C_{\text{carb}}\text{-}H) = -2.12$ |
| $[PtW(\mu\text{-}CC_6H_4Me\text{-}p)(CO)_2(PMe_2Ph)cp]$[9] | 336 | $J(PC)$ 59 Hz, $J(PtC)$: 747 |
| $[Mo(CO)_5C(OEt)CH(SiMe_3)_2]$[19] | 350.6 | 207.2 (CO-*cis*), 213.8 (CO-*trans*) |
| $[Mo(CO)_5C(OEt)CH_2SiMe_3]$[19] | 346.5 | 206.5 (CO-*cis*), 213.5 (CO-*trans*) |
| *trans*-$[PdCl_2(PBu_3^n)L']$*[14,15] | 200.5 | $^2J(CP)$ 180.5 Hz |
| *cis*-$[PdCl_2(PBu_3^n)L]$**[14,15] | 195.1 | $^2J(CP)$ <2.4 Hz |
| *cis*-$[PtCl_2(PBu_3^n)_2L]$[14,15] | 178.0 | $^2J(CP)$ ~60 Hz |
| *trans*-$[PtCl_2(PBu_3^n)L]$[14,15] | 196.5 | $^2J(CP)$ 146.4 Hz |
| $[Fe(CO)_4L]$[14,15] | 213.2 | |
| $[Cr(CO)_4\{C(SMe)_2\}_2]$[14,15] | 141.6 | |
| $[Ta(CHSiMe_3)(CH_2SiMe_3)N(SiMe_3)_2]$[j] | 231 | $J(CH)$ 95 Hz, $\tau^1H(C_{\text{carb}}\text{-}H)$ 3.70 |
| $[Tacp(CHCMe_3)Cl_2]$[k] | 246 | $J(CH)$ 84 Hz, $\tau^1H(C_{\text{carb}}\text{-}H)$ 3.62 |
| $[Nb(C_5H_4Me)(CHCMe_3)Cl_2]$[k] | 254 | $J(CH)$ 95 Hz, $\tau^1H(C_{\text{carb}}\text{-}H)$ 2.70 |
| $[TaCl(\eta\text{-}C_5Me_5)(CCMe_3)(PMe_3)_2]$[l] | 354 | $J(CP)$ 19 Hz |
| $[TacpCl(CCMe_3)(PMe_3)_2]$[l] | 348 | |
| *trans*-$[WCl(CMe)(CO)_4]$[14,15] | 288.8 | $\delta C$, 194.0 (CO) |
| *trans*-$[WBr(CMe)(CO)_4]$[14,15] | 288.1 | $\delta C$, 192.7 (CO) |
| *trans*-$[WI(CMe)(CO)_4]$[14,15] | 286.3 | $\delta C$, 191.7 (CO) |
| *trans*-$[WBr((CC_6H_4\text{-}p)(CO)_4]$[1] | 266.15 | |
| *trans*-$[WBr(CPh)(CO)_4]$[1] | 271.30 | |
| *trans*-$[WBr(CC_6H_4Me\text{-}p)(CO)_4]$[1] | 271.43 | |
| *trans*-$[WBr(CC_6H_4OMe\text{-}p)(CO)_4]$[1] | 273.16 | |
| *trans*-$[WBr(CC_6H_2Me_3\text{-}2,4,6)(CO)_4]$[1] | 275.13 | |
| $[(OC)_5ReMo(CO)_4(CPh)]$[e] | 293.98 | $\delta C$, 215.44 (OC–Mo); 199.40 (OC-Re) |
| $[Mocp(CCH_2Bu^t)\{P(OMe)_3\}_2]$[f] | 299.8 | $^2J(CP)$ 27 Hz |
| $[Mocp(H)(CCH_2Bu^t)\{P(OMe)_3\}_2]^+$[f] | 346.7 | $^2J(CP)$ 24 Hz |
| $[Cr(CNEt_2)(CO)_5][BF_4]$[a] | 282.22 | $\delta C$, 207.68 (CO-*cis*); 201.85 (CO-*trans*) |
| $[WBr(CO)_4(CC_5H_4Fecp)]$[g] | 275.1 | $\delta C$, 193.4 (CO) |
| $[Wcp(CSPh)(CO)PPh_3]$[h] | 254.9 | |
| $[Wcp(CC_6H_4Me\text{-}p)(CO)_2]$[i] | 300 | |

* L' = $-\overline{CN(Ph)CH_2CH_2}NPh$;  ** L = $-\overline{CN(Me)CH_2CH_2}NMe$.

[a] See Ref. *e*, Table 5.2.
[b] See Ref. *b*, Table 5.1.
[c] R. A. Andersen, A. L. Galyer, and G. Wilkinson, *Angew. Chem., Int. Ed. Engl.*, **15**, 609 (1976).
[d] See Ref. *d*, Table 5.2.
[e] See Ref. *c*, Table 5.2.
[f] M. Bottrill and M. Green, *J. Am. Chem. Soc.*, **99**, 5795 (1977).
[g] See Ref. *g*, Table 5.2.
[h] See Ref. *j*, Table 5.2.
[i] E. P. Fischer, T. L. Lindner, G. Huttner, P. Friedrich, F. R. Kreissl, and J. O. Bosenhard, *Chem. Ber.*, **110**, 3397 (1977).
[j] R. A. Andersen, *Inorg. Chem.*, **18**, 3622 (1979).
[k] See Ref. *k*, Table 5.2.
[l] S. J. McLain, C. D. Wood, L. Messerle, R. R. Schrock, F. J. Hollander, W. J. Youngs, and M. R. Churchill, *J. Am. Chem. Soc.*, **100**, 5962 (1978).
[m] R. R. Schrock, L. W. Messerle, C. D. Wood, and L. J. Guggenberger, *J. Am. Chem. Soc.*, **100**, 3793 (1978).

*Table 5.4. Photoelectron Spectra of Carbene Complexes*[a]

| Compound | | Orbital ionization energy (eV) | | | |
|---|---|---|---|---|---|
| | | | $(n-1)d$ | | $(M-CXR)$ |
| [Cr(CO)₅(CXR)] | | | | | |
| X | R | | | | |
| OMe | Me | 7.47 | 7.89 | | 9.89 |
| SMe | Me | 7.35 | 7.59 | 7.79 | 9.91 |
| NMe₂ | Me | 7.12 | 7.35 | 7.61 | 9.72 |
| NH₂ | Me | 7.45 | 7.80 | | 10.31 |
| OMe | 2-C₄H₃O | 7.37 | 7.68 | | 9.92 |
| NH₂ | 2-C₄H₃O | 7.22 | 7.52 | | 10.30 |
| OMe | Ph | 7.39 | 7.78 | | 9.26 |
| NH₂ | Ph | 7.25 | 7.52 | 7.73 | 9.80 |
| NMe₂ | Ph | 7.02 | 7.26 | 7.54 | 9.49 |
| OEt | CH(SiMe₃)₂ | 7.57 | | | |
| [Cr(CO)₅L]* | | 7.12 | | | |
| fac-[Cr(CO)₃(DMPE)L] | | 6.70 | | | |
| [Mo(CO)₅L] | | 6.90 | | | |
| [Mo(CO)₅{C(OEt)CH(SiMe₃)₂}] | | 7.27 | | | |
| [W(CO)₅L] | | 7.02 | | | |
| [W(CO)₅{C(OEt)CH(SiMe₃)₂}] | | 7.55 | | | |
| [Fe(CO)₄(COCMe₂CMe₂O)] | | 7.3 | 8.3 | | 10.8 |
| [Fe(CO)₄(CNMeCH=CHNMe)] | | 7.0 | 7.5 | 8.2 | |
| [Fe(CO)₄(CNMeCH₂CH₂NMe)] | | 7.1 | 7.3 | 8.3 | |

[a] According to References 19, 71, and 72.

* L = − CN(Et)CH₂CH₂NEt

## 6. ELECTRONIC SPECTRA[20,21]

In electronic spectra of carbene complexes (Table 5.5), the bands due to $d$–$d$ transitions are less commonly observed compared to other complexes. The reason for this phenomenon is low energy of charge-transfer transitions from the metal to the

*Table 5.5. Electronic Spectra of Carbene Complexes*[a]

| Compound | $\nu$ (cm⁻¹)($\varepsilon$) |
|---|---|
| [Ni(C₆Cl₅){C(NHMe)₂}(PMe₂Ph)₂][PF₆] | 28 500 (798) |
| [Ni(C₆Cl₅){C(NHMe)NMe₂}(PMe₂Ph)₂][PF₆] | 28 150 (757) |
| [Ni(C₆Cl₅){C(NHMe)(OMe)}(PMe₂Ph)₂][ClO₄] | 28 800 (974) |
| [Ni(C₆Cl₅){C(NHMe)OEt}(PMe₂Ph)₂][ClO₄] | 28 800 (927) |
| [Ni(C₆Cl₅){C(NMe₂)OMe}(PMe₂Ph)₂][ClO₄] | 28 400 (903) |
| [Ni(C₆Cl₅){C(OMe)₂}(PMe₂Ph)₂][PF₆] | 29 150 (1614) |
| [Ni(C₆Cl₅){C(OMe)Me}(PMe₂Ph)₂]⁺ | 30 100 |
| [W(CO)₅(CPh₂)][b] | 43 100 (46 500), 35 100 sh, 20 600 (10 400) |
| [Mncp(CO)₂(CPh₂)][c] | 26 300 (7000), $d_\pi$–$p_\pi^{carbene}$, 16 000 (140) |

[a] According to References 20 and 21.
[b] See Ref. d, Table 5.2.
[c] R. E. Wright and A. Volger, *J. Organomet. Chem.*, **160**, 197 (1978).

alkylidene ligands as well as transitions within the ligands. In spectra of carbene complexes of Ni(II), such as $[Ni(C_6Cl_5)(CXR)(PMe_2Ph)]X'$, the $d$–$d$ band occurs at higher frequencies than in spectra of aryl compounds $[Ni(C_6Cl_5)(aryl)(PMe_2Ph)]$ because the $\pi$ bonding in the former complexes is stronger. This is also evidenced by the fact that the lower the $d$–$d$ band energy of alkylidene nickel compounds, the greater the electron density of the carbene. (This causes weakening of the $\pi$ metal–carbene bond.)

# 7. ESR AND MAGNETIC PROPERTIES[18,19]

Very few paramagnetic alkylidene complexes are known; among them are 1,3-dimethylimidazolidyne-2-ylidene Fe(I) as well as carbene complexes of Cr(I) (Table 5.6). In compounds of Fe(I), the unpaired electron is mainly localized on the iron metal. The $A$ values are, for such complexes, greater than for corresponding $[Fe(CO)_3(PR_3)_2]^+$ and $[Fe(CO)_3\{P(OR)_3\}_2]^+$ compounds, probably due to the higher ratio of $\sigma$-donor to $\pi$-acceptor properties of the carbene ligand compared to $PR_3$ or $P(OR)_3$ groups. The $g$ factors for chromium carbene complexes quite considerably exceed the free electron value of 2.0023, and may show that relatively strong delocalization of the unpaired electron occurs. This is also confirmed by the small value of the isotropic, hyperfine coupling constant $A_{53_{Cr}}$. From the ESR and IR measurements, it is found that the Fe(I) complexes $[Fe(CO)_3L_2]^+$ and $[FeL(CO)_xZ_{4-x}]^+$ have distorted trigonal–bipyramidal structure (Table 5.6).

Table 5.6. *EPR Parameters and Magnetic Properties of Carbene Complexes of Chromium(I) and Iron(I)*[a]

| Compound | $g$ | $A_p$ (mT) | $A_{Cr}$ (mT) | $\mu_{eff}$ (BM) |
|---|---|---|---|---|
| $[FeL_2(CO)_3]^+[BF_4]^-$ | 2.044 | | | |
| $[FeL^*(CO)_3(PPh_3)]^+[BF_4]^-$ | 2.047 | 2.26 | | |
| $[FeL(CO)_3(PEt_3)]^+[BF_4]^-$ | 2.046 | 2.39 | | |
| $[FeL_2(CO)_2(PPh_3)]^+[BF_4]^-$ | 2.045 | 1.85 | | |
| $[FeL(CO)_2(PPh_3)_2]^+]BF_4]^-$ | 2.047 | 2.03 | | 1.68 |
| $[FeL(CO)_2(PEt_3)_2]^+[BF_4]^-$ | 2.046 | 2.22 | | |
| $[FeL(CO)_2(P(OPh)_3)_2]^+[BF_4]^-$ | 2.042 | 3.15 | | |
| $[(OC)_3LFe(\mu\text{-}Ph_2PC_2H_4PPh_2)FeL(CO)_3]^{2+}(BF_4^-)_2$ | 2.045 | 2.38 | | 3.5 |
| $[Cr(CO)_3\{C(OEt)CH(SiMe_3)_2\}(DMPE)][BF_4]$ | 2.0221 | 2.00 | 1.34 | |
| $[Cr(CO)_3\{C(OEt)CH(SiMe_3)_2\}(DPPE)][BF_4]$ | 2.0154 | 1.93 | 1.24 | |
| $[Cr(CO)_3L'^{**}(DMPE)][BF_4]$ | 2.0199 | 2.23 | 1.34 | |
| $[Cr(CO)_3L'DPPE][BF_4]$ | 2.0205 | 2.16 | 1.35 | |
| $[Cr(CO)_4L_2']BF_4$ | 2.018 | | | |
| $[Cr(CO)_3L_2'PPh_3][BF_4]$ | 2.012 | 1.56 | | |

[a] According to References 18 and 19.

* $L = -\overline{CN(Me)CH_2CH_2N}Me$; ** $L' = -\overline{CN(Et)CH_2CH_2N}Et$.

# 8. METHODS OF PREPARATION OF CARBENE COMPLEXES[1-7, 18, 19]

## a. From metal carbonyls

The most general method of synthesis of carbene complexes is the action of nucleophiles on coordinated carbon monoxide leading to acyl derivatives followed by their alkylation. In this way, Fischer synthesized the first carbene complex.

$$[W(CO)_6] + LiPh \xrightarrow{Et_2O} Li[(OC)_5W-\overset{\overset{O}{\|}}{C}-Ph] \xrightarrow{H^+}_{H_2O} (OC)_5W = C \overset{OH}{\underset{Ph}{\diagdown}} \tag{5.6}$$

$$\xrightarrow{CH_2N_2}_{Et_2O} [W(CO)_5\{C(OMe)Ph\}]$$

$$Li[W(CO)_5(COPh)] \xrightarrow{(Me_3O)BF_4} [W(CO)_5\{C(OMe)Ph\}] \tag{5.7}$$

Also, hydroxycarbene complexes may be obtained.

$$[Cr(CO)_6] \xrightarrow[2.\ HBr(H_2O)pentane]{1.\ LiMe} (OC)_5Cr = O \overset{OH}{\underset{Me}{\diagdown}} \tag{5.8}$$

In addition to diazomethane and trialkyloxonium salts, methyl fluorosulfonate, $MeOSO_2F$, may be used as an alkylating agent. Similarly, instead of LiR, many other nucleophilic lithium compounds may be utilized.

$$[Cr(CO)_6] \xrightarrow[2.\ Et_3O^+BF_4^-]{1.\ LiNMe_2} [Cr(CO)_5\{C(NMe_2)OEt\}] + LiBF_4 \tag{5.9}$$

$$[Mo(CO)_6] \xrightarrow[2.\ Et_3O^+BF_4^-]{1.\ LiSiPh_3} [Mo(CO)_5\{C(SiPh_3)OEt\}] \tag{5.10}$$

$$[W(CO)_6] \xrightarrow[2.\ Et_3O^+BF_4^-]{1.\ LiPMe_2} cis\text{-}[W(CO)_4\{C(PMe_2)OEt\}_2] \tag{5.11}$$

This type of reaction may also involve the action of anionic carbonyl complexes on metal alkyl compounds:

$$[MnR(CO)_5] + Na[Re(CO)_5]$$

$$\xrightarrow[10h]{THF} [(OC)_5ReMn(CO)_4(COR)]^- \xrightarrow{MeOSO_2F} [(OC)_5ReMn(CO)_4\{C(OMe)R\} \tag{5.12}$$

## b. Nucleophilic addition to isocyanide ligands

Alcohols, thiols, and amines undergo addition to coordinated isocyanide molecules forming carbene complexes:

$$(Et_3P)Br_2Pt\text{-}C \equiv NMe \xrightarrow{RXH} (Et_3P)Br_2Pt = C \overset{XR}{\underset{NHMe}{\diagdown}} \tag{5.13}$$

$$(RXH = MeOH, n\text{-}PrOH, PhNH_2, s\text{-}BuNH_2)$$

$$[Pt(CNMe)_4]^{2+} \xrightarrow{\text{EtSH}} [(MeNC)_2 Pt\{C(SEt)NHMe\}_2]^{2+} \qquad (5.14)$$

$$[M(CNMe)_4]^{2+} \xrightarrow{\text{MeNH}_2} [M\{C(NHMe)_2\}_4]^{2+} \qquad (5.15)$$

(M = Pd, Pt)

$$[Mo(CNR)_5(NO)]^+ \xrightarrow{\text{R'NH}_2} [Mo(CNR)_4\{C(NHR)NHR'\}(NO)]^+ \qquad (5.16)$$

This group of reactions also encompasses reactions of 2-hydroxyisocyanides and 3-hydroxyisocyanides with transition metal halides often leading to the formation of homoleptic carbene coordination compounds[48,49] (Figure 5.5).

$$MCl_n \cdot mH_2O + CNCH_2CHROH \xrightarrow[H_2O, \Delta]{O_2} \left[ M\left( C \underset{O}{\overset{NH}{\underset{\quad}{\diagup}}} \underset{CHR}{\overset{CH_2}{\diagdown}} \right)_6 \right] Cl_3 \qquad (5.17)$$

M = Co, n = 2, m = 0; M = Rh, n = 3, m = 3; R = H, Me

$$MX_2L_n + CNCHRCHR'OH \longrightarrow \left[ M\left( C \underset{O}{\overset{NH}{\underset{\quad}{\diagup}}} \underset{CHR'}{\overset{CHR}{\diagdown}} \right)_4 \right] X_2 \qquad (5.18)$$

M = Ni, Pd, Pt; X = Cl, Br, I, ClO$_4$; L = MeCN, PhCN, PPh$_3$;
n = 0, 2; R = R' = H; R = Et, R' = H; R = H, R' = Me;

$$MX_2L_2 + CNCH_2CH_2CH_2OH \longrightarrow \left[ M\left( C \underset{O-CH_2}{\overset{NH-CH_2}{\diagup\diagdown}} CH_2 \right)_4 \right] X_2 \qquad (5.19)$$

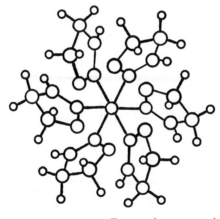

Figure 5.5.  The structure of $\left[ Co^{III}\left( C\underset{O}{\overset{NH}{\diagup\diagdown}} \right)_6 \right] Cl_3$.

### c. *Hydrogen atom abstraction from σ bonded carbon ligands*

This method is often utilized for preparation of carbene compounds which do not contain heteroatoms. Trityl salts and $Me_3P=CH_2$ are utilized as hydrogen atom trapping agents:

$$Tacp_2Me_3 \xrightarrow[-Me^-]{Ph_3C^+BF_4^-} [Tacp_2Me_2]^+BF_4^- \xrightarrow{Me_3P=CH_2} [Tacp_2Me(CH_2)] \quad (5.20)$$

Also, one of the ligands may react with a hydrogen atom of another ligand leading to the formation of a carbene compound and evolution of a hydrocarbon.[6, 69]

In the case of $[Ta(2, 4, 6\text{-}C_6H_2Me_3)(Me)X_3(PMe_3)_2]$ $(X = Br, Cl)$ $\gamma$-elimination of hydrogen from the methylene group of mesitylene takes place to give a carbene complex and methane[69]:

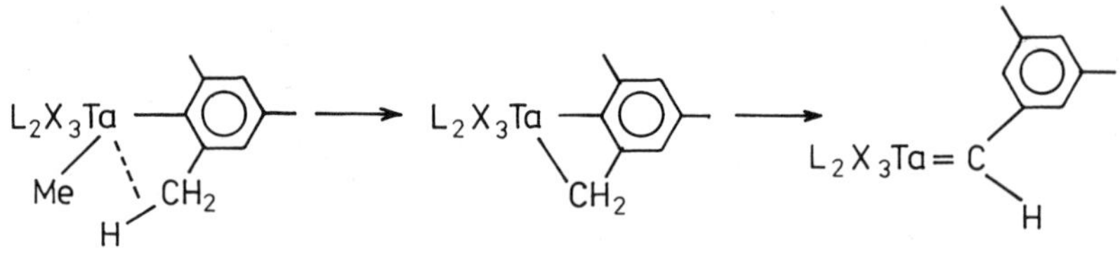

$$ \quad (5.21)$$

$$[M(CH_2CMe_3)_3Cl_2] \xrightarrow[-LiCl]{LiCH_2CMe_3} [M(CH_2CMe_3)_4Cl]$$

$$\xrightarrow{-CMe_4} [M(CH_2CMe_3)_2Cl(CHCMe_3)]$$

$$\xrightarrow[-LiCl]{LiCH_2CMe_3} [M(CH_2CMe_3)_3(CHCMe_3)] \quad (5.22)$$

$$(M = Nb, Ta)$$

$$[Ta(CH_2Ph)_3Cl_2] \xrightarrow{LiC_5Me_5} [Ta(\eta\text{-}C_5Me_5)(CH_2Ph)Cl(CHPh)]$$

$$\xdownarrow[2Tlcp]{-PhMe}$$

$$[Tacp_2(CH_2Ph)(CHPh)] \quad (5.23)$$

$$[Recp(Me)(NO)PPh_3] + Ph_3C^+ \xrightarrow[203\,K]{CH_2Cl_2} [Recp(CH_2)(NO)(PPh_3)]^+ \quad (5.24)$$

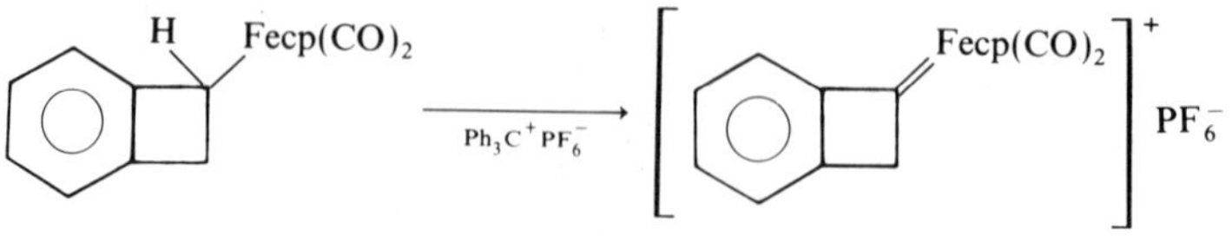

$$ \quad (5.25)$$

In some syntheses, the addition of a nucleophile to a carbene atom is first carried out followed by elimination of another group to regenerate carbene complex:

$$[W(CO)_5\{C(OMe)Ph\}] \xrightarrow[195\,K]{LiPh} [W(CO)_5\{CPh_2(OMe)\}]^- \xrightarrow[195\,K]{HCl} [W(CO_5(CPh_2)] \quad (5.26)$$

$$[W(CO)_5\{C(OMe)Ph\}] \xrightarrow[195\,K]{LiMe} [W(CO_5\{CMePh(OMe)\}]$$

$$\xrightarrow[195\,K]{HCl} [W(CO)_5\{C(Ph)Me\}] \quad (5.27)$$

### d. *Alkylation of complexes with some* $M-C\sigma$ *bonds*

Many complexes with $M-C$ bonds react with alkylating agents to give ionic carbene compounds:

$$[\text{Fecp(DPPE)(COPh)}] \xrightarrow{\text{Me}_3\text{O}^+\text{BF}_4^-} [\text{Fecp(DPPE)}\{\text{C(OMe)Ph}\}]^+ \, [\text{BF}_4]^- \qquad (5.28)$$

$$trans\text{-}[\text{PtCl}(\overset{\displaystyle S}{\overset{\parallel}{\text{C}}}-\text{NMe}_2)(\text{PPh}_3)_2] \xrightarrow{\text{MeSO}_3\text{F}} trans\text{-}[\text{PtCl}\{\text{C(SMe)NMe}_2\}(\text{PPh}_3)_2]^+ + \text{SO}_3\text{F}^- \quad (5.29)$$

### e. *Electrophilic addition to coordinated ligands*

Electrophilic reagents react with $\sigma$-bonded imino group to afford carbenes:

$$(\text{RNC})\text{cpNi}-\overset{\displaystyle \overset{NR}{\parallel}}{\underset{\displaystyle R'}{\text{C}}} \xrightarrow{\text{HBF}_4} [\text{cp(RNC)Ni}\{\text{C(NHR)R}'\}]^+ \, \text{BF}_4^- \qquad (5.30)$$

$$trans\text{-}(\text{Et}_3\text{P})_2\text{ClPt}-\overset{\displaystyle \overset{NPh}{\parallel}}{\underset{\displaystyle H}{\text{C}}} \xrightarrow[\text{NaClO}_4]{\text{Me}_2\text{SO}_4} [(\text{Et}_3\text{P})_2\text{ClPt}\{\text{C(NMePh)H}\}]^+ + \text{ClO}_4^- \qquad (5.31)$$

Also, complexes containing $\sigma$-bonded vinyl ligands undergo protonation to give carbene complexes[50]:

$$[(\text{Ph}_3\text{P})(\text{OC})\,\text{cpFeCMe}=\text{CH}_2] + \text{HBF}_4 \xrightarrow[273\text{K}]{\text{Et}_2\text{O}} [(\text{Ph}_3\text{P})(\text{OC})\,\text{cpFe}=\text{CMe}_2]^+\text{BF}_4^- \quad (5.32)$$

### f. *Reactions of complexes with mixtures of acetylenes and alcohols*

Alkyl and arylacetylenes in presence of alcohols react with platinum compounds to furnish carbene complexes:

$$[\text{PtMeClL}_2] + \text{RC}\equiv\text{CH} + \text{R}'\text{OH} \xrightarrow{\text{AgPF}_6} trans\text{-}[\text{PtMe}\{\text{C(OR}')\text{CH}_2\text{R}\}\text{L}_2] \qquad (5.33)$$

$$(\text{L} = \text{PMe}_2\text{Ph, AsMe}_3)$$

$$[\text{PtMe}_2(\text{CF}_3)\text{I}(\text{PMe}_2\text{Ph})_2] + \text{CH}\equiv\text{CCH}_2\text{CH}_2\text{OH}$$

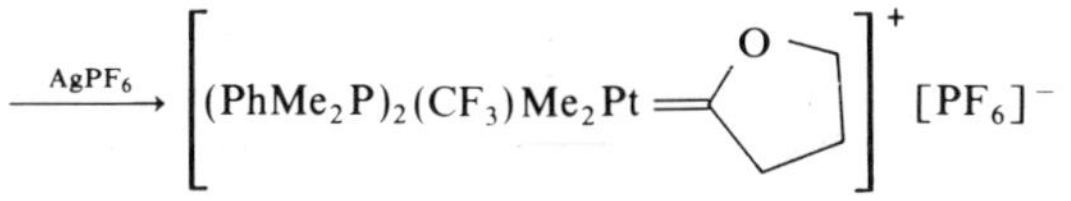

$$\xrightarrow{\text{AgPF}_6} \left[(\text{PhMe}_2\text{P})_2(\text{CF}_3)\text{Me}_2\text{Pt}=\!\!\!\!\bigg\langle\!\!\overset{\displaystyle O}{\phantom{x}}\!\bigg\rangle\right]^+ [\text{PF}_6]^- \qquad (5.34)$$

### g. *Synthesis from ionic complexes*

Anionic metal complexes may react with organic dihalogeno compounds to form carbene complexes. Similarly, metal carbenes may be obtained by reactions of hydrido complexes with compounds possessing active hydrogen.

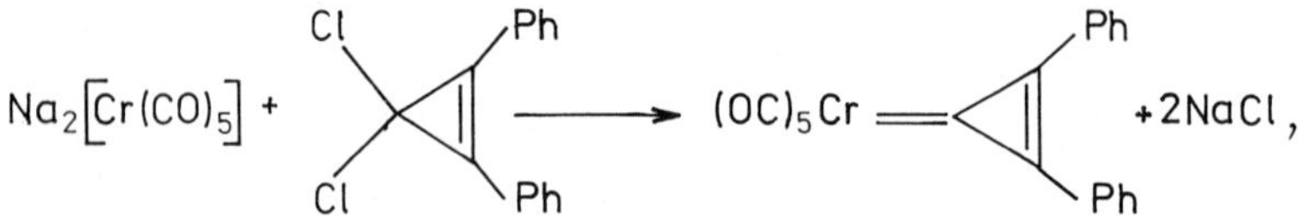

$$Na_2[Cr(CO)_5] + \underset{Cl\quad Ph}{\overset{Cl\quad Ph}{\times}} \longrightarrow (OC)_5Cr = \underset{Ph}{\overset{Ph}{<}} +2NaCl, \qquad (5.35)$$

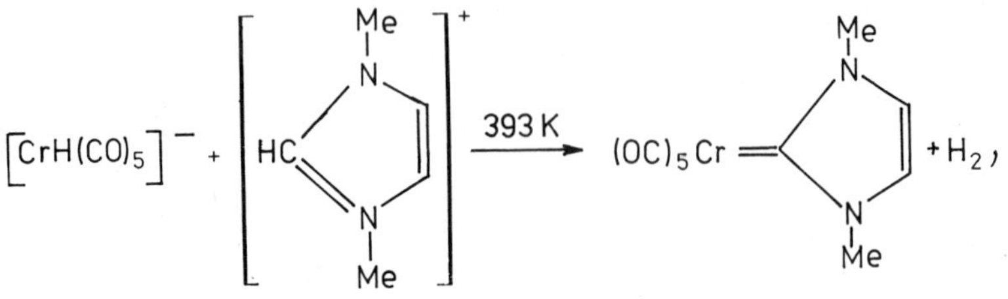

$$[CrH(CO)_5]^- + \left[HC\underset{\underset{Me}{N}}{\overset{\overset{Me}{N}}{\diamondsuit}}\right]^+ \xrightarrow{393\ K} (OC)_5Cr = \underset{\underset{Me}{N}}{\overset{\overset{Me}{N}}{<}} + H_2, \qquad (5.36)$$

$$[FeH(CO)_4]^- + \left[HC\underset{\underset{Me}{N}}{\overset{\overset{Me}{N}}{\diamondsuit}}\right]^+ \longrightarrow (OC)_4Fe = C\underset{\underset{Me}{N}}{\overset{\overset{Me}{N}}{<}} + H_2. \qquad (5.37)$$

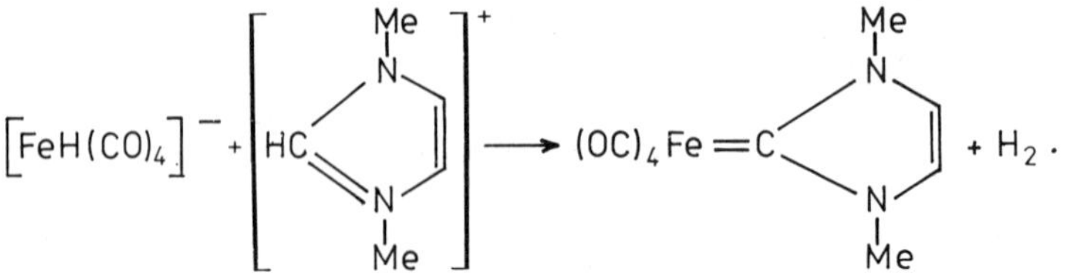

Like the organic dihalides, dialkyl(chloromethylene)ammonium chlorides react with many transition metal complexes. This method allows preparation of carbene complexes in which central atoms possess high oxidation states.

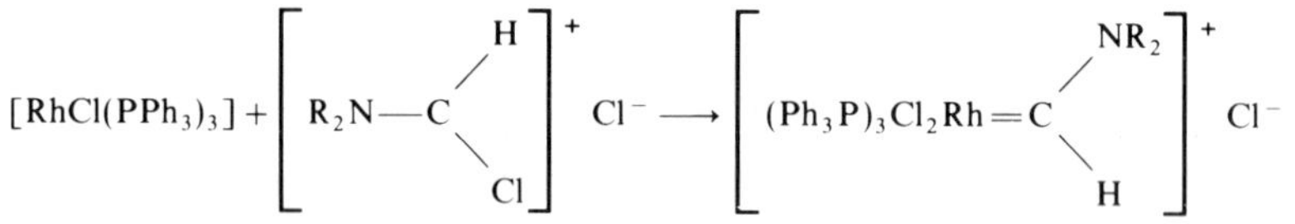

$$[RhCl(PPh_3)_3] + \left[R_2N-\underset{Cl}{\overset{H}{C}}\right]^+ Cl^- \longrightarrow \left[(Ph_3P)_3Cl_2Rh=\underset{H}{\overset{NR_2}{C}}\right]^+ Cl^- \qquad (5.38)$$

(R = Me, i-Pr, n-Bu)

$$Na_2[Fe(CO)_4] + [ClCHNMe_2]^+ Cl^- \longrightarrow [Fe(CO)_4C(NMe_2)H] \qquad (5.39)$$

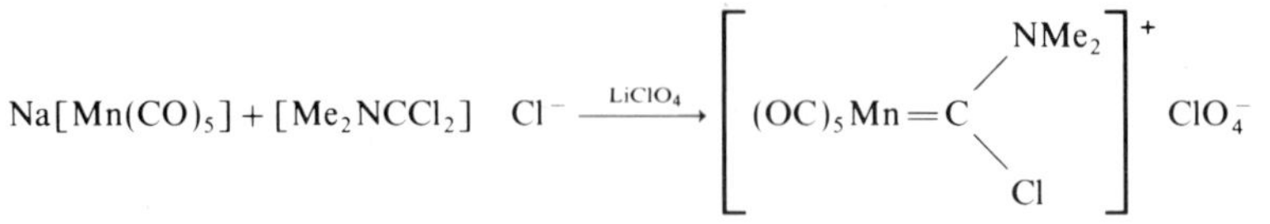

$$Na[Mn(CO)_5] + [Me_2NCCl_2]\ Cl^- \xrightarrow{LiClO_4} \left[(OC)_5Mn=\underset{Cl}{\overset{NMe_2}{C}}\right]^+ ClO_4^- \qquad (5.40)$$

## h. Oxidative addition reactions

Imidoyl chloride $Cl(Ph)C=NMe$ forms carbene complexes in the course of oxidative addition reactions to metal compounds in their low oxidation states.

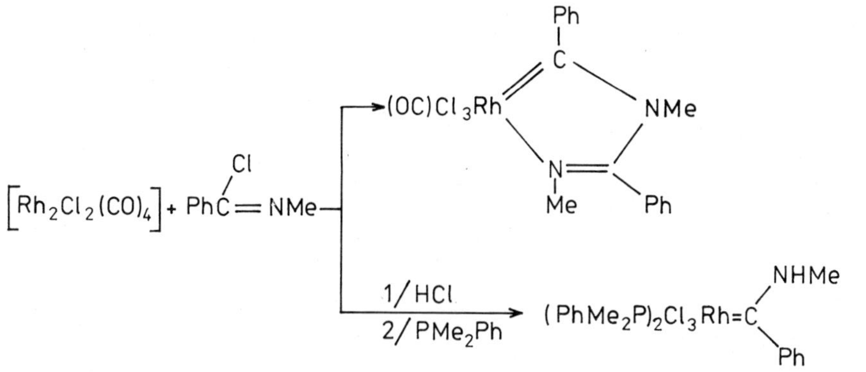

$$\left[Rh_2Cl_2(CO)_4\right] + PhC\overset{Cl}{\underset{}{=}}NMe \Big\{$$

(5.41)

## i. Reactions of complexes with neutral carbene precursors

In order to explain reactions of diazomethane and its derivatives with coordination compounds, it has been proposed that the formation of intermediate carbene complexes takes place. In the reaction of bis(trifluoromethyl)diazomethane with $[IrCl(CO)(PMePh_2)_2]$ the carbene complex $[IrCl(CO)(PMePh_2)_2\{C(CF_3)_2\}]$ is formed rather than the chlorobis(trifluoromethyl)methyl iridium complex $Ir\{CCl(CF_3)_2\}(CO)(PMePh_2)_2$. In many cases, such complexes were isolated and investigated[42,51,52]:

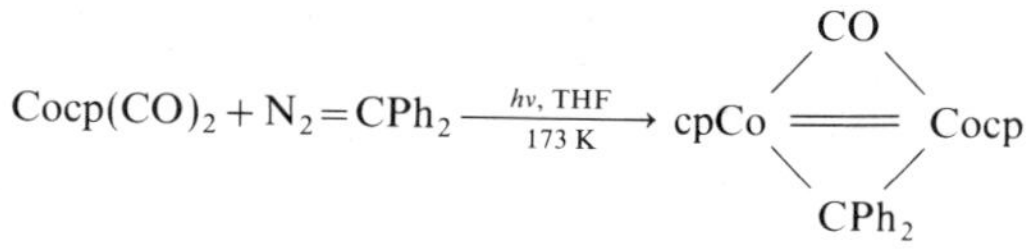

$$Cocp(CO)_2 + N_2=CPh_2 \xrightarrow{h\nu, THF \atop 173\,K} \text{ cpCo} \underset{CPh_2}{\overset{CO}{=\!=}} Cocp \tag{5.42}$$

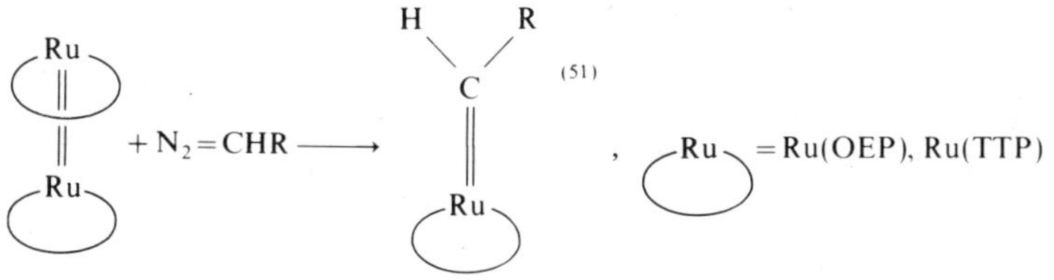

(5.43)

$$Ru_3(CO)_8(np\text{-}DAB) \xrightarrow{CH_2N_2} Ru_3(CO)_8(\mu\text{-}CH_2)(np\text{-}DAB)^{(52)} \tag{5.44}$$

Electron-rich alkenes react with transition metal compounds with cleavage of the carbon–carbon double bond to afford carbene compounds.

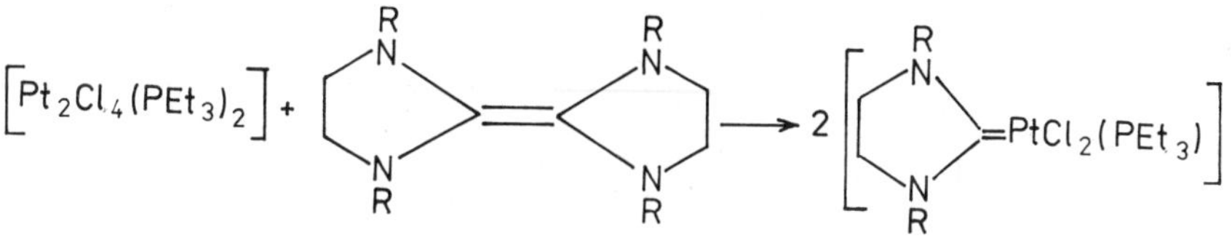

(5.45)

$$2[RhCl(PPh_3)_3] + R\,\overline{NCH_2CH_2N(R)C} = \overline{CN(R)\,CH_2CH_2N}R$$

$$\longrightarrow 2[RhCl(\overline{CN(R)\,CH_2CH_2N}R)(PPh_3)_2] + 2PPh_3 \tag{5.46}$$

## j. *Nucleophilic addition to carbyne complexes*

This method is usually utilized when other methods cannot be used.[53,54]

$$[Mcp(CO)_2(CPh)]^+[BCl_4]^- \xrightarrow{R^-} [Mcp(CO)_2(CRPh)] \qquad (5.47)$$

$$(M = Mn, Re; R = CN, NCS)$$

$$[Cr(CO)_5(CNMe_2)]^+ \xrightarrow{LiNMe_2} [Cr(CO)_5\{C(NMe_2)_2\}] \qquad (5.48)$$

$$[Cr(CO)_5(CNEt_2)]^+BF_4^- \xrightarrow{[Bu_4N]^+F^-} [Cr(CO)_5\{C(NEt_2)F\}] \qquad (5.49)$$

$$[L_nM\equiv CR]^+ + X^- \longrightarrow \left[ L_nM=C\begin{smallmatrix} R \\ \\ X \end{smallmatrix} \right] \qquad (5.50)$$

$M = Cr, Mo, W; L_n = (C_6H_6)(CO)_2, (CO)_5, (PR_3)(CO)_4),$ etc.;
$X = F, Cl, Br, I, NCO, NCSe, CN, SPh, SeR, TePh,$
$OEt, NMe_2, AsPh_2, SnPh_3, PbPh_3, Me, [(OC)_5MC(O)R]$
$(M = Cr, W),$ etc.

## k. *Oxidation reactions*

Paramagnetic carbene complexes of Fe(I) and Cr(I) are obtained by oxidation of Fe(O) and Cr(O) complexes, respectively, using silver tetrafluoroborate.[18,19]

$$[Fe(CO)_3L_2] + AgBF_4 \xrightarrow[293K]{THF} [Fe(CO)_3L_2]^+[BF_4]^- + Ag \qquad (5.51)$$

$$[Cr(CO)_3L(dppe)] + AgBF_4 \xrightarrow{CH_2Cl_2} [Cr(CO)_3L(dppe)]^+[BF_4]^- + Ag \qquad (5.52)$$

$$[Cr(CO)_3L(dmpe)] + AgBF_4 \xrightarrow{THF} [Cr(CO)_3L(dmpe)]^+[BF_4]^- + Ag \qquad (5.53)$$

$$(L = - \overline{CN(Et) CH_2CH_2N}Et)$$

## l. *Addition to carbonyl ligands*

Cationic carbonyls, some neutral carbonyls, and thiocarbonyls react with heterocyclic compounds such as $\overline{YCH_2CH_2}$ (Y = O, NH) in the presence of halides as catalysts to furnish cyclic aminoxy- and dioxycarbene complexes[56]:

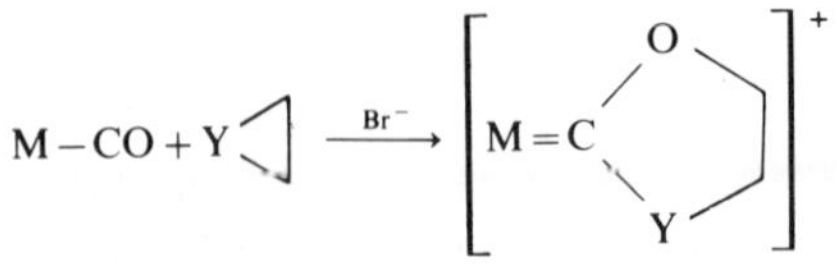

$$M-CO+Y \overset{Br^-}{\longrightarrow} \left[ M=C \begin{smallmatrix} O \\ \\ Y \end{smallmatrix} \right]^+ \qquad (5.54)$$

$M = Fecp(CO)_2^+, Rucp(CO)_2^+, Mncp(CO)(NO)^+,$
$Fecp(CO)(PPh_3)^+, Mn(CO)_4X, Re(CO)_4X$
$(X = Cl, Br, I), Y = NH, O$

Reaction (5.54) depends on the electrophilicity of CO groups; $\overline{YCH_2CH_2}$ reacts at room temperature only with those complexes in which the force constant $k(CO)$ is higher than $1700\ Nm^{-1}$. Complexes in which CO groups are more nucleophilic [the force constant $k(CO)$ is lower] react in this way at higher temperatures, or they do not react at all. The reaction of ethylene oxide with $Mn_2(CO)_{10}$, $Re_2(CO)_{10}$, $Ru_3(CO)_{12}$, and Fe $(CO)_5$ takes place at 373 K.

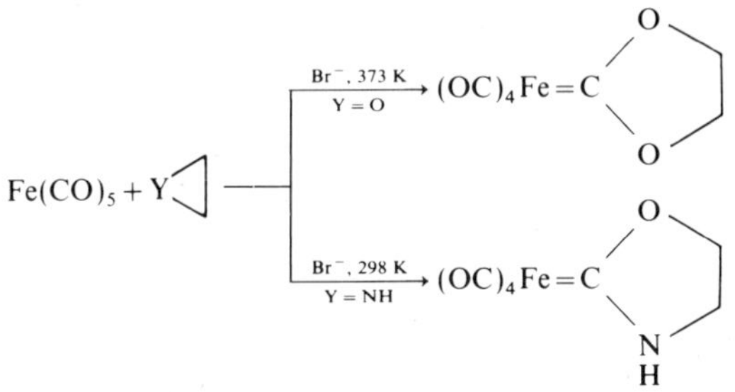

$$Fe(CO)_5 + Y \overset{\displaystyle \triangleleft}{} \quad \begin{cases} \xrightarrow[Y=O]{Br^-,\ 373\ K} (OC)_4Fe=C\diagdown^{O}_{O}\diagup \\ \\ \xrightarrow[Y=NH]{Br^-,\ 298\ K} (OC)_4Fe=C\diagdown^{O}_{\underset{H}{N}}\diagup \end{cases} \qquad (5.55)$$

$$Re_2(CO)_{10} + O\triangleleft \longrightarrow Re_2(CO)_9(=\overline{COCH_2CH_2O}) + Re_2(CO)_8(=\overline{COCH_2CH_2O})_2 \qquad (5.56)$$

## m. Synthesis of $\mu$-methylene complexes[42]

These complexes of the general type $M(\mu\text{-}CH_2)M$ containing direct $M-M$ bonds may be obtained by reactions of various complexes with the following:

1. Diazomethane and its derivatives [reactions (5.42) and (5.44)].
2. Geminal dihaloalkanes and their derivatives.
3. Wittig reagents, $R_3P=CH_2$.

$$Na_2[Fe_2(CO)_8] + CH_2I_2 \longrightarrow Fe_2(\mu\text{-}CH_2)(CO)_8 \qquad (5.57)$$

$$Fe_2cp_2(CO)_4 + Ph_3P=CH_2 \longrightarrow (OC)cpFe(\mu\text{-}CH_2)(\mu\text{-}CO)Fecp(CO) \qquad (5.58)$$

$$Li^+[(ON)_2Fe(\mu\text{-}PPh_2)(\mu\text{-}NO)Fe(NO)(PPh_2H)]^-$$

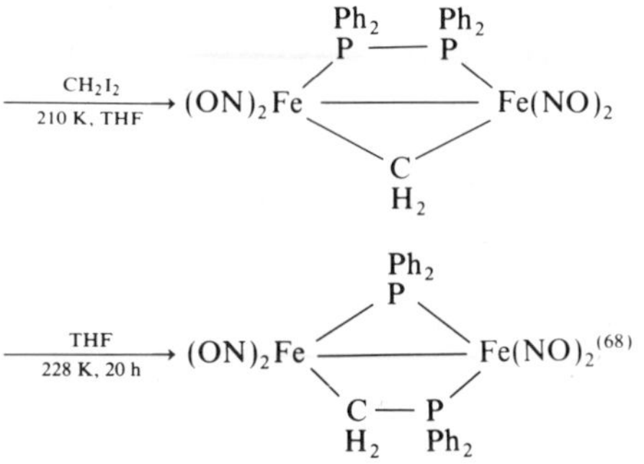

$$\xrightarrow[210\ K,\ THF]{CH_2I_2} (ON)_2Fe \underset{\underset{H_2}{C}}{\overset{\overset{Ph_2}{P}\text{---}\overset{Ph_2}{P}}{\diagup\diagdown}} Fe(NO)_2$$

$$\xrightarrow[228\ K,\ 20\ h]{THF} (ON)_2Fe \underset{\underset{H_2\ \ Ph_2}{C-P}}{\overset{\overset{Ph_2}{P}}{\diagup\diagdown}} Fe(NO)_2{}^{(68)} \qquad (5.59)$$

Complexes with $\mu$-methylene bridges are also formed in reactions of metal carbene compounds with systems such as $M(PR_3)_2$ ($M=Ni$, Pd, Pt) which are isolobal to the $CH_2$ group [reactions (3.17)–(3.21)].

## 9. METHODS OF PREPARATION OF CARBYNE COMPLEXES[1,2,67]

### a. Removal of alkoxy, alkylthio, and amino groups from the carbene ligand[53,54]

Fischer first prepared carbyne complexes by the action of boron halides on carbene compounds of chromium, molybdenum, and tungsten.

$$[M(CO)_5\{C(OMe)R\}] + BX_3 \xrightarrow[-BX_2(OMe)]{\text{pentane}} [MX(CO)_4(CR)] \qquad (5.60)$$

$$(M = Cr, Mo, W; X = Cl, Br, I; R = Me, Et, Ph)$$

$$[(OC)_5Cr\{C(OEt)(NEt_2)\}] + 2BF_3 \xrightarrow[CH_2Cl_2]{173\,K} [(OC)_5Cr \equiv CNEt_2]^+ + [BF_4]^- \qquad (5.61)$$

Elimination of alkoxy group may be also affected by using aluminum and gallium halides. In the case of elimination of halides from halocarbene complexes, silver salts may be used.

$$[W(CO)_5\{C(OEt)(C_5H_4Fecp)\}] +Al_2Br_6$$

$$\longrightarrow [WBr(CO)_4(CC_5H_4Fecp)] + CO + Al_2Br_5(OEt) \qquad (5.62)$$

$$[Cr(CO)_5\{C(Cl)NMe_2\}] + AgBF_4 \xrightarrow[243\,K]{PhMe} [Cr(CO)_5(CNMe_2)]^+[BF_4]^- + AgCl \qquad (5.63)$$

$$[(OC)_5WC(OMe)Ph] \xrightarrow{GaCl_3} [ClW(CO)_4(CPh)] + GaCl_2(OMe) + CO \qquad (5.64)$$

### b. Hydrogen elimination from the carbene ligand

In some carbene complexes of tantalum possessing less than 18 electrons, the $C-H$ bond in the carbene ligand is weak, and therefore hydrogen splitting readily takes place

$$[TaCl(CH_2Ph)\,cp(CHPh)] + 2PMe_3 \xrightarrow{-PhMe} [TaClcp(CPh)(PMe_3)_2] \qquad (5.65)$$

$$[Ta(CH_2CMe_3)_3(CHCMe_3)] + LiBu + L \longrightarrow (LiL)[Ta(CH_2CMe_3)_3\{C(CMe_3)\}] \qquad (5.66)$$

$$(L = N, N'\text{-dimethylpiperazine})$$

### c. Cleavage of $C \equiv C$ bonds in acetylenes and $C \equiv N$ bonds in cyanides

The dinuclear tungsten complex $W_2(OBu')_6$ reacts with acetylenes and cyanides to give metathesis products of $W \equiv W$, $C \equiv C$, and $C \equiv N$ bonds.[57-59,61-63]

$$W_2(OCMe_3)_6 + C_2R_2 \xrightarrow{298\,K} W(CR)(OCMe_3)_3 \qquad (5.67)$$

$$R = alkyl, aryl, H, etc.$$

$$W_2(OCMe_3)_6 + RCN \longrightarrow W(CR)(OCMe_3)_3 + [W(N)(OCMe)_3]_x \qquad (5.68)$$

$$W_2(OCMe_3)_6 + C_2Ph_2 \xrightarrow{343\,K} (Me_3CO)_2W(\mu\text{-CPh})_2W(OCMe_3)_2 \qquad (5.69)$$

The analogous molybdenum compound, $Mo_2(OCMe_3)_6$, reacts in this way with $RC\equiv CH$ (R = Ph, Pr).[58]

$$Mo_2(OCMe_3)_6 + PhCCH \longrightarrow Mo(CPh)(OCMe_3)_3 \qquad (5.70)$$

Studies on $W_2(OCMe_3)_6(\mu\text{-}C_2H_2)(py)$ utilizing $H^{13}C^{13}CH$ and $D^{12}C^{12}CD$ show that this complex exists in equilibrium with carbyne complexes of the type $(Me_3CO)_3W\equiv CH$.[62]

### d. *Reactions of carbyne complexes with anionic transition metal complexes*

These reactions lead to the formation of dinuclear carbyne complexes containing $M'-M$ bonds.

$$Na[M'(CO)_5] + [BrM(CO)_4(CPh)] \xrightarrow[248\,K]{THF} [(OC)_5M'M(CO)_4(CPh)] + NaBr \qquad (5.71)$$

$$(M' = Mn, Re; M = Cr, Mo, W)$$

### e. *Spontaneous carbene ligand rearrangement*[30, 53–54]

Fischer obtained tetracarbonyl(diethylaminocarbyne)triphenylstannylchromium by spontaneous rearrangement of the carbene ligand within the following chromium complex:

$$(OC)_5Cr=C\begin{smallmatrix}SnPh_3\\ \\NEt_2\end{smallmatrix} \longrightarrow [Ph_3SnCr(CO)_4(\equiv CNEt_2)] + CO \qquad (5.72)$$

Similar rearrangement occurs within halocarbene complexes in solution and even in the solid state:

$$(OC)_5Cr=C\begin{smallmatrix}X\\ \\NEt_2\end{smallmatrix} \xrightarrow[Br:253\,K]{Cl:303\,K} [XCr(CO)_4(CNEt_2)] + CO \qquad (5.73)$$

### f. *Reactions of acyl complexes with dibromotriphenylphosphorus(V)*

Dibromotriphenylphosphorus(V) reacts with acyl complexes to give metal carbyne compounds. The first stage of the reaction probably involves the formation of a carbene complex containing a $C_{carbene}-O-P$ bond:

$$(OC)_5WC\begin{smallmatrix}OLi\\ \\Ph\end{smallmatrix} + PBr_2Ph_3 \xrightarrow[223\,K]{Et_2O} [(OC)_4BrW\equiv CPh] + CO + POPh_3 + \cdots. \qquad (5.74)$$

Table 5.7. Carbene Complexes

| Compound | Color | Mp (K) | Remarks |
|---|---|---|---|
| $[Er(CHSiMe_3)(CH_2SiMe_3)]^{(70)}$ | Yellow-pink | 658 dec | |
| $[Nb_2(\mu\text{-}CSiMe_3)_2(CH_2SiMe_3)_4]^{(23)}$ | Red | 425 dec | |
| $[Nb(C_5H_4Me)(CHCMe_3)Cl_2]^a$ | Purple | | |
| $[NbCl(CHCMe_3)_2(PMe_3)_2]^{(22)}$ | Orange-red | | |
| $[Nbcp(CHCMe_3)Cl_2]^a$ | Purple | | |
| $[Nb(CH_2CMe_3)_3(CHCMe_3)]^b$ | Orange | | |
| $[Nbcp_2(CHCMe_3)Cl]^c$ | Orange-gold | | |
| $[Tacp_2Me(CH_2)]^d$ | Light-yellow | | $\tau(H)$: $-0.22$ ($CH_2$); 4.82 (cp); 9.88 (Me) |
| $[Ta(CH_2CMe_3)_3(CHCMe_3)]^b$ | Orange | | |
| $[Tacp_2(CH_2Ph)CHPh]^c$ | Brown-orange | | |
| $[Ta(CHC_6H_3Me_2)Br_3(PMe_3)_2]^{(69)}$ | Green | | |
| $[Tacp_2(CHCMe_3)Cl]^c$ | Orange-gold | | |
| $[Ta(CH_2CMe_3)(CHCMe_3)_2(PMe_3)_2]^{(22)}$ | Orange-red | | |
| $[Ta(CH_2SiMe_3)\{N(SiMe_3)_2\}(CHSiMe_3)]^e$ | Green | 407–408 | |
| $[TacpCl_2(CHCMe_3)]^a$ | Black | | |
| $[Cr(CO)_5\{C(OMe)Me\}]^{(5)}$ | Yellow | 307 | |
| $[Cr(CO)_5\{C(Cl)NMe_2\}]^m$ | Yellow | 322–324 dec | |
| $[Cr(CO)_5\{C(OMe)Ph\}]^{(5)}$ | Orange | 319 | |
| $[Cr(CO)_5\{C(OMe)C_4H_4NMe_2\text{-}p\}]^{(5)}$ | Red | 404 dec | |
| $[Cr(CO)_5\{C(NH_2)Me\}]^{(5)}$ | Light-yellow | 349–350 | |
| $[Cr(CO)_5\{C(NEt_2)F\}]^f$ | Yellow | 313 | |
| $[Cr(CO)_3L'(DMPE)][BF_4]^{**(19)}$ | Dark-orange | | |
| $cis\text{-}[Cr(CO)_4\{\overline{CN(Me)CH=CHNMe}\}_2]^i$ | Yellow | 503 | $v(CO)$ 1980 m, 1886 sh, 1850 s, 1832 sh |
| $[Cr(CO)_5\{C(NH_2)Ph\}]^{(5)}$ | Yellow | 352 | |
| $[Cr(CO)_5\{C(NMe_2)Me\}]^{(5)}$ | Light-yellow | 347 | |
| $[Cr(CO)_5\{C(NMe_2)Ph\}]^{(5)}$ | Light-yellow | 354 | |
| $[Cr(CO)_5\{\overline{CC(Ph)}=C(Ph)\}]^{(5)}$ | Dark-yellow | 472–473 dec | |
| $[Mo(CO)_5\{C(OMe)Me\}]^{(5)}$ | Yellow | | Decomposes rapidly |
| $[Mo(CO)_5\{C(OMe)Ph\}]^{(5)}$ | Orange-red | | |
| $cis\text{-}[Mo(CO)_4\{C(OMe)Me\}(PPh_3)]^{(5)}$ | Yellow-orange | 377–379 | Stable |
| $[Mo(CO)_5\{\overline{CC(Ph)CPh}\}]^g$ | Yellow | 453 dec | |
| $cis\text{-}[Mo(CO)_4\{\overline{CN(Me)CH=CHNMe}\}_2]^i$ | Yellow | 508 | |
| $[Mo(CO)_5\{C(OEt)CH(SiMe_3)_2\}]^{(19)}$ | Yellow | 317 | |
| $[Mo(CO)_5\{C(OEt)(C_6H_5\text{-}\eta^6)Cr(CO)_3\}]^{(29)}$ | Red | 390 | |
| $[Mocp(CO)_3\{C(H)NMe_2\}][Mocp(CO)_3]^m$ | Yellow | 413 dec | |
| $[W(CO)_5\{C(OMe)Me\}]^{(5,1)}$ | Yellow | 325 | |
| $[W(CO)_5\{C(OMe)Ph\}]^{(5,1)}$ | Orange | 332 | |
| $[W(CO)_5(CPh_2)]^h$ | Dark-red | 338–339 | |
| $cis\text{-}[W(CO)_5\{C(NHMe)Me\}]^{(5)}$ | Light-yellow | 354 | $v(CO)$ 2066 s, 1976 sh, 1949 sh, 1904 vs |
| $trans\text{-}[W(CO)_5\{C(NHMe)Me\}]^{(5)}$ | Light-yellow | 362 | $v(CO)$ 2070 s, 1980 sh, 1930 vs |

*(Table continued)*

*Table 5.7. (continued)*

| Compound | Color | Mp (K) | Remarks |
|---|---|---|---|
| *cis*-[W(CO)$_4${C(OMe)Me}(Ph$_3$)][5] | Orange-red | 384.5–386 | $v$(CO) 2024 vs, 1925 s, 1914 s, 1897 vs |
| *fac*-[MnBr(CO)$_3$L$_2$]*[ii] | Light-yellow | | |
| [Mncp(CO)$_2${C(OEt)Ph}][hh] | Yellow | 311–312 | |
| [Mncp(CO)$_2${C(OMe)Me}][5)j] | Red-brown | liquid | $v$(CO) 1965 vs, 1894 vs |
| [Mn$_2$(CO)$_9${C(OMe)Me}][k] | Orange-red | 354–356 | |
| [Mn$_2$(CO)$_9${C(OMe)Ph}][k] | Orange-yellow | 341 | |
| [Mn$_2$(CO)$_9${C(OEt)Ph}][k] | Orange-yellow | 245 | |
| *cis*-[MnCl(CO)$_4${C(H)NMe$_2$}][m(27)] | Yellow | 364–366 | |
| [Mn(CO)$_5${C(Cl)NMe$_2$}]ClO$_4$[m(13)] | Light-yellow | 394–395 dec | |
| [Mn$_2$(CO)$_9${C(NH$_2$)Me}][k] | Yellow | 403 | |
| [Mncp(CO)$_2$(CPh$_2$)][24] | Green-brown | 364–365 | |
| [Recp(CO)$_2${C(OH)Me}][l] | Yellow | 357 dec | |
| [Recp(CO)$_2${C(OMe)Me}][l] | Yellow | 374 | $v$(CO) 1972 vs, 1894 vs |
| [Recp(CO)$_2${C(OMe)Ph}][l] | Yellow | 379 | $v$(CO) 1976 vs, 1892 vs |
| *cis*-[ReCl(CO)$_4${C(H)NMe$_2$}][m(27)] | Fawn | 369–371 | |
| [Recp(CH$_2$)(NO)(PPh$_3$)][PF$_6$][60] | | | Unstable |
| [ReCl{C(OH)Me}(CO)(N$_2$)(PPh$_3$)$_2$][ff] | Orange | | $v$(N$\equiv$N) 2050 s, $v$(CO) 1895 s |
| *fac*-[ReBr(CO)$_3${C(OH)Me}$_2$][gg] | Yellow | 353–355 | $v$(CO) 2060 m, 2001 s, 1914 s |
| [Fe(CO)$_4${C(OMe)Ph}][5] | Red-brown | | Liquid |
| [Fecp(CO){C(OEt)Me}(PPh$_3$)]$^+$[BF$_4$]$^-$[n] | Yellow | 384 | $v$(CO) 1990 |
| [Fecp(CO){C(OEt)Me}(PCy$_3$)]$^+$[BF$_4$]$^-$[n] | Yellow | 448 | $v$(CO) 1965 |
| [Fecp(CO){C(OH)Me}(PCy$_3$)]$^+$[BF$_4$]$^-$[n] | Green | 393 | $v$(CO) 1961 |
| [Fe(CO)$_4$L][ii] | Yellow | 321–323 | |
| [FeL$_2$(CO)$_3$]$^+$[BF$_4$]$^-$[18] | Dark-green | | |
| [Fe(CO)$_3$L$_2$][ii] | Dark-yellow | 413 | |
| [Fe(CCl$_2$)(H$_2$O)(tetraphenylporphyrin)][26] | Purple-red | | |
| [Fecp{C(OMe)Ph}(DPPE)][BF$_4$][l] | Yellow | | |
| [Fecp(CO)$_2$(CHPh)]$^+$[25] | Dark-orange | | |
| [RuCl{ $\overline{\text{CN(C}_6\text{H}_4\text{Me-P)CH}_2\text{CH}_2\text{N}}$(C$_6$H$_3$Me-*p*)} (PPh$_3$)$_2$][o] | Red | >613 | |
| [RuCl$_2$L$_4$][p] | Yellow | 509 | |
| [Rucp(CO){C(OEt)Me}PCy$_3$][BF$_4$][n] | White | 473 | $v$(CO) 1968 |
| [Ru$_3$(CO)$_{11}$L'][ii] | Purple-red | 386 | |
| [OsBr(CO)$_2$( $\overline{\text{CSCH}_2\text{CH}_2\text{S}}$)(PPh$_3$)$_2$]$^+$ClO$_4^-$[q] | | | $v$(CO) 2060, 1985 |
| [OsClH(CO){ $\overline{\text{CNHCH:C(Ph)O}}$ }(PPh$_3$)$_2$][r] | | | $v$(CO) 1920 |
| [Os(CO)$_4${ $\overline{\text{CSHCH=C(Me)NMe}}$ }][s] | Yellow | 370–373 | |
| [OsI{C(SMe)$_2$}(CO)$_2$(PPh$_3$)$_2$]$^+$[r] | Yellow | | $v$(CO) 2055, 1980 |
| [Os(CO)$_3${ $\overline{\text{CN(Me)C(Me)=CHS}}$ }(PPh$_3$)][s] | Yellow | 473 dec | |
| [Co(CO)$_3${C(OEt)Ph}(SnPh$_3$)][5] | Yellow | 395–397 | $v$(CO) 1967, 1955 |
| [Co(CO)$_2${C(Et)NMe}(NO)][u] | Orange | liquid | |

*(Table continued)*

*Table 5.7. (continued)*

| Compound | Color | Mp (K) | Remarks |
|---|---|---|---|
| $[CoCl(CO)_2\{CH(NMe_2)\}(PPh_3)]^{m\,(27)}$ | Yellow | 451–453 | |
| $[Co(CO)_2L(NO)]^{ii}$ | Scarlet | 268 | |
| $[Co(CO)L_2(NO)]^{ii}$ | Purple-brown | 409 dec | |
| $[RhCl\{\overline{CN(Ph)CH_2CH_2NPh}\}(PPh_3)_2]^{(5)}$ | Orange | 482 dec | $v(RhCl)$ 287 |
| $[RhCl\{\overline{CN(Ph)CH_2CH_2NPh}\}(CO)(PPh_3)]^{(5)}$ | Yellow | 471 | $v(CO)$ 1973, $v(RhCl)$ 299 |
| $[Rh_2Cl_6(CO)_2\{C(Cl)NMe_2\}_2]^m$ | Orange | >573 | |
| $[RhCl_3\{C(Cl)NMe_2\}(PPh_3)_2]^m$ | Yellow | 511–515 | |
| $[RhCl_3\{(CO)C(Ph)NHMe\}_n]^{(5)}$ | Yellow | 463 dec | $v(CO)$ 2120, $v(RhCl)$ 355, 340 |
| $[RhCl_3(CO)\{C(Ph)NHEt\}(PPh_3)]^{(5)}$ | White | 437–440 | $v(CO)$ 2105, $v(RhCl)$ 330, 299 |
| $[RhCl_3(CHNMe_2)(PEt_3)_2]^v$ | | 342–346 | |
| $[IrCl_3\{C(H)NMe_2\}(PPh_3)_2]^v$ | | 558–563 | |
| $[IrCl(CO)\{C(CF_3)_2\}(PPh_2Me)_2]^{(5)}$ | Cream | 427 | $v(CO)$ 2056 |
| $[\overline{IrCl(Bu_2^tPCH_2CH_2CCH_2CH_2PBu_2^t)}]^w$ | Red-brown | | $v(IrCl)$ 279, $\delta^{13}C$(carbene) 66.6 ppm |
| $cis$-$[NiCl_2L_2]^{ii}$ | Orange-yellow | 503 dec | |
| $[Ni(C_6Cl_5)\{C(NHMe)_2\}(PMe_2Ph)_2][PF_6]^{(20)}$ | Yellow-cream | 485–498 dec | |
| $[Ni(C_6Cl_5)\{C(NHMe)NMe_2\}(PMe_2Ph)_2][PF_6]^{(20)}$ | Yellow-cream | 474–477 dec | |
| $[Ni(C_6Cl_5)\{C(OMe)_2\}(PMe_2Ph)_2][PF_6]^{(20)}$ | Yellow | 449–451 dec | |
| $[\overline{Ni(CF_2CF_2CF_2CF_2)}\{C(NHBu^t)NMe_2\}CNBu^+]^x$ | | 424–425 | |
| $[Nicp\{C(NHCy)C_6H_4Cl\text{-}p\}CNCy][BF_4]^y$ | Dark-yellow | 424–425 | |
| $[NiClL_2'(NO)]^{(27b)}$ | Dark-blue | 357 dec | |
| $trans$-$[Pd_2Cl_4\{\overline{CC(Ph)CPh}\}_2]^{(5)}$ | Light-yellow | 488 dec | |
| $cis$-$[PdCl_2L(PEt_3)]^{(5)}$ | White | 518 dec | $v(PdCl)$ 309, 285 |
| $cis$-$[PdCl_2L''(PEt_3)]$***$^{(5)}$ | White | 548–563 | $v(PdCl)$ 302, 272, $v(PdP)$ 435 |
| $trans$-$[PdCl_2L''PEt_3]^{(5)}$ | Yellow | 470–473 | $v(PdCl)$ 351, 318, $v(PdP)$ 415 w |
| $[\{\overline{HON=C(Me)C_6H_4}\}PdClL'']^z$ | Light-yellow | 503–541 dec | $v(PdCl)$ 295 |
| $cis$-$[PtCl_2\{C(OMe)NHPh\}(PEt_3)]^{(5)}$ | White | 467–468 | $v(PtCl)$ 305, 275 |
| $trans$-$[PtMe\{C(OMe)Me\}(AsMe_3)_2]^+ PF_6^{-\,aa}$ | White | 443–445 | $v(Pt\text{-}C)$ 514 |
| $cis$-$[PtCl_2L(PEt_3)]^{(5)}$ | White | 553 dec | $v(PtCl)$ 312, 288 $v(PtP)$ 435 |
| $trans$-$[PtCl_2L(PEt_3)]^{(5)}$ | Yellow | 456–458 | $v(PtCl)$ 339, $v(PtP)$ 421 |
| $cis$-$[PtCl_2L''(PEt_3)]^{(5)}$ | White | 581–583 dec | $v(PtCl)$ 308, 277, $v(PtP)$ 439 |

*(Table continued)*

*Table 5.7. (continued)*

| Compound | Color | Mp (K) | Remarks |
|---|---|---|---|
| *trans*-[PtCl$_2$L″(PEt$_3$)][(5)] | Yellow | 474 | $\nu$(PtCl) 341, 326 sh, $\nu$(PtP) 422 |
| *trans*-[PtCl{C(NHMe)(NHC$_6$H$_4$Cl-*p*} (PEt$_3$)$_2$][ClO$_4$]$^{-bb}$ | White | 421–426 | $\nu$(PtCl) 303 |
| *cis*-[PtCl$_2${C(NHMe)$_2$}(PEt$_3$)]$^{bb}$ | White | | $\nu$(PtCl) 276 s, 297 s |
| [PtCl$_4${C(NHMe)$_2$}PEt$_3$]$^{bb}$ | Light-yellow | 430–433 | $\nu$(PtCl) 342 s, 297 s, 263 s |
| [Au{C(NHMe)NMe$_2$}$_2$][PF$_6$]$^{cc,dd}$ | White | | |
| [AuI$_2${C(NHMe)$_2$}$_2$][PF$_6$]$^{cc,dd}$ | Yellow | | |
| [AuBr$_2${C(p-C$_6$H$_4$Me)$_2$}][ClO$_4$] · $\frac{1}{2}$Et$_2$O$^{ee}$ | Yellow | 408 dec | |

* L = $\overline{\text{CN(Me)CH}_2\text{CH}_2\text{N}}$Me

** L′ = $\overline{\text{CN(Et)CH}_2\text{CH}_2\text{N}}$Et

*** L″ = $\overline{\text{CN(Ph)CH}_2\text{CH}_2\text{N}}$Ph

[a] See Ref. *k*, Table 5.2.
[b] R. R. Schrock and J. D. Fellman, *J. Am. Chem. Soc.*, **100**, 3359 (1978); R. R. Schrock, *J. Am. Chem. Soc.*, **96**, 6796 (1974).
[c] See Ref. *n*, Table 5.3.
[d] R. R. Schrock, *J. Am. Chem. Soc.*, **97**, 6577 (1975).
[e] See Ref. *j*, Table 5.3.
[f] See Ref. *e*, Table 5.2.
[g] C. W. Rees and E. von Angerer, *Chem. Commun.*, 420 (1972).
[h] See Ref. *d*, Table 5.2.
[i] C. G. Kreiter, K. Oefele, and G. W. Wieser, *Chem. Ber.*, **109**, 1749 (1976).
[j] E. O. Fischer and A. Maasboel, *Chem. Ber.*, **100**, 2445 (1967).
[k] E. O. Fischer and E. Offhaus, *Chem. Ber.*, **102**, 2449 (1969).
[l] E. O. Fischer and A. Riedel, *Chem. Ber.*, **101**, 156 (1968).
[m] See Ref. *b*, Table 5.2.
[n] M. L. H. Green, L. C. Mitchard, and M. G. Swanwick, *J. Chem. Soc. (A)*, 794 (1971).
[o] See Ref. *c*, Table 5.1.
[p] P. B. Hitchcock, M. F. Lappert, and P. L. Pye, *Chem. Commun.*, 644 (1976); *J. Chem. Soc., Dalton Trans.*, 826 (1978).
[q] T. J. Collins, K. R. Grundy, W. R. Roper, and S. F. Wong, *Organomet. Chem.*, **107**, C37 (1976).
[r] K. R. Grundy and W. R. Roper, *J. Organomet. Chem.*, **91**, C61 (1975); K. R. Grundy, R. O. Harris, and W. R. Roper, *J. Organomet. Chem.*, **90**, C34 (1975).
[s] M. Green, F. G. A. Stone, and M. Underhill, *J. Chem. Soc., Dalton Trans.*, 939 (1975).
[t] H. Felkin, B. Meunier, C. Pascard, and T. Prange, *J. Organomet. Chem.*, **135**, 361 (1977).
[u] E. P. Fischer, F. R. Kreissl, E. Winkler, and C. G. Kreiter, *Chem. Ber.*, **105**, 588 (1972).
[v] B. Cetinkaya, M. F. Lappert, and K. Turner, *Chem. Commun.*, 851 (1972).
[w] See Ref. *d*, Table 5.1.
[x] C. H. Davies, C. H. Game, M. Green, and F. G. A. Stone, *J. Chem. Soc., Dalton Trans.*, 357 (1974).
[y] Y. Yamamoto and H. Yamazaki, *Bull. Chem. Soc. Jpn.*, **48**, 3691 (1975).
[z] K. Hiraki, M. Onishi, K. Sewaki, and K. Sugino, *Bull. Chem. Soc. Jpn.*, **51**, 2548 (1978).
[aa] M. H. Chisholm, H. C. Clark, D. H. Hunter, *Chem. Commun.*, 809 (1971); M. H. Chisholm and H. C. Clark, *Chem. Commun.*, 763 (1970); *Inorg. Chem.*, **10**, 1711 (1971).
[bb] J. Chatt, R. L. Richards, and G. H. D. Royston, *J. Chem. Soc., Dalton Trans.*, 599 (1976).
[cc] G. Minghetti and F. Bonati, *J. Organomet. Chem.*, **54**, C62 (1973).
[dd] J. E. Parks and A. L. Balch, *J. Organomet. Chem.*, **57**, C103 (1973); **71**, 453 (1974).
[ee] G. Minghetti, F. Bonati, and G. Banditelli, *Inorg. Chem.*, **15**, 1718 (1976).
[ff] J. Chatt, G. J. Leigh, C. J. Pickett, and D. R. Stanley, *J. Organomet. Chem.*, **184**, C64 (1980).
[gg] K. P. Darst, P. G. Lenhert, C. M. Lukehart, and L. T. Warfield, *J. Organomet. Chem.*, **195**, 317 (1980).
[hh] See Ref. *m*, Table 5.8.
[ii] M. F. Lappert and P. L. Pye, *J. Chem. Soc., Dalton Trans.*, 2172 (1977).

## g. *Oxygen abstraction from acyl ligands*

Stable carbyne complexes of the type *trans*-$L_2(OC)_2XM \equiv CR$ [$M = Cr, Mo, W$; $R = Ph, Me$; $X = Cl, Br, CF_3COO$; $L_2 = (py)_2$, TMED] are formed from corresponding acyl complexes involving the following reagents: $COCl_2$, $C_2O_2Cl_2$, $ClC(O)OC-Cl_3$, $C_2O_2Br_2$, or $(CF_3CO)_2O$. Nitrogen containing ligands are added after the reaction has taken place.[64]

$$[NMe_4][OC)_5W-C(O)R] \xrightarrow[-NMe_4Cl, -CO_2, -CO]{CH_2Cl_2, COCl_2} [(OC)_4ClW \equiv CR] \tag{5.75}$$

$$R = Ph, 203\,K; R = Me, 183\,K$$

$$[(OC)_4ClWCR] \xrightarrow[-2CO]{py, 298\,K} [(py)_2(OC)_2ClWCR] \tag{5.76}$$

## h. *Protonation of isocyanide ligands*[65]

Electron-rich isocyanide complexes react with acids from which the addition of a proton to the nitrogen atom takes place leading to the formation of carbyne complexes. Additional $H^+$ ions may be bound by the metal atom or by the nitrogen atom.

$$\begin{array}{c} \textit{trans-}[M(CNMe)_2(dppe)_2] \xrightarrow[+CO_3^{2-}]{Et_2O, +HBF_4} \textit{trans-}[M(CNHMe)(CNMe)(dppe)_2]^+ \\ \Big\downarrow \text{solvent} \\ \textit{trans-}[MH(CNHMe)(CNMe)(dppe)_2]^{2+} \xleftarrow{HBF_4} [MH(CNMe)_2(dppe)_2]^+ \\ (M = Mo, W) \end{array} \tag{5.77}$$

## i. *Miscellaneous methods of synthesis of carbyne complexes*

The first anionic carbyne complex was prepared in 1985.[55] Until that time, only neutral and cationic carbyne complexes were known. The anionic complex is formed in the following reaction:

$$\textit{trans-}[WX(CNEt_2)(CO)_2L_2] + \textit{cis-}[Mo(PPh_2K)_2(CO)_4]$$

$$\xrightarrow[-L_2, -KX]{THF} K^+[Et_2NC \equiv W(CO)_2(\mu\text{-}PPh_2)_2Mo(CO)_4]^- \tag{5.78}$$

$$X = Br, I; L_2 = bipy, phen$$

Iron carbonyls $Fe_2(CO)_9$ and $Fe(CO)_5$ cleave the $C-Br$ bond of tribromofluoromethane to give the following carbyne complex.[66]

$$Fe_2(CO)_9 \xrightarrow[333-353\,K]{CFBr_3, \textit{n-}heptane} [Fe_3(\mu_3\text{-}CF)_2(CO)_9] \tag{5.79}$$

In many cases, complexes containing bridging carbyne ligands are formed via electrophilic addition to the oxygen atom of coordinated carbonyl group or to the carbon atom of carbide clusters [reactions (3.71)–(3.83), Section 3.10].

# 10. PROPERTIES OF CARBENE AND CARBYNE COMPLEXES

Among presently known carbene complexes of group 5–10 transition metals possessing electronic configuration $d^2$–$d^8$ and $d^{10}$, some are thermally and air stable. However, many of them are readily oxidized, and they decompose even at low temperatures. Exceptionally stable are some compounds containing chelating carbene ligands. Carbene complexes are usually neutral; nevertheless, many are cationic and few are anionic. Because the $M=CXY$ and $M\equiv CR$ bonds are greatly covalent, the majority of known Fischer-type compounds containing CXY and CR ligands have 18 valence

*Table 5.8. Carbyne Complexes*

| Compound | Color | $T_t$ (K) | Remarks |
|---|---|---|---|
| $[Ta(\eta\text{-}C_5Me_5)Cl(CCMe_3)(PMe_3)_2]^a$ | Light-yellow | | |
| $[TacpCl(CCMe_3)(PMe_3)_2]^a$ | Light-yellow | | |
| $[Li(Me_2piperazine)][Ta(CH_2CMe_3)_3CCMe_3]^b$ | Yellow-orange | | |
| $[Ta(\eta\text{-}C_5Me_5)Cl(CPh)(PMe_3)_2]^a$ | Orange | | |
| $[Cr(CO)_5(CNEt_2)][BF_4]^c$ | Dark-red | | |
| $[CrI(CO)_4(CPh)]^d$ | Yellow | | $\nu(CO)$ 2106 m, 2046 vs |
| $[CrI(CO)_4(C-\overline{C=CHCH_2CH_2CH_2})]^e$ | Yellow | 321 | |
| $[CrBr(CO)_4(C\text{-}SiPh_3)]^f$ | Yellow | 313 dec | |
| $[(OC)_5ReCr(CO)_4CPh]^g$ | Orange | | |
| $[(OC)_5ReMo(CO)_4CPh]^g$ | Orange | 358–360 | |
| $[Wcp(CO)_2(CMe)]^h$ | Orange | | |
| $[WCl(CO)_4(C\text{-}C_6H_5\text{-}\eta^6)Cr(CO)_3]^{(29)}$ | Light-red | 316 dec | |
| $[WCl(CO)_4(C\text{-}C_5H_4Rucp)]^{(28)}$ | Light-yellow | >253 dec | |
| $[W(CH_2CMe_3)(CHCMe_3)(CCMe_3)(DMPE)]^i$ | Yellow | | |
| $[W(CH_2CMe_3)(CHCMe_3)(CCMe_3)(PMe_3)_2]^i$ | Yellow | | |
| $[WBr(CO)_4(CSiMe_2Ph)]^f$ | Yellow | 314–315 dec | |
| $[Wcp(CO)(CSPh)(PPh_3)]^j$ | Orange | | $\nu(CO)$ 1886 |
| $[WBr(CO)_4(CC_5H_4Fecp)]^k$ | Red | >343 dec | |
| $[Mncp(CO)_2CPh][BCl_4]^m$ | Yellow | | $\nu(CO)$ 2088 vs, 2047 vs, $\delta^{13}C(Mn\equiv C)$ 356.9 |
| $[(Me_3P)(OC)_2Fe(\mu\text{-}SMe)_2Fe(CCF_3)(CO)(PMe_3)][BF_4]^l$ | Dark-red | 455 | $\nu(CO)$ 2042 vs, 2015 s, 1984 w |

$^a$ See Ref. *l*, Table 5.3.
$^b$ R. R. Schrock and J. D. Fellmann, *J. Am. Chem. Soc.*, **100**, 3359 (1978); L. J. Guggenberger and R. R. Schrock, *J. Am. Chem. Soc.*, **97**, 2935 (1975).
$^c$ See Ref. *e*, Table 5.2.
$^d$ E. O. Fischer, G. Kreis, C. G. Kreiter, J. Mueller, G. Huttner, and H. Lorenz, *Angew. Chem., Int. Ed. Engl.*, **12**, 564 (1973).
$^e$ E. O. Fischer, W. R. Wagner, F. R. Kreissl, and D. Neugebauer, *Chem. Ber.*, **112**, 1320 (1979).
$^f$ E. O. Fischer, H. Hollfelder, and F. R. Kreissl, *Chem. Ber.*, **112**, 2177 (1979).
$^g$ See Ref. *c*, Table 5.2.
$^h$ W. Vedelhoven, K. Eberl, and F. R. Kreissl, *Chem. Ber.*, **112**, 3376 (1979).
$^i$ M. R. Churchill and W. J. Youngs, *Chem. Commun.*, 321 (1979); *Inorg. Chem.*, **18**, 2454 (1979); D. N. Clark and R. R. Schrock, *J. Am. Chem. Soc.*, **100**, 6774 (1978).
$^j$ See Ref. *j*, Table 5.2.
$^k$ See Ref. *g*, Table 5.2.
$^l$ J. J. Bonnet, R. Mathieu, R. Poilblanc, and J. A. Ibers, *J. Am. Chem. Soc.*, **101**, 7487 (1979).
$^m$ E. O. Fischer, W. Schambeck, and F. R. Kreissl, *J. Organomet. Chem.*, **169**, C27 (1979); E. O. Fischer, E. W. Meineke, and F. R. Kreissl, *Chem. Ber.*, **110**, 1140 (1977).

electrons. Exceptions are paramagnetic carbene complexes of Cr(I) and Fe(I) possessing electronic configuration $d^5$ and $d^7$, respectively, as well as compounds containing $1e$ ligands or other ligands which do not show $\pi$ acceptor properties, and also square-planar $16e$ complexes which contain a $d^8$ central atom. In general, the Schrock carbene complexes are electron-deficient compounds. Niobium and tantalum even form $10e$ complexes such as $[M(CH_2CMe_3)_3(CHCMe_3)]$.

There are considerably fewer metal carbyne compounds than carbene compounds. Among them are compounds known to be relatively resistant to oxidation and thermally quite stable. Carbenes and carbynes constitute "soft" ligands. Therefore, they are found in complexes in which the central atom adopts a low oxidation state; nevertheless, carbene complexes of the following higher oxidation state metals are known: M(III) (M = Ta, Nb, W, Mo, Rh) and Pt(IV). Also known are carbyne complexes such as Mo(III), Mo(IV), W(III), Ta(III), etc. (Tables 5.7 and 5.8).

## 11. *REACTIONS OF CARBENE COMPLEXES*[1–7,37,38]

In Fischer complexes, the carbene carbon atom constitutes an electrophilic center. Therefore, this atom is readily attacked by nucleophiles. Probably, in many reactions, the addition of a nucleophile to the carbon carbene atom takes place. In the case of phosphines and amines, such intermediate addition products may be isolated and identified. In the subsequent step, the nucleophile may substitute the ligand bonded to the metal or the OR group of the carbene ligand. Therefore, the OR group may easily be displaced by $NR_2$, SR, SeR, etc. Therefore, Fischer[1–7] considers carbene complexes as ester-type compounds, $X = C(OR)R'$, in which the oxygen atom was replaced by $ML_n$, $X = ML_n$, instead of $X = O$. Substitution reactions of hydrogen atoms are also possible in alkyl radicals situated in the $\alpha$ position with respect to the carbene carbon atom. This is possible due to the acidity of these hydrogen atoms which is similar to the acidity of hydrogen atoms in nitromethane because of electron-accepting properties of the $ML_n$ group [for example, $M(CO)_5$]. The heteroatom of the carbene ligand $C(XR_n)R'$ in Fischer-type complexes may react with electrophiles which, after the $C-X$ bond breaking, often leads to the formation of carbyne complexes. Substitution reactions involving other ligands in carbene complexes or reactions of those ligands with electrophiles (or nucleophiles) represent yet another type of reaction which carbene complexes undergo. Also, migration rearrangement or dissociation of the carbene ligand from the complex may occur, leading to possible application in organic synthesis. In Schrock complexes, the carbene carbon atom has nucleophilic properties; thus, it readily reacts with electrophiles. Strong bases cause abstraction of the hydrogen atom bonded to the carbene carbon atom. This is a convenient method for synthesizing some carbyne complexes. As in the case of Fischer complexes, Schrock-type compounds may undergo substitution, deprotonation, and other such reactions involving other ligands.

### a. *Nucleophilic substitution reactions*

In Fischer-type alkylidene complexes, the carbene carbon atom very readily undergoes nucleophilic attack. Unlike the acyl complexes such as $Mn(COMe)(CO)_5$, in pentacarbonylcarbene complexes nucleophilic attack occurs at the carbene carbon atom and not at carbonyl ligands. This fact cannot be explained on the basis of positive charge distribution, because calculations as well as ESCA data show that carbonyl groups in these pentacarbonylcarbene complexes have greater positive charge than the carbene carbon atoms.[18a,b] The reactivity of carbenes toward nucleophiles is well explained by calculations of molecular orbital energies which show that, in $[Cr(CO)_5\{C(OMe)Me\}]$, the lowest unoccupied orbital (LUMO) is mainly localized on the carbene carbon atom. The following nucleophiles react with carbenes: $NH_3$, $RNH_2$, $R_2NH$, $RSH$, $R_2NNH_2$, $PhCH=NNH_2$, $R^-$, $Ar^-$, etc.

On the basis of kinetic and spectroscopic studies, mechanism (5.80) for nucleophilic

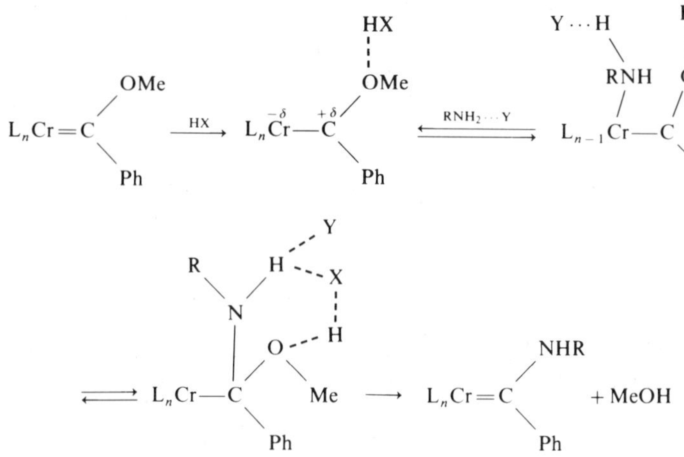

$$
\begin{array}{c}
L_nCr\!=\!C\!\!\begin{array}{c}\nearrow OMe\\ \searrow Ph\end{array}
\xrightarrow{HX}
L_nCr\!-\!C\!\!\begin{array}{c} \vdots HX \\ \nearrow OMe \\ \searrow Ph\end{array}
\underset{RNH_2\cdots Y}{\rightleftharpoons}
L_{n-1}Cr\!-\!C\!\!\begin{array}{c} Y\cdots H \quad HX \\[2pt] \searrow\; \vdots \\ RNH \;\; OMe \\ \nearrow \quad \nearrow \\ \\ \searrow Ph\end{array}
\end{array}
$$

$$
\rightleftharpoons \; L_nCr\!-\!C\!\!\begin{array}{c} R\;\; H\cdots_{\cdot\cdot}Y \\ \searrow \nearrow \quad \vdots X \\ N \qquad \vdots \\ | \qquad \vdots \\ O\text{---}H \\ \searrow \\ Me \\ \\ Ph \end{array}
\longrightarrow L_nCr\!=\!C\!\!\begin{array}{c}\nearrow NHR\\ \searrow Ph\end{array} + MeOH \qquad (5.80)
$$

substitution has been proposed, where HX is a donor and Y is an acceptor of a proton. The rate law equation for this reaction is

$$
\frac{d[L_nCrC(NHR)\,Ph]}{dt} = k[L_nCrC(OMe(Ph)][RNH_2][HX][Y] \qquad (5.81)
$$

For reactions carried out in saturated hydrocarbons, the amine itself must play the role of HX and Y, and therefore the rate law equation becomes

$$
\frac{d[L_nCrC(NH_2)\,Ph]}{dt} = k_1[L_nCrC(OMe)\,Ph][RNH_2]^3 \qquad (5.82)
$$

The formation of $\sigma$-carbyl complexes as an intermediate probably also occurs in reactions with other nucleophiles.

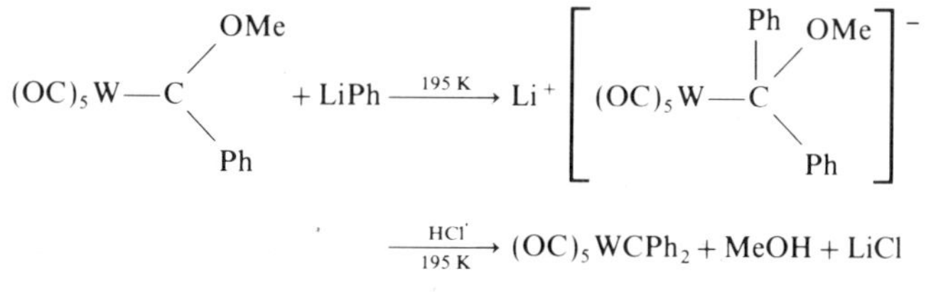

$$
(OC)_5W\!-\!C\!\!\begin{array}{c}\nearrow OMe\\ \searrow Ph\end{array} + LiPh \xrightarrow{195\,K} Li^+ \left[ (OC)_5W\!-\!C\!\!\begin{array}{c} Ph \quad OMe \\ \backslash \; / \\ \\ \searrow Ph\end{array} \right]^-
$$

$$
\xrightarrow[195\,K]{HCl} (OC)_5WCPh_2 + MeOH + LiCl \qquad (5.83)
$$

$$[(OC)_5MC(OMe)Ph] \xrightarrow{p\text{-}RC_6H_4S^-} (OC)_5M-\underset{\underset{Ph}{|}}{\overset{\overset{OMe}{|}}{C}}-SC_6H_4R$$

$$\xrightarrow{H^+} [(OC)_5MC(SC_6H_4R)Ph] + MeOH \qquad (5.84)$$

$$[Recp(CH_2)(NO)(PPh_3)]^+ + LiPh \longrightarrow [Recp(CH_2Ph)(NO)(PPh_3)] + Li^+ \qquad (5.85)$$

Compared to alkoxycarbene complexes, the siloxycarbene compounds behave differently. In siloxycarbene complexes, the nucleophilic substitution of the alkylsiloxy group takes place with great difficulty because the $d\pi$–$p\pi$ interaction between silicon and oxygen increases the donor properties of oxygen leading to the increase in stability of the carbon–oxygen bond.

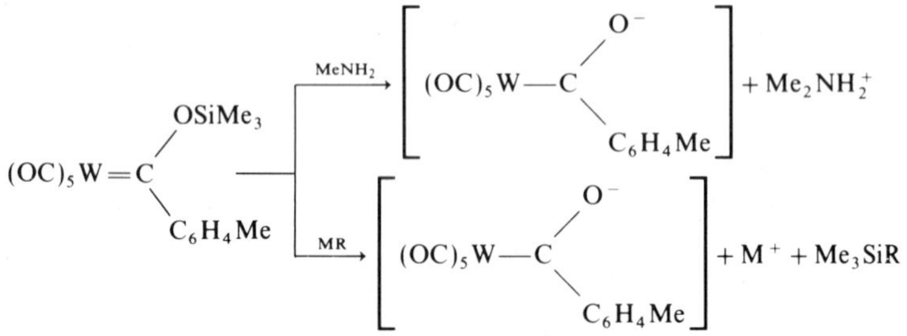

$$(OC)_5W=\underset{\underset{C_6H_4Me}{\diagdown}}{\overset{\diagup OSiMe_3}{}}
\begin{cases}
\xrightarrow{MeNH_2} & \left[ (OC)_5W-\underset{\underset{C_6H_4Me}{\diagdown}}{\overset{\diagup O^-}{}}C \right] + Me_2NH_2^+ \\[3em]
\xrightarrow{MR} & \left[ (OC)_5W-\underset{\underset{C_6H_4Me}{\diagdown}}{\overset{\diagup O^-}{}}C \right] + M^+ + Me_3SiR
\end{cases} \qquad (5.86)$$

Reactions of carbene complexes with tertiary amines, di-, or trialkylphosphines give ylid-type compounds:

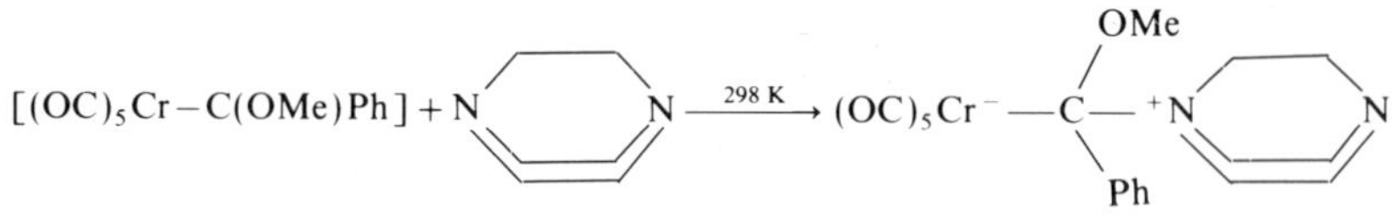

$$[(OC)_5Cr-C(OMe)Ph] + N\!\!\diagup\!\!\diagdown\!\!N \xrightarrow{298\ K} (OC)_5Cr^- -\underset{\underset{Ph}{|}}{\overset{\overset{OMe}{|}}{C}}-^+N\!\!\diagup\!\!\diagdown\!\!N \qquad (5.87)$$

$$[(OC)_5M-C(X)R] + PR'_3 \xrightleftharpoons{213\ K} (OC)_5M^- -\underset{\underset{R}{|}}{\overset{\overset{X}{|}}{C}}-P^+R'_3$$

$$(5.88)$$

$$(M = Cr, W; X = OMe, SMe, SeMe, OPh, OSiMe_3; R = Ph,$$
$$Me, p\text{-}C_6H_4Me; R'_3 = Me_3, Et_3, n\text{-}Bu_3, Me_2H)$$

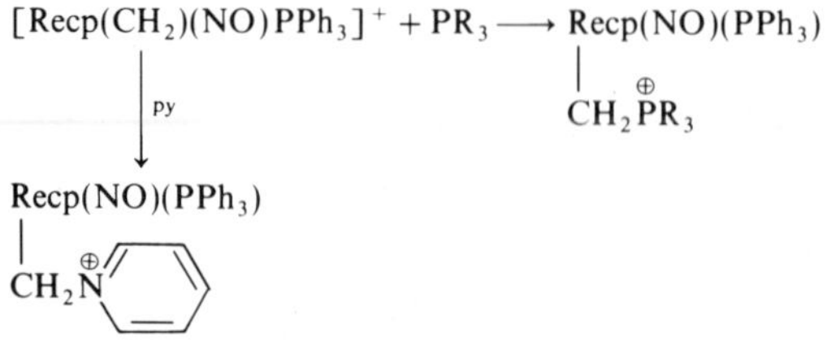

$$[Recp(CH_2)(NO)PPh_3]^+ + PR_3 \longrightarrow \underset{\underset{CH_2\overset{\oplus}{P}R_3}{|}}{Recp(NO)(PPh_3)}$$

$$\Big\downarrow py$$

$$\underset{\underset{CH_2\overset{\oplus}{N}\!\!\diagup\!\!\diagdown}{|}}{Recp(NO)(PPh_3)} \qquad (5.89)$$

## b. *Rearrangement reactions caused by addition*

In certain cases, carbene complexes and proton compounds give products in which the carbene ligand undergoes rearrangement and bonds through the heteroatom. It is assumed that, in the first step of the reaction, protonation of the heteroatom of the carbene ligand takes place followed by nucleophilic addition to the strongly electrophilic carbene carbon atom and hydrogen transfer to $C_{carb}$.

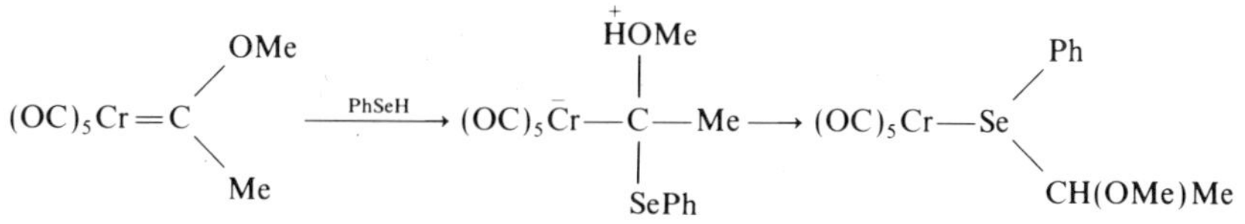

$$(OC)_5Cr=C \overset{OMe}{\underset{Me}{\diagdown}} \xrightarrow{\text{PhSeH}} (OC)_5\bar{Cr}-\overset{\overset{+}{H}OMe}{\underset{SePh}{C}}-Me \longrightarrow (OC)_5Cr-Se \overset{Ph}{\underset{CH(OMe)Me}{\diagup}} \qquad (5.90)$$

$$(OC)_5Cr=C \overset{OMe}{\underset{Ph}{\diagdown}} \xrightarrow{\text{Ca(CN)}_2} (OC)_5CrN\equiv CCH \overset{OMe}{\underset{Ph}{\diagup}} \qquad (5.91)$$

$$[(OC)_5CrC(SMe)Me] \xrightarrow[\text{243 K}]{\text{HBr}} (OC)_5Cr-S \overset{Me}{\underset{CH(Me)Br}{\diagup}} \qquad (5.92)$$

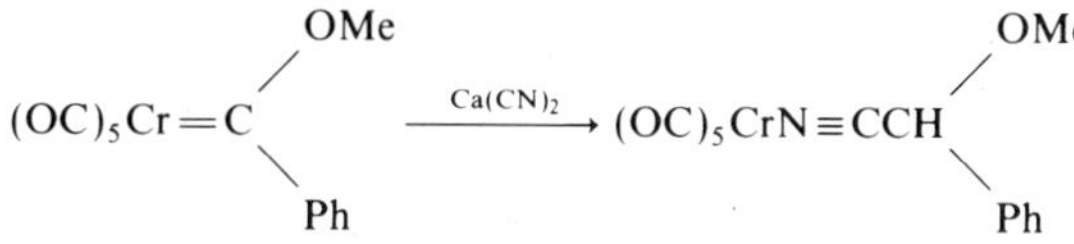

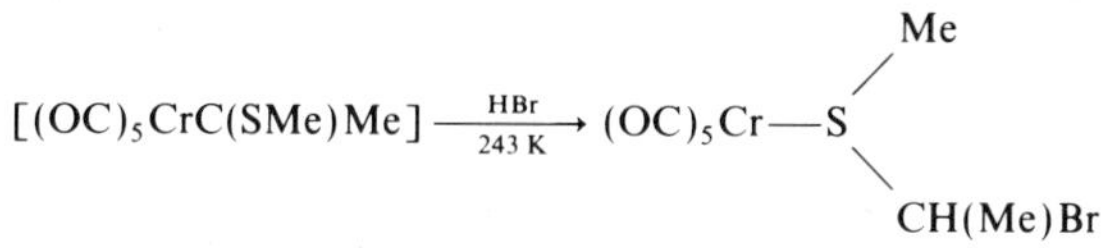

$$[(OC)_5CrC(NR_2)Ph] \xrightarrow[\text{233 K}]{\text{HX}} [(OC)_5Cr-\overset{R}{\underset{R}{N}}-CH(Ph)X] \longrightarrow [R_2\overset{+}{N}=CHPh][CrX(CO)_5]^- \qquad (5.93)$$

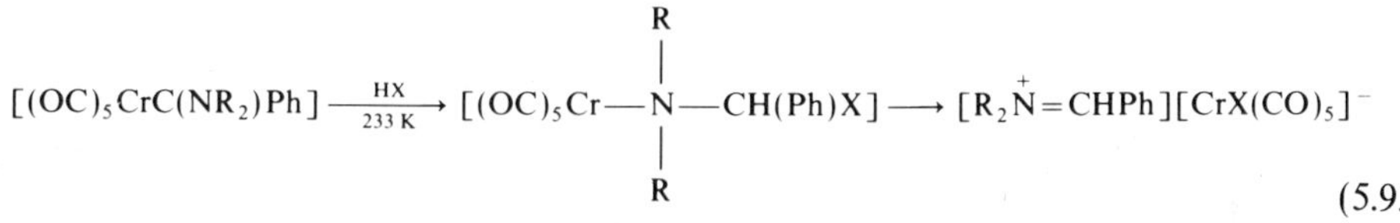

Reaction (5.93) involves decomposition of the ligand which is bonded to the metal through nitrogen to give ammonium salt.

Pentacarbonyl(methoxymethylcarbene)chromium, $(OC)_5CrC(OMe)Me$, reacts with cyclohexylisocyanide to form the aziridine complex which, when reacting with methanol, affords a carbene complex with the opening of the three-member ring.

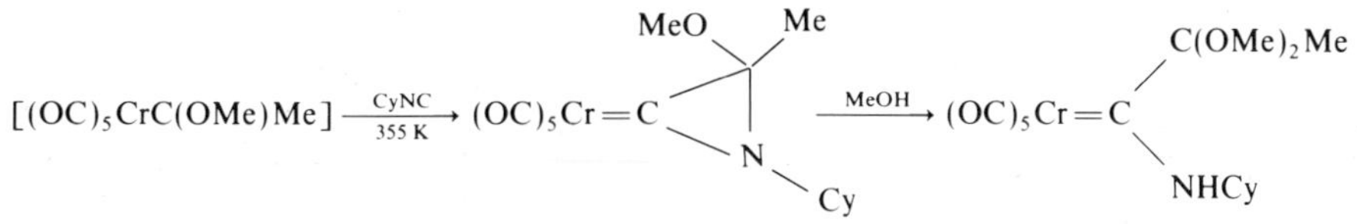

$$[(OC)_5CrC(OMe)Me] \xrightarrow[\text{355 K}]{\text{CyNC}} (OC)_5Cr=C\overset{MeO\diagdown \diagup Me}{\underset{N\diagdown Cy}{\bigtriangleup}} \xrightarrow{\text{MeOH}} (OC)_5Cr=C \overset{C(OMe)_2Me}{\underset{NHCy}{\diagup}} \qquad (5.94)$$

Fischer-type complexes react with ynamines to give carbene compounds, which formally are products of the carbene ligand migration; this process probably begins with nucleophilic attack of the ynamine on $C_{carb}$.

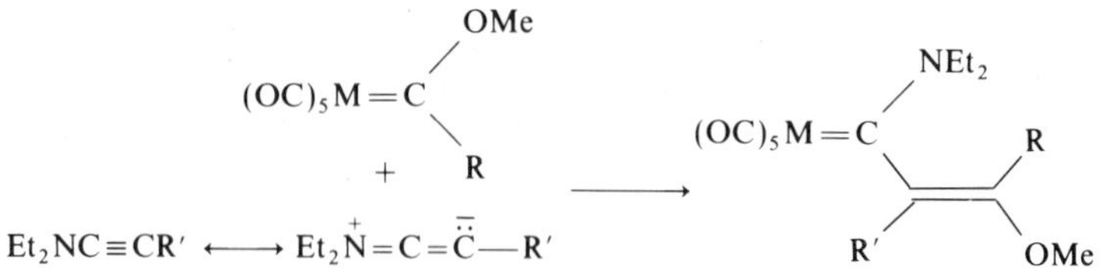

$$(OC)_5M{=}C\diagup^{OMe}_{\diagdown R}$$

$$+$$

$$Et_2NC{\equiv}CR' \longleftrightarrow Et_2\overset{+}{N}{=}C{=}\overset{..}{\underset{..}{C}}{-}R'$$

$$\longrightarrow \qquad (OC)_5M{=}C\diagup^{NEt_2}\diagup R \diagdown_{R'}\diagdown^{OMe}$$

$$(5.95)$$

$$(M = Cr, Mo, W; R = Ph, Me; R' = H, Me)$$

This reaction is stereospecific; the vinylcarbene complex possessing $E$ configuration is formed in which the R and R' are *trans* with respect to each other.

Also, insertion reaction product is formed when diphenylacetylene reacts with $[TacpCl_2(CHMe_3)]$:

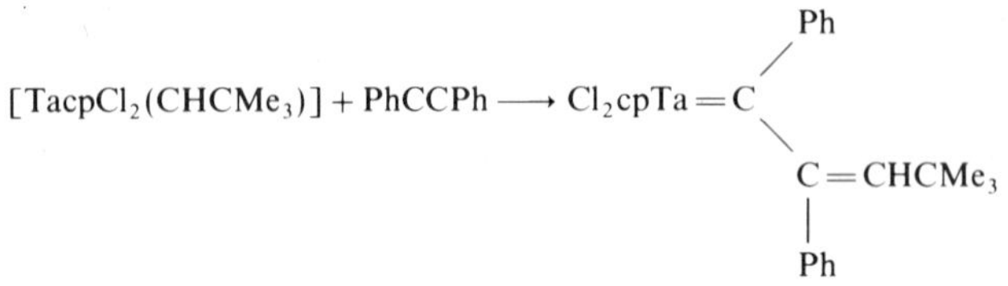

$$[TacpCl_2(CHMe_3)] + PhCCPh \longrightarrow Cl_2cpTa{=}C\diagup^{Ph}_{\diagdown}$$
$$C{=}CHCMe_3$$
$$|$$
$$Ph$$

$$(5.96)$$

### c. Hydrogen substitution at α-C

In MeOD solutions, in the presence of catalytic amounts of sodium methanolate, pentacarbonyl[methoxy(methyl)carbene]chromium(0) exchanges all hydrogen atoms of the methyl group for deuterium. Analogous molybdenum and tungsten complexes react in the same manner.

$$[(OC)_5MC(OMe)CH_3] + 3MeOD \xrightarrow[\text{NaOMe, cat.}]{} [(OC)_5MC(OMe)CD_3] + 3MeOH \quad (5.97)$$

Hydrogen atoms may also be replaced by other groups, for example, methyl groups may be introduced to the following carbene complex.

$$[(OC)_5CrC(OMe)CH_3] \xrightarrow[\text{2. Me}_3\text{OBF}_4]{\text{1. THF, 1 NaOMe}} [(OC)_5CrC(OMe)CH_nMe_{3-n}]$$
$$(n = 1, 2, 3)$$

$$(5.98)$$

Because of proton abstraction, anionic carbene ligands are formed as intermediates (Z denotes a base):

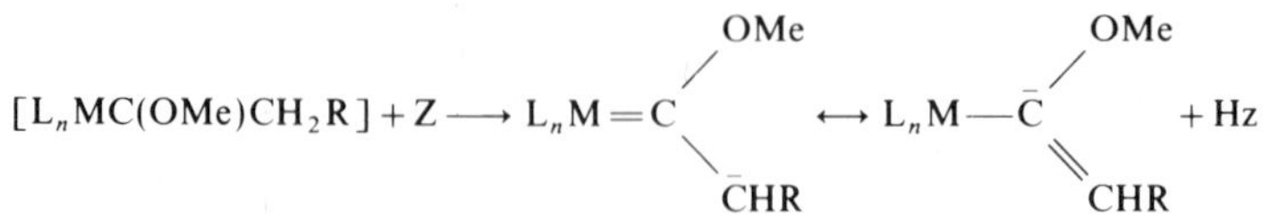

$$[L_nMC(OMe)CH_2R] + Z \longrightarrow L_nM{=}C\diagup^{OMe}_{\diagdown CHR} \longleftrightarrow L_nM{-}C\diagup^{OMe}_{\diagdown\!\!\!\backslash CHR} + Hz$$

$$(5.99)$$

### d. Reduction of carbene complexes

Reduction of carbene complexes of group 6 metals leads to their decomposition, or does not proceed at all because of unfavorable changes in oxidation states. Nevertheless, reduction reactions are known for carbene complexes of group 8 elements.

$$[(Ph_3P)(OC)\,cpFeC(OEt)\,Me]^+ \xrightarrow[\text{EtOH}]{\text{NaBH}_4} [(Ph_3P)(OC)\,cpFeCH(OEt)\,Me] \qquad (5.100)$$

$$[(DPPE)\,cpFeC(OMe)\,Ph]^+ \xrightarrow{\text{NaBH}_4} [(DPPE)\,cpFeCH(OMe)\,Ph] \qquad (5.101)$$

In these compounds, the carbene carbon atom is prochiral. However, during reduction, no asymmetric induction was observed; enantiomers are formed in equal quantities (the Fe atom may also be asymmetric).

The following iridium carbene complex may be reduced by hydrogen as recently described:

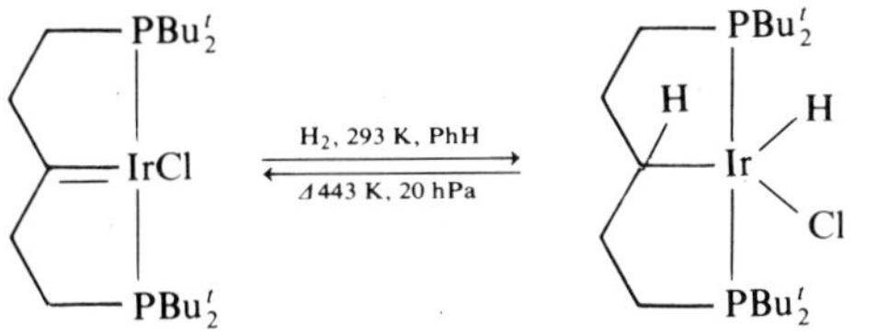

$$(5.102)$$

The reductive removal of carbene ligands by means of lithium aluminum hydride does not proceed in a satisfactory manner; good results may be obtained by hydrogen reduction.

$$[W(CO)_5CPh_2] \xrightarrow[\text{decaline, 373 K}]{6.9\ \text{MPa H}_2} Ph_2CH_2 + [W(CO)_6] \qquad (5.103)$$

$$[W(CO)_5C(OMe)Ph] \xrightarrow[413\ \text{K}]{0.18\ \text{MPa H}_2} PhCH_2OMe + PhMe + \underset{MeO}{\overset{Ph}{\diagdown}}C=C\underset{Ph}{\overset{OMe}{\diagup}} \qquad (5.104)$$

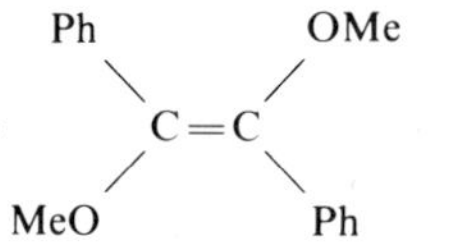

$$[Cr(CO)_5(\overline{CCH_2CH_2CH_2O})] \xrightarrow[443\ \text{K}]{6.9\ \text{MPa H}_2} \underset{O}{\text{\Pentagon}} + [Cr(CO)_6] \qquad (5.105)$$

### e. *Oxidation reactions*

Oxidation of carbene complexes leads either to their decomposition with evolution of organic molecules or to formation of paramagnetic carbene complexes:

$$[(OC)_5WCPh_2] \xrightarrow{(NH_4)_2[Ce(NO_3)_6]} Ph_2CO \qquad (5.106)$$

$$[(OC)_5WCPh_2] \xrightarrow[\text{or Me}_2SO]{O_2} Ph_2CO \qquad (5.107)$$

$$[(OC)_5CrC(OMe)Ph] \xrightarrow[341\ \text{K}]{O_2} PhCOOMe$$
$$\xrightarrow[308\ \text{K}]{S_8} PhCS(OMe) \qquad (5.108)$$
$$\xrightarrow[374\ \text{K}]{Se} PhCSe(OMe)$$

During preparation of paramagnetic carbene complexes of Fe(I) and Cr(I), silver salts are utilized as oxidizing reagents [reactions (5.51)–(5.53)].[18,19]

## f. Reactions with electrophiles

In contrast to reactions with nucleophiles, only few reactions of Fischer-type carbene complexes with electrophiles are known.

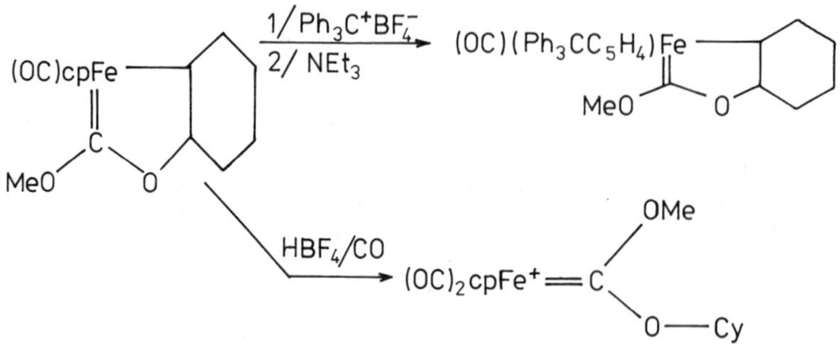

$$(5.109)$$

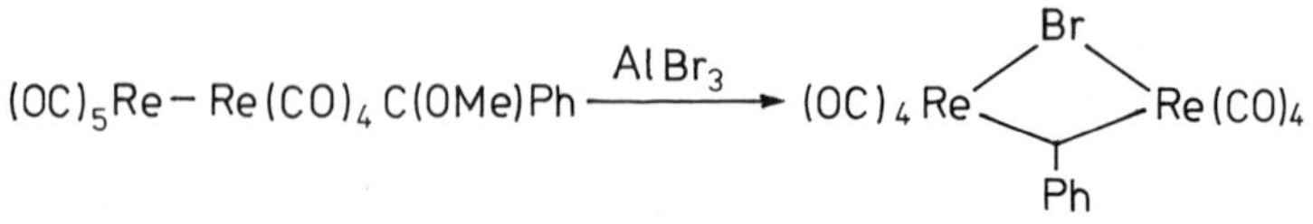

$$(5.110)$$

Reactions with electrophiles also give alkenylidenecarbene complexes:

$$(5.111)$$

$$(M = Cr, W)$$

Unlike Fischer-type compounds, alkylidene derivatives of niobium and tantalum have nucleophilic carbene carbon atoms. Therefore, they react with Lewis acids.

$$[Tacp_2(Me)\,CH_2] \xrightarrow{\text{AlMe}_3} [Tacp_2(Me)(CH_2AlMe)] \qquad (5.112)$$

$$[M(CH_2CMe_3)_3(CHCMe_3)] + HCl \longrightarrow [M(CH_2CMe_3)_4Cl] \qquad (5.113)$$

$$(M = Nb, Ta)$$

## g. Substitution reactions of noncarbene ligands

These reactions, like most substitution reactions, proceed readily. In 18e complexes, dissociation of a ligand usually occurs; thus, the substitution proceeds according to an $S_N1$ mechanism. These reactions are therefore sources of information concerning coordinately unsaturated carbene complexes. In carbene complexes of chromium, the dissociation of the carbonyl ligand occurs much more easily than in other compounds of the type $Cr(CO)_5L$. The rate of the reaction

$$[(OC)_5CrL] + PCy_3 \longrightarrow [(Cy_3P)(OC)_4CrL] + CO \qquad (5.114)$$

is 21,800 times greater for $L = C(OMe)$ than for $L = CO$.

Substitution reactions involving noncarbene ligands also occur for many other complexes.

$$\textit{trans-}[RuCl_2L_4] \xrightarrow{\ PF_3\ } \textit{trans-}[RuCl(PF_3)L_4]^+ \qquad (5.115)$$

$$\textit{trans-}[PtCl_2(PEt_3)L] \xrightarrow{\ LiMe\ } \textit{trans-}[PtMe_2(PEt_3)L] \qquad (5.116)$$

$$[MoBr_2(CO)_2L_2] \xrightarrow{\ NO\ } [MoBr_2(NO)_2L_2] \qquad (5.117)$$

$$[RhCl(PPh_3)_2L'] \xrightarrow{\ CO\ } [RhCl(CO)(PPh_3)L'] \qquad (5.118)$$

$$(L = \overline{CN(Me)\,CH_2CH_2N}Me, \ L' = \overline{CN(Ph)\,CH_2CH_2N}Ph)$$

Carbene complexes possessing metal atoms in low oxidation states undergo oxidative addition reactions:

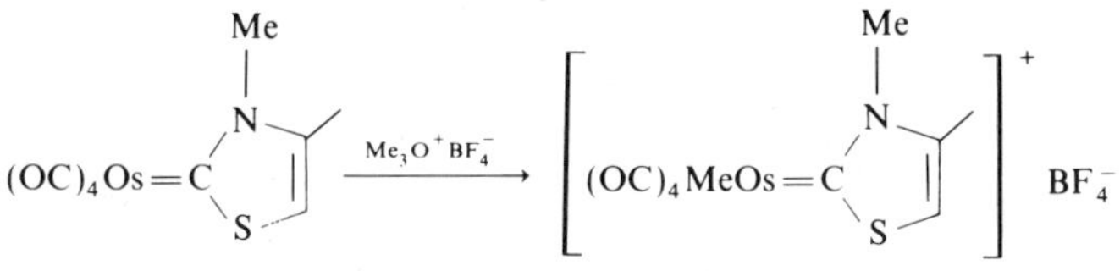

$$(5.119)$$

$$\textit{trans-}[Pt(Me)(PMe_2Ph)_2(\overline{CCH_2CH_2CH_2O})]^+$$

$$\xrightarrow{\ MeI\ } [PtMe_2I(PMe_2Ph)_2(\overline{CCH_2CH_2CH_2O})]^+ \qquad (5.120)$$

## h. *Release of carbene ligands*

In order to release carbene ligands and prepare a variety of organic compounds, the following methods may be utilized: oxidation [reactions (5.106)–(5.108)], reduction [reactions (5.100)–(5.105)], decomposition by acids and bases, thermal decomposition, as well as substitution of carbene ligands.

$$[W(CO)_5CPh(OMe) \xrightarrow[\leqslant 343\,K]{\ HX\ } [WX(H)(CO)_5] + PhCHO + \cdots \qquad (5.121)$$

$$(X = Cl, Br, I)$$

$$[Cr(CO)_5C(OR)R'] \xrightarrow{\ >403\,K\ } R'(RO)C = C(OR)R' + [Cr(CO)_6] + \cdots \qquad (5.122)$$

$$[W(CO)_5(CPh_2)] \xrightarrow{\ 373\,K\ } Ph_2C = CPh_2 + Ph_2 + Ph_2CH_2 + [W(CO)_6] + \cdots \qquad (5.123)$$

$$[Cr(CO)_5C(OMe)Me] + 2PH_3 \xrightarrow{\ h\nu\ } \textit{cis-}[Cr(CO)_4(PH_3)_2] + CH_2 = CHOMe \qquad (5.124)$$

$$[Cr(CO)_5C(OMe)Ph] + Ph_2SiH_2 \xrightarrow{328\,K} Ph_2HSiCH(OMe)Ph \qquad (5.125)$$

$$[Cr(CO)_5C(OMe)Ph] + MeOH \xrightarrow[373\,K]{17\,MPa\,CO} [Cr(CO)_6] + (MeO)_2CHPh \qquad (5.126)$$

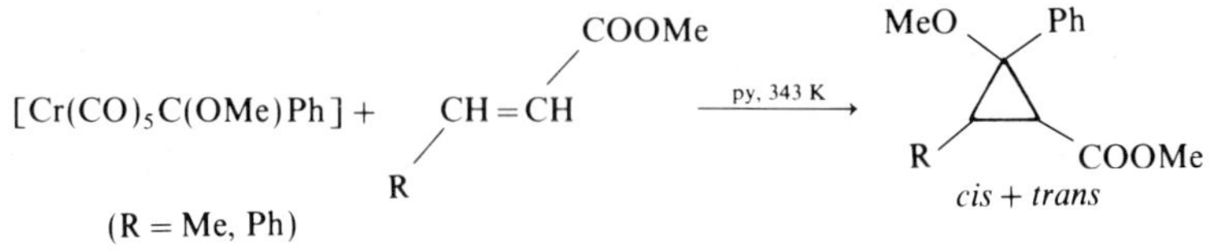

$$[Cr(CO)_5C(OMe)Ph] + \underset{R}{\overset{COOMe}{CH=CH}} \xrightarrow{py,\,343\,K} \underset{cis\,+\,trans}{\overset{MeO\quad Ph}{\triangle}} \qquad (5.127)$$

(R = Me, Ph)

$$[Cr(CO)_5C(OMe)Ph] + CH_2=CHOEt \longrightarrow CH_2=C(OMe)Ph + \cdots \qquad (5.128)$$

Fischer-type complexes react only with selected activated olefins. More reactive carbene complexes which do not contain heteroatoms in the alkylidene ligand react with a greater number of olefins.

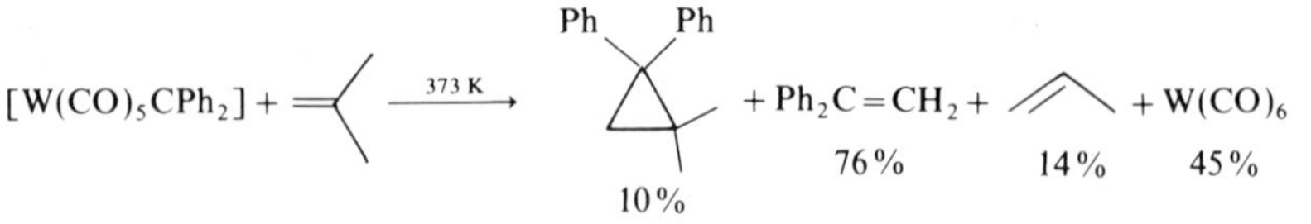

$$[W(CO)_5CPh_2] + {} \xrightarrow{373\,K} \underset{10\%}{\overset{Ph\quad Ph}{\triangle}} + Ph_2C=CH_2 + {} + W(CO)_6 \qquad (5.129)$$

76%   14%   45%

$$[(OC)_5W=CPh_2] + MeCH=CHMe \xrightarrow[4h]{323\,K} MeCH=CPh_2 + \cdots \qquad (5.130)$$

$$[Cl_2cpTa=CHCMe_3] + PhCH=CD_2 \longrightarrow PhCH=CDCHDCMe_3 \qquad (5.131)$$

Acetylenes also react with carbene compounds [cf. reaction (5.96)].[38,79]

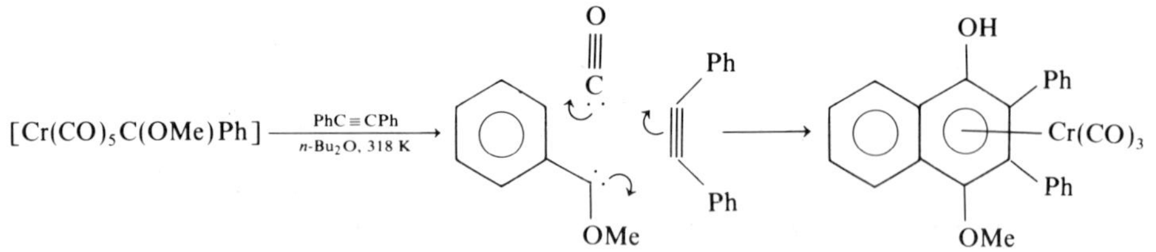

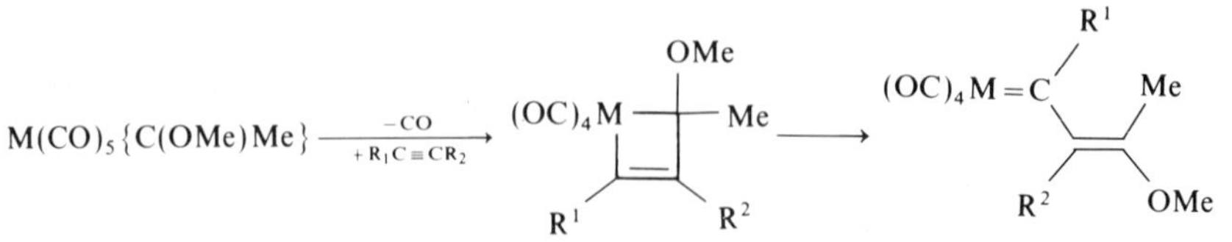

$$[Cr(CO)_5C(OMe)Ph] \xrightarrow[n\text{-}Bu_2O,\,318\,K]{PhC\equiv CPh} \cdots \qquad$$

$$M(CO)_5\{C(OMe)Me\} \xrightarrow[+\,R_1C\equiv CR_2]{-CO} \cdots \qquad (5.132)$$

$$\xrightarrow{R^1CCR^2} \cdots \qquad$$

M = Cr, W

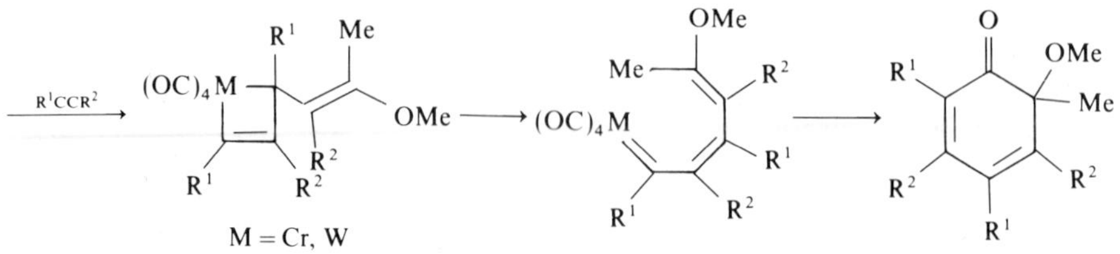

$$(5.133)$$

Reactions of acetylenes with metal carbenes may also lead to linear products.

$$[(OC)_5W=C(OMe)CH_2R^1] + PhC \equiv CR^2 \longrightarrow \quad (80)$$

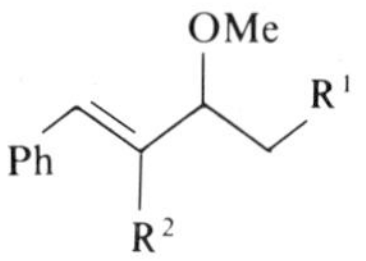

Nucleophilic alkylidene complexes of niobium and tantalum react with ketones and aldehydes to afford olefins.

$$[(Me_3CCH_2)_3M=CHCMe_3] + RCOR'$$

$$\xrightarrow[298K]{pentane} \; cis, \; trans\text{-}RR'C=CHCMe_3 + [(Me_3CCH_2)_3TaO]_n \qquad (5.134)$$

### i. *Migration reactions*

Migration of an alkyl group to the carbene ligand was observed in the case of cationic complexes[81] such as $[Wcp_2(CH_2R)(CH_2)]^+$. This process leads to the formation of $[Wcp_2(CH_2CH_2R)]^+$, which subsequently rearranges to the olefin–hydrido complex $[Wcp_2(CH_2=CHR)H]^+$.

In the case of $[Wcp_2Ph(CH_2)]^+PF_6^-$ in acetonitrile solution, the benzyl complex $[Wcp_2(CH_2Ph)(NCMe)]^+PF_6^-$ is formed because of phenyl migration. No olefin is formed in this reaction from the carbon ligand due to the lack of the hydrogen atom in $\beta$ position.[81]

One of the characteristic reactions of free carbenes is their insertion into the C−H bond. So far, examples of insertions into the C−H bonds of carbenes which are coordinated to the metal are not known. However, such carbenes may insert into more reactive M−H bonds where M = Si, Ge, or Sn.[38]

Diphenyldiazomethane reacts with excess of Ni(CO)$_4$ to give diphenylketene. It is assumed that a carbene complex is formed as an intermediate.

$$Ph_2CN_2 + [Ni(CO)_4] \xrightarrow[-N_2]{-CO} [(OC)_3NiCPh_2] \longrightarrow Ph_2C=C=O \qquad (5.135)$$

The reaction of carbon monoxide with free carbene is not known.

The complex $[Mncp(CO)_2(CPh_2)]$ reacts with carbon monoxide to form a diphenylketene compound.[24]

$$[Mncp(CO)_2(CPh_2)] + CO \xrightarrow{65\,MPa} Ph_2C-Mncp(CO)_2 \qquad (5.136)$$

Reactions (5.95), (5.96), and (5.132) may also be considered as migration reactions.

## 12. *REACTIONS OF CARBYNE COMPLEXES*[1,2a,2b,6,31,32,33]

### a. *Reactions with nucleophiles*

Fischer-type carbyne complexes react with nucleophiles to form metal carbene compounds.

$$[Cr(OC)_5(CNMe_2)]^+ + CN^- \longrightarrow [Cr(CO)_5\{C(CN)NMe_2\}] \qquad (5.137)$$

$$[Cr(CO)_5(CNEt_2)][BF_4] + [NBu_4]F$$

$$\xrightarrow[CH_2Cl_2, THF]{195K} [Cr(CO)_5C(NEt_2)F] + [NBu_4][BF_4] \qquad (5.138)$$

$$[Mncp(CO)_2(CPh)]^+[BCl_4]^- + MOR \longrightarrow [Mncp(CO)_2\{C(OR)Ph\}] + M(BCl_4) \quad (5.139)$$

$$(M = Li, Na, K; R = alkyl, aryl)$$

### b. *Reactions leading to bridging carbyne complexes*

Some carbyne compounds may give μ-carbyne complexes as a result of the existence of an equilibrium between monomeric and dimeric species or due to reactions with
other complexes. This makes it possible to prepare heteronuclear compounds.

$$2[W(CH_2CMe_3)_3(CCMe_3)] \longrightarrow [(Me_3CCH_2)_3W(\mu\text{-}CCMe_3)_2W(CH_2CMe_3)_3] \quad (5.140)$$

$$[(OC)_4BrW(CC_6H_4Me\text{-}p)$$

$$\xrightarrow[hv, CH_2Cl_2(Et_2O)]{F_2PN(Me)PF_2} [W_2(\mu\text{-}Br)(\mu\text{-}CC_6H_4Me\text{-}p)(\mu\text{-}F_2PNMePF_2)_2Br_2(CO)_2^{(31)} \quad (5.141)$$

$$[Wcp(CC_6H_4Me\text{-}p)(CO)_2] + [Pt(C_2H_4)(PMe_2Ph)_2]$$

$$\longrightarrow [PtW(\mu\text{-}CC_6H_4Me\text{-}p)cp(CO)_2(PMe_2Ph)_2]^{(32)} \qquad (5.142)$$

### c. *Reactions of carbyne complexes with CO and $CO_2$[(33)]*

Reactions of carbonyl carbyne complexes with phosphine ligands or CO lead to the
addition of a carbonyl group to the carbyne carbon atom affording ketenyl compounds
[scheme (5.143)]. As a result of dissociation of a phosphine or CO molecule, σ-ketenyl

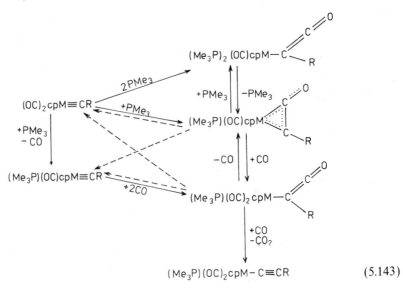

(5.143)

(---- reactions in mass spectrometer; M = Mo, W; R = Me, 1-cyclopentyl, Ph, $C_6H_4Me\text{-}p$, $SiPh_3$,
$C_5H_4Fecp$, $C_6H_2Me_3$-2, 4, 6)

complexes rearrange to $\pi$-ketenyl coordination compounds in which $RC{=}C{=}O$ is a $\eta^2$ three-electron ligand. The high $v(CO)$ value for the ketenyl ligand (2025–2041 cm$^{-1}$), higher than for carbonyl groups, as well as low values of $\delta$ $^{13}C$ for ketenyl carbon atoms, show that the most important contributing representation of the metal–ketenyl bond has the following ionic structure:

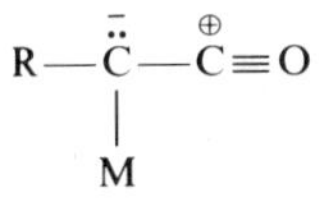

$$R-\overset{\displaystyle..}{\underset{\displaystyle |}{C}}-\overset{\displaystyle\oplus}{C}{\equiv}O$$
$$M$$

In the presence of bases MZ, a reaction occurs involving metal-coordinated CO, the carbyne ligand, and the corresponding anion:

$$Br(OC)_4Cr{\equiv}CC_6H_4Me\text{-}p + KOH \xrightarrow[\text{2. } H_2O/HCl, 293\ K]{\text{1. EtOH/}H_2O, 253\ K} p\text{-}MeC_6H_4CH_2COOH + \cdots \quad (5.144)$$

$$Br(OC)_4W{\equiv}CPh + LiPh \xrightarrow[\text{2. } H_2O/HCl, 293\ K]{\text{1. ether, 253 K}} PhCH_2COPh + \cdots \quad (5.145)$$

$$Br(OC)_4W{\equiv}CPh + NaOEt \xrightarrow[\text{2. } H_2O/HCl, 293\ K]{\text{1. EtOH, 253 K}} PhCH_2COOEt + \cdots \quad (5.146)$$

In these reactions, the carbyne carbon atom undergoes protonation. In complexes of the type $[(OC)_2cpW{\equiv}CR]$, the carbyne ligand also undergoes protonation when reacted with a solution of HCl in ether to give benzyl compounds, and subsequently acyl compounds, due to migratory insertion reaction.[82]

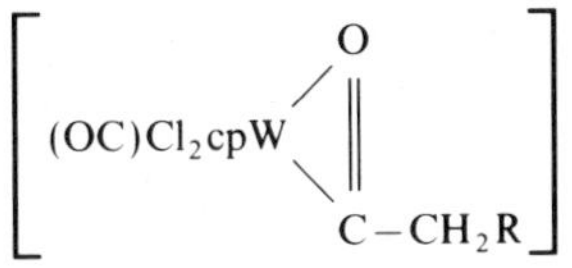

$(R = Me, p\text{-}C_6H_4Me)$

Carbon dioxide reacts with an anionic carbyne compound to furnish a cyclo-addition product.[55]

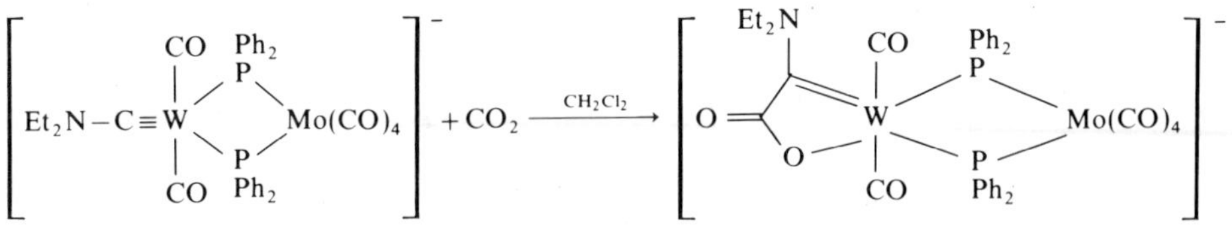

$$(5.147)$$

### d. Removal of carbyne ligands

In less reactive solvents at relatively high temperatures, the carbyne ligand dissociates off and undergoes dimerization to form alkynes.

$$[(OC)_4BrCr{\equiv}CPh] \xrightarrow[\text{303 K, 1.5h}]{\text{hexane}} PhC{\equiv}CPh + \cdots \quad (5.148)$$

$$[(OC)_4BrCr{\equiv}CMe] \xrightarrow[\substack{\text{1. 303 K, 2h}\\ \text{2. 323 K, 3h}}]{\text{n-heptane}} MeC{\equiv}CMe + \cdots \quad (5.149)$$

### e. Exchange reactions involving noncarbyne ligands

These reactions occur quite readily [reaction (5.143)]. The carbonyl groups much more easily undergo substitution reactions in 18e carbyne complexes of chromium group metals than in $M(CO)_6$:

$$[(OC)_5Cr\equiv CNMe_2]BF_4 \xrightarrow[\text{PEt}_3,\ 263\text{K}]{\text{Me}_2\text{CO},\ hv} trans\text{-}[Et_3P(OC)_4Cr\equiv CNMe_2]^+ + [BF_4]^- \quad (5.150)$$

$$[(OC)_5Cr\equiv CNMe_2][BF_4] \xrightarrow[\text{[NBu}_4\text{]I}]{\text{Me}_2\text{CO},\ 263\text{K}} trans\text{-}[(OC)_4ICr\equiv CNMe_2] \quad (5.151)$$

These reactions indicate strong *trans* effect of the carbyne ligand.

### f. Reactions with acids

The tantalum carbyne complex $[Ta(CH_2CMe_3)_3(CCMe_3)]$ reacts with $CF_3COOD$ to give $[Ta(CH_2CMe_3)_3(CDCMe_3)]$.

### g. Miscellaneous reactions

The triple bond $M\equiv C$ in *trans*-bromotetracarbonyl(phenylethynylcarbyne)tungsten(I) is less reactive than the $C\equiv C$ bond because the addition of diethylamine takes place only in the case of $C\equiv C$.

$$[(OC)_4BrW\equiv C-C\equiv CPh]$$

$$\xrightarrow[\text{233 K, Et}_2\text{O}]{\text{Me}_2\text{NH}} [(CO)_4BrW\equiv C-CH=C(NMe_2)Ph] \quad (5.152)$$

$[Mncp(CO)_2(CPh)]^+[BCl_4^-]$ reacts with *tert*-butylisonitrile to form a paramagnetic carbene complex (cf. Ref. *m*, Table 5.8):

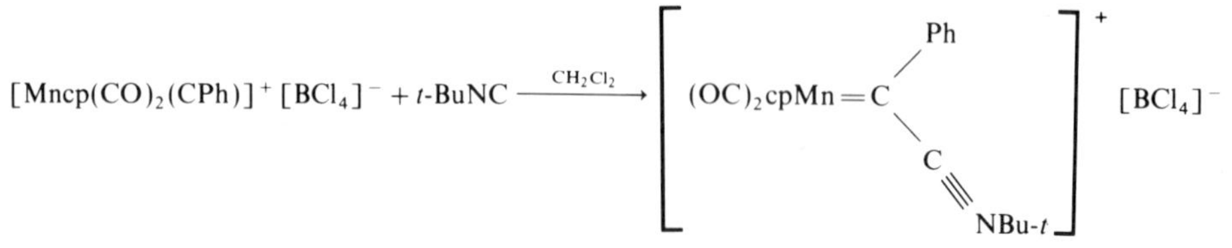

$$(5.153)$$

## 13. REACTIONS OF COMPLEXES CONTAINING METHYLENE BRIDGES

The $\mu$-methylene complexes of the type $L_n\overline{M(\mu\text{-}CH_2)}ML_n$ are quite stable. Photolytic and thermal stability of dimetallacyclopropane compounds is exceptional. For example, the complex $(OC)cp\overline{Rh(\mu\text{-}CH_2)}Rhcp(CO)$ does not decompose at reflux in benzene solution for two days; it also resists long-term UV irradiation. The methylene complex is more stable than the analogous carbonyl complex $(OC)cpRh(\mu\text{-}CO)Rhcp(CO)$. There are many examples showing that $\mu$-methylene compounds are at least as stable as their $\mu$-carbonyl analogs, e.g., stable $[\{Cocp(CO)\}_2(\mu\text{-}CH_2)]$; unstable $[\{Cocp(CO)\}_2(\mu\text{-}CO)]$; stable $[\{Rhcp(CO)\}_2 (\mu\text{-}CH_2)]$; photo- and thermally labile $[\{Rhcp(CO)\}_2(\mu\text{-}CO)]$. Despite their thermodynamic stability, the dimetallacyclopropane complexes readily react, especially with acids and unsaturated hydrocarbons. Complexes $[M_2cp_2(CO)_2(\mu\text{-}CHMe)(\mu\text{-}CO)]$ $(M = Fe, Ru)^{(73)}$ react with acetylenes upon irradiation to give insertion products into the $M-CHMe$ bond. Reaction products may be described as $\eta^3$-vinylcarbene or $\pi$-allyl complexes:

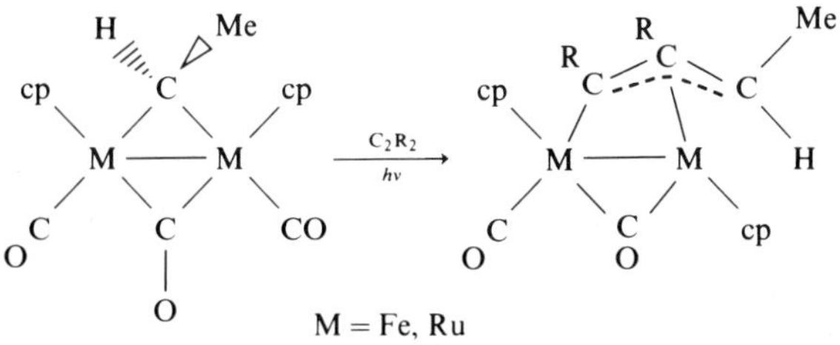

$$ (5.154) $$

$$ M = Fe, Ru $$

Tebbe's complex $[cp_2Ti(\mu\text{-}CH_2)(\mu\text{-}Cl) AlMe_2]$ readily reacts with alkenes and alkynes to form metallacyclobutanes (unstable) and metallacyclobutenes. It was found that there is a $CH_2$ exchange between metallacyclobutene and $^{13}CH_2=CMe_2$ [equation (5.155)].$^{(74)}$ The $\mu$-methylene compounds of the first series of the transition

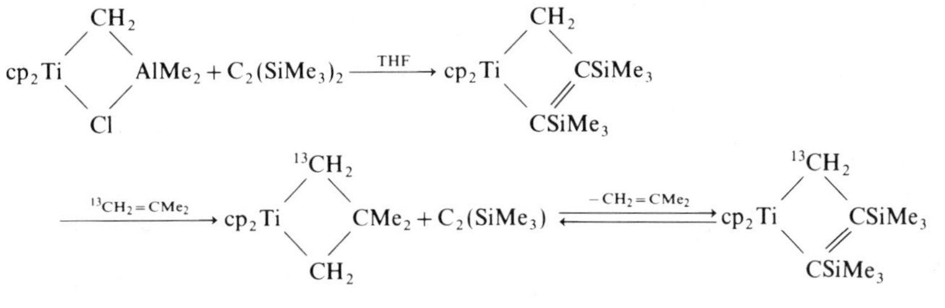

$$ (5.155) $$

metals are resistant toward protic acids (for example, $HBF_4$), while the complexes of the second and third series undergo interesting reactions with protons. The dinuclear dirhodacyclopropane $(OC)cpRh(\mu\text{-}CH_2)Rhcp(CO)$ reacts with $HBF_4$ in tetrahydrofuran, unexpectedly giving a trinuclear cluster containing $\mu_3$-carbyne bridges [equation (5.156)].$^{(75)}$ However, the ruthenium complex containing three methylene

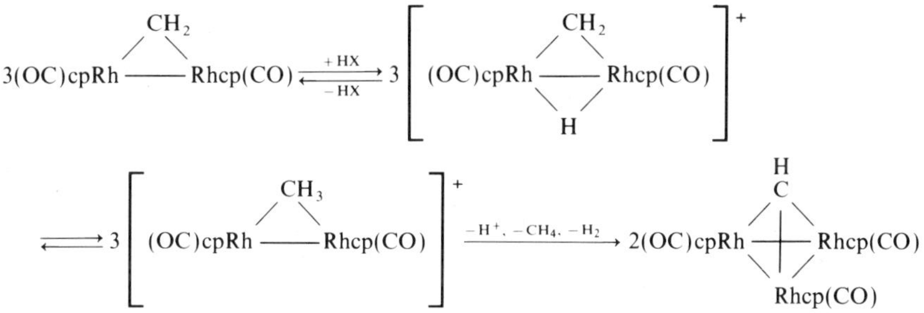

$$ (5.156) $$

bridges $(Me_3P)_3Ru(\mu\text{-}CH_2)_3Ru(PMe_3)_3$ reacts with acids according to expectations to give $\mu$-methyl complexes [equation (5.157)].[76] It is surprising that trityl tetrafluoro-

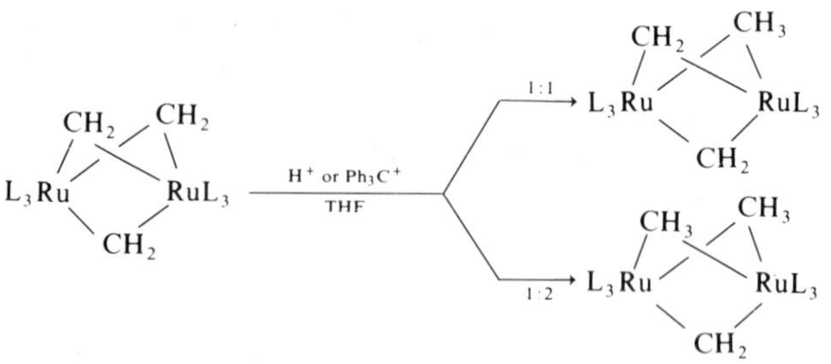

(5.157)

borate $[Ph_3C]^+[BF_4]^-$ does not abstract $H^-$ from the methylene groups of the complex in THF. Conversely, there is protonation of the $CH_2$ groups.

## 14. APPLICATIONS OF CARBENE AND CARBYNE COMPLEXES

The discussed reactions of carbene and carbyne complexes show that they have essential significance as catalysts or unstable transient intermediate compounds in such catalytic processes as metathesis of olefins and other unsaturated compounds, Fischer–Tropsch synthesis, syntheses of cyclopropanes from diazoalkanes and olefins, and polymerization of olefins and alkynes[77,78] as well as in organic synthesis. Except for alkynes [reaction (5.132)[38,79]] some compounds containing double bonds react with carbon monoxide and carbene ligands to form bonds with those groups. Examples of such compounds are enamines, ynamines, and Schiff bases. The N-vinylpyrrolidone (enamine), methoxyphenylcarbene, and excess of CO (higher pressure) react to furnish enaminoketone.

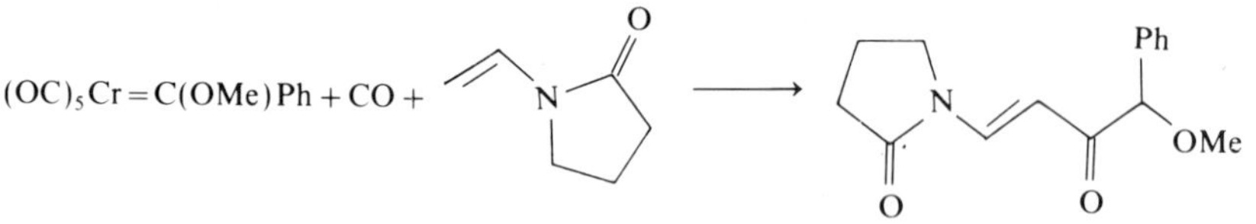

(5.158)

Schiff bases give linear as well as cyclic products in reactions of carbene with CO.

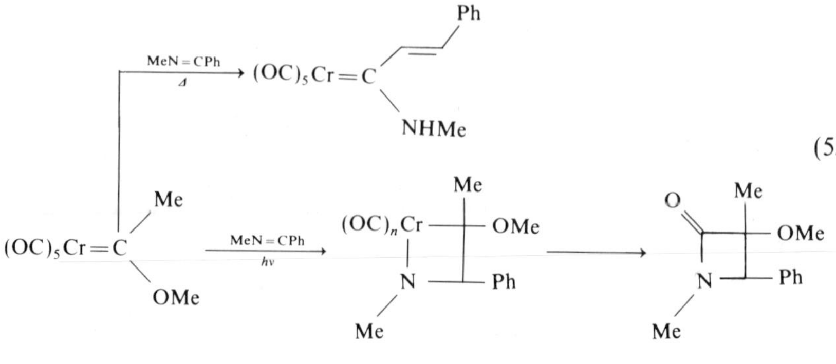

(5.159)

Perspectives of application of carbene complexes for natural product synthesis seem very promising. Coordination carbene compounds are utilized in the synthesis of peptides, vitamins K and E, and antibiotics.[38,79]

Owing to activation of $\alpha$ hydrogen atoms by heteroatoms, the enol forms of carbene complexes are relatively stable. Such carbonyl carbenes may react with carbonyl compounds to give aldols [equation (5.160)].[83] The reaction of the "enol" form with

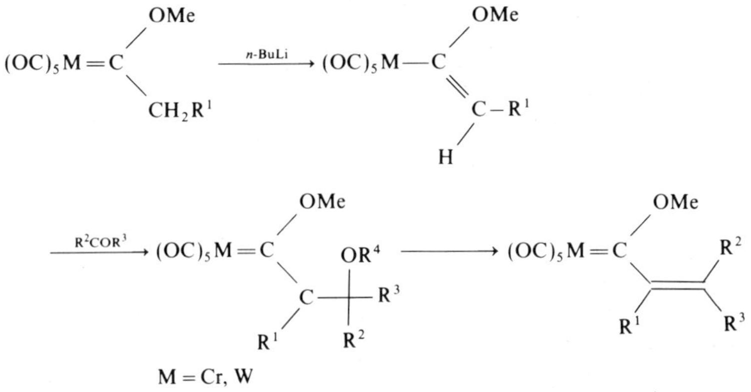

$$(OC)_5 M = C \begin{array}{c} OMe \\ \\ CH_2 R^1 \end{array} \xrightarrow{\text{n-BuLi}} (OC)_5 M - C \begin{array}{c} OMe \\ \\ C - R^1 \\ H \end{array}$$

$$\xrightarrow{R^2 COR^3} (OC)_5 M = C \begin{array}{c} OMe \\ \\ C \begin{array}{c} OR^4 \\ \\ R^3 \\ R^1 \quad R^2 \end{array} \end{array} \longrightarrow (OC)_5 M = C \begin{array}{c} OMe \\ \\ \\ R^1 \quad R^3 \end{array} \quad (5.160)$$

$$M = Cr, W$$

a carbonyl compound proceeds with high yield only in the presence of strong Lewis acids, because the enol forms of carbene complexes are relatively unreactive.

Metathesis reactions occurring between carbene complexes and compounds containing the C=N or N=N double bonds are interesting. Dicyclohexylcarbodiimide react with $(OC)_5 W = CPhR^1$ according to equation (5.161).[84]

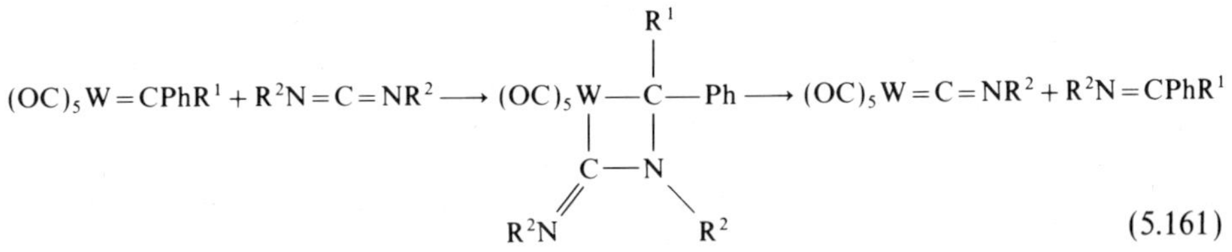

$$(OC)_5 W = CPhR^1 + R^2 N = C = NR^2 \longrightarrow (OC)_5 W - \overset{R^1}{\underset{\underset{R^2N}{C}}{\overset{|}{C}}} - Ph \longrightarrow (OC)_5 W = C = NR^2 + R^2 N = CPhR^1$$

$$(5.161)$$

In addition to metathesis products, photolytic reaction of azobenzenes with carbene complexes affords cyclic compounds such as 1,2-diazetidinone and 1,3-diazetidinone [equation (5.162)].[85] Photolytic reactions of imines with carbene com-

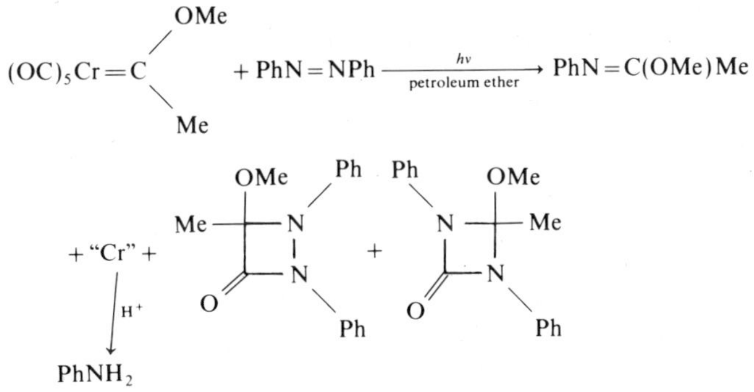

$$(OC)_5 Cr = C \begin{array}{c} OMe \\ \\ Me \end{array} + PhN = NPh \xrightarrow[\text{petroleum ether}]{hv} PhN = C(OMe)Me$$

$$+ \text{"Cr"} + \begin{array}{c} OMe \quad Ph \\ Me - \phantom{x} - N \\ \phantom{x} \phantom{x} | \\ O \phantom{x} N \\ \phantom{xxxx} Ph \end{array} + \begin{array}{c} Ph \quad OMe \\ N - \phantom{x} - Me \\ | \phantom{x} \phantom{x} \\ O \phantom{x} N \\ \phantom{xxxx} Ph \end{array} \quad (5.162)$$

$$\downarrow H^+$$

$$PhNH_2$$

plexes lead to $\beta$-lactams [equation (5.163)].[86] The CO molecules were incorporated into diazetidinones and $\beta$-lactams.

$$(OC)_5Cr = C \begin{array}{c} OMe \\ \diagdown \\ R^1 \end{array} + \begin{array}{c} R^3 \\ \diagup \\ C = N \\ \diagup \\ R^2 \end{array} R^4 \xrightarrow[\text{Et}_2\text{O}]{h\nu} \quad \text{(5.163)}$$

The compound $Nb_2Cl_6(SMe_2)_3$ containing a niobium–niobium double bond reacts readily with azobenzene to afford $[NbCl_2(Me_2S)(NC_6H_5)]_2(\mu\text{-Cl})_2$. This dinuclear molecule can be seen as the product of an inorganic double-bond metathesis reaction in which Nb=Nb and N=N bonds react to give two Nb=N bonds[87]:

$$\cdots + PhN = NPh \longrightarrow \cdots + R_2S \quad \text{(5.164)}$$

$$Nb = Nb + -N = N - \longrightarrow 2Nb = N-$$

## REFERENCES

1. E. O. Fischer, *Adv. Organomet. Chem.*, **14**, 1 (1976).
2. a. E. O. Fischer, U. Schubert, and H. Fischer, *Pure Appl. Chem.*, **50**, 857 (1978); b. E. O. Fischer and U. Schubert, *J. Organomet. Chem.*, **100**, 59 (1975).
3. F. J. Brown, *Prog. Inorg. Chem.*, **27**, 1 (1980).
4. D. J. Cardin, B. Cetinkaya, M. J. Doyle, and M. F. Lappert, *Chem. Soc. Rev.*, **2**, 99 (1973).
5. a. D. J. Cardin, B. Cetinkaya, and M. F. Lappert, *Chem. Rev.*, **72**, 545 (1972); b. F. A. Cotton and C. M. Lukenhart, *Prog. Inorg. Chem.*, **16**, 487 (1972).
6. R. R. Schrock, *Acc. Chem. Res.*, **12**, 98 (1979).
7. B. L. Shaw and N. I. Tucker, in: *Comprehensive Inorganic Chemistry* (J. C. Bailar, ed.), Vol. 4, p. 781, Pergamon, Oxford (1973).
8. P. Jutzi and W. Steiner, *Angew. Chem., Int. Ed. Engl.*, **15**, 684 (1976).
9. T. V. Ashworth, J. A. K. Howard, and F. G. A. Stone, *J. Chem. Soc., Dalton Trans.*, 1609 (1980).
10. R. R. Schrock, L. W. Messerle, C. D. Wood, and L. J. Guggenberger, *J. Am. Chem. Soc.*, **100**, 3793 (1978); L. J. Guggenberger and R. R. Schrock, *J. Am. Chem. Soc.*, **97**, 6578 (1975).
11. C. G. Kreiter, K. Oefele, and G. W. Wieser, *Chem. Ber.*, **109**, 1749 (1976).
12. E. O. Fischer, C. G. Kreiter, J. Kollmeier, J. Muller, and R. D. Fischer, *J. Organomet. Chem.* **28**, 237 (1971). b. E. O. Fischer, N.-Q. Dao, and W. R. Wagner, *Angew. Chem., Int. Ed. Engl.*, **17**, 50 (1978); E. O. Fischer, N.-Q. Dao, W. R. Wagner, and D. Neugebauer, *Chem. Ber.*, **112**, 2552 (1979). c. E. O. Fischer, C. Kappenstein, and N.-Q. Dao, *Proceedings*, XXI ICCC, p. 342, Tolouse, France (1980).
13. E. Moser and E. O. Fischer, *J. Organomet. Chem.*, **13**, 387 (1968).
14. M. H. Chisholm and S. Godleski, *Prog. Inorg. Chem.*, **20**, 299 (1976).
15. B. E. Mann, *Adv. Organomet. Chem.*, **12**, 135 (1974).
16. A. H. Cowley, *Prog. Inorg. Chem.*, **26**, 45 (1979).
17. C. Furlani and C. Cauletti, *Structure and Bonding*, **35**, 119 (1978).
18. a. M. F. Lappert, J. J. MacQuitty, and P. L. Pye, *Chem. Commun.*, 411 (1977). b. W. B. Perry, T. F.

Schaaf, and W. L. Jolly, *Inorg. Chem.*, **13**, 2038 (1974). c. T. F. Block, R. F. Fenske, and C. P. Casey, *J. Am. Chem. Soc.*, **98**, 441 (1976).

19. M. F. Lappert, R. W. McCabe, J. J. MacQuitty, P. L. Pye, and P. I. Riley, *J. Chem. Soc., Dalton Trans.*, 90 (1980).
20. M. Wada, S.-I. Kanai, R. Maeda, M. Kinoshita, and K. Oguro, *Inorg. Chem.*, **18**, 417 (1979).
21. E. O. Fischer, C. G. Kreiter, H. J. Kollmeier, J. Muller, and R. D. Fischer, *J. Organomet. Chem.*, **22**, C 39, (1970).
22. J. D. Fellmann, G. A. Ruprecht, C. D. Wood, and R. R. Schrock, *J. Am. Chem. Soc.*, **100**, 5964 (1978).
23. W. Mowat and G. Wilkinson, *J. Chem. Soc., Dalton Trans.*, 1120 (1973).
24. W. A. Herrmann and J. Plank, *Angew. Chem., Int. Ed. Engl.*, **17**, 525 (1978); W. A. Herrmann, *Chem. Ber.*, **108**, 486 (1975).
25. M. Brookhart and G. O. Nelson, *J. Am. Chem. Soc.*, **99**, 6099 (1977).
26. D. Mansuy, M. Lange, J. C. Chottard, J. F. Bartoli, B. Chevrier, and R. Weiss, *Angew. Chem., Int. Ed. Engl.*, **17**, 781 (1978); D. Mansuy, M. Lange, J. C. Chottard, P. Guerin, P. Morliere, D. Brault, and M. Rougee, *Chem. Commun.*, 648 (1977).
27. a. A. J. Hartshorn, M. F. Lappert, and K. Turner, *J. Chem. Soc., Dalton. Trans.*, 348 (1978). b. M. F. Lappert and P. L. Pye, *J. Chem. Soc., Dalton Trans.*, 837 (1978).
28. E. O. Fischer, F. J. Gammel, J. O. Besenhard, A. Frank, and D. Neugebauer, *J. Organomet. Chem.*, **191**, 261 (1980).
29. E. O. Fischer, F. J. Gammel, and D. Neugebauer, *Chem. Ber.*, **113**, 1010 (1980).
30. E. O. Fischer, H. Fischer, U. Schubert, and R. B. A. Pardy, *Angew. Chem., Int. Ed. Engl.*, **18**, 871 (1979).
31. E. O. Fischer, W. Kellerer, B. Zimmer-Gasser, and U. Schubert, *J. Organomet. Chem.*, **199**, C 24 (1980).
32. T. V. Ashworth, J. A. K. Howard, and F. G. A. Stone, *J. Chem. Soc., Dalton Trans.*, 1609 (1980).
33. W. Uedelhoven, K. Eberl, and F. R. Kreissl, *Chem. Ber.*, **112**, 3376 (1979).
34. A. A. Bagaturianc and W. B. Kazanskij, *Khimicheskaya Svyaz i Stroyeniye Molekul* (V. I. Nefedow, ed.), p. 51, Nauka, Moscow (1984).
35. D. M. P. Mingos, in: *Comprehensive Organometallic Chemistry* (G. Wilkinson, F. G. A. Stone, and E. W. Abel, eds.), Vol. 3, Chapter 19, Pergamon, Oxford (1982).
36. U. Schubert, *Coord. Chem. Rev.*, **55**, 261 (1984).
37. A. I. Svatkovskii and B. D. Babitski, *Usp. Khim.*, **53**, 1152 (1984).
38. K. H. Doetz, *Angew. Chem., Int. Ed. Engl.*, **23**, 587 (1984).
39. R. J. Goddard, R. Hoffmann, and E. D. Jemmis, *J. Am. Chem. Soc.*, **102**, 7667 (1980).
40. J. Ushio, H. Nakatsuji, and T. Yoneszawa, *J. Am. Chem. Soc.*, **106**, 5892 (1984).
41. A. R. Gregory and E. A. Mintz, *J. Am. Chem. Soc.*, **107**, 2179 (1985).
42. W. A. Herrmann, *Adv. Organomet. Chem.*, **20**, 159 (1982).
43. B. E. Mann, in: *Comprehensive Organometallic Chemistry* (G. Wilkinson, F. G. A. Stone, and E. W. Abel, eds.), Vol. 3, Chapter 20, Pergamon, Oxford (1982).
44. A. R. Pinhas, T. A. Albright, P. Hofmann, and R. Hoffmann, *Helv. Chim. Acta*, **63**, 29 (1980).
45. D. C. Calabro, D. L. Lichtenberger, and W. A. Herrmann, *J. Am. Chem. Soc.*, **103**, 6852 (1981).
46. J. C. Green, *Structure and Bonding*, **43**, 37 (1981).
47. J. Evans and G. S. McNulty, *J. Chem. Soc., Dalton Trans.*, 79 (1984); J. Evans and G. S. McNulty, *J. Chem. Soc., Dalton Trans.*, 639 (1983).
48. U. Plaia, H. Stolzenberg, and W. P. Fehlhammer, *J. Am. Chem. Soc.*, **107**, 2171 (1985).
49. W. P. Fehlhammer, K. Bartel, U. Plaia, A. Voelkel, and A. T. Liu, *Chem. Ber.*, **118**, 2235 (1985).
50. C. P. Casey, W. H. Miles, and H. Tukada, *J. Am. Chem. Soc.*, **107**, 2924 (1985).
51. J. P. Collman, P. J. Brothers, L. M. Elwee-White, E. Rose, and L. J. Wright, *J. Am. Chem. Soc.*, **107**, 4570 (1985).
52. J. Keijsper, L. H. Polm, G. van Koten, K. Vrieze, K. Goubitz, and C. H. Stam, *Organometallics*, **4**, 1876 (1985).
53. H. Fischer and E. O. Fischer, *J. Mol. Catal.*, **28**, 85 (1985).
54. H. Fischer, A. Motsch, R. Markl, and K. Ackermann, *Organometallics*, **4**, 726 (1985).
55. E. O. Fischer, A. C. Filippou, H. G. Alt, and U. Thewalt, *Angew. Chem., Int. Ed. Engl.*, **24**, 203 (1985).

56. M. M. Singh and R. J. Angelici, *Inorg. Chem.*, **23**, 2691 (1984); *Inorg. Chem.*, **23**, 2699 (1984); *Inorg. Chim. Acta*, **100**, 57 (1985).
57. R. R. Schrock, M. L. Listemann, and L. G. Sturgeoff, *J. Am. Chem. Soc.*, **104**, 4291 (1982).
58. H. Strutz and R. R. Schrock, *Organometallics*, **3**, 1600 (1984).
59. M. L. Listemann and R. R. Schrock, *Organometallics*, **4**, 74 (1985).
60. W. K. Wong, W. Tam, C. E. Strouse, and J. A. Gladysz, *Chem. Commun.*, 530 (1979); W. K. Wong, W. Tam, and W. E. Strouse, *J. Am. Chem. Soc.*, **101**, 5440 (1979); J. A. Gladysz and W. Tam, *J. Am. Chem. Soc.*, **100**, 2545 (1978).
61. F. A. Cotton, W. Schwotzer, and E. S. Shamshoum, *J. Am. Chem. Soc.*, **2**, 1167 (1983); F. A. Cotton, W. Schwotzer, and E. S. Shamshoum, *J. Am. Chem. Soc.*, **2**, 1340 (1983).
62. M. H. Chisholm, K. Folting, D. M. Hoffman, and J. C. Huffman, *J. Am. Chem. Soc.*, **106**, 6794 (1984).
63. M. H. Chisholm, D. M. Hoffman, and J. C. Huffman, *J. Am. Chem. Soc.*, **106**, 6806 (1984).
64. A. Mayr, G. A. McDermott, and A. M. Dorries, *Organometallics*, **4**, 608 (1985).
65. A. J. L. Pombeiro and R. L. Richards, *Transition Met. Chem.*, **5**, 55 (1980).
66. D. Lentz, I. Brudgam, and H. Hartl, *Angew. Chem., Int. Ed. Engl.*, **24**, 119 (1985).
67. J. A. Roth and M. Orchin, *J. Organomet. Chem.*, **172**, C 27 (1979).
68. Y.-F. Yu, C.-N. Chau, A. Wojcicki, M. Calligaris, G. Nardin, and G. Balducci, *J. Am. Chem. Soc.*, **106**, 3704 (1984).
69. P. R. Sharp, D. Astruc, and R. R. Schrock, *J. Organomet. Chem.*, **182**, 477 (1979).
70. H. Schumann and J. Mueller, *J. Organomet. Chem.*, **169**, C 1 (1979).
71. T. F. Block and R. F. Fenske, *J. Am. Chem. Soc.*, **99**, 4321 (1977).
72. M. C. Boehm, J. Daub, R. Gleiter, P. Hofmann, and K. Oefele, *Chem. Ber.*, **113**, 3629 (1980).
73. A. F. Dyke, S. A. R. Knox, P. J. Naish, and G. E. Taylor, *J. Chem. Soc., Chem. Commun.*, 803 (1980).
74. F. N. Tebbe, *J. Am. Chem. Soc.*, **100**, 3611 (1978); F. N. Tebbe, G. W. Parshall, and G. S. Reddy, *J. Am. Chem. Soc.*, **100**, 3611 (1978); F. N. Tebbe and R. L. Harlow, *J. Am. Chem. Soc.*, **102**, 6149 (1980).
75. W. A. Herrman, J. Plank, E. Guggolz, and M. L. Ziegler, *Angew. Chem., Int. Ed. Engl.*, **19**, 651 (1980); W. A. Herrmann, J. Plank, M. L. Ziegler, and B. Balbach, *J. Am. Chem. Soc.*, **102**, 5906 (1980); W. A. Herrmann, J. Plank, D. Riedel, M. L. Ziegler, K. Weidenhammer, E. Guggolz, and B. Balbach, *J. Am. Chem. Soc.*, **103**, 63 (1981).
76. R. A. Anderson, R. A. Jones, G. Wilkinson, M. B. Hursthouse, and K. M. A. Malik, *J. Chem. Soc., Chem. Commun.*, 865 (1977); M. B. Hursthouse, R. A. Jones, K. M. A. Malik, and G. Wilkinson, *J. Am. Chem. Soc.*, **101**, 4128 (1979); R. A. Jones, G. Wilkinson, A. M. R. Galas, M. B. Hursthouse, and K. M. A. Malik, *J. Chem. Soc., Dalton Trans.*, 1771 (1980).
77. B. A. Dolgoplosk and Ju. W. Korshak, *Usp. Khim.*, **53**, 65 (1984); B. A. Dolgoplosk and E. I. Tinyakova, in: *Metalloorganicheskii Kataliz v Processakh Polimerizatsii*, Izd. Nauka, Moscow (1982).
78. K. Weiss, in: *Transition Metal Carbene Complexes* ( K. H. Doetz, ed.), Verlag Chemie, Weinheim (1983).
79. P.-C. Tang and W. D. Wulff, *J. Am. Chem. Soc.*, **106**, 1132 (1984); W. D. Wulff and D. C. Yang, *J. Am. Chem. Soc.*, **106**, 7565 (1984); W. D. Wulff and P.-C. Tang, *J. Am. Chem. Soc.*, **106**, 434, (1984); W. D. Wulff, R. W. Kaesler, G. A. Peterson, and P.-C. Tang, *J. Am. Chem. Soc.*, **107**, 1060 (1985).
80. D. W. Macomber, *Organometallics*, **3**, 1589 (1984).
81. P. Jernakoff and N. J. Cooper, *J. Am. Chem. Soc.*, **106**, 3026 (1984); J. C. Hayes, G. D. N. Pearson, and N. J. Cooper, *J. Am. Chem. Soc.*, **103**, 4648 (1981); J. C. Hayes and N. J. Cooper, *J. Am. Chem. Soc.*, **104**, 5570 (1982).
82. F. R. Kreissl, W. J. Sieber, M. Wolfgruber, and J. Riede, *Angew. Chem., Int. Ed. Engl.*, **23**, 640 (1984).
83. W. D. Wulff and S. R. Gilbertson, *J. Am. Chem. Soc.*, **107**, 503 (1985).
84. K. Weiss and P. Kindl, *Angew. Chem., Int. Ed. Engl.*, **23**, 629 (1984).
85. L. S. Hegedus and A. Kramer, *Organometallics*, **3**, 1263 (1984).
86. M. A. McGuire and L. S. Hegedus, *J. Am. Chem. Soc.*, **104**, 5538 (1982); L. S. Hegedus, M. A. McGuire, L. M. Schultze, C. Yijun, and O. P. Anderson, *J. Am. Chem. Soc.*, **106**, 2680 (1984).
87. F. A. Cotton, S. A. Duraj, and W. J. Roth, *J. Am. Chem. Soc.*, **106**, 4749 (1984).

*Chapter 6*

# Compounds Containing Two-Electron π-Ligands

## A. *OLEFIN COMPLEXES*

### 1. *BONDING AND STRUCTURE*

The first olefin complex, $K[PtCl_3C_2H_4]$, was obtained in 1827 by the Danish chemist W. O. Zeise by the reaction of $K_2[PtCl_4]$ with aqueous ethyl alcohol. However, the metal–olefin bond was correctly described only in 1951 by Dewar,[1] and subsequently by Chatt and Duncanson.[2]

A π-bonding olefin orbital forms a bond with $ns$, $np$, $(n-1)\,d_{z^2}$, and $(n-1)\,d_{x^2-y^2}$ orbitals or with corresponding hybridized orbitals which are formed from these orbitals of the metal (Figure 6.1). This causes an increase of electron density on the metal. As a result of the interaction between a $d_\pi$ metal orbital ($d_{xy}, d_{xz}, d_{yz}$) and anti-bonding π* alkene orbital, π metal–olefin back-bonding occurs and leads to an increase of electron density on the π* orbital and a decrease of electron density on the central atom.[1-5, 56] The formation of these two bonds decreases the multiplicity of the carbon–carbon bond. As in the carbonyl group, the stability increase of the σ olefin–metal bond causes an increase in the strength of π back-bonding. Molecular orbital calculations utilizing SCF-Xα-SW for the Zeise's salt $K[PtCl_3(C_2H_4)]$[3] and *ab initio* calculations for $[Ag(C_2H_4)]^+$ [4] confirmed main characteristics of the Dewar–Chatt–Duncanson model. It was found that in $K[PtCl_3(C_2H_4)]$, the most important contribution to the metal–olefin bonding comes from the π ethylene orbital which overlaps with $d$ orbitals possessing appropriate symmetry, in this case $d_{z^2}$ and $d_{x^2-y^2}$ (Figure 6.2). These two orbitals undergo mixing to a considerable degree as in many other transition metal complexes having $C_{2v}$ symmetry. Considerably less is the contribution of $6s$ and $6p$

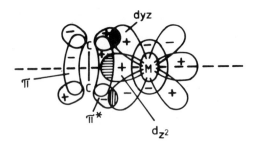

Figure 6.1.   The olefin–metal bond.

orbitals in the formation of the alkene–platinum bonds than estimated by Chatt and Duncanson. This is due to a greater energy difference between $5d$ orbitals and $6s$ and $6p$ orbitals. Therefore pure, hybridized $d\mathrm{sp}^2$ orbitals are not formed. The calculations confirmed the significance of the $\pi$ back-bonding in the Dewar–Chatt–Duncanson model. They show that the contribution of $\pi$ back-bonding to the platinum–ethylene bond energy reaches 25%. Thus, stable $\pi$-olefin complexes are formed by metals possessing a large number of $d$ electrons and low oxidation states, and therefore those which may form strong $\pi$ back-bonding.[1–7, 9–11, 56, 57] The Hartree–Fock–Slater calculations concerning stabilities of the complexes $[M(PH_3)_2L]$ and $[M'(PH_3)_4L]^+$ ($L = O_2$, $C_2H_4$, and $C_2H_2$; $M$ = Ni, Pd, Pt, and $M'$ = Co, Rh, Ir) show that the $M-L$ bond energies change as a function of L according to $O_2 > C_2H_2 \backsim C_2H_4$ and for different metals M according to $3d > 5d > 4d$.[58]

Stable olefin complexes of metals containing fewer than two $d$ electrons are not known. Complexes possessing $d^2$ and $d^4$ electronic structures are numerous. In the case of $[Ag(C_2H_4)]^+$, it can be expected that the $\pi$ back-bonding is not as important as in the Pt(II) complexes. This was confirmed by *ab initio* calculations carried out for the silver compound.[4] They show that the most essential changes in the electronic structure of ethylene in the silver complex are caused by polarization effects which are induced by the positive charge of the silver atom. The olefin–metal bonding orbital has only 6.5% contribution of the $5s$ Ag orbital. This result was confirmed by ESR measurements carried out for olefin silver complexes irradiated with $\gamma$-rays. Based on

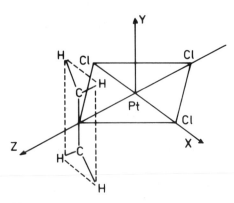

Figure 6.2.   The structure of $[PtCl_3(C_2H_4)]^-$.

this experiment, the spin density of the 5$s$ orbital was estimated; it equals 4.6% for the $\sigma$ bonding orbital of the complex.[5, 6] The formation of both $\sigma$ and $\pi$ redonor metal–olefin bonds decreases the multiplicity of the carbon–carbon bond. The experimentally observed elongation of the C—C bond of the coordinated olefin with respect to the free molecule reaches 20 pm values and depends on the metal, its oxidation state, and other ligands coordinated to the metal. The substituents of the complexed alkene are tilted away from the complex; they are located in the plane which is farther from the metal than olefin carbon atoms themselves. This shift is particularly large in the case of olefins bearing electron-withdrawing substituents (F, CN). These data show that there is partial change in hybridization of carbon atoms from sp$^2$ to sp$^3$ in olefin complexes.

In extreme cases, the carbon atoms of tetracyanoethylene have pseudotetrahedral hybridization. The properties of complexes containing C$_2$(CN)$_4$ may also be explained utilizing the valence bond theory, because the olefin molecule coordinated to the metal may be regarded as a cyclopropane derivative possessing two $\sigma$ M—C bonds. For example, in [Pt{C$_2$(CN)$_4$}(PPh$_3$)$_2$], the carbon–carbon distance equals 149 pm, and is therefore almost equal to the typical single C—C bond distance (in a free olefin, this distance equals 134 pm). The C—C axis forms an angle of 10° with the P–Pt–P plane (Figure 6.3). The coordinated olefin show tendencies to specific conformations with respect to the plane created by the metal and ligands. In complexes having $d^8$ electronic structure of the central atom, possessing formula [M(C$_2$R$_4$)L$_3$] where M = Pt(II), Pd(II), Rh(I), Ir(I), etc., the C—C bond axis is perpendicular to the ML$_3$ plane, while in $d^{10}$ trigonal complexes and $d^8$ coordination compounds possessing trigonal-bipyramidal structures, the alkene axis is coplanar with the trigonal coordination plane within the limits of several degrees (Figure 6.3, Tables 6.1 and 6.2).

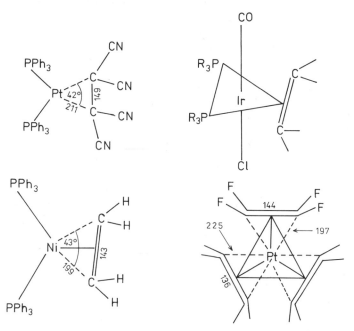

Figure 6.3.   Structures of trigonal–bipyramidal and trigonal olefin complexes. The distances are in pm units.

The $C-C$ bond lengthening in olefins depends on the relative effectiveness of the following bonds: $\sigma$ olefin $\rightarrow$ metal and $\pi$ metal $\rightarrow$ olefin, which in turn depends on the formal oxidation state of the metal. The olefin, with respect to the metal possessing a higher oxidation state [Pt(II)], may be a Lewis base. However, with respect to the same reduced central atom [Pt(0)], this olefin may be a Lewis acid.

The average Pt(II)–olefin bond distance is higher by 10 pm than the analogous average distance for the Pt(0)–olefin compound, despite the fact that the covalent Pt(II) radius is smaller than that of Pt(0). Therefore, $\pi$ back-bonding is stronger in the Pt(0) alkene compound. This is also caused by strong $\pi$ basicity of the $L_2Pt$ fragment. The strength of the $\pi$ redonor bond for trigonal compounds also depends on the LML angle because the L ligands influence the degree of $d$–$p$ hybridization, which in turn influences the effectiveness of the $\pi$ bond (Figure 6.4).

The sign of mixing of $p_x$ and $d_{xz}$ orbitals is determined by the following matrix element:

$$\frac{\langle d_{xz}|L\rangle\langle L|p_x\rangle}{(E_{xz}-E_L)(E_{xz}-E_x)} \tag{6.1}$$

where $L=(\tfrac{1}{2})^{1/2}(\Phi_1-\Phi_2)$ and represents asymmetric combination of ligand orbitals. If it is assumed that the overlap integrals are positive, and therefore the numerator is positive, then the sign of the entire expression will depend on the orbital energy.

From the energy values $E_L<E_{xz}<E_x$ it follows that expression (6.1) is negative for the donor orbital of the $ML_2$ fragment; therefore, the $p_x$ orbital must undergo hybridization with the $d_{xz}$ orbital with the opposite sign with respect to the combination $\Phi_1-\Phi_2$ (Figure 6.4a).

The degree of hybridization, and therefore the strength of the redonor bond, is a function of the LML angle since the expression $\langle xz|L\rangle$ also depends on it. This expression equals zero for 180° (Figure 6.4b), which represents the linear LML grouping. For the T-shaped $ML_3$ fragment in $[M(C_2R_4)L_3]$ complexes, the degree of $d$–$p$ hybridization is also 0.

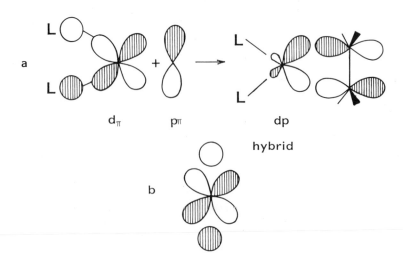

Figure 6.4.  Influence of ligands on $dp$ hybridization and $\pi$ back-bonding.

These considerations show that the angular $ML_2$ grouping causes an increase in bond strength of the $\pi$ back-bonding of the metal olefins in the plane occupied by ligands and the metal. The influence of ligands on $d$–$p$ hybridization explains the planar structure of the $[M(C_2R_4)L_2]$ grouping in $d^{10}$ trigonal and $d^8$ trigonal–bipyramidal complexes. The $d_{xz}$ orbital energy is considerably higher than the $d_{yz}$ orbital energy (the energy difference between the $d_{xz}$ metal orbital and $\pi^*$ olefin orbital is small). Thus, the $d_{xz}$ orbital is a more effective donor with respect to the $\pi^*$ olefin orbital than is the $d_{yz}$ orbital.

In $[M(C_2R_4)L_3]$ complexes, there is no specific olefin conformation decisively favored by electronic factors. The perpendicular orientation of the $C=C$ bond with respect to the $ML_3$ coordination plane results from the steric effects. This explains the low energetic barrier for olefin rotation in such complexes.

Analogous considerations for $M(C_2H_4)_3$ complexes possessing $d^{10}$ central ions predicted that the structure of such complexes should be planar.[7] Indeed, this was proven for $[Pt(C_2F_4)(C_2H_4)_2]$.[8] Energy calculations for $[PtCl_2(C_2H_4)NH_3]$ as a function of the angle between the $C-C$ bond and the $ML_x$ coordination plane show that there are two minima for $0°$ and $90°$, the latter being deeper. However, for $[Pt(C_2Me_2)(PPh_3)_2]$ and $[Pt(C_2H_4)(PPh_3)_2]$, there is only one minimum at zero, and it is considerably deeper.[9] The calculated energies of activation for orientation changes of the unsaturated hydrocarbons in the above-mentioned complexes are 100, 330, and $310\ kJ\ mol^{-1}$, respectively, with the highest value for the acetylene trigonal complex. These calculations are in agreement with the experimental data obtained for olefin rotation about the metal–olefin bond. It was also found that olefins in octahedral and square-planar complexes rotate more easily than they do in pentacoordinate $d^8$ $[M(C_2R_4)L_4]$ compounds possessing trigonal–bipyramidal structures. However, this rotation is also difficult for trigonal complexes (Tables 6.1 and 6.7).[5, 9–11, 56–59, 64]

Table 6.1. *Properties of Metal–Olefin Complexes*

| Properties | Structures of complexes | | | |
| --- | --- | --- | --- | --- |
| | Square, octahedron | Triangle | Trigonal bipyramid | Tetrahedron |
| Coordination number | 4, 6 | 3 | 5 | 4 |
| Rotation of the olefin about the M–olefin axis | yes | difficult | difficult | — |
| Angle between $C=C$ and coordination plane | 75–90° | 0–24° | 0–20° | |
| Elongation of $C=C$ bond due to coordination | <10 pm | 10–20 pm | 10–20 pm | <10 pm |
| Angle of tilting of substituents with respect to the $C=C$ axis | 20° | 50° | 35° | 42° |
| Electronic structures of metals | $d^8$, $d^6$, $d^5$ | $d^{10}$ | $d^8$ | $d^{10}$ |
| Examples of typical central metal atoms | Pt(II), Pd(II), Rh(I), Ir(I), Re(I), Mn(I), Ir(III), Cr(O), Mo(O), W(O), V(O), Rh(III) | Ni(O), Pd(O), Pt(O) | Fe(O), Ir(I), Rh(I) | Ni(O), Cu(I) |

*Table 6.2. Structural Data for Metal–Olefin Complexes*

| Compound | Structure* | Distances* (pm) | | |
|---|---|---|---|---|
| | | C–C | M–C | M–L |
| $[Ni(C_2H_4)(PPh_3)_2]$[a] | T | 143 | 199 | 215.2 |
| $[Ni(PhHC{=}CHPh)\{P(p\text{-tol})_3\}_2]$[b] | T | 147.1 | 201.9 | 218.1 |
| $[Ni(C_2H_4)_2(PCy_3)]$[c] | T | 140.1 | 201.4 | 219.6 |
| $[Ni(CDT)]$[d] | T | 137.2 | 202.4 | |
| $[Ni\{C_2(CN)_4\}(t\text{-BuNC})_2]$[e] | T | 147.6 | 195.4 | 186.6 |
| $[Ni(COD)_2]$[f] | $T_d$ | 139 | 212 | |
| $[Ni(COD)(DUR)]$[g] | $T_d$ COD | 132.5 | 210 | |
| | DUR | 140 | 222 | |
| $[Ni(C_2F_4)(TDPME)]$[h] | $T_d$ | 137 | 186 | 225 |
| $[Pt(C_2H_4)(PPh_3)_2]$[i] | T | 143 | 211 | 226.8 |
| $[Pt\{(NC)HC{=}CN(CN)\}(PPh_3)_2]$[j] | T | 153 | 211 | 228.7 |
| $[Pt\{C_2(CN)_4\}(PPh_3)_2]$[k] | T | 149 | 211 | 229 |
| $[Nbcp_2(Et)(C_2H_4)]$[l] | | 140.3 | 232 | cp 240.2 |
| | | | 227.7 | Et 231.6 |
| $K[PtCl_3(C_2H_4)]$[m] | S | 137 | 212.7 | 232.7 |
| $[Rh(acac)(C_2F_4)(C_2H_4)]$[n] | S $C_2H_4$ | 142 | 219 | 202.7 |
| | $C_2F_4$ | 140 | 201 | 204.7 |
| $[Rh(acac)(C_2H_4)_2]$[n] | S | 141 | 214 | 203 |
| $[Rh_2Cl_2(COD)_2]$[o] | S | 144 | 212 | |
| $[IrBr(CO)\{C_2(CN)_4\}(PPh_3)_2]$[p] | TBP | 150.6 | 214.8 | 240 (Ir–P) |
| $[IrH(CO)\{(NC)HC{=}CH(CN)\}(PPh_3)_2]$[q] | TBP | 143.1 | 211 | 231.7 (Ir–P) |
| $[Ir(Me)(COD)(PMe_2Ph)_2]$[r] | TBP E | 138.6 | 219.1 | 232.3 (Ir–P) |
| | A | 136.2 | 222.4 | |
| $[Ir(ME)(COD)(DPPE)]$[s] | TBP E | 145.9 | 213.9 | 230.8 (Ir–P) |
| | A | 137.4 | 221.5 | |
| $[Fe(C_2H_4)(CO)_4]$[t] | TBP | 146 | | |
| $[Cu_2Cl_2(COD)_2]$[u] | $T_d$ | 140 | 217; 221 | |
| | | 141 | 229; 238 | |

* C=C distances in free olefins: $C_2H_4$—133.7; $C_2(CN)_4$—134; $C_2F_4$—131 pm; T—triangle; S—square plane; $T_d$—tetrahedron; TBP—trigonal bipyramid; E—equatorial; A—axial.

[a] P. T. Cheng, C. D. Cook, C. H. Koo, S. C. Nyburg, and M. T. Shiomi, *Acta Crystallogr.*, **B27**, 1904 (1971).

[b] S. D. Ittel and J. A. Ibers, *J. Organomet. Chem.*, **74**, 121 (1974).

[c] C. Krüger and Y. H. Tsay, *J. Organometal. Chem.*, **34**, 387 (1972).

[d] D. J. Brauer and C. Krüger, *J. Organomet. Chem.*, **44**, 397 (1972).

[e] J. K. Stalick and J. A. Ibers, *J. Am. Chem. Soc.*, **92**, 5333 (1970).

[f] H. Dierks and H. Dietrich, *Z. Kristallogr.*, **122**, 1 (1965).

[g] M. D. Glick and L. F. Dahl, *J. Organomet. Chem.*, **3**, 200 (1965).

[h] J. Browning and B. R. Penfold, *Chem. Commun.*, 198 (1973).

[i] P. T. Cheng and S. C. Nyburg, *Can. J. Chem.*, **50**, 912 (1972).

[j] C. Panattoni, R. Graziani, G. Bandoli, D. A. Clemente, and U. Belluco, *J. Chem. Soc. (B)*, 371 (1970).

[k] G. Bombieri, E. Forsellini, C. Panattoni, R. Graziani, and G. Bandoli, *J. Chem. Soc. (A)*, 1313 (1970).

[l] J. Guggenberger, P. Meakin, and F. N. Tebbe, *J. Am. Chem. Soc.*, **96**, 5420 (1974).

[m] J. A. J. Jarvis, B. T. Kilbourn, and P. G. Owston, *Acta Crystallogr.*, **B27**, 326 (1971).

[n] J. A. Evans and D. R. Russell, *Chem. Commun.*, 197 (1971).

[o] J. A. Ibers and R. G. Snyder, *J. Am. Chem. Soc.*, **84**, 495 (1962); J. A. Ibers and R. G. Snyder, *Acta Crystallogr.*, **15**, 923 (1962).

[p] L. M. Muir, K. W. Muir, and J. A. Ibers, *Discuss. Faraday Soc.*, **47**, 84 (1969).

[q] K. W. Muir and J. A. Ibers, *J. Organomet. Chem.*, **18**, 175 (1969).

[r] M. R. Churchill and S. A. Bezman, *Inorg. Chem.*, **11**, 2243 (1972).

[s] M. R. Churchill and S. A. Bezman, *Inorg. Chem.*, **12**, 260 (1973).

[t] M. I. Davis and C. S. Speed, *J. Organomet. Chem.*, **21**, 401 (1970).

[u] J. H. van den Hende and W. C. Baird, *J. Am. Chem. Soc.*, **85**, 1009 (1963).

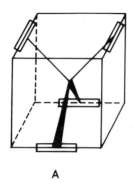   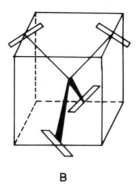

      A                                   B

Figure 6.5. "Cubic" (a) and "dodecahedral" (b) modification of tetrahedral olefin complexes.

Complexes of the composition, $[M(olefin)_4]^{n-}$ where $M = Fe$, $n = 2$; $M = Co$, $n = 1$; and $M = Ni$, $n = 0$, have tetrahedral structures (Figure 6.5).[59-63] Compounds containing cyclic olefins such as 1,5-cyclooctadiene adopt a "cubic" structure (Figure 6.5). However, calculations show that, in the case of monoolefins which do not force "cubic" configuration, "dodecahedral" modification of tetrahedral complexes should be more stable (Figure 6.5).[7] The $^{13}C$ NMR spectra show that this type of structure is observed for $[Co(C_2H_4)_4]^-$ in solution at low temperatures.[60] Structural data for some olefin complexes are given in Table 6.2.

## 2. IR SPECTRA OF OLEFIN COMPLEXES

Coordinated olefins have decreased multiplicity and therefore lowered strength of the double bond. However, the difference between vibrational frequencies of a free olefin and a complexed olefin cannot be a quantitative measure of the metal–olefin bond strength, because the stretching $v(C=C)$ vibration often interacts with other vibrations such as $CH_2$ scissor-type oscillation.[12-14] Usually, the lowering of the $v(C=C)$ band of an olefin bonded to the metal is in the 70–170 cm$^{-1}$ range compared to the free alkene. Most commonly, the band caused by $v(C=C)$ bond vibrations in complexes occurs in the 1460–1560 cm$^{-1}$ range, and has weak or medium intensity.

J. Hiraishi believes that the lowering of the stretching frequencies $\Delta v(C=C)$ with respect to the free olefin is considerably larger.[14] In the case of $[PtCl_3(C_2H_4)]$, Hiraishi assigned the band due to $v(C=C)$ vibration at *ca* 1240 cm$^{-1}$. The validity of Hiraishi's calculations are confirmed by R. Rericha[15] who, based on the assumption that the coordinated alkene molecule is equivalent to the free olefin in its excited state, showed that for $CH_2=CH_2$, the $v(C=C)$ band occurs at 1240 cm$^{-1}$. The energy of this band increases for olefins possessing substituents at the double bond. For these reasons, the $v(metal–olefin)$ frequency is considered to be a better measure of the metal–olefin bond strength.

In many cases, it was found that the lowering of the $v(C=C)$ frequency is also accompanied by lowering of the $v(CO)$ or $v(CN)$ in olefin–carbonyl or olefin–isocyanide complexes. Therefore, better $\pi$ redonor properties of the metal with respect to CO groups cause an increase of $\pi$ interaction with olefins. The $\Delta v(C=C)$ value is not always

Table 6.3. The $v(C=C)$ and $v(M-Olefin)$ Stretching Frequencies[a]

| Complex | Free olefin $v(C=C)$ $(cm^{-1})$ | Coordinated olefin $v(C=C)$ $(cm^{-1})$ | $\Delta v(C=C)$ $(cm^{-1})$ | $v_s(M-\|)$ $(cm^{-1})$ |
|---|---|---|---|---|
| $K[PtCl_3(C_2H_4)] \cdot H_2O$ | 1623 | 1526 | 97 | 407 |
| $K[PtCl_3(C_2D_4)] \cdot H_2O$ | 1515 | 1428 | 87 | 387 |
| $trans\text{-}[PtCl_2(NH_3)(C_2H_4)]$ | 1623 | 1521 | 102 | 383 |
| $trans\text{-}[PtBr_2(NH_3)(C_2H_4)]$ | 1623 | 1517 | 106 | 383 |
| $K[PtBr_3(C_2H_4)] \cdot H_2O$ | 1623 | 1511 | 112 | 395 |
| $trans\text{-}[PtCl_2(C_2H_4)]_2$ | 1623 | 1516 | 107 | 410 |
| $trans\text{-}[PtCl_2(C_2D_4)]_2$ | 1515 | 1428 | 87 | 395 |
| $trans\text{-}[PdCl_2(C_2H_4)]_2$ | 1623 | 1527 | 96 | 427 |
| $K[PtCl_3(cis\text{-}C_4H_8)]$ | 1672 | 1505 | 167 | 405 |
| $K[PtCl_3(C_3H_6)]$ | 1649 | 1505 | 144 | 393 |
| $K[PtCl_3(trans\text{-}C_4H_8)]$ | 1681 | 1522 | 159 | 387 |
| $trans\text{-}[PtCl_2(C_2H_4)py]$[b] | | | | 387 $\{v_{as}(Pt-\|)-470\}$ |
| $trans\text{-}[PtCl_2(cis\text{-}C_4H_8)py]$ | | | | 380 $\{v_{as}(Pt-\|)-487\}$ |

[a] According to References 13 and 14.
[b] M. A. Meester, H. vanDam, D. J. Stufkens, and A. Oskam, *Inorg. Chim. Acta*, **20**, 155 (1976).

Table 6.4. IR Spectra $(cm^{-1})$ of Zeise's Salt: $K[PtCl_3(C_2H_4)] \cdot H_2O$[a]

| $C_2H_4$ | $K[PtCl_3(C_2H_4)] \cdot H_2O$ | $C_2D_4$ | $K[PtCl_3(C_2D_4)] \cdot H_2O$ | Assignment |
|---|---|---|---|---|
| 3019 | 2920 | 2251 | 2115 | $v(CH)$ |
| 1623 | 1526 | 1515 | 1428 | $v(C=C) + \delta_s(CH_2)$[b] |
| 1342 | 1418 | 981 | 978 | $\delta_s(CH_2) + v(C=C)$[b] |
| 3108 | 2975 | 2304 | 2219 | $v(CH)$ |
| 1236 | 1251 | 1009 | 1021 | $\rho_r(CH_2)$ |
| 3106 | 3098 | 2345 | 2335 | $v(CH)$ |
| 810 | 844 | 586 | 536 | $\rho_r(CH_2)$ |
| 2990 | 3010 | 2200 | 2185 | $v(CH)$ |
| 1444 | 1428 | 1078 | 1067 | $\delta_a(CH_2)$ |
| 1007 | 730 | 726 | 450 | $\rho_t(CH_2)$ |
| 943 | 1023 | 780 | 811 | $\rho_w(CH_2)$ |
| 949 | 1023 | 721 | 818 | $\rho_w(CH_2)$ |
| | 331 | | 329 | $v_s(Pt-Cl)$ |
| | 407 | | 387 | $v(Pt-C_2H_4)$ |
| | 310 | | 305 | $v(Pt-Cl_{trans})$ |
| | 183 | | 185 | $\delta(Cl-Pt-Cl)$ |
| | 339 | | 339 | $v_{as}(Pt-Cl)$ |
| | 210 | | 198 | $\delta(ClPtC_2H_4) + \delta(ClPtCl)$ |
| | 161 | | 160 | $\delta(ClPtCl_{trans}) + \delta(ClPtC_2H_4)$ |
| | 121 | | 117 | $\pi(C_2H_4PtCl_t)$ |
| | 92 | | 92 | $\pi(ClPtCl)$ |

[a] According to Reference 110.
[b] This interaction does not take place for the $C_2D_4$ compound.

Table 6.5. *Far-Infrared Spectra of* $K_2[(PtCl_3)_2 C_4H_6]$ *and*
$K[PtCl_3(C_2H_4)] \cdot H_2O$ $(cm^{-1})^a$

| $K_2[(PtCl_3)_2(C_4H_6)]$ | $K[PtCl_3(C_2H_4)]$ | Assignment |
|---|---|---|
| 415 | 407 | $v_s$(Pt–olefin) |
| 339, 332, 308 | 339, 331, 310 | $v$(Pt–Cl) |
| 231 | 210 | $\delta$(ClPt–‖) |
| 203 | 183 | $\delta$(ClPtCl) |
| 155 | 161 | $\delta$(ClPtCl) |
| 128 | 121 | $\pi$(ClPt–‖) |
| 101 | 92 | $\pi$(ClPtCl) |

$^a$ According to Reference 111.

proportional to the energy of the $v$(metal–olefin) band (Table 6.3). The $v$(metal–alkene) bands most commonly occur in the 350–450 cm$^{-1}$ range, and are usually of medium intensity.

Pt(II) and Pd(II) complexes containing acyclic dienes of the general formula $K_2[(MCl_3)_2$ (diene)] are dimeric; each metal atom is bonded to only one carbon–carbon double bond. This was shown by the similarity of IR spectra of dimeric complexes containing dienes and those of monomeric complexes such as K[PtCl$_3$(alkene)] (Tables 6.3–6.5). The IR spectra of $[Co_2(\eta$-$C_4H_6)_2 (CO)_4]$ and $[Fe(C_4H_6)(CO)_3]$, in which the butadiene molecule is cisoidal, differ considerably from that of $K_2[(PtCl_3)_2 C_4H_6]$ which contains a transoidal butadiene molecule.

## 3. $^1H$ NMR SPECTRA

The $\tau$ values of olefin hydrogen atoms of coordinated alkenes are sometimes similar to chemical shifts of the free olefin. However, generally, for olefins bonded in a complex, these shifts occur at higher (sometimes considerably higher) fields[16–19] (Table 6.6).

In complexes, the olefins may undergo exchange reactions with free alkenes or they may rotate about the metal–olefin axis. For example, in the complex $[RhCl(C_2H_4)(PPh_3)_2]$, rapid alkene exchange takes place, and the $^1H$ NMR spectrum of this compound shows a broad resonance at $\tau = 5.35$. The subsequent addition of ethylene causes the shift of this signal to lower fields. At high concentrations of $C_2H_4$, there is only one sharp signal at $\tau = 4.68$, which is characteristic of free ethylene. The presence of ethylene dissociation in this complex, even without an excess of the alkene, confirms exceptional kinetic lability of $[RhCl(C_2H_4)(PPh_3)_2]$, although this is not evidence of its thermodynamic instability. In contrast to chloroethylenebis(triphenylphosphine)rhodium(I), in the $[Rh(acac)(C_2H_4)_2]$ complex, the dissociation of the ethylene molecule does not occur in chloroform despite the fact that the olefin excess causes olefin proton shifts to lower $\tau$ values as a result of rapid exchange. The NMR spectra enable investigations of alkene rotation about the metal–olefin bond. The energy of activation for olefin rotation in the case of square-planar complexes possessing $d^8$ electronic configuration is 40–70 kJ mol$^{-1}$. The energy of activation for octahedral com-

*Table 6.6.* $^1$H-NMR Spectra of Olefin Complexes

| Compound | Solvent | τ, Olefin–H | Coupling constants (Hz) |
|---|---|---|---|
| $C_2H_4$[a,b] | CHCl$_3$ | 4.68 | |
| | CDCl$_3$ | 4.59 | |
| $[Re(C_2H_4)_2(CO)_4]^+$[c] | D$_2$O | 6.6 | |
| $[RhCl(C_2H_4)(PPh_3)_2]$[a] | CDCl$_3$ | 5.35 | |
| $[RhCl(C_2H_4)(AsPPh_3)_2]$[d] | CHCl$_3$ | 7.26 | |
| $[Rhcp(C_2H_4)_2]$[e] | CHCl$_3$ | 8.07 (330 K) | J(Rh–H): 1.8; 2.8 |
| | | 7.23; 8.88 (253 K) | J(H–H): 8.8; |
| | | | 12.2; −0.06 |
| $[Rh(acac)(C_2H_4)_2]$[e] | CHCl$_3$ | 7.04 (298 K) | J(Rh–H): 1.8; 2.4 |
| | | 6.24; 7.49 (215 K) | J(H–H): 8.8; |
| | | | 14.4; −0.07 |
| $[Rhcp(C_2H_4)(SO_2)]$[e,f] | SO$_2$ | 6.68 (271 K) | |
| | | 7.13; 7.33 (223 K) | |
| $[RhClcpEt(C_2H_4)]$[g] | | 6.12 | |
| $[PtCl_3(C_2H_4)]^-$[b] | MeOH + HCl | 5.59 | J($^{195}$Pt–H) 66.0 |
| $[PtCl_2(C_2H_4)(ONC_5H_5)]$[b] | CDCl$_3$ | 5.62 | J($^{195}$Pt–H) 68.8 |
| $[PtCl_2(C_2H_4)(ONC_5H_4NO_2)]$[b] | CDCl$_3$ | 5.34 | |
| $[Rh(Me_2NCS_2)(p\text{-tol}NC)_2 \cdot (MA)]$[h] | CH$_2$Cl$_2$ | 5.61 (298 K) | |
| $[Rh(Me_2NCS_2)(p\text{-tol}NC)_2 \cdot (MN)]$[h] | CH$_2$Cl$_2$ | 6.96 | J($^{103}$Rh–H) 2.0 (213 K) |
| $[Rh(Me_2NCS_2)(p\text{-tol}NC)_2 \cdot (FN)]$[h] | CH$_2$Cl$_2$ | 6.78 (298 K) | |
| $[NiMe(C_2H_4)_2]^-$[i] | D$_8$-THF | 8.26 | At 193–293 K there is no |
| $[NiEt(C_2H_4)_2]^-$[i] | D$_8$-THF | 8.42 | trigonal complexes |
| $[NiPh(C_2H_4)_2]^-$[i] | D$_8$-THF | 8.34 | |
| $[PtCl_2(py)(C_2H_4)]$[j] | CDCl$_3$ | 5.1 | J($^{195}$Pt–H) 61 |
| $[PtCl_2(py)_2(C_2H_4)]$[j] | CDCl$_3$ | 6.1 | J($^{195}$Pt–H) 70 |

[a] J. A. Osborne, F. H. Jardine, J. F. Young, and G. Wilkinson, *J. Chem. Soc. (A)*, 1711 (1966).
[b] P. D. Kaplan and M. Orchin, *Inorg. Chem.*, **4**, 1393 (1965).
[c] E. O. Fischer and K. Ofele, *Angew. Chem.*, **74**, 76 (1962).
[d] J. T. Mague and G. Wilkinson, *J. Chem. Soc. (A)*, 1736 (1966).
[e] R. Cramer, *J. Am. Chem. Soc.*, **86**, 217 (1964); R. Cramer, J. B. Kline, and J. D. Roberts, *J. Am. Chem. Soc.*, **91**, 2519 (1969).
[f] R. Cramer, *J. Am. Chem. Soc.*, **89**, 5377 (1967).
[g] R. Cramer, *J. Am. Chem. Soc.*, **87**, 4717 (1965).
[h] T. Kaneshima, Y. Yumoto, K. Kawakami, and T. Tanaka, *Inorg. Chim. Acta*, **18**, 29 (1976).
[i] K. Jonas, K. R. Pörschke, C. Crüger, and Y.-H. Tsai, *Angew. Chem., Int. Ed. Engl.*, **15**, 621 (1976).
[j] I. Al-Najjar and M. Green, *J. Chem. Soc., Chem. Commun.*, 212 (1977).

plexes (electron configuration $d^6$) is relatively low: 20–50 kJ mol$^{-1}$. In this case, like in complexes possessing other structures, the activation energy depends strongly on both electronic and steric factors which are evidenced by $\Delta G^+$ values for olefin rotation in the following complexes: cis-[W($\eta$-cis-MeOOCCHCHCOOMe)(CO)$_4$(PMe$_3$)], [W-*abf*-(CO)$_3$-ce-(PMe$_3$)$_2$ ($\eta$-cis-MeOOCCHCHCOOMe)], and [W-*abf*-(CO)$_3$-ce-(PMe$_3$)$_2$ ($\eta$-trans-MeOOCCHCHCOOMe)]. These values are 33.5, 58.1, and 80.0 kJ mol$^{-1}$, respectively.[64] The energy of activation for complexes having trigonal–bipyramidal structures is usually low if the olefin rotation is accompanied by Berry's pseudorotation process (see Chapter 7). In the case of trigonal complexes of the type [ML$_2$(olefin)], the energy of activation is 70–105 kJ mol$^{-1}$, but may be considerably lower for diolefin complexes [ML(olefin)$_2$] having $d^{10}$ electron configuration; it equals 42–55 kJ mol$^{-1}$ (Table 6.7).

*Table 6.7. Thermodynamic Parameters for the Rotation of Olefins*

| Compound | $\Delta G_{T_c}^{\neq}$ (kJ mol$^{-1}$) | $E_a$ (kJ mol$^{-1}$) | $\Delta S^{\neq}$ [J(mol·K)$^{-1}$] |
|---|---|---|---|
| $[Rhcp(C_2H_4)_2]^a$ | | $62.7 \pm 0.8$ | |
| $[Rhcp(C_2H_4)(C_2F_4)]^a$ | | $56.8 \pm 2.4$ | |
| $[Rhcp(C_2H_4)SO_2]^a$ | | $51.0 \pm 3.2$ | |
| $[Rh(Me_2NCS_2)(MeC_6H_4NC)_2(MA)]^b$ | 49.3 | | |
| $[Rh(Me_2NCS_2)(MeC_6H_4NC)_2(MN)]^b$ | 56.0 | | |
| $[Rh(Me_2NCS_2)(Me_3C_6H_2NC)_2(MN)]^b$ | 55.6 | | |
| $[Rh(Me_2NCS_2)(MeC_6H_4NC)_2(FN)]^b$ | 59.0 | | |
| $[Rh(Me_2NCS_2)(MeC_6H_4NC)_2(TCNE)]^b$ | 76.1 | | |
| $[Rh(acac)(MeC_6H_4NC)_2(TCNE)]^c$ | 61.9 | | |
| $[Pt(acac)Cl(C_2H_4)]^c$ | $52.3 \pm 2.5$ | | |
| $[Pt(acac)Cl(C_3H_6)]^c$ | $56.4 \pm 0.85$ | | |
| $[Pt(acac)Cl(cis\text{-}C_4H_8)]^c$ | $55.7 \pm 0.85$ | | |
| $[Pt(acac)Cl(trans\text{-}C_4H_8)]^c$ | 66.0 | $56.4 \pm 5.9$ | $-35.5 \pm 21$ |
| $[Pt(acac)Cl(C_2Me_4)]^c$ | 45.6 | $36.4 \pm 2.9$ | $-50.1 \pm 17.8$ |
| $[Pt(C_2H_4)_2PPh_3]^d$ | 42.7 | | |
| $[Pt(C_2H_4)_2PMe_3]^d$ | 54.4 | | |

[a] R. Cramer, J. B. Kline, and J. D. Roberts, *J. Am. Chem. Soc.*, **91**, 2519 (1969); R. Cramer, *J. Am. Chem. Soc.*, **89**, 5377 (1967); R. Cramer, *J. Am. Chem. Soc.*, **87**, 4717 (1965); R. Cramer, *J. Am. Chem. Soc.*, **86**, 217 (1964).
[b] T. Kaneshima, Y. Yumoto, K. Kawakami, and T. Tanaka, *Inorg. Chim. Acta*, **18**, 29 (1976).
[c] C. E. Holloway, G. Hulley, B. F. G. Johnson, and J. Lewis, *J. Chem. Soc. (A)*, 53 (1969).
[d] A. J. Campbell, C. E. Cottrell, C. A. Fyfe, and K. R. Jeffrey, *Inorg. Chem.*, **15**, 1321 (1976); J. R. Lyerla, C. A. Fyfe, and C. S. Yannoni, *J. Am. Chem. Soc.*, **101**, 1351 (1979).

## 4. $^{13}C$ *NMR SPECTRA*

The $^{13}$C chemical shifts of olefin carbon atoms change as a result of olefin coordination to the metal (coordination shift $\Delta\delta$) in the *ca* $+30$ ppm to *ca* $-115$ ppm range. The signs $+$ and $-$ represent changes in chemical shift in the direction of a weaker and a stronger field, respectively. The value of the coordination shift approximately corresponds to the strength of the metal–olefin interaction. The smallest coordination shifts were found for weak Ag(I) compounds, and the largest shifts for stable complexes of Pt(0), Pt(II), and Rh(I).[20, 21]

The increase in the screening of the carbon atom in platinum olefin complexes corresponds to the lowering of the $\nu(C{=}C)$ frequency. Current theories of $^{13}$C screening do not give satisfactory explanations for chemical shifts of coordinated olefin atoms. However, at least for some series of analogous complexes, the most important factor is paramagnetic screening $\sigma_p$,[22] because there is a straight-line dependence of $\Delta\delta$ on the reciprocal of the cube of the M–C distances. This is also confirmed by the straight-line dependence between chemical shifts of carbon atoms in π-olefin complexes and the carbon atoms which create a $\sigma$ bond with the metal, for example, in alkyl complexes. Therefore, for these series of compounds, the influence of the π bond on the chemical shift of olefin carbon atoms is small. However, the chemical shifts obtained by utilizing only paramagnetic screening factors are too small. $^{13}$C NMR spectra of some olefin complexes are given in Table 6.8.

Table 6.8. $^{13}C$ NMR Spectra of Metal Olefin Complexes

| Compound | $\delta$ (ppm) | $\Delta\delta$ (ppm) | Coupling constants* (Hz) |
|---|---|---|---|
| trans-[PtMe(C$_2$H$_4$)(PMe$_2$Ph)$_2$]PF$_6$[a] | 84.4 | −37.2 | $^1J$(Pt−C): 50 |
| [Pt(C$_2$H$_4$)(PPh$_3$)$_2$][a] | 39.6 | −82.0 | $^1J$(Pt−C): 194 |
| K[PtCl$_3$(C$_2$H$_4$)][a] | 67.1 to 75.1 | −56.2 to −48.2 | $^1J$(Pt−C): 195 |
| [PtMe$_2$(COD)][a] | 98.8 | −29.0 | $^1J$(Pt−C): 55 |
| [Pt(CF$_3$)$_2$(COD)][a] | 111 | −16.8 | $^1J$(Pt−C): 56 |
| [PtI$_2$(COD)][a] | 103.2 | −24.6 | $^1J$(Pt−C): 124 |
| K[PtCl$_3$(cis-2-butene)][20] | 89.3 | −34.4 | |
| K[PtCl$_3$(trans-2-butene)][20] | 87.8 | −37.3 | |
| [PtCl$_2$(C$_{10}$H$_{12}$)][20] | 116.2 | −19.3 | |
| [PtCl$_2$(PhCHCH$_2$)(py)][b,A] | 98.2 (C$_\alpha$) | −39.5 (C$_\alpha$) | $^1J$(Pt−C$_\alpha$): 136, |
| | 62.1 (C$_\beta$) | −50.2 (C$_\beta$) | $^1J$(Pt−C$_\beta$): 167 |
| [PtCl$_2$(cis-2-butene)(py)][b,A] | 91.0 | −32.3 | $^1J$(Pt−C): 150 |
| [PtCl$_2$(C$_3$H$_6$)(py)][b,A] | 99.5 (C$_\alpha$) | −33.6 (C$_\alpha$) | $^1J$(Pt−C$_\alpha$): 152, |
| | 71.4 (C$_\beta$) | −43.6 (C$_\beta$) | $^1J$(Pt−C$_\beta$): 159 |
| [PtCl$_2$(C$_2$H$_4$)(py)][b,A] | 75.3 | −47.5 | $^1J$(Pt−C): 165 |
| [PtCl$_2$(MeCOOCHCH$_2$)(py)][b,A] | 111.6 (C$_\alpha$) | −30.1 (C$_\alpha$) | $^1J$(Pt−C$_\alpha$): 171; |
| | 52.9 (C$_\beta$) | −43.5 (C$_\beta$) | $^1J$(Pt−C$_\beta$): 166 |
| [PtCl$_2$(C$_2$H$_4$)(p-MeC$_5$H$_4$N)][21a] | 75.05 | −47.35 | $^1J$(Pt−C): 164.4 |
| [PtCl$_2$(C$_3$H$_6$)(p-MeC$_5$H$_4$N)][21a] | 71.54 (C1) | −43.46 | $^1J$(Pt−Cl): 158.9 |
| | 99.20 (C2) | −36.10 | $^1J$(Pt−C2): 151.3 |
| [PtCl$_2$(C$_8$H$_{14}$)(p-MeC$_5$H$_4$N)][21a] | 93.84 | −36.13 | $^1J$(Pt−C): 161.8 |
| [PtCl$_2$(CH$_2$:CHCH$_2$Cl)(p-MeC$_5$H$_4$N)][21a] | 66.09 (C1) | −52.45 | $^1J$(Pt−C1): 145.6 |
| | 90.29 (C2) | −43.68 | $^1J$(Pt−C2): 175.1 |
| [PdCl$_2$(COD)][20] | 117.2 | −10.6 | |
| [Rhcp(C$_2$H$_4$)$_2$][c] | 60.2 | −61.6 | $^1J$(Rh−C): 10; |
| | | | $J$(C−H): 160 |
| [Rh(acac)(C$_2$H$_4$)$_2$][c] | 36.3 | −85.3 | $^1J$(Rh−C): 14; |
| | | | $J$(C−H): 158 |
| [Rh(acac)(C$_3$H$_6$)][d] | 65.0 (C1) | −53.6 | |
| | 75.8 (C2) | −61.0 | |
| | 23.3 (C3) | +0.9 | |
| [Rhcp(COD)][e] | 62.4 | −65.4 | $J$(C−H): 152 |
| [Rh(acac)(COD)][c] | 76.0 | −51.8 | |
| [Rh(acac-F$_3$)(COD)][c] | 76.3 | −51.5 | |
| [Rh(acac-F$_6$)(COD)][c] | 78.2 | −49.6 | |
| [Rh$_2$Cl$_2$(COD)$_2$][e] | 78.5 | −49.3 | $J$(Rh−C): 13.9 |
| [RhCl(NOR)]$_2$[B(22)] | 55.9 (C1) | −86.5 | $J$(Rh−C): 10.1 |
| | 50.3 (C3) | −102.15 | $J$(Rh−C): 12.1 |
| [Rh(acac)(NOR)][B(22)] | 57.1 (C1) | −85.3 | $J$(Rh−C): 11.0 |
| | 50.9 (C3) | −101.55 | $J$(Rh−C): 12.5 |
| [Rh(acac-F$_6$)(NOR)(py)][B(22)] | 62.29 (C1) | −80.12 | $J$(Rh−C): 8.0 |
| | 28.99 (C3) | −123.46 | $J$(Rh−C): 14.0 |
| [Ir$_2$Cl$_2$(COD)$_2$][20] | 62.1 | −65.7 | |
| [Fecp(CO)$_2$(C$_3$H$_6$)]CF$_3$COO[d] | 58.0 (C1) | −60.6 | |
| | 88.8 (C2) | −48.0 | |
| | 24.4 (C3) | +2.0 | |
| [Cr(NBD)(CO)$_4$][f] | 76.0 | −68.1 | |
| [Mo(NBD)(CO)$_4$][f] | 79.6 | −64.5 | |

*(Table continued)*

*Table 6.8. (continued)*

| Compound | $\delta$ (ppm) | $\Delta\delta$ (ppm) | Coupling constants* (Hz) |
|---|---|---|---|
| [Ag(cyclopentene)]NO₃[g] | 126.2 | −4.4 | |
| [Ag(cyclohexene)]NO₃[g] | 122.8 | −4.4 | |
| [Ag(cyclohexene)]⁺[h] | 126.5[C] | −0.7 | |
| [Ag(COD)]⁺[h] | 127.4 | −0.4[C] | |
| [Ag(1-hexene)]⁺[h] | | −1.7 (C1) | |
| | | −0.4 (C2)[C] | |
| [Ag(CH₂CEt₂)]⁺[h] | | −2.0 (C1) | |
| | | +1.1 (C2)[C] | |
| [Ag(CH₂CHCN)]⁺[i] (1 2 3) | | +4.4 (C1) | |
| | | −1.6 (C2) | |
| | | +1.0 (C3) | |
| [Hg(C₂H₄)]⁺[j] | 134.8 | +13.2 | |
| [Hg(cyclohexene)]²⁺[j] | 157.8 | +30.5 | |
| [Mocp₂(C₂H₄)]ᵏ | 11.8 | −109.8 | |
| [Nbcp₂H(C₂H₄)][l,D] | 13.44; | −108.16 | |
| | 7.96 | −113.64 | |

[A] Data calculated from the CDCl₃ reference: $\delta_{TMS} = \delta_{CDCl_3} - 77.1$ ppm.

[B]NOR: 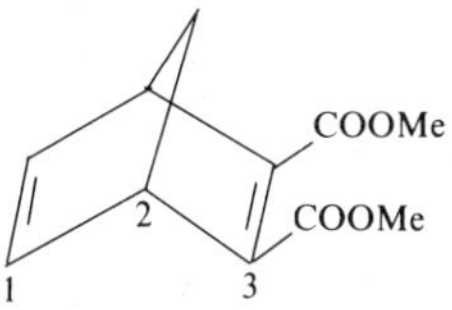   Rh(acac)(NOR):

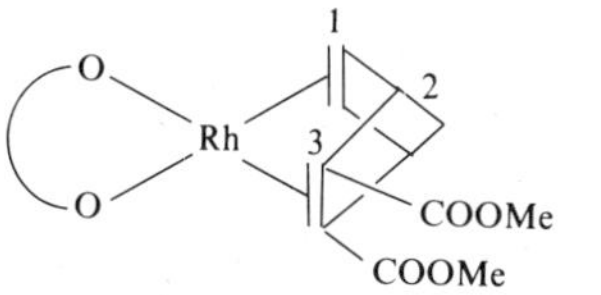

[C] 1.0 M in MeOH, [alkene]/[AgNo₃] = 8/1.
[D] Nonequivalent carbon atoms of the olefin.
* Coupling constants: $J(^{195}Pt - ^{13}C)$, $J(^{103}Rh - ^{13}C)$, or $J(^{13}C - ^1H)$.
[a] M. H. Chisholm, H. C. Clark, L. E. Manzer, and J. B. Stothers, *J. Am. Chem. Soc.*, **94**, 5087 (1972); D. G. Cooper and J. Powell, *Inorg. Chem.*, **15**, 1959 (1976).
[b] M. A. M. Meester, H. Van Dam, D. J. Stufkens, and A. Oskam, *Inorg. Chim. Acta*, **20**, 155 (1976).
[c] G. M. Bodner, B. N. Storhoff, D. Doddrell, and L. J. Todd, *Chem. Commun.*, 1530 (1970); A. N. Nesmeyanov, E. I. Fedin, L. A. Federov, and P. V. Petrovskii, *J. Strukt. Chem.*, **13**, 964 (1972).
[d] K. R. Aris, V. Aris, and J. M. Brown, *J. Organomet. Chem.*, **42**, C67 (1972).
[e] D. E. Axelson, C. E. Holloway, and A. J. Oliver, *Inorg. Nucl. Chem. Lett.*, **9**, 885 (1973).
[f] B. E. Mann, *Chem. Commun.*, 976 (1971); B. E. Mann, *J. Chem. Soc., Dalton Trans.*, 2012 (1973).
[g] R. G. Parker and J. D. Roberts, *J. Am. Chem. Soc.*, **92**, 743 (1970).
[h] C. D. M. Beverwijk and J. P. C. M. van Dongen, *Tetrahedron Lett.*, **42**, 4291 (1972).
[i] J. P. C. M. van Dongen and C. D. M. Beverwijk, *J. Organomet. Chem.*, **51**, C36 (1973).
[j] G. A. Olah and P. R. Clifford, *J. Am. Chem. Soc.*, **95**, 6067 (1973); G. A. Olah and P. R. Clifford, *J. Am. Chem. Soc.*, **93**, 2320 (1971).
[k] J. L. Thomas, *Inorg. Chem.*, **17**, 1507 (1978).
[l] J. Guggenberger, P. Meakin, and F. N. Tebbe, *J. Am. Chem. Soc.*, **96**, 5420 (1974).

## 5. PHOTOELECTRON SPECTRA

In alkenes, the highest occupied orbitals are bonding π orbitals as well as free pair electron orbitals of oxygen and nitrogen atoms in corresponding olefin derivatives. The π bonding orbital energy is approximately equal to the energy of the transition metal *d* orbitals.

*Table 6.9. Photoelectron and ESCA Spectra of Some Olefins and Metal Olefin Complexes*[24,34,65,112]

| Compound | Ionization energy (eV) | Energy of the transition $\pi \to \pi^*$ (eV) |
|---|---|---|
| $C_2H_4$ | 10.51 | 7.60 |
| PHCHCH$_2$ | 8.50 | 4.33 |
| *cis*-2-butene | 9.36 | 7.10 |
| propylene | 9.73 | 7.10 |
| CH$_2$CHOOCMe | 9.77 | 6.42 |
| CH$_2$CHCOOMe | 10.52 | 6.39 |
| *trans*-NCCHCHCN | 11.15 | 5.64 |
| CH$_2$CHCN | 10.91 | 6.43 |
| MeCOCHCH$_2$ | 10.10 | 5.95 |
| 1-butene | 9.58 | 7.10 |
| 1-hexene | 9.46 | 7.00 |
| *trans*-2-hexene | 9.13 | 6.84 |
| *cis*-2-hexene | 9.13 | 6.98 |
| 2-Me-1-pentene | 9.12 | 6.58 |
| MeCH$=$CMe$_2$ | 8.67 | 6.98 |
| $C_2F_4$ | 10.52 | 8.88 |
| CH$_2=$CHF | 10.58 | 7.44 |
| CH$_2=$CHCl | 10.00 | 6.4 |
| CHF$=$CF$_2$ | 10.53 | 7.61 |
| *trans*-CHF$=$CHF | 10.38 | 7.28 |
| *trans*-PhCH$=$CHPh | 7.95 | 3.71 |
| CH$_2=$CHO(CH$_2$)$_3$Me | 8.93 | 6.43 |
| [Fe(C$_2$H$_4$)(CO)$_4$]$^a$ | 8.38 ($d_{xy}, d_{x^2-y^2}$); 9.23 ($d_{xz}, d_{yz}$); 10.56 ($\pi$C$=$C); 12.48 ($\sigma$CH); 14–16 (CO, olefin) | |
| [Fe(CH$_2=$CHCHO)(CO)$_4$]$^a$ | 8.69 ($d_{xy}, d_{x^2-y^2}$); 9.42 ($d_{xz}, d_{yz}$); (9.67; 10.76) ($\pi$C$=$C); 12.9 ($\sigma$CH) | |
| [Ir(acac)(C$_2$H$_4$)$_2$]$^b$ | 7.36; 7.83; 8.37; 8.86 ($d$ orbitals); 10.41 ($\pi$C$=$C); 9.35; 9.51; 11.24 (acac) | |
| [Rh(acac)(C$_2$H$_4$)$_2$]$^b$ | 7.54; 8.11 ($d$ orbitals); 10.22 ($\pi$C$=$C); 8.94; 9.33; 10.76 (acac) | |
| [Mocp$_2$C$_2$H$_4$]$^c$ | 6.0; 6.9 ($d$ orbitals); 11.3 ($\pi$C$=$C); 8.8; 9.2 (cp) | |
| [Wcp$_2$C$_2$H$_4$]$^c$ | 6.0; 7.1 ($d$ orbitals); 11.3 ($\pi$C$=$C); 9.0; 9.3; 9.5 (cp) | |
| [Ni(CH$_2$CHCN)$_2$]$^d$ | 856.0 (Ni2p$_{3/2}$) | |
| [Ni(C$_2$H$_4$)(PPh$_3$)$_2$]$^d$ | 855.6 (Ni2p$_{3/2}$) | |
| [Ni(PhCHCH$_2$){P(O-$o$-tolyl)$_3$}$_2$]$^d$ | 855.4 (Ni2p$_{3/2}$) | |
| [Ni(MA){P(O-$o$-tolyl)$_3$}$_2$]$^d$ | 855.1 (Ni2p$_{3/2}$) | |
| [Ni(C$_2$H$_4$)(PCy$_3$)$_2$]$^d$ | 855.0 (Ni2p$_{3/2}$) | |
| [Ni(COD){P(O-$o$-tolyl)$_3$}$_2$]$^d$ | 855.0 (Ni2p$_{3/2}$) | |
| [Ni(C$_2$H$_4$){P(O-$o$-tolyl)$_3$}$_2$]$^d$ | 854.7 (Ni2p$_{3/2}$) | |
| [Ni(COD)$_2$]$^d$ | 853.7 (Ni2p$_{3/2}$) | |

$^a$ H. van Dam and A. Oskam, *J. Electron Spectrosc. Relat. Phenom.*, **17**, 357 (1979); H. van Dam and A. Oskam, *J. Electron Spectrosc. Relat. Phenom.*, **16**, 307 (1979).
$^b$ H. van Dam, A. Terpstra, D. J. Stufkens, and A. Oskam, *Inorg. Chem.*, **19**, 3448 (1980).
$^c$ J. C. Green, S. E. Jackson, and B. Higginson, *J. Chem. Soc., Dalton Trans.*, 403 (1975).
$^d$ C. A. Tolman, W. M. Riggs, W. J. Linn, C. M. King, and R. C. Wendt, *Inorg. Chem.*, **12**, 2770 (1973).

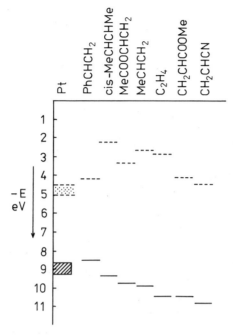

Figure 6.6. Orbital energy levels of platinum and olefins for *trans*-[PtCl₂(olefin) py]: - - - free orbitals, —— occupied orbitals.

The ionization potentials of free olefins[23–28, 65] are in the 8–11 eV range and are therefore analogous to the ionization potentials of $d$ orbitals in many transition metal complexes (Figure 6.6, Table 6.9).

The platinum–olefin bond strength depends on the overlap between orbitals of the central atom and those of the alkene and the energy difference between these orbitals. Figure 6.6 shows that the $\pi$ platinum–olefin bond strength is greater for olefins containing electron–accepting substituents (vinyl acetate, methylacrylate, and acrylnitrile) compared to the ethylene complex, and it is smaller for alkenes possessing donor substituents such as propylene and *cis*-2-butene because, for the former olefins, the $\pi^*$ orbital energy decreases, while in the latter olefins this energy becomes higher. The change of the $\sigma$ platinum–olefin bond strength is not analogous, however. The relative electron density on $\sigma$ and $\pi$ orbitals in $\pi$ olefin complexes may be evaluated more precisely on the basis of data obtained by means of other investigative methods, such as ¹³C NMR, IR, electronic spectra, etc.

## 6. ELECTRONIC SPECTRA

Electronic spectra of olefin complexes (Table 6.10) exhibit, as in all other complexes, three kinds of bands:

1.  Bands due to $d$–$d$ transitions (with the exception of metal complexes containing closed $d^{10}$ electronic structures).

2. Bands due to metal–ligand charge transfer (MLCT).

3. Bands due to transitions within ligands themselves (olefins).

In the case of the presence of ligands other than olefins in complexes, there may be other bands present in the electronic spectra due to transition within these ligands as well as charge transfer ligand → metal and metal → ligand. The charge transfer metal → olefin bands occur in the 30,000–50,000 cm$^{-1}$ range.[29–33] The $\pi \to \pi^*$ bands in olefins occur in far ultraviolet usually above 50,000 cm$^{-1}$. For substituted olefins, these bands may occur at lower energies, particularly if these substituents contain sulfur and nitrogen atoms. As a result of alkene coordination, the $\pi$ to $\pi^*$ bands shift to higher energies and therefore these bands are not generally observed.

Table 6.10. Electronic Spectra of Metal Olefin Complexes

| Compound | $(v)\,(10^3\ \mathrm{cm}^{-1})\,(\varepsilon)$ | Transition |
|---|---|---|
| $[PtCl_3(C_2H_4)]^{-\ (30)}$ | 41.6 (3220) | |
| | 37.7 (1360) | $d \to \pi^*$ |
| | 34.4 (750) | |
| | 30.0 (250) | $d \to d$ |
| $[PtCl_2(H_2O)(C_2H_4)]^{(30)}$ | 41.3 (1360) | $d \to \pi^*$ |
| | 38.5 (1150) | |
| | 35.2 (710) | $d \to \pi^*$ |
| | 30.0 (150) | $d \to d$ |
| $[PtCl_3(CH_2CHCH_2OH)]^{-\ (30)}$ | 41.2 (3325) | |
| | 37.4 (1780) | $d \to \pi^*$ |
| | 33.8 (850) | |
| | 29.7 (181) | $d \to d$ |
| | 23.0 (25) | |
| $[PtCl_3(CH_2CHCH_2NH_3)]^{(30)}$ | 40.9 (3300) | |
| | 37.9 (1800) | $d \to \pi^*$ |
| | 34.7 (800) | |
| | 30.0 (200) | $d \to d$ |
| $[PtCl_3CH_2CHCH_2PEt_3]^{(30)}$ | 40.6 (4300) | |
| | 37.6 (2000) | $d \to \pi^*$ |
| | 34.4 (920) | |
| | 30.0 (240) | $d \to d$ |
| $[Ag(C_6H_6)]^+ ClO_4^{-\ (32)}$ | 43.0 (1500) | $d \to \pi^*$ |
| | 39.3 | within $C_6H_6$ |
| $[Cu(\text{norbornene})Br]^{(31)}$ | 41.9 | $d \to \pi^*$ |
| $[Ni(C_2H_4)]^{(33)}$ | 31.35 | |
| $[Ni(C_2H_4)_2]^{(33)}$ | 35.71 | |
| $[Ni(C_2H_4)_3]^{(33)}$ | 42.37 | |
| $[Ni(MeCHCH_2)]^{(33b)}$ | 30.86 | $d \to \pi^*$ |
| $[Ni(MeCHCH_2)_2]^{(33b)}$ | 34.84 | |
| $[Ni(MeCHCH_2)_3]^{(33b)}$ | 41.15 | |
| $[Ni(C_2F_4)]^{(33b)}$ | 30.67 | |
| $[Ni(C_2F_4)_2]^{(33b)}$ | 35.71 | |
| $[Ni(C_2F_4)_3]^{(33b)}$ | 41.32 | |
| $[Pd(C_2H_4)]^{(33b)}$ | 41.7 | $d \to \pi^*$ |
| $[Pd(C_2H_4)_2]^{(33b)}$ | 45.3 | |
| $[Pd(C_2H_4)_3]^{(33b)}$ | 49.0 | |

## 7. STABILITY OF OLEFIN COMPLEXES

Stability constants for many olefin complexes are currently known.[10, 34] A variety of experimental methods are used to determine these constants: potentiometry, polarography, thermometric titration, and others.[10]

The metal–olefin bond theory shows that the stability of the metal–olefin bond depends on the following factors: $\pi$ bonding and $\pi^*$ antibonding alkene orbital energy, metal valence orbital energy, symmetry of orbitals involved and their interactions. Substitution of hydrogen atoms at the $C=C$ bond by donor substituents which increase the electron density between olefin carbon atoms causes the lowering of stability constants of complexes. This is influenced not only by the changes of $\pi$ and $\pi^*$ orbital energy (Figure 6.6), but also by the increase of steric hindrance. The steric factor is more important because the donor substituents increase the $\sigma$ metal–olefin bond strength and decrease the $\pi$ bond strength. Therefore, to a significant degree these trends are mutually compensated. Substitution of hydrogen atoms at the carbon atom which neighbors the unsaturated carbon atom also decreases the stability of the complex.

In olefins possessing acceptor substituents (CN, $CF_3$, COOMe, etc.), the changes in electronic structure are more pronounced. However, even in those complexes, steric factors play an essential role. Usually, for this type of olefin complex, the $\pi$ redonor bond strength increases and, to a certain degree, the $\sigma$ bond strength decreases (Figure 6.6) while the stability of the metal–olefin bond increases. However, this is not a simple relationship, because the equilibrium constant of the reaction

$$[Rh(acac)(C_2H_4)_2] + C_2H_nF_{4-n} \underset{}{\overset{K_1}{\rightleftharpoons}} [Rh(acac)(C_2H_4)(C_2H_4F_{4-n})] + C_2H_4$$

is smaller than 1 for vinyl fluoride ($K_1 = 0.32$) and vinylidene fluoride ($K_1 = 0.10$). Therefore, complexes containing these olefins are less stable than the ethylene complex. However, complexes containing $C_2HF_3$ ($K_1 = 88$) and $C_2F_4$ ($K_1 = 59$) are more stable than the ethylene complex.

Generally, the $\pi^*$ olefin orbital energy is more sensitive to the influence of substituents than the $\pi$ orbital energy (Table 6.9).[34] Therefore, the dependence of the equilibrium constant on the lowest $\pi^*$ unoccupied orbitals is often linear. This type of relationship is observed for the following reaction[34]:

$$NiL_3 + olefin \longrightarrow Ni(olefin)L_2 + L \qquad [L = P(O\text{-}o\text{-tolyl})_3]$$

These considerations do not involve Ag(I) and Cu(I) complexes with olefins containing both electron-accepting and electron-donating substituents; such complexes are less stable compared to the ethylene complex (Table 6.11).

The stability of Ag(I) complexes with cyclic olefins (Table 6.11) decreases according to the series $C_5 > C_7 > C_6 > C_8$, which is caused almost exclusively by enthalpy changes; the entropy changes are almost the same for all olefins. The stability of Ag(I) coordination compounds with diolefins of the type $CH_2CH(CH_2)_nCHCH_2$ is maximum for $n = 2$. Therefore, the chain containing six carbon atoms is probably optimal for the formation of a chelate with silver.

The metal–olefin bond energy data are scarce. Such energies for group 8 transition metals are usually about 20–30% lower than bond energies of terminal carbonyl groups.[34, 35] The metal–olefin bond energies for some iron, nickel, and rhodium complexes are given in Table 6.12.

Table 6.11. Stability Constants of Metal Olefin and Acetylene Complexes

| Compound | $K_s$ | $\Delta H$ (kJ mol$^{-1}$) | $\Delta S$ (J(mol·K)$^{-1}$) | Remarks |
|---|---|---|---|---|
| | $K_s = \dfrac{[\text{Rh(acac)}(C_2H_4)(\text{olefin})][C_2H_4]^a}{[\text{Rh(acac)}(C_2H_4)_2][\text{olefin}]}$ | | | |
| | $10^3 K_s$ ($T = 298$ K) | | | |
| Propylene | $78 \pm 7$ | $5.9 \pm 3.8$ | $-2.1 \pm 2.1$ | |
| 1-Butene | $92 \pm 18$ | $4.2 \pm 5.9$ | $-7.1 \pm 21$ | Solvent: toluene |
| cis-2-Butene | $4.1 \pm 0.3$ | $7.5 \pm 3.4$ | $-20.5 \pm 12.6$ | |
| trans-2-Butene | $2.0 \pm 0.3$ | $8.0 \pm 2.9$ | $-25.5 \pm 10.4$ | |
| Me$_2$CCH$_2$ | $0.35 \pm 0.02$ | $16.3 \pm 2.9$ | $-10.4 \pm 10.2$ | |
| CH$_2$CHCl | $170 \pm 19$ | $3.35 \pm 3.35$ | $-3.35 \pm 12.1$ | |
| CH$_2$CHF | $320 \pm 22$ | $-6.7 \pm 4.6$ | $-31.5 \pm 10.9$ | |
| trans-FHCCHF | $1240 \pm 360$ | | | |
| cis-FHCCHF | $1590 \pm 330$ | | | |
| | $K_s = \dfrac{[\text{IrX(CO)}(PPh_3)_2(\text{olefin})]^b}{[\text{IrX(CO)}(PPh_3)_2][\text{olefin}]}$ | | | |
| | $K_s$ ($T = 303$ K) | | | |
| Ethylene | $<1$ | | | |
| CH$_2$CHCN | 1.2 | | | PhCl, X = Cl |
| NCCHCHCN | 1500 | | | |
| (NC)$_2$CC(CN)$_2$ | 140000 | | | THF, X = Cl |
| Ethylene | 1.1 | | | |
| CH$_2$CHCN | 0.4 | | | |
| NCCHCHCN | 87 | | | Chlorobenzene, X = I |
| C$_2$F$_4$ | 100 | | | |
| C$_2$H$_2$ | 1.2 | | | |
| | $K_s = \dfrac{[\text{Ni(dipy)(olefin)}]^c}{[\text{Ni(dipy)(solvent)(olefin)}]}$ | | | |
| Maleic anhydride | 47000 | $-53.5$ | $-92$ | 295.35 K |
| | 17000 | | | 313.25 K |
| Acroleine | 5400 | $-79.5$ | $-196.5$ | 299.95 K |
| | 1400 | | | 315.65 K |
| Acrylonitrile | 1300 | $-66.9$ | $-169$ | 293.55 K |
| | 350 | | | THF    314.95 K |
| Methyl acrylate | 790 | $-57.2$ | $-193.5$ | 291.45 K |
| | 140 | | | 314.75 K |
| Methacrylonitrile | 510 | $-59.0$ | $-151$ | 295.0 K |
| | 95 | | | 316.45 K |

[a] R. Cramer, J. Am. Chem. Soc., **89**, 4621 (1967).
[b] R. Vaska, Acc. Chem. Res., **1**, 335 (1968).
[c] T. Yamamoto, A. Yamamoto, and S. Ikeda, J. Am. Chem. Soc., **93**, 3360 (1971).

(Table continued)

*Table 6.11. (continued)*

| Compound | $K_s$ | $\Delta H$ (kJ mol$^{-1}$) | $\Delta S$ (J(mol·K)$^{-1}$) | Remarks |
|---|---|---|---|---|
| $$K_s = \frac{[PdCl_3(olefine)^-][Cl^-]^d}{[PdCl_4^{2-}][olefin]}$$ | | | | |
| Ethylene | 15.6 | 6.3 | 0 | 281.15 K |
| | 16.3 | | | 286.55 K |
| | 15.2 | | | 293.15 K |
| | 13.1 | | | 298.15 K |
| Propylene | 8.4 | 0 | 16.7 | 283.45 K |
| | 7.9 | | | 293.25 K |
| 1-Butene | 13.9 | 0 | 20.9 | 278.15 K |
| | 12.4 | | | 293.15 K |
| $$K_s = \frac{[PtCl_3(olefin)^-][X^-]^e}{[PtX_4^{2-}][olefin]}$$ | | | | |
| $CH_2CHCH_2NH_3^+$ | 2829 | −29.6 | −31.7 | Solvent 0.1 M HCl + 1.9 M NaCl; 303.35 K |
| $CH_2CHCH_2NEt_3^+$ | 260 | −20.4 | −19.2 | 0.1 M HCl + 1.9 M NaCl; 298.20 K |
| $CH_2CHCH_2PEt_3^+$ | 505 | | | 332.2 K |
| $CH_2CHCH_2AsEt_3^+$ | 917 | −24.6 | −17.9 | 331.2 K |
| $K_s = [ML_2(C_2H_4)]\cdot[L]/[ML_3]\cdot[C_2H_4]^f$ | | | | |
| Ni(PPh$_3$)$_3$ | 300 ± 40 | | | |
| Pt{P($p$-C$_6$H$_4$Me)$_3$}$_3$ | 0.21 ± 0.002 | | | |
| Pt(PPh$_3$)$_3$ | 0.122 ± 0.003 | | | |
| Pt{P($m$-C$_6$H$_4$Me)$_3$}$_3$ | 0.07 ± 0.02 | | | |
| Pd(PPh$_3$)$_3$ | 0.013 ± 0.002 | | | |
| Pd{P($p$-C$_6$H$_4$Me)$_3$}$_3$ | 0.016 ± 0.002 | | | |
| $K_s = [Cu(olefin)^+]/[Cu^+]\cdot[olefin]^g$ | | | | |
| Cyclohexadien-1,4 | 339 | −29.2 | | |
| Cyclohexene | 123 | −34.5 | | 1 M LiClO$_4$ in propanol-2 acetone |
| NBD | 1288 | −62.6 | | |
| 1,5-COD | 32000 | | | |
| $K_s = [Ag(olefin)^+]/[Ag^+][olefin]^h$ | | | | |
| Ethylene | 17.5 | −14.6 | −25.1 | Ethylene |
| Propylene | 7.5 | −14.6 | −31.3 | Glycol |
| Butene-1 | 8.8 | −15.4 | −33.4 | H$_2$O, |
| Pentene-1 | 6.7 | −15.1 | −33.8 | 298 K |
| C$_2$H$_2$ | 42.7 | | | Ethylene |
| Hexene-2 | 2.0 | | | Glycol, 313 K |

$^d$ I. I. Moiseev, M. N. Vargaftik, and K. Syrkin, *Dokl. Akad. Nauk SSSR*, **152**, 147 (1963); S. V. Pestrikov, I. I. Moiseev, and B. A. Tsvilikjovskaya, *Zh. Neorg. Khim.*, **11**, 1742 (1966); S. V. Pestrikov and I. I. Moiseev, *Izv. Akad. Nauk SSSR, Ser. Khim.*, 349 (1964); S. V. Pestrikov, I. I. Moiseev, and A. M. Sverzh, *Zh. Neorg. Khim.*, **11**, 2081 (1966); S. V. Pestrikov, I. I. Moiseev, and T. N. Romanova, *Zh. Neorg. Khim.*, **10**, 2203 (1965).
$^e$ R. G. Denning, F. R. Hartley, and L. M. Venanzi, *J. Chem. Soc. (A)*, 324 (1967).
$^f$ C. A. Tolman, W. C. Seidel, and D. H. Gerlach, *J. Am. Chem. Soc.*, **94**, 2669 (1972).
$^g$ J. M. Harvilchuck, D. A. Alkens, and R. C. Murray, *Inorg. Chem.*, **8**, 539 (1969); S. E. Manakan, *Inorg. Chem.*, **5**, 2063 (1966).
$^h$ R. J. Cvetanovic, R. J. Duncan, W. E. Falconer, and R. S. Irvin, *J. Am. Chem. Soc.*, **87**, 1827 (1965).

*(Table continued)*

Table 6.11. (continued)

| Compound | $K_s$ | $\Delta H$ (kJ mol$^{-1}$) | $\Delta S$ (J(mol·K)$^{-1}$) | Remarks |
|---|---|---|---|---|

$$K_s = [NiL_2(olefin)] \cdot [L]/[NiL_3] \cdot [olefin]^{(34)} \quad [L = P(O\text{-}o\text{-tolyl})_3]$$

| Compound | $K_s$ |
|---|---|
| MA | $4 \cdot 10^8$ |
| trans-NCCHCHCN | $1.6 \cdot 10^8$ |
| cis-MeOOCCHCHCOOMe | $4.8 \cdot 10^4$ |
| CH$_2$CHCN | $4 \cdot 10^4$ |
| CH$_2$CHCOOMe | $6.3 \cdot 10^3$ |
| CH$_2$CHCOMe | $2.5 \cdot 10^3$ |
| PhCHCH$_2$ | $1.0 \cdot 10$ |
| CH$_2$CHCH$_2$OMe | 1.4 |
| 1-Hexene | $5 \cdot 10^{-1}$ |
| CH$_2$CHCH$_2$SiMe$_3$ | $2 \cdot 10^{-1}$ |
| trans-PhCHCHPh | $8 \cdot 10^{-2}$ |
| CH$_2$CHCMe$_3$ | $1.6 \cdot 10^{-2}$ |
| CH$_2$CHO(CH$_2$)$_3$Me | $3.1 \cdot 10^{-3}$ |
| trans-2-hexene | $2.7 \cdot 10^{-3}$ |
| cis-2-hexene | $2.3 \cdot 10^{-3}$ |
| 2-Me-1-pentene | $1.2 \cdot 10^{-3}$ |
| MeCHCMe$_2$ | $3 \cdot 10^{-4}$ |
| CH$_2$CHCF$_3$ | $1.0 \cdot 10^3$ |
| Ethylene | $2.5 \cdot 10^2$ |
| C$_2$F$_4$ | $\sim 10^2$ |
| CH$_2$CHF | $9 \cdot 10^1$ |
| CH$_2$CHCl | $1.6 \cdot 10^1$ |
| Propene | $5.3 \cdot 10^{-1}$ |
| 1-Butene | $5.2 \cdot 10^{-1}$ |

Table 6.12. Thermodynamic Data for Metal Olefin Complexes

| Compound | Enthalpy of decomposition of the complex $\Delta H_D$ (kJ mol$^{-1}$) | $E$ (M–olefin) (kJ mol$^{-1}$) | $T$ (MCO) (kJ mol$^{-1}$) | $T/E$ |
|---|---|---|---|---|
| Fe(CO)$_4$C$_2$H$_4$[a] | 567 | 99 | 117 | 1.18 |
| Fe(CO)$_3$(BD)[a] | 554 | 203 | 117 | 1.15[e] |
| Fe(CO)(BD)$_2$[a] | 482 | 183 | 117 | 1.27[e] |
| Fe(CO)(cyclohexadiene)$_2$[a] | 500 | 192 | 117 | 1.22[e] |
| Ni(COD)$_2$[b] | | 206 | 147 | 1.43[e] |
| Ni{P(O-o-tolyl)$_3$}$_2$(C$_2$H$_4$)[(34)] | | 138 | 147 | 1.07 |
| Ni{P(O-o-tolyl)$_3$}$_2$(trans-Bu$^t$CHCHBu$^t$)[(34)] | | 104 | 147 | 1.41 |
| Ni{P(O-o-tolyl)$_3$}$_2$(trans-NCCHCHCN)[(34)] | | 176 | 147 | 0.84 |
| Ni{P(O-o-tolyl)$_3$}$_2$(CH$_2$=CHPr$^n$)[c] | | 150 | 147 | 0.98 |
| Ni{P(O-o-tolyl)$_3$}$_2$(CH$_2$=CHPr$^n$)[c] | | 155 | 147 | 1.05 |
| Rhcp(C$_2$H$_4$)$_2$[d] | | 130 | 166 | 1.28 |

[a] D. L. S. Brown, J. A. Connor, M. L. Leung, M. I. Paz-Andrade, and H. A. Skinner, *J. Organomet. Chem.*, **110**, 79 (1976).
[b] C. A. Tolman, D. W. Reutter, and W. C. Seidel, *J. Organomet. Chem.*, **117**, C30 (1976).
[c] C. A. Tolman, W. C. Seidel, and L. W. Gosser, *Organometallics*, **2**, 1391 (1983).
[d] R. Cramer, *J. Am. Chem. Soc.*, **94**, 5681 (1972).
[e] Calculated assuming that $E = E$(M-diene)/2.

# 8. *METHODS OF SYNTHESIS OF OLEFIN COMPOUNDS*[9, 17, 18, 36, 37]

## a. *Olefin addition to the transition metal salts*

The simplest method for preparation of $\pi$-olefin coordination compounds is the reaction of addition of alkenes to inorganic salts in the solid state or in solution. Such reactions readily take place with group 11 metal compounds, e.g., Cu(I) and Ag(I):

$$CuCl + C_2H_4 \xrightarrow{\text{6 MPa}} [CuCl(C_2H_4)] \qquad (6.2)$$

$$CuCl + CH_2CHCN \xrightarrow{\text{Ar}} [CuCl(CH_2CHCN)] \qquad (6.3)$$

$$AgNO_3 + C_2H_4 \xrightarrow[\text{H}_2\text{O}]{\text{273 K}} AgNO_3 \cdot \tfrac{1}{2}C_2H_4 \qquad (6.4)$$

$$AgX + 2\ \text{olefin} \longrightarrow AgX \cdot 2\ \text{olefin} \qquad (X = NO_3^-, ClO_4^-, BF_4^-) \qquad (6.5)$$

$$AuCl + \text{olefin} \xrightarrow{\text{Et}_2\text{O}} AuCl \cdot \text{olefin} \qquad (6.6)$$

In the case of Cu(I) compounds, substituted olefins serve simultaneously as solvents. Silver compounds, AgX, react to give complexes containing two alkene molecules with the exception of ethylene. $PtCl_2$ reacts with ethylene in the presence of HCl to afford Zeise's anion $PtCl_3(C_2H_4)^-$. However, this reaction does not proceed to completion even at higher pressures of $C_2H_4$.

The reaction of Pd(II) chloride with ethylene affords a dimer:

$$2PdCl_2 + 2C_2H_4 \xrightarrow[\text{CCl}_4 \text{ or CHCl}_3]{\text{1 MPa}} [Pd_2Cl_4(C_2H_4)_2] \qquad (6.7)$$

This reaction is strongly solvent dependent.

## b. *Addition of olefins to coordination compounds of the transition metals*

Transition-metal coordinatively unsaturated 16$e$ complexes, most commonly of the square-planar type, readily react with olefins to form 18$e$ compounds possessing trigonal–bipyramidal structures:

$$\textit{trans-}[IrCl(CO)(PPh_3)_2] + \text{olefin} \xrightarrow{\text{solvent}} [IrCl(CO)(PPh_3)_2 \cdot (\text{olefin})] \qquad (6.8)$$

$$\textit{trans-}[RhCl(CO)(PPh_3)_2] + C_2(CN)_4 \xrightarrow{\text{PhH}} [RhCl(CO)(PPh_3)_2 C_2(CN)_4] \qquad (6.9)$$

$$\textit{trans-}[IrCl(N_2)(PPh_3)_2] + EtOOCCHCHCOOEt \xrightarrow[\text{CHCl}_3]{\text{275 K}} [IrCl(N_2)(PPh_3)_2(EtOOCCHCHCOOEt)]$$
$$(6.10)$$

## c. *Substitution of coordinated ligands with olefins*

These reactions are most commonly utilized for preparation of olefin coordination compounds (Table 6.11). Usually, the following ligands are substituted: halogens, nitriles, isonitriles, carbonyls, phosphines, etc.

$$[PtCl_4]^{2-} + C_2H_4 \longrightarrow [PtCl_3(C_2H_4)]^- + Cl^- \qquad (6.11)$$

Other alkenes such as substituted and chelating alkenes, for example, 2-allylpyridine or *ortho*-allylphenyldiphenylphosphine, react with $[PtCl_4]^{2-}$ in the same manner.[17–19, 36, 37, 66, 68]

$$[PtCl_4]^{2-} + C_5H_4NCH_2CHCH_2 \longrightarrow \quad (6.12)$$

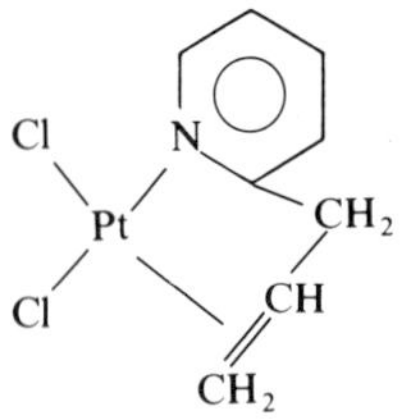

In the case of complexes also containing ligands other than Cl, the geometry of the resulting olefin compound is dictated by the *trans* effect.

$$[PtCl_3(NH_3)]^- + C_2H_4 \longrightarrow cis\text{-}[PtCl_2(C_2H_4)(NH_3)] + Cl^- \quad (6.13)$$

Commonly, aluminum halides, $AlX_3$, are used to bind the departing halogen ion:

$$[FecpCl(CO)_2] + C_6H_{10} + AlCl_3 \longrightarrow [Fecp(CO)_2(C_6H_{10})]AlCl_4 \quad (6.14)$$

$$[Mn(CO)_5Cl] + C_2H_4 + AlCl_3 \longrightarrow [Mn(CO)_5(C_2H_4)]AlCl_4 \quad (6.15)$$

Two ligands may be substituted in these reactions:

$$[Re(CO)_5Cl] + 2C_2H_4 + AlCl_3 \xrightarrow[C_2H_4]{21\ MPa} [Re(CO)_4(C_2H_4)_2]AlCl_4 \quad (6.16)$$

The best results are obtained if a threefold excess of aluminum chloride is used.

Phosphines are readily substituted, particularly in the case of metals of groups 8–10.

$$[RhCl(PPh_3)_3] + C_2H_4 \longrightarrow [RhCl(C_2H_4)(PPh_3)_2] + PPh_3 \quad (6.17)$$

$$[RuCl_2(PPh_3)_3] + CH_2CHCN \longrightarrow [RuCl_2(CH_2CHCN)(PPh_3)_2] + PPh_3 \quad (6.18)$$

Phosphine complexes of platinum, palladium, and nickel also react in a similar manner:

$$[Ni(PPh_3)_3] + C_2H_4 \longrightarrow [Ni(PPh_3)_2(C_2H_4)] + PPh_3 \quad (6.19)$$

$$[Pd(PPh_3)_3] + C_2H_4 \longrightarrow [Pd(PPh_3)_2(C_2H_4)] + PPh_3 \quad (6.20)$$

$$[Pd(PPh_3)_3] + C_2(CN)_4 \longrightarrow [Pt(PPh_3)_2C_2(CN)_4] + PPh_3 \quad (6.21)$$

$$[Pt(PPh_3)_3] + C_2H_4 \longrightarrow [Pt(PPh_3)_2(C_2H_4)] + PPh_3 \quad (6.22)$$

These reactions occur most easily with activated olefins containing electron-accepting substituents.

In many complexes, nitrile ligands may readily be substituted by olefin ligands:

$$fac\text{-}[M(CO)_3(NCMe)_3] + 3CH_2CHCN$$
$$\longrightarrow mer\text{-}[M(CO)_3(CH_2CHCN)_3] + 3MeCN \quad (M = W, Mo) \quad (6.23)$$

$$2[PdCl_2(NCPh)_2] + 2C_2H_4 \xrightarrow[295\ K]{PhH} [Pd_2Cl_4(C_2H_4)_2] + 4PhCN \quad (6.24)$$

Carbonyl groups in metal carbonyls also may be substituted, particularly during photochemical processes. The first step involves dissociation of the CO molecule:

$$[M(CO)_6] + C_2(CN)_4 \xrightarrow[\text{PhH}]{hv} [M(CO)_5 C_2(CN)_4] + CO \tag{6.25}$$

$$(M = Cr, Mo, W)$$

$$[Mncp(CO)_3] + C_2H_4 \xrightarrow[\text{THF}]{hv} [Mncp(CO)_2(C_2H_4)] + CO \tag{6.26}$$

Many other carbonyls and their derivatives react similarly:

$$[Ni(CO)_4] + 2CH_2CHCN \xrightarrow{\text{reflux}} 1/n[Ni(CH_2CHCN)_2]_n + 4CO \tag{6.27}$$

$$[Fe(CO)_5] + CH_2CHCN \xrightarrow{hv} [Fe(CO)_4(CH_2CHCN)] + CO \tag{6.28}$$

$$[Rh_2Cl_2(CO)_4] + 2COD \longrightarrow [Rh_2Cl_2(COD)_2] + 4CO \tag{6.29}$$

Substitution reactions also involve splitting of polynuclear coordination compounds containing bridging ligands by means of olefins:

$$[Pt_2Cl_4(C_2H_4)_2] + 2C_2H_4 \longrightarrow 2[PtCl_2(C_2H_4)_2] \tag{6.30}$$

$$[Pt_2Cl_4(PR_3)_2] + 2C_2H_4 \longrightarrow 2cis\text{-}[PtCl_2(C_2H_4)(PR_3)] \tag{6.31}$$

$$[Pt_2Cl_4(PR_3)_2] + 2C_2H_4 \longrightarrow 2trans\text{-}[PtCl_2(C_2H_4)(PR_3)] \tag{6.32}$$

The *trans* phosphine–olefin compound forms reversibly after a short reaction period, while the *cis* compound forms irreversibly after a long time period.

Complexes which contain metal–metal bonds also undergo cleavage reactions:

$$[Fe_2(CO)_9] + C_2H_4 \longrightarrow [Fe(CO)_4(C_2H_4)] + Fe(CO)_5 \tag{6.33}$$

## d. *Reduction in the presence of olefins*

The formation of olefin complexes may occur during reduction of transition metal compounds in the presence of alkenes. Reducing agents may be compounds introduced to the reaction medium, the solvent, or the olefin itself. Preparation of complexes possessing the composition $Rh_2Cl_2(olefin)_4$ may serve as an example of reductions by alkenes:

$$2RhCl_3 + 2H_2O + 6C_2H_4 \longrightarrow [Rh_2Cl_2(C_2H_4)_4] + 2CH_3CHO + 4HCl \tag{6.34}$$

Reactions with propene and allyl alcohol occur in a similar way. Analogous complexes containing cycloheptene, cyclooctene, 1,5-COD, norbornene, and norbornadiene are also formed from rhodium trichloride in aqueous ethyl alcohol; however, the reaction stoichiometry for these olefins was not established. It is possible that the oxidation of alcohol occurs. For the preparation of rhodium and iridium complexes with cycloolefins of the formula $[M_2Cl_2(olefin)_4]$, the best results are obtained if aqueous ethyl or isopropyl alcohol is used as the solvent:

$$RhCl_3 + cyclooctene \xrightarrow{80\% \text{ EtOH}} [Rh_2Cl_2(cyclooctene)_4] \tag{6.35}$$

$$IrCl_3(IrCl_4) + cyclooctene \xrightarrow[\text{295 K}]{80\% \text{ }i\text{-PrOH}} [Ir_2Cl_2(cyclooctene)_4] \tag{6.36}$$

The complexes AuCl(olefin) and $Au_2Cl_4(olefin)_n$ ($n = 1-3$) are formed from $AuCl_3$, $H[AuCl_4]$, or $Na[AuCl_4]$ as a result of oxidation of excess olefin. Ketones, 1,2-dioles, $\beta$-chloroalcohols, or chloroalkanes are the oxidation products.

The following reducing agents are also utilized: $N_2H_4$, $SO_2$, CO, and organometallic aluminum compounds, such as $AlR_3$ ($R = Me$, Et, $Bu^i$, etc.) and $Al(OR)_xR_{3-x}$.

$$[Ni(acac)_2] + 2PPh_3 + C_2H_4 \xrightarrow[PhH]{Al(OEt)Et_2} [Ni(C_2H_4)(PPh_3)_2] \tag{6.37}$$

$$2[PtCl_2(PPh_3)_2] + 2PhCHCHPh + N_2H_4 \xrightarrow[333\,K]{EtOH} 2[Pt(PhCHCHPh)(PPh_3)_2] + 4HCl + N_2 \tag{6.38}$$

$$[Fe(acac)_3] + 2\,diphos + C_2H_4 \xrightarrow[Et_2O]{Al(OEt)Et_2} [Fe(C_2H_4)(diphos)_2] \tag{6.39}$$

### e. Preparation of olefin complexes from hydrocarbon ligands coordinated to the metal

Some olefin complexes may be obtained by reduction of diolefins coordinated to the metal.

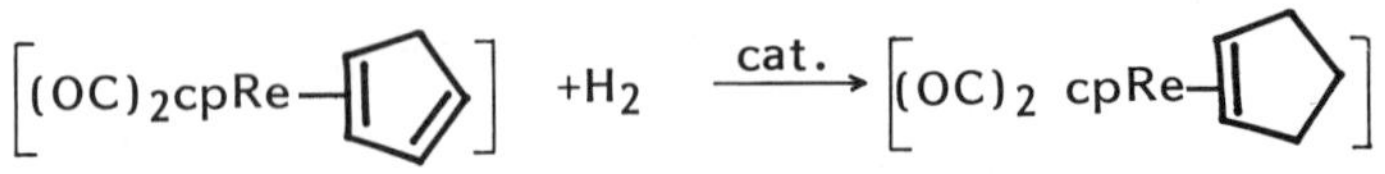

$$\left[(OC)_2cpRe{-}\!\!\bigodot\right] + H_2 \xrightarrow{cat.} \left[(OC)_2\ cpRe{-}\!\!\bigodot\right] \tag{6.40}$$

Metal olefin compounds are also formed by protonation reactions of $\sigma$-allyl complexes and $H^-$ eliminations from alkyl compounds by means of trityl salts $[CPh_3]^+ X^-$:

$$[Fecp(CO)_2(C_2H_5) \xrightarrow[THF]{[CPh_3]ClO_4} [Fecp(CO)_2(C_2H_4)]^+ ClO_4^- + CHPh_3 \tag{6.41}$$

$$[Mn(CO)_5(C_2H_5)] + [CPh_3]BF_4 \longrightarrow [Mn(CO)_5(C_2H_4)]^+ BF_4^- + CHPh_3 \tag{6.42}$$

$$[Fecp(CO)_2(CH_2CH{=}CHR)] + HCl \longrightarrow [Fecp(CO)_2(CH_2{=}CHCH_2R)]^+ Cl^- \tag{6.43}$$

### f. Exchange reactions

Exchange reactions of one olefin for another is a particular case of substitution reaction. The possibility of obtaining a new olefin complex in this way depends on the ratio of the formation constants of two alkene complexes remaining in the equilibrium

$$[PtCl_3(C_2H_4)]^- + PhCHCH_2 \xrightarrow[ETOH]{vacuum} [PtCl_3(PhCHCH_2)]^- + C_2H_4 \tag{6.44}$$

Application of vacuum which removes $C_2H_4$ allows for the increase of reaction yield.

$$[Ni(PPh_3)_2(C_2H_4)] + PhCHCH_2 \longrightarrow [Ni(PPh_3)_2(PhCHCH_2)] + C_2H_4 \tag{6.45}$$

$$[Rh(acac)(C_2H_4)_2] + 2MeCH{=}CH_2 \longrightarrow [Rh(acac)(CH_2CHMe)_2] + 2C_2H_4 \tag{6.46}$$

## g. *Dehydration of alcohols*

Dehydration of ethyl alcohol as a result of the reaction with $PtCl_4$ or $[PtCl_6]^{2-}$ leads to the formation of Zeise's complexes: $[PtCl_3(C_2H_4)]^-$ and $[Pt_2Cl_4(C_2H_4)_2]$. Alcohol also serves as a reducing agent.

$$2[PtCl_4] + 4C_2H_5OH \longrightarrow [Pt_2Cl_4(C_2H_4)_2] + 2CH_3CHO + 2H_2O + 4HCl \qquad (6.47)$$

$$[PtCl_6]^{2-} + 2C_2H_5OH \longrightarrow [PtCl_3(C_2H_4)]^- + 3Cl^- + H_2O + CH_3CHO + 2H^+ \qquad (6.48)$$

However, yields of these reactions are small. The dehydration of ethanol, n-propanol, and n-butanol by means of $Na_2[PtCl_4]$ allows preparation of olefin compounds $[PtCl_3(C_nH_{2n})]^-$ with 20–40% yields.

$$[PtCl_4]^{2-} + CH_3CH_2CH_2CH_2OH \longrightarrow [PtCl_3(MeCH_2CHCH_2)]^- + H_2O + Cl^- \qquad (6.49)$$
$$(40\%)$$

$$[PtCl_4]^{2-} + [MeCH_2CH_2OH] \longrightarrow [PtCl_3(MeCH=CH_2)]^- + H_2O + Cl^- \qquad (6.50)$$
$$(32\%)$$

The reaction of $[PtCl_4]^{2-}$ with n-alcohols proceeds with formation of considerable amounts of metallic platinum. Branched alcohols do not react in this manner and do not form corresponding olefin complexes.

## h. *Electrochemical syntheses*[69, 70]

Some olefin complexes may conveniently be obtained by electrochemical reduction methods. Cyclooctadiene and cyclododecatriene nickel(0) complexes such as $Ni(COD)_2$ and $Ni(CDT)$ may be prepared by electrochemical reduction of nickel acetylacetonate in the presence of butadiene in acetonitrile, dimethoxymethane, or DMF solutions.[69] Complexes $[Fe(CO)_4(\text{alkene})]$ are reduced electrochemically to $17e$ compounds $[Fe(CO)_3(\text{alkene})]^-$. In the case of complexes $[Fe(CO)_3(\eta^4\text{-diene})]$ (diene = BD, cyclohexadiene), during reduction, $17e$ anions $[Fe(CO)_3(\text{diene})]^-$ are formed in which one olefin bond is decomplexed.[70]

## 9. *PROPERTIES OF TRANSITION METAL OLEFIN COMPLEXES*

Elements of groups 3 and 4 do not form stable π-olefin complexes. However, such complexes are formed as unstable compounds during olefin polymerization reactions in the presence of Ziegler–Natta-type catalysts which contain titanium and vanadium compounds. Exceptions are cyclopentadienyl complexes of titanium such as $Ti(pmcp)_2(C_2H_4)$.[123, 124] Few vanadium compounds which contain chelating olefin ligands possessing C=C bonds as well as phosphine phosphorus atoms are known, such as *ortho*-allylphenyldiphenylphosphine–vanadium compounds. Transition metal olefin complexes of groups 4–10 are known. The most stable compounds are those in which the central atom contains 8 or 10 electrons.

Compounds containing $4d$ and $5d$ electron elements are generally more stable than those of $3d$ electron metals. The former compounds are characterized by greater thermal stability and by greater resistance to water and oxidizing agents. Many platinum group olefin complexes are air stable in the presence of water. This is a result of the character of the metal–olefin bond.

## a. *Group 4 metal complexes*

Titanium olefin compounds are unstable and are formed only as intermediate complexes during oligomerization and polymerization of olefins. The following titanium(II) complex is relatively stable: $Ti(pmcp)_2 (C_2H_4)$.[123, 124] In the presence of excess ethylene,

this compound is formed in equilibrium with $(pmcp)_2 Ti$ ⌐⌐. The equilibrium is shifted in the direction of the olefin complex.

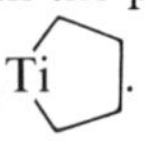

The ethylene molecule readily undergoes substitution by ligands exhibiting good $\pi$-acceptor properties, e.g., CO, MeNC, $MeC \equiv CMe$. Bis(pentamethylcyclopentadienyl)(ethylene)titanium(II) catalyzes isomerization of 1-butene, H/D exchange between propene $C_3H_6$ and deuterated propene $C_3D_6$. In the presence of this complex, ethylene affords 1,3-butadiene and ethane.

Reactions of $[Ti(pmcp)_2 (C_2H_4)]$ with alkynes, nitriles, $CO_2$, and aldehydes give the metallacyclic compounds:

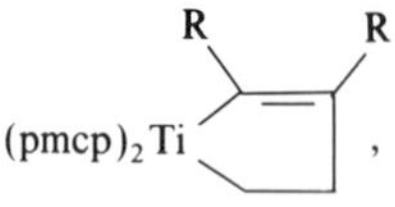

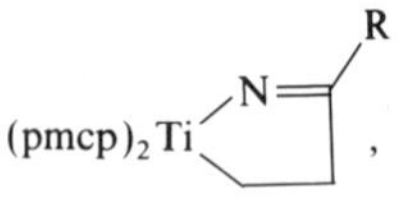

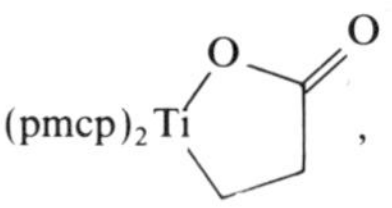

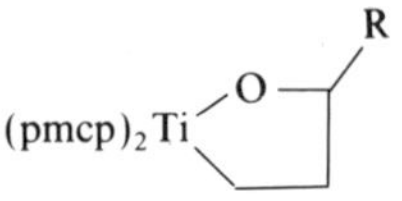

## b. *Group 5 metal complexes*

Not many olefin compounds of elements of this group are known. Very unstable $\pi$ olefin complexes are formed as intermediates during polymerization of alkenes which require compounds of this group as components of Ziegler–Natta catalysts.

Sometimes at low temperatures, it is possible to isolate alkene complexes. At 195 K, $VCl_4$ forms complexes of the type $[VCl_3(alkene)_n]$ ($n = 1,2$, alkene = 1-heptene, 1-decene; $n = 2$, alkene = acrylonitrile, 1-heptene, 4-methyl-1-pentene) in pentane solution.[71] The formation of colored complexes is also observed in solution of $VOCl_3$ in carbon tetrachloride containing olefins. These complexes decompose immediately after evaporation of the solvent. Stable vanadium(0) complexes containing *o-cis*-propenylphenyldiphenylphosphine and *ortho*-allylphenyldiphenylphosphine are known.[38] These complexes are formed from stoichiometric amounts of hexacarbonylvanadium(0).

$$Ph_2PC_6H_4CHCHCH_3 + [V(CO)_6] \longrightarrow \qquad (6.51)$$

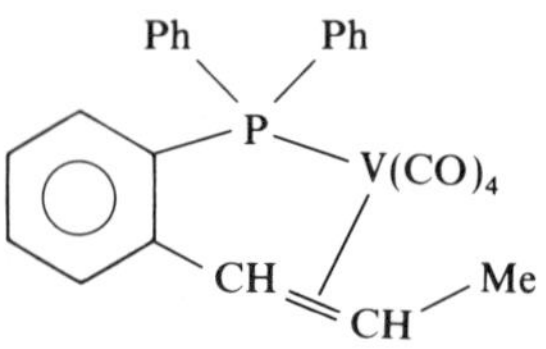

$$Ph_2PC_6H_4CH_2CH=CH_2 + [V(CO)_6] \longrightarrow \qquad (6.52)$$

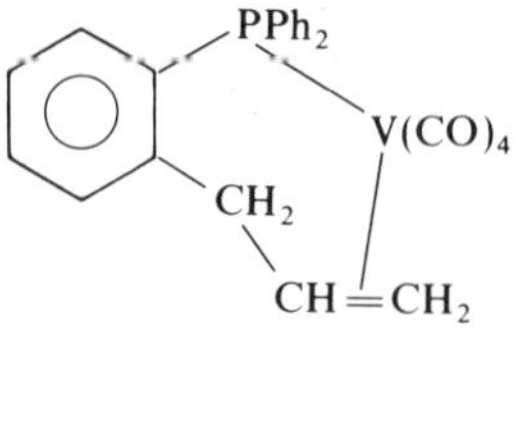

These compounds are monomeric and paramagnetic. The magnetic moment of the allylphosphine vanadium(0) derivative equals 1.98 BM at room temperature. The $v(C=C)$ shifts from 1637 cm$^{-1}$ in the free ligand to 1513 cm$^{-1}$ in the complex. Reduction by sodium amalgam leads to the anion $[V(CO)_5L]^-$ in which the ligand is coordinated only through the phosphorus atom.

Diphenylketene reacts with biscyclopentadienylvanadium to give the dark-green paramagnetic compound $[Vcp_2(Ph_2C=C=O)]$ ($g = 1.9976$, mp $= 432$–$436$ K).[38a]

$$[Vcp_2] + Ph_2C=C=O \xrightarrow[\text{PhMe}]{333\ K} \begin{array}{c} \text{cp} \\ \diagdown \\ \diagup \\ \text{cp} \end{array} V \begin{array}{c} \text{O} \\ \| \\ \text{C} \\ \| \\ \text{C} \\ \diagup \diagdown \\ \text{Ph} \quad \text{Ph} \end{array} \qquad (6.53)$$

Niobium forms the complex $[Nbcp_2(Et)(C_2H_4)]$ in which the $C=C$ distance is 140.6 pm. It is formed in the reaction of dicyclopentadienyltrihydridoniobium(V) with ethylene. When the reaction is carried out with a limited amount of ethylene, the dicyclopentadienylethylenehydridoniobium(III) complex $[Nbcp_2H(C_2H_4)]$ is isolated (Table 6.8). Niobium and tantalum form many similar alkylalkene hydridoalkene, etc., complexes, for example: $Mcp_2H(alkene)$, $Mcp_2R(alkene)$, $Mcp_2X(alkene)$ (alkene = ethylene, propylene, 1-butene, 1-pentene, cyclopropene, cyclopentene, R = Me, Et, Pr, X = Cl, I).[72] Complexes which do not contain cyclopentadienyl ligands or which contain only one $C_5H_5$ are also known, for example: $TacpX_2(alkene)$ (X = Cl, Br, alkene = cyclooctene, $CH_2=CHR$, R = Me, Pr, np, Ph), $[MCl_{3-x}(OBu')_x (C_2H_4)(PMe_3)_2]$ ($x = 1, 2$), $[MEt(C_2H_4)_2 (PMe_3)_2]$, etc.[72]

### c. Group 6 metal complexes

Almost all stable olefin complexes of the chromium group metals are derivatives of carbonyl and cyclopentadienyl compounds (Table 6.13). Most of the metal alkene compounds may be classified as one of the following groups:

1. Substitution products of the carbonyl group in $M(CO)_6$: $[M(CO)_5 (olefin)]$ $[M(CO)_4 (olefin)_2]$, $[M(CO)_4 (CMO)]$, and $[M(CO)_4 (CDO)]$.
2. Complexes containing arene $6e$ ligands: $[M(arene)(CO)_2 (olefin)]$.
3. Anionic halogen complexes: $[M(CO)_2 (olefin)_3X]^-$.
4. Cationic cyclopentadienylcarbonyl complexes: $[Mcp(CO)_3 (olefin)]^+$.
5. Derivatives of cyclopentadienyl compounds: $[Mcp_2(H)(olefin)]^+$ and $[Mcp_2(olefin)]$.
6. Complexes containing quinones as chelating olefins: $[Et_4N][MX(benzoquinone)_3]$ and $M(benzoquinone)_3$.

Complexes of the general formula $[M(CO)_5 C_2(CN)_4]$ (M = Cr, Mo, W) are obtained by reactions of carbonyls with tetracyanoethylene in benzene solutions irradiated by ultraviolet light. IR, electronic, and mass spectra show that the chromium compound has $C_{4v}$ symmetry with symmetrically bonded olefin. The remaining

Table 6.13. Olefin Complexes of Chromium Group Metals

| Compound | Color | Mp (K) | IR spectra |
|---|---|---|---|
| $[Cr(CO)_5(TCNE)]^a$ | Blue-violet | 375 dec | $\nu(CN)$ 2203 air-stable |
| $[Cr(C_6Me_6)(CO)_2(AN)]^b$ | Orange | 433 dec | $\nu(CN)$ 2195, $\tau(CH=CH_2)$ 9.0–9.2 |
| $[Cr(1,3,5\text{-}C_6H_3Me_3)(CO)_2(AN)]^b$ | Red | 382 dec | $\nu(CN)$ 2195 |
| $[Cr(C_6Me_6)(CO)_2(C_2H_4)]^c$ | Red | 353 dec | $\nu(CO)$ 1890, 1835 |
| $[Cr(C_6H_3Me_3)(CO)_2(C_2H_4)]^{c,d}$ | Red | >100 dec | $\nu(CO)$ 1901, 1852 |
| $[Mo(CO)_5(BD)]^e$ | | | $\nu(CO)$ 2087 $(A_1)$, 1977 $(A_1)$, 1961 $(E)$ |
| $[Mo(CO)_5(C_2H_4)]^e$ | | | $\nu(CO)$ 2087 $(A_1)$, 1976 $(A_1)$, 1960 $(E)$ |
| $[Mo(CO)_4(C_2H_4)_2]^e$ | | | $\nu(CO)$ 1964 $(E_u)$ |
| $[Mo(CO)_2(BD)_2]^f$ | | | $\nu(CO)$ 1980 $(A_1)$, 1938 $(B_1)$ |
| $[Mocp(CO)_3(CH_2CHMe)]PF_6{}^g$ | Yellow | | $\nu(CO)$ 2110, 2062, 2008, 1980 sh |
| $[Mo(CO)_3(AN)_3]^h$ | | | $\nu(CO)$ 1975 $(A_1)$, 1900 $(B_1)$, 1832 $(A_1)$; $\nu(C=C)$ 1456 |
| $[Mo(CO)_4(CH_2CHC_6H_4PPh_2]^i$ | Yellow | 428–438 dec | $\nu(C=C)$ 1505 |
| $[Mo(CO)_4(cis\text{-}MeCHCHC_6H_4PPh_2)]^j$ | Yellow | 457–458 dec | $\nu(C=C)$ 1524 |
| $[W(CO)_5BD]^e$ | | | $\nu(CO)$ 2087 $(A_1)$, 1975 $(A_1)$, 1955 $(E)$ |
| $[W(CO)_5(MeCHCH_2)]^e$ | | | $\nu(CO)$ 2085 $(A_1)$, 1966 $(A_1)$, 1951 $(E)$ |
| $trans\text{-}[W(CO)_4(C_2H_4)_2]^e$ | | | $\nu(CO)$ 1964 $(E_u)$ |
| $[Wcp(CO)_3(CH_2CHMe)]PF_6{}^k$ | Light-yellow | | |
| $[W(CO)_3(AN)_3]^h$ | Red | | $\nu(C=C)$ 1440; $\nu(CO)$ 1979 $(A_1)$, 1901 $(B_1)$, 1833 $(A_1)$ |
| $[W(CO)_4(CH_2CHC_6H_4PPh_2)]^i$ | Yellow | 433–448 dec | $\nu(C=C)$ 1490 |
| $[W(CO)_4(cis\text{-}MeCHCHC_6H_4PPh_2)]^j$ | Yellow | 452–453 dec | $\nu(C=C)$ 1502 |
| $[Et_4N][WCl(benzoquinone)_3]^l$ | | | |
| $[MoHcp_2(C_2H_4)]PF_6{}^m$ | White | | |
| $[WHcp_2(C_2H_4)]PF_6{}^m$ | White | | |
| $[WHcp_2(C_3H_6)]PF_6{}^m$ | Light-yellow | | |
| $[Mocp_2(C_2H_4)]^{m,n}$ | Orange-brown | 423 | |
| $[Wcp_2(C_2H_4)]^{m,n}$ | Orange | 449 | |
| $[Cr(CO)_4(COD)]^o$ | Yellow | >343 dec | |
| $[Cr(CO)_4(NBD)]^p$ | Orange | 365 | $\nu(CO)$ 2033, 2959, 1944, 1913 |
| $[Mo(CO)_4(COD)]^o$ | Light-yellow | 392–393 | |
| $[Mo(CO)_4(NBD)]^p$ | Yellow | 350–351 | |
| $[W(CO)_4(COD)]^o$ | Yellow | 420 dec | |

[a] M. Herberhold, *Angew. Chem., Int. Ed. Engl.*, **7**, 305 (1968).

[b] J. F. Guttenberger and W. Strohmeier, *Chem. Ber.*, **100**, 2807 (1967).

[c] W. Strohmeier and H. Hellmann, *Chem. Ber.*, **98**, 1598 (1965).

[d] W. Strohmeier and H. Hellmann, *Ber. Bunsenges. Phys. Chem.*, **69**, 178 (1965).

[e] J. W. Stolz, G. R. Dobson, and R. K. Sheline, *Inorg. Chem.*, **2**, 1264 (1963).

[f] E. O. Fischer, H. P. Kögler, and P. Kuzel, *Chem. Ber.*, **93**, 3006 (1960).

[g] M. Cousins and M. L. H. Green, *J. Chem. Soc.*, 889 (1963).

[h] B. L. Ross, J. G. Grosselli, W. M. Ritchey, and H. D. Kaesz, *Inorg. Chem.*, **2**, 1023 (1963).

[i] M. A. Bennet, R. S. Nyholm, and J. D. Saxby, *J. Organomet. Chem.*, **10**, 301 (1967).

[j] L. V. Interrante, M. A. Bennet, and S. Nyholm, *Inorg. Chem.*, **5**, 2212 (1966).

[k] M. L. H. Green and A. N. Stear, *J. Organomet. Chem.*, **1**, 230 (1964).

[l] F. Calderazzo and R. Henzi, *J. Organomet. Chem.*, **10**, 483 (1967).

[m] F. W. Benfield, B. R. Francis, and M. L. H. Green, *J. Organomet. Chem.*, **44**, C13 (1972); F. W. S. Benfield and M. L. H. Green, *J. Chem. Soc., Dalton Trans.*, 1324 (1974).

[n] J. L. Thomas, *J. Am. Chem. Soc.*, **95**, 1838 (1973).

[o] M. A. Bennet and G. Wilkinson, *Chem. Ind.*, 1516 (1959); E. O. Fischer and W. Fröhlich, *Chem. Ber.*, **92**, 2995 (1959); T. A. Manuel and F. G. A. Stone, *Chem. Ind.*, 1349 (1959).

[p] M. A. Bennet, L. Pratt, and G. Wilkinson, *J. Chem. Soc.*, 2037 (1961).

[M(CO)$_5$(olefin)] complexes have similar structures. In complexes with dienes of the composition [M(CO)$_5$(diene)], [M(CO)$_4$(diene)$_2$], the formation of the metal–olefin bond involves only one C=C.

Compounds [M(CO)$_4$(BD)$_2$] are formed during prolonged irradiation of [M(CO)$_6$] in the presence of butadiene, and have *trans* structures. However, butadiene is a 4$e$ ligand in the complex [Mo(CO)$_2$(BD)$_2$] in which the carbonyl groups are in *cis* position. In all these compounds, the central atom has a $d^6$ electronic configuration.

Complexes in which metals have higher oxidation state are also known. In cationic [Mcp(CO)$_3$(olefin)]$^+$ (M = Mo, W) complexes, the central atom has a $d^4$ electronic configuration. Such compounds are obtained by protonation of the corresponding allyl complexes, e.g., [Mocp(CO)$_3$($\sigma$-CH$_2$CHCH$_2$)], or by H$^-$ elimination from the metal alkyl compounds:

$$[Mocp(CO)_3(C_2H_5)] + [Ph_3C]BF_4 \longrightarrow [Mocp(CO)_3(C_2H_4)]^+ + BF_4^- + CHPh_3 \quad (6.54)$$

In aqueous solution, cationic compounds are unstable. Therefore, during their preparation by protonation of allyl complexes, a large anion such as [PF$_6$]$^-$, [PtCl$_6$]$^{2-}$, or [BPh$_4$]$^-$ should be present in the solution causing immediate precipitation of the reaction product. Molybdenum and tungsten form stable compounds in which the central atom has a $d^2$ electronic configuration, [MHcp$_2$(C$_2$H$_4$)]PF$_6$ (M = Mo, W) and [WHcp$_2$(C$_3$H$_6$)]PF$_6$. The presence of two cyclopentadienyl rings in the complex considerably increases the electron density on the metal atom, allowing formation of a more stable olefin–metal bond. These complexes are rare examples of hydrido–olefin compounds. They are prepared by reactions of [MCl$_2$cp$_2$] (M = Mo, W) with an excess of dichloroethylaluminum.

The reaction of [WCl$_2$cp$_2$] with excess of MgBr($i$-Pr) affords [WHcp$_2$(C$_3$H$_6$)]PF$_6$. The olefins are formed from alkylaluminum or alkylmagnesium compounds. Cationic complexes are reversibly deprotonated by means of 1 M KOH solution:

$$[MHcp_2(C_2H_4)]^+ + OH^- \rightleftharpoons [Mcp_2C_2H_4] + H_2O \quad (6.55)$$

Olefins of the resulting M(II) and W(II) complexes are resistant to substitution, even by molten triphenylphosphine. However, migration of a hydrido ligand in cations takes place with exceptional ease.

$$[MHcp_2(C_2H_4)]^+ + PPh_3 \longrightarrow [Mcp_2Et(PPh_3)]^+ \quad (6.56)$$

The complexes Mcp$_2$(C$_2$H$_4$) (M = Mo, W) and Mocp$_2$(1,2-$\eta$-C$_4$H$_6$) may be prepared by reduction of MCl$_2$cp$_2$ by means of sodium amalgam in the presence of olefins.

$$[MCl_2cp_2] + C_2H_4 + Na/Hg \xrightarrow[\text{15 MPa}]{\text{ethylene}} [Mcp_2(C_2H_4)] \quad (6.57)$$

$$[MCl_2cp_2] + C_4H_6 + Na/Hg \xrightarrow[\text{0.1 MPa}]{\text{BD}} [Mcp_2(1,2-\eta-C_4H_6)] \quad (6.58)$$

Compounds containing chelating diolefins (NBD and COD) and monoolefins (*ortho*-vinylphenyldiphenylphosphine, *ortho*-allylphenyldiphenylphosphine, etc.), similarly to other complexes of this group possessing zero oxidation state of the metal, have octahedral structures. Complexes possessing *cis* arrangements ($C_{2v}$ symmetry) must be formed with chelating ligands. Therefore, the IR spectra of $[M(CO)_4 (CMO)]$ and $[M(CO)_4 (CDO)]$ should exhibit four $v(CO)$ bands.

Phosphine–olefin compounds $[M(CO)_4 (CMO)]$ are air stable. The stability of complexes containing $d^6$ central atoms becomes greater as the $\pi$-acidic properties of alkenes increase. The lowering of frequencies $\Delta v(C=C)$ most commonly equals $100–150$ cm$^{-1}$. Nonionic olefin compounds of these metals have properties like corresponding carbonyl complexes from which they were obtained. Thus, they are volatile, soluble in organic solvents such as benzene and acetone, but less soluble in saturated hydrocarbons.

### d. *Group 7 metal complexes*

Olefin complexes of these metals may be classified to one of the following groups:

1.  Substitution products of $[M(CO)_6]^+$ and $[M(CO)_5 X]$: $[M(CO)_5 (olefin)]^+$, $[M(CO)_4 (olefin)_2]^+$ (olefin $= C_2H_4$, $C_3H_6$, TCNE, MA), $[M(CO)_3 (CMO)X]$, and $[M(CO)_4 (CMO)]^+$, where

CMO =

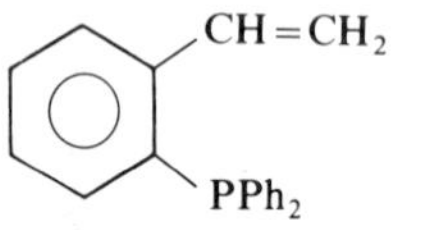

**a.** *o*-vinylphenyldiphenyl        **b.** *o*-(*cis*-propenylphenyl)        **c.** *o*-allylphenyldiphenylphosphine
phosphine                                 diphenylphosphine

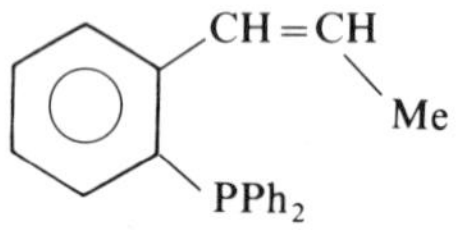

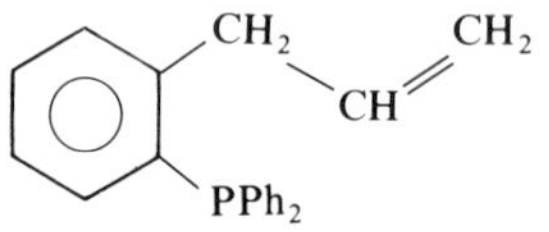

2.  Substitution products of $[MR(CO)_5]$ and $[Mcp(CO)_3]$: $[MR(CO)_4 (olefin)]$ and $[Mcp(CO)_2(olefin)]$.

3.  Complexes which do not contain carbonyl groups.

Manganese group olefin compounds have $d^6$ electronic configuration (Table 6.14). Cationic complexes such as $[Mn(CO)_5 (olefin)]^+$ are not formed from the unstable $Mn(CO)_6^+$ but from the chlorocarbonyl compound $Mn(CO)_5 Cl$ and olefin in the presence of $AlCl_3$, or by protonation reaction of $\sigma$-allyl compounds:

$$[Mn(CO)_5 Cl] + AlCl_3 + C_2H_4 \xrightarrow[14 \text{ MPa}]{C_2H_4} [Mn(CO)_5 (C_2H_4)]^+ AlCl_4^-$$

$$[Mn(CO)_5 (\sigma\text{-}CH_2CHCH_2)] + H^+ \longrightarrow [Mn(CO)_5 (CH_3CH=CH_2)]^+ \qquad (6.59)$$

Under more forcing conditions at higher pressures of $C_2H_4$, it is also possible to obtain complexes containing two olefin molecules, e.g., $[Tc(CO)_4 (C_2H_4)_2]^+$ and $[Re(CO)_4 (C_2H_4)_2]^+$.

Under the influence of water, cationic manganese compounds decompose immediately, as in organic donor solvents such as acetone and tetrahydrofuran. In contrast to isoelectronic compounds of the chromium group, diolefin technetium and rhenium complexes have *cis* structures. In aqueous solution, the cation $[Re(CO)_4 (C_2H_4)_2]^+$ is stable; the $Re-C_2H_4$ bond in $[Re(CO)_5 (C_2H_4)]^+$ is also stable. There is no exchange between free ethylene and the ethylene in the complex.

*Table 6.14. Olefin Complexes of the Manganese Group Metals*

| Compound | Color | Mp (K) | IR and NMR spectra |
|---|---|---|---|
| $[Mn(CO)_5(C_2H_4)]AlCl_4$[a] | White | | $v(CO)$ 2165, 2083; $v(C=C)$ 1541 |
| $[Mncp(CO)_2(C_2H_4)]$[b] | Yellow-brown | 398–400 | $v(CO)$ 1976, 1916 |
| $[Mncp(CO)_2(NBD)]$[b] | Yellow-brown | 366–367 | $v(CO)$ 1972, 1912 |
| $[Mncp(CO)_2(norbornene)]$[b] | Dark-yellow | 410 dec | $v(CO)$ 1972, 1912 |
| $[Mncp(CO)_2(TCNE)]$[c] | | | $v(CN)$ 2203, $\tau = 4.44$ |
| $[Mncp(CO)_2MA]$[d] | Yellow | 459 dec | $\tau(CH=CH) = 5.59$; $v(CO)$ 2016, 1957 |
| $[Mn(CO)_4(CH_2CHC_6H_4PPh_2)]AlCl_4$[e] | Light-yellow | | $v(C=C)$ 1519 |
| $[Tc(CO)_4)(C_2H_4)_2]AlCl_4$[(18)] | | | $v(CO)$ 2137, 2058, 2033; $v(C=C)$ 1559 |
| $[Re(CO)_5(C_2H_4)_2]PF_6$[f (18)] | White | | $v(CO)$ 2178, 2075; $v(C=C)$ 1582 |
| $[Re(CO)_4(C_2H_4)_2]PF_6$[f] | | | $v(CO)$ 2146, 2053, 2016; $v(C=C)$ 1539 |
| $[Recp(CO)_2(C_5H_6)]$[g] | Yellow | | |
| $[Recp(CO)_2(C_5H_8)]$[g] | White | 368–369 | |
| $[Re(CO)_4(o\text{-}CH_2CHC_6H_4PPh_2)][PF_6]$[e] | White | | $v(C=C)$ 1515 |
| $[Re(CO)_4(cis\text{-}MeCHCHC_6H_4PPh_2)]PF_6$[e] | White | | $v(C=C)$ 1525 |
| $[ReH(C_2H_4)(diphos)_2]$[h] | | | $\tau(H)$ 7.64 $(C_2H_4)$; 16.92 (ReH) |

[a] E. O. Fischer and K. Oefele, *Angew. Chem.*, **73**, 581 (1961); E. O. Fischer and K. Oefele, *Angew. Chem., Int. Ed. Engl.*, **1**, 52 (1962).
[b] E. O. Fischer and M. Herberhold, in: *Essays in Coordination Chemistry*, Exper. Suppl. IX (W. Schneider, G. Anderegg, and R. Gut, eds.), pp. 259–305, Birkhauser Verlag, Basel (1964).
[c] M. Herberhold, *Angew. Chem., Int. Ed. Engl.*, **7**, 305 (1968); H. Herberhold and H. Brabetz, *Angew. Chem., Int. Ed. Engl.*, **8**, 902 (1964); M. Herberhold and H. Brabetz, *Chem. Ber.*, **103**, 3896, 3909 (1970).
[d] M. Herberhold and C. R. Jablonski, *Chem. Ber.*, **102**, 767, 768 (1969).
[e] L. V. Interrante and G. V. Nelson, *Inorg. Chem.*, **7**, 2059 (1968).
[f] E. O. Fischer and K. Oefele, *Angew. Chem., Int. Ed. Engl.*, **1**, 52 (1962).
[g] M. L. H. Green and G. Wilkinson, *J. Chem. Soc.*, 4314 (1958).
[h] M. E. Tully and A. P. Ginsberg, *J. Am. Chem. Soc.*, **95**, 2042 (1973).

Compounds containing phosphine–olefin ligands have *cis*-octahedral structures as evidenced by IR spectra. The lowering of the $v(C=C)$ frequencies with respect to the free ligand is in the 100–120 cm$^{-1}$ range. A variety of monoolefins as well as diolefins may form complexes possessing the formula $Mcp(CO)_2$ (olefin). In compounds with diolefins, the metal–olefin bond contains only one $C=C$. The lowering of $v(C=C)$ of the coordinating olefin bond is in the 100–170 cm$^{-1}$ range. The electron density on the metal depends on the π acidity of the alkene. Alkene coordination causes shifts of the olefin protons by 1.5–3 ppm to higher fields as well as deshielding of the cyclopentadienyl protons. Such manganese group olefin compounds are quite stable, and some of them may be sublimed. The σ-donor ligands expel olefins from cyclopentadienyl complexes.

$$[Mncp(CO)_2 \text{ (olefin)}] + PR_3 \longrightarrow [Mncp(CO)_2 (PR_3)] + \text{olefin} \qquad (6.60)$$

In the case of triphenylphosphine, the reaction proceeds according to an $S_N1$ dissociative mechanism involving cleavage of the manganese–olefin bond. The rate of dissociation increases according to $C_2H_4 \ll C_3H_6 < 1$-pentene, NBD < norbornene $\ll$ cyclooctene < cycloheptene < cyclohexene.

Cyclooctene complexes [Mcp(CO)$_2$(cyclooctene)] are utilized for preparation of thiocarbonyl and selenocarbonyl complexes [see reactions (2.176) and (2.177):

$$[Mncp(CO)_2(cyclooctene)] + CS_2 \longrightarrow [Mncp(CO)_2(CS)] + S \hspace{3cm} (6.61)$$

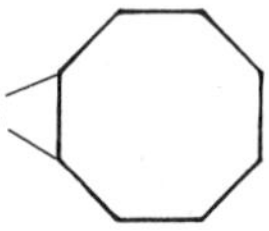

$$[Recp(CO)_2(cyclooctene)] + CS_2 + PPh_3 \longrightarrow [Recp(CO)_2CS] + PPh_3S + C_8H_{14} \qquad (6.62)$$

## e. Group 8 metal complexes

In most known stable complexes of these elements, central atoms have $d^{10}$, $d^8$, or $d^6$ electronic configurations (Table 6.15). Like compounds of the chromium and manganese groups, they are $18e$ complexes. The iron group olefin compounds may belong to one of the following classes of compounds:

1.  [M(CO)$_4$(olefin)]   (olefin: ethylene, $C_3H_6$, $C_4H_8$, PhCHCH$_2$, BD, cyclohexene, cyclooctene, MA, AN, maleic acid, fumaric acid, vinyl chloride, etc.

2.  [Mcp(CO)$_2$(olefin)]$^+$ PF$_6^-$.

3.  [M(CO)$_3$COD)], [M(CO)$_3$(NBD)], and [M(CO)$_3$(CMO)].

4.  [MCl$_2$(CDO)]$_n$, [MCl$_2$(COD)(PEt$_2$Ph)$_2$] (M = Ru, Os), [RuHCl(NBD) (PPh$_3$)$_2$], and [Fe(diphos)$_2$(C$_2$H$_4$)].

5.  Li$_2$(Fe(alkene)$_4$], Li[Fecp(alkene)$_2$] etc. (2 alkene = 2C$_2$H$_4$, COD).[60]

Iron compounds [Fe(CO)$_4$(olefin)] are prepared by irradiation of solutions containing [Fe(CO)$_5$] and the olefin or by reactions of olefins with [Fe$_2$(CO)$_9$] or [Fe$_3$(CO)$_{12}$]:

$$[Fe_3(CO)_{12}] + 3 \text{ olefin} \longrightarrow 3[Fe(CO)_4(\text{olefin})] \hspace{3cm} (6.63)$$

In all cases, a mixture of olefin compounds and [Fe(CO)$_5$] or other complexes is formed. During isolation of [Fe(CO)$_4$(olefin)], substantial losses of the product take place. These compounds are thermally unstable and with carbonyl groups or any other electron-accepting (withdrawing) substituent in the $\alpha$, $\beta$ position to the double bond are relatively stable and may be stored without decomposition for long periods. At room temperature, compounds with ethylene, propene, and styrene are liquids.

Complexes Fe(CO)$_4$(olefin) have trigonal–bipyramidal structures with the olefin occupying the equatorial position (Figure 6.7). In the compound containing acrylonitrile, Fe(CO)$_3$(AN)$_2$, the olefin ligand constitutes a bridge and is bonded with iron atoms through the C=C double bond as well as through the nitrogen atom of the nitrile group.

The relative iron–olefin bond strength is evidenced by the ease of olefin substitution by phosphines. For complexes which contain olefins possessing $\alpha,\beta$ carbonyl substituents, the rate of olefin and carbonyl group substitution by PPh$_3$ increases according

*Table 6.15. Olefin Complexes of Fe, Ru, and Os*

| Compound | Color | Mp (K) | IR and NMR spectra |
|---|---|---|---|
| $[Fe(CO)_4(C_2H_4)]^a$ | Orange-yellow | 254.3 | $v(C=C)$ 1551; $v(CO)$ 2088, 2013 sh, 2007, 1986 |
| $[Fe(CO)_4(PhCHCH_2)]^b$ | Orange-red | | $v(C=C)$ 1505; $v(CO)$ 2088, 2002, 1968 |
| $[Fe(CO)_4(C_4H_6)]^{a,c}$ | | | $v(C=C)$ 1487, 1620*; $v(CO)$ 2082, 2010 sh, 2002 |
| $[Fe(CO)_4]_2(C_4H_6)^c$ | | | $v(C=C)$ 1480; $v(CO)$ 2071, 2004, 1980 |
| $[Fecp(CO)_2(C_2H_4)]PF_6{}^d$ | Light-yellow | 438 dec | $v(C=C)$ 1527; $v(CO)$ 2083, 2049 |
| $[Fe(CO)_4MA]^e$ | Light-yellow | 421 dec | |
| $[Fe(diphos)_2(C_2H_4)]^f$ | Red | 437 | |
| | Violet | 443–444 dec | |
| $[Fe(PF_3)_4(PhCHCH_2)]^g$ | Yellow | 301 302 | |
| $[Fe(PF_3)_3(NBD)]^h$ | Yellow | 325–326 | |
| $[Fe(CO)_3(CH_2:CHC_6H_4PPh_2)]^{(39)}$ | Yellow-brown | 318–320 | $v(CO)$ 2039, 1972, 1945 |
| $[Rucp(CO)_2(C_2H_4)]PF_6{}^i$ | White | 488 dec | $v(C=C)$ 1523; $v(CO)$ 2090, 2040 |
| $[RuCl_2(COD)]_n{}^j$ | Brown | >473 dec | |
| $[RuCl_2(NBD)]_n{}^j$ | Red | | |
| $[RuHCl(NBD)(PPh_3)_2]^k$ | Brown | 409–411 | |
| $[Ru(CO)_3(o\text{-}CH_2CHC_6H_4PPh_2)]^l$ | Orange | 336–337 | $v(CO)$ 2060, 1990, 1968 |
| $[Ru(CO)_2(o\text{-}CH_2CHC_6H_4PPh_2)_2]^l$ | Yellow | 443 | $v(CO)$ 1978, 1917 |
| $[Ru(C_2H_4)(CO)_2(PPh_3)_2]^m$ | | | $v(CO)$ 1955, 1900 |
| $[OsCl_2(COD)]_n{}^n$ | Yellow-brown | | |
| $[Oscp(CO)_2(C_2H_4)]PF_6{}^{(18)}$ | White | | $v(C=C)$ 1546; $v(CO)$ 2080, 2033 |
| $[Os(CO)(NO)(C_2H_4)(PPh_3)_2]^{+o}$ | Yellow | | |
| $[Os(CO)_2(C_2F_4)\{P(OMe)_3\}_2]^p$ | White | 401 | $v(CO)$ 2010 sh, 2015 s |

* Uncoordinated $C=C$ bond.
[a] H. D. Mudroch and E. Weiss, *Helv. Chim. Acta*, **46**, 1588 (1963).
[b] E. Koerner von Gustorf, M. C. Henry, and C. Di Pietro, *Z. Naturforsch.*, **21b**, 42 (1966).
[c] H. D. Mudroch and E. Weiss, *Helv. Chim. Acta*, **45**, 1156 (1962).
[d] E. O. Fischer and K. Fichtel, *Chem. Ber.*, **94**, 1200 (1961); E. O. Fischer and K. Fichtel, *Chem. Ber.*, **95**. 2063 (1962).
[e] E. Weiss, K. Stark, J. E. Lancaster, and H. D. Mudroch, *Helv. Chim. Acta*, **46**, 288 (1963).
[f] G. Hata, H. Kondo, and A. Miyake, *J. Am. Chem. Soc.*, **90**, 2278 (1968).
[g] Th. Kruck and L. Knoll, *Chem. Ber.*, **106**, 3578 (1973).
[h] Th. Kruck, L. Knoll, and J. Laufenberg, *Chem. Ber.*, **106**, 697 (1973).
[i] E. O. Fischer and A. Vogler, *Z. Naturforsch.*, **17b**, 421 (1962).
[j] E. W. Abel, M. A. Bennett, and G. Wilkinson, *J. Am. Chem. Soc.*, **81**, 3178 (1959).
[k] P. S. Hallman, B. R. McGarvey, and G. Wilkinson, *J. Chem. Soc. (A)*, 3143 (1968).
[l] M. A. Bennett, G. B. Robertson, J. B. Tomkins, and P. O. Whimp, *Chem. Commun.*, 341 (1971); M. A. Bennett, G. B. Robertson, and J. B. Tomkins, *J. Organomet. Chem.*, **32**, C19 (1971).
[m] B. E. Cavit, K. R. Grundy, and W. R. Roper, *Chem. Commun.*, 60 (1972).
[n] G. Winkhaus, H. Singer, and M. Kricke, *Z. Naturforsch.*, **21b**, 1109 (1966); R. R. Schrock, B. F. G. Johnson, and J. Lewis, *J. Chem. Soc., Dalton Trans.*, 951 (1974).
[o] J. A. Segal and F. G. A. Johnson, *J. Chem. Soc., Dalton Trans.*, 677 (1975); B. F. G. Johnson and J. A. Segal, *J. Chem. Soc., Chem. Commun.*, 1312 (1972).
[p] M. Cooke, M. Green, and T. A. Kuc, *J. Chem. Soc. (A)*, 1200 (1971).

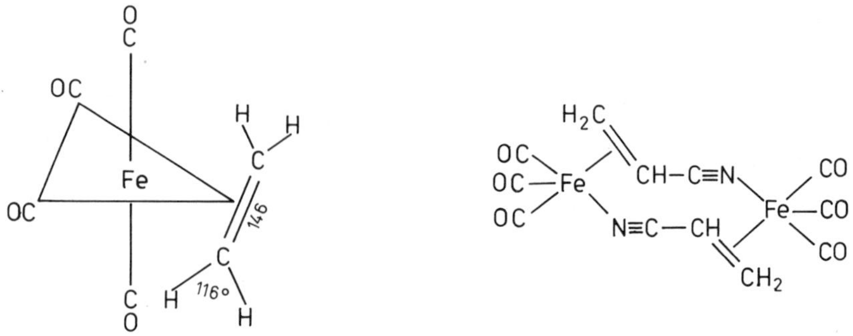

Figure 6.7. Structures of iron olefin complexes.

to the series $-COCH=CHCO < CO < -COCH=CH-$. Cationic complexes $[Fecp(CO)_2][(olefin)]^+$ are more stable than corresponding cationic compounds of chromium and manganese groups. The complex $[Fecp(CO)_2(CH_2=CHMe)]^+$ may be obtained according to one of the methods (6.64) which are also utilized for the prepara-

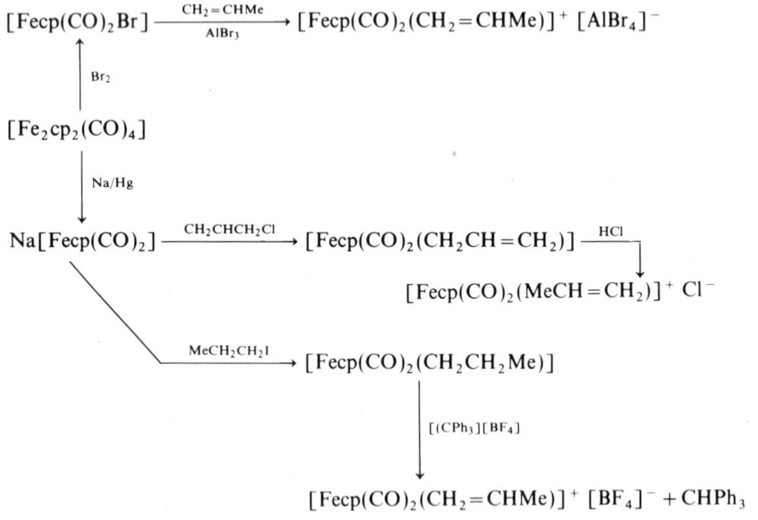

$$[Fecp(CO)_2Br] \xrightarrow[\text{AlBr}_3]{\text{CH}_2=\text{CHMe}} [Fecp(CO)_2(CH_2=CHMe)]^+ [AlBr_4]^-$$

$$\uparrow \text{Br}_2$$

$$[Fe_2cp_2(CO)_4]$$

$$\downarrow \text{Na/Hg}$$

$$Na[Fecp(CO)_2] \xrightarrow[]{\text{CH}_2\text{CHCH}_2\text{Cl}} [Fecp(CO)_2(CH_2CH=CH_2)] \xrightarrow[]{\text{HCl}} $$

$$[Fecp(CO)_2(MeCH=CH_2)]^+ Cl^-$$

$$\xrightarrow[]{\text{MeCH}_2\text{CH}_2\text{I}} [Fecp(CO)_2(CH_2CH_2Me)]$$

$$\downarrow {\scriptstyle[(\text{CPh}_3][\text{BF}_4]}$$

$$[Fecp(CO)_2(CH_2=CHMe)]^+ [BF_4]^- + CHPh_3 \qquad (6.64)$$

tion of other compounds. Hydrogen abstraction from the coordinated alkyl takes place from the carbon atom, which is situated in the $\beta$ position with respect to the metal. In the case of isopropyl, the hydrogen abstraction is reversible, while protonation of the $\sigma$-allyl complex is irreversible. According to the above reaction which requires aluminum bromide, ruthenium and osmium complexes are also obtained.

$$[Rucp(CO)_2Br] + C_2H_4 + AlBr_3 \xrightarrow[\text{Al/Ru}=3/1]{\text{C}_2\text{H}_4,\ 15\ \text{MPa}} [Rucp(CO)_2C_2H_4]^+ AlBr_4^- \qquad (6.65)$$

$$[Oscp(CO)_2Br] + C_2H_4 + AlBr_3 \xrightarrow[\text{Al/Os}=6/1]{\text{C}_2\text{H}_4,\ 30\ \text{MPa}} [Oscp(CO)_2C_2H_4]^+ AlBr_4^- \qquad (6.66)$$

The alkyl complex $[(OC)_2cpFe\overline{CHCH_2N(SO_2Me)\ SN(SO_2Me)}CH_2]$ reacts with $HPF_6 \cdot Et_2O$ to give the olefin compound $[(OC)_2cpFe(\eta^2\text{-}CH_2=CHCH_2N(SO_2Me)$

SNH(SO$_2$Me)]$^+$ PF$_6^-$ which results from cleavage of the C$-$N bond by protonation of the nitrogen atom.[125]

Chelating monoolefins such as *o*-styryldiphenylphosphine or *o*-propenylphenyl-diphenylphosphine react with [M$_3$(CO)$_{12}$] (M = Fe, Ru) to form mononuclear compounds [M(CO)$_3$(CHRCHC$_6$H$_4$PPh$_2$)] which, based on spectroscopic data, possess a structure containing the phosphorus atom in the axial position and the olefin grouping in the equatorial position.[39]

In complexes [M(CO)$_4$(olefin)] and [Mcp(CO)$_2$(olefin)]$^+$, the alkene molecules rotate as in the corresponding chromium and manganese group compounds. Iron, ruthenium, and osmium also form olefin coordination compounds possessing $\sigma$ and $\pi$ bonds (Figure 6.8).

$$[RCH=CHFecp(CO)_2] \xrightarrow[\text{313 K, benzene}]{[Fe_2(CO)_9]} RCH=CHFecp(CO) \tag{6.67}$$

$$\begin{array}{c} | \diagup \diagdown \\ (OC)_3Fe \text{———} CO \end{array}$$

Elements of this triad also afford complexes with fluoroolefins, for example: [Fe(CO)$_4$(olefin)] (olefin = C$_2$F$_4$, C$_3$F$_6$, CCl$_2$=CF$_2$, etc.), [RuL$_2$(CO)$_2$(olefin)] (L = PR$_3$, P(OR)$_3$; olefin = C$_2$F$_4$, C$_3$F$_6$, ...).

## f. *Group 9 metal complexes*

Olefin complexes of these elements possessing zero oxidation states ($d^9$ electronic configuration of the central atom) are scarce. Interaction of cobalt atoms with ethylene at low temperatures gives mono- and polynuclear complexes of the type CO$-$(C$_2$H$_4$), (C$_2$H$_4$)$-$Co$-$(C$_2$H$_4$), Co$-$Co$-$(C$_2$H$_4$), (C$_2$H$_4$)$-$Co$-$Co$-$(C$_2$H$_4$), as well as Co$_4$(C$_2$H$_4$)$_4$. The complex Co$_4$(C$_2$H$_4$)$_4$ is formed by allowing Co$_2$(C$_2$H$_4$)$_2$ to warm to 113 K and probably is a cluster in which cobalt atoms lie at the vertices of a tetrahedron.[73] Some heteroleptic cobalt(0) olefin complexes are stable at room temperature. Reduction of anhydrous Co(II) salts in nitrile solutions

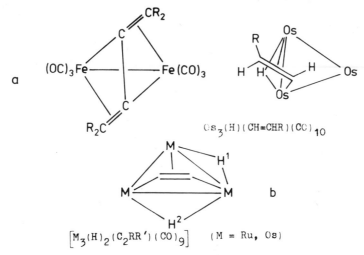

Figure 6.8. The structure of $\sigma$, $\pi$-complexes of iron group metals.[40]

in the presence of olefins leads to the formation of complexes of the type $Co(RCH=CHR)_2 (R'CN)_2$ $(R = Pr^n, Bu^n, COOEt, R' = Me, Et)$.[74] The compound $[Co(EtOOCCH=CHCOOEt)_2 (NCMe)_2]$ as shown by X-ray analysis has a trigonal–pyramidal structure with olefin molecules occupying positions at the pyramid base. Arylfulvenes react with octacarbonyldicobalt to give dinuclear complexes of the formula

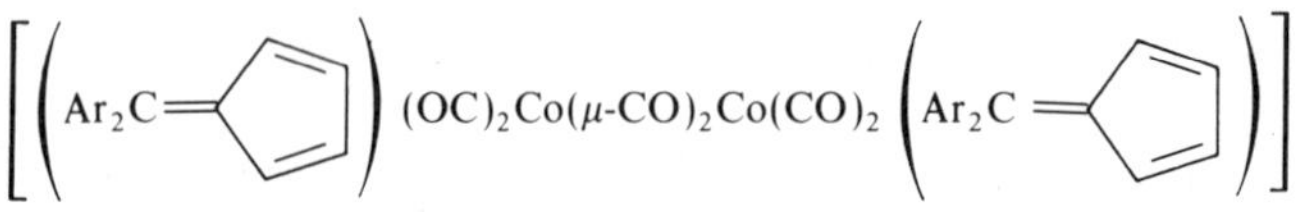

$(Ar = Ph, p\text{-}C_6H_4Cl, p\text{-}C_6H_4Me)$

Compounds of group 9 metals in $+1$ oxidation states possess square or trigonal–bipyramidal structures. In the case of Co(I), 18$e$ five-coordinate complexes are generally formed. However, rhodium(I) and iridium(I) give both types of compounds. Often, square-planar complexes are more stable than compounds possessing the coordination number five. Typical complexes are: $[Mcp(alkene)_2]$, $[Mcp(alkene)L]$, $[ML_4(alkene)]^+$ $(M = Co, Rh, Ir, L = 2e$ ligand), $[M_2X_2(alkene)_4]$, $[(MXL_3(alkene)]$ $(M = Rh, Ir, X = Cl, Br, I)$, $[M(acac)(alkene)_2]$ $(M = Rh, Ir)$, etc.

Compounds containing metals in their $-1$ oxidation states are also known, for example: $[M(alkene)(NO)(PR_3)_2]$, $[Co(CNR)_2 (TCNE)(NO)]$, $M[Co(alkene) (PMe_3)_3]$, and $M[Co(alkene)_4]$.[60]

The dimeric rhodium(II) compound $Rh_2(CF_3COO)_4$ reacts with olefins to give $1:1$ adducts in which the alkene molecules are coordinated along the Rh–Rh axis: $[Rh_2(\mu\text{-}CF_3COO)_4 (alkene)]$ (alkene = styrene, 1-hexene, cyclohexene, $trans$-3-hexene, 2,5-dimethyl-2,4-hexadiene, $Bu^nOCH=CH_2$, etc.).[117]

Ionic complexes $[Rh(L–L)(NH_3)_2]^+ [PF_6]^-$ $(L–L = COD, NBD)$ form stable adducts with crown ethers and crown ether-like molecule $NH(CH_2CH_2O)_2 CH_2CH_2NH(CH_2CH_2O)_2 CH_2CH_2$: $[Rh(L–L)(NH_3)_2 \cdot$ (crown ether)$]^+ PF_6^-$ and $[\{Rh(COD)(NH_3)\}_2\{(OCH_2CH_2)_2NHCH_2CH_2\}_2]^+ PF_6^-$.[118–120]

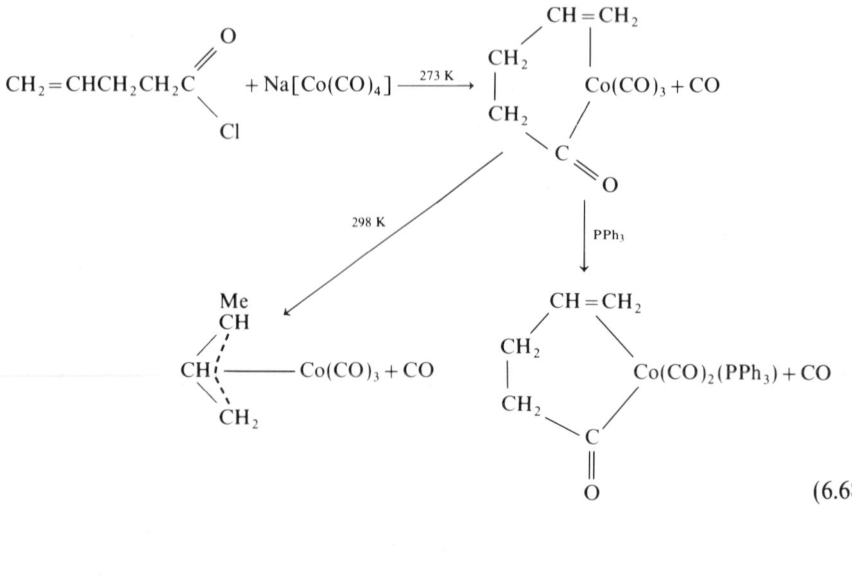

(6.68)

The stability of these adducts depends on the formation of hydrogen bonds between the complex molecules and the host molecules. The stability of the latter complex also depends on the coordination which takes place between nitrogen atoms of the macrocyclic ring and the rhodium atoms.

Dialkyl cobalt complexes $CocpR_2(PPh_3)$ react with methyl esters of maleic and fumaric acids to furnish compounds of the formula $[Cocp(MeOOCCH=CHCOOMe)(PPh_3)]$. Chlorides of unsaturated carboxylic acids react with $NaCo(CO)_4$ in the manner shown in scheme (6.68). The stability of complexes of the type

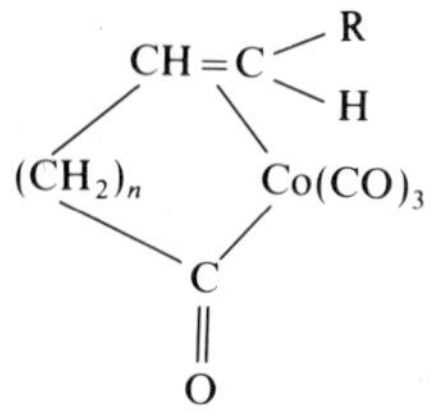

$$
\begin{array}{c}
CH=C{<}^{R}_{H}\\
(CH_2)_n \qquad Co(CO)_3\\
C\\
\parallel\\
O
\end{array}
$$

depends on R and on the size of the ring, that is, on $n$. The most stable compounds are formed when $n=0$ and 2. For $n>4$, compounds in which the olefin bond is not coor-

dinated are formed: $CH_2=CH(CH_2)_nC{\overset{O}{\underset{Co(CO)_4^-}{\diagdown}}}$

In the compound containing a *trans* methyl group ($R=CH_3$) both forms exist in equilibrium:

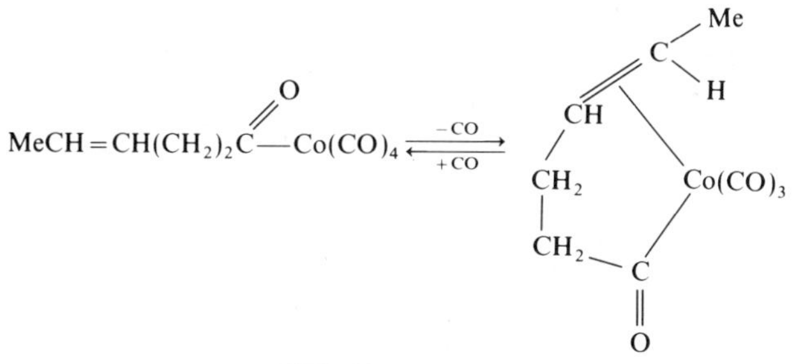

$$
MeCH=CH(CH_2)_2\overset{O}{\overset{\parallel}{C}}{-}Co(CO)_4 \underset{+CO}{\overset{-CO}{\rightleftharpoons}}
\begin{array}{c}
\quad C{<}^{Me}_{H}\\
CH\\
CH_2 \qquad Co(CO)_3\\
CH_2\\
C\\
\parallel\\
O
\end{array}
\tag{6.69}
$$

Reduction of $CoCl_2$ with $NaBH_4$ in ethanol in the presence of butadiene affords $CoC_{12}H_{19}$ which has the following structure:

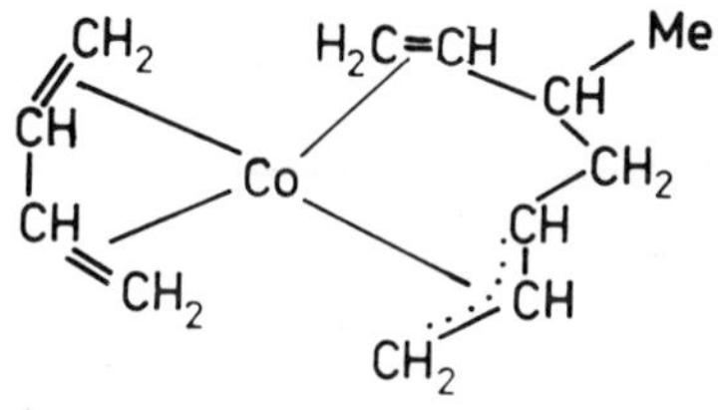

$$
\begin{array}{c}
CH_2 \qquad H_2C=CH \qquad Me\\
CH \qquad\qquad CH\\
\qquad Co \qquad CH_2\\
CH \qquad\qquad CH\\
CH_2 \qquad\qquad CH\\
\qquad CH_2
\end{array}
$$

In the presence of phosphines, this complex catalyzes oligomerization of butadiene. For Rh(I) and Ir(I), square-planar complexes are typical. Very characteristic are dinuclear compounds containing halogen bridges: $[M_2Cl_2(\text{olefin})_4]$ $(M = Rh, Ir)$. $[Rh_2Cl_2(C_2H_4)_4]$ is formed by reaction of $RhCl_3$ with ethylene in hydrated methanol:

$$2RhCl_3 + 2H_2O + 6C_2H_4 \longrightarrow [Rh_2Cl_2(C_2H_4)_4] + 2CH_3CHO + 4HCl \qquad (6.70)$$

Rhodium and iridium complexes with cyclooctadiene, cyclooctene, and norbornadiene such as $[M_2Cl_2(\text{olefin})_4]$ and $[M_2Cl_2(CDO)_2]$ are also formed according to analogous reaction in aqueous ethanol $(EtOH:H_2O = 5:1)$ at room temperature. As a result convenient synthesis and greater reactivity of these complexes in solutions, they are commonly utilized for preparation of other rhodium and iridium derivatives; particularly useful are those containing cyclooctene such as $[M_2Cl_2(C_8H_{14})_4]$ $(M = Rh, Ir)$ as well as $[Rh_2Cl_2(C_2H_4)_4]$. One of the most important applications of these compounds is the "in situ" preparation of catalysts for hydrogenation, isomerization, oligomerization, etc. Figure 6.9 shows examples of $[Rh_2Cl_2(C_2H_4)_4]$ reactivity.

Similar properties are also exhibited by di-$\mu$-chloro-tetracyclooctenedirhodium and di-$\mu$-chloro-dicyclooctadienedirhodium $[Rh_2Cl_2(C_8H_{14})_4]$ and $[Rh_2Cl_2(C_8H_{12})_2]$. The first compound is utilized for synthesis of many other rhodium complexes because it is easily prepared. Moreover, its cyclooctene ligands are readily substituted and, as in other complexes of this type, the $Rh-Cl-Rh$ bridges may easily be broken. However, this compound is relatively unstable; it is slowly oxidized by air in the solid state. In cyclooctadiene complexes, substitutions of the chelating hydrocarbon molecules are more difficult. Some reactions of $[Rh_2Cl_2(C_8H_{14})_4]$ are represented in Figure 6.10.

As in other $16e$ complexes, the rhodium atom in $[Rh_2Cl_2(C_2H_4)_4]$ has square-planar coordination with olefin $C=C$ double bonds perpendicular to the

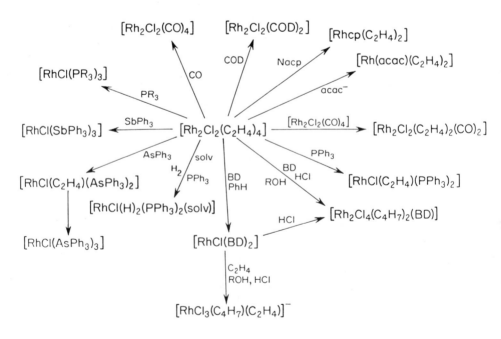

Figure 6.9.  Reactions of $[Rh_2Cl_2(C_2H_4)_4]$.

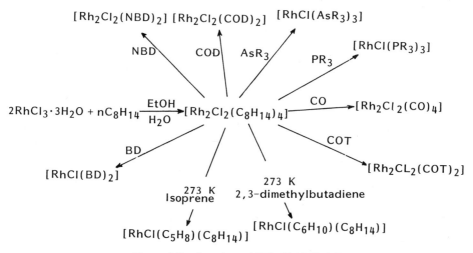

Figure 6.10. Reactions of $[Rh_2Cl_2(C_8H_{14})_4]$.

coordination plane (Figure 6.11). As in $Rh_2Cl_2(CO)_4$, both coordination planes are situated at a certain angle due to a weak $Rh-Rh$ bond in the dimer. Therefore, the ethylene complex readily reacts with di-$\mu$-chlorotetracarbonyldirhodium to give $[Rh_2CL_2(C_2H_4)_2(CO)_2]$. The compound $[Rh_2Cl_2(C_2H_4)_4]$ also reacts with dilute hydrochloric acid to afford the anionic complex cis-$[RhCl_2(C_2H_4)_2]^{2-}$. This compound may also be obtained by the action of HCl on bis(ethylene)rhodium(I) acetylacetonate. Wilkinson's complex $[RhCl(PPh_3)_3]$ reacts with ethylene to give $[RhCl(C_2H_4)(PPh_3)_2]$. The stability of the analogous bromo derivative is lower while the iodo compound cannot be isolated from the solution. The stability of corresponding complexes containing higher hydrocarbon olefins is also lower.

Many 18$e$ rhodium(I) compounds having trigonal–bipyramidal structure are known, including $Rhcp(C_2H_4)_2$ because the cyclopentadienyl ring may be treated as a ligand which occupies three coordination sites. This compound may be obtained by careful reaction of $Rh_2Cl_2(C_2H_4)_4$ with sodium cyclopentadienide.

The same group also comprises the following compounds: $[RhX(CO)(PPh_3)_2(TCNE)]$, $[Rh(SnCl_3)(NBD)_2]$, $[Rh(acac)(RNC)_2(TCNE)]$, and $[Rh(Me_2NCS_2)(RNC)_2(olefin)]$ ($R = p\text{-}MeC_6H_4$, $2,4,6\text{-}Me_3C_6H_2$; olefin = TCNE, MN, FN, MA) (Figure 6.11, Table 6.16).

Much less stable are coordination compounds involving rhodium(III) and alkenes rather than rhodium(I) olefin complexes. These rhodium(III)–alkene compounds are formed as unstable derivatives during reactions of Rh(III) allyl complexes such as $[RhCl_2(C_4H_7)(PPh_3)_2]$ and $[RhCl_2(C_4H_7)(AsPh_3)_2]$ with ethylene. In the phosphine complex, chlorine undergoes substitution, while in the arsine compound, triphenylarsine is replaced.

The reaction of $[Rh_2Cl_2(C_2H_4)_4]$ with substantial excess of hydrochloric acid ($HCl/Rh > 4.3$) gives an oxidative addition product:

$$[Rh_2Cl_2(C_2H_4)_4] + 2Cl^- \longrightarrow 2[RhCl_2(C_2H_4)_2]^- \tag{6.71}$$

$$[RhCl_2(C_2H_4)_2]^- + HCl + solvent \longrightarrow [RhCl_3(C_2H_5)(C_2H_4)(solvent)]^- \tag{6.72}$$

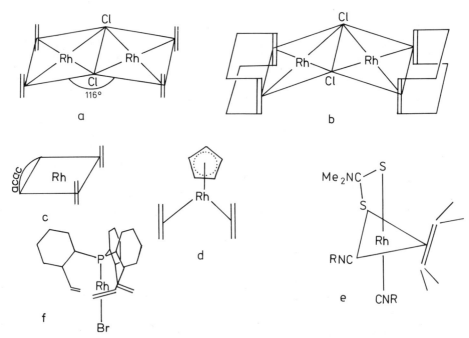

Figure 6.11. Structures of (a) $[Rh_2Cl_2(C_2H_4)_4]$, (b) $[Rh_2Cl_2(COD)_2]$, (c) $[Rh(acac)(C_2H_4)_2]$, (d) $[Rhcp(C_2H_4)_2]$, (e) $[Rh(Me_2NCSS)(p\text{-}MeC_6H_4NC)_2(C_2R_4)]$, (f) $[RhBr(o\text{-}CH_2{=}CHC_6H_4)_3P]$.

The Rh(III) ethyl complex is an active catalyst for ethylene dimerization. Properties and structures of olefin complexes of iridium(I) are similar to those of corresponding rhodium(I) complexes. Ir(I) compounds have square-planar (16e four-coordinate compounds) or trigonal–bipyramidal (18e five-coordinate) structures. $[Ir_2Cl_2(CDO)_2]$ (CDO = COD, NBD) and $[Ir_2Cl_2(C_8H_{14})_4]$ readily react with various ligands to give olefin substitution products or cleavage of the bridging groupings. The cyclooctene complex reacts at low temperatures with ethylene according to this equation:

$$[Ir_2Cl_2(C_8H_{14})_4] + 8C_2H_4 \longrightarrow 2[IrCl(C_2H_4)_4] + 4C_8H_{14} \qquad (6.73)$$

The above reaction furnishes a coordination compound containing four olefin molecules; this is one of the few examples of such complexes. This shows that Ir(I) possesses better coordination abilities than Rh(I) with respect to the olefin. In the case of rhodium, 18e complexes containing chelating olefins which have three coordinated olefin bonds are known: $[RhX\{P(CH_2CH_2CH{=}CH_2)_3\}]$ and $[RhX\{P(C_6H_4CH{=}CH_2)_3\}]$. These compounds have trigonal–bipyramidal structures with phosphorus and halogen atoms occupying the axial position and the three olefin bonds coordinating in the plane perpendicular to the $P{-}Rh{-}X$ axis (Figure 6.11). Based on $^1H$ NMR spectra, the compound $[IrCl(C_2H_4)_4]$ was found to have a trigonal–bipyramidal structure with three ethylene molecules occupying equatorial positions. It is stable at 223 K; however, it readily undergoes decomposition with ethylene evolution.

$$2[IrCl(C_2H_4)_4] \longrightarrow [Ir_2Cl_2(C_2H_4)_4] + 4C_2H_4 \qquad (6.74)$$

The structure of the dimeric compound is analogous to that of the corresponding rhodium complex. Some reactions of $[Ir_2Cl_2(C_8H_{14})_4]$ are presented in Figure 6.12.

Chloroiridic acid, $H_2IrCl_6 \cdot 6H_2O$, and iridium(III) chloride, $IrCl_3 \cdot 4H_2O$, react with cyclic monoolefins and cyclooctadiene in aqueous ethyl alcohol to give $18e$ olefin–carbonyl complexes: $[IrCl(C_7H_{12})_3(CO)]$ and $[IrCl(C_8H_{14})_3(CO)]$. At higher temperatures at which syntheses of these compounds take place, decomposition of the

*Table 6.16. Olefin Complexes of Co, Rh, and Ir*

| Compound | Color | Mp (K) | IR spectra |
|---|---|---|---|
| $[Cocp(FN)(PPh_3)]^a$ | Dark red | | |
| $[Co(MA)(NO)(PPh_3)_2]^b$ | | | |
| $[Co(BD)(CH_2CHCHCH_2CH(Me)CH=CH_2)]^c$ | Red | 309.5–310 | |
| $[Rh_2Cl_2(C_2H_4)_4]^d$ | Orange-red | 388 dec | |
| $[Rh_2Cl_2(COD)_2]^e$ | Yellow | 529–531 dec | |
| $[Rh_2Cl_2(C_8H_{14})_4]^f$ | Yellow | 423 dec | $v(C=C)$ 1555; $\Delta v(C=C)$ 95 |
| $[Rh(acac)(C_2H_4)_2]^g$ | Orange-yellow | 417–419 | |
| $[Rhcp(C_2H_4)_2]^h$ | Yellow | 345–346 | |
| $[Rh(Me_2NCS_2)(2,4,6-Me_3C_6H_2NC)_2(TCNE)]^i$ | Yellow | 483–485 | |
| $[RhCl(CO)(PPh_3)_2TCNE]^j$ | Gold | | |
| $[Rh_2Cl_2(C_2H_4)_2(CO)_2]^k$ | Dark red | 386–387 dec | |
| $[RhCl(C_2H_4)(PPh_3)_2]^l$ | Light-yellow | | |
| $[Rh_2Cl_2(norbornene)_4]^f$ | Yellow-brown | 413–418 dec | |
| $[Ir_2Cl_2(C_8H_{14})_4]^{m,n}$ | Yellow | 428 dec | |
| $[Ir_2Cl_2(COD)_2]^p$ | Orange-red | >473 dec | |
| $[IrCl(C_2H_4)_4]^o$ | White | | |
| $[IrCl(C_7H_{12})_3(CO)]^p$ | White | 441–443 dec | $v(CO)$ 1990 |
| $[IrCl(C_8H_{14})_3(CO)]^p$ | White | 443–448 dec | $v(CO)$ 1996 |
| $[IrCl(C_8H_{14})_2(CO)]_2^p$ | Lemon-yellow | 418–423 dec | $v(CO)$ 1996 |
| $[IrCl(CO)(PPh_3)_2(TCNE)]^q$ | White | 538–543 dec | $v(CO)$ 2060; $v(CN)$ 2235 |
| $[IrBr(CO)(PPh_3)_2(TCNE)]^{q,r}$ | Light-yellow | 497–503 dec | $v(CO)$ 2056; $v(CN)$ 2233 |
| $[Ir_2Cl_4(COD)_2H_2]^s$ | Cream | >473 dec | |

[a] H. Yamazaki and N. Hagihara, *Bull. Chem. Soc. Jpn.*, **44**, 2260 (1971).
[b] G. LaMonica, G. Navazio, P. Sandrini, and S. Cenini, *J. Organomet. Chem.*, **31**, 89 (1971).
[c] G. Natta, U. Giannini, P. Pino, and A. Cassata, *Chim. Int. Milan*, **47**, 524 (1965); G. Allegra, F. LoGiudice, G. Natta, U. Giannini, G. Fagherazzi, and P. Pino, *Chem. Commun.*, 1263 (1967).
[d] R. Cramer, *Inorg. Chem.*, **1**. 722 (1962).
[e] J. Chatt and L. M. Venanzi, *J. Chem. Soc.*, 4735 (1957).
[f] G. Winkhaus and H. Singer, *Chem. Ber.* **99**, 3602 (1966); L. Porri and A. Lionetti, *J. Organomet. Chem.*, **6**, 422 (1966).
[g] R. Cramer, *J. Am. Chem. Soc.*, **86**, 217 (1964).
[h] R. B. King, *Inorg. Chem.*, **2**, 528 (1963).
[i] T. Kaneshima, Y. Yumoto, K. Kawakami, and T. Tanaka, *Inorg. Chim. Acta*, **18**, 29 (1976).
[j] W. H. Baddley, *J. Am. Chem. Soc.*, **88**, 4545 (1966).
[k] J. Powell and B. L. Shaw, *J. Chem. Soc. (A)*, 211 (1968).
[l] J. A. Osborn, F. H. Jardine, J. F. Young, and G. Wilkinson, *J. Chem. Soc. (A)*, 1711 (1966).
[m] B. L. Shaw and E. Singleton, *J. Chem. Soc. (A)*, 1683 (1967).
[n] M. Zuber and F. Pruchnik, *React. Kinet. Catal. Lett.*, **4**, 281 (1976).
[o] M. A. Bennett and D. L. Milner, *J. Am. Chem. Soc.*, **91**, 6983 (1969).
[p] G. Winkhaus and H. Singer, *Chem. Ber.*, **99**, 3610 (1966).
[q] W. H. Baddley, *J. Am. Chem. Soc.*, **90**, 3705 (1968).
[r] J. A. McGinnety and J. A. Ibers, *Chem. Commun.*, 235 (1968).
[s] S. D. Robinson and B. L. Shaw, *J. Chem. Soc.*, 4997 (1965).

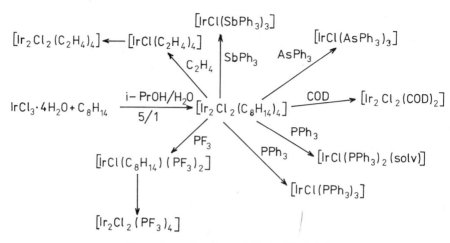

Figure 6.12.   Reactions of $[Ir_2Cl_2(C_8H_{14})_4]$.

alcohol occurs followed by the addition of a carbonyl group (arising from alcohol) to iridium. Reactions leading to formation of metal carbonyls from organic compounds are characteristic for the platinum group, particularly for Ru, Os, Rh, and Ir. Such reactions most easily occur for iridium. Coordination of the CO group, which comes from solvents or organic compounds added to the reaction mixture as reagents, easily occurs in the case of aldehydes, ketones, formamide, dimethylformamide, formic acid, as well as primary alcohols. The CO group extraction from secondary alcohols is considerably more difficult and practically does not take place in the case of tertiary alcohols. Consequently, it is convenenient to prepare $[Ir_2Cl_2(C_8H_{14})_4]$ in isopropyl alcohol instead of in primary alcohols. The major products of ethanol decomposition are CO and $CH_4$.

Monomeric carbonyl complexes rapidly undergo dimerization:

$$2[IrCl(C_8H_{14})_3(CO)] \rightleftharpoons [Ir_2Cl_2(C_8H_{14})_4(CO)_2] + 2C_8H_{14} \qquad (6.75)$$

This reaction is reversible. The dimer is formed during washings of the monomer with chloroform or ethyl ether. As in $[M_2Cl_2(C_8H_{14})_4]$ (M = Rh, Ir) complexes, cycloolefin molecules may be substituted by various ligands such as dienes, phosphines, arsenes, etc. In contrast to rhodium, iridium in addition to $Ir_2Cl_2(COD)_2$ also forms the hydrido complex $[Ir_2Cl_4H_2(C_8H_{12})_2]$ (Figure 6.13).

Olefin Ir(I) complexes containing phosphine ligands are also known: $[Ir_2Cl_2(C_2H_4)_2(PPh_3)_2]$, $[IrCl(C_2H_4)(PPh_3)_2]$, $[IrCl(C_2H_4)_2(PPh_3)_2]$,[42] and $[Ir(C_2H_4)_2(C_6H_4PPh_2)(PPh_3)]$.[43] Chelating ligands such as 2,2-bipyridine and 1,10-phenanthroline form cationic four- and five-coordinate olefin complexes with iridium[44]: $[Ir(olefin)(COD)(chel)]^+$ (chel = bipy, phen; olefin = $C_2H_4$, 1-hexene, AN, FN, TCNE, MA, BD), and $[Ir(COD)(chel)]^+$.

### g. *Group 10 metal complexes*

Alkene complexes of these metals may have the following structures: square-planar, trigonal–bipyramidal, tetrahedral, and trigonal. The first two types are encountered for

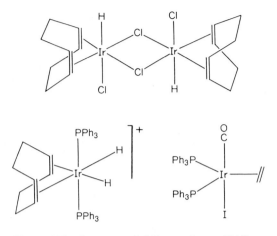

Figure 6.13. Structures of olefin complexes of iridium.

metals in their $+2$ oxidation states with $d^8$ central atom electronic configuration. The remaining two structural types are formed by $d^{10}$ metals possessing zero oxidation states. Not long ago, the following Pt(IV) and Pt(II) olefin complex was obtained and studied: $[\text{Cl(cyclonon-6-ene-1,3-diyl)} \, \text{Pt}^{IV}(\mu\text{-Cl})_2 \, \text{Pt}^{II}\text{Cl}(C_2H_4)]$ (Figure 6.18 below). It is

formed by the reaction of $\text{Pt}_2(\mu\text{-Cl})_2\text{Cl}_2$ with $[\text{Pt}_2\text{Cl}_4(C_2H_4)_2]$ in

chloroform at 378 K.[113] In this complex, the coordination number of platinum(IV) is 6 and that of platinum(II) is 4. Only nickel and dienes form tetrahedral complexes, for example, Ni(COD)$_2$. Nickel(0) olefin compounds are quite stable and numerous, while nickel(I) and nickel(II) alkene complexes are rare, i.e., $[\text{Ni(acac)(COD)}]$ and $[\text{NiR}_2(\text{bipy})(\text{AN})]$.

Nickel(II) forms many relatively stable allyl complexes. In the case of Pt(II), the situation is reversed; numerous, stable olefin compounds are known, while allyl complexes are quite scarce. Intermediate properties are exhibited by Pd(II) which forms many of both olefin and allyl complexes. However, allyl complexes are more stable because they may be obtained from olefin complexes at higher temperatures. This suggests that in π-complexes, the chemical properties of Pd(II) more closely resemble those of Ni(II) than Pt(II).

Nickel(0) complexes containing phosphines, phosphites, isonitriles, etc., react with olefins in solutions to give trigonal compounds, $[\text{NiL}_2(\text{olefin})]$ ($L = PR_3$, $P(OR)_3$, bipy, CNR; olefin $= C_2H_4$, MA, $C_nH_{2n}$, FN, MN, etc.) (Tables 6.11 and 6.17; Figures 6.3 and 6.14).

The same type of complex is also represented by $[\text{Ni}(C_2H_4)_3]$, which is formed by reaction of ethylene with $[\text{Ni}(t,t,t\text{-CDT})]$:

$$[\text{Ni}(t,t,t\text{-CDT})] + 3C_2H_4 \xrightarrow{\ 273 \text{ K}\ } [\text{Ni}(C_2H_4)_3] + t,t,t\text{-CDT} \qquad (6.76)$$

This is a thermally unstable compound in which the coordinated molecules rapidly exchange with free ethylene.[45]

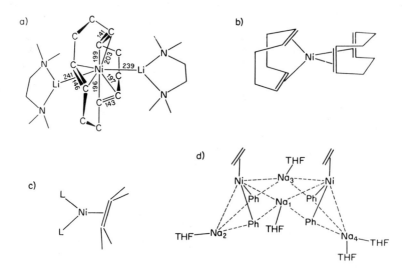

Figure 6.14. Structures of (a) $Li_2(Me_2NCH_2CH_2NMe_2)_2Ni(CDT)$, (b) $Ni(COD)_2$, (c) $Ni(C_2R_4)L_2$, (d) $[(NaPh)_4(THF)_5(NiC_2H_4)_2]$.

The nickel complex with $t,t,t$-cyclododecatriene is also trigonal; it may be obtained by reaction of $[Ni(C_3H_5)_2]$ with butadiene first affording an allyl complex, which subsequently undergoes cyclization to form the olefin compound.

$$[Ni(C_3H_5)_2] + 2\,BD \longrightarrow \quad \cdots \text{Ni} \cdots \xrightarrow{\;BD\;}$$

$$\longrightarrow \quad \cdots \text{Ni} \cdots \longrightarrow \quad \text{Ni} \qquad (6.77)$$

Such processes are essential in catalytic reactions of oligomerization and polymerization of butadiene. Nickel ethylene complexes are also formed during condensation of nickel vapors with ethylene in argon matrices. Compounds including the following compositions are formed: $[Ni(C_2H_4)]$, $[Ni(C_2H_4)_2]$, and $[Ni(C_2H_4)_3]$ (Table 6.10). In contrast to $[PdL_4]$ and $[PtL_4]$ the complex $[Ni(PPh_3)_4]$ reacts with $C_2F_4$ to give an oxidative coupling cycloaddition product (Table 1.4):

$$[Ni(PPh_3)_4] + 2C_2F_4 \longrightarrow (Ph_3P)_2Ni \overset{CF_2-CF_2}{\underset{CF_2-CF_2}{\Big|}} \qquad (6.78)$$

*Table 6.17. Olefin Complexes of Ni, Pd, and Pt*

| Compound | Color | Mp (K) | Remarks |
|---|---|---|---|
| $[Ni(AN)_2]^a$ | Red | 378 dec | $v(CN)$ 2200; $v(C=C)$ 1446; $v_{as}(Ni-AN)$ 451 |
| $[Ni(CH_2CHCHO)_2]^b$ | Violet | 363–368 dec | $v(C=C)$ 1605; $v(C=O)$ 1518 |
| $[Ni(COD)_2]^c$ | Yellow | | |
| $[Ni(C_2H_4)(PPh_3)_2]^d$ | Yellow | | $v(C=C)$ 1195; $\tau(C_2H_4)$ 7.27 |
| $[Ni(C_2H_4)_3]^{(45)}$ | White | 273 dec | |
| $[Ni(C_{12}H_{18})]^c$ | Dark-red | 413–423 dec | |
| $[Pd_2Cl_4(C_2H_4)_2]^e$ | Yellow | >445 dec | Stable <403 K |
| $[PdCl_2(C_2H_4)(ONC_5H_5)]^f$ | Gold-brown | 380–382 | |
| $[Pd(TCNE)(PPh_3)_2]^g$ | | 529–530 dec | |
| $[Pd(MA)(PPh_3)_2]^g$ | Light-yellow | 420–424 dec | |
| $[PdCl_2(CH_2=CHCH_2C_6H_4PPh_2)]^h$ | Yellow | 463–464 dec | $v(C=C)$ 1530 |
| $[PdCl_2(COD)]^i$ | Light-orange | 478–483 dec | |
| $[Pd(C_2H_4)(PPh_3)_2]^j$ | White | | |
| $[Pt(C_2H_4)(PPh_3)_2]^k$ | White | 395–398 dec | |
| $[Pt(PhCHCHPh)(PPh_3)_2]^l$ | White | 404–414 dec | |
| $[Pt(TCNE)(PPh_3)_2]^{(47)}$ | White | 541–542 dec | |
| $[Pt_2Cl_4(C_2H_4)_2]^m$ | Orange | 463–483 dec | |
| $K[PtCl_3(C_2H_4)]\cdot H_2O^n$ | Yellow | 473 dec | |
| $[Pt_2Cl_4(cyclohexene)_2]^m$ | Light-orange | 418–419 dec | |
| $[Pt_2Cl_4(CH_2CHPh)]^m$ | Orange | 442–444 | |
| $t\text{-}[PtCl_2(C_2H_4)(NH_3)]^o$ | Yellow | 396–405 dec | |
| $cis\text{-}[PtCl_2(C_2H_4)(NH_3)]^o$ | Green-yellow | 386–392 dec | |
| $[PtCl_2(CH_2=CHCH_2C_6H_4PPh_2)]^h$ | White | 429–430 | $v(C=C)$ 1497; $\Delta v(C=C)$ 143 |

[a] G. N. Schrauzer, *J. Am. Chem. Soc.*, **81**, 5310 (1959); G. N. Schrauzer, *Adv. Organomet. Chem.*, **2**, 2 (1964); H. P. Fritz and G. N. Schrauzer, *Chem. Ber.*, **94**, 650 (1961).

[b] R. J. Cvetanovic, F. J. Duncan, W. E. Falconer, and R. S. Irvin, *J. Am. Chem. Soc.*, **87**, 1827 (1965).

[c] G. Wilke, *Angew. Chem.*, **72**, 581 (1969); G. Wilke, *Angew. Chem.*, **75**, 10 (1963).

[d] G. Wilke and G. Herrmann, *Angew. Chem., Int. Ed. Engl.*, **1**, 549 (1962).

[e] A. D. Ketley, L. P. Fischer, A. J. Berlin, C. R. Morgan, E. H. Gorman, and T. R. Steadman, *Inorg. Chem.*, **6**, 657 (1967).

[f] W. H. Clement, *J. Organomet. Chem.*, **10**, P19 (1967).

[g] P. Fitton and J. E. McKeon, *Chem. Commun.*, 4 (1968).

[h] M. A. Bennett, W. R. Kneen, and R. S. Nyholm, *Inorg. Chem.*, **7**, 556 (1968).

[i] J. Chatt, L. M. Vallarino, and L. M. Vananzi, *J. Chem. Soc.*, 3413 (1957).

[j] R. Van der Linde and R. O. Jongh, *Chem. Commun.*, 653 (1971).

[k] C. D. Cook and G. S. Jauhal, *Inorg. Nucl. Chem. Lett.*, **3**, 31 (1967); C. D. Cook and G. S. Jauhal, *J. Am. Chem. Soc.*, **90**, 1464 (1968).

[l] J. Chatt, B. L. Shaw, and A. A. Williams, *J. Chem. Soc.*, 3269 (1962).

[m] M. S. Kharasch and T. A. Ashford, *J. Am. Chem. Soc.*, **58**, 1733 (1936); J. Chatt and L. A. Duncanson, *J. Chem. Soc.*, 2939 (1953).

[n] W. C. Zeise, *Pogg. Ann.*, **21**, 497 (1831); J. Jira, J. Sedlemeier, and J. Smidt, *Ann. Chem.*, **693**, 99 (1966); E. W. Abel, M. A. Bennett, and G. Wilkinson, *Proc. Chem. Soc.*, 152 (1958).

[o] I. I. Chernyaev and A. D. Gelman, *Dokl. Akad. Nauk SSSR*, **4**, 181 (1936); A. N. Shidlovskaya and Y. K. Syrkin, *Dokl. Akad. Nauk SSSR*, **55**, 231 (1947).

As mentioned earlier, nickel forms tetrahedral olefin complexes, viz. $[Ni(COD)_2]$. It is obtained by reduction of nickel(II) acetylacetonate by means of trialkylaluminum in the presence of 1,5-cyclooctadiene. Analogous nickel complexes containing other dienes are also known.

Nickel(0) compounds of the type $[(MR)_x (n\text{-donor–ligand})_y (Ni(0)(alkene)_z]$ where M = Li, Na, K, ..., constitute an interesting group. Such complexes are formed by reactions of Ni(0) olefin complexes with lithium organometallic compounds.[60] If the ratio MR/Ni = 1, ionic compounds $M^+[NiR(alkene)_n]^-$ are formed, while if the ratio MR/Ni = 2 or 3, nonionic compounds are obtained. Nonionic complexes exhibit characteristic phenyl bridging ligands between M and Ni as well as bonds between alkali metal cations and olefins coordinated to nickel. Typical examples of complexes of the first group are: $[Li(THF)_4]^+[NiPh(CDT)]^-$ and $[Li(TMED)_2]^+[NiR(C_2H_4)_2]^-$ (R = Me, Et, Ph). Ionic compounds are also formed by reactions of olefin nickel complexes with phosphides and hydrides.

$$Ni(CDT) + LiPh \xrightarrow[233\,K]{THF} [Li(THF)_4]^+[PhNi(CDT)]^- \tag{6.79}$$

$$Ni(CDT) + LiR + 2TMED + 2C_2H_4 \xrightarrow{Et_2O} [Li(TMED)_2]^+[NiR(C_2H_4)_2]^- \tag{6.80}$$

$$2Ni(C_2H_4)_3 + Li[AlBu_2^i H_2] + 2TMED$$

$$\xrightarrow[233\,K]{Et_2O} [Li(TMED)_2]^+[(C_2H_4)_2Ni\text{-}H\text{-}Ni(C_2H_4)_2]^- + C_2H_4 + AlEtBu_2^i \tag{6.81}$$

$$2Ni(COD)_2 + LiPPh_2 + 4C_2H_4$$

$$\xrightarrow{THF} [Li(THF)_4]^+[(C_2H_4)_2Ni(\mu\text{-}PPh_2)Ni(C_2H_4)_2]^- \tag{6.82}$$

$$Ni(CDT) + 2(m\,NaPh + n\,LiPh) \xrightarrow[C_2H_4]{Et_2O} m[\{NaPh\}_2(Et_2O)_2\}NiC_2H_4]$$

$$+ n[\{(LiPh)_2(Et_2O)\}NiC_2H_4] \tag{6.83}$$

The compound $[(NaPh)_4(THF)_5(NiC_2H_4)_2]$, which forms as a result of an exchange of ethyl ether molecules for tetrahydrofuran ligands in the complex according to reaction (6.83), has its structure represented in Figure 6.14d. Sodium and α-carbon atoms of the phenyl rings form an arc-like, eight-membered ring. The α-carbon atoms of the phenyl group are almost coplanar with the plane in which the nickel atom and ethylene carbon atoms are located. In contrast to $Na_4$, $Na_2$ coordinates only one tetrahydrofuran molecule; thus, the free coordination site is used for the formation of intermolecular bridging bonds with a partially negative neighboring ethylene molecule of $[(NaPh)_4(THF)_5(NiC_2H_4)_2]$. This situation leads to the formation of a chain-like polymeric structure.

Different types of compounds are formed by reactions of nickel olefin complexes with alkali metals. In these compounds, both alkene molecules and alkali metal atoms are bonded to nickel (Figure 6.14a).[60, 75–77]

$$Ni(CDT) + 2Li + 2TMED \xrightarrow[273\,K]{THF} [(LiTMED)_2(NiCDT)] \tag{6.84}$$

$$Ni(COD)_2 + 2Li \xrightarrow[243-273\,K]{THF} [Li(THF)_2]_2 Ni(COD)_2 \tag{6.85}$$

$$[Li(THF)_2]_2 Ni(COD)_2 + z\,alkene + 2y\,L \longrightarrow [(LiL_y)_2 Ni(alkene)_z] \tag{6.86}$$

The olefin bond distance in $(LiTMED)_2NiCDT$ equals 143 pm and is greater than in NiCDT (137 pm).[75] As shown by NMR spectra in complexes $[Li(THF)_4]_2 Ni(COD)_2$, only three olefin bonds are coordinated to the nickel atom.

Nickel(I) and nickel(II) olefin compounds are rare. Examples are [Ni(acac) (COD)], [NiX(COD)] (X = Br, I),[49] and [NiR(olefin)(bipy)].[46] Alkyl complexes most probably have trigonal–bipyramidal structures. They are formed as unstable intermediate compounds during reactions of [NiR$_2$(bipy)] with olefins. Compounds containing acrolein and acrylonitrile[46] were also isolated:

$$[NiR_2(bipy)] + (olefin) \longrightarrow [NiR_2(olefin)(bipy)] \longrightarrow [Ni(olefin)_n(bipy)] \quad (n = 1, 2) \quad (6.87)$$

Nickel complexes containing acrylonitrile and acrolein, i.e., [Ni(AN)$_2$] and [Ni(CH$_2$CHCHO)$_2$], are rapidly oxidized in air. Depending upon dispersion, their color may be light yellow to red for the weakly dispersed forms. IR spectra, as well as chemical and physical properties such as nonsolubility, nonvolatility of [Ni(AN)$_2$], and readiness to react with Lewis bases, show that this compound is polymeric in the solid state. It is not possible to exclude that acrylonitrile constitutes a bridge which coordinates through the double bond and the nitrile group.

Compounds Ni(olefin)$_2$ have significant application in catalysis. Olefin–carbonyl complexes of the type [Ni(CO)$_x$(olefin)$_{n-x}$] are not known. Their existence as unstable intermediate compounds in many reactions results from catalytic properties of tetracarbonylnickel in olefin oligomerization processes, oxo synthesis, olefin isomerization, etc.

Palladium, as the other metals of this triad, forms trigonal complexes of the PdL$_2$(olefin) type. Such complexes are formed by reactions of PdL$_4$ with olefins:

$$[PdL_4] + olefin \longrightarrow [Pd(olefin)L_2] + 2L \quad (6.88)$$

$$[L = PR_3, P(OR)_3, AsR_3, RNC]$$

Palladium(II) complexes are considerably less stable than those of Pt(II). Stable complexes such as [PdCl$_2$(olefin)$_2$] and [PdCl$_3$(olefin)]$^-$ are not known. The unstable yellow complex [PdCl$_2$(C$_2$H$_4$)$_2$] is obtained by action of ethylene (at 1 MPa pressure) on PdCl$_2$ solutions in organic solvents containing small amounts of ethanol. This compound may be isolated at 230 K; however, after warming to room temperature, it is transformed to the dimer [Pd$_2$Cl$_4$(C$_2$H$_4$)$_2$]. Analogous unstable complexes with other olefins are also known. The compound *cis*-[PdCl$_2$(C$_2$H$_4$)$_2$] is most probably an unstable intermediate which is formed during ethylene dimerization catalyzed by palladium(II) chloride.

Neutral complexes containing pyridine *N*-oxide were obtained: [PdCl$_2$(olefin) (ONC$_5$H$_4$Z)] (Z = H, Me, OMe, NO$_2$). These complexes are formed by reactions involving cleavage of the chloride bridges:

$$[Pd_2Cl_4(C_2H_4)_2] + 2ONC_5H_5 \xrightarrow[C_2H_4]{273 \text{ K}} 2t\text{-}[PdCl_2(C_2H_4)(ONC_5H_5)] \quad (6.89)$$

Complexes [PdCl$_3$(olefin)]$^-$ are also unstable. They are also formed as unstable intermediate compounds during catalytic oxidation of olefins, particularly during oxidation of ethylene to acetaldehyde in the Wacker process. Stability constants for [PdCl$_3$(olefin)]$^-$ complexes are given in Table 6.11.

The formation of [PdCl$_2$(C$_2$H$_4$)$_2$] in the presence of Cl$^-$ ions does not occur, because chloride ions bond more strongly to palladium. However, compounds containing one or two ethylene molecules are formed in aqueous solutions of palladium phosphate, sulfate, and nitrate which do not contain $Cl^-$ ions because these solutions absorb 1.5 mol of ethylene per one mole of palladium. The most stable Pd(II) olefin complexes are those which are dinuclear and contain bridging and terminal chlorine atoms (Figure 6.15). The other palladium halides do not form analogous complexes. The compound $Pd_2Cl_4(olefin)_2$ may form three isomers. In the case of ethylene complexes, a and b were isolated, while for *cis*-pent-2-ene all three isomers were obtained.

Compounds $Pd_2Cl_4(olefin)_2$ are most commonly prepared from bis(benzonitrile)dichloropalladium(II) and corresponding olefin. The benzonitrile ligand can easily be displaced:

$$2[PdCl_2(PhCN)_2] + 2C_2H_4 \longrightarrow [Pd_2Cl_4(C_2H_4)_2] + 2PhCN \qquad (6.90)$$

Chelating monoolefin ligands form complexes of the composition $PdCl_2(CMO)$, for example: $[PdX_2(CH_2{=}CHCH_2C_6H_4PPh_2)]$ (X = Cl, Br, I, NCS). Compounds possessing similar compositions are also formed by chelating dienes, viz. $PdCl_2(CDO)$, where CDO = COD, NBD, COT, and DHMB. Olefin complexes containing cyclopentadienyl ligands are also known, such as $[Pdcp(alken)L]^+$ (L = phosphine).

Platinum forms trigonal, square-planar, and pentacoordinate olefin complexes. All these structural types are relatively stable. Most Pt(II) compounds are stable in air, and their decomposition temperatures are high. Trigonal complexes are most commonly obtained by ligand substitution reactions of $PtL_4$ or by olefin exchanges. Complexes with alkenes and alkenes containing withdrawing substituents of the following compositions are known: $[PtL_2(olefin)]$ and $[Pt(olefin)_3]$ where L = $PR_3$, $AsR_3$, CNR, $P(OR)_3$. The stability of olefin coordination compounds of the type $[PtL_2(olefin)]$ increases if the $\pi$ acidity of the olefin becomes greater, i.e., with the increase in electron-withdrawing properties of substituents. Olefins possessing strong $\pi$ acidic properties displace less acidic alkenes from a given complex. It is believed that, in most cases, the olefin exchange proceeds according to an associative mechanism which involves 18*e* compounds of the type $PtL_2L'L''$ where L' and L'' are olefin molecules. Molecules $[ML_2(olefin)]$ (M = Ni, Pd, Pt) react with alkyl, aryl, and acyl halides to give complexes in which the oxidation state of the metal is +2 by oxidative addition.

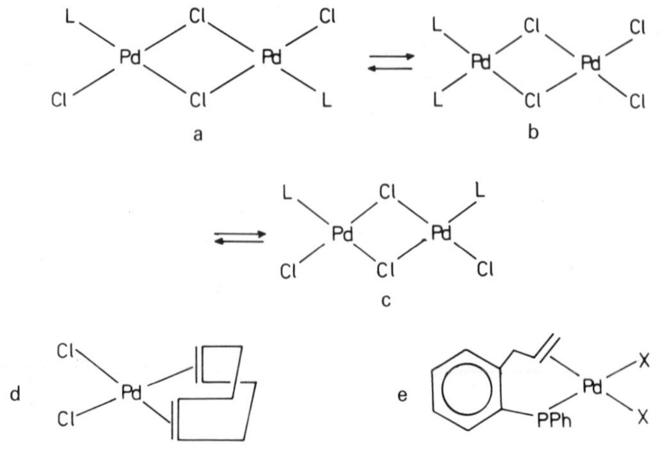

Figure 6.15.  Structures of olefin complexes of palladium.

The olefin is expelled from the coordination sphere according to a dissociative mechanism:

$$[Pt(C_2H_4)(PPh_3)_2] \xrightarrow{\text{fast}} [Pt(PPh_3)_2] \tag{6.91}$$

$$[Pt(PPh_3)_2] + RX \xrightarrow{\text{slow}} [PtX(R)(PPh_3)_2] \tag{6.92}$$

Ethylene complexes $[M(C_2H_4)(PPh_3)_2]$ (M = Pt, Ni) react with peroxides ROOR, disulfides RSSR, and $Sn_2Me_6$ to form $[M(ER)_2(PPh_3)_2]$ (E = O, S) and $[M(SnMe_3)_2(PPh_3)_2]$; these reactions involve O−O, S−S, and Sn−Sn bond breaking.

Platinum(II) forms 16*e* square-planar complexes and 18*e* compounds possessing trigonal–bipyramidal structures. Such compounds are prepared by ligand substitution, olefin addition, dehydration of alcohols, olefin exchange reactions, etc. Platinum(II) and platinum(IV) halides as well as halogenoplatinates(II) and halogenoplatinates(IV) are most commonly utilized for preparation of olefin platinum complexes. Olefin compounds of platinum(IV) are known. Platinum(II) forms the following types of olefin compounds: $[PtX_3(\text{olefin})]^-$, *cis*-$[PtX_2(\text{olefin})_2]$, *trans*-$[PtX_2(\text{olefin})_2]$, *cis*-$[PtX_2$ (olefin)L], *trans*-$[PtX_2(\text{olefin})L]$, $[PtX_2(CDO)]$, $[PtX_2(CMO)]$ (CDO = COD, NBD, hexa-1,5-diene, DHMB, COT; CMO = $CH_2$=$CHCH_2C_6H_4PPh_2$, $CH_2$=$CHCH_2C_6H_4AsPh_2$, $MeCH$=$CHC_6H_4PPh_2$, etc.). Platinum(II) square-planar complexes are diamagnetic and range in color from yellow to red in the case of dinuclear compounds $[Pt_2Cl_4(\text{olefin})_2]$. Most complexes containing coordinated ammonia are light green. Olefin complexes react with various ligands; thus, reactivity of these complexes allows for their use in diverse syntheses.

Typical reactions of $[Pt_2Cl_4(C_2H_4)_2]$ and $[PtCl_3(C_2H_4)]^-$ are given in Figures 6.16 and 6.17.

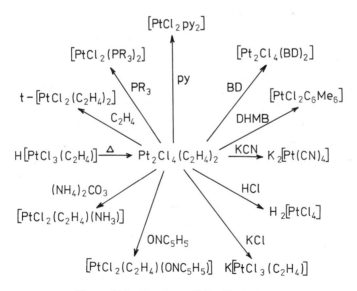

Figure 6.16.   Reactions of $[Pt_2Cl_4(C_2H_4)_2]$.

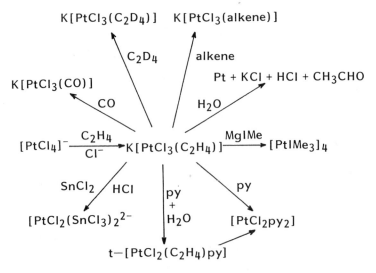

Figure 6.17.   Reactions of $K[PtCl_3(C_2H_4)]$.

Complexes such as $t\text{-}[PtX_2(\text{alkene})_2]$ are more difficult to prepare because the following equilibrium exists:

$$2[PtCl_2L_2] \rightleftharpoons [Pt_2Cl_4L_2] + 2L \qquad (6.93)$$

This equilibrium is strongly shifted to the right in the case of ligands possessing strong $\pi$ acceptor properties, for example, $C_2H_4$, CO, and $PF_3$, and which show strong *trans* effect. Nevertheless, the complex $[PtCl_2(CO)_2]$ is transformed into the dinuclear compound above 370 K. For $t\text{-}[PtCl_2(C_2H_4)_2]$, the equilibrium is shifted completely in the direction of the dinuclear compound at 267 K. Therefore, preparation of this compound in pure state, as well as preparation of $cis\text{-}[PtCl_2(C_2H_4)_2]$, is very difficult. Complexes of the type $[PtCl_2(\text{alkene})_2]$ with olefins possessing higher molecular masses are relatively more stable.

As far as acyclic hydrocarbon olefins are concerned, the most stable compound is formed by 1-octene.

$$[Pt_2Cl_4(C_8H_{16})_2] + 2C_8H_{16} \longrightarrow 2cis\text{-}[PtCl_2(C_8H_{16})] \qquad (6.94)$$

Olefins such as 1-heptene and 1-nonene do not form analogous compounds with platinum; however, the cyclooctene compound $PtCl_2(C_8H_{14})_2$ is stable. Even more stable are compounds containing chelating monoolefins and diolefins, e.g., $[PtX_2(CMO)]$ and $[PtX_2(CDO)]$.

Platinum(II) 18e pentacoordinate complexes are less stable than square-planar compounds due to a relatively great difference between $6p$ and $6s$ orbital energies (see Chapter 1 and discussion of stability of Ni, Pd, and Pt carbonyls). The stable complex $[Pt(CN)H(PEt_3)_2(TCNE)]$ possessing CN and H ligands which are in *trans* position to each other was isolated.[47] Addition of TCNE to $trans\text{-}[PtX(H)(PR_3)_2]$ leads to reductive elimination of HX where $X = Cl$ or Br and to the formation of the Pt(0) complex $Pt(TCNE)(PR_3)_2$. Unstable olefin compounds possessing trigonal–bipyramidal

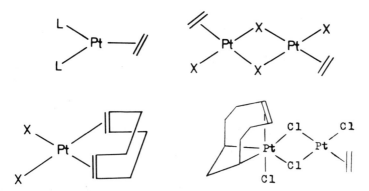

Figure 6.18. Structures of olefin complexes of platinum.

structures are also known: $[PtCl_2(C_2H_4)py_2]$, $[PtCl_2(C_2H_4)(bipy)]$, and $[PtCl_2(C_2H_4)(acac)]$.[48]

## h. *Group 11 metal complexes*

Copper and silver form olefin complexes only if the metal has $+1$ oxidation state. Gold, in addition to Au(I) compounds, also forms polynuclear coordination compounds containing Au(I) and Au(III) (Figures 6.19 and 6.20, Table 6.18). The IR, [1]H

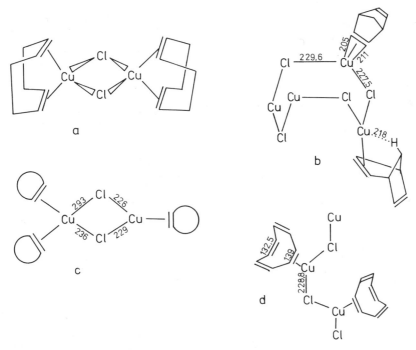

Figure 6.19. Structures of (a) $Cu_2Cl_2(COD)_2$, (b) $[CuCl(NBD)]_4$, (c) $Cu_2Cl_2(C_8H_{14})_3$, (d) $[CuCl(COT)]_n$.

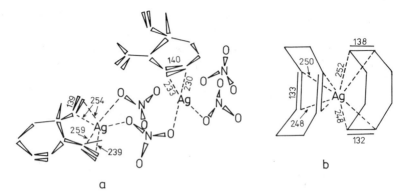

Figure 6.20. Structures of (a) $C_{12}H_8(AgNO_3)_2$, *J. Chem. Soc.* (B), 965 (1970); (b) $Ag(COD)_2BF_4$, *J. Organomet. Chem.*, **182**, 269 (1979).

NMR, $^{13}C$ NMR, etc. spectra show that the Cu(I)–alkene and Ag(I)–alkene bonds are weak, rendering Cu(I) and Ag(I) complexes unstable. Gold(I) olefin compounds, however, are characterized by somewhat greater stability comparable with that of Pd(II) compounds.

Copper, silver, and gold form complexes with olefins of the following composition: $[CuX(olefin)]$ (X = Cl, Br; olefin = ethylene, propylene, butenes, COT, NBD, etc.; dienes in these complexes are monodentate ligands), $[Cu_2X_2(COD)_2]$, $[Cu_2X_2(CMO)_2]$ $[CMO = CH_2CHCH_2C_6H_4PPh_2$, $CH_2C(Me)CH_2C_6H_4PPh_2$; X = Cl, Br, I], $[Cu_2X_2(diene)]$ (diene = cyclopolyolefins), $[CuCl(NBD)]_4$, $[CuCl(COT)]_n$, $[Cu_2Cl_2(cyclooctene)_3]$, $[Ag(alkene)_2]X$, $[Ag(alkene)_3]BF_4$, $[Ag(diene)]X$ (alkene = ethylene, propene, cyclohexene, etc.; diene = cyclopentadiene,

Table 6.18. *Olefin Complexes of Cu, Ag, and Au*

| Compound | Color | Mp (K) | IR spectra |
|---|---|---|---|
| $[CuCl(AN)]^a$ | White | | $v(C=C)$ 1502; $\Delta v(C=C)$ 101 |
| $[CuCl(CH_2=CHCHO)]^b$ | Yellow | 313 dec | $v(C=C)$ 1530; $\Delta v(C=C)$ 90 |
| $[Cu_2Cl_2(CH_2CHCH_2C_6H_4PPh_2)_2]^c$ | White | 530–531 | $v(C=C)$ 1562; $\Delta v(C=C)$ 77 |
| $[Cu_2I_2(CH_2=C(Me)CH_2C_6H_4PPh_2)_2]^c$ | | 456–458 | $v(C=C)$ 1575; $\Delta v(C=C)$ 78 |
| $[Ag(C_2H_4)_2]BF_4^d$ | White | | $\Delta v(C=C)$ 38 |
| $[Ag(C_3H_6)_2]BF_4^d$ | White | | $\Delta v(C=C)$ 53 |
| $[Ag(CH_2CHEt)_2]BF_4^d$ | White | 310 dec | $\Delta v(C=C)$ 55 |
| $[Ag(CH_2CHCH_2C_6H_4AsPh_2)]NO_3^c$ | White | | $\Delta v(C=C)$ 50; $v(C=C)$ 1590 |
| $[AuCl(cyclooctene)]^e$ | White | 366–369 dec | $\Delta v(C=C)$ 136; $v(C=C)$ 1512 |
| $[AuCl(norbornene)]^e$ | White | 368 dec | |
| $[AuCl(1-hexadecene)]^e$ | White | 316–318 dec | $\Delta v(C=C)$ 115; $v(C=C)$ 1525 |
| $[AuCl(C_{18}H_{36})]^e$ | | 319–321 dec | $\Delta v(C=C)$ 115; $v(C=C)$ 1525 |

$^a$ G. N. Schrauzer, *Chem. Ber.*, **94**, 1891 (1961).
$^b$ S. Kawaguchi and T. Ogura, *Inorg. Chem.*, **5**, 844 (1966).
$^c$ M. A. Bennett, W. R. Kneen, and R. S. Nyholm, *Inorg. Chem.*, **7**, 552 (1968).
$^d$ H. W. Quinn and D. N. Glew, *Can. J. Chem.*, **40**, 1103 (1962).
$^e$ R. Hüttel, H. Reinheimer, and H. Dietl, *Chem. Ber.*, **99**, 462 (1966); R. Hüttel and H. Reinheimer, *Chem. Ber.*, **99**, 2778 (1966).

COD, NBD, COT; $X = NO_3^-$, $BF_4^-$, $ClO_4^-$), [Ag$_2$diene)]X (diene = NBD, 1,4-COD), [Ag$_2$(NBD)$_3$][BF$_4$]$_2$, [Ag(CMO)]NO$_3$, [AuCl(alkene)], [Au$_2$Cl$_2$(diene)] (alkene = cyclopentene, cyclohexene, cyclooctene, long chain terminal olefins, diene = 1,4-hexadiene, 1,5-hexadiene, COD, NBD), and [Au$_2$Cl$_4$(NBD)$_x$] ($x = 1, 2, 3$). Some of the above-mentioned compounds may be prepared in solid state. Often, alkenes react with metal salts in the solid state to give complexes as a result of incorporating alkene molecules into the crystal lattice of the salt. In such cases, the ease of complex formation depends not only on the electronic structure of the olefin, but also on its geometry. The ease of formation of silver complexes also depends on the silver salt anion; the ease of this formation decreases in the series: $AgSbF_6 > AgBF_4 > Ag_2SiF_6 > AgClO_4 \backsim AgOOCCF_3 > AgNO_3 \gg AgBrO_3$. In the complex [Cu$_2$Cl$_2$(COD)$_2$], the copper atom has tetrahedral coordination.

## 10. *COMPLEXES CONTAINING HETEROOLEFINS*

The formation of multiple bonds is characteristic for the second row of the Periodic Table. The stability of the $E=E$ and $E\equiv E$ bonds in the case of the remaining elements is considerably lower. An example is provided by energies of single and multiple bonds in nitrogen and phosphorus compounds (in kJ mol$^{-1}$): $N-N \backsim 170$, $N=N \backsim 420$, $N\equiv N$ 944, $P-P \backsim 200$, $P=P \backsim 310$, and $P\equiv P \backsim 490$.

The relative increase of bond energies when the bond multiplicity becomes higher is considerably greater for the second-row elements than for heavier homologs.[78] Therefore, in the case of heavier elements, it is more favorable to form two $\sigma$ bonds than one $\sigma$ bond and one $\pi$ bond. Thus, compounds (PR)$_n$ and (SiR$_2$)$_n$ usually have cyclic structures.

$$
\begin{array}{ccc}
\text{R---P------P---R} & & \begin{array}{c} \text{R}_2 \\ \text{Si} \\ \diagup \quad \diagdown \end{array} \\
\text{R---P \quad P---R} & \text{R}_2\text{Si} \qquad \text{SiR}_2 \\
\diagdown \diagup & | \qquad\qquad | \\
\text{P} & \text{R}_2\text{Si} \qquad \text{SiR}_2 \\
| & \diagdown \qquad \diagup \\
\text{R} & \begin{array}{c} \text{Si} \\ \text{R}_2 \end{array}
\end{array}
\qquad (6.95)
$$

Silenes R$_2$Si=SiR$_2$ (for example, R = 2,4,6-Me$_3$C$_6$H$_2$, Bu$^t$, etc.) and diphosphenes RP=PR [for example, R = 2,4,6-Bu$_3'$C$_6$H$_2$, C(SiMe$_3$)$_3$, CH(SiMe$_3$)$_2$] are formed only if the substituents R create so much steric hindrance that the formation of oligomers containing single $E-E$ bonds is precluded.[78–86] The $E=E$ double bond distances ($E=$ Si, Ge, Sn, P, and As) are considerably shorter than in corresponding single bonds. At present, the following examples of transition metal complexes containing RE=ER compounds (R = aryl, alkyl, H; E = N, P, As, Sb,[78, 79, 103, 104] CR$_2$=PR,[82–84] CH$_2$=O,[80] and CH$_2$=E (E = S, Se), are known.[81] In contrast to olefins, diphosphenes (RP=PR) possess free electron pairs localized on phosphorus atoms as do corresponding arsenic and antimony compounds; therefore, these compounds may form the types of complexes shown in structures (6.96). Examples of such complexes are shown in structures (6.97). Owing to the presence of free electron

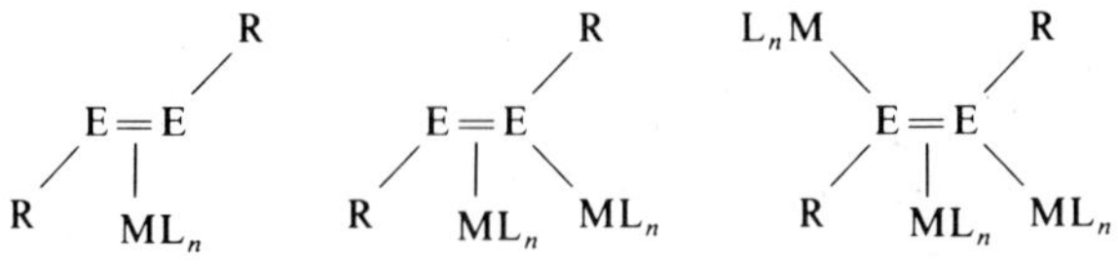

$$(6.96)$$

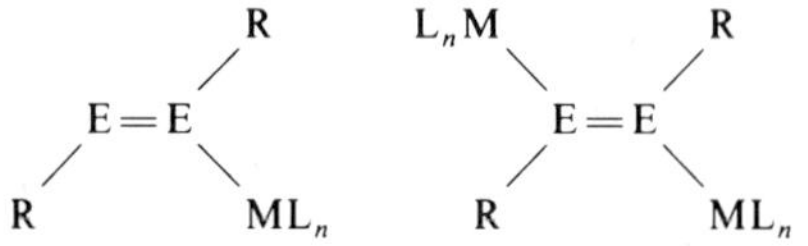

$$E = P, As, Sb$$

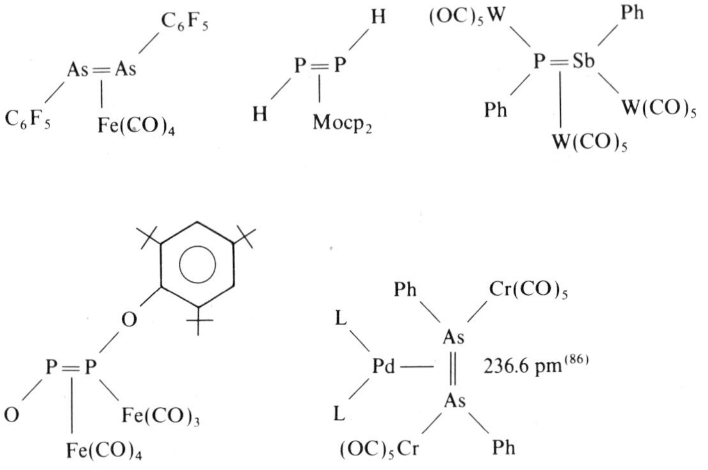

$$(6.97)$$

$$L = P(OMe)_3$$

pairs on the heteroatoms, ligands such as $CR_2=PR'$, $CR_2=O$, and $CR_2=S$ may coordinate in a similar manner.

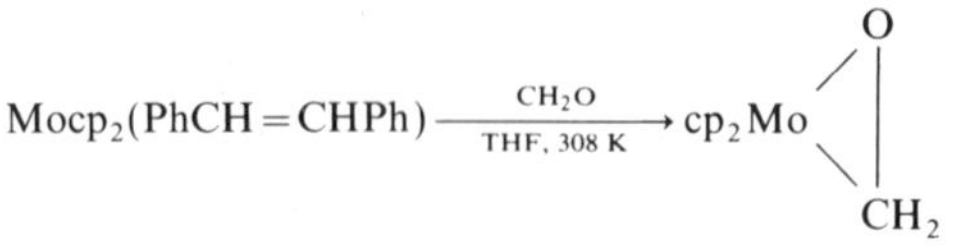

$$Mocp_2(PhCH=CHPh) \xrightarrow[\text{THF, 308 K}]{CH_2O} cp_2Mo \begin{array}{c} O \\ \diagup \\ \diagdown \\ CH_2 \end{array} \qquad (6.98)$$

$$[(Me_3P)cpRh(CH_2I)I] + NaSeH \longrightarrow [Me_3P)cpRh \begin{array}{c} Se \\ \diagup \\ \diagdown \\ CH_2 \end{array} \qquad (6.99)$$

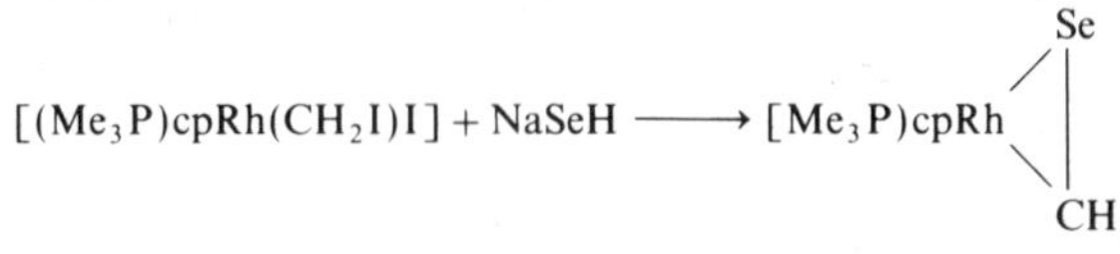

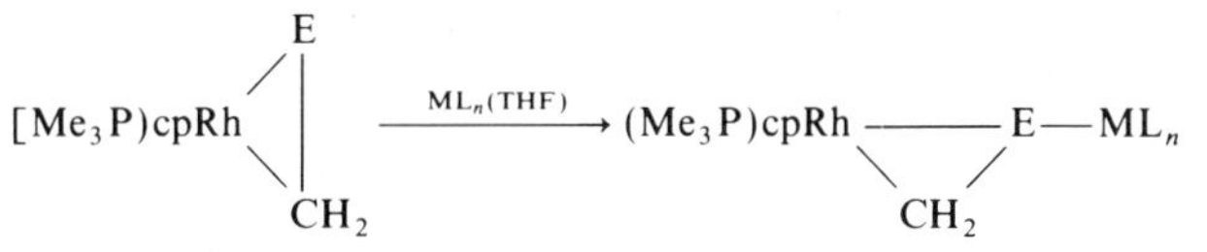

$$[Me_3P)cpRh \overset{E}{\underset{CH_2}{<}} \xrightarrow{ML_n(THF)} (Me_3P)cpRh \overset{}{\underset{CH_2}{<}} E—ML_n \qquad (6.100)$$

1. E = Se    $ML_n = Cr(CO)_5, W(CO)_5, Mncp(CO)_2$
2. E = S    $ML_n = Cr(CO)_5$

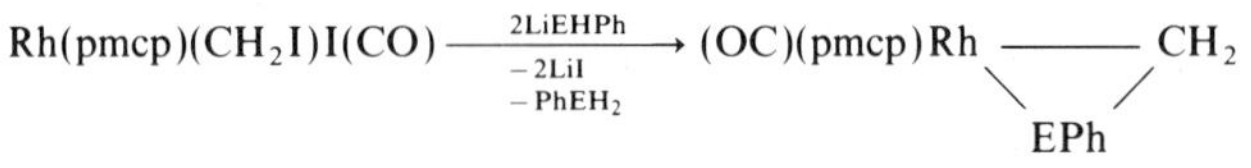

$$Rh(pmcp)(CH_2I)I(CO) \xrightarrow[\substack{-2LiI \\ -PhEH_2}]{2LiEHPh} (OC)(pmcp)Rh \overset{}{\underset{EPh}{<}} CH_2 \qquad (6.101)$$

$$E = P, As^{[84]}$$

## 11. APPLICATIONS OF π-OLEFIN COMPLEXES

### a. Reactions of olefin complexes with nucleophiles

Coordination of an olefin molecule to the metal leads to its activation. This facilitates reactions of alkenes with various compounds. The coordinated olefin may have nucleophilic as well as electrophilic properties, and therefore may react in the first case with electrophiles, and in the second case with nucleophiles.[17, 18, 37, 50] Thus, either the central atom or the alkene may undergo nucleophilic attack. In the case of weak π-olefin Cu(I) and Ag(I) complexes, even weak nucleophiles attack metal ions.

In copper and silver olefin compounds, alkenes readily undergo exchange. Gold compounds are more stable, and their olefins react with nucleophiles. The compound $Au_2Cl_2(cis$-2-butene$)$ reacts with $D_2O$ to give 2-butanone. The following $d^8$ and $d^6$ electron complexes readily react with nucleophiles: Pd(II), Pt(II), Fe(II), etc.

Reactions involving nucleophilic attack on Pd(II) and Pt(II) complexes are best known. For example, cyclooctadiene complexes react according to equation (6.102). Analogous reactions involving diene Pd(II) and Pt(II) complexes also take place with other nucleophiles.

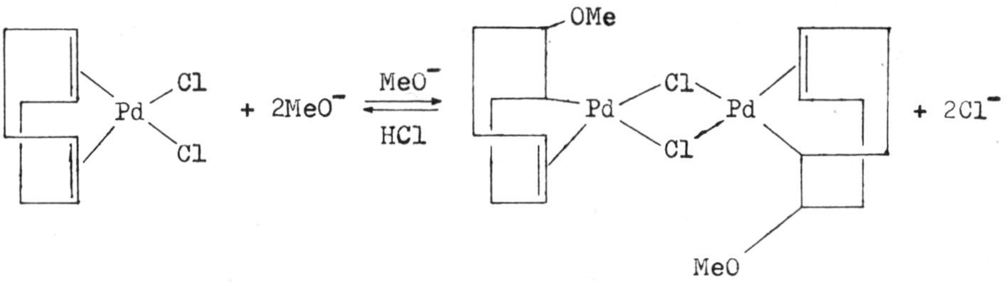

$$(6.102)$$

Products of reactions with $R_2NH$, β-diketones, ethyl acetylacetate $MeCOCH_2COOEt$, and diethyl malonate $CH_2(COOEt)_2$ are as follows:

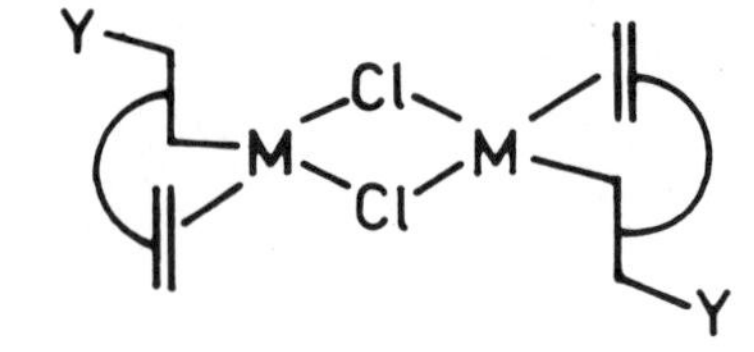

$$(6.103)$$

$$Y = NR_2, CR(COR)_2, MeCOCHCOOEt, CH(COOEt)_2$$

The nucleophile may coordinate to the metal as in the Pt(II) acetylacetonate complex, which reacts with triphenylphosphine according to equation (6.104).

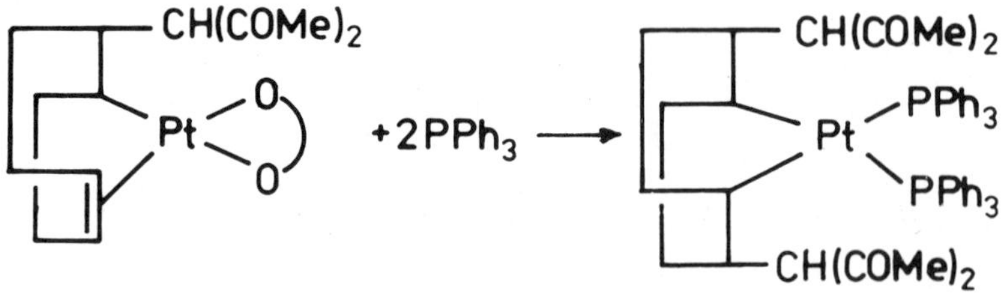

$$(6.104)$$

These types of reactions of monoolefin transition metal complexes have very significant practical application, because they allow for preparation of many compounds from inexpensive olefin hydrocarbons. Examples are provided by olefin Pd(II) complexes, which show electrophilic properties and react with water and other nucleophiles.

$$[PdCl_4]^{2-} + C_2H_4 \longrightarrow [PdCl_3(C_2H_4)]^- + Cl^- \qquad (6.105)$$

$$[PdCl_3(C_2H_4)]^- + H_2O \longrightarrow CH_3CHO + 3Cl^- + 2H^+ + Pd \qquad (6.106)$$

Alkenes such as $RCH=CH_2$ form ketones in these reactions:

$$MeCH=CH_2 \xrightarrow[\text{303 K}]{\text{H}_2\text{O, PdCl}_2} MeCOMe \qquad (6.107)$$

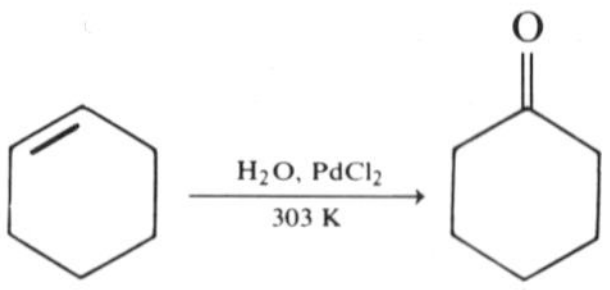

$$(6.108)$$

The oxidation reaction of ethylene utilizing $PdCl_4^{2-}$ as a catalyst found very broad industrial applications. In the United States in 1975, 410,000 tons of acetaldehyde was produced by oxidation of ethylene.

In order to prevent precipitation of metallic palladium, the following oxidizing agents are used: Cu(II), Fe(III), benzoquinone, and 1,2-naphthoquinone-4-sulfonic acid. Utilization of these compounds is necessary because oxygen does not oxidize Pd(0) to Pd(II). However, this oxidation takes place if the above-mentioned oxidation agents used have their reduced forms oxidized by air or oxygen:

$$Pd + 2Cu^{2+} \longrightarrow Pd(II) + 2Cu^+ \qquad (6.109)$$

$$4Cu^+ + O_2 + 4H^+ \longrightarrow 4Cu^{2+} + 2H_2O \qquad (6.110)$$

In summary, this oxidation reaction may be represented as follows:

$$2C_2H_4 + O_2 \xrightarrow[\text{HCl}]{\text{CuCl}_2,\ [\text{PdCl}_4]^{2-}} 2CH_3CHO \qquad (6.111)$$

Studies on $C_2H_4$ oxidation show that the formation of the $\beta$-hydroxyethyl palladium complex takes place rather as a result of attack on the coordinated ethylene molecule by an external nucleophile, i.e., a water molecule,[51] than as a result of migration (insertion) of the OH group bonded to palladium, and therefore by nucleophilic attack of the group occupying the Pd(II) coordination sphere as was assumed earlier.[50, 52]

Oxidation of $C_2D_4$ leads to $CD_3CDO$, and therefore, most probably in the transition state, hydrido palladium complex containing coordinated vinyl alcohol molecule is formed [equation (6.112)].

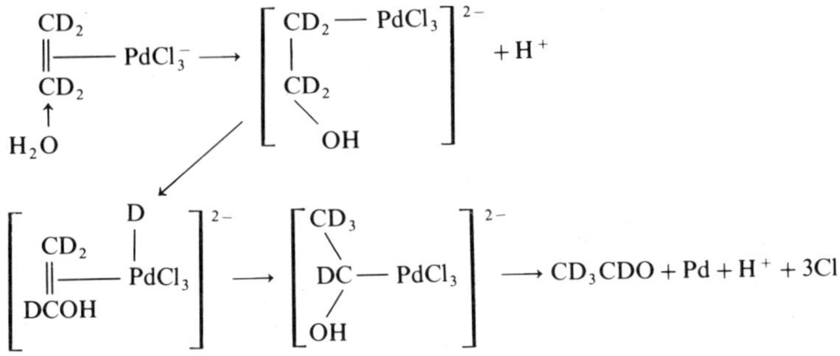

$$
\begin{array}{c}
CD_2 \\
\| \text{—— PdCl}_3^- \\
CD_2 \\
\uparrow \\
H_2O
\end{array}
\longrightarrow
\begin{bmatrix}
CD_2 \text{—— PdCl}_3 \\
| \\
CD_2 \\
\backslash \\
OH
\end{bmatrix}^{2-}
+ H^+
$$

$$
\begin{bmatrix}
\quad\ \ D \\
CD_2 \ | \\
\| \text{—— PdCl}_3 \\
DCOH
\end{bmatrix}^{2-}
\longrightarrow
\begin{bmatrix}
CD_3 \\
\backslash \\
DC \text{—— PdCl}_3 \\
/ \\
OH
\end{bmatrix}^{2-}
\longrightarrow CD_3CDO + Pd + H^+ + 3Cl^- \qquad (6.112)
$$

The rates of oxidation of ethylene, propylene, 1-butene, and 2-butene are described by the following expression:

$$\frac{d[\text{alkene}]}{dt} = \frac{k[\text{PdCl}_4^{2-}][\text{alkene}]}{[\text{Cl}^-][\text{H}^+]} \qquad (6.113)$$

The process of ethylene oxidation may be represented by scheme (6.114).

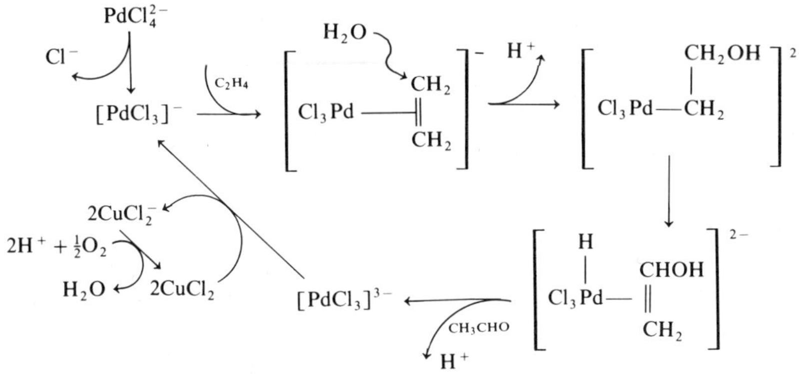

$$(6.114)$$

Olefin complexes, particularly those of palladium, are also attacked by nucleophiles in nonaqueous solvents. In many cases, vinyl compounds are formed:

$$[Pd_2Cl_4(C_2H_4)_2] + 4ROH \longrightarrow 2CH_3CH(OR)_2 + Pd + 4HCl \tag{6.115}$$

$$[Pd_2Cl_4(C_2H_4)_2] + 2ROH \longrightarrow 2CH_2=CHOR + Pd + 4HCl \tag{6.116}$$

$$C_2H_4 + 2CH_3COO^- + Pd^{2+} \longrightarrow CH_2=CHOOCCH_3 + Pd + CH_3COOH \tag{6.117}$$

Mechanisms of formation of vinyl ethers and vinyl acetate are analogous to that of the oxidation reaction (6.118) of ethylene.

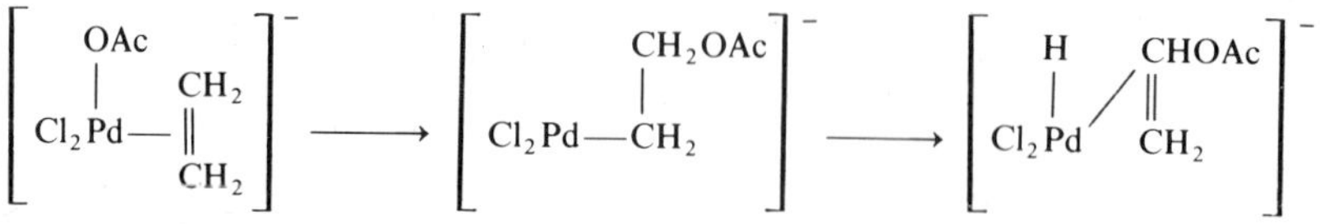

$$\longrightarrow CH_2=CHOAc + Pd + H^+ + 2Cl^- \tag{6.118}$$

Ethylene is also a feed stock for the production of ethylene glycol:

$$C_2H_4 + \tfrac{1}{2}O_2 + H_2O \longrightarrow HOCH_2CH_2OH \tag{6.119}$$

The first step involves formation of ethylene oxide, which subsequently reacts with water. The overall reaction yield is low (65–75%), because during ethylene oxide formation a considerable amount of ethylene is oxidized to $CO_2$.

Homogeneous oxidation of ethylene was introduced with application of Te(VI), Pd(II), or Tl(III) compounds as catalysts.[53]

$$CH_2=CH_2 + Br_2 \longrightarrow BrCH_2CH_2Br \xrightarrow{\text{2HOAc}} AcOCH_2CH_2OAc + 2HBr \tag{6.120}$$

$$TeO_3 + 2HBr \longrightarrow H_2TeO_3 + Br_2$$
$$\underset{\tfrac{1}{2}O_2}{\big\downarrow} \quad TeO_3 + H_2O \tag{6.121}$$

$$AcOCH_2CH_2OAc + 2H_2O \longrightarrow HOCH_2CH_2OH + 2HOAc \tag{6.122}$$

$$CH_2=CH_2 + TlCl_3 + H_2O \longrightarrow ClCH_2CH_2OH + TlCl + HCl$$
$$\underset{H_2O}{\big\downarrow} \quad HOCH_2CH_2OH + HCl \tag{6.123}$$

$$TlCl + 2CuCl_2 \longrightarrow TlCl_3 + 2CuCl$$
$$\underset{O_2}{\overset{HCl}{\big\downarrow}} \quad CuCl_2 \tag{6.124}$$

In the case of palladium catalysts, nitrogen oxide catalysts are utilized for oxidation of Pd(0) to Pd(II).

## b. Reactions of olefin complexes with electrophiles

Strong π back-bonding increases negative charge on olefin carbon atoms and increases their nucleophilicity; this facilitates attack by electrophiles. The coordination of olefins by Pt(0) so greatly increases their nucleophilicity that even such strong π acids as $C_2F_4$ may undergo protonation in the platinum complex and give the Pt(II) alkyl compound.

Rhodium(I) complexes react similarly.

$$[Rh_2Cl_2(C_2H_4)_2(CO)_2] + HCl \longrightarrow [RhCl_2(C_2H_5)(CO)_2]_x \qquad (6.125)$$

$$[Rhcp(C_2H_4)_2] + HCl \longrightarrow [Rhcp(C_2H_5)Cl(C_2H_4)] \qquad (6.126)$$

Protonation of the olefin probably takes place during formation of $[RhCl(CO)L_2]$ $[L = PPh_{3-x}(C_6F_5)_x, x = 1{-}3]$ from $[RhClL_3]$ in the presence of $C_2F_4$ and water.

$$[RhClL_3] + C_2F_4 \longrightarrow [L_xRhCl(C_2F_4)] \xrightarrow{H^+} [L_xRhCl(CF_2CF_2H)]^+$$

$$\xrightarrow{-HF} [L_xRhCl(CF=CF_2)]^+ \xrightarrow{H_2O} [L_xRhCl(CF(OH)CF_2H)]^+$$

$$\xrightarrow{-HF} [L_xRhCl(COCF_2H)]^+ \longrightarrow [L_xRhCl(CO)] + \tfrac{1}{2}CF_2HCF_2H + \cdots \qquad (6.127)$$

Owing to protonation of ethylene, $[Rh_2Cl_2(C_2H_4)]_4$ catalyzes ethylene dimerization [equation (6.128)].

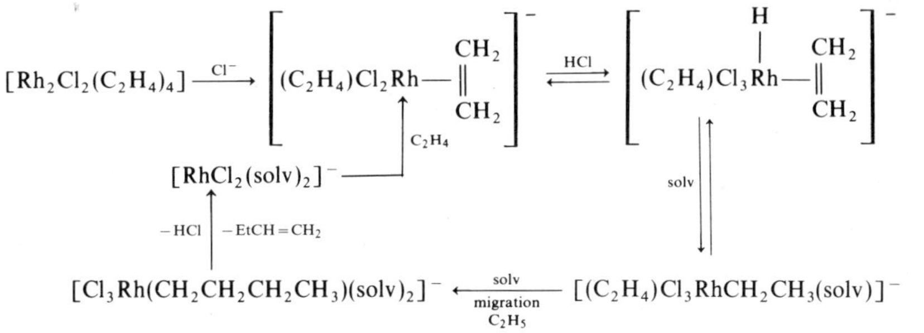

$$[Cl_3Rh(CH_2CH_2CH_2CH_3)(solv)_2]^- \xleftarrow[\substack{\text{migration} \\ C_2H_5}]{\text{solv}} [(C_2H_4)Cl_3RhCH_2CH_3(solv)]^- \qquad (6.128)$$

Some olefin coordination compounds may react with hydrogen acceptors to give allyl complexes. The ligand itself may serve as the acceptor:

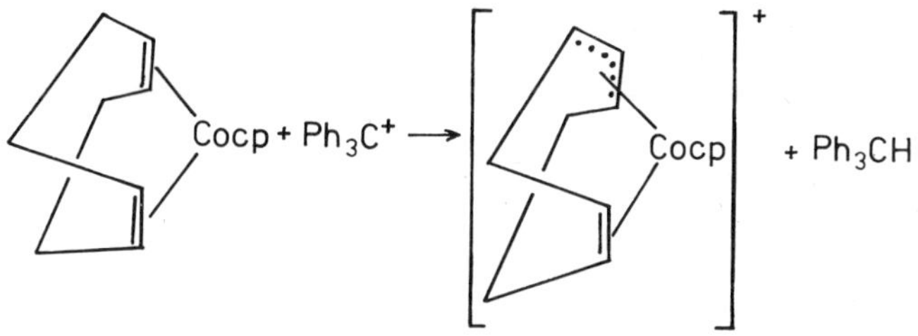

$$+ Ph_3CH \qquad (6.129)$$

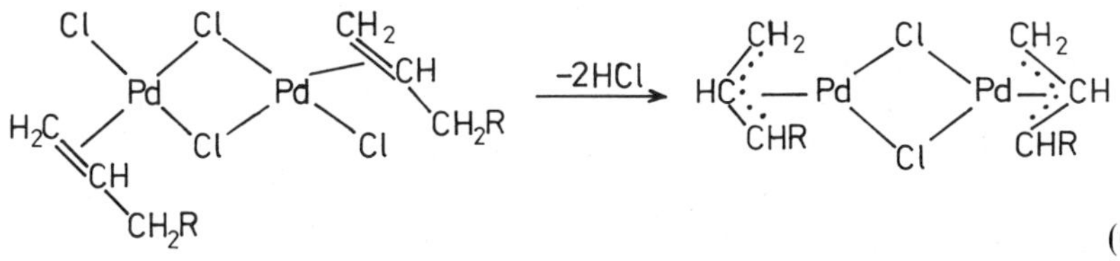

$$(6.130)$$

### c. *Amination of olefins*

Amination of olefins represents interesting reactions (involving nucleophiles) which may have essential practical applications:

$$RCH=CH_2 + NH_3 \longrightarrow RCH_2CH_2NH_2 \tag{6.131}$$

This reaction is thermodynamically favorable (for ethylene $K_{298} = 230$). However, it proceeds under unusually drastic conditions. The formation of ethylamine from ethylene and ammonia occurs in the presence of metallic sodium at 473 K and under 40 MPa pressure. At the same temperature, this reaction is also catalyzed by molybdenum oxide. However, the reaction rate is very slow, and at higher temperatures the equilibrium of this reaction is not favorable.

The platinum(II) ethylene complex stoichiometrically reacts with $NH_3$:

$$[C_2H_4PtCl_2PR_3] + NH_3 \longrightarrow [H_3^+NCH_2CH_2PtCl_2PR_3]^-$$

$$\xrightarrow{\ NH_3\ } [H_2NCH_2CH_2PtCl_2PR_3]^- + NH_4^+$$

$$\xrightarrow{\ HCl\ } EtNH_3^+Cl^- + NH_4Cl + [PtCl_4]^{2-} + [PR_3H]^+Cl^- \tag{6.132}$$

Reactions of secondary amines with alkenes and dienes were also carried out.[54, 114–116] Such reactions are significant in organic synthesis, and are catalyzed mainly by palladium, platinum, nickel, and rhodium complexes.

$$RCH=CHR' + R_2''NH \longrightarrow RCH_2CHR'NR_2'' \tag{6.133}$$

$$C_4H_6 + R_2NH \xrightarrow[BH_4^-]{Ni(II)} CH_2=CHCH_2CH_2NR_2 \tag{6.134}$$

### d. *Olefin exchange reactions*

Substitution of olefins by other alkenes takes place in many complexes [see methods of preparation of alkene complexes: reactions (6.44)–(6.46), Table 6.11, Section 6.7]. Sometimes, substitution of olefins may be carried out even if the equilibrium constant of the reaction is not favorable by applying excess of alkene, removing more volatile alkene, etc.

### e. *Substitution of olefin ligands*

Olefins are often readily displaced by Lewis bases, particularly soft ones such as phosphines, CO, acetylenes, and so on.

### f. *Transformation of olefin complexes into vinyl compounds*

Platinum(0) haloolefin complexes are transformed into Pt(II) vinyl complexes. This reaction proceeds during heating in nucleophilic solvents such as alcohols, 1,5-COD, and acetonitrile. However, it does not take place in less polar solvents, for example, benzene, acetone, and dichloromethane. The transformation may be induced by $SnCl_4$ or ligands such as CO, $SCN^-$, $NO_2^-$, $CF_3COO^-$. During measurement of ESCA spectra of $[Pt(C_2Cl_4)(PPh_3)_2]$, formation of the vinyl complex occurs on the crystal surface.

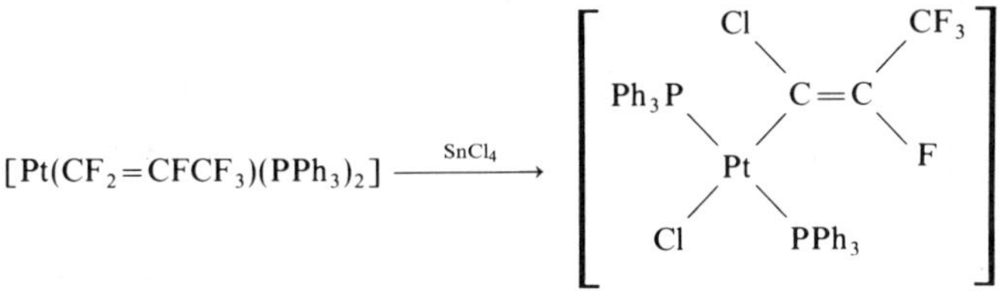

$$[Pt(CF_2=CFCF_3)(PPh_3)_2] \xrightarrow{SnCl_4} \left[ \begin{array}{c} \text{Ph}_3\text{P} \quad \overset{\text{Cl}}{\diagdown} \quad \overset{\text{CF}_3}{\diagup} \\ \underset{\text{Cl}}{\diagdown} \text{Pt} \overset{\diagup}{\underset{\text{PPh}_3}{}} \text{C}=\text{C} \diagdown \text{F} \end{array} \right] \qquad (6.135)$$

$$[Pt(CCl_2=CCl_2)(PPh_3)_2] \longrightarrow cis\text{-}[Pt(CCl=CCl_2)Cl(PPh_3)_2] \qquad (6.136)$$

### g. *Hydrogen/deuterium exchange in olefin compounds*

Isotope H/D exchange between a coordinated olefin molecule and $D_2$ or $C_nD_{2n}$ may take place as a result of formation of allyl complexes (via hydrogen elimination) or $\sigma$ carbyl complexes, if hydrido compounds are present and allow formation of the metal–alkyl bond by migration of the hydrogen atom to the olefin.

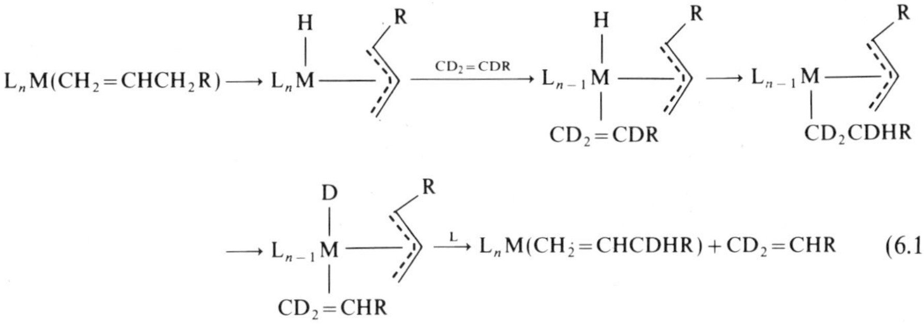

$$L_nM(CH_2=CHCH_2R) \longrightarrow L_nM \overset{\text{H}}{\underset{}{|}} \diagdown^R \xrightarrow{CD_2=CDR} L_{n-1}M \overset{\text{H}}{\underset{CD_2=CDR}{|}} \diagdown^R \longrightarrow L_{n-1}M \overset{\text{H}}{\underset{CD_2CDHR}{|}} \diagdown^R$$

$$\longrightarrow L_{n-1}M \overset{\text{D}}{\underset{CD_2=CHR}{|}} \diagdown^R \xrightarrow{L} L_nM(CH_2=CHCDHR) + CD_2=CHR \qquad (6.137)$$

$$L_nM(CH_2=CHR) \xrightarrow{\text{D}} L_nM(CH_2CHDR) \longrightarrow L_nM(CH_2=CDR) \qquad (6.138)$$

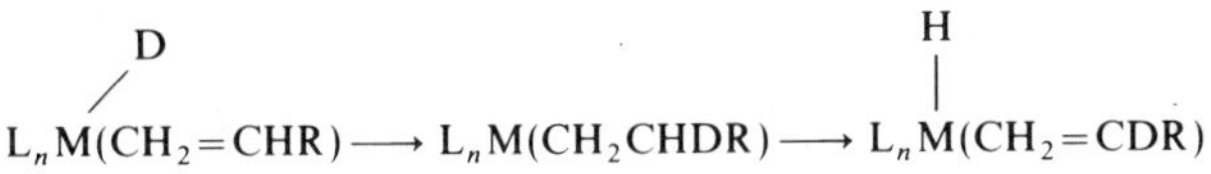

The H/D exchange within the propene molecule coordinated to $Fe^+$, $Co^+$, and $Ni^+$ takes place in the case of $C_2D_4$, and does not proceed with $D_2$.[121] The mechanism of the reaction is shown in equation (6.139).

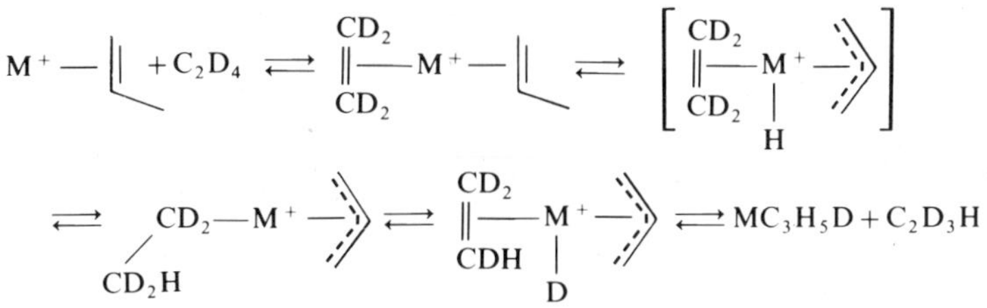

$$M^+ - \| + C_2D_4 \rightleftarrows \overset{CD_2}{\underset{CD_2}{\|}} - M^+ - \| \rightleftarrows \left[ \overset{CD_2}{\underset{CD_2}{\|}} - M^+ \overset{\diagdown}{\underset{H}{|}} \right]$$

$$\rightleftarrows \overset{CD_2}{\underset{CD_2H}{\diagup}} - M^+ \diagdown \rightleftarrows \overset{CD_2}{\underset{CDH}{\|}} - M^+ \overset{\diagdown}{\underset{D}{|}} \rightleftarrows MC_3H_5D + C_2D_3H \qquad (6.139)$$

Migratory insertion reactions involving olefins, that is, reactions of H/D exchange, isomerization, *cis–trans* isomerization, hydroformylation and other oxo syntheses, hydrosilylation, hydrocyanation, hydrogermylation, hydrogenation, polymerization, and others will be discussed in Chapter 13.


## h. *Photochemical reactions*

Photochemical studies of olefin complexes are more concentrated on diene compounds because they are more stable. Photochemical reactions of hydrogenation, isomerization, and hydrosilation of olefins were carried out.[122]

The process of isomerization of norbornadiene to quadricyclane[126–128] taking place in the presence of Cu(I) compounds appears to be very interesting. The reverse reaction which is catalyzed by Rh(I), Pd(II), and Pt(II) compounds furnishes norbornadiene from quadricyclane. About 84 kJ mol$^{-1}$ of heat energy is given off.

Photolysis of $Fe(CO)_5$ in the presence of olefins leads to formation of olefin complexes which exhibit catalytic properties. The complex $[Fe(cyclooctene)_2 (CO)_3]$ catalyzes isomerization of olefins[129] in alkene solutions. The compound $Fe(C_2H_4)(CO)_4$ which is formed in the gas phase by photolysis of $Fe(CO)_5$ and ethylene catalyzes hydrogenation of $C_2H_4$.[130] Photochemistry of the following dinuclear complex of iridium(I) is interesting:

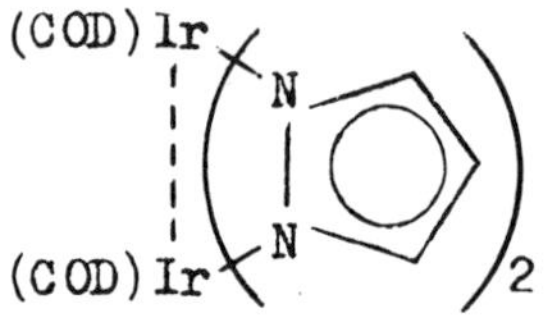

$$(6.140)$$

In dichloroethane and methylene chloride solutions, photoinduced two-center oxidative addition of chloroalkane takes place, leading to formation of the iridium(II) coordination compound. In the case of $C_2H_4Cl_2$, ethylene elimination takes place[131] [scheme (6.141)].

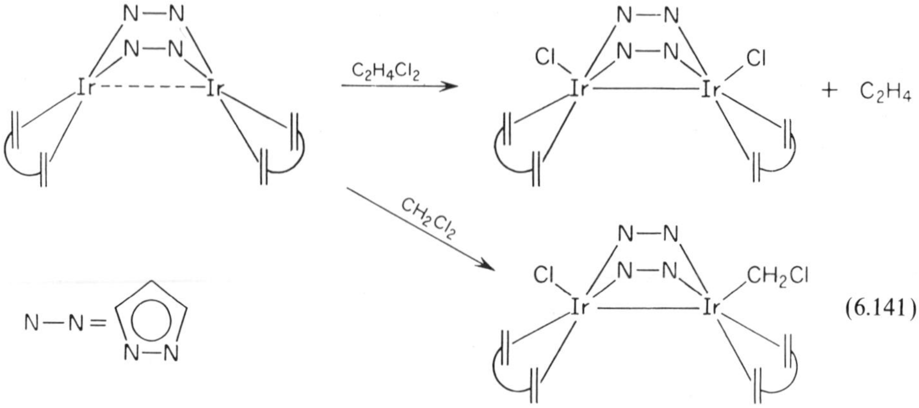

$$(6.141)$$

## B. *ACETYLENE COMPLEXES*

### 1. *BONDING AND STRUCTURE*

The metal–alkyne bond character is similar to that of the metal–olefin bond. The $\sigma$ bond is formed as a result of overlap between $\pi$ bonding acetylene orbital and metal orbitals possessing appropriate symmetry. Also, two $\pi$ molecular orbitals and one $\delta$ orbital may be formed. Acetylenes have two $\pi$-bonding and two $\pi^*$-antibonding orbitals which are mutually perpendicular and therefore may constitute bridging ligands bonding two or more metal atoms, for example, $[Co_2(C_2Ph_2)(CO)_6]$ and $[Co_4(C_2Et_2)(CO)_{10}]$. In the tetranuclear complex, the carbon atoms which form triple bond may be treated as skeleton atoms of the cluster (see Chapter 3). Molecular orbitals which are formed as a result of interaction between metal and acetylene are represented in Figure 6.21. The character of the bond is decided by the effectiveness of orbital overlap and orbital energy, and therefore by the effective nuclear charge of the metal, its oxidation state, types of ligands connected to the central atom, and substituents bonded to acetylene carbon atoms.

The orbital overlap decreases in the order of $a > b > c > d$. In complexes in which the central atom has empty $d\pi$ orbitals such as ($[Nbcp(CO)(PhCCPh)_2]$, $[Ticp_2(PhCCPh)(CO)]$, $[W(O)\,cp_2(RCCR)]$, etc.) orbital c is bonding, while in compounds of metals ending a given transition series (groups 8–10), this orbital has antibonding character. Interaction d may be neglected because of weak orbital overlap.

As in olefin coordination compounds, acetylene complexes may have trigonal, square-planar, octahedral, trigonal–bipyramidal, and tetrahedral structures. In four-coordinate square-planar complexes, the acetylene molecule is perpendicular to the plane which comprises the remaining ligands and the central atom, while in trigonal and

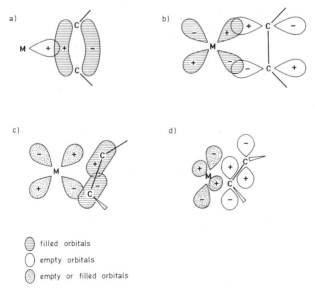

filled orbitals

empty orbitals

empty or filled orbitals

Figure 6.21.  The interaction of the metal with acetylenes: (a) $\sigma$ bond, (b) and (c) $\pi$ bond, (d) $\delta$ bond.

five-coordinate compounds possessing trigonal–bipyramidal structure, the alkyne molecule forms only a small angle with the triangular coordination plane (Figure 6.22).

This phenomenon, like in olefin complexes, is caused by the influence of $ML_2$ ligands on the degree of $d$–$p$ hybridization of the central atom, and therefore on the back-bonding interaction [equation (6.1), Figure 6.4] which is the strongest if the acetylene molecule lies in the $ML_2$ plane. A stable, two-coordinate $14e$ complex of platinum with diphenylacetylene $Pt(C_2Ph_2)_2$ is also known (Table 6.19, Figure 6.22). Due to the existence of a $\pi$ metal–alkyne bond, the coordinated alkyne molecule may be treated as a molecule which is in the excited state. It is deformed, and the substituents at the acetylene carbon atoms are tilted away from the complex. Usually, the $C-C-R$ angle assumes values of 170–140°.

The C–C distance in the complex may be longer than that of the free acetylene by 20 pm. The greatest elongation of the $C-C$ bond is observed in complexes containing bridging acetylenes. The increase of the $C-C$ bond distance is considerably greater for compounds possessing trigonal–bipyramidal or trigonal structures than for square-planar compounds (Figure 6.22, Table 6.19). The M–C distances in acetylene complexes are shorter by about 7 pm compared to corresponding olefin complexes. This is partially due to the decrease by 4 pm of a single bond radius while changing from hybridization $sp^2$ to sp.

Therefore, comparison of interatomic distances leads to the conclusion that acetylenes are more strongly bonded to the metal than olefins. This is also confirmed by theoretical considerations.[9, 11, 17, 29, 85–87] Acetylene complexes containing a small number of d electrons such as $[Nbcp(CO)(C_2Ph_2)_2]$ show very pronounced elongation of the $C-C$ bond distance. This can be explained by interaction between $\pi$ donor alkyne and the central atom (Figure 6.21c).

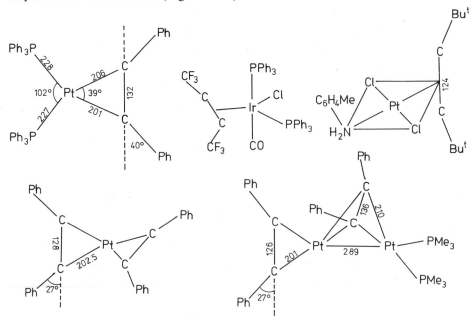

Figure 6.22. Structures of acetylene complexes: $[Pt(PhCCPh)(PPh_3)_2]$, $[IrCl(CF_3CCCF_3)CO(PPh_3)_2]$, $[PtCl_2(Bu^tCCBu^t)(NH_2C_6H_4Me)]$, $[Pt(PhCCPh)_2]$, and $[Pt_2(PhCCPh)_2(PMe_3)_2]$.

*Table 6.19. Structures of Metal Acetylene Complexes*

| Compound | Structure* | Distance (pm) | | | C−C−R Angle |
|---|---|---|---|---|---|
| | | C−C | M−C | M−L | |
| [Co$_2$(C$_2$Ph$_2$)(CO)$_6$]$^a$ | Dimer | 137 | 196 | 175 | 138° |
| [Co$_2$(C$_2$t-Bu$_2$)(CO)$_6$]$^b$ | | 133.5 | 200 | | 144.5° |
| [Ni(C$_2$Ph$_2$)(t-BuNC)$_2$]$^c$ | T | 128 | | | 149° |
| [Ni$_2$(COD)$_2$(μ-C$_2$Ph$_2$)]$^d$ | Dimer | 138.6 | 193 | | |
| [Pd{C$_2$(COOMe)$_2$}(PPh$_3$)$_2$]$^e$ | T | 128 | | | 145° |
| [Pd$_2$(C$_5$Ph$_5$)$_2$(μ-C$_2$Ph$_2$)]$^f$ | Dimer | 133 | 205 | | |
| [Pt(C$_2$Ph$_2$)(PPh$_3$)$_2$]$^g$ | T | 132 | 203 | | 140° |
| [Pt{C$_2$(CF$_3$)$_2$}(PPh$_3$)$_2$]$^h$ | | 125.5 | | | 140° |
| [PtClMe{C$_2$(CF$_3$)$_2$}(AsMe$_3$)$_2$]$^h$ | TBP | 132 | | | |
| [PtCl$_2$(p-toluidine)(C$_2$t-Bu)$_2$]$^i$ | S | 124 | | | 165° |
| [Pt(C$_2$Ph$_2$)$_2$]$^j$ | | 128 | 202.5 | | 153° |
| [Pt$_2$(μ-C$_2$Ph$_2$)(C$_2$Ph$_2$)(PMe$_3$)$_2$]$^j$ | Dimer | 136 (μ) | 210 (μ) | | |
| | | 126 | 201 | | 153° |
| [Rh$_2$(PPh$_3$)$_2$(PF$_3$)$_4$(μ-C$_2$Ph$_2$)]$^k$ | | 136.9 | 211 | | |
| [Ir(t-C(CN)=CHCN)(CO)(PPh$_3$)(NCC≡CCN)]$^l$ | TBP | 129 | 208 | | 139° |
| [Fe(CO)$_6$(C$_2$t-Bu$_2$)]$^b$ | Dimer | 131.1 | 209 | | 145° |
| [Mo$_2$cp$_2$(CO)$_4$(C$_2$Et$_2$)]$^m$ | Dimer | 133.5 | 196.1 | | |
| [WOcp(C$_2$Ph$_2$)]$^n$ | | 129 | | | 143.5° |
| [W(C$_2$Ph$_2$)$_3$(CO)]$^o$ | T$_d$ | 130 | 206 | | 140° |
| [Nbcp(CO)(C$_2$Ph$_2$)$_2$]$^p$ | | 135 | 219 | | 138° |
| [Nbcp(CO)(C$_2$Ph$_2$)(C$_4$Ph$_4$)]$^q$ | | 126 | | | 141° |
| [Ticp$_2$(C$_2$Ph$_2$)(CO)]$^{(96)}$ | | 128.5 | 210.7 | 205 (CO) | 145.8° |
| | | | 223.0 | 239.6 (cp) | 138.8° |
| [Rh$_3$cp$_3$(μ$_3$-CO)(μ$_3$-C$_2$Ph$_2$)]$^r$ | | 138.5 | 213 | | 128.3° |
| | | | 203 | | 125.1° |
| [Co$_2$(HCCH)(CO)$_4$(PMe$_3$)$_2$]$^s$ | Dimer | 132.7 | 195.9 | M–CO 178 | |

* TBP—trigonal bipyramid, S—square planar, T—trigonal, T$_d$—tetrahedron.
$^a$ W. G. Sly, *J. Am. Chem. Soc.*, **81**, 18 (1959); D. A. Brown, *J. Chem. Phys.*, **33**, 1037 (1960).
$^b$ F. A. Cotton, J. D. Jamerson, and B. R. Stults, *J. Am. Chem. Soc.*, **98**, 1774 (1976).
$^c$ R. S. Dickson and J. A. Ibers, *J. Organomet. Chem.*, **36**, 191 (1972).
$^d$ V. W. Day, S. A. Abdel-Meguid, S. Dalestani, G. Thomas, W. R. Pretzer, and E. L. Muetterties, *J. Am. Chem. Soc.*, **98**, 8289 (1976).
$^e$ J. A. McGinnety, *J. Chem. Soc., Dalton Trans.*, 1038 (1974).
$^f$ E. Ban, P. T. Cheng, T. Jack, S. C. Nyburg, and J. Powell, *Chem. Commun.*, 368 (1973).
$^g$ J. O. Glanville, J. M. Stewart, and S. O. Grim, *J. Organomet. Chem.*, **7**, P9 (1967).
$^h$ B. W. Davies, R. J. Puddephatt, and N. C. Payne, *Can. J. Chem.*, **50**, 2276 (1972).
$^i$ G. R. Davies, W. Hewerston, R. H. B. Mais, P. G. Owston, and C. G. Patel, *J. Chem. Soc. (A)*, 1873 (1970).
$^j$ M. Green, D. M. Grove, J. A. K. Howard, J. L. Spencer, and F. G. A. Stone, *Chem. Commun.*, 759 (1976).
$^k$ M. A. Bennett, R. N. Johnson, G. B. Robertson, T. W. Turney, and P. O. Whimp, *Inorg. Chem.*, **15**, 97 (1976).
$^l$ R. M. Kirchner and J. A. Ibers, *J. Am. Chem. Soc.*, **95**, 1095 (1973).
$^m$ W. I. Bailey, Jr., F. A. Cotton, J. D. Jamerson, and J. R. Kolb, *Organomet. Chem.*, **121**, C23 (1976).
$^n$ N. G. Bokiy, Y. V. Galitov, Y. T. Struchkov, and N. A. Ustynyuk, *J. Organomet. Chem.*, **54**, 213 (1973).
$^o$ M. R. Laine, R. E. Moriarty, and R. Bau, *J. Am. Chem. Soc.*, **94**, 1402 (1972).
$^p$ A. N. Nesmeyanov, A. I. Gusev, A. S. Pasynskii, K. N. Anisimov, N. E. Kolobova, and Y. T. Struchkov, *Chem. Commun.*, 277 (1969).
$^q$ A. N. Nesmeyanov, A. I. Gusev, A. S. Pasynskii, K. N. Anisimov, N. E. Kolobova, and Y. T. Struchkov, *Chem. Commun.*, 739 (1969).
$^r$ Trinh-Toan, R. W. Broach, S. A. Gardner, M. D. Rausch, and L. F. Dahl, *Inorg. Chem.*, **16**, 279 (1977).
$^s$ J. J. Bonnet and R. Mathieu, *Inorg. Chem.*, **17**, 1973 (1978).

## 2. IR SPECTRA

There are three vibrations for the triatomic group $MC_2$, namely $a_1$, $a_1$, and $b_1$ (local symmetry $C_2$).

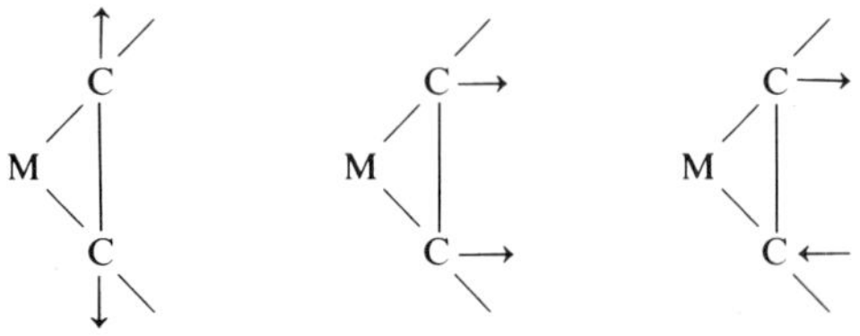

$$
\nu_1(a_1)(\nu_{C\equiv C}) \qquad \nu_2(a_1)(\nu_{M-\|}^s) \qquad \nu_3(b_1)(\nu_{M-\|}^{as}) \tag{6.142}
$$

The bands due to metal–acetylene vibrations $\nu_2$ and $\nu_3$ in complexes have not yet been identified because of a strong interaction between them. The frequency $\nu_1$ is due to the $C\equiv C$ stretching vibration. The position of $\nu(C\equiv C)$ bands of coordinated acetylenes is influenced by the strength of $\pi$ back-bonding. The stronger the metal–acetylene interaction, the greater the lowering of the $\nu(C\equiv C)$ frequency of bonded acetylene compared to the free acetylene molecule (Table 6.20).

Table 6.20. $\nu(C\equiv C)$ Frequencies for Metal Acetylene Complexes

| Compound* | | $\nu(C\equiv C)$ (cm$^{-1}$) |
|---|---|---|
| $[Cu(OAc)(C_2Ph_2)]^a$ | | 1985, 1971 |
| $[Pt_2Cl_4(t\text{-}BuCCR)]^b$ | | |
| R: $t$-Bu | | 2023, 2005 |
| $i$-Pr | | 2011 |
| Et | | 2023 |
| $CMe_2Ph$ | | 2001 |
| $Na[PtCl_3(t\text{-}BuCCMe)]^b$ | | 2028 |
| $[PtMe(RCCR')L_2]^+PF_6^{-c}$ | | |
| RCCR' | L | |
| $C_2Me_2$ | $PMe_2Ph$ | 2114 |
| MeCCEt | $PMe_2Ph$ | 2116 |
| $C_2Et_2$ | $PMe_2Ph$ | 2101 |
| $C_2Ph_2$ | $PMe_2Ph$ | 2087 |
| $C_2Ph_2$ | $AsMe_3$ | 2024 |
| $[Ni(C_2Ph_2)(PPh_3)_2]^d$ | | 1800 |
| $[Pd\{C_2(CO_2Me)_2\}(PPh_3)_2]^d$ | | 1830, 1845 |
| $[Pt(C_2Ph_2)(PPh_3)_2]^d$ | | 1740, 1768 |
| $[Ni(C_2Ph_2)(t\text{-}BuNC)_2]^e$ | | 1810 |
| $[Pd(C_2Ph_2)(t\text{-}BuNC)_2]^e$ | | 1825 |
| $[Vcp(CO)_2(C_2Ph_2)]^f$ | | 1700 |
| $[Nbcp(CO)_2(C_2Ph_2)]^f$ | | 1680 |
| $[Nb_2cp_2(CO)_4(\mu\text{-}C_2Ph_2)_2]^f$ | | 1500 |
| $[Crcp(CO)(NO)(C_2H_2)]^g$ | | 1720 |
| $[IrCl(CO)(PPh_3)_2\{C_2(COOMe)_2\}]^h$ | | 1770 |

*(Table continued)*

*Table 6.20. (continued)*

| Compound* | $v(C \equiv C)$ (cm$^{-1}$) |
|---|---|
| [IrCl(CO)(PPh$_3$)$_2$(HOOCCCCOOH)]$^h$ | 1600 |
| [Ir(CO)$_2$(PPh$_3$)$_2$(MeOOCCCCOOMe)]$^{+i}$ | 1814 |
| [RhCl(CO)(PPh$_3$)$_2$(HOOCCCCOOH)]$^h$ | 1600 |
| [RhCl(CO)(PPh$_3$)$_2$(NCCCCN)]$^j$ | 1775 |
| [IrCl(PPh$_3$)$_2$(MeOOCCCCCOOMe)]$^h$ | 1885 |
| [IrCl(PPh$_3$)$_2$(C$_2$Ph$_2$)]$^h$ | 1850 |
| [RhCl(PPh$_3$)$_2$(C$_2$Ph$_2$)]$^k$ | 1916 |
| [RhCl(AsPh$_3$)$_2$(C$_2$Ph$_2$)]$^k$ | 1883 |
| [Pt$_2$(PhCCPh)$_2$(PMe$_3$)$_2$]$^l$ | 1851 |
| [Pt(C$_2$Ph$_2$)$_2$]$^l$ | 1881 |
| [Co$_2$(C$_2$H$_2$)(CO)$_6$]$^m$ | 1402.5 |
| [Mocp$_2$(C$_2$H$_2$)]$^n$ | 1613 |
| [Wcp(Me)(O)(HCCH)]$^o$ | 1585 |
| [Ni(CHCH)(PPh$_3$)$_2$]$^p$ | 1630 |

* $v(C \equiv C)$ in free alkynes: C$_2$H$_2$—1973.8 cm$^{-1}$; C$_2$Ph$_2$—2223 cm$^{-1}$; C$_2$(COOMe)$_2$—2256 cm$^{-1}$; C$_2$(CN)$_2$—2218 cm$^{-1}$; C$_2$(CF$_3$)$_2$—2300 cm$^{-1}$.

$^a$ R. G. Guy and J. Chatt, *J. Adv. Inorg. Chem. Radiochem.*, **4**, 77 (1962).
$^b$ J. Chatt, R. G. Guy, and L. A. Duncanson, *J. Chem. Soc.*, 827 (1961).
$^c$ M. H. Chisholm and H. C. Clark, *Inorg. Chem.*, **10**, 2557 (1971).
$^d$ G. Davidson, *Organomet. Chem. Rev.*, **8**, 342 (1972); G. Wilke and G. Herrmann, *Angew. Chem.*, **74**, 693 (1962).
$^e$ S. Otsuka, T. Yoshida, and Y. Tatsuno, *J. Am. Chem. Soc.*, **93**, 6462 (1971).
$^f$ A. N. Nesmeyanov, K. N. Anisimov, N. E. Kolobova, and A. A. Pasynskii, *Izv. Akad. Nauk SSSR, Ser. Khim.*, 100 (1969).
$^g$ M. Herberhold, H. Alt, and C. G. Kreiter, *J. Organomet. Chem.*, **42**, 413 (1972).
$^h$ J. P. Collman and J. W. Kang, *J. Am. Chem. Soc.*, **89**, 844 (1967).
$^i$ M. J. Church, M. J. Mays, R. N. F. Simpson, and F. D. S. Stefanini, *J. Chem. Soc. (A)*, 2909 (1970).
$^j$ C. L. McClure and W. H. Baddley, *J. Organomet. Chem.*, **27**, 155 (1971).
$^k$ J. T. Mague and G. Wilkinson, *J. Chem. Soc. (A)*, 1736 (1966).
$^l$ M. Green, D. M. Grove, J. A. K. Howard, J. L. Spencer, and F. G. A. Stone, *J. Chem. Soc., Chem. Commun.*, 759 (1976).
$^m$ Y. Iwashita, F. Tamura, and A. Nakamura, *Inorg. Chem.*, **8**, 1179 (1969).
$^n$ K. L. Tang Wong, J. L. Thomas, and H. H. Brintzinger, *J. Am. Chem. Soc.*, **96**, 3694 (1974).
$^o$ See Reference *j*, Table 6.21.
$^p$ See Reference *k*, Table 6.21.

For complexes possessing a weak metal–alkyne bond, the lowering of the $v(C \equiv C)$ frequency is 150–250 cm$^{-1}$. Examples are provided by Cu(I), Ag(I), and Hg(II) acetylene complexes as well as square-planar Pt(II) compounds for which $v(C \equiv C)$ occurs most commonly in the 1950–2120 cm$^{-1}$ range. In trigonal, three-coordinate [M(C$_2$R$_2$)L$_2$] and pentacoordinate [M(C$_2$R$_2$)L$_4$] complexes possessing trigonal–bipyramidal structure as well as in alkyne compounds of vanadium, chromium, and manganese group metals, the $v(C \equiv C)$ band occurs in the 1600–1900 cm$^{-1}$, and therefore the lowering of the frequency is even 400–500 cm$^{-1}$. Intermediate values are exhibited by Rh(I) and Ir(I) square-planar (isoelectronic with Pt$^{2+}$ compounds), which usually have their $v(C \equiv C)$ bands at *ca* 1850–1950 cm$^{-1}$. The lowest values of $v(C \equiv C)$, which occur in the 1400–1600 cm$^{-1}$ region, are observed for polynuclear complexes in which the alkyne constitutes the bridging ligand. These bands occur even at lower energies than $v(C = C)$ bands for free olefins and for olefin coordination compounds.

## 3. NMR SPECTRA

Relatively few data are available concerning $^1$H NMR spectra of protons connected to the alkyne carbon atoms in acetylene metal complexes. In contrast to olefins, coordinated acetylenes have their proton signals shifted to lower $\tau$ values. The shift to lower fields generally equals 2.5–4 ppm for coordinated acetylenes. Therefore, acetylene protons in alkynes bonded to the central atom have chemical shifts which are typical for olefin hydrogen atoms. This is in agreement with theoretical predictions, X-ray data, IR spectra, and the metal–alkyne bond model. The chemical shift of protons for some alkyne complexes are given in Table 6.21.

The $^{13}$C NMR chemical shift change for acetylene complexes caused by coordination and expressed as $\Delta\delta$ is also reversed compared to the analogous change for olefins (Table 6.22). In contrast to olefin compounds, in metal acetylene complexes an increase of the redonor bond strength causes a decrease in screening.

*Table 6.21. $^1$H NMR Spectra of Metal Acetylene Compounds*

| Compound | $\tau(C\equiv CH)$ | Coupling constants (Hz) |
|---|---|---|
| $[Pt\{HC\equiv CC(OH)Me_2\}(PPh_3)_2]^a$ | 3.76 | $J(t\text{-}P-H)$ 22, $J(c\text{-}P-H)$ 10, $J(Pt-H)$ 59.4 |
| $[Pt\{HC\equiv CC(OH)Ph_2\}(PPh_3)_2]^a$ | 3.55 | $J(t\text{-}P-H)$ 22.3, $J(c\text{-}P-H)$ 10, $J(Pt-H)$ 58 |
| $\left[Pt\left(HC\equiv C-\!\!\!\left\langle\begin{array}{c}\\[-2pt]\bigcirc\\[-2pt]OH\end{array}\right\rangle\right)(PPh_3)_2\right]^{a\,(29)}$ | 3.74 | $J(t\text{-}P-H)$ 22.8, $J(c\text{-}P-H)$ 10, $J(Pt-H)$ 63 |
| $[Pt(C_2H_2)\{P(C_6D_5)_3\}_2]^b$ | 2.91 | |
| $[Ir(SnCl_3)(C_2H_2)(CO)(PPh_3)_2]^c$ | 5.4 | |
| $[Vcp(C_2H_2)(CO)_2]^d$ | 4.92 (CCl$_4$), 5.63 (PhH) | |
| $[Vcp(HCCPr)(CO)_2]^d$ | 4.93 (CHCl$_3$) | |
| $[Vcp(HCCBu)(CO)_2]^d$ | 5.60 (PhH) | |
| $[Co_2(CO)_6(C_2H_2)]^e$ | 6.0 | |
| $[Co_2(C_2H_2)(CO)_4(PMe_3)_2]^f$ | 5.17 | $J\{P-H(C\equiv CH)\}$ 3.5, $\tau(CH_3)$ 8.72, $J\{P-H(CH_3)\}$ 8.4 |
| $[Mocp_2(C_2H_2)]^g$ | 2.32 | $\tau_{cp}$ 5.70 |
| $[Mocp_2(HCCMe)]^h$ | 2.95 | $\tau_{CH_3}$ 7.25, $\tau_{cp}$ 5.53 |
| $[Ticp_2(C_2H_2)PMe_3]^i$ | 1.18 | $J(P-H)$ 9.0 |
| | 3.01 | $J(P-H)$ 5.4 |
| $[Wcp(Me)(O)(C_2H_2)]^j$ | $-0.35$ (d, q) (A) | q: $^4J(HH)$ 1.3, $\tau_{cp}$ 3.93, $\tau_{Me}$ 8.29 ($^4J_{H-H}=1.3$) |
| | 1.1 (d) (B) | $J(AB)$ 0.5; $^2J(W-H)$ 9.7 and 7.0 |
| $[Ni(HCCH)(PPh_3)_2]^k$ | 3.59 | |

$^a$ J. H. Nelson, H. B. Jonassen, and D. M. Roundhill, *Inorg. Chem.*, **8**, 2591 (1969).
$^b$ C. D. Cook and K. Y. Wan, *J. Am. Chem. Soc.*, **92**, 2595 (1970).
$^c$ M. Camia, M. P. Lachi, L. Benzoni, L. Zanzottera, and M. Tacci Venturi, *Inorg. Chem.*, **9**, 251 (1970).
$^d$ R. Tsumura and N. Hagihara, *Bull. Chem. Soc. Jpn.*, **38**, 1901 (1965).
$^e$ Y. Tashita, F. Tamura, and A. Nakamura, *Inorg. Chem.*, **8**, 1179 (1969).
$^f$ J. J. Bonnet and R. Mathieu, *Inorg. Chem.*, **17**, 1973 (1978).
$^g$ K. L. Tang Wong, J. L. Thomas, and H. H. Brintzinger, *J. Am. Chem. Soc.*, **96**, 3694 (1974).
$^h$ J. L. Thomas, *Inorg. Chem.*, **17**, 1507 (1978).
$^i$ H. G. Alt and H. E. Engelhardt, *J. Am. Chem. Soc.*, **107**, 3717 (1985).
$^j$ H. G. Alt and H. I. Hayen, *Angew. Chem.*, **24**, 497 (1985).
$^k$ K. R. Poerschke, Y.-H. Tsay, and C. Krueger, *Angew. Chem.*, **24**, 323 (1985).

*Table 6.22.* $^{13}C$ *NMR Spectra of Metal Acetylene Compounds*

| Compound | $\delta(\equiv CR)$ (ppm) | Other data |
|---|---|---|
| $[PtMe(MeCCMe)(PMe_2Ph)_2]PF_6{}^a$ | 69.5 | |
| $[Pt(MeCCMe)(PPh_3)_2]^a$ | 112.8 | |
| $[Pt(C_2Ph_2)_2]^b$ | 124.8 | $J(^{195}Pt-{}^{13}C)$ 311 Hz |
| $[Mocp_2(C_2H_2)]^c$ | 117.7 | $\delta$(cp) 84.8 |
| $[Mocp_2(HC^1\equiv C^2Me)]^c$ | $C^1$ 106.0; $C^2$ 126.1 | $\delta$cp 84.1; $\delta$Me 21.6 |
| $[Co_2\{EtCCCH(OH)Me\}(CO)_2]^d$ | 102.2; 100.7 | |
| $[Co_2(EtCCCOMe)(CO)_6]^d$ | 102.6; 87.2 | |
| $[PtCl_2(PhCCPh)(p\text{-}MeC_5H_4N)]^{(21b)}$ | 72.9 | $-16.62\,\Delta\delta$; $^1J(Pt-C)$ 165 Hz |
| $[PtCl_2(PhCCMe)(p\text{-}MeC_5H_4N)]^{(21b)}$ | $\equiv$CPh 72.90 | $-12.69\,\Delta\delta$; $^1J(Pt-C)$ 132 |
| | $\equiv$CMe 65.79 | $-14.84\,\Delta\delta$ |
| $[PtCl_2(MeCCBu')(p\text{-}MeC_5H_4N)]^{(21b)}$ | $\equiv$CMe 65.83 | $-3.93\,\Delta\delta$; $^1J(Pt-C)$ 141.1 |
| | $\equiv$CBu' 75.37 | $-12.55\,\Delta\delta$ |
| $[PtCl_2(Bu'CCBu')(p\text{-}MeC_5H_4N)]^{(21b)}$ | 76.34 | $-10.91\,\Delta\delta$; $^1J(Pt-C)$ 183.4 |
| $[Wcp(Me)(O)(HCCH)]^e$ | 152.6; 148.4 | $\delta$(cp) 105.3; $\delta$(Me) 2.3 |
| $[W(PhCCPh)_3SnPh_3]^{-(132)}$ | 197.3; 183.3 | |
| $[Ni(HCCH)(PPh_3)_2]^f$ | 122.1 | |

$^a$ M. H. Chisholm, H. C. Clark, L. E. Manzer, and J. B. Stothers, *J. Am. Chem. Soc.*, **94**, 5087 (1972).
$^b$ Green, D. M. Grove, J. A. K. Howard, J. L. Spencer, and F. G. A. Stone, *J. Chem. Soc., Chem. Commun.*, 759 (1976).
$^c$ J. L. Thomas, *Inorg. Chem.*, **17**, 1507 (1978).
$^d$ S. Aime, L. Milone, and D. Osella, *J. Chem. Soc., Chem. Commun.*, 704 (1979).
$^e$ See Reference *j*, Table 6.21.
$^f$ See Reference *k*, Table 6.21.

For 2-butyne, this screening decreases according to the following series: $trans\text{-}[PtMe(MeC\equiv CMe)(PMe_2Ph)_2]^+ > MeC\equiv CMe > [Pt(MeC\equiv CMe)(PPh_3)_2]$. For the Pt(II) complex, the shift is 69.5 ppm, while for the Pt(0) compound, it equals 112.8 ppm. The direction of these changes may be explained by assuming formation of metallacyclic compounds with olefins and acetylenes and therefore by required sp–sp$^2$ rehybridization of acetylenes (shifts to lower fields) and sp$^2$–sp$^3$ rehybridization of olefins (shifts to higher fields).

Alternatively, the $^{13}C$ NMR chemical shift change may be explained by π metal–ligand interaction which causes the following: (1) an increase in density of total charge on the olefin or acetylene carbon atoms, (2) a lowering of the multiplicity of the π carbon–carbon bond due to formation of the π-donor bond, (3) a lowering of the π character of carbon atoms due to formation of the σ hydrocarbon–metal bond (overlap of π orbitals with the $d$sp hybridized metal orbital), and (4) a decrease of electron density on olefin or acetylene carbon atoms due to formation of the σ hydrocarbon–metal bond. For olefins, the first three effects are responsible for an increase of screening, and only the fourth factor operates in the opposite direction. However, in the case of coordinated alkynes, factors (2), (3), and (4) cause a lowering of the screening, and only factor (1) increases the screening. Therefore, these factors explain the increase of the screening for olefins and its lowering for olefins and its lowering for alkynes.

The coupling constants $J(^{195}Pt-C)$ are greater for olefin carbon atoms than for acetylene carbon atoms. This argues against the metallacyclic model of the metal-unsaturated hydrocarbon bond which suggests that $J(^{195}Pt-Csp^2)$ for the acetylene complex should be greater than $J(^{195}Pt-Csp^3)$ for the olefin compound.

Table 6.23. Enthalpy of Thermal Decomposition of
[IrX(CO)(PPh₃)₂L]

| X | L = CF₃C≡CCF₃<br>$\Delta H_{dec}$ (kJ mol$^{-1}$) | L = CF₂=CF₂<br>$\Delta H_{dec}$ (kJ mol$^{-1}$) |
|---|---|---|
| F | 99 | 79 |
| Cl | 96 | 67 |
| Br | 79 | 41 |
| I | 82 | 57 |

The metal–alkyne bond energy is higher than that of the metal–olefin bond. Calculations show that this bond energy dependence on the angle between the carbon–carbon bond and the MP₂ plane for the trigonal complex [Pt(MeCCMe)(PH₃)₂] has only one minimum for the angle 0°.[9] The calculated depth of the minimum, and therefore the energy of activation for the process of acetylene rotation in this complex, is high and equals 3.6 eV, that is, *ca* 330 kJ mol$^{-1}$. Consequently, for trigonal complexes, the rotation of alkynes about the metal–acetylene bond has not been observed. Rotation of coordinated alkynes takes place only in the case of some polynuclear complexes or weak mononuclear complexes containing a small number of $d$ electrons. Examples are provided by trinuclear rhodium complexes such as [Rh₃cp₃(CO)(C₂Ph₂)] and [Rh₃cp₃(C₂Ph₂)], in which diphenylacetylene is bonded with all rhodium atoms.

In the first complex, the alkyne does not rotate at 186 K but shows dynamic properties at room temperature. In the second compound, the diphenylacetylene undergoes rotation even at 146 K.[89] The first compound for which the rotation of coordinated acetylene was observed is [Crcp(C₂H₂)(CO)(NO)].[90] The free energy of activation $\Delta G^{\neq}$ varies for this compound for different solvents in the range of 50–60 kJ mol$^{-1}$; it assumes lower values for more polar solvents. However, there is no direct relationship between the dipole moment of the solvent and $\Delta G^{\neq}$. Alkyne complexes have somewhat lower stability constants than analogous metal olefin compounds (Table 6.11), despite the fact that X-ray studies, IR, and NMR spectra show that the metal–alkyne bond is stronger. This is also indicated by enthalpy changes of thermal decomposition of Ir(I) complexes [IrX(CO)(PPh₃)₂L][91] (Table 6.23).

ESCA studies show that alkyne carbon atoms in [Pt(C₂Ph₂)(PPh₃)₂], like olefin carbon atoms in [Pt(C₂H₄)(PPh₃)₂], possess a negative charge. The positive charge on the platinum atom for complexes of the type PtL(PPh₃)₂ increases when L changes according to the series: (PPh₃)₂ < C₂Ph₂ ⌐ C₂H₄ < CS₂ < O₂ < Cl₂.[92]

## 4. METHODS OF PREPARATION OF ACETYLENE COMPLEXES[93, 86, 85]

### a. *Alkyne addition to the transition metal salts and complexes*

Reactions of Cu(I) and Ag(I) salts with 1-alkynes usually lead to formation of acetylide complexes which possess both $\sigma$ metal–acetylide and $\pi$ metal–acetylene bonds. However, under particular conditions such as alcohol or acidic aqueous solutions, acidic properties of alkynes RCCH do not play an essential role, and these acetylenes react with CuCl and $AgNO_3$ to give alkyne complexes.

Solid state copper(I) chloride reacts also with gaseous acetylene or $RC \equiv CR'$ to give complexes of the following composition:

$$[(CuCl)_n C_2 H_2] \ (n = 1,2,3,6), \ [(CuCl)_n (RCCH)] \ (n = 1,2) \tag{6.143}$$

$$3CuCl_{(s)} + C_2 H_2 \longrightarrow [(CuCl)_3 (C_2 H_2)] \ (\varDelta G = -2400 \ J) \tag{6.144}$$

$$2CuCl_{(s)} + C_2 H_2 \longrightarrow [(CuCl)_2 (C_2 H_2)] \ (\varDelta G = -1380 \ J) \tag{6.145}$$

$$2[(CuCl)_3 (C_2 H_2)] + C_2 H_2 \longrightarrow 3[(CuCl)_2 (C_2 H_2)] \ (\varDelta G = 660 \ J) \tag{6.146}$$

$$Ag^+ + C_2 H_2 \longrightarrow [Ag(C_2 H_2)]^+ \tag{6.147}$$

$$Ag^+ + RCCR' + NO_3^- \longrightarrow [Ag(RCCR')]NO_3 \tag{6.148}$$

$$Ag^+ + RCCR' \longrightarrow [Ag(RCCR')_2]^+ \tag{6.149}$$

$$AuCl + C_2 R_2 \longrightarrow [AuCl(C_2 R_2)] \tag{6.150}$$

Acetylene complexes may also be obtained from salts containing metals in their higher oxidation states. In such cases, the alkyne plays the role of a reducing agent.

$$4MeCCMe + 2AuCl_3 \longrightarrow [Au(C_2 Me_2)_2][AuCl_4] + \quad \quad \tag{6.151}$$

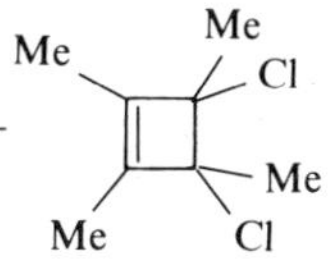

Metal alkyne compounds are also generally formed as a result of acetylene addition to unsaturated coordination complexes, most commonly to 16e compounds. An example is provided by the formation of pentacoordinate Ir(I) alkyne complexes:

$$[IrCl(CO)(PPh_3)_2] + RCCR' \longrightarrow [IrCl(RCCR')(CO)(PPh_3)_2] \tag{6.152}$$

Other examples are:

$$[RhCl(CO)(PPh_3)_2] + RCCR' \longrightarrow [RhCl(RCCR')(CO)(PPh_3)_2] \tag{6.153}$$

$$[Vcp_2] + MeOOCC \equiv CCOOMe \longrightarrow [Vcp_2(MeOOCC \equiv CCOOMe)] \tag{6.154}$$

## b. *Ligand substitution in coordination compounds*

Some labile complexes react with alkynes to afford products in which at least one ligand was displaced by an acetylene molecule. Substitution reactions most commonly occur for complexes containing the following ligands: CO, halides, phosphines, water, $N_2$, etc.

$$[W(CO)_6] + 3RCCR' \xrightarrow{hv} [W(RCCR')_3 (CO)] + 5CO \tag{6.155}$$

$$[Co_2(CO)_8] + RCCR' \longrightarrow [Co_2(RCCR')(CO)_6] + 2CO \tag{6.156}$$

$$3[Fe(CO)_5] + 2RCCR' \longrightarrow [Fe(RCCR')(CO)_4] + [Fe_2(RCCR')(CO)_6] \tag{6.157}$$

Phosphine Pd(0) and Pt(0) complexes react with acetylenes according to the following equations:

$$[M(PPh_3)_4] + RCCR' \longrightarrow [M(RCCR')(PPh_3)_2] + 2PPh_3 \qquad (6.158)$$

Other compounds react similarly.

$$[Rh_2(PF_3)_8] + RCCR' \longrightarrow [Rh_2(RCCR')(PF_3)_6] + 2PF_3 \qquad (6.159)$$

Nitriles and $N_2$ readily undergo substitution:

$$[W(MeCN)_3(CO)_3] + RCCR \longrightarrow [W(RCCR)_3(CO)] \qquad (6.160)$$

$$[IrCl(N_2)(PPh_3)_2] + RCCR \longrightarrow [IrCl(RCCR)(PPh_3)_2] \qquad (6.161)$$

Platinum(II) halides readily form alkyne complexes via halogen substitution reactions:

$$M_2[PtCl_4] + RCCR' \longrightarrow M[PtCl_3(RCCR')] + MCl \qquad (6.162)$$

$$trans\text{-}[PtCl(Me)(PMe_2Ph)_2] + RCCR \xrightarrow[\text{MeOH}]{\text{AgPF}_6} trans\text{-}[Pt(Me)(RCCR)(PMe_2Ph)_2]PF_6 \qquad (6.163)$$

Olefin ligands may also be replaced:

$$K[PtCl_3(C_2H_4)] + RCCR' \longrightarrow K[PtCl_3(RCCR')] + C_2H_4 \qquad (6.164)$$

$$[Pt_2Cl_4(C_2H_4)_2] + 2RCCR' \longrightarrow [Pt_2Cl_4(RCCR')_2] + 2C_2H_4 \qquad (6.165)$$

The only stable Pd(II) compound with bis(*tert*-butyl)acetylene may be prepared by reaction of Bu′CCBu′ with bis(benzonitrile)dichloropalladium(II) or di-$\mu$-chloro-dichlorodi(ethylene)dipalladium(II):

$$2[PdCl_2(PhCN)_2] + 2\,t\text{-BuCCBu-}t \longrightarrow [Pd_2Cl_4(t\text{-BuCCBu-}t)_2] + 4PhCN \qquad (6.166)$$

$$[Pd_2Cl_4(C_2H_4)_2] + 2\,t\text{-BuCCBu-}t \longrightarrow [Pd_2Cl_4(t\text{-BuCCBu-}t)_2] + 2C_2H_4 \qquad (6.167)$$

Other alkyne palladium(II) complexes decompose and the acetylenes undergo oligomerization and polymerization with the formation of organometallic and organic compounds.

Many acetylene compounds are obtained by reactions involving reduction of transition metal compounds carried out in the presence of alkynes:

$$[PtCl_2(PPh_3)_2] + RCCR \xrightarrow[\text{EtOH}]{\text{N}_2\text{H}_4} [Pt(RCCR)(PPh_3)_2] \qquad (6.168)$$

## 5. PROPERTIES OF ACETYLENE COMPLEXES

Titanium group metals do not form stable alkyne complexes. The complex Ticp$_2$(PhCCPh)(CO) is an exception. It is prepared by reaction of Ticp$_2$(CO)$_2$ with a diphenylacetylene under vacuum which is necessary for removal of CO. Additional exceptions are Ticp$_2$(RCCR′)[122, 123] and Ticp$_2$(HCCH)(PMe$_3$) (Reference *i*, Table 6.21). The yellow acetylene carbonyl complex readily changes back to dicyclopentadienyldicarbonyltitanium(II) in the atmosphere of CO. During these decompositions, diphenylacetylene is given off.[96] Vanadium forms paramagnetic complexes of the type $[Vcp_2(C_2R_2)]$ as well as compounds such as $[Vcp(C_2R_2)(CO)_2]$. Niobium and tantalum also form carbonyl complexes as follows: $[Mcp(C_2R_2)(CO)_2]$, $[Mcp(C_2R_2)_2(CO)]$, and $[Nbcp(C_4Ph_4)(C_2Ph_2)CO]$. In the third complex, the tetraphenylcyclobutadiene ligand is formed from diphenylacetylene.

In contrast to vanadium compounds, niobium and tantalum complexes of the formula $[Mcp(C_2R_2)(CO)_2]$ exist in solution in monomer–dimer equilibria:

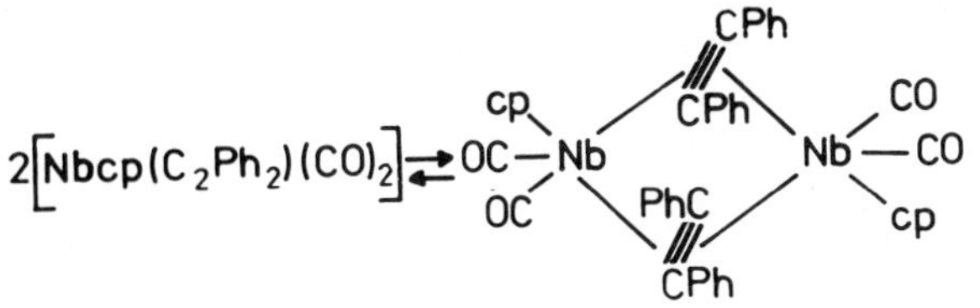

$$2\left[Nbcp(C_2Ph_2)(CO)_2\right] \rightleftharpoons$$

(6.169)

Owing to its smaller atomic radius, vanadium does not form dimers.

Complexes $[Mcp(C_2Ph_2)(CO)]_2$ are formed as a result of decomposition of $[Mcp(C_2Ph_2)(CO)_2]$. The dimeric compounds possess a metal–metal double bond.

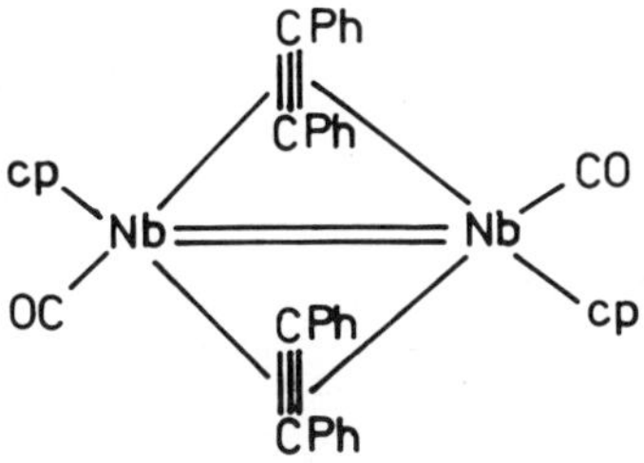

(6.170)

The following complexes are also known: $[Nbcp_2X(RCCR)]$ (X = Cl, I, H, Me, SMe), $[Ta(pmcp)_2 Cl(RCCR)]$, $[Ta(C_5H_4Me)_2 X(RCCR)]$ (X = H, I), $[Ta(pmcp)_2 Cl_2(RCCR)]$, $[TaCl_4(PhCCPh)py]^-$,[133, 134] $[Ta_2Cl_6(\mu\text{-}Bu^tCCBu^t)(THF)]$.[133]

For all acetylene compounds of these elements, substantial lowering of the carbon–carbon bond multiplicity is observed despite a small number of $d$ electrons, as evidenced by a considerable lowering of the $v(C\equiv C)$ frequency. This frequency equals $ca$ 1700 cm$^{-1}$ for nonbridging alkynes and $ca$ 1500 cm$^{-1}$ for bridging acetylenes.

Chromium group elements form alkyne complexes of the type $[M(C_2R_2)(CO)_5]$ (R = H, Me, Et). However, such compounds are unstable and are formed only during irradiation of carbonyls $[M(CO)_6]$ in the presence of alkynes. The compound $[W(C_2H_2)_2(CO)_4]$ is obtained in a similar fashion. These complexes were not isolated in the solid state.

As in olefin complexes, the presence of arene ligands in metal alkyne compounds increases the alkyne–metal bond stability. Stability of complexes of the type Cr(alkyne)(arene)(CO)$_2$ increases according to the following series: benzene < trimethylbenzene < hexamethylbenzene. Complexes containing cyclopentadienyl ligands are also known: $[Crcp(C_2R_2)(CO)(NO)]$ and $[Mocp_2(C_2R_2)]$.

The molybdenum and tungsten acetylene complexes $[M(RCCR)_3(CO)]$ and $[W(RCCR)_3(MeCN)]$ are very interesting. All diphenylacetylene molecules in

[W(PhCCPh)$_3$ (CO)] are equivalent; however, the two Ph—C≡ fragments are not equivalent and the complex possesses $C_{3v}$ symmetry (Figure 6.23). Therefore, the ligands lie at the vertices of a deformed tetrahedron. As a result, for asymmetric alkynes, $R^1CCR^2$, the existence of isomeric compounds such as [M(R$^1$CCR$^2$)$_3$L] is possible.

These complexes provide the classic examples of compounds in which nonbridging alkyne ligands use the electrons in both of their π-bonds to bind to a central atom (4e ligands).

The reduction of [W(PhCCPh)$_3$CO] by LiC$_{10}$H$_8$ probably first forms [W(PhCCPh)$_3$]$^{2-}$, which subsequently reacts with SnPh$_3$Cl to give [W(PhCCPh)$_3$SnPh$_3$]$^-$.[132] Triphenylchlorotin is commonly utilized for preparation of crystalline compounds from solutions of strongly reduced anionic complexes such as M(CO)$_n^{x-}$ which are otherwise difficult to isolate. Molybdenum also forms the acetylene–tetraphenylcyclobutadiene compound:

$$ \tag{6.171} $$

Only a few acetylene complexes possessing $d^2$ central atoms are known, for example, previously discussed [Ticp$_2$(PhCCPh)(CO)], Ti(II), and complexes of the type [MX$_4$(RCCR)] where M = Mo(IV), X = Cl, R = Me or Ph, and M = W(IV), X = Br,

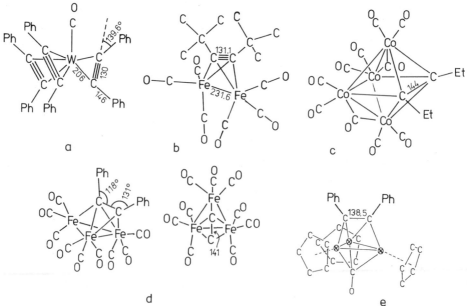

Figure 6.23. Structures of (a) [W(PhCCPh)$_3$(CO)], (b) [Fe$_2$(*t*-BuCCBu-*t*)(CO)$_6$], (c) [Co$_4$(EtCCEt)(CO)$_{10}$], (d) [Fe$_3$(PhCCPh)(CO)$_9$], (e) [Rh$_3$cp$_3$(PhCCPh)(CO)].

$R = Me$ or $Ph$.[94] The acetylene complex is even formed by tungsten($V$) ($d^1$ electron configuration): $[WOCl_3(MeCCMe)]$.[94] In dimers such as $[Mo_2cp_2(RCCR)(CO)_4]$ ($R = H$, Et, Ph) alkynes form bridges and the central atoms are bonded via metal–metal single bonds. The Mo–Mo distances are 298.4 and 295.6 pm, respectively.

Manganese group metals form the acetylene complexes $[Mncp(RCCR)(CO)_2]$, which are analogous to olefin compounds, as well as $[ReCl(RCCR)_2]$ and $[ReCl(RCCR)(PPh_3)]$.[97]

Iron forms mononuclear complexes $[Fe(RCCR)(CO)_4]$ ($R = Ph$, $t$-Bu, $SiMe_3$), dinuclear compounds $[Fe_2(RCCR)(CO)_6]$, and trinuclear derivatives, $[Fe_3(RCCR)(CO)_9]$ and $[Fe_3(RCCR)_2(CO)_8]$. In the dinuclear compound, there is an iron–iron double bond (Table 6.19). Mononuclear and polynuclear ruthenium and osmium complexes are also known: $[Ru(PhCCPh)(CO)_2(PPh_3)_2]$, $[RuCl(CF_3CCCF_3)(NO)(PPh_3)]$, $[Os(PhCCPh)(CO)(NO)(PPh_3)_2]^+$, $[Os(CF_3CCCF_3)(CO)_2\{P(OMe)_3\}_2]$, $[Os_3H_2(C_2H_2)(CO)_9]$, and $[Ru_3H_2(PhCCPh)(CO)_9]$.

The diphenylacetylene molecule in $[Cocp(PhCCPh)(PPh_3)]$ is easily substituted by olefins containing electron-withdrawing substituents to give $[Cocp(RCH=CHR)(PPh_3)]$ ($R = CN$, COOMe). The ion $[Co(CNMe)_5]^{2+}$ forms substitution products with acetylenes, e.g., $[Co(CNMe)_4(RCCR)]^{2+}$.

Acetylenes undergo oxidative addition with $[Co(CN)_5]^{3-}$ to afford dinuclear compounds containing a vinylene bridge.[99]

$$K_6 \left[ \begin{array}{c} (NC)_5Co \diagdown \qquad\qquad \diagup H \\ C = C \\ H \diagup \qquad\qquad \diagdown CO(CN)_5 \end{array} \right] \qquad (6.172)$$

Dinuclear cobalt acetylene complexes $[Co_2(RCCR)(CO)_6]$ in which the alkyne molecule constitutes a bridge are very stable. In these compounds, as in other similar complexes, acetylenes are $4e$ ligands. In complexes of the type $[Co_4(RCCR)(CO)_{10}]$, the alkyne molecule is bonded to all cobalt atoms. In dimeric cobalt complexes, acetylene molecules may undergo exchange reactions:

$$[Co_2(RCCR)(CO)_6] + R'CCR' \longrightarrow [Co_2(R'CCR')(CO)_6] + RCCR$$

The stability of complexes changes according to the following series: $C_2(CH_2NEt_2)_2 < C_2H_2 < MeCCH < PhCC < C_2Me_2 < PhCCMe < C_2Ph_2 < C_2(COOMe)_2 < C_2(CF_3)_2$. Alkyne cobalt complexes are stabilized by electron-withdrawing substituents in the same way as Pt(0) alkyne complexes.

For rhodium(I) and iridium(I) complexes, the most characteristic are square-planar complexes of the type $MX(RCCR)L_2$. Examples are provided by the following compounds: $[IrX(RCCR')(CO)(ER''_3)_2]$ ($X = Cl$, Br, I, SCN, $SnCl_3$; $E = P$, As, Sb), $[IrX(RCCR')(PR_3)_2]$, $[RhCl(RCCR)(PR_3)_2]$, and $[RhCl(RCCR')(CO)(PPh_3)]$. Other acetylene Rh(I) and Ir(III) complexes are also known: $[Rhcp(MeOOCCCCOOMe)(PPh_3)]$ and $[IrCl_2\{C(COOMe)=CH(COOMe)\}(MeOOCC \equiv CCOOMe)(PEt_3)_2]$. The latter compound is octahedral, with the alkyne molecule occupying one coordination site.

Rhodium also forms the dimeric compound $[Rh_2(PhCCPh)_2(CO)_3(PPh_3)]$ in which each alkyne molecule is probably bonded to one rhodium atom.[100]

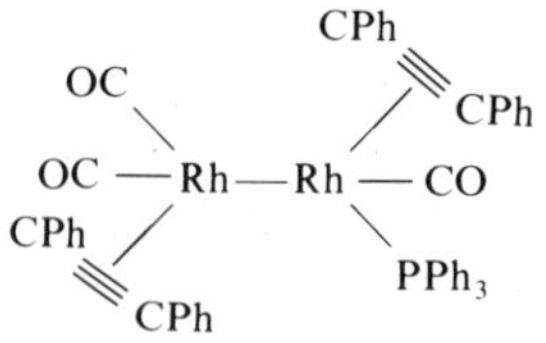

$$(6.173)$$

Polynuclear acetylene rhodium complexes are also encountered: $[Rh_3cp_3(\mu_3\text{-}PhCCPh)(\mu_3\text{-}CO)]$, $[Rh_3cp_3(\mu_3\text{-}C_6F_5CCC_6F_5)(\mu\text{-}CO)]$, and $[Rh_2(\mu\text{-}C_2Ph_2)(PPh_3)_2(PF_3)_4]$ (Tables 6.24 and 6.19).

Alkyne nickel(I) compounds such as $[Ni_2cp_2(PhCCPh)]$ and $[Ni_2cp_2(HCCH)]$ have structures analogous to $[Co_2(PhCCPh)(CO)_6]$. The Ni–Ni distances are 232.9

*Table 6.24. Metal Acetylene Complexes*

| Compound | $T_t$ (K) | Color | Other data |
|---|---|---|---|
| $[Vcp_2(MeOOCC_2COOMe)]^a$ | 419–420 | Dark-green | $\nu(C\equiv C)$ 1821, $g=1.9976$, $A=4.4$ mT |
| $[Vcp_2(CF_3CCCF_3)]^a$ | 397–399 | Brown | $\nu(C\equiv C)$ 1800, $g=1.9971$, $A=4.6$ mT |
| $[Vcp(PhCCPh)(CO)_2]^b$ | 331–333 | Green | |
| $[Nbcp(PhCCPh)(CO)_2]^c$ | 413 dec | Violet | |
| $[Nbcp(PhCCPh)_2(CO)]^d$ | 408–410 | Light-yellow | |
| $[Tacp(PhCCPh)_2(CO)]^e$ | 412–413 | Light-yellow | $\nu(C\equiv C)$ 1700 |
| $[Crcp(C_2H_2)(CO)(NO)]^{(90)}$ | | | |
| $[Mocp_2(C_2H_2)]^f$ | | Orange | |
| $[Mocp_2(HCCMe)]^g$ | | Red | |
| $[MoCl_4(MeCCMe)]^{(94)}$ | | Brown | $\nu(C\equiv C)$ 1755, $M=288$ (mass spectrum) |
| $[W(PhCCPh)_3(CO)]^h$ | 466 | Light-yellow | |
| $[NEt_4][W(PhCCPh)_3SnPh_3]^{(132)}$ | | White | Light-sensitive |
| $[Mncp(PhCCPh)(CO)_2]^i$ | 377–378 | Brown | |
| $[Mncp(CF_3CCCF_3)(CO)_2]^j$ | 341 | Orange-yellow | |
| $[Fecp(SnPh_3)(C_2Ph_2)(CO)]^k$ | 457 | Red | |
| $[Fe_2(t\text{-}BuCCBu\text{-}t)_2(CO)_4]^m$ | 448 (sublimation) | Black-violet | |
| $[Fe_2(t\text{-}BuCCBu\text{-}t)(CO)_6]^l$ | | Black | $\nu(CO)$ 2050, 2005, 1984, 1973 (sh), 1968 sh |
| $[Ru(PhCCPh)(CO)_2(PPh_3)_2]^n$ | | | $\nu(C\equiv C)$ 1776 |
| $[RuCl(CF_3CCCF_3)(NO)(PPh_3)]^o$ | 482–485 dec | Red | |
| $[Os(HCCH)(CO)(NO)(PPh_3)_2]^{+\,p}$ | | | $\tau(H)$ 1.62, 4.25 (243 K) |
| $[Os(CF_3CCCF_3)(CO)_2\{P(OMe)_3\}_2]^q$ | | | $\nu(C\equiv C)$ 1753 |
| $[Cocp(PhCCPh)(PPh_3)]^r$ | | | $\nu(C\equiv C)$ 1818 |
| $[Co_2(PhCCPh)(CO)_6]^s$ | 383 | Dark-red | |
| $[Co_2(C_2H_2)(CO)_5(PPh_3)]^t$ | 403 | Violet | |
| $[Co_4(MeCCH)(CO)_{10}]^u$ | 442 dec | Dark-blue | |
| $[RhCl(CO)(HOOCCCCOOH)(PPh_3)_2]^v$ | 478–480 | Light-yellow | |
| $[Rh_3cp_3(PhCCPh)(CO)]^w$ | | Dark-purple | |

*(Table continued)*

*Table 6.24. (continued)*

| Compound | $T_t$ (K) | Color | Other data |
|---|---|---|---|
| $[Rhcp\{C_2(COOMe)_2\}(PPh_3)]^r$ | | | |
| $[IrCl(MeOOCCCCOOMe)(PPh_3)_2]^v$ | 455–456 dec | Yellow | |
| $[IrCl(PhCCPh)(PPh_3)_2]^v$ | 466–468 dec | Yellow | |
| $[Ir(C_2H_2)(COD)(phen)]^{+x}$ | | | |
| $[Ir(CNC_6H_4Me)_2(MeOOCCCCOOMe)$ | | | |
| $\quad (PPh_3)_2]^+ClO_4^{-y}$ | | White | $v(C\equiv C)$ 1772 |
| $[Ni(COD)(p\text{-}MeC_6H_4CCC_6H_4\text{-}Me\text{-}p)]^z$ | 406–408 | Orange-yellow | |
| $[Ni_2(COD)_2(PhCCPh)]^z$ | 446–448 | Dark-red | |
| $[Ni_4(PhCCPh)_3(CNCMe_3)_4]^{(95)}$ | | | $d(Ni-Ni)$ 237.4, |
| | | | $d(C\equiv C)$ 134.0 |
| $[Pd_2Cl_4(t\text{-}BuCCBu\text{-}t)_2]^{aa}$ | 437–438 dec | Red | $v(C\equiv C)$ 2000 |
| $[Pd_2(C_5Ph_5)_2(PhCCPh)]^{bb}$ | 513 dec | Dark-green | |
| $[Pt(CF_3CCCF_3)(PPh_3)_2]^j$ | 488–489 | White | $v(C\equiv C)$ 1775 |
| $[PtCl_2\{C_2(CMe_2OH)_2\}py]^{cc}$ | 383–385 dec | Light-yellow | |
| $[Pt\{C_2(CEtMeOH)_2\}_2]^{dd}$ | 437–446 dec | White | |
| $[CuBr(cyclooctyne)]^{ee}$ | 404 | | |
| $[CuBr(cyclooctyne)_2]^{ee}$ | 383–393 | | |
| $[AuBr(cyclooctyne)_2]^{ee}$ | 408–416 dec | | |
| $[Au_2Cl_6(MeCCMe)]^{ff}$ | | Red | $\tau(Me)$ 7.56 (233 K) |
| $[Au_2Cl_6(MeCCMe)_2]^{ff}$ | | Red | $\tau(Me)$ 7.36 (233 K) |

[a] R. Tsumura and N. Hagihara, *Bull. Chem. Soc. Jpn.*, **38**, 861 (1965); H. J. de Liefde Meijer and F. Jellinek, *Inorg. Chim. Acta*, **4**, 651 (1970).

[b] A. N. Nesmeyanov, K. N. Anisimov, N. E. Kolobova, and A. A. Pasynskii, *Dokl. Akad. Nauk SSSR*, **182**, 112 (1968).

[c] A. N. Nesmeyanov, K. N. Anisimov, N. E. Kolobova, and A. A. Pasynskii, *Izv. Akad. Nauk SSSR, Ser. Khim.*, 774 (1966).

[d] A. N. Nesmeyanov, K. N. Anisimov, N. E. Kolobowa, and A. A. Pasynskii, *Izv. Akad. Nauk SSSR, Ser. Khim.*, 100 (1969).

[e] A. A. Pasynskii, K. N. Anisimov, N. E. Kolobova, and A. N. Nesmeyanov, *Izv. Akad. Nauk SSSR, Ser. Khim.*, 183 (1969).

[f] K. L. Tang Wong, J. L. Thomas, and H. H. Brintzinger, *J. Am. Chem. Soc.*, **96**, 3694 (1974).

[g] J. L. Thomas, *Inorg. Chem.*, **17**, 1507 (1978).

[h] D. P. Tate, I. M. Augl, W. M. Ritchey, B. W. Ross, and J. G. Grasselli, *J. Am. Chem. Soc.*, **86**, 3261 (1964).

[i] W. Strohmeier and D. von Hobe, *Z. Naturforsch.*, **16b**, 402 (1961).

[j] J. T. Boston, S. O. Grim, and G. Wilkinson, *J. Chem. Soc.*, 3468 (1963).

[k] A. N. Nesmeyanov, N. E. Kolobowa, V. V. Skripkin, and K. N. Anisimov, *Dokl. Akad. Nauk SSSR*, **196**, 606 (1971).

[l] F. A. Cotton, J. D. Jamerson, and B. R. Stults, *J. Am. Chem. Soc.*, **98**, 1774 (1976).

[m] K. Nicholas, L. S. Bray, R. E. Davis, and R. Pettit, *Chem. Commun.*, 608 (1971).

[n] B. E. Cavit, K. R. Grundy, and W. R. Roper, *J. Chem. Soc., Chem. Commun.*, 60 (1972).

[o] J. Clemens, M. Green, and F. G. A. Stone, *J. Chem. Soc., Dalton Trans.*, 375 (1973).

[p] J. Ashley-Smith, B. F. G. Johnson, and J. A. Segal, *J. Organomet. Chem.*, **49**, C38 (1973).

[q] R. Burt, M. Cooke, and M. Green, *J. Chem. Soc. (A)*, 2981 (1970).

[r] Y. Wakatsuki and H. Yamazaki, *Chem. Commun.*, 280 (1973); Y. Wakatsuki, H. Yamazaki, and H. Iwasaki, *J. Am. Chem. Soc.*, **95**, 5781 (1973).

[s] H. W. Sternberg, H. Greenfield, R. A. Freidel, J. Wotiz, R. Markby, and L. Wender, *J. Am. Chem. Soc.*, **76**, 1457 (1954).

[t] Y. Iwashita, F. Tamura, and A. Nakamura, *Inorg. Chem.*, **8**, 1179 (1969).

[u] R. S. Dickson and G. R. Tailby, *Aust. J. Chem.*, **23**, 229 (1970).

[v] J. P. Collman and J. W. Kang, *J. Am. Chem. Soc.*, **89**, 844 (1967).

[w] S. A. Gardner, P. S. Andrews, and M. D. Rausch, *Inorg. Chem.*, **12**, 2396 (1973).

[x] G. Mestroni and A. Camus, *Inorg. Nucl. Chem. Lett.*, **9**, 261 (1973).

[y] K. Kawakami, Masa-AkiHaga, and T. Tanaka, *J. Organomet. Chem.*, **60**, 363 (1973).

[z] E. L. Muetterties, W. R. Pretzer, M. G. Thomas, B. F. Beier, D. L. Thorn, V. W. Day, and A. B. Anderson, *J. Am. Chem. Soc.*, **100**, 2090 (1978).

[aa] T. Hosokawa, I. Moritani, and S. Nishioka, *Tetrahedron Lett.*, 3833 (1969).

[bb] E. Ban, P.-T. Cheng, T. Jack, S. C. Nyburg, and J. Powell, *Chem. Commun.*, 368 (1973); T. R. Jack, C. J. May, and J. Powell, *J. Am. Chem. Soc.*, **99**, 4707 (1977).

[cc] A. D. Allen and T. Theophanides, *Can. J. Chem.*, **42**, 1551 (1964); A. D. Allen and T. Theophanides, *Can. J. Chem.*, **44**, 2703 (1966).

[dd] F. D. Rochon and T. Theophanides, *Can. J. Chem.*, **46**, 2973 (1968).

[ee] G. Wittig and S. Fischer, *Chem. Ber.*, **105**, 3542 (1972).

[ff] R. Huttel and H. Forkl, *Chem. Ber.*, **105**, 1664 (1972); R. Huttel and H. Forkl, *Chem. Ber.*, **105**, 2913 (1972).

and 234.5 pm, respectively. The Pd–Pd distance in the palladium(I) compound $[Pd_2(C_5Ph_5)_2(PhCCPh)]$ equals 263.9 pm. These compounds contain $M-M$ single bonds.

The compound $[Ni_4(PhCCPh)_3(CNMe_3)_4]$ represents a tetranuclear cluster possessing direct $Ni-Ni$ bonds and $\mu_3$ acetylene bridging ligands (Table 6.24). The nickel group elements characteristically form trigonal complexes $[M(RCCR)L_2]$ ($L = PR_3$, $AsR_3$, RNC, etc.) while Pd(II) and Pt(II) ions give square-planar compounds such as $[M_2Cl_4(RCCR)_2]$, $[PtCl_3(RCCR)]^-$, and $[PtCl_2(RCCR)L]$ ($L =$ amines, phosphines).

In the case of platinum, $14e$ complexes of the following composition are also known: $[Pt(RCCR)_2]$ $[R = Ph, C(OH)R^1R^2; R^1, R^2 = Me, Et, Pr, Bu]$ (Tables 6.20 and 6.24). The dimer $[(PhCCPh)Pt(\mu\text{-}PhCCPh)Pt(PMe_3)_2]$ was also prepared (Tables 6.20 and 6.24). The Pt–Pt distance (289 pm) is greater than the distance of most platinum dimeric complexes possessing bridges. In $[Pt(RCCR)_2]$ the $C\equiv C$ bonds are perpendicular, and therefore the alkyne ligands occupy corners of a deformed tetrahedron. Platinum(II) pentacoordinate compounds having trigonal–bipyramidal structures are also known, for example, $[PtCl(Me)(CF_3CCCF_3)(AsMe_2Ph)_2]$.

The copper group metals form relatively weak complexes with alkynes in which the ratio M/acetylene is $\geqslant 1$: $[CuClC_2H_2]$, $[(CuCl)_2C_2H_2]$, $[(CuCl)_3C_2H_2]$, $[(CuCl)_6C_2H_2]$, $[CuClRCCH]$, $[(CuCl)_2(RCCH)]$, $[CuCl_2(RCCR)]^-$, $[Ag(RCCR)_n]^+$ ($n = 1\text{–}3$), $[Ag_2(RCCR)]^{2+}$, $[AuCl(RCCR)]$, and $[Au(MeCCMe)_2][AuCl_4]$ (see preparation of acetylene complexes). In acetylene compounds containing $\sigma$ metal–carbon bonds, e.g., $CuC\equiv CH$, the ethynide anion constitutes a bridge and forms a $\pi$-acetylene bond with the adjacent copper atom:

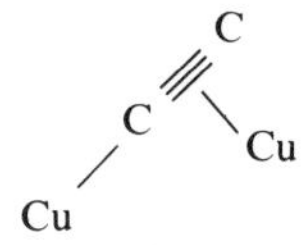

## 6. REACTIONS OF ACETYLENE COMPLEXES

The acetylene molecule, particularly in the case of weak alkyne complexes, may undergo dissociation reaction. This process occurs quite readily in copper group compounds. In pentacoordinate complexes, many dissociation reactions are known only in solution.

$$[CuCl(HCCH)] \xrightarrow{\Delta} CuCl + HCCH \qquad (6.174)$$

$$[AgNO_3(EtCCEt)_3] \longrightarrow [AgNO_3(EtCCEt)] + 2EtCCEt \qquad (6.175)$$

$$[PtCl(Me)(CF_3CCCF_3)(AsMe_2Ph)_2] \xrightarrow{350\ K} [PtClMe(AsMe_2Ph)_2] + CF_3CCCF_3 \qquad (6.176)$$

Thermal decompositions may lead to release of the alkyne or, most frequently, to formation of oligomers, cyclooligomers, or cooligomers. Free acetylene may be given off under the influence of oxidizing agents. Oxidation of $Ni_2cp_2(RCCR)$ by air or oxidation of $Co_2(F_5C_6CCC_6F_5)(CO)_6$, $Cocp(PhCCPh)(PPh_3)$, and $Fe_3(PhCCPh)(CO)_9$ by iodine causes alkyne liberation. The reaction of $Mcp(PhCCPh)(CO)_2$ (M = Nb, Ta) with HCl occurs in a similar way. Alkyne complexes decompose under the influence of acids, bromine, and $LiAlH_4$, and the acetylenes form corresponding addition products: olefins, bromoolefins, or alkanes. Complexes containing $\sigma$-vinyl groups are formed in intermediate stages. Such vinyl complexes may constitute main reaction products under

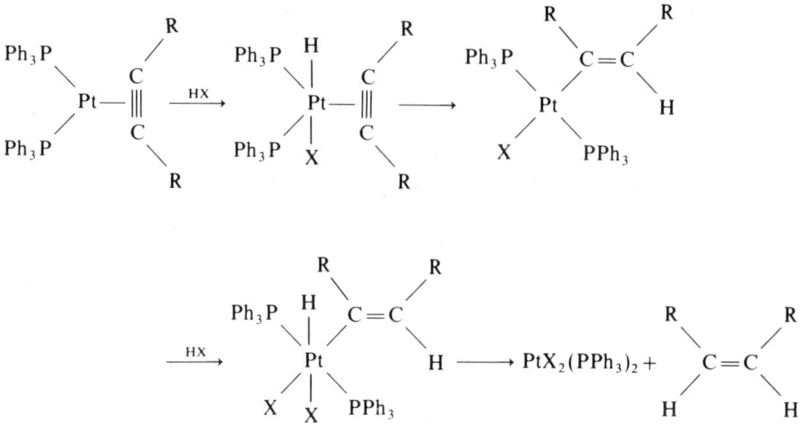

mild conditions [equation (6.177)]. Olefin complexes may be formed in the presence of mildly brominating compounds such as $pyH^+Br_3^-$:

$$[Pt(NCC\equiv CCN)(PPh_3)_2] + pyH^+Br_3^- \longrightarrow [Pt\{\eta^2\text{-}BrC(CN)=C(CN)Br\}(PPh_3)_2]$$

Acetylenes undergo exchange reactions or they may be displaced by other ligands which are present in excess in the reaction mixture or exhibit considerably stronger complex formation properties. For cationic coordination compounds of the type $[PtMe(RCCR)L_2]^+$ the ease of alkyne substitution by other ligands decreases as follows:

$$RNC > PMe_2Ph > PPh_3 > CO > AsPh_3 > SbPh_3 > py$$

$$> RCN > CH_2=CH_2 \backsim MeCCMe > MeCOMe \backsim MeOH \qquad (6.178)$$

Alkyne substitution reactions by Lewis bases also take place for many other complexes. In such reactions, the following ligands are most commonly utilized: phosphines, arsines, phosphites, $P(OR)_3$, CO, etc.

In neutral and anionic platinum(II) complexes the alkyne may be displaced by thiourea, KI, KSCN, $KNO_2$, and $Na_2S_2O_3$. Acetylenes exhibit a relatively strong *trans* effect. In $[PtCl_3(RCCR)]^-$, the chlorine atom which is *trans* to acetylene is weakly bonded. The $\nu(Pt-Cl_{trans})$ frequency is lower than that of $\nu(Pt-Cl_{cis})$ by $20\ cm^{-1}$. Consequently, ligands that are *trans* to alkyne readily undergo substitution.

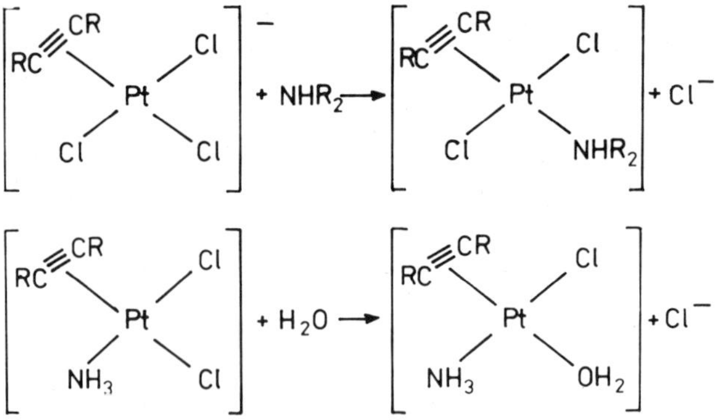

$$\left[\begin{array}{c} RC{\equiv}CR \\ Cl \quad\quad Cl \\ Pt \\ Cl \end{array}\right]^{-} + NHR_2 \longrightarrow \left[\begin{array}{c} RC{\equiv}CR \\ Cl \quad\quad Cl \\ Pt \\ Cl \quad NHR_2 \end{array}\right] + Cl^{-}$$

$$\left[\begin{array}{c} RC{\equiv}CR \\ Cl \quad\quad Cl \\ Pt \\ NH_3 \quad Cl \end{array}\right] + H_2O \longrightarrow \left[\begin{array}{c} RC{\equiv}CR \\ Cl \quad\quad Cl \\ Pt \\ NH_3 \quad OH_2 \end{array}\right] + Cl^{-} \qquad (6.179)$$

During reactions with chelating ligands such as en and bipy, pentacoordinate complexes are formed as intermediates:

$$[PtCl_3(RCCR)]^- + bipy \longrightarrow [PtCl_2(RCCR)(bipy)] + Cl^-$$
$$\downarrow \qquad\qquad (6.180)$$
$$[PtCl(RCCR)(bipy)]^+ Cl^-$$

It is possible to conduct an acylation reaction on the substituent phenyl of coordinated diphenylacetylene in the complex $Co_2(PhCCPh)(CO)_6$:

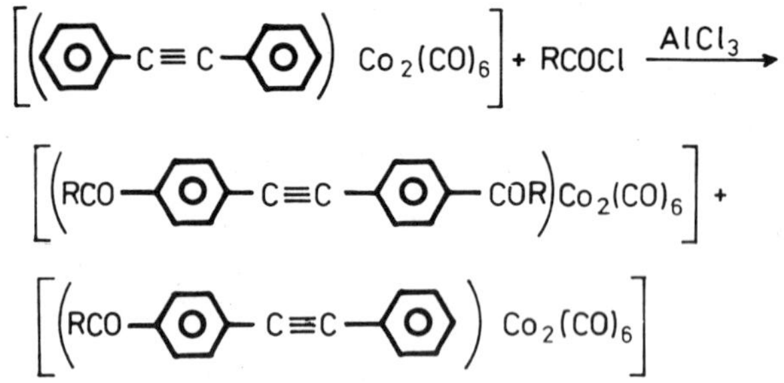

$$\left[\left(\langle\bigcirc\rangle{-}C{\equiv}C{-}\langle\bigcirc\rangle\right)Co_2(CO)_6\right] + RCOCl \xrightarrow{AlCl_3}$$

$$\left[\left(RCO{-}\langle\bigcirc\rangle{-}C{\equiv}C{-}\langle\bigcirc\rangle{-}COR\right)Co_2(CO)_6\right] +$$

$$\left[\left(RCO{-}\langle\bigcirc\rangle{-}C{\equiv}C{-}\langle\bigcirc\rangle\right)Co_2(CO)_6\right] \qquad (6.181)$$

The migration (insertion) reaction of coordinated ligands to alkynes constitutes (as in many other complexes) an important elementary reaction involved in catalytic hydrogenation, oligomerization, or polymerization. Owing to the presence of two mutually perpendicular $\pi$ orbitals, the migration reaction is complicated and may lead to formation of *cis* or *trans* products dependent upon the metal, its oxidation state, the character of the remaining ligands, and substituents on the alkyne molecule. The structure of these $\sigma$-alkenyl complexes cannot be unequivocally determined on the basis of the structure of the olefin that results from reaction of these complexes with hydrogen or water, because isomerization may occur during their decomposition. Fluoroacetylenes such as $CF_3C{\equiv}CCF_3$ and $CF_3C{\equiv}CH$ form stable $\sigma$-alkenyl complexes which may be studied by $^1H$ and $^{19}F$ NMR methods. The coupling constants unequivocally allow determination of the stereochemistry of investigated complexes. In the case of compounds involving diphenylacetylene and $MeOOCC{\equiv}CCOOMe$, structures of metal alkenyl compounds prepared by these ligands are inferred from $^{13}C$ NMR spectra.

Migration reactions of the hydrido ligand to $CF_3CCCF_3$ and $PhCCPh$ give the products shown in equations (6.182)–(6.185).[87]

$CF_3C \equiv CCF_3$

*trans* migration:

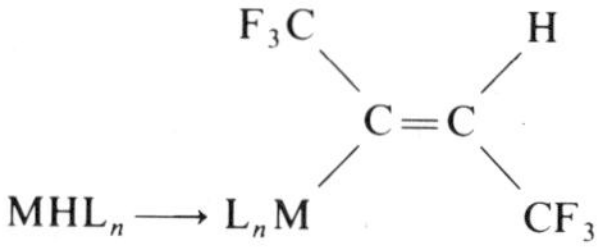

$$MHL_n \longrightarrow L_nM \quad \begin{array}{c} F_3C \\ \diagdown \\ \diagup \\ \end{array} C = C \begin{array}{c} H \\ \diagup \\ \diagdown \\ CF_3 \end{array} \qquad (6.182)$$

$ML_n = Mn(CO)_5, Re(CO)_5, MoHcp_2, WHcp_2, Recp_2$

*cis* migration:

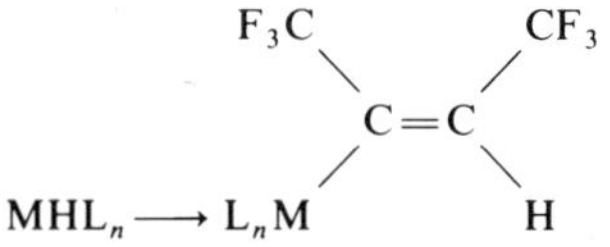

$$MHL_n \longrightarrow L_nM \quad \begin{array}{c} F_3C \\ \diagdown \\ \diagup \\ \end{array} C = C \begin{array}{c} CF_3 \\ \diagup \\ \diagdown \\ H \end{array} \qquad (6.183)$$

$ML_n = PtCl(PEt_3)_2, IrCl_2(PMe_2Ph)_3, Rucp(PPh_3)_2, RhCO(PPh_3)_3$

$PhC \equiv CPh$

*trans* migration:

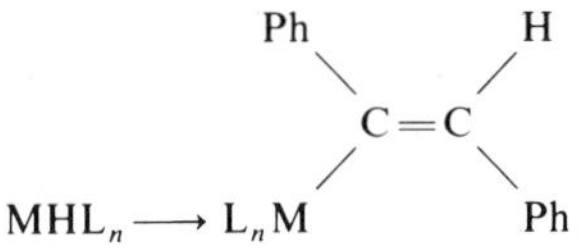

$$MHL_n \longrightarrow L_nM \quad \begin{array}{c} Ph \\ \diagdown \\ \diagup \\ \end{array} C = C \begin{array}{c} H \\ \diagup \\ \diagdown \\ Ph \end{array} \qquad (6.184)$$

$ML_n = Rh(CO)(PPh_3)_3, Cocp(PPh_3)$

*cis* migration:

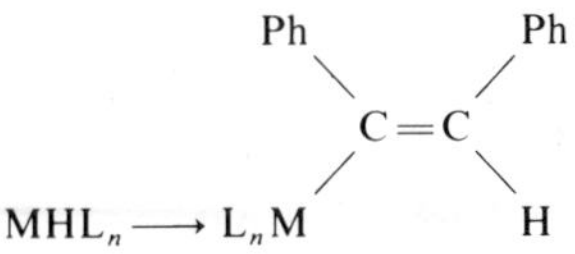

$$MHL_n \longrightarrow L_nM \quad \begin{array}{c} Ph \\ \diagdown \\ \diagup \\ \end{array} C = C \begin{array}{c} Ph \\ \diagup \\ \diagdown \\ H \end{array} \qquad (6.185)$$

$ML_n = MoHcp_2, IrCl_2(MeSOMe)_2, PtCl(PEt_3)_2, Co(dmgH)_2$

The stereochemistry of the *cis* migration in reactions of acetylenes with $Mocp_2H_2$ may best be described by a four-center transition state which is polarized to a certain degree.

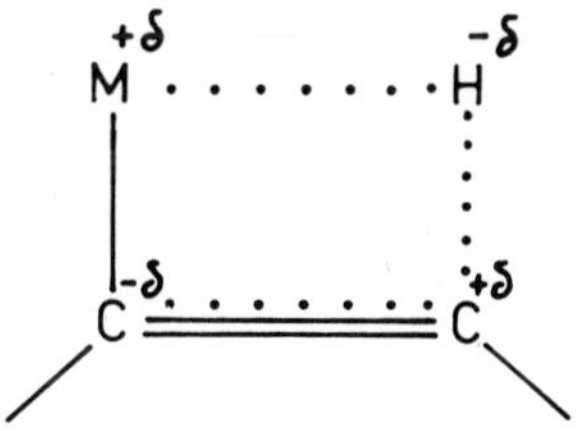

$$ (6.186) $$

It is assumed that the reaction proceeds through a thermally excited, strongly π-basic complex containing parallel cyclopentadienyl ligands because this reaction should be very slow for weakly σ-basic dicyclopentadienyldihydridomolybdenum(IV) in the ground state. An example is provided by *cis* migration to the asymmetric alkyne $CF_3 \equiv CH$.

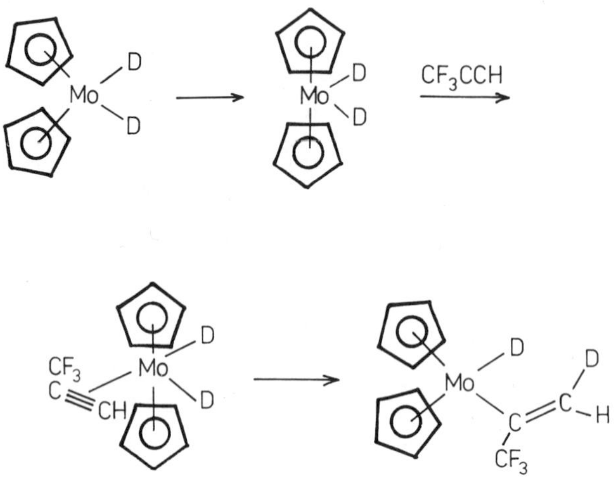

$$(6.187)$$

The *trans* migration to $CF_3C \equiv CCF_3$ in complexes of the type $Mcp_2H_2$ proceeds differently. The lack of the influence of compounds which readily interact with radicals on the reaction course, as well as the fact that the reaction rate does not depend on the polarity of the solvent, indicate that the reaction is neither radical nor ionic. It is assumed that, in the transition state, a four-membered nonpolar ring containing a twisted (slanted, skewed) arrangement (6.188) of σ and π bonds is formed. It is assumed

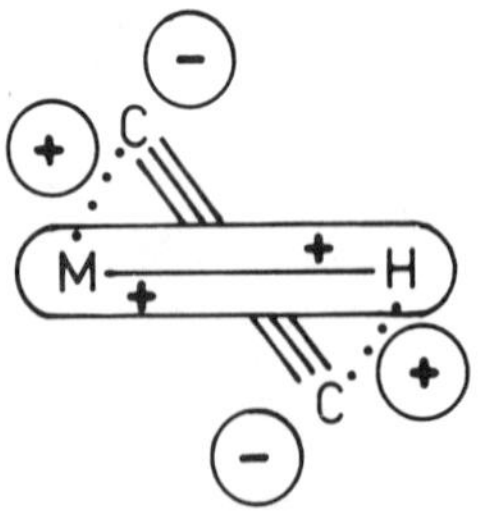

$$(6.188)$$

that in the case of the reaction of $[Mo(C_5H_5)_2H_2)]$ with bistrifluoromethylacetylene, complexes in which the acetylene molecule is located either between hydrido ligands or between the hydrido ligand and the cyclopentadienyl group may be formed as intermediates. Such structures of intermediate complexes explain that there is more of isomer a than b in the reaction products [scheme (6.189)]. The isomers are fairly stable, for a ⇌ b isomerization $E_a = 70.3$ kJ mol$^{-1}$ and $\Delta S^* = 3.5$ JK$^{-1}$ mol$^{-1}$.

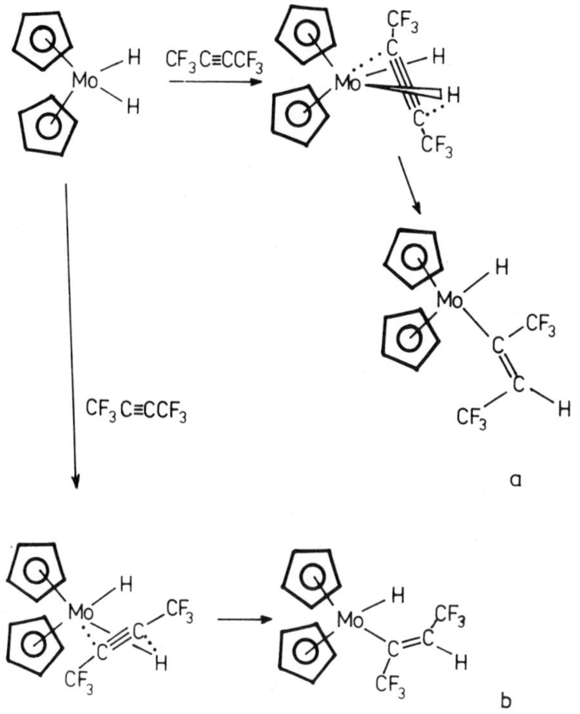

(6.189)

a

b

## 7. METALLACYCLIZATION REACTIONS

Alkynes may react with acetylene complexes to form metallacyclic compounds, i.e., compounds containing a higher number of coordinated acetylene molecules, or compounds possessing other alkynes (substitution reaction); see scheme (6.190).

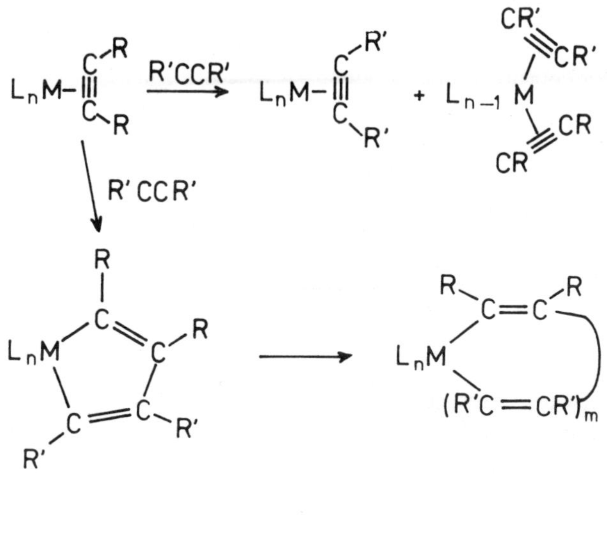

(6.190)

Metallacyclization reactions occur most readily in the case of complexes in which metals possess $d^8$–$d^{10}$ electron configurations (Fe, Co, Ni, Ru, Rh, Pd, and Ir); see equations (6.191) and (6.192).

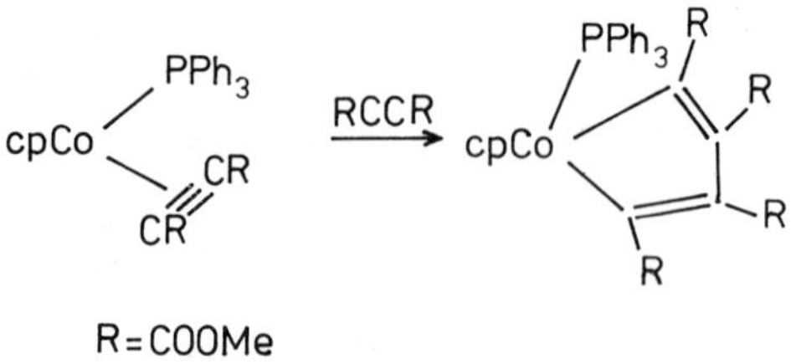

$$\text{cpCo} \underset{\text{CR}}{\overset{\text{PPh}_3}{<}} \overset{\text{RCCR}}{\longrightarrow} \quad (6.191)$$

R = COOMe

$$\text{L}_2\text{Pd} \overset{\text{RCCR}}{\longrightarrow} \text{L}_2\text{Pd} \qquad (6.192)$$

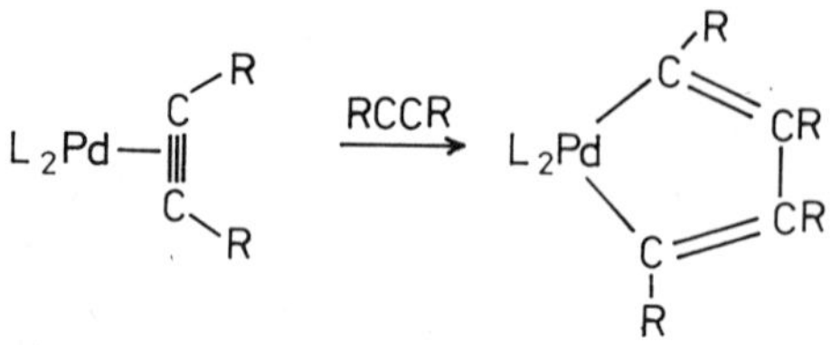

As a result of the antibonding character of the orbital c (Figure 6.21) (in complexes which have occupied $d_{\pi_\perp}$ orbitals) it is possible to excite thermally the acetylene molecule and to form a monohapto coordination compound or to dissociate the alkyne [scheme (6.193)].

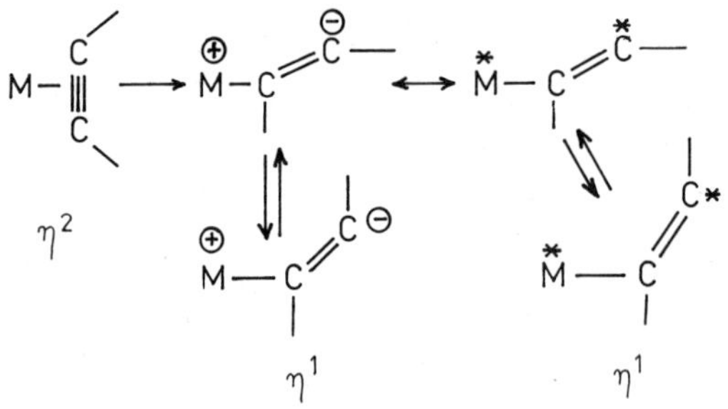

$$ \eta^2 \qquad\qquad \eta^1 \qquad\qquad \eta^1 \qquad (6.193)$$

In the case of coordination compounds with central metals possessing $d^1$ and $d^2$ electron configurations, the metallacyclic compounds are also readily formed. Because of a very weak $\pi$ back-bonding interaction, the thermal excitation of acetylene leads to the reverse polarization of the metal–acetylene grouping compared to that in previous complexes.

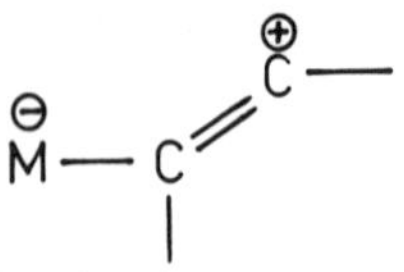

$$\overset{\ominus}{M} - C \overset{\overset{\oplus}{C}-}{=} $$

Decreased activity of alkyne complexes containing 4–6 $d$ electrons in the metallacyclic reaction may be explained by the bonding character of the c orbital (Figure 6.21). Examples are provided by niobium and molybdenum complexes ($d^4$–$d^6$ electron configurations). The ionic or radical character of the intermediate complex depends on the metal, its oxidation state, the remaining ligands and substituents on the alkyne carbon atoms. Therefore, all these factors influence the stereochemistry and selectivity of the migration (insertion) reaction (6.194). The decisive excess of isomer b in the reaction products (and the lack of isomer c) for (1) indicates ionic character of the intermediate complex $[(PPh_3)cpCo^+ - CMe = C^- - COOMe]$. In case (2) the content of isomers a and b may be accounted for by a less polar intermediate complex and by a certain contribution of radical character.

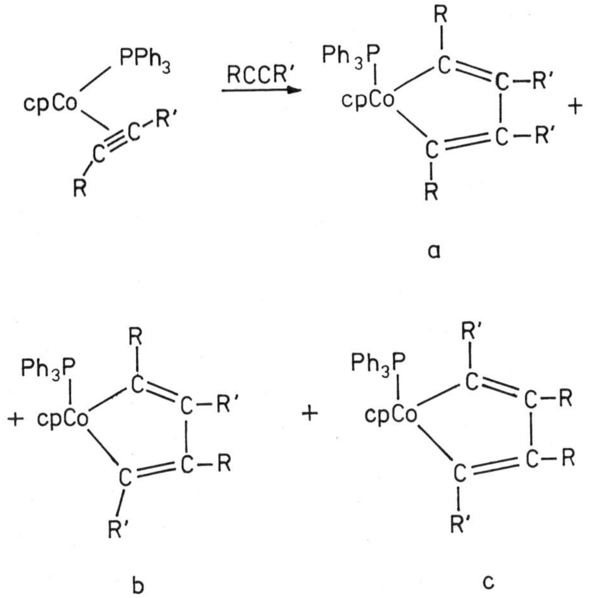

$$(6.194)$$

(1)   R = COOMe, R′ = Me: a—9%, b—50%
(2)   R = Ph, R′ = COOMe: a—13%, b—20%

Platinum(II) complexes may form polarized intermediate compounds which have an excess of positive charge on the carbon atom[101]; see equation (6.195). Metalla-

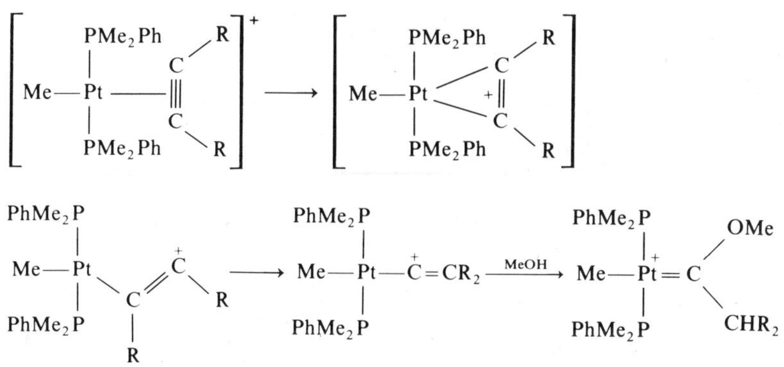

$$(6.195)$$

cyclopentadienyl complexes may react with an additional alkyne molecule to give benzene derivatives or metallacycloheptatrienyl complexes; see scheme (6.196).

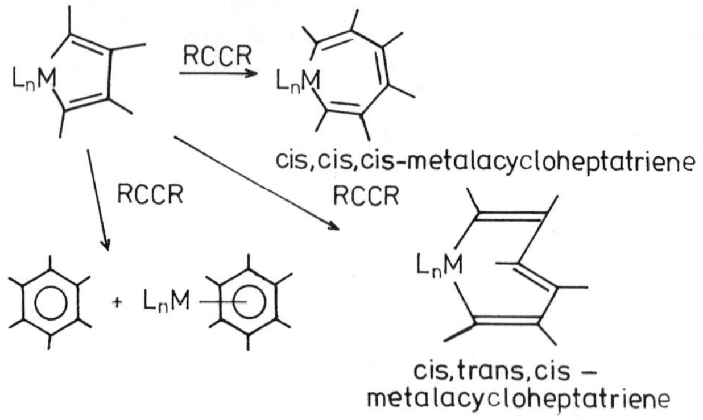

cis,cis,cis-metalacycloheptatriene

cis,trans,cis –
metalacycloheptatriene

(6.196)

The metallacycloheptatrienyl complex may undergo transformation to a benzene derivative as witnessed by the reverse reaction involving cleavage of the $C-C$ bond in hexakis(trifluoromethyl)benzene in the presence of $[Pt_3(t\text{-BuNC})_6]$ or $[Pt(trans\text{-PhCH}=CHPh)(PMe_3)_2]$.[102]

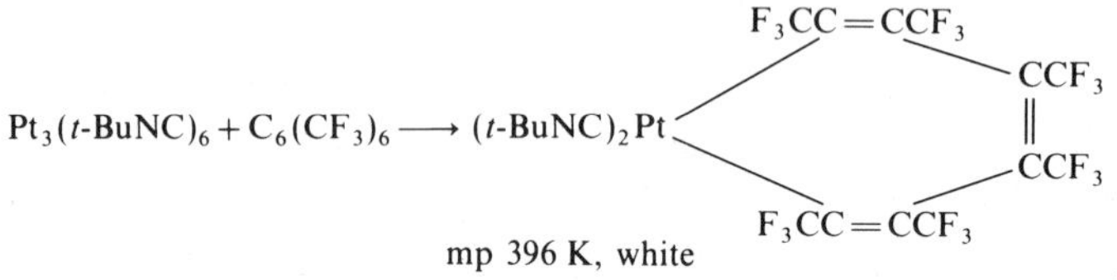

$$Pt_3(t\text{-BuNC})_6 + C_6(CF_3)_6 \longrightarrow (t\text{-BuNC})_2 Pt \cdots$$

(6.197)

mp 396 K, white

## 8. COMPLEXES CONTAINING $R-C\equiv Y$ AND $R-Y\equiv C$ LIGANDS

Ligands containing triple bonds $-C\equiv Y$ and $-Y\equiv C$ are nitriles $RC\equiv N$, phosphaalkynes $RC\equiv P$, and isocyanides $R-N\equiv C$. All these ligands form mono- and polynuclear $\pi$ complexes with transition metals.[78, 135–137] The reaction of $[Pt(trans\text{-PhCH}=CHPh)(PPh_3)_2]$ with trifluoroacetonitrile affords a $\pi$ complex possessing the following structure.[138]

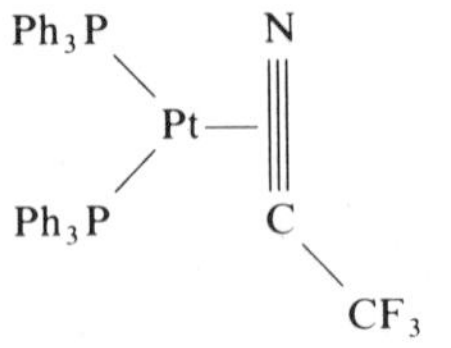

(6.198)

The compound $Pt(PPh_3)_4$ reacts with $CF_3C\equiv N$ to give the complex

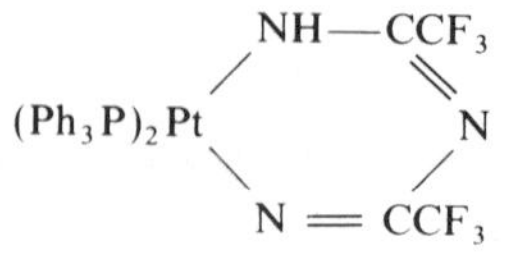

(6.199)

The NH group is formed as a result of hydrolytic decomposition of $CF_3CN$ which is caused by traces of water absorbed on glass.[138] The stability of complexes of the type

$[PtL(PPh_3)_2]$ decreases in the order $L = (CF_3)_2CO$, $PhC \equiv CPh > CF_3C \equiv N > CF_2 = CH_2 > PhCH = CHPh$.

Excess 3,3-dimethyl-1-phosphabutyne in benzene solution causes ethylene substitution in $[Pt(C_2H_4)(PPh_3)_2]$ resulting in the following compound:

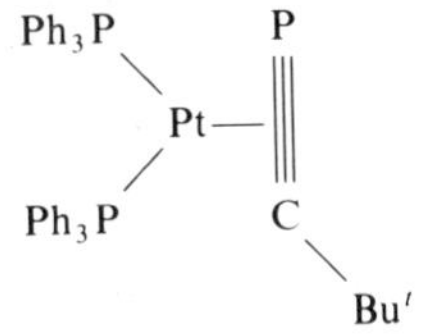

$$\begin{array}{c} \text{Ph}_3\text{P} \\ \diagdown \\ \qquad \quad \text{Pt}\!-\!\Big\Vert \\ \diagup \\ \text{Ph}_3\text{P} \end{array} \quad \begin{array}{c} \text{P} \\ \\ \\ \text{C} \\ \diagdown \\ \quad \text{Bu}' \end{array} \tag{6.200}$$

The C–P distance in the complex equals 167.2 pm and is considerably greater than in the free ligand (154.4 pm).[139]

Bis(1,5-cyclopentadiene)platinum(0) reacts with a mixture of the phosphaalkyne $Bu'C \equiv P$ and the phosphaalkene $Ph_2C = P(C_6H_2Me_3\text{-}2,4,6)$ to give the complex[140]

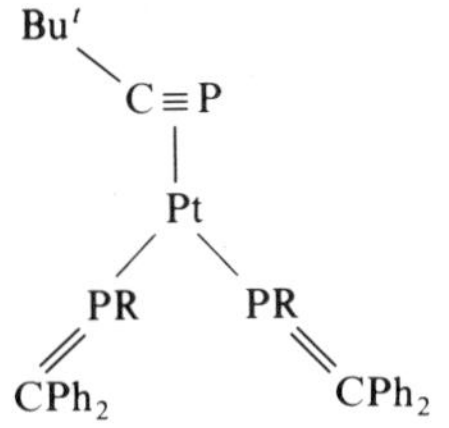

$$\begin{array}{c} \text{Bu}' \\ \diagdown \\ \quad \text{C} \equiv \text{P} \\ | \\ \text{Pt} \\ \diagup \quad \diagdown \\ \text{PR} \quad \text{PR} \\ \diagup\diagup \qquad \diagdown \\ \text{CPh}_2 \qquad \text{CPh}_2 \end{array} \tag{6.201}$$

In this complex, the phosphaalkene is coordinated only through the phosphorus atom. Equal molar quantities of $Co_2(CO)_8$ and $Bu'CP$ give $[(OC)_3Co(\mu\text{-}\eta^2\text{-}Bu'CP)Co(CO)_3]$ possessing a structure analogous to that of corresponding alkyne complexes.[140] This compound interacts with $[W(CO)_5(solvent)]$ to afford the complex[141]:

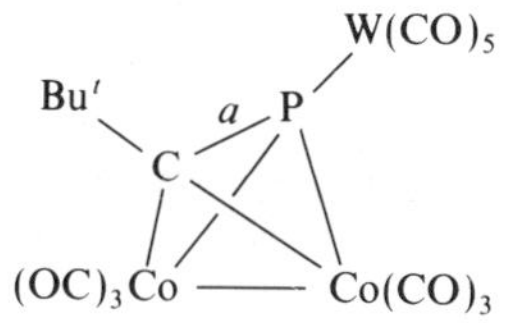

$$\begin{array}{c} \qquad\qquad \text{W(CO)}_5 \\ \qquad\qquad \diagup \\ \text{Bu}' \qquad a \quad \text{P} \\ \diagdown \quad \diagup \diagdown \\ \quad \text{C} \diagup \\ \diagup \diagdown \\ (\text{OC})_3\text{Co} \!-\!\!-\!\!-\! \text{Co(CO)}_3 \end{array} \qquad a = 169.5 \text{ pm} \tag{6.202}$$

Rhodium and molybdenum dinuclear complexes containing $Bu'CP$ are also known.[142]

$$Bu'C \equiv P + [(OC)_2cpMo \equiv Mocp(CO)_2] \longrightarrow [(OC)_2cpMo(\mu\text{-}\eta^2\text{-}Bu'C \equiv P)Mocp(CO)_2] \tag{6.203}$$

$$Bu'C \equiv P + [(pmcp)Rh(\mu\text{-}CO)_2Rh(pmcp)] \longrightarrow \tag{6.204}$$

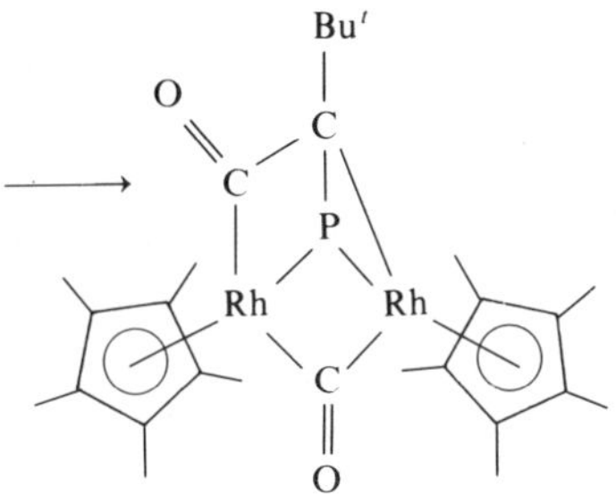

The carbon–phosphorus distance in the molybdenum complex equals 171.3 pm.

In contrast to the nitriles which are capable of coordinating to the metal as $\pi$ ligands (through carbon and nitrogen atoms) only if they contain a strongly electron-withdrawing group such as $CF_3CN$, phosphaalkynes form stable $\pi$ complexes in which the $RC\equiv P$ molecule is coordinated through the $C\equiv P$ triple bond. This is caused by a considerably greater difference between $\pi$ and $\sigma$ orbital energy in phosphaalkynes RCP than in nitriles RCN as shown by photoelectron spectra of these compounds.[143]

Usually, isonitriles are coordinated to the metal through the carbon atom and therefore analogously to the isoelectronic carbon monoxide molecule. Like the CO ligand, nitriles may also constitute $\eta^1$ terminal ligands as well as $\mu$-$\eta^1$ and $\mu_3$-$\eta^1$ bridging ligands. However, in many complexes, nitriles are coordinated as 4e ligands through free electron pairs of both carbon and nitrogen atoms:

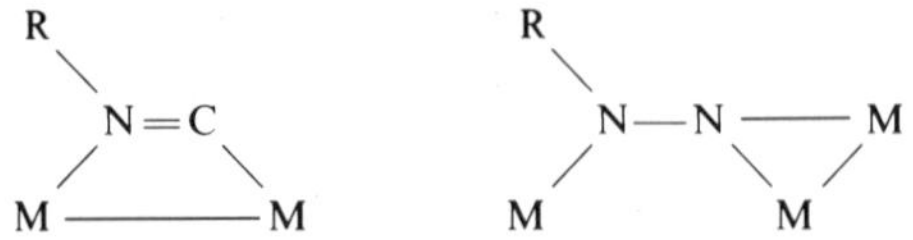

(6.205)

Nitriles may also coordinate through an electron pair of the carbon atom and $\pi$ orbitals of the $C\equiv N$ bond, thus behaving as a 4e ligand. This coordination is analogous to the CO molecule as shown in Figure 2.2e:

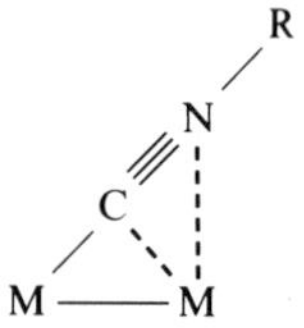

(6.206)

The compound $CF_3NC$ reacts with $(OC)_2cpMo\equiv Mocp(CO)_2$ to give the dinuclear complex $(OC)_2cpMo(\mu\text{-}CF_3NC)Mocp(CO)_2$ possessing structure (6.205).[144]

The reaction of $Mn_2(\mu\text{-}CO)(CO)_4(dppm)_2$ (cf. Sections 2.8.e and 12.1) with p-tolyl isocyanide leads to the formation of the complex in which the group $\mu$-$\eta^1$-$\eta^2$-CO is displaced by $RNC$[145]; see structure (6.207).

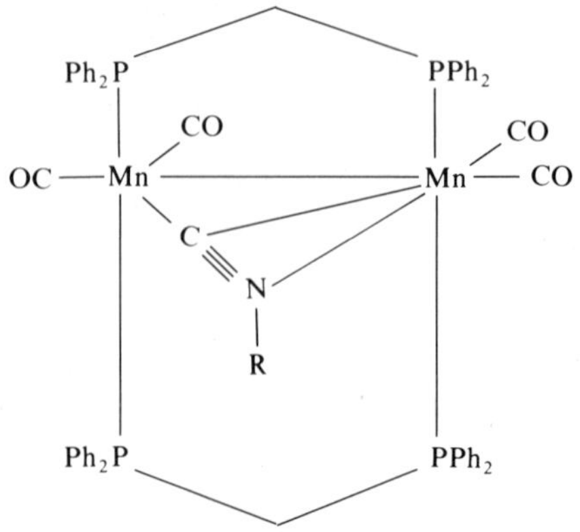

(6.207)

$R = p\text{-}C_6H_4Me$

In the cluster compound $[Ru_5(CO)_{14}(\mu_5\text{-}\eta^2\text{-CNBu}')(CNBu')]$ one of the *tert*-butyl isocyanide molecules constitutes a bridge bonded to five ruthenium atoms[146] (cf. Section 12.1).

## C. ALLENE COMPLEXES

### 1. BONDING AND STRUCTURE

Complexes containing allene, $H_2C=C=CH_2$, are relatively less known because allene undergoes oligomerization during storage. Also, transition metal complexes catalyzes oligomerization and polymerization. Compounds containing derivatives of allene such as $C_3R_4$ are more stable.

Allenes may behave as monodentate or as bridging ligands connecting two central atoms.[87, 105]

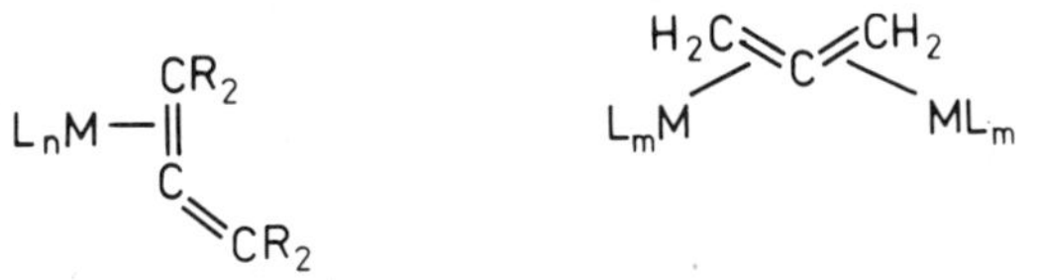

$$(6.208)$$

Mononuclear allene complexes possessing trigonal, square-planar, as well as trigonal–bipyramidal structures are known; see structures (6.209).

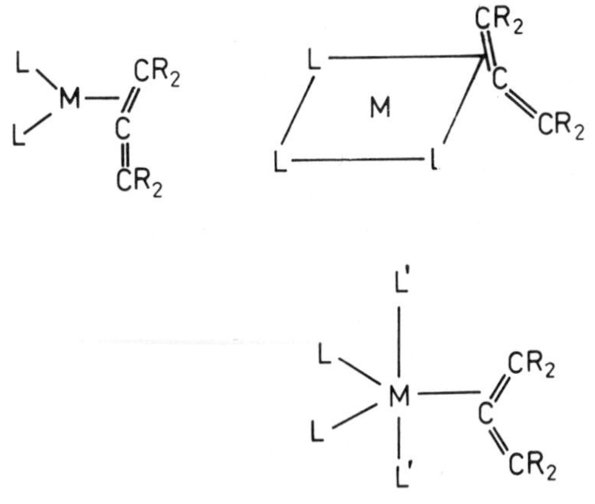

$$(6.209)$$

Structures of these compounds are analogous to those of olefin and alkyne complexes. In square-planar compounds, the allene molecule is approximately perpendicular to the $ML_3$ plane, while in the remaining types of compounds, the allene molecule lies almost in the $ML_2$ plane. The angle between this plane and the plane containing the allene carbon atoms equals about $10°$. The character of the metal–allene bond may be described by means of the Dewar–Chatt–Duncanson model as in the case of alkene and acetylene complexes. The formation of the $\sigma(d_\sigma\text{-}\pi)$ and $\pi(d_\pi\text{-}\pi^*)$ bonds causes an increase in the carbon–carbon distance and tilting of the allene substituents away from the complex. Therefore, the coordinated allene molecule may be regarded as a molecule which is in the excited state, because in this state it is bent. The allene molecule is linear in the ground state.

The distance $a$ between the middle carbon and the metal is shorter than the distance $b$ between the central atom and the terminal allene carbon atom.

$$(6.210)$$

This situation may be explained by stronger interaction between the middle carbon atom and the metal atom. Such interaction is caused by the presence of two mutually perpendicular bonding and antibonding $\pi$ orbitals. The terminal carbon atoms have only $\pi$ and $\pi^*$ orbitals. This is equivalent to the interaction of the metal with the uncoordinated double bond of the allene which also leads to the change of distance $d$ between carbon atoms. The smaller distance $a$ is a result of a smaller atomic radii of the middle allene carbon atom which forms two double bonds. The distance and angles in the allene–metal complexes are given in Table 6.25. Structures of some allene complexes are depicted in Figure 6.24.

*Table 6.25. Structures of Metal Allene Complexes*

| Compound | $a$ (pm) | $b$ (pm) | $c$ (pm) | $d$ (pm) | $\theta$ | $\phi^*$ |
|---|---|---|---|---|---|---|
| $[Fecp(CO)_2(Me_2C=C=CMe_2)]^+ BF_4^-$ [a] | 206 | 224 | 137 | 133.5 | 145.7° | |
| $[Rh(acac)(Me_2C=C=CMe_2)_2]$ [b] | 203.3 | 217.6 | 137.7 | 132.1 | 148.9° | 97.9° |
| | 202.7 | 217.7 | 137.3 | 132.5 | 147.2° | 97.8° |
| $[RhI(CH_2=C=CH_2)(PPh_3)_2]$ [c] | 204 | 217 | 135 | 134 | 158° | |
| $[Pd(CH_2=C=CH_2)(PPh_3)_2]$ [d] | 206.8 | 211.8 | 140 | 130 | 148.3° | 8.6° |
| $[Pt(CH_2=C=CH_2)(PPh_3)_2]$ [e] | 203 | 213 | 148 | 131 | 142° | 9° |
| $[Pt_2Cl_4(Me_2C=C=CMe_2)_2]$ [b,f] | 207 | 225 | 137 | 136 | 151.4° | 95.4° |
| $[Rh_2(acac)_2(\mu\text{-}CH_2=C=CH_2)(CO)_2]$ [g] | 205 | 212 | 137 | 141 | 144.5° | 95.5° |
| | 206 | 214 | | | | |
| $[Mo_2cp_2(CO)_4(C_3H_4)]$ [h] | 223 | 213 | 144 | 141 | 146° | |
| | 223 | 211 | | | | |
| $R_2C=C=CR_2$ | | | 130.9–131.2 | | | |

* $\phi$—dihedral angle $ML_2-MC_1C_2$.
[a] B. M. Foxman, *Chem. Commun.*, 221 (1975).
[b] T. G. Hewitt and J. J. De Boer, *J. Chem. Soc. (A)*, 817 (1971).
[c] N. Kasai, T. Kashiwagi, M. Kukudo, and N. Yasuoka, *Chem. Commun.*, 317 (1969).
[d] K. Okamoto, Y. Kai, N. Yasuoka, and N. Kasai, *J. Organomet. Chem.*, **65**, 427 (1974).
[e] M. Kadonaga, N. Yasuoka, and N. Kasai, *Chem. Commun.*, 1597 (1971).
[f] T. D. Hewitt, K. Anzehoffer, and J. J. de Boer, *J. Organomet. Chem.*, 3, 19 (1969).
[g] R. Racanelli, G. Pantini, A. Immirzi, G. Allegra, and L. Porri, *Chem. Commun.*, 361 (1969).
[h] W. J. Bailey, Jr., M. H. Chisholm, F. A. Cotton, C. A. Murillo, and L. A. Rankel, *J. Am. Chem. Soc.*, **100**, 802 (1978).

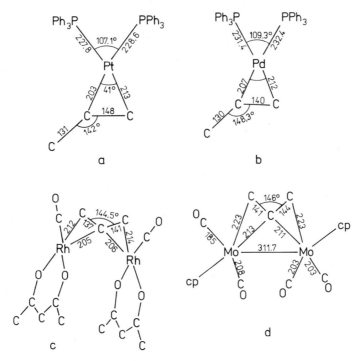

Figure 6.24. Structures of allene complexes: (a) [Pt(C$_3$H$_4$)(PPh$_3$)$_2$], (b)[Pd(C$_3$H$_4$)(PPh$_3$)$_2$], (c) [Rh$_2$(acac)$_2$(C$_3$H$_4$)(CO)$_2$], (d) [Mo$_2$cp$_2$(C$_3$H$_4$)(CO)$_4$].

## 2. IR AND NMR SPECTRA
## REACTIONS OF ALLENE COMPLEXES

In free allene, the $v_{as}$ (C=C=C) band occurs at 1940 cm$^{-1}$. Because of coordination, this band is shifted by 180–260 cm$^{-1}$ in the direction of lower energies. Therefore, the lowering of this frequency is greater than $\Delta v$(C=C) for olefin complexes and smaller than $\Delta v$(C≡C) for alkyne complexes. The $v_{as}$(C=C=C) band for [Pt(C$_3$H$_4$)(PPh$_3$)$_2$], [RhCl(C$_3$H$_4$)(PPh$_3$)$_2$], [IrCl(C$_3$H$_4$)(PPh$_3$)$_2$], and [IrCl(O$_2$)(C$_3$H$_4$)(PPh$_3$)$_2$] occurs at 1680, 1730, 1719, and 1760 cm$^{-1}$,[105] respectively.

The $^{13}$C NMR chemical shift for C$_3$H$_4$ in [Mo$_2$cp$_2$(C$_3$H$_4$)(CO)$_4$] equals 36.66 ppm for the terminal carbon atoms CH$_2$= and 196.96 ppm for the middle carbon atom =C=, while in [W$_2$cp$_2$(C$_3$H$_4$)(CO)$_4$], the chemical shift of the group CH$_2$= equals 31.58 ppm. For the free allene molecule, these shifts equal 73.67 and 212.58 ppm, respectively. Considerable lowering of the chemical shift $\delta$ of the coordinated allene compared to the free hydrocarbon molecule is in agreement with the Dewar–Chatt–Duncanson model of the metal–allene bond (reference *h*, Table 6.25).

Like in the case of olefin complexes, the chemical shift $\tau$ of the coordinated CHR group increases by 2–3 ppm compared to the free allene. This is also in agreement with the Dewar–Chatt–Duncanson model of the metal–allene bond (rehybridization sp$^2$ → sp$^3$) according to the valence bond theory.

Three signals are observed in the $^1H$ NMR spectrum. The highest $\tau$ values are seen for a and b protons, while the lowest value is for c; see structure (6.211). The coupling

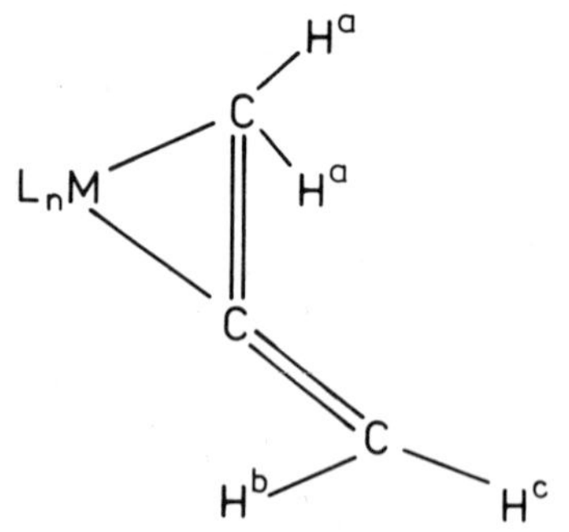

(6.211)

constant $J(H^a - H^b)$ is smaller for the coordinated allene than for the free allene. If the symmetry is lower, four signals may be observed. Some allene complexes show dynamic properties. An example may be provided by $[Fe(C_3Me_4)(CO)_4]$, which exhibits three signals ($\tau = 7.93$—1H, 8.00—1H, 8.23—2H) in $CS_2$ at 213 K in the spectrum. This indicates that the iron atom is bonded only to one double bond. At room temperature, these peaks become a singlet at $\tau$ 8.16 resulting from coordination of the allene to the iron atom in all possible ways by both double bonds alternatively. The coordinated allene complex $[Fecp(C_3R_4)(CO)_2]^+$ rotates about the metal–allene axis as well as undergoes 1,2 shift; see equation (6.212).

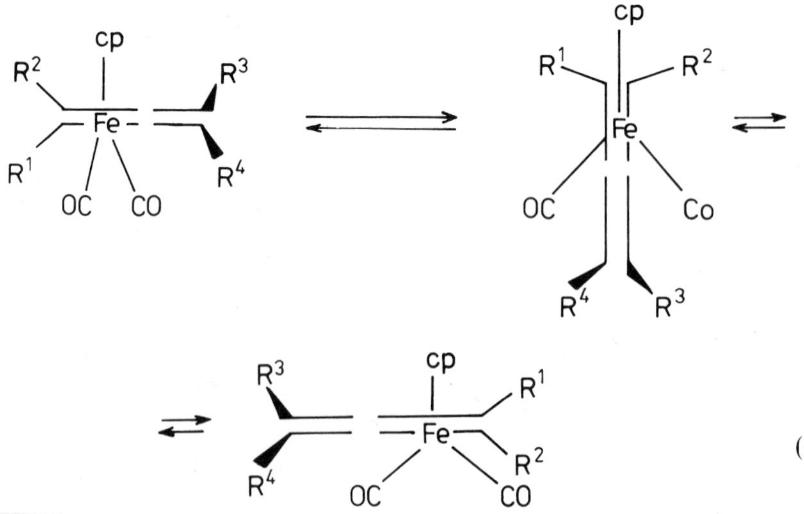

(6.212)

(R = H, Me)

The energy of activation for the rotation is lower than $32.5 \text{ kJ mol}^{-1}$ for $R^1 = R^2 = R^3 = R^4 = H$ and is equal to $45.0 \text{ kJ mol}^{-1}$ for $R^1 = R^2 = R^3 = H$, $R^4 = Me$.

The energy of activation of the 1,2 shift is considerably higher and varies in the 68.2–96.5 kJ mol$^{-1}$ range for various methylallenes.[108] The lowering of electron density on the central atom decreases the activation energy of migration and therefore facilitates this process. Examples are provided by stable, rigid Pt(0) compounds such as [Pt(allene)(PPh$_3$)$_2$] and dynamic Pt(II) complexes [Pt$_2$Cl$_4$(C$_3$Me$_4$)$_2$] and [PtCl$_2$(C$_3$Me$_4$)L] (L = $p$-XC$_5$H$_4$N). The rate of migration increases in the series X = NH$_2$ < Me < Et < H < Br < CN because the electron density on the central atom decreases in the same direction.[106] It is assumed that in the transition state in the migration reaction, a compound is formed in which there is a bond formed as a result of an overlap of the metal $d_{xy}$ orbitals with perpendicular π* orbitals of the middle allene carbon atom (Figure 6.25).

The allene molecule dissociates from the complex [Pt(C$_3$Me$_4$)(PPh$_3$)$_2$] at room temperature. The dissociation constant equals $1.5 \times 10^{-3}$ at 298 K. For analogous platinum complexes containing other allenes, the dissociation of the hydrocarbon does not take place. The exchange of allenes also occurs in the case of Pd(0), e.g., [Pd(allene)(PPh$_3$)$_2$]. In contrast to Pd(0) and Pt(0) complexes, Ni(0) compounds have dynamic properties. The complex [Ni(C$_3$H$_4$)(PPh$_3$)$_2$] is unstable and cannot be isolated, while the compound [Ni(PhHC=C=CHPh)(PPh$_3$)$_2$] is stable and the dissociation of the allene does not take place before this compound's decomposition. An excess of phosphine causes its rapid exchange according to a dissociative mechanism, while an excess of the allene leads to oligomerization. The thermal stability of allene compounds containing substituents is greater than that of compounds with C$_3$H$_4$. The lability of allene ligands is utilized for the preparation of other coordination compounds by means of substitution reactions using such ligands as phosphines, CS$_2$, acetylenes, CO, etc. For allene rhodium compounds, the hydrogenation of C$_3$R$_4$ occurs only at higher temperatures (353 K).

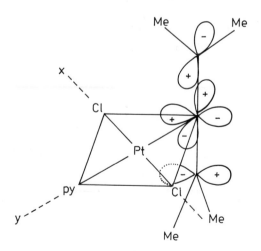

Figure 6.25. The proposed transition state in the exchange reaction of methyl groups in PtCl$_2$(C$_3$Me$_4$) Py.

Protonation of allene coordination compounds leads to formation of allyl, alkenyl, and alkyl complexes; see equations (6.213) and (6.214).

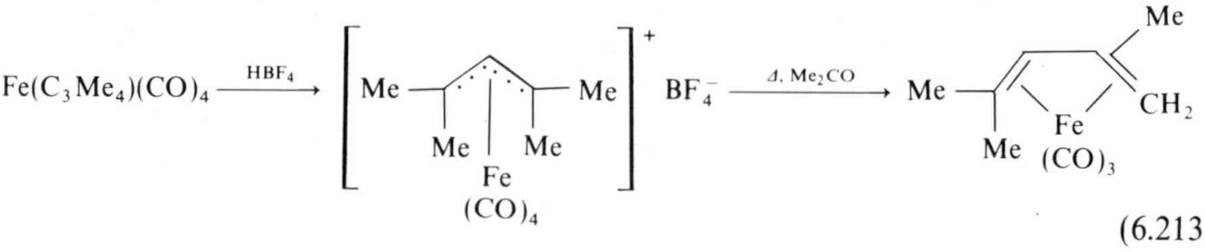

Fe(C3Me4)(CO)4
HBF4
Me
Me
Me
Me
Fe
(CO)4
BF4−
Δ, Me2CO
Me
Me
CH2
Fe
Me
(CO)3


(6.213)

[Pt(C3H4)(PPh3)2]
CF3COOH
Me
Ph3P
C=CH2
Pt
Ph3P
OCOCF3


(6.214)

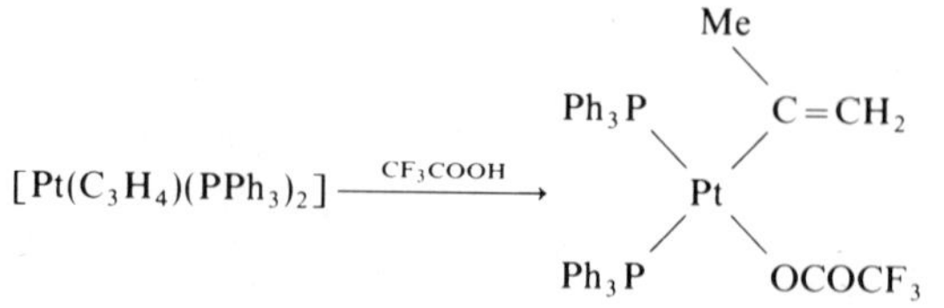

Some metal allene compounds undergo migration (insertion) reactions with $SO_2$ and $CF_3C{\equiv}CCF_3$ to give compounds containing $\sigma$ M—C bonds; see equations (6.215) and (6.216).

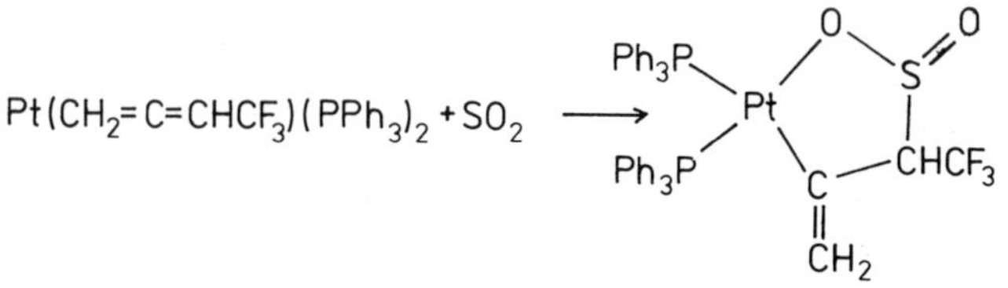

Pt(CH2=C=CHCF3)(PPh3)2 + SO2
Ph3P
Pt
Ph3P
O
S
O
CHCF3
C
CH2


(6.215)

[RhCl(C3H4)(PPh3)2] + CF3C≡CCF3
CF3
CF3
Ph3P
C=C
Cl
Rh
Ph3P
C−CH2
CH2


(6.216)

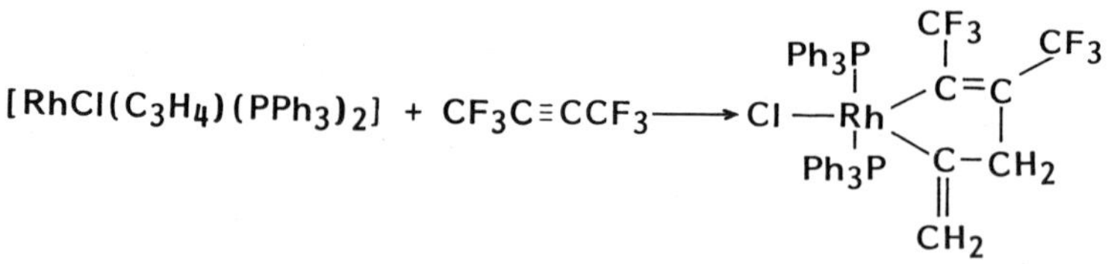

Many complexes react with allene to afford metal allyl compounds [equations (6.217)–(6.219)].

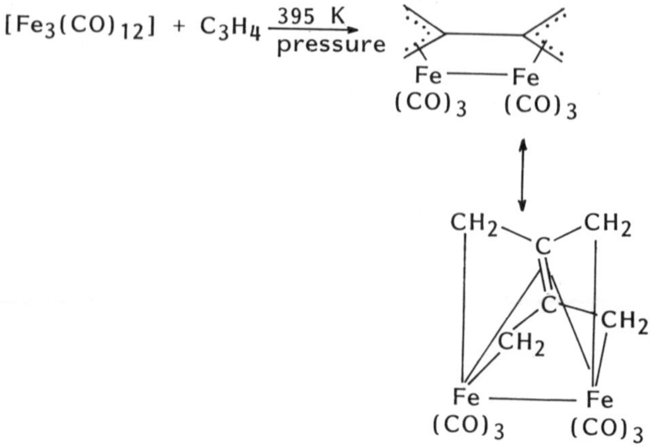

[Fe3(CO)12] + C3H4
395 K
pressure
Fe
Fe
(CO)3
(CO)3
CH2
C
CH2
C
CH2
CH2
Fe
Fe
(CO)3
(CO)3


(6.217)

$$[PdCl_2(PhCN)_2] + C_3H_4 \xrightarrow{PhH} \left[ \begin{matrix} CH_2 \\ Cl-C \\ CH_2 \end{matrix} Pd \begin{matrix} Cl \end{matrix} \right]_2 \qquad (6.218)$$

$$[PdCl(Me)(PR_3)_2] + C_3H_4 \xrightarrow{AgBF_4} \left[ Me \text{---} C \text{--} Pd(PPh_3)_2 \right]^+ BF_4^- \qquad (6.219)$$

Allene complexes and coordination compounds that readily bind allenes often catalyze oligomerization and polymerization of allenes.[87, 105, 107, 147] Allene, phosphine, phosphine–carbonyl, phosphite, olefin, carbonyl, etc. complexes of metals of groups 8–10 exemplify such behavior, for instance, $[Ni\{P(OR)_3\}_4]$, $[Ni(PR_3)_4]$, $[Ni(CO)_2(PR_3)_2]$, $[Ni(COD)_2]$, $[Rh_2Cl_2(CO)_4]$, $[RhCl(PPh_3)_3]$, $[Rh(acac)(CH_2=C=CH_2)_2]$, etc.

Oligomerization and polymerization reactions of allenes are discussed in Chapter 13.

## REFERENCES

1. M. J. S. Dewar, *Bull. Soc. Chim. Fr.*, **18**, C71 (1951).
2. J. Chatt and L. A. Duncanson, *J. Chem. Soc.*, 2939 (1953).
3. N. Rösch, R. P. Messmer, and K. H. Johnson, *J. Am. Chem. Soc.*, **96**, 3855 (1974).
4. H. Basch, *J. Chem. Phys.*, **56**, 441 (1972).
5. D. M. P. Mingos, *Adv. Organomet. Chem.*, **15**, 1 (1971).
6. D. R. Gee and J. K. S. Wan, *Chem. Commun.*, 641 (1970).
7. N. Rösch and R. Hoffmann, *Inorg. Chem.*, **13**, 2656 (1974).
8. F. G. A. Stone, *J. Organomet. Chem.*, **100**, 257 (1975); M. Green, J. A. K. Howard, J. L. Spencer, and F. G. A. Stone, *J. Chem. Soc. Chem. Commun.*, 449 (1975).
9. L. D. Pettit and D. S. Barnes, *Top. Curr. Chem.*, **28**, 85 (1972).
10. F. R. Hartley, *Chem. Rev.*, **73**, 163 (1973).
11. S. D. Ittel and J. A. Ibers, *Adv. Organomet. Chem.*, **14**, 33 (1976).
12. K. Nakamoto, in: *Characterization of Organometallic Compounds* (M. Tsutsui, ed.), p. 73, Interscience, New York (1969).
13. K. Nakamoto, in: *Coordination Chemistry* (A. E. Martell, ed.), Vol. 1, p. 134, Van Nostrand Reinhold, New York (1971).
14. J. Hiraishi, *Spectrochim. Acta A*, **25**, 749 (1969).
15. R. Rericha and J. Hetflejs, *Collect. Czech. Chem. Commun.*, **40**, 1811 (1975); R. Rericha, *Collect. Czech. Chem. Commun.*, **42**, 3518 (1977); R. Rericha, *Collect. Czech. Chem. Commun.*, **42**, 3530 (1977).
16. R. G. Kidd, in: *Characterization of Organometallic Compounds, Part II*, p. 373, Wiley, New York (1971).
17. B. L. Shaw and N. I. Tucker, in: *Comprehensive Inorganic Chemistry* (J. C. Bailar, Jr., ed.), Vol. 4, p. 781, Pergamon, Oxford (1973).
18. M. Herberhold, in: *Metal π-Complexes, Part I*, Vol. 2, Elsevier, Amsterdam (1972).
19. M. Herberhold, in: *Metal π-Complexes, Part II*, Vol. 2, Elsevier, Amsterdam (1974).
20. B. E. Mann, *Adv. Organomet. Chem.*, **12**, 135 (1974).
21. a. M. H. Chisholm and S. Godleski, *Prog. Inorg. Chem.*, **20**, 299 (1976); b. D. G. Cooper and J. Powell, *Inorg. Chem.*, **15**, 1959 (1976).
22. R. P. Hughes, N. Krishnamachari, C. J. L. Lock, J. Powell, and G. Turner, *Inorg. Chem.*, **16**, 314 (1977).

23. C. Furlani and C. Cauletti, *Structure and Bonding*, **35**, 119 (1978).
24. M. A. Meester, H. van Dam, J. Stufkens, and A. Oskam, *Inorg. Chim. Acta*, **20**, 155 (1976).
25. K. Kimura, S. Katsumata, T. Yamazaki, and H. Wakabayashi, *J. Electron Spectrosc. Relat. Phenom.*, **6**, 41 (1975).
26. J. P. Maier and D. W. Turner, *J. Chem. Soc., Faraday Trans. 2*, **69**, 196 (1973).
27. A. Katrib and J. W. Rabalais, *J. Phys. Chem.*, **77**, 2358 (1973).
28. M. J. S. Dewar and S. D. Worley, *J. Chem. Phys.*, **50**, 654 (1969).
29. J. H. Nelson and H. B. Jonassen, *Coord. Chem. Rev.*, **6**, 27 (1971).
30. R. G. Denning, F. R. Hartley, and L. M. Venanzi, *J. Chem. Soc. (A)*, 1322 (1967).
31. D. J. Trecker, J. P. Henry, and J. E. McKeon, *J. Am. Chem. Soc.*, **87**, 3261 (1965).
32. T. N. Murrel and S. Carter, *J. Chem. Soc.*, 6185 (1964).
33. a. H. Huber, G. A. Ozin, and W. J. Power, *J. Am. Chem. Soc.*, **98**, 6508 (1976). b. G. A. Ozin and W. J. Power, *Inorg. Chem.*, **17**, 2836 (1978).
34. C. A. Tolman, *J. Am. Chem. Soc.*, **96**, 2780 (1974).
35. J. A. Connor, *Top. Curr. Chem.*, **71**, 71 (1977).
36. M. L. H. Green, in: *Organometallic Compounds, The Transition Elements*, Methuen, London (1968).
37. M. I. Rybinskaya and D. A. Bochvar, in: *Metody Elementoorganicheskoi Khimii*, p. 217, Izd. Nauka, Moscow (1975).
38. a. L. V. Interrante and G. V. Nelson, *J. Organomet. Chem.*, **25**, 153 (1970). b. P. Hong, K. Sonogashira, and N. Hagihara, *Bull. Chem. Soc. Jpn.*, **39**, 1821 (1966).
39. M. A. Bennett, G. B. Robertson, J. B. Tomkins, and P. O. Whimp, *Chem. Commun.*, 341 (1971); M. A. Bennett, G. B. Robertson, J. B. Tomkins, and P. O. Whimp, *J. Organomet. Chem.*, **32**, C 19 (1971).
40. J. R. Shapley, S. I. Richter, M. Tachikawa, and J. B. Keister, *J. Organomet. Chem.*, **94**, C 43 (1975).
41. J. Evans, F. G. Johnson, J. Lewis, and T. W. Matheson, *J. Organomet. Chem.*, **97**, C 16 (1975).
42. A. van der Ent and A. L. Onderdelinden, *Inorg. Chim. Acta*, **7**, 203 (1973).
43. G. Perego, G. del Piero, M. Cesari, M. G. Clerici, and E. Perrotti, *J. Organomet. Chem.*, **54**, C 51 (1973).
44. G. Mestroni, G. Zassinovich, and A. Camus, *Inorg. Nucl. Chem. Lett.*, **11**, 359 (1975); G. Mestroni and A. Camus, *Inorg. Nucl. Chem. Lett.*, **9**, 261 (1973).
45. K. Fischer, K. Jonas, and G. Wilke, *Angew. Chem., Int. Ed. Engl.*, **12**, 565 (1973).
46. T. Yamamoto, A. Yamamoto, and S. Ikeda, *J. Am. Chem. Soc.*, **93**, 3360 (1971).
47. W. H. Baddley and L. M. Venanzi, *Inorg. Chem.*, **5**, 33 (1966); P. Uguagliati and W. H. Baddley, *J. Am. Chem. Soc.*, **90**, 5446 (1968).
48. I. Al-Najjar and M. Green, *J. Chem. Soc., Chem. Commun.*, 212 (1977).
49. L. Porri, G. Vitulli, and M. C. Gallazzi, *Angew. Chem., Int. Ed. Engl.*, **6**, 452 (1967).
50. I. I. Moiseev, *π-Kompleksy v Zhidkofaznom Okislenii Olefinov*, Izd. Nauka, Moskow (1970).
51. J.-E. Backvall, B. Akermark, and S. O. Lijunggren, *J. Chem. Soc., Chem. Commun.*, 264 (1977).
52. D. R. Amstrong, R. Fortune, and P. G. Perkins, *J. Catal.*, **45**, 339 (1976).
53. G. P. Pez and S. T. Bustany, in: *Fundamental Research in Homogeneous Catalysis* (M. Tsutsui and R. Ugo, eds.), Plenum, New York (1977).
54. D. R. Coulson, *Tetrahedron Lett.*, 429 (1971).
55. R. Baker, D. E. Halliday, and T. N. Smith, *J. Chem. Soc., Chem. Commun.*, 1583 (1971).
56. D. M. P. Mingos, in: *Comprehensive Organometallic Chemistry* (G. Wilkinson, F. G. A. Stone, and E. W. Abel, eds.), Vol. 3, Chapter 19, Pergamon Press, Oxford (1982).
57. T. A. Albright, R. Hoffmann, J. C. Thibeault, and D. L. Thorn, *J. Am. Chem. Soc.*, **101**, 3801 (1979).
58. T. Ziegler, *Inorg. Chem.*, **24**, 1547 (1985).
59. P. W. Yolly and E. Mynolt, *Adv. Organomet. Chem.*, **19**, 257 (1981).
60. K. Jonas, *Adv. Organomet. Chem.*, **19**, 97 (1981).
61. K. Jonas, R. Mynott, C. Krueger, J. C. Sekutowski, and Y.-H. Tsay, *Angew. Chem., Int. Ed. Engl.*, **15**, 767 (1976).
62. J. Jonas, L. Schieferstein, C. Krüger, and Y.-H. Tsay, *Angew. Chem., Int. Ed. Engl.*, **18**, 550 (1979).
63. K. Jonas and L. Schieferstein, *Angew. Chem., Int. Ed. Engl.*, **18**, 549 (1979).

64. B. E. Mann, in: *Comprehensive Organometallic Chemistry* (G. Wilkinson, F. G. A. Stone, and E. W. Abel, eds.), Vol. 3, Chapter 20, Pergamon Press, Oxford (1982).
65. J. C. Green, *Structure and Bonding*, **43**, 37 (1981).
66. M. I. Rybinskaya and W. W. Krivykh, *Usp. Khim.*, **53**, 825 (1984).
67. E. Maslowsky, Jr., *Vibrational Spectra of Organometallic Compounds*, Wiley, New York (1977).
68. I. Omae, *Coord. Chem. Rev.*, **32**, 235 (1980); I. Omae, *Angew. Chem., Int. Ed. Engl.*, **21**, 889 (1982).
69. A. P. Tomilov, I. N. Chernykh, and Yu. M. Kargin, *Electrokhimia Elementoorganicheskikh Soedinenii*, Izd. Nauka, Moscow (1985).
70. N. G. Connelly, and W. E. Geiger, *Adv. Organomet. Chem.*, **23**, 1 (1984).
71. D. R. Burfield, *J. Organomet. Chem.*, **150**, 321 (1978); D. R. Burfield, *Makromol. Chem.*, **180**, 1565 (1979).
72. J. A. Labinger, in: *Comprehensive Organometallic Chemistry* (G. Wilkinson, F. G. A. Stone, and E. W. Abel, eds.), Vol. 3, Chapter 25, Pergamon Press, Oxford (1982).
73. G. A. Ozin, *Coord. Chem. Rev.*, **28**, 117 (1979).
74. G. Agnes, I. W. Bassi, C. Benedicenti, R. Intrito, M. Calcaterra, and C. Santini, *J. Organomet. Chem.*, **129**, 401 (1977).
75. D. J. Brauer, C. Krüger, and J. C. Sekutowski, *J. Organomet. Chem.*, **178**, 249 (1979); K. Jonas, *Angew. Chem., Int. Ed. Engl.*, **14**, 752 (1975).
76. K. Jonas and C. Krüger, *Angew. Chem., Int. Ed. Engl.*, **19**, 520 (1980).
77. P. W. Jolly, in: *Comprehensive Organometallic Chemistry* (G. Wilkinson, F. G. A. Stone, and E. W. Abel, eds.), Chapter 37.5, Pergamon Press, Oxford (1982).
78. A. H. Cowley, *Acc. Chem. Res.*, **17**, 386 (1984); A. H. Cowley, *Polyhedron*, **3**, 389 (1984); O. J. Scherer, *Angew. Chem., Int. Ed. Engl.*, **24**, 924 (1985).
79. K. A. Schugart and R. F. Fenske, *J. Am. Chem. Soc.*, **105**, 3384 (1985).
80. G. E. Herberich and J. Okuda, *Angew. Chem., Int. Ed. Engl.*, **24**, 402 (1985).
81. H. Werner and W. Paul, *Angew. Chem., Int. Ed. Engl.*, **23**, 58 (1984).
82. K. Knoll, G. Huttner, M. Wasiucionek, and L. Zsolnai, *Angew. Chem., Int. Ed. Engl.*, **23**, 739 (1984).
83. T. A. VanderKnaap, F. Bickelhaupt, J. G. Kraaykamp, G. vanKoten, J. P. C. Bernards, H. T. Edzes, W. S. Veeman, E. deBoer, and E. J. Baerends, *Organometallics*, **3**, 1804 (1984).
84. W. Werner, W. Paul, and R. Zolk, *Angew. Chem., Int. Ed. Engl.*, **23**, 626 (1984).
85. O. N. Tiomkin and R. M. Flid, *Kataliticheskie Prevrashchemia Acetilenowykh Soedinenii v Rastvorakh Kompleksov Metallov*, Izd. Nauka, Moscow (1968).
86. L. P. Jureva, in: *Metody Elementoorganicheskoi Khimii* (D. A. Bochvar, ed.), p. 384, Izd. Nauka, Moscow (1975).
87. S. Otsuka and A. Nakamura, *Adv. Organomet. Chem.*, **14**, 245 (1976).
88. R. S. Dickson and P. J. Fraser, *Adv. Organomet. Chem.*, **12**, 323 (1974).
89. L. J. Todd, J. R. Wilkinson, M. D. Rausch, S. A. Gardner, and R. S. Dickson, *J. Organomet. Chem.*, **101**, 133 (1975).
90. M. Herberhold, H. Alt, and C. G. Kreiter, *J. Organomet. Chem.*, **42**, 413 (1972).
91. J. L. McNaughton, C. T. Mortimer, J. Burgess, M. J. Hacker, and R. D. W. Kemmitt, *J. Organomet. Chem.*, **71**, 287 (1974).
92. C. D. Cook, K. Y. Wan, U. Gelius, K. Hamrin, G. Johansson, E. Olsson, H. Siegbahn, C. Nordling, and K. Siegbahn, *J. Am. Chem. Soc.*, **93**, 1904 (1971).
93. R. D. W. Kemitt, *MTP International Review of Science*, in: *Inorg. Chem. Ser. 1*, Vol. 6 (M. J. Mays, ed.), p. 227, Butterworths, London (1972).
94. A. Greco, F. Pirinoli, and G. Dall'Asta, *J. Organomet. Chem.*, **60**, 115 (1973).
95. M. G. Thomas, E. L. Muetterties, R. O. Day, V. W. Day, *J. Am. Chem. Soc.*, **98**, 4645 (1976).
96. G. Fachinetti and C. Floriani, *Chem. Commun.*, 66 (1974); G. Fachinetti, C. Floriani, F. Marchetti, and M. Mellini, *J. Chem. Soc.*, 1398 (1978).
97. R. Colton, R. Lewitus, and G. Wilkinson, *Nature*, **186**, 233 (1960).
98. W. C. Kaska and M. E. Kimball, *Inorg. Nucl. Chem. Lett.*, **4**, 719 (1968).
99. M. E. Kimball, J. P. Martella, and W. C. Kaska, *Inorg. Chem.*, **6**, 414 (1967); W. A. McAllister and L. T. Farias, *Inorg. Chem.*, **8**, 2806 (1969).

100. Y. Iwashita and F. Tamura, *Bull. Chem. Soc. Jpn.*, **43**, 1517 (1970).
101. M. H. Chisholm and H. C. Clark, *J. Am. Chem. Soc.*, **94**, 1532 (1972).
102. J. Browning, M. Green, A. Laguna, L. E. Smart, J. L. Spencer, and F. G. A. Stone, *Chem. Commun.*, 723 (1975).
103. G. Huttner, U. Weber, B. Sigwart, and O. Scheidsteger, *Chem. Comm.*, **21**, 215 (1982); G. Huttner, H.-G. Schmid, A. Frank, and O. Orama, *Chem. Commun.*, **15**, 234 (1976); J. Born, L. Zsolnai, and G. Huttner, *Chem. Commun.*, **22**, 974 (1983).
104. G. Huttner and I. Jibril, *Chem. Commun.*, **23**, 740 (1984).
105. F. L. Bowden and R. Giles, *Coord. Chem. Rev.*, **20**, 81 (1976).
106. H. C. Volger, K. Vrieze, and A. P. Praat, *J. Organomet. Chem.*, **21**, 467 (1970); K. Vrieze, H. C. Volger, M. Gronert, and A. P. Praat, *J. Organomet. Chem.*, **16**, P19 (1969).
107. R. Baker, *Chem. Rev.*, **73**, 487 (1973).
108. B. Foxman, D. Marten, A. Rosan, S. Raghu, and M. Rosenblum, *J. Am. Chem. Soc.*, **99**, 2160 (1977).
109. H. Boennemann, R. Brinkmann, and H. Schenkluhn, *Synthesis*, 575 (1974); H. Boennemann and R. Brinkmann, *Synthesis*, 600 (1975); H. Boennemann and H. Schenkluhn, Patent 2416295 (1975); H. Boennemann, *VIII Polish—GDR Colloquy on Organometallic Chemistry*, Mausz (1977).
110. M. J. Grogan and K. Nakamoto, *J. Am. Chem. Soc.*, **88**, 5454 (1966).
111. M. J. Grogan and K. Nakamoto, *Inorg. Chim. Acta*, **1**, 228 (1967).
112. C. D. Cook, K. J. Wan, K. Gelius, K. Hamrin, G. Johanson, E. Olson, M. Siegbahn, C. Nordling, and K. Siegbahn, *J. Am. Chem. Soc.*, **93**, 1904 (1971).
113. E. J. Parsons, R. D. Larsen, and P. W. Jennings, *J. Am. Chem. Soc.*, **107**, 1793 (1985).
114. B. Akermark and K. Zetterberg, *J. Am. Chem. Soc.*, **106**, 5560 (1984).
115. L. S. Hegedus, B. Akermark, K. Zetterberg, and L. F. Olsson, *J. Am. Chem. Soc.*, **106**, 7122 (1984).
116. B. M. Trost and T. R. Verhoeven, in: *Comprehensive Organometallic Chemistry* (G. Wilkinson, F. G. A. Stone, and E. W. Abel, eds.), Vol. 8, Chapter 57, Pergamon Press, Oxford (1982).
117. M. P. Doyle, M. R. Colsman, and M. S. Chinn, *Inorg. Chem.*, **23**, 3684 (1984).
118. H. M. Colquhoun, S. M. Doughty, J. F. Stoddort, and D. J. Williams, *Angew. Chem., Int. Ed. Engl.*, **23**, 235 (1984).
119. D. R. Alston, A. M. Z. Slawin, J. F. Stoddart, and D. J. Williams, *Angew. Chem., Int. Ed. Engl.*, **23**, 821 (1984).
120. H. M. Colquhoun, S. M. Doughty, A. M. Z. Slawin, J. F. Stoddart, and D. J. Williams, *Angew. Chem., Int. Ed. Engl.*, **24**, 135 (1985).
121. D. B. Jacobson and B. S. Freiser, *J. Am. Chem. Soc.*, **107**, 72 (1985).
122. G. L. Geoffroy and M. S. Wrighton, *Organometallic Photochemistry*, Academic Press, New York (1979).
123. S. A. Cohen, P. R. Auburn, and J. E. Bercaw, *J. Am. Chem. Soc.*, **105**, 1136 (1983).
124. S. A. Cohen and J. E. Bercaw, *Organometallics*, **4**, 1006 (1985).
125. Y.-R. Hu, T. W. Leung, S.-C. H. Su, A. Wojcicki, M. Calligaris, and G. Nardin, *Organometallics*, **4**, 1001 (1985).
126. N. Kamiya and T. Ohta, in: *Solar-Hydrogen Energy Systems* (T. Ohta, ed.), Pergamon Press, Oxford (1979).
127. K. I. Zamaraev and W. N. Parmon, *Fotokataliticheskoe Preobrazovanie Solnechnoi Energii*, Izd. Nauka, Sibirskoje Otdelenie, Nowosibirsk (1985).
128. D. P. Schwendman and C. Kutal, *Inorg. Chem.*, **16**, 719 (1977); D. P. Schwendman, *J Am. Chem. Soc.*, **99**, 5677 (1977).
129. H. Fleckner, F.-W. Grevels, and D. Hess, *J. Am. Chem. Soc.*, **106**, 2027 (1984).
130. M. E. Miller and E. R. Grant, *J. Am. Chem. Soc.*, **106**, 4635 (1984); M. E. Miller and E. R. Grant, *J. Am. Chem. Soc.*, **107**, 3386 (1985).
131. J. L. Marshall, S. R. Stobart, and H. B. Gray, *J. Am. Chem. Soc.*, **106**, 3027 (1984); J. V. Caspar and H. B. Gray, *J. Am. Chem. Soc.*, **106**, 3029 (1984).
132. J. M. Maher, J. R. Fox, B. M. Foxman, and N. J. Cooper, *J. Am. Chem. Soc.*, **106**, 2347 (1984).
133. F. A. Cotton and W. T. Hall, *J. Am. Chem. Soc.*, **101**, 5094 (1979); F. A. Cotton and W. T. Hall, *Inorg. Chem.*, **19**, 2352 (1980); F. A. Cotton and W. T. Hall, *Inorg. Chem.*, **19**, 2354 (1980).
134. E. A. Robinson, *J. Chem. Soc., Dalton Trans.*, 2373 (1981).

135. F. R. Hartley, in: *Comprehensive Organometallic Chemistry* (G. Wilkinson, F. G. A. Stone, and E. W. Abel, eds.), Vol. 6, Chapter 39, p. 713, Pergamon Press, Oxford (1982).
136. E. A. Ishmaeva and I. I. Pacanovski, *Usp. Khim.*, **54**, 418 (1985).
137. E. Singleton and H. E. Oosthuizen, *Adv. Organomet. Chem.*, **22**, 209 (1983).
138. W. J. Bland, R. D. W. Kemmitt, and R. D. Moore, *J. Chem. Soc., Dalton Trans.*, 1292 (1973).
139. J. C. T. R. Burckett-St. Laurent, P. B. Hitchcock, H. W. Kroto, and J. E. Nixon, *Chem. Commun.*, 1141 (1981).
140. S. I. Al-Ressayes, S. I. Klein, H. W. Kroto, M. F. Meidine, and J. F. Nixon, *Chem. Commun.*, 930 (1983).
141. J. C. T. R. Burckett-St. Laurent, P. B. Hitchcock, H. W. Kroto, M. F. Meidine, and J. F. Nixon, *J. Organomet. Chem.*, **238**, C82 (1982).
142. G. Becker, W. E. Herrmann, W. Kalcher, G. W. Kriechbaum, C. Pahl, C. T. Wagner, and M. L. Ziegler, *Angew. Chem., Int. Ed. Engl.*, **22**, 413 (1983).
143. J. C. T. R. Burckett-St. Laurent, M. A. King, H. W. Kroto, J. F. Nixon, and R. J. Suffolk, *J. Chem. Soc., Dalton Trans.*, 755 (1983).
144. D. Lentz, I. Bruedgam, and H. Hartl, *Angew. Chem., Int. Ed. Engl.*, **23**, 525 (1984).
145. L. S. Benner, M. M. Olmstead, and A. L. Balch, *J. Organometal. Chem.*, **159**, 289 (1978); A. L. Balch and L. S. Benner, *J. Organometal. Chem.*, **135**, 339 (1977).
146. M. I. Bruce, J. G. Matisons, J. R. Rodgers, and R. C. Wallis, *Chem. Commun.*, 1070 (1981).
147. D. J. Pasto, N.-Z. Huang, and C. W. Eigenbrot, *J. Am. Chem. Soc.*, **107**, 3160 (1985).

# Chapter 7

# Complexes Containing Three-Electron π-Ligands

## 1. THE METAL–ALLYL BOND

The allyl group $C_3R_5$ (R = H, alkyl, or aryl) may behave as an $\eta^1$ ligand which forms a $\sigma$ bond with the metal (a) or it may behave as an $\eta^3$ ($\pi$) ligand. The $\eta^3$ allyl group may be bonded to the metal in a symmetrical fashion (b), with equivalent terminal carbon atoms, or it may be bonded in an asymmetrical way (c).[1–7,32–36]

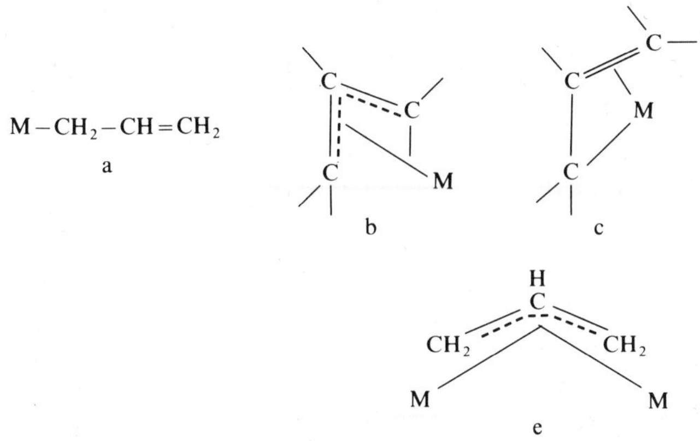

$$(7.1)$$

In some polynuclear compounds, the allyl ligand may play the role of an asymmetrical bridge (d); it forms a $\sigma$ bond with one metal atom, and a $\pi$ olefin bond with the other metal atom. Allyl complexes containing a symmetrical bridging allyl group are also known (e).

In $\pi$-allyl complexes, the CCC angle is *ca* 120°; all carbon atoms have sp² hybridization. Therefore, the multiplicity of the $C-C$ bond is higher than one, and rotation about the carbon–carbon axis does not occur. Thus, the allyl groups containing substituents at the terminal carbon atoms may form the following isomers: *syn-*, *syn-*;

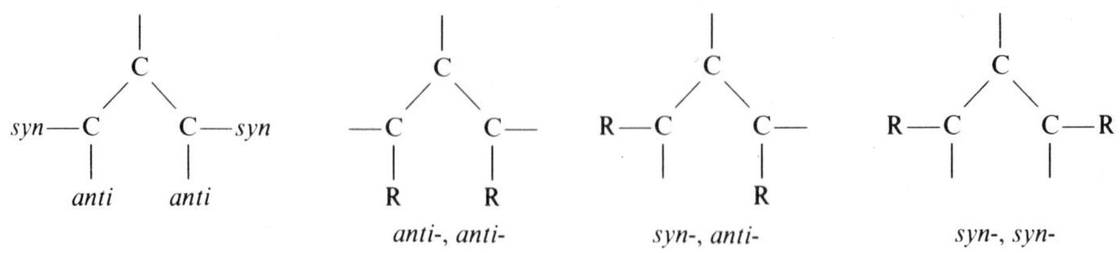

$$(7.2)$$

*syn-*, *anti-*; and *anti-*, *anti-* [see structures (7.2)]. The three $p\pi$ orbitals of the allyl group form three molecular orbitals:

$$\psi_1 = \frac{1}{2}(\varphi_1 + \sqrt{2}\varphi_2 + \varphi_3), \qquad \psi_2 = \frac{1}{\sqrt{2}}(\varphi_1 - \varphi_3), \qquad \psi_3 = \frac{1}{2}(\varphi_1 - \sqrt{2}\varphi_2 + \varphi_3)$$

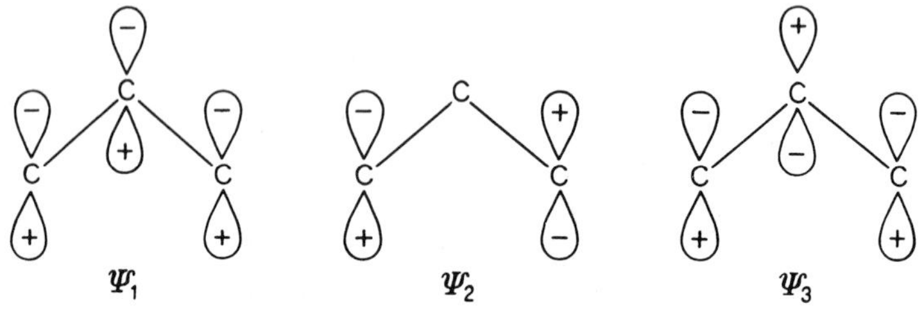

$$(7.3)$$

The energy of these orbitals varies in the following series: $\psi_1 < \psi_2 < \psi_3$. Therefore, the electronic structures of the $C_3H_5^-$ anion, the $C_3H_5$ radical, and the $C_3H_5^+$ cation are $\psi_1^2\psi_2^2$, $\psi_1^2\psi_2^1$, and $\psi_1^2$, respectively.

All carbon atoms of the allyl group are bonded to the metal. The anion $C_3H_5^-$ formally resembles the acetylene molecule [cf. structures (7.3) and Figure 6.21], because $\psi_2$ is a $\pi$-donor orbital, and $\psi_3$ is a $\pi$-acceptor orbital. Therefore, like acetylene, the ion $C_3H_5^-$ interacts in the most effective way with fragments possessing both occupied and empty $d\pi$ orbitals. On the other hand, the cation $C_3H_5^+$ is formally similar to ethylene and therefore $\psi_2$ constitutes a good $\pi$ acceptor. In complexes, all allyl orbitals may play an important role in the formation of the metal–allyl bond. Preferred conformations in allyl complexes may be explained by considering interactions of frontier orbital fragments $ML_n$ (see Figures 3.7–3.9) and $C_3H_5$. In the case of compounds such as $L_2M(C_3H_5)$, $L_3M(C_3H_5)$, and $L_4M(C_3H_5)$ conformations (7.4) are possible. In the 16e complex $[Pt(C_3H_5)(PPh_3)_2]^+$, $\psi_2$ of the allyl cation constitutes the LUMO orbital

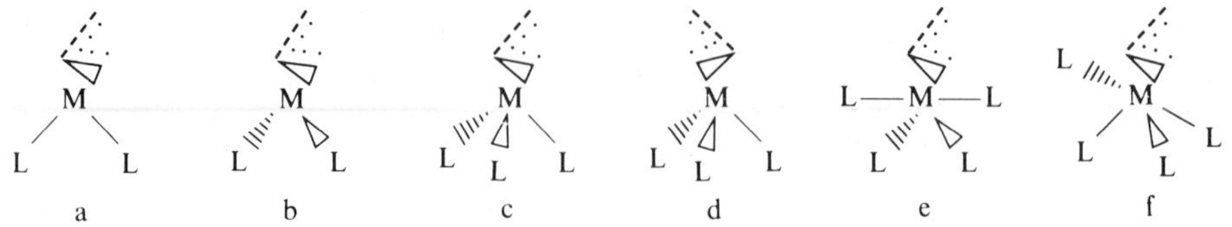

$$(7.4)$$

which may interact with the orbital of the $d^{10}$ Pt(PPh$_3$)$_2$ fragment possessing $b_2$ symmetry [equation (7.5)] to give a relatively strong $\pi$ bond. The interaction of $\psi_2$ with

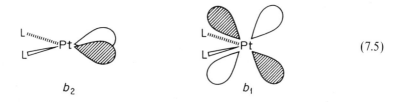

$$b_2 \qquad\qquad b_1 \qquad\qquad (7.5)$$

the $b_1$ orbital which possesses lower energy is less effective. Therefore, conformation (7.4b) is strongly preferred over conformation (7.4a) in which the ligands of the ML$_2$ fragments lie in the plane passing between the carbon atoms which do not form a bond. Conformation (7.4a) is more stable in the case of 18$e$ complexes such as M(C$_3$H$_5$)L$_2$. This conformation corresponds to 18$e$ tetrahedral complexes, while 16$e$ compounds are more stable if they have structures (7.4b) corresponding to square-planar complexes. Similarly, conformations (7.4c) and (7.4e) (pseudooctahedral) are more stable for 18$e$ complexes, and conformations (7.4d) and (7.4f) (bicapped tetrahedron) for 16$e$ complexes.

Hydrogen atoms or other groups bonded to the carbon atoms do not lie in the plane of the allyl ligand. Based on the neutron diffraction studies, it was found that in *trans*-[Ni(C$_3$H$_5$)$_2$], the *anti* hydrogen atoms are "shifted" from the plane of the allyl group away from the complex by 29.4°, while the *syn* and *meso* atoms are moved in the direction of the nickel atom by 8.9° and 15.8°, respectively (for *syn* and *anti* hydrogen atoms, the average values are given).[34] The LCAO–SCF calculations show that the total energy of *trans*-bis($\eta$-allyl)nickel is lower by 187 kJ mol$^{-1}$ if the hydrogen atoms do not lie in the plane of the allyl group compared to the complex possessing flat C$_3$H$_5$ ligands. Very good agreement of the calculated values[34] with those obtained by experiment was achieved. The lowering of the total energy was explained by repulsion between the *anti*-hydrogen atoms and the nickel atom, destabilization of the allyl ligands, and the increase in the interaction of the allyl groups with nickel which is caused by rehybridization of the carbon atoms, especially the terminal ones.

The dihedral angle between the plane containing other ligands as well as the metal atom and the plane created by the carbon atoms of the allyl group fluctuates in the 110–125° range. At these geometries, the overlap of all ligand orbitals with metal orbitals is possible. Kettle and Mason[6] considered the overlap of ligand orbitals with the $d$ metal orbitals in allyl–palladium complexes. Their calculations show that the $\psi_1$ bonding orbital and $\psi_2$ antibonding orbital of the allyl group are essential for the formation of the Pd–allyl bond, while the contribution of the $\psi_3$ antibonding orbital is not significant. Van Leeuwen and Pratt[7] also considered 5$s$ and 5$p$ palladium orbitals in their qualitative analysis. The $\psi_1$ orbital may overlap with the 5$s$ palladium orbital as well as with a combination of the 5$p_x$ and 5$p_y$ orbitals, while the $\psi_2$ orbital interacts with the $4d_{x^2-y^2}$ orbital as well as with the combination of the $4d_{xz}$ and $4d_{yz}$ orbitals. The $\psi_3$ orbital overlaps with the $4d_{xz}$ and $4d_{yz}$ combination. The overlap of the $\psi_2$ orbital with the metal orbitals is greatest if the allyl group is perpendicular to the MLL' plane, and the terminal carbon atoms are located in this plane (Figure 7.1a). On the other hand, the interaction between the $\psi_3$ orbital and the palladium orbitals is more

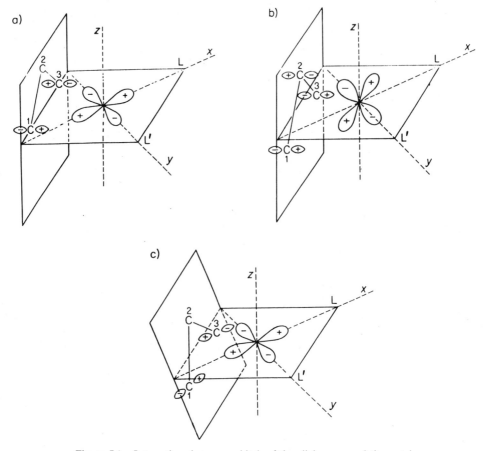

Figure 7.1. Interactions between orbitals of the allyl group and the metal.

effective if the nodal planes of the orbital of the allyl group and palladium also overlap [Figure 7.1b). Consequently, the allyl group tilts with respect to the MLL′ plane at an angle which is larger than 90°, and the terminal carbon atoms are located underneath this plane. The plane bisects the C−C bond, causing effective overlap of the $\psi_2$ orbital and yet sufficiently good overlap of the $\psi_1$ and $\psi_3$ orbitals with corresponding metal orbitals (Figure 7.1c). This bonding scheme is very useful for an explanation of the structure of asymmetric allyl complexes, such as $PdCl(C_3H_5)PPh_3$ [see scheme (7.6)].

$$
\begin{array}{c}
^3CH_2 \\
147 \diagup \quad \diagdown 210 \qquad PPh_3 \\
H^2C \underline{\qquad 212 \qquad} Pd \\
128 \diagdown\!\diagdown \quad \diagup 217 \qquad Cl \\
_1CH_2
\end{array}
\qquad (7.6)
$$

If L = PPh$_3$ and L' = Cl (Figure 7.1) the contribution of the $p_x$ orbital in the formation of the Pd$-$C$_1$ bond is smaller than that of the $p_y$ orbital in the formation of the Pd$-$C$_3$ bond. Therefore, the electron density on the $\psi_2$ orbital is greater on the C$_3$ carbon atom located in the *trans* position to the chlorine atom. This leads to an increase in stability of the Pd$-$C$_3$ bond (*trans* to Cl) and weakening of the C$_2-$C$_3$ bond.

## 2. IR SPECTRA OF ALLYL COMPLEXES

IR spectra of allyl complexes exhibit bands due to vibrations of the ligand itself as well as bands resulting from sekeletal metal–ligand and metal–allyl vibrations. The spec-

*Table 7.1. IR Spectra of Allyl Complexes (cm$^{-1}$)[a]*

| [Pd$_2$Cl$_2$(C$_3$H$_5$)$_2$] | [Pd$_2$Br$_2$(C$_3$H$_5$)$_2$] | [Pd(C$_3$H$_5$)$_2$] | [Rh(C$_3$H$_5$)$_3$] | Band |
|---|---|---|---|---|
| 3079 sh | | 3090 w | 3061 | $\nu$(CH$_2$) |
| 3065 w | 3070 | | | $\nu$(CH$_2$) |
| | | 2915 m | | |
| 3055 w | | 2842 m | | |
| 3030 vw | 3038 | | | $\nu$(CH) |
| 2997 vvw | 2997 | | 3000 m, | $\nu$(CH$_2$) |
| | | | 2960 w | $\nu$(CH$_2$) |
| | | 2330 m | 2350 w, | |
| | | | 2928 w | |
| | | | 2850 vw | |
| 1491 w | 1490 w | 1499 m* | 1483 w* | $\nu_{as}$(CCC) |
| 1461 s | 1459 s | | 1453 m | $\delta_{as}$(CH$_2$) |
| 1383 s | 1382 vs | 1268 vvw* | 1380 m* | $\delta_s$(CH$_2$) |
| 1230 w | 1229 w | 1208 vvw* | 1259 m* | $\rho_r$(CH$_2$) |
| | | | 1220 m* | $\rho_r$(CH$_2$) |
| 1193 w | 1197 w | | 1198 ms | $\delta$(CH) |
| 1024 s | 1020 s | 1010 w | 1010 s | $\nu_s$(CCC) |
| 998 m | 999 m | 991 w | 982 m | $\rho_t$(CH$_2$) |
| 968 s | 961 s | 910 w | 922 | $\rho_w$(CH$_2$) |
| 943 vs | 941 s | 878 m | 897 s | $\rho_w$(CH$_2$) |
| 913 w | 913 w | 863 w* | 886 m* | $\pi$(CH) |
| | | | 862 m* | $\pi$(CH) |
| 767 m | 763 m | 800 w | 805 w | $\rho_r$(CH$_2$)$^+$ + $\rho_t$(CH$_2$) |
| | | 752 w | | |
| 512 s | 505 vs | 500 w | 556 | $\delta$(CCC) |
| | | | (CS$_2$) | |
| 403 m | 398 m | 350 w | 357 | $\nu_{as}$ |
| | | | (Raman) | (M–allyl) |
| 379 sh | 373 m | 329 | 329 | $\nu_s$ |
| 369 w | 361 sh | (Raman) | (Raman) | (M–allyl) |
| 254 s | 187 vs | | | $\nu_{as}$(Pd–X) |
| 245 s | 187 vs | | | $\nu_s$(Pd–X) |
| | | 468 w | 533 s | (M–allyl) |
| | | 402 w | 393 | Tilts |
| | | | (Raman) | |

[a] According to References 9–11.

\* There is a lack of agreement with regard to band assignment.

tra of allyl complexes in the range of 500–3100 cm$^{-1}$ are very similar because they are due to the 18 vibrations of the $C_3H_5$ group.[8–11] Interactions between allyl ligands which are bonded to the same central atom may be neglected. The similarity of spectra of the compounds $[Pd_2Cl_2(C_3H_5)_2]$ and $[Pd(C_3H_5)_2]$ may serve as an example. Practically all bands are observed in both the IR spectrum and in the Raman spectrum. This is due to the very low local symmetry of the allyl group ($C_s$). There is no agreement with regard to the assignment of the $v_{as}(CCC)$. Nakamoto assigned a very weak band at *ca* 1470–1490 cm$^{-1}$ to this vibration. However, Fritz assigned a strong band at *ca* 1450–1460 cm$^{-1}$ to this vibration.[1–8] Still other authors assigned vibrations at *ca* 1380 cm$^{-1}$ to this mode.[10,11] The $\delta(CCC)$ band which is medium or weakly intense occurs for the $C_3H_5$ group at *ca* 510 cm$^{-1}$, while for the 2-methylallyl ligand, this band occurs at 570 cm$^{-1}$.

The $v(M–allyl)$ bands as well as bands due to M–allyl deformation vibrations occur in the 300–550 cm$^{-1}$ range. The spectra of the complexes $Pd_2X_2(allyl)_2$, $Pd(C_3H_5)_2$, and $Rh(C_3H_5)_3$ are given in Table 7.1.

## 3. NMR SPECTRA

Many conclusions concerning the structure and properties of allyl complexes in solution may be obtained from $^1H$ NMR spectra (Table 7.2).[1–5,12,37,38,40] The spectra of $\sigma$-allyl derivatives, such as $[(\sigma–C_3H_5)\,Mn(CO)_5]$, are of the ABCX$_2$ type. For complexes containing substituted allyl groups, it is possible to assign positions of these substituents. For compounds of transition metals possessing a $\pi$-allyl $C_3H_5$ group which is bonded in a symmetrical fashion, the spectrum of the AM$_2$X$_2$ type is characteristic if the complexes are rigid. However, the spectrum is of AX$_4$ type if the complexes have dynamic properties. For rigid complexes containing an asymmetrically bonded allyl ligand, the AGMPX spectra are characteristic. The AM$_2$X$_2$ spectrum show doublets for the M and X protons as well as a multiplet for the A proton, while the AGMPX spectrum exhibits four doublets of the 1, 2, 3, 4 protons, and a multiplet for the proton number 5 consisting of nine lines (Figure 7.2).

These spectra are simple owing to small values of the $J(MX)$ and $J(MX')$ coupling constants. However, in high resolution spectrometers, the M and X resonance signals may split into doublets or triplets, while the A signal may split into fifteen lines. The $J(AM)$ coupling constant is smaller than the $J(AX)$ coupling constant.

$$
\begin{array}{c}
H_A \\
| \\
C \\
\diagup \quad \diagdown \\
H_M{-}C \qquad\qquad C{-}H_M \\
| \qquad\qquad\qquad | \\
H_x \qquad\qquad\quad H_x
\end{array}
\qquad\qquad (7.7)
$$

At higher temperatures, allyl complexes are no longer rigid; they become dynamic, fluxional compounds. Therefore, AM$_2$X$_2$ spectra become AX$_4$ spectra. Temperatures of such transitions for various compounds range from 200 to 450 K. The hafnium allyl compound $[Hf(\eta^3\text{-}C_3H_5)_4]$ exhibits an AX$_4$-type spectrum even at 199 K. For dynamic

Table 7.2. $^1H$ NMR Spectra of Some Allyl Complexes

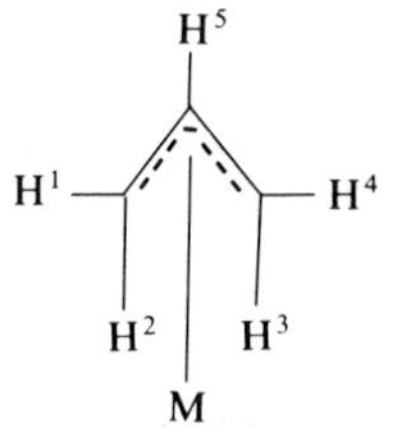

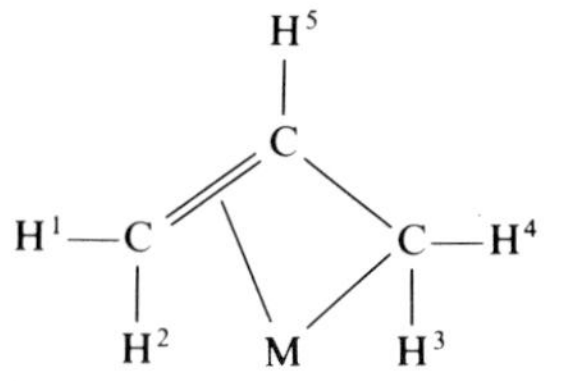

| Compound | $\tau_1-\tau_4$ | $\tau_5$ | Temperature (K) | Type of spectrum |
|---|---|---|---|---|
| $[Zr(C_3H_5)_4]^a$ | $\tau_1 = 6.72$; $\tau_2 = 8.10$; $\tau_3 = 8.10$; $\tau_4 = 6.72$; | 4.82 | 199 | $AM_2X_2$ |
| | $\tau_{1,2,3,4} = 7.37$ | 4.81 | 263 | $AX_4$ |
| $[Th(C_3H_5)_4]^a$ | $\tau_{1,4} = 6.46$; $\tau_{2,3} = 7.61$ | 3.97 | 283 | $AM_2X_2$ |
| | $\tau_{1,2,3,4} = 7.09$ | 3.95 | 353 | $AX_4$ |
| $[Mo(C_3H_5)_4]^b$ | $\tau_1 = 6.89$; $\tau_2 = 7.12$; $\tau_3 = 10.20$; $\tau_4 = 8.07$ $J(1,5) = 9.1$; $J(2,5) = 13.3$; $J(3,5) = 11.5$; $J(4,5) = 8.5$; $J(1,4) = 1.6$; $J(3,4) = 0.8$ Hz | 5.92 | 303 | AGMPX |
| $[Mn(C_3H_5)(CO)_4]^c$ | $\tau_{1,4} = 7.31$; $\tau_{2,3} = 8.28$ | 5.30 | 298 | $AM_2X_2$ |
| $[Pd_2Cl_2(C_3H_5)_2]^d$ | $\tau_{1,4} = 5.89$; $\tau_{2,3} = 6.94$ | 4.58 | 308 | $AM_2X_2$ |
| $[Rh_2Cl_2(C_3H_5)_4]^{c,e}$ | $\tau_1 = 4.92$; $\tau_2 = 5.98$; $\tau_3 = 8.18$; $\tau_4 = 7.50$ $J(1,5) = 6.8$; $J(2,5) = 12.0$; $J(3,5) = 10.8$; $J(4,5) = 6.0$ Hz | 5.24 | 253 | AGMPX |
| | $\tau_{1,2,3,4} = 6.78$; $J(5, X) = 9.0$; $J(Rh, 5) = 2.3$ Hz | 5.40 | 423 | $AX_4$ |
| $[Rh_2Br_2(C_3H_5)_4]^f$ | $\tau_1 = 4.70$; $\tau_2 = 5.65$; $\tau_3 = 7.95$; $\tau_4 = 7.38$ $J(1,5) = 8.0$; $J(2,5) = 14.0$; $J(3,5) = 11.0$; $J(4,5) = 7.0$ Hz | 5.20 | 297 | AGMPX |
| $[Ir(C_3H_5)_2py_2][BF_4]^g$ | $\tau_1 = 5.62$; $\tau_2 = 7.63$; $\tau_3 = 8.67$; $\tau_4 = 7.16$ $J(1,5) = 6.8$; $J(2,5) = 6.3$; $J(3,5) = 10.2$; $J(4,5) = 9.8$; $J(1,4) = 1.6$; $J(3, 4) = 1.9$ Hz | 4.90 | 300 | AGMPX |
| $[Re_2(C_3H_5)_4]^h$ | $\tau_{1,4} = 5.88$; $\tau_{2,3} = 10.53$ $J(1,5) = J(4,5) = 5$ Hz; $J(2,5) = J(3,5) = 9$ Hz $J(1,3) \approx 5$ Hz | 4.23 | | $AM_2X_2$ |

[a] J. K. Becconsall and S. O. O'Brien, *Chem. Commun.*, 302 (1966).
[b] K. C. Ramey, D. C. Lini, and W. B. Wise, *J. Am. Chem. Soc.*, **90**, 4275 (1968).
[c] W. R. McClellan, H. H. Hoehn, H. N. Cripps, E. L. Muetterties, and B. W. Hawk, *J. Am. Chem. Soc.*, **83**, 1601 (1961).
[d] K. C. Ramey and G. L. Statton, *J. Am. Chem. Soc.*, **88**, 4387 (1966).
[e] J. Powell and B. L. Shaw, *J. Chem. Soc.*, 583 (1968).
[f] H. Pasternak, T. Glowiak, and F. Pruchnik, *Inorg. Chim. Acta*, **19**, 11 (1976); F. Pruchnik and H. Pasternak, *Rocz. Chem.*, **51**, 1581 (1977).
[g] M. Green and G. J. Parker, *J. Chem. Soc., Dalton Trans.*, 333 (1974).
[h] A. F. Masters, K. Mertis, J. F. Gibson, and G. Wilkinson, *Nouv. J. Chim.*, **1**, 389 (1977); F. A. Cotton and M. W. Etine, *J. Am. Chem. Soc.*, **100**, 3788 (1978).

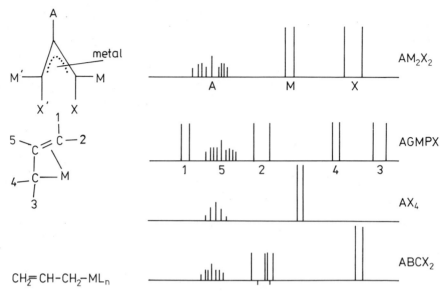

Figure 7.2.   NMR spectra of some allyl complexes.

allyl compounds, the fluxionality may be caused either by the movement of the allyl group with respect to the remaining part of the coordination compound, or by dynamic properties of the entire complex. The precise mechanism of such transformations is not known for the majority of complexes.

Three types of transformations may be distinguished:

1.   The *syn* ↔ *anti* proton exchange at one or at both terminal carbon atoms [equation (7.8)].

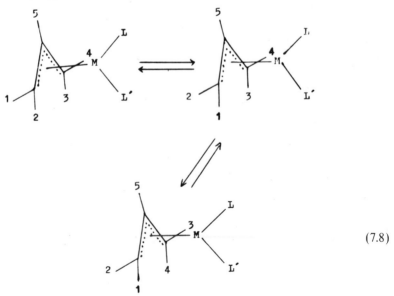

(7.8)

2.  The *syn ↔ syn* and *anti ↔ anti* proton exchange [equation (7.9)]. This process may be observed only for complexes containing an asymmetrically bonded allyl group.

$$\tag{7.9}$$

3.  The equilibrium between conformers in which the allyl groups have different environments. In such cases, the allyl group may be bonded to the metal both in symmetric as well as asymmetric fashions [equation (7.10)].

$$\tag{7.10}$$

The *syn ↔ anti* exchange takes place owing to the formation of the intermediate σ-allyl complex, followed by reformation of the π complex in which the allyl group is bonded to the metal through the other side (Figure 7.3a). The rotation around the C−C bond

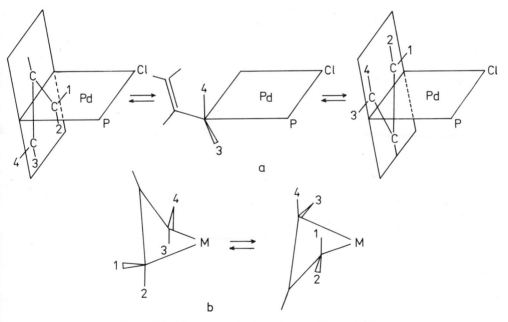

Figure 7.3. The π–σ mechanism of *syn–anti* isomerization.

in the $\pi$-allyl complex as well as the isomerization occurring due to formation in the transition state of a species in which the allyl group is bonded via two $\sigma$ bonds between terminal carbon atoms and the metal is less probable. The energy of activation of the $syn \leftrightarrow anti$ exchange most often assumes values in the 34–80 kJ mol$^{-1}$ range.

The $syn$–$syn$, $anti$–$anti$ exchange in square-planar complexes often depends on the ligand exchange process. This process may be explained by the "angular twist" mechanism in the pentacoordinate intermediate complex (Figure 7.4a) or by Berry's pseudorotation (Figure 7.4b).

The isomerization occurring according to the above-described mechanism also sometimes takes place in pentacoordinate compounds without exchange of the ligands.

The $syn \rightleftarrows syn$, $anti \rightleftarrows anti$ exchange may also proceed according to a dissociative mechanism via a 14$e$ complex, e.g., Pd(1-MeC$_3$H$_4$)(C$_6$Cl$_5$)(PPh$_3$)[22] [equation (7.11)].

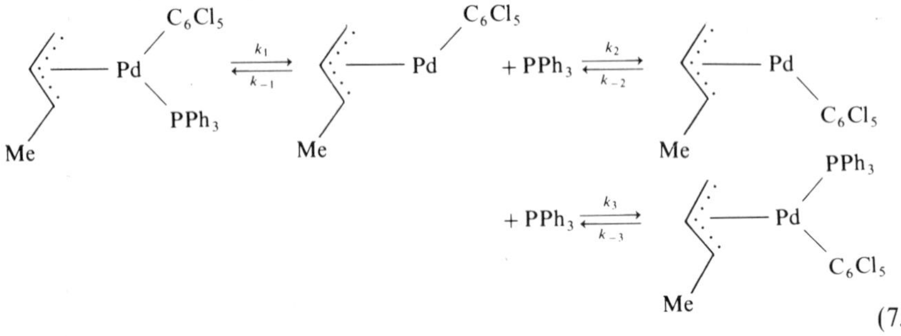

$$(7.11)$$

Examples of complexes whose dynamic properties are caused by the existence of temperature-dependent equilibrium between different conformers are compounds of the type [Mcp($\eta^3$-C$_3$H$_4$R)(CO)$_2$] (M = Mo, W; R = H, Me) (Figure 7.5).[3,40] This equi-

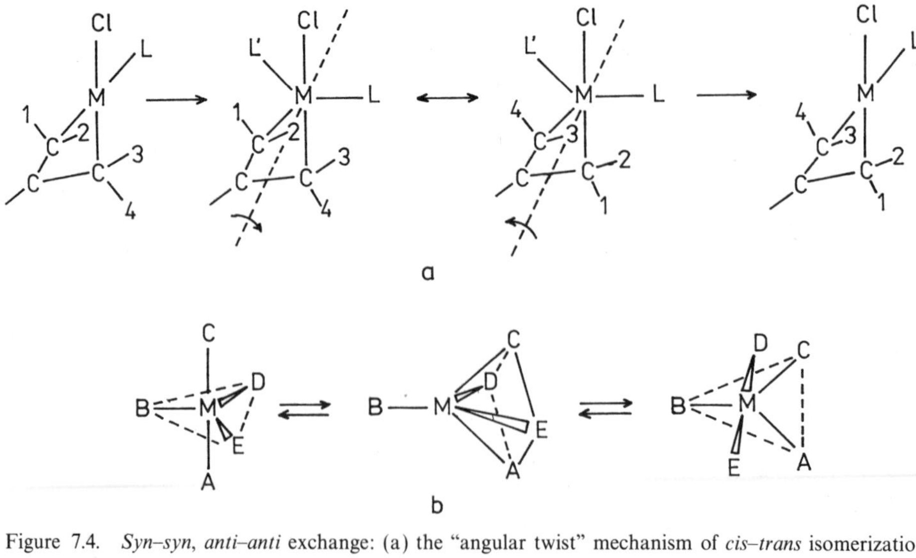

Figure 7.4. *Syn–syn, anti–anti* exchange: (a) the "angular twist" mechanism of *cis–trans* isomerization and (b) Berry's pseudorotation mechanism of the exchange of equatorial and axial ligands in the trigonal bipyramid.

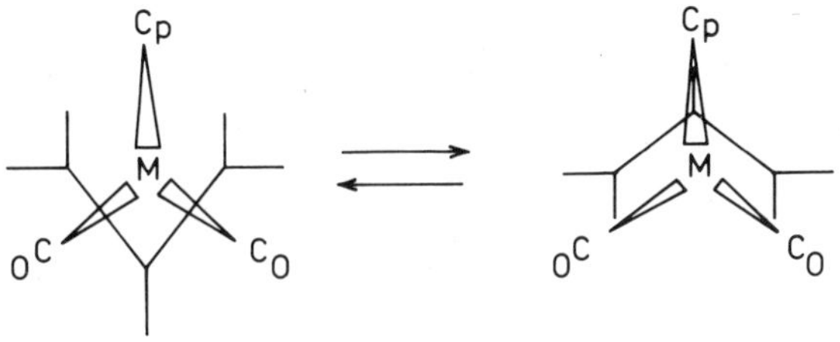

Figure 7.5. Equilibrium between the conformers $Mcp(\eta^3\text{-}C_3H_4R)(CO)_2$ where M = Mo, W; R = H, Me.

librium is achieved by the rotation of the allyl ligand in the $C^1C^2C^3$ plane about the allyl–metal axis.

The rotation of the allyl group is also probably reponsible for the dynamic properties of the triallylrhodium(III) compounds. At 194 K, all allyl groups in this compound, as in $Ir(C_3H_5)_3$, are not equivalent. At 307 K, two allyl ligands become equivalent owing to the rotation of the third one in its own plane. The energy of activation of this process is 39.8 kJ $mol^{-1}$.

$$\text{(7.12)}$$

At higher temperatures, all allyl group begin to rotate; however, the π–σ–π rearrangement begins to occur simultaneously.

The $^{13}C$ NMR spectra of allyl complexes of transition metals are distinctive. The terminal carbon atoms have chemical shifts in the 35–80 ppm range, while the middle cabon atom has chemical shifts in the 80–140 ppm range[37,38] (see Table 7.3). These values may be explained on the basis of the valence bond theory. The chemical shift of terminal carbon atoms have values which are intermediate between shifts of carbon atoms creating the σ bond with the metal and those of free olefins. This is in agreement with the following canonical forms:

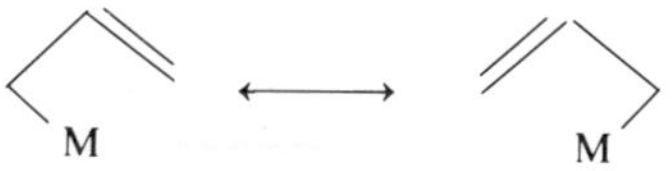

$$\text{(7.13)}$$

For a full explanation of the $^{13}C$ NMR shifts, it is also necessary to take into consideration other forms, for example, ionic. The $^1J(CC)$ coupling constants of the allyl group depend on the character of the metal–allyl bond. These constants equal *ca* 59 Hz in the case of ionic compounds such as $LiC_3H_5$ and $KC_3H_5$, while for allyl complexes of transition metals coupling constants within the range of 40–50 Hz are characteristic. The $^1J(CC)$ constant is influenced mainly by the metal–allyl bond, while the $^1J(^{13}CH)$ constant may also be influenced by the tilting of hydrogen atoms away from the skeletal plane of the allyl group. The $^1J(CC)$ constant decreases as the atomic number of the metal increases both in the group as well as in the period of the Periodic Table. This indicates the increasing contribution of the $\psi_2$ and $\psi_3$ orbitals in the metal–allyl bond.

The $^{13}C$ NMR spectra also allow investigation of isomers of allyl complexes, e.g., *cis*- and *trans*-$M(C_3H_5)_2$ (M = Ni, Pd, Pt), $M(allyl)_n$ (M = Rh, Ir, $n = 3$;

Table 7.3. $^{13}C$ NMR Spectra of Allyl Complexes

| Compound | $\delta(^{13}C)$ (ppm) | | | $\delta(^{13}C)$ (ppm) for orher atoms |
|---|---|---|---|---|
| | C1 | C2 | C3 | |
| $[Ni_2Cl_2(1\text{-}MeC_3H_4)_2]^a$ | 70 | 106.9 | 48.0 | |
| $[Ni_2Br_2(1\text{-}MeC_3H_4)_2]^a$ | 71.2 | 105.6 | 49.6 | |
| $[Pd_2Cl_2(C_3H_5)_2]^b$ | 63.2 | 111.9 | 63.2 | |
| $[Pd(acac)(C_3H_5)]^b$ | 55.8 | 113.5 | 55.8 | |
| $[Pdcp(C_3H_5)]^b$ | 45.8 | 95.0 | 45.8 | |
| $[Pd_2Cl_2(2\text{-}MeC_3H_4)_2]^b$ | 61.7 | 127.9 | 61.7 | Me = 23.1 |
| $[Pdcp(2\text{-}MeC_3H_4)]^b$ | 47.0 | 112.2 | 47.0 | Me = 23.6 |
| $[Pd(acac)(2\text{-}MeC_3H_4)]^b$ | 54.8 | 129.2 | 54.8 | Me = 23.4 |
| 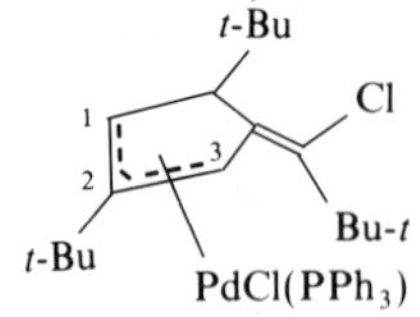 (structure c: t-Bu substituted allyl PdCl(PPh₃)) | 77.6 | 132.4 | 96.8 | |
| 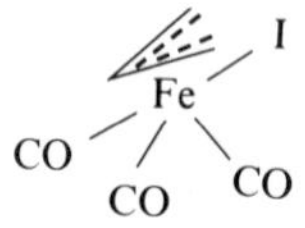 (structure d,e: allyl Fe(CO)₃I) | 52.6 | 106.6 | 52.6 | |
| 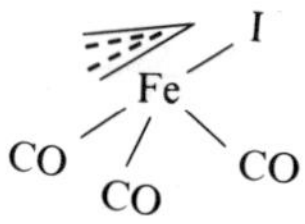 (structure d,e: allyl Fe(CO)₃I) | 59.7 | 100.9 | 59.7 | |
| $cis\text{-}Pt(C_3H_5)_2^{(37)}$ | 48.5 | 104.8 | 48.5 | $^1J(CC) = 40.4$ Hz |
| $Ni(1\text{-}MeCHCHCH_2)cp^{(37)}$ | 58.37 | 92.32 | 36.33 | $^1J(C^1C^2) = 43.5$ Hz; $^1J(C^2C^3) = 42.5$ Hz; $^1J(C^1 - Me) = 41.4$ Hz |
| $[Fe(C_3H_5)(OAc)(CO)_3]^e$ | 68.07 | 104.13 | 68.07 | $CO_{eq} = 204.59$; $CO_{ax} = 206.08$ |
| $[Fe(C_3H_5)(CF_3COO)(CO)_3]^e$ | 69.05 | 104.26 | 69.05 | $CO_{eq} = 203.22$; $CO_{ax} = 204.72$ |
| $[Rhcp(1\text{-}MeC_3H_4)(PPh_3)]^+BF_4^{-f}$ | 76.5 | 84.6 | 48.1 | Me = 22.5 |
| $[Rhcp(2\text{-}MeC_3H_4)(PPh_3)]^+BF_4^{-f}$ | 51.8 | 108.2 | — | Me = 26.6; cp = 94.4 |
| $[Cr_2(C_3H_5)_4]^{(37)}$ | 70.9 | 130.6 | 76.5 | $^1J(CC) = 46.4; 47.9$ Hz |
| | 58.6 | 101.5 | 58.6 | $^1J(CC) = 43.6$ Hz |
| | 55.8 | 106.0 | 55.8 | $^1J(CC) = 44.2$ Hz |

[a] L. A. Churlyaeva, M. I. Lobach, G. P. Kondratenkov, and V. A. Kormer, *J. Organomet. Chem.*, **39**, C23 (1972).
[b] B. E. Mann, R. Pietropaolo, and B. L. Shaw, *Chem. Commun.*, 790 (1971).
[c] P. M. Bailey, B. E. Mann, A. Segnitz, K. L. Kaiser, and P. M. Maitlis, *J. Chem. Soc., Chem. Commun.*, 567 (1974).
[d] E. W. Randall, E. Rosenberg, and L. Milone, *J. Chem. Soc., Dalton Trans.*, 1672 (1973).
[e] A. N. Nesmeyanov, L. A. Fedorov, N. P. Avakyan, P. V. Petrovskii, E. I. Fedin, E. V. Arshavskaya, and I. I. Kritskaya, *J. Organomet. Chem.*, **101**, 121 (1975).
[f] P. Powell and L. J. Russell, *J. Organomet. Chem.*, **129**, 415 (1977).

$M = Mo$, $W$, $n = 4$), etc., because the $^{13}C$ chemical shifts differ for different isomers.[37,38] The $^1J(M - {}^{13}C)$, $J(^{31}P - {}^{13}C)$ coupling constants are also utilized in structural studies.[38]

## 4. ELECTRONIC SPECTRA

The electronic spectra of allyl complexes (Table 7.4) exhibit relatively low intensity bands due to the $d$–$d$ ytansitions as well as intense bands resulting from ligand → metal and metal → ligand charge transfer transitions. In palladium complexes of the type $Pd_2Cl_2(allyl)_2$, the first three low intensity bands at 29,500 ($\varepsilon = 543$), 34,000 ($\varepsilon = 618$), and 40,500 cm$^{-1}$ ($\varepsilon = 2006$) were assigned to the $d$–$d$ transitions. The two more intense bands at 44,000 cm$^{-1}$ ($\varepsilon = 5222$) and 46,500 cm$^{-1}$ ($\varepsilon = 2286$) are due to Cl → Pd and Pd → allyl electron transfer transitions, respectively.[13]

The spectra of $[Rh_2X_2(allyl)_4]$ ($X = Cl$, Br, J) show broad shoulders in the 27,800–29,400 cm$^{-1}$ range which are assigned to the following $d$–$d$ transitions: $^1A_1 \to {}^1A_2$, $^1B_2$, $^1B_2$ (in the field possessing $C_{2v}$ symmetry). They correspond to the $^1A_{1g} \to {}^1T_{1g}$ transition in octahedral complexes in which the electronic structure of the central ion is $d^6$. In addition, a shoulder in the spectra of these rhodium compounds occurs in the 33,300–37,000 cm$^{-1}$ range due to the $^1A_1 \to {}^1A_1$, $^1B_1$, $^1B_2$ transitions ($C_{2v}$ symmetry) which correspond to the $^1A_{1g} \to {}^1T_{2g}$ transition for $O_h$ symmetry.[4,15] The bands due to electron transfer X → Rh should appear at higher energies. The Cl → Rh CT band in the complex $[RhCl_6]^{3-}$ occurs at 39,200 cm$^{-1}$, in the compound $[RhCl_2(en)_2]^+$ at 42,200 cm$^{-1}$, and in the allyl complex $[Rh_2Cl_2(allyl)_4]$ at even higher energies, since the allyl ligands create a stronger field than the ethylenediamine groups. In complexes in which the central ion has a $d^6$ electronic configuration, the electron transfer must take place into antibonding orbitals of the metal whose energy is greater, the higher the splitting parameter of the ligands.

*Table 7.4. Electronic Spectra of Some Allyl Complexes*

| Compound | Band assignment $v$ (extinction coefficient) (transition) (cm$^{-1}$) |
|---|---|
| $[Ni_2Br_2(C_3H_5)_2]^{(8)}$ | 40,200, 34,500, 22,800 |
| $[Ni(C_3H_5)cp]^{(8)}$ | 46,000, 36,800, 31,800, 26,500, 18,200 |
| $[Pd_2Cl_2(C_3H_5)_2]^{(13)}$ | 46,500 (2286) (CT M → allyl), 44,000 (CT Cl → Pd), 40,500 (2006) ($d$–$d$), 34,000 (618) ($d$–$d$), 29,500 (543) ($d$–$d$) |
| $[Pd_2Cl_2(2\text{-}MeC_3H_4)_2]^a$ | 49,000, 44,700, 29,500 |
| $[Rh_2Cl_2(C_3H_5)_4]^{(14)}$ | 37,000$^a$ (10,000) ($^1A_1 \to {}^1A_1$, $^1B_1$, $^1B_2$), 29,400$^a$ (2500) ($^1A_1 \to {}^1A_2$, $^1B_1$, $^1B_2$) |
| $[Rh_2Br_2(C_3H_5)_4]^{(14)}$ | 40,000$^a$ (14,900) (CT Br→Rh), 35,700$^a$ (12,100) ($^1A_1 \to {}^1A_1$, $^1B_1$, $^1B_2$), 28,600$^a$ (3100) ($^1A_1 \to {}^1A_2$, $^1B_1$, $^2B_2$) |
| $[Rh_2I_2(C_3H_5)_4]^{(14)}$ | 37,700$^a$ (17,300) (CT J→Rh), 33,300$^a$ (14,200) ($^1A_1 \to {}^1A_1$, $^1B_1$, $^1B_2$), 27,800$^a$ (5400) ($^1A_1 \to {}^1A_2$, $^1B_1$, $^1B_2$) |
| $[RhCl_2(2\text{-}MeC_3H_4)]_n^{(14,15)}$ | 41,400 (CT Cl → Rh), 31,500 ($^1A_1 \to {}^1A_1$, $^1B_1$, $^1B_2$), 21,500 ($^1A_1 \to {}^1A_2$, $^1B_1$, $^1B_2$) |

$^a$ S. D. Robinson and B. L. Shaw, *J. Chem. Soc.*, 4806 (1963).

*Table 7.5. Interatomic Distances and Structures of Allyl Complexes*

| Compound | Distances (pm) | | | | | Angle $C^1C^2C^3$ |
|---|---|---|---|---|---|---|
| | $C^1$–$C^2$ | $C^2$–$C^3$ | M–$C^1$ | M–$C^2$ | M–$C^3$ | |
| $[Cr_2(C_3H_5)_4]^a$ | | | | | | |
| $\mu$-$C_3H_5$ | 140 | 140 | 217 (Cr1) | 245 (Cr1) | | |
| | | | | 248 (Cr2) | 217 (Cr2) | 125° |
| $\mu$-$C_3H_5$ | 157 | 155 | 215 (Cr1) | 240 (Cr1) | | 117° |
| | | | | 237 (Cr2) | 216 (Cr2) | |
| $C_3H_5$ | 145 | 146 | 224 (Cr1) | 231 (Cr1) | 222 (Cr1) | 121° |
| $C_3H_5$ | 131 | 145 | 227 (Cr2) | 230 (Cr2) | 219 (Cr2) | 126° |
| $[Mo_2(C_3H_5)_4]^b$ | | | | | | |
| $\mu$-$C_3H_5$ | 135 | 136 | | 252 (Mo1) | 224 (Mo1) | 144° |
| | | | 224 (Mo2) | 244 (Mo2) | | |
| $\mu$-$C_3H_5$ | 140 | 136 | 222 (Mo2) | 253 (Mo1) | | 128° |
| | | | | 215 (Mo2) | 222 (Mo2) | |
| $C_3H_5$ | 136 | 140 | 229 (Mo1) | 231 (Mo1) | 230 (Mo1) | 120° |
| $C_3H_5$ | 138 | 134 | 235 | 235 (Mo2) | 228 (Mo2) | 122° |
| $[Re_2(C_3H_5)_4]^c$ | | | | | | |
| Re1 | 138 | 137 | 219 | 219 | 214 | 119° |
| | 145 | 161 | 220 | 221 | 226 | 113° |
| Re2 | 151 | 141 | 231 | 228 | 229 | 122° |
| | 137 | 147 | 219 | 219 | 222 | 118° |
| $[Ni(C_3H_5)\{SC(NH_2)_2\}_2]^+Cl^{-\,d}$ | 146 | 131 | 207 | 199 | 204 | |
| $[Pd_2Cl_2(C_3H_5)_2]^e$ | 135 | 137 | 214 | 202 | 217 | 129° |
| $[PdCl(2\text{-}MeC_3H_4)PPh_3]^f$ | 140 | 147 | 228 | 222 | 214 | |
| $[Pd_2Cl_2(2\text{-}MeC_3H_4)_2]^g$ | 136.9 | 134.6 | 208.2 | 210.2 | 206.2 | |
| $[Pd_2Cl_2(Me_2CCHCMe_2)_2]^h$ | 141 | 142 | 212 | 214 | 214 | 118.4° |
| $[Rh_2Cl_2(C_3H_5)_4]^i$ | 148 | 140 | 212 | 215 | 225 | |
| | 145 | 141 | 212 | 217 | 226 | |
| $[Rh_2Br_2(C_3H_5)_4]^{(20)}$ | 148.2 | 140.7 | 215.6 | 214.9 | 226.7 | 119.39 |
| | (25) | (25) | (17) | (16) | | (148)° |
| | 143.8 | 141.1 | 213.8 | 216.7 | 225.9 | 116.03 |
| | (23) | (23) | (16) | (17) | (17) | (147)° |
| $[IrClH(1\text{-}PhC_3H_4)(PPh_3)_2]^{(16)}$ | 142.5 | 138 | 228.3 | 217.7 | 220.2 | |
| $[Pt(C_3H_5)(PPh_3)_2]^+[PF_6]^{-\,j}$ | 136 | 147 | 224 | 219 | 224 | 125.7° |
| $[Ru(C_3H_5)(NO)(PPh_3)_2]^k$ | 138 | 141 | 221.4 | 213.0 | 225.8 | |
| $[FeI(C_3H_5)(CO)_3]^l$ | 138 | 138 | 230 | 209 | 230 | 131° |

$^a$ T. Aoki, A. Furusaki, Y. Tomiie, K. Ono, and K. Tanaka, *Bull. Chem. Soc. Jpn.*, **42**, 545 (1969).
$^b$ F. A. Cotton and J. R. Pipal, *J. Am. Chem. Soc.*, **93**, 5441 (1971).
$^c$ F. A. Cotton and M. W. Extine, *J. Am. Chem. Soc.*, **100**, 3788 (1978).
$^d$ A. Sirigu, *Chem. Commun.*, 256 (1969).
$^e$ W. E. Oberhansli and L. F. Dahl, *J. Organomet. Chem.* **3**, 43 (1965); A. E. Smith, *Acta Crystallogr.*, **18**, 331 (1965); V. F. Levdik and M. A. Poraj-Koszyc, *Zh. Strukt. Khim.*, **3**, 472 (1962).
$^f$ R. Mason and D. R. Russell, *Chem. Commun.*, 26 (1966).
$^g$ R. Mason and A. G. Wheller, *J. Chem. Soc. (A)*, 2549 (1968).
$^h$ R. Mason and A. G. Wheller, *J. Chem. Soc. (A)*, 2543 (1968).
$^i$ M. McPartlin and R. Mason, *Chem. Commun.*, 16 (1967).
$^j$ J. D. Smith and J. D. Oliver, *Inorg. Chem.*, **17**, 2585 (1978).
$^k$ M. W. Schoonover, C. P. Kubiak, and R. Eisenberg, *Inorg. Chem.*, **17**, 3050 (1978).
$^l$ M. H. Minasyants, Yu. T. Struchkov, I. I. Kritskaya, and R. A. Avoyan, *Zh. Strukt. Khim.*, **7**, 903 (1966); M. H. Minasyants and J. T. Struchkov, *Zh. Strukt. Khim.*, **9**, 665 (1968).

## 5. STRUCTURAL PROPERTIES

In π-allyl complexes, the carbon–carbon distances (Table 7.5) usually assume higher values than in olefins; however, they are smaller than in compounds containing $sp^3$ hybridized carbon atoms. In the case of asymmetrically bonded groups

$$
\begin{array}{c}
\quad\quad C_3 \\
\diagup \quad \diagdown \\
M \quad\quad\quad C_2 \\
\diagdown \quad \diagup \\
\quad\quad C_1
\end{array}
$$

very large differences between the $C_1$–$C_2$ and $C_2$–$C_3$ distances are sometimes observed. The following complexes serve as examples: $[Ni(C_3H_5)\{SC(NH_2)_2\}_2]^+Cl^-$ 131 pm ($C_1$–$C_2$) and 146 pm ($C_2$–$C_3$), $[Cr_2(C_3H_5)_4]$ 131 pm ($C_1$–$C_2$) and 145 pm ($C_2$–$C_3$) for one of the allyl group, and $[Re_2(C_3H_5)_4]$ 145 pm ($C_1$–$C_2$) and 161 pm ($C_2$–$C_3$). In palladium complexes of the type $[Pd_2Cl_2(allyl)_2]$, the $C_1$–$C_2$ and $C_2$–$C_3$ distances are generally almost equal. The M–C distances for terminal allyl groups vary in the 200–230 pm range, and for the bridging allyl groups they are most commonly in the 220–255 pm range.

In complexes containing substituted allyl groups the substituents at the middle carbon atom are tilted away from the $C^1C^2C^3$ plane in the direction of the metal. The tilting reaches up to 50 pm. The *anti* substituents are also generally tilted in the same direction, in contrast to the *syn* groups (Figures 7.6 and 7.7).

The nickel and palladium complexes containing phenyl- and methylcyclobutenyl ligands are exceptions; in these compounds, the substituents are tilted in the opposite direction (away from the complex). Neutron diffraction studies showed that the *anti* hydrogen atoms in the complex *trans*-$[Ni(C_3H_5)_2]$ are tilted away from the plane of the allyl group by 29.4°, and the remaining hydrogens are tilted in the direction of the metal by 15.8° for the *meso* atom and 8.9° for the *syn* atoms.[34]

## 6. PHOTOELECTRON SPECTRA[16b, 34, 41–43, 50]

The $[M(allyl)_2]$ allyl complexes may be regarded as the simplest sandwich compounds. The interpretation of the photoelectron spectra of allyl complexes of nickel and palladium is not unequivocal because the energies of $d$ orbitals as well as the energy of the $\pi_2(\psi_2 - d_\pi)$ molecular orbital are similar. At first, it was suggested that the first three bands in the spectrum of $[Ni(C_3H_5)_2]$ possessing the 2:2:1 intensity ratio appear owing to ionization of the $3d$ metal orbitals whose energy changes according to the series: $d_{z^2} < d_{xz}, d_{yz} < d_{x^2-y^2}, d_{xy}$. However, subsequent calculations showed that the ionization of molecular orbitals possessing predominant $d$ metal orbital character is associated with very large energies of relaxation. The ligand orbitals exhibit minimal relaxation effect. The PE spectra of some allyl compounds of Ni, Cr, and Mo are given in Table 7.6. In dinuclear chromium(II) and molybdenum(II) complexes the lowest ionization energy is possessed by the $\delta$(M–M) orbital.

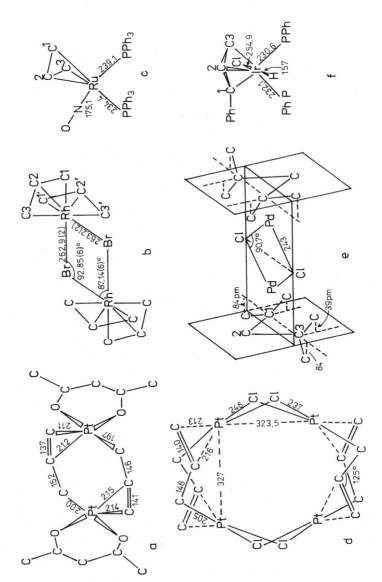

Figure 7.6. Structures of (a) [Pt₂(acac)₂(C₃H₅)₂], (b) [Rh₂Br₂(C₃H₅)₄], (c) [Ru(C₃H₅)(NO)(PPh₃)₂], (d) [Pt₄Cl₄(C₃H₅)₄], (e) [Pd₂Cl₂(Me₂CCHCHCMe₂)₂], (f) [IrClH(1-PhC₃H₄)(PPh₃)₂]. (the C–C and M–C distances are given in Table 4.5; distances are in pm units).

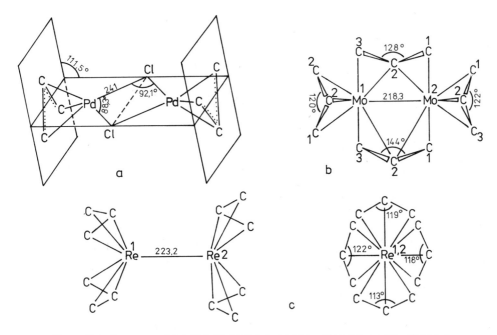

Figure 7.7. Structures of (a) $[Pd_2Cl_2(C_3H_5)_2]$, (b) $[Mo_2(C_3H_5)_4]$, (c) $[Re_2(C_3H_5)_4]$.

Table 7.6. *Experimental Ionization Energies (eV) for Allyl Complexes*[50]

| $[Ni(C_3H_5)_2]$ | $[Ni(2\text{-}MeC_3H_4)_2]$ | $[Ni(1\text{-}MeC_3H_4)_2]$ | $[Ni(1,3\text{-}Me_2C_3H_3)_2]$ |
|---|---|---|---|
| 7.76 | 7.53 | 7.53 | 7.22 |
| 8.19 | 7.91 | 8.00 | 7.68 |
| 8.58 | 8.32 | 8.40 | 8.10 |
| 9.40 | 9.22 | 9.13 | 8.78 |
| 10.38 | 9.86 | 10.10 | 9.73 |
| 11.55 | 10.93 | 11.15 | 10.70 |
| 12.7 | 12.2 | 12.3 | 12.2 |
| 14.2 | 12.7 | 12.8 | 13.4 |
| 15.6 | 15.0 | 15.0 | 14.0 |

| $[Cr(C_3H_5)_3]$ | $[Cr_2(C_3H_5)_4]$ | $[Mo_2(C_3H_5)_4]$ | Band assignment |
|---|---|---|---|
|  | 6.90 | 6.72 | $(M-M)\delta$ |
| 7.20 | 7.94 | 7.61 | $M_d + 1a_2$ |
| 7.84 |  | 7.94 |  |
| 8.80 |  | 8.46 |  |
| 10.54 | 10.73 | 10.04 | $1b_2$ |
| 11.79 |  |  |  |
| 12.86 | 12.72 | 12.46 | (allyl)$\sigma$ |
| 14.36 | 14.27 | 13.36 |  |
| 15.48 | 15.63 | 14.52 |  |
| 17.9 |  | 15.98 |  |
| 21.2 |  | 18.2 |  |

## 7. METHODS OF PREPARATION OF
## $\pi$-ALLYL COMPLEXES[1,2,5,8,17,33,35,44,45]

### a. *Reactions of transition metal complexes with allyl compounds of main group metals*

Allyl complexes of the transition metals are most commonly synthesized by reactions of Grignard compounds or organolithium derivatives with halides or acetylacetonates of the transition metals. Dependent upon the starting material, complexes containing exclusively allyl groups (homoleptic complexes) may be obtained, or derivatives also possessing other ligands (heteroleptic complexes) may be prepared:

$$NiBr_2 + 2CH_2 = CHCH_2MgBr \longrightarrow [Ni(C_3H_5)_2 + 2MgBr_2 \qquad (7.14)$$

$$RhCl_3 + 3CH_2 = CHCH_2MgCl \longrightarrow [Rh(C_3H_5)_3] + 3MgCl_2 \qquad (7.15)$$

$$Rh_2Cl_2(C_3H_5)_4 + 2CH_2 = CHCH_2MgCl \longrightarrow 2[Rh(C_3H_5)_3] + 2MgCl_2 \qquad (7.16)$$

$$RuCl_2(COD) + 2CH_2 = CHCH_2MgCl \longrightarrow [Ru(C_3H_5)_2(COD)] + 2MgCl_2 \qquad (7.17)$$

$$mer\text{-}[IrCl_3(PPh_3)_3] + 2CH_2 = CHCH_2MgCl \longrightarrow [Ir(C_3H_5)_2(PPh_3)_2]^+Cl^- + 2MgCl_2 \qquad (7.18)$$

$$mer\text{-}[IrCl_3(PPh_3)_3] + CH_2 = CHCH_2MgCl \longrightarrow [IrCl_2(C_3H_5)(PPh_3)_2] \qquad (7.19)$$

The following complexes may be prepared by utilizing these methods: $[M(allyl)_2]$ $(M = Ni, Pd, and Pt)$, $[M(allyl)_3]$ $(M = V, Cr, Fe, Co, Rh, Ir, Sc, lanthanides, and actinides)$, $[M(allyl)_4]$ $(M = Zr, Hf, Nb, Ta, Mo, W, Th, and U)$, and $[M_2(allyl)_4]$ $(M = Cr, Mo, and Re)$.

In addition to allylhalogenomagnesium and allyllithium, the synthesis of allyl coordination compounds of the transition metals may involve allyl derivatives of sodium, zinc, tin, boron, and other main group elements. The synthesis is carried out at 190–295 K in ethers or hydrocarbon solvents.

### b. *Reactions of allyl halides with transition metal complexes in the presence of reducing agents*

Some carbonyl complexes may react with allyl halides to give metal allyl compounds. The metal as well as carbon monoxide undergo oxidation:

$$[Rh_2Cl_2(CO)_4] + 4H_2O + 6C_3H_5Cl \longrightarrow [Rh_2Cl_2(C_3H_5)_4] + 4CO_2 + 2C_3H_6 + 6HCl \qquad (7.20)$$

Many allyl palladium complexes may be prepared in an analogous manner. The starting materials for this reaction are allyl halides and sodium tetrachloropalladate(II) in the presence of CO. Instead of carbon monoxide, $SnCl_2$ may be used as a reducing agent.

$$2[PdCl_4]^{2-} + 2C_3H_5Cl + 2CO + 2H_2O \longrightarrow [Pd_2Cl_2(C_3H_5)_2] + 4HCl + 2CO_2 + 4Cl^- \qquad (7.21)$$

The mechanism of this reaction is shown in equation (7.22).

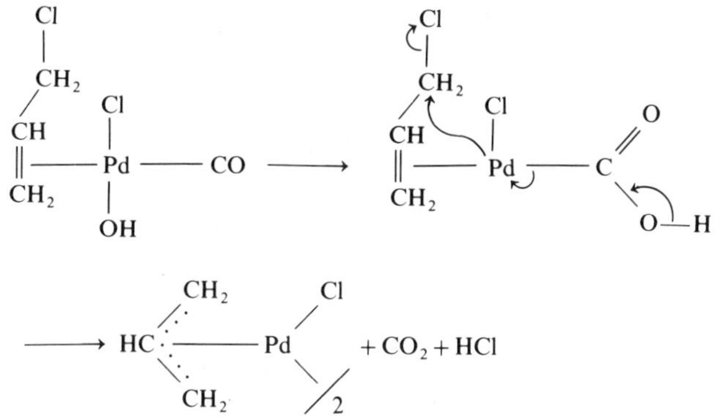

$$2[PdCl_4]^{2-} + 2C_3H_5Cl + 2SnCl_2 \longrightarrow [Pd_2Cl_2(C_3H_5)_2] + 2[SnCl_6]^{2-} \tag{7.23}$$

## c. Oxidative addition of allyl halides

This type of addition takes place in the case of complexes in which the metal possesses a low oxidation state. Coordinatively saturated complexes (18e) or compounds with a coordination number lower than the maximum (e.g., 16e) may be utilized:

$$[RhCl(PPh_3)_3] + C_3H_5Cl \longrightarrow [RhCl_2(C_3H_5)(PPh_3)_2] + PPh_3 \tag{7.24}$$

$$\qquad 16e \qquad\qquad\qquad\qquad 18e$$

$$[Rh\{P(OR)_3\}_4]^+ + C_3H_5Cl \longrightarrow [RhCl(C_3H_5)\{P(OR)_3\}_3]^+ + P(OR)_3 \tag{7.25}$$

$$2[Ni(CO)_4] + 2C_3H_5X \longrightarrow [Ni_2X_2(C_3H_5)_2] + 8CO \tag{7.26}$$

$$2[Ni(COD)_2] + 2C_3H_5X \longrightarrow [Ni_2X_2(C_3H_5)_2] + 2COD \tag{7.27}$$

$$[Fe_2(CO)_9] + C_3H_5X \longrightarrow [FeX(C_3H_5)(CO)_3] + CO + [Fe(CO)_5] \tag{7.28}$$

$$2[Pd\{(PhCH=CH)_2CO\}] + 2CH_2=CMeCH_2Cl \longrightarrow [Pd_2Cl_2(2\text{-}MeC_3H_4)_2] \tag{7.29}$$

## d. Reactions of olefins with metal complexes

An important method of preparation of palladium allyl compounds is the reaction of $PdX_4^{2-}$ or $PdX_2$ with olefins:

$$2RCH_2CH:CH_2 + 2PdX_2 \longrightarrow [Pd_2X_4(CH_2:CHCH_2R)_2] \tag{7.30}$$

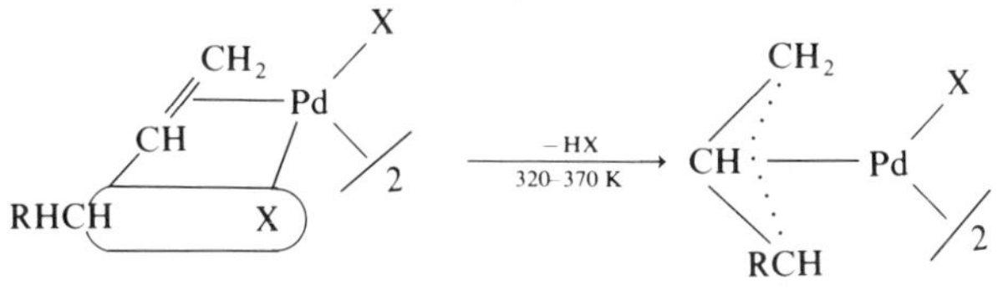

Cyclic olefins, Dewar hexamethylbenzene, steroids, alkenes containing other functional groups, etc., may also react in an analogous fashion:

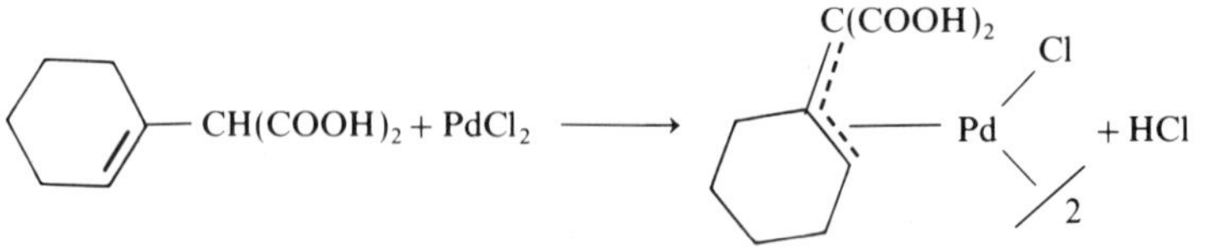

$$\text{(cyclohexenyl)}-CH(COOH)_2 + PdCl_2 \longrightarrow \text{(cyclohexyl-}C(COOH)_2)-Pd\begin{pmatrix}Cl\\ \end{pmatrix}_2 + HCl \qquad (7.32)$$

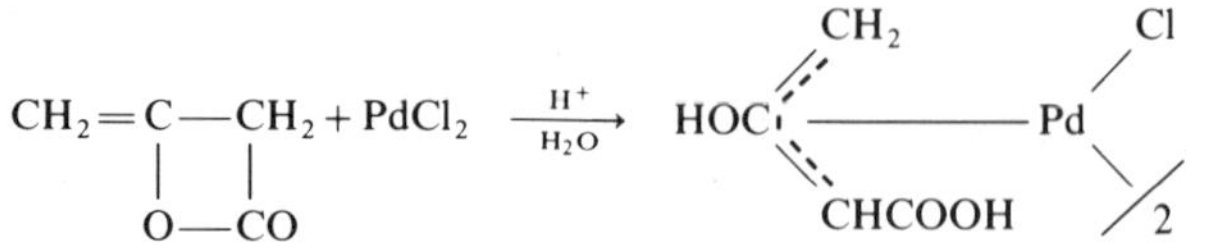

$$CH_2{=}C{-}CH_2 + PdCl_2 \xrightarrow[H_2O]{H^+} HOC\begin{pmatrix}CH_2\\ \\ CHCOOH\end{pmatrix}-Pd\begin{pmatrix}Cl\\ \end{pmatrix}_2 \qquad (7.33)$$
$$\quad\;\; |\;\;\; |$$
$$\quad\;\; O{-}CO$$

### e. Reactions of 1, 3-dienes with metal compounds

These reactions occur via attack of various atoms or groups on the terminal carbon atom. The following entities may be utilized for the attack: H, RCO, RO, Cl, and AcO. The most commonly utilized complexes for this purpose are the transition metal hydrido complexes.

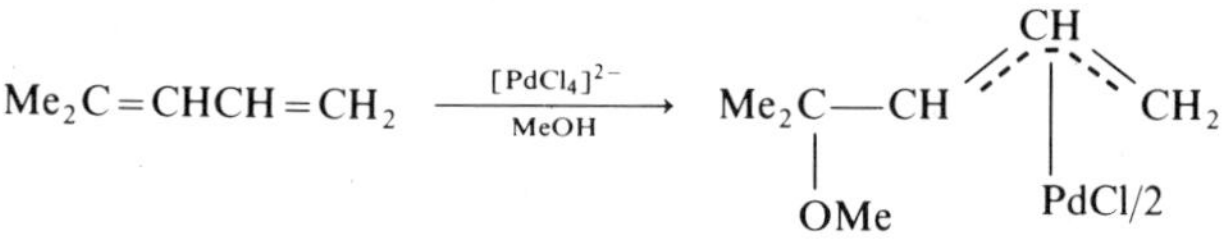

$$Me_2C{=}CHCH{=}CH_2 \xrightarrow[MeOH]{[PdCl_4]^{2-}} Me_2C{-}CH\overset{CH}{\underset{PdCl/2}{\diagup\diagdown}}CH_2 \qquad (7.34)$$
$$\qquad\qquad\qquad\qquad\qquad\qquad |$$
$$\qquad\qquad\qquad\qquad\qquad\;\; OMe$$

$$[CoH(CO)_4] + CH_2{=}CHCH{=}CH_2 \longrightarrow [Co(1\text{-}MeC_3H_4)(CO)_3] + CO \qquad (7.35)$$

$$[CoH(CN)_5]^{3-} + CH_2{=}CHCH{=}CH_2 \longrightarrow [Co(CN)_4(1\text{-}MeC_3H_5)]^{2-} \qquad (7.36)$$

$$[RhH(PPh_3)_4] + CH_2{=}CHCH{=}CHR \longrightarrow Rh(CH_3CHCHCHR)(PPh_3)_2 \qquad (7.37)$$

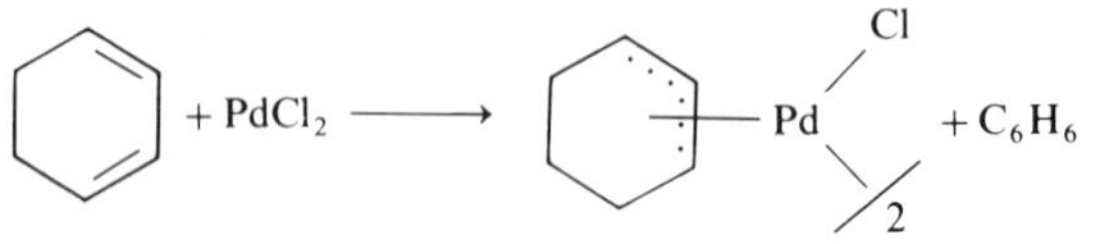

$$\text{(cyclohexadiene)} + PdCl_2 \longrightarrow \text{(cyclohexadienyl)}-Pd\begin{pmatrix}Cl\\ \end{pmatrix}_2 + C_6H_6 \qquad (7.38)$$

Coordinated dienes may undergo protonation reactions.

$$[Fe(C_4H_6)(CO)_3] + HCl \longrightarrow anti\text{-}[FeCl(1\text{-}MeC_3H_4)(CO)_3]$$
$$\qquad\quad |$$
$$\qquad\quad \underset{HClO_4}{\longrightarrow} syn\text{-}[Fe(1\text{-}MeC_3H_4)(CO)_3]^+ + ClO_4^- \qquad (7.39)$$

Diene oligomerization reactions may also lead to the formation of allyl compounds. During reduction of Ni(acac)$_2$ in the presence of trans, trans, trans-cyclododecatriene, olefin complexes of Ni(0) are formed which react with butadiene at 233 K to give the allyl complex:

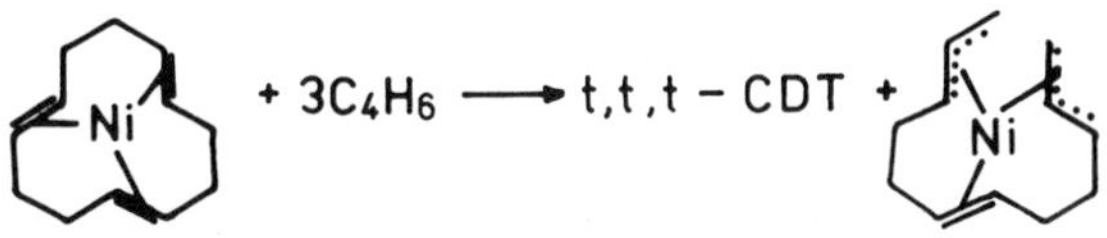

   $+\ 3C_4H_6\ \longrightarrow\ t,t,t - CDT\ +$     (7.40)

### f. Rearrangements of σ-allyl compounds to π-allyl compounds

The σ-allyl complexes may be prepared by reactions of anionic metal carbonyl complexes with allyl halides. These σ compounds subsequently rearrange to π-allyl complexes by thermal or photochemical reactions.

$$Na[Mn(CO)_5] + C_3H_5Cl \xrightarrow{\ -NaCl\ } [Mn(\sigma\text{-}C_3H_5)(CO)_5]$$

$$\xrightarrow[\text{or } 353\,\text{K}]{hv} [Mn(\pi C_3H_5)(CO)_4] + CO \qquad (7.41)$$

The σ–π-allyl rearrangement may easily be conducted for other complexes as well, e.g., $[Mo(\sigma\text{-}C_3H_5)\,cp(CO)_3]$ and $[Fe(\sigma\text{-}CH_2CH=CHMe)\,cp(CO)_2]$.

### g. Synthesis of metal complexes containing chelated allyl ligands[44]

Diolefins, monoolefins, and transition metal compounds give 1-σ-alkyl-4, 5, 6-allyl chelated complexes during irradiation with ultraviolet light [equation (7.42)].

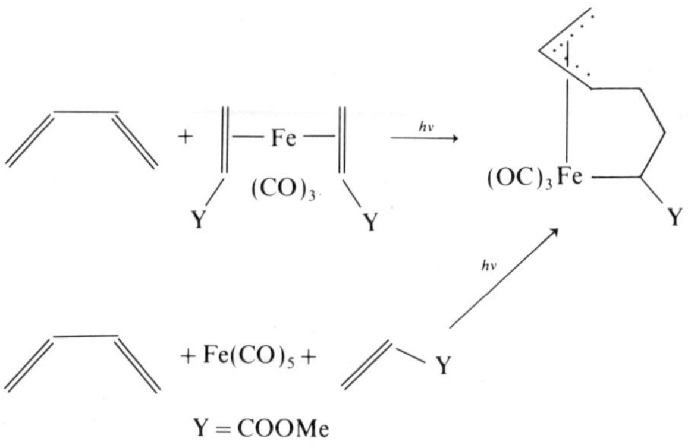

$$Y = COOMe$$

Similar compounds are formed as a result of ring opening of cyclopropane in bicyclic compounds during their reactions with metal carbonyls [equation (7.43)]. In

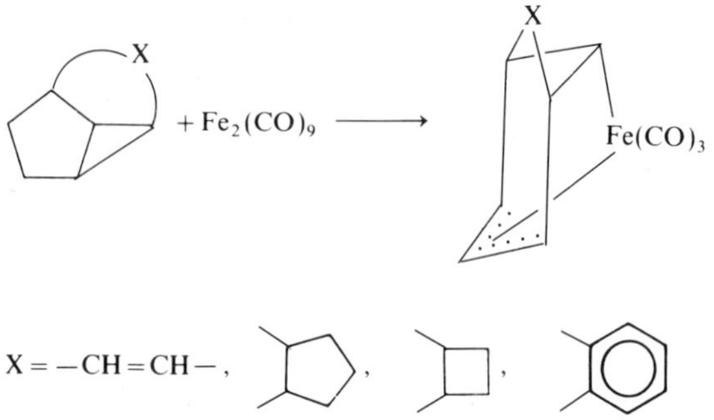

$$X = -CH=CH-, \quad \text{(image)} , \quad \text{(image)} , \quad \text{(image)} \tag{7.43}$$

the presence of Lewis acids, $\eta^4$-cyclodiolefin carbonyl complexes also form carbonyl allyl compounds [equation (7.44)].

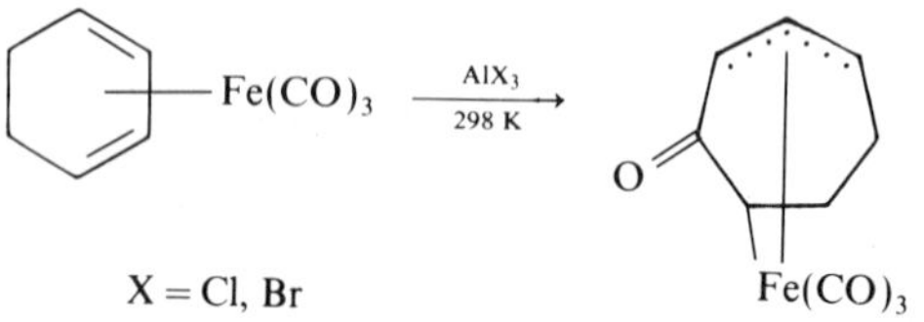

$$X = Cl, Br$$ 

$$\tag{7.44}$$

### h. *Miscellaneous methods*

Allenes may react with $PdCl_2$ to furnish $\pi$-allyl derivatives (see allene complexes) [equations (7.45) and (7.46)].

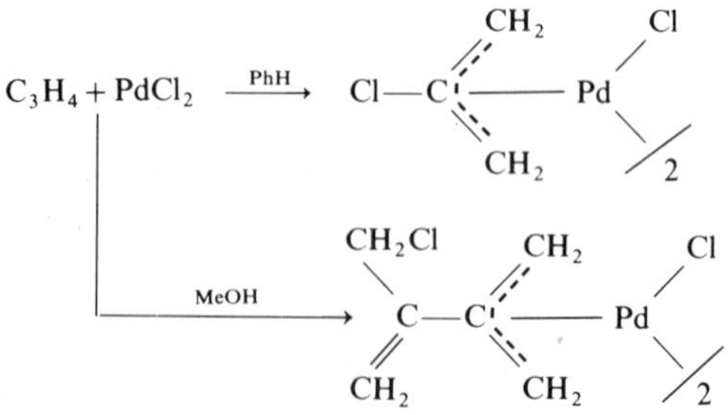

$$\tag{7.45}$$

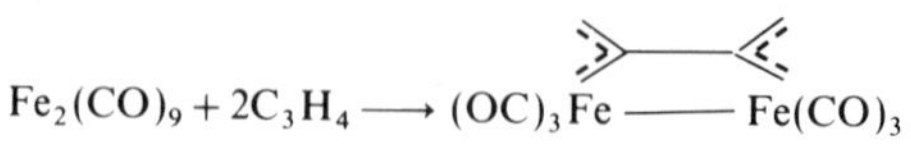

$$Fe_2(CO)_9 + 2C_3H_4 \longrightarrow (OC)_3Fe \longrightarrow Fe(CO)_3 \tag{7.46}$$

The complex $[RhCl_2(2\text{-}MeC_3H_4)]_n$ may be prepared by heating a saturated solution of rhodium(III) chloride in *tert*-butyl alcohol.[15]

$$n RhCl_3 + 2n(CH_3)_3COH \longrightarrow [RhCl_2(CH_2CMeCH_2)]_n + 2nH_2O + n(CH_3)_3CCl \tag{7.47}$$

Metal allyl halogen compounds may be synthesized by reactions of $M(allyl)_n$ with hydrogen halides, allyl halides, halogens, etc.; for instance,

$$4[Pt(C_3H_5)_2] + 4HCl \longrightarrow [Pt_4Cl_4(C_3H_5)_4] + 4C_3H_6 \qquad (7.48)$$

$$2[Cr(C_3H_5)_3] + I_2 \longrightarrow [Cr_2I_2(C_3H_5)_4] + C_6H_{10} \qquad (7.49)$$

$$2[Ni(C_3H_5)_2] + 2C_3H_5X \longrightarrow [Ni_2X_2(C_3H_5)_2] + 2C_6H_{10} \qquad (7.50)$$

The allyl group exchange also takes place between atoms of various metals:

$$[Rh(C_3H_5)_3] + [PdCl_4]^{--} \longrightarrow [Pd_2Cl_2(C_3H_5)_2] \qquad (7.51)$$

A common allylating reagent is tetraallyltin:

$$RuCl_3 + Sn(C_3H_5)_4 + PR_3 \longrightarrow [RuCl(C_3H_5)(PR_3)_3] \qquad (7.52)$$

The stable complex $[IrClH(1\text{-}PhC_3H_4)(PPh_3)_2]$ is formed as a result of ring opening and hydrogen atom abstraction from phenylcyclopropane,[16a]

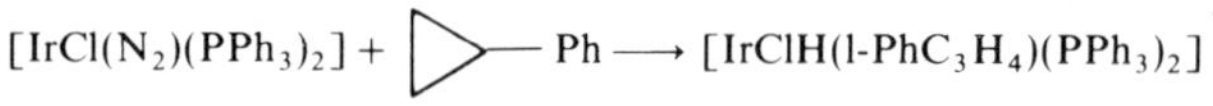

$$[IrCl(N_2)(PPh_3)_2] + \triangleright\!-Ph \longrightarrow [IrClH(1\text{-}PhC_3H_4)(PPh_3)_2] \qquad (7.53)$$

Also, in the presence of Pd(II) complexes, three-membered rings of the following compounds open easily: methylenecyclopropanes, spiro-2, 2′-pentanes, and cyclo-propenes. The position of the ring opening depends upon substituents:

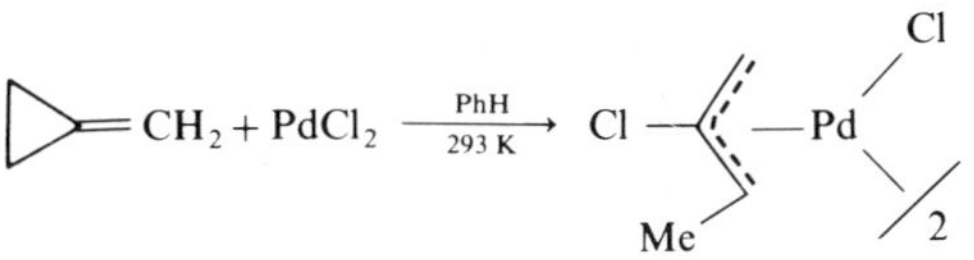

$$(7.54)$$

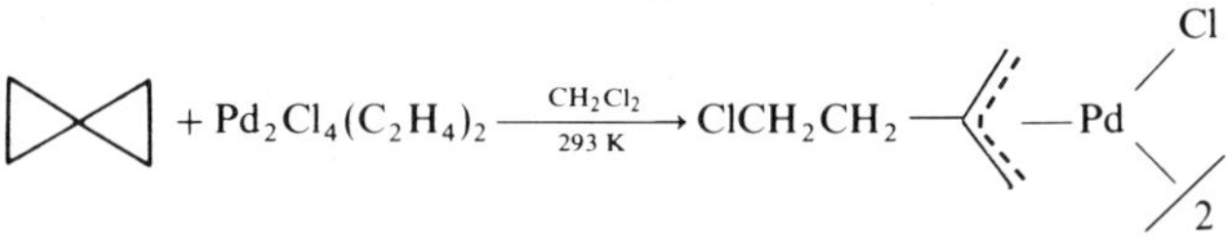

$$(7.55)$$

The reaction of $[Pd(C_3H_5)cp]$ with $PdL_2$ $[L = PR_3, P(OR)_3]$ leads to the formation of dinuclear sandwich complexes[18, 35] [equations (7.56) and (7.57)].

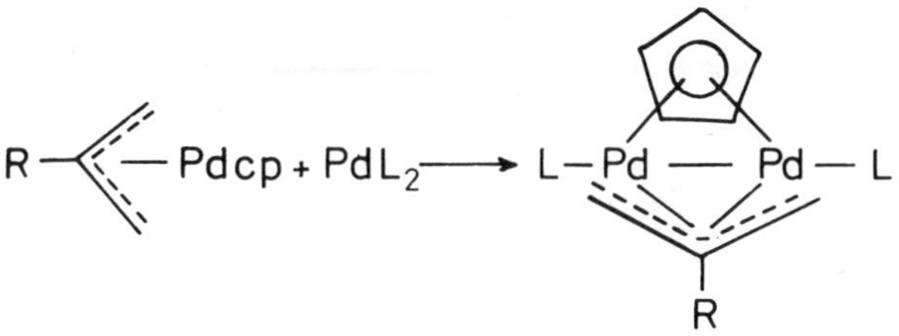

$$(7.56)$$

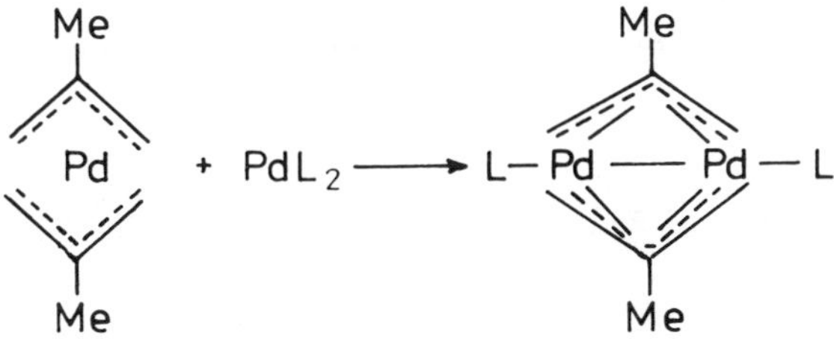

$$(7.57)$$

## 8. *PROPERTIES OF π-ALLYL METAL COMPOUNDS*

π-Allyl complexes differ considerably with respect to thermostability, resistance to air, water, and other protic solvents. Usually, homoleptic allyl complexes, $M(C_3H_5)_n$, are less stable than compounds which also contain other ligands, for example, halogens. As for many other complexes, e.g., olefin complexes, the stability of allyl compounds increases when the atomic number becomes greater, and therefore also with the increase in the number of *d* electrons in a given series of the transition metals as well as with the increase of the atomic number of the central ion in a given group. For example, rhodium allyl complexes are considerably more stable than cobalt allyl compounds.

### a. *Complexes of group 4 metals*

These metals form the following π-allyl complexes: $[Ticp_2(allyl)]$, $[Ti(\eta^3\text{-}1\text{-}MeC_3H_4)(\eta^8\text{-}C_8H_8)]$, $[Ti(C_4H_7)_4]$, $[Ti(allyl)_xCl_{4-x}]$ ($x = 1, 2, 3$), $[Ti(C_3H_5)Cl_3(bipy)]$, $[Zr(C_3H_5)_4]$, $[Hf(C_3H_5)_4]$, $[Zr(\eta^3\text{-}C_3H_5)(\eta^1\text{-}C_3H_5)cp_2]$, $[ZrCl(C_3H_5)cp_2]$, and $[Zr_2(C_3H_5)_3cp_4Cl]$.[51,52]

Such compounds are usually obtained by reaction of titanium group metal halides with Mg(allyl)X:

$$MCl_4 + 4Mg(allyl)X \longrightarrow M(allyl)_4 + 4MgClX \qquad (7.58)$$

Titanium group allyl compounds are sensitive to air, water, and alcohols. Tetrallyl derivatives of zirconium and hafnium are dynamic compounds. The complex $Hf(allyl)_4$ gives the $AX_4$-type NMR spectrum even at very low temperatures (199 K). The Ti(III) complexes are paramagnetic; magnetic moments correspond to one unpaired electron and, for various allyl groups, vary in the 1.47–1.75 BM range. The properties of allyl

*Table 7.7. Allyl Complexes of Titanium Group Metals*

| Compound | Color | Mp (K) | Electronic spectra and magnetic properties |
|---|---|---|---|
| $[Ti(C_3H_5)cp_2]^{a,b,c}$ | Purple | 384–385 dec | $\mu_{eff} = 1.65$ BM<br>UV 19,500 cm$^{-1}$ |
| $[Ti(1\text{-}MeC_3H_4)cp_2]^{a,b,c}$ | Violet | 370 dec | $\mu_{eff} = 1.67$ BM<br>UV 18,600 cm$^{-1}$ |
| $[Ti(2\text{-}MeC_3H_4)cp_2]^{a,b,c}$ | Purple | 387 dec | $\mu_{eff} = 1.72$ BM<br>UV 20,100, 16,800 cm$^{-1}$ |
| $[Zr(C_3H_5)_4]^{d(19)}$ | Red | 273 dec | |
| $[Zr(\eta^3\text{-}C_3H_5)(\eta^1\text{-}C_3H_5)cp_2]^e$ | Yellow | 352 dec | |
| $[Zr(\eta^3\text{-}2\text{-}MeC_3H_4)][(\eta^1\text{-}2\text{-}MeC_3H_4)cp_2]^e$ | Yellow | 258 | |
| $[Hf(C_3H_5)_4]^d$ | Orange-red | | |
| $[Hf(\eta^3\text{-}C_3H_5)_2(\eta^8\text{-}C_8H_8)]^f$ | Yellow | | |

[a] H. A. Martin and F. Jellinek, *J. Organomet. Chem.*, **8**, 115 (1967).
[b] H. A. Martin and F. Jellinek, *J. Organomet. Chem.*, **12**, 149 (1968).
[c] R. B. Helmholdt, F. Jellinek, H. A. Martin, and A. Vos, *Recl. Trav. Chim. Pays-Bas*, **86**, 1263 (1967).
[d] J. K. Becconsall, B. E. Job, and S. O'Brien, *J. Chem. Soc. (A)*, 423 (1967).
[e] H. A. Martin, P. J. Lemaire, and F. Jellinek, *J. Organomet. Chem.*, **14**, 149 (1968).
[f] H. J. Kablitz, R. Kallweit, and G. Wilke, *J. Organomet. Chem.*, **44**, C49 (1972).

complexes of titanium group metals are given in Table 7.7. The stability of allyl complexes of group 4 elements decreases as follows: Hf > Zr > Ti.

### b. Complexes of group 5 metals

These elements form homoleptic allyl complexes as well as alkenyl compounds with mixed coordination spheres. These compounds are also sensitive to air and proton containing solvents (Table 7.8). They decompose at low temperatures, for example, the compound $[Ta(C_3H_5)_4]$ is stable only in saturated hydrocarbon solutions below 273 K. The complexes $[Nb(allyl)_4]$ and $[Ta(allyl)_4]$ are paramagnetic. The complex $[V(C_3H_5)_3]$ has pyrophoric properties. Halogen and cyclopentadienyl allyl complexes are more stable. In the compound $[V(C_3H_5)cp_2]$ the allyl forms a $\sigma$ bond to the central atom; thus, it acts as an $\eta^1$ ligand.

The complex $[V(\eta^3\text{-}C_3H_5)Cl_3]$ is formed in the following reaction:

$$3VCl_4 + B(C_3H_5)_3 \xrightarrow{Et_2O} 3[VCl_3(\eta^3\text{-}C_3H_5)] \cdot Et_2O + BCl_3 \qquad (7.59)$$

Tetraallylniobium reacts with $VOCl_3$. The reaction mixture resulting from these compounds catalyzes polymerization of olefins.

The following complexes are typical for the group 5 metals: $[V(allyl)_3]$, $[V(allyl)Cl_3] \cdot OEt_2$, $[V(allyl)(CO)_5]$, $[V(allyl)(CO)_{5-x}L_x]$, $[Nb(allyl)_3)]$, $[M(allyl)_4]$, $[Mcp_2(allyl)]$ (M = Nb, Ta), and $Ta(allyl)(PF_3)_5$.

### c. Complexes of group 6 metals

Many allyl complexes of chromium group elements are known, particularly those possessing mixed cyclopentadienyl and carbonyl ligands. Chromium forms compounds of the composition $[Cr(allyl)_3]$, while molybdenum and tungsten, like metals of the second and third transition metal series of the preceding groups, coordinate a larger

*Table 7.8. Allyl Complexes of Vanadium Group Metals*

| Compound | Color | Mp (K) |
|---|---|---|
| $[V(C_3H_5)_3]$[19] | Brown | 243 dec |
| $[VCl_3(C_3H_5)] \cdot Et_2O$[a] | Red | |
| $[V(C_3H_5)(CO)_3(diphos)]*$[b] | Orange-red | 365–371 dec |
| $[V(C_3H_5)(CO)_5]$[b] | Dark-red | 369 dec |
| $[V(\eta^3\text{-}C_3Ph_3)(CO)_5]$[b] | Dark-red | 503 dec |
| $[Nb(C_3H_5)_4]$[c] | Black-green | 273 dec |
| $[Nb(C_3H_5)cp_2]$[d] | Black-green | |
| $[Nbcp_2(\eta^3\text{-}C_8H_9)]$[e] | | |
| $[Ta(C_3H_5)_4]$[19] | Green | 273 dec |

* $\tau_s = 4.96$; $\tau_{syn} = 7.15$; $\tau_{anti} = 8.07$.
[a] K. H. Thiele and S. Wagner, *J. Organomet. Chem.*, **20**, 25 (1969).
[b] U. Franke and E. Weiss, *J. Organomet. Chem.*, **153**, 39 (1978); U. Franke and E. Weiss, *J. Organomet. Chem.*, **121**, 345 (1976); U. Franke and E. Weiss, *J. Organomet. Chem.*, **121**, 355 (1976).
[c] J. K. Becconsal, B. E. Job, and S. O'Brien, *J. Chem. Soc. (A)*, 423 (1967).
[d] F. W. Siegert and H. J. de Liefde Meijer, *J. Organomet. Chem.*, **23**, 177 (1970).
[e] C. P. Verkade, A. Westerhof, and H. J. de Liefde Meijer, *J. Organomet. Chem.*, **154**, 317 (1978).

number of allyl groups to afford tetraallyl derivatives of the type $[M(allyl)_4]$. For chromium and molybdenum, dinuclear complexes such as $[M_2(allyl)_4]$ are known in which there are two bridging allyl groups. These compounds possess quadruple metal–metal bonds; the Cr–Cr and Mo–Mo distances are 197 and 218.3 pm, respectively. Therefore, these distances are considerably shorter than interatomic distances in metallic chromium and molybdenum (Table 7.9, Figure 7.7). Heating $[Cr(C_3H_5)_3]$ in dioxane leads to the formation of $[Cr_2(C_3H_5)_4]$.

The decomposition of mononuclear chromium complexes with HX leads to the formation of halogen–allyl compounds:

$$[Cr(C_3H_5)_3] + HCl \longrightarrow [Cr_2Cl_2(C_3H_5)_4] \tag{7.60}$$

$$[W(C_3H_5)_4] + HCl \longrightarrow [WCl(C_3H_5)_3] \tag{7.61}$$

$$2Mo(C_3H_5)_4 \xrightarrow{\text{HCl}} [Mo(C_3H_5)_3Cl]_2 \xrightarrow{\text{HgOAc}} 2Mo(C_3H_5)_3(OAc)^{(53)} \tag{7.62}$$

Table 7.9. Allyl Complexes of Chromium Group Metals

| Compound | Color | Mp (K) |
|---|---|---|
| $[Cr(C_3H_5)_3]^{(19)}$ | Red | 343–353 dec |
| $[Cr_2(C_3H_5)_4]^{(19)}$ | Red-brown | 394–402 dec |
| $[Cr_2Cl_2(C_3H_5)_4]^{(19)}$ | Red-brown | |
| $[Cr(1\text{-}MeC_3H_4)cp(CO)_2]^a$ | Yellow-green | 391–394 dec |
| $[Mo(C_3H_5)_4]^{(19)}$ | Green | 398–403 dec |
| $[Mo_2(C_3H_5)_4]^{(19)}$ | Black-green | 383–388 dec |
| $[Et_4N][Mo_2Cl_3(C_3H_5)_2(CO)_4]^b$ | Yellow | 424–425 dec |
| $[MoCl(2\text{-}MeC_3H_4)(CO)_2bipy]^c$ | Dark-orange | 507–511 dec |
| $[Mo(C_3H_5)cp(CO)_2]^d$ | Yellow | 407 dec |
| $[MoCl(C_3H_5)(CO)_2\{P(OMe)_3\}_2]^{*\,e}$ | | |
| $[Mocp_2(1\text{-}MeC_3H_4)]^+[PF_6]^-$ | Red | |
| $[Mo_2Cl_2(C_3H_5)_2(C_6H_5Me)_2]^g$ | Purple | 433 dec |
| $[W(C_3H_5)_4]^{(19)}$ | Light-brown | 368 dec |
| $[WBr(C_3H_5)(CO)_4]^h$ | Orange-red | |
| $[Et_4N][W_2Cl_3(C_3H_5)_2(CO)_4]^b$ | Yellow | 410–411 |
| $[W(C_3H_5)cp(CO)_2]^i$ | Yellow | |
| $[W(2\text{-}MeC_3H_4)cp_2]^+[PF_6]^{-\,f}$ | | |

* $\delta(^{13}C)$ 76.02 (C1,3), 114.3 ppm (C2).
$^a$ E. O. Fischer, H. P. Kogler, and P. Kuzel, Chem. Ber., 93, 3006 (1960).
$^b$ H. D. Murdoch, J. Organomet. Chem., 4, 119 (1965); H. T. Dieck and H. Friedel, J. Organomet. Chem., 14, 375 (1968).
$^c$ H. Murdoch and R. Henzi, J. Organomet. Chem., 5, 552 (1966); C. G. Hull and M. H. B. Stiddard, J. Organomet. Chem., 9, 519 (1967).
$^d$ M. Cousins and M. L. Green, J. Chem. Soc., 899 (1963); R. B. King and M. Ishaq, Inorg. Chim. Acta, 4, 258 (1970).
$^e$ M. G. B. Drew, B. J. Brisdon, D. A. Edwards, and E. Paddock, Inorg. Chim. Acta, 35, L381 (1979).
$^f$ M. Ephritikhine, M. L. H. Green, and R. E. MacKenzie, J. Chem. Soc., Chem. Commun., 619 (1976).
$^g$ M. L. H. Green, L. C. Mitchard, and W. E. Silverthorn, J. Chem. Soc. (A), 2929 (1971); M. L. H. Green, L. C. Mitchard, and W. E. Silverthorn, J. Chem. Soc., Dalton Trans., 1403, (1973); M. L. H. Green, J. Knight, L. C. Mitchard, G. G. Roberts, and W. E. Silverthorn, Chem. Commun., 1619 (1971); M. L. H. Green and W. E. Silverthorn, J. Chem. Soc., Dalton Trans., 301 (1973).
$^h$ E. Holloway, J. D. Kelhy, and M. H. Stiddard, J. Chem. Soc. (A), 931 (1969).
$^i$ M. L. H. Green and A. N. Stear, J. Organomet. Chem., 1, 230 (1963).

The reaction of $[Mo(C_3H_5)_3Cl]_2$ with sodium cyclopentadienide affords the paramagnetic (1.44 BM) orange complex $[Mo(C_3H_5)_2cp]$.[53] This compound has a pseudotrigonal structure; the ligand centers occupy the vertices of an isosceles triangle.

Many arene–allyl molybdenum and tungsten complexes are also known:

$$[Mocp_2H_3]^+[PF_6]^- + C_4H_6 \longrightarrow [Mocp_2(\eta^3\text{-MeC}_3H_4)]^+[PF_6]^- \qquad (7.63)$$

$$[Mo(C_6H_5Me)_2] + C_3H_5Cl \longrightarrow [Mo_2Cl_2(C_3H_5)_2(C_6H_5Me)_2] \qquad (7.64)$$

Molybdenum and tungsten form a large number of allyl compounds, such as $[MX(C_3H_5)(CO)_2L_2]$, $[M(allyl) cp(CO)_2]$, $[Mo(allyl) cp(NO)(CO)]^+$, and $[W(allyl) cp(I)(NO)]$. The complex $[MoCl(C_3H_5)(CO)_2\{P(OMe)_3\}_2]$ also belongs to this group; it possesses a pentagonal bipyramidal structure. The phosphite molecules, the carbonyl and allyl groups are coordinated in the pentagonal plane, while the axial position is occupied by CO and Cl.

## d. *Complexes of group 7 metals*

A relatively small number of allyl compounds of manganese group metals exists (Table 7.10). Manganese forms complexes of the formula $Mn(allyl)(CO)_4$ (allyl $= C_3H_5$, $1\text{-MeC}_3H_4$, $2\text{-MeC}_3H_4$, $1,1\text{-Me}_2C_3H_3$, etc.). These complexes are obtained by the reaction of allyl halides with $Na[Mn(CO)_5]$. First, the $\sigma$-allyl compound is formed which, after heating or UV irradiation, gives off a CO group to afford the $\pi$-allyl compound. This method cannot be utilized for the preparation of analogous rhenium compounds. Ferra-, mangana-, or rhena-diketonates of boron (which may be regarded as carbene compounds of these transition metals) react with KH to give allyl complexes [54] [equation (7.65)].

$$ML_n = cp(OC)Fe,\ cis = (OC)_4Mn,\ and\ cis\text{-}(OC)_4Re \qquad (7.65)$$

*Table 7.10. Allyl Complexes of Manganese Group Metals*

| Compound | Color | Mp (K) | Remarks |
|---|---|---|---|
| $[Mn(C_3H_5)(CO)_4]^{a,b}$ | Yellow | 328–329 | |
| $[Mn(2\text{-MeC}_3H_4)(CO)_4]^a$ | Yellow | | Liquid, boiling point 323 K (2.5 mm Hg) |
| $[Mn(Me_2CCHCH_2)(CO)_4]^a$ | Yellow | 285–287 | |
| $[Re(C_3H_5)(CO)_4]^{b,c}$ | | | |
| $[Re_2(C_3H_5)_4]^{d,e}$ | Orange | | $\tau_5 = 4.23$; $\tau(syn)$ 5.88; $\tau(anti)$ 10.53 $J(5, syn)$ 5 Hz; $J(5, anti)$ 9 Hz; $J(syn, anti) \approx 5$ Hz |

[a] W. R. McClellan, H. H. Hoehn, H. N. Cripps, E. L. Muetterties, and B. W. Howk, *J. Am. Chem. Soc.*, **83**, 1601 (1961); G. Davidson and D. C. Andrews, *J. Chem. Soc., Dalton Trans.*, 126 (1972).
[b] E. W. Abel and S. Moorhouse, *Angew. Chem., Int. Ed. Engl.*, **10**, 339 (1971).
[c] B. J. Brisdon, D. A. Edwards, and J. W. White, *J. Organomet. Chem.*, **175**, 113 (1979).
[d] A. F. Masters, K. Mertis, J. F. Gibson, and G. Wilkinson, *Nouv. J. Chim.*, **1**, 389 (1977).
[e] F. A. Cotton and M. W. Extine, *J. Am. Chem. Soc.*, **100**, 3788 (1978).

Allyltetracarbonylrhenium(I) was prepared by reaction (7.66).

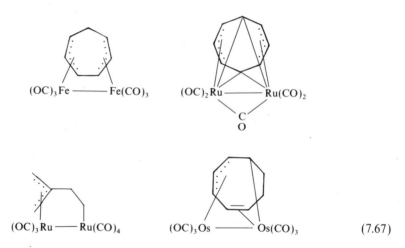

$$Sn(C_3H_5)Me_3 + [ReCl(CO)_5] \xrightarrow{-CO} (OC)_4Re \xrightarrow{-SnClMe_3} [Re(\eta^3\text{-}C_3H_5)(CO)_4] \qquad (7.66)$$

Recently, the stable compound $Re_2(C_3H_5)_4$ was prepared. It is a dimer in which in which there is a rhenium–rhenium triple bond. None of the allyl groups is bridging in contrast to molybdenum and chromium complexes possessing analogous composition (Table 7.5, Figure 7.7). The complex $Re_2(C_3H_5)_4$ may be prepared by the reaction of $ReCl_5$ with $MgCl(C_3H_5)$. The $^1H$ NMR spectrum of $Re_2(C_3H_5)_4$ is of the $AM_2X_2$ type, and it does not change in the 188–358 K range.

### e. Complexes of group 8 metals

Iron forms the homoallyl complex $Fe(C_3H_5)_3$ which is very unstable, decomposing at 233 K. Like the complex $V(C_3H_5)_3$, the compound $Fe(C_3H_5)_3$ is also pyrophoric. A relatively large number of stable complexes of the composition $[MX(allyl)(CO)_3]$ (M = Fe and Ru) are known (Table 7.11). In the case of iron, 18e anionic allyl complexes as well as 16e cationic allyl compounds are known, for instance, $[Fe(allyl)(CO)_3]^-$ and $[Fe(allyl)(CO)_3]^+$. The anionic derivative $[Fe(C_3H_5)(CO)_3]^-$ is formed during reduction of $[FeBr(C_3H_5)(CO)_3]$ with zinc. The anionic compound reacts with the excess of $[FeBr(C_3H_5)(CO)_3]$ to give the pyrophoric red dimer $[(OC)_3(C_3H_5)Fe-Fe(C_3H_5)(CO)_3]$. All group 8 metals form di- and polynuclear allyl omplexes, particularly with cyclic allyl ligands[55–58]; an example is shown in scheme (7.67). Also 16e, 17e, an 18e iron compounds containing phosphorus ligands

were obtained: $[Fe(C_3H_5)(PF_3)_3]$, $[Fe(\eta^3\text{-}C_8H_{13})\{P(OMe)_3\}_3]$, $[Fe(\eta^3\text{-}C_8H_{13})\{P(OMe)_3\}_3]^+ BF_4^-$, $[FeBr(C_3H_5)(PF_3)_3]$, $[Fe(C_3H_5)I(CO)_2(PPh_3)]$. Tetracarbonyl and carbonyl–nitrosyl iron complexes also exist, e.g., $[Fe(allyl)(CO)_4]^+ BF_4^-$, $[Fe(allyl)(CO)_2NO]$, and $[Fe(allyl)_2(CO)_2]$.

*Table 7.11. Allyl Complexes of Fe, Ru, and Os*

| Compound | Color | Mp (K) |
|---|---|---|
| $[Fe(C_3H_5)_3]$[(19)] | Gold-red | 233 dec |
| $[FeCl(C_3H_5)(CO)_3]^a$ | Yellow | 361–362 dec |
| $[Fe(C_3H_5)cp(CO)]^b$ | Yellow | 338 dec |
| $[Fe(1\text{-}MeC_3H_4)cp(CO)]^b$ | Yellow | Liquid |
| $[Fe(acac)_2(C_3H_5)]^c$ | Red | 435 dec |
| $[Fe(1\text{-}MeC_3H_4)(CO)_3]^+ BF_4^{-\,d}$ | Red | |

| Compound | Color | Mp (K) |
|---|---|---|
| (structure with $Fe(CO)_3$ and $Fe(CO)_4$) $^e$ | Orange | Liquid |
| $[RuCl(C_3H_5)(CO)_3]^f$ | Light-yellow | 348–350 |
| $[Ru(C_3H_5)_2(CO)_2]^g$ | Light-yellow | <298 |
| $[Ru(C_3H_5)(NO)(PPh_3)_2]^h$ | Red | |

| Compound | Color | Mp (K) |
|---|---|---|
| (structure with $(OC)_3Ru$ — $Ru(CO)_4$ and $CH_2$) $^i$ | Light-green | 363 dec |
| $[RuCl_2(C_{12}H_{18})]^j$ | Yellow-brown | 483–493 dec |
| $[RuCl_2(C_{10}H_{16})]$ (prepared from isoprene)$^k$ | Red-brown | |
| $[Ru_2Cl_2(C_3H_5)_2(COD)_2]^l$ | Gold-yellow | 408–413 dec |
| $[Ru(C_3H_5)_2(COD)]^l$ | White | 403–413 dec |
| $[Ru(C_3H_5)cp(CO)]^m$ | Light-yellow | 317–318 |
| $[Os(C_3H_5)_2(COD)]^n$ | | |
| $[Os(C_3H_5)(COD)(MeCN)_2]^+[BF_4]^{-\,n}$ | White | |
| $[Os(C_3H_5)(COD)(bipy)]^+[BF_4]^{-\,n}$ | Gold | |
| $[Os(C_8H_8)(CO)_3]^o$ | Light-yellow | 363.5 |

$^a$ A. N. Nesmeyanov and I. I. Kritskaya, *J. Organomet. Chem.*, **14**, 387 (1968); R. E. Heck and C. R. Boss, *J. Am. Chem. Soc.*, **86**, 2580 (1964); A. N. Nesmeyanov, S. P. Gubin, and A. Z. Rubezhov, *J. Organomet. Chem.*, **16**, 163 (1969).

$^b$ M. L. H. Green and P. L. I. Nagy, *J. Chem. Soc.*, 189 (1963).

$^c$ G. A. Razuvaev, G. A. Domrachev, O. N. Suvorova, and L. G. Abakumova, *J. Organomet. Chem.*, **32**, 113 (1971); G. A. Domrachev, J. A. Sorokin, G. A. Razuvaev, and O. N. Suvorova, *Dokl. Akad. Nauk SSSR*, **183**, 1085 (1968).

$^d$ G. F. Emerson and R. Pettit, *J. Am. Chem. Soc.*, **84**, 4591 (1962).

$^e$ R. Ben-Shoshan and R. Pettit, *Chem. Commun.*, 247 (1968).

$^f$ G. Sbrana, G. Braca, F. Placenti, and P. Pino, *J. Organomet. Chem.*, **13**, 240 (1968).

$^g$ M. Cooke, R. J. Goodfellow, M. Green, and G. Parker, *J. Chem. Soc. (A)*, 16 (1971); E. W. Abel and S. Moorhouse, *Angew. Chem., Int. Ed. Engl.*, **10**, 339 (1971).

$^h$ M. W. Schoonover, C. P. Kubiak, and R. Eisenberg, *Inorg. Chem.*, **17**, 3050 (1978).

$^i$ J. S. Ward and R. Pettit, *Chem. Commun.*, 1419 (1970).

$^j$ J. E. Lydon, J. K. Nicholson, B. L. Shaw, and M. R. Truter, *Proc. Chem. Soc.*, 421 (1964); J. K. Nicholson and B. L. Shaw, *J. Chem. Soc. (A)*, 807 (1966).

$^k$ A. Colombo and G. Allegra, *Acta Crystallogr.*, **B27**, 1653 (1971); L. Porri, M. C. Galazzi, A. Colombo, and G. Allegra, *Tetrahedron Lett.*, 4187 (1965).

$^l$ J. Powell and B. L. Shaw, *J. Chem. Soc. (A)*, 159 (1968).

$^m$ R. B. King and M. Ishaq, *Inorg. Chim. Acta*, **4**, 258 (1970).

$^n$ R. R. Schrock, B. F. G. Johnson, and J. Lewis, *J. Chem. Soc., Dalton Trans.*, 951 (1974).

$^o$ M. I. Bruce, M. Cooke, M. Green, and G. J. Westlake, *J. Chem. Soc. (A)*, 987 (1969).

In the complex $[Ru(\eta^3\text{-}C_8H_{13})\{PPh_2(OMe)\}_3]^+PF_6^-$,[59] as in the compounds $[Fe(\eta^3\text{-}C_8H_{13})\{P(OMe)_3\}_3]^+$[60] and $[Mn(\eta^3\text{-}C_7H_{11})(CO)_3]$,[61] there is a $Ru-H-C$ agostic bond (7.68).

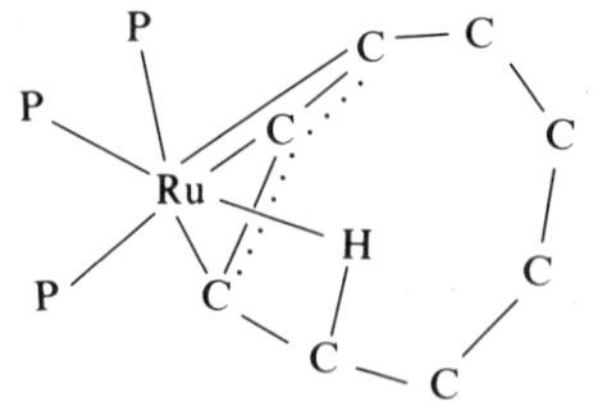

$$(7.68)$$

Ruthenium and osmium form clusters containing a bridging ligand of the allyl type bonded to three metal atoms[64]; see scheme (7.69).

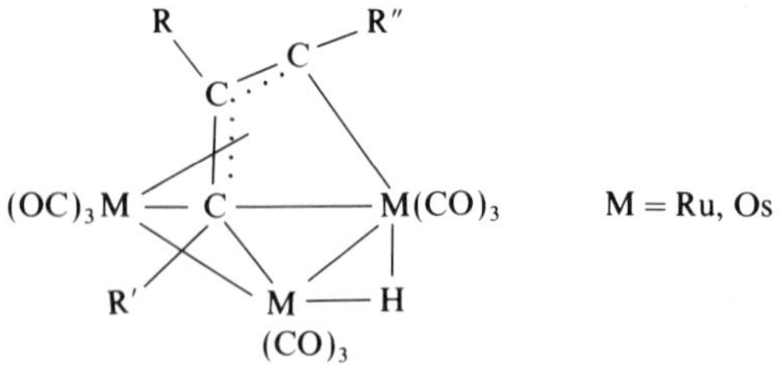

$$M = Ru, Os \qquad (7.69)$$

Iron and ruthenium also form compounds of the type $[M(\text{allyl})\,cp(CO)]$ and $[Fe(\text{allyl})\,cp(PPh_3)]$. Ruthenium furnishes the typical derivative $[Ru(\text{allyl})_2(COD)]$, $[Ru(\text{allyl})_2(PR_3)_2]$, and $[Ru_2Cl_2(\text{allyl})_2(COD)_2]$. The reaction of ruthenium trichloride with butadiene gives $[RuCl_2(C_{12}H_{18})]$, while the reaction of $RuCl_3$ with isoprene affords the dinuclear complex with the structure (7.70).

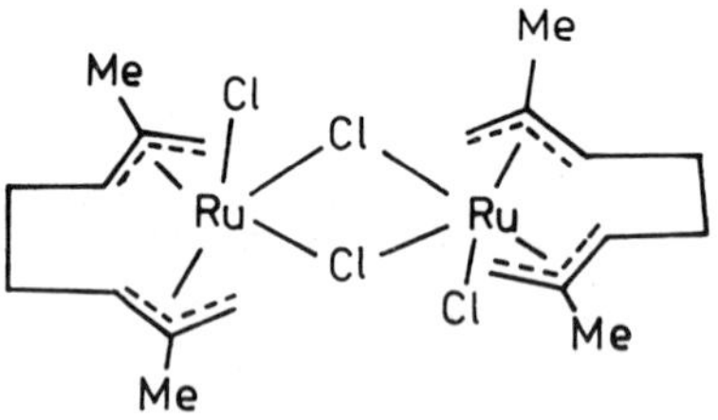

$$(7.70)$$

The reactions of allenes and dienes with iron and ruthenium compounds form a large number of various allyl complexes (see allene complexes). Osmium forms allyl compounds of the formulas $[OsCl(C_3H_5)(PPh_3)_3]$ and $[Os(C_8H_8)(CO)_3]$.

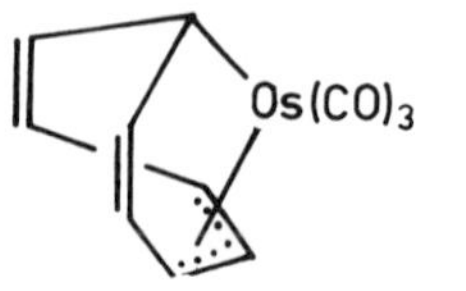

$$(7.71)$$

Known olefin-allyl derivatives include $[Os(C_3H_5)_2(COD)]$ and $[Os(C_3H_5)(COD)L_2]^+$, as well as complexes such as $[Os\{CH_2C(R)CH_2\}Cl(\eta\text{-}C_6H_6)]$ (R = H, Me).

*Table 7.12. Allyl Complexes of Co, Rh, and Ir*

| Compound | Color | Mp (K) |
|---|---|---|
| $[Co(C_3H_5)_3]^{(10,19)}$ | Gold-red | 233 dec |
| $[CoCl(C_3H_5)_2]^{(19)}$ | Dark-red | |
| $[Co(C_3H_5)(CO)_3]^a$ | Orange-red | Liquid |
| $[Co(1\text{-}MeC_3H_4)(CO)_2PPh_3]^b$ | Orange | 366–369 |
| $[Co(anti\text{-}1\text{-}MeC_3H_4)(BD)PPh_3]^c$ | Orange-red | 383 dec |
| $[CoBr(C_3H_5)cp]^d$ | Dark-red | 369–371 dec |
| $[Co(C_3H_5)\{P(OPr^i)_3\}_3]^f$ | Yellow | |
| $[Co(C_3H_5)(PF_3)_3]^e$ | Orange | Liquid |
| $[Co(C_3H_5)(PMe_3)_3]^f$ | Red-orange | 383 decoloration >413 melting |
| $[Rh(C_3H_5)_3]^g$ | Yellow | 353–358 |
| $[Rh_2Cl_2(C_3H_5)_4]^g$ | Yellow | 453–458 dec |
| $[Rh_2Br_2(C_3H_5)_4]^{g(20)}$ | Yellow | 463–473 dec |
| $[RhCl_2(2\text{-}Me-C_3H_4)]_n^{g(15)}$ | Red | 493–503 dec |
| $[Rh_2(\mu\text{-}Cl)_2Cl_2(1\text{-}MeC_3H_4)_2BD]^h$ | Dark-orange | 423 dec |
| $[Rh(C_3H_5)(PF_3)_3]^i$ | Light-yellow | Liquid |
| $[Rh(C_3H_5)\{P(OMe)_3\}_3]^j$ | Yellow | |
| $[Rh(C_3H_5)(CO)_2]^k$ | Yellow | 328–333 dec |
| $[RhCl(C_3H_5)cp]^g$ | Orange-red | 448–458 dec |
| $[Rh(C_3H_5)(COD)]^l$ | Light-yellow | 381–385 |
| $[Ir(C_3H_5)_3]^m$ | White | 338 |
| $[IrCl_2(C_3H_5)(PPh_3)_2]^n$ | Yellow | 348–356 dec |
| $[Ir_2Cl_4(1\text{-}MeC_3H_4)_2BD]^o$ | Yellow-orange | 478 dec |
| $[IrCl_2(C_3H_5)(CO)(C_8H_{14})]^p$ | White | 387 dec |
| $[Ir(C_3H_5)(CO)(PPh_3)_2]^q$ | Light-yellow | 418 dec |
| $[IrCl(1\text{-}MeC_3H_4)(C_5Me_5)]^r$ | | 399 dec |
| $[Ir(C_3H_5)(PPh_3)_3]^{(1)}$ | Yellow | |
| $[Ir(2\text{-}MeC_3H_4)Cl_2(PEt_3)_2]^n$ | Yellow | 348–356 |
| $[IrCl(C_3H_5)(CO)(PMe_2Ph)_2]^+[PF_6]^{-s}$ | White | 453–454 |
| $[Ir(C_3H_5)_2(PPh_3)]^+[BF_4]^{-t}$ | White | 468–469 |

$^a$ R. F. Heck and D. S. Breslow, *J. Am. Chem. Soc.*, **82**, 750 (1960); D. C. Andrews and G. Davidson, *J. Chem. Soc., Dalton Trans.*, 1381 (1972).
$^b$ W. W. Spooncer, A. C. Jones, and L. H. Slaugh, *J. Organomet. Chem.*, **18**, 327 (1969); P. V. Rinze and H. Nöth, *J. Organomet. Chem.*, **30**, 115 (1971).
$^c$ G. Vitulli, L. Porri, and A. L. Segre, *J. Chem. Soc. (A)*, 3246 (1971).
$^d$ R. F. Heck, *J. Org. Chem.*, **28**, 604 (1963).
$^e$ M. A. Cairns and J. F. Nixon, *J. Chem. Soc., Dalton Trans.*, 2001 (1974).
$^f$ M. C. Rakowski, F. J. Hirsekora, L. S. Stuhl, and E. L. Muetterties, *Inorg. Chem.*, **15**, 2379 (1976).
$^g$ J. Powell and B. L. Shaw, *J. Chem. Soc. (A)*, 583 (1968).
$^h$ J. Powell and B. Shaw, *J. Chem. Soc. (A)*, 597 (1968).
$^i$ J. F. Nixon, B. Wilkins, and D. A. Clement, *J. Chem. Soc., Dalton Trans.*, 1993 (1974).
$^j$ M. Bottrill and M. Green, *J. Organomet. Chem.*, **111**, C6 (1976).
$^k$ S. O'Brien, *Chem. Commun.*, 757 (1968); E. W. Abel and S. M. Moorehouse, *J. Chem. Soc., Dalton Trans.*, 1706 (1973).
$^l$ A. Kasahara and K. Tanaka, *Bull. Soc. Chem. Jpn.*, **39**, 634 (1966).
$^m$ P. Chini and S. Martinego, *Inorg. Chem.*, **6**, 837 (1967).
$^n$ J. Powell and B. L. Shaw, *J. Chem. Soc. (A)*, 780 (1968).
$^o$ B. L. Shaw and E. Singleton, *Chem. Soc. (A)*, 1972 (1967).
$^p$ B. L. Shaw and E. Singleton, *J. Chem. Soc. (A)*, 1683 (1967).
$^q$ C. K. Brown, W. Mowat, G. Yagupsky, and G. Wilkinson, *J. Chem. Soc. (A)*, 850 (1971).
$^r$ K. Moseley, J. W. Kang, and P. M. Maitlis, *J. Chem. Soc. (A)*, 2875 (1970).
$^s$ A. J. Deeming and B. L. Shaw, *J. Chem. Soc. (A)*, 1562 (1969).
$^t$ M. Green and G. J. Parker, *J. Chem. Soc., Dalton Trans.*, 333 (1974).

## f. Complexes of group 9 metals

Elements of this triad form allyl compounds of the type [M(allyl)₃] (Table 7.12). The cobalt complex is very unstable. It decomposes above 233 K and easily reacts with water, alcohol, and air. However, rhodium and iridium compounds are stable and may be stored in air. They do not decompose under the influence of water. Many carbonyl, cyclopentadienyl, and halogen allyl complexes of the cobalt triad are known: $[Co(allyl)(CO)_3]$, $[CoX(allyl)cp]$, $(Co(C_3H_5)I_2]$, $K_2[Co(C_3H_5)(CN)_4]$, $[CoX(C_3H_5)_2]_2$, $[Co(allyl)\,cp(CO)]X$, $[Co(allyl)(COD)]$, $[Co(allyl)_2(L-L)]^+X^-$ (L–L = bipy, phen), $[Co(BD)(C_8H_{13})]$, $[Co(allyl)(PF_3)_3]$, $[M(allyl)\{P(OR)_3\}_3]$, $[M(allyl)(PR_3)_3]$ (M = Co, RhIr), $[Rh_2(C_3H_5)_2(CO)_4]$, $[Rh_2(CH_2C(Me)CH_2)_2$ $(dppm)_2]$,$^{(63)}$ $[Rh_2X_2(allyl)_4]$, $[RhX(allyl)_2\,L]$, $[Rh(allyl)_2\,L_2]^+$ (L = PR₃, AsR₃, py, etc.), $[Rh(allyl)_2\,cp]$, $[Rh_2Cl_4(allyl)_2\,(BD)]$, $[RhX_2(allyl)]_n$, $[RhX_2(allyl)\,L_2]$, $[RhCl_2(C_{12}H_{19})\,L_2]$, $[RhCl(allyl)\,cp]$, $[Rh(allyl)(COD)]$, $[IrCl_2(allyl)(PR_3)_2]$, $[Ir_2X_4(allyl)_2\,(BD)]$, $[IrX_2(allyl)(CO)L]$ (X = Cl, Br; L = C₈H₁₄, PR₃, AsR₃, and py), $[IrCl(allyl)(C_5Me_5)]$, $[Ir(allyl)_2L_2]^+$, $[Ir(allyl)L_2]$, and $[Ir(allyl)(CO)_nL_{3-n}]$ (L = PR₃, etc.).

Iridium also forms the stable hydrido allyl complex $[IrCl(1-PhC_3H_4)(H)(PPh_3)_2]$. This compound is formed by the reaction of $[IrCl(N_2)(PPh_3)_2]$ with phenylcyclopropane. During the reaction, opening of the ring and migration of

Table 7.13. Allyl Complexes of Ni, Pd, and Pt

| Compound | Color | Mp (K) |
|---|---|---|
| $[Ni(C_3H_5)_2]^{a\,(19)}$ | Light-orange | 274 |
| $[Ni_2Cl_2(C_3H_5)_2]^{a,b,c}$ | Brown | 352–355 dec |
| $[Ni(C_3H_5)\{SC(NH_2)_2\}_2]^+Cl^{-d}$ | Orange-red | |
| $[NiBr(C_3H_5)(CO)(PPh_3)]^e$ | Orange-red | |
| $[Ni(C_3H_5)cp]^f$ | Dark-violet | Liquid |
| $[Ni_2(C_3H_4)_2cp_2]^g$ | Dark-red | |
| $[Ni_2(OAc)_2(C_3H_5)_2]^h$ | Orange-brown | |
| $[Ni(C_3H_5)(H)(PPh_3)]^i$ | Red-brown | |
| $[Ni(C_{12}H_{18})]^{(19)}$ | | |
| $[Pd(C_3H_5)_2]^{a\,(19)}$ | Light-yellow | 303 |
| $[Pd_2Cl_2(C_3H_5)_2]^{f\,(9-11)}$ | Yellow | 425–428 dec |
| $[Pd(C_3H_5)cp]^f$ | Dark-red | 336 |
| $[Pd_2(OAc)_2(C_3H_5)_2]^k$ | Yellow | 373–400 dec |
| $[Pd_2Br_2(C_3H_5)_2]^{l,m}$ | Yellow-green | 408 dec |
| $[PdCl(C_3H_5)PPh_3]^n$ | Yellow | 445–451 dec |
| $[Pd(C_3H_5)(acac)]^{m,o}$ | Light-yellow | 344–347 |
| $[Pd_2(\mu\text{-}C_3H_5)(\mu\text{-}C_5H_5)\{P(i\text{-}Pr)_3\}_2]^{(18)}$ | Orange-red | 337 |
| $[Pd_4(\mu\text{-}CN)_4(1,2,3\text{-}Me_3C_3H_2)_4]^p$ | White | 467–469 |
| (structure below) | | 362 dec |

$$\begin{array}{c} \text{CHMePh} \quad \text{Me}\;^{(21)} \\ \diagdown \qquad \diagup \\ N = C \\ \diagup \qquad \diagdown \\ 2-\text{MeC}_3\text{H}_4-\text{Pd} \qquad \text{CH} \\ \diagdown \qquad \diagup\!\!\diagup \\ O - C \\ \diagdown \\ \text{Me} \end{array}$$

*(Table continued)*

*Table 7.13. (continued)*

| Compound | Color | Mp (K) |
|---|---|---|
| $[Pt(C_3H_5)_2]^{q\,(19)}$ | White | 317 |
| $[Pt_2(C_3H_5)_2(acac)_2]^q$ | White | 428–434 dec |
| $[Pt_2Cl_2(2\text{-}MeC_3H_4)_2]^q$ | Light-yellow | 423–427 dec |
| $[Pt_4Cl_4(C_3H_5)_4]^{q,r}$ | Yellow-orange | 447–457 dec |
| $[Pt(C_3H_5)cp]^q$ | Yellow | 336–337 |
| $[Pt(2\text{-}MeC_3H_4)(acac)]^q$ | White | 356–358 |
| $[PtCl(C_3H_5)(PPh_3)]^q$ | White | 455–461 dec |
| $[Pt(2\text{-}MeC_3H_4)_2(PPh_3)_2]^q$ | Light-yellow | 354–358 |
| $[Pt(2\text{-}MeC_3H_4)(PPh_3)_2]^+[BPh_4]^{-\,q}$ | White | 447–452 |
| $[Pt(C_3H_5)(PPh_3)_2]^+Cl^{-\,s}$ | White | 465–473 |
| $[Pt(C_3H_5)(PCy_3)_2]^+[PF_6]^{-\,t}$ | White | |

[a] D. A. Brown and A. Owens, *Inorg. Chim. Acta*, **5**, 675 (1971).

[b] V. A. Vashkevich, A. M. Lazutkin, and B. I. Mintser, *Z. Org. Khim.*, **37**, 1926 (1967).

[c] E. O. Fischer and H. Werner, *Z. Chem.*, **2**, 174 (1962).

[d] A. Sirigu, *Chem. Commun.*, 256 (1969); A. Sirigu, *Inorg. Chem.*, **9**, 2245 (1970).

[e] F. Guerrieri and G. P. Chiusoli, *J. Organomet. Chem.*, **15**, 209 (1968).

[f] W. R. McClellan, H. H. Hochn, H. N. Cripps, E. L. Muetterties, and B. W. Howck, *J. Am. Chem. Soc.*, **83**, 1601 (1961).

[g] W. Keim, *Angew. Chem.*, **80**, 968 (1968).

[h] F. Dawans, J. C. Marechal, and P. Teyssie, *J. Organomet. Chem.*, **21**, 259 (1970).

[i] H. Boennemann, *Angew. Chem.*, **82**, 699 (1970).

[j] I. I. Moyseev, E. A. Fedorovskaya, and Ya. K. Syrkin, *Zh. Neorg. Khim.*, **4**, 2641 (1959); R. F. Heck and D. S. Breslow, *J. Am. Chem. Soc.*, **82**, 750 (1960).

[k] S. D. Robinson and B. L. Shaw, *Organomet. Chem.*, **3**, 367 (1965).

[l] M. S. Lupin, J. Powell, and B. L. Shaw, *J. Chem. Soc. (A)*, 1410 (1966).

[m] B. E. Mann, R. Pietropaolo, and B. L. Shaw, *Chem. Commun.*, 790 (1971).

[n] J. Powell and B. L. Shaw, *J. Chem. Soc. (A)*, 1839 (1967).

[o] S. D. Robinson and B. L. Shaw, *J. Chem. Soc.*, 4806 (1963); G. A. Razuvaev, G. A. Domrachev, O. M. Suvrova, and L. G. Abakumova, *J. Organomet. Chem.*, **32**, 113 (1971).

[p] B. L. Shaw and G. Shaw, *J. Chem. Soc. (A)*, 3533 (1971).

[q] B. E. Mann, B. L. Shaw, and G. Shaw, *J. Chem. Soc. (A)*, 3536 (1971).

[r] G. Raper and W. S. McDonals, *J. Chem. Soc., Dalton Trans.*, 265 (1972).

[s] H. C. Volger and K. Vrieze, *Organomet. Chem.*, **9**, 527 (1967); **9**, 524 (1967).

[t] B. F. Mann and A. Musco, *J. Organomet. Chem.*, **181**, 439 (1979); Y. D. Smith and J. D. Oliver, *Inorg. Chem.*, **17**, 2585 (1978).

hydrogen from the C2 carbon atom to iridium take place [Figure 7.6, reaction (7.53)]. The complex $[Ir(\eta^3\text{-}C_3H_5)\,(pmcp)H]$ is also quite stable.[62]

Polymeric rhodium compounds such as $[RhX_2(allyl)]_n$ contain both bridging as well as terminal chlorine atoms. These complexes may also contain bridging allyl groups as rhodium(III) exhibits a tendency to achieve the coordination number six.

### g. *Complexes of group 10 metals*

Allyl complexes of this group have been more thoroughly investigated than those of other elements, because these complexes are relatively stable and have broad applications in various catalytic reactions. Compounds of the type $[M(allyl)_2]$ (M = Ni, Pd, and Pt) and $[M_2X_2(allyl)_2]$ (M = Ni, Pd) are known (Table 7.13).

The platinum atom also gives analogous derivatives $[Pt_2X_2(allyl)_2]$; however, these complexes contain exclusively allyl groups with substituents, for example, 2-methylallyl, 1, 2-dimethylallyl, 1, 2, 3-trimethylallyl, etc. The allyl complex

$[PtCl(C_3H_5)]_4$ is tetrameric and contains both bridging chlorine atoms as well as bridging allyl groups (Figure 7.6). The compound $[Ni(C_3H_5)_2]$ inflames in air and the complex $[Ni_2Cl_2(C_3H_5)_2]$ is unstable. The corresponding palladium and platinum compounds are considerably more stable. Also, $18e$ compounds such as $[M(allyl)XL_2]$ are known.

All metals of this triad additionally form complexes of the type $[MX(allyl)L]$ ($X = Cl$, Br and I; $L = PR_3$, amines, $AsR_3$, etc.), $[M(allyl)cp]$, and $[M(acac)(allyl)]$. In the case of platinum, the acetylacetonate complexes are dimeric with bridging allyl groups. For $[2\text{-}(R, S)\text{-}\alpha\text{-phenylethylimino-3-penten--olate}](2\text{-methylallyl})palladium(II)$ a mixture of diastereoisomers is formed because the imino ligand is asymmetric. Moreover, the plane which is perpendicular to the coordination plane and passes through the $C_2$ carbon atom of the allyl group and palladium atom is chiral. The conformation of the complex in which the methyl group of the allyl ligand and the phenyl group of the imino ligand are located on the same side of the coordination plane is favored (Figure 7.8). This is probably due to electronic rather than steric factors.[21]

Dimeric palladium(I) and platinum(I) compounds are also known [see equations (7.56) and (7.57)]. They contain the $M-M$ bond.[18,35]

$$L-M-M'-L \qquad L-M-M-L \qquad (7.72)$$

$$M = M' = Pd, Pt; L = PR_3$$

## h. Complexes of group 3 metals

Scandium group elements as well as the lanthanides form allyl complexes in which the central atom has $+3$ oxidation state, while the actinides give compounds possessing $+4$ oxidation states (Table 7.14). These complexes are formed by reactions of metal halides with allyl compounds of magnesium, lithium, tin, etc.[45,46]

$$[ScClcp_2] + MgCl(CH_2CH=CH_2) \longrightarrow [Sccp_2(\eta^3\text{-}C_3H_5)] \qquad (7.73)$$

$$[LnClcp_2] + MgBr(CH_2CH=CH_2) \xrightarrow{\text{THF + Et}_2\text{O}}_{195\,\text{K}} [Lncp_2(\eta^3\text{-}C_3H_5)] \qquad (7.74)$$

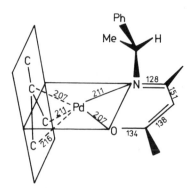

Figure 7.8. The structure of $[2\text{-}(R, S)\text{-phenylethylimino-3-pentene-4-olate}](2\text{-methylallyl})$palladium(II) (distances are in pm units).

*Table 7.14. Allyl Complexes of the Group 3 Metals,
the Lanthanides, and the Actinides*

| Compound | Color | Mp (K) |
|----------|-------|--------|
| $[Sc(\eta^3\text{-}C_3H_5)cp_2]^a$ | Orange | 424 dec |
| $[Sm(\eta^3\text{-}C_3H_5)cp_2]^b$ | Yellow | >473 dec |
| $[Ho(\eta^3\text{-}C_3H_5)cp_2]^b$ | Bright-yellow | >473 dec |
| $[Er(\eta^3\text{-}C_3H_5)cp_2]^b$ | Pink | >473 dec |
| $[Th(C_3H_5)_4]^{(19)}$ | Dark-yellow | 273 dec |
| $[U(C_3H_5)_4]^{*c,d}$ | Dark-red | 253 dec |

\* $\mu_{eff} = 2.6$ BM, $\theta = 100$ K.
[a] R. S. P. Coutts and P. C. Wailes, *J. Organomet. Chem.*, **25**, 117 (1970).
[b] M. Tsutsui and N. Ely, *J. Am. Chem. Soc.*, **197**, 3551 (1975).
[c] G. Lugli, M. Marconi, A. Mazzei, N. Paladino, and U. Pedretti, *Inorg. Chim. Acta*, **3**, 253 (1969).
[d] N. Paladino, G. Lugli, U. Pedretti, and M. Brunelli, *Phys. Lett.*, **5**, 15 (1970).

For the lanthanides, the following complexes are typical: $LiLn(allyl)_4$ (Ln = Ce, Nd, Sm, Gd, Dy), $Lncp_2(allyl)$ (Ln = Sm, Er, Ho, Lu), and for actinides: $M(allyl)_4$ (M = Th, U), $U(allyl)_3X$ (X = Cl, Br, I), $[U(C_3H_5)_2(OR)_2]_2$, and $[U(pmcp)(allyl)_3]$. The $^1H$ NMR spectra show that in the compound $Th(C_3H_5)_4$ the π-allyl ligand is bonded symmetrically to the central atom because at 283 K these spectra are of the $AM_2X_2$ type (Table 7.2). The complex $U(pmcp)(CH_2CMeCH_2)_3$ has pseudotetrahedral structure

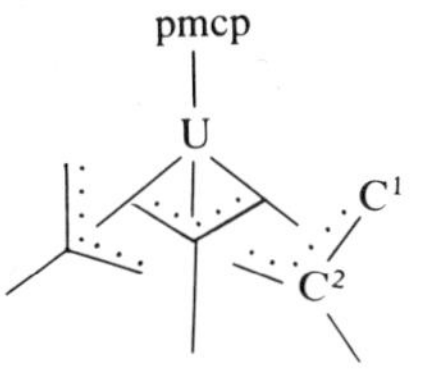

$$U\text{-}C^1 = 266 \text{ pm}$$
$$U\text{-}C^2 = 280 \text{ pm}$$

$$(7.75)$$

Compounds of the type $Lncp_2(allyl)$ and $Sc(pmcp)_2(allyl)$ have fluxional allyl groups as shown by NMR studies.

Homoleptic allyl compounds of the actinides such as $Th(C_3H_5)_4$ and $U(C_3H_5)_4$, like other derivatives of the same type, are considerably less thermally stable. Also, they very easily react with water and other protic solvents as well as with oxygen; $U(C_3H_5)_4$ inflames in air.

## 9. CYCLOPROPENYL COMPLEXES

The cyclopropenyl group $C_3H_3$ and its derivatives such as $C_3R_3$ are also considered to be 3e ligands. Complexes with triphenylcyclopropenyl and tris(*tert*-butyl)cyclopropenyl ligands are known.

These complexes are usually obtained from $C_3R_3^+X^-$ (X = halogen, $BF_4^-$) and metal compounds in their low oxidation states[47,48]:

$$[Ni(CO)_4] + C_3Ph_3^+Cl^- \xrightarrow{MeOH} [NiCl(C_3Ph_3)(CO)] \qquad (7.76)$$

$$[Ni(CO)_4] + C_3Bu_3^t + NaBr \longrightarrow [NiBr(C_3Bu_3^t)(CO)] \qquad (7.77)$$

In the nickel complex containing pyridine, the cyclopropenyl ring is symmetrically coordinated to the nickel atom[48] [structure (7.78)]. Symmetrically bonded cyclo-

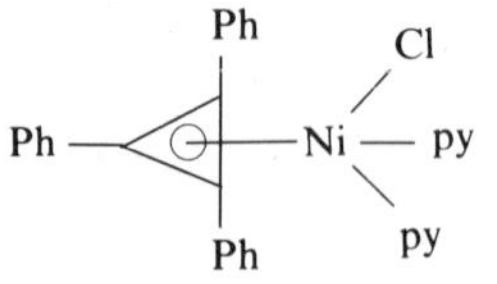

$$(7.78)$$

propenyl ligands are found in these complexes: $[NiBr(C_3Ph_3)(CO)]_2$,[65] $[Ni(C_3Ph_3)cp]$, $Co(C_3Ph_3)(CO)_3]$, $[MoBr(C_3Ph_3)(CO)_2$ (bipy)], and $[Wcp\{C_3(CMe_3)_2 Me\}$ $Cl_2(PMe_3)]$, while cationic compounds such as $[M(C_3Ph_3)(PPh_3)_2]^+$ (M = Ni, Pd, Pt) have asymmetrically bonded cyclopropenyl groups.[66,67]

Some metal cyclopropenyl compounds are very stable. Nickel complexes with $C_3Bu_3^t$ in the solid state are stable in the air for short periods in the solid state, and the vanadium compound $[V(C_3Ph_3)(CO)_5]$ decomposes only at 503 K (Table 7.8).

## 10. *HETEROALLYL COMPLEXES*

Transition metal complexes containing heteroallyl ligands, such as those in structures (7.79), have been less studied than metal allyl compounds.[68–70,72] Complexes con-

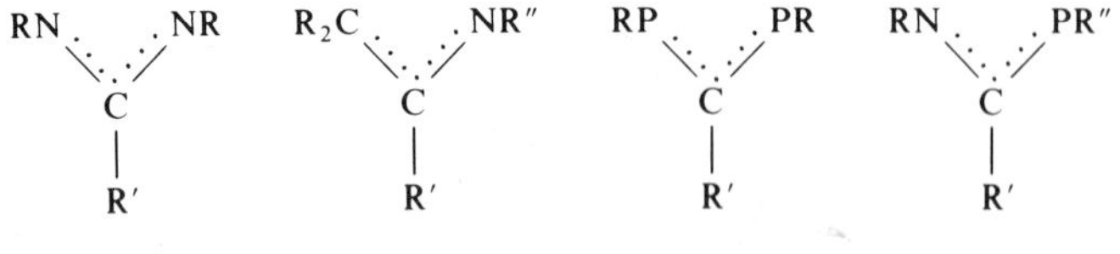

$$(7.79)$$

taining a 1-azaallyl ligand are obtained by reactions of substituted (in the 2 position) azirines with transition metal complexes[68] [equation (7.80)]. The reaction of

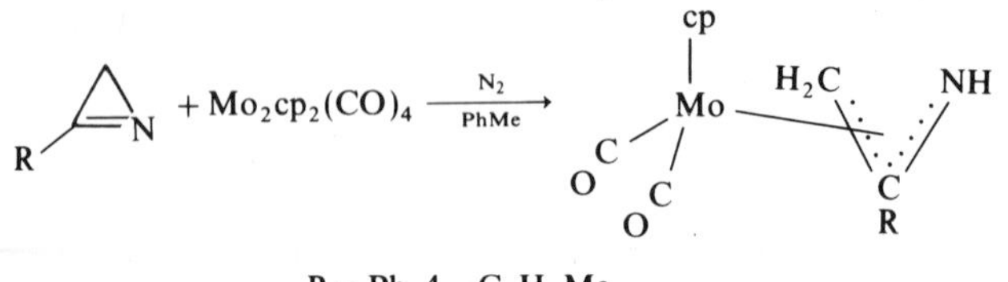

$$(7.80)$$

$$R = Ph, 4 - C_6H_4Me$$

RP = CHPClR with sodium tetracarbonylcobaltate(−I) gives the 1, 3-diphosphaallyl cobalt complex[69] [equation (7.81)]. The C–P distances are 176.9(7) and 179.1(7) pm.

$$RP = CHP\begin{matrix} Cl \\ \diagup \\ \diagdown \\ R \end{matrix} + Na[Co(CO)_4] \xrightarrow[\substack{-NaCl \\ -CO}]{THF} (OC)_3Co \overset{RP}{\underset{CH}{\diagdown\diagup}} PR \qquad (7.81)$$

Therefore, these distances are longer than for the $P=C$ double bond and shorter than in the case of the $P-C$ single bond. The compound has a square-pyramidal structure. The base contains two CO ligands and the phosphorus atoms while the third CO group is located at the pyramid vertex. The distances for Co–C(allyl), Co–$P^1$, and Co–$P^2$ are 203.4(7), 238.2(2), and 239.5(2) pm, respectively.

Many complexes of transition metals possessing metallatriphosphatetrahedrane or metallatriarsenatetrahedrane structures were recently prepared and investigated[71-76] [structures (7.82)]. Formally, these complexes are analogous to cyclopropenyl com-

$$\begin{matrix} L_nM \\ \diagup \quad \diagdown \\ E \text{---}\!\!|\text{---} E \\ \diagdown \quad \diagup \\ E \end{matrix} \qquad \begin{matrix} L_nM \\ \diagup \quad \diagdown \\ E \text{---}\!\!|\text{---} E \\ \diagdown \quad \diagup \\ E \\ | \\ ML_n \end{matrix} \qquad (7.82)$$

$$E = P, As$$

plexes because the CH group is isolobal with the P and As atoms. The preparation of these complexes may be achieved by reactions of white phosphorus or yellow arsenic with hydrated transition metal salts in the presence of polydentate phosphine ligands or by reactions of $PX_3$ with anionic complexes.

$$[Co(H_2O)_6]^{2+} + P_4 + MeC(CH_2PPh_2)_3 \xrightarrow{EtOH/THF} [Co(\eta^3\text{-}P_3)\{(Ph_2PCH_2)_3CMe\}] \quad (7.83)$$

$$PX_3 + Co(CO)_4^- \xrightarrow[Co, 0.1\,MPa]{THF, 298\,K} [Co(CO)_3(\eta^3\text{-}P_3)] \qquad (7.84)$$

Examples of such complexes are $[M(\eta^3\text{-}P_3)L_3]$ (M = Co, Rh, Ir), $[M(\eta^3\text{-}P_3)L_3]^+$ (M = Pd, Pt), and $[L_3M(\mu\text{-}\eta^3\text{-}P_3)ML_3]X_2$ (M = Co, Ni) where $L_3 = MeC(CH_2PPh_2)_3$, $N(CH_2CH_2PPh_2)_3$, and $(CO)_3$; $X = BF_4^-$.

## 11. REACTIONS OF ALLYL COMPLEXES

### a. Thermal decomposition

Thermal stability of transition metal allyl complexes varies greatly. Therefore, derivatives which melt at temperatures higher than 470 K without decomposition (e.g., some palladium and molybdenum compounds with mixed coordination spheres) and very unstable complexes containing exclusively allyl ligands which decompose above 230 K are known. In neutral solvents many coordination compounds decompose with the evolution of 1,5-hexadiene (diallyl), for example, $[Pd(C_3H_5)(acac)]$ and $[FeX(C_3H_5)(CO)_3]$. The same compounds decompose to give propylene and other products in the presence of alcohols. The allylchloropalladium complexes $[Pd_2Cl_2(allyl)_2]$ give off allyl halides and olefins when heated to 473 K.

Palladium complexes containing allyls with substituents evolve HCl and the corresponding diene [equation (7.85)].

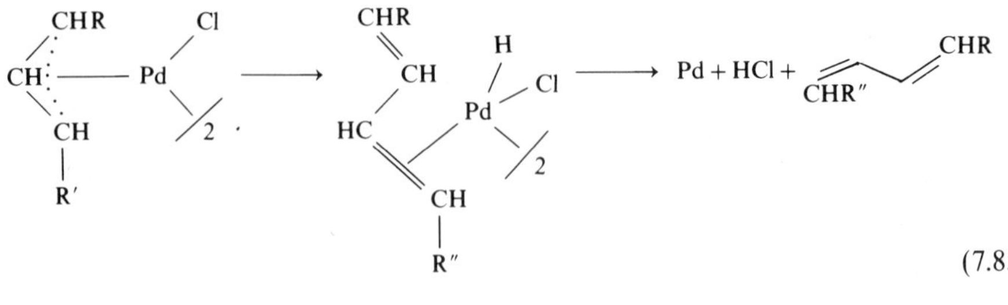

$$(7.85)$$

## b. *Interactions with nucleophilic reagents; oxidative hydrolysis*

Halogen–allyl complexes are relatively resistant to air oxidation. However, "pure" allyl compounds (homoleptic) $[M(allyl)_n]$ in many cases rapidly react with oxygen from air. In the presence of nucleophilic reagents, oxidation reaction of the allyl group or olefin from which the allyl complex was obtained may take place.

Oxidative hydrolysis may serve as an example [see equation (7.86)].

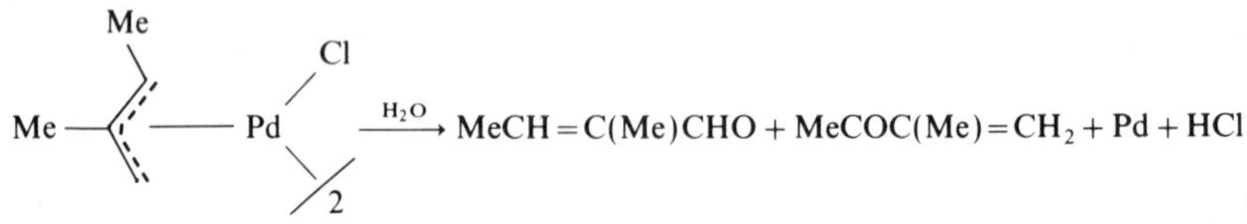

$$\xrightarrow{H_2O} MeCH{=}C(Me)CHO + MeCOC(Me){=}CH_2 + Pd + HCl$$

$$(7.86)$$

The mechanism of this reaction is shown in equation (7.87). The mechanism of the reaction is analogous to the oxidation reaction of ethylene to acetaldehyde in the presence of palladium chloride.

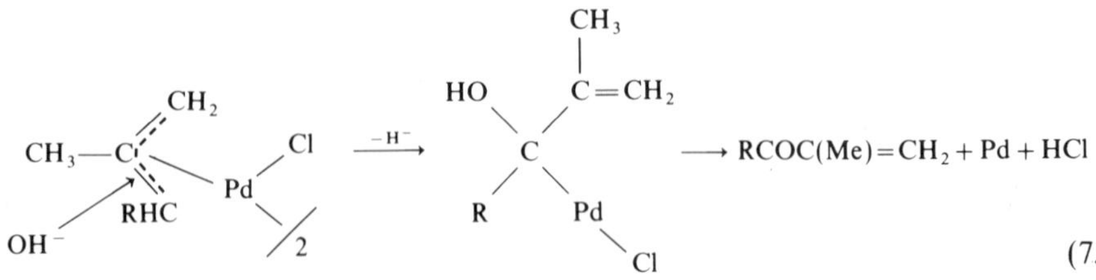

$$\longrightarrow RCOC(Me){=}CH_2 + Pd + HCl$$

$$(7.87)$$

Sometimes, oxidation of the alkene substituents occurs:

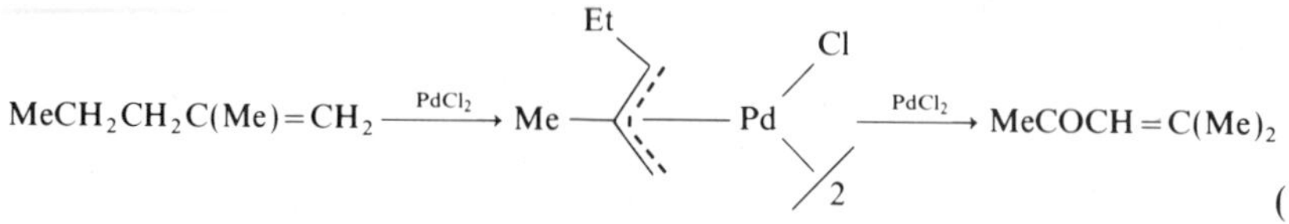

$$MeCH_2CH_2C(Me){=}CH_2 \xrightarrow{PdCl_2} \cdots \xrightarrow{PdCl_2} MeCOCH{=}C(Me)_2$$

$$(7.88)$$

Hydrolysis of a cationic iron complex proceeds similarly [see equation (7.89)].

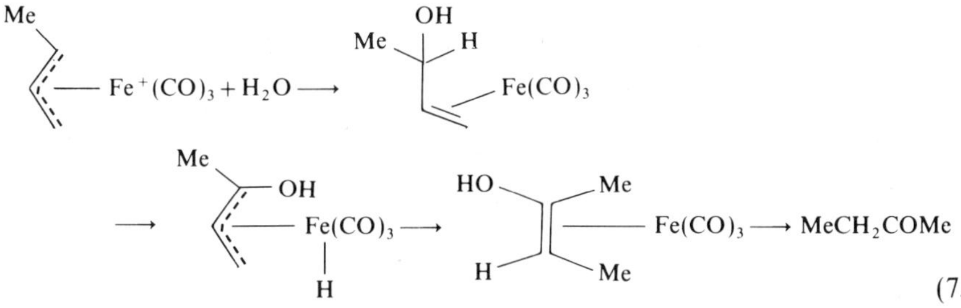

$$(7.89)$$

Some organometallic compounds of the main group elements may play the role of nucleophilic reagents. They react with coordinated allyl groups to give olefin derivatives[39] [see equation (7.90)].

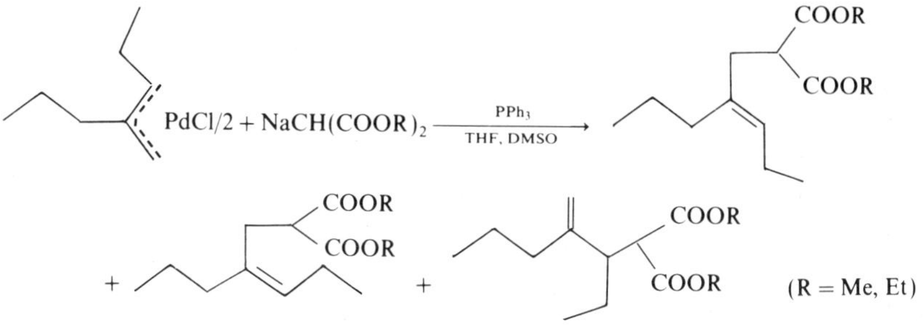

$$(R = Me, Et)$$

$$(7.90)$$

Hydrolysis of allyl complexes often furnishes olefins in addition to the oxidation products. Sometimes, these olefins may constitute main products of the reaction, particularly if the reaction is carried out in basic medium in water or alcohols. It is assumed that the first stage of the reaction is attack of the $H^-$ or the $CH_3O^-$ ion on the allyl complex, which subsequently decomposes with olefin evolution.[23]

$$\tfrac{3}{2}[Pd_2Cl_2(C_3H_5)_2] + 2CH_3O^- + OH^- \longrightarrow 3C_3H_6 + 3Pd + 3Cl^- + CO_2 + CH_3OH \qquad (7.91)$$

$$[Pd_2Cl_2(C_3H_5)_2] + 2CH_3O^- \longrightarrow 2[PdCl(OCH_3)(C_3H_5)] \longrightarrow 2[PdHCl(C_3H_5)]^- + 2CH_2O \qquad (7.92)$$

$$2CH_2O + OH^- \longrightarrow CH_3OH + HCOO^- \qquad (7.93)$$

$$\tfrac{1}{2}[Pd_2Cl_2(C_3H_5)_2] + HCOO^- \longrightarrow [PdHCl(C_3H_5)]^- + CO_2 \qquad (7.94)$$

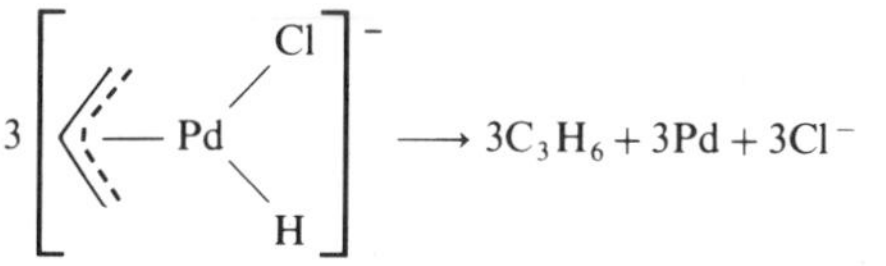

$$3C_3H_6 + 3Pd + 3Cl^- \qquad (7.95)$$

Halogen–allyl rhodium or iridium complexes react in a similar manner with olefin evolution.

In aqueous solutions, the decomposition of $[Pd_2Cl_2(C_3H_5)_2]$ leads to the formation of acrolein and propylene or acetone and propylene, dependent upon conditions.[25–27] This indicates a complicated mechanism of decomposition of allyl coordination compounds and the possibility of preparation of various products depending upon reaction conditions. The hydrogen exchange between allyl groups is possible because it was found that during decomposition of diallyl-di-$\mu$-chloro-dipalladium in a solution of NaOD in $D_2O$, propylene which does not contain deuterium is given off.[28]

### c. *Electrophilic attack on allyl groups*

Electrophilic reagents such as $H^+$ and $Ph_3C^+$ cause allyl–metal bond splitting, leading to the formation of corresponding complexes and organic derivatives[19,24,29]:

$$[M(C_3H_5)_n + nH^+ \longrightarrow nC_3H_6 + M^{n+} \tag{7.96}$$

$$[Ru(2\text{-}MeC_3H_4)_2(NBD)] + 2H^+ \xrightarrow[\text{H}^+]{\text{MeCN}} [Ru(NBD)(MeCN)_4]^{2+} + C_4H_8 \tag{7.97}$$

$$[Ru(2\text{-}MeC_3H_4)_2(NBD)] + [Ph_3C]^+[BF_4]^-$$

$$\xrightarrow{\text{MeCN}} [Ru(2\text{-}MeC_3H_4)(NBD)(MeCN)_2]^+[BF_4]^- + Ph_3CCH_2CMe=CH_2 \tag{7.98}$$

The action of acids on allyl palladium compounds also causes olefin evolution. However, the reaction conditions must be very carefully selected because slight changes or a small excess of acid may lead to various side reactions owing to lability of olefin palladium complexes.

### d. *Formation of $\sigma$-allyl complexes*

Some allyl complexes may react with $2e$ ligands to give $\sigma$-allyl compounds.

$$[RhCl_2(\eta^3\text{-}C_3H_5)(PPh_3)_2] + SO_2 \longrightarrow [RhCl_2(\eta^1\text{-}C_3H_5)SO_2(PPh_3)_2] \tag{7.99}$$

$$[IrCl_2(\eta^3\text{-}C_4H_7)(PMe_2Ph)_2] + CO \longrightarrow [(rCl_2(\eta^1\text{-}C_4H_7)(CO)(PMe_2Ph)_2] \tag{7.100}$$

$$[Re(\eta^3\text{-}C_3H_5)(CO)_4] + PPh_3 \longrightarrow [Re(\eta^1\text{-}C_3H_5)(CO)_4)(PPh_3)] \tag{7.101}$$

Many $\pi$-allyl complexes may spontaneously rearrange to $\sigma$-allyl compounds (transformation from $18e$ to $16e$ complexes). This is an important step in many catalytic reactions involving allyl complexes.

### e. *Reduction*

Some allyl complxes may be reduced to alkene and the metal. The complex dodeca-2, 6, 10-triene-1,12-diylnickel is reduced by hydrogen with the formation of nickel and dodecane. Many other metal allyl compounds react in a similar manner.

## f. *Reductive elimination*

Allyl complexes may react with ligands which stabilize low oxidation states with the formation of dienes and transition metal complexes.

$$[Ni(C_3H_5)_2] + 4CO \longrightarrow [Ni(CO)_4] + \text{\hspace{3cm}} \qquad (7.102)$$

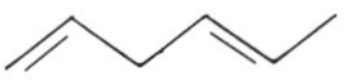

$$[Co(C_3H_5)_3] + 3CO \longrightarrow [Co(C_3H_5)(CO)_3] + C_6H_{10} \qquad (7.103)$$

Reductive elimination also occurs in the case of unstable hydrido allyl complexes.

$$[L_nM(C_3H_5)H] \longrightarrow [L_nM(C_3H_6)] \qquad (7.104)$$

Such reactions have considerable significance in oligomerization and diene polymerization:

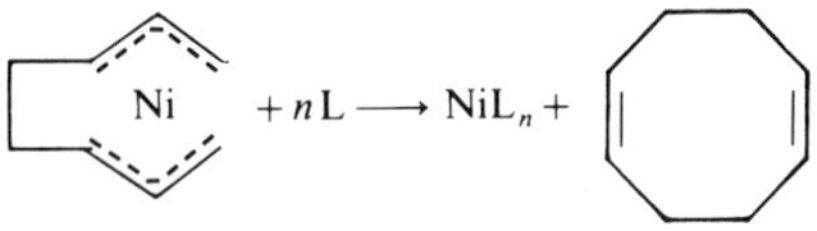

$$\text{\hspace{5cm}} \qquad (7.105)$$

## g. *Migration (insertion) reactions of ligands*

The described reactions show that, with the influence of new ligands and an increase in temperature, often the equilibrium betwen σ- and π-allyl compounds is established. Owing to attack of ligands, complete removal of allyl groups as dienes or as migration (insertion) products of the allyl group to the ligand may take place. Therefore, allyl complexes are often utilized as catalysts for oligomerization, polymerization, cooligomerization, and copolymerization of dienes, olefins, alkynes, as well as diene carbonylation, telomerization, etc. Allyl complexes are formed as intermediate compounds in many oligomerization and polymerization reactions.

Carbonylation reactions take place via coordination of the carbon monoxide molecule to the complex, formation of the σ-allyl compounds, and subsequent allyl migration to the CO group[1,5,30,31] [scheme (7.106)].

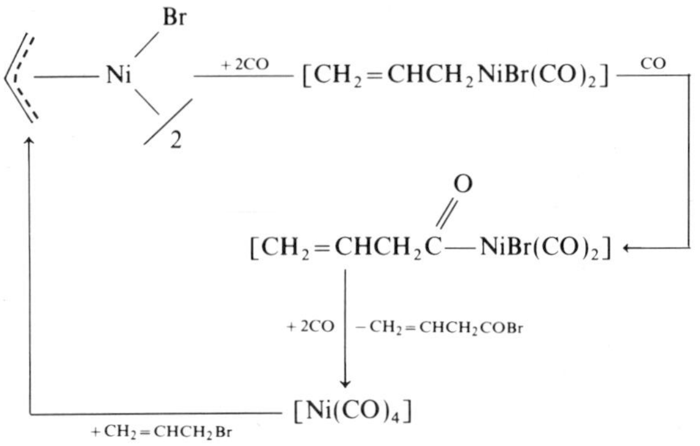

$$\qquad (7.106)$$

Metal complexes in low oxidation states are often used as catalysts for reactions of allyl halides with CO, acetylenes, and olefins as well as their derivatives. In the first step of the reaction, there is an oxidative addition of $CH_2CHCH_2X$ and the formation of the allyl complex:

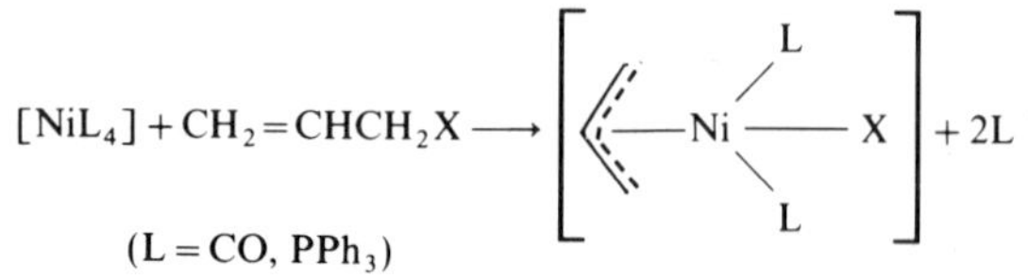
[NiL4] + CH2=CHCH2X
Ni
L
X
L
+ 2L
(L = CO, PPh3)

(7.107)

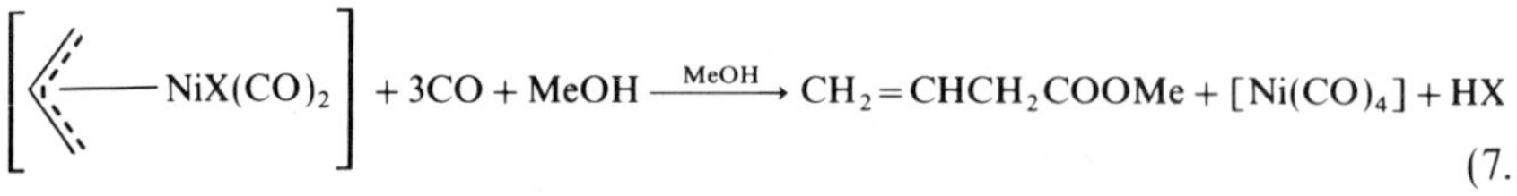
NiX(CO)2
+ 3CO + MeOH
MeOH
CH2=CHCH2COOMe + [Ni(CO)4] + HX

(7.108)

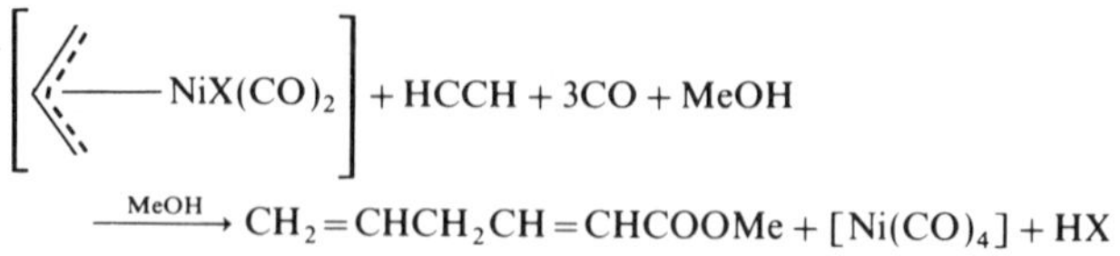
NiX(CO)2
+ HCCH + 3CO + MeOH
MeOH
CH2=CHCH2CH=CHCOOMe + [Ni(CO)4] + HX

(7.109)

Acetylene becomes coordinated in the vacated site because of the $\pi$ to $\sigma$ rearrangement of the complex.

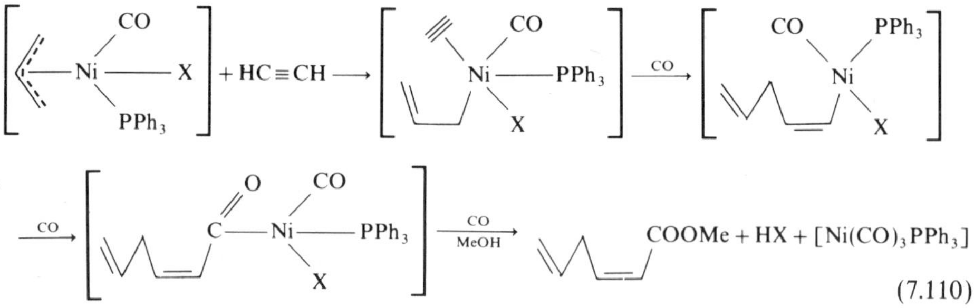
CO
Ni
X
PPh3
+ HC≡CH
CO
Ni
PPh3
X
CO
CO
PPh3
Ni
X
CO
O
CO
C
Ni
PPh3
X
CO
MeOH
COOMe + HX + [Ni(CO)3PPh3]

(7.110)

## REFERENCES


1. I. I. Kritskaya and D. A. Bochwar, in: *Metody Elementoorganicheskoy Khimii*, p. 734, Izd. Nauka, Moscow (1975).
2. M. L. H. Green, *Organometallic Compounds, The Transition Elements*, Vol. 2, Methuen, London (1968).
3. K. Vrieze, in: *Dynamic Nuclear Magnetic Resonance Spectroscopy* (L. M. Jackman and F. A. Cotton, eds.), p. 441, Academic Press, New York (1975).
4. H. L. Clarke, *J. Organomet. Chem.*, **80**, 155 (1974).
5. B. L. Shaw and N. I. Tucker, in: *Comprehensive Inorganic Chemistry* (J. C. Bailar, Jr., ed.), p. 781, Pergamon Press, Oxford (1973).
6. S. F. A. Kettle and R. Mason, *J. Organomet. Chem.*, **5**, 573 (1966).
7. P. W. N. M. Leeuwen and A. P. Pratt, *J. Organomet. Chem.*, **21**, 501 (1970).

8. E. O. Fischer and H. Werner, *Metal π-Complexes*, Elsevier, Amsterdam (1966).
9. K. Shabotake and K. Nakamoto, *J. Am. Chem. Soc.*, **92**, 3339 (1970); H. P. Fritz, *Chem. Ber.*, **94**, 1217 (1961).
10. D. C. Andrews and G. Davidson, *J. Organomet. Chem.*, **55**, 383 (1973).
11. D. M. Adams and A. Squire, *J. Chem. Soc. (A)*, 1808 (1970).
12. R. G. Kidd, in: *Characterization of Organometallic Compounds, Part II* (M. Tsutsui, ed.), p. 373, Wiley–Interscience, New York (1971).
13. F. R. Hartley, *J. Organomet. Chem.*, **21**, 227 (1970).
14. F. Pruchnik and H. Pasternak, *Rocz. Chem.* **51**, 1581 (1977).
15. F. Pruchnik, *Inorg. Nucl. Chem. Lett.*, **9**, 1229 (1973).
16. a. T. H. Tulip and J. A. Ibers, *J. Am. Chem. Soc.*, **100**, 3252 (1978). b. A. H. Cowley, *Prog. Inorg. Chem.*, **26**, 45 (1979).
17. J. Powell, *M.T.P. International Review of Science, Inorganic Chemistry, Ser. I*, Vol. 5, p. 273, Butterworths, London (1972).
18. H. Werner and A. Kuhn, *Angew. Chem., Int. Ed. Engl.*, **16**, 412 (1977).
19. G. Wilke, B. Bogadnovic, P. Hardt, P. Heimbach, W. Keim, M. Kroener, W. Oberkirch, K. Tanaka, E. Steinruecke, D. Walter, and H. Zimmerman, *Angew. Chem., Int. Ed. Engl.*, **5**, 151 (1966).
20. H. Pasternak, T. Głowiak and F. Pruchnik, *Inorg. Chim. Acta*, **19**, 11 (1976).
21. A. Musco and P. F. Swinton, *J. Organomet. Chem.*, **50**, 333 (1973); R. Claverini, P. Ganis and C. Pedone, *J. Organomet. Chem.*, **50**, 327 (1973).
22. H. Kurosawa and S. Numata, *J. Organomet. Chem.*, **175**, 143 (1979).
23. A. P. Belov, I. I. Mojseev, N. G. Satsko, and J. K. Syrkin, *Izv. Akad. Nauk SSSR, Ser. Khim.*, 265 (1971); A. P. Belov, A. I. Zakhariev, I. I. Mojseev, N. G. Satsko, and J. K. Syrkin, *Izv. Akad. Nauk SSSR, Ser. Khim.*, 2681 (1970); A. P. Belov, I. I. Mojseev, and J. K. Syrkin, *Izv. Akad. Nauk SSSR, Ser. Khim.*, **40** (1970).
24. R. R. Schrock, B. F. G. Johnson, and J. Lewis, *J. Chem. Soc., Dalton Trans.*, 951 (1974).
25. H. Hrist and R. Huettel, *Angew. Chem., Int. Ed. Engl.*, **2**, 626 (1963).
26. R. Huettel and P. Kochs, *Chem. Ber.*, **101**, 1043 (1968).
27. F. R. Hartley and M. D. Higs, *J. Organomet. Chem.*, **44**, 197 (1972).
28. T. A. Schenach and F. F. Caserio, *J. Organomet. Chem.*, **18**, P17 (1969).
29. P. M. Maitlis, *The Organic Chemistry of Palladium*, Academic Press, New York (1971).
30. R. Baker, *Chem. Rev.*, **73**, 487 (1973).
31. G. P. Chiusoli, *Acc. Chem. Res.*, **6**, 422 (1973).
32. D. M. P. Mingos, in: *Comprehensive Organometallic Chemistry* (G. Wilkinson, F. G. A. Stone, and E. W. Abel, eds.), Vol. 3, Chapter 19, Pergamon Press, Oxford (1982).
33. P. W. Jolly, *Angew. Chem., Int. Ed. Engl.*, **24**, 283 (1985).
34. R. Goddard, C. Kruger, F. Mark, R. Stansfield, and X. Zhang, *Organometallics*, **4**, 285 (1985).
35. H. Werner, *Adv. Organomet. Chem.*, **19**, 155 (1981).
36. S. P. Gubin and G. B. Shulpin, *Khimia Kmpleksov so Svyazyami Metall-uglerod*, Izdatielstvo Nauka, Sibirskoye Otdelenie, Novosibirsk (1984).
37. R. Benn and A. Rufinska, *Organometallics*, **4**, 209 (1985).
38. P. W. Jolly and R. Mynot, *Adv. Organomet. Chem.*, **19**, 257 (1981).
39. B. M. Trost, L. Weber, P. E. Strege, T. J. Fullerton, and T. J. Dietsche, *J. Am. Chem. Soc.*, **100**, 3416 (1978).
40. B. E. Mann, in: *Comprehensive Organometallic Chemistry* (G. Wilkinson, F. G. A. Stone, and E. W. Abel, eds.), Vol. 3, Chapter 20, Pergamon Press, Oxford (1982).
41. J. C. Green, *Structure and Bonding*, **43**, 37 (1981).
42. M. C. Boehm, R. Gleiter, and C. D. Batich, *Helv. Chim. Acta*, **63**, 990 (1980).
43. M. C. Boehm and R. Gleiter, *Theor. Chim. Acta*, **57**, 315 (1980).
44. I. Omae, *Coord. Chem. Rev.*, **53**, 261 (1984).
45. M. Schumann, *Angew. Chem., Int. Ed. Engl.*, **23**, 474 (1984).
46. T. J. Marks, in: *Comprehensive Organometallic Chemistry* (G. Wilkinson, F. G. A. Stone, and E. W. Abel, eds.), Vol. 3, Chapter 21, Pergamon Press, Oxford (1982).
47. E. W. Gawling and S. F. A. Kettle, *Inorg. Chem.*, **3**, 604 (1964).

48. D. L. Weaver and R. M. Tuggle, *J. Am. Chem. Soc.*, **91**, 6506 (1969); M. D. Rausch, R. M. Tuggle, and D. L. Weaver, *J. Am. Chem. Soc.*, **92**, 4981 (1970).
49. D. H. Richards, *Chem. Soc. Rev.*, **6**, 235 (1977).
50. C. D. Batich, *J. Am. Chem. Soc.*, **96**, 7585 (1976); J. C. Green and E. A. Seddon, *J. Organomet. Chem.*, **198**, C61 (1980).
51. M. Bottrill, P. D. Gavens, J. W. Kelland, and J. McMeeking, in: *Comprehensive Organometallic Chemistry* (G. Wilkinson, F. G. A. Stone, and E. W. Abel, eds.), Vol. 3, Section 22.3, Pergamon Press, Oxford (1982).
52. D. J. Cardin, M. F. Lappert, C. L. Raston, and P. I. Riley, in: *Comprehensive Organometallic Chemistry* (G. Wilkinson, F. G. A. Stone, and E. W. Abel, eds.), Vol. 3, Section 23.2, Pergamon Press, Oxford (1982).
53. P. W. Jolly, C. Kruger, C. C. Ramao, and M. J. Ramao, *Organometallics*, **3**, 936 (1984).
54. P. G. Lenhert, C. M. Lukehart, and K. Srinivasan, *J. Am. Chem. Soc.*, **106**, 124 (1984).
55. A. J. Deeming, in: *Comprehensive Organometallic Chemistry* (G. Wilkinson, F. G. A. Stone, and E. W. Abel, eds.), Vol. 4, Section 31.3, Pergamon Press, Oxford (1982).
56. W. P. Fehlhammer and H. Stolzenberg, in: *Comprehensive Organometallic Chemistry* (G. Wilkinson, F. G. A. Stone, and E. W. Abel, eds.), Vol. 4, Section 31.4, Pergamon Press, Oxford (1982).
57. M. A. Bennett, M. I. Bruce, and T. W. Matheson, in: *Comprehensive Organometallic Chemistry* (G. Wilkinson, F. G. A. Stone, and E. W. Abel, eds.), Vol. 4, Sections 32.3 and 32.4, Pergamon Press, Oxford (1982).
58. R. D. Adams and J. P. Selegue, in: *Comprehensive Organometallic Chemistry* (G. Wilkinson, F. G. A. Stone, and E. W. Abel, eds.), Vol. 4, Chapter 33, Pergamon Press, Oxford (1982).
59. T. V. Ashworth, D. C. Liles, and E. Singleton, *Organometallics*, **3**, 1851 (1984).
60. R. K. Brown, J. M. Williams, A. J. Schultz, G. D. Stucky, S. D. Ittel, and R. L. Harlow, *J. Am. Chem. Soc.*, **102**, 981 (1980).
61. M. Brookhart, W. Lamanna, and M. B. Humphrey, *J. Am. Chem. Soc.*, **104**, 2117 (1982); A. J. Schultz, R. G. Teller, M. A. Beno, J. M. Williams, M. Brookhart, W. Lamanna, and M. B. Humphrey, *Science*, **220**, 197 (1983).
62. W. D. McGhee and R. G. Bergman, *J. Am. Chem. Soc.*, **107**, 3388 (1985).
63. M. D. Fryzuk, *Inorg. Chim. Acta*, **54**, L265 (1981).
64. G. Granozzi, E. Tondello, R. Bertoncello, S. Aime, and D. Osella, *Inorg. Chem.*, **24**, 570 (1985).
65. C. A. Ghilardi, S. Midollini, and A. Orlandini, *J. Organomet. Chem.*, **295**, 377 (1985).
66. M. R. Churchill, J. C. Fettinger, L. G. McCullough, and R. R. Schrock, *J. Am. Chem. Soc.*, **106**, 3356 (1984).
67. C. Mealli, S. Midollini, S. Moneti, L. Sacconi, J. Silvestre, and T. A. Albright, *J. Am. Chem. Soc.*, **104**, 95 (1982); M. D. McClure and D. L. Weaver, *J. Organomet. Chem.*, **54**, C59 (1974).
68. M. Green, R. J. Mercer, C. E. Morton, and A. G. Orpen, *Angew. Chem., Int. Ed. Engl.*, **24**, 422 (1985).
69. R. Appel, W. Schuhn, and F. Knoch, *Angew. Chem., Int. Ed. Engl.*, **24**, 420 (1985).
70. H. R. Keable and M. Kilner, *J. Chem. Soc., Dalton Trans.*, 153 (1972); T. Inglis, M. Kilner, T. Reynoldson, and E. E. Robertson, *J. Chem. Soc., Dalton Trans.*, 924 (1975); T. Inglis and M. Kilner, *J. Chem. Soc., Dalton Trans.*, 930 (1975).
71. M. DiVaira and L. Sacconi, *Angew. Chem., Int. Ed. Engl.*, **21**, 330 (1982).
72. O. J. Scherer, *Angew. Chem., Int. Ed. Engl.*, **24**, 924 (1985).
73. M. DiVaira, C. A. Ghilardi, S. Midollini, and L. Sacconi, *J. Am. Chem. Soc.*, **100**, 2550 (1978).
74. M. DiVaira, S. Midollini, and L. Sacconi, *J. Am. Chem. Soc.*, **101**, 1757 (1979).
75. L. Fabbrizzi and L. Sacconi, *Inorg. Chim. Acta*, **36**, L407 (1979).
76. C. Bianchini, M. DiVaira, A. Meli, and L. Sacconi, *Inorg. Chem.*, **20**, 1169 (1981); *Angew. Chem., Int. Ed. Engl.*, **19**, 405 (1980); *J. Am. Chem. Soc.*, **103**, 1448 (1981).

# Chapter 8

# Compounds Containing Four-Electron π-Ligands

## 1. THE BONDING

The following compounds are classified as $4e$ ligands: 1,3-dienes, cyclobutadiene, and trimethylenemethane, and their derivatives.

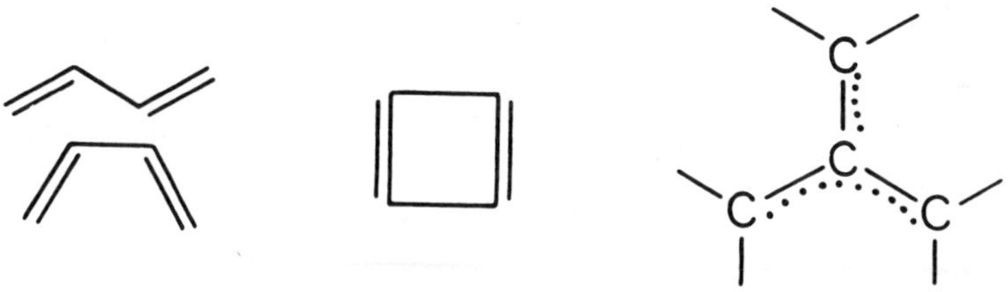

$$(8.1)$$

In most complexes, acyclic 1,3-dienes are coordinated as *cis* planar ligands. However, compounds in which such acyclic groups are coordinated to the metal atom in their *trans* forms are also known.[38–41] Conjugated dienes may also function as bridging ligands in polynuclear compounds in which the $C-C$ double bonds are attached to two central atoms. Each $4e$ ligand has four molecular orbitals (Figure 8.1).[1–5, 31–34]

The two lowest orbitals are occupied in the 1,3-diene molecule, therefore the electron configuration is $\psi_1^2\psi_2^2$. Cyclobutadiene and trimethylenemethane have electron configuration $\psi_1^2\psi_2^1\psi_3^1$ and thus are biradicals.[6] Table 8.1 gives symmetries of orbitals of the ligands and the central atom as well as the type of metal–ligand bonding for the M(butadiene), M(cyclobutadiene), and M(trimethylenemethane) fragments, which have local symmetries $C_s$, $C_{4v}$, and $C_{3v}$, respectively. In metal compounds containing

471

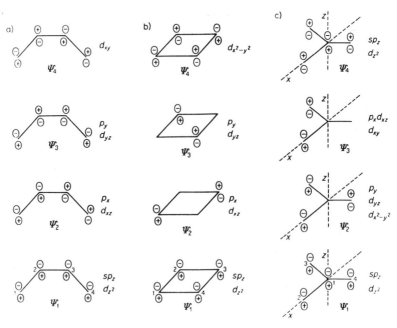

Figure 8.1. Molecular orbitals of (a) 1,3-dienes, (b) cyclobutadiene, (c) trimethylenemethane. Metal orbitals which may form bonds with the given orbital of the ligand are shown.

1,3-dienes, the overlap of occupied $\psi_1$ and $\psi_2$ orbitals with corresponding orbitals of the metal causes charge transfer from the ligand to the central atom. An essential role is also played by $\pi$ back-bonding. Structural data show that, owing to coordination of the butadiene molecule, there is a shortening of the central $C-C$ bond and elongation of the terminal carbon–carbon double bonds. In contrast to the free molecule, the coordinated butadiene possesses a central bond which is somewhat shorter than the terminal bonds. The *ab initio* calculations for $Fe(C_4H_6)(CO)_3$[8] are in agreement with these observations. They show that the butadiene ligand possesses a considerably negative charge $(-0.88)$.[8] Therefore, this situation more resembles the anion than the diene molecule in the excited state.

The accumulation of the negative charge on the ligand occurs owing to the formation of the effective iron–butadiene back-bonding which results from the overlap of the occupied $d$ orbital and the lowest antibonding diene $\psi_3$ orbital. As a result of the position of the nodal planes of this orbital (Figure 8.1), the charge transfer from the central atom to the ligand leads to a shortening of the central carbon–carbon bond and elongation of the terminal bonds. For cyclobutadiene complexes, the most essential interaction takes place between the $\psi_2$ and $\psi_3$ ligand orbitals and the metal. The overlap between the $\psi_4$ and $d_{x^2-y^2}$ orbitals is not great, and therefore their contribution to the metal–cyclobutadiene bond is small. Figure 8.2 represents molecular orbital schemes for the $MC_4R_4$ (for $C_4$ symmetry) and $ML_3$ (for $C_3$ symmetry) fragments of complexes possessing $[M(C_4R_4)L_3]$ formula; in the case of $ML_3$, only $\sigma$ ligand orbitals are considered. Separate consideration of molecular orbitals allows a discussion of fundamental characteristics of the metal–ligand bond for simple systems possessing

Table 8.1. *Orbital Symmetry and Types of Bonds for the $M(C_4H_6)$, $M(C_4H_4)$, and $M\{C(CH_2)_3\}$ Fragments*[a]

| Symmetry | Orbitals of the ligand | Orbitals of the metal | Type of bond |
|---|---|---|---|
| | $M(CH_2{=}CHCH{=}CH_2)$, group $C_s$ | | |
| $A'$ | $\psi_1 = 0.37\,\psi_1 + 0.60\,\psi_2 + 0.60\,\psi_3 + 0.37\,\psi_4$ | $s, p_z, d_{z^2}$ | $\sigma$ |
| $A'$ | $\psi_3 = 0.60\,\psi_1 - 0.37\,\psi_2 - 0.37\,\psi_3 + 0.60\,\psi_4$ | $p_y, d_{yz}$ | $\pi$ |
| $A''$ | $\psi_2 = 0.60\,\psi_1 + 0.37\,\psi_2 - 0.37\,\psi_3 - 0.60\,\psi_4$ | $p_x, d_{xz}$ | $\pi$ |
| $A''$ | $\psi_4 = 0.37\,\psi_1 - 0.60\,\psi_2 + 0.60\,\psi_3 - 0.37\,\psi_4$ | $d_{xy}$ | $\delta$ |

$\text{M}\!-\!\diamondsuit\!\bigcirc$ , group $C_{4v}$

| | | | |
|---|---|---|---|
| $A_1$ | $\psi_1 = \dfrac{1}{2}(\psi_1 + \psi_2 + \psi_3 + \psi_4)$ | $s, p_z, d_{z^2}$ | $\sigma$ |
| $E$ | $\psi_2 = \dfrac{1}{\sqrt{2}}(\psi_1 - \psi_3);\ \psi_3 = \dfrac{1}{\sqrt{2}}(\psi_2 - \psi_4)$ | $p_x, d_{xz},\ p_y d_{yz}$ | $\pi$ |
| $B_1$ | $\psi_4 = \dfrac{1}{2}(\psi_1 - \psi_2 + \psi_3 - \psi_4)$ | $d_{x^2-y^2}$ | $\delta$ |
| $B_2$ | | $d_{xy}$ | |

| | $M\{C(CH_2)_3\}$, group $C_{3v}$ | | |
|---|---|---|---|
| $A_1$ | $\psi_1 = \dfrac{1}{\sqrt{10}}\psi_1 + \sqrt{\dfrac{3}{10}}(\psi_2 + \psi_3 + \psi_4)$ | $s, p_z, d_{z^2}$ | |
| $E$ | $\psi_2 = \dfrac{1}{\sqrt{6}}(\psi_2 + \psi_3 - 2\psi_4);\ \psi_3 = \dfrac{1}{2}(\psi_2 - \psi_3)$ | $p_x, p_y, d_{xy}, d_{xz}, d_{yz}, d_{x^2-y^2}$ | |
| $A_1$ | $\psi_4 = \dfrac{1}{\sqrt{10}}\psi_1 - \sqrt{\dfrac{3}{10}}(\psi_2 + \psi_3 + \psi_4)$ | $s, p_z, d_{z^2}$ | |

[a] According to References 1, 2, and 5–7.

higher symmetry. The symmetry of the entire $M(C_4R_4)L_3$ molecule, for example, $[Fe(C_4R_4)(CO)_3]$ and $[Ni_2Cl_4(C_4R_4)_2]$, is considerably lower $(C_s)$.[2, 5, 7, 9]

The molecular orbital diagram (Figure 8.2) shows that these compounds are 18*e* systems. The calculations show that in the complexes $Fe(\eta\text{-}C_4H_4)(CO)_3$, $[Fe(\eta^4\text{-}C_4H_4CO)(CO)_3]$, $Cocp(\eta\text{-}C_4H_4)$, and $[Cocp(\eta^4\text{-}C_4H_4CO)]$ $(C_4H_4CO =$ cyclopentadienone) the Cocp fragment forms a stronger bond with cyclobutadiene than with cyclopentadienone.[47] In the case of the $Fe(CO)_3$ fragment, the situation is reversed. One of the essential reasons for this is the difference in electron repulsion in these two fragments. In both cobalt complexes the diene–M bonds are more covalent than in the iron compounds.

The analysis of the diagram representing the interaction of $Fe(CO)_3$ with 1,3-butadiene leads to the conclusion that conformation (8.2) of $Fe(C_4H_6)(CO)_3$ is the most stable (Figure 8.3). Experiments show that in the case of trimethylenemethane complexes of the type $Fe(tmm)(CO)_3$ (tmm = trimethylenemethane and its derivatives),

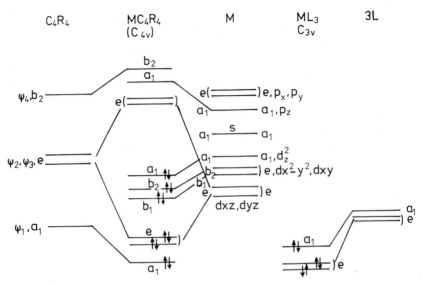

Figure 8.2. Molecular orbital diagram for complexes $M(C_4R_4)L_3$, $M(C_4R_4)(CO)_3$, and $Ni_2Cl_4(C_4R_4)_2$.

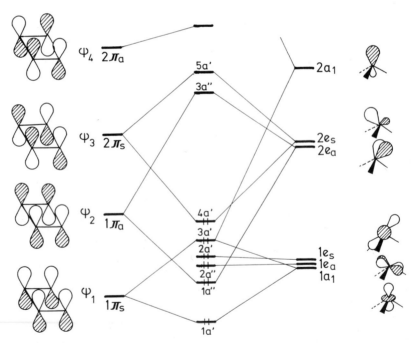

Figure 8.3. Molecular orbital diagram for the complex $[Fe(C_4H_6)(CO)_3]$ possessing the conformation presented in (8.2). Three Fe–CO bonding orbitals of the $Fe(CO)_3$ fragment are not shown (*cf.* Figure 8.2).

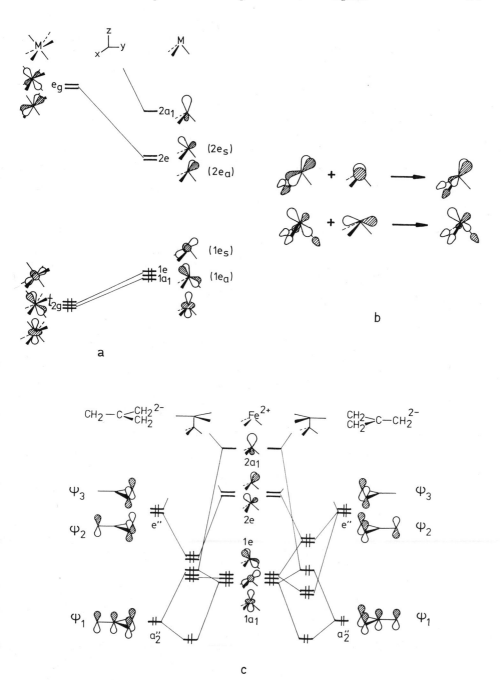

Figure 8.4. (a) Valence orbitals for $ML_6$ and $ML_3$, (b) hybridization of $d$ and $p$ orbitals for the $ML_3$ fragment, and (c) molecular orbital diagram for the complex tmm–$Fe(CO)_3$ for staggered and eclipsed conformations.

the staggered conformation is considerably more stable than the eclipsed one. The $M(CO)_3$ fragment has three "$t_{2g}$" orbitals, that is, one $1a_1(d_{z^2})$ and two $1e$ orbitals: $d_{xy}$ (with a certain contribution of the $d_{xz}$ orbital) and $d_{x^2-y^2}$ (with a $d_{yz}$ contribution). The

(8.2)

"$e_g$" orbitals which possess higher energy are $d_{xz}$ (with a $d_{xy}$ contribution) and $d_{yz}$ (with a certain contribution of $d_{x^2-y^2}$). This different composition of $t_{2g}$ and $e_g$ orbitals results from a different choice of coordinates (Figure 8.4). The threefold axis of the octahedron represents the $z$ axis. The lowering of the symmetry to $C_{3v}$ in the case of the $M(CO)_3$ fragment enables hybridization of the $2e$ ($d_{xz}$ and $d_{yz}$) orbitals with the $p_x$ and $p_y$ orbitals. The mixing of these orbitals causes their shifting away from the carbonyl groups and affords possibilities for a more effective interaction with the hydrocarbon. In addition to $1a_1$, $1e$, and $2e$ orbitals, a certain role may also be played by the $2a_1$ orbital which, in the starting octahedron, had been as $s$ orbital. In $M(CO)_3$, this orbital is mixed with the $p_z$ and $d_{z^2}$ orbitals to a considerable degree. Figure 8.4 represents the interaction between orbitals of $[C(CH_2)_3]^{2-}$ and $[Fe(CO)_3]^{2+}$ for two extreme conformations: staggered and eclipsed. Trimethylenemethane is a ligand in which the $\pi$ orbitals are tilted. Therefore, considerably more effective interaction between the $2e_s$ ($d_{yz}$) and $e_s''$ orbitals as well as between the $2e_a$ ($d_{xz}$) and $e_a''$ orbitals takes place in the staggered complex than in the eclipsed compound[42] ($e_s$ and $e_a$ designate corresponding symmetrical and asymmetrical orbitals with respect to the plane of the page). This explains high activation energy of the rotation of trimethylenemethane about the Fe–tmm axis.

## 2. STRUCTURE OF COMPOUNDS WITH 4e LIGANDS

In cyclobutadiene, and trimethylenemethane complexes, all metal–carbon distances are equal. Exceptions were found only in coordination compounds in which an essential role is played by steric factors. In compounds containing 1,3-dienes, the distance of terminal C–C bonds is larger, and the distance of the middle C–C bond is shorter than for the free dienes. This argues for very strong back-bonding between the metal and the diene. (See discussion of the M–diene bond.) The above-described changes are greater in cyclopentadienyl complexes [Mcp(diene)] than in carbonyl derivatives such as [Fe(diene)(CO)_3] because the CO groups possess stronger electron-acceptor properties than the cyclopentadienyl ligand. Therefore, in the complex [Mcp(diene)], the $C^1$–$C^2$ and $C^3$–$C^4$ distances are longer than the $C^2$–$C^3$ distances, while in [Fe(diene)(CO)_3], all C–C distances are almost equal. Hydrogen atoms at the middle carbon atoms of the diene grouping in tricarbonyl[2-{nitrophenyl)amine}-3,5-heptadiene] iron are tilted away from the plane of the diene carbon atoms toward the iron atom by 6°. The *syn*-atoms are shifted in the same direction (by 20°) whereas the *anti*-atoms are tilted to the opposite side by 30°.

In the compound [Mn(C_4H_6)_2(CO)], the hydrogen atoms bonded to the middle

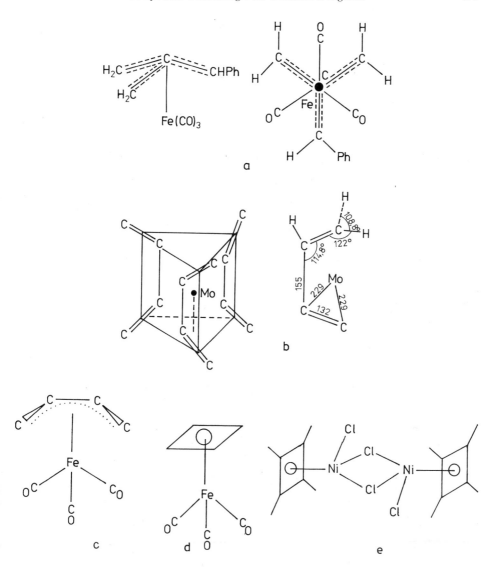

Figure 8.5  Structures of (a) [Fe{C(CH₂)₂CHPh}(CO)₃], (b) [Mo(C₄H₆)₃], (c) [Fe(C₄H₆)(CO)₃], (d) [Fe(C₄R₄)(CO)₃], (e) [Ni₂Cl₄(C₄Me₄)₂] (distances are in pm units).

carbon atoms lie in the plane which is closer to the manganese atom than the carbon atoms (tilting by 24 pm). The *syn*-protons are located on the outer side of the C¹C²C³C⁴ plane (tilting by +72 pm), and the *anti*-protons lie in the ligand plane.[13]

In most complexes, 1,3-dienes are coordinated in the s-*cis* form. However, compounds in which such ligands are bonded only to one metal atom in the s-*trans* form are known, for example, [Mocp(s-*trans*-η⁴-diene)]  (diene = 2,3-dimethylbutadiene, 2-methylbutadiene, and 2,5-dimethyl-2,4-hexadiene),[38] [Zrcp₂(CR¹Me=CR²CR³= =CR⁴H),[39,41] and Zrcp₂(*trans*-η⁴-PhCH=CHCH=CHPh).[40] In the complex

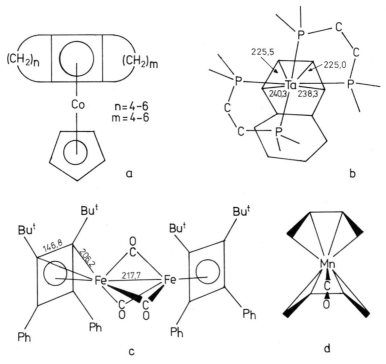

Figure 8.6. Structures of (a) $[Cocp(CH_2)_nC_4(CH_2)_m]$, (b) $Ta(\eta^4\text{-}C_{10}H_8)(Me_2PC_2H_4PMe_2)_2]$, (c) $[Fe_2(C_4Bu_2^tPh_2)_2(CO)_3]$, (d) $[Mn(C_4H_6)_2(CO)]$ (distances are in pm units).

Table 8.2. *Structures of Metal Diene, Cyclobutadiene, and Trimethylenemethane Complexes*

| Compound | Interatomic distances (pm) | | | | | | |
|---|---|---|---|---|---|---|---|
| | $M\text{-}C^1$ | $M\text{-}C^2$ | $M\text{-}C^3$ | $M\text{-}C^4$ | $C^1\text{-}C^2$ | $C^2\text{-}C^3$ | $C^3\text{-}C^4$ |
| $[Fe(C_4H_6)(CO)_3]^a$ | 214 | 206 | 206 | 214 | 146 | 145 | 146 |
| $[Fe\{C_5(CF_3)_4O\}(CO)_3]^b$ | 204 | 199 | 201 | 212 | 136 | 139 | 142 |
| $[Fe(PhCH=CHCH=CHPh)(CO)_3]^c$ | 215 | 213 | 213 | 215 | 142 | 142 | 142 |
| $[Fe(C_4H_6)(C_8H_8)(CO)]^d$ | 214 | 205 | 205 | 214 | BD | | |
| | 214 | 204 | 204 | 214 | COT | | |
| $[Fecp(GeCl_2Me)(C_4H_6)]^e$ | 213 | 202 | 202 | 213 | 149 | 136 | 149 |
| $[Fe(1,3\text{-cyclohexadiene})_2(CO)]^f$ | 211.9 | 203.3 | 203.7 | 211.5 | 140.4 | 140.8 | 139.6 |
| | 212.7 | 203.8 | 203.7 | 212.4 | 138.4 | 143.2 | 139.1 |
| $[Fe_2(\mu\text{-COT})(CO)_6]^g$ | 214 | 206 | 206 | 214 | 144 | 140 | 144 |
| $[Fe(COT)(CO)_3]^{g,h}$ | 218 | 205 | 205 | 218 | 142 | 142 | 142 |
| $[Mo(C_4H_6)_3]^{(25c)}$ | 229 | 229 | 229 | 229 | 132 | 155 | 132 |
| $[Co_2(C_4H_6)_2(CO)_4]^i$ | 210 | 215 | 215 | 210 | 137 | 164 | 137 |
| $[Cocp(C_5H_5Ph)]^j$ | 202 | 203 | 197 | 198 | 149 | 136 | 154 |
| $[Rhcp\{C_6(CF_3)_6\}]^k$ | 215 | 206 | 211 | 212 | 148 | 142 | 153 |
| $[RhCl(C_4H_6)_2]^l$ | 221 | 215 | 215 | 221 | 138 | 145 | 138 |
| $[Mn(C_4H_6)_2(CO)]^m$ | 215 | 206 | 206 | 215 | 139 | 146 | 139 |

*(Table continued)*

*Table 8.2. (continued)*

| Compound | Interatomic distances (pm) | | | | | | |
|---|---|---|---|---|---|---|---|
| | M–C$^1$ | M–C$^2$ | M–C$^3$ | M–C$^4$ | C$^1$–C$^2$ | C$^2$–C$^3$ | C$^3$–C$^4$ |
| Cyclobutadiene and trimethylmethane complexes | | | | | | | |
| [Ni$_2$Cl$_4$(C$_4$Me$_4$)$_2$]$^m$ | Ni–C: 199.7–204.7; C–C: 140–145 | | | | | | |
| [Fe(C$_4$Ph$_4$)(CO)$_3$]$^n$ | Fe–C: 206.7 (av); C–C: 146 (av), ∢CCC: 90.2° | | | | | | |
| [Co(C$_5$H$_4$SiMe$_3$)(C$_4$Ph$_4$)]$^o$ | Co–C: 198.5 (av); C–C: 147 (av) | | | | | | |
| [Nbcp(C$_2$Ph$_2$)(C$_4$Ph$_4$)(CO)]$^p$ | Nb–C: 238; C–C: 146, ∢CCC: 90.0° | | | | | | |
| [(OC)Fe{Ph$_2$C$_4$(C$_6$H$_4$)$_2$C$_4$Ph$_2$}Fe(CO)$_3$]$^q$ | Fe–C: 205; C–C: 145, ∢CCC: 89.6° | | | | | | |
| [see equation (8.3)] | Fe–Fe: 249.4 | | | | | | |
| [Me$_4$C$_4$Ni(MeCCMe)$_2$Fe(CO)$_3$]$^r$ | Ni–C: 202; C–C: 146, ∢CCC: 90.0° | | | | | | |
| | Fe–Ni: 244.9 | | | | | | |
| [Mo(C$_4$Ph$_4$)$_2$(CO)$_2$]$^{(9)}$ | Mo–C: 231; C–C: 147, ∢CCC: 89.9° | | | | | | |
| [Mo$_2$Br$_2$(C$_4$Ph$_4$)$_2$(CO)$_4$]$^s$ | Mo–C: 225; C–C: 146, ∢CCC: 90.0° | | | | | | |
| | Mo–Mo: 295.4 | | | | | | |
| [Cocp(C$_4$H$_4$)]$^t$ | Co–C: 196; C–C: 144, ∢CCC: 90.1° | | | | | | |
| [Rhcp(C$_4$Ph$_4$)]$^u$ | Rh–C: 210; C–C: 147, ∢CCC: 90.0° | | | | | | |
| [Vcp(C$_4$Ph$_4$)(CO)$_2$]$^v$ | V–C: 226; C–C: 147, ∢C: 90.0° | | | | | | |
| [Fe{(PhC$^4$H)C$^1$(C$^{2,3}$H$_2$)$_2$}(CO)$_3$]$^w$ | Fe–C$^3$: 211.8, Fe–C$^4$: 216.2, C$^1$–C$^2$: 140.5 | | | | | | |
| | C$^1$–C$^3$: 140.6, C$^1$–C$^4$: 143.6 | | | | | | |

$^a$ O. S. Mills and G. Robinson, *Acta Crystallogr.*, **16**, 758 (1963); M. J. Davis and C. S. Speed, *J. Organomet. Chem.*, **21**, 401 (1970).

$^b$ N. A. Bailey and R. Mason, *Acta Crystallogr.*, **21**, 652 (1966).

$^c$ L. G. Kuzmina, J. T. Struchkov, and A. I. Nekhaev, *Zh. Strukt. Khim.*, **13**, 1115 (1972).

$^d$ I. W. Bassi and R. Scordamaglia, *J. Organomet. Chem.*, **37**, 353 (1972).

$^e$ F. M. Chaudhari and P. L. Pauson, *J. Organomet. Chem.*, **5**, 73 (1966); V. G. Andrianov, V. P. Martynov, and J. T. Struchkov, *Zh. Strukt. Khim.*, **12**, 866 (1971); V. G. Andrianov, V. P. Martynov, K. N. Anisimov, N. E. Kolobova, and V. V. Skripkin, *Chem. Commun.*, 1252 (1970).

$^f$ C. Kruger and Y.-H. Tsay, *J. Organomet. Chem.*, **33**, 59 (1971); C. Kruger and Y.-H. Tsay, *Angew. Chem., Int. Ed. Engl.*, **10**, 261 (1971).

$^g$ B. Dickens and W. N. Lipscomb, *J. Chem. Phys.*, **37**, 2084 (1962); B. Dickens and W. N. Lipscomb, *J. Am. Chem. Soc.*, **83**, 489 (1961).

$^h$ B. Dickens and W. N. Lipscomb, *J. Am. Chem. Soc.*, **83**, 4862 (1964).

$^i$ R. O. Jones and E. N. Maslen, *Z. Kristallogr.*, **123**, 330 (1966).

$^j$ M. R. Churchill and R. Mason, *Proc. R. Soc. London, Ser. A*, **279**, 191 (1964).

$^k$ M. R. Churchill and R. Mason, *Proc. R. Soc. London, Ser. A*, **292**, 61 (1966).

$^l$ A. Immirzi and G. Allegra, *Acta Crystallogr.*, **B25**, 120 (1969).

$^m$ J. D. Dunitz, H. C. Mez, O. S. Mills, and H. M. M. Shearer, *Helv. Chim. Acta*, **45**, 647 (1962).

$^n$ R. P. Dodge and V. Schomaker, *Acta Crystallogr.*, **18**, 614 (1965).

$^o$ M. Calligaris and K. Venkatasubramanian, *J. Organomet. Chem.*, **175**, 95 (1979).

$^p$ A. I. Gusew and J. T. Struchkov, *Zh. Strukt. Chim.*, **10**, 515 (1969).

$^q$ E. F. Epstein and L. F. Dahl, *J. Am. Chem. Soc.*, **92**, 493 (1970).

$^r$ E. F. Epstein and L. F. Dahl, *J. Am. Chem. Soc.*, **92**, 502 (1970).

$^s$ M. Mathew and G. L. Palenik, *J. Organomet. Chem.*, **61**, 301 (1973).

$^t$ P. E. Riley and R. E. Davis, *J. Organomet. Chem.*, **113**, 157 (1976).

$^u$ G. G. Cash, J. F. Helling, M. Mathew, and G. J. Palenik, *J. Organomet. Chem.*, **50**, 227 (1973).

$^v$ A. I. Gusev, G. G. Aleksandrov, and Yu. T. Struchkov, *Zh. Strukt. Khim.*, **10**, 665 (1969).

$^w$ M. R. Churchill and K. Gold, *Inorg. Chem.*, **8**, 401 (1969).

[Mocp(s-*trans*-$\eta^4$-Me$_2$C=CH–CH=CMe$_2$)] the diene is a nonplanar, transoidal ligand [the torsional angle is 124.8(4°)]. The C–C distances in the diene fragment are similar, ranging from 140.1 to 141.8 pm, and the Mo–C$_{\text{diene}}$ distances are: Mo–C$^2$ 239.0 pm, Mo–C$^3$ 220.9 pm, Mo–C$^4$ 223.4 pm, and Mo–C$^5$ 236.5 pm. In the complex Zrcp$_2$(*trans*-$\eta^4$-PhCH=CHCH=CHPh) the structure of the diene is different; the C=C distances are 140 pm, and the C–C distance is 148 pm.[40] Dimeric complexes containing bridging *trans*-1,3-dienes bonded to two central atoms are also known, i.e., $L_nM$ ⟋⟍ $ML_n$. The following complexes serve as examples: $[\{Mncp(CO)_2\}_2$ $(\mu\text{-}\eta^4\text{-}C_4H_6)]$,[43] $[\{Mn(CO)_4\}_2$ $(\mu\text{-}\eta^4\text{-}C_4H_6)]$,[44] $[Os_3(CO)_{10}$ $(\mu\text{-}\eta^4\text{-}C_4H_6]$,[45] $[M_2(allyl)_2$ $(\mu\text{-}Cl)_2Cl_2(\mu\text{-}\eta^4\text{-}C_4H_6)]$ $(M = Rh, Ir)$ (see Table 7.12).

The C–C distances in cyclobutadiene complexes are the same; generally the substituents are located away from the metal with respect to the plane of the ligand. Exceptions are encountered in some complexes possessing substituents which force their tilting toward the metal. The dinuclear compound which is prepared by reaction (8.3) may

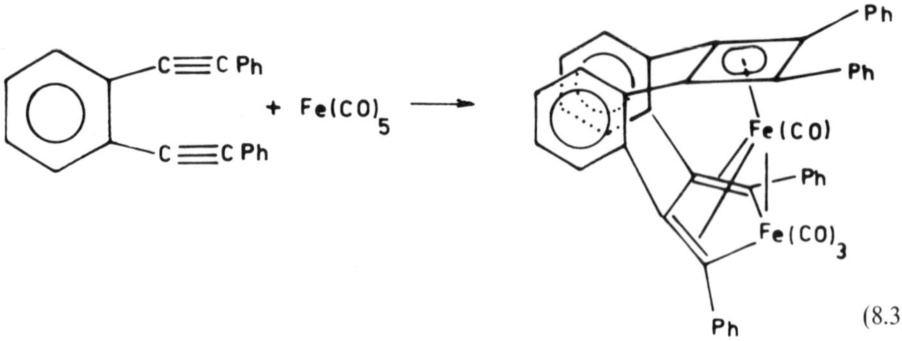

$$(8.3)$$

serve as an example. The *o*-phenyl substituents are tilted toward the iron atom, while the phenyl substituents are tilted in the opposite direction.

In trimethylenemethane complexes, the metal atom is located under the central carbon atom. The terminal carbon atoms of the ligand are located in a plane which is closer to the metal atom than the central carbon, and therefore all metal–carbon distances are the same. Thus, the coordinated trimethylenemethane molecule is bent and possesses $C_{3v}$ symmetry. The structures of complexes containing $4e$ ligands are represented in Figures 8.5 and 8.6, and the interatomic distances are given in Table 8.2.

## 3. NMR SPECTRA

The proton nuclear magnetic resonance spectrum of free *trans*-butadiene consists of two groups of signals. The signal located at lower fields ($\tau$ *ca* 3.8) corresponds to protons connected to $C^2$ and $C^3$ atoms, while the doubly intense signal at $\tau = 4.9$ corresponds to protons at $C^1$ and $C^4$.

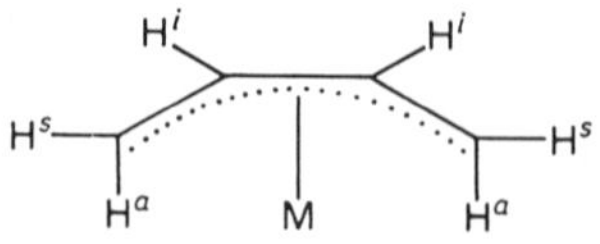

$$(8.4)$$

The spectrum of the iron complex $[Fe(C_4H_6)(CO)_3]$ is of $A_2M_2X_2$ type. Owing to coordination, the bands are shifted in the direction of stronger fields. The internal protons give a signal at $\tau = 4.7$, the *syn* $H^s$ protons at $\tau = 8.3$, and the *anti* $H^a$ hydrogens at $\tau = 9.78$. Therefore, the spectra of butadiene complexes are similar to those of allyl compounds.

In 1,3-diene compounds of iron that do not contain substituents at $C^1$ or $C^2$, the $\tau$ values are as follows: $\tau(H^a) = 9.5$–$9.8$ and $\tau(H^s) = 8.0$–$8.4$. Substituents somewhat decrease the $\tau$ values. Based on the $^1H$ NMR spectra it is possible to determine if the diene is coordinated in the transoidal or cisoidal form.[39] In complexes (8.5), the $J(2,3)$

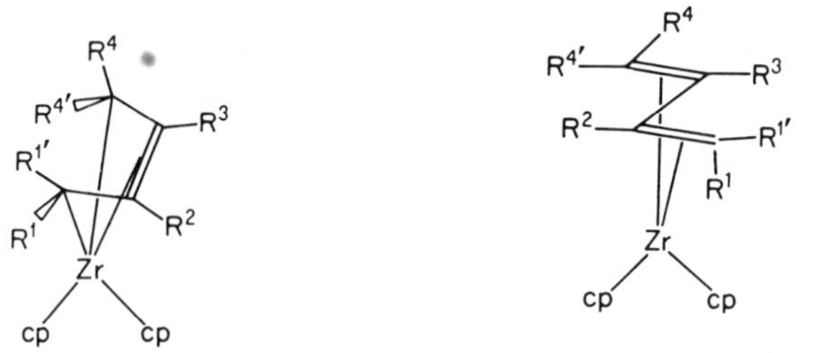

(8.5)

a. $R^{1' \ 3} = R^{4'} = H$, $R^4 = Me$
b. $R^{2,3} = R^{1'} = R^{4'} = H$
   $R^1 = R^4 = Me$
c. $R^{1' \ 3} = R^{4'} = H$, $R^4 = Et$

d. $R^{1' \ 3} = R^{4'} = H$, $R^4 = Me$
e. $R^{1'} = R^{2,3} = R^{4'} = H$
   $R^1 = R^4 = CH_3$
f. $R^{1' \ 3} = R^{4'} = H$, $R^4 = Et$

coupling constants for cisoidal complexes (8.5a–c) are in the 9.5–10.5 Hz range, while for transoidal compounds these constants are 14.5–15.2 Hz. The values of chemical shifts of the $H^1$ proton (the syn proton with respect to $H^2$) and those of $H^{1'}$ (the *anti*)proton may also serve to determine structures of diene complexes. The $H^1$ signals in *cis* isomers (8.5a–c) occur at $6.76 \pm 0.23$, and the $H^{1'}$ resonances are at $10.66 \pm 0.27$. These values are in complete agreement with shifts observed for $[Zrcp_2(s\text{-}cis\text{-}C_4H_6)]$ and $[Zrcp_2(s\text{-}cis\text{-}CH_2{=}CMeCH{=}CH_2)]$. The $H^{1'}$ signals in transoidal compounds (8.5d–f) appear at lower fields by 1.6–1.7 ppm.[39] Therefore, based on the $^1H$ NMR spectra, it is possible to determine the ratio of cisoidal to transoidal forms. The rotation of butadiene occurs relatively easily. The free enthalpy of activation $\Delta G^{\neq}$ for the rotation of butadiene and other 1,3-dienes is in the 25–65 kJ mol$^{-1}$ range. In the solid state all 18$e$ allyl, butadiene, pentadienyl, and hexatriene compounds possessing the $M(C_nH_{n+2})(CO)_3$ formula adopt the conformation (8.6a) [cf. structures (7.4)]. In the case of $Fe(C_4H_6)(CO)_3$, the conformation (8.6a) is preferred due to strong interaction

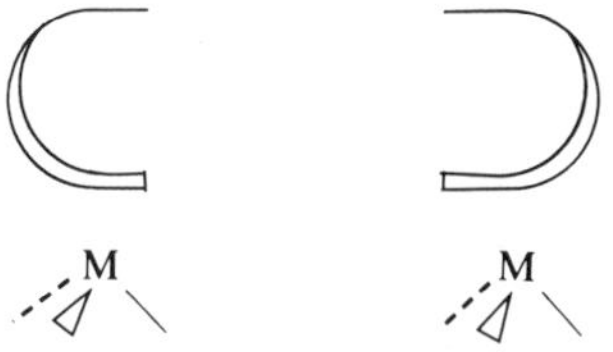

(8.6)

of the $\psi_2$ and $\psi_3$ orbitals of butadiene with the 2$e$ ($d_{xz}$ and $d_{yz}$) orbitals of the $Fe(CO)_3$ fragment (see Figure 8.3). For 16$e$ complexes, the conformation (8.6b) is more stable. The rearrangement of the 18$e$ butadiene complex (8.7a) into the form (8.7b) requires energy of activation. The rotation of the butadiene molecule may take place according to the mechanism of Berry's pseudorotation or angular twist (Figure 7.4) for pentacoordinate complexes, or according to the mechanism of trigonal twist for hexa-coordinate complexes.[34] The $^1$H NMR spectra of cyclobutadiene metal compounds or

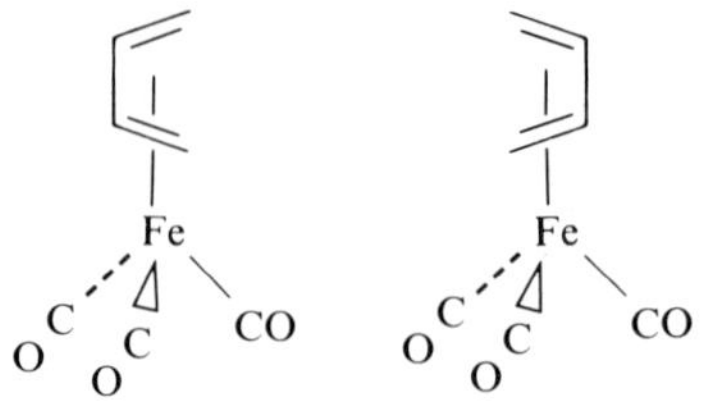

$$(8.7)$$

their totally and symmetrically substituted congeners show that all four positions of the ring are magnetically equivalent. This can be explained by a low-energy barrier to rotation about the metal–cyclobutadiene axis. NMR investigations of tetrakis(p-tolyl) cyclobutadiene complexes showed that the tolyl substituents undergo rapid rotation about the bond with the four-membered ring.[10]

*Table 8.3. NMR Spectra of Complexes Containing 4e Ligands*

| Compound | Chemical shifts $\tau$ | Coupling constant (Hz) |
|---|---|---|
| Butadiene complexes 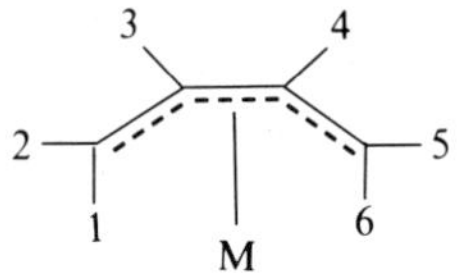 | | |
| | | |
| $[Fe(C_4H_6)(CO)_3]^a$ | 9.78 (H 1,6), 8.26 (H 2,5), 4.69 (H 3,4) | $J(1,6)=0$, $J(3,4)=4.4$, $J(1,3)=J(4,6)=8.4$, $J(1,4)=J(3,6)=-1.7$, $J(1,2)=J(5,6)=2.5$, $J(2,3)=J(4,5)=6.9$ |
| $[Vcp(C_4H_6)(CO)_2]$ | $\tau(1,6)=10.2$; $\tau(2,5)=7.8$; $\tau(3,4)=5.6$ | $\tau(cp)=6.4$ |
| $[Mncp(C_4H_6)(CO)]$ | $\tau(1,6)=11.6$; $\tau(2,5)=8.2$; $\tau(3,4)=5.3$ | $\tau(cp)=6.4$ |
| $[Mo(C_4H_6)_2(CO)_2]^{(11)}$ | $\tau(1,6)=9.8$; $\tau(2,5)=8.6$; $\tau(3,4)=5.8$ | |
| $[Co_2(C_4H_6)_2(CO)_4]^{(11)}$ | $\tau(1,6)=9.3$; $\tau(2,5)=8.3$; $\tau(3,4)=6.0$ | |
| $[Co_2(CH_2CHCMeCH_2)_2(CO)_4]^{(11)b}$ | $\tau(1,6)=9.2$; $\tau(2,5)=8.4$; $\tau(3)=6.2$ | $\tau(Me)=8.9$ |
| $[Co_2(CH_2CMeCMeCH_2)_2(CO)_4]^{(11)b}$ | $\tau(1,6)=8.4$; $\tau(2,5)=7.84$ | $\tau(Me)=8.2$ |
| $[Mn(C_4H_6)(CO)_2(NO)]^{(12)}$ | $\tau(1,6)=9.95$; $\tau(2,5)=8.53$; $\tau(3,4)=5.08$ | |

*(Table continued)*

*Table 8.3. (continued)*

| Compound | Chemical shifts $\tau$ | Coupling constant (Hz) |
|---|---|---|
| Trimethylene and cyclobutadiene complexes | | |
| $[Fe(C_4H_4)(CO)_3]^{c,d}$ | 6.09 | |
| $[Fe(C_4H_4)(CO)_2(NO)]^+[PF_6]^{-e}$ | 3.86 | |
| $[Cocp(C_4H_4)]^f$ | 6.39 | |
| $[Rhcp(C_4H_4)]^g$ | 5.95(d) | $J(Rh-H)=1.6$ |
| $[Ru(C_4H_4)(CO)_3]^d$ | 5.87 | |
| $[Cr(C_4H_4)(CO)_4]^h$ | 6.12 | |
| $[Mo(C_4H_4)(CO)_4]^d$ | 5.94 | |
| $[W(C_4H_4)(CO)_4]^d$ | 5.98 | |
| $[Fe(1\text{-}HOOCC_4H_3)(CO)_3]^i$ | $\tau(2,4)=5.45(s);$ $\tau(3)=5.65(s)$ | |
| $[Fe(1\text{-}HOOCCH_2C_4H_3)(CO)_3]^i$ | $\tau(2,4)=5.66(s);$ $\tau(3)=5.73(s)$ | |
| $[Fe\{1,2\text{-}(HOOC)_2C_4H_2\}(CO)_3]^j$ | 4.82(s) | |
| $]Fe(1,2\text{-}Bu_2^tC_4H_2)(CO)_3]^j$ | 5.88(s) | |
| $[Fe\{1,2\text{-}(MeOOC)(NH_2)C_4H_2\}(CO)_3]^j$ | 5.78(s) 6.11(s) | |
| $[Fe\{C(CH_2)_3\}(CO)_3]^k$ | 8.0 | |
| $[Cr\{C(CH_2)_3\}(CO)_4]^l$ | 8.42 | |
| $[Mo\{C(CH_2)_3\}(CO)_4]^l$ | 8.40 | |

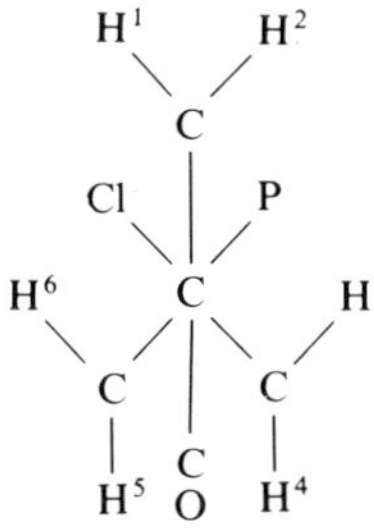

H¹ H² diagram = $[Ir\{C(CH_2)_3\}Cl(CO)PPh_3]^{(46)}$

$\tau(2)=9.30; \tau(1)=8.34;$
$\tau(3)=7.60$
$\tau(6)=7.06; \tau(4)=6.97;$
$\tau(5)=6.84$

[a] A. N. Nesmeyanov, N. E. Kolobova, V. V. Skripkin, K. N. Anisimov, and L. A. Fedorov, *Dokl. Akad. Nauk SSSR*, **195**, 368 (1970).
[b] G. Winkhaus and G. Wilkinson, *J. Chem. Soc.*, 602 (1961).
[c] G. F. Emerson, L. Watts, and R. Pettit, *J. Am. Chem. Soc.*, **87**, 131 (1965).
[d] R. G. Amiet, P. C. Reeves, and R. Pettit, *Chem. Commun.*, 1208 (1967).
[e] A. Efraty, R. Bystrek, J. A. Geaman, S. S. Sandhu, Jr., M. H. A. Huangg, and R. H. Herber, *Inorg. Chem.*, **13**, 1269 (1974).
[f] R. G. Amiet and R. Pettit, *J. Am. Chem. Soc.*, **90**, 1059 (1968).
[g] S. A. Gardner and M. D. Rausch, *J. Organomet. Chem.*, **56**, 365 (1973).
[h] J. S. Ward and R. Pettit, *Chem. Commun.*, 1419 (1970).
[i] D. Stierle, E. R. Biehl, and R. C. Reeves, *J. Organomet. Chem.*, **72**, 221 (1974).
[j] G. Berens, F. Kaplan, R. Rimerman, B. W. Roberts, and A. Wissner, *J. Am. Chem. Soc.*, **97**, 7076 (1975).
[k] G. F. Emerson, K. Ehrlich, W. P. Giering, and P. C. Lauterbur, *J. Am. Chem. Soc.*, **88**, 3172 (1966).
[l] J. S. Ward and R. Pettit, *Chem. Commun.*, 1419 (1970).

The lack of noticeable interaction between neighboring, vicinal, and diagonal hydrogen atoms is characteristic for disubstituted metal cyclobutadiene complexes. In spectra of almost all cyclobutadiene compounds, there are singlets whose intensity depends on the number of magnetically equivalent protons of each kind. Also, in coordination complexes with the $ML_n$ fragments possessing axial symmetry ($C_{3v}$ and $C_{4v}$)

and containing trimethylenemethane, all protons are magnetically equivalent. The NMR spectra of the complexes $[Fe\{C(CH_2)_3\}(CO)_3]$, $[Mo\{C(CH_2)_3\}(CO)_4]$, and $[Cr\{C(CH_2)_3\}(CO)_4]$ show singlets in the $\tau = 8.0$–8.5 range. In compounds containing nonsymmetrical trimethylenemethane derivatives, all protons are magnetically non-equivalent. The NMR spectra of some compounds containing $4e$ ligands are given in Table 8.3.

The $^{13}C$ NMR spectra of complexes with $\eta^4$ ligands have also been investigated.[36, 37] They show that the M–diene bond may be described by formulas (8.8). Better agreement with experiment gives model $\pi(a)$ rather than model $\sigma, \pi(b)$. The

$$\text{a} \qquad \text{or} \qquad \text{b} \tag{8.8}$$

coupling constants $J(C-H)$ for $[Fe(C_4H_4)(CO)_3]$ suggest that the carbon atoms have intermediate hybridization between sp and sp$^2$.

For cyclobutadiene complexes, the energy of activation of rotation of ligands about the M–$C_4R_4$ axis is very small, while for the trimethylenemethane iron derivative the $\Delta G^{\neq}$ of rotation of trimethylenemethane is high, equaling 71–75 kJ mol$^{-1}$ (Table 8.4).

Table 8.4. $^{13}C$ NMR Spectra of Complexes Containing $\eta^4$ Ligands

| Compound | $\delta$ (ppm) (SiMe$_4$, $\delta = 0$) | Coupling constants (Hz) |
|---|---|---|
| $[Fe(C_4H_6)(CO)_3]^{a,b}$ | 41.1 (C1,4); 85.8 (C2,3) | $^1J(C1-H)$ 151.7; $^1J(C2-H)$ 170 |
| $[Fe(C_4H_4)(CO)_3]^{a,b}$ | 61.0 | $^1J(C-H)$ 191 |
| $[Fe(C_8H_8)(CO)_3]^c$ | 62.0; 89.2; 120.5; 129.1 | $J(CH)$ 158 |
| $[Fe\{C^1(C^2H_2)_3\}(CO)_3]^d$ | 105.3 (C1); 53.0 (C2) | $J(CH)$ 158 |
| (structure) $^e$ | 39.2 (C1); 85.8 (C2); 81.3 (C3) | |
| | 62.1 (C4); 138.6 (C5); 114.4 (C6) | |
| $[Mo(C_4H_4)(CO)_4]^a$ | 60.7 | $J(CH)$ 198 |
| $[Fe(C_4H_4)(CO)_3]^f$ | | $^1J(^{13}C-^{57}Fe)$ 3.62; $^1J(^{13}CO-^{57}Fe)$ 28.73 |
| $[Fe(HOCH_2C_4H_3)(CO)_3]^g$ | 84.5 (C1); 64.6 (C2,4); 63.1 (C3) | |
| $[(OC)_3Fe(CH_2C_4H_3)]^+HSO_4^{-\,g}$ | 98.0 (C1); 84.6 (C2,4); 109 (C3) | |
| (structure) $^h$ | 69.4 (C1); 107.3 (C2); 54.3 (C3); 58.4 (C4) | $\Delta G^{\neq}$ for ligand rotation 71–75 kJ mol$^{-1}$ |

$^a$ A. N. Nesmeyanov, E. I. Fedin, L. A. Fedorov, and P. V. Petrovskii, Zh. Strukt. Khim., 13, 964 (1972).
$^b$ H. G. Preston and J. C. Davis, J. Am. Chem. Soc., 88, 1585 (1966).
$^c$ G. Rigatti, G. Boccalon, A. Ceccon, and G. Giacometti, Chem. Commun., 1165 (1972).
$^d$ See Reference k, Table 8.3.
$^e$ M. Anderson, A. D. H. Clague, L. P. Blaauw, and D. A. Couperus, J. Organomet. Chem., 56, 307 (1973).
$^f$ P. S. Nielsen, R. S. Hansen, and H. J. Jakobsen, J. Organomet. Chem., 114, 145 (1976).
$^g$ C. S. Eschhbach, D. Seyferth, and P. C. Reeves, J. Organomet. Chem., 104, 363 (1976).
$^h$ E. S. Magyar and C. P. Lillya, J. Organomet. Chem., 116, 99 (1976).

## 4. *PHOTOELECTRON SPECTRA*

In the photoelectron spectrum of $[Fe(C_4H_6)(CO)_3]$ in the region of ionization potentials of up to 14 eV (before the ionization of CO occurs), there are bands which are assigned to ionizations from nine orbitals of the complex, that is, four $d$ orbitals and five orbitals of the ligand ($2\pi$ and $3\sigma$).[14, 15] This indicates strong mixing of orbitals of the ligand and the metal. Also, in the spectrum of $[Fe(C_4H_4)(CO)_3]$, the first band at 8.45 eV corresponds to the ionization of $d$ orbitals of iron, and the subsequent one at 9.21 eV corresponds to the orbitals of the ligand.[16] The spectrum of $[Fe\{C(CH_2)_3\}(CO)_3]$ is similarly interpreted.[17, 48] In the spectra of compounds of the type $[M(BD)_3]$ (M = Mo, W) in the low-energy region, three bands occur whose relative intensity varies only slightly with the increase of the photon energy. This implies that in this region, no $d$ bands occur, and all molecular orbitals possessing high energies have the same contribution of ligand orbitals. Thus, these bands concern the $\psi_1$, $\psi_2$, and $\psi_3$ orbitals of butadiene.[48] The considerable population of the $\psi_3$ orbital indicates much charge transfer to the butadiene ligand.

The first band in the spectra of $[Mcp(C_4H_6)]$ (M = Rh, Ir) was also assigned to the molecular orbital which possesses a considerable contribution to the $\psi_3$ orbital.[48] Spectra of the complexes $[\{M(CO)_3\}_n diene]$ (diene = 5,6,7,8-tetramethylidene-bicyclo-2,2,2-oct-2-ene, M = Fe, Ru) differ for *exo* and *endo* isomers because of the difference in energies of molecular orbitals which are caused by long-range electrostatic effects.[49]

## 5. *IR SPECTRA*

The infrared spectra of butadiene complexes show three characteristic bands at *ca* 1400 cm$^{-1}$ which are assigned to stretching vibrations of coordinated double bonds as well as to the $\delta(CH)$ asymmetric oscillation.[18, 20]

Carbonyl complexes exhibit the presence of $\nu(CO)$ bands in the IR spectrum. Their position is dependent, among other factors, on the donor properties of ligands. For $[Fe(C_4H_{4-x}Me_x)(CO)_3]$ complexes, the $\nu(CO)$ frequency becomes lower when X increases because the donor properties of cyclobutadiene are also increased. The spectra of the cyclobutadiene complex $[FeC_4H_4(CO)_3]^{[21, 22]}$ as well as those of the trimethylenemethane complex $[Fe\{C(CH_2)_3\}(CO)_3]^{[23]}$ were also interpreted.

## 6. *METHODS OF PREPARATION OF 1,3-DIENE COMPLEXES*[2, 4, 5, 11, 24, 31, 32, 35]

### a. *Reactions of coordination compounds with dienes*

Diene complexes are most commonly obtained by reactions of coordination compounds with conjugated dienes. However, they can also be synthesized by reactions involving nucleophilic attack on metal dienyl complexes.

Metal carbonyls are commonly utilized in ligand substitution reactions:

$$[Fe(CO)_5] + C_4H_6 \longrightarrow [Fe(C_4H_6)(CO)_3] + 2CO \qquad (8.9)$$

*Table 8.5. 1,3-Diene Metal Complexes*

| Compound | Color | Mp (K) | Remarks |
|---|---|---|---|
| $[Th(pmcp)_2(C_4H_6)]$[55] | Yellow-orange | | |
| $[Ti(C_4H_6)_2(depe)]$[52] | Blue | | |
| $[Hf(C_4H_6)_2(dmpe)]$[52] | Red | | |
| $[Hfcp_2(C_4H_6)]$[39,54] | Orange | | |
| $[Hfcp_2(CH_2CMeCMeCH_2)]$[54] | Dark-red | 435–437 | |
| $[NbCl(C_4H_6)(dmpe)_2]$[52] | Orange | | |
| $[Vcp(C_4H_6)(CO)_2]^a$ | Red | 408–413 dec | |
| $[Vcp(CH_2=CMeCMe=CH_2)(CO)_2]^a$ | Red | 408–409 dec | |
| $K^+[Nb(\eta^3-C_8H_8)_2\eta^4-C_8H_8]^{-b}$ | Red-black | | |
| $[TaCl(\eta^4-C_{10}H_8)(Me_2PC_2H_4PMe_2)_2]^c$ | Dark-red | | |
| $[Cr(cyclohexadiene)_2(CO)_2]$[11] | Yellow | | |
| $[Cr(C_4H_6)(CO)_4]^d$ | | 273 dec | |
| $[Mo(C_4H_6)_2(CO)_2]^a$ | Amber | 383 dec | |
| $[Mo(C_4H_6)_3]$[26] | Yellow | 403 dec | air-stable; H 5.27 m; 8.38 m; 9.46 m |
| $[W(C_4H_6)_3]$[26,25] | White | 408 dec | air-stable; H 5.72 m; 8.50 m; 9.62 m |
| $[Mocp(t-Me_2CCHCHCMe_2)(NO)]$[38] | Yellow | | |
| $[Mn(C_4H_6)_2(PMe_3)]$[56] | Green | | $\mu_{eff} = 1.74$ BM |
| $[Mncp(C_4H_6)(CO)]^a$ | Red | 407–409 dec | |
| $[Mn(C_4H_6)(CO)_2(NO)]^e$ | Yellow | 288 | |
| $[Mn(C_4H_6)_2(CO)]^e$ | Blue-green | 423 | $\mu_{eff} = 1.79$ BM; 298 K |
| $[Fe(C_4H_6)(CO)_3]^f$ | Light-yellow | 292 | |
| $[Fe(C_4H_6)_2(CO)]^g$ | | 403–410 dec | |
| $[Fe(PhCH=CHCH=CHPh)(CO)_3]^h$ | Yellow-orange | 438–439 | |
| $[Co(C_4H_7)(C_4H_6)(PPh_3)]^i$ | Orange-red | 383 dec | |
| $[Co_2(C_4H_6)_2(CO)_4]^j$ | Red | 391 dec | |
| $[Co_2(CH_2=CMeCH=CH_2)_2(CO)_4]$[11] | Brown | | |
| $[RhCl(C_4H_6)_2]^{k,l}$ | Yellow | 313 dec | |
| $[Rhcp(C_4H_6)]^l$ | Yellow | 356–357 dec | |
| $[Rh_2Cl_2(t,t-MeCH=CHCH=CHMe)_2]^l$ | Red | 405–406 dec | |
| $[Rh(acac)(t,t-MeCH=CHCH=CHMe)]^l$ | Orange | 401–403 dec | |
| $[RhCl(C_4H_6)(PPh_3)_2]^m$ | Yellow | 358 dec | |
| $[IrCl(CH_2=CMeCMe=CH_2)_2]^n$ | White | 418–423 | |
| $[Ir(C_4H_6)(CO)(PMe_2Ph)_2]^+[BF_4]^{-o}$ | White | 403–407 | $v(CO)$ 2024 |
| $[IrCl(C_4H_6(PPh_3)_2]^p$ | Light-yellow | 437–440 dec | |

[a] E. O. Fischer, H. P. Kogler, and P. Kuzel, *Chem. Ber.*, **93**, 3006 (1960).
[b] L. Y. Guggenberger and R. R. Schrock, *J. Am. Chem. Soc.*, **97**, 6693 (1975).
[c] J. O. Albright, L. D. Brown, S. Datta, J. K. Kouba, S. S. Wreford, and B. M. Foxman, *J. Am. Chem. Soc.*, **99**, 5518 (1977).
[d] E. Koerner von Gustorf, O. Jaenicke, and O. E. Polansky, *Angew. Chem., Int. Ed. Engl.*, **11**, 533 (1972).
[e] M. Herberhold and A. Razavi, *Angew. Chem., Int. Ed. Engl.*, **14**, 351 (1975).
[f] B. F. Hallman and P. L. Pauson, *J. Chem. Soc.*, **642** (1958).
[g] D. L. Williams-Smith, L. R. Wolf, and P. S. Skell, *J. Am. Chem. Soc.*, **94**, 4042 (1972).
[h] M. Cais and M. Feldkimel, *Tetrahedron Lett.*, 444 (1961).
[i] G. Vitulli, L. Porri, and A. L. Segre, *J. Chem. Soc. (A)*, 3246 (1971).
[j] E. O. Fischer, P. Kuzel, and H. P. Fritz, *Z. Naturforsch.*, **16b**, 138 (1961).
[k] L. Porri, A. Lionetti, G. Allegra, and A. Immirzi, *Chem. Commun.*, 336 (1965).
[l] S. M. Nelson, M. Sloan, and M. G. B. Drew, *J. Chem. Soc., Dalton Trans.*, 2195 (1973).
[m] D. M. Roundhill, D. N. Lawson, and G. Wilkinson, *J. Chem. Soc. (A)*, 845 (1968).
[n] G. Winkhaus and H. Singer, *Chem. Ber.*, **99**, 3610 (1966).
[o] A. J. Deeming and B. L. Shaw, *J. Chem. Soc. (A)*, 376 (1971).
[p] A. van der Ent and A. L. Onderdelinden, *Inorg. Chim. Acta*, **7**, 203 (1973).

Often, hydrogen transfer processes take place leading to the rearrangement of the diene [equation (8.10)]. Also, other metal carbonyls react with dienes:

$$\text{Me} {\overset{\displaystyle \diagup}{\underset{\displaystyle \text{Me}}{}}}{=}{\cdots}{-}\text{Me} + [\text{Fe(CO)}_5] \longrightarrow \qquad \qquad {-}\text{CHMe}_2 \qquad (8.10)$$

$$[\text{Co}_2(\text{CO})_8] + \text{C}_4\text{H}_6 \longrightarrow [\text{Co}_2(\text{C}_4\text{H}_6)(\text{CO})_6] \qquad (8.11)$$

$$[\text{Co}_2(\text{CO})_8] + 2\text{C}_4\text{H}_6 \longrightarrow [\text{Co}_2(\text{C}_4\text{H}_6)_2 (\text{CO})_4] \qquad (8.12)$$

$$[\text{Mo(CO)}_6] + \text{C}_6\text{H}_8 \longrightarrow [\text{Mo}(\text{C}_6\text{H}_8)_2 (\text{CO})_2] + 4\text{CO} \qquad (8.13)$$

$$(\text{C}_6\text{H}_8 = \text{cyclohexadiene})$$

Often, other ligands may be substituted:

$$[\text{Mo}(\text{C}_6\text{H}_3\text{Me}_3)(\text{CO})_3] + \text{C}_4\text{H}_6 \longrightarrow [\text{Mo}(\text{C}_4\text{H}_6)_2 (\text{CO})_2] + \text{CO} + \text{C}_6\text{H}_3\text{Me}_3 \qquad (8.14)$$

$$[\text{Co(bipy)}(\text{PR}_3)\text{H}_2]\text{ClO}_4 \underset{\text{H}_2}{\overset{\text{diene}}{\rightleftharpoons}} [\text{Co(bipy)}(\text{PR}_3)(\text{diene})]\text{ClO}_4 \qquad (8.15)$$

Butadiene complexes are also formed by reduction reactions of transitions metal compounds by $\text{Al}_2\text{R}_6$, $\text{BH}_4^-$, etc.

$$[\text{CoCl}_2(\text{PPh}_3)_2] + \text{C}_4\text{H}_6 \xrightarrow[273\ \text{K}]{\text{Zn}} [\text{Co}(\text{C}_4\text{H}_7)(\text{C}_4\text{H}_6)(\text{PPh}_3)] \qquad (8.16)$$

$$\text{RhCl}_3 + \text{C}_4\text{H}_6 \xrightarrow{\text{EtOH}} [\text{RhCl}(\text{C}_4\text{H}_6)_2] \qquad (8.17)$$

In reaction (8.17), rhodium(III) is reduced by alcohol. The compound $[\text{IrCl}(\text{CH}_2{=}\text{CMeCMe}{=}\text{CH}_2)_2]$ is formed in an analogous way.

Sometimes, 1,3-diene complexes are formed from nonconjugated dienes:

$$[\text{Fe}_3(\text{CO})_{12}] + \text{CH}_2{=}\text{CHCH}_2\text{CH}{=}\text{CH}_2 \longrightarrow [\text{Fe}(\text{CH}_2{=}\text{CHCH}{=}\text{CHMe})(\text{CO})_3] \qquad (8.18)$$

Some compounds containing 1,3-dienes may be obtained by substitution reactions of other ligands or as a result of changes involving the functional group of the diene.

$$[\text{Fe(diene)}(\text{CO})_3] \xrightarrow{\text{L}} [\text{Fe(diene)}(\text{CO})_2\text{L}] + \text{CO} \xrightarrow{\text{L}} [\text{Fe(diene)}(\text{CO})\text{L}_2] + \text{CO} \qquad (8.19)$$

$$[\text{Fe}(\text{RCH}{=}\text{CR}^1\text{CR}^2{=}\text{CHCOR}^3)(\text{CO})_3] \left\{ \begin{array}{l} \xrightarrow{\text{LiAlH}_4/\text{AlCl}_3} [\text{Fe}(\text{RCH}_3{=}\text{CR}^1\text{CR}^2{=}\text{CHCH}_2\text{R}^3)(\text{CO})_3] \\[2ex] \xrightarrow{\text{NaBH}_4} [\text{Fe}(\text{RCH}{=}\text{CR}^1\text{CR}^2{=}\text{CHCH(OH)}\text{R}^3(\text{CO})_3] \end{array} \right. \qquad (8.20)$$

The compound $[\text{Mg}(\text{C}_4\text{H}_6)(\text{THF})_2]_n$ is often used for the preparation of butadiene complexes.[52, 55]

$$HfCl_4 + dmpe + MgC_4H_6 \cdot 2THF \xrightarrow{\text{THF}} Hf(C_4H_6)_2 \, (dmpe) \tag{8.21}$$

$$FeCl_2 3PMe_3 + MgC_4H_6 \cdot 2THF \xrightarrow{\text{THF}} Fe(BD)(PMe_3)_3 \tag{8.22}$$

$$Th(pmcp)_2Cl_2 + MgC_4H_6 \cdot 2THF \xrightarrow{\text{Et}_2\text{O}} Th(pmcp)_2 \, (BD) \tag{8.23}$$

### b. *Reactions involving nucleophilic attack on diene complexes*

Nucleophiles attack many dienyl compounds to give diene derivatives. Most such reactions are known for iron compounds. Thus, the reaction of (1,3-cyclohexadiene)tri-carbonyliron(0) with trityl tetrafluoroborate affords (cyclohexadienyl)tricarbonyliron cation, which reacts with a nucleophile forming a diene complex [equation (8.24)]. The

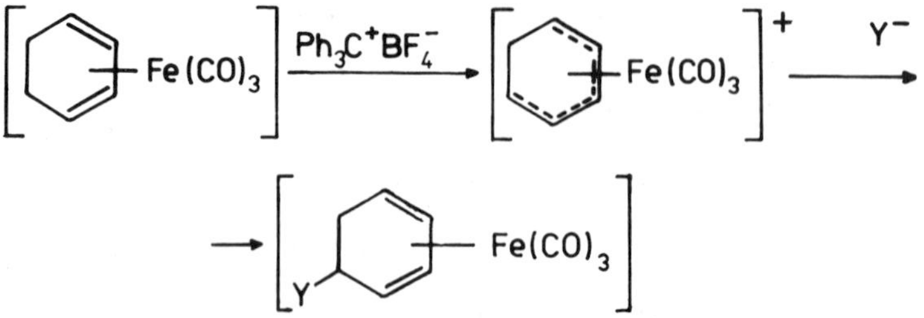

$$\tag{8.24}$$

following reagents may serve as nucleophiles: water, $CN^-$, $MeO^-$, pyrrolidine, morpholine, diethyl malonate, acetylacetonate, etc. [equation (8.25)]. The $Cocp_2^+$ ions react with nucleophiles in the manner shown in equation (8.26).

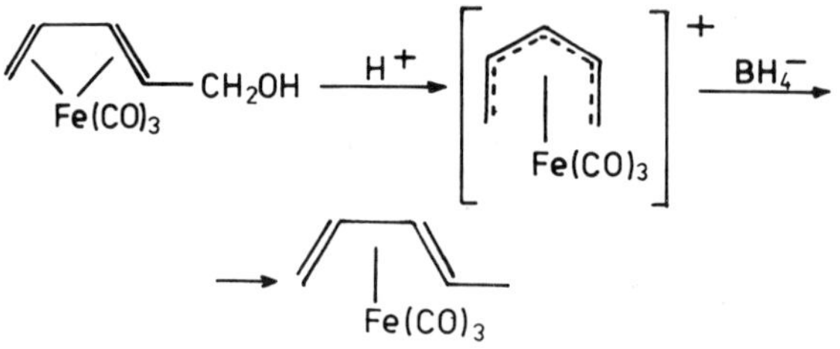

$$\tag{8.25}$$

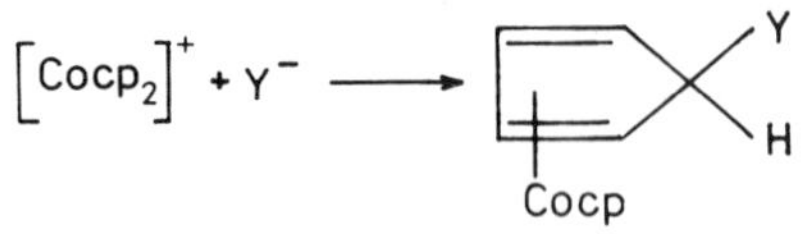

$$\tag{8.26}$$

$$(Y = H^-, Ph^-)$$

A very important method of preparation of diene compounds is the condensation reaction of dienes with metal vapors. The reaction is carried out at low temperature and at low pressure in a specifically designed flask containing the diene either neat or in solution. The metal atoms vaporize from the electrically heated metal wire and react with dienes which are present in the liquid phase[25] (Figure 8.7). Often, such condensations of metal vapors with dienes are carried out in matrices at liquid nitrogen temperature. In this manner, the air-stable compounds $[M(C_4H_6)_3]$ $(M = Mo, W)$[26] as well as the complexes $[Cr(C_4H_6)(CO)_4]$ and $[Fe(C_4H_6)_2]$ have been obtained.[27]

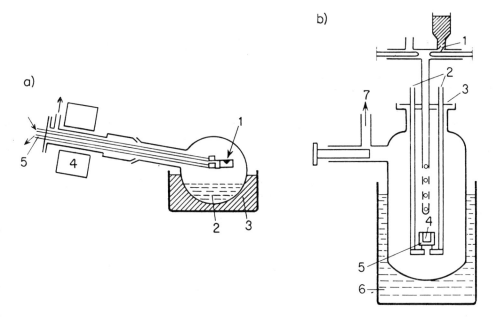

Figure 8.7. Apparatus for preparation of compounds from metal vapors: (a) 1—metal, 2—substrate solution, 3—cooling bath, 4—rotor, 5—inlet for electricity and water; (b) 1—inlet for the liquid or gas, 2—water cooling and electrical supply, 3—seal, 4—heated metal, 5—Mo wire coiled $Al_2O_3$ crucible, 6—liquid nitrogen, 7—vacuum.

## 7. METHODS OF PREPARATION OF CYCLOBUTADIENE COMPLEXES[2, 3, 5, 9, 11, 24]

The most important methods for the preparation of cyclobutadiene complexes are oxidative addition reactions of dihalogenocyclobutenes to low-valent transition metal compounds [equations (8.27) and (8.28)], as well as reactions of acetylenes with coordination compounds.

$$(8.27)$$

$$(8.28)$$

Anionic carbonyl complexes of iron, ruthenium, chromium, molybdenum, tungsten, etc., also react with 3,4-dichlorocyclobutene or its derivatives. In order to obtain anionic derivatives, metal carbonyls are reduced by means of sodium amalgam [equations (8.29)–(8.31)].

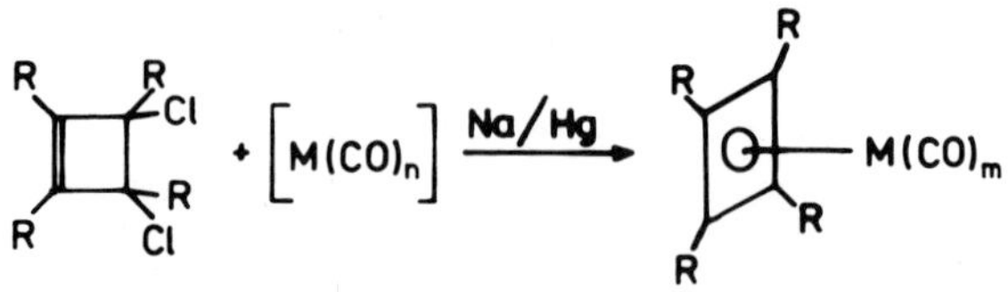

$$(m = 3 \text{ for Fe, Ru}; m = 4 \text{ for Cr, Mo, W}) \qquad (8.29)$$

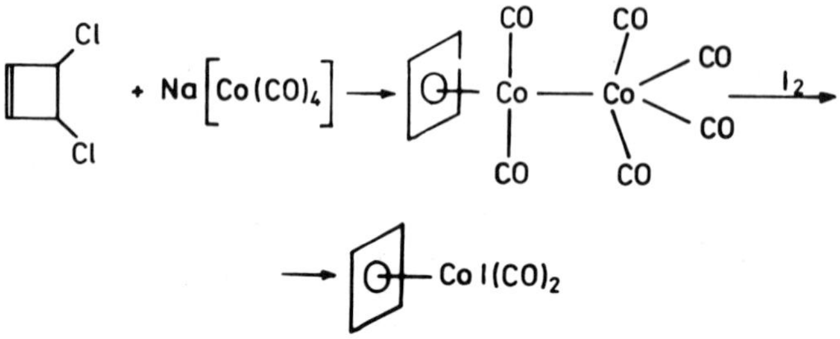

$$(8.30)$$

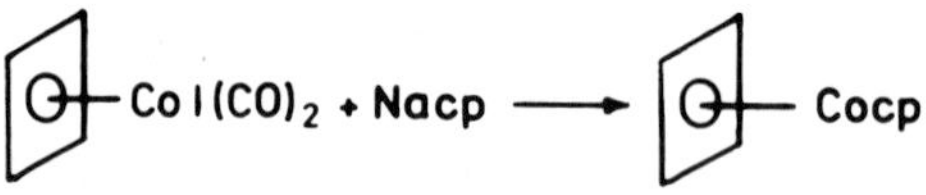

$$(8.31)$$

The tetrahalogenocyclobutane derivatives may also undergo dehalogenation reactions [equation (8.32)]. Vinyl carbonate reacts with alkynes to give *cis*-3,4-carbonyl-

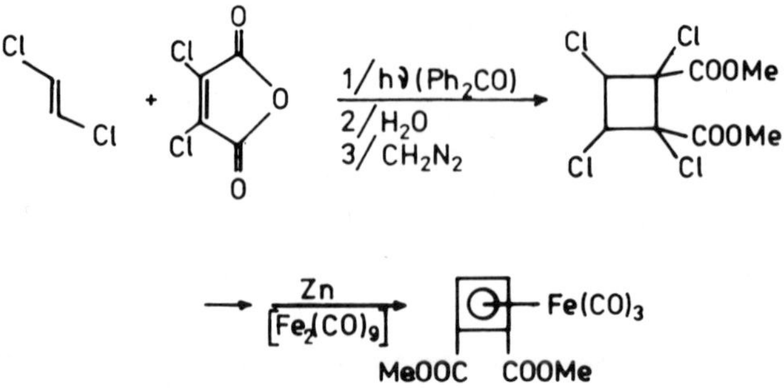

$$(8.32)$$

dioxycyclobutenes, which further react with $[Fe_2(CO)_9]$ or $Na_2[Fe(CO)_4]$ to afford cyclobutadiene compounds [equation (8.33)].

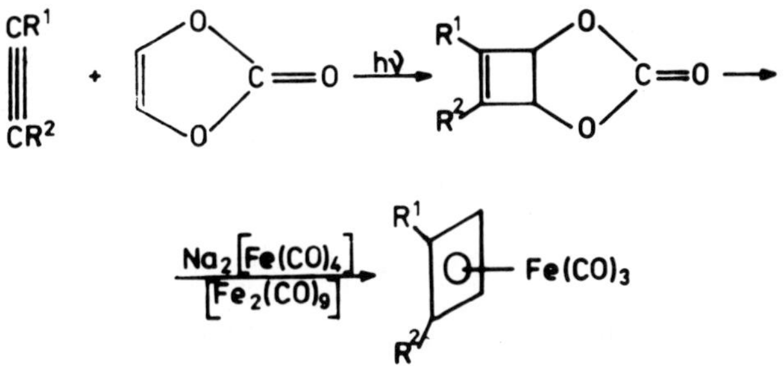

$$(8.33)$$

$$(R^1 = H, Me; R^2 = H, Me, Bu, CH_2OMe)$$

Acetylenes easily undergo dimerization to give cyclobutadiene compounds [equations (8.34)–(8.36)]. Cyclobutadiene coordination compounds may be obtained [equation (8.37)] by reductive decyclization reaction of metallacyclopentadiene, which is

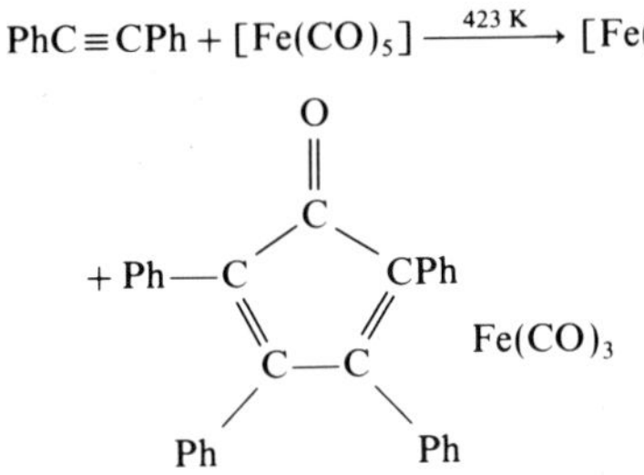

$$PhC \equiv CPh + [Fe(CO)_5] \xrightarrow{423 \text{ K}} [Fe(C_4Ph_4)(CO)_3]$$

$$(8.34)$$

$$PhC \equiv CPh + [Cocp_2] \longrightarrow [Cocp(C_4Ph_4)]$$  $$(8.35)$$

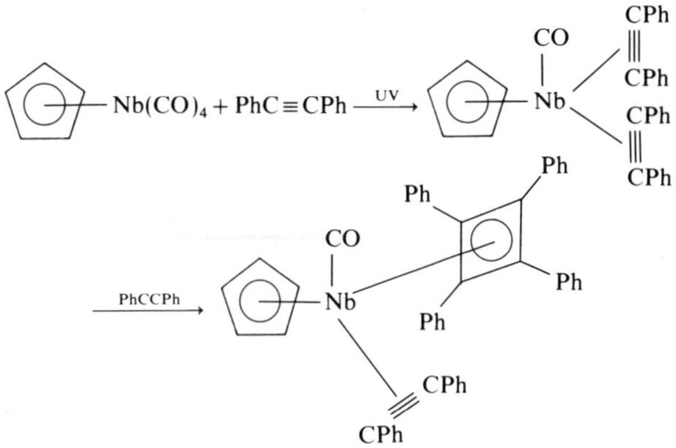

$$(8.36)$$

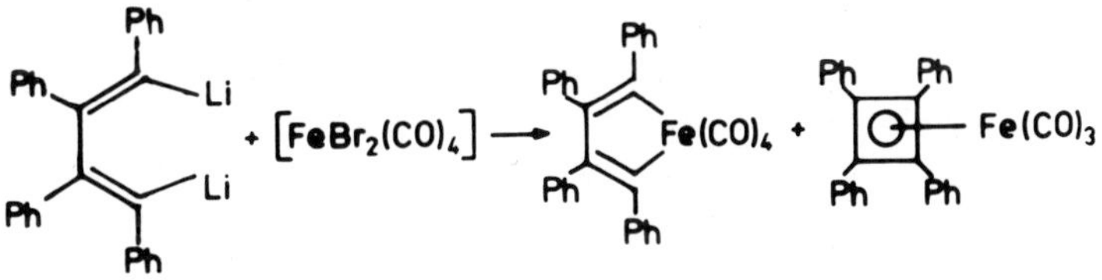

$$(8.37)$$

formed by dimerization of acetylene or from butadiene derivatives [mechanisms of these reactions were discussed in Chapter 6, Equations (6.194)–(6.196)].

Acetylenes may form various organometallic compounds with metal complexes [equations (8.38)–(8.40)].

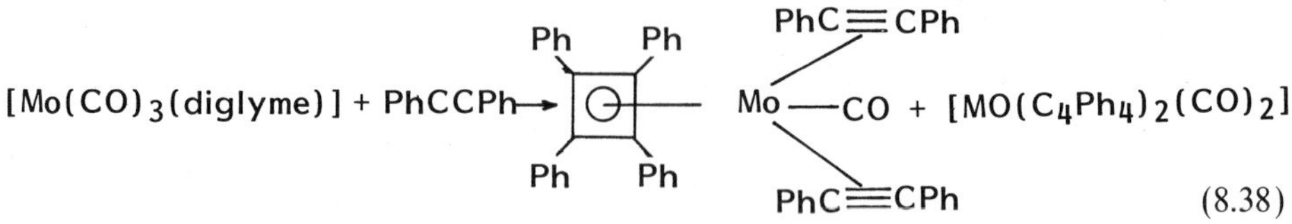

$$[Mo(CO)_3(diglyme)] + PhCCPh \longrightarrow$$

(8.38)

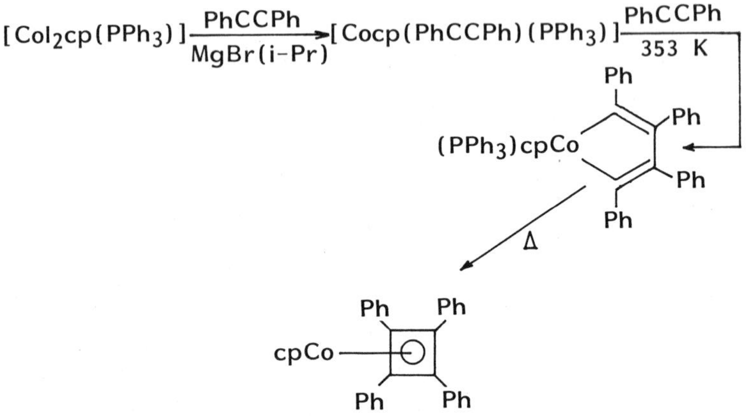

(8.39)

$$ArC \equiv CAr + [PdCl_2(PhCN)_2] \xrightarrow{PhH}$$

$$\xrightarrow{HCl} [Pd_2Cl_4(C_4Ar_4)_2]$$

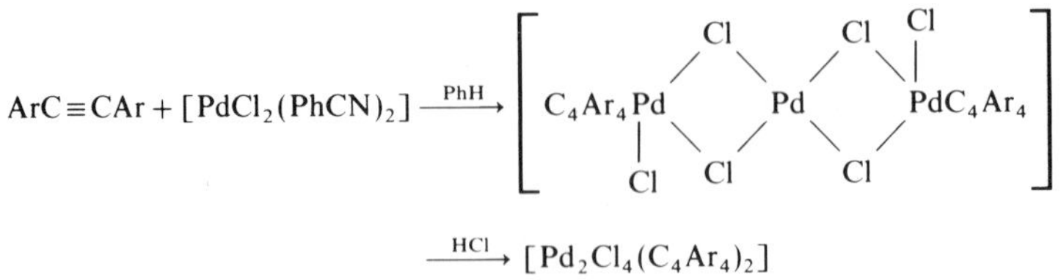

(8.40)

The reaction of molybdenum carbonyl with $PhC \equiv CPh$ furnishes one of the few known compounds containing two cyclobutadiene ligands coordinated to only one central atom [equation (8.41), Table 8.6].

1,3-Dienes as well as cyclobutadienes may undergo exchange reactions between various complexes [equation (8.42)]. During flash vacuum pyrolysis of cyclopentadienyl(1,1-dioxo-2,5-dimethylthiophene)cobalt(I), $SO_2$ elimination takes place and

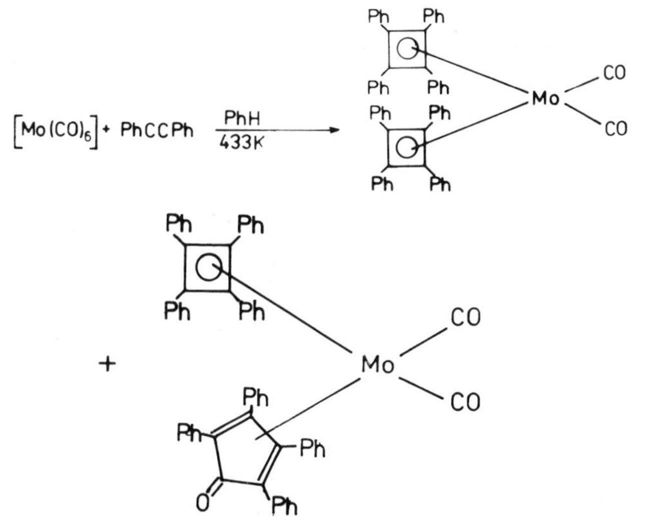
[Mo(CO)6] + PhCCPh
PhH
433K
Ph
Mo
CO
O

(8.41)

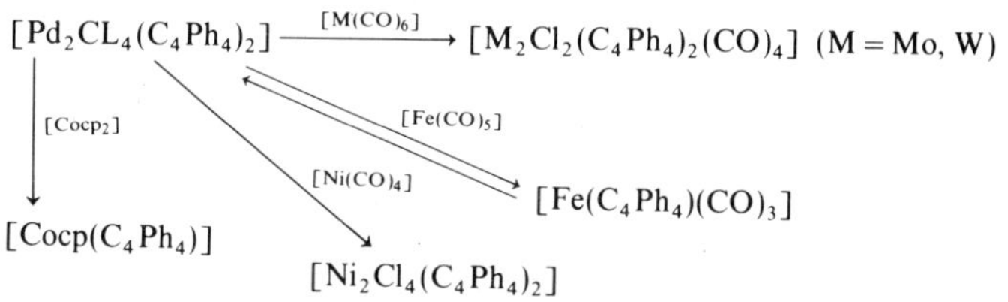
[Pd2CL4(C4Ph4)2]
[M(CO)6]
[M2Cl2(C4Ph4)2(CO)4] (M = Mo, W)
[Cocp2]
[Fe(CO)5]
[Ni(CO)4]
[Fe(C4Ph4)(CO)3]
[Cocp(C4Ph4)]
[Ni2Cl4(C4Ph4)2]

(8.42)

the 1,2-dimethylcyclobutadiene cobalt(I) complex is formed[50] [equation (8.43)]. Molybdenum and tungsten oxocyclobutenyl complexes give cationic cyclobutadiene compounds via reaction with titanacyclobutane followed by reaction with ether solution of tetrafluoroboric acid[51] [equation (8.44)].

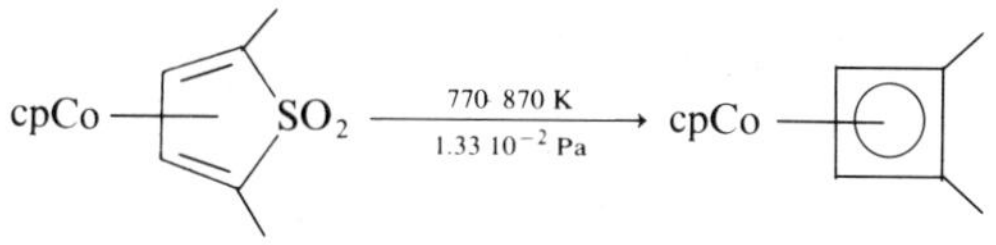
cpCo
SO2
770 870 K
1.33 10−2 Pa
cpCo

(8.43)

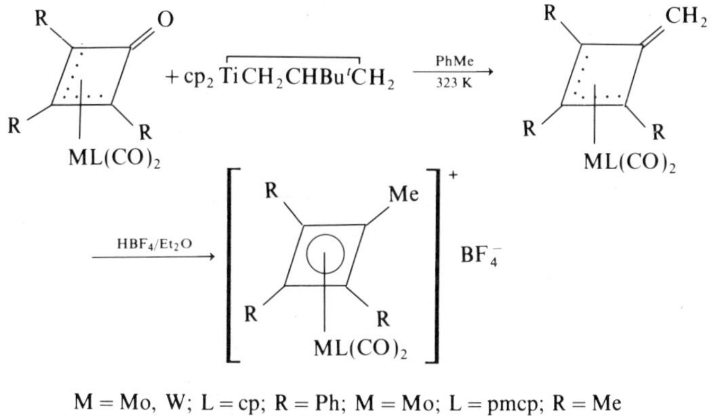
R
O
+ cp2 TiCH2CHBu'CH2
PhMe
323 K
CH2
ML(CO)2
HBF4/Et2O
Me
BF4−

(8.44)

M = Mo, W; L = cp; R = Ph; M = Mo; L = pmcp; R = Me

Table 8.6. Cyclobutadiene and Trimethylmethane Metal Complexes

| Compound | Color | Mp (K) | Remarks |
|---|---|---|---|
| $[Ti(C_4Ph_4)(C_8H_8)]^a$ | Green | | Stable to 573 K, air-stable in solid state |
| $[Vcp(C_4Ph_4)(CO)_2]^b$ | Orange | 508–513 dec | |
| $[Nbcp(C_4Ph_4)(PhCCPh)(CO)]^c$ | Red | 393 dec | $v(CO)$ 2000; $v(C\equiv C)$ 1780 |
| $[Tacp(C_4Ph_4)(PhCCPh)(CO)]^d$ | Red | | |
| $[Cr(C_4H_4)(CO)_4]^e$ | Yellow | 319 | $\tau(H)$ 3.88; $v(CO)$ 1970, 2080 |
| 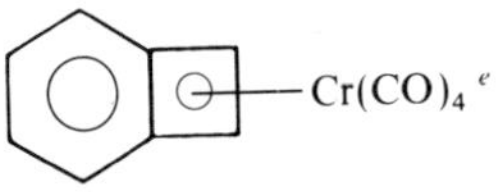 —Cr(CO)$_4$ $^e$ | Orange | | $\tau(H)$ 4.35 (s, 1H); 6.92 (m, 2H); $v(CO)$ 1945, 1980, 2080 |
| $[Mo(C_4H_4)(CO)_4]^f$ | Orange-red | 290 | |
| $[MoHCp(C_4Ph_4)(CO)]^g$ | Yellow | 447–448 dec | $\tau(HMe)$ 16.83; $v(CO)$ 1933 |
| $[Mo(C_4Ph_4)_2(CO)_2]^h$ | Light-yellow | 528–535 dec | |
| $[Mo(C_4Ph_4)(PhCCPh)_2(CO)]^h$ | Violet | 473–475 dec | |
| $[Mo_2Br_2(CO)_4(C_4Ph_4)_2]^{i,j}$ | Red | 553 dec | $v(CO)$ 1974, 2021 |
| $[W(C_4H_4)(CO)_4]^f$ | Orange-red | | |
| $[WBrcp(C_4Ph_4)(CO)]^k$ | | 500–501 dec | $v(CO)$ 1958 |
| $[Fe(C_4H_4)(CO)_3]^l$ | Light-yellow | 299 | $v(CO)$ 1985, 2055 |
| $[Fe(C_4Ph_4)(CO)_3]^m$ | Light-yellow | 507 | |
| $[Fe_2(C_4Bu_2^tPh_2)_2(\mu\text{-}CO)_3]^n$ | Violet | 477–478 dec | $d(Fe–Fe)$ 217.7 pm; $d(C–C)$ 146.8 |
| $[Ru(C_4H_4)(CO)_3]^f$ | | | |
| $[Cocp(C_4H_4)]^{o,p}$ | Yellow | 361.5–362 | $\tau(cp)$ 5.14; $\tau(C_4H_4)$ 6.39 |
| $[CoBr(C_4Ph_4)(CO)_2]^q$ | Red-brown | 528–535 dec | $v(CO)$ 2037, 2072 |
| 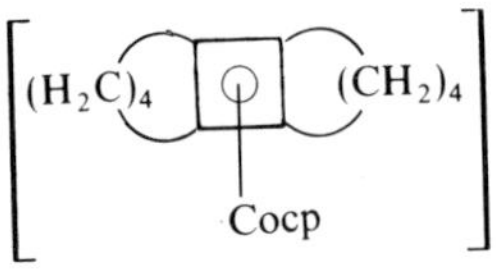 $(H_2C)_4$ $(CH_2)_4$ Cocp | Yellow | 358.5 | |
| $[Cocp(C_4Ph_4)]^s$ | Yellow-orange | 537 | |
| $[Co(C_6H_6)(C_4Ph_4)]^+ Br^{-q}$ | Yellow | 443–453 dec | |
| $[Rhcp(C_4Ph_4)]^t$ | Yellow | 500–501 | |
| $[Rh_2Cl_2(C_4Ph_4)_2]^u$ | Red | 473 dec | |
| $[Rhcp(C_4H_4)]^v$ | White | 402.5–403 | |
| $[Ni_2Cl_4(C_4Me_4)_2]^w$ | Black-violet | 523 dec | |
| $[Ni_2Br_4(C_4Me_4)_2]^x$ | Blue-black | 573 | Air-stable |
| $[Ni(C_4Ph_4)_2]^{(29,30)}$ | | | |
| 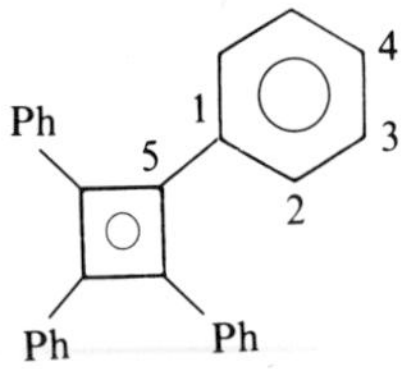 | Blue | 677 dec | $\delta(^{13}C)$ 114.5 (C5); 133.8 (C1); 128.9 (C2); 128.4 (C3); 127.0 (C4) |
| $[Pd\{(3)\text{-}1,2\text{-}B_9C_2H_{11}\}(C_4Ph_4)]^y$ | Red | 583 dec | |
| $[Pt_2Cl_4(C_4Ph_4)_2]^z$ | Orange | 573 dec | |
| $[Pt_2Br_4(C_4Ph_4)_2]^z$ | Red-violet | 571 | |
| $[PtCl_2(C_4Et_4)]_n^{aa}$ | yellow | | |

*(Table continued)*

*Table 8.6. (continued)*

| Compound | Color | Mp (K) | Remarks |
|---|---|---|---|
| $[(C_4Et_4)ClPt(\mu\text{-}Cl)_2Pt(CO)$ $(\mu\text{-}Cl)_2PtCl(C_4Et_4)]^{aa}$ | Light-yellow | 440 | |
| $[Ta(pmcp)Me_2\{\eta^4\text{-}C(CH_2)_3\}]^{(58)}$ | Yellow | | |
| $[Cr\{C(CH_2)_3\}(CO)_4]^{bb}$ | Yellow | 356 | $\nu(CO)$ 1955, 2020 |
| $[Mo\{C(CH_2)_3\}(CO)_4]^{bb}$ | Yellow | 323 | $\nu(CO)$ 1958, 2062 |
| $[Fe\{C(CH_2)_3\}(CO)_3]^{cc,dd}$ | Yellow | 305.5–306 | $\nu(CO)$ 2061 m, 1995 s, 1966 w |
| $[Fe\{(PhCH)C(CH_2)_2\}(CO)_3]^{dd}$ | Yellow | 336–337 | $\nu(CO)$ 2058 m, 1996 s, 1966 w |

| Compound | Color | Mp (K) | Remarks |
|---|---|---|---|
| (structure above) $^{ee,ff}$ | Yellow | 375 | |
| $[\{Fe(CO)_3\}_2(H_2C \cdots \bigcirc \cdots CH_2)]^{(57)}$ | | >423 dec | |
| $[Os\{C(CH_2)_3\}(CO)_2(PPh_3)]^{(46)}$ | | | NMR, IR |
| $[IrCl\{C(CH_2)_3\}(CO)(PPh_3)]^{(46)}$ | | | NMR, IR, structure |

[a] H. O. Van Owen, *J. Organomet. Chem.*, **55**, 309 (1973).

[b] A. N. Nesmeyanov, K. N. Anisimov, N. E. Kolobova, and A. A. Pasynskii, *Dokl. Akad. Nauk SSSR*, **182**, 112 (1968).

[c] A. N. Nesmeyanov, K. N. Anisimov, N. E. Kolobova, and A. A. Pasynskii, *Izv. Akad. Nauk SSSR*, 100 (1969); A. N. Nesmeyanov, A. I. Gusev, A. A. Pasynskii, K. N. Anisimov, N. E. Kolobova, and Yu. T. Struchkov, *Chem. Commun.*, 739 (1969).

[d] A. A. Pasynskii, K. N. Anisimov, N. E. Kolobowa, and A. N. Nesmejanov, *Izv. Akad. Nauk SSSR*, 183 (1969).

[e] J. S. Ward and R. Pettit, *Chem. Commun.*, 1419 (1970).

[f] R. G. Amiet, P. C. Reeves, and R. Pettit, *Chem. Commun.*, 1208 (1967).

[g] R. B. King and A. Efraty, *Chem. Commun.*, 1370 (1970).

[h] W. Hubel and R. Merenyi, *J. Organomet. Chem.*, **2**, 213 (1964).

[i] M. Mathew and G. L. Palenik, *Can. J. Chem.*, **47**, 705 (1969).

[j] P. M. Maitlis and M. L. Games, *Chem. Ind.*, 1624 (1963).

[k] P. M. Maitlis and A. Efraty, *J. Organomet. Chem.*, **4**, 172 (1965).

[l] G. F. Emerson, L. Watts, and R. Pettit, *J. Am. Chem. Soc.*, **87**, 131 (1965).

[m] W. Huebel and E. H. Braye, *J. Inorg. Nucl. Chem.*, **10**, 250 (1959); W. Huebel, E. H. Braye, A. Clauss, E. Weiss, U. Kruerke, D. A. Brown, G. S. D. King, and C. Hoogzand, *J. Inorg. Nucl. Chem.*, **9**, 204 (1959).

[n] S.-I. Murahashi, T. Mizoguchi, T. Hosokawa, and I. Moritani, *Chem. Commun.*, 563 (1974).

[o] R. G. Amiet and R. Pettit, *J. Am. Chem. Soc.*, **90**, 1059 (1968).

[p] M. Rosenblum and B. North, *J. Am. Chem. Soc.*, **90**, 1060 (1968).

[q] P. M. Maitlis and A. Efraty, *J. Organomet. Chem.*, **4**, 175 (1965).

[r] R. B. King and A. Efraty, *J. Am. Chem. Soc.*, **94**, 3021 (1972).

[s] J. L. Boston, D. W. A. Aharp, and G. Wilkinson, *J. Chem. Soc.*, 3488 (1962).

[t] G. G. Cash, J. F. Helling, M. Mathew, and G. J. Palenik, *J. Organomet. Chem.*, **50**, 277 (1973).

[u] J. F. Nixon and M. Kooti, *J. Organomet. Chem.*, **104**, 231 (1976); M. Kooti and J. F. Nixon, *Inorg. Nucl. Chem. Lett.*, **9**, 1031 (1973).

[v] S. A. Gardner and M. D. Rausch, *J. Organomet. Chem.*, **56**, 365 (1973).

[w] R. Criegee and G. Schroeder, *Angew. Chem.*, **71**, 70 (1959); R. Criegee and G. Schroeder, *Ann. Chem.*, **623**, 1 (1959).

[x] H. H. Freedman, *J. Am. Chem. Soc.*, **83**, 2194 (1961).

[y] M. F. Hawthorne, D. C. Young, T. D. Andrews, D. V. Howe, R. P. Pilling, A. D. Pitts, M. Reintjes, L. F. Warren, Jr., and P. A. Wegner, *J. Am. Chem. Soc.*, **90**, 879 (1968).

[z] F. Canziani, P. Chini, A. Quarta, and A. DiMartino, *J. Organomet. Chem.*, **26**, 285 (1971).

[aa] F. Canziani and M. C. Malatesta, *J. Organomet. Chem.*, **90**, 235 (1975).

[bb] J. S. Ward and R. Pettit, *Chem. Commun.*, 1419 (1970).

[cc] G. F. Emerson, K. Ehrlich, W. P. Giering, and P. C. Lauterbur, *J. Am. Chem. Soc.*, **88**, 3172 (1966).

[dd] R. Noyori, T. Nishimura, and H. Takaya, *Chem. Commun.*, 89 (1969).

[ee] A. N. Nesmeyanov, I. I. Kritskaya, G. P. Zolnikova, Yu. A. Ustinyuk, G. M. Babakhina, and A. M. Wajnberg, *Dokl. Akad. Nauk SSSR*, **182**, 1091 (1968); I. I. Kritskaya, G. P. Zolnikova, I. F. Leshcheva, Yu. A. Ustinyuk, and A. N. Nesmeyanov, *J. Organomet. Chem.*, **30**, 103 (1971); A. N. Nesmeyanov, I. S. Astakhova, G. P. Zolnikova, I. I. Kritskaya, and Yu. T. Struchkov, *Chem. Commun.*, 85 (1970).

[ff] I. S. Astakhova and Yu. T. Struchkov, *Zh. Strukt. Khim.*, **11**, 473 (1970).

## 8. METHODS OF PREPARATION OF TRIMETHYLENEMETHANE COMPLEXES[2, 5, 24, 28]

The first trimethylenemethane compound $[Fe\{C(CH_2)_3\}(CO)_3]$ was obtained by reaction of 3-chloro-2-chloromethylpropene with nonacarbonyldiiron [equation (8.45)]. The carbonyl $[Cr(CO)_6]$ also reacts in an analogous manner [equation (8.46)]. A

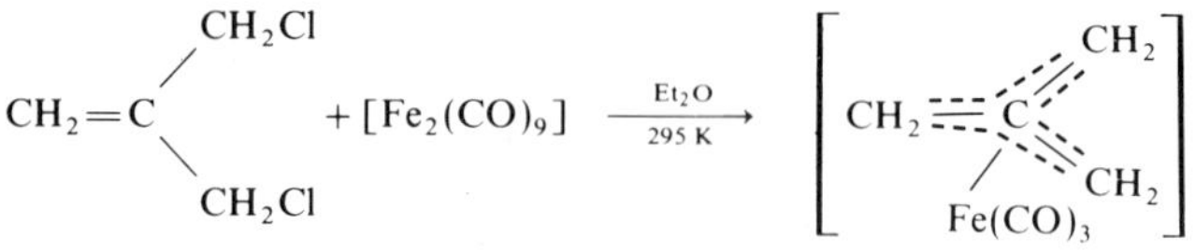

$$CH_2=C\begin{array}{c}CH_2Cl\\[4pt]CH_2Cl\end{array} + [Fe_2(CO)_9] \xrightarrow[295\ K]{Et_2O} \left[CH_2\overset{CH_2}{\underset{Fe(CO)_3}{\cdots C\cdots CH_2}}\right] \qquad (8.45)$$

$$[Cr(CO)_6] + CH_2=C(CH_2Cl)_2 \xrightarrow{hv} [Cr\{C(CH_2)_3\}(CO)_4] \qquad (8.46)$$

similar reaction of $[Fe_2(CO)_9]$ with 1,4-bis(bromomethyl)benzene and 1,4-bis(bromo-methyl)-2,5-dimethylbenzene takes place[57] [equation (8.47)]. Also, other derivatives of isobutene are utilized successfully for synthesis of trimethylenemethane complexes[46]

$$C_6H_2R_2(CH_2Br)_2 + Fe_2(CO)_9 \xrightarrow[413\text{-}8\ K]{PhH} \qquad (8.47)$$

[equations (8.48) and (8.49)]. However, a more general method of synthesis of tri-methylenemethane compounds is the reaction of anionic metal carbonyls with $CH_2=C(CH_2Cl)_2$ [equation (8.50)].

$$CH_2=C(CH_2SiMe_3)(CH_2Cl) + IrCl(CO)(PPh_3)_2 \begin{array}{c} \nearrow [IrCl\{\eta^4\text{-}C(CH_2)_3\}(CO)(PPh_3)] \\[10pt] \searrow \underset{KPF_6}{} [Ir\{\eta^4\text{-}C(CH_2)_3\}(CO)(PPh_3)_2]^+ \\ PF_6^- \end{array} \qquad (8.48)$$

$$CH_2=C(CH_2SiMe_3)(CH_2OAc) + Os(CO)_2(PPh_3)_3 \longrightarrow [Os\{\eta^4\text{-}C(CH_2)_3\}(CO)_2(PPh_3)] \qquad (8.49)$$

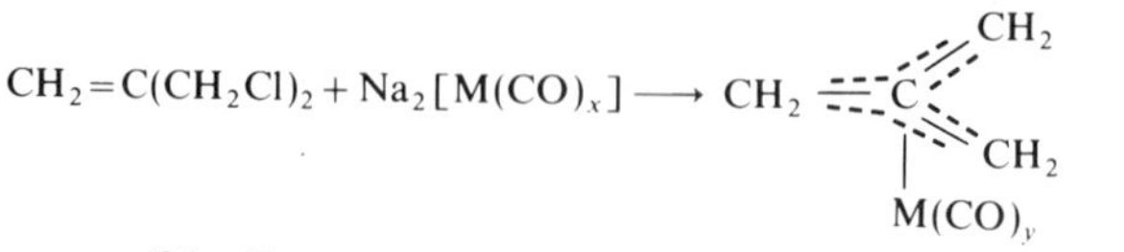

$$CH_2=C(CH_2Cl)_2 + Na_2[M(CO)_x] \longrightarrow CH_2\overset{CH_2}{\underset{M(CO)_y}{\cdots C\cdots CH_2}} \qquad (8.50)$$

$$(M = Fe;\ x = 4;\ y = 3;\ M = Mo;\ x = 5;\ y = 4)$$

Another relatively general method of preparation of trimethylenemethane com-pounds is the breaking of the methylenecyclopropane ring [equations (8.51) and (8.52)].

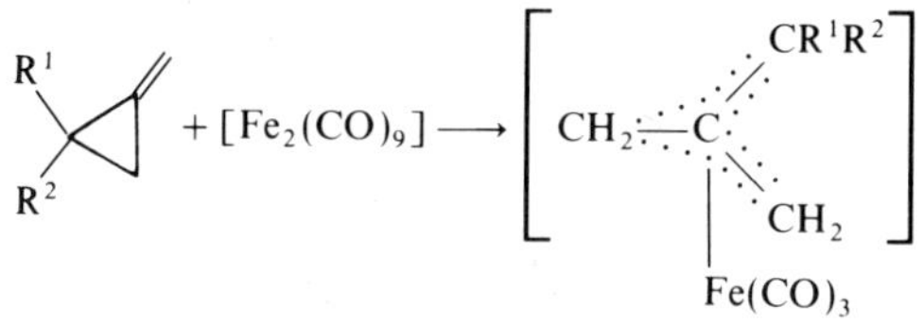

$$R^1 \diagdown \triangle \!\!=\!\! + [Fe_2(CO)_9] \longrightarrow \left[ CH_2 \!\cdots\! \overset{CR^1R^2}{\underset{CH_2}{C}} \right] \quad (8.51)$$

$$R^2 \qquad\qquad\qquad\qquad Fe(CO)_3$$

$$(R^1 = H, Me, Ph; R^2 = Ph)$$

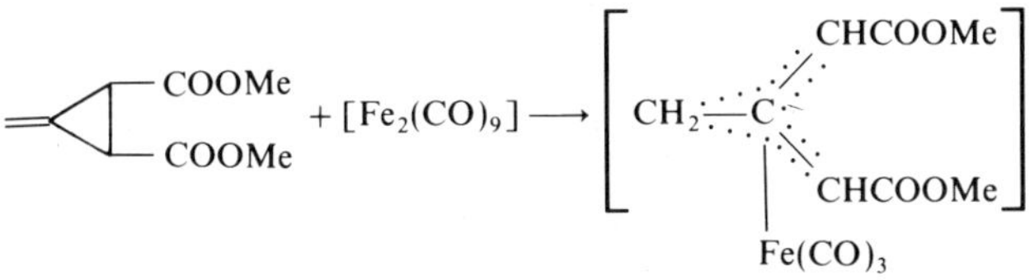

$$\triangle \!\!\!\! \overset{COOMe}{\underset{COOMe}{}} + [Fe_2(CO)_9] \longrightarrow \left[ CH_2 \!\cdots\! \overset{CHCOOMe}{\underset{CHCOOMe}{C}} \right] \quad (8.52)$$

$$Fe(CO)_3$$

The elimination of HCl from tricarbonylchloro(2-methylallyl)iron may also take place [equation (8.53)].

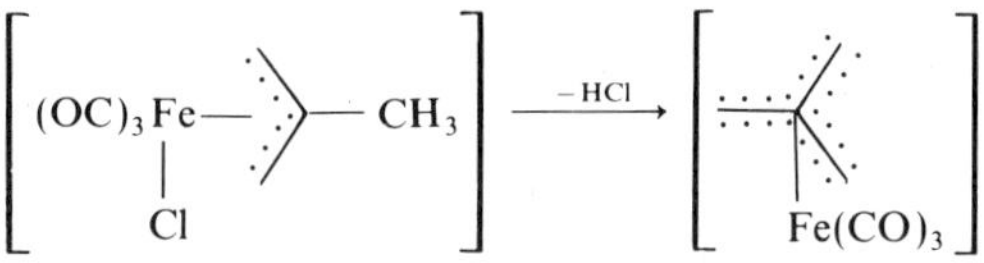

$$\left[ (OC)_3Fe \!\!-\!\! \underset{Cl}{\overset{\diagup}{\diagdown}}\!\!-\! CH_3 \right] \overset{-HCl}{\longrightarrow} \left[ \underset{Fe(CO)_3}{\diagup\!\!\diagdown} \right] \quad (8.53)$$

In the case of (2-acetylallyl)tricarbonylcobalt, there exists an equilibrium between the allyl complex and the trimethylenemethane complex which depends on the acidity of the solution [equation (8.54)].

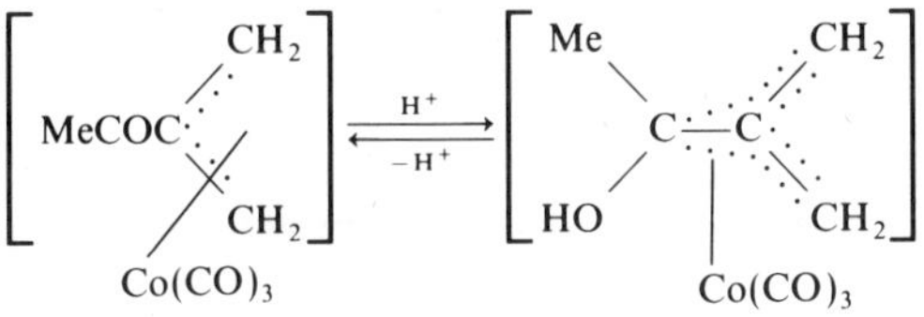

$$\left[ MeCOC \!\!\overset{CH_2}{\underset{CH_2}{\diagup\!\!\diagdown}} \right] \underset{-H^+}{\overset{H^+}{\rightleftarrows}} \left[ \overset{Me}{\underset{HO}{}}C\!\!=\!\!\overset{CH_2}{\underset{CH_2}{C}} \right] \quad (8.54)$$

$$Co(CO)_3 \qquad\qquad\qquad Co(CO)_3$$

The carbonyl $[Fe_2(CO)_9]$ reacts with 2-bromomethylnaphthalene to give a trimethylenemethane complex [equation (8.55)].

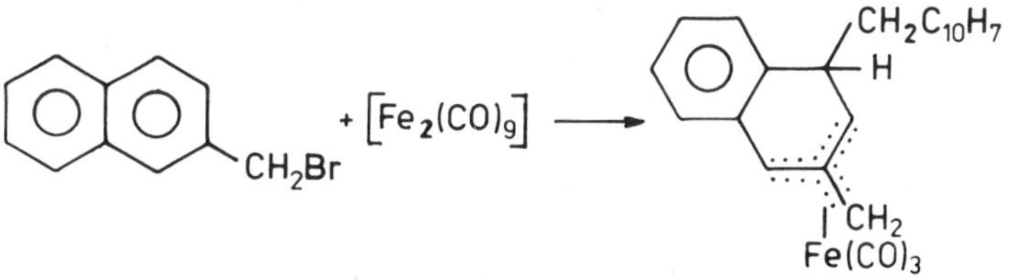

$$\cdots CH_2Br + [Fe_2(CO)_9] \longrightarrow \overset{CH_2C_{10}H_7}{\underset{H}{}} \cdots \overset{}{\underset{Fe(CO)_3}{CH_2}} \quad (8.55)$$

## 9. REACTIONS OF 1,3-DIENE COMPLEXES

Dienes may react with nucleophiles to give allyl complexes or they may react with electrophiles to afford dienyl compounds [equations (8.24) and (8.25)]. Some diene reactions were already discussed in Chapter 7. The nucleophiles which may attack the terminal carbon atom to give allyl compounds are $H^-$, $Cl^-$, $RCO^-$, $RO^-$, $OAc^-$, etc. In complexes, the diene molecules often undergo oxidative cycloaddition reaction, and allyl compounds are subsequently formed [equation (7.40), Table 1.4].

Trityl tetrafluoroborate eliminates $H^-$ from tricarbonyl(cis-penta-1,3-diene)iron to furnish a dienyl complex [equation (8.56)].

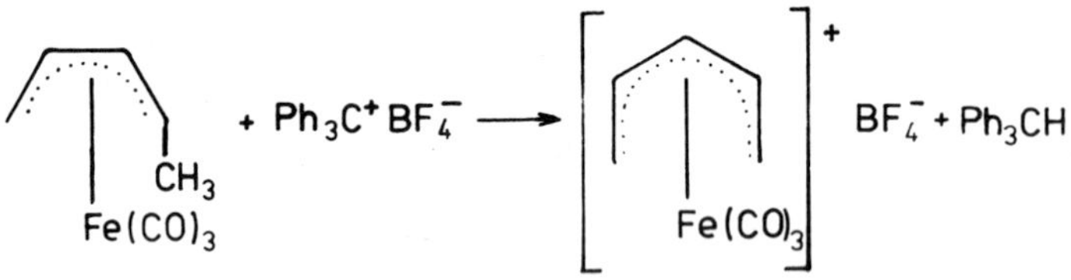

$$\text{(structure)} + Ph_3C^+ BF_4^- \longrightarrow [\text{(structure)}]^+ \quad BF_4^- + Ph_3CH \qquad (8.56)$$

The analogous *trans* compound does not react in the same manner because the methyl group is not favorably situated with respect to the iron atom.

Strong acids, such as $HBF_4$, $HClO_4$, and $HSbF_6$, react with dienes to give allyl complexes [equation (8.57)].

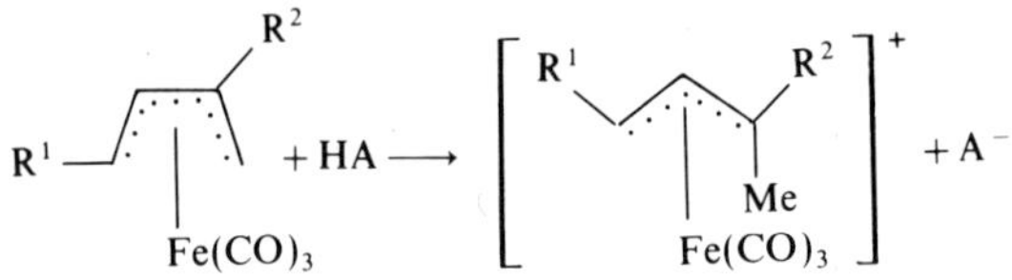

$$\text{(structure)} + HA \longrightarrow [\text{(structure)}]^+ + A^- \qquad (8.57)$$

Hydrogen chloride forms an iron allyl chloride complex when reacting with diene iron compounds [equation (8.58)].

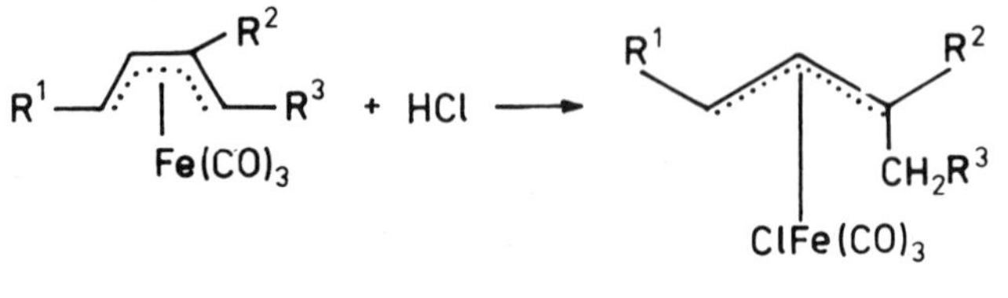

$$\text{(structure)} + HCl \longrightarrow \text{(structure)} \qquad (8.58)$$

Under Friedel–Crafts reaction conditions, acetyl chloride acylates butadiene and 2,3-dimethylbutadiene in iron derivatives in the presence of $AlCl_3$ in $CHCl_3$ or $CCl_4$ solution [equation (8.59)].

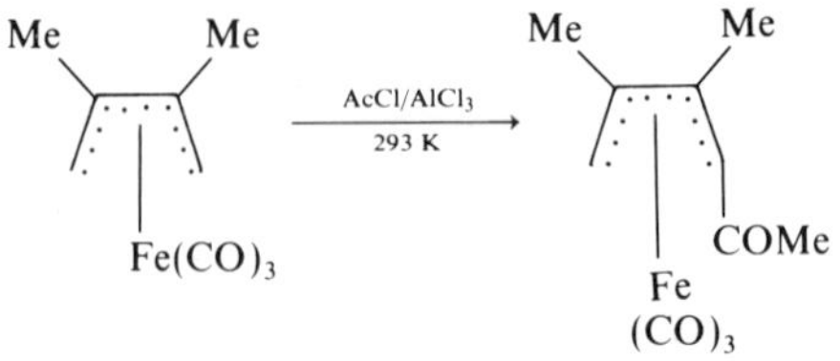

$$\text{(structure)} \xrightarrow[293 \text{ K}]{AcCl/AlCl_3} \text{(structure)} \qquad (8.59)$$

In compounds containing dienes with substituents at $C^1$ which possess easily removable hydrogen, dienyl derivatives in addition to acetylation products are formed [equation (8.60)].

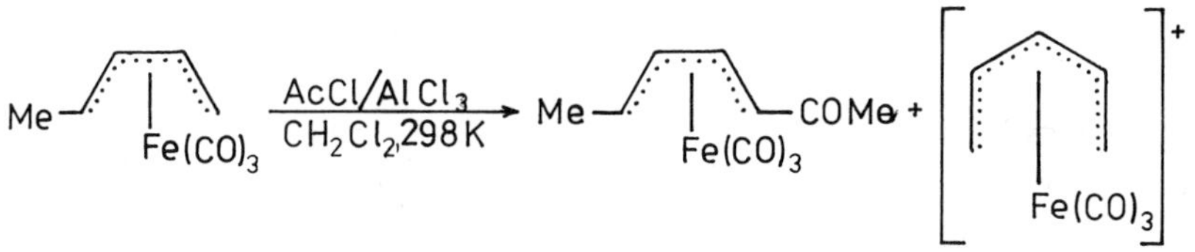

$$Me\diagup\!\!\!\diagdown\text{---}[Fe(CO)_3] \xrightarrow[\text{CH}_2\text{Cl}_2, 298\,\text{K}]{\text{AcCl/AlCl}_3} Me\diagup\!\!\!\diagdown\text{---COMe}[Fe(CO)_3] + \left[\diagup\!\!\!\diagdown[Fe(CO)_3]\right]^+$$

$$(8.60)$$

In the presence of bases, in complexes containing electron-withdrawing constituents, an exchange of hydrogen for deuterium may take place [equation (8.61)]. The rate of this reaction decreases as the electronegativity of the substituent becomes smaller: $X = CN > CHO > COPh > COOMe$.

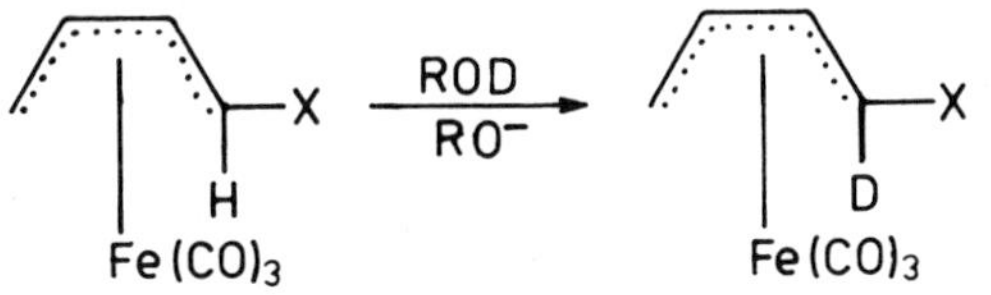

$$\underset{\underset{Fe(CO)_3}{H}}{\diagup\!\!\!\diagdown}\text{---X} \xrightarrow[\text{RO}^-]{\text{ROD}} \underset{\underset{Fe(CO)_3}{D}}{\diagup\!\!\!\diagdown}\text{---X} \qquad (8.61)$$

$(R = Me, Et, Bu')$

Coordinated dienes containing CHO groups have properties of aromatic aldehydes; therefore, they differ from $\alpha,\beta$-unsaturated aldehydes. The reduction by means of $Na[BH_4]$ leads to a primary alcohol, and the reaction with MeMgI gives a secondary alcohol [equation (8.62)].

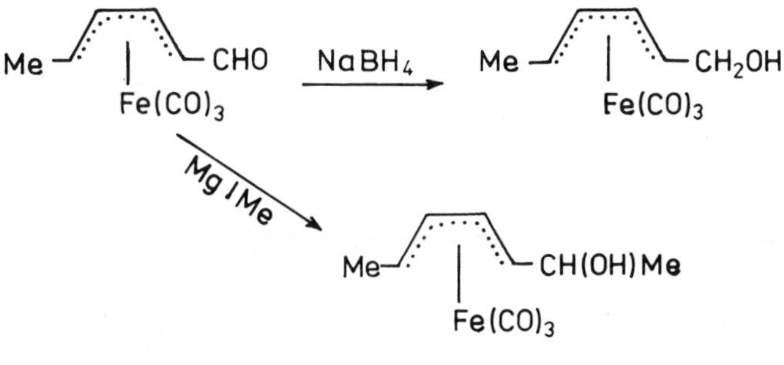

$$Me\diagup\!\!\!\diagdown\text{---CHO}[Fe(CO)_3] \xrightarrow{\text{Na}BH_4} Me\diagup\!\!\!\diagdown\text{---CH}_2\text{OH}[Fe(CO)_3]$$

$$\xrightarrow{Mg/Me} Me\diagup\!\!\!\diagdown\text{---CH(OH)Me}[Fe(CO)_3]$$

$$(8.62)$$

Reductions to alcohols by means of $Na[BH_4]$ also occur in the case of dienes that contain ketone groups.

In complexes with 1,3-dienes, exchange reactions of dienes as well as other ligands also take place.

$$[M(C_4H_6)(CO)_n] + L \longrightarrow [M(C_4H_6)(CO)_{n-1}L] + CO \qquad (8.63)$$

$$[M(C_4H_6)(CO)_n] + C_4H_5R \longrightarrow [M(C_4H_5R)(CO)_n] + C_4H_6 \qquad (8.64)$$

Oxidation of diene derivatives by means of $Fe^{3+}$ or $Ce^{4+}$ leads to their decomposition and the evolution of the diolefin molecule.

## 10. REACTIONS OF CYCLOBUTADIENE COMPLEXES

### a. Thermal decomposition

Thermal decomposition of cyclobutadiene complexes leads to various organic compounds. Heating $[Ni_2Cl_4(C_4Me_4)_2]$ forms, among others, the products shown in scheme (8.65).

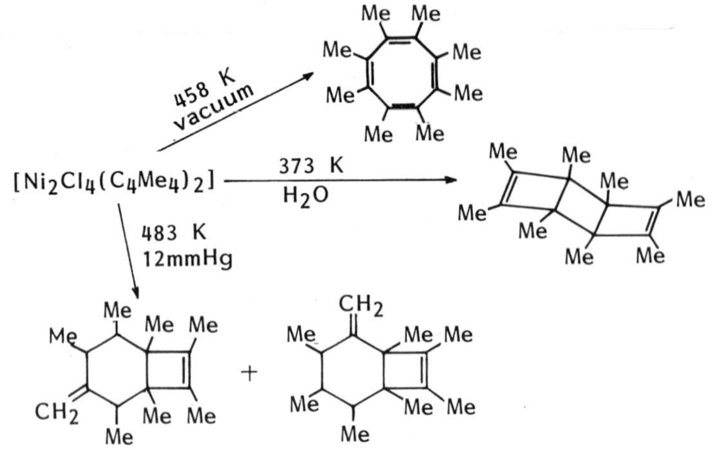

$$(8.65)$$

The pyrolysis of $[Pd_2Cl_4(C_4Ph_4)_2]$ mainly gives 1,4-dichlorotetraphenylbutadiene. Cyclobutadiene–acetylene niobium and tantalum compounds give off hexaphenyl-benzene at higher temperatures [equation (8.66)].

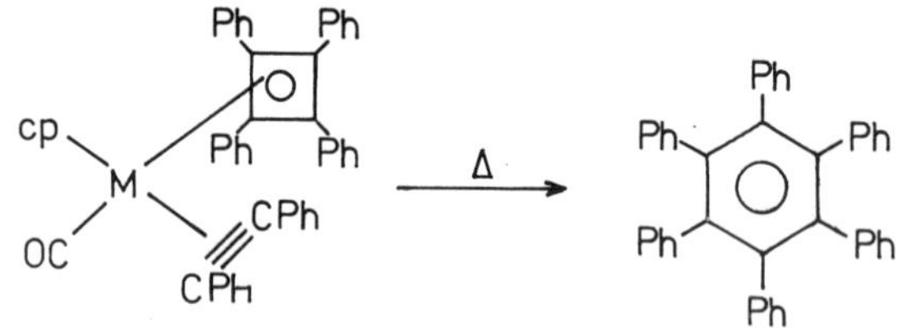

$$(8.66)$$

### b. Ligand substitution reactions

Many complexes undergo ligand substitution reactions.

$$[(Ph_4C_4)Co(CO)_2Br] + CH_2{=}CHCH_2MgCl \longrightarrow [(Ph_4C_4)Co{-}CO]$$

$$(8.67)$$

$$\xrightarrow[\text{2. } Br^-, I^-, PF_6^-]{\text{1. PhH/AlCl}_3} [(Ph_4C_4)Co(\eta\text{-}C_6H_6)]^+\, X^-$$

$$[(C_4R_4)CoX(CO)_2] + [Fe_2cp_2(CO)_4] \longrightarrow [Cocp(C_4R_4)] \qquad (8.68)$$

$$(R = Ph,\ p\text{-}MeC_6H_4,\ Me)$$

$$[M_2Br_4(C_4R_4)_2] + [FeBrcp(CO)_2] \longrightarrow [Mcp(C_4R_4)]^+ Br^- \tag{8.69}$$

$$(M = Ni, Pd; R = Ph, p\text{-}MeC_6H_4)$$

$$[Fe(C_4R_4)(CO)_3] + NO^+ PF_6^- \longrightarrow [Fe(C_4R_4)(CO)_2(NO)]^+ PF_6^- \tag{8.70}$$

$$(R = H, Me, Ph)$$

$$[Fe(C_4R_4)(CO)_3] + PPh_3 \longrightarrow [Fe(C_4R_4)(CO)_2(PPh_3)] + CO \tag{8.71}$$

$$[Fe(C_4H_4)(CO)_3] + MeOOCCH=CHCOOMe \xrightarrow{h\nu} [Fe(C_4H_4)(CO)_2]$$

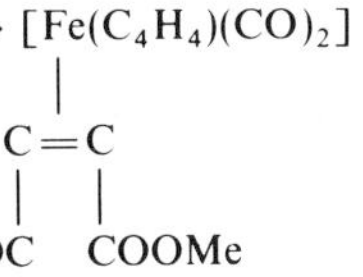

$$\tag{8.72}$$

## c. Reactions involving the cyclobutadiene ligand

*Nucleophilic attack.* The complex $[Ni_2Cl_4(C_4Me_4)_2]$ reacts with sodium cyclopentadienide to give a cyclobutenyl complex. Other derivatives also react in the same manner.

$$[Ni_2Cl_4(C_4Me_4)_2] + Nacp \longrightarrow \tag{8.73}$$

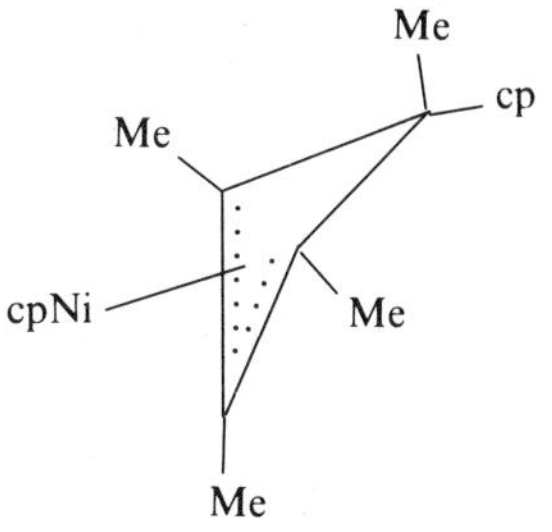

$$[Mcp(C_4Ph_4)]^+ FeBr_4^- \xrightarrow{RO^-} \qquad (M = Ni, Pd) \tag{8.74}$$

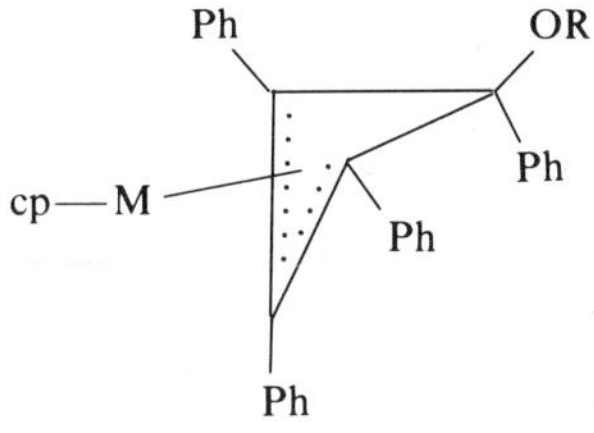

$$[M_2Cl_4(C_4R_4)_2] \xrightarrow{R'OH}$$

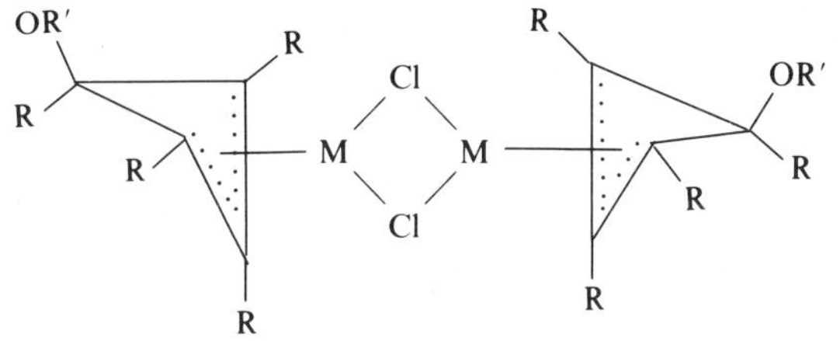

$$(M = Pd; R = Ph, p\text{-}MeC_6H_4; M = Pt; R = Me, Et) \tag{8.75}$$

*Electrophilic attack.*   Reaction of cyclobutadienetricarbonyliron with electrophilic reagents leads to the substitution of hydrogen in the ring. This is an analogous reaction to hydrogen substitution in cyclopentadienyl complexes. Reactivities of $Fe(C_4H_4)(CO)_3$ and $Fecp_2$ are similar.

Acetylation and benzoylation give acetyl and benzoyl derivatives, respectively [equation (8.76)]. Mercuration reactions occur unusually easily, leading to the formation of an equilibrium mixture of all possible complexes [equation (8.77)]. The reaction

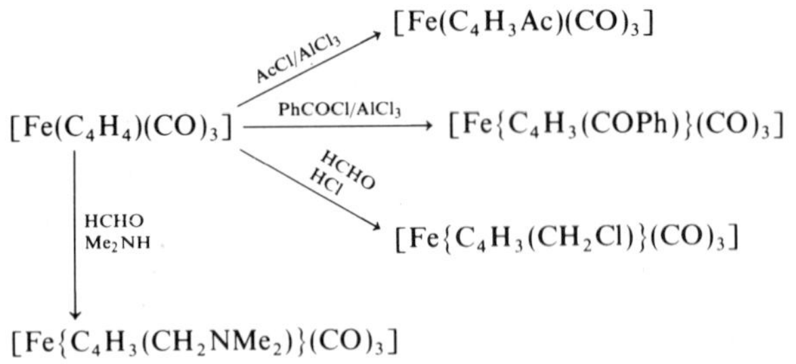

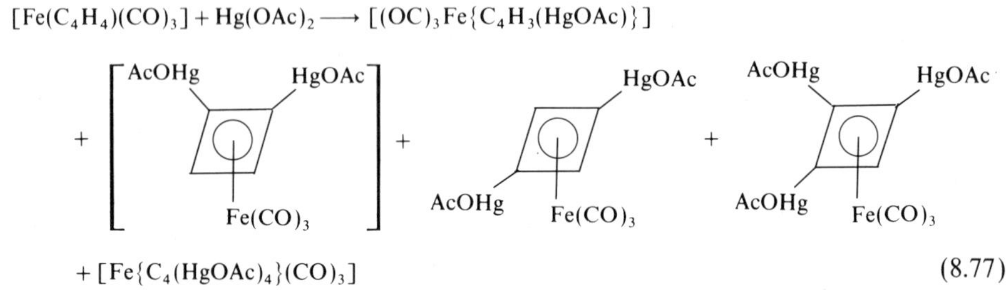

$$[Fe(C_4H_4)(CO)_3] + Hg(OAc)_2 \longrightarrow [(OC)_3Fe\{C_4H_3(HgOAc)\}]$$

of these compounds with $KI_3$ allows preparation of all iodocyclobutadiene derivatives which are inaccessible by other methods, for example, $[Fe(C_4H_2I_2)(CO)_3]$. It is assumed that the mechanism of the substitution reaction in cyclobutadiene complexes is analogous to that in other aromatic compounds. First, the addition of the $R^+$ cation to the ring takes place to give the cationic intermediate, which subsequently gives off the proton and forms the products [equation (8.78)].

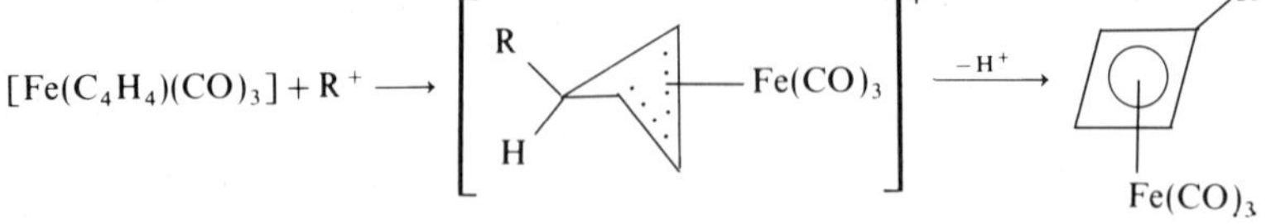

### d. *Oxidation reactions*

Iron(III) and Ce(IV) compounds readily oxidize cyclobutadiene complexes. In this way it is possible to obtain many organic compounds which are otherwise difficult to prepare.

Most commonly, oxidation reactions are carried out in the presence of various dienophilic compounds.

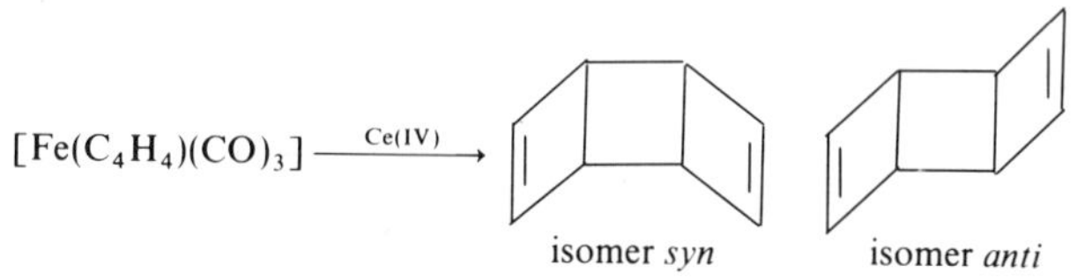

$$[Fe(C_4H_4)(CO)_3] \xrightarrow{Ce(IV)} \qquad \text{isomer } syn \qquad \text{isomer } anti \qquad (8.79)$$

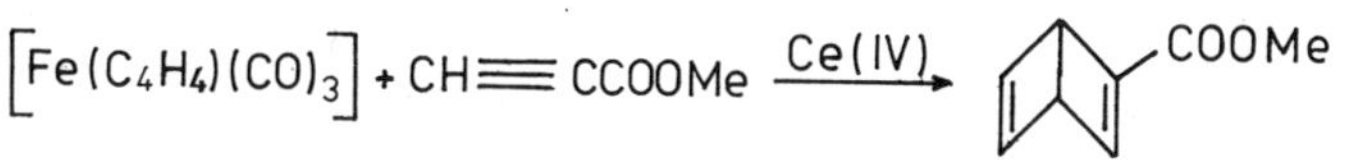

$$\left[Fe(C_4H_4)(CO)_3\right] + CH\equiv CCOOMe \xrightarrow{Ce(IV)} \qquad COOMe \qquad (8.80)$$

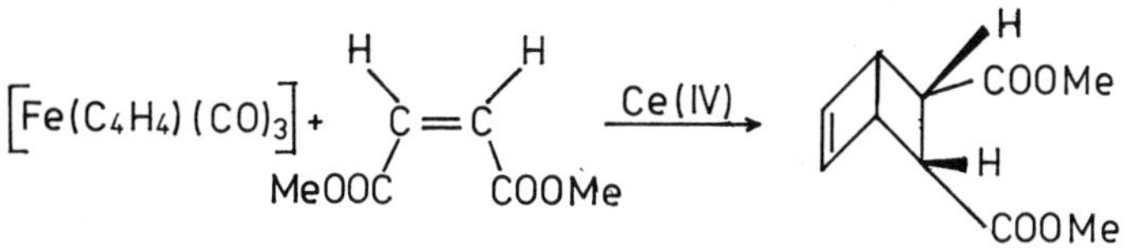

$$\left[Fe(C_4H_4)(CO)_3\right] + \underset{MeOOC}{\overset{H}{\underset{\phantom{a}}{C}}} = \underset{COOMe}{\overset{H}{\underset{\phantom{a}}{C}}} \xrightarrow{Ce(IV)} \qquad (8.81)$$

$$\left[\phantom{xx} O + Fe(CO)_3 \right] \xrightarrow{Pb(OAc)_4} \qquad (8.82)$$

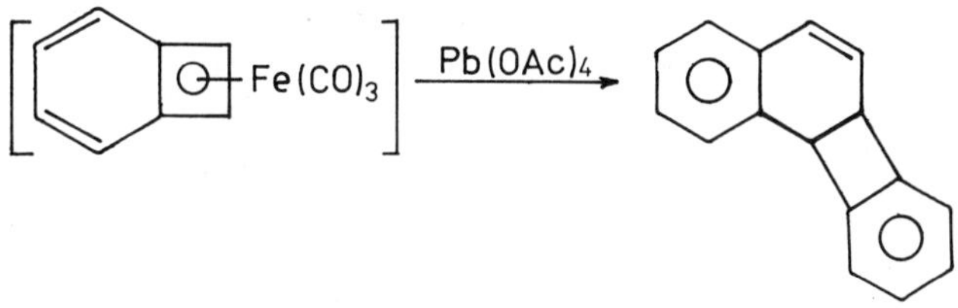

$$\left[Fe(C_4H_4)(CO)_3\right] + O = \left\langle \phantom{x} \right\rangle = O \xrightarrow{Ce(IV)}$$

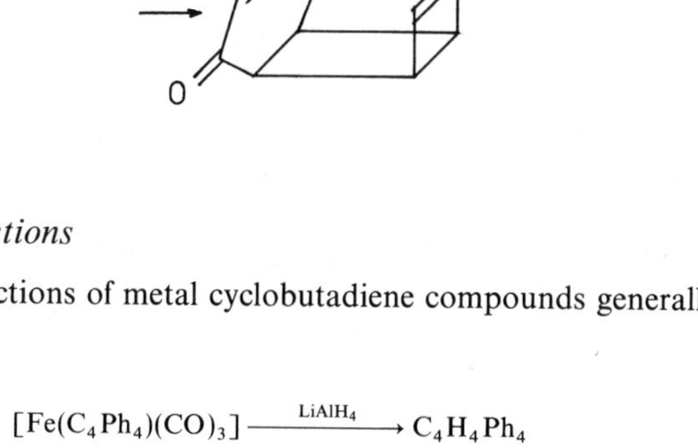

$$\longrightarrow \qquad (8.83)$$

## e. *Reduction reactions*

Reduction reactions of metal cyclobutadiene compounds generally proceed without ring breaking.

$$[Fe(C_4Ph_4)(CO)_3] \xrightarrow{LiAlH_4} C_4H_4Ph_4 \qquad (8.84)$$

$$[Ni_2Cl_4(C_4Me_4)_2] \xrightarrow[\text{or Zn/HCl}]{H_2/Pd} C_4H_4Me_4 \qquad (8.85)$$

$$[Ni_2Br_4(C_4Ph_4)_2] \xrightarrow{\text{LiAlH}_4} \text{cis-}C_4H_2Ph_4 \qquad (8.86)$$

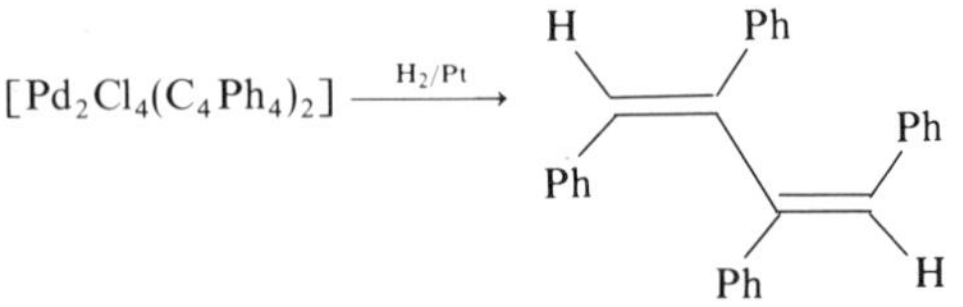

$$[Pd_2Cl_4(C_4Ph_4)_2] \xrightarrow{\text{H}_2/\text{Pt}} \qquad (8.87)$$

## 11. REACTIONS OF TRIMETHYLENEMETHANE COMPLEXES

The complex $[Fe\{C(CH_2)_3\}(CO)_3]$ reacts with bromine to afford a 2-bromo-methylallyl iron complex [equation (8.88)]. The addition of HCl may also take place [equation (8.89)]. Some other trimethylenemethane complexes may also react in a similar manner[46] [equation (8.90)].

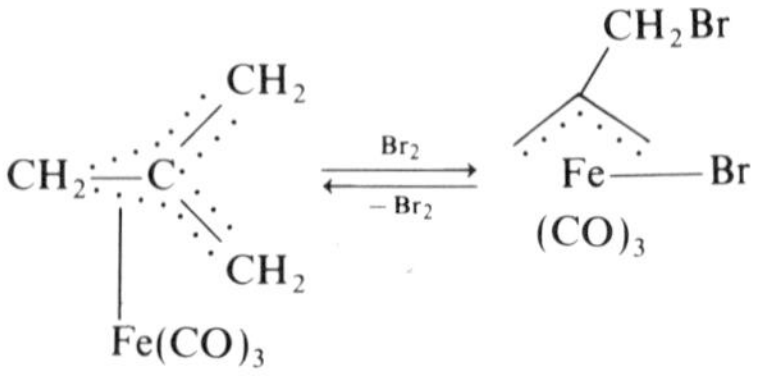

$$(8.88)$$

$$[Fe\{C(CH_2)_3\}(CO)_3] \xrightleftharpoons[-HCl]{HCl} [FeCl(2\text{-MeC}_3H_4)(CO)_3] \qquad (8.89)$$

$$[Os\{C(CH_2)_3\}(CO)_2(PPh_3)] + HCl \longrightarrow [Os(CH_2CMeCH_2)Cl(CO)_2(PPh_3)] \qquad (8.90)$$

Oxidation of the iron compound $[Fe\{C(CH_2)_3\}(CO)_3]$ by Ce(IV) causes loss of trimethylenemethane which gives an adduct with tetracyanoethylene. The HCl addition, similarly to other electrophilic substitution reactions (e.g., acylation):

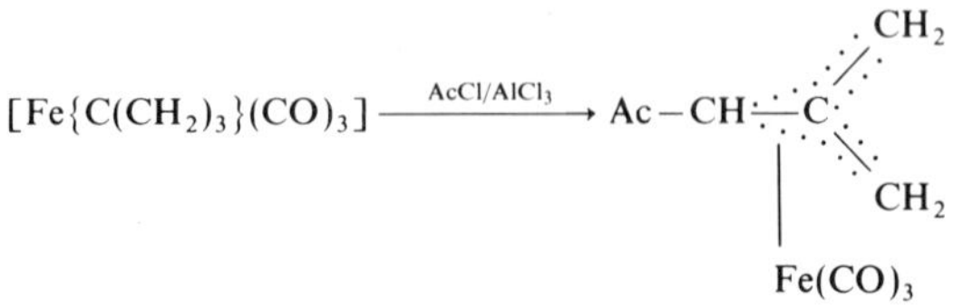

$$[Fe\{C(CH_2)_3\}(CO)_3] \xrightarrow{\text{AcCl/AlCl}_3} Ac-CH \qquad (8.91)$$

proceeds with a formation of the cationic allyl complex as an intermediate.

## 12. PROPERTIES OF DIENE COMPLEXES

Stable complexes containing 1,3-dienes are known for many transition metals of groups 3–10. The central atoms have $d^2$–$d^8$ electronic structures. Generally, these compounds are quite air stable, are easily soluble in organic solvents, and, after heating,

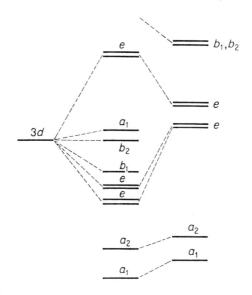

Figure 8.8.  The molecular orbital scheme for $Ni(C_4R_4)_2$.

they decompose with the formation of the metal. Table 8.5 gives properties of some complexes with 1,3-dienes.

Usually, 1,3-diene compounds constitute $18e$ complexes. However, exceptions are known; these comprise square-planar $16e$ complexes possessing $d^8$ electronic configuration of the central atom.

Bis(butadiene)carbonylmanganese, $[Mn(C_4H_6)_2(CO)]$, is a $17e$ paramagnetic compound with the magnetic moment corresponding to one unpaired electron.

Cyclobutadiene complexes are also relatively stable, but many of them undergo decomposition in air. Such complexes are formed by metals of groups 4–10 containing central atoms of $d^4$–$d^8$ electronic structures. Recently, the first biscyclobutadiene compound was obtained; it is the nickel complex $[Ni(C_4Ph_4)_2]^{(29,\ 30)}$ which is isoelectronic with ferrocene, $Fecp_2$. The molecular orbital scheme for this compound is presented in Figure 8.8.

## REFERENCES

1. D. M. P. Mingos, *Adv. Organomét. Chem.*, **15**, 1 (1977).
2. M. L. H. Green, *Organometallic Compounds, The Transition Elements*, Methuen, London (1968).
3. L. W. Rybin, in: *Metody Elementoorganicheskoy Khimii* (A. N. Nesmeyanov and K. A. Kocheskhov, eds.), Vol. 2, p. 542, Izd. Nauka, Moscow (1976).
4. S. P. Gubin, in: *Metody Elementoorganicheskoy Khimii* (A. N. Nesmeyanov and K. A. Kocheskhov, eds.), Vol. 3, p. 7, Izd. Nauka, Moscow (1976).
5. B. L. Shaw and N. I. Tucker, in: *Comprehensive Inorganic Chemistry* (J. C. Bailar, Jr., ed.), Vol. 4, p. 781, Pergamon Press, Oxford (1973).
6. A. Streitwieser, Jr., *Molecular Orbital Theory for Organic Chemists*, Wiley, New York (1961).

7. F. A. Cotton, *Teoria Grup, Zastosowania w Chemii*, PWN, Warsaw (1973).
8. J. A. Connor, L. M. R. Derrick, B. Hall, I. H. Hilier, M. F. Guest, B. R. Higginson, and D. R. Lloyd, *Mol. Phys.* **28**, 1193 (1974).
9. A. Efraty, *Chem. Rev.*, **77**, 691 (1977).
10. A. Efraty and P. M. Maitlis, *J. Am. Chem. Soc.*, **89**, 3744 (1967).
11. E. O. Fischer and H. Werner, *Metal Pi-Complexes*, Elsevier, Amsterdam (1966).
12. M. Herberhold and A. Razavi, *Angew. Chem., Int. Ed. Engl.*, **14**, 351 (1975).
13. G. Huttner, D. Neugebauer, and A. Razavi, *Angew. Chem., Int. Ed. Engl.*, **14**, 352 (1975).
14. C. Furlani and C. Cauletti, *Structure and Bonding*, **35**, 119 (1978).
15. J. C. Green, P. Powell, and J. van Tilborg, *J. Chem. Soc., Dalton Trans.*, 1974 (1976).
16. M. B. Hall, I. H. Hillier, J. A. Connor, M. F. Guest, and D. R. Lloyd, *Mol. Phys.*, **30**, 839 (1975).
17. J. A. Connor, L. M. R. Derrick, I. H. Hillier, M. F. Guest, and D. R. Lloyd, *Mol. Phys.*, **31**, 23 (1976); M. J. S. Dewar and S. D. Worley, *J. Chem. Phys.*, **51**, 1672 (1969).
18. H. P. Fritz, *Advanced Chemistry of the Coordination Compounds*, McMillan, New York (1961); H. P. Fritz, *Adv. Organomet. Chem.*, **1**, 239 (1964).
19. G. Davidson, *Inorg. Chim. Acta*, **3**, 596 (1969).
20. G. Davidson and D. A. Duce, *J. Organomet. Chem.*, **44**, 365 (1972).
21. W. T. Aleksanian and M. N. Nefedova, *Zh. Strukt. Khim.*, **14**, 839 (1973).
22. D. C. Andrews and G. Davidson, *J. Organomet. Chem.*, **76**, 373 (1974).
23. D. C. Andrews, G. Davidson, and D. A. Duce, *J. Organomet. Chem.*, **97**, 95 (1975).
24. J. Powell, in: *M.T.P. International Review of Inorganic Chemistry Ser. 1* (M. H. Mays, ed.), Vol. 6, p. 309.
25. a. P. L. Timms, *Angew. Chem., Int. Ed. Engl.*, **14**, 273 (1975). b. K. H. Klabunde, *Acc. Chem. Res.*, **8**, 393 (1975). c. P. S. Skell and M. J. McGlinchey, *Angew. Chem., Int. Ed. Engl.*, **14**, 195 (1975).
26. P. S. Skell, E. M. Van Dam, and M. P. Silvon, *J. Am. Chem. Soc.*, **96**, 626 (1974).
27. R. E. Mackenzie and P. L. Timms, *Chem. Commun.*, 650 (1974).
28. I. I. Kritskaya, D. A. Bochvar, and A. N. Nesmeyanov, *Metody Elementoorganicheskoy Khimii*, Izd. Nauka, Moscow (1975).
29. H. Hoberg, R. Krause-Goring, and R. Mynott, *Angew. Chem., Int. Ed. Engl.*, **17**, 123 (1978).
30. D. W. Clack and K. D. Warren, *J. Organomet. Chem.*, **161**, C55 (1978).
31. S. P. Gubin and G. B. Shulpin, *Khimia Kompleksov so Svyazyami Metall-Uglerod*, Izd. Nauka, Sibirskoye Otdielenie, Novosibirsk (1984).
32. S. P. Gubin and A. V. Gołounin, *Dieny i ich π-Kompleksy*, Izd. Nauka, Sibirskoye Otdielenie, Novosibirsk (1983).
33. D. M. P. Mingos, in: *Comprehensive Organometallic Chemistry* (G. Wilkinson, F. G. A. Stone, and E. W. Abel, eds.), Vol. 3, Chapter 19, Pergamon Press, Oxford (1982).
34. B. E. Mann, in: *Comprehensive Organometallic Chemistry* (G. Wilkinson, F. G. A. Stone, and E. W. Abel, eds.), Vol. 3, p. 20, Pergamon Press, Oxford (1982).
35. A. J. Pearson, *Transition Met. Chem.*, **6**, 67 (1981).
36. P. W. Jolly and R. Mynott, *Adv. Organomet. Chem.*, **19**, 257 (1981).
37. B. E. Mann, *Adv. Organomet. Chem.*, **12**, 135 (1974); M. H. Chisholm and S. Godleski, *Progr. Inorg. Chem.*, **20**, 299 (1976).
38. A. D. Hunter, P. Legzdins, C. R. Nurse, F. W. B. Einstein, and A. C. Willis, *J. Am. Chem. Soc.*, **108**, 1791 (1985).
39. H. Yasuda, K. Nagasuna, M. Akita, K. Lee, and A. Nakamura, *Organometallics*, **3**, 1470 (1984); H. Yasuda, Y. Kajihara, K. Mashima, K. Nagasuna, K. Lee, and A. Nakamura, *Organometallics*, **1**, 388 (1982); H. Yasuda, Y. Kajihara, K. Mashima, K. Nagasuna, K. Lee, and A. Nakamura, *Organometallics*, **2**, 478 (1983).
40. Y. Kai, N. Kanehisa, K. Miki, N. Kasai, K. Mashima, K. Nagasuna, H. Yasuda, and A. Nakamura, *Chem. Commun.*, 191 (1982).
41. G. Erker, J. Wicher, K. Engel, F. Rosenfeldt, W. Dietrich, and C. Krueger, *J. Am. Chem. Soc.*, **102**, 6346 (1980); P. Czisch, G. Erker, H.-G. Korth and R. Sustmann, *Organometallics*, **3**, 945 (1984).
42. T. A. Albright, *Acc. Chem. Res.*, **15**, 149 (1982).
43. M. Ziegler, *Z. Anorg. Allg. Chem.*, **355**, 12 (1967).
44. H. E. Sasse and M. L. Ziegler, *Z. Anorg. Allg. Chem.*, **392**, 167 (1972).

45. C. G. Pierpont, *Inorg. Chem.*, **17**, 1976 (1978).
46. M. D. Jones, R. D. W. Kemmitt, A. W. G. Platt, D. R. Russell, and L. J. S. Sherry, *Chem. Commun.*, 673 (1984).
47. J. W. Chinn, Jr. and M. B. Hall, *Organometallics*, **3**, 284 (1984).
48. J. C. Green, *Structure and Bonding*, **43**, 37 (1981).
49. G. Granozzi, E. Lorenzoni, R. Roulet, J.-P. Daudey, and D. Ajo, *Organometallics*, **4**, 836 (1985).
50. J. S. Drage and K. P. C. Vollhardt, *Organometallics*, **4**, 389 (1985).
51. R. P. Hughes, J. W. Reisch, and A. L. Rheingold, *Organometallics*, **3**, 1761 (1984).
52. S. S. Wreford and J. F. Whitney, *Inorg. Chem.*, **20**, 3918 (1981).
53. G. Erker, K. Berg, C. Krueger, G. Mueller, K. Argermund, R. Benn, and G. Schroth, *Angew. Chem., Int. Ed. Engl.*, **23**, 455 (1984).
54. C. Krueger, G. Mueller, G. Erker, U. Dorf, and K. Engel, *Organometallics*, **4**, 215 (1985).
55. G. Erker, T. Muehlenbernd, R. Benn, and A. Rufinska, *Organometallics*, **5**, 402 (1986).
56. R. L. Harlow, P. J. Krusic, R. J. McKinney, and S. S. Wreford, *Organometallics*, **1**, 1506 (1982).
57. A. R. Koray, C. Krieger, and H. A. Staab, *Angew. Chem., Int. Ed. Engl.*, **24**, 521 (1985).
58. J. M. Mayer, C. J. Curtis, and J. E. Bercaw, *J. Am. Chem. Soc.*, **105**, 2651 (1983).

# Chapter 9

# *Compounds Containing Five-Electron π-Ligands*

The most typical $5e$ dienyl ligands are five-, six-, and seven-membered cyclic and some acyclic hydrocarbons.[1-7,59,60] The simplest ligand of this type is cyclopentadienyl in which all carbon atoms are equivalent and lie in the same plane.

   The cyclopentadienyl ligand may be bonded to only one metal atom (9.1a) or it may form two kinds of bridges between two central metal atoms (9.1b, c). In the case of (9.1a) and (9.1c), all metal–carbon distances are the same.

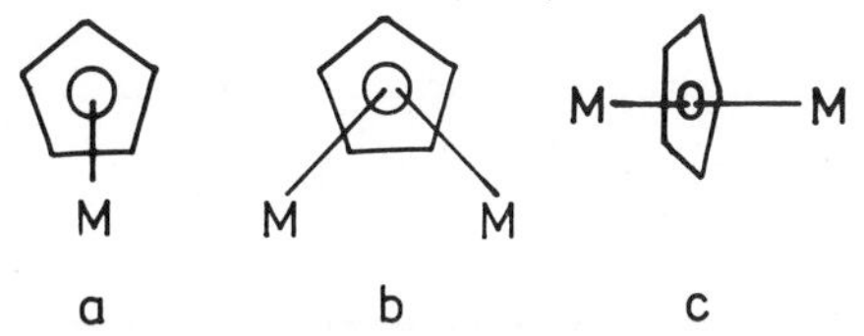

$$\text{(9.1)}$$

a         b         c

   The $\eta^1:\eta^5\text{-}C_5H_4$ group may also be considered a $5e$ ligand toward one metal[7b] [structures (9.2)].

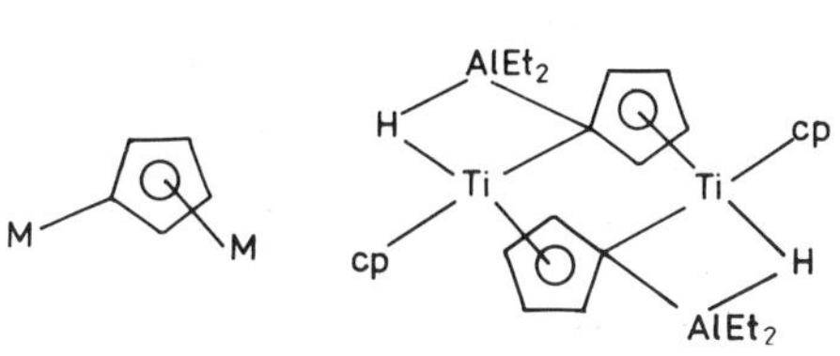

$$\text{(9.2)}$$

In most typical dicyclopentadienyl compounds, the parallel $C_5H_5$ groups may have eclipsed (9.3a) or staggered (9.3b) configuration. Acyclic dienyl ligands may form

509

analogous compounds (9.4). Bispentadienyl complexes usually have synclinal (9.4d) or perpendicular (9.4e) conformation.

$$(9.3)$$

a            b

$$(9.4)$$

a            b            c            d            e

## 1. *THE METAL–CYCLOPENTADIENYL BOND*

The $p_z$ orbitals of the five-membered ring form five $\pi$ orbitals of the cyclopentadienyl ligand (Figure 9.1). In the $C_{5v}$ group, they have the $a_1$, $e_1$, and $e_2$ symmetries. The energy of these orbitals varies according to the series: $a_1 < e_1 < e_2$. The interaction between the $e_1$ orbitals and the metal orbitals is the strongest. In the case of biscyclopentadienyl metal compounds with parallel rings possessing staggered configuration (the $D_{5d}$ group) the linear combination of $p$ orbitals gives ten molecular orbitals of the two $C_5H_5$ groups (Figure 9.2). These orbitals are as follows: $a_{1g}$, $a_{2u}$, $e_{1g}$, $e_{1u}$, $e_{2u}$. As for the $MC_5H_5$-type complexes, in $M(C_5H_5)_2$ compounds, the

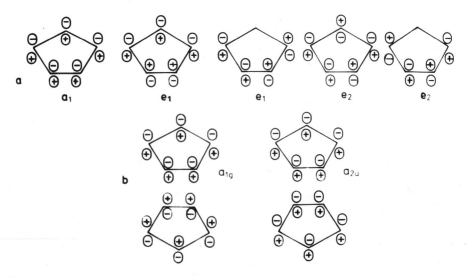

Figure 9.1.   (a) Molecular orbitals of the cyclopentadienyl ring. (b) The $a_{1g}$ and $a_{2u}$ molecular orbitals of two cyclopentadienyl rings.

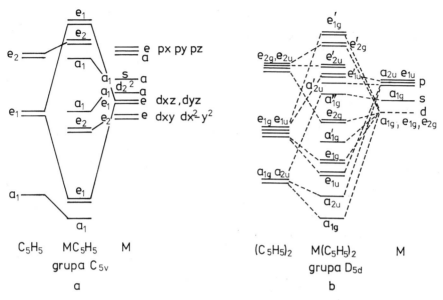

Figure 9.2 The molecular orbital diagram for (a) $MC_5H_5$ and (b) $M(C_5H_5)_2$ (by Shustorovich and Diatkina).

interaction between the orbitals of the ligands and the $e_{1g}$ $(d_{xz}, d_{yz})$ orbitals of the metal is the strongest. The remaining three $d$ orbitals of the metal are practically nonbonding. Therefore, the splitting of the $d$ level is as follows: $e_{2g}$ $(d_{x^2-y^2}, d_{xy}) \leqslant a_{1g}(d_{z^2}) < e_{1g}$ $(d_{xz}, d_{yz})$.

There is controversy with regard to the order of the $e_{2g}$ and $a_{1g}$ levels. The intensity of bands in the photoelectron spectra show that the $a_{1g}$ orbital has lower energy. This conclusion is in agreement with the SCF–LCAO calculations for ferrocene.[8,61,62] The reverse order $e_{2g} < a_{1g} < e_{1g}$ comes from electron absorption spectra and the $X\alpha$ calculations.[9,10] The disagreement of the intensity of bands in the photoelectron spectra may be due to a small contribution of the $s$ orbital in the $a_{1g}$ orbital (this phenomenon may influence the relative intensity of bands). Earlier calculations[3a,16] show that the $e_{1g}$ $(d_{xy}, d_{yz})$ orbital has very high energy. It lies $ca$ 15 eV above the non-bonding $d$ orbitals of the metal. The orbitals of the ligands are located in between the nonbonding and antibonding $d$ orbitals of the metal. Experimental data (electronic, photoelectron, and EPR spectra as well as magnetic measurements) show that the energy difference between the $d$ orbitals of the metal are relatively small. Properties of sandwich compounds of the type $[Mcp_2]^{n+}$ $(n=0, 1, ...)$, $[Mcp(arene)]^{m+}$ $(m=0, 1, ...)$, $[M(arene)_2]^{p+}$ $(p=0, 1, ...)$, etc., may be well explained on the basis of the crystal field theory.[26,27] However, a better description of properties of cyclopentadienyl complexes may be made by employing the scheme depicted in Figure 9.3, which incorporates the results of physicochemical studies.

The relative energy of the $d$ orbitals in various sandwich compounds depends partly on the ring size of the ligand, because the energy of the orbitals interacting with the metal depends on the size of the ring, as illustrated in Figure 9.4.

Generally, the hydrogen atoms or the substituents of the sandwich or half-sandwich

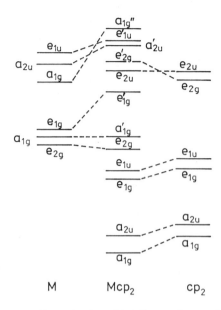

Figure 9.3.   The molecular orbital diagram for metallocenes.

compounds do not lie in the plane defined by the carbon atoms of the polyene. The magnitude of this phenomenon may be measured qualitatively by the angle $\theta$ which is formed by the $C-H$ or $C-R$ bond and the plane of the ring [diagram (9.5)].

$$(9.5)$$

Experiments show that the substituents are tilted away from the complex in the case of small rings (negative $\theta$) while they are tilted inward, toward the metal, in complexes possessing large rings (positive $\theta$). The value of the angle $\theta$ is close to zero for cyclopentadienyl and arene complexes, and therefore for compounds containing five- and six-membered rings. This is confirmed by calculations of the optimal angle $\theta$ for a series of the $M(CH)_n$ fragments employing the Hückel method.

For isolectronic fragments, the calculated values of angle $\theta$ are given below in (9.6).[11]

| Fragment | Optimal angle |
|---|---|
| $Co(C_3H_3)$ | $-26$ |
| $Fe(C_4H_4)$ | $-6$ |
| $Mn(C_5H_5)$ | $0$ |
| $Cr(C_6H_6)$ | $3$ |
| $V(C_7H_7)$ | $9$ |
| $Ti(C_8H_8)$ | $13$ |

$$(9.6)$$

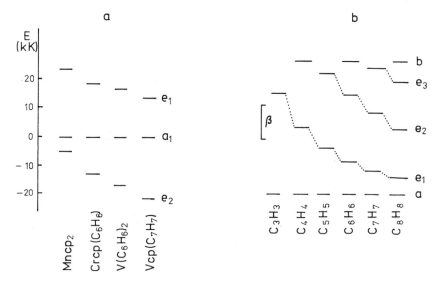

Figure 9.4. (a) Splitting of the $3d$ orbitals in sandwich compounds containing various rings. (b) Dependence of the orbital energy of the ligand on the size of the ring.[27]

This phenomenon is a result of the interaction of the orbitals of the ligand possessing the $e_1$ symmetry and the $d_{xz}$ and $d_{yz}(e_1)$ orbitals of the metal.

In the case of benzene, the optimal overlap between the $p\pi$ orbitals of the ring with the $d$ orbitals of the metal occurs when the hydrogen atoms are tilted toward the metal (Figure 9.5a). The situation is reversed for small rings (Figure 9.5b).

For cyclopentadienyl complexes with $d^0$–$d^4$ electronic configurations of the central atom, some nonbonding $e_{2g}$ and $a_{1g}$ orbitals are not occupied and therefore these metals may form many compounds of the type $Mcp_2X$, $Mcp_2X_2$, and $Mcp_2X_3$. Owing to deformation of the $Mcp_2$ fragment as the angle $\theta$ becomes smaller in going from the linear molecule $Mcp_2$ having $D_{5d}$ (or $D_{5h}$) symmetry to the molecule possessing the $C_{2v}$ symmetry, the orbitals $e_{2g}$ $(e'_2)$, $a_{1g}$ $(a'_1)$, and $e_{1g}$ $(e''_1)$ become $1a_1, b_2, 2a_1, b_1$, and $a_2$. The first three orbitals have a predominantly metal $d$ orbital character, and they lie in

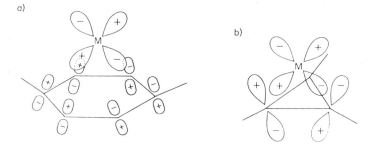

Figure 9.5. Interaction of the ring's $p\pi$ orbitals with the $d\pi$ orbitals: (a) $M(C_6H_6)$ and (b) $M(C_3H_3)$.

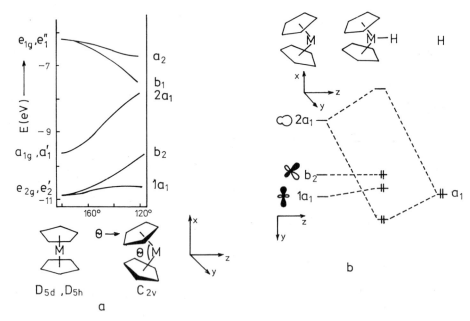

Figure 9.6. (a) Dependence of the molecular orbital energy of the bent fragment $Mcp_2$ on the cp—M—cp angle. (b) Scheme of the frontier molecular orbitals for $[cp_2MH]$, for example, $[Recp_2H]$ and $[Fecp_2H]^+$.

the $yz$ plane (Figure 9.6). Therefore, these orbitals may form bonds with additional ligands[7a, 12] (Figure 9.6).

In the case of the $Mcp_2H$ complexes in which the metal atom has pseudotrigonal coordination (assuming that the cyclopentadienyl ligand occupies one coordination site), the hydrogen atom is located on the $z$ axis. This structure is also the most favorable one because of steric reasons. In this compound, the only $\sigma$ orbital ($a_1$ symmetry) of the $H^-$ anion interacts strongly with the $2a_1$ orbital of the $Mcp_2$ fragment, interacts weakly with the $1a_1$ orbital, and does not interact at all with the $b_2$ orbital. The interaction of the $\sigma$ orbital of the hydride with the $2a_1$ orbital causes the formation of the M—H bond, while the $b_2$ orbital is nonbonding and the $1a_1$ orbital is slightly destabilized. Therefore, the most stable complexes possessing such structures are formed by metals with $d^4$ electronic structure because all orbitals possessing low energies are occupied (Figure 9.6b). The compounds $[Recp_2H]$ and $[Fecp_2H]^+$ are examples of such complexes.

For compounds containing 0–2$d$ electrons it is more favorable to place the ligand in the $yz$ plane, but on the axis which forms an angle with the $z$ axis. This enables the interactions of the $H^-$ ion with the $1a_1$ and $b_2$ orbitals (for the $C_{2v}$ symmetry) which become the orbitals possessing $a'$ symmetry in the complex belonging to the $C_s$ group [diagram (9.7)].

$$
\begin{array}{c}
\text{cp} \\
\diagdown \\
\text{----} \quad M \text{----} z \\
\diagup \quad \big)\alpha \\
\text{cp} \qquad H
\end{array}
\qquad (9.7)
$$

The energy of these orbitals is comparable with the energy of the $\sigma$ orbital of the $H^-$ ligand, and therefore their interaction is stronger than for the complex possessing $C_{2v}$ symmetry in which $H^-$ is located on the $z$ axis. These considerations show that for $d^0$, $d^1$, and $d^3$ electronic structures of the central ion, the X ligand in the $[Mcp_2X]$-type complexes should not be located on the $z$ axis. The $M-X$ bond should form a certain angle with the $z$ axis in low-spin $d^2$-electron compounds. So far unequivocal experimental confirmation of the predicted deformation is lacking.

In pseudotetrahedral complexes $[Mcp_2X_2]$ the XMX angle $\varphi$ depends on the number of $d$ electrons of the metal [diagram (9.8)]. This angle decreases in going from

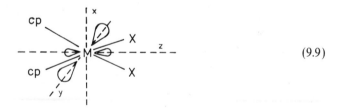

$$(9.8)$$

the $d^0$ through $d^1$ to $d^2$ structures. Calculations show[12] that in the case of $d^0$ electron complexes, two $\sigma$ orbitals of the X ligands possessing $a_1$ and $b_2$ symmetry interact strongly (the XMX angles are about $110°$) with the $1a_1$, $b_2$, and $2a_1$ orbitals. This interaction leads to only two bonding orbitals ($a_1$ and $b_2$). The remaining three orbitals are antibonding (two $a_1^*$ and one $b_2^*$ orbitals). If the $1a_1$ orbital of the $Mcp_2$ fragment is occupied ($d^1$ or $d^2$ structures), then its interaction with the corresponding orbital of the ligands decreases. The minimal energy is attained for the XMX angle of *ca* $75°$. At this angle, the X ligands are approximately located in the nodal planes of the $1a_1$ orbital (Figure 9.6).

The spatial arrangement of the $1a_1$ orbital shows that the free (nonbonding) electron pair is localized mainly outside the X ligands [diagram (9.9)].

$$(9.9)$$

It was confirmed by experiments that the largest XMX angles ($94–97°$) are encountered in complexes possessing the $d^0$ electron configurations ($[Zrcp_2Cl_2]$, $[Zrcp_2F_2]$, $[Ticp_2(CPh)_2]$, etc.), smaller ($85–88°$) in $d^1$ compounds ($[Nbcp_2Cl_2]$, $[Mocp_2Cl_2]^+$, $[Vcp_2(SPh)_2]$, etc.), and smallest ($76–82°$) in $d^2$ complexes such as $[Mocp_2Cl_2]$, $[Recp_2Br_2]^+$, $[Mocp_2Br(SnBr_3)]$, etc.[13]

The $d^0$ central atoms may form compounds of the composition $Mcp_2X_3$, such as $Tacp_2H_3$. All X ligands lie in the $yz$ plane. Calculations predict that the angle between bonds which are formed between the metal and the outside X atoms should equal *ca* $120°$. This angle is $122°$ for $[Tacp_2H_3]$.

In addition to biscyclopentadienyl sandwich-type complexes, triple-decker and multidecker compounds are known.[6,63–66] [structures (9.10)].

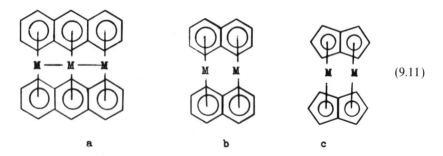

$$R^1 = Me, Et; R^2 = Me, Et; R^3 = H, Me \qquad\qquad (9.10)$$

In 1954, Wilkinson first pointed out the possibility of the existence of triple-decker compounds.[67] The first triple-decker complex $[Ni_2cp_3]^+$ was prepared in 1972 by Salzer and Werner.[6] The triple-decker complexes have 26–34 valence electrons. Based on the analysis of molecular orbitals of the Mcp and cp fragments, Hoffmann and coworkers[68] explained the formation of 30 and 34 electron triple-decker complexes. The complexes $[cpV(\mu-\eta-C_6H_6)Vcp]$ and bis(cyclopentadienyl)$(\mu-\eta$-mesitylene)divanadium $[cpV(\mu-\eta-C_6H_3Me_3)Vcp]$ have 26 valence electrons.[69,70] Photoelectron spectra and molecular orbital calculations show that these compounds have four unpaired electrons.[69] Chesky and Hall pointed out that analogous anions containing 27 and 28 electrons and perhaps cations possessing 25 and 24 electrons should be stable.[69] Calculations by Burdett and Canadell[63] show that sandwich complexes with polycyclic hydrocarbons containin M−M bonds should be stable [see structures (9.11)]. As

**a**      **b**      **c**    (9.11)

predicted by cluster theory (see Chapter 3), the calculations show that a relatively large difference between HOMO and LUMO exists if the $M_3$ and $M_2$ fragments in stuctures (9.11a), (9.11b), and (9.11c) have 22, 14, and 18 electrons, respectively.[63] In these cases, it is possible to form linear 50 and 34e clusters.

Many bispentadienyl $M(C_5R_{7-x}H_x)_2$ and pentadienyl $M(C_5R_{7-x}H_x)L_n$[59–61,71,72] complexes were recently prepared and investigated. Calculations for $Fe(\eta^5-C_5H_7)_2$ show that compounds with antiperiplanar (9.12a) and synclinal (9.12b) conformation are the most stable, noting that for the first case the potential minimum is slightly lower. The synperiplanar (9.12c) and anticlinal (9.12d) conformers have high energies.

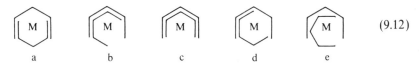

$$(9.12)$$

a           b           c           d           e

The complex $Fe(\eta^5\text{-}CH_2CMeCHCMeCH_2)_2$ has the synclinal (9.12b) conformation in the solid state, as in $Ru(\eta^5\text{-}CH_2CMeCMeCMeCH_2)_2$, despite the repulsions of the methyl groups which are located on the pentadienyl ligand, particularly in the case of the ruthenium compound. However, chromium and vanadium complexes $M(CH_2CMeCHCMeCH_2)_2$ have perpendicular conformation (9.12e). The fact that the perpendicular confirmation in vanadium and chromium compounds and synclinal conformation in iron and ruthenium pentadienyl complexes is preferred may be due to more effective interactions of the $\delta$ orbitals of the metal with the orbitals of the ligands than in the case of metallocenes, $Mcp_2$.[73] The reason for this is the greater dimensions of the open dienyl ligand compared to the cyclopentadienyl group. This leads to a lowering of the energy of orbitals which are directed along the $z$ axis ("$d_{z^2}$") and an increase in energy of the "$d_{xy}$" and "$d_{x^2-y^2}$" orbitals. As a result of lower symmetry of the dienyl complexes compared to $Mcp_2$ compounds, the $d$ metal orbitals and the orbitals of the ligands mix to a considerably greater degree.[73] Thus, the energy differences between the "$d$" metal orbitals become even greater. Therefore, the titanium and vanadium dienyl complexes $M(CH_2CMeCHCMeCH_2)_2$ are low-spin (0 and 1 unpaired electron).

In the case of metals of group 8, the $d_{xy}$ and $d_{x^2-y^2}$ orbitals are completely occupied. Therefore, they may form stronger $\delta$ bonds with ligands and force antiperiplanar or synclinal conformations which correspond to the octahedral coordination of the iron atom.

The compounds $M(C_nH_{n+2})L_2$, $M(C_nH_{n+2})L_3$, and $M(C_nH_{n+2})L_n$ show tendencies to the formation of some more stable conformers depending on the number of valence electrons in the complex. The 18$e$ complexes prefer conformations in which one of the L ligands is directed to the "open" edge of the polyene and therefore the "tetrahedral" and octahedral conformations, (9.13a) and (9.13e), and the conformation (9.13c) result. In the case of 16$e$ compounds, the square-planar coordination (9.13b), the bicapped tetrahedron structure (9.13f), and the (9.13d) conformation for complexes

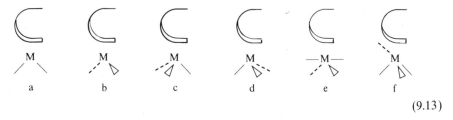

a          b          c          d          e          f

$$(9.13)$$

containing the $ML_3$ fragment are more stable [cf. discussion of stabilities of allyl complexes, Chapter 7, equation (7.4) and metal dienyl compounds, Chapter 8, equations (8.6) and (8.7), Figure 8.3].

## 2. STRUCTURE OF CYCLOPENTADIENYL COMPLEXES

Most sandwich compounds of the composition $Mcp_2$ involving the first series of the transition metals have staggered structures of the ferrocene type, while $[Rucp_2]$ and $[Oscp_2]$ compounds possess prismatic structures. However, in the gaseous state, ferrocene, like other analogous compounds of the $3d$ electron elements, has eclipsed $D_{5h}$ symmetry. The interchange (eclipsed–staggered) occurs easily because the barrier to the rotation of the cyclopentadienyl group in its own plane is very small and usually reaches $10 \, kJ \, mol^{-1}$. The C–C distance in the $C_5H_5$ ring varies in the 130–145 pm range.

Table 9.1. Structures of Cyclopentadienyl and Dienyl Metal Complexes

| Compound | $d_{M-C}$ (pm) | $d_{C-C}$ (pm) | Remarks |
|---|---|---|---|
| $[Vcp_2]^a$ | 230 | | $D_{5d}$ |
| $[Vcp(C_7H_7)]^b$ | 223 | 142 | V–C ($C_7H_7$) 225, C–C ($C_7H_7$) 140 |
| $[Crcp_2]^a$ | 222 | | |
| $[Mncp_2]^c$ | 238 | | |
| $[Fecp_2]$ | $205.6^d$ | | |
| | $204.5^e$ | 140.3 | |
| $[Cocp_2]^a$ | 213 | | |
| $[Nicp_2]$ | $220^a$ | | |
| | $218^f$ | | |
| $[Rucp_2]^g$ | 221 | 143 | |
| $[Oscp_2]^h$ | 222 | | |
| $[Ni_2cp_3]^{+i}$ | Ni–C (cp) 209 and 208 pm, Ni–C ($\mu$-cp) 213 and 216 pm | | |
| $[Pd_2(\mu\text{-}2\text{-}MeC_3H_4)(\mu\text{-}C_5H_5)(PPh_3)_2]^g$ | | | Pd–Pd 267.9 |
| $[Ti_2(\mu\text{-}Cl)_2cp_4]^j$ | 235 | 133 | |
| | | 136 | Ti$^1$–Cl 253.6; |
| | | 135 | 254.5 |
| | | | Ti$^2$–Cl 255.8 |
| $[Co(C_5H_4SiMe_3)(C_4Ph_4)]^k$ | 208 | 142.5 | |
| $[V(CH_2CMeCHCMeCH_2)_2]^{(74)}$ | 217.9 (C1,5)–223.6 (C3) | 140.2* | |
| $[Fe(CH_2CMeCHCMeCH_2)_2]^{(75)}$ | 210.8 (C1,5)–207.3 (C2,4) | 141.1* | |
| $[Ru\{CH_2(CMe)_3CH_2\}_2]^{(76)}$ | 216.2 (C1,5)–225.8 (C3) | 142.8* | |

* Averaged values.
[a] E. Weiss and E. O. Fischer, Z. Naturforsch., 10b, 58 (1955); E. Weiss and E. O. Fischer, Z. Naturforsch., 14b, 737 (1959); E. Weiss and E. O. Fischer, Z. Anorg. Allg. Chem., 278, 219 (1955); R. Schneider and E. O. Fischer, Naturwiss., 50, 349 (1963).
[b] G. Engebretson and R. Rundle, J. Am. Chem. Soc., 85, 481 (1963).
[c] A. Almenningen, A. Haaland, and S. Sandal, J. Organomet. Chem., 149, 219 (1978); A. Haaland, VIII Polish–GDR Colloquy on Organometallic Compounds, Mausz (1977); A. Haaland, Acc. Chem. Res., 12, 415 (1979).
[d] R. K. Boehn and A. Haaland, J. Organomet. Chem., 5, 470 (1966).
[e] J. D. Dunitz, L. E. Orgel, and A. Rich, Acta Crystallogr., 9, 373 (1956).
[f] I. A. Ronowa and N. V. Alekseev, Zh. Strukt. Khim., 7, 886 (1966).
[g] L. G. Herdgrove and D. H. Templeton, Acta Crystallogr., 12, 28 (1959).
[h] F. Jellinek, Z. Naturforsch., 14b, 737 (1959).
[i] E. Dubler, M. Textor, R. Oswald, and A. Salzer, Angew. Chem., Int. Ed. Engl., 13, 135 (1974).
[j] R. Jungst, D. Sekutowski, J. Davis, M. Luly, and G. Stucky, Inorg. Chem., 16, 1645 (1977).
[k] M. Calligaris and K. Venkatasubramanian, J. Organomet. Chem., 175, 195 (1979).

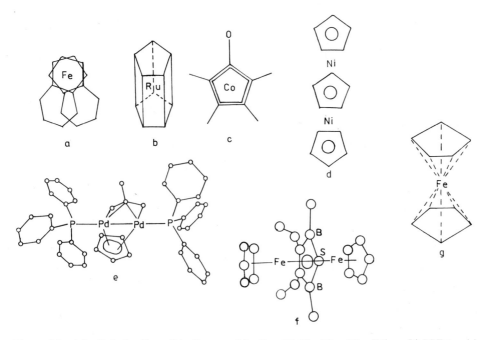

Figure 9.7.   (a)   Fe(indenyl)$_2$,   (b)   Rucp$_2$,   (c)   Cocp(C$_5$Me$_4$O),   (d)   [Ni$_2$cp$_3$]$^+$ BBF$_4^-$,   (e)

Pd$_2$($\mu$-2-MeC$_3$H$_4$)($\mu$-cp)(PPh$_3$)$_2$,   (f) Fe$_2$ $\left( \mu - \begin{array}{c} \text{CEt} - \text{BMe} \\ \| \\ \text{CEt} - \text{BMe} \end{array} \right\rangle \text{S} \right)$ cp$_2$, (g) Fecp$_2$.

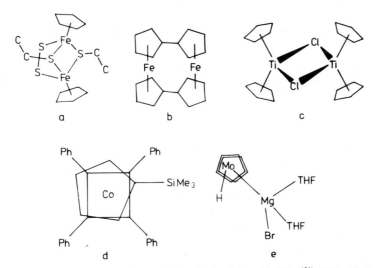

Figure 9.8.   Structures of dienyl complexes: (a) [Fe$_2$(SEt)$_2$ (S$_2$) cp$_2$]$^+$ PF$_6^-$,[54] (b) Fe$_2$(C$_5$H$_4$-C$_5$H$_4$)$_2$],
(c) [Ti$_2$Cl$_2$(cp)$_4$], (d) [Co(C$_4$H$_4$SiMe$_3$)(C$_4$Ph$_4$), (e) [Mocp$_2$HMgBr(THF)$_2$].[55]

Usually, all M–C distances are approximately the same. Exceptions are $C_5H_5$ groups which do not constitute $\eta^5$ ligands, and some $C_5H_5$ bridges. In "open" metallocenes M(dienyl)$_2$, the average M–C distance is usually shorter than in the cyclopentadienyl Mcp$_2$ complexes (Table 9.1). The situation is reversed in the case of the iron compounds Fecp$_2$ and Fe(CH$_2$CMeCHCMeCH$_2$)$_2$.[59,60,74–76] The C–C distances of the internal bonds are longer than those of the external ones. However, these differences are small and reach up to 4 pm. An exception is the complex Mn$_3$(CH$_2$CHCMeCHCH$_2$)$_4$ which possesses structure (9.14).[77]

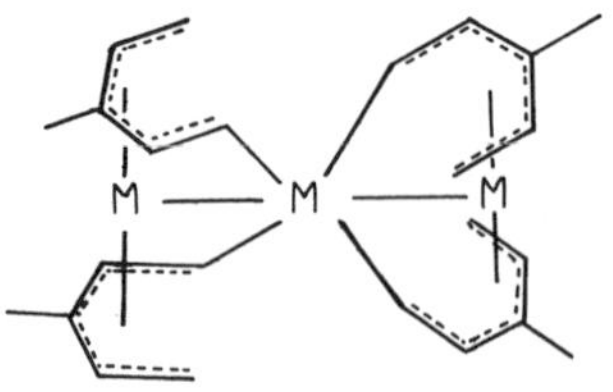

$$(9.14)$$

Structural data of some cyclopentadienyl complexes are given in Table 9.1. Examples of structures of complexes containing dienyl ligands are presented in Figures 9.7 and 9.8.

## 3. NMR SPECTRA[14–16]

The $^1$H NMR spectra of cyclopentadienyl groups for diamagnetic compounds contain only one narrow line. The ligand easily undergoes rotation about the fivefold M–C$_5$H$_5$ axis because the activation energy of this process is very small. Therefore, all cyclopentadienyl protons are equivalent. The chemical shifts are in the $\tau = 3.5$–7.0 range. The ferrocene molecule containing one substituent may show a considerable difference in the chemical shifts of protons in the 2 and 3 positions. For compounds with electronegative substituents, the resonances for protons in the 2 position appear at lower fields. The four protons of the ring give an AA'BB'-type spectrum in which there are often two triplets because the $\Delta\delta$ difference of the chemical shifts of protons A and B is much greater than the $J(AB)$–$J(AB')$ coupling constant. The spectra of paramagnetic complexes exhibit characteristic, unusually broad $^1$H NMR resonances. The determined second moment for ferrocene shows that there is also a rotation of the cyclopentadienyl rings in the solid state. Table 9.2 gives typical $\tau$ values for $C_5H_5$ groups bonded to various metals.

Table 9.2. Chemical Shift Ranges for the $\pi$-$C_5H_5$ Group Bonded to Transition Metals[a]

| Metal | $\tau$ Range | Metal | $\tau$ Range | Metal | $\tau$ Range |
|-------|--------------|-------|--------------|-------|--------------|
| Ti | 4.0–4.2 | Mn | 3.8–6.6 | Co | 4.0–5.8 |
| V  | 4.1 | Tc | 4.6–5.6 | Rh | 4.0–5.2 |
| Ta | 5.2 | Re | 3.9–6.0 | Ir | 5.8 |
| Cr | 4.2–6.9 | Fe | 4.4–6.0 | Ni | 4.2–5.2 |
| Mo | 3.4–5.6 | Ru | 4.5–6.0 | Pd | 3.5–4.9 |
| W  | 4.0–5.7 | Os | 4.7–5.3 | Pt | 4.2–4.9 |

[a] According to Reference 14.

The $^{13}C$ chemical shifts in cyclopentadienyl complexes (Table 9.3) occur in the 64–125 ppm range for unsubstituted $C_5H_5$ groups and up to *ca* 140 ppm for substituted $CR_5$ ligands.[17-22] There is a linear relationship between the $^1H$ and $^{13}C$ chemical shifts for the $\eta$-$C_5H_5$ ligands.[19-21]

The complexation causes a considerable shift of the signal to stronger fields and the increase in the $^1J(CH)$ coupling constant. The increase of screening is due to greater electron density on the carbon atoms of the ligand compared to the free $C_5H_5^-$ ion. The increase of the electron density is due to the formation of the poly-center covalent

Table 9.3. $^{13}C$ *NMR Spectra of Cyclopentadienyl Metal Complexes*[a]

| Compound | $\delta\ ^{13}C$ (ppm) | Coupling constants $J(CH)$ (Hz) |
|---|---|---|
| 1 | 2 | 3 |
| $[Fecp_2]$ | 67.9 | 175 |
| $[Rucp_2]$ | 70.6 | |
| $[Oscp_2]$ | 63.9 | |
| $[Fecp(CO)_2(COPh)]$ | 87.1 | |
| $[Fecp(C_6H_6)]^+[BF_4]^-$ | 72.8 | |
| $[Fecp(C_5H_4OAc)]$ | 69.5; 70.0 (C2); 71.1 (C3) | |
| $[Fecp(C_5H_4Ph)]$ | 69.5; 85.7 (C1); 68.8 (C2); 66.5 (C3) | |
| $[Fecp(C_5H_4COPh)]$ | 69.6; 77.9 (C1); 70.9 (C2); 72.0 (C3) | |
| $[Fecp(C_5H_4Me)]$ | 68.7; 84.0 (C1); 69.2 (C2); 67.2 (C3) | |
| $[Fecp(C_5H_4Et)]$ | 67.7; 90.3 (C1); 66.4 (C2); 66.8 (C3) | |
| Nacp | 102.1 | |
| Mgcp$_2$ | 108.0 | |
| $[TicpCl_3]$ | 123.1 | |
| $[Ticp(OEt)_3]$ | 112.3 | 173.5 |
| $[Ti(C_5Me_5)Br_3]$ | 138.2 | |
| $[Ticp_2Cl_2]$ | 119.7 | |
| $[Zrcp_2Cl_2]$ | 115.7 | |
| $[Zrcp(acac)_2Cl]$ | 116.5 | 185 |
| $[Hfcp_2Cl_2]$ | 114.4 | |
| $[Vcp(CO)_4]$ | 90.7 | |
| $[Nbcp(CO)(PhCCPh)_2]$ | 98.7 | |
| $[MocpCl(CO)_3]$ | 94.9 | |
| $[Mocp(CO)_2NO]$ | 93.6 | |
| $[WcpPh(CO)_3]$ | 92.2 | |
| $[Wcp(CH_2Ph)(CO)_3]$ | 92.4 | 173 |
| $[Mncp(CO)_3]$ | 83.1 | 175 |
| $[Mn(C_5H_4Me)(CO)_3]$ | 103.0 (C1); 83.1 (C2); 82.5 (C3) | |
| $[Mncp(CO)(Ph_2PCH_2PPh_2)]$ | 76.7 | |
| $[Recp(CO)_3]$ | 84.5 | |
| $[Rhcp(COD)]$ | 86.2 | 167; 3.8 $J(Rh-C)$ |
| $[Rhcp(C_2H_4)_2]$ | 86.9 | 194 |
| $[Rhcp(CO)_2]$ | 87.6 | 176; 3.0 $J(Rh-C)$ |
| $[Rhcp(NBD)]$ | 84.8 | 172 |
| $[Nicp(C_3H_5)]$ | 89.2 | |
| $[Pdcp(C_3H_5)]$ | 94.3 | 171 |

[a] According to References 19–22.

metal–ligand bond. The formation of this bond involves participation of the $p\pi$ orbitals of the ligand, and therefore the contribution of the $s$ orbitals in the $C-H$ bond becomes greater. This leads to an increase in the $J(CH)$ coupling constant. Only in cyclopentadienyl titanium complexes is the screening weaker than for the $C_5H_5^-$ anion.

The strongest screening is observed for iron cyclopentadienyl compounds. It is greater for ferrocene derivatives than for monocyclopentadienyl complexes. The greatest changes in chemical shifts, which depend on the substituent R in the ligand $C_5H_4R$, occur for the $C_1$ carbon (bonded to the substituent), while the change of R has less influence on the $C_2$ and $C_3$ shifts [structure (9.15)].

$$
\begin{array}{l}
C^3 \!-\! C^2 \\
| \qquad\quad \diagdown \\
| \qquad\qquad\; C^1 \!-\! R \\
| \qquad\quad \diagup \\
C^3 \!-\! C^2
\end{array}
\qquad\qquad (9.15)
$$

There is a linear relationship between $\delta(C_1)$ for complexes containing benzene drivatives $C_6H_5R$ and $\delta(C_1)$ for derivatives of ferrocene. A similar relationship was found for $\delta(C_2)$ of ferrocene derivatives and $\delta(C_2)$ of benzene derivatives and for $\delta(C_3)$ in $C_5H_4R$ and $\delta(C_4)C_6H_5R'$ [there is no such correlation with $\delta(C_3)$ for benzene compounds].[19,20] For complexes of the type $Fecp(CO)_2X$, there is no dependence of $\delta\,{}^{13}C(cp)$ on the character of X despite the existence of such a dependence for the carbonyl groups for which $\delta({}^{13}C)$ depends on Taft's $\sigma_I$ constant for the ligands X ($X=CN$, Cl, Br, I, $SnR_3$, COPh, COMe, Ph etc.). Dienyl complexes of titanium, iron, and ruthenium $[M(dienyl)_2]$ show dynamic properties.[59,60] This shows that ligands may oscillate about the M–dienyl axis. The barrier ($\Delta G^+$) of oscillation for iron complexes containing $C_5H_7$, $3\text{-}C_5H_6Me$, and $2,4\text{-}C_5H_5Me_2$ equals 35–38 kJ mol$^{-1}$, for ruthenium complexes with $2,4\text{-}C_5H_5Me_2$ and $2,3,4\text{-}C_5H_4Me_3$ it is 40–43 kJ mol$^{-1}$, and for the titanium complex $Ti(2,4\text{-}C_5H_5Me_2)_2$ it equals 64 kJ mol$^{-1}$. Therefore, this barrier is considerably greater than for cyclopentadienyl complexes for which the activation energy of rotation of the $C_5H_5$ group in solution equals up to 10 kJ mol$^{-1}$. In the solid state, this value ranges from 2 to 50 kJ mol$^{-1}$. The highest $E_a$ values were found for substituted rings.[78] The $^1H$ NMR spectra of iron, ruthenium, and titanium complexes $M(2,4\text{-}C_5H_5Me_2)_2$ at low temperature show 7 lines which suggest that these compounds have nonsymmetrical conformations. Thus, the synclinal conformation which is observed in the case of the iron complex in the solid state is indeed the ground state.

## 4. IR SPECTRA[23,24]

Cyclopentadienyl complexes may be divided into the following four groups, each of which shows a characteristic IR spectrum:

1. Compounds with ionic bonds. The $C_5H_5$ ring in these compounds gives a spectrum which is analogous to that of the $C_5H_5^-$ ion ($D_{5h}$ symmetry).

2. Compounds with central $\sigma M - C_5H_5$ bonds. These compounds have local $C_{5v}$ symmetry. The metal forms a $\sigma$ bond with the middle of the ring. The far-infrared spectra show a band due to the stretching metal-ring vibrations.

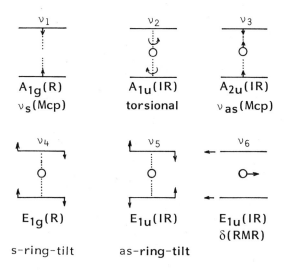

Figure 9.9.   Skeletal vibrations of metal dicyclopentadienyl complexes.

3.  Complexes in which the ring is bonded to the metal via a π bond (local symmetry $C_{5v}$). The far-infrared spectra show the stretching metal-ring vibrations and other skeletal oscillations.

4.  Compounds with a σ-metal–carbon bond (monohapto $M(\eta^1\text{-}C_5H_5)$. The IR spectrum is analogous to the cyclopentadiene molecule $C_5H_6$ and differs completely from IR spectra of the above-mentioned groups of compounds.

Ionic compounds, such as $M(C_5H_5)$ (M = K, Rb, Cs), $[M(C_5H_5)_2]$ (M = Ca, Sr, and Ba), and $[Eu(C_5H_5)_3]$, show four bands in the IR spectra: $3000\text{–}3100 \text{ cm}^{-1}$—$v(CH)$, $1400\text{–}1500 \text{ cm}^{-1}$—$v(CC)$; $1000\text{–}1010 \text{ cm}^{-1}$—$\delta(CH)$ and $660\text{–}770$—$\pi(CH)$. The IR spectra of complexes of group 2 metals such as $M(C_5H_5)$ (M = Li, Na, and Tl) and $M(C_5H_5)_2$ (M = Be, Mg, Sn, and Pb) exhibit eight bands. The frequency $v(M-cp)$ occurs in the $300\text{–}550 \text{ cm}^{-1}$ range. The IR spectra of π-com-

Table 9.4.   *IR Spectra of Cyclopentadienyl Metal Complexes* $(cm^{-1})^a$

| Complex | $v_{CH}$ | $v_{CH}$ | $v_{CC}$ | $v_{CC}$ | $\delta_{CH}$ | $\pi_{CH}$ | $\pi_{CH}$ | $E_{1u}$ as tilting | $v_{Mcp}$ | $k_{Mcp}$ (N m$^{-1}$) |
|---|---|---|---|---|---|---|---|---|---|---|
| Crcp$_2^+$ | 3012 | | 1421 | 1112 | 1007 | 835 | | 479 | 381 | |
| Crcp$_2$ | 3076 | | 1404 | 1089 | 987 | 829 | 766 | 429 | 408 | 160 |
| Fecp$_2^+$ | 3106 | 2933 | 1412 | 1107 | 1003 | 852 | 817 | 510 | 423 | 200 |
| Fecp$_2$ | 3086 | 2909 | 1408 | 1104 | 1001 | 854 | 814 | 490 | 478 | 270 |
| Rucp$_2$ | 3097 | | 1410 | 1101 | 1002 | 866 | 821 | 446 | 379 | 240 |
| Oscp$_2$ | 3061 | 2907 | 1400 | 1098 | 998 | 831 | 823 | 428 | 353 | 280 |
| Cocp$_2^+$ | 3106 | 2916 | 1414 | 1103 | 1008 | 867 | 820 | 500 | 461 | 250 |
| Cocp$_2$ | 3041 | | 1412 | 1101 | 995 | 828 | 778 | 464 | 355 | 150 |
| Nicp$_2$ | 3052 | 2891 | 1421 | 1109 | 1002 | 839 | 772 | | 355 | 150 |

$^a$ According to Reference 24.

plexes are similar to those of compounds with a central $\sigma$ metal–cyclopentadienyl bond. The metallocenes $Mcp_2$ have in addition six skeletal vibrations, four of which are IR active (Figure 9.9).

As far as skeletal vibrations are concerned, the following frequencies are usually observed: $v_3-v_{as}$ (M−cp) and $v_5$ (asymmetric tilting vibration). The IR spectra and the M–cp force constants of some metallocenes are given in Table 9.4. The M − cp bonds are the strongest in complexes possessing closed $18e$ configurations. The IR spectroscopy is very useful for studies of substituted cyclopentadienyl complexes.

The $9-10\mu$ rule is important for cyclopentadienyl compounds. The IR spectra of complexes containing insubstituted $C_5H_5$ groups show two bands at ca $9\mu$ ($1100\ cm^{-1}$) and at $10\mu$ ($1000\ cm^{-1}$). These bands do not occur in spectra of complexes possessing substituents in the ring. Therefore, two isomeric compounds in which only one possesses the unsubstituted $C_5H_5$ ring may be distinguished from each other because the unsubstituted one will show two bands at 1100 and $1000\ cm^{-1}$: $[Fe(C_5H_5)(C_5H_3R_2)]$ and $[Fe(C_5H_4R)_2]$.

# 5. MAGNETIC PROPERTIES

Many cyclopentadienyl complexes contain partially filled electron shells. Therefore, such compounds are paramagnetic. The magnetic susceptibility may also be influenced by the orbital momentum which may increase the magnetic moment. Table 9.5 gives magnetic properties of some cyclopentadienyl compounds.

Table 9.5. Magnetic Properties of Cyclopentadienyl Complexes of Transition Metals[a]

| Compound | Electronic configuration of the metallocene | (BM) "spin only" | (BM) expected | (BM) experimental |
|---|---|---|---|---|
| $Ticp_2^+$ | $e_{2g}^1$ | 1.73 | >1.73 | 2.29 ± 0.05 |
| $Vcp_2^+$ | $e_{2g}^2$ | 2.83 | 2.83 | 2.86 ± 0.06 |
| $Vcp_2$ | | 3.87 | 3.87 | 3.78 ± 0.19 |
| | $e_{2g}^2(a'_{1g})^1$ | | | 3.84 ± 0.04 |
| $Crcp_2^+$ | | 3.87 | 3.87 | 3.73; 3.81 |
| $Crcp_2$ | $e_{2g}^3(a'_{1g})^1$ or $e_{2g}^2(a'_{1g})^2$ | 2.83 | 2.83 | 3.02 ± 0.15 |
| $Mncp_2$ (antiferromagnetic) | $e_{2g}^2(a'_{1g})^1(e'_{1g})^2$ | 5.91 | 5.91 | 5.8–5.9 |
| $Fecp_2^+$ | $e_{2g}^3(a'_{1g})^2$ | 1.73 | >1.73 | 2.34 ± 0.12 |
| $Fecp_2$ | $e_{2g}^4(a_{1g})^2$ | 0 | 0 | 0 |
| $Cocp_2$ | $e_{2g}^4(a'_{1g})^2(a''_{1g})^1$ | 1.73 | 1.73 | 1.76 ± 0.07 |
| $Nicp_2^+$ | $e_{2g}^4(a'_{1g})^2(e'_{1g})$ | 1.73 | >1.73 | 1.82 ± 0.09 |
| $Nicp_2$ | $e_{2g}^4(a'_{1g})^2(e'_{1g})^2$ | 2.83 | 2.83 | 2.86 |
| $[Ucp_3Cl]$ | | 2.83 | | 3.16 |
| $[Vcp_2Cl_2]$ | | 1.73 | >1.73 | 1.95 |
| $[Mocp_2Cl]^+$ | $a_1^2(a'_1)^2 b_2^2$ | 0 | 0 | 0 |

[a] According to References 3a, 4, 16, 25–27, 58.

## 6. *EPR SPECTRA*[26-28,93]

Cyclopentadienyl complexes may possess the following electronic states: (a) orbitally nondegenerate and (b) orbitally degenerate. The first category includes compounds in which orbitally degenerate levels are occupied by an even number of electrons and the $a'_{1g}$ orbital which mainly has a $d_{z^2}$ character contains one electron. Examples are Vcp$_2$ and [Crcp$_2$]$^+$ possessing the $e^2_{2g}(a'_{1g})^1$ electronic structure (ground state term $^4A_{2g}$) as well as Mncp$_2$ which has the high-spin $e^2_{2g}(a'_{1g})^1(e'_{1g})^2$ configuration (term $^6A_{1g}$). Therefore, the information obtained from the ESR spectra concerns the properties of the $a_{1g}$ orbital. Because the spin–orbit interaction with the excited states is relatively weak, both $g_{\parallel}$ and $g_{\perp}$ are close to 2.

The vanadium and chromium compounds show superfine structures in the spectra. This enables estimation of mixing of the $4s$ orbital with the $3d_{z^2}$ orbital (2–5%). The distance between the high-spin ground term $^6A_{1g}$ and the low-spin term $^2E_g$ $(e^3_{2g}a'^2_{1g})$ is small for Mncp$_2$. The first term is the ground term for manganocene in organic solvents and in the solid solutions in Mgcp$_2$, while the second term is for Mncp$_2$ in Fecp$_2$ and Rucp$_2$ solutions. Therefore, manganocene is located in the state which is close to the intersection of the $^6A_{1g}$ and $^2E_{2g}$ terms in the Tanabe–Sugano diagram. The temperature dependence of the magnetic moment for Mncp$_2$ shows that $E(^6A_{1g}) - E(^2E_{2g}) \sim$ 175 cm$^{-1}$, with the higher $^6A_{1g}$ level being thermally populated.

In the case of the methylcyclopentadienyl ligand C$_5$H$_4$Me which creates a stronger field, the complex Mn(C$_5$H$_4$Me)$_2$ is low-spin at low temperatures. The magnetic moment of Mn(C$_5$H$_4$Me)$_2$ in toluene solution between 214 and 371 K was found to increase from 3.23 to 4.77 BM. The observed temperature dependence of the moment may be explained by assuming $E(^6A_{1g}) - E(^2E_g) \sim 630$ cm$^{-1}$. These results indicate that Mn(C$_5$H$_4$Me)$_2$ should lie further to the low-spin side of the Tanabe–Sugano diagram, although still close to the spin crossover point.[26,27] In addition to the manganese complex, three more paramagnetic compounds possessing orbitally degenerate states are known: [Nicp$_2$]$^+$, [Cocp$_2$] ($^2E_{1g}$), and [Fecp$_2$]$^+$ ($^2E_{2g}$). For all complexes possessing orbitally degenerate ground states, the g values are strongly anisotropic and differ considerably from 2. These compounds quite easily undergo Jahn–Teller distortion, although experimental data show that the stabilization energy which is caused by this effect, $E(JT)$, is always smaller than the energy of the corresponding vibration (oscillation). Therefore, the distortion of rings is completely dynamic. It was found that the mixing of the ligand orbitals with the degenerate metal orbital occupied by one electron considerably increases in the series: Fecp$_2^+$ < Mncp$_2$ < Cocp$_2$ < Nicp$_2^+$. Quantity $E(JT)$ increases in the same direction. Thus, the data show that the value of $E(JT)$ is considerably greater for $d^7$ electron complexes with the unpaired electron occupying degenerate $e_{1g}$ orbitals than for $d^5$ electron compounds containing an odd number of electrons occupying the $e_{2g}$ level.

## 7. *ELECTRONIC SPECTRA*

Electronic spectra of cyclopentadienyl complexes may be explained on the basis of the ligand field theory. The cyclopentadienyl ligands create a strong field. The values of

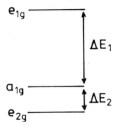

Figure 9.10.  Splitting of the $d$ orbitals in the field with pseudoaxial symmetry.

Racah's parameter $B$ for ions in complexes are considerably lower than for free ions, and therefore the nephelauxetic ratios are considerably smaller than one:

$$\beta = B_{complex}/B_{free\ ion}$$

This proves that the M—cp bond is considerably covalent. The structure of sandwich compounds with ligands possessing fivefold axis $C_5$ (or higher) may be considered by assuming the $C_{\infty v}$ symmetry.[26,27]

Thus, the energies (one-electron) of the $a_{1g}$ $(d_{z^2})$, $e_{1g}$ $(d_{xz}, d_{yz})$, and $e_{2g}$ $(d_{x^2-y^2}, d_{xy})$ orbitals are given by the following formulas:

$$E(e_{1g}) = Ds + 4Dt, \qquad E(a_{1g}) = 2Ds - 6Dt, \qquad E(e_{2g}) = -2Ds - Dt \qquad (9.16)$$

where $Ds$ and $Dt$ represent the splitting parameters.

Experimental data show that the order of these orbitals is as follows: $E(e_{2g}) < E(a_{1g}) < E(e_{1g})$. For the Mcp$_2$ compounds, the splitting parameter $Dt \approx 0.55\ Ds$. Often, it is convenient to use the energy difference between these orbitals, i.e., $\Delta E_1 = E(e_{1g}) - E(a_{1g})$ and $\Delta E_2 = E(a_{1g}) - E(e_{2g})$. From the value of the $Dt/Ds$ ratio, it follows that $\Delta E_1 \gg \Delta E_2$ ($\Delta E_1/\Delta E_2 = 18/5$) (Figure 9.10, Table 9.6).

Table 9.6.  Electronic Spectra of Cyclopentadienyl Complexes of Transition Metals[a]

| Transition | Energy of the transition (kK) | | | |
|---|---|---|---|---|
| | Fecp$_2$ | Rucp$_2$ | Cocp$_2^+$ | Fecp($C_5H_4$Ph) |
| $^1A_{1g} \to a^3E_{1g}$ | 18.9 | 26.0 | 21.8 | 17.5 |
| $^1A_{1g} \to a^1E_{1g}$ | 21.80 | 29.5 | 24.3 | 20.8 |
| $^1A_{1g} \to {}^1E_{2g}$ | 24.0 | 32.5 | 26.4 | 23.0 |
| $^1A_{1g} \to b^1E_{1g}$ | 30.8 | 36.6 | 33.3 | 29.5 |
| CT | 37.7 sh | 42.0 | | |
| CT | 41.7 sh | 46.1 | | |
| CT | 50.0 | >51.3 | 38.0 | >50.0 |
| $\Delta E_1$ | 22.0 | 29.8 | 24.4 | 21.0 |
| $\Delta E_2$ | 7.1 | 6.6 | 7.2 | 6.9 |
| $Ds$ | 5.17 | 6.14 | 5.54 | |
| $Dt$ | 2.72 | 3.59 | 2.99 | |
| $B$ | 0.390 | 0.260 | 0.400 | 0.370 |
| $\beta^b$ | 0.42 | 0.42 | 0.36 | 0.40 |

[a] According to References 9 and 26.
[b] $\beta = B(Mcp_2^{x+})/B(M^{n+})$ ($n = 2, 3; x = 0, 1$).

## 8. PHOTOELECTRON SPECTRA

Photoelectron spectra of cyclopentadienyl complexes, $Mcp_2$, show that the order of $d$ metal orbital energy is as follows: $e_{1g}(d_{xz}, d_{yz}) > e_{2g}(d_{xy}, d_{x^2-y^2}) > a_{1g}(d_{z^2}) > e_{1u}(\pi-cp) > e_{1g}(\pi-cp)$; therefore, this order is not in agreement with the series resulting from the electronic spectra and calculations. The ionization potentials for metallocenes and compounds of the type $Mcp_2 L_n$, $Mcp_m L_n$ and $Mcp_m$ ($m = 3, 4$) are low and vary from *ca* 5.5 to 18 eV.[26, 27, 29, 30, 62, 69, 79] The spectra of chromocene, $Crcp_2$, show that the ground term is rather $^3E_{2g}[(e_{2g})^3(a_{1g})^1]$ than $^3A_{2g}(e_{2g})^2(a_{1g})^2$. The spectra of manganocene, $Mncp_2$, show unequivocally that the energy of the terms $^6A_{1g}[(e_{2g})^2(a_{1g})^1(e_{1g})^2]$ and $^2E_{2g}[(e_{2g})^3(a_{1g})^2]$ are close for the compound in the gas phase.

This behavior is even more pronounced in the complex $Mn(C_5H_4Me)_2$, for which a small energy difference between these terms was observed in solution based on the measurements of magnetic susceptibility as a function of temperature. They proved that there is an equilibrium between the high- and low-spin form of the compound which is temperature dependent. Photoelectron spectra of some cyclopentadienyl complexes are given in Table 9.7.

*Table 9.7. Ionization Potentials (eV) of Cyclopentadienyl Metal Complexes[a]*

### a. Complexes with Closed Electron Shells

| Compound | Resulting term and orbital which undergoes ionization | | | |
|---|---|---|---|---|
| | $^2E_{2g}[e_{2g}(nd)]$ | $^2A_{1g}[a_{1g}(nd)]$ | $^2E_{1u}(\pi\text{-cp})$ | $^2E_{1g}[e_{1g}(\pi\text{-cp})]$ |
| $Fecp_2$ | 6.858 | 7.234 | 8.715 | 9.38 |
| $[Fe(C_5H_4Me)_2]$ | 6.72 | 7.06 | 8.53 | 9.17 |
| $[Fe(C_5H_4Cl)_2]$ | 7.03 | 7.37 | 8.71 | 9.49 |
| $Rucp_2$ | 7.45 | 7.63 | 8.51 | 9.93 |
| $[Ru(C_5H_4Me)_2]$ | 7.25 | 7.25 | 8.24 | 9.76 |
| $Oscp_2$ | 6.93; 7.55 | 7.21 | 8.23 | 9.90 |
| $Mgcp_2$ | | | 9.03; 9.26 | 8.23; 8.44 |

### b. Complexes Possessing Open Electron Shells

| Compound | Ionization energy (eV) | Ground term | Electronic configuration of the ion | Term of the resulting ion |
|---|---|---|---|---|
| $Vcp_2$ | 6.78 | $^4A_{1g}[(e_{2g})^2(a_{1g})^1]$ | $(e_{2g})^1(a_{1g})^1$ | $^3E_{2g}$ |
| | 6.78 | | $(e_{2g})^2$ | $^3A_{2g}$ |
| $Cocp_2$ | 5.56 | $^2E_{1g}[(e_{2g})^4(a_{1g})^2(e_{1g})^1]$ | $(e_{2g})^4(a_{1g})^2$ | $^1A_{1g}$ |
| | 7.18 | | | $^3E_{1g}$ |
| | 7.18 | | $(e_{2g})^3(a_{1g})^2(e_{1g})^1$ | $^3E_{2g}$ |
| $Nicp_2$ | 6.51 | $^3A_{2g}[(e_{2g})^4(a_{1g})^2(e_{1g})^2]$ | $(e_{2g})^4(a_{1g})^2(e_{1g})^1;$ | $^2E_{1g}$ |
| | 8.0–10.5 | | | $^4E_{2g}$ |
| | 8.0–10.5 | | $(e_{2g})^3(a_{1g})^2(e_{1g})^2$ | $^2E_{2g}$ |

[a] According to References 26, 27, 29, 30, and 62.

## 9. THERMOCHEMISTRY OF CYCLOPENTADIENYL COMPLEXES[31]

The $M-cp$ bond is very stable. The energy of the dissociation of this bond varies from *ca* 200 to almost 400 kJ mol$^{-1}$. This explains the exceptional thermal stability of these compounds. The energy of dissociation of the metal–cyclopentadienyl bond is a function of electronic structure, oxidation state of the metal, and properties of other ligands coordinated to the metal. For metallocenes of the first series of the transition metals, the energy of dissociation decreases linearly as the metal–cyclopentadienyl distance becomes greater (Table 9.1 and 9.8). The dissociation energy of the first $M-cp$ bond in metallocene may be considerably higher than the dissociation energy of the second bond, and higher than the average dissociation energy. The pyrolysis method at very low pressures shows that the dissociation energy of the first $Fe-cp$ bond equals $381 \pm 13$ kJ mol$^{-1}$, while this energy for the second bond is 213.5 kJ mol$^{-1}$.[80] The $Fe-C_5H_5$ bond energy of the *5e* ligand is almost precisely 2.5 times greater than the $Fe-C_2H_4$ bond energy of the *2e* ligand in the $Fe(C_2H_4)(CO)_4$ complex which is equal to 155.8 kJ mol$^{-1}$.[80]

Table 9.8. *Thermodynamic Properties of Cyclopentadienyl Metal Complexes*[a]

| Compound | Enthalpy of formation in solid state $\Delta H_s^\circ$ (kJ mol$^{-1}$) | Enthalpy of formation in gaseous state $\Delta H_g^\circ$ (kJ mol$^{-1}$) | Enthalpy of sublimation (kJ mol$^{-1}$) $\Delta H_{subl.}^\circ$ | Energy of dissociation of bonds $M-cp$ $D(M-cp)$ (kJ mol$^{-1}$) | Energy of dissociation of other bonds: $D(M-X)$ (kJ mol$^{-1}$) | Force constant $k(M-cp)$ (N m$^{-1}$) |
|---|---|---|---|---|---|---|
| Vcp$_2$ | $142.3 \pm 8.4$ | $200.8 \pm 8.4$ | $58.6 \pm 4.2$ | $368.2 \pm 4.2$ | | |
| Crcp$_2$ | $178.2 \pm 2.1$ | $248.1 \pm 2.5$ | $69.9 \pm 1.7$ | $283.7 \pm 2.5$ | | 160 |
| Mncp$_2$ | $201.3 \pm 2.1$ | $277.0 \pm 2.5$ | $75.7 \pm 1.7$ | $213.0 \pm 1.7$ | | |
| Fecp$_2$ | $154.8 \pm 4.2$ | $228.0 \pm 4.2$ | $73.2 \pm 0.8$ | $303.8 \pm 2.5$ | | 270 |
| Cocp$_2$ | $235.6 \pm 2.5$ | $305.6 \pm 3.3$ | $70.3 \pm 2.1$ | $270.3 \pm 2.5$ | | 150 |
| Nicp$_2$ | $274.5 \pm 4.2$ | $346.9 \pm 4.2$ | $72.4 \pm 1.3$ | $250.2 \pm 2.5$ | | 150 |
| Sccp$_3$ | $-13.4 \pm 4.2$ | $96.2 \pm 6.3$ | $109.6 \pm 4.2$ | $304.2 \pm 2.1$ | | |
| Ycp$_3$ | $-45.2 \pm 4.2$ | $66.5 \pm 6.3$ | $111.7 \pm 4.2$ | $328.4 \pm 2.1$ | | |
| Lacp$_3$ | $-26.8 \pm 4.2$ | $87.9 \pm 6.3$ | $114.6 \pm 4.2$ | $323.4 \pm 2.1$ | | |
| Prcp$_3$ | $-28.0 \pm 8.4$ | $97.5 \pm 8.4$ | $125.5 \pm 4.2$ | $300.8 \pm 4.2$ | | |
| Tmcp$_3$ | $-49.4 \pm 4.2$ | $61.9 \pm 6.3$ | $111.3 \pm 4.2$ | $271.1 \pm 2.1$ | | |
| Ybcp$_3$ | $29.3 \pm 4.2$ | $138.0 \pm 6.3$ | $108.8 \pm 4.2$ | $213.8 \pm 2.1$ | | |
| Ticp$_2$Cl$_2$ | $-383.3 \pm 4.2$ | $-264.4 \pm 4.2$ | $118.8 \pm 2.1$ | $269.0 \pm 4.2$ | | |
| Ticp$_2$Me$_2$ | $54.4 \pm 8.4$ | $133.9 \pm 8.2$ | $79.5 \pm 4.2$ | | $D(Ti-Me)\ 251 \pm 8.4$ | |
| Zrcp$_2$Cl$_2$ | $-538.1 \pm 2.9$ | $-433.0 \pm 3.8$ | $105.1 \pm 2.1$ | $358.6 \pm 4.2$ | | |
| Zrcp$_2$Me$_2$ | $-44.4 \pm 2.1$ | $36.8 \pm 2.9$ | $81.2 \pm 2.1$ | | $D(Zr-Me)\ 274.5 \pm 4.6$ | |
| Hfcp$_2$Cl$_2$ | $-536.0 \pm 2.5$ | $-430.1 \pm 3.3$ | $106.7 \pm 2.1$ | $359.0 \pm 2.5$ | | |
| Mocp$_2$H$_2$ | $195.0 \pm 2.5$ | $287.4 \pm 3.3$ | $92.5 \pm 2.1$ | $352.7 \pm 7.5$ | $D(Mo-H)\ 259.0 \pm 3.8$ | |
| Wcp$_2$H$_2$ | $250.6 \pm 2.5$ | $346.9 \pm 3.3$ | $96.2 \pm 2.1$ | $395.0 \pm 3.3$ | $D(W-H)\ 287.0 \pm 4.2$ | |

[a] According to References 31a, 31b, and 57.

## 10. *MASS SPECTRA*[32,33]

Mass spectra of cyclopentadienyl metal complexes contain peaks of parent ions $Mcp_2^+$ which are the most intense. Exceptions are compounds possessing predominantly ionic character of the $M-cp$ bond, for instance, $Mncp_2$ and $Mgcp_2$, which are relatively unstable and undergo fragmentation to a greater extent to give higher yields of $Mcp^+$ and $M^+$ ions. The main peaks in the spectra of $Mcp_2$ complexes correspond to the $Mcp_n^+$ ions where $n = 0, 1, 2$.

Other types of the fragmentation pattern take place to a lesser extent. This is especially true for compounds of the first series of the transition metals for which the intensity of ions, owing to elimination of various neutral hydrocarbon fragments, is small. Such fragmentation occurs to a greater degree in the case of the second and third series of the transition metals. The intensity of the $Rucp_2^+$ ion is particularly high compared to that of $Rucp^+$ and $Ru^+$. This indicates greater stability of $Rucp_2^+$ compared to $Fecp_2^+$. For complexes of the type $Mcp_3$ and $Mcp_4$ (M = titanium group metals or lanthanide and actinide), the main peaks are those of $Mcp_{n-1}^+$. The subsequent splitting of the cyclopentadienyl groups was demonstrated by the presence of metastable ions. Small amounts of the ions $MC_3H_3^+$, $MC_3H_2^+$, and $MC_2H^+$ (M = Fe, Ni) show that there is fragmentation of the ring.

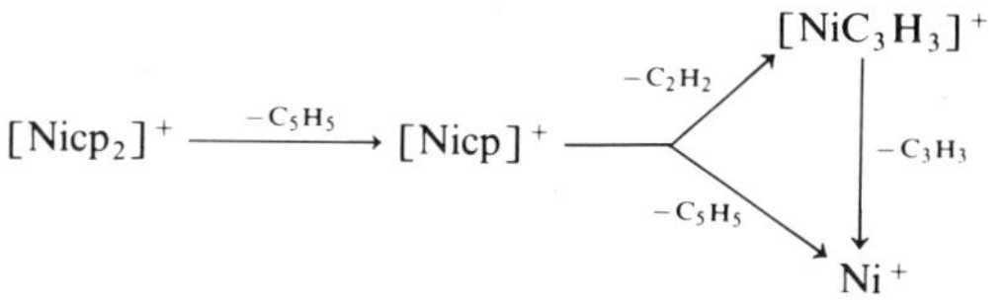

$$[Nicp_2]^+ \xrightarrow{-C_5H_5} [Nicp]^+ \xrightarrow{\begin{subarray}{l} -C_2H_2 \\ -C_3H_3 \\ -C_5H_5 \end{subarray}} \begin{array}{l} [NiC_3H_3]^+ \\ Ni^+ \end{array} \qquad (9.17)$$

The fragmentation of $[Ticp_3]$ proceeds as shown in equations (9.18).

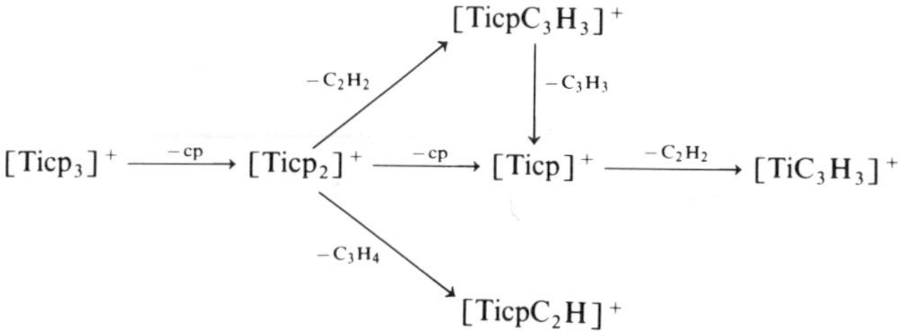

$$[Ticp_3]^+ \xrightarrow{-cp} [Ticp_2]^+ \xrightarrow{-cp} [Ticp]^+ \xrightarrow{-C_2H_2} [TiC_3H_3]^+ \qquad (9.18)$$

(with branches: $[Ticp_2]^+ \xrightarrow{-C_2H_2} [TicpC_3H_3]^+$; $[TicpC_3H_3]^+ \xrightarrow{-C_3H_3} [Ticp]^+$; $[Ticp_2]^+ \xrightarrow{-C_3H_4} [TicpC_2H]^+$)

Mass spectra of ferrocene and mixtures of ferrocene and nickelocene also show polynuclear ions such as $[Fe_2cp_2]^+$, $[Fe_2cp_2C_3H_3]^+$, $[Fe_2cpC_3H_3]^+$, $[Fe_2cp_3]^+$, $[Ni_2cp_3]^+$, and $[FeNicp_3]^+$. For the $M_2cp_3^+$ ions, a triple-decker structure has been proposed (see also preparation of cp complexes and Figure 9.7d).

## 11. METHODS OF PREPARATION OF CYCLOPENTADIENYL COMPLEXES[1-6, 16, 63-66, 71, 72]

a. *Reactions utilizing cyclopentadiene in the presence of bases or cyclopentadienyl compounds of the main group elements*

This method is utilized for the preparation of ferrocene and its derivatives:

$$C_5H_6 + base \longrightarrow base\ H^+C_5H_5^- \xrightarrow{FeCl_2} Fecp_2 \qquad (9.19)$$

Both weak bases such as diethylamine and pyridine and strong bases, for example, alkali metals (most commonly sodium), may be utilized in this reaction.

$$NiBr_2 + 2Et_2NH + 2C_5H_6 \longrightarrow Nicp_2 + 2[Et_2NH_2]Br \qquad (9.20)$$

In the case of $PdCl_2$ the role of the base is played by the chloride ligand:

$$PdCl_2 + C_5H_6 \longrightarrow PdcpCl + HCl \qquad (9.21)$$

Sodium metal reacts with cyclopentadiene to give sodium cyclopentadienide, $Na^+C_5H_5^-$.

$$Na + C_5H_6 \longrightarrow NaC_5H_5 + \tfrac{1}{2}H_2 \qquad (9.22)$$

$$MCl_n + nNaC_5H_5 \longrightarrow Mcp_n + nNaCl \qquad (9.23)$$

Ferrocene is usually obtained by the reaction of cyclopentadiene with $FeCl_2$ in the presence of diethylamine or by the action of $FeCl_2$ on Nacp in tetrahydrofuran or 1, 2-dimethoxyethane. Cyclopentadienyl derivatives also react with sodium:

$$C_5H_5R \xrightarrow[\text{2. FeCl}_2]{\text{1. Na}} [Fe(C_5H_4R)_2]$$

$$(R = Bu^t, allyl, acetyl, SiMe_3, etc.) \qquad (9.24)$$

Metallocenes containing two substituents in the ring may be obtained in a similar manner:

$$C_5H_4R_2 \xrightarrow{\text{Na}/\text{FeCl}_2} [Fe(C_5H_3R_2)_2] \qquad (9.25)$$

$$(R = Me, Ph)$$

Often, these syntheses also involve organomagnesium compounds:

$$[Ru(acac)_3] + MgBrcp \longrightarrow Rucp_2 \qquad (9.26)$$

$$TiCl_4 + MgBrcp \longrightarrow [TiBr_2cp_2] \qquad (9.27)$$

$$VCl_4 + MgBrcp \longrightarrow Vcp_2 \qquad (9.28)$$

In some cases complexes also containing σ-cyclopentadienyl groups may be formed in addition to π-cyclopentadienyl ligands:

$$MoCl_5 + 8Nacp \xrightarrow{Et_2O} [Mo(\eta\text{-}C_5H_5)(\sigma\text{-}C_5H_5)_3] \tag{9.29}$$

$$NbCl_5 + Nacp \longrightarrow [Nb(\eta\text{-}C_5H_5)_2(\sigma\text{-}C_5H_5)_2] \tag{9.30}$$

Some reactions with Nacp give compounds with other coordinated groups, for instance, hydrido and halide ligands:

$$ReCl_5 + Nacp(\text{excess}) \xrightarrow{THF} [Recp_2H] \tag{9.31}$$

$$MCl_5 + Nacp(\text{excess}) \xrightarrow{THF} [Mcp_2H_2] \tag{9.32}$$
$$(M = Mo, W)$$

$$TiCl_4 + 2Nacp \xrightarrow{THF} [Ticp_2Cl_2] \tag{9.33}$$

$$[Ticp_2Cl_2] + Nacp \longrightarrow Ticp_3 \tag{9.34}$$

$$NiCl_2 + Nacp \longrightarrow Nicp_2 + [Nicp(C_5H_7)] \tag{9.35}$$

Cyclopentadienyl–carbonyl compounds are conveniently prepared by the reaction of metal carbonyl halides with sodium cyclopentadienide, or by the reaction of Nacp with metal halides in the atmosphere of carbon monoxide (sometimes under pressure):

$$[MX(CO)_5] + Nacp \longrightarrow [Mcp(CO)_3] + NaX + 2CO \tag{9.36}$$
$$(M = Mn, Tc, Re)$$

$$[FeI_2(CO)_4] + Nacp \longrightarrow [FecpI(CO)_2] + NaI + 2CO \tag{9.37}$$

$$[Ru_2Cl_2(CO)_6] + Nacp \longrightarrow [Ru_2cp_2(CO)_4] \tag{9.38}$$

Often, in the synthesis of cyclopentadienyl compounds, the following cp salts are used: Tlcp, Kcp, Licp, and Mgcp$_2$. The application of the light-yellow Tlcp is especially convenient, because this compound is air stable and it is easily prepared by the reaction of Tl(I) salt with cyclopentadiene in the presence of a base. Potassium cyclopentadienide is obtained in a manner analogous to Nacp, while lithium cyclopentadienide is prepared by the reaction of alkyllithium with cyclopentadiene. The pyrophoric Mgcp$_2$ is formed from $C_5H_6$ and magnesium at high temperatures.

### b. *Syntheses utilizing fulvenes*

Fulvenes may readily be obtained from cyclopentadiene and aldehydes or ketones:

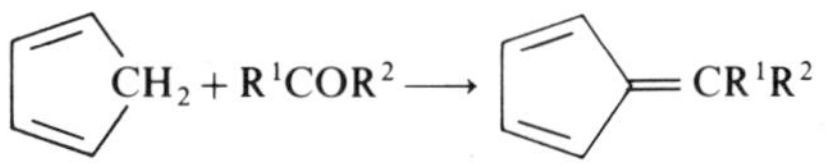

$$\text{CH}_2 + R^1COR^2 \longrightarrow {=}CR^1R^2 \tag{9.39}$$

Sodium, potassium, organolithium compounds, and LiAlH$_4$ react with fulvenes to give ionic cyclopentadienides, which subsequently form cyclopentadienyl complexes with transition metal compounds [equations (9.40) and (9.41).

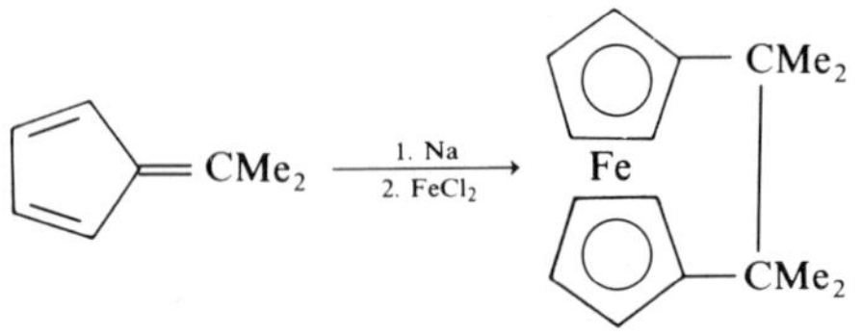

$$(9.40)$$

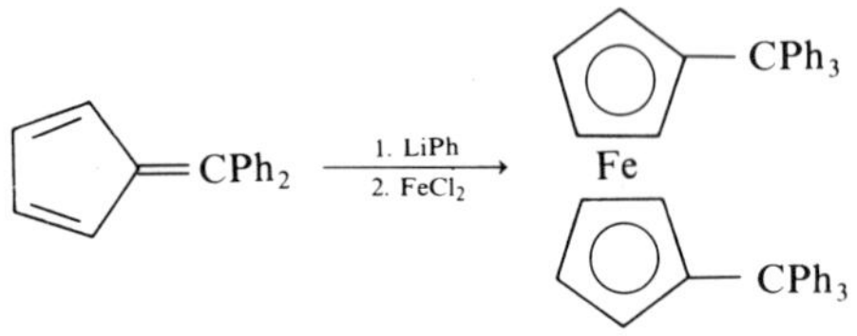

$$(9.41)$$

Some transition metal carbonyls may also form cyclopentadienyl derivatives with fulvenes. Most commonly, the solvent is the source of hydrogen. In the absence of the solvent, disproportionation reactions occur [equations (9.42) and (9.43)].

$$(M = Mo, W)$$

$$(9.42)$$

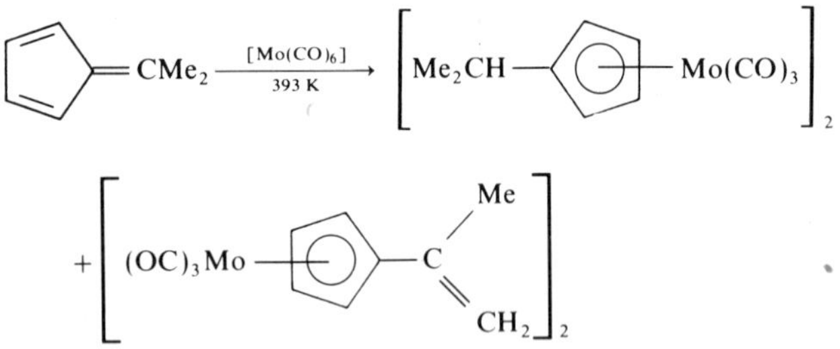

$$(9.43)$$

c. *Reactions of transition metal compounds with unsaturated hydrocarbons*

Acetylenes and olefins may react with metal carbonyls, metal halides, or other transition metal compounds to give cyclopentadienyl complexes [equations (9.44)–(9.46)].

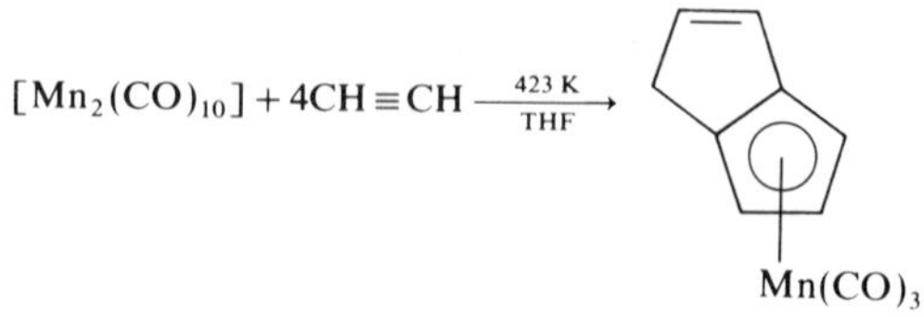

$$[Mn_2(CO)_{10}] + 4CH \equiv CH \xrightarrow[\text{THF}]{423 \text{ K}}$$

$$(9.44)$$

$$[Fe_3(CO)_{12}] + HCCH \longrightarrow \qquad\qquad\qquad\qquad\qquad (9.45)$$

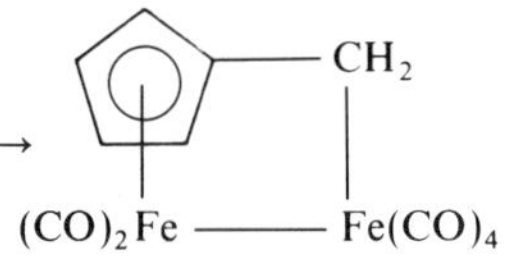

$$TiCl_4 + CH_2{=}CMe_2 \xrightarrow[\text{3–6 MPa}]{573\,K} [TiCl_3(C_5Me_5)] + H_2 + CH_4 + C_2H_6 + C_3H_8 + \cdots \qquad (9.46)$$

### d. Syntheses utilizing cyclopentadiene

Halides and other salts as well as metal carbonyls sometimes react with cyclopentadiene to afford cyclopentadienyl complexes with concomitant evolution of hydrogen, hydrogen chloride, or other reduction products, such as $C_5H_8$. These reactions take place under relatively severe conditions at high temperatures and often under elevated pressures:

$$MX_3 + C_5H_6 \longrightarrow [McpX_3] + \tfrac{1}{2}H_2 \qquad\qquad (9.47)$$
$$(M = Ti, X = Cl; M = Zr, X = Cl, Br, I)$$

$$[Co(C_{17}H_{33}COO)_2] + C_5H_6 + CO \xrightarrow{453–563\,K} [Cocp(CO)_2] \qquad (9.48)$$

$$[Co_2(CO)_8] + 3C_5H_6 \longrightarrow 2[Cocp(CO)_2] + 4CO + C_5H_8 \qquad (9.49)$$

$$[Ni(CO)_4] + 2C_5H_6 \xrightarrow[\text{hexane}]{343\,K} \left[ cpNi \right] \qquad (9.50)$$

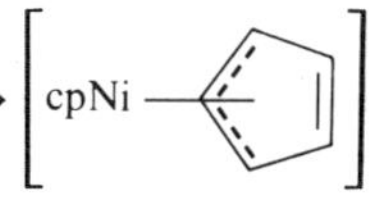

During reactions with metal carbonyls, cyclopentadiene complexes are probably first formed which subsequently undergo rearrangement to hydridocyclopentadienyl compounds followed by decomposition with hydrogen evolution:

$$[Fe(CO)_5] + C_5H_6 \longrightarrow \qquad \longrightarrow [Fecp(H)(CO)_2] + [Fe_2cp_2(CO)_4] + H_2 \qquad (9.51)$$

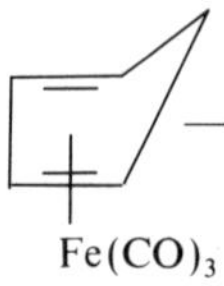

It is possible that in some syntheses, the ligands may play the role of a base abstracting the proton.

### e. Reactions of dienes with metal vapors[34]

Ferrocene and chromocene are formed during condensation of iron or chromium atoms with cyclopentadiene:

$$Fe + 2C_5H_6 \longrightarrow Fecp_2 + H_2 \qquad\qquad (9.52)$$
$$Cr + 2C_5H_6 \longrightarrow Crcp_2 + H_2 \qquad\qquad (9.53)$$

Vapors of nickel and cobalt react to give compounds containing allyl, cyclopentadiene, and cyclopentadienyl ligands:

$$Ni + 2C_5H_6 \longrightarrow cpNi \qquad (9.54)$$

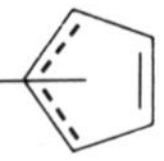

$$Co + 2C_5H_6 \longrightarrow Cocp(C_5H_6) \qquad (9.55)$$

The molybdenum and tungsten metals react with $C_5H_6$ to afford hydrido complexes:

$$M + 2C_5H_6 \longrightarrow [Mcp_2H_2] \qquad (M = Mo, W) \qquad (9.56)$$

### f. Preparation of triple-decker cyclopentadienyl complexes[6,63–66,69,71]

The compound $[Ni_2cp_3]^+X^-$ (Figure 9.7d) possessing triple-decker structure with the bridging cyclopentadienyl ligand is formed from biscyclopentadienylnickel and a trityl salt:

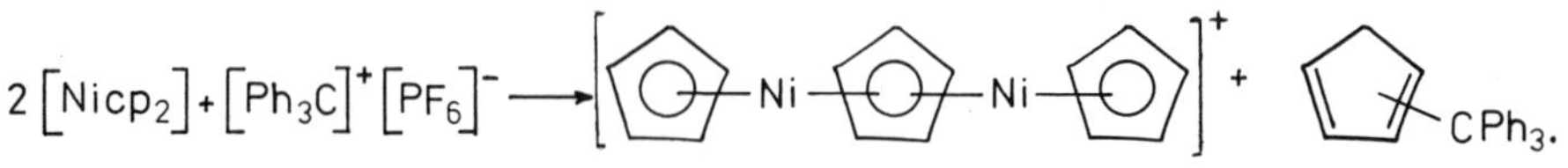

$$2[Nicp_2] + [Ph_3C]^+[PF_6]^- \longrightarrow [\quad -Ni- \quad -Ni- \quad]^+ + \quad_{CPh_3.} \qquad (9.57)$$

The same cationic complex may be obtained by utilizing the following starting materials in place of $[Ph_3C][PF_6]$: $[Ph_3C][BF_4]$, $[Ph_2CH][BF_4]$, $[C_7H_7][BF_4]$, $[Me_3O][BF_4]$, and $HBF_4$. The first step of the reaction which occurs by electrophilic attack probably gives a 14e cationic complex which interacts with the electron-rich nickelocene to furnish the dinuclear compound:

$$Nicp_2 + H^+ \longrightarrow [Nicp(C_5H_6)]^+ \xrightarrow{-C_5H_6} [Nicp]^+ \xrightarrow{Nicp_2} [Ni_2cp_3]^+ \qquad (9.58)$$

The cationic complexes of the type $[Nicp]^+X^-$ were isolated and investigated.

An attempt at synthesis of an analogous palladium–nickel compound was unsuccessful; the reaction of $Pdcp^+$ with $Nicp_2$ leads to $Ni_2cp_3^+$ [equation (9.59)].

$$2[Pdcp(PR_3)X] \xrightarrow[-2AgX]{+2Ag^+} 2[PdcpPR_3]^+ \xrightarrow{-[Pdcp(PR_3)_2]^+} [Pdcp]^+ \xrightarrow{Nicp_2} [Ni_2cp_3]^+$$
$$\xrightarrow{Nicp_2} \big/\!\!\big/ \quad [PdNicp_3]^+ \qquad (9.59)$$

Another triple-decker compound is obtained by the reaction (9.60) of 3,4-diethyl-2,5-dimethyl-1,2,5-thiadiborole with $Fe_2cp_2(CO)_4$ (Figure 9.7f).

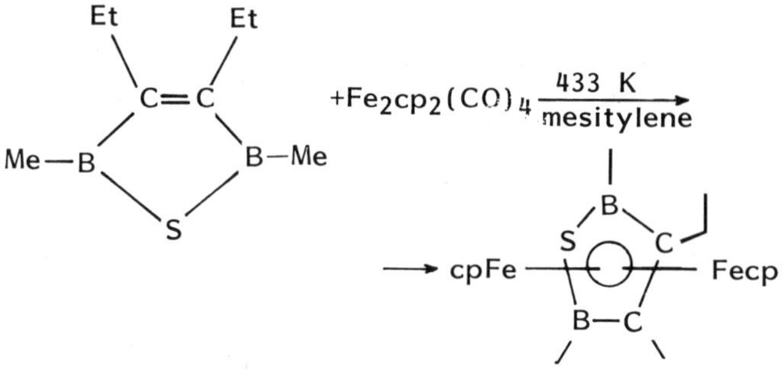

$$(9.60)$$

Both 1,2,5-thiadiborole and 2,3-dihydro-1,3-diborole derivatives (9.61) are utilized for the preparation of multidecker sandwich complexes[64] [equations (9.62)–(9.64)].

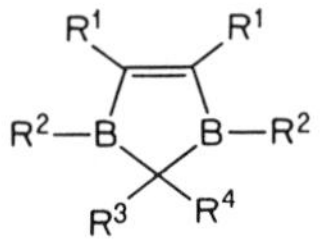

$R^1$=Me, Et; $R^2$=Me, Et; $R^3$=H, Me; $R^4$=H, Me          (9.61)

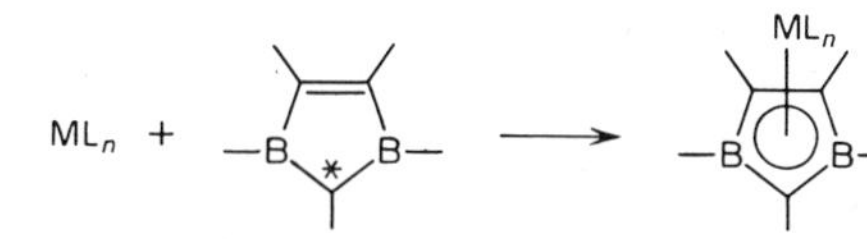

$ML_n$=Nicp, Co(CO) itp.          (9.62)

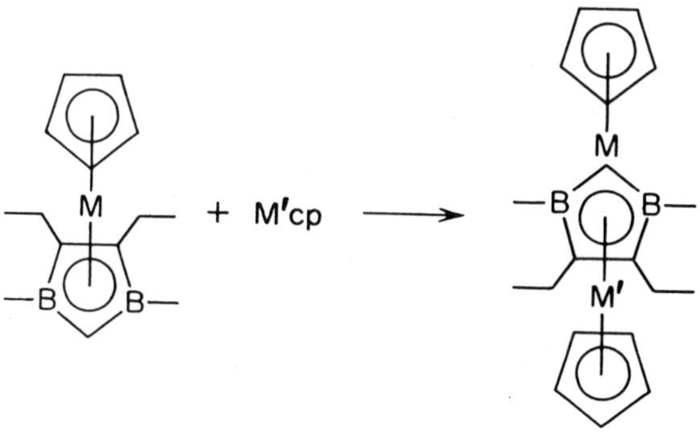

M = Ni, Co, Fe
M′ = Ni, Co, Fe

$$(9.63)$$

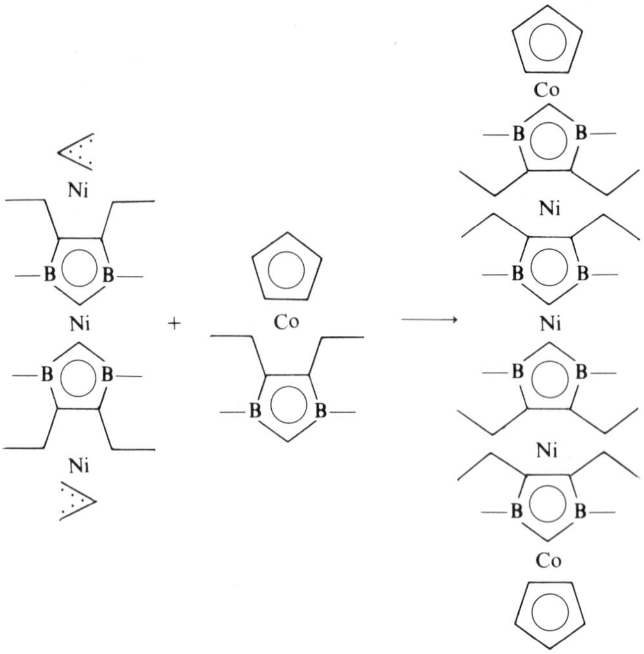

$$(9.64)$$

### g. Preparation of compounds containing bridging cyclopentadienyl ligands

The compound $[\text{Pdcp}(2\text{-RC}_3\text{H}_4)]$ ($R = H$, Me, Bu$^t$, Cl) reacts with PdL$_2$ ($L = PR_3$, $R = i\text{-Pr}$, Ph, Cy) (Figure 9.7e) to give dinuclear complexes containing cyclopentadienyl and allyl bridges:

$$[\text{Pdcp}(2\text{-RC}_3\text{H}_4)] + \text{PdL}_2 \longrightarrow [\text{LPd}(\mu\text{-cp})(\mu\text{-allyl})\,\text{PdL}] \qquad (9.65)$$

Platinum compounds react in a similar way:

$$[\text{Ptcp}(2\text{-MeC}_3\text{H}_4)] + \text{PtL}_2 \longrightarrow [\text{LPt}(\mu\text{-cp})(\mu\text{-allyl})\,\text{PtL}] \qquad (9.66)$$

$$[L = P(i\text{-Pr})_3]$$

### h. Synthesis of trinuclear M-P(O)(OR)$_2$-cyclopentadienyl complexes[6]

Owing to prolonged reaction time to high temperatures, the interaction of Cocp$_2$ with phosphites (P(OR)$_3$) furnishes trinuclear analogues of multidecker compounds [equation (9.67)].

$$\text{Cocp}_2 \xrightarrow[+\,\text{P(OR)}_3]{363\ \text{K}} [\text{Cocp}\{\text{P(OR)}_3\}_2] \xrightarrow{403\ \text{K}}$$

$$[\text{cpCo}\{\mu\text{-P(O)(OR)}_2\}_3\,\text{Co}\{\mu\text{-P(O)(OR)}_2\}\text{Cocp}] \qquad (9.67)$$

363–403 K
+ HP(O)(OR)$_2$

## i. *Electrochemical synthesis*[81,82]

Electrochemical methods may be successfully utilized for the preparation of both mono- and poly-cyclopentadienyl complexes of many transition metals. During the dissolution of manganese at the anode in DMF which contains methylcyclopentadiene in the atmosphere of carbon monoxide, (methylcyclopentadienyl)tricarbonylmangenese is obtained:

$$C_5H_5Me + Mn + 3CO \longrightarrow Mn(C_5H_4Me)(CO)_3 \qquad (9.68)$$

Ferrocene may be obtained in a similar fashion:

$$Fe + C_5H_6 \xrightarrow[\text{[NEt}_4\text{]Cl}]{\text{MeCN}} Fecp_2 \qquad (9.69)$$

Many cyclopentadienyl complexes may be prepared by electrochemical oxidation or reduction of metallocenes and other compounds possessing coordinated cyclopentadienyl ligands.

## j. *Synthesis of dienyl complexes*

Elimination of the proton from 1, 4-diene or 1, 3-diene compounds is often used for the preparation of metal dienyl complexes. The following compounds are utilized as deprotonating reagents: alkyllithium compounds, (trimethylsilylmethyl)potassium, alkali metal amides, etc. Also useful are alkali metals as sands in the presence of triethylamine, which prevents anionic polymerization of dienes.[59,60]

$$MCl_2 + Li(2, 4\text{-}C_5H_5Me_2) \longrightarrow M(2, 4\text{-}C_5H_5Me_2)_2 \qquad (9.70)$$

$$(M = Fe, Cr, V)$$

Pentadienyl complexes are also formed by reactions of metal alkylbutadiene complexes with reagents which accept the hydride ion, for instance, with trityl tetrafluoroborate [equations (9.71) and (9.72)]. Such complexes may also be obtained

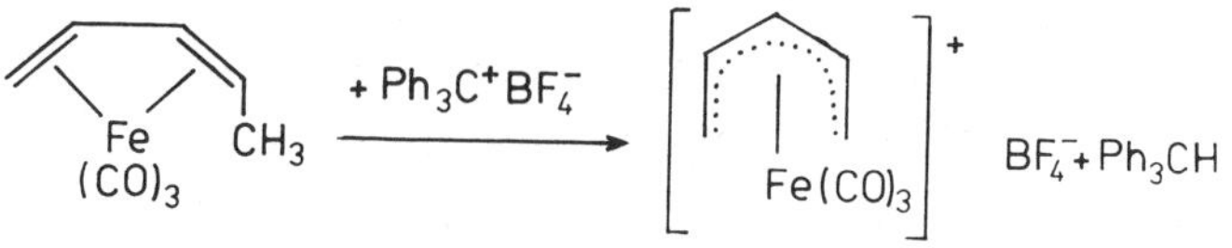

$$(9.71)$$

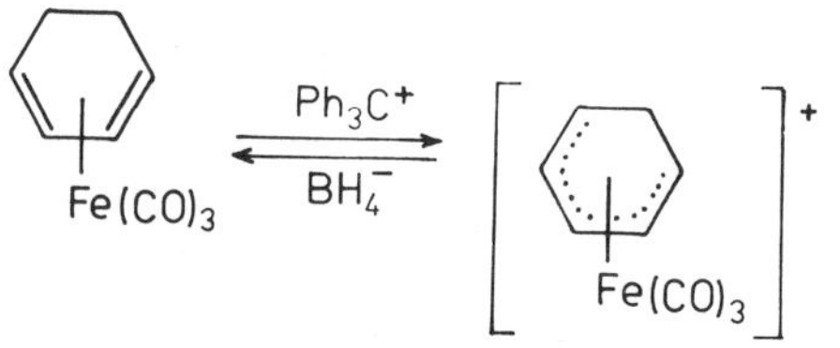

$$(9.72)$$

not only by hydride abstraction from the diene molecule, but also through the addition of a proton to the triene compound [equation (9.73)].

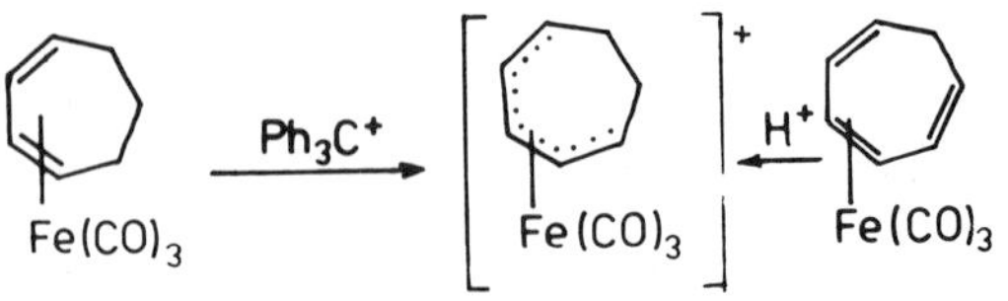

$$\text{(structure)} \xrightarrow{\text{Ph}_3\text{C}^+} \left[\text{structure}\right]^+ \xleftarrow{\text{H}^+} \text{structure} \qquad (9.73)$$

## 12. PROPERTIES OF CYCLOPENTADIENYL COMPLEXES

Sandwich complexes $\text{Mcp}_2$ are thermally very stable. They melt at relatively high temperatures, usually without decomposition. The resistance of metallocenes to oxidation varies. Some are completely air stable, for example, ferrocene and ruthenocene. However, some are pyropophoric, such as chromocene and vanadocene which spontaneously inflame in air. Often during controlled oxidation, oxo $\pi$-cyclopentadienyl complexes are obtained.

All metallocenes are oxidized to the $\text{Mcp}_2^+$ cations in acidic media. Sometimes this reaction is reversible. Many complexes become protonated to give $\text{Mcp}_2\text{H}^+$ in acidic solutions. Numerous cyclopentadienyl compounds which also possess other ligands coordinated to the metal are known, for instance, halogens, CO, H, OH, O, etc. Also, the formation of cyclopentadienyl polynuclear and cluster complexes is well known.

### a. Group 4 metal complexes

These metals form cyclopentadienyl derivatives possessing the following formulas: $[\text{McpX}_3]$, $[\text{Mcp}_2\text{X}_2]$ (M = Ti, Zr, Hf), $[\text{Mcp}_2(\text{CO})_2]$ (M = Ti, Zr, Hf), $[\text{M}_2\text{cp}_4(\mu\text{-Cl})_2]$ (M = Ti, Zr, Hf), $[\text{Ticp}_3]$, $[\text{Mcp}_4]$ (M = Ti, Zr, Hf), $[\text{Mcp}_2\text{R}_2]$, $\text{McpR}_3$ (M = Ti, Zr, Hf, R = alkyl, aryl). The tetrakiscyclopentadienyl compounds $[\text{M}(\eta^5\text{-C}_5\text{H}_5)_2(\sigma\text{-C}_5\text{H}_5)_2]$ contain two $\pi$ and two $\sigma$ ligands. Thus, these compounds are analogs of other complexes $\text{Mcp}_2\text{X}_2$. An exception is the zirconium compound $[\text{Zr}(\eta^5\text{-cp})_3(\eta^1\text{-cp})]$. The complex $[\text{Ti}_2\text{cp}_4]$ is diamagnetic and therefore probably possesses a titanium–titanium bond. Based on the IR, $^1\text{H NMR}$, $^{13}\text{C NMR}$, UV, and mass spectra as well as chemical properties, it is assumed that the formula of this compound is $[\text{Ti}_2\text{cp}_2(\text{C}_5\text{H}_4)_2(\text{H})_2]$ (Table 9.9). Structure (9.74) is proposed for this

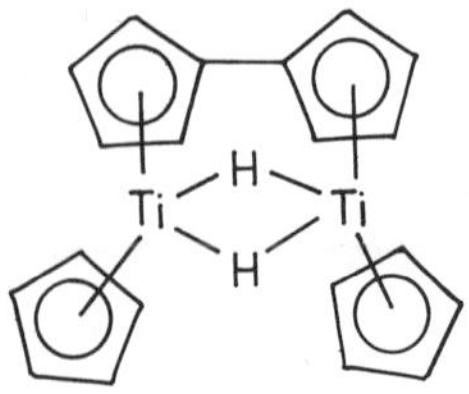

$$\text{(structure)} \qquad (9.74)$$

compound.[83,84,90] The presence of the fulvalene ligand $\text{C}_5\text{H}_4\text{-C}_5\text{H}_4$ is confirmed by

*Table 9.9. Properties of Titanium Group Cyclopentadienyl Complexes*

| Compound | Color | Mp (K) |
|---|---|---|
| $[TiBr_3cp]^a$ | Light-orange | 447.5–448.5 dec |
| $[TicpCl_3]^b$ | Orange | 490 |
| $[Ticp_2Cl_2]^a$ | Red | 560 dec |
| $[Ti_2cp_2(C_5H_4C_5H_4)(H)_2]^c$ | Dark-green | >473 dec |
| $[Ti(C_5Me_5)_2]^c$ | Orange | |
| $[Ticp_2(\sigma\text{-}C_5H_5)]^d$ | Dark-green | 411–413 dec |
| $[Ticp_2(\sigma\text{-}C_5H_5)_2]^d$ | Violet-black | 401 |
| $[Ti_2cp_4Cl_2]^e$ | Green, violet-brown | 554–555 |
| $[ZrcpCl_3]^b$ | Cream | 511 dec |
| $[Zr(C_5H_5)_2]^f$ | Purple-black | >573 dec |
| $[Zrcp_2Cl_2]^g$ | White | 517 |
| $[Zr(C_5H_4Me)_3Cl]^h$ | Yellow | 411–413 dec |
| $[Zrcp_3(\sigma\text{-}C_5H_5)]^i$ | Yellow | 533 dec |
| $[Hfcp_2Cl_2]^{j,k}$ | White | 508 dec |
| $[Hfcp_2(\sigma\text{-}C_5H_5)_2]^k$ | | 480–481 dec |
| $[Hf(C_5H_5)_2]^l$ | Red | >573 dec |

[a] C. L. Sloan and W. A. Barber, *J. Am. Chem. Soc.*, **81**, 1364 (1959).
[b] A. F. Reid and P. C. Wailes, *J. Organomet. Chem.*, **2**, 329 (1964).
[c] H. H. Brintzinger and J. E. Bercaw, *J. Am. Chem. Soc.*, **92**, 6182 (1970); J. E. Bercaw, R. H. Marvich, L. G. Bell, and H. H. Brintzinger, *J. Am. Chem. Soc.*, **94**, 1219 (1972); J. E. Bercaw, *J. Am. Chem. Soc.*, **96**, 5087 (1974); J. M. Manriquez and J. E. Bercaw, *J. Am. Chem. Soc.*, **96**, 6229 (1974).
[d] F. W. Siegert and H. J. de Liefde Meijer, *J. Organomet. Chem.*, **20**, 141 (1969).
[e] A. F. Reid and P. C. Wailes, *Aust. J. Chem.*, **18**, 9 (1965).
[f] G. W. Watt and F. O. Drummond, Jr., *J. Am. Chem. Soc.*, **88**, 5926 (1966); G. W. Watt and F. O. Drummond, Jr., *J. Am. Chem. Soc.*, **92**, 826 (1970).
[g] E. M. Brainina, M. C. Minacheva, and R. C. Freidlina, *Izv. Akad. Nauk SSSR, Ser. Khim.*, 1716 (1961); R. C. Freidlina, E. M. Brainina, and A. N. Nesmeyanov, *Dokl. Akad. Nauk SSSR*, **138**, 1369 (1961).
[h] E. M. Brainina, E. I. Mortikova, L. A. Petrashkevich, and R. C. Freidlina, *Dokl. Akad. Nauk SSSR*, **169**, 335 (1966).
[i] E. M. Brainina and G. G. Dworancewa, *Izv. Akad. Nauk SSSR, Ser. Khim.*, 442 (1967).
[j] A. N. Nesmeyanov, R. B. Materikova, E. M. Brainina, and N. S. Kochetkova, *Izv. Akad. Nauk SSSR, Ser. Khim.*, 1323 (1969); E. M. Brainina, M. C. Minacheva, B. W. Lokshin, E. I. Fedin, and P. W. Petrovskii, *Izv. Akad. Nauk SSSR*, 2492 (1969).
[k] M. Minacheva, E. M. Brainina, and R. C. Freidlina, *Dokl. Akad. Nauk SSSR*, **173**, 581 (1967).
[l] G. E. Watt and E. O. Drummond, *J. Am. Chem. Soc.*, **92**, 826 (1970).

$^{13}$C NMR spectra and X-ray structures (9.75) for analogous complexes containing OH and Cl bridges.

$$X = OH, Cl \qquad (9.75)$$

The reduction of $TiMe_2cp_2$ by means of hydrogen leads to the dimeric compound, which contains hydrido bridges and in which all cyclopentadienyl groups are pentahapto terminal ligands.

$$TiMe_2cp_2 + 3H_2 \longrightarrow cp_2Ti(\mu\text{-}H)_2Ticp_2 + 4CH_4 \qquad (9.76)$$

Bis(cyclopentadienyl)dimethyltitanium reacts with $R_3SiH$ [$R_3 = PhH_2$, $(RO)_3$, $Me(RO)_2$] to give the hydride [$(cp_2HTi)_2(\mu\text{-}H)$] which, based on UV, ESR and IR spectra, possesses the formula (9.77).[86]

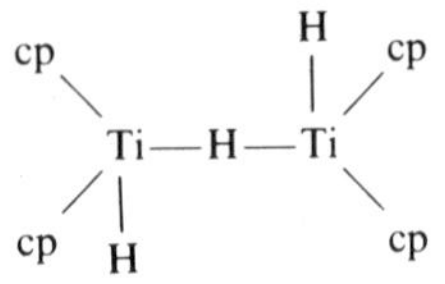

$$(9.77)$$

The compound Ti(pmcp)$_2$ is more stable. It may be obtained from the complex [$\{(pmcp)_2Ti\}_2(\mu\text{-}N_2)$] by heating it to room temperature. The decomposition of Ti(pmcp)$_2$H$_2$ gives a mixture of Ti(pmcp)$_2$ and Ti(pmcp)$_2$H. Solutions of permethylitanocene at room temperature are slowly converted to an equilibrium mixture containing the complex [Ti($C_5Me_4CH_2$)($C_5Me_5$)H]; see equation (9.78). It has not

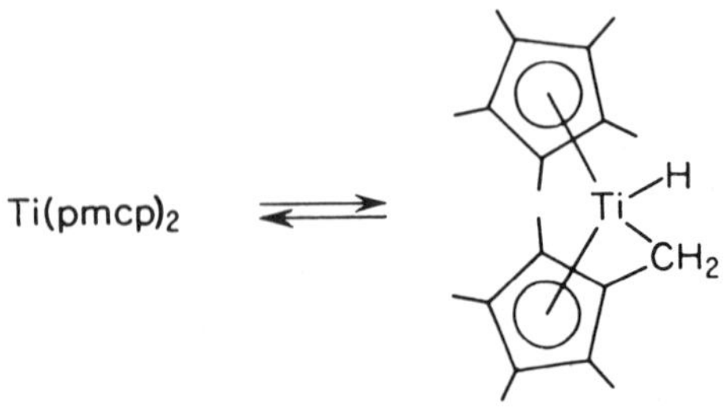

$$(9.78)$$

been ruled out that the $C_5Me_4CH_2$ group constitutes a bridging ligand as in the complex (9.79), [$(pmcp)Ti(\mu\text{-}C_5Me_4CH_2)(\mu\text{-}O)_2Ti(pmcp)$].[92] The chemistry of pen-

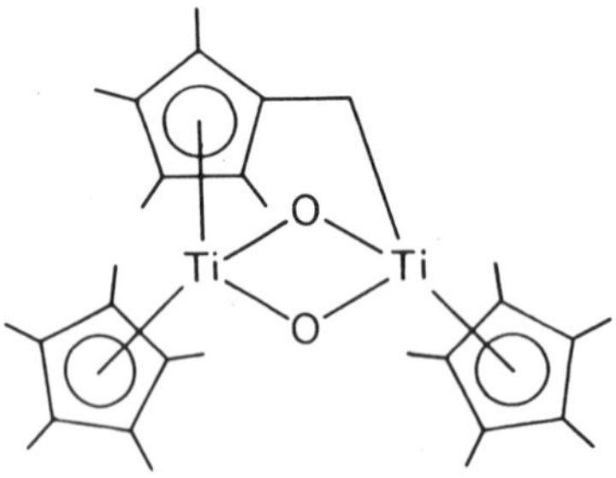

$$(9.79)$$

tamethylcyclopentadienyl complexes of zirconium and hafnium is similar. Like titanium, these metals form compounds of the following composition: M(pmcp)$_2$X$_2$ (X = Cl, Br, I, H), M(pmcp)$_2$(CO)$_2$, [$\{M(pmcp)_2(\mu\text{-}N_2)\}$].[83-85,90,91] The compounds of zirconium and hafnium of the formula $MC_{10}H_{10}$ are isomorphic with $TiC_{10}H_{10}$. The sandwich complex Ticp$_2$ is unstable compared to the dimeric derivative. However, it is probably formed as an intermediate product in some reactions.[35] On the other hand, titanium and zirconium complexes such as M($C_5Me_5$)$_2$ are relatively stable (Table 9.9). For the Zr(pmcp)$_2$ complex, there is also hydrogen atom abstraction from the methyl group of the ligand [see equation (9.78)]. Zirconium(III) and hafnium(III) compounds have been considerably less investigated than titanium(III) complexes. Compounds of the type $[M_2cp_4(\mu\text{-}Cl_2)]$ (M = Zr, Hf)[87] are also known. The reduction of $Zrcp_2Cl_2$ with sodium amalgam gives the fulvalene coordination compound[88] [equation (9.80)].

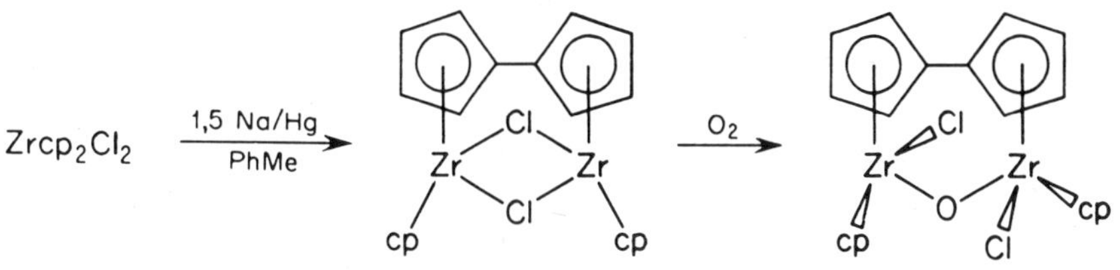

$$Zrcp_2Cl_2 \xrightarrow[\text{PhMe}]{\text{1,5 Na/Hg}} \qquad \xrightarrow{O_2} \qquad (9.80)$$

Cyclopentadienyl di- and polynuclear as well as cluster-type compounds have also been prepared, for example, $[cpTi(\mu\text{-}OOCR)_4Ticp]$, $[Ti(C_5H_4Me)\,Cl(\mu\text{-}O)]_4$, and $[Ti_6cp_6(\mu_3\text{-}O)]_8$. Titanium group compounds containing cyclopentadienyl ligands are quite strong, relatively hard, Lewis acids and therefore they form heteronuclear complexes with metals which behave as softer acids and in which the CO ligands are bonded through oxygen with the titanium group metal, for instance, $[(\mu_3\text{-}cp_2Ti\text{-}OC)_2(Co_3cp_3)]$.[89]

## b. *Group 5 metal complexes*[1–3,71,90,93–98]

These metals form the following cyclopentadienyl compounds: $[VcpX(CO)_3]^-$ (X = CN, H, etc.), $[V_2cp_2(CO)_5]$, $[Mcp(CO)_4]$ (M = V, Nb, Ta), $Vcp_2$, $Nbcp_2$, $Vcp_2X$, $[Mcp_2R]$, $[Mcp_2XL]$ (M = V, Nb, Ta), $[Nb_2(mcp)_2(\mu\text{-}Cl)_2Cl_2(CO)_4]$,[95] $[VcpX_2(PR_3)]$, $[McpXL_2]^+$ (M = Nb, Ta), $[VcpX_2]_2$, $[VcpX_2L_2]$, $[Mcp_2X_2]$, $[Mcp_2R_2]$, $[McpX_3]$ (M = V, Nb, Ta), $Vcp_2E_5$ (E = S, Se), $[NbcpX_3(dmpe)]$, $[Mcp_2X_3]$, $[Mcp_2H_3]$ (M = Nb, Ta), $[\{Nbcp_2X_2\}_2O]$, $[McpX_4]$ (M = V, Nb, Ta), $[\{\eta^5\text{-}C_5Bu'(CH_2CMe_3)_2(CH_2CMe_2CH_2)_2\}\,TaCl_2]$,[96] $[M(\eta\text{-}cp)_2(\sigma\text{-}cp)_2]$ (M = Nb, Ta). Halogen complexes are more resistant to air, and some are quite air stable.

Cyclopentadienyl compounds of this group are most commonly obtained by the reaction of Nacp with metal halides $MCl_n$. Some properties of cyclopentadienyl complexes of the vanadium group are presented in Table 9.10. The properties of $[Mcp(CO)_4]$ are given in Chapter 2.

Biscyclopentadienylniobium is considerably more reactive than $Vcp_2$ and therefore this niobium complex is more similar to the titanium group compounds. It is relatively stable below 213 K. At higher temperatures it rapidly forms the dinuclear complex $[cpHNb(\mu\text{-}\eta^1:\eta^5\text{-}C_5H_4)_2NbcpH]$.[90] This compound is also formed during the decomposition of $[Nbcp_2H_3]$ in benzene. It possesses a Nb–Nb bond[36] [structure (9.81)].

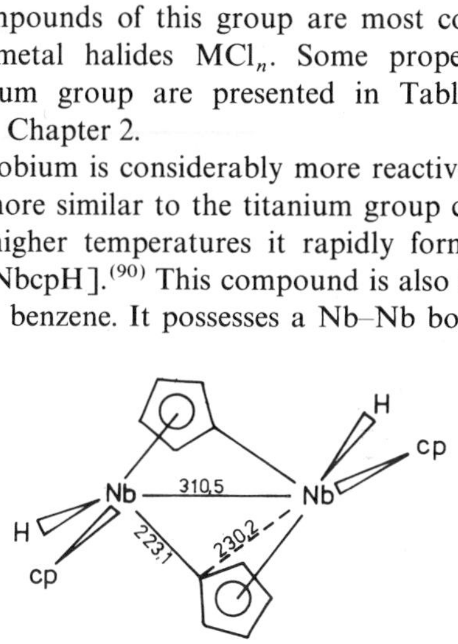

$$(9.81)$$

Table 9.10. *Properties of Vanadium Group Cyclopentadienyl Complexes*

| Compound | Color | Mp (K) | Remarks |
|---|---|---|---|
| $Vcp_2{}^a$ | Purple | 440–441 | |
| $[Vcp_2py][BPh_4]^b$ | Blue-green | | $\mu_{eff} = 2.76$ BM (291 K) |
| $[Vcp_2Cl]^c$ | Blue | 479 dec | |
| $[Vcp_2Cl_2]^{a,d}$ | Light-green | 523 dec | |
| $[Vcp_2Cl_3]^e$ | Violet | 433 dec | |
| $[Nb(\eta\text{-}cp)_2(\sigma\text{-}cp)_2]^f$ | Black-blue | | $\mu_{eff} = 1.41$ BM |
| $[Nbcp_2Ph_2]^f$ | | >423 dec | |
| $[Nbcp_2Br_3]^d$ | Red-brown | 533 | |
| $[Nbcp_2Cl_2]^f$ | Black | 473 | $\mu_{eff} = 1.61$ BM |
| $[Nbcp_2H_3]^g$ | | | $\nu(Nb-H)$ 1720 cm$^{-1}$; $\tau(cp) = 5.20$; $\tau(H) = 12.73$ (1H); 13.72 (2H) |
| $[Ta(\eta\text{-}cp)_2(\sigma\text{-}cp)_2]^h$ | Red-violet | | $\mu_{eff} = 1.73$ BM |
| $[Tacp_2H_3]^{i,j}$ | White | 460–462 dec | |
| $[Tacp_2Br_3]^d$ | Red | 553 | |

[a] R. B. King, *Organometallic Synthesis*, Vol. 1, Academic Press, New York (1965); G. Wilkinson, F. A. Cotton, and J. Birmingham, *Inorg. Nucl. Chem.*, **2**, 95 (1956).
[b] G. Fachinetti and C. Floriani, *Chem. Commun.*, 578 (1975).
[c] H. J. de Liefde Meijer, M. J. Janssen, and G. J. M. van der Kerk, *Recl. Trav. Chim. Pays-Bas*, **80**, 831 (1961).
[d] G. Wilkinson and J. M. Birmingham, *J. Am. Chem. Soc.*, **76**, 4281 (1954).
[e] E. O. Fischer, S. Vigoureux, and P. Kuzel, *Chem. Ber.*, **93**, 701 (1960).
[f] F. W. Siegert and H. J. de Liefde Meijer, *Recl. Trav. Chim. Pays-Bas*, **87**, 1445 (1968).
[g] F. N. Tebbe and G. W. Parshall, *J. Am. Chem. Soc.*, **93**, 3793 (1971).
[h] E. O. Fischer and A. Treiber, *Chem. Ber.*, **94**, 2193 (1961).
[i] M. L. H. Green, J. A. McCleverty, L. Pratt, and G. Wilkinson, *J. Chem. Soc.*, 4854 (1961); R. B. King, *J. Am. Chem. Soc.*, **91**, 7217 (1969).
[j] F. N. Tebbe, *J. Am. Chem. Soc.*, **95**, 5412 (1973).

In contrast to $Vcp_2$, fulvalene vanadium complexes (9.82) are diamagnetic and have vanadium-to-vanadium bonds. Niobium fulvalene complexes may be obtained

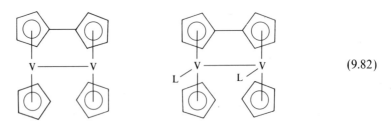

                                                                              (9.82)

from the compound (9.81) in the presence of ligands which easily form bridges[90]; see structures (9.83).

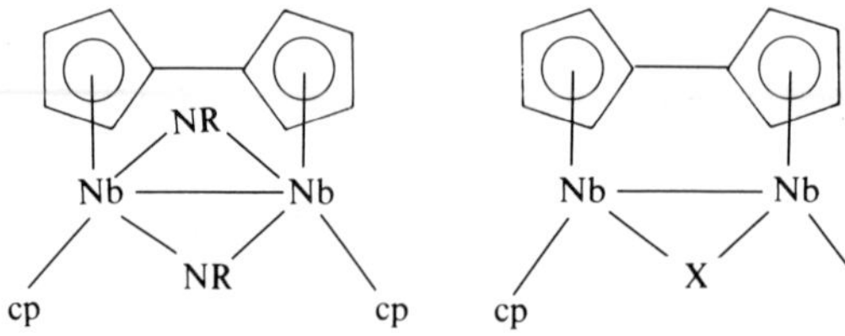

X = Cl, NR     (9.83)

## c. *Group 6 metal complexes*[1–3,5,16,90,99,100]

These metals readily form complexes of the formula $Mcp_2$ ($M = Cr$, Mo, W) which may be easily oxidized. The compounds bis(cyclopentadienyl)molybdenum and bis(cyclopentadienyl)tungsten are obtained by photolytic decomposition of $[Mcp_2H_2]$ and $[Mcp_2CO]$. $Mcp_2$ ($M = Mo$, W) are stable at 10 K, but they undergo association when the temperature is raised[90,100] to give most likely dimers which are analogous to compounds obtained by reduction of $Mcp_2Cl_2$ by sodium amalgam. The heaviest ions in the mass spectra of products of the reduction are $[M_2C_{20}H_{20}]^+$. Because the $^1H$ NMR spectra show only one signal resulting from the unsubstituted $C_5H_5$ rings, it is therefore probable that the association occurs owing to the formation of metal–metal bonds (9.84). From these complexes, various other compounds such as $Mcp_2L$ and

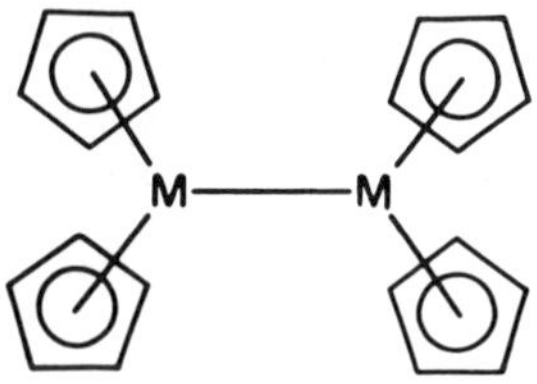

(9.84)

$Mcp_2X_2$ ($X = 1e$ ligand) may be obtained, although complexes containing $\eta^1 : \eta^5\text{-}C_5H_4$ and $\eta^5 : \eta^5\text{-}C_5H_4\text{-}C_5H_4$ have not been prepared. However, these latter complexes may be synthesized from $Mcp_2H_2$ [equations (9.85) and (9.86)]. Electron-deficient cyclopentadienyl complexes of molybdenum and tungsten, like those of zirconium and hafnium, activate $C-H$ bonds. Very typical for elements of the chromium group (Table 9.11) are dinuclear compounds of the type $[M_2cp_2(CO)_n]$ ($n = 4, 6$) which possess the $M \equiv M$ triple bonds for $n = 4$ and $M-M$ single bonds for $n = 6$, as well as derivatives $[MXcp(CO)_3]$ ($X = H$, alkyl, aryl, acyl, etc.); see structure (9.87).

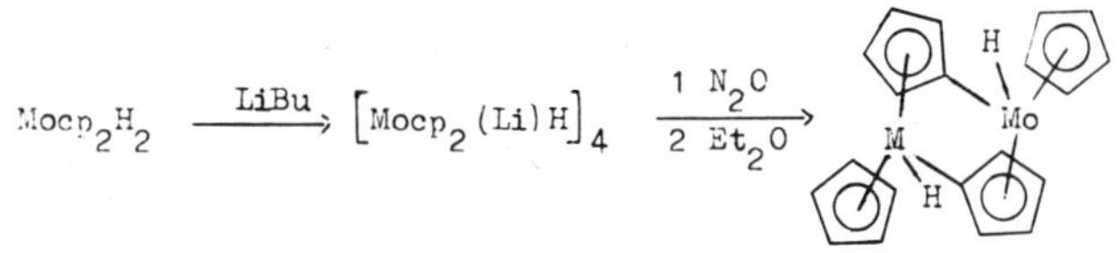

(9.85)

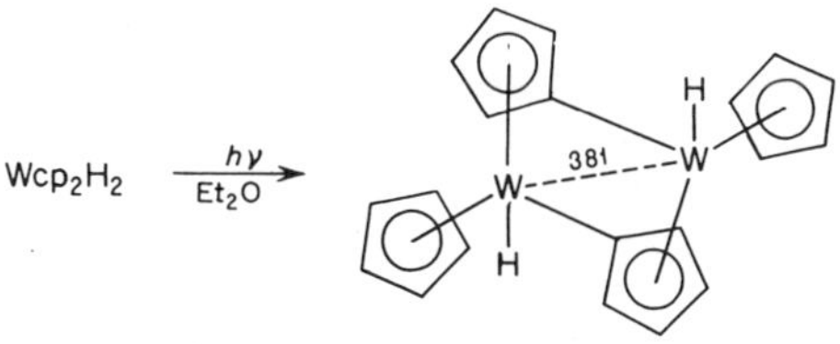

(9.86)

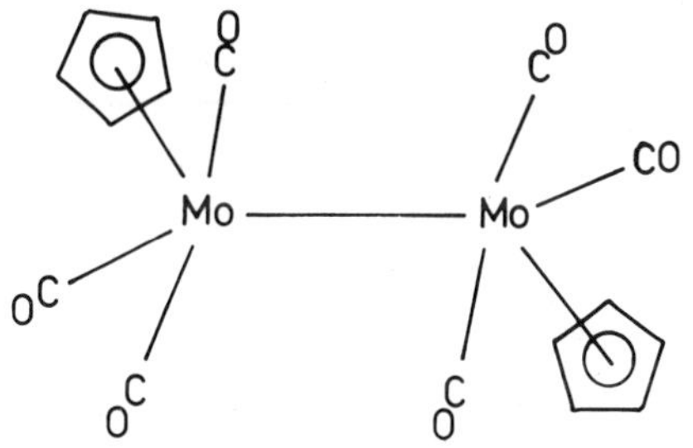

(9.87)

Table 9.11. Properties of Chromium Group Cyclopentadienyl Complexes

| Compound | Color | Mp (K) |
|---|---|---|
| 1 | 2 | 3 |
| $Crcp_2{}^a$ | Light-red | 446 |
| $[Cr_2cp_2(\mu\text{-}NO)_2(NO)_2]^b$ | Red-violet | 431–432 dec |
| $[CrcpCl(NO)_2]^c$ | Green-yellow | 413 dec |
| $[Cr_2cp_2(CO)_6]^{d,e}$ | Dark-green | 436–441 dec |
| $[CrcpH(CO)_3]^{d,e}$ | Yellow | 330–331 |
| $[Crcp(CO)_2(NO)]^f$ | Orange-red | 1340–1341 |
| $[(OC)_3cpCrVcp_2]^g$ | Dark-brown | 420–422 |
| $[Mocp_2]_x{}^h$ | Red-brown | |
| $[Mocp_2H_2]^i$ | Yellow | 456–458 |
| $[cp(H)Mo(\mu\text{-}C_5H_4)_2Mocp(H)]^j$ | Yellow | |
| $[Mocp_2Cl_2]^k$ | Green | 543 dec |
| $[Mocp_2Br_2]^k$ | Green | 543 dec |
| $[Mo(\eta\text{-}cp)(\sigma\text{-}cp)_3]^l$ | Red-violet | 393 dec |
| $[Mo_2cp_2(CO)_6]^{m,e}$ | Purple-red | 488–490 dec |
| $[MocpH(CO)_3]^{e,d,n}$ | Yellow | 323–325 |
| $[Mocp(C_6H_6)Cl]^o$ | Purple | |
| $[Mocp(CO)_2(NO)]^f$ | Orange-red | 358–359 |
| $[Mocp_2HMgBr(THF)_2]^p$ | Orange | |
| $[Wcp_2]_x{}^h$ | Red | |
| $[Wcp_2H_2]^i$ | Yellow | 436–438 |
| $[Wcp_2Cl_2]^k$ | Green | 523 dec |
| $[Wcp_2Br_2]^k$ | Green | 523 dec |
| $[W_2cp_2(CO)_6]^q$ | Purple-red | 513–515 dec |
| $[WcpH(CO)_3]^d$ | Yellow | 341–342 |
| $[Wcp(CO)_2(NO)]^f$ | Orange | 378–380 |

[a] G. Wilkinson, F. A. Cotton, and J. M. Birmingham, *J. Inorg. Nucl. Chem.*, **2**, 95 (1956); G. Wilkinson, *J. Am. Chem. Soc.*, **76**, 209 (1954).

[b] R. B. King and M. B. Bisnette, *Inorg. Chem.*, **3**, 791 (1964).

[c] T. S. Piper and G. Wilkinson, *J. Inorg. Nucl. Chem.*, **2**, 136 (1956).

[d] E. O. Fischer, W. Hafner, and H. O. Stahl, *Z. Anorg. Allg. Chem.*, **282**, 47 (1955); E. O. Fischer, W. Hafner, and H. O. Stahl, *Inorg. Synth.*, **7**, 136 (1963).

[e] R. B. King and F. G. A. Stone, *Inorg. Synth.*, **7**, 99 (1963).

[f] E. O. Fischer, O. Beckert, W. Hafner, and H. Stahl, *Z. Naturforsch.*, **10b**, 598 (1955).

[g] A. Mijake, H. Kondo, and M. Arojama, *Angew. Chem., Int. Ed. Engl.*, **8**, 520 (1969).

[h] J. L. Thomas, *J. Am. Chem. Soc.*, **95**, 1838 (1973).

[i] M. L. H. Green, J. A. McCleverty, L. Pratt, and G. Wilkinson, *J. Chem. Soc.*, 4854 (1961).

[j] N. J. Cooper, M. L. H. Green, C. Couldwell, and K. Prout, *J. Chem. Soc., Chem. Commun.*, 145 (1977).

[k] R. L. Cooper and M. L. H. Green, *Z. Naturforsch.*, **19b**, 652 (1964); R. L. Cooper and M. L. H. Green, *J. Chem. Soc. (A)*, 1155 (1967).

[l] E. O. Fischer and J. Hristidu, *Chem. Ber.*, **95**, 253 (1962).

[m] R. G. Hayter, *Inorg. Chem.*, **2**, 1031 (1963).

[n] S. A. Keppie and M. F. Lappert, *J. Organomet. Chem.*, **19**, 5 (1969).

[o] M. L. H. Green, J. Knight, and J. A. Segel, *J. Chem. Soc., Chem. Commun.*, 283 (1975).

[p] S. G. Davies, M. L. H. Green, K. Prout, A. Coda, and V. Tazzoli, *J. Chem. Soc., Chem. Commun.*, 135 (1977).

[q] G. Wilkinson, *J. Am. Chem. Soc.*, **76**, 209 (1954).

Molybdenum and tungsten also form tetracoordinate complexes such as $[Mcp_2X_2]$ (X = H, Cl, Br, etc.) and $[Mcp_2X_2]^+$. In the case of molybdenum, many mixed sandwich compounds of the type $Mocp(arene)^+$ and $Mocp(arene)X$ are known. The unusually reactive complex $[Mocp_2HMgBr(THF)_2]$ possesses a Mo—Mg bond (Figure 9.8e). This complex represents one of few examples of complexes with a M—Mg bond. This shows that the reaction of Grignard compounds with transition metal complexes may give intermediates containing direct metal–magnesium interactions which were postulated earlier. The most characteristic cyclopentadienyl compounds of the chromium group metals are $[Mcp_2]$, $[M_2cp_2(CO)_n]$ ($n = 4, 6$), $[Mcp(CO)_3]^-$, $[McpX(CO)_3]$ [X = H, alkyl, aryl, $ER_3$ (E = Si, Ge, Sn, Pb), halide, pseudohalide], $[Mcp(CO)_2NO]$, $[Mcp(X)(NO)_2]$ (M = Cr, Mo, W), $[CrcpCl_2]$, $[CrcpX_2L]$, $[CrcpX_3]^-$, $[MocpX_2(O)]$, $[MocpX_4]$, $[Mcp_2X_2]$ (M = Mo, W; X = halide, NCS, $N_3$), $[Mcp_2H_2]$ (M = Mo, W), $[Mo(R)cp_2(PR_3)]^+$, and $[Wcp_2R_2]$.

### d. Group 7 metal complexes[1-3,5,16,90,101-111]

These elements form the following dienyl complexes: $[Mcp(CO)_3]$, $[Mcp(CO)_{3-x}L_x]$ (M = Mn, Tc, Re), $[Mcp_2]^-$ (M = Mn, Re), $[(OC)_2cpMn(\mu\text{-SO})Mncp(CO)_2]$,[103] $[Mncp(CO)_2(NS)]^+$,[104] $[Mncp_2]$, $[Re_2cp_4]$,[105] $[Recp_2(CH_2=CH_2)]$, $[Re_2(pmcp)_2(\mu\text{-O})(CO)_4]$,[106] $[Mcp_2L]$,[107] $[Mcp_2H]$ (M = Mn, Tc, Re), $[Mcp_2X]$ (X = halide, M = Mn, Re), $[Recp_2R]$, $[Recp_2(SnMe_3)]$,[108] $[Mncp(\eta^1\text{-}C_5H_5)(TMEDA)]$,[109] $[Recp_2X_2]^+$, $[Mcp_2H_2]^+$ (M = Tc, Re), $[Recp_2R_2]^+$, $[ReBr_2(pmcp)O]$,[106] $[Re_2(pmcp)_2(\mu\text{-O})_2(O)_2]$,[106] $[Re(pmcp)Me_2(O)]$,[106] $[Re_3(pmcp)_3(O)_6]^{2+}$,[110] $[Re(pmcp)(O)_3]$[111] (Table 9.12). Manganocene is pyrophoric. It is a brown solid in an inert atmosphere up to 431 K. Above this temperature, it turns light pink. It melts at 445 K. After cooling, it again becomes brown. Manganocene is monometic in the gaseous state and shows sandwich structure; however, in the solid state at room temperature the brown isomorphous form of $Mncp_2$ is a polymer in which the Mncp units are connected by the bridging $C_5H_5$ ligands. The compound $[Mn(pmcp)_2]$ has a ferrocene type structure and adopts a low-spin $^2E_{2g}$ state. Rhenium forms dimeric cyclopentadienyl complex in which all $C_5H_5$ groups are equivalent (only one signal in the $^1H$ NMR spectrum). Therefore, this dimer possesses a Re—Re bond.[105] Like 4d and 5d metals of groups 4 and 5, rhenium forms cyclopentadienyl compounds in high oxidation states, for examples, $[Re_3(pmcp)_3(O)_6]^{2+}$,[110] $[Re(pmcp)(O)_3]$,[111] $[Re(pmcp)Me_2(O)]$,[106] and $[Recp_2X_2]^+$. The compounds $Mcp_2H$ in acidic solutions undergo protonation to give the $[Mcp_2H_2]^+$ derivatives.

### e. Group 8 metal complexes[1-5,16,112-118]

Metallocenes of these metals are exceptionally stable because they achieve 18e configuration of the valence shell. They are stable in air in the solid state. They undergo reduction with difficulty; by contrast, in acidic media, they are easily oxidized to the $[Mcp_2]^+$ cations. Dilute, acidic solutions of $Fecp_2^+$ are blue-green, while more concentrated solutions are blood red. Neutral and even more basic solutions of the ferrocenium cation $Fecp_2^+$ undergo decomposition to give ferrocene and iron hydroxide. Ferrocene forms charge transfer complexes with nitrobenzene, tetracyanoethylene, and 2, 3-dichloro-5, 6-dicyanoquinone.

*Table 9.12. Properties of Manganese Group Cyclopentadienyl Complexes*

| Compound | Color | Mp (K) |
|---|---|---|
| $Mncp_2^{a,b}$ | Brown | 445–446 |
| $[Mncp(CO)_3]^c$ | Yellow | 350 |
| $[Mn_2cp_2(CO)_2(NO)_2]^d$ | Purple-brown | 473 |
| $[Mn_3cp_3(NO)_4]^{d,k}$ | Black | 493–498 dec |
| $[Tc_2cp_4]^e$ | Yellow | 428 |
| $[Tccp_2H]^f$ | Gold-yellow | 423–428 |
| $[Tccp(CO)_3]^g$ | | 360.5 |
| $[Recp_2H]^h$ | Light-yellow | 434–435 |
| $[RecpMe_2(2\text{-}4\text{-}\eta C_5H_4Me)]^i$ | Yellow | 346–347 |
| $[Recp(CO)_3]^j$ | White | 384–387 |

[a] R. B. King, *Organometallic Synthesis*, Vol. 1, p. 67, Academic Press, New York (1965).
[b] G. Wilkinson, F. A. Cotton, and J. M. Birmingham, *J. Inorg. Nucl. Chem.*, **2**, 95 (1956).
[c] T. S. Piper, F. A. Cotton, and G. Wilkinson, *J. Inorg. Nucl. Chem.*, **1**, 165 (1955).
[d] R. B. King and M. B. Bisnette, *Inorg. Chem.*, **3**, 791 (1964).
[e] D. K. Huggins and H. D. Kaesz, *J. Am. Chem. Soc.*, **83**, 4474 (1961).
[f] E. O. Fischer and M. W. Schmidt, *Chem. Ber.*, **102**, 1954 (1969).
[g] C. Palm, E. O. Fischer, and F. Baumgaertner, *Naturwiss.*, **49**, 279 (1962).
[h] M. L. H. Green, L. Pratt, and G. Wilkinson, *J. Chem. Soc.*, 3916 (1958).
[i] R. L. Cooper, M. L. H. Green, and J. T. Moelwyn-Hughes, *J. Organomet. Chem.*, **3**, 261 (1965).
[j] M. L. H. Green and G. Wilkinson, *J. Chem. Soc.*, 4314 (1958); K. H. Anisimov, N. Kolobova, and A. J. Joganson, *Izv. Akad. Nauk SSSR, Ser. Khim.*, 1749 (1969).
[k] T. S. Piper and G. Wilkinson, *J. Inorg. Nucl. Chem.*, **2**, 38 (1956); R. C. Elder, *Inorg. Chem.*, **13**, 1037 (1974).

The charge transfer tetracyanoethylene complex has a structure analogous to those of charge transfer complexes of aromatic hydrocarbons. The tetracyanoethylene is located above the plane of the cyclopentadienyl ring. Ruthenocene may be oxidized by one-electron oxidation to the ruthenocenium cation $Rucp_2^+$ which is light yellow. On the other hand, chronopotentiometric oxidation occurs as a two-electron process to give the dipositive cation $Rucp_2^{++}$. Osmocene may be oxidized chronopotentiometrically or by means of Fe(III) compounds to give also the Os(IV) complex, $Oscp_2^{++}$. In acidic solutions (for example, $BF_3$, $H_2O$) ferrocene and ruthenocene undergo protonation to give cationic hydrides such as $[Fecp_2H]^+$ and $[Rucp_2H]^+$. Their presence was confirmed on the basis of the $^1H$ NMR spectra. The signal of the proton which is bonded to the metal occurs at higher magnetic fields than for the tetramethylsilane. Osmocene is a weaker base and does not undergo protonation under these conditions. Many coordination compounds containing rings with substituents and complexes possessing 5e polycyclic ligands of these metals are known. These metals also form carbonyl, halogen, etc. derivatives: $[M_2cp_2(CO)_4]$ (M = Fe, Ru, Os), $[Fecp(CO)_3]^+$, $[Fecp(CO)_2]^-$, $[Mcp(CO)_2X]$ (M = Fe, Ru; X = Cl, Br, I, Ph, CN, etc.), $[McpX(PPh_3)_2]$ (M = Ru, Os; X = Cl, Br) (Table 9.13). Thousands of derivatives, both sandwich and monocyclopentadienyl of group 8, are known, particularly those of iron. These are mainly obtained either from the cyclopentadiene derivatives by their reaction with metal

*Table 9.13. Fe, Ru, and Os Cyclopentadienyl Complexes*

| Compound | Color | Mp(K) |
|---|---|---|
| Fecp$_2^a$ | Orange | 446 |
| [Fe$_2$cp$_2$(CO)$_4$]$^{b,c}$ | Purple-red | 467 |
| [Fe$_4$cp$_4$S$_4$]$^e$ | Black | 503–513 dec |
| [Fe$_4$cp$_4$(CO)$_4$]$^d$ | Dark-green | 493 dec |
| [Fe$_2$cp$_2$H(CO)$_4$]PF$_6^f$ | Red-brown | |
| Rucp$_2^g$ | Light-yellow | 472–473 dec |
| [Ru$_2$cp$_2$(CO)$_4$]$^{h,i}$ | Orange-red | 458 dec |
| [Rucp(CO)$_2$Cl]$^j$ | Yellow | |
| [RucpCl(PPh$_3$)$_2$]$^k$ | Orange | 404–408 |
| Oscp$_2^l$ | White | 502–503 |
| [Os$_2$cp$_2$(CO)$_4$]$^m$ | Yellow | 471 |
| [OsBrcp(PPh$_3$)$_2$]$^k$ | Light-yellow-orange | 455–458 dec |

[a] J. M. Birmingham, D. Seyferth, and G. Wilkinson, *J. Am. Chem. Soc.*, **76**, 4179 (1954); T. J. Kealy and P. L. Pauson, *Nature*, **168**, 1039 (1961).

[b] S. Monastyrskyj, R. E. Maginn, and M. Dubeck, *Inorg. Chem.*, **2**, 904 (1963).

[c] E. O. Fischer and R. Boettcher, *Z. Naturforsch.*, **10b**, 600 (1955).

[d] R. B. King, *Inorg. Chem.*, **5**, 2227 (1966).

[e] R. G. Hayter, *J. Am. Chem. Soc.*, **85**, 3120 (1963); R. A. Schunn, C. J. Fritchie, and C. T. Prewitt, *Inorg. Chem.*, **5**, 892 (1966).

[f] D. A. Symon and T. C. Waddington, *J. Chem. Soc. (A)*, 953 (1971).

[g] G. Wilkinson, *J. Am. Chem. Soc.*, **74**, 6146 (1952).

[h] E. O. Fischer and A. Vogler, *Z. Naturforsch.*, **17b**, 421 (1962).

[i] F. A. Cotton and G. Yagupsky, *Inorg. Chem.*, **6**, 15 (1967).

[j] R. J. Haines and A. L. duPreez, *J. Chem. Soc., Dalton Trans.*, 944 (1972).

[k] T. Blackmore, M. I. Bruce, and F. G. A. Stone, *J. Chem. Soc. (A)*, 2376 (1971).

[l] E. O. Fischer and H. Grubert, *Chem. Ber.*, **92**, 2302 (1959).

[m] E. O. Fischer and K. Bittler, *Z. Naturforsch.*, **17b**, 274 (1962).

compounds or by substitution of hydrogen atoms in complexes of the type [M(C$_5$H$_5$)$_2$] or [MC$_5$H$_5$L$_n$], etc. Such compounds include $n$ metallocenophanes and, for instance, [1 : 1]metallocenophanes[112–115,119,120] [structures (9.88)]. Based on the ${}^1$H and

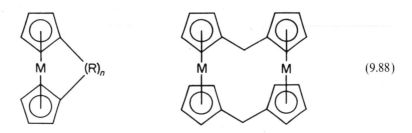

(9.88)

${}^{13}$C NMR spectra, it was found that in the carboanions (9.89) there is intramolecular hydrogen bond of the type C – H – C.[120] Also, clusters such as Fe$_4$cp$_4$(CO)$_4$, Fe$_4$cp$_4$S$_4$, etc., constitute a large group of cyclopentadienyl compounds. Iron also forms dinuclear and polynuclear fulvalene complexes, for example, 1, 1′-biferrocenylene (9.90a) and polyferrocenylene (9.90b).

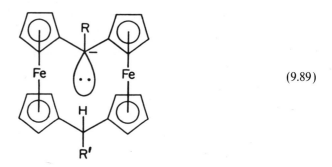

(9.89)

a)                                    b)

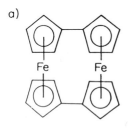

(9.90)

## f. Group 9 metal complexes[1-3, 5, 16, 121-128]

Cobalt and its heavier homologs form paramagnetic metallocenes $Mcp_2$ (Tabele 9.14). Biscyclopentadienylcobalt(II) is relatively stable, while compounds of rhodium and iridium are stable only at low (liquid nitrogen) or high (above 393 K) temperatures in gaseous state. At room temperature, they dimerize to afford the diene–dienyl complexes (9.91).

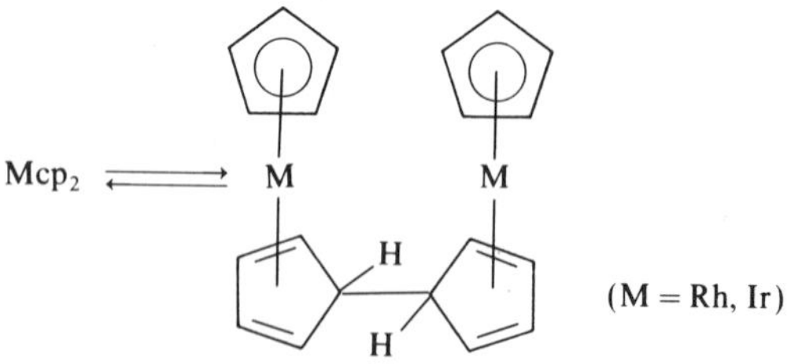

$$Mcp_2 \rightleftharpoons \qquad\qquad\qquad (9.91)$$

$$(M = Rh, Ir)$$

All $Mcp_2$ complexes have 19 valence electrons, and therefore easily undergo oxidation to the $Mcp_2^+$ cations:

$$4Cocp_2 + O_2 + 4H^+ \longrightarrow 4[Cocp_2]^+ + 2H_2O \qquad (9.92)$$

The cations $[Rhcp_2]^+$ and $[Ircp_2]^+$ undergo reduction to $Mcp_2$ with difficulty, therefore the $Mcp_2$ compounds may be obtained only by the reaction of molten sodium with $[Mcp_2]^+[PF_6]^-$. The isolation of the products from the reaction mixture is done

*Table 9.14. Properties of Cyclopentadienyl Complexes of Co, Rh, and Ir*

| Compound | Color | Mp (K) | Remarks |
|---|---|---|---|
| $Cocp_2{}^a$ | Purple | 446–474 | |
| $[Cocp(CO)_2]^b$ | Red | 251 | |
| $[Co_3cp_3(CO)_3]^c$ | Black | | |
| $[Co_2cp_2(\mu\text{-}PPh_2)_2]^d$ | Dark-brown | 560–565 dec | |
| $[Co_4cp_4H_4]^o$ | Black-violet | 573 | |
| $Rhcp_2{}^e$ | Brown-black | | |
| $[Rh_2(\mu\text{-}Cl)_2Cl_2(C_5Me_5)_2]^{f,g}$ | Red | >573 | |
| $[Rh_2(\mu\text{-}Cl)(\mu\text{-}H)Cl_2(C_5Me_5)_2]^h$ | Purple | | $\tau(RhH)$ 21.37; $\tau(Me)$ 8.46 $J(RhH)$ 23 Hz |
| $[Rhcp(CO)_2]^i$ | Orange | 262 | |
| $[Rh_3cp_3(CO)_3]^j$ | Black | | |
| $[Rh_3cp_3(CO)_3]^k$ | Dark-green | | |
| $[Rh_2cp_2(CO)_3]^l$ | Dark-red | | |
| $[Ircp_2]^e$ | White | | |
| $[Ir_2(\mu\text{-}Cl)_2Cl_2(C_5Me_5)_2]^f$ | Orange | >503 | |
| $[Ir_2(\mu\text{-}Cl)(\mu\text{-}H)Cl_2(C_5Me_5)_2]^h$ | Red | | $\tau(IrH)$ 23.55; $\tau(Me)$ 8.38 |
| $[Ircp(CO)_2]^m$ | Yellow liquid | | $\tau(H) = 5.81$ |
| $[Ir(C_5Me_5)(C_4H_4N)][PF_6]^n$ | White | >523 | |

$^a$ See Reference *b*, Table 9.12.
$^b$ See Reference *c*, Table 9.12.
$^c$ See Reference *d*, Table 9.13.
$^d$ J. M. Coleman and L. F. Dahl, *J. Am. Chem. Soc.*, **89**, 542 (1967).
$^e$ E. O. Fischer and H. Wawersik, *J. Organomet. Chem.*, **5**, 559 (1966).
$^f$ J. W. Kang and P. M. Maitlis, *J. Am. Chem. Soc.*, **90**, 3259 (1968); J. W. Kang, K. Moseley, and P. M. Maitlis, *J. Am. Chem. Soc.*, **91**, 5970 (1969).
$^g$ B. L. Booth, R. N. Haszeldine, and M. Hill, *J. Chem. Soc. (A)*, 1299 (1969).
$^h$ C. White, D. S. Gill, J. W. Kang, H. B. Lee, and P. M. Maitlis, *Chem. Commun.*, 734 (1971).
$^i$ E. O. Fischer and K. Bittler, *Z. Naturforsch.*, **16b**, 225 (1961).
$^j$ O. S. Mills and E. E. Paulus, *J. Organomet. Chem.*, **10**, 331 (1967).
$^k$ E. F. Paulus, E. O. Fischer, H. P. Fritz, and H. Schuster-Woldan, *J. Organomet. Chem.*, **10**, P3 (1967).
$^l$ O. S. Mills and J. P. Nice, *J. Organomet. Chem.*, **10**, 337 (1967).
$^m$ E. O. Fischer and K. S. Brenner, *Z. Naturforsch.*, **17b**, 774 (1962).
$^n$ C. White and P. M. Maitlis, *J. Chem. Soc. (A)*, 3322 (1971).
$^o$ J. Müller and H. Dorner, *Angew. Chem., Int. Ed. Engl.*, **12**, 843 (1973); G. Huttner and H. Lorenz, *Chem. Ber.*, **108**, 973 (1975).

by condensation on the surface cooled with liquid nitrogen. The reduction in the presence of solvents leads to $[Mcp(C_5H_6)]$ or $[Mcp(C_5H_5R)]$ dependent on the reducing agent. The cation $[Cocp_2]^+$ is reduced to $[Cocp_2]^-$ in dimethoxyethane solution. Complexes possessing the following composition are also known: $[Co_2cp_2(\mu\text{-}NO)_2]$, $Na[Cocp(NO)]$,[124] $[Mcp(CO)_2]$, $[McpL_2]$, $[Mcp(diene)]$ $(M = Co,$ $Rh, Ir)$, $[M_2cp_2(\mu\text{-}CO)(CO)_2]$, $[M_2cp_2(\mu\text{-}CO)_2]$, $[Co_2(fulvalene)_2]$, $[Co(pmcp)I]$, $[RhcpCl_2]_n$, $[Mcp_2]^+$, $[McpX_2L]$ $(M = Co, Rh, Ir)$, $[M_2(pmcp)_2(\mu\text{-}Cl)_2Cl_2]$, $[M_2(pmcp)_2(\mu\text{-}OH)_3]^+$, $[M_2(pmcp)_2(\mu\text{-}Cl)(\mu\text{-}H)Cl_2]^{[127]}$ $(M = Rh, Ir)$, $[Ircp(R)_2(PPh_3)]$, $[Ir(pmcp)H_2(PPh_3)]$,[122] $[Ir(pmcp)H_4]$,[123] $[Rh(pmcp)(H)_2(SiEt_3)_2]$,[126] $[(pmcpRh(cis\text{-}Nb_2W_4O_{19})]^{2-}$,[125] and $[cpCo\{P(OR)_2O\}_3Rh(\mu\text{-}CO)_3Rh\{OP(OR)_2\}_3 Cocp]$.[128]

In the case of rhodium and iridium, cyclopentadienyl compounds for high oxidation states $(+5)$ of the metals are known.[123,126] These complexes, like rhodium(III) and iridium(III) compounds, with the pentamethylcyclopentadienyl ligand are catalysts for hydrogenation of arenes and for hydrosilylation. Also, many clusters of metals of group 9 containing cyclopentadienyl ligands are known, for instance, $[Co_4(pmcp)_2(CO)_7]$, $[Co_4cp_4(\mu_3\text{-S})_2(\mu_3\text{-S}_2)_2]$, $[Co_4cp_4(\mu_3\text{-S})_4]$, $[M_3cp_3(CO)_3]$ (M = Co, Rh), $[Rh_4cp_4(\mu_3\text{-CO})_2]$, etc. (Figure 9.11).

## g. Group 10 metal complexes[1–3, 5, 16, 129–132]

Nickel forms the 20e complex $Nicp_2$ which is oxidized to $[Nicp_2]^+$ in acidic media. Under these conditions the dication $[Nicp_2]^{2+}$ is not formed. However, $[Nicp_2]^{2+}$ may be obtained by electrochemical oxidation. In contrast to the isoelectronic compounds $Fecp_2$ and $[Cocp_2]^+$, the dication $[Nicp_2]^{2+}$ is very unstable.

The following cyclopentadienyl compounds of the nickel triad are known: $[M_2cp_2(\mu\text{-CO})_2]$ (M = Ni, Pt), $[M_2cp_2(\mu\text{-RC} \equiv CR)]$ (M = Ni, Pd), $[Ni_2cp_2(\mu\text{-NCPh})_2]$, $[Nicp(PBu_2Ph)_2]$, $[M_2(\mu\text{-cp})_2L_2]$, $[M_2(\mu\text{-cp})(\mu\text{-C}_3H_5)L_2]$ (M = Pd, Pt), $[Mcp(NO)]$ (M = Ni, Pd, Pt), $Nicp_2$, $[Pd_n(C_5H_5)_{2n}]$, $[cpPt(\mu\text{-}\eta^4\text{-C}_5H_5\text{-C}_5H_5) Ptcp]$, $[Mcp(X)L]$, $[Mcp(R)L]$, $[McpL_2]^+$, $[Mcp(diene)]^+$ (M = Ni, Pd, Pt), $[Mcp(allyl)]$ (M = Ni, Pd), $[Ni_2cp_2(\mu\text{-Y})_2]$ [Y = SPh, $P(CF_3)_2$], $[cpNi\{P(OR)_2\text{-O}\}_2 M\{O-P(OR)_2\}_2Nicp]$ (M = Co, Zn), and $[Ptcp(Me)_3]$. The complex $[Pt_2(C_5H_5)_4]$ was prepared by the reaction of $PtCl_2$ with sodium cyclopentadienide in hexane. The $^1H$ NMR spectrum shows a singlet with satellites resulting from the interaction with both $^{195}Pt$ atoms; this indicates the presence of a Pt—Pt bond. However, the IR and UV spectra show that the $C_5H_5$ groups are bonded as $\sigma$ and $\pi$ ligands. These data suggest that in the solution the complex is not rigid, but fluxional. In the solid state, this compound has structure (9.93).

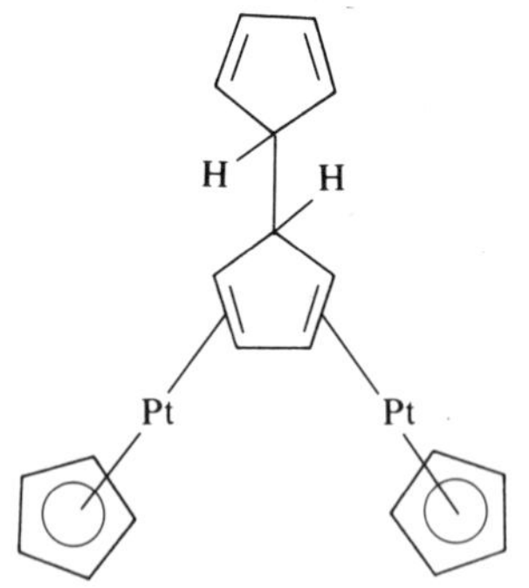

(9.93)

The palladium compound $[Pd_n(C_5H_5)_{2n}]$ is very unstable. Palladium and platinum, like other metals of the second and third series of the transition metals, do not show tendencies toward the formation of stable metallocenes, in contrast to the 3d metals.

The complex $[(pmcp) Pd(\mu\text{-PhCCPh}) Pd(pmcp)]$ ($= M_2$) undergoes two one-electron reversible oxidation steps to give mono- and dipositive ions. It may also be reduced reversible in a single electron process. Therefore, in the case of this dimer, the formation of four quite stable oxidation states is indicated.[129]

$$[M_2]^- \rightleftharpoons [M_2] \rightleftharpoons [M_2]^+ \rightleftharpoons [M_2]^{2+} \qquad (9.94)$$

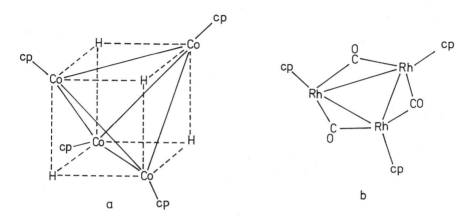

Figure 9.11. Structures of cluster cyclopentadienyl complexes: (a) $Co_4cp_4H_4^{(56)}$ and (b) $Rh_3cp_3(CO)_3$ (the black isomer).

Many stable cyclopentadienyl clusters of metals of the nickel group are also known: $[Ni_3cp_3(CO)_2]$, $[Ni_3cp_3(NBu')]$, $[Ni_4cp_4H_3]$, $[Ni_2Pt_2W_4(\mu_3\text{-}CPh)_4(CO)_8cp_4]$,[133] $[Ni_2Zn_4cp_6]$, $[Ni_6cp_6]$, and $[Ni_6cp_6]^+$ (Table 9.15).

### h. *Group 11 metal complexes*

In cyclopentadienyl compounds of copper such as $[Cucp(CO)]$ and $[Cucp(PR_3)]$ $(R = Et, Bu, Ph)^{(37)}$ the $C_5H_5$ group is a pentahapto ligand. Formally these compounds have 18 valence electrons. However, because of a considerable lowering of the energy of $3d$ copper orbitals and a substantial increase in the energy difference between the $4p$ and $4s$ orbitals, the $e_1(p_x, p_y)$ orbitals of copper probably interact weakly with the orbitals of the cyclopentadienyl ligand. Therefore, the complex CucpL may be regarded as a 14$e$ compound with the sp hybridization of the copper atom. One of the sp orbitals interacts with the carbonyl or phosphine ligand, while the other forms the central $\sigma$ bond with the cyclopentadienyl ring. In the case of gold, the compounds Aucp and $[Aucp(PPh_3)]$ are known. In the phosphine complex, the cyclopentadienyl group is a pentahapto ligand.[38,39] Cyclopentadienyl compounds of silver are not known.

### i. *Group 3 metal complexes including compounds of the lanthanides and the actinides*[40–42,134–146,159–162]

Cyclopentadienyl complexes of these elements most commonly contain pentahapto and monohapto cyclopentadienyl ligands. The $(n-1)d$ orbital energy and particularly the $np$ orbital energy of metals which begin a given series are relatively high (see Chapter 1). Therefore, for elements of groups 3, 4, and 5, the energy difference between the $M-(\eta^5\text{-cp})$ and $M-(\eta^1\text{-cp})$ bonds is smaller in contrast to the elements of group 7 and some of groups 8, 9, and 10. Thus, for metals which begin the transition metal series, all kinds of Mcp complexes are known: $[Mo(\eta^5\text{-cp})(\eta^1\text{-cp})_3]$, $[M(\eta^5\text{-cp})_2(\eta^1\text{-cp})_2]$ $(M = Ti, Hf, Nb, Ta)$, $[Zr(\eta^5\text{-cp})_3(\eta^1\text{-cp})]$, and $[M(\eta^5\text{-cp})_4]$ $(M = U, Th)$. The $M-(\eta\text{-cp})$ bonds in these complexes are also central $\sigma$ bonds as in the case of

*Table 9.15. Cyclopentadienyl Complexes of Ni, Pd, and Pt*

| Compound | Color | Mp (K) | Remarks |
|---|---|---|---|
| Nicp$_2$[a,b] | Green | 446–447 dec | |
| [Ni$_2$cp$_2$($\mu$-CO)$_2$][c] | Red | 419–420 dec | |
| [Ni$_3$cp$_3$($\mu_3$-CO)$_2$][c,d] | Green | 473 dec | $d$(Ni–Ni) 239 pm |
| [Ni$_3$cp$_3$($\mu_3$-NBu$^t$)][e] | Black | | $\mu_{eff}$ = 1.68 BM, $d$(Ni–Ni) 221, 234, 227 pm |
| [NicpCl(PPh$_3$)][f,g] | Red-brown | 413 dec in air 439–441 under N$_2$ | |
| [Ni$_2$cp$_3$][PF$_6$][h] | Black-brown | | |
| [Ni$_4$cp$_4$H$_3$][s] | Black-violet | > 593 | |
| [PdCpBr(PEt$_3$)][i] | Dark-green | 338–339 | |
| [Pdcp(NO)][j] | Red-brown | 243 | |
| [Pd$_2$cp$_2$($\mu$-Cl)$_2$][k] | Red-brown | | |
| [Pd$_2$($\mu$-cp)($\mu$-C$_3$H$_5$)(PPr$^i_3$)$_2$][l] | Orange-yellow | 337 | |
| [Pt$_2$cp$_4$][m] | Green | | |
| [Pt$_2$cp$_2$(CO)$_2$][n] | Red | 376 dec | |
| [PtcpNO][o] | Orange | 337 | |
| [PtcpI(PEt$_3$)][i] | Dark-orange | 341–342 | |
| [PtcpMe$_3$][p,q,r] | White | 338 | |

[a] F. Engelman, *Z. Naturforsch.*, **8b**, 775 (1953).
[b] G. Wilkinson, P. L. Pauson, and F. A. Cotton, *J. Am. Chem. Soc.*, **76**, 1970 (1954).
[c] E. O. Fischer and C. Palm, *Chem. Ber.*, **91**, 1725 (1958).
[d] A. A. Hock and O. S. Nills, in: *Advances in the Chemistry of the Coordination Compounds*, Proceedings 6th ICCC (S. Kirschner, ed.), p. 640, McMillan, London (1961).
[e] S. Otsuka, A. Nakamura, and T. Yoshida, *Inorg. Chem.*, **7**, 261 (1968).
[f] M. D. Rausch, Y. F. Chang, and H. B. Gordon, *Inorg. Chem.*, **8**, 1355 (1969).
[g] H. Yamazaki, T. Nishido, Y. Matsumoto, S. Sumida, and N. Hagihara, *J. Organomet. Chem.*, **6**, 86 (1966).
[h] H. Werner and A. Salzer, *Synth. React. Inorg. Met.-Org. Chem.*, **2**, 239 (1972).
[i] R. J. Gross and R. Wardle, *J. Chem. Soc. (A)*, 2000 (1971).
[j] E. O. Fischer and A. Vogler, *Z. Naturforsch.*, **18b**, 771 (1963).
[k] J. Smidt and R. Jira, *Angew. Chem.*, **71**, 651 (1959).
[l] H. Werner and A. Kuehn, *Angew. Chem., Int. Ed. Engl.*, **16**, 412 (1977).
[m] E. O. Fischer and H. Schuster-Woldan, *Chem. Ber.*, **100**, 705 (1967).
[n] E. O. Fischer, H. Schuster-Woldan, and K. Bittler, *Z. Naturforsch.*, **18b**, 429 (1963).
[o] E. O. Fischer and H. Schuster-Woldan, *Z. Naturforsch.*, **19b**, 766 (1964).
[p] S. D. Robinson and B. L. Shaw, *J. Chem. Soc.*, 1529 (1965).
[q] W. A. Semion, A. Z. Rubezhov, Ju. T. Struchkov, and S. P. Gubin, *Zh. Strukt. Khim.*, **10**, 151 (1969).
[r] G. W. Adamson, J. C. J. Bart, and J. J. Dally, *J. Chem. Soc. (A)*, 2616 (1971).
[s] J. Müller, H. Dorner, G. Huttner, and H. Lorenz, *Angew. Chem., Int. Ed. Engl.*, **12**, 1005 (1973); G. Huttner and H. Lorenz, *Chem. Ber.*, **107**, 996 (1974).

compounds of copper and gold. The scandium group metals and the lanthanides form characteristic compounds: Mcp$_2$X (X = Cl, alkyl, aryl, H, C≡CPh, $\pi$-C$_3$H$_5$),[40–42,134–138,142,144] McpCl$_2$(THF)$_3$,[134] [(Mcp$_2$)$_3$($\mu_3$-H)($\mu$-H)$_3$]$^-$,[136,137] and [Smcp$_2$(SiMe$_3$)$_2$]$^-$.[143] Also, compounds containing other halogens, cyanide, oxygen, etc., are known: [Ercp$_2$I], [Ndcp$_2$CN], [Ybcp$_2$CN],[40–42,134] [{Sm(pmcp)$_2$}$_2$ ($\mu$-O)],[140] and *cis*- and *trans*-[{(Ph$_3$PO)(pmcp)$_2$Sm}$_2$($\mu$-OCH=CHO)] (*cf.* p. 63).[141]

The scandium complex Sccp$_3$ has polymeric structure with the Sc($\eta^5$-C$_5$H$_5$)$_2$ fragments bonded via $\eta^1$:$\eta^1$-C$_5$H$_5$ bridges. In tetrahydrofuran solution, Sccp$_3$ has dynamic properties.[138]

Tris(cylopentadienyl)praseodymium is a polymer in which the Pr($\eta^5$-C$_5$H$_5$)$_2$ units are bonded by $\eta^2$:$\eta^5$-C$_5$H$_5$ bridges [structure (9.95)]. The Ln—cp bond is decisively

$$(9.95)$$

ionic. The complexes $Mcp_3$ ($M = Sc$, $Y$, $Ln$) are very reactive. They are easily oxidized by air and react with $CO_2$, maleic anhydride, aldehydes, and ketones. Magnetic properties of $Lncp_3$ are analogous to those of other lanthanide complexes under magnetically diluted solutions. All lanthanides form $Mcp_3$-type compounds. Compounds containing alkylcyclopentadienyl ligands are known only in the case of $C_5H_4Me$ (practically for all the lanthanides) and $C_5H_4(i\text{-}C_3H_7)$ (for La, Pr, and Nd). With larger ligands L such as pentamethylcyclopentadienyl, trimethylsilylcyclopentadienyl, or bis(trimethylsilyl)-cyclopentadienyl, only bis-L compounds of the type $[M(L)_2X]$ are formed.

The compounds $Mcp_3$ show characteristic tendencies to the formation of complexes with donor ligands, such as $NH_3$, THF, $PPh_3$, and RNC. Some of them are so stable that they decompose with the evolution of cyclopentadiene:

$$[Ybcp_3(NH_3)] \longrightarrow [Ybcp_2(NH_2)] + C_5H_6 \qquad (9.96)$$

The complexes $Mcp_2X$ ($X =$ halogen, H, Me) are dimers in which the metal atoms are bonded by the X bridges. Other compounds possessing analogous composition $Mcp_2X$ have probably similar structures ($X = NH_2$, MeO, PhO, HCOO, AcO, PhCOO). These complexes easily react with alkali metal halides to give compounds containing halogen bridges which bond the lanthanide atoms with alkali metals. Complexes of the type $McpCl_2(THF)_n$ ($n = 3$; $n = 4$ for La, Sm, Eu, Tm, Yb) are monomeric.

Some lanthanides such as samarium, europium, and ytterbium form compounds in which the lanthanide has +2 oxidation state: $[Mcp_2]$, $[Mcp_2(THF)_n]$, $[M(pmcp)_2L]$ ($M = Sm$, Eu, Yb; $L = THF$, $Et_2O$),[40–42, 134] $[Sm(pmcp)_2(THF)_2]$,[139] $[Sm(pmcp)(\mu\text{-}I)(THF)_2]_2$,[139] and $[Sm(pmcp)_2]$.[145] In the last samarium compound, the metal atom is coordinated only to two pentamethylcyclopentadienyl ligands. However, in contrast to metallocenes, the $C_5Me_5$ rings are not parallel. The angle pmcp center–Sm–pmcp center equals $140.1°$ ($136.7°$ in the compound $Sm(pmcp)_2(THF)_2$).[139,145] The shortest distance between the samarium atom and the hydrogen atom of the adjacent molecule equals 280 pm, suggesting that in this complex there is no agostic bond $Sm \cdots H - C$. The $M - cp$ bond in $Mcp_2$ complexes is ionic. Therefore, these compounds are very oxygen sensitive and immediately react with water. They are not soluble in hydrocarbons and ethers, although they are soluble in DMF and liquid ammonia.

As far as the lanthanides are concerned, only for cerium is the complex possessing the +4 oxidation state known, $Cecp_4$. Its physical and chemical properties differ considerably from compounds of the type $Mcp_3$. It does not react with water and dilute acids. It decomposes under the action of concentrated acids and hot, dilute hydroxides. It reacts with hydrogen chloride to give $CecpCl_3$, and therefore it behaves analogously to Ti, Zr, and Nb complexes. The chemical properties and the IR spectra show that the Ce–cp bond is covalent to a considerable degree. Thorium, protactinium, uranium, and neptunium form $Mcp_4$ complexes. Heavier actinides (from plutonium to californium) afford cyclopentadienyl compounds in which the metal adopts only a $+3$ oxidation state.[42,146] Compounds of the following types are also known: $[Mcp_3X]$ (M = Th, Pa, U, Np; X = Cl, Br, I, F, OR, SCN etc.), $[Mcp_3]$ (M = Th, U, Np, Pu, Am, Cm, Bk, Cf), $[Mcp_2X]$ (M = U, X = CN; M = Th, Bk, X = Cl), $[Ucp_3Cl_2]$, and $[McpX_3]$ (M = Th, U, X = Cl, Br). All these compounds are very easily oxidized by air, and are decomposed by water and acids. They have a decisively ionic character. Compounds of actinide(III), like those of lanthanide(III), may add neutral ligands, for example, $[Ucp_3(THF)]$ and $[Ucp_3(CNC_6H_{11})]$. Complexes possessing U=C or U=N double bonds are interesting: $[cp_3U=CHPMePhR]$ (R = Me, Ph),[159,160] $[cp_3U=NC(Me)$

$$CHPPh_2Me],^{[160]} \quad \begin{bmatrix} & NCy \\ & \parallel \\ cp_3U & | \\ & \diagdown \\ & C=CHPPh_2Me \end{bmatrix},^{[160]} \quad and \quad [(C_5H_4Me)_3U=NPh].^{[161]}$$

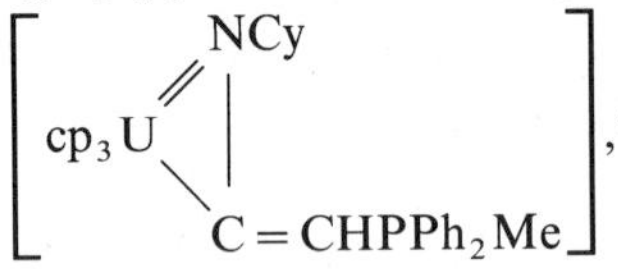

In the compound $[(pmcp)_2Th(\mu\text{-}PPh_2)_2Ni(CO)_2]$,[162] there is a Th–Ni interaction, because the distance between the metal atoms equals 320.6 pm and is shorter by *ca* 50 pm than the nonbonding distance.

## 13. POLYNUCLEAR AND CLUSTER CYCLOPENTADIENYL COMPLEXES

Some *d*-electron metals show characteristic tendencies to form polynuclear and cluster-type cyclopentadienyl complexes. Almost always, the $C_5H_5$ group constitutes a terminal ligand in these compounds. The $\mu$ and $\mu_3$ bridges which link metal atoms very often include the following groups and atoms: OR, OH, O, NR, $NR_2$, SR, $PR_2$, CO, CS, H, S, NO, halogen, and acetylenes. Examples of cluster-type cyclopentadienyl compounds are given in Tables 9.12–9.15. The $M_3$ clusters in which the metal atoms are located at the corners of a triangle and the $M_4X_4$ clusters in which the metal atoms and the $\mu_3$ bridging ligands are located at every other vertex of a cube are the most commonly encountered clusters (see Chapter 3).

## 14. PENTADIENYL COMPLEXES

The transition metals form many "open" (9.97a) and "closed" (9.97b) dienyl complexes. "Open" complexes comprise those containing the pentadienyl $C_5H_7$ ligand and its derivatives $C_5H_{7-x}R_x$, while "closed" compounds possess cyclohexadienyl, cycloheptadienyl, cyclooctatrienyl, etc. Pentadienyl and its derivatives $C_5H_{7-x}R_x$ as 5*e* ligands are usually coordinated in the "U" form (9.97a),[59,60] although complexes are

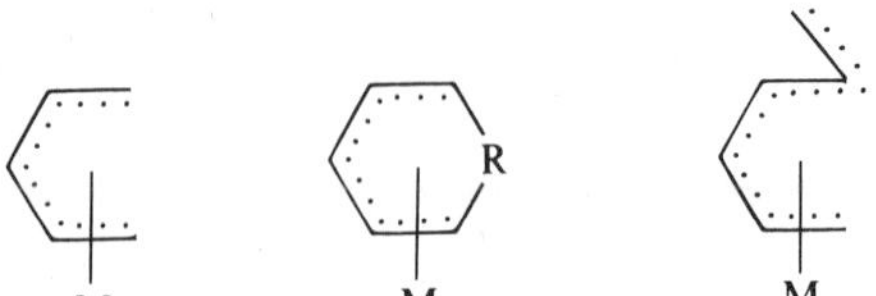

$$(9.97)$$

a                    b                    c

$$R = (CH_2)_n, etc.$$

known in which this ligand adopts the "S" form (9.97c).[147] So far, the cause of the existence of the "S" form is not known. In the case of $Mo(C_5H_5Me_2)_2(PEt_3)$, both steric and electronic factors are influential.[147] In this complex, in contrast to the "U" ligand, the "S" ligand is not planar. The $C^1C^2C^3$ plane is twisted toward the central atom by 52° with respect to the $C^2C^3C^4C^5$ plane [structure (9.98)]. The complexes

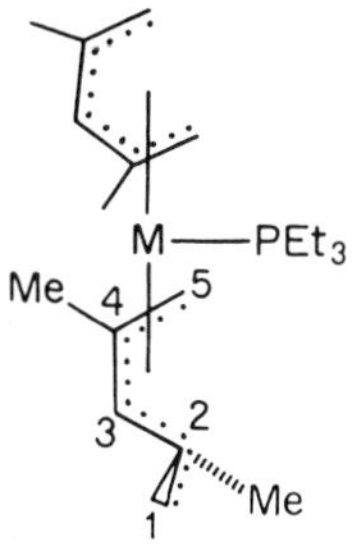

$$(9.98)$$

$M(2, 4\text{-}C_5H_5Me_2)_2(PEt_3)$ (M = Zr, Nb) have synperiplanar conformation, [147] as do titanium and vanadium compounds $M(2, 4\text{-}C_5H_5Me_2)_2L$ (L = CO, $PF_3$).[59,60] Pentadienyl complexes of the type $M(2, 4\text{-}C_5H_5Me_2)_2$ (M = Ti, V, Cr, Fe, Ru) are known. "Open" metallocenes containing many other dienyl ligands are also stable, for example, $C_5H_7$, $2\text{-}C_5H_6Me$, $3\text{-}C_5H_6Me$, $2, 3, 4\text{-}C_5H_4Me_3$, etc. In contrast to Ticp$_2$, the "open" titanocene $[Ti(2, 4\text{-}C_5H_5Me_2)_2]$ is stable as shown by NMR.[59,60]

The reaction of $2, 4\text{-}C_5H_4Me_2^-$ with $CoCl_2$ leads to the dimer (9.99). Many

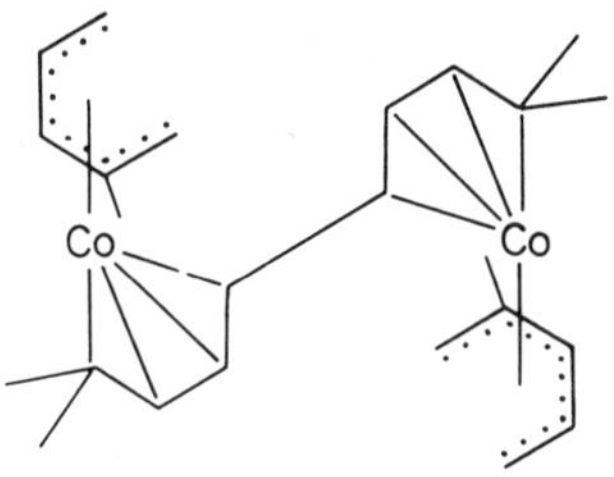

$$(9.99)$$

monodienyl complexes are known: $[Fe(C_5H_7)(CO_3)]^+$, $[Mn(\text{cyclohexadienyl})(CO)_3]$, $[Mn(CHT)(CO)_3]$, $[Fe(\text{cyclooctatrienyl})(CO)_3]^+$, etc. Nickel forms a dimeric complex (9.100) containing the "W" form of the pentadienyl.

$$(9.100)$$

## 15. PYRROLE COMPLEXES

Azacyclopentadienyl complexes may be obtained by the reaction of the anion of pyrrole (sodium or potassium salt) with transition metal compounds [equations (9.101) and (9.102)].

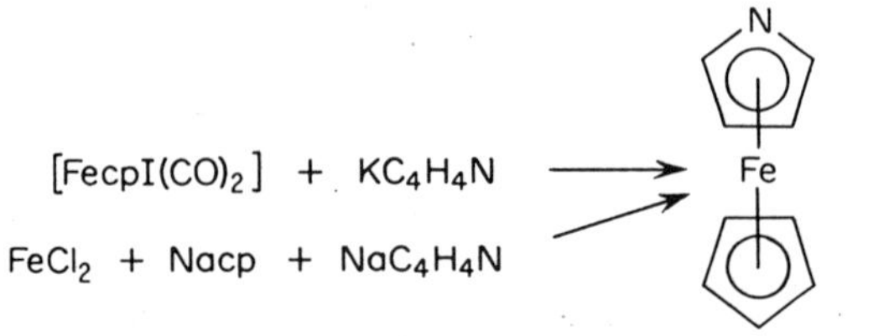

$$[FecpI(CO)_2] + KC_4H_4N \longrightarrow \quad \text{Fe} \qquad (9.101)$$

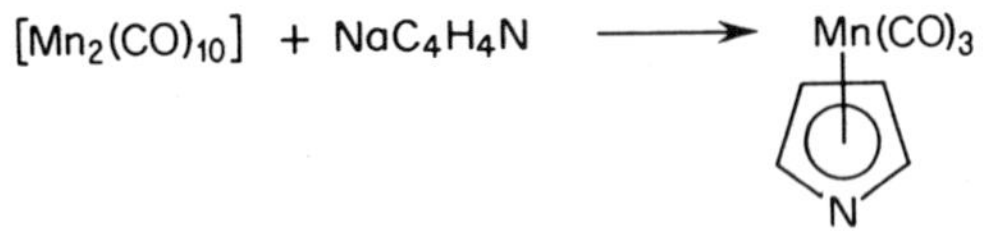

$$FeCl_2 + Nacp + NaC_4H_4N \longrightarrow$$

$$[Mn_2(CO)_{10}] + NaC_4H_4N \longrightarrow \quad Mn(CO)_3 \qquad (9.102)$$

## 16. MULTIDECKER COMPLEXES

Recently, numerous multidecker complexes of transition metals, particularly those involving heterocyclic ligands with boron and sulfur, were prepared.[64-66] Triple-, quadruple-, pentuple-, and hexuple-decker compounds were characterized by X-ray crystallography, mass spectrometry, and NMR [see equations (9.67)–(9.69)].[64] The composition of the hexuple-decker complex, which has the greatest known number of layers, was studied by mass spectrometry and NMR. Recently, a polymeric multideckered sandwich compound of nickel (9.103) was also prepared.[148] In addition to cyclic, five-membered rings, arenes may also act as bridging ligands. Benzene and mesitylene

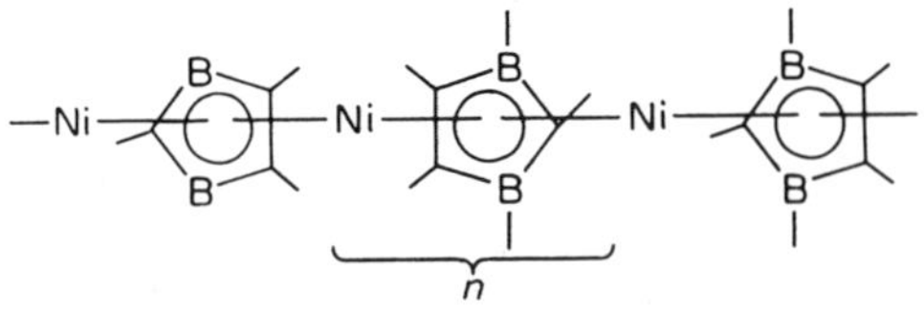

$$\qquad (9.103)$$

constitute bridging ligands in the triple-decker complexes $[cpV(C_6H_6)Vcp]$ and $[cpV(1,3,5-C_6H_3Me_3)Vcp]$,[69,70] and $[cpCo(C_6R_6)Cocp]$ (R = $\overset{O}{\overset{\|}{C}}$-OMe).[150a] Multidecker compounds containing the ligands $\mu$-$P_n$ ($n$ = 5, 6), $\mu$-$As_5$,[65,66] $\mu$-$P_3$, and $\mu$-$As_3$ are also stable [see Chapter 7, equation (7.82)]:

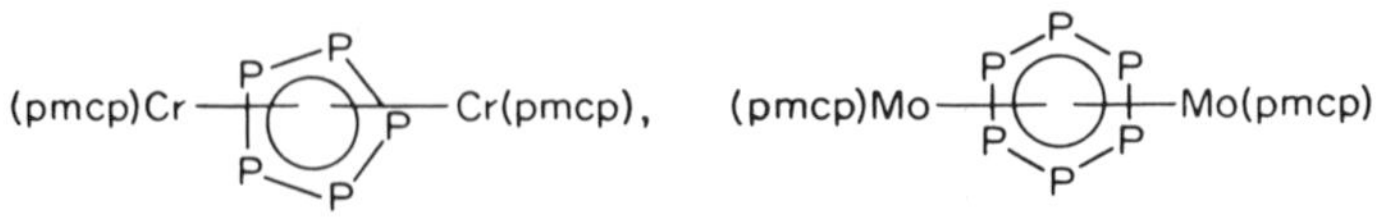

$$(pmcp)Cr \text{---} \quad \text{---} Cr(pmcp), \quad (pmcp)Mo \text{---} \quad \text{---} Mo(pmcp) \qquad (9.104)$$

# 17. REACTIONS OF CYCLOPENTADIENYL COMPOUNDS[1–6, 150b–d]

## a. *Oxidation of metallocenes and electron transfer reactions*

All metallocenes of the transition metals may be oxidized to the $[Mcp_2]^+$ or $[Mcp_2]^{2+}$ cations. Sometimes, the oxidation is reversible. Ferrocene may be oxidized by $HNO_3$, $H_2SO_4$, $Ce(SO)_4)_2$, $H[AuCl_4]$, $Fe(III)$, $CuCl_2$, quinone, etc. Other metallocenes are oxidized in a similar way:

$$Rucp_2 \underset{Na_2S_2O_3}{\overset{FeCl_3,\ Et_2O}{\rightleftharpoons}} [Rucp_2Cl]^+ FeCl_4^- \tag{9.105}$$

$$Oscp_2 \xrightarrow{I_2,\ H_2SO_4} [Oscp_2]^+ \tag{9.106}$$

$$Oscp_2 \xrightarrow{NH_4Fe(SO_4)_2} [Oscp_2OH]^+ \tag{9.107}$$

Cobaltocene is a strong reducing agent; it reduces $FeCl_2$ and $H^+$:

$$2Cocp_2 + FeCl_2 \longrightarrow 2[Cocp_2]^+Cl^- + Fe \tag{9.108}$$

$$Cocp_2 + H^+ \longrightarrow [Cocp_2]^+ + \tfrac{1}{2}H_2 \tag{9.109}$$

The cations $[Cocp_2]^+$, $[Rhcp_2]^+$, and $[Ircp_2]^+$ are very stable. They do not decompose even on boiling with concentrated sulfuric and nitric acids or aqua regia. Therefore, this enables the oxidation of substituents in the ring:

$$[Co(C_5H_4Me)_2]^+ \xrightarrow{KMnO_4} [Co(C_5H_4COOH)_2]^+ \tag{9.110}$$

Between the oxidized $Mcp_2^+$ cation and the metallocenes, there is often rapid electron transfer. The half-time for the electron transfer reaction between the ferrocenium ion and ferrocene is several milliseconds at 198 K. The exchange between $Cocp_2^+$ and $Cocp_2$ is also fast.

Metallocenes may also be oxidized by halogens ($Cl_2$, $Br_2$ $I_2$) and some halogenated organic compounds.

$$Recp_2H \underset{LiAlH_4}{\overset{X_2}{\rightleftharpoons}} [Recp_2X_2]^+X^- \tag{9.111}$$

$$Crcp_2 + I_2 \xrightarrow{Et_2O} [Crcp_2I] \tag{9.112}$$

$$Crcp_2 \xrightarrow{CH_2=CHCH_2I} [Crcp_2I] \tag{9.113}$$

$$[Mo(C_5Ph_5)_2] \xrightarrow{Br_2} [Mo(C_5Ph_5)_2]^+(Br_3)^- \tag{9.114}$$

$$2Vcp_2 + I_2 \xrightarrow{Et_2O} 2[Vcp_2I] \tag{9.115}$$

$$Vcp_2 + RX \longrightarrow [Vcp_2RX] \xrightarrow{Vcp_2} [Vcp_2R] + [Vcp_2X] \tag{9.116}$$

$$[Vcp_2R] + RX \longrightarrow [Vcp_2X] + R-R \tag{9.117}$$

The same oxidizing agents are also utilized for oxidation of cyclopentadienyl complexes of the titanium group metals:

$$Ticp_2 + 2CCl_4 \longrightarrow [Ticp_2Cl_2] + C_2Cl_6 \tag{9.118}$$

$$Ticp_2Cl + O_2 \xrightarrow{\text{HCl}} [Ticp_2Cl_2]$$

$$[Ticp_2Cl] + O_2 \longrightarrow [Ticp_2Cl]_2O \tag{9.119}$$

Vanadocene and decamethylvanadocene undergo an interesting oxidation reaction with benzoic acid to give $(\eta^5\text{-}C_5R_5)V(\mu\text{-}O_2CPh)_4 V(\eta^5\text{-}C_5R_5)$ compounds (R = H or $CH_3$) in which the V–V distance is *ca* 360 pm, suggesting that little if any direct bonding occurs between the two metal atoms[150a] [equation (9.120)]. Both metallocenes and

$$V(pmcp)_2 + C_6H_5CO_2H \longrightarrow \tag{9.120}$$

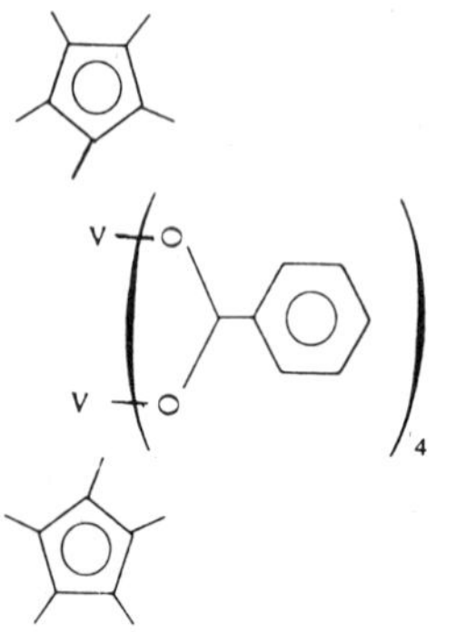

other cyclopentadienyl complexes may be oxidized and reduced electrochemically.[81,82] Many of these oxidation–reduction processes are reversible. Ferrocene was proposed to be an internal reference for measurement of the potential, because it is assumed that the couple $Fecp_2$–$Fecp_2^+$ ($E° = 0.31$ V, 0.2 M $LiClO_4$ in MeCN) is basically reversible. However, this has not yet been definitely proven. As an internal reference, the $[Cr(C_6H_6)_2]$–$[Cr(C_6H_6)_2]^+$ couple may be utilized because it is reversible.

## b. *Reduction of cyclopentadienyl complexes*

Cyclopentadienyl complexes in higher oxidation states undergo reduction to metallocenes in the presence of various reducing agents such as $LiAlH_4$, Zn, Nacp, MgXR, Na, etc. The reduction of the cyclopentadienyl rings in ferrocene does not occur even in the presence of catalysts, which ordinarily hydrogenate benzene within several minutes. However, alkali metals in amine solutions reduce $Fecp_2$ to metallic iron and the cyclopentadienyl ion. Partial reduction of the ring easily occurs in the case of some other cyclopentadienyl complexes which do not have an 18$e$ valence electron structure:

$$[Rhcp_2]^+ + LiAlH_4 \longrightarrow [Rhcp(C_5H_6)] \tag{9.121}$$

$$[Cocp_2]^+ + LiAlH_4 \longrightarrow [Cocp(C_5H_6)] \tag{9.122}$$

$$[Cocp_2]^+ + LiPh \longrightarrow [Cocp(C_5H_5Ph)] \tag{9.123}$$

$$Nicp_2 + Na/Hg \longrightarrow CpNi \tag{9.124}$$

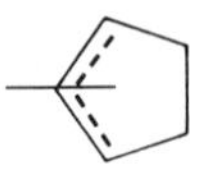

## c. *Protonation*

In the strongly acidic solution of boron trifluoride hydrate, ferrocene and ruthenocene undergo protonation to give hydrido complexes $[Mcp_2H]^+$. This was proved by $^1H$ NMR measurements, because signals at $\tau > 10$ for hydrido protons were found. Many other cyclopentadienyl complexes also undergo protonation, for example, $[Mocp_2H_2]$, $[Wcp_2H_2]$, and $[Recp_2H]$ give $[Mcp_2H_3]^+$ and $[Recp_2H_2]^+$ in acidic media.

## d. *The M—cp bond rupture*

The rupture of the $M-cp$ bond in compounds of the iron group and in the cobaltocenium cation $[Cocp_2]^+$ is difficult. It occurs considerably more easily in other complexes in which the character of the metal–cyclopentadienyl bond is to a greater degree ionic. This is especially true for $Mncp_2$, although in other compounds such reactions take place quite readily. $Mncp_2$ easily hydrolyzes in water and it reacts with $FeCl_2$ to give ferrocene. Hydrogen chloride causes splitting of the cyclopentadienyl ring in $Pd(C_3H_5)cp$.

$$[Pd(C_3H_5)cp] + HCl \longrightarrow [Pd_2Cl_2(C_3H_5)_2] \tag{9.125}$$

Other examples of $M-cp$ bond rupture are

$$[Pd(C_3H_5)cp] + FeCl_2 \longrightarrow Fecp_2 \tag{9.126}$$

$$[Crcp_2] + FeCl_2 \longrightarrow Fecp_2 \tag{9.127}$$

$$[Zrcp_2Cl_2] + Zr(acac)_4 \longrightarrow [ZrcpCl(acac)_2] \tag{9.128}$$

$$Zrcp_4 + FeCl_2 \longrightarrow Fecp_2 + [Zrcp_2Cl_2] \tag{9.129}$$

$$[Ticp_2(OOCMe)_2] \xrightarrow{H_2O} [Ticp(OOCMe)_2]_2O \tag{9.130}$$

$$[Ticp_2Cl_2] + TiCl_4 \longrightarrow 2[TicpCl_3] \tag{9.131}$$

$$Crcp_2 + CO \xrightarrow[10\,MPa]{373\,K} [Crcp_2]^+[Crcp(CO)_3]^- \tag{9.132}$$

$$Crcp_2 + CO \xrightarrow{420-440\,K} [Cr_2cp_2(CO)_6] \tag{9.133}$$

$$Vcp_2 + CO \xrightarrow[7-25\,MPa]{373\,K} [Vcp(CO)_4] \tag{9.134}$$

$$Nicp_2 + 4L \longrightarrow [NiL_4] \tag{9.135}$$

$$L = PR_3, P(OR)_3, PF_3, CNPh$$

Reactions in which the $+2$ oxidation state is retained may also occur:

$$Nicp_2 + py \longrightarrow [Ni(py)_4]^{2+} \tag{9.136}$$

$$Nicp_2 + dmgH_2 \longrightarrow [Ni(dmgH)_2] \tag{9.137}$$

$$Nicp_2 + KCN \longrightarrow K_2[Ni(CN)_4] \tag{9.138}$$

Nickelocene undergoes reductive substitution with tetracyclene to furnish bis(tetracyclene)nickel(0), $\eta^3$-1,2,3,4-tetraphenyl-5-ketocyclopentenylcyclopentadienyl-nickel, and trans-2,3,4,5-tetraphenyl-2-cyclopenten-1-one[150d]:

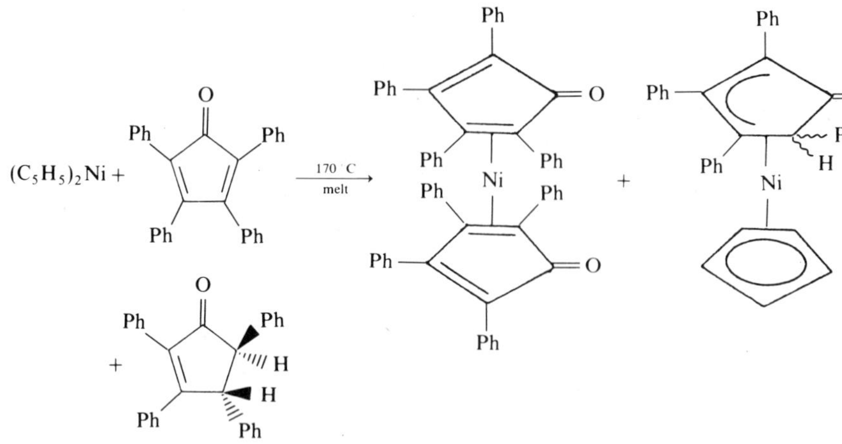

Similarly, reductive substitution of a cyclopentadienyl ring of cobaltocene by diene, quinone, and cyclopentadienone moieties serves as a synthetic procedure for a series of ($\eta^4$-ligand)(cyclopentadienyl)cobalt compounds[150d]:

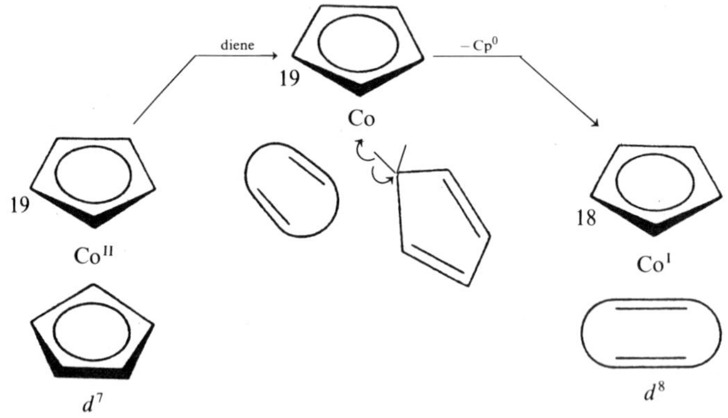

Nickelocene reacts with acetylenes to give dinuclear complexes. On the other hand, the reaction of nickelocene with $PhCH_2MgCl$ [see equation (9.140)] affords the tri-nuclear cluster, which is isoelectronic with the $Co_3(\mu_3\text{-}CR)(CO)_9$ carbonyl:

$$Nicp_2 + PhC\equiv CPh \longrightarrow [CpNi(\mu\text{-}PhC\equiv CPh)\,Nicp] \qquad (9.139)$$

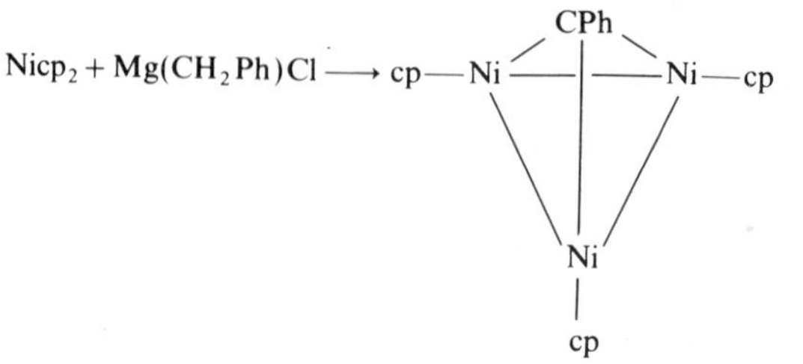

$$Nicp_2 + Mg(CH_2Ph)Cl \longrightarrow \text{cp—Ni} \cdots \text{Ni—cp} \qquad (9.140)$$

## e. *Metallation reactions*[165]

Metallocenes of the iron triad and some other cuclopentadienyl complexes undergo metallation reactions of the ring and therefore show aromatic properties as in the acylation reactions. Both metallation and acylation occur more easily than in the case of benzene.

The most commonly utilized metallation reagents are butyllithium and mercury acetate [equation (9.141)].

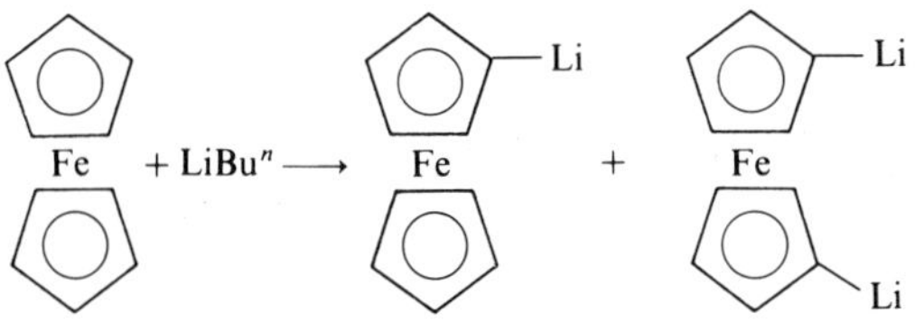

Fe
+ LiBu^n
Fe
Li
+
Fe
Li
Li

(9.141)

Dependent upon the conditions, 1, 1′-dilithioferrocene or lithioferrocene may be obtained in tetramethylethylenediamine. Lithioferrocene and 1, 1′-dilithioferrocene are utilized for the preparation of various ferrocene derivatives [schemes (9.142) and (9.143)]. Lithioferrocenes, mercurioferrocenes, and derivatives of ferrocene containing various functional groups such as amines, carboxyls, phosphines, etc. are utilized for the

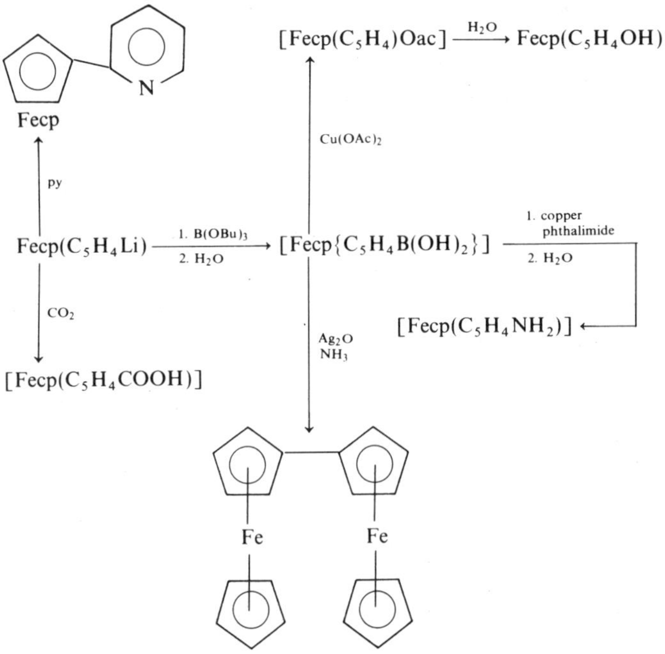

N
Fecp
py
[Fecp(C5H4)Oac]
H2O
Fecp(C5H4OH)
Cu(OAc)2
Fecp(C5H4Li)
1. B(OBu)3
2. H2O
[Fecp{C5H4B(OH)2}]
1. copper phthalimide
2. H2O
CO2
[Fecp(C5H4COOH)]
Ag2O
NH3
[Fecp(C5H4NH2)]
Fe
Fe

(9.142)

preparation of ferrocenyl complexes,[149] for example, $[M(OOCC_5H_4Fecp)_n]$, $[(OC)_3Cr(\eta^6\text{-}C_6H_5\text{-}C_5H_4Fecp)]$, $[cp_2Ti(\sigma\text{-}C_5H_4Fecp)_2]$, $[MX_n(H_2NC_5H_4Fecp)_m]$, and $[NiCl_2(Me_2PC_5H_4Fecp)_2]$.

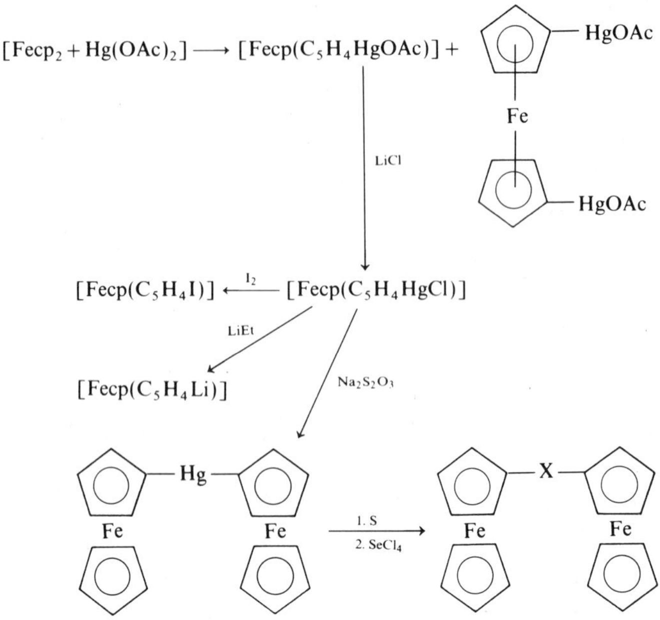

$$[Fecp_2 + Hg(OAc)_2] \longrightarrow [Fecp(C_5H_4HgOAc)] +$$

$$[Fecp(C_5H_4I)] \xleftarrow{\;I_2\;} [Fecp(C_5H_4HgCl)]$$

(9.143)

$$[Fecp(C_5H_4Li)]$$

(for 1, X = S; for 2, X = Se)

Acetoxymercururuthenocene, chloromercururuthenocene, and lithioruthenocene and analogous compounds of osmium may be obtained in the same manner as the ferrocene compounds.

$$Mcp_2 + LiBu^n \longrightarrow [Mcp(C_5H_4Li)] + [M(C_5H_4Li)_2] \qquad (9.144)$$

$$Rucp_2 + Hg(OAc)_2 \longrightarrow [Rucp(C_5H_4HgOAc)] + [Ru(C_5H_4HgOAc)_2] \qquad (9.145)$$

However, if the reaction of ruthenocene is carried out with mercury(II) cyanide followed by the addition of boric acid, then the diamagnetic complex containing the $Ru-Hg-Ru$ group is obtained.

$$Rucp_2 \xrightarrow[\text{2. HCl}]{\text{1. Hg(CN)}_2} [cp_2Ru-Hg-Rucp_2]^{2+}[BF_4]_2^- \qquad (9.146)$$

The cyclopentadienyl complexes $[Recp_2H]$, $[Mncp(CO)_3]$, $[Recp(CO)_3]$, and $[Cocp(C_4Ph_4)]$ also undergo metallation of this ring [scheme (9.147)].

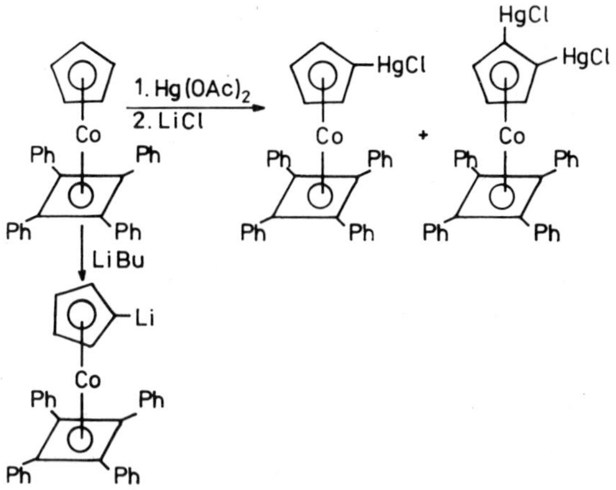

(9.147)

## f. Aromatic substitution reactions

Cyclopentadienyl complexes show aromatic properties. This is the origin of the names metallocenes, ferrocene, etc. For example, the Friedel–Crafts acylation proceeds very easily. Ferrocene is acylated $3.3 \times 10^6$ times faster than benzene. The acylating reagents are the organic acid chlorides and acid anhydrides. The reaction is catalyzed by the typical Friedel–Crafts catalysts $BF_3$, $AlCl_3$, etc.

$$Fecp_2 + RCOCl \longrightarrow [Fe(C_5H_4COR)_2] + [Fecp(C_5H_4COR)] \tag{9.148}$$

$$Rucp_2 + 3RCOCl \xrightarrow{\ 3AlCl_3\ } [Rucp(C_5H_4COR)] + [Ru(C_5H_4COR)_2] \tag{9.149}$$
$$(R = Me, Ph)$$

In the case of ruthenocene, in addition to dibenzoyl heteroannular derivatives (substituents are located at both rings), also small amounts of homoannular dibenzoylruthenocene are obtained in which both substituents are bonded to the same ring. Ferrocene derivatives containing electron-withdrawing substituents give heteroannular compounds in the acylation reaction while the complex $[Fecp(C_5H_4R)]$ where $R = $ alkyl gives mainly compounds in which the acyl is located in the 3 position with respect to the alkyl, although all other derivatives are also formed. Alkylation of ferrocene under the Friedel–Crafts conditions leads to a mixture of polyalkylferrocenes.

Monocyclopentadienyl complexes also undergo acylation. Studies of competitive acylation showed that the rate of acylation increases in the series

$$[Recp(CO)_3] < C_6H_6 < Vcp(CO)_4 < Oscp_2 < Mncp(CO)_3$$

$$< Rucp_2 < PhOMe < Fecp^2 \tag{9.150}$$

Derivatives of metallocenes may be used as acylating agents [scheme (9.151)].

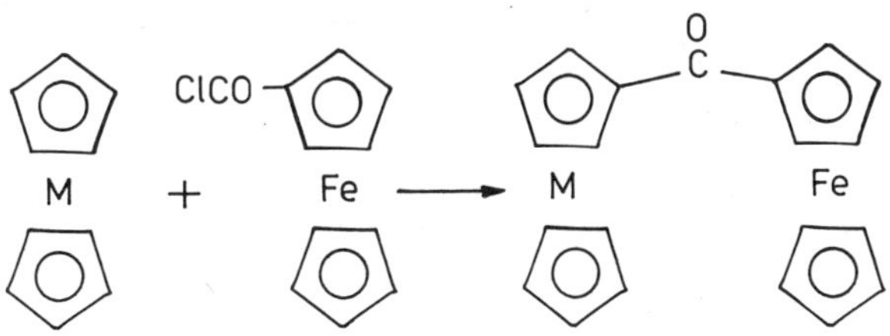

(9.151)

## g. Preparation of polymetallocenes

Biferrocenyl, terferrocenyl, and polyferrocenyl are obtained by the reaction of a mixture of bromo and dibromoferrocene with copper [equation (9.152)].

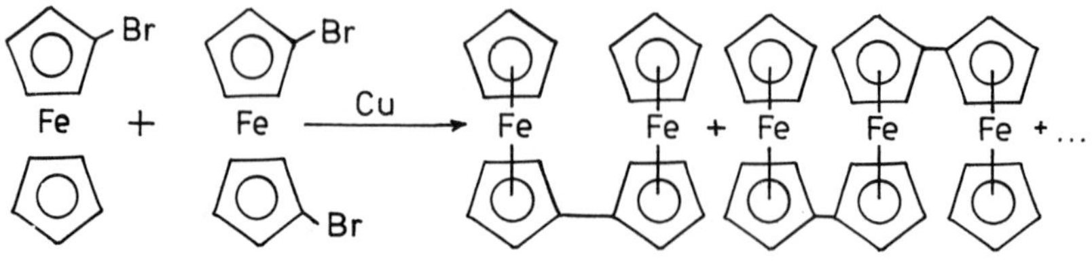

$$\tag{9.152}$$

## h. Ligand substitution reactions

Complexes containing other ligands in addition to cyclopentadienyl groups may also undergo substitution reactions involving the non-cp ligands. In this way, a variety of coordination compounds may be obtained. Most commonly, these displacement reactions are carried out in the case of cyclopentadienyl metal carbonyls:

$$[Mncp(CO)_3] + PPh_3 \longrightarrow [Mncp(CO)_2PPh_3] \tag{9.153}$$

$$[Fecp(CO)_2]_2 + R_2S_2 \longrightarrow [Fecp(CO)(SR)]_2 \tag{9.154}$$

$$[Mncp(CO)_3] + NaCN \xrightarrow[\text{MeOH}]{hv} Na[Mncp(CO)_2(CN)] + CO \tag{9.155}$$

$$[Mncp(CO)_3] + NO^+ \longrightarrow [Mncp(CO)_2(NO)]^+ \tag{9.156}$$

$$[Crcp(CO)_3]_2 + NO \xrightarrow{PhH} [Crcp(CO)_2(NO)] \tag{9.157}$$

## i. Reactions with substituents containing various functional groups

Ferrocene derivatives may be used for the preparation of many compounds containing different substituents by utilizing reactions between functional groups bonded to the substituents of the ferrocene ring and organic or inorganic compounds. Ferrocenylmethanol and 1-ferrocenylethanol react with dimercaptans HSRSH $[R = C_2H_4, CH_2CH_2CH_2, CH_2CHMe, (CH_2CH_2)_2O, (CH_2CH_2)_2S]$ to give corresponding sulfides[43,44]; see equations (9.158)–(9.160).

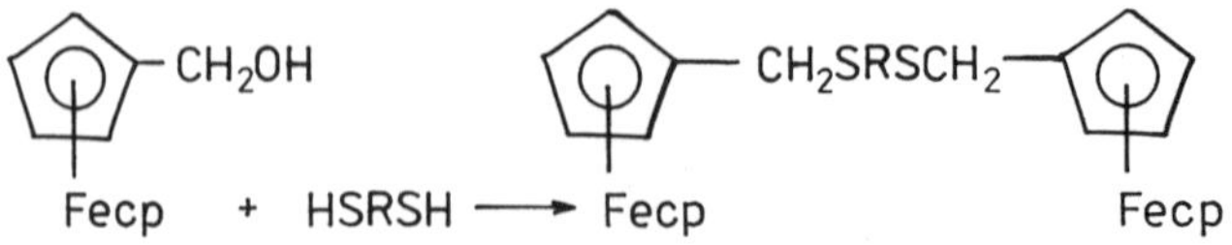

$$\tag{9.158}$$

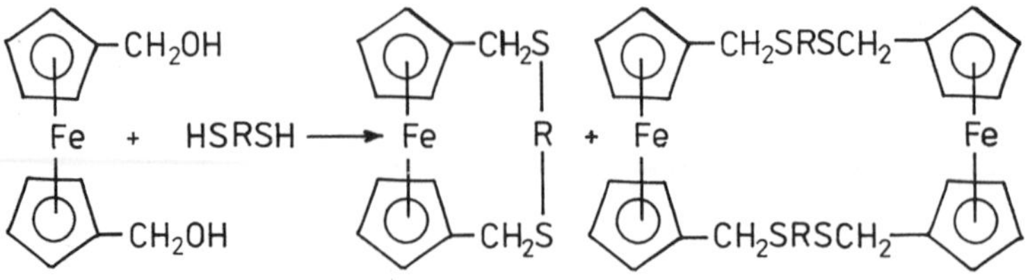

$$\tag{9.159}$$

$$[Fecp(C_5H_4CH_2OH)] + HSCH_2COOH \longrightarrow [Fecp(C_5H_4CH_2SCH_2COOH)] \tag{9.160}$$

The $(\pm)$-*S*-(1-ferrocenylethyl)thioglycolic acid racts with the optically active ephedrine to give the $(-)$-*S*-(1-ferrocenylythyl)thioglycolic acid, which gives optically active 1-ferrocenylethanol upon treatment with $HgCl_2$.[45]

## j. *Activation of C−H bonds*

Electron-deficient cyclopentadienyl complexes are the most effective activators of C−H bonds of alkanes, alkenes, and arenes (see Section 4.13). In such compounds, for example, $Mcp_2$ (M = Ti, Zr, Hf, Nb, Ta), the C−H bond in the cyclopentadienyl ligand is broken very easily [equations (9.74), (9.78), (9.80), (9.81), (9.85), (9.86)].

## k. *Photochemistry of cyclopentadienyl complexes*[154]

At one time it was thought that ferrocene does not undergo photochemical reactions. However, later studies showed that ferrocene and its derivatives are susceptible to light and give products of photodecomposition. Under the influence of light, in the first step, there is electron transfer from the orbital, which is mainly localized on the metal, to the orbital, which is mainly located on the ligand, followed by electron removal. Dependent upon the conditions of the reaction, solvent, and other reagents present in solution, the outcome may be decomposition of ferrocene, formation of its derivatives, or oxidation to $Fecp_2^+$ [equations (9.161) and (9.162)]. The

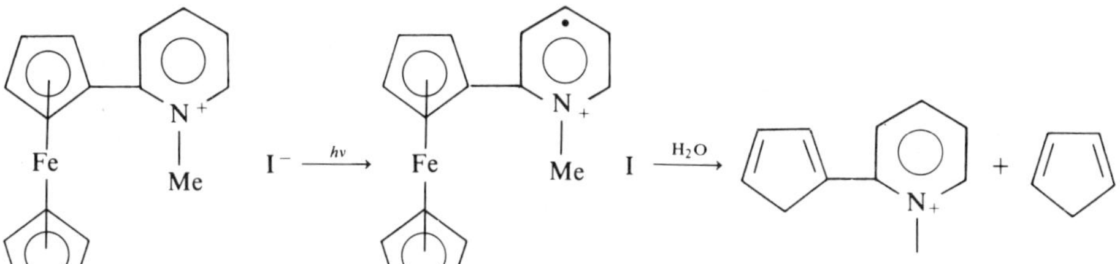

$$\text{(9.161)}$$

$$Fecp_2 \xrightarrow[CCl_4]{hv} [Fecp_2]Cl + {}^*CCl_3 \longrightarrow$$

$$Fecp(C_5H_4CCl_3) + H^+ \left\langle \begin{array}{l} \xrightarrow{EtOH} Fecp(C_5H_4COOEt) \\ \\ \longrightarrow FeCl_3 \end{array} \right. \qquad \text{(9.162)}$$

photolysis of ruthenocene proceeds similarly to give $Rucp(C_5H_4R)$ and $[Rucp_2]^+$. Cobaltocene is oxidized to $[Cocp_2]^+$, while nickelocene, under the influence of light, first gives $[Nicp_2]^+$ which subsequently decomposes to give a polymer and $Ni^{2+}$.

## 18. *APPLICATION OF CYCLOPENTADIENYL COMPLEXES*

Cyclopentadienyl compounds belong to the most commonly utilized organometallic derivatives in various areas of chemistry and technology. They are commonly utilized in organic synthesis.[150a] Ferrocene derivatives containing asymmetric substituents are utilized as ligands for preparation of asymmetric hydrogenation catalysts (Chapter 13). Cyclopentadienyl complexes are also used as catalysts for polymerization and oligomerization of olefins and dienes (Chapter 13). They also serve in the preparation of metals and deposition of thin metallic layers in electronics (Table 2.31).

Tricarbonylcyclopentadienylmanganese and tricarbonylmethylcyclopentadienylmanganese are utilized as antiknock agents in gasoline.[46,163] Other derivatives may also be utilized for the same purpose: $[Mn(C_5H_4R)(CO)_3]$ [R = Ac, Me−CH−OH, Me−CH−(OMe), Me−CH(OEt), Et, $CH_2Cl$, $CH_2OR$, PhCO, Bu′], $[Mn(C_5H_{5-x}R_x)(CO)_3]$ ($x = 1$–5; R = Et). Methyl derivative $Mn(mcp)(CO)_3$ may also be used as an additive to rocket and diesel fuel. These compounds improve the burning process and decrease the amount of soot during the burning of dispersed fuel, aromatic oils, and coal. In addition, they decrease the amount of $SO_3$ in fumes by 45%. The complexes $[Mncp(CO)_2L]$ [L = $PR_3$, $P(OR)_3$, $AsR_3$, and $SbR_3$] may be utilized as herbicides.

The compounds $[Mocp(CO)_3]$ and $[Fecp(CO)_2]_2$ catalyze the reaction of addition of carbon tetrachloride to olefins[47]:

$$CCl_4 + RCH=CH_2 \longrightarrow RCHCl-CH_2-CCl_3 \tag{9.163}$$

Polymerization reaction of isocyanides RNC is catalyzed by $[Fecp(CH_2Ph)(CO)]$. It proceeds as a result of multistep migration to coordinated ligands and is evidenced by the isolation and identification of the tris(imino) complex (9.164).[48]

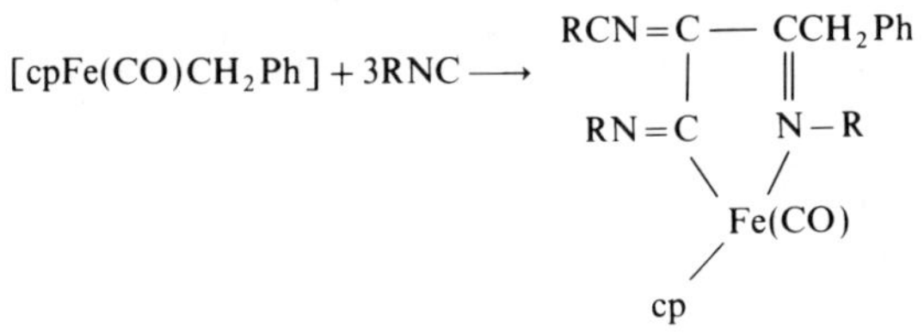

Cyclopentadienylbis(triphenylphosphine)cobalt catalyzes the oligomerization of acetylenes and cooligomerization of acetylenes and nitriles or acetylenes and isothiocyanates, $RNCS$[49] [scheme (9.165)].

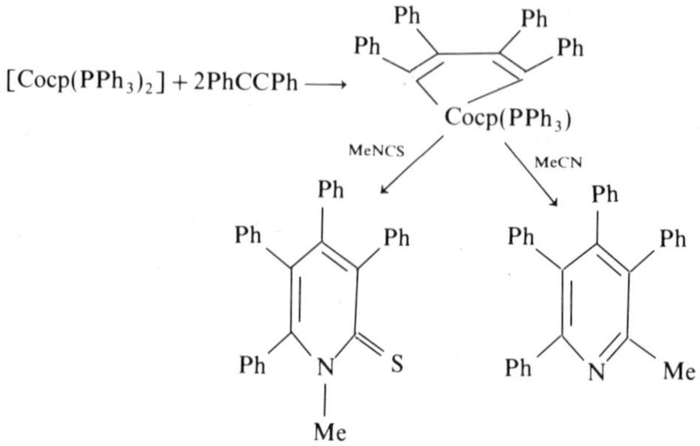

In the presence of oxygen, cobaltocene oxidizes α-diketones and *o*-quinones[50]:

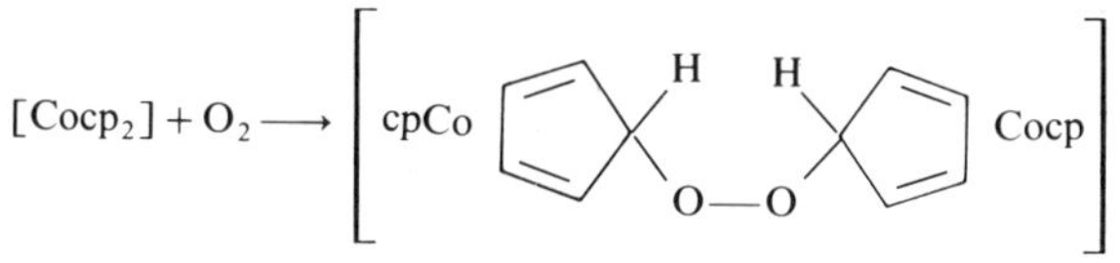

$$[Cocp_2] + O_2 \longrightarrow \left[ cpCo \underset{O-O}{\overset{H \quad H}{\diagdown \diagup}} Cocp \right]$$

$$\xrightarrow{RCOCOR} 2[Cocp_2]^+ RCOO^- \xrightarrow{HCl} [Cocp_2]^+ Cl^- + RCOOH$$

(9.166)

$$(R = Me, Ph)$$

The compound $[Mocp(CO)_3]_2$ catalyzes oxidation of substituted olefins to epoxides and allyl alcohols, while in the presence of $[Vcp(CO)_4]$ epoxy alcohols are formed[50] [equation (9.167)].

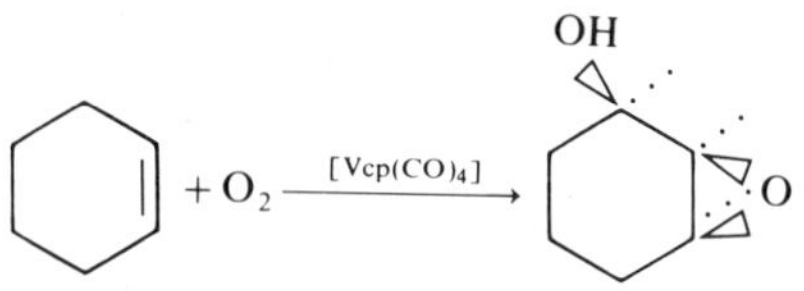

$$\bigcirc \; | \; + O_2 \xrightarrow{[Vcp(CO)_4]} \bigcirc O$$

(9.167)

Dichlorobis(pentamethylcyclopentadienyl)zirconium(IV) during reduction with sodium amalgam in nitrogen atmosphere forms a dinuclear complex containing a dinitrogen bridge.[51,151,152] This compound reacts with hydrochloric acid to give dinitrogen and hydrazinehydrochloride [equation (9.168)]. Dicyclopentadienyl-

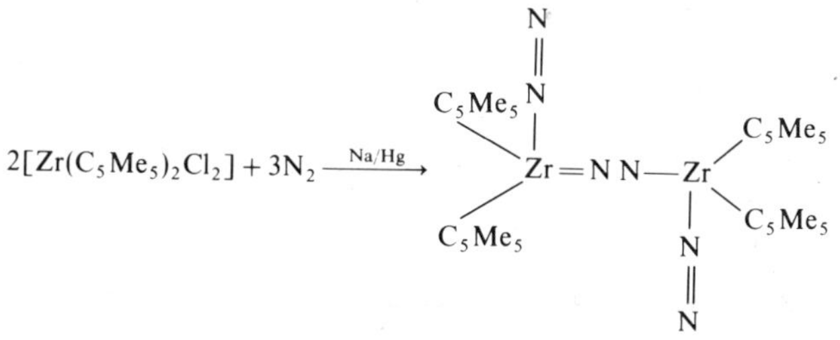

$$2[Zr(C_5Me_5)_2Cl_2] + 3N_2 \xrightarrow{Na/Hg} \begin{array}{c} N \\ \| \\ C_5Me_5 \quad N \\ \diagdown \quad | \qquad \qquad C_5Me_5 \\ Zr = N \; N - Zr \\ \diagup \qquad \qquad | \quad \diagdown C_5Me_5 \\ C_5Me_5 \qquad \quad N \\ \| \\ N \end{array}$$

(9.168)

$$\xrightarrow{HCl} [Zr(C_5Me_5)_2Cl_2] + 2N_2 + N_2H_4 \cdot 2HCl$$

dichlorotitanium(IV) after reduction with Grignard compounds or other reducing agents (Li, LiR, Na, K, Mg, Ca, Ce, La, naphthalene–sodium naphthalene–lithium, etc.) forms compounds that bond dinitrogen in a particularly active manner.[52,151,152] In addition to ammonia, the reaction also furnishes certain amounts of amines. At 173 K, the dimeric compound $[Ticp_2R]_2N_2$ is formed and undergoes decomposition at 203 K (see Section 9.12a). The reaction of $[Ti_2(\mu\text{-}\eta^1 : \eta^5\text{-}C_5H_4)cp_3]$ with nitrogen in the mixture of dimethoxyethane and diglyme gives the nitrogen titanium complex of the formula $[cp_2Ti(\mu\text{-}\eta^1 : \eta^5\text{-}C_5H_4) \; Ticp(\mu_3\text{-}N_2) \; Ti_2(\mu\text{-}\eta^5 : \eta^5\text{-}C_5H_4\text{-}C_5H_4)cp_2]$ · $[Ticp_2\{O(C_2H_4OMe)_2\}]$ · $O(C_2H_4OMe)_2$, which is composed of two units, one of which contains nitrogen[151,164] [scheme (9.169)].

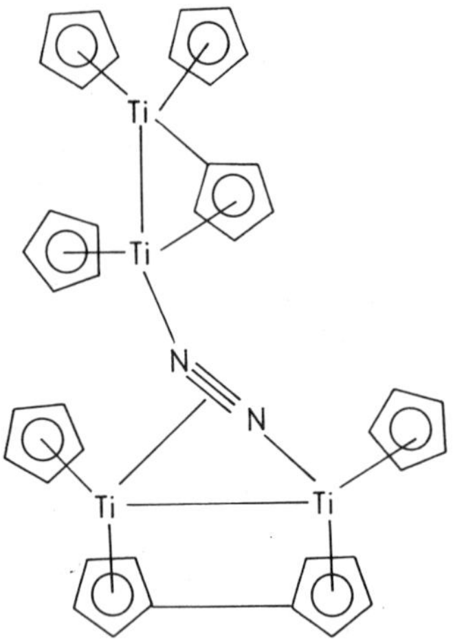

(9.169)

Optically active cyclopentadienyl iron complexes [Fecp(RO)(CO)L] were utilized for investigating the reaction mechanism of the alkyl ligand with the carbonyl group. It was found that migration of the alkyl group and the CO group both take place [see Section 4.12.b, equations (4.173) and (4.174)].

Polymers containing ferrocene and other cyclopentadienyl compounds were investigated intensively for the last twenty years because of their valuable properties. Polymers that contain ferrocene are highly thermally stable. They have low toxicity, and exhibit high absorption of ultraviolet and gamma radiation. Polymers that contain partially oxidized iron (ferrocene–ferrocenium) are semiconductors. The following monomers are most commonly utilized for polymerization[53]: vinylferrocene [$Fe(C_5H_4CH=CH_2)cp$], divinylferrocene [$Fe(C_5H_4CH=CH_2)_2$], ferrocenenylacetylene [$Fecp(C_5H_4C \equiv CH)$], 2-ferrocenylbutadiene [$Fecp(C_5H_4\text{-}\overset{\displaystyle \|}{\underset{\displaystyle CH_2}{C}}\text{-}CH=CH_2)$],

[$MnC_5H_4CH=CH_2)(CO)_3$], [$Cr(C_5H_4CH=CH_2)(CO)_2(NO)$], etc.[53,153]

Divinylferrocene affords a "net-like" polymer [equation (9.170)].

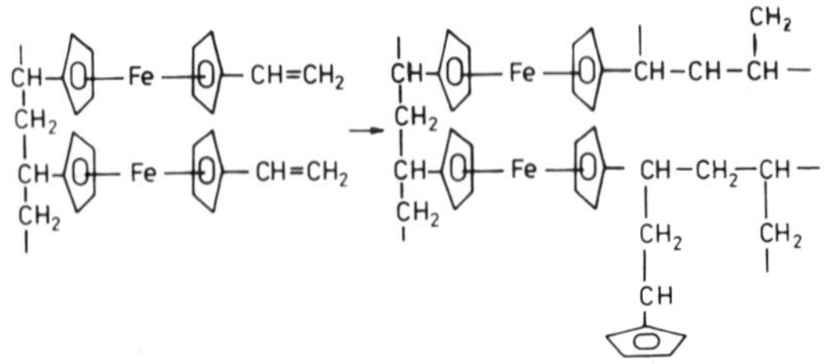

(9.170)

Ferrocenylacetylene gives unsaturated polymer which reacts with iodine [equation (9.171)].

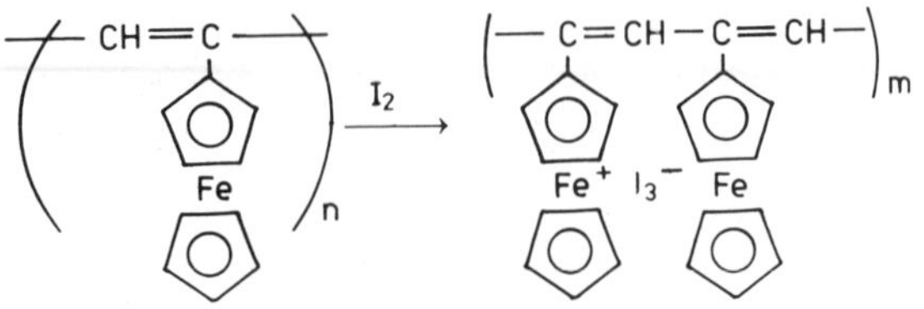

(9.171)

Polymers of cyclopentadienyl metal complexes find various applications; poly ($\eta^5$-vinylcyclopentadienyl)tricarbonylmanganese is utilized as a fungicide. Polymers arising from ferrocene are used as catalysts for dehydrogenation and dehydration.[53,153]

Complexes of Cr, Mo, W, Co, Fe, Ti, and Zr possessing the pentaphenylcyclopentadienyl ligand have been synthesized directly from the ligand itself; these compounds may serve as models for highly phenylated transition metal containing polymers; see, for example, equation (9.172).[166]

$$C_5Ph_5H \xrightarrow{\text{\textit{n}-BuLi}} C_5Ph_5Li \xrightarrow{Mo(CO)_6} [(C_5Ph_5)Mo(CO)_3]Li \xrightarrow{CH_3I}$$

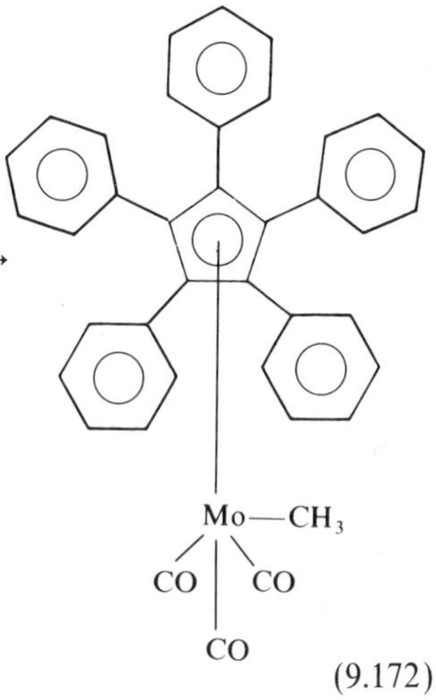

$$\text{(9.172)}$$

Ferrocene is utilized as an ingredient in light-sensitive materials.[154] Ferrocene, cobaltocene, and nickelocene serve as photoconductors in electrophotography. Many other ferrocene derivatives have photostabilizing polymer properties.[154] Ferrocene, ($Fecp_2/Fecp_2^+$), $Fe(pmcp)_2$, (1-hydroxyethyl)ferrocene, and $Cocp_2(Cocp_2/Cocp_2^+)$ are utilized in semiconducting photoelectrochemical couples.[155]

The complexes $Mcp_2X_2$ (M = Ti, V, Nb, Mo, X = F, Cl, Br, I, NCS, $N_3$) have strong anticancer properties.[156,157] Such properties are also shown by the salts $[Fecp_2]^+X^-$ [$X^- = Cl_3CCOO^-$, 2, 4, 6-$(O_2N)_3C_6H_2O^-$, $FeCl_4^-$].[158]

## REFERENCES

1. E. G. Perevalova and T. W. Nikitina, on: *Metody Elementoorganicheskoy Khimii* (A. N. Nesmeyanov and K. A. Kocheshkov, eds.), p. 687, Izd. Nauka, Moscow (1975).
2. T. W. Nikitina, in: *Metody Elementoorganicheskoy Khimii* (A. N. Nesmeyanov and K. A. Kocheshkov, eds.), p. 585, Izd. Nauka, Moscow (1975).
3. a. M. L. H. Green, *Organometallic Compounds, The Transition Metals*, Methuen, London (1969). b. A. Haaland, *Acc. Chem. Res.*, **12**, 415 (1979).
4. M. Rosenblum, *Chemistry of the Iron Group Metallocenes*, Wiley, New York (1965).
5. E. P. Fischer and H. Werner, *Metal Pi-Complexes*, Elsevier, Amsterdam (1966).
6. H. Werner, *Angew. Chem., Int. Ed. Engl.*, **16**, 1 (1977); R. N. Grimes, *Coord. Chem. Rev.*, **28**, 47 (1979).
7. a. D. M. P. Mingos, *Adv. Organomet. Chem.*, **15**, 1 (1977). b. L. H. Guggenberger and F. N. Tebbe, *J. Am. Chem. Soc.*, **95**, 7870 (1973).

8. M. M. Coutiere, J. Demuynck, and A. Veillard, *Theor. Chim. Acta*, **27**, 281 (1972).
9. Y. S. Sohn, D. N. Hendrickson, and H. B. Gray, *J. Am. Chem. Soc.*, **93**, 3603 (1971).
10. N. Roesch and K. H. Johnson, *Chem. Phys. Lett.*, **24**, 179 (1974).
11. M. Elian, M. M. L. Chen, D. M. P. Mingos, and R. Hoffmann, *Inorg. Chem.*, **15**, 1148 (1976).
12. J. W. Lauher and R. Hoffmann, *J. Am. Chem. Soc.*, **98**, 1729 (1976).
13. J. C. Green, M. L. H. Green, and C. K. Prout, *Chem. Commun.*, 421 (1972).
14. M. L. Maddox, S. L. Stafford, and H. D. Kaesz, *Adv. Organomet. Chem.*, **3**, 1 (1965).
15. R. G. Kidd, in: *Characterization of Organometallic Compounds, Part II* (M. Tsutsui, ed.), Wiley, New York (1971).
16. B. L. Shaw and N. I. Tucker, in: *Comprehensive Inorganic Chemistry I* (I. C. Bailar, Jr., ed.), Vol. 4, p. 781, Pergamon Press, Oxford (1973).
17. M. H. Chisholm and S. Godleski, *Prog. Inorg. Chem.*, **20**, 299 (1976).
18. B. E. Mann, *Adv. Organomet. Chem.*, **12**, 135 (1974).
19. A. N. Nesmeyanov, P. W. Petrovskii, L. A. Federov, W. I. Robas, and E. I. Fedin, *Zh. Strukt. Khim.*, **14**, 49 (1973).
20. A. N. Nesmeyanov, O. W. Nogina, E. I. Fedin, W. A. Dubovitskii, B. A. Kvasov, and P. W. Petrovskii, *Dokl. Akad. Nauk SSSR*, **205**, 857 (1972).
21. A. N. Nesmeyanov, E. I. Fedin, L. A. Federov, and P. W. Petrovskii, *Zh. Strukt. Khim.*, **13**, 964 (1972).
22. P. C. Lauterbur and R. B. King, *J. Am. Chem. Soc.*, **87**, 3266 (1965).
23. H. P. Fritz, *Adv. Organomet. Chem.*, **1**, 240 (1964).
24. K. Nakamoto, in: *Characterization of Organometallic Compounds* (M. Tsutsui, ed.), Part I, p. 73, Interscience, New York (1969); E. Maslowsky, Jr., *Vibrational Spectra of Organometallic Compounds*, Wiley, New York (1977).
25. L. M. Mulay and J. R. Dehn, in: *Characterization of Organometallic Compounds* (M. Tsutsui, ed.), Part II, p. 439, Wiley, New York (1971).
26. K. D. Warren, *Structure and Bonding*, **27**, 45 (1976).
27. D. W. Clark and K. D. Warren, *Structure and Bonding*, **39**, 1 (1980).
28. J. J. Ammeter, *J. Magn. Reson.*, **30**, 299 (1978).
29. A. H. Cowley, *Prog. Inorg. Chem.*, **26**, 45 (1979).
30. C. Furlani and C. Cauletti, *Structure and Bonding*, **35**, 119 (1978).
31. a. W. I. Telnoy and I. B. Rabinovich, *Usp. Khim.*, **46**, 1337 (1977). b. J. A. Connor, *Top. Curr. Chem.*, **71**, 71 (1977).
32. M. R. Litzow and T. R. Spalding, *Mass Spectrometry of Inorganic and Organometallic Compounds*, Elsevier, Amsterdam (1973).
33. R. W. Kiser, in: *Characterization of Organometallic Compounds* (M. Tsutsui, ed.), p. 137, Interscience, New York (1969).
34. K. J. Klabunde, *Acc. Chem. Res.*, **8**, 393 (1975).
35. O. W. Nogina, in: *Metody Elementoorganicheskoy Khimii* (A. N. Nesmeyanov and K. A. Kocheshkov, eds.), p. 145, Izd. Nauka, Moscow (1974).
36. L. J. Guggenberger and F. N. Tebbe, *J. Am. Chem. Soc.*, **93**, 5924 (1971).
37. F. A. Cotton and T. J. Marks, *J. Am. Chem. Soc.*, **92**, 5114 (1970); F. A. Cotton and J. Takats, *J. Am. Chem. Soc.*, **92**, 2353 (1970).
38. R. Huettel, U. Raffay, and H. Reinheimer, *Angew. Chem.*, **79**, 859 (1967).
39. F. A. Cotton and T. J. Marks, *J. Am. Chem. Soc.*, **91**, 7281 (1969).
40. T. W. Nikitina, in: *Metody Elementoorganicheskoy Khimii* (A. N. Nesmeyanov and K. A. Kocheshkov, eds.), pp. 138, 143, Izd. Nauka, Moscow (1974).
41. N. S. Vyazankin, R. N. Shchelokov, and O. A. Kruglaya, in: *Metody Elementoorganicheskoy Khimii* (A. N. Nesmeyanov and K. A. Kocheshkov, eds.), p. 905, Izd. Nauka, Moscow (1974).
42. R. N. Shchelokov, G. T. Bolotova, and N. S. Vyazankin, in: *Metody Elementoorganicheskoy Khimii* (A. N. Nesmeyanov and K. A. Kocheshkov, eds.), p. 915, Izd. Nauka, Moscow (1974).
43. A. Ratajczak and B. Czech, *Polish J. Chem.*, **52**, 837 (1978); A. Ratajczak and B.Czech, *Bull. Acad. Polon. Sci., Ser. Chim.*, **25**, 635 (1977); A. Ratajczak, B. Misterkiewicz, and B. Czech, *Bull. Acad. Polon. Sci., Ser. Chim.*, **25**, 27 (1977).

44. A. Ratajczak and B. Czech, *Polish J. Chem.*, **54**, 57 (1980).
45. A. Ratajczak, B. Misterkiewicz, and B. Czech, Pol. Patent 89736; *Chem. Abstr.*, **89**, 109979 (1978).
46. A. A. Joganson, K. H. Anisimov, N. E. Kolobova, and Z. P. Valuyeva, in: *Metody Elementoorganicheskoy Khimii* (A. N. Nesmeyanov and K. A. Kocheshkov, eds.), p. 839, Izd. Nauka, Moscow (1974).
47. J. Tsuji, *Organic Synthesis by Means of Transition Metal Complexes*, Springer-Verlag, Berlin (1975).
48. Y. Yamamoto and H. Yamazaki, *Inorg. Chem.*, **11**, 211 (1972).
49. Y. Wakatsuki and H. Yamazaki, *Chem. Commun.*, 280 (1973); Y. Wakatsuki and H. Yamazaki, *Tetrahedron Lett.*, 3383 (1973).
50. J. E. Lyons, in: *Aspects of Homogeneous Catalysis* (R. Ugo, ed.), Vol. 3, p. 1, Reidel, Dordrecht (1977).
51. J. M. Manriquez and J. E. Bercaw, *J. Am. Chem. Soc.*, **96**, 6229 (1974); J. E. Bercaw, *Fundamental Research in Homogeneous Catalysis* (M. Tsutsui and R. Ugo, eds.), p. 129, Plenum, New York (1977).
52. M. E. Volpin and V. B. Shur, *Organomet. React.*, **1**, 55 (1970).
53. C. E. Carraher, Jr., J. E. Sheats, and C. U. Pittman, Jr. (eds.), *Organometallic Polymers*, Academic Press, New York (1978).
54. P. J. Vergamini, R. R. Ryan, and G. J. Kubas, *J. Am. Chem. Soc.*, **98**, 1980 (1976).
55. S. G. Davies, M. L. H. Green, K. Prout, A. Coda, and V. Tazzoli, *Chem. Commun.*, 135 (1977).
56. G. Huttner and H. Lorenz, *Chem. Ber.*, **108**, 973 (1975).
57. L. M. Dyagileva, V. P. Marin, E. I. Tsyganova, and G. A. Razuvaev, *J. Organomet. Chem.*, **175**, 63 (1979); A. M. Mulokozi, *J. Organomet. Chem.*, **187**, 107 (1980); J. R. Chipperfield, J. C. R. Sneyd, and D. E. Webster, *J. Organomet. Chem.*, **178**, 177 (1979).
58. E. Koenig, V. P. Desai, B. Kannellakopulos, and E. Dornberger, *J. Organomet. Chem.*, **187**, 61 (1980).
59. R. D. Ernst, *Structure and Bonding*, **57**, 1 (1984).
60. R. D. Ernst, *Acc. Chem. Res.*, **18**, 56 (1985).
61. D. M. P. Mingos, in: *Comprehensive Organometallic Chemistry* (G. Wilkinson, F. G. A. Stone, and E. W. Abel, eds.), Pergamon Press, Oxford (1982).
62. J. C. Green, *Structure and Bonding*, **43**, 37 (1981).
63. J. K. Burdett and E. Canadell, *Organometallics*, **4**, 805 (1985).
64. W. Siebert, *Angew. Chem., Int. Ed. Engl.*, **24**, 943 (1985); W. Siebert, *Adv. Organomet. Chem.*, **18**, 301 (1980).
65. O. J. Scherer, *Adv. Organomet. Chem.*, **24**, 924 (1985).
66. O. J. Scherer, J. Schwalb, G. Wolmerahauser, W. Keim, and R. Gross, *Angew. Chem., Int. Ed. Engl.*, **25**, 363 (1986).
67. G. Wilkinson, *J. Am. Chem. Soc.*, **76**, 209 (1954).
68. J. Lauher, M. Elian, R. Summerville, and R. Hoffman, *J. Am. Chem. Soc.*, **98**, 3219 (1976).
69. P. T. Chesky and M. B. Hall, *J. Am. Chem. Soc.*, **106**, 5186 (1984).
70. A. W. Duff, K. Jonas, R. Goddard, H.-J. Kraus, and C. Krueger, *J. Am. Chem. Soc.*, **105**, 5479 (1983).
71. S. P. Gubin and G. B. Shulpin, *Khimia Kompleksov so Svyazyami Metall-Uglerod*, Izd. Nauka, Sibirskoye Otdelenie, Novosibirsk (1984).
72. T. W. Nikitina, in: *Metody Elementoorganicheskoy Khimii, Metalloorganicheskiye Soyedineniya Zheleza* (A. N. Niesmiejanow and K. A. Kocheshkov, eds.), Izd. Nauka, Moscow (1985).
73. M. C. Boehm, M. Ecker-Maksic, R. D. Ernst, D. R. Wilson, and R. Gleiter, *J. Am. Chem. Soc.*, **104**, 2699 (1982).
74. C. F. Campana, R. D. Ernst, D. R. Wilson, and J.-Z. Liu, *Inorg. Chem.*, **23**, 2732 (1984).
75. D. R. Wilson, R. D. Ernst, and T. H. Cymbaluk, *Organometallics*, **2**, 1220 (1983).
76. L. Stahl and R. D. Ernst, *Organometallics*, **2**, 1229 (1983).
77. M. C. Boehm, R. D. Ernst, R. Gleiter, and D. R. Wilson, *Inorg. Chem.*, **22**, 3815 (1983).
78. B. E. Mann, in: *Comprehensive Organometallic Chemistry* (G. Wilkinson, F. G. A. Stone and E. W. Abel, eds.), Pergamon Press, Oxford (1982).
79. B. E. Bursten, M. Casarin, S. DiBella, A. Fang, and I. Fragala, *Inorg. Chem.*, **24**, 2169 (1985).

80. K. E. Lewis and G. P. Smith, *J. Am. Chem. Soc.*, **106**, 4650 (1984).
81. A. P. Tomilow, I. N. Chernykh, and J. M. Kargin, *Elektrokhimia Elementoorganicheskikh Soyedinenii*, Izd. Nauka, Moscow (1985).
82. N. G. Connelly and W. E. Geiger, *Adv. Organomet. Chem.*, **23**, 1 (1984).
83. G. P. Pez and J. N. Armor, *Adv. Organomet. Chem.*, **19**, 1 (1981).
84. M. Bottrill, P. D. Gavens, J. W. Kelland, and J. McMeeking, in: *Comprehensive Organometallic Chemistry* (G. Wilkinson, F. G. A. Stone, and E. W. Abel, eds.), Sections 22.1–22.3, Pergamon Press, Oxford (1982).
85. D. J. Cardin, M. F. Lappert, C. L. Raston, and P. I. Riley, in: *Comprehensive Organometallic Chemistry* (G. Wilkinson, F. G. A. Stone, and E. W. Abel, eds.), Sections 23.1–23.2, Pergamon Press, Oxford (1982).
86. E. Samuel and J. F. Harrod, *J. Am. Chem. Soc.*, **106**, 1859 (1984).
87. T. Cuenca and P. Royo, *J. Organomet. Chem.*, **293**, 61 (1985).
88. T. V. Ashworth, T. C. Agreda, E. Herdtweek, and W. A. Herrmann, *Angew. Chem., Int. Ed. Engl.*, **25**, 289 (1986).
89. S. Gambarotta, S. Stella, C. Floriani, A. Chiesi-Villa, and C. Guastini, *Angew. Chem., Int. Ed. Engl.*, **25**, 254 (1986).
90. D. A. Lemenovskii and W. P. Fedin, *Usp. Khim.*, **55**, 303 (1986).
91. D. M. Roddick, M. D. Fryzuk, P. F. Seidler, G. L. Hillhouse, and J. E. Bercaw, *Organometallics*, **4**, 97 (1985).
92. F. Bottomley, G. O. Egharevba, I. J. B. Lin, and P. S. White, *Organometallics*, **4**, 550 (1985).
93. G. A. Razuwajew, G. A. Abakumov, and W. K. Czerkasow, *Usp. Khim.*, **54**, 1235 (1985).
94. C. J. Curtis, J. C. Smart, and J. L. Robbins, *Organometallics*, **4**, 1283 (1985).
95. M. D. Curtis and J. Real, *Organometallics*, **4**, 940 (1985).
96. H. vanderHeijden, A. W. Gal, P. Pasman, and A. G. Orpen, *Organometallics*, **4**, 1847 (1985).
97. N. G. Connelly, in: *Comprehensive Organometallic Chemistry* (G. Wilkinson, F. G. A. Stone, and E. W. Abel, eds.), Vol. 3, Chapter 24, Pergamon Press, Oxford (1982).
98. J. A. Labinger, in: *Comprehensive Organometallic Chemistry* (G. Wilkinson, F. G. A. Stone, and E. W. Abel, eds.), Vol. 3, Chapter 25, Pergamon Press, Oxford (1982).
99. R. Davis and L. A. P. Kane-Maguire, *Comprehensive Organometallic Chemistry* (G. Wilkinson, F. G. A. Stone, and E. W. Abel, eds.), Vol. 3, Sections 26.2, 27.2, and 28.2, Pergamon Press, Oxford (1982).
100. R. B. Hitam, K. A. Mahmoud, and A. J. Rest, *Coord. Chem. Rev.*, **55**, 1 (1984).
101. P. M. Treichel, in: *Comprehensive Organometallic Chemistry* (G. Wilkinson, F. G. A. Stone, and E. W. Abel, eds.), Vol. 4, Chapter 29, Pergamon Press, Oxford (1982).
102. N. M. Boag and H. D. Kaesz, in: *Comprehensive Organometallic Chemistry* (G. Wilkinson, F. G. A. Stone, and E. W. Abel, eds.), Vol. 4, Chapter 30, Pergamon Press, Oxford (1982).
103. I.-P. Lorenz, J. Messelhauser, W. Hiller, and K. Haung, *Angew. Chem., Int. Ed. Engl.*, **24**, 228 (1985).
105. P. Pasman and J. J. M. Snel, *J. Organomet. Chem.*, **276**, 387 (1984).
106. W. A. Herrmann, R. Serrano, U. Kuesthardt, M. L. Ziegler, E. Guggolz, and T. Zahn, *Angew. Chem., Int. Ed. Engl.*, **23**, 515 (1984).
107. C. G. Howard, G. S. Girolami, G. Wilkinson, M. Thornton-Pett, and M. B. Hursthouse, *J. Am. Chem. Soc.*, **106**, 2033 (1984).
108. V. K. Belsky, A. N. Protsky, I. V. Molodnitskaya, B. M. Bulychev, and G. L. Soloveichik, *J. Organomet. Chem.*, **293**, 69 (1985).
109. J. Heck, W. Massa, and P. Weing, *Angew. Chem., Int. Ed. Engl.*, **23**, 722 (1984).
110. W. A. Herrmann, R. Serrano, M. L. Ziegler, H. Pfisterer, and B. Nuber, *Angew. Chem., Int. Ed. Engl.*, **24**, 50 (1985).
111. W. A. Herrmann, R. Serrano, and H. Bock, *Angew. Chem., Int. Ed. Engl.*, **23**, 383 (1984); A. H. Klahn-Oliva and D. Sutton, *Organometallics*, **3**, 1313 (1984).
112. E. G. Perevalova, M. D. Reshotova, and K. I. Grandberg, in: *Metody Elementoorganicheskoy Khimii. Zhelezoorganicheskiye Soyedineniya. Ferrocen* (A. N. Nesmeyanov and K. A. Kocheshkov, eds.), Izd. Nauka, Moscow (1983).
113. T. W. Nikitina, *Metody Elementoorganicheskoy Khimii. Metalloorganicheskiye Soyedineniya Zeleza* (A. N. Nesmeyanov and K. A. Kocheshkov, eds.), Izd. Nauka, Moscow (1985).

114. A. A. Koridze, in: *Metody Elementoorganicheskoy Khimii. Kobalt, Nikel. Platinovye Metally* (A. N. Nesmeyanov and K. A. Kocheshkov, eds.), Chapters 3 and 4, Izd. Nauka, Moscow (1978).
115. A. J. Deeming, in: *Comprehensive Organometallic Chemistry* (G. Wilkinson, F. G. A. Stone, and E. W. Abel, eds.), Vol. 4, Section 31.3, Pergamon Press, Oxford (1982).
116. W. P. Fehlhammer and H. Stolzenberg, in: *Comprehensive Organometallic Chemistry* (G. Wilkinson, F. G. A. Stone, and E. W. Abel, eds.), Vol. 4, Section 31.4, Pergamon Press, Oxford (1982).
117. M. A. Bennett, M. I. Bruce, and T. W. Matheson, in: *Comprehensive Organometallic Chemistry* (G. Wilkinson, F. G. A. Stone, and E. W. Abel, eds.), Vol. 4, Sections 32.3 and 32.4, Pergamon Press, Oxford (1982).
118. M. I. Bruce, in: *Comprehensive Organometallic Chemistry* (G. Wilkinson, F. G. A. Stone, and E. W. Abel, eds.), Vol. 4, Section 32.5, Pergamon Press, Oxford (1982).
119. U. T. Mueller-Westerhoff and A. Nazzal, *J. Am. Chem. Soc.*, **106**, 5381 (1984); A. Waleh, G. H. Loew, and U. T. Mueller-Westerhoff, *Inorg. Chem.*, **23**, 2859 (1984).
120. U. T. Mueller-Westerhoff, A. Nazzal, and W. Proessdorf, *J. Am. Chem. Soc.*, **103**, 7678 (1981).
121. W. S. Khandkarova, in: *Metody Elementoorganicheskoy Khimii. Kobalt. Nikel. Platinowye Metally* (A. N. Nesmeyanov and K. A. Kocheshkov, eds.), Chapters 5 and 6, Izd. Nauka, Moscow (1978).
122. T. M. Gilbert and R. G. Bergman, *J. Am. Chem. Soc.*, **107**, 3502 (1985).
123. T. M. Gilbert, F. J. Hollander, and R. G. Bergman, *J. Am. Chem. Soc.*, **107**, 3508 (1985).
124. W. P. Weiner, F. J. Hollander, and R. G. Bergman, *J. Am. Chem. Soc.*, **106**, 7462 (1984).
125. C. J. Beseker, V. W. Day, W. G. Klemperer, and M. R. Thompson, *J. Am. Chem. Soc.*, **106**, 4125 (1984).
126. M.-J. Fernandez, P. M. Bailey, P. O. Bentz, J. S. Ricci, T. F. Koetzle, and P. M. Maitlis, *J. Am. Chem. Soc.*, **106**, 5458 (1984).
127. P. M. Maitlis, *Coord. Chem. Rev.*, **43**, 377 (1982).
128. W. Klaeui, M. Scotti, M. Valderrama, S. Rojas, G. M. Sheldrick, P. G. Jones, and T. Schroeder, *Angew. Chem., Int. Ed. Engl.*, **24**, 683 (1985).
129. K. Broadley, G. A. Lance, N. G. Connelly, and W. E. Geiger, *J. Am. Chem. Soc.*, **105**, 2486 (1983).
130. P. W. Jolly, in: *Comprehensive Organometallic Chemistry* (G. Wilkinson, F. G. A. Stone, and E. W. Abel, eds.), Vol. 2, Section 37.8, Pergamon Press, Oxford (1982).
131. P. M. Maitlis, P. Espinet, and M. J. H. Russel, in: *Comprehensive Organometallic Chemistry* (G. Wilkinson, F. G. A. Stone, and E. W. Abel, eds.), Vol. 2, Sections 38.3 and 38.8, Pergamon Press, Oxford (1982).
132. F. R. Hartley, in: *Comprehensive Organometallic Chemistry* (G. Wilkinson, F. G. A. Stone, and E. W. Abel, eds.), Vol. 2, Chapter 39, Pergamon Press, Oxford (1982).
133. G. P. Elliott, J. A. K. Howard, T. Mise, C. M. Nunn, and F. G. A. Stone, *Angew. Chem., Int. Ed. Engl.*, **25**, 190 (1986).
134. H. Schumann, *Angew. Chem., Int. Ed. Engl.*, **23**, 474 (1984).
135. P. L. Watson and G. W. Parshall, *Acc. Chem. Res.*, **18**, 51 (1985).
136. J. V. Ortiz and R. Hoffmann, *Inorg. Chem.*, **24**, 2095 (1985).
137. W. J. Evans, J. H. Meadows, and T. P. Hanusa, *J. Am. Chem. Soc.*, **106**, 4454 (1984).
138. P. Bougeard, M. Mancini, B. F. Sayer, and M. J. McGlinchey, *Inorg. Chem.*, **24**, 93 (1985).
139. W. J. Evans, J. W. Grate, H. W. Choi, I. Bloom, W. E. Hunter, and J. L. Atwood, *J. Am. Chem. Soc.*, **107**, 941 (1985).
140. W. J. Evans, J. W. Grate, I. Bloom, W. E. Hunter, and J. L. Atwood, *J. Am. Chem. Soc.*, **107**, 405 (1985).
141. W. J. Evans, J. W. Grate, and R. J. Doedens, *J. Am. Chem. Soc.*, **107**, 1671 (1985).
142. W. J. Evans, J. H. Meadows, A. G. Kostka, and G. L. Closs, *Organometallics*, **4**, 324 (1985).
143. H. Schumann, S. Nickel, E. Hahn, and M. J. Heeg, *Organometallics*, **4**, 800 (1985).
144. K. N. Raymond and C. W. Eigenbrot, Jr., *Acc. Chem. Res.*, **13**, 276 (1980).
145. W. J. Evans, L. A. Hughes, and T. P. Hanusa, *J. Am. Chem. Soc.*, **106**, 4270 (1984).
146. T. J. Marks and R. D. Ernst, in: *Comprehensive Organometallic Chemistry* (G. Wilkinson, F. G. A. Stone, and E. W. Abel, eds.), Chapter 21, Pergamon Press, Oxford (1982).
147. L. Stahl, J. P. Hutchinson, D R. Wilson, and R. D. Ernst, *J. Am. Chem. Soc.*, **107**, 5016 (1985).
148. T. Kuhlmann, S. Roth, J. Roziere, and W. Siebert, *Angew. Chem., Int. Ed. Engl.*, **25**, 105 (1986).
149. P. N. Gaponik, A. I. Lesnikovich, and J. G. Orlik, *Usp. Khim.*, **52**, 294 (1983).

150. a. K. Jonas, *Angew. Chem., Int. Ed. Engl.*, **24**, 295 (1985). b. F. A. Cotton, S. A. Duraj, and W. J. Roth, *Organometallics*, **4**, 1174 (1985). c. D. W. Slocum, T. R. Engelmann, R. L. Fellows, M. Moronski, and S. A. Duraj, *J. Organomet. Chem.*, **260**, C21 (1984). d. D. W. Slocum, M. Moronski, R. Gooding, and S. A. Duraj, *J. Organomet. Chem.*, **260**, C26 (1984).
151. A. J. L. Pombeiro, in: *New Trends in the Chemistry of Nitrogen Fixation* (J. Chatt, L. M. daCamara Pina, and R. L. Richards, eds.), Chapter 6, Academic Press, London (1980); A. J. L. Pombeiro, *Rev. Port. Quim.*, **26**, 30 (1984).
152. M. E. Volpin and V. B. Shur, in: *New Trends in the Chemistry of Nitrogen Fixation* (J. Chatt, L. M. daCamara Pina, and R. L. Richards, eds.), Chapter 3, Academic Press, London (1980).
153. A. D. Pomogajlo and W. S. Savostyanov, *Usp. Khim.*, **52**, 1698 (1983).
154. N. S. Kochetkova, M.-G. A. Shvekhgeimer, and L. W. Balabanova, *Usp. Khim.*, **53**, 2009 (1984).
155. B. Parkinson, *Acc. Chem. Res.*, **17**, 431 (1984).
156. H. Koepf and P. Koepf-Maier, *Acc. Chem. Res.*, **18**, 477 (1979); H. Koepf and P. Koepf-Maier, *ACS Symp. Ser.*, **209**, 315 (1983).
157. J. H. Toney and T. J. Marks, *J. Am. Chem. Soc.*, **107**, 947 (1985).
158. P. Koepf-Maier, H. Koepf, and E. W. Neuse, *Angew. Chem., Int. Ed. Engl.*, **23**, 456 (1984).
159. R. E. Cramer, K. T. Higa, and J. W. Gilje, *J. Am. Chem. Soc.*, **106**, 7245 (1984).
160. R. E. Cramer, K. Panchanatheswaran, and J. W. Gilje, *J. Am. Chem. Soc.*, **106**, 1853 (1984); R. E. Cramer, K. Panchanatheswaran, and J. W. Gilje, *Angew. Chem., Int. Ed. Engl.*, **23**, 912 (1984).
161. J. G. Brennan and R. A. Andersen, *J. Am. Chem. Soc.*, **107**, 514 (1985).
162. J. M. Ritchey, A. J. Zozulin, D. A. Wrobleski, R. R. Ryan, H. J. Wasserman, D. C. Moody, and R. T. Paine, *J. Am. Chem. Soc.*, **107**, 501 (1985).
163. W. G. Syrkin, *Karbonily Metallov*, Khimia, Moscow (1983).
164. G. P. Pez, P. Apgar, and R. K. Crissey, *J. Am. Chem. Soc.*, **104**, 482 (1982).
165. D. W. Slocum, T. R. Engelmann, C. Ernst, C. A. Jennings, W. Jones, B. Koonsvitsky, J. Lewis, and P. Shenkin, *J. Chem. Educ.*, **46**, 144 (1969).
166. D. W. Slocum, S. A. Duraj, M. Matusz, J. L. Cmarik, K. M. Simpson, and D. A. Owen, in: *Metal-Containing Polymeric Systems* (J. E. Sheats, C. E. Carraber, Jr., and C. U. Pittman, Jr., eds.), p. 59, Plenum Press, New York (1985).

# Chapter 10

# Complexes Containing Six-Electron π-Ligands

## 1. THE BONDING

The role of $6e$ ligands may be assumed by benzene and its derivatives, cycloheptatriene and other trienes, cyclooctatriene, pyridine, thiophene, etc. Sandwich compounds of the general formula M(arene)$_2$ which possess structures analogous to metallocenes and monoarene derivatives of the composition M(arene)L$_n$ are known. Arenes may be considered as ligands occupying three coordination sites. Usually, arenes are terminal ligands bonded to one metal atom; however, sometimes they behave as bridging ligands and form $\pi$ bonds with two or three metal atoms.[1–3, 29–32, 44] Examples include [Cr(C$_6$H$_6$)$_2$], [Cr(C$_6$H$_6$)(CO)$_3$], [Pd$_2$(AlCl$_4$)$_2$ (C$_6$H$_6$)$_2$], [Pd$_2$(Al$_2$Cl$_7$)$_2$ (C$_6$H$_6$)$_2$], [Os$_3$($\mu_3$-C$_6$H$_6$)(CO)$_9$], and [cpV($\mu$-C$_6$H$_6$)Vcp] (Figure 10.1). In [Pd$_2$(AlCl$_4$)$_2$ (C$_6$H$_6$)$_2$] each benzene molecule appears to be bonded rather as a bridging conjugated diene, the benzene rings are folded (7°), and two carbon atoms of each ring are bent away from the two palladium atoms. The structure of [Pd$_2$(Al$_2$Cl$_7$)$_2$ (C$_6$H$_6$)$_2$] is similar but, due to disorder, the stereochemistry of the arene molecules is not known exactly.[61]

The arene molecule has six $p_\pi$ orbitals which forms six molecular orbitals. These MOs are represented in Figure 10.2. In the case of sandwich compounds possessing $D_{6h}$ symmetry, they may form twelve orbitals which have the following symmetries: $a_{1g}$, $e_{1g}$, $e_{2g}$, $a_{2u}$, $e_{1u}$, $e_{2u}$, $b_{2g}$, and $b_{1u}$ (Table 10.1). Calculations by Dyatkina and Shustorovich for the Cr(C$_6$H$_6$)$_2$ molecule[1–5] show that the antibonding $e_{1g}(d_{xz}, d_{yz})$ orbital has energy which is considerably higher than that of the $4s$ and $4p$ orbitals.

However, this is not confirmed by later calculations[4, 5, 33, 34, 44] predicting relatively small splitting of $d$ orbitals whose order is as follows: $e_{2g}(d_{xy}, d_{x^2-y^2}) < a_{1g}(d_{z^2}) < e_{1g}(d_{xz}, d_{yz})$. The energy difference $\Delta E_2$ (Figure 10.3) is greater for arene complexes than for metallocenes; it increases as the ring becomes larger[4, 5] (Figure 9.4).

575

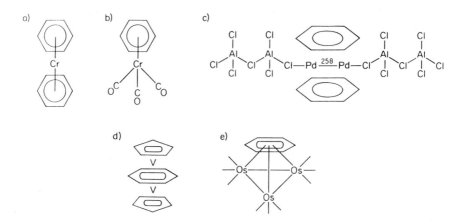

Figure 10.1. Structures of arene compounds: (a) $[Cr(C_6H_6)_2]$, (b) $[Cr(C_6H_6)(CO)_3]$, (c) $[Pd_2(C_6H_6)_2 (Al_2Cl_7)_2]$, (d) $[V_2cp_2(\mu\text{-}C_6H_6)]$, (e) $[Os_3(\mu_3\text{-}C_6H_6)(CO)_9]$.[29]

The structure of arene sandwich compounds (like that of cyclopentadienyl, cycloheptatrienyl, etc., derivatives) and the character of the M–arene bond may also be considered on the basis of ligand field theory assuming $C_{\infty v}$ symmetry of the compound (cf. Chapter 9). The ligands connected to the metal via carbon atoms create a strong field. Therefore, the splitting parameters are large. The nephelauxetic coefficients $\beta = B/B_0$, where $B$ is the Racah parameter for the complex and $B_0$ is the Racah parameter for the free metal ion ($B_0$ and $B$ represent interelectron repulsion), generally assume values $0.5 \pm 0.1$ for metallocenes and arene sandwich complexes. Low values of nephelauxetic coefficients indicate considerable covalent character of the metal–hydrocarbon bond.

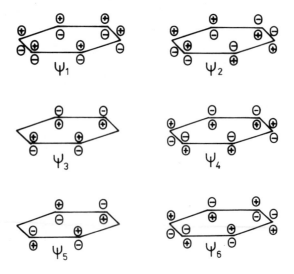

Figure 10.2. The molecular orbitals of arenes.

*Table 10.1. Orbitals of Ligands, the Metal, and Molecular Orbitals for M(arene)$_2$ Complexes Possessing D$_{6h}$ Symmetry*

| Arene orbitals | (Arene)$_2$ orbitals | Metal orbitals | Symmetry $m$ | Bond type |
|---|---|---|---|---|
| $\psi_1 = \dfrac{1}{\sqrt{6}}(\psi_1 + \psi_2 + \psi_3 + \psi_4 + \psi_5 + \psi_6)$ | $\psi_1 + \psi_1'$ | s, d$_{z^2}$ | $a_{1g}$ | σ |
| $\psi_2 = \dfrac{1}{\sqrt{12}}(2\psi_1 + \psi_2 - \psi_3 - 2\psi_4 - \psi_5 + \psi_6)$ | $\psi_2 + \psi_2'; \psi_3 + \psi_3'$ | d$_{xz}$, d$_{yz}$ | $e_{1g}$ | π |
| $\psi_3 = \dfrac{1}{2}(\psi_2 + \psi_3 - \psi_5 - \psi_6)$ | $\psi_4 + \psi_4'; \psi_5 + \psi_5'$ | d$_{xy}$, d$_{x^2-y^2}$ | $e_{2g}$ | δ |
| $\psi_4 = \dfrac{1}{\sqrt{12}}(2\psi_1 - \psi_2 - \psi_3 + 2\psi_4 - \psi_5 - \psi_6)$ | $\psi_1 - \psi_1'$ | p$_z$ | $a_{2u}$ | σ |
| $\psi_5 = \dfrac{1}{2}(\psi_2 - \psi_3 + \psi_5 - \psi_6)$ | $\psi_2 - \psi_2'; \psi_3 - \psi_3'$ | p$_x$, p$_y$ | $e_{1u}$ | π |
| $\psi_6 = \dfrac{1}{\sqrt{6}}(\psi_1 - \psi_2 + \psi_3 - \psi_4 + \psi_5 - \psi_6)$ | $\psi_4 - \psi_4'; \psi_5 - \psi_5'$ | | $e_{2u}$ | |
| | $\psi_6 + \psi_6'; \psi_6 - \psi_6'$ | | $b_{2g}, b_{1u}$ | |

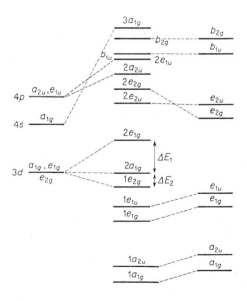

Figure 10.3. The molecular orbital diagram for [Cr(C$_6$H$_6$)$_2$].

## 2. ELECTRONIC SPECTRA OF ARENE COMPLEXES

Electronic spectra of arene complexes have not been widely investigated compared to cyclopentadienyl derivatives. Only in some cases is their interpretation unequivocal. The spectrum of the cation $[Cr(C_6H_6)_2]^{+(6)}$ exhibits bands at 8540 ($\varepsilon = 7$), 10200sh, 29400, 37000, and 42600 cm$^{-1}$. The first two bands are due to the $d$–$d$ transitions. Bis(benzene)chromium exhibits bands at 15600 ($\varepsilon = 25$) and 31250 cm$^{-1}$ ($\varepsilon = 8000$). The $\pi \rightarrow \pi^*$ transitions within coordinated benzene molecules are shifted only slightly toward higher energies. This indicates that benzene ligands preserve their individual character in complexes.[2] In the spectrum of $[Fe(C_6Me_6)_2]^+$ with a $d^7$ structure of the central ion, there are bands at 17200 ($\varepsilon = 604$) and 20900 ($\varepsilon = 416$) cm$^{-1}$, which are due to the $d$–$d$ transitions. The compound $[Fe(C_6Me_6)_2]$ shows bands at 21500 ($\varepsilon = 470$) and 33600 ($\varepsilon = 345$) cm$^{-1}$. In dichloromethane, $[Co(C_6Me_6)_2]^+$ $[PF_6]^-$ was found to absorb at 11910 ($\varepsilon = 16.3$), 13610 ($\varepsilon = 23.9$), 20500sh ($\varepsilon = 186$), 24100sh ($\varepsilon = 1230$), 26650sh ($\varepsilon = 1990$), 30300 ($\varepsilon = 4870$), 32600sh ($\varepsilon = 3800$), and 39200 ($\varepsilon = 3000$) cm$^{-1}$. The first three bands correspond to the $d$–$d$ transitions.[7] From the spectrum of this compound, the following parameters were calculated[4, 5]: $Ds = 2676$ cm$^{-1}$, $Dt = 1460$ cm$^{-1}$, $\Delta E_1 = 11920$ cm$^{-1}$, $\Delta E_2 = 3410$ cm$^{-1}$, $B = 573$ cm$^{-1}$, $\beta = 0.69$ [see also equation (9.16)]. The spectra of complexes containing naphthalene and its derivative, such as $M(C_{10}H_{8-n}R_n)_2$ and $M(C_{10}H_{12})_2$ (M = Ti, V, Cr), were also investigated in matrices.[38]

## 3. MAGNETIC PROPERTIES

Like many cyclopentadienyl complexes, a series of arene derivatives, particularly those of the sandwich type, do not obey Sidgwick's rule. Some of them have

Table 10.2. Magnetic Properties of Arene Metal Complexes[a]

| Compound | Electronic structure | Number of unpaired electrons | Magnetic moment, the "spin-only" values (BM) | Experimental magnetic moment (BM) |
|---|---|---|---|---|
| $[V(PhH)_2]$ | $e_{2g}^4 2a_{1g}^1$ | 1 | 1.73 | $1.68 \pm 0.08$ |
| $[V(mesitylene)_2][AlCl_4]$ | $e_{2g}^3 2a_{1g}^1$ | 2 | 2.83 | $2.80 \pm 0.17$ |
| $[Cr(PhH)_2][AlCl_4]$ | | 1 | 1.73 | $1.73 \pm 0.05$ |
| $[Mo(PhH)_2][AlCl_4]$ | $e_{2g}^4 2a_{1g}^1$ | 1 | 1.73 | $1.74 \pm 0.09$ |
| $[W(PhH)_2][AlCl_4]$ | | 1 | 1.73 | $1.58 \pm 0.08$ |
| $[Re(PhH)_2]$ | $e_{2g}^4 2a_{1g}^2 2e_{1g}^1$ | 1 | 1.73 | |
| $[Fe(C_6Me_6)_2]$ | $e_{2g}^4 2a_{1g}^2 2e_{1g}^2$ | 2 | 2.83 | 3.08 |
| $[Fe(C_6Me_6)_2]^+ [PF_6]^-$ | $e_{2g}^4 2a_{1g}^2 2e_{1g}^1$ | 1 | 1.73 | 1.89 |
| $[Co(C_6Me_6)_2][PtCl_6]$ | | 1 | 1.73 | $1.73 \pm 0.05$ |
| $[Co(C_6Me_6)_2][PF_6]$ | $e_{2g}^4 2a_{1g}^2 2e_{1g}^2$ | 2 | 2.83 | $2.95 \pm 0.08$ |
| $[Co(C_6Me_6)_2]$ | $e_{2g}^4 2a_{1g}^2 2e_{1g}^3$ | 1 | 1.73 | 1.86 |
| $[Rh(C_6Me_6)_2]^{2+}$ | $e_{2g}^4 2a_{1g}^2 2e_{1g}^1$ | 1 | 1.73 | $1.32 \pm 0.08$ |
| $[Ni(C_6Me_6)_2]^{2+}$ | $e_{2g}^4 2a_{1g}^2 2e_{1g}^2$ | 2 | 2.83 | $3.00 \pm 0.09$ |

[a] According to References 1, 2, 4, 5, and 7.

paramagnetic properties. For complexes possessing electronic structure $e_{2g}^x 2a_{1g}^1 2e_{1g}^y$ and $e_{2g}^x 2a_{1g}^2 2e_{1g}^y$ ($x = 1$, 3 and $y = 0$ or $x = 4$ and $y = 1$, 3) the complete quenching of the orbital momentum does not take place, and therefore these complexes should have magnetic moments that are higher than the "spin-only" value (Table 10.2).

## 4. IR SPECTRA

The IR spectra of complexes possessing $D_{6h}$ symmetry show ten bands. Seven of them are due to the vibrations of benzene molecules, while the remainder are due to skeletal oscillations: stretching $v(M-Ar)$, ring tilting, and deformation $\delta(Ar-M-Ar)$[8, 9, 35] (cf. Figure 9.9). These bands occur in the ranges I–X defined in Table (10.1). The spectra of some complexes of the type $[M(C_6H_6)_2]$ are given in Table 10.3.

| | | | |
|---|---|---|---|
| I | 3100–3000 cm$^{-1}$ | $v(CH)$ | |
| II | 2950–2890 cm$^{-1}$ | $v(CH)$ | |
| III | 1450–1400 cm$^{-1}$ | $v(CC)$ or ring deformations | |
| IV | 1000–970 cm$^{-1}$ | $\delta(CH)$ | |
| V | 970–900 cm$^{-1}$ | $v(CC)$ or ring deformations | |
| VI | 880–810 cm$^{-1}$ | $\pi(CH)$ | |
| VII | 800–730 cm$^{-1}$ | $\pi(CH)$ | |
| VIII | 500–300 cm$^{-1}$ | asymmetric metal-ring tilt | |
| IX | 460–330 cm$^{-1}$ | $v(M-Ar)$ | |
| X | ∽140 cm$^{-1}$ | $\delta(Ar-M-Ar)$ | (10.1) |

Investigations of IR spectra of $[Cr(C_6H_6)(CO)_3]$ and $[Cr(C_6D_6)(CO)_3]$ showed that the benzene ligand has $C_{3v}$ symmetry. The bands of the benzene spectrum which lie at *ca* 1600 and 1500 cm$^{-1}$ become shifted to lower frequencies after the benzene molecule is coordinated to the metal.

## 5. NMR SPECTRA

The $^1H$ NMR signal for the free benzene molecule occurs at $\tau = 2.7$. It becomes shifted toward higher fields after coordination to the metal by approximately 3 ppm. For benzene complexes, the singlet resonance usually occurs in the range of $\tau = 4.5$–6,

*Table 10.3. IR Spectra of Benzene Complexes of Transition Metals*

| Compound | Energy of the band (cm$^{-1}$) | | | | | | | | | |
|---|---|---|---|---|---|---|---|---|---|---|
| | I | II | III | IV | V | VI | VII | VIII | IX | X |
| $[Cr(C_6H_6)_2)$ | 3037 | | 1426 | 999 | 971 | 833 | 749 | 490 | 459 | 140 |
| $[Cr(C_6H_6)_2]^+$ | 3040 | | 1430 | 1000 | 972 | 857 | 795 | 466 | 415 | 144 |
| $[Mo(C_6H_6)_2]$ | 3030 | 2916 | 1425 | 995 | 966 | 811 | 773 | 424 | 362 | |
| $[W(C_6H_6)_2]$ | 3012 | 2898 | 1412 | 985 | 963 | 882 | 793 | 386 | 331 | |
| $[V(C_6H_6)_2]$ | 3058 | | 1416 | 985 | 959 | 818 | 739 | 470 | 424 | |

*Table 10.4. Proton Chemical Shifts of*
*Arene Complexes of Transition Metals$^a$*

| Compound | $\tau$ |
|---|---|
| $[Cr(C_6H_6)(CO)_3]$ | 4.63 |
| $[W(C_6H_6)(CO)_3]$ | 4.77 |
| $[Fe(C_6H_6)(C_6H_8)]$ | 5.13 |
| $[Ru(C_6H_6)(C_6H_8)]$ | 5.0 |
| $[Re(C_6H_6)(C_6H_7)]$ | 5.35 |
| $[Cr(C_6H_5Me)(CO)_3H]^+$ | 3.35$^b$ |
| $[Cr(C_6H_3Me_3)(CO)_3H]^+$ | 4.13$^b$ |

$^a$ According to Reference 10.
$^b$ *J. Chem. Soc.*, 3653 (1962).

while compounds containing methyl derivatives of benzene have resonances at some-what lower fields (Table 10.4).

The $\delta^{13}C$ chemical shift of coordinated benzene is lower (the shift of the signal due to coordination toward higher fields) compared to the free ligand. This is in agreement with the lowering of the multiplicity of the $C-C$ bond due to coordination of the arene molecule (Table 10.5).

The rotation of the coordinated benzene molecule occurs readily. The energy of activation of the process of rotation of benzene in $[Cr(C_6H_6)(CO)_3]$ is of the order of 3–4 kJ mol$^{-1}$.[37] However, energy of activation of rotation of hexahapto ligands is considerably higher if they contain bulky substituents. Ligands in bis(1,4-*tert*-butylmethylbenzene)chromium(0) rotate with greater difficulty. The data are $E_a = 35.6 \pm 1.6$ kJ mol$^{-1}$ ($\Delta H^{\neq} = 34.0 \pm 1.6$ kJ mol$^{-1}$, $\Delta S^{\neq} = -22 \pm 8$ J K$^{-1}$ mol$^{-1}$).[37] The energy of activation of rotation of cycloheptatriene in the complex $[Cr(\eta^6\text{-}C_7H_7)(CO)_3]$[60] is of the same order ($\Delta H^{\neq} = 41.5 \pm 0.9$ kJ mol$^{-1}$, $\Delta S^{\neq} = -2.9 \pm 4.6$ J K$^{-1}$ mol$^{-1}$).

A different type of dynamic process may be observed in the compound $Ru(\eta^6\text{-}C_6Me_6)(\eta^4\text{-}C_6Me_6)$[40] in which the carbon atoms of the tetrahapto ligands, uncoordinated to the metal, are tilted away from the compound. At 263 K, the $^1H$ NMR spectra are in agreement with the structure ($\eta^6$-arene) Ru($\eta^4$-arene), while at 323 K, the methyl groups are equivalent on the NMR time scale. A line-shape analysis of the $^1H$ DNMR spectra showed that there is the following rearrangement: $Ru(\eta^6\text{-}C_6Me_6)(\eta^4\text{-}C_6Me_6) \rightleftharpoons Ru(\eta^4\text{-}C_6Me_6)_2$ ($\Delta H^{\neq} = 65.3 \pm 5.5$ kJ mol$^{-1}$, $\Delta S^{\neq} = -12.6 \pm 21.0$ J K$^{-1}$ mol$^{-1}$).[40]

# 6. PHOTOELECTRON SPECTRA

Photoelectron spectra of arene complexes show bands in the 5–7 eV region due to ionization of completely occupied $a_{1g}$ and $e_{2g}$ orbitals which mainly have the character of $d$ orbitals of the metal. The $e_{2g}$ orbitals have lower energy. The ionization of $e_{1u}$ and $e_{1g}$ orbitals of the arene occurs at energies between 8–10 eV.[11–14, 36] Similar spectra are observed below 8 eV for compounds that are isoelectronic with dibenzenechromium,

*Table 10.5. $^{13}C$ NMR Spectra of Arene Metal Complexes*[28]

| Compound | δ $^{13}$C (ppm) free ligand | Complex |
|---|---|---|
| 1 | 2 | 3 |
| [Cr(C$_6$H$_6$)(CO)$_3$] | 128.5 | 93.7 |
| [Cr(mesitylene)(CO)$_3$] | 21.4 (CH$_3$) | 21.0 |
| | 127.6 (CH) | 92.4 |
| | 138.6 (CMe) | 111.5 |
| [Cr(durene)(CO)$_3$] | 19.2 (CH$_3$) | 18.6 |
| | 131.6 (CH) | 99.0 |
| | 134.4 (CMe) | 107.6 |
| [Cr(C$_6$Me$_6$)(CO)$_3$] | 16.9 (CH$_3$) | 17.4 |
| | 132.5 (CMe) | 107.5 |
| [Mo(*m*-xylene)(CO)$_3$] | | 115.6 (C-1.3) |
| | | 96.4 (C-2) |
| | | 98.2 (C-5) |
| | | 92.7 (C-4.6) |
| [Mo(mesitylene)(CO)$_3$] | | 21.5 (CH$_3$) |
| | | 21.2 (CH$_3$) |
| | | 94.7 (CH) |
| | | 111.7 (CMe) |
| [Mo(C$_6$Me$_6$)(CO)$_3$] | | 18.0 (CH$_3$) |
| | | 111.7 (CMe) |
| [Mo(durene)(CO)$_3$] | | 19.1 (CH$_3$) |
| | | 101.4 (CH) |
| | | 118.8 (CMe) |
| [W(mesitylene)(CO)$_3$] | | 20.9 (CH$_3$) |
| | | 90.9 (CH) |
| | | 111.1 (CMe) |
| [W(durene)(CO)$_3$] | | 18.9 (CH$_3$) |
| | | 97.3 (CH) |
| | | 107.9 (CMe) |
| [W(C$_6$Me$_6$)(CO)$_3$] | | 17.9 (CH$_3$) |
| | | 107.9 (CMe) |
| [Ag(PhH)]$^+$ [BF$_4$]$^{-a}$ | | 128.2 |
| [Ag(*p*-xylene)]$^+$ [BF$_4$]$^{-a}$ | | 133.3 (CMe) |
| | | 123.1 (CH) |
| [Cr(PhH)(CO)$_2$(PPh$_3$)]$^b$ | | 89.6 |
| [Cr(PhMe)(CO)$_2$(PPh$_3$)]$^b$ | | 103.5 (CMe) |
| | | 88.7 (C2) |
| | | 89.6 (C3) |
| | | 88.4 (C4) |
| [Cr(mesitylene)(CO)$_2$(PPh$_3$)$_2$]$^b$ | | 105.2 (CMe) |
| | | 90.2 (CH) |

$^a$ J. D. C. M. Von Dongen and C. O. M. Beverwijk, *J. Organomet. Chem.*, **81**, C36 (1973).
$^b$ L. A. Fedorov, P. V. Petrovskij, E. I. Fedin, N. K. Baranetskaya, V. I. Zdanovich, V. N. Setkina, and D. N. Kursanov, *Dokl. Akad. Nauk SSSR*, **209**, 266 (1973).

Table 10.6. *Ionization Energy of Valence Orbitals. Assignment of Bands for Arene Metal Complexes*[a]

| | Ionization energy of the term (eV) | | | | | | |
|---|---|---|---|---|---|---|---|
| Compound | $^2A_{1(g)}$ metal | $^2E_{2(g)}$ metal | $^2E_{2(u)}$ ligand | $^2E_{1(g)}$ ligand | $^3E_{2(g)}$ metal | $^1A_{1(g)}$ metal | $^1E_{2(g)}$ metal |
| [Cr(PhH)$_2$] | 5.4 | 6.4 | 9.6 | 9.6 | | | |
| [Cr(PhMe)$_2$] | 5.24 | 6.19 | 9.16 | 9.53 | | | |
| [Cr(mesitylene)$_2$] | 5.01 | 5.88 | 8.90 | 8.90 | | | |
| [Mo(PhH)$_2$] | 5.52 | 6.59 | 9.47 | 10.15 | | | |
| [Mo(PhMe)$_2$] | 5.32 | 6.33 | 9.05 | 9.75 | | | |
| [Mo(mesitylene)$_2$] | 5.13 | 6.03 | 8.63 | 9.33 | | | |
| [Crcp(C$_7$H$_7$)] | 5.59 | 7.19 | 8.69 ($e_1$, cp) | 9.00 ($e_1$, cp) 10.4 ($e_1$, C$_7$H$_7$) | | | |
| [Mncp(PhH)] | 6.36 | 6.72 | 8.75 9.25 | 9.79 | | | |
| [V(mesitylene)$_2$] | | | 8.75 | | 5.33 | 5.61 | 6.08 |
| [Vcp(C$_7$H$_7$)] | | | 10.2 ($e_1$, C$_7$H$_7$) | 8.66 ($e_1$, cp) 8.99 ($e_1$, cp) | 6.77 | 6.42 | 7.28 |
| [Crcp(PhH)] | | | | | 6.20 | 6.20 | 7.15 |
| [Ticp(C$_8$H$_8$)] | | | 8.63 ($e_1$, cp) 8.93 ($e_1$, cp) | 10.51 ($e_1$, C$_8$H$_8$) | 7.62 | 5.67 | 7.62 |
| [Ticp(C$_7$H$_7$)][b] | | 6.83 | 10.21 ($e_1$, C$_7$H$_7$) | 8.71 ($e_1$, cp) 9.1 ($e_1$, cp) | | | |
| [Zrcp(C$_7$H$_7$)][b] | | 6.94 | 10.2 ($e_1$, C$_7$H$_7$) | 8.89 ($e_1$, cp) 9.26 ($e_1$, cp) | | | |
| [Mocp(C$_7$H$_7$)][b] | 5.87 | 7.55 | 10.4 ($e_1$, C$_7$H$_7$) | 8.93 ($e_1$, cp) 9.28 ($e_1$, cp) | | | |

[a] According to References 11, 14, and 36.
[b] C. J. Groenenboom, H. J. De Liefde Meijer, F. Jellinek, and A. Oskam, *J. Organomet. Chem.*, **97**, 73 (1975).

such as [Cr(C$_5$H$_5$)(C$_7$H$_7$)] and [Mn(C$_5$H$_5$)(C$_6$H$_6$)]. In the case of sandwich compounds possessing open electronic shells, ground terms may appear to be dubious (Table 10.6). The ground term may be $^2E_2$ ($e_2^3 a_1^2$) or $^2A_1$ ($e_2^4 a_1$). The ESR data seem to confirm the second possibility. For the $A_1$ ground term, the ejection of the $a_1$ electron causes the formation of the ion which is in the $^1A_1$ state, while the ionization of the $e_2$ electron leads to ions possessing $^1A_2$ and $^3E_2$ states. Photoelectron spectra and molecular orbital calculations show that 26$e$ triple-decker complexes [cpV($\mu$-arene)Vcp] (arene = benzene, mesitylene) have four unpaired electrons in the ground state.[31]

## 7. ESR SPECTRA[4,5]

Paramagnetic electron resonance spectra of V(PhH)$_2$ and [Cr(PhH)$_2$]$^+$ show that the unpaired electron is located in the $a_{1g}(d_{z^2})$ orbital. The overlap of this orbital with the orbitals of the ligand is small; the contribution of the orbitals of the ligand is merely several percent, and it is somewhat higher for the vanadium compound than for the cation $[Cr(PhH)_2]^+$. Weak overlap of the $a_1(d_{z^2})$ orbital with $a_1$ orbitals of the ligands was also found for $[Vcp(C_7H_7)]$, $[Crcp(C_7H_7)]^+$, and $[Crcp(PhH)]$.

The reduction of $[Cr(PhH)_2]$ by means of potassium affords the anion $[Cr(PhH)_2]^-$ in which one benzene molecule acts as an $\eta^6$ ligand while the other $C_6H_6$ is an $\eta^4$ group as evidenced by the ESR data.[39] The tetrahapto ligand is not planar. The two carbon atoms are tilted away from the complex as in the case of $Ru(C_6Me_6)_2$. The ESR spectrum confirms that the unpaired electron is located in the $\pi$ orbital of the ligand. The anisotropy $\Delta g = |g_\parallel - g_\perp|$, which is higher for this compound than for aromatic radicals, results from the interaction of the $\pi$ orbital of the ligand with the orbitals of the chromium center.

# 8. STRUCTURE OF ARENE COMPLEXES

The carbon–carbon distances in arene rings which are coordinated to the metal are longer than those in free molecules. Chromium complexes of the type $[Cr(arene)(CO)_3]$ may have, depending upon the substituents in the ring, eclipsed or staggered configurations. In the first case the carbon atoms of the ring and those of the CO groups lie in the $\sigma_v$ planes which contain the $C_3$ axis, while in the second case the CO ligands lie in the $\sigma_v'$ planes which dissect the $C-C$ bonds of the arene (Figure 10.4). Examples of eclipsed compounds are $[Cr(PhOMe)(CO)_3]$, $[Cr(o\text{-}MeC_6H_4NH_2)(CO)_3]$, and $[Cr(PhCOOMe)(CO)_3]$, while the following compounds represent staggered configurations: $[Cr(PhH)(CO)_3]$, $[Cr(C_6Me_6)(CO)_3]$, and $[Cr(arene)(CO)_3]$, where arene denotes fused-ring aromatic compounds such as napthalene, phenanthrene, anthracene, and some of their derivatives (Figure 10.4, Table 10.7).

*Table 10.7. Interatomic Distances in Arene Metal Complexes*

| Compound | Distances (pm) | | Remarks |
|---|---|---|---|
| | M–C | C–C | |
| $[V(PhH)_2]^a$ | 217 | | |
| $[Cr(PhH)_2]^b$ | 213.2; 213.8 | 138.7 | |
| $[Cr(PhH)(CO)_3]^c$ | 222.1 | 140 | 184.2 (Cr–CO) |
| $[Cr(PhOMe)(CO)_3]^d$ | 223 | 140 | 179 (Cr–CO) |
| $[Nb_3(\mu\text{-}Cl)_6(C_6Me_6)_3]^{2+e}$ | 228.2–249.2 | 144.7 | 249.4 (Nb–Cl) |
| | | (average) | (average) |
| $[Cr(2,6\text{-}Me_2C_5H_3N)_2]^f$ | 211.6–216.1 | | 215.1 (Cr–N) |
| $[Fecp(\eta^6\text{-}C_{13}H_9)]^g$ | 203.2–230 | | |
| $[Ru_4(OH)_4(PhH)_4](SO_4)_2 \cdot 12H_2O^h$ | 223 | | |
| $[Cu_4(CF_3COO)_4(PhH)_2]^i$ | 270–300 | | |

$^a$ E. O. Fischer, H. P. Fritz, J. Manchet, E. Priebe, and R. Schneider, *Chem. Ber.*, **96**, 1418 (1963).
$^b$ F. A. Cotton, W. A. Dollase, and J. S. Wood, *J. Am. Chem. Soc.*, **85**, 1543 (1963).
$^c$ M. F. Bailey and L. F. Dahl, *Inorg. Chem.*, **4**, 1314 (1965).
$^d$ O. L. Carter, A. T. McPhail, and G. A. Sim, *J. Chem. Soc. (A)*, 822 (1966).
$^e$ S. Z. Goldberg, B. S. Spivack, G. Stanley, R. Eisenberg, D. M. Braitsch, J. S. Miller, and M. Abkowitz, *J. Am. Chem. Soc.*, **99**, 110 (1977).
$^f$ L. H. Simons, P. E. Riley, R. E. Davis, and J. J. Lagowski, *J. Am. Chem. Soc.*, **98**, 1044 (1976).
$^g$ J. W. Johnson and P. M. Treichel, *Chem. Commun.*, 688 (1976).
$^h$ R. O. Gould, C. L. Jones, D. R. Robertson, and T. A. Stephenson, *Chem. Commun.*, 222 (1977).
$^i$ P. F. Rodesiler and E. L. Amma, *Chem. Commun.*, 599 (1974).

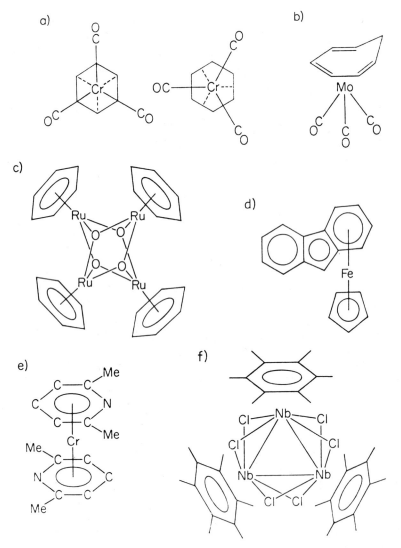

Figure 10.4. Structures of arene complexes: (a) Cr(arene)(CO)$_3$, (b) Mo(C$_7$H$_8$)(CO)$_3$, (c) [Ru$_4$(OH)$_4$ (PhH)$_4$]$^{4+}$, (d) Fecp($\eta^6$-C$_{13}$H$_9$), (e) Cr(C$_5$H$_3$NMe$_2$)$_2$, (f) [Nb$_3$Cl$_6$(C$_6$Me$_6$)$_3$]$^{2+}$.

## 9. *THERMOCHEMISTRY OF ARENE COMPLEXES*[15]

Thermochemical data for arene complexes are less abundant than for metallocenes. This is due in part to incomplete combustion of the compounds. The energies of dissociation of metal–arene bonds were obtained from mass spectra for most of the complexes. However, data obtained by this method are somewhat higher (by *ca* 100 kJ mol$^{-1}$ and even higher). In some cases, even the reverse order of the relative stabilities for a given series of compounds is obtained. Table 10.8 gives energies of dissociation of metal–arene bonds which are obtained by calorimetric methods.

*Table 10.8. Dissociation Energy of the*
*M–Arene Bond*

| Compound | $D$(M–arene) (kJ mol$^{-1}$) |
|---|---|
| $[V(PhH)_2]^a$ | 290 |
| $[Cr(PhH)_2]^{b,c}$ | 163 |
| $[Cr(PhEt)_2]^d$ | 159 |
| $[Cr(o\text{-}C_6H_4Et_2)_2]^d$ | 167 |
| $[Cr(PhPr^i)_2]^d$ | 176 |
| $[Cr(o\text{-}C_6H_4Pr^i_2)_2]^d$ | 176 |
| $[Mo(PhH)_2]^d$ | 211 |

[a] E. P. Fischer and A. Reckziegel, *Chem. Ber.*, **94**, 2204 (1961).
[b] V. I. Telnoi, I. B. Rabinowich, B. G. Gribov, A. S. Pashinkin, B. A. Salamatin, and V. I. Chernova, *Zh. Fiz. Khim.*, **46**, 802 (1972).
[c] J. A. Connor, H. A. Skinner, and Y. Virmani, *J. Chem. Soc., Faraday Trans. 1*, **69**, 1218 (1973).
[d] V. I. Telnoi, I. B. Rabinovich, and V. A. Umilin, *Dokl. Akad. Nauk SSSR, Ser. Khim.*, **203**, 127 (1973).

## 10. MASS SPECTRA[16–18]

The mass spectrum of bisbenzenechromium shows the following ions: $[Cr(PhH)_2]^+$ (100), $[Cr(PhH)]^+$ (92), $[Cr(C_5H_5)]^+$ (6.6), $[Cr(C_3H_3)]^+$ (8.0), $Cr^+$ (95.4), and $[Cr(PhH)_2]^{2+}$ (5.6) (values in parentheses represent relative intensities).

The dipositive ion undergoes unusual decomposition to singly charged ions in the metastable transition:

$$[Cr(PhH)_2]^{2+} \longrightarrow [Cr(PhH)]^+ + C_6H_6^+ \tag{10.2}$$

Analogous reaction is observed for vanadium compounds $[V(PhH)_2]$. The band intensity of the $C_6H_6^+$ ion changes due to thermal decomposition, which is temperature dependent. It was also found that the compound decomposes to chromium and benzene. This decomposition depends on temperature and is catalyzed by metallic mirror deposited on the walls of the ionization chamber. The mass spectrum of $[Cr(PhH)(CO)_3]$ shows the presence of ions which contain benzene and also ions possessing only carbonyl groups: $[Cr(PhH)(CO)_x]^+$ ($x = 0$–3) and $[Cr(CO)_n]^+$ ($n = 0$–2). The ion $Cr^+$ shows the highest intensity. However, the presence of ions containing the metal and fragment of hydrocarbons resulting from the decomposition of the benzene ring are not observed. Mass spectra of complexes of the type $[W(\text{arene})(CO)_3]$, where arene = toluene, *p*-xylene, and mesitylene, are considerably more complex due to extensive fragmentation of arene ligands.

## 11. *PROPERTIES OF ARENE COMPOUNDS*[1–5, 29–32, 34, 38–43]

The following types of arene compounds were obtained: $[M(arene)_2]$ (M = Ti, V, Nb, Ta, Cr, Mo, W, Fe, Ru, Co), $[M(arene)_2]^+$ (M = Ti, V, Cr, Mo, W, Mn, Tc, Re, Fe, Os, Co, Rh), $[M(arene)_2]^{2+}$ (M = Fe, Ru, Os, Co, Rh, Ir, Ni), $[M(arene)(CO)_3]$ (M = Cr, Mo, W), $[M(arene)(CO)_3]^+$ (M = Mn, Re), $[Mcp(arene)]$ (M = Cr, Mo, Mn, Re), $[Mcp(arene)]^{2+}$ (M = Co, Rh, Ir), $[M(PhH)_2]^-$ (M = V, Cr),[20, 39] $[Mcp(arene)(CO)]$ (M = Mo, W), $[Fecp(arene)]^+$, $[M(arene)L_2]^+$ [M = Rh, Ir, L = diene, alkene, $PR_3$, $P(OR)_3$], $[TiCl_2(C_6Me_6)]^+$, $[M(PhMe)_2(PMe_3)]$, $[M(SnMe_3)_2(PhMe)_2]$ (M = Zr, Hf), $[V(arene)(CO)_4]^+$, $[VX(CO)_3(mesitylene)]$ (X = H, I), $[V(CO)_3(mesitylene)]^-$, $[V(CO)_3(C_6Ph_6)]$, $[Fe(arene)L_2]$ [L = CO, $P(OR)_3$, $PR_3$, diene], $[RuCl_3(PhH)]^-$, $[Ru(NH_3)_3(PhH)]^{2+}$, $[MX_2(arene)L]$ (M = Ru, Os, X = halogen), $[MHL_2(arene)]^+$ (M = Ru, Os), $[M(arene)L_2]^+$ [M = Rh, Ir, L = diene, alkene, $P(OR)_3$, $PR_3$], $[Ni(C_6F_5)_2(arene)]$, $[Ni(arene)(diene)]^{2+}$ $[Ni(allyl)(PhH)]^+$, and $[Pd(2\text{-}MeC_3H_4)(C_6Me_6)]^+$.

Many polynuclear and cluster-type arene compounds are also known, for example, $[M_2Cl_4(arene)_2]$ (M = Nb, Ta), $[M_3(\mu\text{-}Cl)_6(arene)_3]^+$ (M = Ti, Zr, Nb, Ta), $[TiCl_2(arene)(AlCl_3)_2]$, $[U(C_6H_6(AlCl_4)_3]$, $[M_3(\mu\text{-}Cl)_6(arene)_3]^{2+}$ (M = Zr, Nb), $[Nb_3Br_6(C_6Me_6)_3]^+$, $[cpV(arene)Vcp]$,[30,31] $[Ru_2(\mu\text{-}X)_2X_2(arene)_2]$, $[OsX_2(arene)]_n$ (X = halogen), $[Ru_2(\mu\text{-}X)_3(C_6Me_6)_2]^+$ (X = Cl, H), $[M_2(\mu\text{-}X)_3(arene)_2]^+$ (M = Ru, Os, X = OR, OOCR, OH, Cl),[41] $[Ru_4(OH)_4(PhH)_4]^{4+}$, octahedro-$[Ru_6(\mu_6\text{-}C)(\mu\text{-}CO)(CO)_{13}(PhH)]$, $[Fe_2(\mu\text{-}CO)_2(CO)_2(arene)_2]$, $[Os_4Cl_4(mesitylene)_3]$, $[cpCo\{\mu\text{-}C_6(COOMe)_6\}Cocp]$,[30] $[Co_3(PhH)_3(CO)_2]^+$, $[Co_3(\mu_3\text{-}CY)(arene)(CO)_6]$, $[Co_4(arene)(CO)_9]$, and $[Cu_4(CF_3COO)_4(PhH)_2$ (Table 10.9).

*Table 10.9. Properties of Arene Metal Complexes*

| Compound | Color | Mp (K) |
|---|---|---|
| 1 | 2 | 3 |
| $[Ti_3(\mu\text{-}Cl)_6(C_6Me_6)_3]Cl^a$ | Violet | 303 dec |
| $[TiCl_2(C_6H_6)(AlCl_3)_2]^{b(19)}$ | Violet | |
| $[Zr_3(\mu\text{-}Cl)_6(C_6Me_6)_3]Cl^a$ | | 293–303 dec |
| $[V(PhH)_2]^{c(7)}$ | Red-brown | 450 |
| $[V(PhH)_2]^{-(20)}$ | Blue | |
| $[V(mesitylene)_2]^d$ | Red-brown | 299–300 |
| $[Nb_3(\mu\text{-}Cl)_6(C_6Me_6)_3]Cl^e$ | Dark-green | |
| $[Nb_2Cl_4(C_6Me_6)_2]^e$ | Brown | |
| $[Ta_3Cl_6(C_6Me_6)_3]PF_6{}^a$ | Brown | |
| $[Cr(PhH)_2)]^f$ | Green-black | 359–360 |
| $[Cr(PhMe)_2]^g$ | Brown | 324–325 |
| $[Cr(PhH)(CO)_3]^h$ | Yellow | 335–336 |
| $[Mo(PhH)_2]^i$ | Green | 388 dec |
| $[Mo(PhH)(CO)_3]^j$ | Yellow | 398 |
| $[W(PhH)_2]^k$ | Dark-green | >433 |
| $[W(PhH)(CO)_3]^j$ | Yellow | 413–418 |
| $[Mn(C_6Me_6)_2]^{+l}$ | Light-pink | |
| $[Mn(PhH)(C_6Me_6)]PF_6{}^n$ | Red | 498 |

*(Table continued)*

*Table 10.9. (continued)*

| Compound | Color | Mp (K) |
|---|---|---|
| 1 | 2 | 3 |
| $[Mn(PhH)(CO)_3]^+ AlCl_4^-$ [m] | Light-yellow | |
| $[Tc(PhH)_2]PF_6$ [o] | Yellow-green | 523 dec |
| $[Re(PhH)_2]^+ PF_6^-$ [p] | Light-yellow | 573 dec |
| $[Fe(C_6Me_6)_2]$ [q] | Black | |
| $[Fe(C_6Me_6)_2]PF_6$ [q] [(7)] | Dark-purple | |
| $[Fe(mesitylene)_2]PF_6$ [(7)] | Dark-blue | |
| $[Fecp(C_{13}H_9)]$ [(21)] | Green | 378 |
| $[Fe(C_6Me_6)_2][PF_6]_2$ [r] | Orange | |
| $[Ru(C_6Me_6)_2][PF_6]_2$ [s] | White | >533 |
| $[Ru_4(OH)_4(PhH)_4](SO_4)_2 \cdot 12H_2O$ [t] | Orange | |
| $[Co(C_6Me_6)_2]$ [u] | Dark-brown | ~353 |
| $[Co(C_6Me_6)_2][PF_6]$ [v] | Dark-yellow | ~443 |
| $[Co(C_6Me_6)_2][PtCl_6]$ [(7)] | Yellow-brown | |
| $[Rh(C_6Me_6)_2][PF_6]$ [v] | Light-yellow | 453 |
| $[Ni(C_6Me_6)_2][PtCl_6]$ | Green-yellow [w] Brown-yellow [(7)] | |
| $[Co_3(PhH)_3(CO)_2]^+$ [x] | Dark-brown | |

[a] E. O. Fischer and F. Rohrscheid, *J. Organomet. Chem.*, **6**, 53 (1966).
[b] H. Martin and F. Vohwinkel, *Chem. Ber.*, **94**, 2416 (1961).
[c] E. O. Fischer and R. Reckziegel, *Chem. Ber.*, **94**, 2204 (1961).
[d] F. Calderazzo, *Inorg. Chem.*, **3**, 810 (1964).
[e] E. O. Fischer and F. Roscheid, *J. Organomet. Chem.*, **7**, 121 (1967).
[f] E. O. Fischer and W. T. Hafner, *Z. Anorg. Allg. Chem.*, **286**, 146 (1956).
[g] H. Behrens, K. Meyer, and A. Müller, *Z. Naturforsch.*, **20b**, 74 (1965).
[h] E. O. Fischer and K. Oefele, *Chem. Ber.*, **90**, 2532 (1957).
[i] E. O. Fischer and H. O. Stahl, *Chem. Ber.*, **89**, 1805 (1956).
[j] E. O. Fischer, K. Oefele, H. Essler, W. Fröhlich, J. P. Mortensen, and W. Semlingez, *Chem. Ber.*, **91**, 2763 (1958).
[k] E. O. Fischer, F. Scherer, and H. O. Stahl, *Chem. Ber.*, **93**, 2065 (1960).
[l] M. Tsutsui and H. Zeiss, *J. Am. Chem. Soc.*, **83**, 825 (1961).
[m] G. Winkhaus, L. Pratt, and G. Wilkinson, *J. Chem. Soc.*, 3807 (1961).
[n] E. O. Fischer and M. W. Schmidt, *Chem. Ber.*, **100**, 3782 (1967).
[o] C. Palm, E. O. Fischer, and F. Baumgaertner, *Tetrahedron Lett.*, 253 (1962).
[p] E. O. Fischer and M. W. Schmidt, *Chem. Ber.*, **99**, 2206 (1966).
[q] E. O. Fischer and F. Roehrscheid, *Z. Naturforsch.*, **17b**, 483 (1962).
[r] J. F. Helling and D. M. Braitsch, *J. Am. Chem. Soc.*, **92**, 7207 (1970).
[s] E. O. Fischer and C. Elschenbroich, *Chem. Ber.*, **103**, 162 (1970).
[t] R. O. Gould, C. L. Jones, D. R. Robertson, and T. A. Stephenson, *Chem. Commun.*, 22 (1977).
[u] E. O. Fischer and H. H. Lindner, *J. Organomet. Chem.*, **2**, 222 (1964).
[v] E. P. Fischer and H. H. Lindner, *J. Organomet. Chem.*, **1**, 307 (1964).
[w] H. H. Lindner and E. O. Fischer, *J. Organomet. Chem.*, **12**, P18 (1968).
[x] E. O. Fischer and O. Beckert, *Angew. Chem.*, **70**, 744 (1958).

Neutral bisarene metal complexes are moderately soluble in organic solvents. They are thermally stable, often up to 570 K. They sublime in vacuum at *ca* 370 K. Most bisarene neutral compounds may be easily oxidized to mono and dipositive cations which, as expected, are considerably more stable. In weakly basic solutions, the cation $[Cr(PhH)_2]^+$ may exist in air without decomposition for several weeks. Compared to metallocene sandwich compounds, arene complexes have considerably weaker basic properties.

## 12. ASYMMETRIC ARENE COMPLEXES[21b, 24, 25]

Arene metal carbonyl complexes containing disubstituted *ortho* or *meta* arenes are asymmetric and may be resolved to give enantiomers [structures (10.3)].

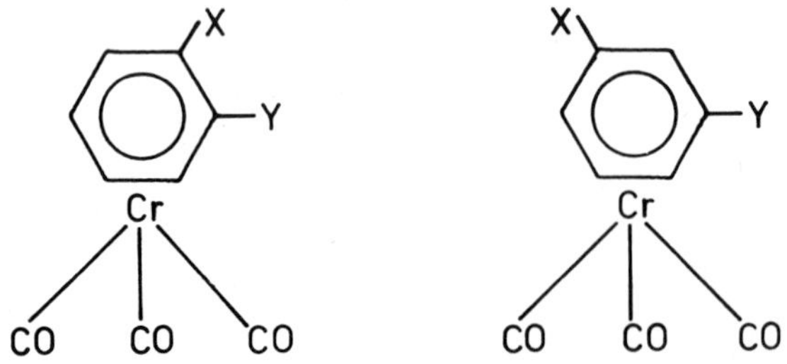

(10.3)

If the metal is bonded to three different ligands in addition to the arene, then the central atom is asymmetric. Chromium, ruthenium, etc., compounds which were obtained as pure enantiomers may serve as examples[25] [structure (10.4)].

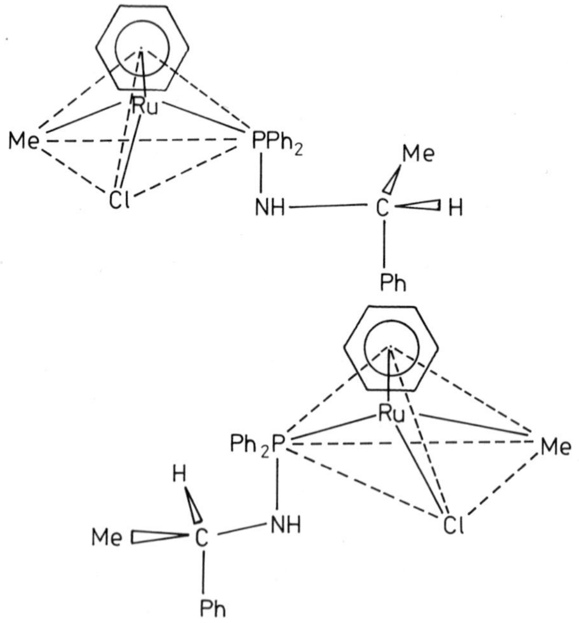

(10.4)

## 13. TRIPLE LAYER AND HETEROARENE COMPLEXES[30–33, 42, 43, 45–47]

Arenes, like the cyclopentadienyl ligand, may serve as bridging ligands in polylayer transition metal complexes,[30, 32] for instance, $cpV(\mu\text{-}C_6H_6)Vcp$, $cpV(\mu\text{-}1,3,5\text{-}C_6H_3Me_3)Vcp$, $[cpCo\{\mu\text{-}C_6(COOMe)_6\}Cocp]$. Heterocyclic, six-mem-

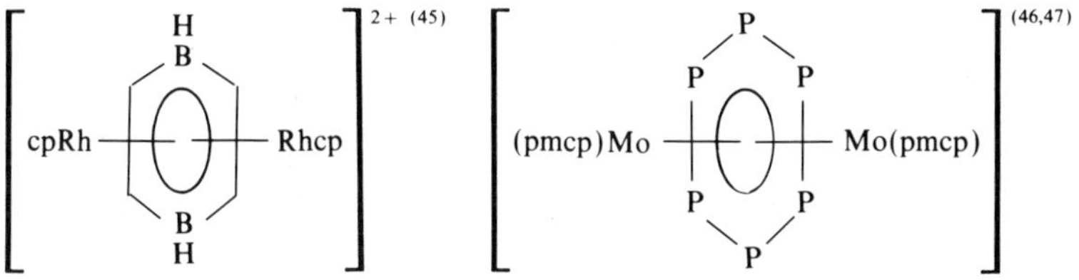

$$(10.5)$$

bered compounds (10.5) may also constitute such ligands. Heteroaromatic compounds such as

$(E = N, P, As, Sb)$

may serve as $2e$ ligands, coordinating through the heteroatom, as $6e$ hexahapto ligands,[47] or as $8e$ $\sigma$, $\pi$ ($\eta^1:\eta^6$) ligands. Chromium forms sandwich compounds with pyridine and its derivatives: $Cr(\eta^6\text{-}C_5H_5N)(PF_3)_3$, $[Cr(\eta^6\text{-}2,6\text{-}C_5H_3Me_2N)(CO)_3]$, and $[Cr(\eta^6\text{-}2,6\text{-}C_5H_3Me_2N)_2]$ (see Table 10.7). The coordination change from $\sigma$ to $\pi$ ($\eta^1 \to \eta^6$) causes large changes of the chemical shift in the direction of stronger fields of protons bonded with the pyridine ring. Very large shift toward stronger fields of the $^{31}P$ NMR signal ($\Delta\delta = 293$ ppm) is observed when the transition from $\sigma$ to $\pi$ occurs for the

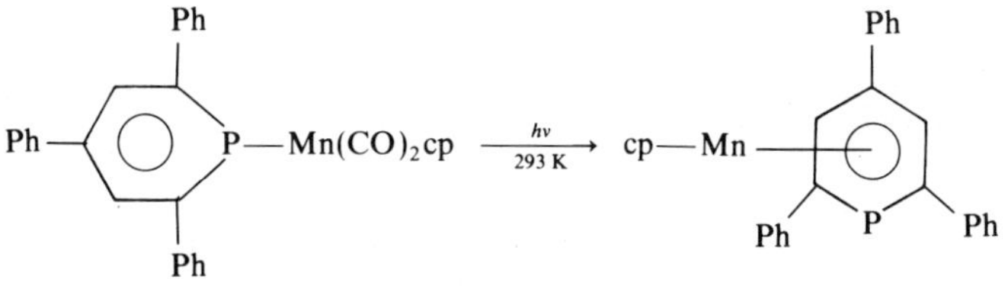

$$(10.6)$$

phosphabenzene compounds [equation (10.6)]. The following complexes, *inter alia*, are also known:

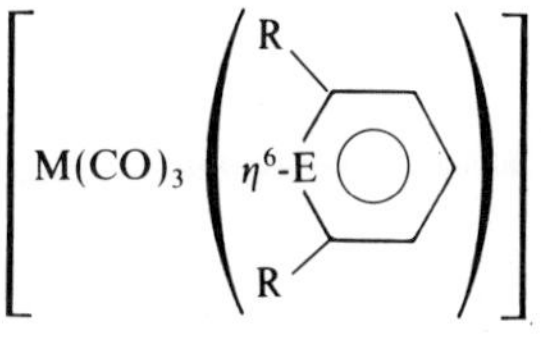

$(M = Cr, Mo, E = P, R = Ph; M = Mo, E = As, Sb, R = H)$

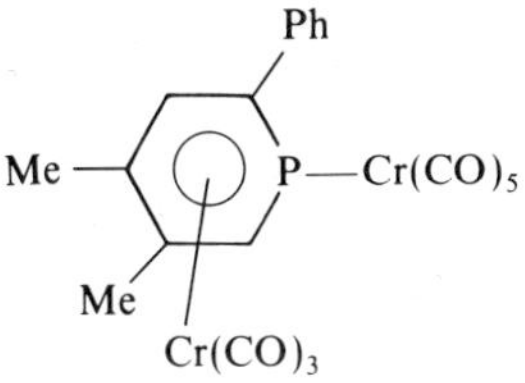

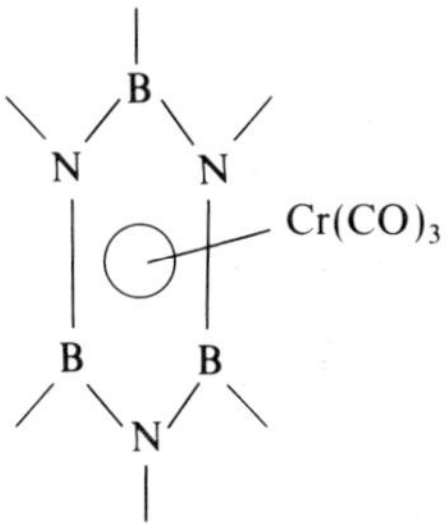

## 14. METHODS OF PREPARATION OF ARENE COMPLEXES[1-3, 7, 22]

### a. Reactions of transition metal compounds with aluminum and aluminum chloride

Fischer and coworkers developed the most important and general method for the preparation of arene complexes. It involves the action of aluminum and its chloride on transition metal compounds in the presence of arene:

$$3CrCl_3 + AlCl_3 + 2Al + 6C_6H_6 \longrightarrow 3[Cr(PhH)_2][AlCl_4] \qquad (10.7)$$

The cation may subsequently be reduced by strongly reducing agents such as the dithionite ion in alkaline solution:

$$2[Cr(PhH)_2]^+ + S_2O_4^{2-} + 4OH^- \xrightarrow[H_2O + MeOH]{} 2[Cr(PhH)_2] + 2SO_3^{2-} + 2H_2O \quad (10.8)$$

This method may be utilized for the preparation of most arene sandwich compounds as well as for many other arene complexes. However, it cannot be used for manganese arene derivatives.

$$3TiCl_4 + 2Al + 4AlCl_3 + 3C_6H_6 \longrightarrow 3[TiCl_2(PhH)Al_2Cl_6] \qquad (10.9)$$

$$3MoCl_5 + 4Al + 6PhH \longrightarrow 3[Mo(PhH)_2]^+ [AlCl_4]^- + AlCl_3 \qquad (10.10)$$

Cationic molybdenum, tungsten, chromium, and vanadium compounds disproportionate to give sandwich complexes $M(arene)_2$:

$$6[M(arene)_2]^+ + 8OH^- \xrightarrow{KOH (30\%)} 5[M(arene)_2] + MO_4^{2-} 4H_2O + 2\,arene \quad (M = Mo, W)$$
$$\qquad (10.11)$$

$$2[Cr(arene)_2]^+ \xrightarrow{OH^-} [Cr(arene)_2] + Cr^{2+} + 2\,arene \qquad (10.12)$$

$$2[V(arene)_2]^+ \xrightarrow{pH\,7} [V(arene)_2] + V^{2+} + 2\,arene \qquad (10.13)$$

In alkaline solution, the cation $V^{2+}$ reduces $[V(arene)_2]^+$ and therefore the disproportionation reaction proceeds according to the equation

$$3[V(arene)_2]^+ \xrightarrow{OH^-} 2[V(arene)_2] + V^{3+} + 2\,arene \qquad (10.14)$$

### b. Reactions of salts of metals with Grignard compounds

The first method for synthesis of arene complexes involved the reaction of chromium chlorides with phenylmagnesium bromide[1-3] which was discovered by Hein as early as 1919. For a long time it was believed that the products of these reactions are phenylchromium compounds ($Ph_nCr$). However, in the 1950s it was established that these derivatives are actually sandwich arene compounds. The tetrahydrofuran solvate of chromium chloride reacts with PhMgBr to form triphenylchromium which, after heating or under the influence of ether, forms a black pyrophoric substance. Hydrolysis of this black substance furnishes benzene and biphenyl sandwich-type complexes [equation (10.15)].

$$CrCl_3(THF)_3 + PhMgBr \longrightarrow [CrPh_3(THF)_3] \xrightarrow{-THF} \text{black pyrophoric substance}$$

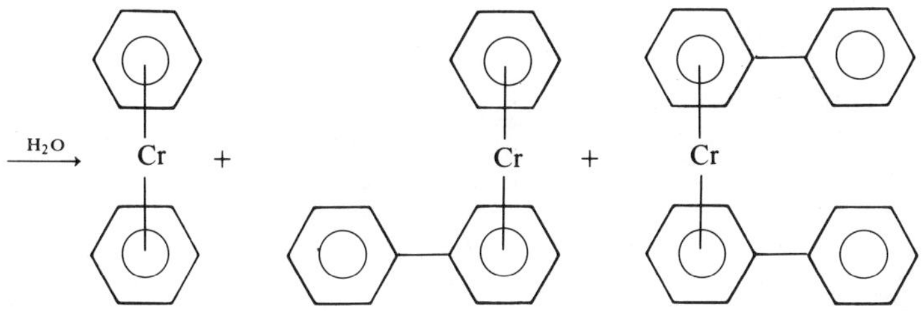

$$\xrightarrow{H_2O} \quad (10.15)$$

Bis(biphenyl)chromium (10.15c) may be extracted from the black substance. However, benzene complexes (10.15a) and (10.15b) are formed during hydrolysis. This is inferred by the presence of deuterium in the benzene ligand in products which are obtained after deuterolysis.

Hein's method may also be utilized for the preparation of heteroleptic arene compounds of manganese and chromium:

$$MnCl_2 + MgBrPh + MgBrcp \xrightarrow{THF} [Mncp(PhH)] \tag{10.16}$$

$$CrCl_3 + MgBrPh + MgBrcp \xrightarrow{THF} [Crcp(PhH)] \tag{10.17}$$

Arene compounds of V, Cr, and Ru are also formed by reactions of salts of these elements with radical anions of aromatic compounds of alkali metals.

## c. *Cyclotrimerization of acetylenes*

Triphenylchromium(III) causes trimerization of acetylenes leading to the formation of arene complexes (cf. reactions of oligomerization of acetylenes, Chapter 6); see equation (10.18).

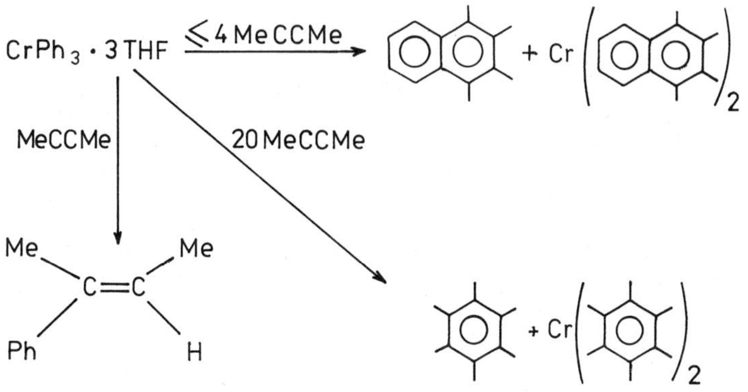

$$\tag{10.18}$$

Other arene derivatives, such as those of manganese, cobalt, nickel, and rhodium, may be prepared in an analogous way.

$$MeCCMe + MnPh_2 \xrightarrow{(n\text{-}Pr)_2O} [Mn(PhH)(C_6Me_6)]^+ \tag{10.19}$$

### d. *Dehydrogenation of 1,3-cyclohexadiene*

During reactions of transition metal salts with cyclohexadiene in the presence of Grignard reagents, dehydrogenation of the diene followed by the formation of arene complexes takes place. Heteroleptic diene–arene compounds may also be synthesized in this manner:

$$CrCl_3 + 1,3\text{-}C_6H_8 + MgBr(i\text{-}Pr) \xrightarrow[hv]{Et_2O} [Cr(PhH)_2] \tag{10.20}$$

$$MCl_3 + 1,3\text{-}C_6H_8 + MgBr(i\text{-}Pr) \xrightarrow[2.\ hv]{1.\ heating} [M(PhH)(C_6H_8) \tag{10.21}$$

$$M = Fe\ (35\%),\ Ru(18\%),\ Os\ (0.2\%)$$

$$ReCl_5 + MgBr(i\text{-}Pr) + MgBrcp + 1,3\text{-}C_6H_8 \xrightarrow[hv]{Et_2O} [Recp(PhH)] + [Recp_2H] \tag{10.22}$$

$$MoCl_5 + MgBr(i\text{-}Pr) + MgBrcp + 1,3\text{-}C_6H_8 \xrightarrow[hv]{Et_2O} [Mocp(PhH)] + [Mocp_2H_2] \tag{10.23}$$

### e. *Substitution of ligands*

The most general method for the preparation of carbonyl arene complexes is substitution of CO groups in metal carbonyls. Vanadium hexacarbonyl undergoes disproportionation in the presence of arenes:

$$[V(CO)_6] + arene \xrightarrow[heptane]{308\ K} [V(CO)_4\ (arene)]^+ [V(CO)_6]^- + V[V(CO)_6]_2 \tag{10.24}$$

This method is most commonly utilized for the preparation of compounds of group 6 metals:

$$[M(CO)_6] + arene \xrightarrow{\Delta} [M(arene)(CO)_3] + 3CO \tag{10.25}$$

$$(M = Cr,\ Mo,\ W)$$

The reaction rate varies according to the series $[Mo(CO)_6] > [W(CO)_6] > [Cr(CO)_6]$. This process is reversible, and therefore the removal of CO causes an increase in yield.

Thiophene and chromium hexacarbonyl react in an analogous manner:

$$[Cr(CO)_6] + C_4H_4S \longrightarrow [Cr(C_4H_4S)(CO)_3] + 3CO \tag{10.26}$$

Carbonyl groups of halogen metal carbonyls may also be substituted in the presence of aluminum chloride:

$$[MCl(CO)_5] + arene + AlCl_3 \longrightarrow [M(arene)(CO)_3]^+ [AlCl_4]^- + 2CO \tag{10.27}$$

$$(M = Mn,\ Re)$$

$$2[McpCl(CO)_3] + PhH + 2AlCl_3 \longrightarrow [Mcp(PhH)(CO)]^+ + [Mcp(CO)_4]^+ + CO + 2AlCl_4^- \quad (M = Mo, W) \qquad (10.28)$$

$$[FecpCl(CO)_2] + arene + AlCl_3 \longrightarrow [Fecp(arene)][AlCl_4] + 2CO \qquad (10.29)$$

Many other ligands also undergo substitution reactions, for instance, cp, RCN, RNC, py, olefins, dienes, $P(OR)_3$, arenes, picolins, etc.

$$Fecp_2 + arene \xrightarrow{Al/AlCl_3} [Fecp(arene)][AlCl_4] + [C_5H_5]_x \qquad (10.30)$$

$$[M(CO)_3(MeCN)_3] + arene \longrightarrow [M(arene)(CO)_3] + 3MeCN \quad (M = Cr, Mo, W) \qquad (10.31)$$

$$[Rhcp(C_2H_4)_2] + Na[BPh_4] \xrightarrow{HClO_4}$$

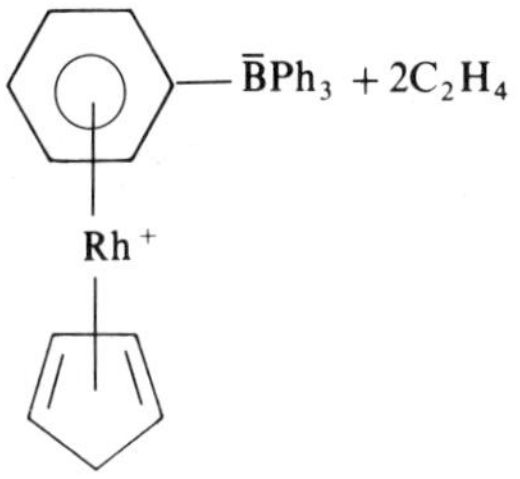

(10.32)

$$[M(arene)(CO)_3] + arene' \longrightarrow [M(arene')(CO)_3] + arene \quad (M = Cr, Mo, W) \qquad (10.33)$$

## f. *Reactions involving rearrangement of ligands*

Cyclic ligands may sometimes undergo ring expansion or contraction leading to the formation of arene complexes. Accordingly, the cation tricarbonyltropylium-chromium(I) reacts with sodium cyclopentadienide to give benzenetricarbonyl-chromium:

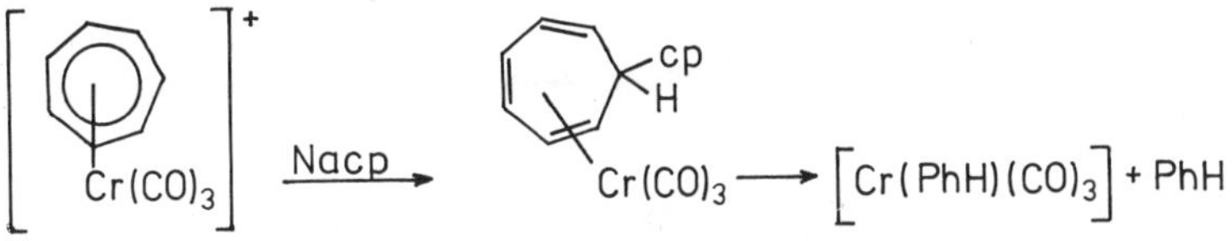

(10.34)

The contraction of the ring occurs also in the case of the following seven-membered heterocyclic compound:

$$+ [Cr(CO)_6] \xrightarrow[413\text{–}423\ K]{diglyme} [Cr(PhH)(CO)_3] + H_2NCOOEt + 3CO \qquad (10.35)$$

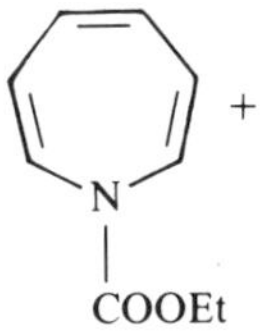

Dibromomethylboron as well as $BBr_2Ph$ and $BBr_3$ add to the cyclopentadienyl ring of cobaltocene to give the complex containing the heterocyclic six-membered ring:

$$3\,Cocp_2 + BBr_2R \longrightarrow RB\!\!\!\!\!\!\!\!\!\!\!\!\!\!\!\!\!\!\underset{Cocp}{\bigcirc} + 2[Cocp_2]^+ \qquad (10.36)$$

### g. Condensation of metal vapors with ligands[22, 23]

Condensation of metal vapors with various $6e$ ligands (cf. preparation of diene complexes, Chapter 8) allows the synthesis of many complexes which would be difficult or impossible to obtain by other methods owing to the reactivity of substituents. The following bisarene complexes were obtained in this manner: $[Cr(arene)_2]$, where arene = PhX (X = F, Cl, $CF_3$, COOMe, Me, $i$-Pr) or $o,m,p$-$C_6H_4XY$ (X = F, Cl, $CF_3$, Me, $i$-Pr; Y = F, Cl, $CF_3$, Me, $i$-Pr) or 2,6-$C_5H_3Me_2N$; $[Mo(PhX)_2]$ (X = H, Me, F, Cl, OMe, $NMe_2$, COOMe); $[W(PhX)_2]$ (X = H, Me, OMe, F); and W($o$-xylene)$_2$.

The condensation of metallic titanium with benzene gave the first sandwich arene compound of titanium(0), $[Ti(PhH)_2]$. This is a diamagnetic red–orange complex.

Several compounds of vanadium of the type $[V(PhX)_2]$ were synthesized (X = H, F, Cl, $CF_3$). The compound $[V(1,4\text{-}C_6H_4F_2)_2]$ was also prepared.

The condensation of metal vapors with hexafluorobenzene and benzene gives unstable arene complexes of V, Cr, Mn, Fe, Co, Ni, and Pd, for instance, $[Cr(C_6H_6)(C_6F_6)]$. The condensation of vapors of chromium or iron with the mixture of vapors of benzene and $PF_3$ or hexafluorobenzene and $PF_3$ leads to the formation, with low yields, of the derivatives $[Cr(C_6F_6)(PF_3)_3]$, $[Cr(PhH)(PF_3)_3]$, and $[Fe(PhH)(PF_3)_2]$. Heteroleptic compounds of chromium, for example, $[Cr(C_6F_6)(PF_3)_3]$, are stable in contrast to $[Cr(C_6F_6)_2]$ which is unstable and explosive. The condensation method was also utilized for the preparation of hexahapto pyridine complexes of chromium.

### h. Miscellaneous reactions

Dibenzenevanadium($-1$) and dibenzenechromium($1-$) may be obtained by reduction of $V(PhH)_2$ by metallic potassium in aprotic chelating solvents such as 1,2-dimethoxyethane or hexamethylphosphoramide.[20, 39] The compound $Ru_2Cl_4(PhH)_2$ reacts with sodium hydroxide to give $[Ru_2(\mu\text{-}OH)_3(PhH)_2]Cl$, while with sodium carbonate it furnishes the tetranuclear complex $[Ru_4(OH)_4(PhH)_4](SO_4)_2 \cdot 12H_2O$ (Figure 10.4c). Many similar reactions in which the arene ligands are not involved may also be carried out.

## 15. REACTIONS OF ARENE COMPLEXES

### a. Redox reactions

Oxidation–reduction reactions which lead to a change in the oxidation state of the metal are the most typical for sandwich arene complexes. However, this is not true for arene carbonyl organometallic compounds. Vanadium compounds V(arene)$_2$ are oxidized immediately in air with complete decomposition. They can be oxidized to other sandwich compounds by means of weak oxidizing agents, for example, $[V(CO)_6]$:

$$[V(arene)_2] + [V(CO)_6] \longrightarrow [V(arene)_2]^+ [V(CO)_6]^- \qquad (10.37)$$

Bis(benzene)vanadium is reduced by potassium to $[V(PhH)_2]^-$. Careful oxidation of the anion allows quantitative recovery of the starting compound:

$$[V(PhH)_2]^- + Ph_2CO \longrightarrow [V(PhH)_2] + Ph_2CO^- \qquad (10.38)$$

Under analogous conditions, bis(benzene)chromium(0) is reduced by potassium to $[Cr(PhH)_2]^-$ in dimethoxyethane solution.[39]

The tendency for oxidation of arene complexes of chromium group metals M(arene)$_2$ decreases in the series W $\geqslant$ Mo > Cr. The chromium compounds are conveniently oxidized to the Cr(arene)$_2^+$ cations by air in the presence of water.

Iodine is a convenient oxidizing agent for molybdenum and tungsten derivatives:

$$2[M(arene)_2] + I_2 \longrightarrow 2[M(arene)_2]I \qquad (10.39)$$

The oxidation–reduction potential for the completely reversible process

$$[Cr(PhH)_2] \underset{+e}{\overset{-e}{\rightleftharpoons}} [Cr(PhH)_2]^+ \qquad (10.40)$$

equals 0.8 V with respect to the calomel electrode in DMF solution.

The reaction of $[Cr(PhH)_2]^+$ with potassium in dimethoxyethane solution leads to the formation of the radical $[Cr(C_6H_6)(C_6H_5)]$ in which rapid exchange of hydrogen between the rings ($\sim 10^7 \, s^{-1}$) takes place as shown by ESR measurements.

The electrochemistry of many sandwich and carbonyl arene complexes was studied.[48, 51]

Stronger oxidizing agents oxidize chromium group compounds $[M(arene)_2]$ leading to their decomposition. The products are inorganic compounds of metals and free arenes.

Arene–carbonyl complexes such as $[M(arene)(CO)_3]$ are very resistant to oxidation. This is due to acceptor properties of the CO groups which cause the increase of the positive charge on the metal. Few cases of oxidation of complexes of the type $[M(arene)(CO)_3]$ are known; however, in these cases, the coordination number increases.

$$[M(arene)(CO)_3] + SbCl_5 \xrightarrow{\text{CHCl}_3} [MCl(arene)(CO)_3]^+ [SbCl_6]^- \qquad (10.41)$$

$$(M = Cr, Mo, W)$$

Standard oxidation potential for complexes $[Cr(arene)(CO)_3]$ in $CF_3COOH + (CF_3CO)_2O$ (93 + 7%) solution is a linear function of the ionization potentials of arene hydrocarbons.[50] The oxidation in trifluoroacetic acid is reversible, while in MeCN and DMF solutions it is irreversible.[51] Further oxidation leads to decomposition of these complexes and the formation of chromium(III) compounds and separation of ligands.

Investigations of electrochemical oxidation of the compounds $[M(arene)(CO)_3]$

(M = Cr, Mo, and W) which were carried out in MeCN solution in the presence of $PR_3$ and $P(OR)_3$ showed that the oxidation products, i.e., complexes possessing metals in +1 oxidation states, react with Lewis bases to give the $[M(arene)(CO)_3L]^+$ cations, the tungsten complexes being the most stable.[51]

The cations of manganese group metals $[M(arene)_2]^+$ are very resistant to both oxidation and reduction. The Re(0) complex $[Re(C_6Me_6)_2]$ is very unstable, however. It easily undergoes dimerization.

In the case of iron, the complexes $[Fe(arene)_2]^{2+}$, $[Fe(arene)_2]^+$, and $[Fe(arene)_2]$ are known. Their resistance to oxidation decreases as the oxidation state of iron becomes lower. The cation $[Ru(C_6Me_6)_2]^{+2}$ undergoes two-electron reduction to give the neutral complex $[Ru(C_6Me_6)_2]$ which is quite stable. This is an 18$e$ compound because one of the hexamethyl benzene molecules is a 4$e$ ligand: $[Ru(\eta^6\text{-}C_6Me_6)(\eta^4\text{-}C_6Me_6)]$. A sandwich complex of ruthenium(I) is not known. The cobalt atom also forms compounds possessing three different oxidation states: $[Co(C_6Me_6)_2]^{2+}$, $[Co(C_6Me_6)_2]^+$, and $[Co(C_6Me_6)_2]$. The bis(hexamethylbenzene)cobalt(I) cation is unexpectedly stable. It is not oxidized by air, while $[Co(C_6Me_6)_2]$ is pyrophoric. The compound $[Ni(C_6Me_6)_2]^{2+}[PtCl_6]^{2-}$, which is stable in air (like the isoelectronic $[Co(arene)_2]^+$ cation), was also prepared.

### b. *Electrophilic substitution reactions*

In arene complexes, there is a charge transfer from the ring to the metal. Therefore, coordinated arenes undergo electrophilic substitutions with greater difficulty than free arenes. Owing to the sensitivity to acids and easy oxidation of bisarene complexes, it has not been possible to perform sulfonation, nitration, and mercuration. However, the acylation reaction of chromium derivatives $[Cr(arene)(CO)_3]$ can be carried out:

$$[Cr(PhR)(CO)_3] + MeCOCl + AlCl_3 \longrightarrow [Cr(RC_6H_4COMe)(CO)_3] \qquad (10.42)$$

$$(R = H, Me, Et, i\text{-}Pr, t\text{-}Bu)$$

### c. *Nucleophilic substitution and addition reactions*

Nucleophilic substitution reactions involving free arenes are very difficult to carry out. The complexation causes a decrease of electron density on the arene molecule and therefore considerably facilitates nucleophilic attack. Thus, the chlorobenzene chromium complex $[Cr(PhCl)(CO)_3]$ easily reacts with sodium methanolate to give anisole chromium tricarbonyl $[Cr(PhOMe)(CO)_3]$. The complex $[Cr(PhF)(CO)_3]$ reacts in a similar way.

The substitution of the halogen (fluorine and chlorine) also occurs very easily in arene cyclopentadienyl iron complexes.

$$[Fecp(PhCl)][BF_4] + MX \xrightarrow{303-333\ K} [Fecp(PhX)][BF_4] + MCl \qquad (10.43)$$

$$(M = H, Na, K; X = OMe, OEt, OPh, SPh, SBu^n, NH_2, CN)$$

Nucleophilic addition to the arene ligand is a specific reaction for the cationic arene complexes. It does not occur for uncoordinated arenes. Dependent on the complex and the conditions of the reaction, addition of neutral nucleophiles such as phosphines, amines, aromatic compounds, or anionic nucleophiles may occur.[52, 53] In some cases, two nucleophiles may be added to the arene to give coordinated dienes.

The decomposition of diene complexes allows the preparation of dienes containing various functional groups.[54, 57] Therefore, this method is significant in organic synthesis. The following complexes are used in such reactions: $[Mn(arene)(CO)_3]^+$, $[Fe(C_6R_6)_2]^{2+}$, $[Fe(C_6R_6H)(C_6R_6)]^+$ $(R = H, Me)$, $[Ru(PhH)_2]^{2+}$, $[Ru(PhH)(C_6Me_6)]^{2+}$, $[Cocp(C_6H_6)]^{2+}$, and $[M(pmcp)(PhH)]^+$ $(M = Rh, Ir)$. In the case of manganese and iron complexes, these reactions are illustrated in equations (10.44) and (10.45). Nucleophilic attack on the arene dienyl iron complex $[Fe(\eta^5\text{-}C_6H_7)(CPhH)]^+$

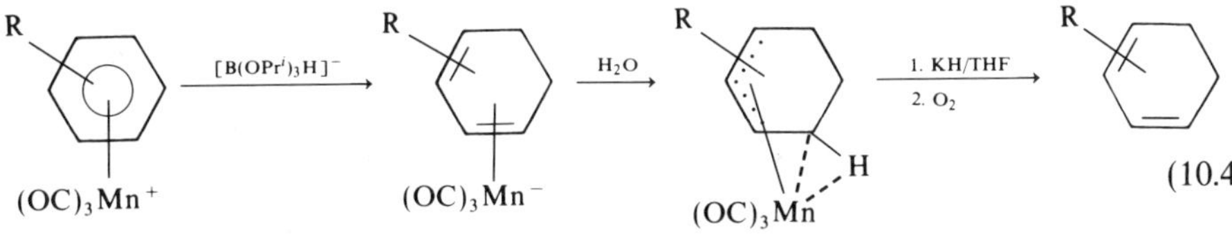

(54)

(10.44)

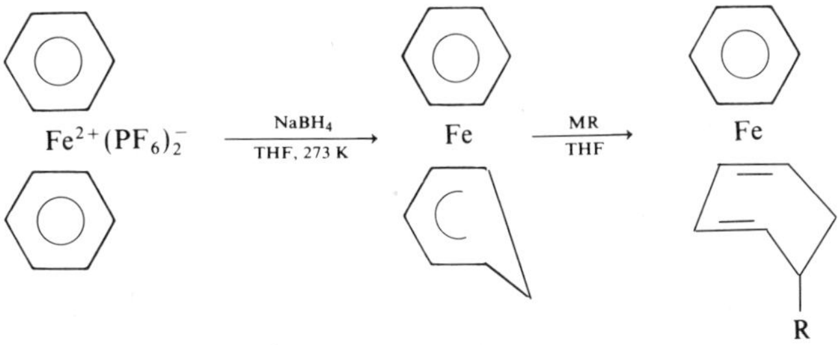

(10.45)

$R = H, CN, CH(COOEt)_2, CH_2Ph, \overline{CHS(CH_2)_3}S, [Fecp(\eta^6\text{-}CH_2C_6Me_5)]^+$ (55-57)

is controlled by frontier orbitals rather than the charge.[57] However, in the mesitylene complex $[Fe(1,3,5\text{-}C_6H_3Me_3)_2]^{2+}$ the attack of the second nucleophile takes place on the arene molecule and not the dienyl ligand, leading to the formation of the bis(dienyl)iron complex; see reaction (10.49).

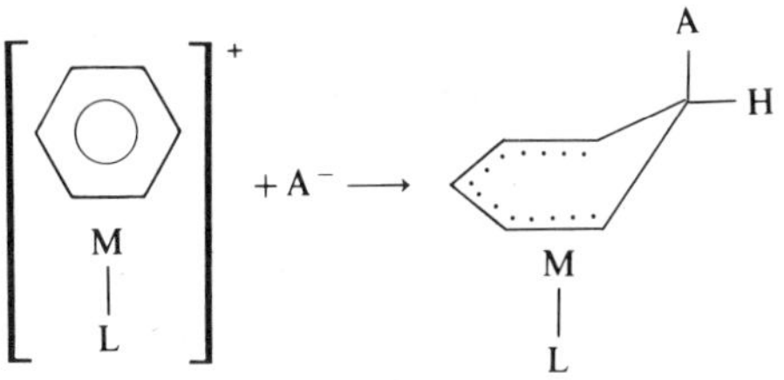

(10.46)

$L = (CO)_n, cp, C_4R_4, arene$

For rhenium and manganese compounds, one molecule of nucleophile is added, while in the case of iron and ruthenium, two molecules are added:

$$[Mn(C_6H_6)(C_6Me_6)]^+ + LiAlH_4 \longrightarrow [Mn(C_6H_7)(C_6Me_6)] \qquad (10.47)$$

$$[Re(C_6H_6)_2]^+ + H^- \longrightarrow [Re(C_6H_6)(C_6H_7)] \qquad (10.48)$$

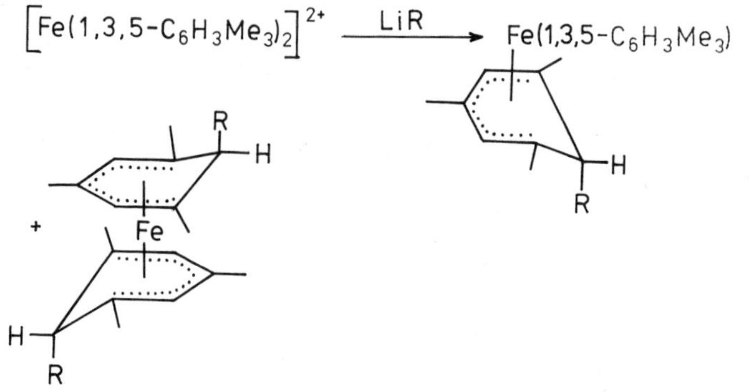

$$\left[Fe(1,3,5\text{-}C_6H_3Me_3)_2\right]^{2+} \xrightarrow{\text{LiR}} Fe(1,3,5\text{-}C_6H_3Me_3)$$

(10.49)

$$[Ru(C_6H_6)_2]^{2+} + LiAlH_4 \longrightarrow [Ru(C_6H_6)(1,3\text{-}C_6H_8)] + [Ru(C_6H_7)_2] \qquad (10.50)$$

$$[V(C_6H_{6-n}Me_n)(CO)_4]^+ + NaBH_4 \xrightarrow{\text{THF}} [V(C_6H_{7-n}Me_n)(CO)_4] \qquad (10.51)$$

$$[Mn(C_6H_6)(CO)_3]^+ + H^- \longrightarrow [Mn(C_6H_7)(CO)_3] + [Mn(C_6H_8)(CO)_3] \qquad (10.52)$$

$$[Mn(C_6H_6)(CO)_3]^+ + LiPh \longrightarrow [Mn(C_6H_6Ph)(CO)_3] \qquad (10.53)$$

### d. Displacement of ligands

The displacement and exchange of arene ligands with $^{14}C$ labeled arenes is a characteristic reaction for arene metal complexes:

$$[M(arene)(CO)_3] + arene\,^{14}C \longrightarrow [M(arene\,^{14}C)(CO)_3] + arene \qquad (10.54)$$

$$(M = Cr, Mo, W; arene = PhH, PhMe, PhCl, C_{10}H_8)$$

The mechanism of the exchange reaction (10.54) is complex. The uncatalyzed reaction is first order with respect to both the complex and the arene.[58] Based on kinetic studies, mechanism (10.55) of the exchange reaction has been proposed.[58] The ligand L may be arene, CO, or catalysts, such as $[Cr(arene)(CO)_3]$, ketones, nitriles, ethers, and other nucleophiles. It is proposed that $[Cr(C_6Me_6)(CO)_3]$ catalyzes the reaction by coordination to the $[Cr(arene)(CO)_3]$ molecule through the oxygen atom of one CO group,[58] $(\eta^4\text{-arene})(OC)_3Cr-O-C-Cr(C_6Me_6)(CO)_2$.

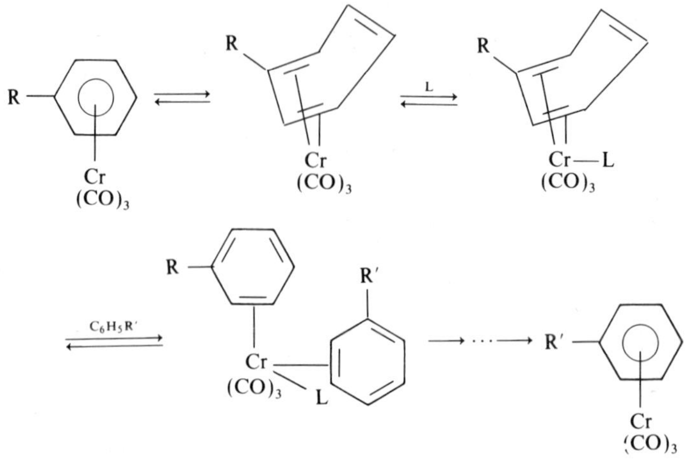

(10.55)

Substitution reactions occur quite easily.

$$[Mo(PhH)_2] + 3L \longrightarrow [Mo(PhH)L_3] + PhH \tag{10.56}$$

$$L = P(OMe)_3, P(OPh)_3, P(OMe)Ph_2, PMe_2Ph, PMePh_2$$

$$[Cr(PhH)_2] + [Cr(CO)_6] \longrightarrow 2[Cr(PhH)(CO)_3] \tag{10.57}$$

$$[V(PhH)_2] + 6CO \longrightarrow [V(CO)_6] + 2PhH \tag{10.58}$$

Arene compounds of molybdenum react with allyl chloride to give oxidative addition products:

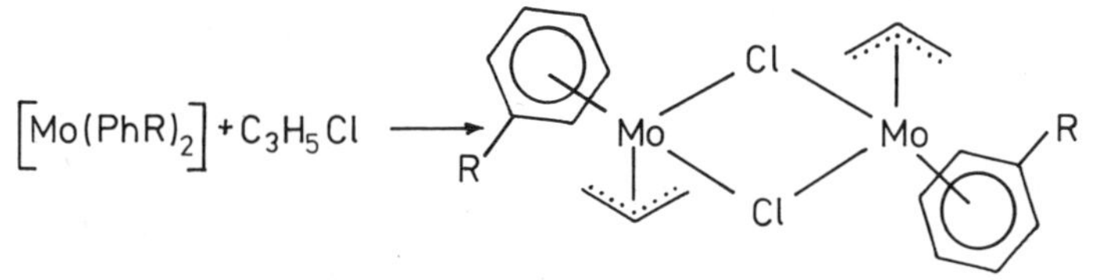

$$\left[Mo(PhR)_2\right] + C_3H_5Cl \longrightarrow \qquad (10.59)$$

In the presence of aluminum chloride, arene displacement by other hydrocarbon arenes with concomitant increase of the oxidation state of the metal may also occur:

$$[Cr(arene)_2] + 2\,arene' \xrightarrow[370\,K]{AlCl_3} [Cr(arene')_2]^+ [AlCl_4]^- + 2\,arene \tag{10.60}$$

In the first step aluminum chloride oxidizes the chromium compound:

$$3[Cr(arene)_2] + 4AlCl_3 \longrightarrow 3[Cr(arene)_2]^+ [AlCl_4]^- + Al \tag{10.61}$$

## 16. APPLICATION OF ARENE COMPLEXES

Arene complexes are utilized for the preparation of metallic layers which are deposited on various surfaces (See Table 2.31). They are also used for the preparation of highly pure metals. This is especially true for chromium group compounds. Pyrolytic decomposition of arene complexes found application for deposition of chromium on magnesium oxide in order to prepare the catalyst for dehydration of isopropyl alcohol.

Titanium(II) compounds such as $[TiCl_2(arene) \cdot Al_2Cl_6]$, where arene $=$ Ph H or $C_6H_{6-n}Me_n$, catalyze oligomerization of butadiene.[19, 26] Tricarbonyl arene complexes of chromium group metals $[M(arene)(CO)_3]$ are effective catalysts for selective hydrogenation of dienes and polyenes to alkenes. The catalytic activity decreases in the series $Mo > W > Cr$, while the selectivity decreases as follows: $Cr > Mo > W$. These compounds also catalyze[27] *cis–trans* isomerization of dienes, migration of double bonds, and olefin metathesis. The compounds $[M(arene)_2]$ and $[M(arene)(CO)_3]$ $(M = Cr, Mo, W)$ may be utilized as catalysts for polymerization of olefins and dienes. This is also true for diarene derivatives of elements of groups 8, 9, and 10. The complexes $[M(arene)(CO)_3]$ $(M = Cr, Mo, W)$ have properties of Lewis acids and may substitute $AlCl_3$ in Friedel–Crafts reactions. They catalyze the following reactions: alkylation, acylation, dehydrohalogenation, polymerization, and sulfonylation. The dinuclear compound $[Mo_2(\mu\text{-}Cl)_2(C_3H_5)_2(PhMe)_2]$ and its derivatives catalyze polymerization of dienes, alkenes, and acetylenes.

Arene complexes are utilized in many organic syntheses [see reactions (10.42)–(10.53)].[1–3, 42, 43, 52–57, 59)

## REFERENCES

1. N. A. Volkenau, in: *Metody Elementoorganicheskoi Khimii* (A. N. Nesmeyanov and K. A. Kocheshkov, eds.), p. 127, Izd. Nauka, Moscow (1976).
2. M. L. H. Green, *Organometallic Compounds, The Transition Elements*, Methuen, London (1968).
3. B. L. Shaw and N. I. Tucker, in: *Comprehensive Inorganic Chemistry* (J. C. Bailar, Jr., ed.), Vol. 4, p. 781, Pergamon Press, Oxford (1973).
4. K. D. Warren, *Structure and Bonding*, **27**, 45 (1976).
5. D. W. Clack and K. D. Warren, *Structure and Bonding*, **39**, 1 (1980).
6. D. R. Scott and R. S. Becker, *J. Phys. Chem.*, **69**, 3207 (1965).
7. S. E. Anderson and R. S. Drago, *J. Am. Chem. Soc.*, **92**, 4244 (1970).
8. K. Nakamoto, in: *Characterization of Organometallic Compounds* (M. Tsutsui, ed.), Part I, p. 73, Interscience, New York (1969).
9. H. P. Fritz, *Adv. Organomet. Chem.*, **1**, 240 (1964).
10. R. G. Kidd, in: *Characterization of Organometallic Compounds* (M. Tsutsui, ed.), Part II, p. 373, Interscience, Nw York (1971).
11. A. H. Cowley, *Prog. Inorg. Chem.*, **26**, 45 (1979).
12. C. Furlani and C. Cauletti, *Structure and Bonding*, **35**, 119 (1978).
13. S. Evans, J. C. Green, S. E. Jackson, and B. Higginson, *J. Chem. Soc., Dalton Trans.*, 304 (1974).
14. S. Evans, J. C. Green, and S. E. Jackson, *J. Chem. Soc., Faraday Trans. 2*, **68**, 249 (1972).
15. J. A. Connor, *Top. Curr. Chem.*, **71**, 71 (1977).
16. R. W. Kiser, in: *Characterization of Organometallic Compounds* (M. Tsutsui, ed.), Part I, p. 137, Intersciene, New York (1969).
17. M. R. Litzow and T. R. Spalding, *Mass Spectrometry of Inorganic and Organometallic Compounds*, Elsevier, Amsterdam (1973).
18. R. B. King, *Appl. Spectrosc.*, **23**, 537 (1969).
19. S. Pasynkiewicz, R. Giezynski, and S. Dzierzgowski, *J. Organomet. Chem.*, **54**, 203 (1973).
20. Ch. Elschenbroich and F. Gerson, *J. Am. Chem. Soc.*, **97**, 3556 (1975).
21. a. J. W. Johnson and P. M. Treichel, *Chem. Commun.*, 688 (1976). b. H. P. Jensen and F. Woldbye, *Coord. Chem. Rev.*, **29**, 213 (1979).
22. K. J. Klabunde, *Acc. Chem. Res.*, **8**, 393 (1975).
23. P. L. Timms, *Angew. Chem.*, **14**, 273 (1975).
24. G. Simonneaux, A. Meyer, and G. Jaouen, *Chem. Commun.*, 69 (1975).
25. H. Brunner and R. G. Gastinger, *Chem. Commun.*, 488 (1977); H. Brunner, *Acc. Chem. Res.*, **12**, 250 (1979).
26. S. Dzierzgowski, R. Giezynski, S. Pasynkiewicz, and N. Nizynska, *J. Mol. Catal.*, **2**, 243 (1977).
27. M. F. Farona, in: *Organometallic Reactions and Syntheses* (E. I. Becker and M. Tsutsui, eds.), Vol. 6, p. 223, Plenum Press, New York (1977).
28. L. F. Farnell, E. W. Randal, and E. Rosenberg, *Chem. Commun.*, 1078 (1971); B. E. Mann, *J. Chem. Soc., Dalton Trans.*, 2012 (1973).
29. M. A. Gallop, A. H. Wright, B. F. G. Johnson, and J. Lewis, *Abstracts*, p. 408, XIIth ICOMC, Vienna (September 8–13, 1985).
30. A. W. Duff, K. Jonas, R. Goddard, H.-J. Kraus, and C. Krueger, *J. Am. Chem. Soc.*, **105**, 5479 (1983); K. Jonas, *Angew. Chem., Int. Ed. Engl.*, **24**, 295 (1985).
31. P. T. Chesky and M. B. Hall, *Angew. Chem., Int. Ed. Engl.*, **106**, 5186 (1984).
32. W. Siebert, *Angew. Chem., Int. Ed. Engl.*, **24**, 943 (1985).
33. D. M. P. Mingos, in: *Comprehensive Organometallic Chemistry* (G. Wilkinson, F. G. A. Stone, and E. W. Abel, eds.), Vol. 3, Chapter 19, Pergamon Press, Oxford (1982).

34. S. P. Gubin and G. B. Shulpin, *Khimia Kompleksov so Svyazyami Metall-Uglerod*, Izd. Nauka, Sibirskoe Otdelenie, Novosibirsk (1984).
35. E. Maslowsky, Jr., *Vibrational Spectra Organometallic Compounds*, Wiley, New York (1977).
36. J. C. Green, *Structure and Bonding*, **43**, 37 (1981).
37. B. E. Mann, in: *Comprehensive Organometallic Chemistry* (G. Wilkinson, F. G. A. Stone, and E. W. Abel, eds.), Vol. 3, Chapter 20, Pergamon Press, Oxford (1982); U. Lenneck, Ch. Elschenbroich, and R. Moeckel, *J. Organomet. Chem.*, **219**, 177 (1981).
38. M. D. Morand and C. G. Francis, *Inorg. Chem.*, **24**, 56 (1985).
39. C. Elschenbroich, E. Bilger, J. Koch, and J. Weber, *J. Am. Chem. Soc.*, **106**, 4297 (1984).
40. M. Y. Darensbourg and E. L. Muetterties, *J. Am. Chem. Soc.*, **100**, 7425 (1978).
41. M. A. Bennett and J. P. Ennett, *Organometallics*, **3**, 1365 (1984).
42. G. Wilkinson, F. G. A. Stone, and E. W. Abel (eds.), *Comprehensive Organometallic Chemistry*, Pergamon Press, Oxford (1982).
43. A. N. Nesmeyanov and K. A. Kocheshkov, (eds.), *Metody Elementoorganicheskoi Khimii*, Vols. 1 and 2, Nauka, Moscow (1974); Vols. 3, 4, and 5, Nauka, Moscow (1975); Vol. 6, Nauka, Moscow (1978); Vol. 7, Nauka, Moscow (1983); Vol. 8, Nauka, Moscow (1985).
44. E. L. Muetterties, J. R. Blecke, E. J. Wucherer, and T. A. Albright, *Chem. Rev.*, **82**, 499 (1982).
45. G. E. Herberich, B. Hessner, G. Huttner, and L. Zsolnai, *Angew. Chem., Int. Ed. Engl.*, **20**, 472 (1981).
46. O. J. Scherer, H. Sitzmann, and G. Wolmershaeusser, *Angew. Chem., Int. Ed. Engl.*, **24**, 351 (1985).
47. O. J. Scherer, *Angew. Chem., Int. Ed. Engl.*, **24**, 924 (1985).
48. N. G. Connelly and W. E. Geiger, *Adv. Organomet. Chem.*, **23**, 1 (1984).
49. A. P. Tomilov, I. N. Chernykh, and Yu. M. Kargin, *Elektrokhimia Elementoorganicheskikh Soedinenii*, Nauka, Moscow (1985).
50. J. O. Howell, J. M. Moncalves, C. Amatore, L. Klasine, R. M. Wrightman, and J. K. Kochi, *J. Am. Chem. Soc.*, **106**, 3968 (1984).
51. K. M. Doxsee, R. H. Grubbs, and F. C. Anson, *J. Am. Chem. Soc.*, **106**, 7819 (1984).
52. L. A. P. Kane-Maguire, E. D. Honig, and D. A. Sweigart, *Chem. Rev.*, **84**, 525 (1984).
53. E. D. Honig, Q.-J. Meng, W. T. Robinson, P. G. Williard, and D. A. Sweigart, *Organometallics*, **4**, 871 (1985).
54. M. Brookhart and A. Lukacs, *Organometallics*, **106**, 4161 (1984).
55. A. M. Madonik, D. Mandon, P. Michaud, C. Lapinte, and D. Astruc, *Organometallics*, **106**, 3381 (1984).
56. A. M. Madonik and D. Astruc, *Organometallics*, **106**, 2437 (1984).
57. D. Astruc, P. Michaund, A. M. Madonik, J.-Y. Saillard, and R. Hoffmann, *Nouv. J. Chim.*, **9**, 41 (1985).
58. T. G. Traylor, K. J. Stewart, and M. J. Goldberg, *J. Am. Chem. Soc.*, **106**, 4445 (1984).
59. S. G. Davies, *Organotransition Metal Chemistry Applications to Organic Synthesis*, Pergamon Press, Oxford (1984).
60. S. D. Reynolds and T. A. Albright, *Organometallics*, **4**, 980 (1985).
61. G. Allegra, G. T. Casagrande, A. Immirzi, L. Porri, and G. Vitulli, *J. Am. Chem. Soc.*, **92**, 289 (1970); G. Nardin, P. Delise, and G. Allegra, *Gazz. Chim. Ital.*, **105**, 1047 (1975).

# Chapter 11

# Complexes Containing Seven- and Eight-Electron π-Ligands

## 1. BONDING, STRUCTURE, SPECTROSCOPY, AND MAGNETIC PROPERTIES

The most commonly encountered $7e$ ligands are the cycloheptatrienyl ligand (CHT), $C_7H_7$, and the azulenium ligand, $C_{10}H_9$. Cyclooctatetraene (COT), $C_8H_8$, represents an $8e$ ligand; it forms sandwich-type compounds with some actinides, for example, $[U(COT)_2]$. Coordinated $\eta^7$-$C_7H_7$ and $\eta^8$-COT molecules are planar, and all C–C distances are equal or differ only slightly.[1-5] The interaction between CHT and COT and the metal is similar to the interaction between cyclopentadienyl or arene and the metal in complexes of the type $Mcp_2$ and $M(arene)_2$. However, due to the lowering of $e_1$ and $e_2$ orbital energy as a result of the ring size increase (Figure 9.4b), there is an increase of overlap between $e_2$ ($d_{x^2-y^2}$ and $d_{xy}$) metal orbitals and ligand orbitals possessing the same symmetry (Figure 11.1). This effect causes an increase in the energy difference between the $e_2$-type molecular orbitals and those of $a_1$ type possessing predominantly $d$ orbital character of the metal (Figure 9.4a).[6, 7]

Complexes with $7e$ and $8e$ ligands ($\eta^7$-$C_7H_7$ and $\eta^8$-$C_8H_8$) contain central atoms which possess a small number of $d$ electrons such as group 4, 5, and 6 metals.

The carbon–metal distances in cycloheptatrienyl molybdenum and vanadium complexes are $ca$ 225 ppm. The distance between the $C_7$ plane and the metal is considerably shorter than the analogous distance for the cyclopentadienyl ligand. In the compound $[Vcp(C_7H_7)]$ the distance between the $C_7$ plane and the vanadium atom is 150 pm, and the distance between the $C_5$ plane and the central atom is 190 pm. In $[Ticp(CHT)]$, the C–C(CHT) distance is 139.7 pm, C–C(cp) 139.6, Ti–C(CHT) 219.4, and the Ti–C(cp)

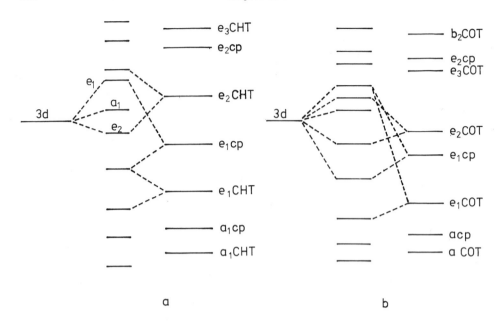

Figure 11.1. Molecular orbital scheme for (a) Crcp(CHT) and (b) Ticp(COT).

distance is 232.1. The titanium–$C_5H_5$ distance is 199.4 pm, and the titanium–$C_7H_7$ is 149 pm.[3b] Thorium, protactinium, uranium, neptunium, and plutonium form complexes of the type $[M(COT)_2]$ possessing sandwich structures. The distances between the rings are 384.7 pm for $[U(COT)_2]$ and 400.7 pm for $[Th(COT)_2]$.[3a]

In the case of uranocene compounds such as $U(C_8H_{8-n}R_n)_2$ (R = H, alkyl, etc.) the metal–ring bond is relatively highly covalent. Theoretical calculations[19] and photoelectron spectra[10, 20, 21] show that the interaction between the $e_{2u}$ and $e_{2g}$ ligand orbitals and the corresponding 5f and 6d uranium orbitals is relatively strong (Figure 11.2). The LCAO semiempirical calculations for $Ce(COT)_2^-$ and $U(COT)_2$[22] show that the ligand orbitals undergo mixing to a greater extent with the $f_{xyz}$ and $f_{z(x^2-y^2)}$ uranium orbitals (ca 22%) than with those of cerium (ca 3%) because the 5f uranium orbitals are larger than the 4f cerium orbitals. However, cyclooctatetraene complexes exhibit considerable ionic character because the $C_8H_8$ ligand easily forms planar $C_8H_8^{2-}$ anions which possess aromatic character ($4n + 2 = 10e$). The structures of some cycloheptatrienyl and cyclooctatraene compounds are represented in Figure 11.3.

Like other sandwich complexes, the CHT and COT compounds are often paramagnetic. The magnetic moment of the compound $Ticp(C_8H_8)$ is 1.60 BM and does not depend on temperature; this strongly indicates the $a_1^1$ electronic configuration.[6, 8] The cycloheptatrienyl complexes $[Ticp(CHT)]$ and $[Vcp(CHT)]^+$ possessing $d^4$ electronic structure of their central atoms are diamagnetic ($e_2^4$ configuration). The magnetic moment of the $d^5$ electron compound Vcp(CHT) is 1.69 BM, which corresponds to the spin-only value. Thus, the complex should have $e_2^4 a_1^1$ electron configuration. The isoelectronic chromium compound $[Crcp(CHT)]^+$ shows a higher

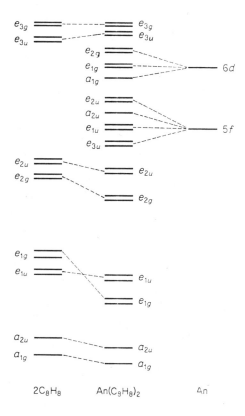

Figure 11.2.  Molecular orbital scheme for cyclooctatetraene complexes of the actinides.[10,20]

magnetic moment than the spin-only value, i.e., 2.0–2.1 BM. Therefore, the following electronic configuration $a_1^2 e_2^3$ is possible for this compound.[6,7] However, the ESR measurements did not confirm such an electronic structure; these measurements show $e_2^4 a_1^1$ configuration.[6] For Ticp(COT) and V(COT)$_2$ complexes, the ESR data indicate the $a_1^1$ configuration[9] and that the overlap of the $a_1$ metal orbitals and ligand orbitals is small.

The photoelectron spectrum of uranocene U(COT)$_2$ shows a band at 6.20 eV corresponding to the ionization of $f$ electrons, a band at 6.90 eV corresponding to the ionization of $e_{2u}$ electrons, and a band at 7.85 eV due to the ionization of $e_{2g}$ electrons. The bands caused by the ejection of electrons from $e_{1u}$ and $e_{1g}$ ligand orbitals appear at 9.90, 10.14, and 10.65 eV $(e_{1u} + e_{1g})$. In the case of the thorium complex [Th(COT)$_2$] in which the central ion has the $d^4 f^0$ configuration, analogous bands occur at the energies 6.97 $(e_{2u})$, 7.91 $(e_{2g})$ as well as at 9.95, 10.28, and 10.56 eV.[10,21] The photoelectron spectra of U(C$_8$H$_7$R)$_2$ (R = Me, Et, Bu$^n$, and Bu$^t$), U(C$_8$H$_6$Me$_2$)$_2$, and U(C$_8$H$_4$Me$_4$)$_2$ show that the contribution of the $6d$ orbitals, and particularly those of the $5f$ orbitals, to the formation of the metal–ring bond increases if the number of alkyl groups in the ring becomes greater.[20] The interaction between the metal orbitals and

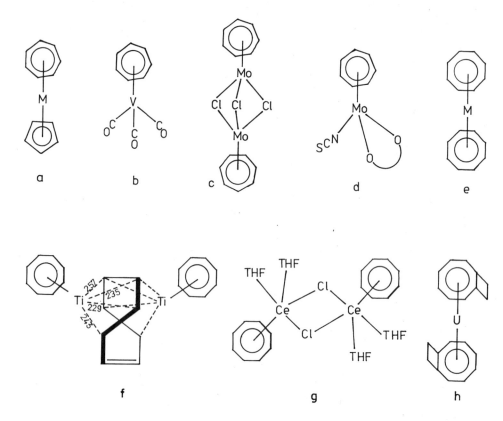

Figure 11.3. Structures of (a) [Vcp(C$_7$H$_7$] and [Ticp(C$_7$H$_7$],[3b] (b) [V(C$_7$H$_7$)(CO)$_3$], (c) [Mo$_2$Cl$_3$($\eta$-C$_7$H$_7$)$_2$]$^+$,[5] (d) [Mo(NCS)(acac)(C$_7$H$_7$)],[4] (e) [M(COT)$_2$] (M = Th, U, Np, Pu) and [M(COT)$_2$]$^-$ (M = Ce, Pr, etc.),[3a, 11] (f) [Ti$_2$(COT)$_3$], (g) [Ce$_2$Cl$_2$(COT)$_2$(THF)$_4$],[11] (h) [U{C$_8$H$_6$(CH$_2$)$_2$}].[12]

the $e_2$ orbitals is stronger than the interaction between the metal orbitals and the $e_1$ orbitals. The photoelectron spectra of some $d$ electron metal complexes with 7e and 8e ligands are given in Table 10.6.

The $^1$H NMR spectra show only one signal for CHT indicating its planar structure. For the compound [Mo$_2$Cl$_3$(CHT)$_2$]$^+$ [MoCl(CF$_3$CCCF$_3$)$_3$]$^-$, the $^1$H NMR signal occurs at $\tau = 3.64$. The spectrum of this compound does not change with temperature.[5] The compound [V(CHT)(CO)$_3$] also shows a $^1$H NMR singlet at 223 K. However, this spectrum changes considerably at higher temperatures, which can be explained by the lowering of the ligand symmetry leading to three delocalized double bonds and one carbonium carbon atom. Such an asymmetrical C$_7$H$_7$ ligand gives a A$_2$B$_2$CD$_2$-type spectrum. These conclusions are also confirmed by IR spectra, which are more complicated in solution than in the solid state.[15] For diamagnetic complexes K[Y(COT)$_2$] and K[La(COT)$_2$], signals at $\tau = 4.25$ and 4.10 (with respect to external TMS) are observed in THF. Similar shifts are given by the free ion C$_8$H$_8^{2-}$ (Table 11.1, ref. $o$).

*Table 11.1. Cycloheptatrienyl and Cyclooctatetraene Metal Complexes*

| Compound | Color | Mp (K) | Remarks |
|---|---|---|---|
| $[Ti_2(COT)_3]^a$ | Yellow | | $d$(TiC) 235 pm |
| $[Ti(COT)_2]^a$ | Violet-red | | |
| $[Ticp(COT)]^{(8,9)}$ | Dark-green | 433 dec | $\mu_{eff} = 1.60$ BM |
| $[Ticp(CHT)]^d$ | Blue | | |
| $[Zrcp(CHT)]^e$ | Violet | | |
| $[Vcp(CHT)]^b$ | Purple | 403 dec | |
| $[V(CHT)(CO)_3]^c$ | Dark-green | 407–410 dec | |
| $[V(COT)_2]^{a(9)}$ | | | |
| $[Nbcp(CHT)]^e$ | | | $\mu_{eff} = 2.00$ BM |
| $[Crcp(CHT)]^{e,f,j}$ | Dark-green | 498 dec | |
| $[Cr(CHT)(CO)_3]^+[PF_6]^{-\,g,h}$ | Orange | | |
| $[Mo_2(\mu\text{-}Cl)_3(CHT)_2]^+$ | | | |
| $[MoCl(CF_3CCCF_3)]^{-\,(5)}$ | Yellow-green | | |
| $[Mocp(CHT)]^{i,e}$ | Brown | 523 dec | |
| $[Mo(PhMe)(CHT)][PF_6]^{(4,14)}$ | Green | | |
| $[Mo_2Cl_3(CHT)_2]^{(14)}$ | Dark-green | | |
| $[W(CHT)(CO)_3][PF_6]^h$ | Orange | | |
| $[Mncp(C_7H_6CH_3)][PF_6]^j$ | Red | | |
| $[U(COT)_2]^{l,k(3)}$ | Green | Pyrophoric | |
| $[Th(COT)_2]^{m(3)}$ | Light-yellow | | |
| $[U(1,3,5,7\text{-}C_8H_4Ph_4)_2]^n$ | Green-black | Air-stable | |
| $[Ce_2Cl_2(COT)_2(THF)_4]^{o(11)}$ | Yellow-green | | |
| $K[Y(COT)_2]^o$ | Yellow | | |
| $K[La(COT)_2]^o$ | Green | | |
| $K[Ce(COT)_2]^o$ | Light-yellow | | |

[a] H. Breil and G. Wilke, *Angew. Chem., Int. Ed. Engl.*, **5**, 898 (1966); H. Dietrich and H. Dierks, *Angew. Chem., Int. Ed. Engl.*, **5**, 899 (1966).
[b] R. B. King and F. G. A. Stone, *J. Am. Chem. Soc.*, **81**, 5263 (1959).
[c] R. P. M. Werner and S. A. Manastyrskyj, *J. Am. Chem. Soc.*, **83**, 2023 (1961).
[d] H. O. van Oven and H. J. de Liefde Meijer, *J. Organomet. Chem.*, **23**, 159 (1970).
[e] H. O. van Oven, C. J. Groenenboom, and H. J. de Liefde Meijer, *J. Organomet. Chem.*, **81**, 379 (1974).
[f] R. B. King and M. B. Bisnette, *Inorg. Chem.*, **3**, 785 (1964).
[g] J. D. Munro and P. L. Pauson, *J. Chem. Soc.*, 3475, 3479, 3484 (1961).
[h] H. J. Dauben, Jr. and L. R. Honnen, *J. Am. Chem. Soc.*, **80**, 5570 (1958).
[i] E. O. Fischer and H. W. Wehner, *J. Organomet. Chem.*, **11**, P29 (1968).
[j] E. O. Fischer and S. Breitschaft, *Angew. Chem., Int. Ed. Engl.*, **2**, 100 (1963); E. P. Fischer and S. Breitschaft, *Chem. Ber.*, **99**, 2213 (1966).
[k] A. Streitwieser, Jr. and U. Muller-Westerhoff, *J. Am. Chem. Soc.*, **90**, 7364 (1968).
[l] A. Streitwieser, Jr., U. Muller-Westerhoff, G. Sonnichsen, F. Mares, D. G. Morrell, K. O. Hodgson, and C. A. Harmon, *J. Am. Chem. Soc.*, **95**, 8644 (1973).
[m] A. Streitwieser, Jr. and N. Yoshida, *J. Am. Chem. Soc.*, **91**, 7528 (1969).
[n] A. Streitwieser, Jr. and R. Walker, *J. Organomet. Chem.*, **97**, C41 (1975).
[o] K. O. Hodgson, F. Mates, D. F. Starks, and A. Streitwieser, Jr., *J. Am. Chem. Soc.*, **95**, 8650 (1973).

## 2. PROPERTIES OF CYCLOHEPTATRIENYL AND CYCLOOCTATETRAENE COMPLEXES[1, 2, 23–27]

The following representative coordination compounds containing cycloheptatrienyl and cyclooctatetraene ligands are known: $[Mcp(L)]$ (M = Ti, Zr, V, Nb, Cr, Mo; L = $C_7H_7$, $C_7H_6Me$), $[Mcp(CHT)]^+$ (M = V, Cr, Mo, Mn), $[M(CHT)(CO)_3]^+$

(M = Cr,   Mo,   W),   [V(CHT)(CO)$_3$],   [Ti(CHT)(C$_7$H$_9$)],   [Zr(CHT)Cl$_2$], [Ta(CHT)$_2$Cl$_3$],   [(OC)(CHT) Mo($\mu$-TePh)$_2$ Mo(CHT)(CO)],   [(CHT) Mo($\mu$-SR)$_3$ Mo(CO)$_3$], [Mo(NCS)(acac)(CHT)], [Mo(acac)(H$_2$O)(CHT)][BF$_4$], [Mo$_2$($\mu$-Cl)$_3$ (CHT)$_2$]$^+$ [MoCl(CF$_3$CCCF$_3$)$_3$]$^-$,   [MoX(CO)$_2$(CHT)]   (X = Cl,   Br,   I), [Mo(CHT)(PhR)]$^+$ (R = H, Me), [V(CHT)$_2$]$^{2+}$ [BF$_4^-$]$_2$,$^{(18)}$ [M(COT)$_2$] (M = Ti, Zr, Hf, V, Ce, Th, Pa, U, Np, Pu), [M$_2$(COT)$_3$] (M = Ti, Ce, V), Ti$_2$(COT)$_3^{2-}$, K[M(COT)$_2$] (M = Sc, Y, La, Ce, Pr, Nd, Sm, Gd, Tb), KM(COT)$_2$ · 2L (M = Ce, Pu, 2L = diglyme;   M = Np,   Pu,   Am,   L = THF),   Th(C$_8$H$_7$R)Cl$_2$ · 2THF,   Th(COT) (BH$_4$)$_2$ · 2THF,   [K(MeOC$_2$H$_4$OMe)]$_2$ [Yb(C$_8$H$_8$)$_2$],$^{(25)}$ [K(MeOC$_2$H$_4$OMe)$^+$]$_2$ [Ce(COT)$_2$]$^{2-}$, [M(COT) Cl(THF)$_2$]$_2$ (M = Ce, Pr, Nd, Sm), [Sc(COT) Cl(THF)]$_n$, [U(C$_8$H$_7$R)$_2$] (R = Me, Et, Bu$^n$, Bu$^t$), [U(C$_8$H$_6$Me$_2$)$_2$], [U(C$_8$H$_4$R$_4$)] (R = Me, Ph), [U{C$_8$H$_6$(CH$_2$)$_2$}$_2$],$^{(12)}$ [Ln(COT)(THF)$_2$]$^+$ [M(COT)$_2$]$^-$ (M = La, Ce, Nd, Er), Ti$_4$($\mu_3$-Cl)$_4$ (COT)$_4$,   [(THF)(COT) Ti($\mu$-Cl)$_2$ Ti(COT)(THF)],   [Mcp(COT)] (M = Ti, Sc, Y, Nd), [ZrCl$_2$(COT)$_2$], [Zr(COT)$_2$ (dmpe)], [ZrCl$_2$(COT)(THF)], [LuR(COT)(THF)$_n$] (R = CH$_2$SiMe$_3$, n = 2; R = 2-C$_6$H$_4$CH$_2$NMe$_2$, n = 1).$^{(26)}$

The   structures   of   cerium   and   ytterbium   complexes   of   the   type [K(dme)]$_2$ [M(COT)$_2$] are not the same because the IR spectrum of the cerium compound does not show bands at 890 and 690 cm$^{-1}$, which are characteristic for the C$_8$H$_8^{2-}$ in K$_2$C$_8$H$_8$. The IR spectra of all cyclooctatetraene complexes of the $f$ block metals exhibit strong bands at 890 ± 20 and 690 ± 20 cm$^1$. The complex [K(dme)]$_2$ [Yb(COT)$_2$] has parallel eclipsed centrosymmetric structure (11.1).$^{(25)}$

$$d(\text{K–C}) = 298.9\text{–}303.6 \text{ pm} \qquad d(\text{Yb–C}) = 271.1\text{–}278.3 \text{ pm}$$

(11.1)

Many mixed intercalation compounds are also known and are obtained by reactions between solutions of sandwich compounds and powdered chalcogenide layered metal compounds such as TaS$_2$, ZrS$_2$, etc. Examples are {[Crcp(CHT)]$_{1/4}$ZrS$_2$} and {[Ticp(COT)]$_{1/4}$ZrS$_2$}.$^{(13)}$ It is assumed that because of low ionization potentials of sandwich complexes, electron transfer takes place from the organometallic compound to the parent metal compound containing the chalcogenide. Similar compounds are formed by arene complexes and metallocenes: {[M(PhH)$_2$]$_{1/6}$ZrS$_2$} (M = Cr, Mo), {[Mo(arene)$_2$]$_x$ZrS$_2$} (x < 1/6, arene = PhMe, mesitylene), {[Crcp(PhH)]$_{1/4}$ZrS$_2$}, {[Crcp$_2$]$_{1/4}$ZrS$_2$}, and {[Cocp$_2$]$_{1/4}$TaS$_2$}.

Cycloheptatrienyl and cyclotatetraene complexes, like other sandwich compounds, and complexes [M(C$_n$H$_n$)L$_x$] containing polyelectron ligands are thermally stable. Often, these compounds melt at higher temperatures without decomposition. Complexes with ligands which do not contain bulky substituents connected to the ring carbon atoms are usually very sensitive to oxidation and hydrolysis. The compounds of the type M(COT)$_2$ (M = U, Pa, Ce, etc.) are easily oxidized but are more resistant to water. Table 8.1 gives properties of some cycloheptatrienyl and cyclooctatetraene metal compounds.

## 3. *METHODS OF PREPARATION OF CYCLOHEPTATRIENYL AND CYCLOOCTATETRAENE COMPLEXES*

Complexes containing CHT may be obtained by reactions of cycloheptatriene compounds with triphenylmethyl tetrafluoroborate:

$$[Mo(C_7H_8)(CO)_3] + [Ph_3C]^+ [BF_4]^- \longrightarrow [Mo(C_7H_7)(CO)_3][BF_4] + Ph_3CH \qquad (11.2)$$

In some cases, hydrogen abstraction from the triene ligand occurs spontaneously:

$$[Vcp(CO)_4] + C_7H_8 \longrightarrow [Vcp(CHT)] \qquad (11.3)$$

$$[V(CO)_6] + C_7H_8 \longrightarrow [V(CHT)(CO)_3] + [V(CHT)(C_7H_8)]^+ + [V(CO)_6]^- \qquad (11.4)$$

The compound Crcp(CHT) is obtained by dehydrogenation of cycloheptatriene-cyclopentadienylchromium using a platinum catalyst,

$$[CrcpCl_2(THF)] + MgBr(i\text{-}Pr) \xrightarrow{Et_2O} [Crcp(i\text{-}Pr)_2 (Et_2O)]$$

$$\xrightarrow[hv]{C_7H_8} [Crcp(C_7H_8)] \xrightarrow{Pt, \text{ benzene}} [Crcp(CHT)] \qquad (11.5)$$

Many cationic cycloheptatrienyl complexes are formed by reactions of various coordination compounds with $C_7H_8$ in the presence of aluminum chloride or aluminum organometallic compounds.[4, 14]

$$[Crcp(PhH)] + C_7H_8 \xrightarrow{AlCl_3} [Crcp(CHT)]^+ \qquad (11.6)$$

$$[MoCl(C_3H_5)(PhMe)]_2 + C_7H_8 \xrightarrow[2. \ [PF_6]^-, H_2O]{1. \ AlEtCl_2} [Mo(CHT)(PhMe)]^+ \qquad (11.7)$$

The compound $[Mo(CHT)(PhMe)]^+$ is utilized for the preparation of other derivatives.[4, 5, 14]

$$[Mo(CHT)(PhMe)]^+ + Na^+acac^- + PPh_3 \xrightarrow{MeOH} [Mo(CHT)(acac)(PPh_3)] \qquad (11.8)$$

$$[Mo(CHT)(PhMe)]^+ + NaOMe \xrightarrow{MeOH} [Mo_2(OMe)_3 (CHT)_2]$$

$$\xrightarrow[PhH]{HCl} [Mo_2Cl_3(CHT)_2] \xrightarrow{O_2} [Mo_2Cl_3(CHT)_2]^+ \qquad (11.9)$$

$$[Mo(CHT)(arene)][BF_4] + SiMe_3X \longrightarrow [Mo_2X_3(CHT)_2] \qquad (11.10)$$

$$(X = Cl, Br, I)$$

Cationic complexes may be reduced by $Na_2S_2O_4$:

$$[Crcp(C_7H_7)]^+ \xrightarrow{Na_2S_2O_4} [Crcp(CHT)] \qquad (11.11)$$

Under Friedel–Crafts conditions, enlargement of the arene ring may take place (see Chapter 10):

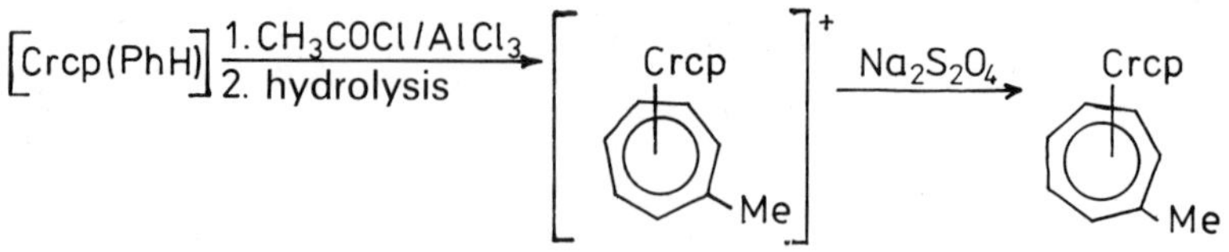

$$\left[Crcp(PhH)\right] \xrightarrow[\text{2. hydrolysis}]{\text{1.} CH_3COCl/AlCl_3} \left[\text{Crcp}\right]^+ \xrightarrow{Na_2S_2O_4} \text{Crcp} \qquad (11.12)$$

Cyclooctatetraene complexes may be obtained by reactions of COT with coordination compounds in the presence of reducing agents.

$$Ti(OBu^n)_4 + COT \xrightarrow{AlEt_3} Ti_2(COT)_3 \qquad (11.13)$$

The most important method of synthesis of compounds with the coordinated $\eta^8$-COT ligand is the reaction of transition metal salts with $K_2C_8H_8$, which in turn can be prepared from dispersed potassium and COT in tetrahydrofuran solution[11] (Table 11.1, references $k$–$o$):

$$2K + C_8H_8 \xrightarrow[\text{210–250 K}]{\text{THF}} 2K^+ + C_8H_8^{2-} \qquad (11.14)$$

$$2K_2C_8H_8 + UCl_4 \longrightarrow [U(COT)_2] + 4KCl \qquad (11.15)$$

The synthesis of cyclooctatetraene complexes may be carried out utilizing many other reducing agents such as metallic sodium and magnesium, aluminum organometallic compounds, etc.

The reaction of chromium trichloride with bromoisopropylmagnesium and azulene gives the chromium complex containing a coordinated $7e$ ligand as well as a coordinated $5e$ ligand:

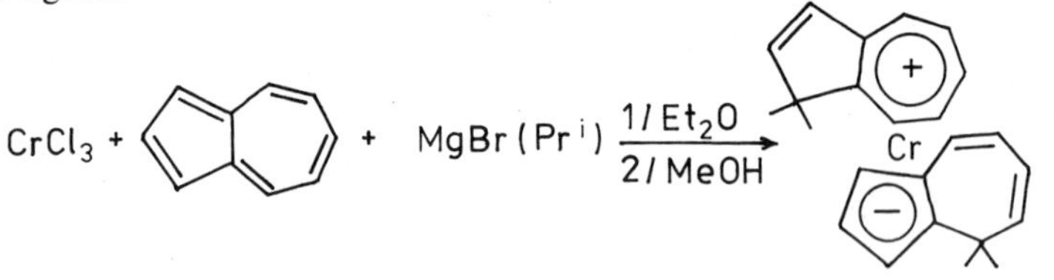

$$CrCl_3 + \quad + \quad MgBr(Pr^i) \xrightarrow[\text{2/ MeOH}]{\text{1/ Et}_2O} \qquad (11.16)$$

In the presence of Raney nickel, the uncoordinated double bond undergoes reduction to afford the complex $[Cr(C_{10}H_{11})(C_{10}H_{13})]$.

The only sandwich complex of molybdenum possessing both a seven-carbon and a six-carbon ring has been prepared by reaction of $[MoI(CHT)(CO)_2]$ with $NaBPh_4$ [cf. reactions (11.7)–(11.10)]. This compound represents a unique zwitterionic sandwich containing the large $\eta^6$-bonding tetraphenylborato ligand[29, 30]:

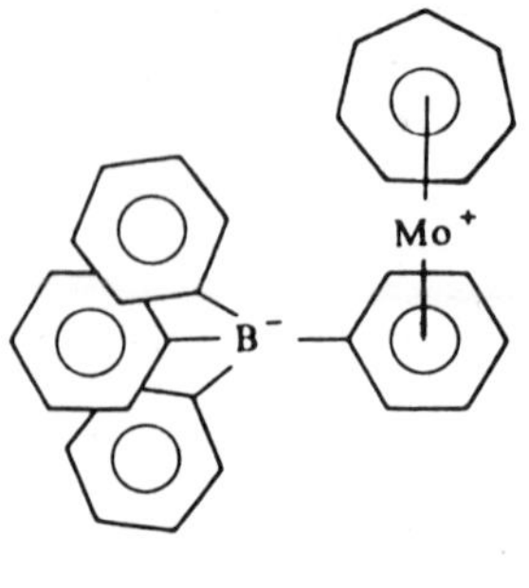

## 4. REACTIONS OF CYCLOHEPTATRIENYL AND CYCLOOCTATETRAENE COMPLEXES

### a. Reactions with nucleophiles

Some nucleophiles react with cationic cycloheptatrienyl complexes to give addition products. Addition of $H^-$, $CN^-$, $OMe^-$, and $Ph^-$ leads to the formation of complexes possessing *exo* structures (11.17).

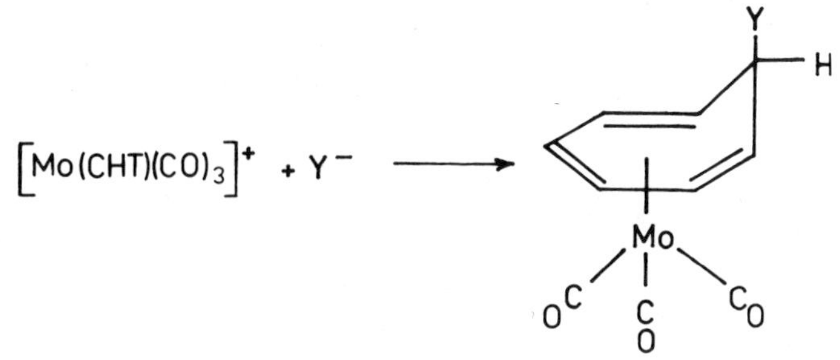

$$\left[Mo(CHT)(CO)_3\right]^+ + Y^- \longrightarrow \qquad\qquad (11.17)$$

However, if $Y^-$ is the cyclopentadienyl ion or malonic ester anion, contraction of the ring takes place, leading to the formation of a neutral arene complex (11.18).

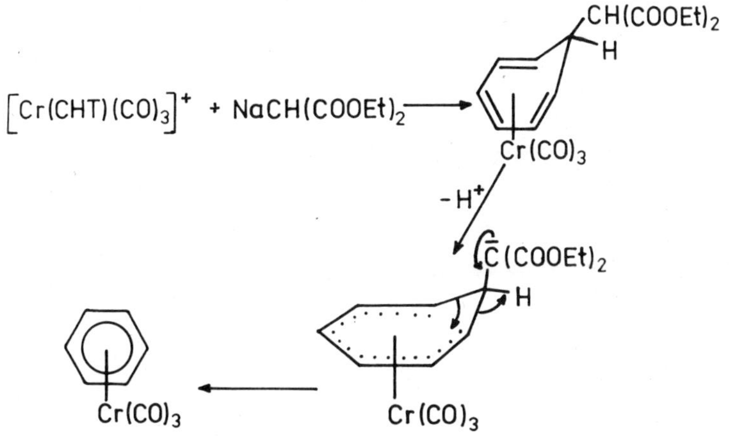

$$\left[Cr(CHT)(CO)_3\right]^+ + NaCH(COOEt)_2 \longrightarrow \qquad\qquad (11.18)$$

$$[Cr(CHT)(CO)_3]^+ + C_5H_4Me^- \longrightarrow [Cr(PhH)(CO)_3] + PhMe \qquad (11.19)$$

The mass spectra of sandwich compounds of the type $[Mcp(CHT)]$ (M = Zr, Nb, Mo) show the presence of $M(C_6H_6)^+$ ions. Therefore, in the mass spectrophotometer, contraction of the cycloheptatrienyl ring and expansion of the cyclopentadienyl ring takes place. The intensity of peaks decreases in the series $Zr > Nb \gg Mo$. The presence of $M(C_6H_6)^+$ ions suggests that in the transition state, the $[M(C_6H_6)_2]^+$ ions are formed, which may result by the internal nucleophilic addition (11.20).

V+
V+
V+
V+

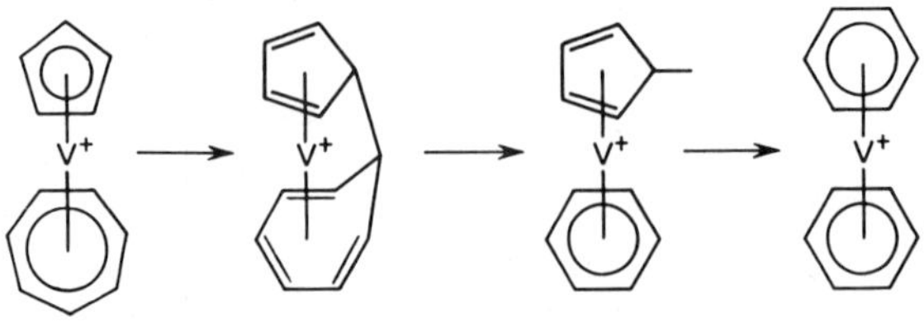

(11.20)

The following nucleophiles may also be added to cycloheptatrienyl ligands: phosphines, phosphites, derivatives of aromatic hydrocarbons, or other complexes[23, 24, 28] [equations (11.21)–(11.23)].

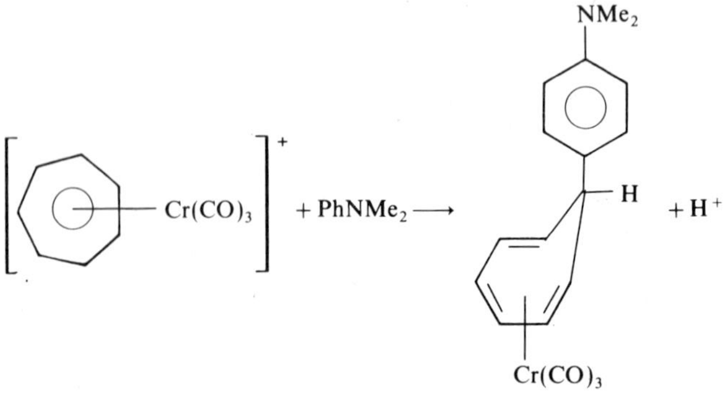

Cr(CO)3
+
+ PhNMe2 ⟶
NMe2
H
Cr(CO)3
+ H+

(11.21)

+
M(CO)3
+ PR3 ⟶
+PR3
H
M(CO)3

(M = Cr, Mo, W) (11.22)

$[M(CHT)(CO)_3]^+ + [Fe(\eta^3\text{-}C_7H_7)(CO)_3]^- \longrightarrow$ 
Fe(CO)3
M(CO)3

(11.23)

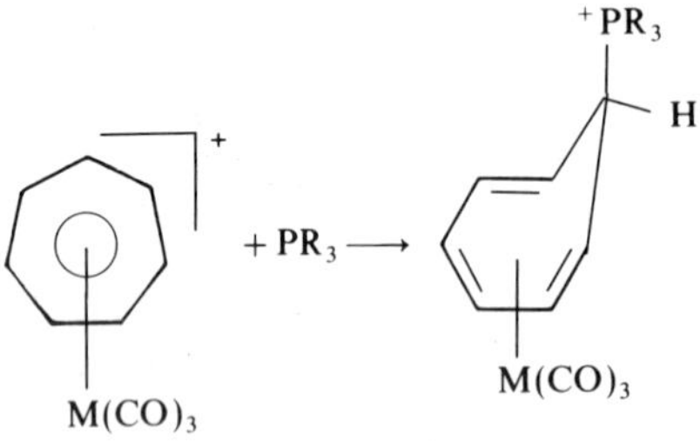

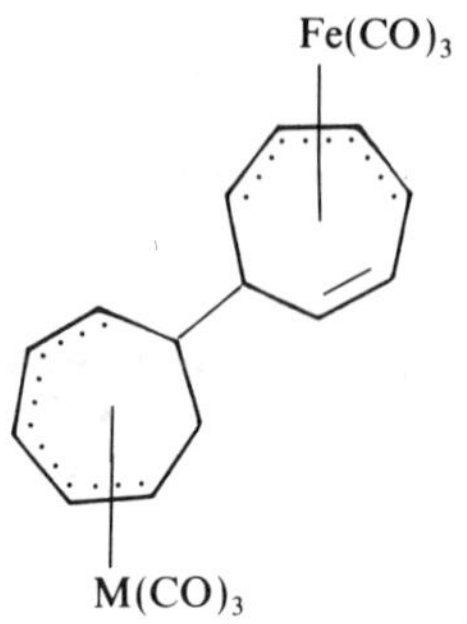

(M = Cr, Mo, W)

## b. *Oxidation and reduction reactions*

Cycloheptatrienyl and cyclooctatetraene complexes may undergo oxidation and reduction reactions. In some cases, these processes are reversible:

$$[Crcp(CHT)]^+ \underset{O_2}{\overset{S_2O_4^{2-}/OH}{\rightleftharpoons}} [Crcp(CHT)] \tag{11.24}$$

$$[Vcp(CHT)]^+ \underset{}{\overset{+0.2\,V}{\rightleftharpoons}} [Vcp(CHT)] \tag{11.25}$$

$$[Mo_2(OMe)_3(CHT)_2] \xrightarrow[NH_4PF_6]{O_2,\ MeOH} [Mo_2(OMe)_3(CHT)_2]^+ [PF_6]^- \tag{11.26}$$

The dinuclear complexes of the type $[Mo_2(\mu\text{-}X)_3(CHT)_2]$ ($X = Cl$, Br, I) undergo reversible one-electron oxidation. The $E_p$ oxidation potentials are $-0.15$ ($X = Cl$), $-0.13$ ($X = Br$), and $-0.11$ V ($X = I$).[4] Is is convenient to utilize phenyldiazonium salts for the preparation of dinuclear cationic complexes containing CHT:

$$[Mo_2X_3(CHT)_2] \xrightarrow{PhN_2^+BF_4^-} [Mo_2X_3(CHT)_2]^+ [BF_4]^- \tag{11.27}$$

The reduction of the dinuclear titanium complex $Ti_2(COT)_3$ leads to the formation of $[Ti_2(COT)_3]^{2-}$, which probably has a triple-decker structure (11.28).[16]

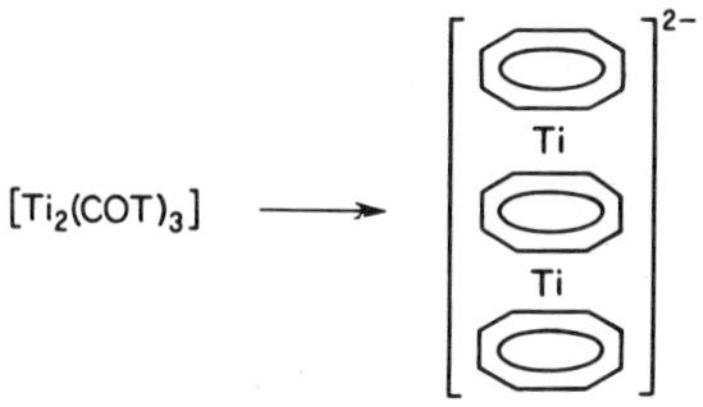

$$[Ti_2(COT)_3] \longrightarrow \tag{11.28}$$

The reduction of $U(COT)_2$ by hydrogen in the presence of the palladium catalyst (Pd/C) does not take place even after long periods, while the reaction of this compound with $LiAlH_4$ in tetrahydrofuran solution for three days gives (after decomposition) a mixture of cyclooctatriene and COT in the ratio of 2/3 (Table 11.1, reference *l*). The oxidation reaction of $[U(COT)_2]$ leads to the formation of inorganic uranium compounds and free COT.

The reaction of bis(cyclooctatetraene)uranium with air occurs explosively; however, the derivatives of $U(COT)_2$ are considerably more stable, for example, $[U(1,3,5,7\text{-}C_8H_4Ph_4)_2]$ is stable in air for several weeks. This happens because of steric hindrance which prevent the attack of oxygen on the uranium atom (Table 11.1). Other cyclooctatetraene complexes of the actinides and lanthanides may be oxidized in a similar way.

### c. *Ligand exchange reactions in complexes containing CHT*

In cationic carbonyl and arene cycloheptatrienyl molybdenum complexes, the arene ligand and the CO group are more easily displaced by other ligands than is cycloheptatrienyl. Such reactivity enables preparation of many molybdenum coordination compounds containing the cycloheptatrienyl ring (corresponding reactions are represented in Figure 11.4[14] and Figure 11.5[14]). Based on the electronic and ESCA spectra,[17] it was shown that, in complexes of the type $[Mo_2X_3(CHT)_2]$, there is a weak bond between the molybdenum atoms; one molybdenum atom has a zero oxidation state, and the other has a $+1$ oxidation state.

### d. *The $M-C_nH_n$ bond breaking*

Cycloheptatrienyl or cyclooctatetraene ligand displacement reactions are often utilized in the synthesis of other complexes.

$$Th(COT)_2 + ThCl_4 \xrightarrow{\text{THF}} 2Th(COT)Cl_2 \cdot 2THF \tag{11.29}$$

$$Ti_2(COT)_3 + HCl_{(g)} \longrightarrow Ti_4(\mu_3\text{-}Cl)_4 (COT)_4 \tag{11.30}$$

$$Zr(COT)_2 + HCl_{(g)} \longrightarrow [ZrCl_2(COT)] \tag{11.31}$$

Acetonitrile may be added stepwise to the $[M(CHT)(CO)_3]^+$ complexes leading to the

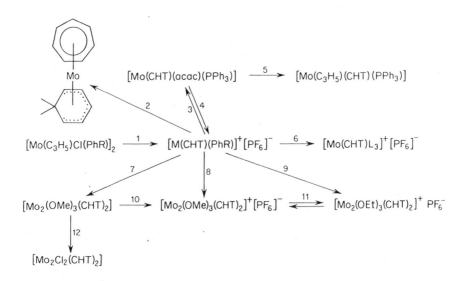

Figure 11.4. Reactions of cycloheptatrienyl molybdenum complexes: (1) $AlEtCl_2 + C_7H_8$, followed by $H_2O$, $[PF_6]^-$, R = H, Me, green crystals; (2) $NaBH_4$, THF, red crystals; (3) $PPh_3$, Na(acac), dark-red crystals; (4) $AlEtCl_2$ in an arene medium; (5) $[MgBr(C_3H_5)]$, $Et_2O$, orange-red crystals; (6) L = MeCN (excess), $PMe_2Ph$, $PMePh_2$ (in EtOH), red-pink crystals; (7) NaOMe, MeOH, dark-green, sublimes in vacuum at 353 K; (8) MeOH, green-brown crystals; (9) EtOH; (10) $O_2$, MeOH, $NH_4[PF_6]$; (11) EtOH, reflux; (12) HCl in benzene, dark-green, sublimes at 353 K, paramagnetic.

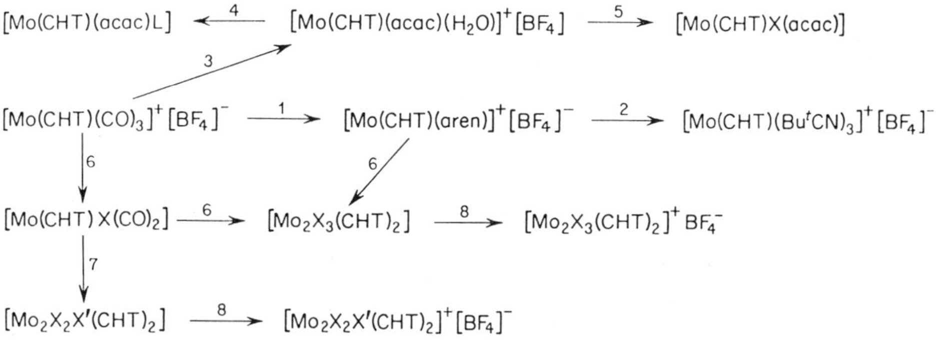

$[Mo(CHT)(acac)L]$ $\xleftarrow{\quad 4 \quad}$ $[Mo(CHT)(acac)(H_2O)]^+[BF_4]$ $\xrightarrow{\quad 5 \quad}$ $[Mo(CHT)X(acac)]$

$\nearrow 3$

$[Mo(CHT)(CO)_3]^+[BF_4]^-$ $\xrightarrow{\quad 1 \quad}$ $[Mo(CHT)(aren)]^+[BF_4]^-$ $\xrightarrow{\quad 2 \quad}$ $[Mo(CHT)(Bu^tCN)_3]^+[BF_4]^-$

$\downarrow 6$           $\downarrow 6$

$[Mo(CHT)X(CO)_2]$ $\xrightarrow{\quad 6 \quad}$ $[Mo_2X_3(CHT)_2]$ $\xrightarrow{\quad 8 \quad}$ $[Mo_2X_3(CHT)_2]^+BF_4^-$

$\downarrow 7$

$[Mo_2X_2X'(CHT)_2]$ $\xrightarrow{\quad 8 \quad}$ $[Mo_2X_2X'(CHT)_2]^+[BF_4]^-$

Figure 11.5. Reactions of cycloheptatrienyl molybdenum complexes: (1) arene (benzene, toluene, mesitylene); (2) MeCN followed by Bu$^t$CN, pink complex; (3) acacH, THF, reflux; (4) L = P(OMe)$_3$, THF, reflux; (5) X = Cl$^-$, Br$^-$, I$^-$, NCS$^-$, acetone, reflux; (6) SiXMe$_3$ (X = Cl, Br, I); (7) X = Cl, Br, SiX'Me$_3$; (8) [PhN$_2$]$^+$ [BF$_4$]$^-$, THF, room temperature.

removal of the cycloheptatrienyl ligand from the coordination sphere [equation (11.32)].

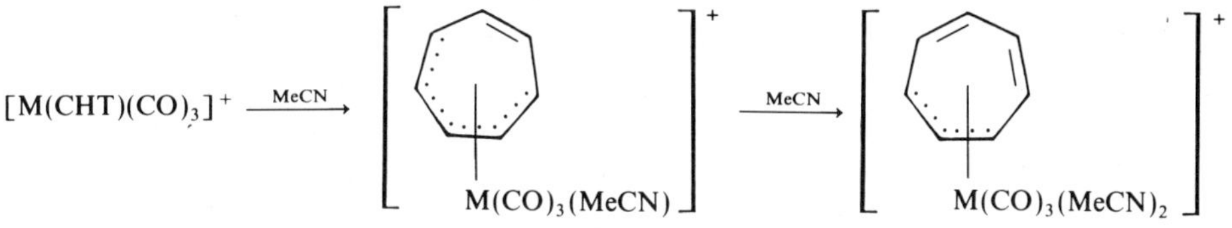

$$[M(CHT)(CO)_3]^+ \xrightarrow{\text{MeCN}} \left[\begin{array}{c} \\ M(CO)_3(MeCN) \end{array}\right]^+ \xrightarrow{\text{MeCN}} \left[\begin{array}{c} \\ M(CO)_3(MeCN)_2 \end{array}\right]^+$$

$$\xrightarrow{\text{MeCN}} M(CO)_3(MeCN)_3 + C_7H_7^+ \qquad (11.32)$$

The $M-C_nH_n$ bond breaking may be accompanied by reactions of the ligand itself:

$$Zr(COT)_2 \xrightarrow{\text{Pr}^i\text{OH}} Zr(OPr^i)_4 + 1,3,5\text{-}C_8H_{10} + 1,3,6\text{-}C_8H_{10} \qquad (11.33)$$

## e. Metallation reactions

The complexes [Mcp(CHT)] undergo metallation reaction with butyllithium. In titanium and zirconium complexes, the lithium addition takes place mainly to the cycloheptatrienyl ring, and in the case of complexes of group 5 and 6 metals the lithium addition occurs to the cyclopentadienyl ligand. This observation indicates that, in titanium and zirconium complexes, the cycloheptatrienyl carbon atoms possess higher negative charges. The situation is reversed for the other compounds. In the complex [Ticp(COT)], the cyclopentadienyl ligand undergoes metallation. The rate of metallation increases in the order:

Crcp(CHT) ≪ Vcp(CHT) ≪ Ticp CHT; Ticp(COT) ≪ Ticp(CHT)

# REFERENCES

1. M. L. H. Green, *Organometallic Compounds. The Transition Elements*, Methuen, London (1964).
2. B. L. Shaw and N. I. Tucker, in: *Comprehensive Inorganic Chemistry* (J. C. Bailer, ed.), Pergamon Press, Vol. 4, p. 781, Oxford (1973).
3. a. A. Avdeef, K. N. Raymond, K. O. Hodgson, and A. Zalkin, *Inorg. Chem.*, **11**, 1083 (1972).
   b. J. Zeinstra and J. L. DeBoer, *J. Organomet. Chem.*, **54**, 207 (1973).
4. M. Bochmann, M. Cooke, M. Green, H. P. Kirsch, F. G. A. Stone, and A. J. Welch, *Chem. Commun.*, 381 (1976).
5. R. Bowerbank, M. Green, H. P. Kirsch, A. Mortreux, L. E. Smart, and F. G. A. Stone, *Chem. Commun.*, 245 (1977).
6. K. D. Warren, *Structure and Bonding*, **27**, 45 (1976).
7. D. W. Clark and K. D. Warren, *Structure and Bonding*, **39**, 1 (1980).
8. H. O. van Oven and H. J. de Liefde Meijer, *J. Organomet. Chem.*, **19**, 373 (1969).
9. J. L. Thomas and R. G. Hayes, *Inorg. Chem.*, **11**, 348 (1972).
10. J. P. Clark and J. G. Green, *J. Organomet. Chem.*, **112**, C14 (1976); J. P. Clark and J. G. Green, *J. Chem. Soc., Dalton Trans.*, 505 (1977); I. Fragala, G. Condorelli, P. Zanella, and E. Tondello, *J. Organomet. Chem.*, **122**, 357 (1976).
11. K. O. Hodgson and K. N. Raymond, *Inorg. Chem.*, **11**, 171 (1972); K. P. Hodgson and K. N. Raymond, *Inorg. Chem.*, **11**, 3030 (1972).
12. A. Zalkin, D. H. Templeton, S. R. Berryhill, and W. D. Luke, *Inorg. Chem.*, **18**, 2287 (1979).
13. W. B. Davies, M. L. H. Green, and A. J. Jacobson, *Chem. Commun.*, 781 (1976).
14. E. F. Ashworth, M. L. H. Green, and J. Knight, *Chem. Commun.*, 5 (1974).
15. E. O. Fischer and H. Werner, *Metal π-Complexes*, Elsevier, Amsterdam (1966).
16. S. P. Kolesnikov, J. E. Dobson, and P. S. Skell, *J. Am. Chem. Soc.*, **100**, 999 (1978).
17. G. C. Allen, M. Green, B. J. Lee, H. P. Kirsch, and F. G. A. Stone, *Chem. Commun.*, 794 (1976).
18. J. Mueller and B. Mertschenk, *Chem. Ber.*, **105**, 3346 (1972); J. Mueller and B. Mertschenk, *J. Organomet. Chem.*, **34**, C41 (1972).
19. N. Roesch and A. Streitwieser, Jr., *J. Organomet. Chem.*, **145**, 195 (1978); N. Roesch and A. Streitwieser, Jr., *J. Am. Chem. Soc.*, **105**, 7237 (1983).
20. J. C. Green, M. P. Payne, and A. Streitwieser, Jr., *Organometallics*, **2**, 1707 (1983).
21. J. C. Green, *Structure and Bonding*, **43**, 37 (1981).
22. K. D. Warren, *Inorg. Chem.*, **14**, 3095 (1975).
23. F. G. A. Stone and E. W. Abel, in: *Comprehensive Organometallic Chemistry* (G. Wilkinson, ed.), Vol. 3, Pergamon Press, Oxford (1982).
24. S. P. Gubin and G. B. Shulpin, *Khimia Kompleksov so Svyazyami Metall-Uglerod*, Izd. Nauka, Sibirskoye Otdelenie, Novosibirsk (1984).
25. S. A. Kinsley, A. Streitwieser, Jr., and A. Zalkin, *Organometallics*, **4**, 52 (1985).
26. A. L. Wayda, *Organometallics*, **2**, 565 (1983).
27. H. Schumann, *Angew. Chem., Int. Ed. Engl.*, **23**, 474 (1984).
28. L. A. P. Kane-Maguire, E. D. Honig, and D. A. Sweigart, *Chem. Rev.*, **84**, 525 (1984).
29. D. A. Owen, A. Siegel, R. Lin, D. W. Slocum, B. Conway, M. Moronski, and S. Duraj, *Ann. N.Y. Acad. Sci.*, **333**, 90 (1980).
30. M. B. Hossain and D. van der Helm, *Inorg. Chem.*, **17**, 2893 (1978).

# Chapter 12

# Isocyanide Complexes

## 1. BONDING AND STRUCTURE

Isocyanides (isonitriles), RNC, are isoelectronic with carbon monoxide, CO. Most commonly, they coordinate to the metal through the carbon atom, although many complexes are also known in which the RNC molecule is bonded to metal atoms through both the carbon and nitrogen atoms.[1-6] Isocyanides may coordinate to metal atoms in the following way:

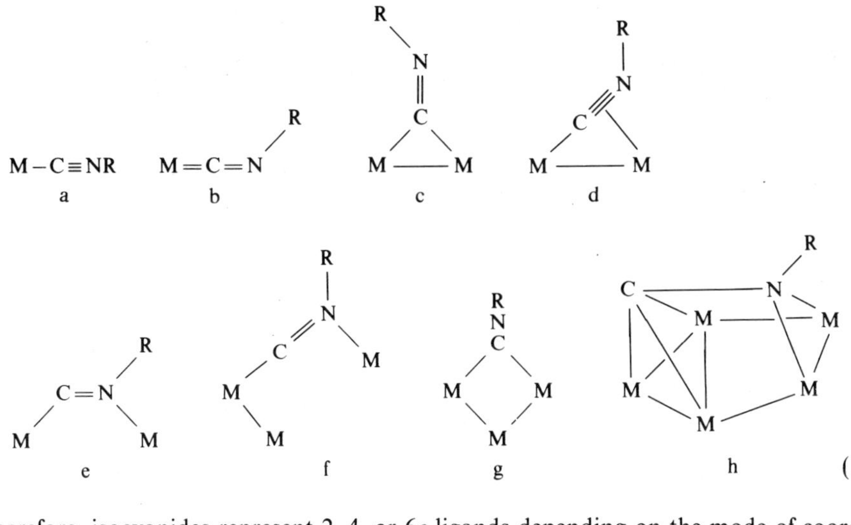

$$M-C\equiv NR \qquad M=C=N \qquad (12.1)$$

Therefore, isocyanides represent 2, 4, or 6$e$ ligands depending on the mode of coordination.

The molecular orbital scheme for alkyl isocyanides is similar to that for the carbon

617

monoxide molecule (Figure 12.1).[1, 3, 5, 7–10] However, in the case of aryl isocyanides one of the $2\pi$ antibonding orbitals interacts with the $\pi$ orbital of the aryl group. This leads to the lowering of energy of the $2\pi_v$ orbital, which is perpendicular to the plane of the aryl group, with respect to the $2\pi_h$ orbital which lies in the same plane (Figure 12.1). The contribution of aryl orbitals to the $2\pi_v$ antibonding orbital is considerable and depends on substituents of the aryl ring.[7–9] For most alkyl isocyanides, the $2\pi$ orbitals are almost degenerate; the compound CNCy is an exception for which the calculated energy difference between the $2\pi_v$ and $2\pi_h$ orbitals is considerable.[8] Terminal isocyanides form both $\sigma$ donor and $\pi$ redonor bonds. The first bond is due to the interaction of the free electron pair localized on the carbon atom with the metal atom, while the second bond results from the overlap between the occupied $d_\pi$ orbitals of the metal with the empty, antibonding $2\pi$ orbitals of the RNC molecule (cf. Section 2.2, Figure 2.5). Isocyanides are particularly interesting ligands because they possess both good $\sigma$ donor and $\pi$ acceptor properties. Aryl isocyanides are considerably better than alkyl isocyanides as $\pi$ acceptors owing to the interaction of the antibonding $2\pi_v$ orbital with the $\pi$ orbital of the aryl group. Thus, isocyanides have intermediate properties between CO and $CN^-$ ligands and may stabilize both high and low oxidation states of the metals.

According to the valence bond theory, the following resonance (canonical) structures of the metal–isocyanide bond are possible:

$$\overset{}{M}-\overset{}{C}\equiv\overset{+}{N}-R \qquad M=C=N\cdot \qquad \overset{}{M}-\overset{}{C}=\overset{\displaystyle R}{\overset{|}{N}}-R \tag{12.2}$$

$$a \qquad\qquad\qquad b \qquad\qquad\qquad c$$

Isocyanide ligands are usually linear; however, in the case of complexes in which the $\pi$ back-bonding is strong, the CNR angle is considerably smaller than 180 °C, often adopting values of 130–140 °C.[1, 2, 8, 11, 28] The ab initio molecular orbital calculations are in agreement with the valence bond theory.[9] They show that the lowering of the CNR angle is directly related to the strength of the $\pi$ back-bonding of the metal isocyanide group. More often, nonlinear CNR ligands are bridges [structure (12.1c)].

Isocyanide ligands containing electron-withdrawing groups form stronger redonor

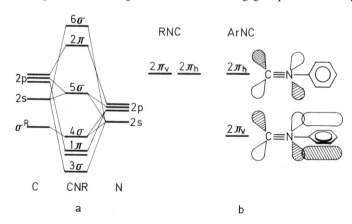

Figure 12.1   (a) Molecular orbital scheme for RNC. (b) $2\pi$ antibonding orbitals of alkyl and aryl isocyanides, R = alkyl, Ar = aryl.

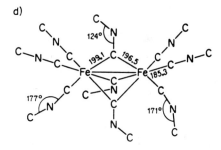
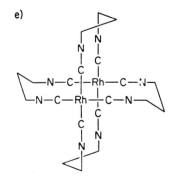
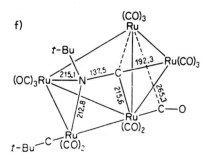

Figure 12.2. Structures: (a) [ReCl(CNMe)(N$_2$){P(OMe)$_3$}$_3$],[12] (b) [Cr(CNCF$_3$)(CO)$_5$],[11] (c) [Mn$_2$(CNC$_6$H$_4$Me)(CO)$_4$(dppm)$_2$],[13] (d) [Fe$_2$($\mu$-CNEt)$_3$(CNEt)$_6$],[14] (e) [Rh$_2${CN(CH$_2$)$_3$NC}$_4$] [BPh$_4$]$_2$·NCMe,[15] (f) [Ru$_5$(CNBu$^t$)$_2$(CO)$_{14}$][16] (the distances are given in pm units).

bonds with the metal than isonitriles with groups possessing weak withdrawing properties. For example, $CNCF_3$ forms complexes possessing quite strong metal–isocyanide
redonor bonds ($<$) $CNC = 142°$), while the coordinated CNMe molecule is usually
linear. The increase in strength of the $\pi$ back-bonding in $M - CNCF_3$ results mainly
from the lowering of the energy of the $2\pi$ ligand orbitals which enables their more
effective interaction with the $d_\pi$ metal orbital.[8–11]

The carbon–nitrogen bond in coordinated isocyanides is usually little changed with
respect to the free ligands. Exceptions are provided by complexes containing 4e and 6e
isocyanide ligands (Figure 12.2c and 12.2f). The metal–carbon distances in isocyanide
complexes are considerably shorter than in $\sigma$-carbyl coordination compounds. In compounds which exhibit strong $\pi$ back-bonding within the $M - CNR$ unit, these distances
are similar to those in metal carbonyls. In compounds which have weak $\pi$ metal–
isocyanide bonds, the metal–carbon distances of the CNR ligands are greater than
the M–C distances in metal carbonyls, and they are comparable with the metal–carbene
distances. Structures of some isocyanide complexes are represented in Figure 12.2.

## 2. ELECTRONIC SPECTRA

The electronic spectra of the isocyanide complexes of the transition metals (Table
12.1) show bands resulting from metal–isocyanide charge transfer (MLCT) and

*Table 12.1. Electronic Spectra of Metal Isocyanide Complexes*

| Compound | $\bar{v}(\varepsilon)$ $\mu m^{-1}$ $(M^{-1}\, cm^{-1})$ | Assigned band |
|---|---|---|
| $[Cr(CNPh)_6]^{(17)}$ | 2.18 (46000) | $d_\pi \to 2\pi_v$ (CNPh) |
| | 2.54 (73000) | $d_\pi \to 2\pi_v$ (CNPh) |
| | 3.23 (36000) | $d_\pi \to 2\pi_h$ (CNPh) |
| $[Mo(CNPh)_6]^{(17)}$ | 2.21; 2.65 | $d_\pi \to 2\pi_h$ (CNPh) |
| | 3.19 | $d_\pi \to 2\pi_h$ (CNPh) |
| | 3.92 | Within the ligand |
| $[Mn(CNPh)_6]Cl^{(17)}$ | 2.94 (61000) | $d_\pi \to 2\pi_v$ (CNPh) |
| | 3.11 (66000) | $d_\pi \to 2\pi_v$ (CNPh) |
| | 4.02 (51000) | Within the ligand |
| | 4.27; 4.44 | Within the ligand |
| $[Mn(CNPh)_6][PF_6]_2^{(17)}$ | 2.04 (4600) | $\sigma(CNPh) \to d_\pi$ |
| $[ReCl(CNBu')(dppe)_2]^{(30)}$ | 3.10 (7950) | MLCT |
| $[Fe(CNMe)_6]^{2+(18)}$ | 3.17 (320) | $^1A_{1g} \to {}^1T_{1g}$ |
| | 3.80 (340) | $^1A_{1g} \to {}^1T_{2g}$ |
| $[Rh(CNEt)_4]^{+(19,20)}$ | 2.30 (260) | $d_{z^2} \to 2\pi$ (CNEt) |
| | 2.63 (8400) | $d_{z^2} \to 2\pi$ (CNEt) |
| | 3.00 (3450) sh | $d_\pi \to 2\pi$ (CNEt) |
| | 3.25 (24350) | $d_\pi \to 2\pi$ (CNEt) |
| | 3.55 (2100) sh | $d_{xy} \to 2\pi$ (CNEt) |
| $[Rh_2\{CN(CH_2)_3NC\}_4]^{2+(21,22)}$ | 1.81 (14500) | $^1A_{1g} \to {}^1A_{2u}$ |
| $[Rh_2\{CN(CH_2)_3NC\}_4Cl_2]^{2+(23)}$ | 2.37 (1880) | $d\sigma \to d\sigma^*$ |
| | 2.97 (59000) | MLCT |
| $[Pd_2(CNMe)_6]^{2+(24)}$ | 3.25 (9900) | $d\sigma \to d\sigma^*$ |
| $[PdPt(CNMe)_6]^{2+(24)}$ | 3.40 (6300) | $d\sigma \to d\sigma^*$ |
| $[Pt_2(CNMe)_6]^{2+(24)}$ | 3.37 (4500) | $d\sigma \to d\sigma^*$ |

isocyanide–metal charge transfer (LMCT), as well as those resulting from transitions within the isocyanide ligands and the ligand field $d$–$d$ transitions. The absorption which occurs below 50,000 cm$^{-1}$ in CNR is due to transitions within the R group, because the $C \equiv N$ fragment does not show electronic transitions possessing lower energies. Therefore, alkyl isocyanides do not show characteristic bands, while aryl isocyanides exhibit typical transitions for arene derivatives.

The bands resulting from the metal-to-ligand charge transfer are often encountered. They are observed at lower energies for aryl isocyanides in which the energy of the $2\pi_v$ orbital is lower than that of $2\pi_h$ (Figure 12.1). If the aryl isocyanide is replaced by the alkyl isocyanide, the energy of the MLCT band increases. This allows the assignment of some bands of isocyanide complexes in their electronic spectra. The MLCT bands are observed for chromium group metal complexes containing aryl isocyanides, for example, [M(CNPh)$_6$], as well as for square-planar isocyanide complexes of elements of groups 9 and 10 which possess the $d^8$ electronic configuration, viz. [M(CNR)$_4$]$^{n+}$.[19–23] The energy of the strong MLCT band in rhodium complexes decreases owing to the formation of polynuclear [Rh(CNR)$_4$]$_n^{n+}$ complexes. This happens because of weak interaction between the rhodium atoms. The lowering of energy of the MLCT transition is greater, the higher the value of $n$.[20–22] The bands due to isocyanide–metal charge transfer (LMCT) are rarely encountered because they may occur only in complexes which have unfilled $d$ orbitals of the central atom, i.e., for complexes possessing the valence electron number smaller than 18. Typical examples of such compounds are [Mn(CNPh)$_6$]$^{2+}$ ($d^5$), [Cr(CNPh)$_6$]$^{n+}$ ($n = 1$, $d^5$; $n = 2$, $d^4$), etc.

## 3. NMR SPECTRA

The $^{13}$C NMR spectra furnish information concerning structures, the character of the isocyanide–metal bond, the reactivity of complexes and their dynamic properties. Sometimes, difficulties arise when $^{13}$C NMR spectra are to be observed. However, application of the relaxing agents and elongation of the accumulation time solve these problems.[25] The $^{13}$C NMR spectrum of the $C \equiv N$ group should show a triplet due to

Table 12.2. $^{13}$C NMR Spectra of Metal Isocyanide Complexes

| Complex | $\delta$ ($^{13}$CNR) (ppm) |
|---|---|
| $t$-[Mo(CNMe$_2$)(CNMe)(dppe)$_2$]FSO$_3$[29] | 160.6 m, br ( = broad) |
| $c$-[Mo(CNMe$_2$)(CNMe)(dppe)$_2$]FSO$_3$[29] | 183.1 m, br |
| [Fe$_2$($\mu$-CNEt)$_3$(CNEt)$_6$][14] | 173.8; 175.7; 176.9; 178.9 (CNEt), |
| | 256.6 ppm ($\mu$-CNEt) (183 K) |
| [Fe(CNBu$'$)$_5$][78] | 209.7 ppm (143 K) |
| [Ru(CNBu$'$)$_5$][78] | 187.5 ppm (singlet at 153–300 K) |
| [Os(CNBu$'$)$_3$(COD)][78] | 160.5 (singlet at 178–300 K) |
| [Co$_2$(CNBu$'$)$_8$](10% $^{13}$C)[79] | 191.4 (295 K); 229.0; 177.7 (183 K) |
| [Pt$_3$($\mu$-CNBu$'$)$_3$(CNBu$'$)$_3$](10% $^{13}$C)[81] | 163.3; 231.3 |
| $trans$-[PtCl(CNMe)(PEt$_3$)$_3$]$^+$[25] | 111.1 |
| $trans$-[PtCl(CNPh)(PEt$_3$)$_3$]$^+$[25] | 120 |
| $cis$-[PtCl$_2$(CNCy)$_2$][25] | 105.8 |
| $cis$-[PdCl$_2$(CNC$_6$H$_4$OMe-4)(PEt$_3$)][25] | 114.3 |

Table 12.3. $^1H$ NMR Spectra of Metal Isocyanide Complexes

| Compound | ($^1$H) (ppm) | (M) (ppm) |
|---|---|---|
| [(Bu$^t$NC)$_4$Cl$_2$Nb($\mu$-Bu$^t$NCCNBu$^t$)NbCl$_4$][31] | 1.821; 1.518 (313 K) | |
| [Vcp(CNMe)(CO)$_3$][32] | 4.57 cp; 2.10 (Me) | |
| [Vcp(CNPr$^i$)(CO)$_3$][32] | 4.58 (cp); 3.1 (CH); 0.71 (Me) | $\delta$($^{51}$V): $-1403$ |
| [Vcp(CNBu$^t$)$_2$(CO)$_2$][32] | 4.84 (cp); 1.08 (Me) | $\delta$($^{51}$V): $-1268$ s; $-1301$ m |
| t-[Mo(CNHBu$^t$)(CNBu$^t$)(dppe)$_2$]$^+$ [33] | 0.64 (CNBu$^t$); 1.16 (CNHBu$^t$); Ca 4.0 (CNHBu$^t$) | |
| t-[ReCl(CNMe)(dppe)$_2$][34] | 7.7–6.9 (Ph); 3.0–2.3 (CH$_2$); 2.42 (CNMe) | |
| [Rh$_2$Cl$_2$($\mu$-CF$_3$CCCF$_3$)(CNMe)(dppm)$_2$][35] | 3.32 (CNMe) | |
| [Ni$_4$($\mu$-CNBu$^t$)$_3$(CNBu$^t$)$_4$][36] | $-0.63$ (s, 3); $-1.41$ (s, 9); $-2.08$ (s, 9) | |
| [Pd$_2$(CNMe)$_6$]$^{2+}$ [24] | 3.47 (s) | |
| [PdPt(CNMe)$_6$]$^{2+}$ [24] | 3.55 (m, J(Pt$-$H) = 18 Hz, 6 H]; 3.44 (s, 9 H); 3.38 (m, 3 H) | $\delta$($^{195}$Pt): $-2833$ |
| [Pt$_2$(CNMe)$_6$]$^{2+}$ [24] | 3.51 [m, J(Pt$-$H) = 16 Hz, 12 H]; 3.33 (m, 6 H) | $\delta$($^{195}$Pt): $-2916$ |

the coupling of the $^{13}$C$^{14}$N spins. However, a broad singlet or a triplet containing a very intense middle peak is usually observed. This is due to the quadrupole moment of $^{14}$N. The $J(^{13}$C$-^{14}$N) coupling constants are 12–30 Hz.[26] The chemical shifts generally are encountered in the range of 100–200 ppm.[14, 25, 26, 29, 78, 79, 81] Such shifts are higher for complexes that show stronger $\pi$ back-bonding of the metal–isocyanide group (Table 12.2) and still higher for the bridging CNR ligands. Free isonitriles give signals in the range of 154–165 ppm.[26, 27] Dinuclear and cluster isocyanide complexes generally show dynamic properties in solution which are mainly investigated by $^1$H NMR. Proton NMR spectroscopy is also utilized for investigations of structures of isocyanide complexes. Selected $^{13}$C and $^1$H NMR spectra are given in Tables 12.2 and 12.3. The chemical shifts of vanadium in the $^{51}$V NMR spectra of compounds of the type [Vcp(CNR)(CO)$_3$] are in good agreement with predicted properties because they occur at ca $-1400$ ppm.[32] This is an intermediate value between the observed values for the two complexes [Vcp(CO)$_3$P(OMe)$_3$] [$-1496$ ppm, a strongly $\pi$ accepting P(OMe)$_3$ ligand] and [Vcp(CO)$_3$PBu$_3^i$] ($-1377$ ppm, an intermediate $\pi$ accepting property of the PBu$_3^i$ ligand). In the case of the complex [Vcp(CNBu$^t$)$_2$(CO)$_2$] the $\delta$ ($^{51}$V) values are $-1268$(s) and $-1301$(m) ppm, and for the compound [Vcp(CNC$_6$H$_3$Me$_2$-2,6)$_2$(CO)$_2$] the $\delta$ ($^{51}$V) data are $-1327$(m) and $-1342$(s) ppm [these data indicate the presence of two isomers for VcpL$_2$(CO)$_2$ complexes]. A still lower value of this shift ($-1235$ ppm) is found in the complex containing three isocyanide ligands, viz., [Vcp(CNC$_6$H$_3$Me$_2$-2,6)$_3$(CO)].[32]

## 4. IR SPECTRA

The number of stretching vibrations $\nu$(C$\equiv$N) and ratios of their intensities furnish information regarding the structure of isocyanide complexes. The selection rules are similar to those for metal carbonyls. For octahedral complexes of the type $M(CNR)_6$ only one band, $v(T_{1u})$, is infrared-active; also for $M(CNR)_4$ complexes which have $T_d$ symmetry only one band, $v(T_2)$, is observed in the infrared spectrum (cf. Chapter 2). In the complex *trans*-$[M(CNR)_4L_2]$, one $v(C\equiv N)$ vibration is infrared-active, while the compound *cis*-$[M(CNR)_4L_2]$ has three infrared-active $v(C\equiv N)$ vibrations. As a result of coordination of the CNR molecule to the central atom, there may be either a decrease or an increase in the $v(C\equiv N)$ frequency. This is in contrast to carbon monoxide, whose $v(CO)$ frequency in complexes is almost always lower than the frequency of the free ligand. Stronger $\sigma$ donor properties of isocyanides account for these differences. Owing to the slightly antibonding character of the $5\sigma$ orbital of the isocyanide molecule (Figure 12.1), the formation of the $\sigma$ bond with the metal leads to an increase in stability of the carbon–nitrogen bond in CNR if the $\pi$ back-bonding is weak. Therefore, lowering of the $v(CN)$ frequency is observed only for complexes with isocyanide ligands possessing strong $\pi$ accepting properties, i.e., in coordination compounds showing strong $\pi$ back-bonding. Thus, lowering of the $v(CN)$ frequency is greater for compounds of metal in low oxidation states. A considerable lowering of the $v(CN)$ frequency is observed in spectra of complexes in which metals adopt zero oxidation states, for instance, $M(CNR)_6$ ($M = Cr$, Mo, W) and $Ni(CNR)_4$ ($\Delta v \gtrsim 100$ cm$^{-1}$).[2] An exceptionally large lowering of this frequency was found for some isonitrile complexes of rhenium(I), viz., *trans*-$[ReCl(CNR)(deppe)_2]$ ($R = Me$, Bu$^t$, $C_6H_4Me$-2, $C_6H_4Cl$-4, $C_6H_3Cl_2$-2,6), $-1$ in which $v(C\equiv N)$ values are 1760–1920 cm$^{-1}$.[34, 37] In the case of $CNC_6H_4Cl$-4 this frequency for the free and coordinated ligand equals 2110 cm$^{-1}$ and 1760 cm$^{-1}$, respectively.[37] In manganese(I) complexes such as $[Mn(CNR)_6]^+$ the $v(CN)$ frequency is slightly lowered with respect to free isocyanide ($\Delta v$ equals from $-12$ to $-40$ cm$^{-1}$),[2, 8] while in isonitrile manganese(II) complexes of the type $[Mn(CNR)_6]^{2+}$ this frequency is strongly increased ($\Delta v \simeq 70$ cm$^{-1}$).[2, 8] Analogous trends are also encountered for other complexes, such as those of rhodium, palladium, etc., which possess various oxidation states. For a series of analogous complexes *trans*-$[MCl(NCR)(PR'_3)_2]^+$, the $\Delta v(C\equiv N)$ values vary as follows: Pt(112–125 cm$^{-1}$) > Pd(85–110 cm$^{-1}$) > Ni(10–70 cm$^{-1}$). Also, the increase in the $v(C\equiv N)$ frequency of the coordinated CNR molecules with respect to the free ones was found for many complexes of metals in oxidation states ranging from 1+ to 4+. The IR spectra of complexes of the type $[M(CNMe)_4][PF_6]_2$ have $v(CN)$ bands which occur at *ca* 2300 cm$^{-1}$. The IR spectra of the compound $[PtCl(CNR)(PEt_3)_2]$ show that the *trans* influence of the isocyanide molecule is somewhat greater than that of CO; however, it is considerably lower than that for phosphines $(PR_3)$. The $v(C\equiv N)$ frequencies are considerably higher for terminal CNR ligands than for the bridging ones. The $v(C\equiv N)$ frequencies usually adopt the following values depending on the bonding of the isocyanide molecule to metals [for free isonitriles, $v(CN)$ occurs in the 2120–2180 cm$^{-1}$ range]:

$$M-C\equiv N-R \qquad 1750\text{–}2300 \text{ cm}^{-1}$$

$$C=N\diagup^{R}_{\diagdown M} \qquad \sim 1600 \text{ cm}^{-1} \qquad (12.3)$$

## 5. *ESR SPECTRA AND MAGNETIC PROPERTIES*

Isocyanide ligands create a strong field and therefore give rise to low-spin complexes. Thus, complexes containing an even number of electrons are diamagnetic. Paramagnetic isonitrile complexes constitute a relatively small group which includes the cobalt(II) compounds $[Co(CNR)_5]^{2+}$, $[Co(CNR)_5(H_2O)]^{2+}$, $[CoX_2(CNR)_4]$, and $[Co(CNR)_4][MX_4]$. The magnetic moment for such complexes corresponds to one unpaired electron.[38–42, 44, 47] The paramagnetic electron resonance spectra of these complexes have also been investigated.[40, 43, 45, 46] Manganese(II) complexes containing type $[Mn(CNR)_6]^{2+}$ are paramagnetic with magnetic moments ranging from 1.95 to 2.1 BM.[48, 49] Chromium group compounds such as $[Cr(CNR)_6]^{2+}$ ($\mu_{eff} = 3.05$–3.14 BM) and $[M(CNR)_6]^+$ (M = Cr, Mo, W)[50, 51] are also paramagnetic.

## 6. *ULTRAVIOLET AND X-RAY PHOTOELECTRON SPECTRA*

Photoelectron spectra of $[Cr(CNMe)(CO)_5]$[88] and $[Mn(CNMe)(CO)_4]$[73] as well as photoelectron and XPS spectra of isonitrile iron complexes such as $[Fe(CNR)(CO)_4]$ (R = Me, Bu$^t$, Ph, SiMe$_3$)[74] have been investigated. Also, XPS spectra of isocyanide complexes of the type $[M(CNR)_6]$ (M = Cr, Mo, W), $[M(CNR)_7]^{2+}$ (M = Cr, Mo, W), $[Cr(CNR)_6]^{2+}$, $[Re(CNR)_6]^+$, $[Rh(CNR)_4]^+$, and those of complexes of the same metals of the composition $[M(CNR)_{n-m}L_m]^{x+}$ ($x = 1$, $n + m = 4$, M = Rh; $x = 2$, $n + m = 7$, M = Mo, W)[89] have been studied.

## 7. *PROPERTIES OF ISOCYANIDE COMPLEXES*[1–6, 55, 56]

Many isocyanide complexes are resistant to oxidation and therefore may be stored for some time in air. In fact, some of them are indefinitely stable in air. Also, isocyanide complexes are characterized by thermal stability. This is true not only for 18e complexes, such as $Cr(CNR)_6$, but also for 17e and 16e compounds; for instance, $Cr(CNR)_6^+$ and $Cr(CNR)_6^{2+}$ are air stable in the solid state. A great number of isocyanide complexes and particularly those which are di- and polynuclear show dynamic properties (Table 12.4). Square-planar complexes of elements of groups 8–10 possessing $d^8$ electronic configuration of the central ion have column structures because there is a weak interaction between the metal atoms. Therefore, these compounds show semiconducting properties (cf. Section 2.14 and Figure 2.32). From this point of view, the complexes of rhodium(I), palladium(II), platinum(III), etc., with aryl diisocyanide such as $C_6H_4(NC)_2$ are interesting.[52–54]

### a. *Group 3 metal complexes*

Metals of this group form complexes in the $+3$ oxidation states possessing the compositions $Mcp_3(CNR)$ (M = Y, Pr, Nd, Tb, Ho, Yb, Th, U), $MX_3(CNR)_n$ (M = U, Np, Am), and $MX_4(CNR)_4$ (M = U, Th). For complexes $Mcp_3(CNR)$, the M$-$CNR bond is mainly formed due to $\sigma$ donor properties of the isocyanide molecule because, as a result of its coordination, the $\nu(C\equiv N)$ frequency increases by *ca* 70 cm$^{-1}$.

Table 12.4. Isocyanide Complexes of Transition Metals

| Compound | Color | Mp (K) | Other data |
|---|---|---|---|
| 1 | 2 | 3 | 4 |
| $[V(CNBu^t)_6](PF_6)_2^{(58)}$ | Yellow | | $\nu(CN)$ 2190 cm$^{-1}$ |
| $[Vcp(CNMe)(CO)_3]^{(32)}$ | Orange-red | 361–363 | $\nu(CN)$ 2130 cm$^{-1}$ |
| $[Cr(CNPh)(CO)_5]^{(56)}$ | Yellow | 338–339 | |
| $[Cr(CNPh)_5(CO)]^{(59)}$ | Orange | 388–390 | |
| $[Cr(CNPh)_6]^{(17,83)}$ | Red | 451–453 | $\nu(CN)$ 2005, 1950 |
| $[Mo(CNEt)(CO)_5]^{(56)}$ | White | 359 | |
| $[Mo(CNPh)_6]^{(17,83)}$ | Red | 438 | $\nu(CN)$ 2008, 1950 |
| $[W(CNCy)(CO)_5]^{(56)}$ | White | 349 | |
| $[W(CNPh)_6]^{(17,83)}$ | Red | 312 | $\nu(CN)$ 2013, 1938 |
| $[Mn(CNPh)_6]Cl^{(17)}$ | White | 450–452 | $\nu(CN)$ 2176 |
| $[Tc(CNMe)_6]PF_6^{(68)}$ | White | >473 | $\nu(CN)$ 2110 s cm$^{-1}$; $\delta(Me)$: 295 ppm |
| $[Tc(CNBu^t)_6Cl][PF_6]_2^{(69)}$ | Bright-yellow | | $\nu(CN)$ 2250 m, 2215 s, 2035 w cm$^{-1}$; $\delta(Me)$ 1.63 ppm |
| $[Re_2(CNBu^t)_3(CO)_7]^{(72)}$ | White | 423–425 | |
| $[Re(CNBu^t)_2(NCMe)_2(PPh_3)_2][BF_4]_2^{(71)}$ | Purple | | |
| $[Re(CNPh)_6]PF_6^{(70)}$ | White | | $\nu(CN)$ 2080 s, 2020 sh |
| $[ReCl(CNMe)(N_2)\{P(OMe)_3\}_3]^{(12)}$ | White | | $\nu(CN)$ 2120; $\nu(NN)$ 2020 s |
| $[Fe_2(CNEt)_9]^{(14)}$ | Orange | | $\nu(CN)$ 2060 vs, 1920 sh, 1701 sh, 1652 vs |
| $[Fe(CNBu^t)_5]^{(78)}$ | Bright-yellow | | $\nu(CN)$ 2110 sh, 2005 s, 1830 s |
| $[Ru(CNBu^t)_5]^{(78)}$ | Yellow | 359–361 | $\nu(CN)$ 2059 sh, 2032 s, 1832 s |
| $[Ru_2(CNPr^i)_9]^{(14)}$ | Dark-red | 358 dec | |
| $[Os(CNBu^t)_3(COD)]^{[78]}$ | Bright-yellow | 375–378 | $\delta(^{13}C)$: 160.5 ppm |
| $[Co_2(CNBu^t)_8]^{(79)}$ | Orange-red | | $\nu(CN)$ 2100 sh, 2010 s, 1680 sh, 1668 s |
| $[Rh_2\{CN(CH_2)_3NC\}_4Cl_2]Cl_2 \cdot 6H_2O^{(23)}$ | Yellow | | |
| $[Ni(CNBu^t)_4]^{(80)}$ | Yellow | 443 dec | $\nu(CN)$ 2000 vs |
| $[Ni_4(CNBu^t)_7]^{(36)}$ | Red | | |
| $[Ni_8(CNBu^t)_{12}]^{(36)}$ | Brown-black | | |
| $[Pd_2(CNMe)_6][PF_6]_2^{(82)}$ | Bright-yellow | | $\nu(CN)$ 2240 |
| $[Pt_3(\mu\text{-}CNBu^t)_3(CNBu^t)_3]PhMe^{(81)}$ | Orange | | $\nu(CN)$ 2155 vs, 2190 sh, 1730 sh, 1714 vs |
| $[Pt_2(CNBu^t)_4(\mu\text{-}dppm)_2]^{(84)}$ | White | >593 dec | |

## b. Group 4 metal complexes

Only a few isonitrile complexes of metals of this group are known. The zirconium complex containing dinitrogen $[\{ZrN_2(pmcp)_2\}_2N_2]$ reacts with $CNC_6H_3Me_2$-2,6 to give the complex $Zr(pmcp)_2(NCC_6H_3Me_2)_2$.$^{(57)}$ During the reaction of $Ticp_2R_2$ with isonitriles at low temperatures, insertion takes place leading to the formation of $Ticp_2R(\eta^2\text{-}CR=NR')$.

## c. Group 5 metal complexes

Vanadium affords homoleptic isocyanide complexes $[V(CNR)_6]^{+}$ [58] as well as compounds containing additional ligands such as cyclopentadienyls, carbonyls, and halogens: $[Vcp(CNR)_n(CO)_{4-n}]$,[32] $[Vcp_2(CNR)]$, $[Vcp_2(CNR)]^{-}$, $[Vcp_2(CNR)_2]^{+}$, $[VCl_3(CNR)_3]$,[58, 62] $[(RNC)_4Cl_2M(\mu\text{-}RNCCNR)MCl_4]$ (M = Nb, Ta, R = Bu$^i$, Pr$^i$, Cy),[31] and $[V(CNR)(CO)_5]^{-}$.[63]

Halogen–isocyanide niobium and tantalum complexes are formed by reaction of isocyanides with $MX_5$ (M = Cl, Br). The coordination of the isonitrile molecule is accompanied by insertion to the $M-X$ bond which leads to compounds of the type $[(RNC)X_3M-CX=NR]$. Insertion also occurs during reactions of alkyl and hydrido carbonyl complexes of niobium and tantalum with CNR. In complexes $[Mcp_2(CH_2CH_2R')(CNR)]$ (R = Me, Et, R' = Me, Cy, Bu$^i$, 2,6-C$_6$H$_3$Me$_2$) the $v(C\equiv N)$ is low which argues for strong $\pi$ back-bonding.

## d. Group 6 metal complexes

All metals of this group form homoleptic isocyanide complexes in 0, 1+, and 2+ oxidation states, e.g., $[M(CNR)_6]$, $[M(CNR)_6]^{+}$, $[M(CNR)_7]^{2+}$ (M = Cr, Mo, W), and $[Cr(CNR)_6]^{2+}$. Chromium complexes $[Cr(CNR)_6]^{n+}$ ($n = 0, 1, 2$) undergo three-step electrochemical oxidation–reduction processes $Cr^0 \rightleftharpoons Cr^{+} \rightleftharpoons Cr^{2+} \rightleftharpoons Cr^{3+}$.[50, 59, 60] Also, many heteroleptic complexes are known, e.g., $[M(CNR)_5py]$, $[M(CNR)_n$ $(CO)_{6-n}]$ (M = Cr, Mo, W), $[M(CN)_4(CNR)_4]$, $[MX(CNR)_6]$ (X = Cl, Br, I; M = Mo, W), $[Cr(arene)(CNR)(CO)_2]$, $[MocpX(CNR)(CO)_2]$, $[Mo_2cp_2(CO)_5$ $(CNR)]$, $[Mocp(CNR)_2(CO)_2]^{+}$, $[Wcp(CNR)(CO)_3]^{+}$, and $[MoX_2(CNR)_5]$. Molybdenum(0) compounds containing the PPh$_3$ and PMePh$_2$ ligands which are coordinated as hexahapto molecules are interesting: $[Mo(CNBu^i)(\eta^6\text{-}C_6H_5PMePh)$ $(PMePh_2)_2]$[66] and $[Ph_3P(RNC) Mo(\mu\text{-}\eta^6\text{-}C_6H_5\text{-}PPh_2)(\mu\text{-}\eta^6\text{-}PPh_2C_6H_5) Mo(CNR)$ $(PPh_3)]$.[67] The bridging group has the following structure:

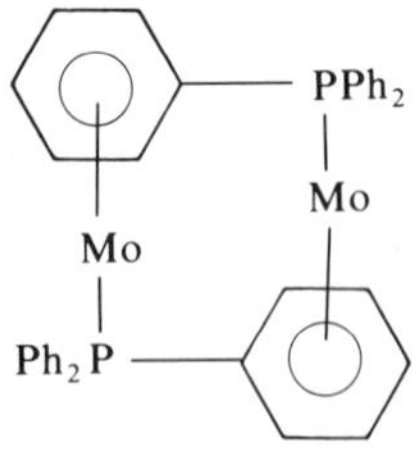

## e. Group 7 metal complexes

Homoleptic complexes of manganese(I), technetium(I), and rhenium(I) $[M(CNR)_6]^{+}$ [17, 68–70] as well as those of manganese(II) and rhenium(II) $[M(CNR)_6]^{2+}$ have been studied. Many heteroleptic complexes of these elements are also known; these include $[MX(CNR)_n(CO)_{5-n}]$ (M = Mn, Re), $[MnR(CNR)_n(CO)_{5-n}]$, $[Mn(CNR)_n(CO)_{5-n}]^{-}$, $[Mncp(CNR)_n(CO)_{3-n}]$ ($n = 1, 3$), $[Mncp(CN)(CNR)_2]^{-}$, $[Mn(arene)(CNR)(CO)_2]^{+}$, $[M(CNR)_6X][PF_6]_2$, $[M(CNR)_5XX'][PF_6]$ (M = Tc, Re),[69] $[Re_2(CNR)_n(CO)_{10-n}]$,[72] $[ReX(CNR)_5]$, $[Re(CNR)_n(CO)_{6-n}]^{+}$ ($n = 4, 5$), $[Re(CNR)_2(dppe)_2]^{+}$, $[ReCl(CNR)(dppe)_2]$,[34]

$[ReCl(CNMe)(N_2)\{P(OR)_3\}_3],^{(12)}$  $[Re(CNR)_2(dppe)_2]^+$,  $[Mn_2(\mu\text{-}CNR)(CO)_4$ $(dppm)_2]$ (Figure 12.2c), and $[Re(CNR)_2(NCMe)L(PPh_3)_2][BF_4]_2.^{(71)}$ Complexes containing metals in $2+$ oxidation states such as $[M(CNR)_6]^{2+}$ and $[Re(CNR)_2$ $(NCMe)L(PPh_3)_2]^{2+}$ $(L = MeCN,$ py$)$ are paramagnetic.[8, 70, 71] Rhenium furnishes isocyanide complexes in high oxidation states $(3+, 4+)$,[70] e.g., $[ReX_3(CNC_6H_4Me\text{-}4)_3]$, $[ReX_3(CNR)_4]$ $(X = Br,$ I$)$, $[ReCl_3(CNR)_3PR'_3]$, $[ReCl_2(CNR)_4(PPh_3)]^+$, and $[ReCl_5(CNMe)]^-.^{(61)}$ Rhenium also forms characteristic complexes such as $[Re_3I_6(CNR)_3]$, $[Re_3I_6(CNR)_6]$, $[Re_3I_9(CNR)_3]$, and $[Re_3Cl_9(CNR)_3]$.

## f. *Group 8 metal complexes*

The following homoleptic isonitrile complexes of the iron group are known: $M(CNR)_5$ $(M = Fe, Ru, Os)$, $M_2(CNR)_9$ $(M = Fe, Ru)$, $[M_2(CNR)_{10}]^{2+}$ $(M = Ru,$ Os$)$, and $[M(CNR)_6]^{2+}$ $(M = Fe, Ru, Os)$. Important heteroleptic complexes include $[MX_2(CNR)_4]$, $[MH(CNR)(dppe)_2]^+$ $(M = Fe, Ru, Os)$, $[FeX(CNR)_5]^+$, $[Fe(CNR)_2(NO)_2]$, $[Fe(CNR)_n(CO)_{5-n}]$, and $[FeX_2(CNR)_n(CO)_{4-n}]$. Halogen iron(III) complexes $[FeCl_3(CNR)_n]$ $(n = 2, 3)$ are very unstable. Ruthenium and osmium form isocyanide–carbene complexes such as $[M(CNR)_{6-n}\{C(NHR)_2\}_n]^{2+}$ $(n = 1, 2)$ $(M = Ru, Os)$, and $[Os(CNR)_3\{C(NHR)_2\}_3]^{2+}$ as well as many isonitrile halogen–carbonyl complexes, for instance, $[RuCl_2(CNR)(CO)_3]$ and $[OsX_2(CNR)(CO)(PR_3)_2]$. The structure of $Fe_2(CNR)_9$ is analogous to that of the carbonyl $Fe_2(CO)_9$ (Figure 12.2d).

## g. *Group 9 metal complexes*

Just like metals of the neighboring groups, elements of group 9 form many homoleptic complexes, for example, $[Co_2(CNR)_8]$, $[Co(CNR)_5]^{n+}$ $(n = 1, 2)$, $[Co_2(CNR)_{10}]^{4+}$, $[M(CNR)_4]^+$ $(M = Rh, Ir)$, $[Rh_2(CNR)_8]^{2+}$, and $[Rh_2\{CN(CH_2)_nNC\}_4]^{2+}$ $(n = 3\text{-}8)$. Rhodium complexes containing $CN(CH_2)_nNC$ $(n = 3\text{-}6)$ are dimeric. They have eclipsed structures in which the Rh-to-Rh interaction is very weak (Figure 12.2e). If $n$ is 7 or 8, monomeric complexes (12.4) containing chelating diisonitrile ligands are

$$(12.4)$$

obtained. The complex $[Rh_2L_4]^{2+}$ $[L = CN(CH_2)_3NC]$ is oxidized in the acidic medium to give $H_2$ and $[Rh_4L_8]^{6+}$, which during photolysis of the solution homolytically splits to give $[Rh_2L_4]^{3+}$. The reduction of $[Rh_4L_8]^{6+}$ by means of $Cr^{2+}$ leads to the polynuclear chain-like compounds $[Rh_6L_{12}]^{8+}$, $[Rh_8L_{16}]^{10+}$, and $[Rh_{12}L_{24}]^{16+}$ [20] Cobalt compounds $Co_2(CNR)_8$ are dinuclear and have structures analogous to $Co_2(CO)_8$ with two bent bridging CNR molecules. For $R = Bu^t$ the CNR angle is $131°$. The Co–Co distance equals 245.7 pm (7 pm lower than in the carbonyl) while the average Co–C distance is 3 pm higher than that in $Co_2(CO)_8$. The complex $Rh_2(CNC_6H_3Me_2\text{-}2,6)_8$ has analogous structure.[64] The compound $[Co_2(CNR)_{10}]^{4+}$ has staggered configuration with terminal, linear CNR groups ($D_{4d}$ symmetry). The Co–Co distance is 274 pm.

Rhodium(I) and iridium(I) furnish two-dimensional polymers[52, 54] with diisocyanide ligands such as 1,4-diisocyanobenzene, 1,3-diisocyanobenzene, 4,4′-diisocyanobiphenyl, 1,4-diisocyanonaphthalene, and 1,5-diisocyanonaphthalene [structure (12.5)]. Typical heteroleptic complexes of these elements include

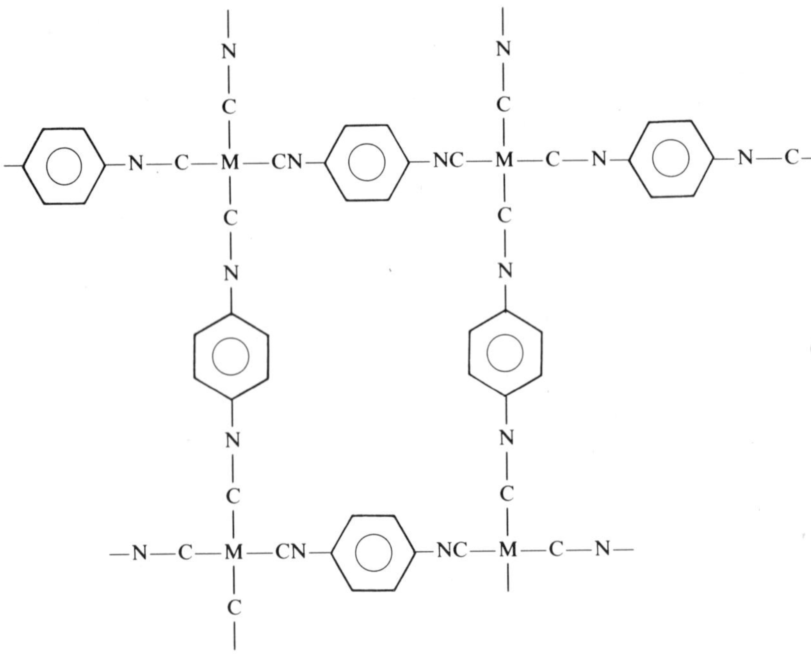

$$(12.5)$$

$[Co_2(CN)_3(CNR)_3]$, $[CoX_2(CNR)_4]^+$, $[CoX(CNR)_5]^{2+}$, $[Cocp(CN)_2(CNR)]$, $[Cocp(CNR)_3]^{2+}$, $[CoX_2(CNR)_2]$, $[CoX_2(CNR)_4]$, $[Co(CNR)_4]^{2+}$, $[(CNR)_4ICo(\mu\text{-}I)CoI(CNR)_4]^+X^-$, $[Co(CNR)_nL_{5-n}]^+$ [$L = PR_3, P(OR)_3, (Ph_2PCH_2CH_2)_3P$, etc.], $[Co_4(CNR)_n(CO)_{12-n}]$ ($n = 1$–$5$), $[RhX_3(CNR)_3]$, $[RhX_2(CNR)_4]^+$, $[RhX_3(CNR)_2L]$, $[(RNC)_4XRhRhX(CNR)_4]^{2+}$, $[RhX(CNR)_3]$, $[Rh_2(CNR)_4(\mu\text{-}dppm)_2]^{2+}$, $[Rh_4(CNR)_n(CO)_{12-n}]$ ($n = 1, 2, 4$), $[Rh_6(CNR)_6(CO)_{10}]$, $[IrX(CNR)_3(PR_3)_2]^{2+}$, $[IrX_2(CNR)_3(PR_3)]^+$, $[IrX(CNR)(diphos)_2]^{2+}$, $[MH(CNR)_3(PR_3)_2]^{2+}$ ($M = Co, Rh, Ir$), $[IrH_3(CNR)(AsR_3)_2]$, $[Ir_2X_4(CNR)_6]^{2+}$, etc.

## h. Group 10 metal complexes

Elements of this group, like metals of groups 8 and 9, form mononuclear and cluster homoleptic complexes in which metals adopt a zero oxidation state. The following homoleptic complexes of these elements are known: $[M(CNR)_4]^{2+}$ ($M = Ni, Pd, Pt$), $[Ni(CNR)_4]$, $[Ni_4(CNR)_7]$, $[Ni_4(CNR)_4]_n$, $[Ni_8(CNR)_{12}]$, $[Ni_4(CNR)_{14}](ClO_4)_8$, and $[Ni_3(CNR)_{11}](ClO_4)_6$. Linear structures were assigned for the last two compounds.[65] Monomeric platinum and palladium compounds of the formula $[M(CNR)_4]$ are unstable compared to triangular clusters $[M_3(\mu\text{-}CNR)_3 (CNR)_3]$. It is assumed that monomers are formed only as intermediate compounds. Platinum also forms the heptanuclear cluster $Pt_7(CNR)_{12}$. Platinum(I) and palladium(I) complexes of the formula $[M_2(CNR)_6]^{2+}$ and $[PdPt(CNR)_6]^{2+}$ are also known. Diisocyanobenzenes and Pd(II) or Pt(II) form two-dimensional polymers[53] which probably have analogous structure to those of the above-mentioned rhodium materials [equation (12.5)]. The nickel group metals furnish the following heteroleptic isocyanide complexes: $[Ni_4(CNR)_6 (RCCR)]$, $[Ni(CNR)_n (CO)_{4-n}]$ $(n = 1\text{–}3)$, $[Ni_2cp_2(\mu\text{-}CNR)_2]$, $[M(CNR)_n L_{4-n}]$ $[L = PR_3P(OR)_3; \ M = Ni, \ Pd, \ Pt]$, $[Ni(CNR)_2O_2]$, $[MX(CNR)_2 (PR_3)]^+$, and $[MX_2(CNR)_2]$ (M = Ni, Pd, Pt), etc.

### i. Group 11 metal complexes

All metals of this group afford homoleptic complexes with the coordination number 4. Silver(I) and gold(I) also give 2-coordinate compounds. Examples are $[M(CNR)_4]^+$, $[Ag(CNR)_2]^+$, and $[Au(CNR)_2]^+$. Heteroleptic isocyanide complexes of these metals are also known, e.g., $[MX(CNR)]$, $[MX(CNR)_2]$, $[CuX(CNR)_3]$, $[Cu(\eta^5\text{-}C_5H_5)(CNR)]$, and $[Cu(\eta^1\text{-}C_5H_5)(CNR)_3]$.

## 8. METHODS OF SYNTHESIS OF ISOCYANIDE COMPLEXES[1–6, 55, 56, 64]

### a. Preparations involving metal cyanides and cyanide metal complexes

Metal cyanides react with organic halides (most commonly with iodine or bromine compounds), olefins, dimethyl sulfate, alkylsulfates $(ROSO_3M)$, diazonium salts, oxonium salts, and compounds of phosphorus(III) or arsenic(III) such as $ER_2Cl$ (E = P or As, R = Et, OEt, Ph) to give isocyanide complexes:

$$Ag_4[Fe(CN)_6] + 6MeI \longrightarrow [Fe(CNMe)_6]I_2 + 4AgI \tag{12.6}$$

$$Ag_2[Pt(CN)_4] + 2MeI \longrightarrow [Pt(CN)_2 (CNR)_2] + 2AgI \tag{12.7}$$

$$K_4[Fe(CN)_6] + 4PhCH_2Br \longrightarrow [Fe(CN)_2 (CNCH_2Ph)_4] + 4KBr \tag{12.8}$$

$$[Fe(CN)_2 (CNCH_2Ph)_4] \underset{\longleftarrow}{\overset{PhCH_2Br}{\longrightarrow}} [Fe(CN)(CNCH_2Ph)_5] Br \underset{\longleftarrow}{\overset{PhCH_2Br}{\longrightarrow}} [Fe(CNCH_2Ph)_6] Br_2 \tag{12.9}$$

$$2AgCN + RI \longrightarrow Ag(CN)(CNR) + AgI \tag{12.10}$$

Complexes containing different types of isocyanide ligand (RNC and R'NC) may be obtained via displacement reaction of R by R':

$$[Fe(CN)(CNCH_2Ph)_5] Br + RBr \rightleftarrows [Fe(CN)(CNCH_2Ph)_{5-n} (CNR)_n] \tag{12.11}$$

Reactions of metal cyanides with various compounds leading to the preparation of isocyanide complexes have less significance at present, because synthetic methods that utilize isocyanides as substrates have been developed. Nevertheless, metal cyanides are still used in the case of synthesis of complexes with unstable or unattainable isocyanides when direct synthesis involving the reaction of the metal compound and the isocyanide ligand cannot be carried out. This method may also be applied for the preparation of chiral isocyanide complexes.

Molybdenum(IV) and tungsten(IV) compounds may be obtained by alkylation utilizing RX, while reactions of $[M(CN)_8]^{4-}$ with isocyanides RNC lead to complexes in lower oxidation state.

$$Ag_4[M(CN)_8] + 4RX \longrightarrow M(CN)_4 (CNR)_4 + 4AgX \qquad (12.12)$$

The anions of the type $[M(CN)_6]^{4-}$ (M = Fe, Ru, Os) react with $[Et_3O][BF_4]$ in acetone solution to give $[M(CNCMe_2CH_2COMe)_6][BF_4]_2$. First, aldol condensation of acetone takes place followed by the attack of carbonium ion on coordinated cyanide ligands. The reaction of $[Fe(CN)_6]^{4-}$ with MeCOEt, cyclohexanone, or acetophenone proceeds analogously.

Complexes containing functionalized isocyanide ligands are formed by reactions of metal cyanides with $R_2ECl$ (E = P, As; R = Et, OEt, Ph)[75] or $R_3ECl$ (E = Si, Ge, Sn, Pb; R = Me, Et).[76]

$$[Mncp(CN)(CO)_2]^- + EClR_2 \longrightarrow [Mncp(CNER_2)(CO)_2] \qquad (12.13)$$

$$E = P, As$$

$$[Fe(CN)(BD)(CO)_2]^- + EClR_3 \longrightarrow [Fe(CNER_3)(BD)(CO)_2] \qquad (12.14)$$

Synthesis of many other functionalized isocyanide coordination compounds have also been developed. Oxidizable anionic cyanides react with diazonium salts to give α-functionalized isocyanide complexes.[77] Arene diazonium chlorides $[4-RC_6H_4N_2]Cl$ (R = H, Cl, Br) react with $[M(CN)(CO)_5]^-$ (M = Cr, Mo, W) in THF, $CH_2Cl_2$, $CHCl_3$, PhCl, or MeCl to afford complexes of the type $[M(CNR)(CO)_5]$, where R = $-\overline{CH(CH_2)_3O}$, $-CHCl_2$, $-CCl_3$, $-COC_6H_4Cl$, and $-COCH_2Cl$.[77] The radical mechanism was proposed for these reactions, for example,

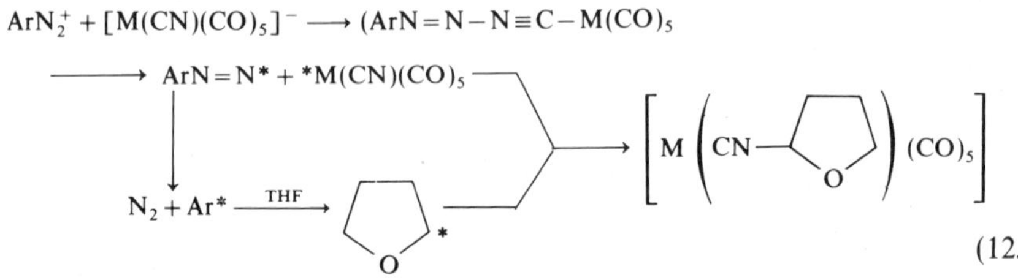

$$ArN_2^+ + [M(CN)(CO)_5]^- \longrightarrow (ArN=N-N\equiv C-M(CO)_5$$

$$(12.15)$$

equation (12.15). Utilizing diazomethane as an alkylating agent, it is possible to obtain complexes which contain isocyanide ligands.

$$H_4[M(CN)_6] + CH_2N_2 \longrightarrow [M(CN)_2 (CNMe)_4] \qquad (12.16)$$

$$M = Fe, Ru$$

The alkylation reaction involving silver cyanide is utilized for the preparation of alkyl isocyanides, because the initially formed isocyanide complex decomposes with RNC evolution while heated with KCN:

$$AgCN + RI \longrightarrow Ag(CN)(CNR) \xrightarrow{\text{KCN}} KAg(CN)_2 + RNC \qquad (12.17)$$

## b. *Reactions of isocyanides with transition metal compounds*

At present, this method is often used for the preparation of isocyanide complexes. During reaction, the addition of the ligand may take place without the change of the oxidation state. However, in many cases, reduction of the metal occurs, and complexes in which the transition element has a lower oxidation state than the starting compound are obtained. Copper(I) salts coordinate one to four RNC molecules to give $[CuX(CNR)_n]$ and $[Cu(CNR)_4]X$. Copper(II) salts are readily reduced by isocyanides to Cu(I) compounds. However, copper(II) chlorate or tetrafluoroborate and *tert*-alkyl isocyanide give the cationic complex $[Cu(CNR)_4]X_2$. This compound must immediately be separated from the reaction mixture in order to avoid its reduction.

Isocyanides themselves, solvents, main group metals and their organometallic compounds, metal iodides, etc., may serve as reducing agents.

$$MnI_2 + CNR \longrightarrow [Mn^I(CNR)_6]I + [Mn^I(CNR)_6][I_3] \qquad (12.18)$$

In the case of synthesis of $Cr(CNR)_6$ from chromium(II) acetate, disproportionation reaction takes place:

$$Cr_2(OOCMe)_4 + 18CNR \longrightarrow [Cr(CNR)_6] + 2[Cr(CNR)_6]^{3+} \qquad (12.19)$$

Isocyanide complexes of the chromium group metals and of rhenium are conveniently obtained in high yields from dinuclear metal compounds in $+2$ or $+3$ oxidation states for the chromium group and for rhenium, respectively, via reductive splitting of the metal–metal bond. The following compounds have been utilized: $M_2(OAc)_4$ (M = Cr, Mo), $[M_2Cl_8]^{4-}$ (M = Mo, W), $W_2(mhp)_4$, $Re_2(OAc)_4Cl_2$, $Re_2X_8^{2-}$ (X = Cl, Br), $[Re_2Cl_6(PR_3)_2]$, and $[Re_2(\mu\text{-}H)_4(PPh_3)_4]$. Under reductive reaction conditions $M(CNR)_6$ (M = Cr, Mo, W) and $[Re(CNR)_6]^+$ are formed, while at low temperatures isocyanide complexes may be obtained which are formed via substitution of ligands and the scission of the $M \equiv M$ bond.[70]

$$[NBu_4]_2[Re_2Cl_8] + CNPh \xrightarrow[\text{MeOH}]{295\ K,\ 4\ h} [Re(CNPh)_6]^+ [ReCl_4(CNPh)_2]^- \qquad (12.20)$$
$$72\%$$

$$[NBu_4]_2[Re_2Cl_8] + CNC_6H_2Me_3\text{-}2,4,6 \xrightarrow[\text{MeOH}]{295\ K,\ 4\ h} [ReCl_3(CNC_6H_2Me_3\text{-}2,4,6)_3]$$
$$+ [Re(CNC_6H_2Me_3\text{-}2,4,6)_6]^+ [ReCl_4(CNC_6H_2Me_3\text{-}2,4,6)]^- \qquad (12.21)$$

$$[NBu_4]_2[Re_2Cl_8] + CNPh \xrightarrow[\text{2. PF}_6^-]{\text{1. MeOH, reflux}} [Re(CNPh)_6][PF_6] \qquad (12.22)$$
$$57\%$$

The substitution reaction of CO ligands is often accelerated by light or catalysts.[32, 72, 85]

$$Mn_2(CO)_{10} + CNR \xrightarrow[\text{PhMe}]{\text{PdO}} [Re_2(CNR)_n(CO)_{10-n}]^{(72)} \qquad (12.23)$$
$$n = 1\text{-}4$$

$$Mncp(CO)_3 + CNR \longrightarrow Mncp(CNR)_3 \qquad (12.24)$$

$$\text{Fe(CO)}_5 + \text{CNR} \xrightarrow[\text{PhH}]{\text{CoCl}_2 \cdot 2\text{H}_2\text{O}} [\text{Fe(CNR)}_n (\text{CO})_{5-n}]^{(85)} \qquad (12.25)$$

$$n = 1\text{--}5$$

The substitution of carbonyl ligands in $\text{Vcp(CO)}_4$ is catalyzed by PdO.[32] Products of these reactions are complexes of the type $[\text{Vcp(CNR)}_n (\text{CO})_{4-n}]$. The substitution of CO groups may also be induced electrochemically.

## c. *Oxidation and reduction reactions*

Many isocyanide complexes, both homoleptic and heteroleptic, may be oxidized and reduced. These reactions are often utilized for the preparation of new isocyanide metal compounds, particularly homoleptic and low valent ones:

$$\text{MCl}_2(\text{CNR})_4 + \text{CNR} \xrightarrow{\text{Na/Hg}} \text{M(CNR)}_5 \qquad (12.26)$$

$$\text{M} = \text{Fe, Ru}$$

$$[\text{Co(CNBu')}_6]^+ + \text{K/Hg} \longrightarrow [\text{Co}_2(\text{CNBu'})_8] \qquad (12.27)$$

$$[\text{Cr(CNR)}_6] + \text{AgPF}_6 \longrightarrow [\text{Cr(CNR)}_6][\text{PF}_6] \qquad (12.28)$$

$$[\text{Mn(CNR)}_6][\text{BF}_4] \xrightarrow[\text{MeCN, HBF}_4]{\text{Cl}_2} [\text{Mn(CNR)}_6][\text{BF}_4]_2^{(8)} \qquad (12.29)$$

$$[\text{Rh}_2\{\text{CN(CH}_2)_3\text{NC}\}_4]\text{Cl}_2 \xrightarrow[\text{6 N HCl}]{\text{Cl}_2} [\text{Rh}_2\{\text{CN(CH}_2)_3\text{NC}\}_4\text{Cl}_2]\text{Cl}_2^{(23)} \qquad (12.30)$$

$$[\text{Re(CNPh)}_6][\text{PF}_6] \xrightarrow[\substack{1.\ \text{Cl}_2 \\ 2.\ \text{KPF}_6}]{\text{acetone}} [\text{Re(CNPh)}_6\text{Cl}][\text{PF}_6]_2^{(70)} \qquad (12.31)$$

$$[\text{Tc(CNR)}_6][\text{PF}_6] \xrightarrow[\text{PF}_6^-]{\text{X}_2} [\text{Tc(CNR)}_6\text{X}][\text{PF}_6]_2^{(69)} \qquad (12.32)$$

## 9. REACTIONS OF ISOCYANIDE COMPLEXES[1–6, 55, 56, 64, 86]

### a. *Cleavage of the M—CNR bond*

Some isocyanide complexes are very stable and, with difficulty, undergo decomposition which involves metal–carbon bond breaking.

Manganese compounds $[\text{Mn(CNR)}_6]^+$ are oxidized to $[\text{Mn(CNR)}_6]^{2+}$ under the influence of fuming nitric acid or bromine. However, loss of ligands does not take place. By the action of the same reagents, $\text{Ni(CNR)}_4$ decomposes completely; yet it is resistant to aqueous solutions of strong acids and bases. Chromium complexes $[\text{Cr(CNAr)}_6]$ are slowly decomposed by aqueous solutions of strong mineral acids. Benzylisocyanide iron complexes are not decomposed by strong acids. The compound $[\text{Fe(CN)(CNCH}_2\text{Ph)}_5]\text{Br}$ hydrolyzes with difficulty while heated at 110 °C in sulfuric acid. However, it furnishes $[\text{Fe(CN)}_6]^{4-}$, $N$-benzylformamide, and polymeric products while boiling in an alcohol solution of KCN. Nickel complexes containing arylisocyanides, $\text{Ni(CNAr)}_4$, do not undergo decomposition when treated with KCN solution.

Consequently the resistance of the complex to $M-CNR$ bond breaking depends upon electronic configuration of the central atom, properties of isocyanide ligands, reagents that react with the isocyanide products, and properties of products.

Isocyanide complexes of copper, silver, and iron decompose to give off free isocyanide ligands while heated, for instance,

$$[M(CNR)_4]Cl \longrightarrow [MCl(CNR)_2] + 2CNR \qquad (12.33)$$

$$M = Cu, Au$$

Tungsten complexes $[W(CN)_4(CNR)_4]$ give off isocyanides when heated at 160–180 °C.

## b. *Oxidation–reduction reactions*

Many reactions of this type are utilized for the preparation of some complexes; see reactions (12.26)–(12.32).

Palladium(II) complexes are reduced to Pd(0) complexes in strongly basic medium containing excess of isocyanide:

$$3[Pd(OH)_2(CNR)_2] + 3CNR \longrightarrow [Pd_3(CNR)_6] + 3RNCO + 3H_2O \qquad (12.34)$$

Cyclopentadienyl and allyl complexes also undergo reduction by means of an excess amount of isocyanide:

$$Nicp_2 \xrightarrow{\ CNR\ } [Nicp(CNR)]_2 \xrightarrow{\ CNR\ } Ni(CNR)_4 \qquad (12.35)$$

$$Pdcp(\eta\text{-}C_3H_5) \xrightarrow{\ CNR\ } Pd_3(CNR)_6 \qquad (12.36)$$

The nickel(I) cyanide complex $[Ni_2(CN)_6]^{4-}$ disproportionates in the presence of isocyanides:

$$[Ni_2(CN)_6]^{4-} + 4RNC \xrightarrow{\ NH_3\ } [Ni(CNR)_4] + Ni(CN)_4^{2-} + 2CN^- \qquad (12.37)$$

Palladium(0) and palladium(II) compounds react to give palladium(I) complexes:

$$Pd_3(CNR)_6 + 3PdX_2(CNR)_2 \longrightarrow 3[PdX(CNR)_2]_2 \qquad (12.38)$$

$$X = Cl, Br, I$$

Compounds of the type $[PdX(CNBu')_2]_2$ are also formed by reaction of $Pd_3(CNBu')_6$ with organic halides, for example, $PhCH_2I$ and $ClCOOPh$.

Many papers devoted to electrochemical processes of oxidation and reduction of metal isocyanide complexes have also appeared.[1, 2, 8, 30, 37, 87] It has been found that the ease of oxidation increases when the number of carbonyl ligands in the coordination sphere decreases and as a result of substitution of aryl isocyanides by alkyl isocyanides. Oxidation of $[Cr(CNR)_6]^{2+}$ to $[Cr(CNR)_6]^{3+}$ proceeds more easily in the case of alkyl isocyanides while reduction of Cr(II) complexes to Cr(I) compounds and Cr(I) complexes to Cr(0) compounds takes place more readily for aryl isocyanide ligands.[50, 59, 60]

The redox potentials of $[MoL_2(dppe)_2]$ and $[ReClL(dppe)_2]$ (L = CNR, CO, and $N_2$) depend on the polarizability of the complexes and on their electron richness.[30, 37] These potentials depend linearly on electron-accepting properties of the isocyanide ligands[30, 37] as in the case of $[Cr(CNR)_6]^{n+}$ complexes.

The electrochemical properties of $[Rh_2(CNR)_4 (dppm)_2][PF_6]_2$ depend strongly upon the solvent. Oxidation of this complex proceeds via two one-electron processes. The stronger the coordinating properties of the solvent or the base present in solution, the lower the positive potential during the second oxidation step.[92]

In some cases, it is possible to isolate oxidized compounds. Thus, electrochemical oxidation of manganese(I) and cobalt(I) compounds has afforded $[Mn(CNR)_6][PF_6]_2$, $[Mn(CNMe)_5 (CO)][PF_6]_2$, and $[Co(CNBu')_3 (PPh_3)_2]$ $[PF_6]_2$. Oxidation of cis-$[Mo(CO)_2 (bipy)_2]$ leads to the formation of $[Mo(CO)_2 (bipy)_2 (NCMe)][BF_4]_2$ and $[\{Mo(CO)_2 (bipy)_2\}_2][BF_4]_2$. These compounds can easily be utilized for the preparation of the metal isocyanide complexes: cis-$[Mo(CNC_6H_4Me-4)_4 (CO)_2]$, $[Mo(CNEt)_2 (CO)_2 (bipy)]$, $[Mo(CNEt)_2 (bipy)_2]$ $[BF_4]_2$, $[Mo(CNEt)_5 (bipy)][BF_4]_2$, and $[Mo(CNEt)_3 (bipy)_2][BF_4]$.[90] Rhenium(I) complexes $[Re(CNR)_2 (NCMe) L(PPh_3)_2][BF_4]$ (L = MeCN, py, CyNH$_2$, t-BuNH$_2$) are oxidized by means of $AgBF_4$ to paramagnetic rhenium(II) compounds $[Re(CNR)_2 (NCMe) L(PPh_3)_2][BF_4]_2$ which may be reduced back to the starting materials by zinc.[71]

## c. Oxidative addition reactions

Low-valent metal isocyanide complexes characteristically undergo oxidative addition reactions which involve XY molecules, where X = Cl, Br, I, H, alkyl, etc., and Y = Cl, Br, I, H, $Mn(CO)_5$, $OsH(CO)_4$, $BF_4$, etc. The complex pentakis(tert-butylisocyanide)iron(0) reacts with RX to give $[FeR(CNBu')_5]X$ (R = Me, X = I; R = $C_6F_5$, X = Br; R = $C_3H_5$, X = Cl, Br) as well as $[Fe(R) X(CNBu')_4]$ (R = PhCH$_2$, X = Br; R = CF(CF$_3$)$_2$, X = I).[91] Many examples of oxidative addition involving halogens, acids, for example, $HBF_4$ and $HClO_4$), hydrides, etc., are known.[1, 2, 55, 56, 64]

## d. Nucleophilic addition

XPS spectra[74, 89] show that coordination of the RNC ligand causes a decrease of the electron density on the isocyanide carbon atom. Therefore, addition of nucleophiles to the carbon atom and addition of electrophiles to the nitrogen atom represent characteristic reactions of the isocyanide ligands. Reactions of metal isocyanide complexes with compounds of the type RXH (RX = RO, RNH, RS, etc.) lead to the formation of metal carbene complexes [Section 5.8.b, reactions (5.13)–(5.19)].

$$[L_nM-CNR] + R'XH \longrightarrow L_nM=C\begin{matrix} XR' \\ \\ NHR \end{matrix} \tag{12.39}$$

This reaction proceeds more readily if the isocyanide R group has electron-accepting properties and the R' group has nucleophilic, electron-donating properties.

The complexes cis-$[PdX_2L(CNC_6H_4Y-4)]$ (L = PPh$_3$, AsPh$_3$) react with amines of the type $4\text{-}ZC_6H_4NH_2$. The more basic the amine, the faster the reaction. For the same amine, e.g., $(4\text{-}MeC_6H_4NH_2)$, the reaction rate decreases depending on Y according to the series: $NO_2 > Cl > H > Me > MeO$. In some cases, the amine molecule may be added to two isocyanide ligands. For instance, in this manner, the complex $[Fe(CNMe)_6]^{2+}$ reacts with methylamine and the compound $[Pt(CNMe)_4]^{2+}$ reacts with hydrazine [equations (12.40) and (12.41)]. Hydroxylamine reacts similarly [equation (12.42)]. It is assumed that hydrolysis of the isocyanide ligand to the carbonyl

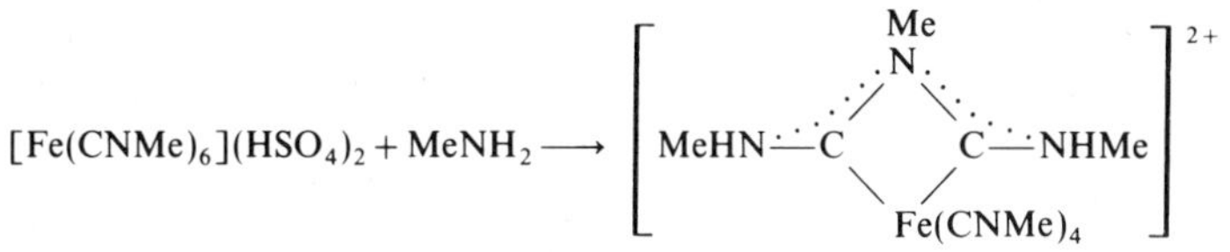

$$[Fe(CNMe)_6](HSO_4)_2 + MeNH_2 \longrightarrow \left[ \begin{array}{c} \overset{Me}{\underset{}{N}} \\ MeHN \text{---} C \qquad C \text{---} NHMe \\ Fe(CNMe)_4 \end{array} \right]^{2+} \qquad (12.40)$$

$$[PtCl_2(CNMe)_2] + H_2NNH_2 \longrightarrow \begin{array}{c} NH\text{---}NH \\ MeHN\text{---}C \qquad C\text{---}NHMe \\ PtCl_2 \end{array} \qquad (12.41)$$

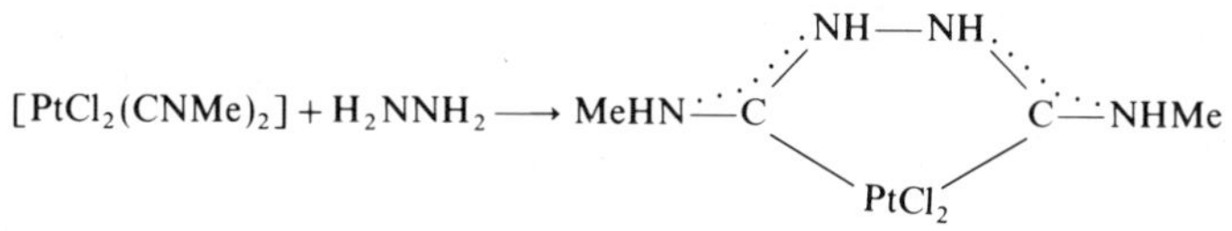

$$[Pt(CNMe)_4]^{2+} + NH_2OH \longrightarrow \left[ \begin{array}{c} N\text{---}O \\ MeHN\text{---}C \qquad C\text{---}NHMe \\ Pt(CNMe)_2 \end{array} \right]^{+}$$

$$\overset{HCl}{\longrightarrow} \begin{array}{c} NH\text{---}O \\ MeHN\text{---}C \qquad C\text{---}NHMe \\ PtCl_2 \end{array} \qquad (12.42)$$

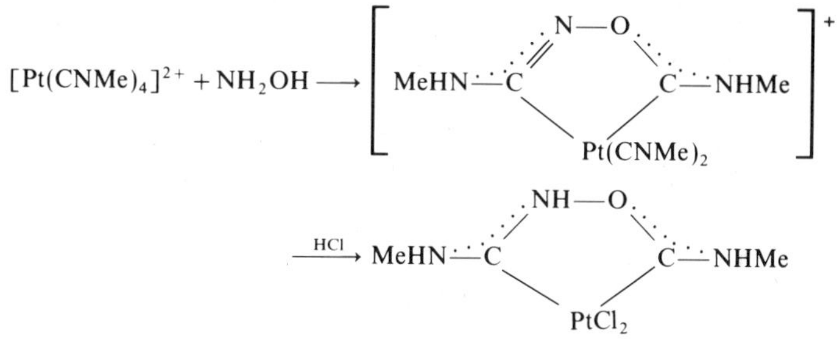

ligand proceeds via a carbene complex [equation (12.43)]. Alcohols react with metal isocyanide complexes less readily than do the amines.

The $C_{1s}$ bond energies show that the carbene carbon atoms are less positive than those of the isocyanide group. Additions of amines, alcohols, and thiols afford many carbene complexes, particularly those of metals of groups 6–10.

$$RhCl(CNMe)(PPh_3)_2 \overset{H_2O}{\longrightarrow} [RhCl\{C(OH)NHMe\}(PPh_3)_2]$$

$$\longrightarrow RhCl(CO)(PPh_3)_2 + MeNH_2 \qquad (12.43)$$

## e. *Formation of metal carbyne complexes*

Electron-rich metal isocyanide complexes are susceptible to electrophilic attack on the nitrogen atom to give metal carbyne complexes. Reacting electrophiles are $R^+$ (R = alkyl) and $H^+$. To carry out these reactions, the following reagents are used: RX (most commonly iodides), $MeFSO_3$, $Me_2SO_4$, $[Et_3O][BF_4]$, and acids such as $HBF_4$, $H_2SO_4$, HCl, $HPF_6$, and $HSFO_3$.[14, 29, 33, 34, 93–95] The $M-CNRR'$ bonding may be described by the two canonical forms (12.44). The C–N distance in carbyne complexes is usually 128–135 pm. Thus, this distance is shorter than that for a single

$$\left[ M \equiv C - N \begin{matrix} H \\ \diagup \\ \diagdown \\ R \end{matrix} \right]^{+} \leftrightarrow Re = C = \overset{+}{N} \begin{matrix} H \\ \diagup \\ \diagdown \\ R \end{matrix} \qquad (12.44)$$

$$\qquad\qquad a \qquad\qquad\qquad\qquad\qquad b$$

N−CH$_3$ bond (142 pm). This argues for a considerable contribution of form (12.44b). Both bridging and terminal isocyanide ligands undergo attack by electrophilic reagents. The reaction of R'I (R' = alkyl) or acids with $[Fe_2(dienyl)_2 (CNR)(CO)_3]$ (dienyl = cp, $C_5H_4Me$, R = alkyl) leads to $[Fe_2(dienyl)_2 (\mu\text{-}CNRR')(CO)_3]I$ and $[Fe_2(dienyl)_2 \{\mu\text{-}CN(H)R\}(\mu\text{-}CO)(CO)_2]^{+}$.[95] The C–N distance in cis-$[Fe_2cp_2(\mu\text{-}CNHMe)(\mu\text{-}CO)(CO)_2][BF_4]$ is 128 pm. Bridging carbyne ligands are also encountered in complexes such as $[Fe_2(\mu\text{-}CNEt)(\mu\text{-}CNREt)_2 (CNEt)_6]I_2$ (R = Me, Et) which are obtained by reaction of $Fe_2(CNEt)_9$ with RI.[14]

Complexes containing terminal carbyne ligands can be obtained by alkylation or protonation of compounds $[M(CNR)_2 (dppe)_2]$ (M = Mo, W),[29, 33, 93, 94] and $[ReCl(CNR)(dppe)_2]$.[34] The reaction products are $[M(CNRR')(CNR)(dppe)_2]^{+}$ (R' = H, alkyl),[29, 33, 93] and $[ReCl(CNHR)(dppe)_2]^{+}$.[34] The protonation of mer-$[W(CNMe)_3 (PMe_2Ph)_3]$ affords $[W_2(\mu\text{-}CNHMe)_2 (CNMe)_4 (PMe_2Ph)_4][BF_4]_2$.[94] Further protonation of molybdenum and tungsten complexes $[M(CNHR)(CNR)(dppe)_2]^{+}$ leads, depending upon conditions, to either the hydrido compounds $[MH(CNHR)(CNR)(dppe)_2]^{2+}$ or the dicarbyne derivatives $[M(CNHR)_2 (dppe)_2]^{2+}$. The intermediate compounds are $[MH(CNR)_2 (dppe)_2]^{+}$, which are formed from $[M(CNHR)(CNR)(dppe)_2]^{+}$ as a result of proton migration from the carbyne ligand to the central atom.[33, 93] By treatment with acids, isocyanide ligands of electron-rich molybdenum(0), tungsten(0), and rhenium(I) complexes are reduced to amines, ammonia, and hydrocarbons[12, 94]; central atoms of these complexes are reducing agents.

Protonation of the manganese(0) compound $[Mn(\mu\text{-}\eta^2\text{-}CNC_6H_4Me\text{-}4)(CO)_4 (dppm)_2]$ (Figure 12.2c) causes proton addition to the Mn−Mn bond with concomitant weakening of the $\mu\text{-}\eta^2$-CNR bridge.[115]

### f. Insertion reactions

Migratory insertion reactions of coordinated isocyanides to $\sigma$-carbyl, $\pi$-acetylene, carbene, carbyne, halogen, and hydrido ligands as well as to groups which are bonded to the central atom through nitrogen, oxygen, and sulfur lead to the preparation of many interesting compounds. Therefore, these reactions have been studied intensively.

Compared to the CO insertion into the M − $\sigma$-carbyl bond, the reaction of this type for isocyanide ligands proceeds more readily, giving iminoacyl groups which are thermodynamically more stable than the acyl ligands. Thus, in contrast to the acyl groups, they do not undergo the reverse reaction, i.e., deinsertion.

The complexes Ticp$_2$RR' react with isocyanides to give compounds possessing the formula Ticp$_2$R($\eta^2$-CR' = NR").[96] Niobium and tantalum halogeno alkyl complexes,[96, 97] ZrR$_4$, and ReMe$_6$[96] react in a similar way. Many other examples of such reactions involving $\sigma$-carbyl compounds (most commonly, those of groups 4–10) are known.[1–6, 55, 56, 64] In some cases, the iminoacyl ligands may become coordinated both through the carbon atom and through the nitrogen atom, e.g., equation (12.45). Hexamethyltungsten reacts with CNBu' to give the complex (12.46).[96]

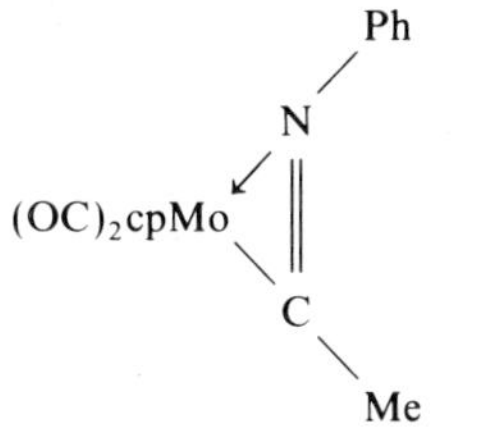

$$(OC)_2cpMo \diagup \overset{\displaystyle Ph}{\underset{\displaystyle C}{\overset{\displaystyle |}{N}}} \diagdown Me \qquad (12.45)$$

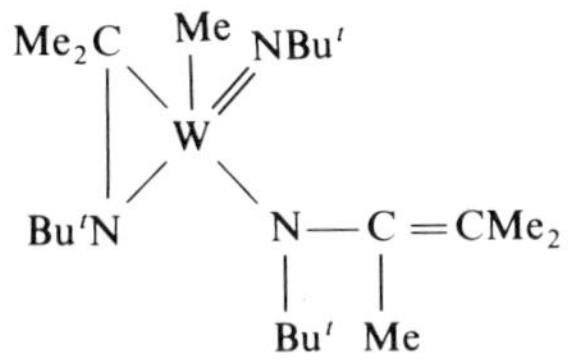

$$
\underset{\displaystyle Bu'N}{\overset{\displaystyle Me_2C \quad Me \quad NBu'}{W}} \diagdown \underset{\displaystyle Bu' \ Me}{N-C=CMe_2} \qquad (12.46)
$$

Also, multiple insertion reactions may occur leading to the formation of polyimine ligands [equation (12.47)]. Alkyl iodides and Ni(CNBu')$_4$ afford pentaimine complexes (12.48). The metal-nitrogen bond of the polyimine complexes may be broken by the

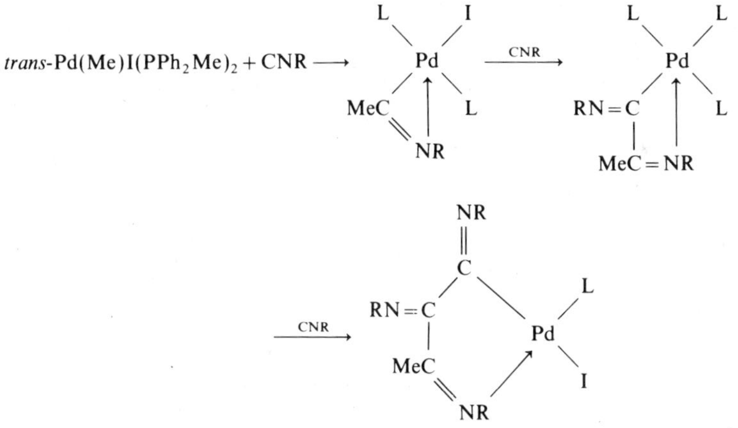

$$
\begin{aligned}
&\textit{trans-}Pd(Me)I(PPh_2Me)_2 + CNR \longrightarrow \quad \underset{MeC\underset{NR}{\Vert}}{\overset{L\quad I}{Pd}}L \xrightarrow{\ CNR\ } \underset{MeC=NR}{\overset{L\quad L}{Pd}}L \\[6pt]
&\xrightarrow{\ CNR\ } \quad \underset{NR}{\overset{NR}{\underset{MeC}{\overset{\Vert}{C}}}}\ RN{=}C\diagup\quad\diagdown Pd \diagup L \diagdown I \qquad (12.47)
\end{aligned}
$$

$$
\underset{I}{\overset{Bu'NC}{\diagdown}}\ \underset{N=C}{\overset{NBu'}{\underset{\Vert}{C}}}\text{—C=NBu'}\ Ni \left(\ C{=}NBu'\ \right)_2 R \qquad (12.48)
$$

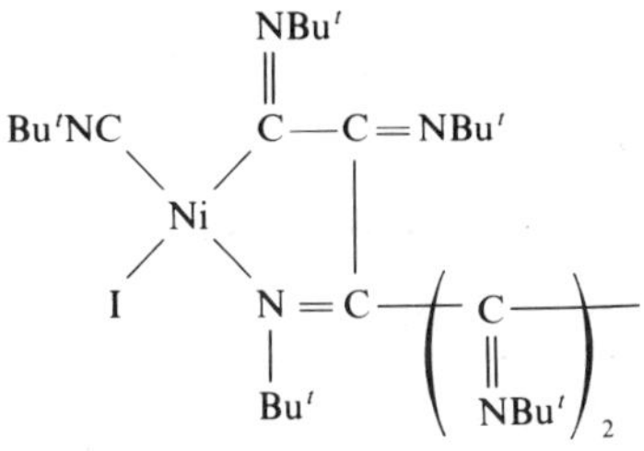

presence of Lewis bases L (for instance, CNR and PR$_3$) and subsequently substituted by the M−L bond. Nitrogen atoms of the polyimine ligands often form bonds with other metals, leading to the formation of polynuclear complexes (12.49). As for $\sigma$-carbyl compounds, the insertion reaction (12.50) also takes place for metal allyl complexes.

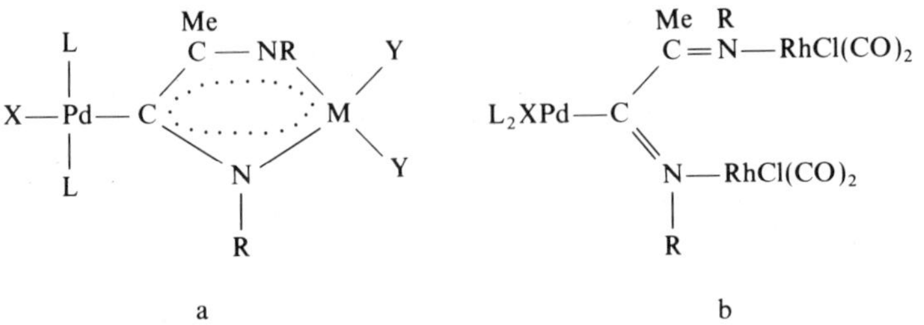

Me
C—NR
L
X—Pd—C
M
Y
Y
N
R
a
Me R
C=N—RhCl(CO)₂
L₂XPd—C
N—RhCl(CO)₂
R
b

(12.49)

$M = Pd, Pt, X = halogen, Y = halogen, S_2CNR_2$

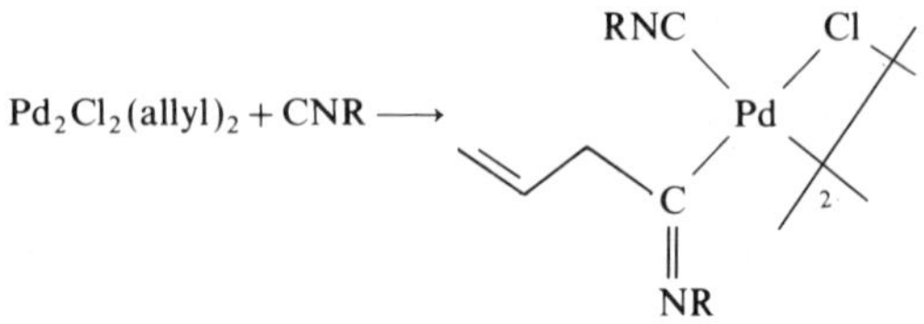

Pd₂Cl₂(allyl)₂ + CNR ⟶
RNC
Cl
Pd
C
NR
2

(12.50)

Niobium and tantalum halides also react with isocyanides to give migration products [equation (12.51)]. Azide complexes react with CNR to afford coordination compounds with tetrazolyl [equation (12.52)]. Codimerization and cooligomerization of

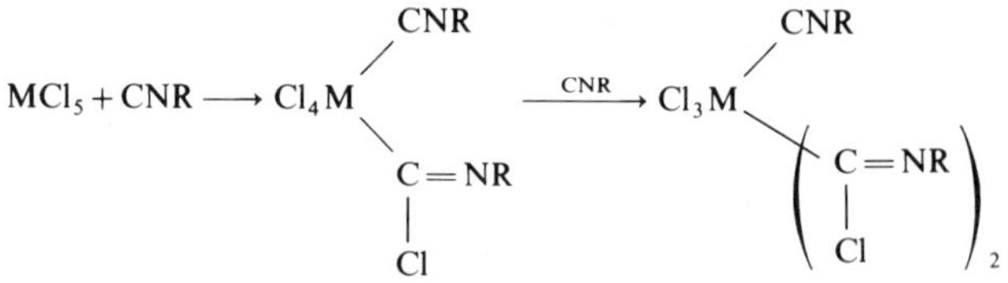

MCl₅ + CNR ⟶ Cl₄M
CNR
C=NR
Cl
CNR
Cl₃M
CNR
C=NR
Cl
2

(12.51)

[Au(N₃)₂]⁻ + CNR ⟶
Au
C
N
R
N
N—N
2

(12.52)

isocyanides and acetylenes proceed via metallacyclic compounds [equations (12.53) and (12.54)]. During pyrolysis of diphenylacetylenebis(arylisocyanide)nickel(0), cyclobutene and cyclopentadiene derivatives are formed [equation (12.55)]. The reactions of metal carbene complexes with isocyanides lead to the formation of azirydynylidene complexes [equation (12.56)]. Upon treatment with methanol, these complexes rearrange to acyclic carbene complexes such as $[(OC)_5W{=}C(NHCy)C(OMe)_2Me]$ and $[\{(MeO)_3P\}_2(OC)_3Mn{=}C(NHR)C(OMe)_2Me][PF_6]$.

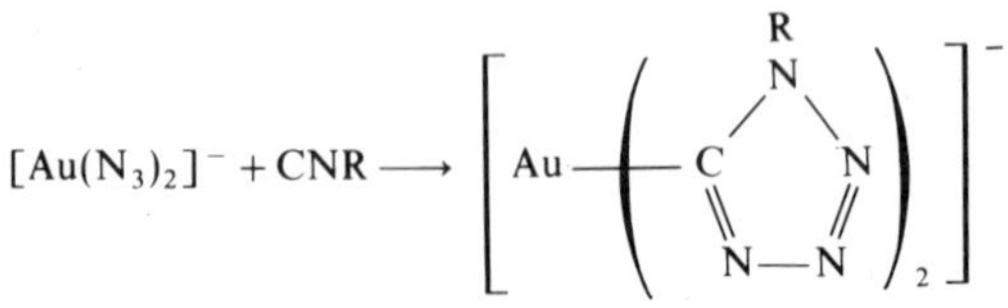

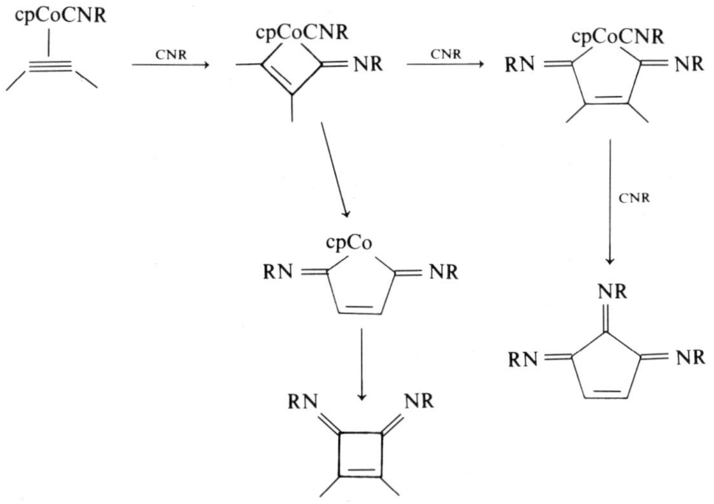

cpCoCNR

$\xrightarrow{\text{CNR}}$   cpCoCNR $=$ NR   $\xrightarrow{\text{CNR}}$   RN $=$ cpCoCNR $=$ NR

cpCo   RN $=$ NR

RN   NR

RN $=$ NR   (NR / RN $=$ NR)

(12.53)

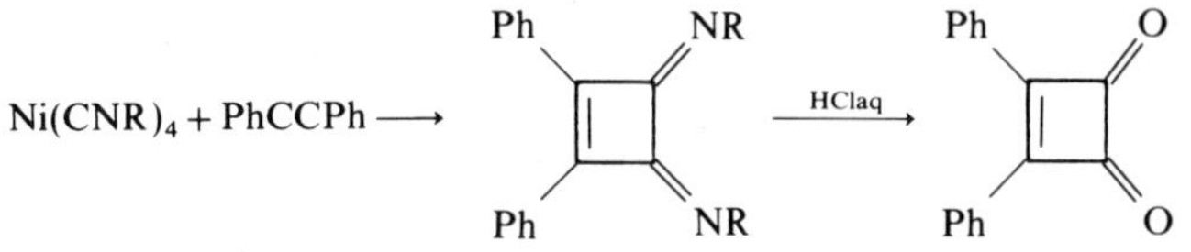

$\text{Ni(CNR)}_4 + \text{PhCCPh} \longrightarrow$   Ph NR / Ph NR   $\xrightarrow{\text{HClaq}}$   Ph O / Ph O   (12.54)

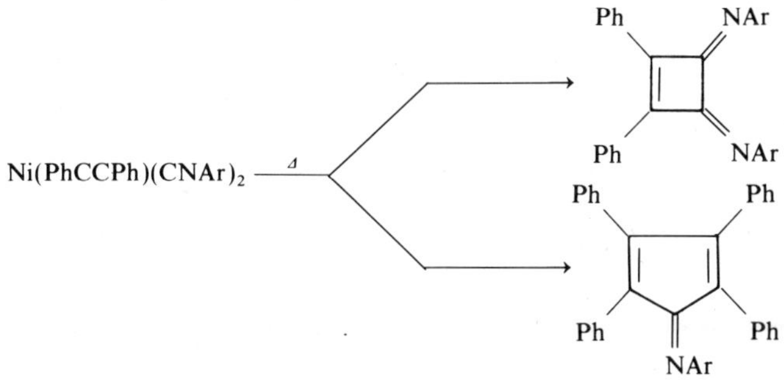

$\text{Ni(PhCCPh)(CNAr)}_2 \xrightarrow{\Delta}$   Ph NAr / Ph NAr   and   Ph Ph / Ph Ph / NAr   (12.55)

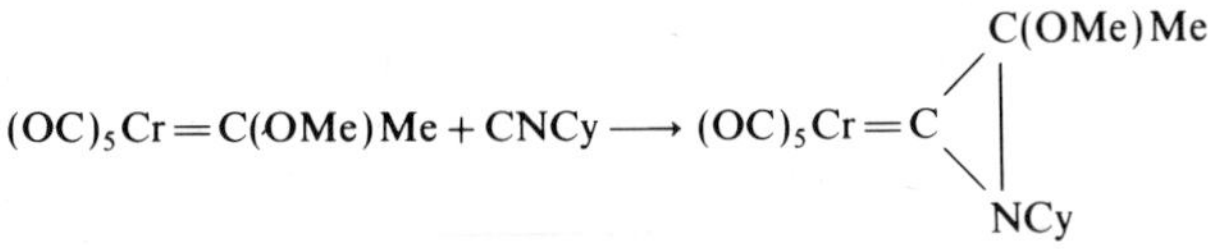

$(\text{OC})_5\text{Cr}=\text{C(OMe)Me} + \text{CNCy} \longrightarrow (\text{OC})_5\text{Cr}=\text{C}\begin{smallmatrix}\text{C(OMe)Me}\\[2pt]\text{NCy}\end{smallmatrix}$   (12.56)

## g. *Dimerization of isocyanides*

Some compounds of transition metals may react with isocyanides to give $RN=C=C=NR$ dimers. Niobium and tantalum compounds of the type $M_2Cl_6(SMe_2)_3$ which possess metal–metal double bonds of the $\sigma^2\pi^2$ configuration ($M=M$ bond distances *ca* 265 pm) react with various isocyanides to furnish the following unusual derivatives (12.57).[31] One of the metal atoms adopts pentagonal bipyramidal configuration with the chlorine atom and the $C-C$ fragment occupying axial positions, while the second metal atom is bonded to four chlorine atoms and to two nitrogen atoms that lie at the vertices of a distorted octahedron.

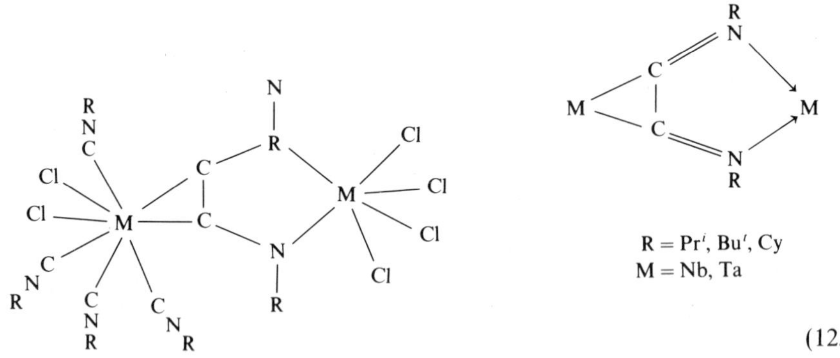

$$R = Pr^i, Bu^i, Cy$$
$$M = Nb, Ta$$

(12.57)

A complex containing another ligand which arises as a result of the formation of a bond between carbon atoms of two isocyanide molecules may be prepared by the reaction of $[Mn(CNMe)(CO)_4]^-$ with methyl iodide [structure (12.58)].[98]

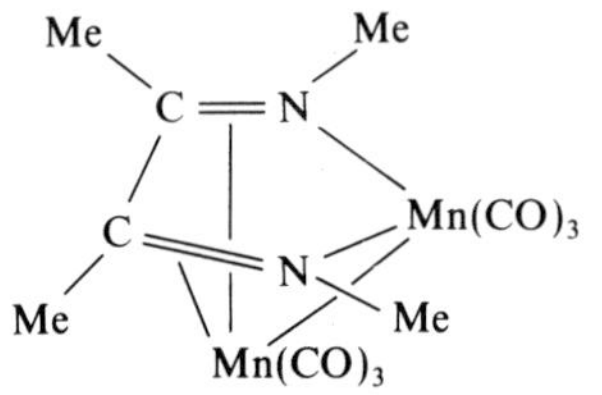

(12.58)

## 10. APPLICATIONS OF ISOCYANIDE COMPLEXES

Isocyanide complexes have found numerous applications in organic synthesis and catalysis. Isocyanides undergo polymerization in the presence of many transition metal complexes, for instance, metal carbonyls, metallocenes, cyclopentadienyl carbonyls, nickel(II), palladium(II), and cobalt(II) complexes. Exceptionally high activity is exhibited by nickel and cobalt carbonyls. The resulting polymers are Schiff bases:

$$\left(\begin{array}{c} +C+ \\ \| \\ NR \end{array}\right)_n \quad \text{[cf. reaction (12.48)]}.$$

Metallic copper and some copper compounds catalyze additions of alcohols to isocyanides. In this way esters of alkylimineformic acid (formimidates) may conveniently be prepared.[99]

$$CNR + R'OH \longrightarrow RN = CHOR'$$

(12.59)

Isocyanides that possess an acidic hydrogen atom in the $\alpha$ position may be added to the $C=C$ bond of the $\alpha$, $\beta$-unsaturated cyanides, ketones, and esters to give heterocyclic compounds [equations (12.60) and (12.61)].[100]

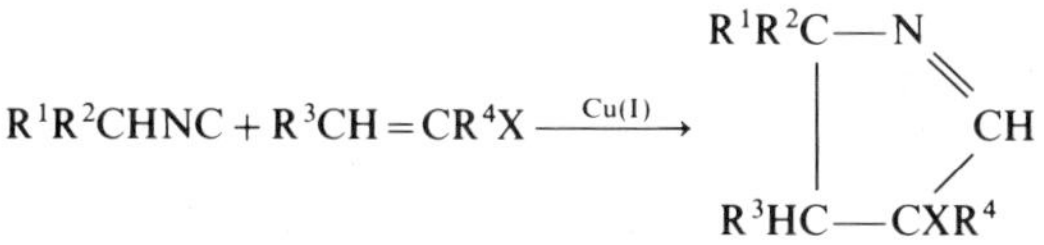

$$R^1R^2CHNC + R^3CH=CR^4X \xrightarrow{Cu(I)} \begin{array}{c} R^1R^2C\!-\!N \\ | \quad\quad\ \ \diagdown \\ \quad\quad\quad CH \\ | \quad\quad\ \ \diagup \\ R^3HC\!-\!CXR^4 \end{array} \qquad (12.60)$$

$$R^1R^2CHNC + R^3R^4CO \xrightarrow{Cu(I)} \begin{array}{c} R^1R^2C\!-\!N \\ | \quad\quad\ \ \diagdown \\ \quad\quad\quad CH \\ | \quad\quad\ \ \diagup \\ R^3R^4C\!-\!O \end{array} \qquad (12.61)$$

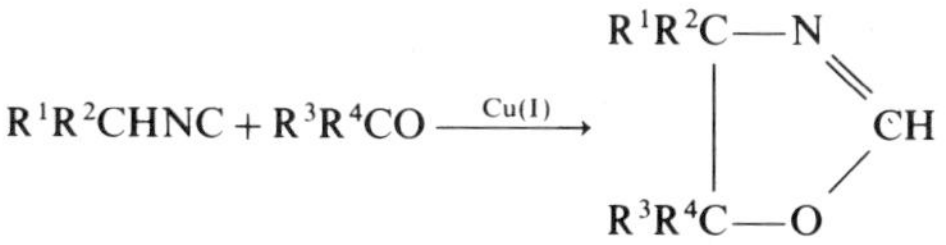

The carbonyls $[Fe(CO)_5]$, $[Fe_2(CO)_9]$, $[Fe(CO)_4](CNPh)]$, and $[M(CO)_6]$ (M = Mo, W) catalyze the formation of carbodiimides from isocyanates. Isocyanide complexes are formed as intermediates:

$$[Fe(CNR)(CO)_4] + RNCO \longrightarrow [Fe(CO)_5] + RN=C=NR \qquad (12.62)$$

$$Fe(CO)_5 + RNCO \longrightarrow [Fe(CNR)(CO)_4] + CO_2 \qquad (12.63)$$

In the presence of cyclooctadiene cobalt complexes as well as compounds such as $[Ni(COD)_2]$, $[Ni(CNR)_4]$, $[RhCl(PPh_3)_3]$, etc., isocyanides are oxidized to isocyanates.[80]

$$[Ni(CNR)_4] \xrightarrow{O_2} [Ni(CNR)_2(O_2)] \xrightarrow{CNR} Ni(CNR)_4 + RNCO \qquad (12.64)$$

The nickel(0) complex $Ni_4(CNBu^t)_7$ catalyzes reduction of acetylenes to olefins, isocyanides and cyanides to amines, as well as cyclotrimerization of acetylene and cyclodimerization of butadiene.[36] Reduction of isocyanides is also catalyzed by other clusters (Chapter 13). Isocyanide copper compounds catalyze[101] addition of $CH_2ClY$ (Y = COOR, COR, CN) to olefins leading to the formation of cyclopropane derivatives

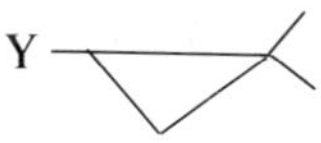

Isocyanides are also utilized for the synthesis of amides,[102] amines,[102, 103] bicyclic compounds,[104, 105] ring compounds of the size $C_{11}$-$C_{15}$,[106] formamides,[107] heterocyclic compounds,[107] serines,[107] products of Michael addition,[108] saturated and unsaturated cyclic derivatives of olefins,[109-113] and heterobicyclic compounds.[114]

# REFERENCES

1. E. Singleton and H. E. Oosthuizen, *Adv. Organomet. Chem.*, **22**, 209 (1983).
2. L. P. Yureva, in: *Metody Elementoorganicheskoi Khimii* (A. N. Nesmeyanov and K. A. Kocheshkov, eds.), Vol. 3, p. 162, Izd. Nauka, Moscow (1975).

3. P. M. Treichel, *Adv. Organomet. Chem.*, **11**, 21 (1973).
4. F. Bonati and G. Minghetti, *Inorg. Chim. Acta*, **9**, 95 (1974).
5. W. P. Griffith, in: *Comprehensive Inorganic Chemistry* (J. C. Bailar, Jr., ed.), Vol. 4, Chapter 46, p. 105, Pergamon Press, Oxford (1973).
6. G. L. Geoffroy and M. S. Wrighton, *Organometallic Photochemistry*, Academic Press, New York (1979).
7. B. E. Bursten and R. F. Fenske, *Inorg. Chem.*, **16**, 963 (1977).
8. R. M. Nielsen and S. Wherland, *Inorg. Chem.*, **24**, 1803 (1985); R. M. Nielsen and S. Wherland, *J. Am. Chem. Soc.*, **107**, 1505 (1985).
9. J. A. S. Howell, J. Y. Saillard, A. LeBeuze, and G. Jaouen, *J. Chem. Soc., Dalton Trans.*, 2533 (1982).
10. J. B. Moffat, *J. Am. Chem. Soc.*, **104**, 3949 (1982); J. B. Moffat, *J. Mol. Struct.*, **108**, 293 (1984); J. B. Moffat, *J. Mol. Struct.*, **44**, 237 (1978).
11. H. Oberhammer and D. Lentz, *Inorg. Chem.*, **24**, 1271 (1985).
12. M. F. N. N. Carvalho, A. J. L. Pombeiro, U. Schubert, O. Orama, C. J. Pickett, and R. L. Richards, *J. Chem. Soc., Dalton Trans.*, 2079 (1985).
13. L. S. Benner, M. M. Omstead, and A. L. Balch, *J. Organomet. Chem.*, **159**, 289 (1978).
14. J.-M. Bassett, G. K. Barker, M. Green, J. A. K. Howard, F. G. A. Stone, and W. C. Wolsey, *J. Chem. Soc., Dalton Trans.*, 219 (1981).
15. K. R. Mann, J. A. Thich, R. A. Bell, C. L. Coyle, and H. B. Gray, *Inorg. Chem.*, **19**, 2462 (1980).
16. M. I. Bruce, J. G. Matison, J. R. Rodgers, and R. C. Wallis, *J. Chem. Soc., Chem. Commun.*, 1070 (1981).
17. K. R. Mann, M. Cimolino, G. L. Geoffroy, G. S. Hammond, A. A. Orio, G. Albertin, and H. B. Gray, *Inorg. Chim. Acta*, **16**, 97 (1976).
18. V. Cerassitti, G. Condorelli, and L. L. Condorelli-Costanzo, *Ann. Chim. (Rome)*, **55**, 329 (1965).
19. H. Isci and W. R. Mason, *Inorg. Chem.*, **14**, 913 (1975).
20. K. R. Mann, J. G. Gordon, II, and H. B. Gray, *J. Am. Chem. Soc.*, **97**, 3553 (1975); H. B. Gray, V. M. Miskowski, S. J. Milder, T. P. Smith, A. W. Maverick, J. D. Buhr, W. L. Gladfelter, I. S. Sigal, and K. R. Mann, *Fundam. Res. Homogeneous Catal.*, **3**, 819 (1979); H. B. Gray, K. R. Mann, N. S. Lewis, J. A. Thich, and R. M. Richman, *Adv. Chem. Ser.*, **168**, 44 (1978); K. R. Mann and H. B. Gray, *Adv. Chem. Ser.*, **173**, 225 (1979); A. W. Maverick and H. B. Gray, *Pure Appl. Chem.*, **52**, 2339 (1980).
21. N. S. Lewis, K. R. Mann, J. G. Gordon, II, and H. B. Gray, *Pure Appl. Chem.*, **98**, 7461 (1976).
22. K. R. Mann, N. S. Lewis, V. M. Miskowski, D. K. Erwin, G. S. Hammond, and H. B. Gray, *Pure Appl. Chem.*, **99**, 5525 (1977).
23. V. M. Miskowski, T. P. Smith, T. M. Loehr, and H. B. Gray, *Pure Appl. Chem.*, **107**, 7925 (1985).
24. M. K. Reinking, M. L. Kullberg, A. R. Cutler, and C. P. Kubiak, *J. Am. Chem. Soc.*, **107**, 3517 (1985).
25. B. Crociani and R. L. Richards, *J. Organomet. Chem.*, **144**, 85 (1978).
26. D. L. Cronin, J. R. Wilkinson, and L. J. Todd, *J. Magn. Reson.*, **17**, 353 (1975).
27. R. W. Stephany, M. J. A. de Bie, and W. Drenth, *Org. Magn. Reson.*, **6**, 45 (1974).
28. A. J. L. Pombeiro, *Rev. Port. Quim.*, **21**, 90 (1979).
29. J. Chatt, A. J. L. Pombeiro, and R. L. Richards, *J. Organomet. Chem.*, **184**, 357 (1980); J. Chatt, A. J. L. Pombeiro, and R. L. Richards, *J. Chem. Soc., Dalton Trans.*, 492 (1980).
30. A. J. L. Pombeiro, *Rev. Port. Quim.*, **23**, 179 (1981).
31. F. A. Cotton, S. A. Duraj, and W. J. Roth, *J. Am. Chem. Soc.*, **106**, 6987 (1984).
32. N. J. Coville, G. W. Harris, and D. Rehder, *J. Organomet. Chem.*, **293**, 365 (1985).
33. M. F. N. N. Carvalho, C. M. C. Laranjeira, A. T. Z. Nobre, A. J. L. Pombeiro, A. C. A. M. Viegas, and R. L. Richards, *Transition Met. Chem.*, **10**, 427 (1985).
34. A. J. L. Pombeiro, M. F. N. N. Carvalho, P. B. Hitchcock, and R. L. Richards, *J. Chem. Soc., Dalton Trans.*, 1629 (1981).
35. M. Cowie, R. S. Dickson, and B. W. Hames, *Organometallics*, **3**, 1879 (1984).
36. V. W. Day, R. O. Day, J. S. Kristoff, F. J. Hirsekorn, and E. L. Muetterties, *J. Am. Chem. Soc.*, **97**, 2571 (1975); M. G. Thomas, W. R. Pretzer, B. F. Beier, F. J. Hirsekorn, and E. L. Muetterties,

*J. Am. Chem. Soc.*, **99**, 743 (1977); E. Band, W. R. Pretzer, M. G. Thomas, and E. L. Muetterties, *J. Am. Chem. Soc.*, **99**, 7380 (1977).

37. A. J. L. Pombeiro, C. J. Pickett, and R. L. Richards, *J. Organomet. Chem.*, **224**, 285 (1982); A. J. L. Pombeiro, *Inorg. Chim. Acta*, **103**, 95 (1985).
38. A. Sacco, *Gazz. Chim. Ital.*, **84**, 370 (1954); L. Malatesta and A. Sacco, *Z. Anorg. Allg. Chem.*, **273**, 247 (1953).
39. J. M. Pratt and P. R. Silverman, *J. Chem. Soc., Chem. Commun.*, 117 (1967); J. M. Pratt and P. R. Silverman, *J. Chem. Soc. (A)*, 1286 (1967).
40. J. P. Maher, *J. Chem. Soc., Chem. Commun.*, 632 (1967); J. P. Maher, *J. Chem. Soc. (A)*, 2918 (1968).
41. F. A. Cotton and R. H. Holm, *J. Am. Chem. Soc.*, **82**, 2983 (1960).
42. C. A. L. Becker, *Inorg. Chim. Acta*, **27**, L105 (1978).
43. E. F. Vansant and J. H. Lunsford, *J. Chem. Soc., Chem. Commun.*, 830 (1972).
44. F. A. Jurnak, D. R. Greig, and K. N. Raymond, *Inorg. Chem.*, **14**, 2585 (1975).
45. N. Kataoka and H. Kon, *J. Am. Chem. Soc.*, **90**, 2978 (1968); N. Kataoka and H. Kon, *J. Phys. Chem.*, **73**, 803 (1969).
46. M. E. Kimball, D. W. Pratt, and W. C. Kaska, *Inorg. Chem.*, **7**, 2006 (1968).
47. A. Sacco and F. A. Cotton, *J. Am. Chem. Soc.*, **84**, 2043 (1962).
48. L. Naldini, *Gazz. Chim. Ital.*, **90**, 871 (1960).
49. D. S. Matteson and R. A. Bailey, *J. Am. Chem. Soc.*, **89**, 6389 (1967); D. S. Matteson and R. A. Bailey, *J. Am. Chem. Soc.*, **91**, 1975 (1969).
50. P. M. Treichel and G. J. Essenmacher, *Inorg. Chem.*, **15**, 146 (1976); P. M. Treichel and G. J. Essenmacher, *Inorg. Chem.*, **16**, 800 (1977).
51. K. R. Mann, H. B. Gray, and G. S. Hammond, *J. Am. Chem. Soc.*, **99**, 306 (1977).
52. A. E. Efraty, I. Feinstein, L. Wackerle, and F. Frolow, *Angew. Chem., Int. Ed. Engl.*, **19**, 633 (1980); A. E. Efraty, I. Feinstein, F. Frolow, and L. Wackerle, *J. Am. Chem. Soc.*, **102**, 6341 (1980); A. Efraty, I. Feinstein, F. Frolow, and A. Goldman, *J. Chem. Soc., Chem. Commun.*, 864 (1980); I. Jaffe and S. Maisuls, XXIV ICCC Athens, August 24–29, 1986, *Abstracts*, p. 239 (1986).
53. J. I. Jaffe, M. Segal, and A. Efraty, *J. Organomet. Chem.*, **244**, C17 (1985).
54. A. Efraty and I. Feinstein, *Inorg. Chem.*, **21**, 3115 (1982); A. Efraty, I. Feinstein, and F. Frolow, *Inorg. Chem.*, **21**, 485 (1982).
55. A. N. Nesmeyanov and K. A. Kocheshkov (eds.), *Metody Elementoorganicheskoi Khimii*, Izd. Nauka, Moscow (1974–1985).
56. G. Wilkinson, F. G. A. Stone, and E. W. Abel (eds.), *Comprehensive Organometallic Chemistry*, Vols. 2–6, Pergamon Press, Oxford (1982).
57. P. T. Wolczanski and J. E. Bercaw, *J. Am. Chem. Soc.*, **101**, 6450 (1979).
58. L. D. Silverman, J. C. Dewan, C. M. Giandomenico, and S. J. Lippard, *Inorg. Chem.*, **19**, 3379 (1980); L. D. Silverman, P. W. R. Corfield, and S. J. Lippard, *Inorg. Chem.*, **20**, 3106 (1981).
59. P. M. Treichel, D. W. Firsich, and G. P. Essenmacher, *Inorg. Chem.*, **18**, 2405 (1979); P. M. Treichel and G. E. Dirren, *J. Organomet. Chem.*, **39**, C20 (1972).
60. W. S. Mialki, D. E. Wigley, T. E. Wood, and R. A. Walton, *Inorg. Chem.*, **21**, 480 (1982).
61. F. A. Cotton, P. E. Fanwick, and P. A. McArdle, *Inorg. Chim. Acta*, **35**, 289 (1979).
62. B. Crociani, M. Nicolini, and R. L. Richards, *J. Organomet. Chem.*, **101**, C1 (1975); M. Benham-Dehkordy, B. Crociani, M. Nicolini, and R. L. Richards, *J. Organomet. Chem.*, **181**, 69 (1979).
63. J. E. Ellis and K. L. Fjare, *J. Organomet. Chem.*, **214**, C33 (1981); J. E. Ellis and K. L. Fjare, *Organometallics*, **1**, 898 (1982); M. Y. Darensbourg and J. M. Hanckel, *J. Organomet. Chem.*, **217**, C9 (1981); M. Y. Darensbourg and J. M. Hanckel, *Organometallics*, **1**, 82 (1982); K. Ihmels and D. Rehder, *Organometallics*, **4**, 1340 (1985).
64. Y. Yamamoto, *Coord. Chem. Rev.*, **32**, 193 (1980).
65. T. J. Weaver and C. A. L. Becker, *J. Inorg. Nucl. Chem.*, **35**, 3739 (1973).
66. R. L. Luck, R. H. Morris, and J. F. Sawyer, *Organometallics*, **3**, 247 (1984).
67. R. L. Luck, R. M. Morris, and J. E. Sawyer, *Organometallics*, **3**, 1009 (1984).
68. M. J. Abrams, A. Davison, A. G. Jones, C. E. Costello, and H. Pang, *Inorg. Chem.*, **22**, 2798 (1983).

69. J. P. Farr, M. J. Abrams, C. E. Costello, A. Davison, S. J. Lippard, and A. G. Jones, *Organometallics*, **4**, 139 (1985).
70. C. J. Cameron, S. M. Tetrick, and R. A. Walton, *Organometallics*, **3**, 240 (1984).
71. J. D. Allison, P. E. Fanwick, and R. A. Walton, *Organometallics*, **3**, 1515 (1984).
72. G. W. Harris and N. J. Coville, *Organometallics*, **4**, 908 (1985); G. W. Harris, J. C. A. Boeyens, and N. J. Coville, *Organometallics*, **4**, 914 (1985).
73. D. L. Lichtenberger, A. C. Sarapu, and R. F. Fenske, *Inorg. Chem.*, **12**, 702 (1973).
74. D. B. Beach, R. Bertoncello, G. Granozzi, and W. P. Jolly, *Organometallics*, **4**, 311 (1985).
75. M. Hoefler and W. Kemp, *Chem. Ber.*, **112**, 1934 (1979).
76. H. Behrens, M. Moll, W. Popp, H. J. Seibold, E. Sepp, and P. Wuerstl, *J. Organomet. Chem.*, **192**, 389 (1980).
77. W. P. Fehlhammer and F. Degel, *Angew. Chem., Int. Ed. Engl.*, **18**, 75 (1979).
78. J.-M. Bassett, D. E. Berry, G. K. Barker, M. Green, J. A. K. Howard, and F. G. A. Stone, *J. Chem. Soc., Dalton Trans.*, 1003 (1979).
79. W. E. Carroll, M. Green, A. M. Galas, M. Murray, T. W. Turney, A. G. Welch, and P. Woodward, *J. Chem. Soc., Dalton Trans.*, 80 (1980).
80. S. Otsuka, A. Nakamura, and Y. Tatsuno, *J. Am. Chem. Soc.*, **91**, 6994 (1969).
81. M. Green, J. A. K. Howard, M. Murray, J. L. Spencer, and F. G. A. Stone, *J. Chem. Soc., Dalton Trans.*, 1509 (1977).
82. J. R. Boehm, D. J. Doonan, and A. L. Balch, *J. Am. Chem. Soc.*, **98**, 4845 (1976).
83. L. Malatesta, A. Sacco, and S. Ghielmi, *Gazz. Chim. Ital.*, **82**, 516 (1952); L. Malatesta, A. Sacco, and M. Gabaglio, *Gazz. Chim. Ital.*, **82**, 548 (1952); L. Malatesta and A. Sacco, *Ann. Chim. (Rome)*, **43**, 622 (1953).
84. C. R. Langrick, P. G. Pringle, and B. L. Shaw, *J. Chem. Soc., Dalton Trans.*, 1233 (1984).
85. M. O. Albers, N. J. Coville, T. V. Ashworth, E. Singleton, and H. E. Swanepoel, *J. Chem. Soc., Chem. Commun.*, 489 (1980); M. O. Albers, N. J. Coville, T. V. Ashworth, E. Singleton, and H. E. Swanepoel, *J. Organomet. Chem.*, **199**, 55 (1980); M. O. Albers, N. J. Coville, and E. Singleton, *J. Chem. Soc., Chem. Commun.*, 96 (1982).
86. O. Temkin and L. G. Bruk, *Usp. Khim.*, **52**, 206 (1983).
87. A. P. Tomilov, I. N. Chernykh, and J. M. Kargin, *Elektrokhimia Elementoorganicheskikh Soedinenii*, Izd. Nauka, Moscow (1985).
88. B. R. Higginson, D. R. Lloyd, D. M. Burroughs, D. M. Gibson, and A. F. Orchard, *J. Chem. Soc., Faraday Trans. 2*, **70**, 1418 (1974).
89. W. S. Mialki and R. A. Walton, *Inorg. Chem.*, **21**, 2791 (1982).
90. J. A. Connor, E. J. James, C. Overton, and N. E. Murr, *J. Organomet. Chem.*, **218**, C31 (1981).
91. J.-M. Bassett, M. Green, J. A. K. Howard, and F. G. A. Stone, *J. Chem. Soc., Dalton Trans.*, 1779 (1980).
92. P. D. Enlow and C. Woods, *Inorg. Chem.*, **24**, 1273 (1985); D. R. Womack, P. D. Enlow, and C. Woods, *Inorg. Chem.*, **22**, 2653 (1983).
93. A. J. L. Pombeiro and R. L. Richards, *Transition Met. Chem.*, **5**, 55 (1980).
94. A. J. L. Pombeiro and R. L. Richards, *Transition Met. Chem.*, **5**, 281 (1980).
95. S. Willis and A. R. Manning, *J. Chem. Soc., Dalton Trans.*, 322 (1981); S. Willis and A. R. Manning, *J. Chem. Soc., Dalton Trans.*, 186 (1980); S. Willis and A. R. Manning, *J. Chem. Res., Synopsis*, 390 (1978); S. Willis and A. R. Manning, *J. Organomet. Chem.*, **97**, C49 (1975).
96. K. W. Chiu, R. A. Jones, G. Wilkinson, A. M. R. Galas, and M. B. Hursthouse, *J. Chem. Soc., Dalton Trans.*, 2088 (1981).
97. J. D. Wilkins, *J. Organomet. Chem.*, **67**, 269 (1974).
98. R. D. Adams, *J. Am. Chem. Soc.*, **102**, 7476 (1980).
99. T. Saegusa, Y. Ito, S. Kobayashi, N. Takeda, and K. Hirota, *Tetrahedron Lett.*, 1273 (1967); T. Saegusa, Y. Ito, S. Kobayashi, N. Takeda, and K. Hirota, *Tetrahedron Lett.*, 521 (1967); T. Saegusa, Y. Ito, S. Kobayashi, N. Takeda, and K. Hirota, *Can. J. Chem.*, **47**, 1217 (1969).
100. T. Saegusa, Y. Ito, H. Kinoshita, and S. Tomita, *J. Org. Chem.*, **36**, 3316 (1971); T. Saegusa, I. Murase, and Y. Ito, *Bull. Chem. Soc. Jpn.*, **45**, 1884 (1972).
101. T. Saegusa, Y. Ito, K. Yonesawa, Y. Inubushi, and S. Tomita, *J. Am. Chem. Soc.*, **93**, 4049 (1971).

102. U. Schoellkopf and W. Frieben, *Justus Liebigs Ann. Chem.*, 1722 (1980).
103. T. Tsuda, H. Habu, S. Horiguchi, and T. Saegusa, *J. Am. Chem. Soc.*, **96**, 5930 (1974).
104. Y. Ito, T. Konoike, and T. Saegusa, *J. Organomet. Chem.*, **85**, 395 (1975).
105. Y. Wakatsuki, O. Nomura, H. Tone, and Y. Yamazaki, *J. Chem. Soc., Perkin Trans. 2*, 1344 (1980).
106. R. Baker, R. C. Cookson, and J. R. Vinson, *J. Chem. Soc., Chem. Commun.*, 515 (1974); H. Breil and G. Wilke, *Angew. Chem., Int. Ed. Engl.*, **9**, 367 (1970).
107. U. Schoellkopf, *Pure Appl. Chem.*, **51**, 1347 (1979).
108. T. Saegusa, Y. Ito, S. Tomita, and H. Kinoshita, *Bull. Chem. Soc. Jpn.*, **45**, 496 (1972); J. Langova and J. Hetflejs, *Collect. Czech, Chem. Commun.*, **40**, 432 (1975).
109. Y. Suzuki and T. Takizawa, *J. Chem. Soc., Chem. Commun.*, 837 (1972).
110. P. B. J. Driessen and H. Hogeveen, *Tetrahedron Lett.*, 271 (1979).
111. T. Saegusa, K. Yonesawa, I. Murase, T. Konnoike, S. Tomita, and Y. Ito, *J. Org. Chem.*, **38**, 2319 (1973).
112. R. Baker and A. H. Copeland, *Tetrahedron Lett.*, 4535 (1976); R. Baker and A. H. Copeland, *J. Chem. Soc., Perkin Trans.*, 2560 (1977).
113. Y. Yamamoto and H. Yamazaki, *Bull. Chem. Soc. Jpn.*, **54**, 787 (1981).
114. H. Sawai and T. Takizawa, *Bull. Chem. Soc. Jpn.*, **49**, 1906 (1976).
115. A. J. Deeming and S. Donovan-Mtunzi, *Organometallics*, **4**, 693 (1985).

# Chapter 13

# Application of Organometallic Compounds in Homogeneous Catalysis

## 1. BASIC PRINCIPLES

Catalysis belongs to older domains of chemistry. The term catalysis was introduced by Berzelius in 1835, and some catalytic processes date back to the beginning of the 19th century, for example, combustion of hydrogen in the presence of platinum or breakdown of many organic compounds by acids.

Chemical compounds A and B react to form the activated complex in the transition state. The formation of this complex requires certain energy. In the presence of catalyst, a different activated compound is formed which generally requires less energy.

Catalyst is a substance that speeds up the chemical reaction because it forms an intermediate compound with reagents. However, a catalyst does not influence the equilibrium of the reaction, and therefore the yield of products is not affected either.

A reaction between compounds A and B [equation (13.1)] forms products C and D and the yield of these products depends on the Gibbs free energy change $\Delta G$ taking place during the reaction. The standard free energy change $\Delta G°$ is related to the equilibrium constant as follows:

$$\Delta G° = -RT \ln K$$

$$A + B \underset{k_{-1}}{\overset{k_1}{\rightleftharpoons}} C + D \tag{13.1}$$

The rate of formation of compounds C and D is proportional to concentrations [A] and [B] and is higher, the higher the temperature, and this rate is also higher, the

647

lower the energy of activation of the reaction, because the rate constant $k$ obeys the Arrhenius equation:

$$\frac{d[C]}{dt} = k_1[A][B] \tag{13.2}$$

where $k_1$ is the reaction rate constant,

$$k_1 = k_0 \exp(-E/RT) \tag{13.3}$$

where $E$ is the activation energy, $T$ the temperature (K), and R the gas constant.
   The constant $k_0$ depends on the entropy of activation,

$$k_0 = \text{const. } \exp(\Delta S^+/R) \tag{13.4}$$

Owing to the formation of the intermediate compound between the reagents and the catalyst, there is a lowering of the energy of activation as well as, almost always, a lowering of entropy. Despite the lowering of entropy (because of the decreased number of molecules in the transition state), the lowering of the activation energy is sufficiently large to increase the rate of the reaction.
   The catalyst increases equally both the rate of the forward reaction and the rate of the reverse reaction, because the equilibrium constant is equal to

$$K = k_1/k_{-1}$$

Figure 13.1 represents the dependence of the free energy change of the noncatalytic and

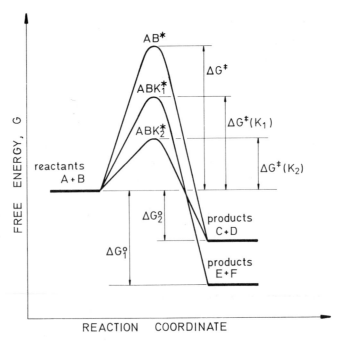

Figure 13.1   Dependence of Gibbs free energy on the reaction coordinate: AB* is the activated complex for a noncatalytic reaction, while $\text{ABK}_1^{\ddagger}$ and $\text{ABK}_2^{\ddagger}$ are the activated complexes for a catalytic reaction.

catalytic reaction on the reaction coordinate. This figure shows that the reaction rate is decided by the Gibbs free energy $\Delta G^{\neq}$ of the formation of the activated complex, while the equilibrium constant is determined by the free energy change $\Delta G$ of the reaction. Figure 13.1 also shows the commonly encountered case whereby the catalyst $K_1$ directs the reaction in different directions, causing the formation of products that are not formed when the reaction proceeds without a catalyst or in the presence of other catalysts. Based on the $\Delta G^{\circ}$ values, it is possible to predict for which reactions searching for a catalyst is a worthwhile endeavor. The yield of products may be considerable even for small positive values of $\Delta G^{\circ}$. When $\Delta G^{\circ} < 0$, the reaction is promising. For $0 < \Delta G^{\circ} < 40 \text{ kJ mol}^{-1}$, searching for the catalyst is still worthwhile, and for $\Delta G^{\circ} > 40 \text{ kJ mol}^{-1}$, no catalyst search need be done.

Catalysis is of tremendous importance in technology and in nature. It is possible to state that without catalysis, life would not be possible because processes such as $CO_2$ assimilation and oxygen binding in animal organisms are catalytic. It is well known that the transition metals are often excellent heterogeneous catalysts of many reactions, both inorganic and organic. In recent years, owing to the development of organometallic chemistry, many interesting homogeneous catalysts have been obtained. They catalyze many important industrial processes. No less important is the role of homogeneous catalysts in explaining the mechanisms of many catalytic reactions, because in these cases it is easier to prepare the intermediate compounds and to investigate their structures, properties, and reactivities. Established reaction mechanisms for mononuclear, polynuclear, or cluster-type homogeneous catalysts were subsequently confirmed more than once for heterogeneous catalysts (Table 13.1).

Many reactions of organic compounds (although not all), such as *cis* addition of dihydrogen (or other molecules) to unsaturated compounds, polymerization, olefin isomerization, etc., could not proceed as concerted processes without the presence of catalysts.

One of the functions of the metal is to provide contact between reagents at specific geometries. In this case, the metal complex acts as a template. However, no less significant is its role in causing the reaction to be allowed by symmetry rules. In this case, the main role is played by the *d* orbitals of the transition metal. The presence of *d* orbitals also causes the transition metal complexes to exhibit some properties which are typical only for this group of compounds and which affect catalytic activity of these compounds. These properties are as follows:

1. The ability to form various types of bonds ranging from ionic bonds or strongly polar bonds to covalent ones which may be single or multiple ($\sigma$, $\pi$, $\delta$).

2. Changeability of oxidation states of the central atom without decomposition of the compound and without significant structural changes.

3. Changeability of the coordination number and lability of ligands.

4. Changeability of $\sigma$ and $\pi$ donor–acceptor properties of a given ligand as a function of properties of other ligands of the complex and mutual influence of ligands.

5. Formation of electron-deficient complexes which are able to activate inert compounds, for example, saturated hydrocarbons.

Moreover, coordination compounds that catalyze reactions of hydrocarbons are characterized by the following properties:

Table 13.1. Homogeneous and Heterogenized Catalysts for Reduction Reactions

| Catalyst (1) | Reduced compounds (2) | Products (3) | Reaction conditions Solvent (4) |
|---|---|---|---|
| $[Nd(pmcp)_2H]_2$[35] | Alkenes | Alkanes | $H_2$; 0.1 MPa, 298 K |
| $[Ticp_2(CO)_2]$[a] | Styrene, 1-alkynes, $PhC_2H$, $PhC_2Ph$ | Ethylbenzene, 1-alkenes, $PhCH_2CH_2Ph$ | $H_2$; 5 MPa, 323–338 K, PhH, heptane |
| $[MCl_2cp_2]$ + $AlR_3$ (or LiR, MgXR) (M = Ti, Zr)[21,23] | Olefins, $PhC_2H$ | Alkanes, ethylbenzene | $H_2$; 0.4 MPa |
| $[M(CH_2Ph)_4]$ (M = Ti, Zr)[b] | Olefins (Ti), aromatic hydrocarbons | Alkanes | $H_2$; 0.1 MPa, 273 K |
| $[M(arene)(CO)_3]$ (M = Cr, Mo, W)[c] | Dienes, trienes | Alkenes | $H_2$; 3–5 MPa, 370–460 K |
| $[Mn_2(CO)_{10}]$[d] | Alkenes | Alkenes | $H_2$; 20 MPa, 353–423 K |
| $[Re_2(CO)_{10}]$[d] | Alkenes | Alkanes | $H_2$; 3–10 MPa, 333–363 K |
| $[ReCl_5]$ + $SnCl_2$[e] | 1-Alkenes | Alkanes, cycloalkanes | $H_2$; <6 MPa, 353–423 K |
| $[Fe(acac)_3]$ + $AlR_3$[f] | Alkenes, arenes | Saturated acids | |
| $[RuCl_6]^{3-}$, $[RuCl_6]^{4-}$ [21] | Unsaturated acids | Alkanes, amines, alcohols | $H_2$; 0.1 MPa, 298 K |
| $[RuClH(PPh_3)_3]$[g] | 1-Alkenes, nitroalkanes, aldehydes | Alkanes | $H_2$; 0.1 MPa, 298 K |
| $[Ru(RCOO)H(PPh_3)_3]$[h] | 1-Alkenes | Alkenes, saturated acids, aldehydes, amines | $H_2$; 5 MPa, 323 K |
| $[RuHCl(\eta\text{-}C_6Me_6)(PPh_3)]$[i] | Arenes, olefins | | $H_2$; 0.1–20 MPa, 273–400 K, Water + alcohols |
| $[Co(CN)_5]^{3-}$ [21,23] | Dienes, polyenes, unsaturated acids, aldehydes, nitriles, nitrocompounds, Schiff bases | Alkenes, saturated acids, aldehydes, amines | |
| $[Co_2(CO)_8]$[21,23] | Arenes, ketones, nitrocompounds, nitriles, imines, epoxides | Hydroarenes, alcohols, alkylarenes, amines | $H_2$; <30 MPa, 400–500 K |
| $[Co(dmgH)_2]$ + L (L = Lewis base, py, $CN^-$, $PR_3$, $AsR_3$)[21,23] | Activated olefins, acetylenes, nitrocompounds, PhN = NPh | Saturated compounds, amines, aniline | $H_2$; 0.1 MPa, 295 K |
| $[CoH(CO)_3(PR_3)]$[21,23] | Alkenes | Alkanes | |
| $[CoH_3\{PhP(C_2H_4PPh_2)_2\}]$[j] | Arenes | | |
| $[Co(\eta\text{-}C_3H_5)\{P(OMe)_3\}_3]$[k][39] | Alkenes | Alkanes | $H_2$; 0.1 MPa, 300 K |
| $[Co(\eta\text{-}C_3H_5)\{P(OPr^i)_3\}_3]$[k][41] | Arenes | | |
| $[RhX(PR_3)_3]$[21-25] | Alkenes, alkynes, steroids | Alkanes | $H_2$; 0.1 MPa, 290 K |

| Complex | Substrate | Product | Conditions |
|---|---|---|---|
| [RhH(CO)(PPh$_3$)$_3$][21-24] | Alkenes | Alkanes | H$_2$; 0.1 MPa, 290 K |
| [Rh$_2$Cl$_4$(C$_5$Me$_5$)$_2$][l] | Arenes | Alkanes | H$_2$; 5 MPa, 323 K |
| [RhH$_2$(PMe$_2$Ph)(solv)$_2$]$^+$[m] | Ketones | Alcohols | H$_2$; 0.1 MPa, 300 K |
| [Rh$_2$Cl$_2$(C$_8$H$_{14}$)$_4$] + PR$_3$ [21-24] | Alkenes, alkynes | Alkanes | H$_2$; 0.1 MPa, 300 K |
| [RhCl$_2$(C$_4$H$_7$)] + PR$_3$, P(OR)$_3$, diphos, SR$_2$, bipy, phen[33] | Alkenes, alkynes | Alkanes | H$_2$; 0.1 MPa, 300 K, alcohols |
| [Rh$_2$Cl$_2$(C$_8$H$_{14}$)$_4$] + 2-aminopyridine[29,32] | | | H$_2$; 0.1 MPa, 280–300 K, alcohols |
| [Ir$_2$Cl$_2$(C$_8$H$_{14}$)$_4$] + PR$_3$ [21-24] | Alkenes, alkynes | Alkanes | H$_2$; 0.1 MPa, 246–300 K, toluene |
| [IrCl(COD)PCy$_3$][n] | Alkenes | | |
| [IrX(CO)(PPh$_3$)$_2$][21-24] | Alkenes | Alkanes | H$_2$; 0.1 MPa, 300–330 K |
| [Ir$_2$Cl$_4$(C$_5$Me$_5$)$_2$][l] | Arenes | | H$_2$; 5 MPa, 323 K |
| [PdX$_2$(PPh$_3$)$_2$] + SnCl$_2$[o] | Alkenes | Alkanes | H$_2$; 0.1 MPa, 300 K |
| [Pd(o-salen)][p,22] | Alkenes | | H$_2$; 0.1 MPa, 300 K |
| [PtCl$_2$(PPh$_3$)$_2$] + SnCl$_2$[q] | Alkenes, dienes | Alkanes, monoenes | H$_2$; 0.1 MPa, 300 K |
| H$_2$[PtCl$_6$] + SnCl$_2$[r] | Alkenes | Alkanes | H$_2$; 0.1 MPa, 300 K |

[a] K. Sonogashira and N. Hagihara, *Bull. Chem. Soc. Jpn.*, **39**, 1178 (1966).

[b] U. Zucchini, U. Giannini, E. Albizzatii, and R. Angelo, *Chem. Commun.*, 1174 (1969).

[c] M. F. Farona, in: *Organometallic Reactions and Syntheses* (E. I. Becker and M. Tsutsui, eds.), Plenum, New York (1977).

[d] T. A. Weil, S. Metlin, and I. Wender, *J. Organomet. Chem.*, **49**, 227 (1973).

[e] A. P. Khrushch and A. E. Shilov, *Kinet. Katal.*, **10**, 466 (1969); A. P. Khrushch and A. E. Szilov, *Kinet. Katal.*, **11**, 86 (1970).

[f] S. Dzierzgowski, R. Giezynski, S. Pasynkiewicz, and A. Korda, *Rocz. Chem.*, **51**, 1127 (1977).

[g] P. S. Hallman, B. R. McGarvey, and G. Wilkinson, *J. Chem. Soc. (A)*, 3143 (1968); J. F. Knifton, *Catalysis in Organic Syntheses*, p. 257, Academic Press, New York (1976).

[h] D. Rose, J. D. Gilbert, R. P. Richardson, and G. Wilkinson, *J. Chem. Soc. (A)*, 2610 (1969).

[i] M. A. Bennett, T.-N. Huang, A. K. Smith, and T. W. Turney, *J. Chem. Soc., Chem. Commun.*, 582 (1978).

[j] D. L. Dubois and D. W. Meek, *Inorg. Chem. Acta*, **19**, L29 (1976).

[k] M. C. Rakowsky, F. J. Hirsekorn, L. S. Stuhl, and E. L. Muetterties, *Inorg. Chem.*, **15**, 2379 (1976); F. J. Hirsekorn, M. C. Rakowsky, and E. L. Muetterties, *J. Am. Chem. Soc.*, **97**, 237 (1975).

[l] P. M. Maitlis, *Acc. Chem. Res*, **11**, 301 (1978).

[m] R. R. Schrock and J. A. Osborn, *Chem. Commun.*, 567 (1970).

[n] W. De Aquino, R. Bonnaire, and C. Potvin, *J. Organomet. Chem.*, **154**, 159 (1978).

[o] J. C. Bailar, Jr. and H. Itatani, *J. Am. Chem. Soc.*, **89**, 1592 (1967).

[p] G. Henrici-Olive and S. Olive, *J. Mol. Catal.*, **1**, 121 (1976).

[q] H. A. Tayim and J. C. Bailar, *J. Am. Chem. Soc.*, **89**, 4330 (1967).

[r] R. D. Cramer, E. L. Jenner, R. V. Lindsey, Jr., and U. G. Stolberg, *J. Am. Chem. Soc.*, **85**, 1691 (1963).

1.  The ability to form $\sigma$ and $\pi$ bonds with hydrocarbons.

2.  The ability to form hydride complexes.

Usually, during catalytic processes, reactions characteristic for a given process and occurring one by one take place; they are known as elementary reactions.

In various stages of a catalytic reaction, the change of oxidation state (OS), the change of coordination number ($N$), and the change of the number of valence electrons (NVE) may take place in a complex. Changes of these quantities in various elementary reactions are given in Table 1.4.[1]

Diamagnetic stable organometallic compounds of metals of groups 4–10 most commonly possess 16 or 18 valence electrons. Tolman[1] broadened this principle to include intermediate compounds formed during chemical reactions. Many stoichiometric and catalytic reactions of organometallic compounds obey Tolman's rule.

In one catalytic process, several elementary organometallic reactions may occur. Very essential are the following reactions: migration, oxidative addition, and oxidative coupling.

a. *Migration reactions*[1–10]

Migration of ligands to coordinated CO or alkene molecules are the best known processes in this class of reactions (see also Sections 2.8, 3.10.b, 4.2, 4.10.c, 4.12.b, 5.11.i, 6.A.11, 7.11.g, 12.9.f). Figure 13.2 shows the mechanism for the migration with orbitals involved in this reaction.[2] The metal–ligand orbitals are occupied, while the $\pi^*$ antibonding orbitals are empty. The transfer of electrons from bonding, occupied molecular orbitals to antibonding, empty orbitals is accompanied by the breaking of the $\sigma$ metal–ligand, the $\sigma$ and $\pi$ metal–carbonyl (or metal–alkene), and the triple carbon–oxygen (or double carbon–carbon) bonds. At the same time, the formation of a $\sigma$ bond between the migrating ligand and the carbonyl group carbon atom (or between the ligand and the closer alkene carbon atom) as well as the formation of a new $\sigma$ metal–carbon bond takes place.

The geometry of orbitals involved in the reactions requires that the migration proceeds to the *cis* position. This phenomenon has been proven for many reactions (Figure 13.2b). However, external addition to alkene of a nucleophilic group which is not present in the coordination sphere may take place. In this case, the product possessing *trans* structure is formed (Figure 13.2c).

The migration reaction is allowed if, in the formation of the metal–carbonyl or metal–alkene $\sigma$ bonds, $d$ or $p$ orbitals are involved, while this reaction is forbidden if the bonding to the ligand is formed by the $s$ orbital of the central atom (Figure 13.2d).

For $Hg^{2+}$ and $Tl^{3+}$ possessing $d^{10}$ electronic structures, addition of the metal and a nucleophile to alkenes is known:

$$Hg^{2+} + C_2H_4 + X^- \longrightarrow [HgC_2H_4X]^+ \tag{13.5}$$

However, these reactions do not proceed by migration of the ligand which is coordinated to the metal. In this case, addition of the external nucleophile occurs. On the other hand, migration reactions are allowed if two $p$ orbitals are involved in them (Figure 13.2e). The reaction of hydroboration of olefins which affords *cis* products may serve as an example.

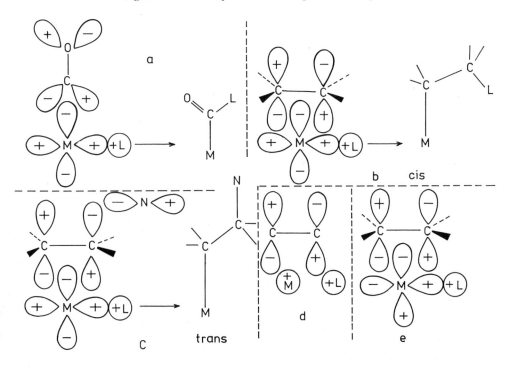

Figure 13.2. Orbitals determining migration reactions: (a) allowed migration to the carbonyl group (the $d$ metal orbital participates in the formation of the $M-L$ bond), (b) allowed migration to alkene, (c) external nucleophilic addition to alkene, (d) forbidden migration to alkene (the $s$ orbital contributes to the formation of the $M-L$ bond), (e) allowed migration to alkene (the $p$ metal orbitals participate in the formation of the $M-L$ bond).

In the formation of the metal–ligand bond, usually hybridized orbitals $(n-1)d^x nsnp^y$ are involved. The greater the contribution of the $s$ orbital in the hybridized orbitals, the higher the activation energy of the migration process. The reaction is forbidden for central atoms whose $d$ electrons are not involved in the formation of bonds, that is, for metals possessing $d^{10}$ electronic structures. Presumably because of this, the elements of group 11 (Cu, Ag, and Au) and their compounds do not catalyze processes in which migration reactions occur. Thus, copper represents a very weak hydrogenation catalyst, while silver and gold do not catalyze the hydrogenation process (an important step in the hydrogenation process is the hydrogen migration to the hydrogenated compound). In contrast to metals of groups 8–10, Cu, Ag, and Au do not catalyze the hydroformylation reaction either. In the case of complexes of the type $MR(CO)L_n$, either the R group or the CO ligand may migrate [see Sections 2.8, 4.12.b, and equation (2.26)]. Migration reactions are usually thermodynamically favorable (Table 4.4). Dedieu's calculations[34] for the reaction of ethylene hydrogenation by means of the $RhCl(PH_3)_3$ complex show that the transfer of the first hydrogen atom to the ethylene molecule in $[RhClH_2(C_2H_4)L_2]$ cannot be regarded as a pure hydride migration process, but rather a complicated process which is probably accompanied by rearrangement of nonreacting ligands.

## b. *Oxidative coupling reactions*[1-3, 8-11]

The oxidative coupling reaction (oxidative cycloaddition) is allowed by symmetry rules if the number of electrons involved in the reaction (two $d$ electrons of the metal and $\pi$ electrons of unsaturated molecules) is equal to $4n + 2$.

In the case of the oxidative cycloaddition of two ethylene molecules ($n = 1$), six electrons in the cyclic compound form two M—C bonds and one C—C bond:

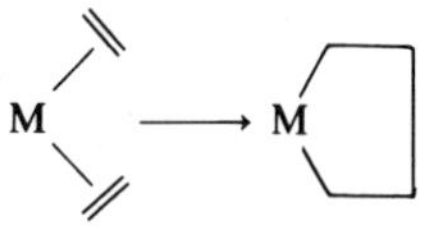

$$(13.6)$$

The metal oxidation state is increased by $+2$. Therefore, it donates two electrons toward the formation of new bonds. A similar situation happens in the case of the cyclization reaction of two butadiene molecules.[2, 3, 11] This reaction involves ten electrons:

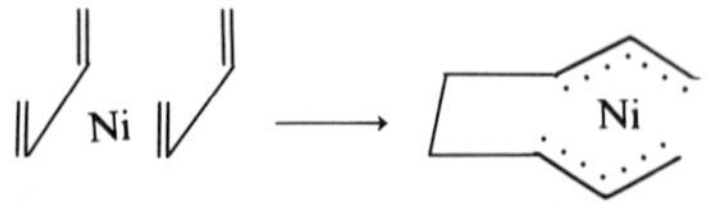

$$(13.7)$$

However, oxidative coupling of butadiene and ethylene is forbidden (8-electron reaction).

## c. *Oxidative addition reactions*[12-17]

The oxidative addition reaction (see also Sections 3.10.e, 4.2, 4.10.b, 4.12.c, 4.13, 5.8.h, 7.7.c, 9.17.j, 12.9.c) in the most general form may be represented as follows:

$$M + X-Y \longrightarrow X-M-Y \qquad (13.8)$$

where M denotes the complex while X and Y represent any atoms or groups connected with a single bond. Therefore, during this reaction, bond breaking of a single bond X—Y occurs, and the formation of two new bonds between the complex and the groups (or atoms) X and Y takes place.

Oxidative addition of dihydrogen, halogens (for example, $Cl_2$), etc., easily takes place, while oxidative addition of the C—H (see Section 4.13) and C—C bonds occurs with more difficulty. Examples of oxidative addition of the latter bonds, and particularly those of C—C bonds, are considerably less numerous than those of the former molecules.

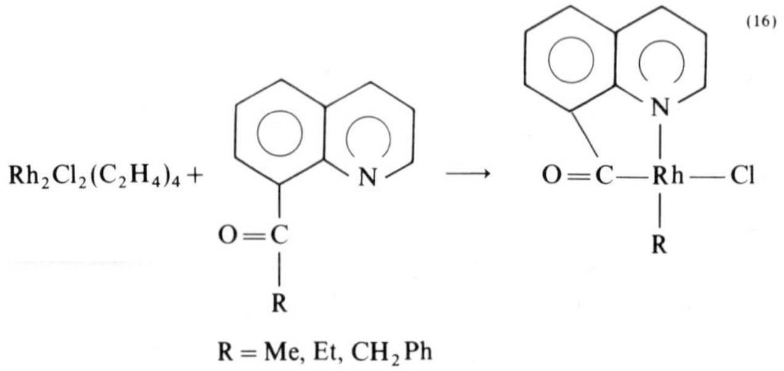

$$Rh_2Cl_2(C_2H_4)_4 + \qquad \longrightarrow \qquad (13.9)$$

$$R = Me, Et, CH_2Ph$$

$$Pt(PEt_3)_3 + PhCN \longrightarrow PtPh(CN)(PEt_3)_2 + PEt_3 \quad [17] \qquad (13.10)$$

However, reductive elimination reactions, that is, the formation of the C$-$H and C$-$C bonds, occur easily.

Three mechanisms of oxidative addition have been proposed[2, 18]:

1.  Concerted addition of X and Y to the metal [equation (13.8)].

2.  A two-step mechanism in which the rate-determining process is described by reaction (13.11).

$$M + XY \longrightarrow M-X^+ + Y^- \tag{13.11}$$

$$M-X^+ + Y^- \longrightarrow Y-M-X \tag{13.12}$$

3.  A radical process,[18]

$$M + XY \longrightarrow X-M^* + Y^* \tag{13.13}$$

$$X-M^* + Y^* \longrightarrow X-M-Y \tag{13.14}$$

The bonds C$-$X where X $=$ I, Br, and Cl most commonly react according to the radical mechanism. Polar solvents favor a two-step mechanism [equations (13.11) and (13.2)], while nonpolar solvents favor a concerted addition and radical mechanism. Reactivity of $d^8$ complexes in oxidative addition processes varies as follows: Os(0) > Ru(0) > Fe(0) > Co(I) > Ir(I) > Rh(I) > Pt(II) > Pd(II) $\gg$ Ni(II). However, this reactivity pattern is changed for $d^{10}$ complexes: Pt(0) $\sim$ Pd(0) $\sim$ Ni(0) > Au(I) > Cu(I) $\gg$ Ag(I).

# 2. HYDROGENATION

## a. Olefin hydrogenation

Activation of olefins due to coordination to the metal allows their hydrogenation under very mild conditions. Homogeneous catalysts are often considerably more active than heterogeneous ones such as the Raney or Urushibara metals.[19] Usually, the complex most commonly activates both reagents, that is, the olefin and dihydrogen. This enables fast reaction of hydrogen with alkene (Figures 13.2)[2, 3, 20] which is otherwise forbidden by orbital symmetry (without the catalyst). Therefore, good catalysts for hydrogenation are hydrido complexes or compounds that easily react with hydrogen to give hydrides and which may coordinate the alkene molecule. At present, homogeneous catalysts for hydrogenation of alkenes, polyolefins, acetylenes, and aromatic hydrocarbons are known.[8–10, 21–25] All these classes of compounds may be reduced under mild conditions such as room temperature and atmospheric pressure with the application of homogeneous catalysts (see Table 13.1).

The activation of the hydrogen molecule occurs as a result of its homolytic or heterolytic splitting:

$$H_2 \longrightarrow 2H \qquad \text{homolytic splitting} \tag{13.15}$$

$$H_2 \longrightarrow H^+ + H^- \qquad \text{heterolytic splitting} \tag{13.16}$$

In the reaction of homolytic splitting of hydrogen, the metal atom formally increases its oxidation state. Thus, during the formation of a $M-H$ bond, the hydrogen atom accepts one electron from the metal.

However, during heterolytic splitting of hydrogen, the oxidation state of the metal does not change:

$$M(I) + H_2 \longrightarrow M(III) \begin{array}{c} \diagup H \\ \diagdown H \end{array} \qquad (13.17)$$

$$2M(II) + H_2 \longrightarrow 2M(III)H \qquad (13.18)$$

$$MA + H_2 \longrightarrow MH + H^+ + A^- \qquad (13.19)$$

In reactions involving heterolytic splitting, an essential role is played by the presence of the base which reacts with the resulting proton. This base may even be the ligand bonded to the central atom which activates hydrogen. Square-planar, $d^8$ complexes, particularly those of Rh(I) and Ir(I), as well as $[Co(CN)_5]^{3-}$, $[CoH(N_2)(PPh_3)_3]$, etc., homolytically split dihydrogen:

$$[RhCl(PPh_3)_3] + H_2 \longrightarrow [RhCl(H)_2 (PPh_3)_3] \qquad (13.20)$$

$$[IrCl(CO)(PPh_3)_2] + H_2 \longrightarrow [IrCl(H)_2 (CO)(PPh_3)_2] \qquad (13.21)$$

$$2[Co(CN)_5]^{3-} + H_2 \longrightarrow 2[Co(CN)_5H]^{3-} \qquad (13.22)$$

$$[CoH(N_2)(PPh_3)_3] + H_2 \longrightarrow [CoH_3(PPh_3)_3] + N_2 \qquad (13.23)$$

The heterolytic splitting most commonly takes place by the influence of complexes of Ru(II), Ru(III), Rh(III), etc.

$$[RuCl_6]^{3-} + H_2 \longrightarrow [RuCl_5H]^{3-} + H^+ + Cl^- \qquad (13.24)$$

$$[RuCl_4(bipy)]^{2-} + H_2 \longrightarrow [RuCl_3H(bipy)]^{2-} + H^+ + Cl^- \qquad (13.25)$$

$$[RhCl_6]^{3-} + H_2 \longrightarrow [RhCl_5H]^{3-} + H^+ + Cl^- \qquad (13.26)$$

Easy isotopic exchange between $H_2$ and deuterated solvents (alcohols, water, amides, etc.) occurs while heterolytic splitting of hydrogen takes place, and this isotopic exchange may be considered to be an important criterion of such splitting. Compounds of the transition metals of all groups (3–12) are utilized as catalysts for the hydrogenation reaction. The most active catalytic complexes which operate under the mildest conditions are those of metals of groups 8–10 and, above all, those of the platinum group. Particularly good catalytic properties are exhibited by rhodium compounds. However, ruthenium and iridium complexes are also known to be unusually catalytically active. Palladium and platinum compounds are also good catalysts. Recently it was found that group 3 elements are also very effective catalysts for reduction of olefins.

The compound $[Nd(pmcp)_2H]_2$ is an unusually active catalyst for olefin reduction (for 1-hexene, the rate of hydrogenation is 77,000 cycles per hour at 298 K and 0.1 MPa).[35] The chloride $UCl_3$ catalyzes hydrogenation of olefins by means of LiH or $LiAlH_4$ in tetrahydrofuran.[36]

Phosphine or phosphite, $P(OR)_3$, complexes of transition metals constitute the most effective and stable catalysts for hydrogenation reactions. This is due to quite strong $\sigma$ donor and $\pi$-acceptor properties of these ligands which stabilize low oxidation states of metals, prevent reduction of complexes to metals, and favor the formation of hydrido complexes. The majority of hydrido complexes contain phosphine or phosphite ligands. Metal complexes show good catalytic properties for hydrogenation reactions if the hydrido olefin complexes formed from these are not too stable, and thus if the energy of the $M-H$ and the M–olefin bonds is not too great.

The energy change $\Delta E$ for the reaction

$$L_xM(I) + H_2 \longrightarrow L_xM(III) \begin{array}{c} H \\ \diagup \\ \diagdown \\ H \end{array}$$

is given by the following expression:

$$\Delta E = 2E_{M-H} - (E_{H-H} + P) \tag{13.27}$$

where $E_{M-H}$ denotes the $M-H$ bond energy, $E_{H-H}$ represents the dissociation energy of the hydrogen molecule, and $P$ is the excitation energy in the ligand field which is related to the change in the bond energy of all $M-L$ bonds and to the electron transfer from the $(n-1) d_{z^2}$ orbital to the $np_z$ orbital. The quantity $\Delta E$ depends chiefly on $P$ because $E_{H-H}$ is constant, and $E_{M-H}$ varies in a narrow range.

Ligands possessing electron-accepting properties will increase $\Delta E$ because they increase the positive charge on the central atom, and therefore they increase the bonding energy of the $d$ electrons, while ligands with electron-donating properties exercise the opposite influence. The validity of these conclusions is confirmed by $\Delta G^+$ values for the reaction of hydrogen with $IrX(CO)(PPh_3)_2$ (X = Cl or Br) which are lower for the bromide complex, because bromine possesses weak electron-accepting properties ($\Delta G^+$ is proportional to $\Delta E$). For complexes exhibiting good catalytic properties, $\Delta E$ should have small positive values. If large positive values are reached, then the energy of the metal–hydrogen bond, $E_{M-H}$, is large and a too stable hydrido complex is formed and the hydrogen atoms cannot be transferred to the olefin molecule. If, however, $\Delta E$ is negative, the complex will not activate dihydrogen.

The solvent may have an essential influence on the activity of the catalyst because it may substitute ligands. The solvent molecules may also change, to a certain degree, the stability of the complex owing to their interaction with ligands. Both of these effects will also influence $\Delta E$.

Kinetics of the olefin hydrogenation reaction were investigated most accurately in the case of cyclohexcne and styrene with the application of Wilkinson's catalyst $[RhCl(PPh_3)_3]$.[25–27]

The mechanism (13.28) of the hydrogenation reaction has been proposed by Halpern.[26]

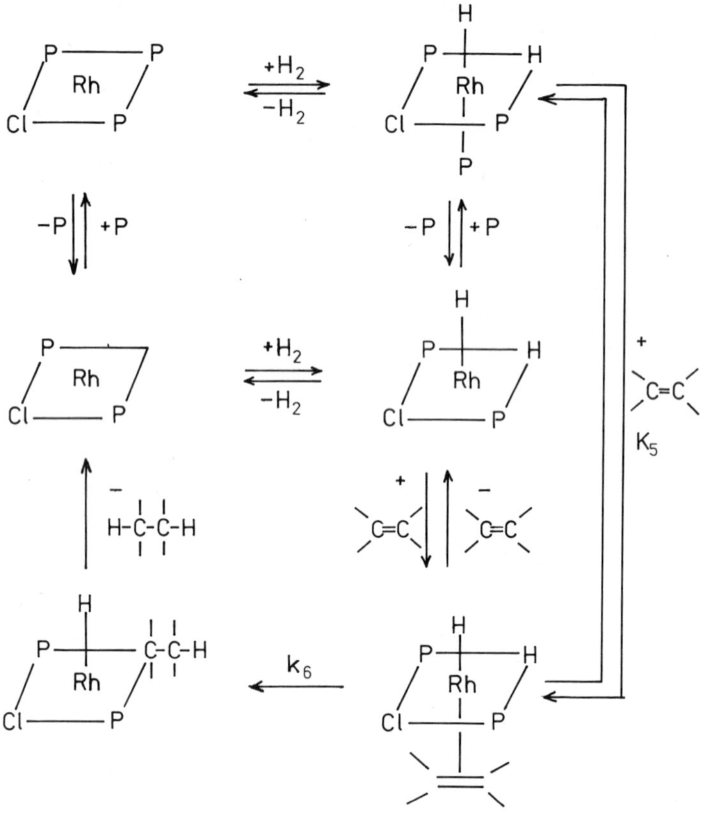

$$(13.28)$$

The rate of the olefin hydrogenation reactions depends on concentrations of reagents according to expression (13.29).[26]

$$-\frac{d[\text{alkene}]}{dt} = \frac{k_6 K_5[\text{alkene}][\text{Rh}]}{K_5[\text{alkene}] + [\text{L}]} \tag{13.29}$$

The reaction scheme represents the most important steps of the catalytic process: the activation of dihydrogen and olefin. Wilkinson[25] assumed that total dissociation of one phosphine ligand in the complex $[\text{RhCl}(\text{PPh}_3)_3]$ occurs. However, later investigations by $^{31}\text{P}$ NMR led to the conclusion that the dissociation of the phosphine ligand is minimal.

Based on the electronic and NMR spectra, the equilibrium constants were determined for reactions (13.30)–(13.32). The constants $K_1'$ and $K_1''$ were determined experimentally; $K_1 = (K_1''/K_1')^{1/2}$.

$$[RhCl(PPh_3)_3] \underset{\leqslant 10^{-5}}{\overset{K_1}{\rightleftharpoons}} [RhCl(PPh_3)_2] + PPh_3 \qquad (13.30)$$

$$2[RhCl(PPh_3)_2] \underset{\geqslant 10^6}{\overset{K_1'}{\rightleftharpoons}} [Rh_2Cl_2(PPh_3)_4] \qquad (13.31)$$

$$2[RhCl(PPh_3)_3] \underset{10^{-4}}{\overset{K_1''}{\rightleftharpoons}} [Rh_2Cl_2(PPh_3)_4] + 2PPh_3 \qquad (13.32)$$

Despite a small concentration of $[RhCl(PPh_3)_2]$ or $[RhCl(PPh_3)_2$ (solvent)] (the formation of $14e$ complexes is less probable), these intermediate complexes still play an important role in the hydrogenation reactions as suggested by Wilkinson, because they react with dihydrogen at least $10^4$ times faster than does $[RhCl(PPh_3)_3]$.

Halpern's proposed mechanism[26] (13.28) is similar in nature. Although his main steps are analogous to the those in the mechanism proposed by Wilkinson,[25] Petit,[27] and de Croon,[28] the kinetic equations for the olefin hydrogenation reaction are considerably different for the latter authors:

$$-\frac{d[alkene]}{dt} = -\frac{d[H_2]}{dt} = \frac{k_e[Rh][alkene][H_2]}{1 + K_1[H_2] + K_2[alkene]} \qquad (13.33)^{[25]}$$

$$-\frac{d[alkene]}{dt} = k\frac{K_1K_D[alkene][H_2]}{2(1 + k_H[H_2])}\left\{\left(1 + \frac{4[Rh](1 + K_H[H_2])}{k_D(1 + K_{el}[alkene] + K_1[H_2])}\right)^{1/2} - 1\right\} \qquad (13.34)^{[27]}$$

In contrast to Halpern's equations, (13.33) and (13.34) take into consideration the dependence of the rate of the reaction on hydrogen pressure.

The complex $[RuHCl(PPh_3)_3]$, isoelectronic with $[RhCl(PPh_3)_3]$, represents an unusually active catalyst. $[RuHCl(PPh_3)_3]$ selectively reduces 1-alkenes *ca* 10 times faster than chlorotriphenylphosphinerhodium(I). The rate of hydrogenation of terminal alkenes is $10^3$–$10^4$ times greater than that of internal alkenes. Recently obtained rhodium catalysts containing 2-aminopyridine[29] and its *N*-alkyl derivatives as well as diamines such as $NH_2(CH_2)_nNH_2$ ($n = 2, 3, 4, 5$)[30] are considerably more active than Wilkinson's complex. These rhodium catalysts are formed *in situ* by the reaction of $[Rh_2Cl_2(C_8H_{14})_4]$ with amines. The catalysts containing diamines are very unstable and undergo reduction to metallic rhodium. The rate of hydrogenation of olefins reaches a maximum for the catalyst which is prepared by the reaction utilizing 1,3-diaminopropane; for the remaining diamines, this rate is considerably lower. An analogous relationship was found for phosphines of the type $Ph_2P(CH_2)_nPPh_2$.[31] Consequently, catalytic activity of the complex depends on steric factors and the chelating effect.

Complexes containing chelating ligands are more stable than coordination compounds with analogous monodentate ligands:

$$M\underset{L}{\overset{L}{\diagup\kern-0.5em\diagdown}}\;\; > \;\; M\underset{L}{\overset{L}{\diagup}}$$

The most stable are complexes possessing five-membered chelating rings. The lowering of the rate of hydrogenation in the presence of compounds containing the groups

$$
\begin{array}{ccc}
\text{Rh}\underset{\diagdown}{\overset{\diagup}{\phantom{x}}}\begin{array}{c}\text{NH}_2\text{—CH}_2\\|\\\text{NH}_2\text{—CH}_2\end{array} & \text{and} & \text{Rh}\underset{\diagdown}{\overset{\diagup}{\phantom{x}}}\begin{array}{c}\text{PR}_2\text{—CH}_2\\|\\\text{PR}_2\text{—CH}_2\end{array}
\end{array}
$$

is caused by the overstability of the ring, leading to the increase of excitation energy $P$ [equation (13.27)] and, at the same time, to the lowering of the ability to activate dihydrogen. The catalytic system which is prepared *in situ* by the reaction of $[Rh_2Cl_2(C_8H_{14})_4]$ with 2-aminopyridine in ethanol belongs to the most active olefin hydrogenation catalysts.[29] It is over a dozen times more active than Wilkinson's complex. Based on kinetic data, mechanism (13.35) and (13.36) for hydrogenation of cyclohexene was proposed.

$$
Rh_2Cl_2(C_8H_{14})_4 + 4ampy \underset{}{\overset{H_2}{\rightleftharpoons}} 2[Rh(ampy)_x(solv)_y]^+ + 2Cl^-
$$

(ampy = 2-aminopyridine)                                        (13.35)

$$
[Rh(ampy)_x H_2(solv)_y]^+
$$

$$
[Rh(ampy)_x(solv)_y]^+ \quad\quad\quad k \;\Big|\; C_6H_{10}
$$

(13.36)

$$
\begin{array}{c}+C_6H_{10}\; \Big\updownarrow\; K_2\end{array} \quad \overset{-C_6H_{12}}{\searrow} \quad [Rh(ampy)_x H_2(C_6H_{10})(solv)_{y-1}]^+
$$

$$
[Rh(ampy)_x(C_6H_{10})(solv)_{y-1}]^+
$$

The dependence of the rate of cyclohexene hydrogenation on reagent concentrations and the catalyst is as follows[32]:

$$
-\frac{d[C_6H_{10}]}{dt} = -\frac{d[H_2]}{dt} = \frac{kK_1[H_2][C_6H_{10}][Rh]}{1 + K_1[H_2] + K_2[C_6H_{10}]}
$$

(13.37)

The constants assume the values $k = 14.0 \text{ mol}^{-1}\text{ dm}^3\text{ s}^{-1}$, $K_1 = 860 \text{ mol}^{-1}\text{ dm}^3$, and $K_2 = 4.9 \text{ mol}^{-1}\text{ dm}^3$.

The formation constant of the hydrido complex is considerably greater than the constant for $[Rh(ampy)_x(C_6H_{10})(solv)_{y-1}]^+$ as in the case of Wilkinson's complex and other catalysts. The activation energy is very low, viz. $26.9 \text{ kJ mol}^{-1}$ ($\Delta H^{\neq} = 24.4 \text{ kJ mol}^{-1}$, $\Delta S^{\neq} = -135.9 \text{ J K}^{-1}$). Very active catalysts for hydrogenation reactions were also obtained by reactions of $[RhCl_2(C_4H_7)]$ and $[Rh_2X_2(allyl)_4]$ ($X = Cl$, Br, and I) with phosphines, phosphites, thioethers, phenanthroline, 2,2-bipyridine, and 2-aminopyridine.[33] The reactions of $[RhCl_2(C_4H_7)]$ with $Rh_2PC_2H_4PPh_2$ in ethanol also afford effective catalysts for olefin hydrogenation. The reaction mixture consists of several compounds, among others, $[RhClH(diphos)_2]Cl$ and $RhCl_2(diphos)$.

The transition metal complexes may also be utilized for selective hydrogenation of dienes and polyenes to alkenes. The simplest catalyst used for this goal is $[Co(CN)_5]^{3-}$, which is obtained by the reaction of $CoCl_2$ with $KCN$.

In the atmosphere of hydrogen, the hydrido complex $[Co(CN)_5H]^{3-}$ is formed. This compound catalyzes hydrogenation of 1,3-dines and polyenes to alkenes as well as reduction of compounds containing activated double bonds, for example, $\alpha,\beta$-unsaturated aldehydes, acids, etc., in such compounds as $\overset{\displaystyle \diagup\!\!\!\diagdown}{\underset{\displaystyle CHO}{}}$ , $\overset{\displaystyle \diagup\!\!\!\diagdown}{\underset{\displaystyle COOH}{}}$ ,

$\overset{\displaystyle \diagup\!\!\!\diagdown}{\underset{\displaystyle CO-R}{}}$ , and styrenes. In the presence of $[Co(CN)_5H]^{3-}$, reduction of nitro and nitroso compounds, nitriles, oximes, and hydrazones also takes place.

Conjugated dienes, as a result of the addition to the hydrido complex, form allyl complexes which subsequently undergo hydrogenation.

$$[Co(CN)_5H]^{3-} + C_4H_6 \rightleftharpoons [Co(CN)_5\,CH(CH_3)CH=CH_2]^{3-} \tag{13.38}$$

$$[Co(CN)_5\,(C_4H_7)]^{3-} + [Co(CN)_5H]^{3-} \rightleftharpoons 2[Co(CN)_5]^{3-} + C_4H_8 \tag{13.39}$$

$$2[Co(CN)_5]^{3-} + H_2 \rightleftharpoons 2[Co(CN)_5H]^{3-} \tag{13.40}$$

At ratios of $CN/Co \leqslant 5$, the equilibrium between $\sigma$- and $\pi$-allyl complexes is possible [equation (13.41)]. Therefore, at high ratios, the main product of the hydrogenation is 1-butene, while at lower ratios, *cis*-2-butene and *trans*-2-butene are also formed.

$$(NC)_4Co-\!\!\!\overset{\displaystyle CH_2}{\underset{\displaystyle CH}{\Big\rangle}}\!CH \xrightarrow[\phantom{xx}]{+CN^-} CH_2=CH-\underset{\displaystyle \underset{\displaystyle Co(CN)_5}{|}}{CH}-CH_3 \tag{13.41}$$

The reduction of some unsaturated acids proceeds according to a radical mechanism:

$$[Co(CN)_5H]^{3-} + S \longrightarrow S^*H + [Co(CN)_5]^{3-} \tag{13.42}$$

$$[Co(CN)_5H]^{3-} + S^*H \rightleftharpoons SH_2 + [Co(CN)_5]^{3-} \tag{13.43}$$

where $S = PhCH = CHCOO^-$, $MeCH = CHCH = CHCOO^-$, and $PhCH = CH_2$. Although the radical $S^*H$ reacts with the cobalt complex according to the reaction

$$[Co(CN)_5]^{3-} + S^*H \rightleftharpoons [Co(CN)_5S^*H]^{3-} \cdots \tag{13.44}$$

the generated organometallic compound does not undergo hydrogenation as shown by kinetic data. The radical mechanism for the hydrogenation reaction is characteristic for its lack of formation of a compound between the metal and the olefin, and therefore the complex does not activate the substrate to be reduced, but only hydrogen.

Also, other transition metal compounds may be utilized for selective hydrogenation of dienes and polyenes to alkenes. Most commonly utilized are the following catalysts: $[M(\eta\text{-arene})(CO)_3]$ $(M = Cr, Mo, W)$, $[McpH(CO)_3]$ $(M = Cr, Mo, W)$, $[Fe(CO)_5]$, $[Fe_2cp_2(CO)_4]$, $[Fe(\eta\text{-diene})(CO)_3]$, $[RuHCl(PPh_3)_3]$, $[Co_2(CO)_8]$, $[Co_2(CO)_6\,(PR_3)_2]$, $[RhH(dmgH)_2PPh_3]$, $[RhX(PR_3)_3]$, $[PtCl_2(PR_3)_2] + SnCl_2$, $H_2(PtCl_6) + SnCl_2$, $[Ni(acac)_2] + AlR_3$, $[Co(acac)_3] + AlR_3$, $CoCl_2 + AlR_3$, $[VCl_2cp] + LiBu$, etc. Many homogeneous and heterogeneous catalysts catalyze hydrogenation of olefins, alkynes, the $C=O$, $C=N$, $C\equiv N$, and $N=N$ bonds, nitroalkanes, nitroarenes, and arenes, by hydrogen transfer of appropriate organic compounds.[59, 62] The following reducing agents are used: alcohols (chiefly $R_2CHOH$), formic acid, cyclohexene, tetrahydronaphthalene, dihydrofuran, dioxane, indoline, piperidine, pyrrolidine, tetrahydroquinoline, etc.

## b. *Hydrogenation of aromatic compounds*[37-42]

Recently, intensive investigations of homogeneous catalysts for hydrogenation of aromatic hydrocarbons have been carried out (see Table 13.1). Their goal is to develop methods for preparation of fuels from aromatic hydrocarbons and coal. The following compounds catalyze reduction of aromatic hydrocarbons: $H[Rh_2A_2Cl]$ where $A$ = phenyl-acetate, $N$-phenylanthranilate, and $L$-tyrosinate, $[Rh_2Cl_4(C_5Me_5)_2]$, $[Ru_2(\mu\text{-}Cl)(\mu\text{-}H)_2 (C_6Me_6)_2]Cl$, $[Ru(\eta\text{-}C_6Me_6)(1\text{-}4\text{-}\eta\text{-}C_6Me_6)$, $[RuHCl(\eta\text{-}C_6Me_6)PPh_3]$, $[Co(\eta\text{-}C_3H_5) \{P(OPr^i)_3\}_3]$, as well as heterogeneous catalysts which are obtained by deposition of compounds of iridium, rhodium, palladium, and platinum on nylon-66 and similar polymers. The following compounds are utilized for this purpose: $RhCl_3 \cdot 3H_2O$, $H_3[RhCl_6]$, $[Rh_2Cl_2(CO)_4]$, $[Rh(acac)(CO)_2]$, $H_2[IrCl_6]$, $[IrCl(CO)_2(PPh_3)]$, $PdCl_2$, $K_2[PdCl_4]$, $[Pd_2Cl_2(allyl)_2]$, $K[PtCl_3(C_2H_4)]$, $H_2[PtCl_6]$, and $K_2[PtCl_4]$. Also, hydrogenation of aromatic hydrocarbons is effectively catalyzed by metallic iridium, palladium, and platinum in the presence of superacids[43] such as HF, $HF + TaF_5$, and $HF + BF_3$. The superacids cause complete protonation of arenes leading to the formation of carbocations, which are quite easily reduced. Metallic rhodium and platinum reduce arenes very rapidly in the presence of aqueous $H_2SO_4$ solution (1.5 M), while hydrogenation of aromatic hydrocarbons proceeds slowly if rhodium phosphine complexes in 1.5 M $H_2SO_4$ are utilized (5–15 cycles $h^{-1}$, 303 K, 0.1 MPa).[44] Reduction of arenes in the presence of $Co_2(CO)_8$ proceeds according to a radical mechanism.

## c. *Asymmetric hydrogenation*[8, 10, 45-51]

Transition metal complexes containing chiral ligands catalyze asymmetric reduction of double bonds. For asymmetric hydrogenation, most commonly complexes of rhodium, ruthenium, and other metals of groups 8–10 possessing chiral ligands, e.g., phosphines, sulphoxides, aminophosphines, derivatives of ferrocene, phosphites $P(OR)_3$, amines, etc., have been employed (Table 13.2). Very good optical yields, and therefore a considerable excess of one of the optical isomers, are obtained during reductions of various substrates in the presence of complexes containing ligands (13.45) and (13.46). The resulting optical yields with the application of complexes of these and

diop                                    dios                        (13.45)

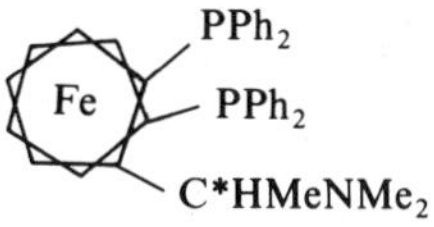

bppfa

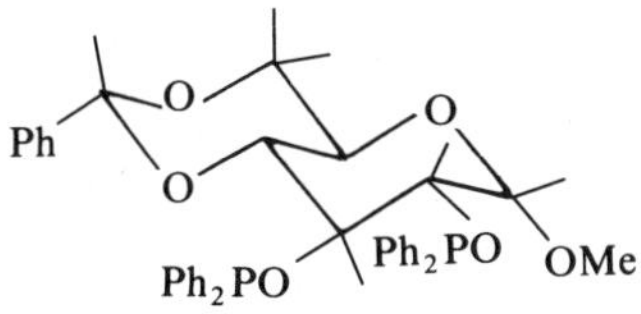

bdpg

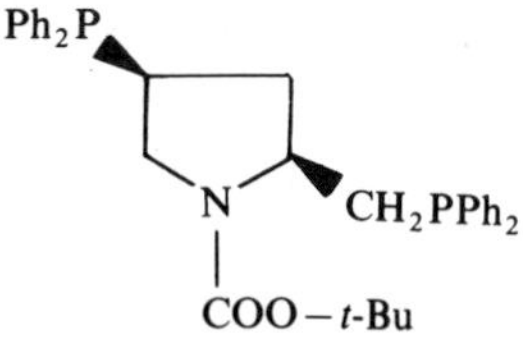

bppm

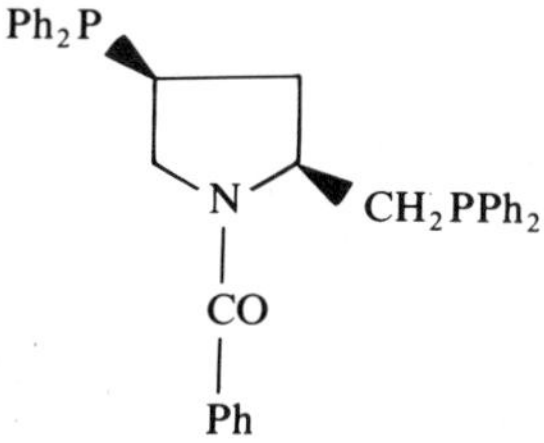

bzpm

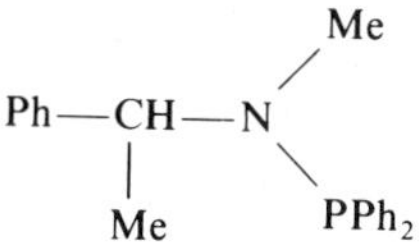

mbap

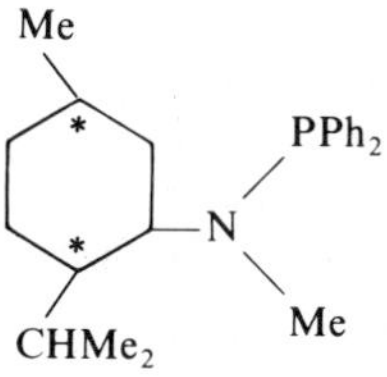

nap

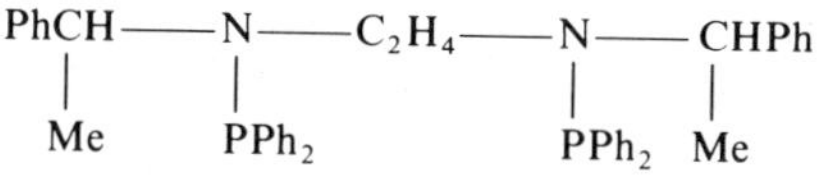

PNNP

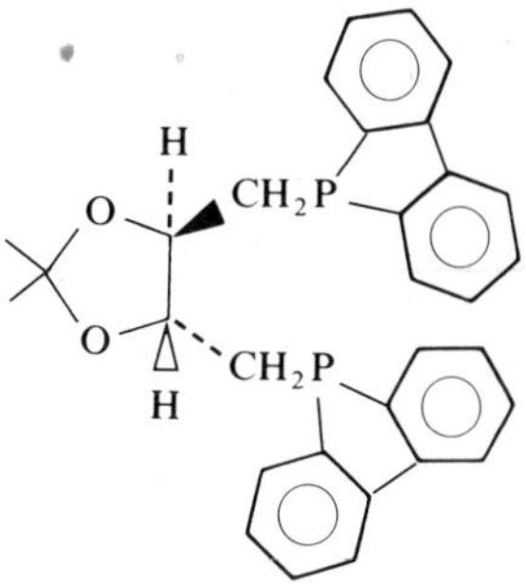

dbp–diop

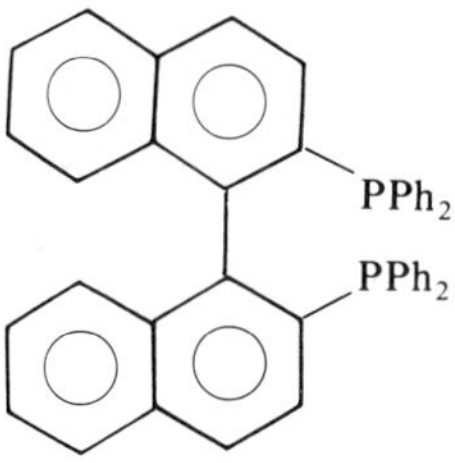

binap

(13.46)

Table 13.2. Catalysts for Asymmetric Hydrogenation

| Catalyst | Substrates | Products | Reaction conditions Solvent |
|---|---|---|---|
| $[RhH(diop)_2]$[46] * | 2-Acetamidoacrylic acid and other unsaturated acids | (S)-2-Acetamidopropionic acid 56% ee | $H_2$; 0.1 MPa, 303 K, BuOH + toluene 2:1 |
| $[Ru_2Cl_4(diop)_3]$[72a,46] | 2-Acetamidoacrylic acid and other unsaturated acids | (S)-2-Acetamidopropionic acid 59% ee | $H_2$; 0.1 MPa, 333 K DMA |
| $[Rh(COD)(PNNP)]^+ClO_4^-$ ᵃ | PhCHC(NHAc)COOH | PhCH₂CH(NHAc)COOH, optical yield 77% | $H_2$; 0.1 MPa, 298 K EtOH |
| $[Rh-(S,S)-bppm(solv)_2]^+$ ᵇ | HOOC—C—CH₂COOH, ‖ CH₂ | HOOCCHCH₂COOH, —CH₃, optical yield 71% | |
| $[Rh-(S,S)-bzpm(solv)_2]^+$ ᵇ | HOOC—C—CH₂COOH, ‖ CH₂ | HOOCCHCH₂COOH, —CH₃, optical yield 83.5% | |
| RhCl(solv)(diop)ᶜ heterogenized | PhCHC(NHCOMe)COOH | PhCH₂CHCOOH, NHCOMe | $H_2$; 0.1–0.2 MPa, 298 K, EtOH, EtOH–PhH |
| $[Rh_2Cl_2(hexadiene)_2]$ + bppfa[47] | PhCHC(NHAc)COOH | PhCH₂CHCOOH, 93% ee, NHCOMe | $H_2$; 5 MPa, 295 K MeOH |

| Catalyst | PhCHC(NHAc)COOH | PhCH₂CHCOOH, NHAC, optical yield 75% | H₂; 0.1 MPa, 298 K, PhH + MeOH |
|---|---|---|---|
| [Rh₂Cl₂(olefin)₄] + bdpg + AgBF₄ [48] | | | |
| [Rh{(R)-Ph₂PCH(CH₂Ph)CH₂PPh₂}L₂]⁺ [52] | PhCH=C(NHAc)COOH | (S)-PhCH₂CH(NHAc)COOH, 99% ee | |
| [Rh(chiraphos)L₂]⁺ [53] | PhCH=C(NHAc)COOH | (R)-PhCH₂CH(NHAc)COOH, 99% ee | |
| [Rh(COD){Ph₂PCH(CH₂OC₂H₄OMe)CH₂PPh₂}]⁺ ClO₄⁻ [54] | PhCH=C(NHAc)COOH | (S)-PhCH₂CH(NHAc)COOH, 86% ee | |
| [Rh₂Cl₂(CH₂=CHCH₂CH₂CH=CH₂)₂] + (R,S)bppfa [55] | PhCH=C(NHAc)COOH | (S)-PhCH₂CH(NHAc)COOH, 93% ee | |
| [Rh(NBD)₂]⁺ + Ph₂PCH{(CHMeO(NHCOBuⁱ)}CH₂PPh₂ [56] | PhCH=C(NHAc)COOH | (S)-PhCH₂CH(NHAc)COOH, 88% ee | |
| [Rh(NBD){(S,R)-bppfa}]⁺ [57] | PhCH=CMeCOOH | (S)-PhCH₂CHMeCOOH, 51% ee | |
| [Rh(NBD)L]⁺ [BF₄]⁻ [58] | PhCH=CH(NHAc)COOH | (R)-PhCH₂CH(NHAc)COOH, 83% ee | |

$L =$ (1,3-dioxolane ring with two substituents: $-CH_2PPh_2$ and $-CH_2P(C_6H_4Me\text{-}4)_2$, with H atoms shown)

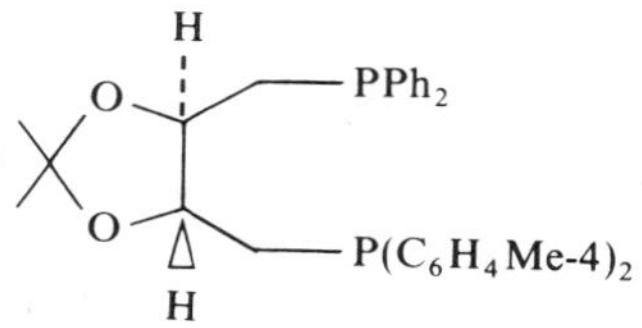

[a] M. Fiorini, F. Marcati, and G. M. Giongo, *J. Mol. Catal.*, **4**, 125 (1978).

[b] K. Achiwa, *Tetrahedron Lett.*, 1475 (1978); K. Achiwa, *Chem. Lett.*, 561 (1978).

[c] N. Takaishi, H. Imai, C. A. Bertelo, and J. K. Stille, *J. Am. Chem. Soc.*, **100**, 264 (1978); T. Masuda and J. K. Stille, *J. Am. Chem. Soc.*, **100**, 268 (1978).

* Abbreviations of chiral ligands are given in equations (13.45) and (13.46).

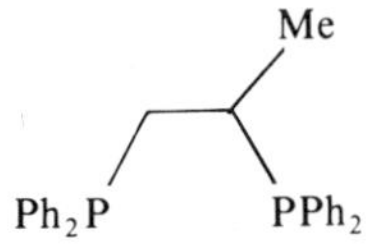

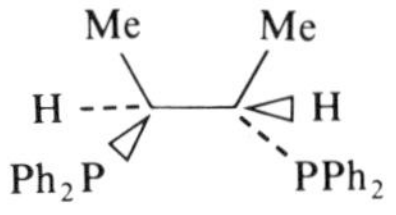

Me

Ph₂P          PPh₂

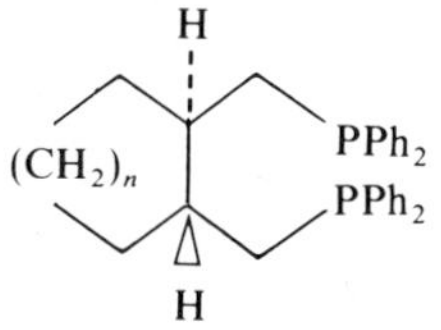

(S, S)-chiraphos

similar ligands often reach values in the 70–99 % range. The optical yield is defined as the ratio of the specific rotation of the product to the specific rotation of a pure enantiomer of that product and it is equal to the excess of one of the enantiomers only if there is a linear relationship between the rotation and the composition of the mixture.

Therefore, the optical yield is given by the formula

$$p = \frac{[\alpha]_x^t}{[\alpha]^t} \, 100\% \tag{13.47}$$

where $[\alpha]_x^t$ is the specific rotation of the product at temperature $t$ and $[\alpha]^t$ is the specific rotation of the optically pure enantiomer of that product at the same temperature.

Complexes containing asymmetric ligands are also utilized in other catalytic processes, such as hydroformylation, hydrosilation, isomerization, oligomerization, polymerization, etc.

The Monsanto Company produces 3,4-dihydroxyphenylalanine (L-DOPA) from substituted acetamidocinnamic acid 3,4-[AcO(MeO)C₆H₃CH=C(NHAc)COOH] by means of its asymmetric reduction. The product of the hydrogenation reaction is subsequently transformed into the final compound, which is used to treat Parkinson's disease. Cationic rhodium(I) complexes containing asymmetric ligands are utilized as catalysts in this synthesis.[10]

The mechanism of asymmetric hydrogenation of prochiral olefins in many cases has not been explained. High optical yields have often been achieved in these cases where chelating phosphines were used and also for hydrogenation of compounds which represent bidentate ligands, such as PhCH=C(NHCOMe)COOH. This compound is coordinated by the double bond and the oxygen of the amide group.[60] Based on the X-ray studies of olefin–phosphine rhodium complexes, the assumption was advanced that the type of resulting enantiomer in the case of 1,2-diphosphine ligands such as chiraphos depends on the confirmation of the five-membered chelating ring. It seems well documented that the chelating ring $\lambda$ induces the formation of (S)-N-acylaminoacids [equation (13.48)]. This principle is usually obeyed also in the case of 1,4-diphosphines.[61]

conformation $\delta$          conformation $\lambda$          (13.48)

(S)-N-acetylphenylalanine

However, based on structural, NMR, and kinetic studies, Halpern found that at room temperature the ratio of diastereoisomeric olefin–phosphine complexes does not influence the optical yield of the product. Formation of an excess of one of the enantiomeric reaction products depends on the rate of hydrogenation of olefin–phosphine complexes A and B. The rate of reaction of isomer A, which occurs in the reaction mixture in a considerably lesser amount, is so great that the main product is formed from this isomer. The mechanism of asymmetric hydrogenation of (Z)-PhCH=C(NHAc)COOEt may be represented by scheme (13.49).

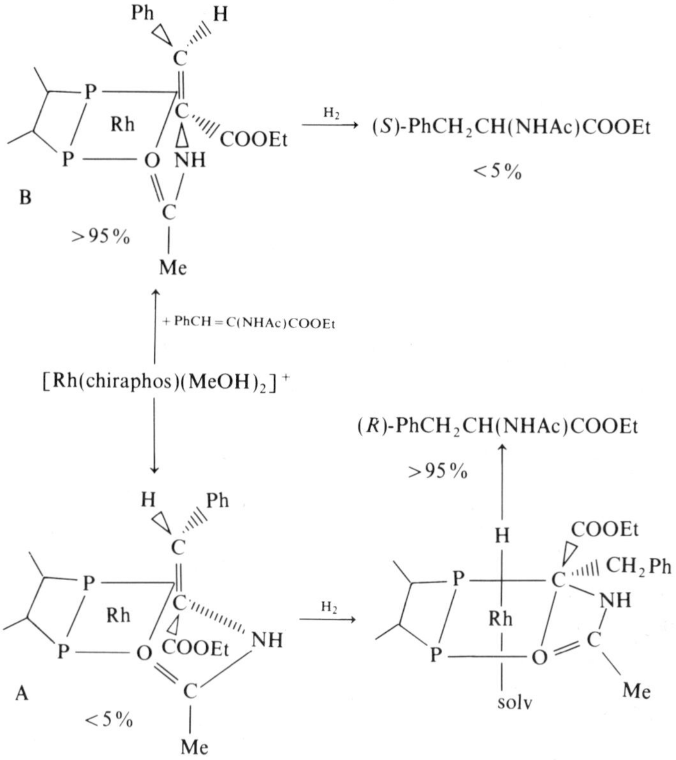

$$(13.49)$$

### d. *Hydrogenation of the C=O, C=N, and C≡N bonds*[8, 21, 63]

Hydrogenation of ketones generally proceeds with more difficulty than reduction of olefins. This is due in part to lesser stability of complexes with ketones (which activate hydrogen) compared to olefin complexes and to the ability to coordinate of the resulting alcohols in contrast to the alkanes. Moreover, the resulting secondary alcohols show a tendency to oxidize themselves back to ketones. Some rhodium complexes oxidize secondary alcohols to ketones. Wilkinson's complexes $RhX(PR_3)_3$ do not represent suitable catalysts for reduction of aldehydes, because decarbonylation of the substrate and the formation of compounds of the type $RhCl(CO)(PR_3)_2$ takes place; these compounds are not catalytically active. However, hydrogenation of ketones is effectively accelerated by complexes such as $[Rh(NBD)(PR_3)_n]^+ ClO_4^-$ ($n = 2, 3$). The catalytic activity decreases as follows[64]: $PEt_3 > PPh_3 \backsim PMe_3 \gg dppe$. Phosphine complexes of ruthenium[65, 67] and iridium[64, 66] are very active catalysts for hydrogenation of aldehydes. The compounds $[RuCl_2(PPh_3)_3]$ and $[RuH(CO)(PPh_3)_3]$, like $[RhCl(PPh_3)_3]$, undergo carbonylation. However, metal carbonyls $[RuClH(CO)(PPh_3)_3]$ and $[RuCl_2(CO)(PPh_3)_3]$ catalyze reduction of RCHO. Very active catalysts are formed by reaction of $[IrH_3(PPh_3)_3]$ with acetic acid; probably the active intermediate compound is $[IrH_2(OAc)(PPh_3)_3]$.

The reduction of ketones in neutral media probably proceeds via an intermediate in which the ketone molecule is coordinated to the metal atom through oxygen. Mechanism (13.50) of hydrogenation of ketones in the presence of cationic complexes

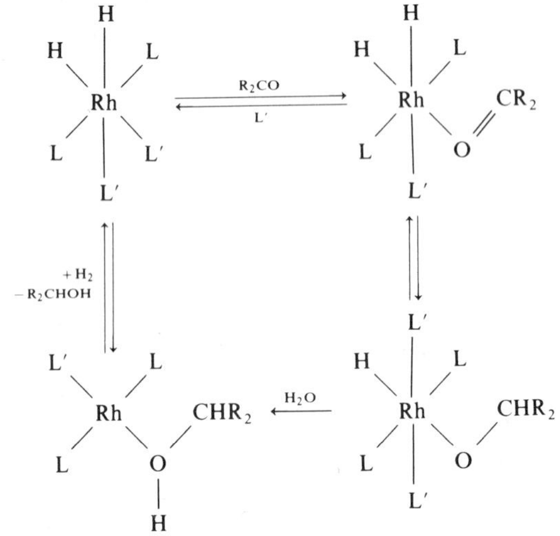

(13.50)

of the type $[RhH_2L_2L'_2]^+$, where $L = PPh_2Me$, $PPhMe_2$, $PMe_3$, and $L' = $ solvent, was proposed. In basic medium, the mechanism of the reaction is different. The ketone molecules are coordinated in the enol form, (13.51) and (13.52). Phosphine rhodium(I) complexes,[63] such as $[RhCl(COD)(PR_3)]$, and rhodium compounds containing 2,2'-dipyridyl and 1,10-phenanthroline and their derivatives[68, 69] are catalysts for reduction of ketones in basic media. Exceptionally active are catalytic systems obtained by reactions of dinuclear rhodium(II) complexes $[Rh_2Cl_2(\mu\text{-}OOCH)_2(N-N)]$ containing $N-N$ ligands in alkaline methanol solution $(N-N = 2,2'\text{-bipyridyl}$, 1,10-phenanthroline).[69] In this case, reduction may be catalyzed not only through mononuclear rhodium complexes, but also through polynuclear ones. The presence of polynuclear rhodium complexes was confirmed by electronic spectra. These catalysts show considerable selectivity in the process of reduction of ketones in the presence of olefins because, when the ratio $Rh/N-N \geqslant 2$, the rate of reduction of the $C=C$ bond

$$RCH_2\overset{\overset{\text{O}}{\|}}{C}CH_2R + OH^- \rightleftarrows RCH_2\overset{\overset{\text{O}^-}{|}}{C}=CHR + H_2O \qquad (13.51)$$

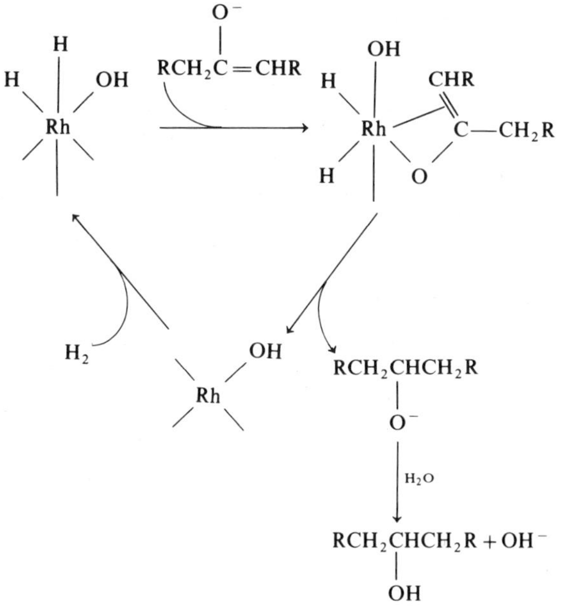

$$
\begin{array}{c}
\text{(reaction scheme)}
\end{array}
$$

(13.52)

is considerably greater than that of the $C=O$ bond. However, if $Rh/N-N=1/1$, the situation is reversed, and olefins are reduced faster than ketones.[68, 69]

Iridium(I) complexes possessing 2,2'-bipyridyl and 1,10-phenanthroline[70] as well as phosphines[71] are also good catalysts for reduction of ketones. Catalysts containing phosphines selectively reduce $\alpha,\beta$-unsaturated aldehydes to saturated alcohols. Relatively selective are rhodium(I) and ruthenium(II) phosphine catalysts such as $Rh_2Cl_2(CO)_4 + PR_3$ and $RuX_2(CO)(PPh_3)_3$ in the presence of which the $C=C$ bond is reduced faster or simultaneously with the $C=O$ bonds in unsaturated aldehydes.[72] Ketones may readily be reduced by means of organic hydrogen donors, mainly secondary alcohols. Selective reduction of the $C=O$ or $C=C$ bonds in unsaturated carbonyl compounds has great preparative significance[63] like the preparation of optically active alcohols from prochiral ketones, RR'CO.[50, 63] The latter reactions utilize ligands which are analogous to those used in asymmetric reduction of olefins.[50, 63, 73] High optical yields ($>90\%$) were obtained utilizing the catalytic system $[RhCl(nbd)]_2 + DIOP.$[73] Phosphine complexes, mainly those of rhodium and ruthenium, are also utilized as catalysts for reduction of the $C=N$ bonds in Schiff bases and oximes.[63] Benzylideneaniline, $PhCH=NPh$, is reduced by hydrogen in the presence of $Co_2(CO)_8$ as catalyst.[74]

Reduction of the ketones cyclohexanone, *p*-chloroacetophenone, acetophenone, butanone, and *p*-methylacetophenone by phase transfer catalysis with the application of $[NMe(C_8H_{17})_3]Cl$, $[NEt_3(CH_2Ph)]Cl$, and 18-crown-6 as phase transfer agents and isopropanol and 1-phenylethanol as hydrogen donors as well as iron carbonyls as catalysts, represents a very interesting reaction.[75]

## 3. ISOMERIZATION OF OLEFINS

### a. Migration of the double bond

Isomerization of olefins which takes place as a result of the transfer of hydrogen atoms with concomitant migration of the double bond is catalyzed by transition metal complexes which may react with olefins to form organometallic compounds. Often, it is necessary that cocatalysts be present as sources of hydrogen. Most commonly utilized cocatalysts are acids, water, alcohol, hydrogen, silanes, etc. The formation of hydrido complexes during isomerization reactions is crucial. The following mechanisms of olefin isomerization reactions are known[8–10, 22, 24, 76, 77]:

   1.  Addition and elimination of the hydride

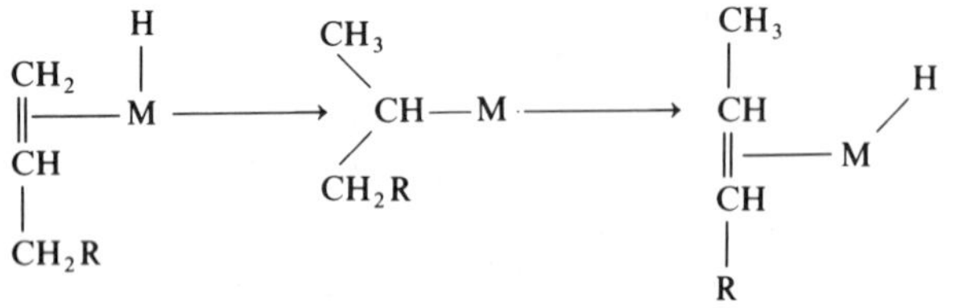

$$
\begin{array}{c}
CH_2 \\
\parallel \quad \text{—M} \\
CH \\
| \\
CH_2R
\end{array}
\quad \overset{H}{\underset{CH_3}{}}
\longrightarrow
\begin{array}{c}
CH_3 \\
\diagdown \\
CH\text{—M} \\
\diagup \\
CH_2R
\end{array}
\longrightarrow
\begin{array}{c}
CH_3 \\
| \\
CH \\
\parallel \text{—M} \\
CH \\
| \\
R
\end{array}\overset{H}{\diagup}
\qquad (13.53)
$$

   2.  Formation of a π-allyl complex

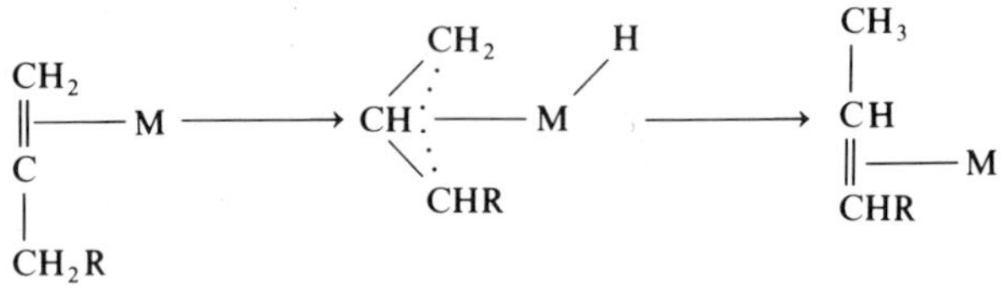

$$
\begin{array}{c}
CH_2 \\
\parallel \text{—M} \\
C \\
| \\
CH_2R
\end{array}
\longrightarrow
\begin{array}{c}
CH_2 \\
\diagup \vdots \\
CH \vdots \text{—M} \\
\diagdown \vdots \\
CHR
\end{array}\overset{H}{\diagup}
\longrightarrow
\begin{array}{c}
CH_3 \\
| \\
CH \\
\parallel \text{—M} \\
CHR
\end{array}
\qquad (13.54)
$$

   3.  Formation of a carbene complex (1,2-hydrogen shift)

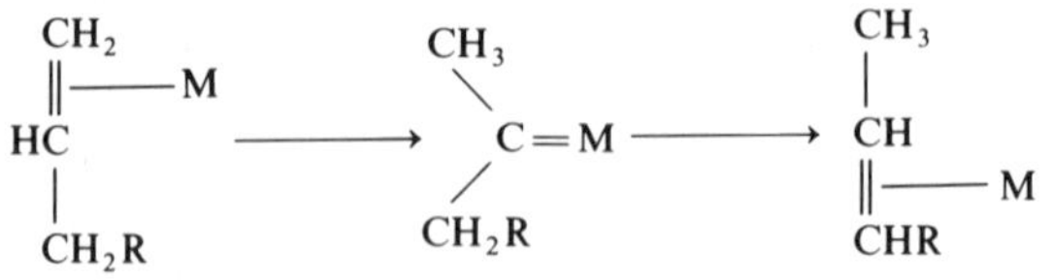

$$
\begin{array}{c}
CH_2 \\
\parallel \text{—M} \\
HC \\
| \\
CH_2R
\end{array}
\longrightarrow
\begin{array}{c}
CH_3 \\
\diagdown \\
C\text{=M} \\
\diagup \\
CH_2R
\end{array}
\longrightarrow
\begin{array}{c}
CH_3 \\
| \\
CH \\
\parallel \text{—M} \\
CHR
\end{array}
\qquad (13.55)
$$

Isomerization most commonly proceeds according to the first two mechanisms. The carbene mechanism was proposed for the isomerization of 1-alkenes in the presence of $[Pd_2Cl_6]^{2-}$ in acetic acid. In the products of isomerization of 3-$d_2$-1-octene with the application of this catalyst neither migration of deuterium to the terminal carbon atom nor deuterium transfer from the solvent to the hydrocarbon was observed which support the carbene mechanism. The absence of deuterium at the first carbon excludes 1,3-hydrogen migration and therefore the π-allyl mechanism. The fact that there is no deuterium transfer from the solvent to the olefin indicates that the formation of hydrido complexes involving the solvent also does not take place, which eliminates the first mechanism. It was proposed that the reaction proceeds via intermediate (13.56) as a result of 1,2-suprafacial hydrogen shift.[79]

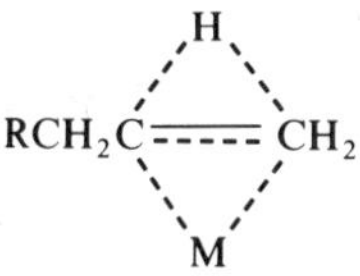

$$RCH_2C \overset{\displaystyle H}{\diamond} CH_2 \qquad (13.56)$$
$$M$$

The role of the catalyst is very essential in the hydride mechanism. The presence of acids, bases, alcohols, hydrogen, etc., facilitates or makes possible the formation of hydrido complexes which govern the reaction course (13.57).

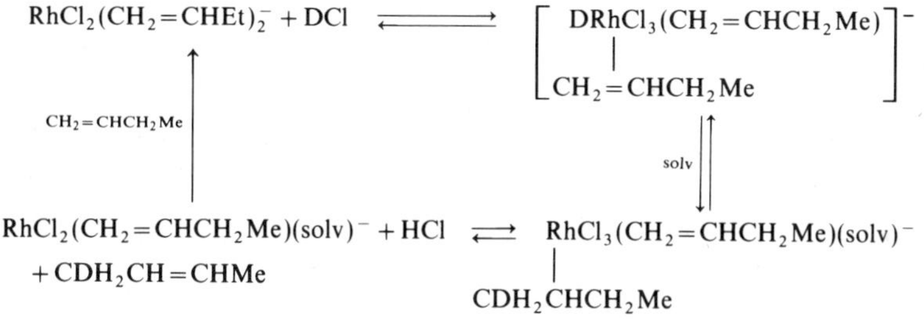

$$RhCl_2(CH_2=CHEt)_2^- + DCl \rightleftharpoons \begin{bmatrix} DRhCl_3(CH_2=CHCH_2Me) \\ | \\ CH_2=CHCH_2Me \end{bmatrix}^-$$

$$CH_2=CHCH_2Me \uparrow \qquad\qquad solv \updownarrow \qquad (13.57)$$

$$RhCl_2(CH_2=CHCH_2Me)(solv)^- + HCl \rightleftharpoons RhCl_3(CH_2=CHCH_2Me)(solv)^-$$
$$+ CDH_2CH=CHMe \qquad\qquad\qquad | \\ CDH_2CHCH_2Me$$

The addition of the hydrido ligand proceeds according to the Markovnikov rule, which requires that the part of the molecule possessing a partial positive charge be added to the carbon atom of the olefin bond containing more hydrogen atoms, and therefore to the carbon atom having a partial negative charge [equation (13.53)]. The anti-Markovnikov addition also takes place; however, it does not lead to isomerization. This is evidenced by the presence of deuterated 1-butene which is formed by reaction (13.58).

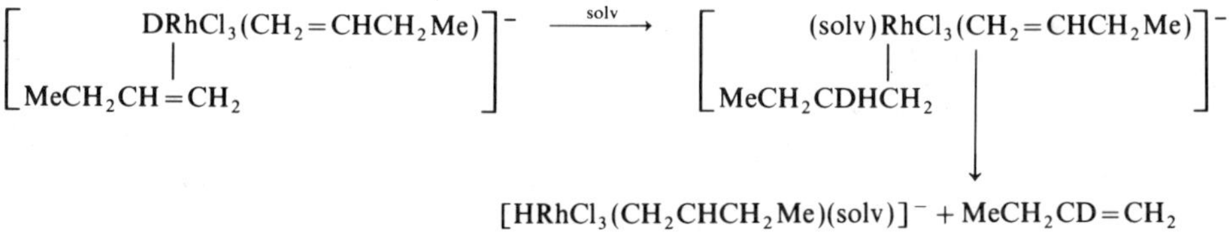

$$\begin{bmatrix} DRhCl_3(CH_2=CHCH_2Me) \\ | \\ MeCH_2CH=CH_2 \end{bmatrix}^- \xrightarrow{solv} \begin{bmatrix} (solv)RhCl_3(CH_2=CHCH_2Me) \\ | \\ MeCH_2CDHCH_2 \end{bmatrix}^-$$

$$\downarrow$$

$$[HRhCl_3(CH_2CHCH_2Me)(solv)]^- + MeCH_2CD=CH_2$$

$$(13.58)$$

The hydrido phosphine rhodium complexes possess a relatively large negative charge on the hydrido ligand, and as such these compounds are not good catalysts for isomerization. Also, the rate of this reaction is influenced negatively by steric repulsion between phosphine molecules and the isoalkyl ligand in the intermediate compound [structure (13.59)].

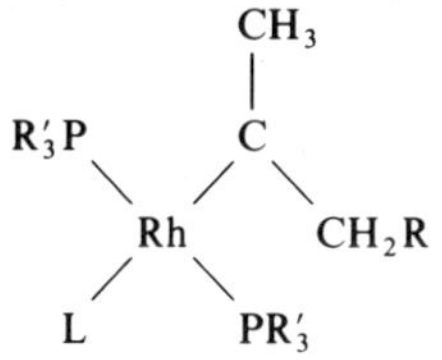

$$\begin{array}{c} CH_3 \\ | \\ R_3'P \diagdown \quad C \diagdown \\ Rh \quad CH_2R \\ \diagup \quad \diagdown \\ L \quad PR_3' \end{array} \qquad (13.59)$$

The catalytic system $Co_2(CO)_8 + H_2$ forms the complex $CoH(CO)_4$, which has strongly acidic character in polar solvents. Therefore, the complex $CoH(CO)_4$ is more likely to be added to 1-alkenes according to the Markovnikov rule than $RhH(CO)(PPh_3)_3$. Based on the studies of hydroformylation of deuterated olefins, it was proposed that the addition–elimination process proceeds according to the concerted mechanism (13.60)

$$
\begin{array}{ccccc}
RCH_2CH=CH_2 & & RCH \text{ --- } CH \text{ --- } CH_2 & & RCH=CHCH_3 \\
| & \rightleftharpoons & \vdots \qquad\qquad \vdots & \longrightarrow & | \\
(OC)_3Co—H & & H\text{-----}Co\text{-----}H & & H—Co(CO)_3 \\
& & (CO)_3 & &
\end{array}
\tag{13.60}
$$

Hydrido complexes or coordination compounds from which hydrido complexes may be formed are found to be the most effective catalysts for isomerization which proceeds via the addition and elimination of the hydride. These are: $[RuClH(PPh_3)_3]$, $[RuCl_2(PPh_3)_3]$, $[RuH_2(N_2)(PPh_3)_3]$, $[RuH_4(PPh_3)_3]$, $[CoH(N_2)(PPh_3)_3]$, $[CoH(CO)_4]$, $[Rh_2Cl_2(olefin)_4] + HCl$, $[RhH(CO)(PPh_3)_3]$, $[Ni\{P(OEt)_3\}_4] + HCl$, $[Ni(PPh_3)_4] + HCl$, $[IrCl(CO)(PPh_3)_2]$, $[Ir_2Cl_4H_2(COD)(PPh_3)_2]$, $[IrH(CO)(PPh_3)_3]$, $[OsClH(CO)(PPh_3)_3]$, $[PtClH(PPh_3)_2]$, $H_2[PtCl_6] + SnCl_2$, etc.

The isomerization reaction proceeds according to the $\pi$-allyl mechanism if the cyclic course of the oxidative addition and reductive elimination is possible. This is confirmed by the existence of an equilibrium between the $\pi$-allyl and olefin complexes. This phenomenon was studied by the $^1H$ NMR method.[78]

$$
\begin{array}{ccc}
F_3P \quad\; CH_2 & & CH_2 \\
\diagdown \quad\;\; \diagdown & \xrightleftharpoons[223\ K]{233\ K} & \| \\
Ni\text{ ------ }CH & & F_3P—Ni— \\
\diagup \quad\;\; \diagup & & CHCH_3 \\
H \qquad CH_2 & &
\end{array}
\tag{13.61}
$$

In the case of isomerization which proceeds according to the $\pi$-allyl mechanism, 1,3-hydrogen transfer takes place. These reactions are catalyzed by palladium(II) complexes which easily form $\pi$-allyl complexes from $\pi$-olefin compounds. Also, compounds of nickel, rhodium, iron, etc., are utilized as catalysts. Effective isomerization is possible if the hydrogen addition to both terminal carbon atoms of the $\pi$-allyl asymmetric grouping takes place.

In isomerization catalyzed by $Fe(CO)_5$ and $Fe_3(CO)_{12}$, first olefin complexes of the type $Fe(CO)_4(CH_2=CHR)$ are formed which split off one CO group and subsequently afford a $\pi$-allyl complex [equation (13.62)].

$$
[Fe(CO)_4(CH_2=CHCH_2R)] \underset{}{\overset{-CO}{\rightleftharpoons}} [Fe(CO)_3(CH_2=CHCH_2R)]
$$

$$
\rightleftharpoons \left[
\begin{array}{cc}
H \quad\; CH_2 \\
| \quad\;\; \diagdown \\
(OC)_3Fe\text{ ------ }CH \\
\diagup \\
CHR
\end{array}
\right] \rightleftharpoons [Fe(CO)_3(CH_3CH=CHR)]
\tag{13.62}
$$

The mechanism of isomerization of 1,4-dichloro-2-butene to 3,4-dichloro-1-butene is analogous. However, in this case, breaking of the C—Cl bond occurs followed by formation of the chloride $\pi$-allyl iron complex, which facilitates migration of the double bond and 1,3-chloride transfer. The compound $[PdCl_2(PhCN)_2]$ also catalyzes isomerization according to the $\pi$-allyl mechanism. The characteristic property of this mechanism is oxidative addition and the formation of a hydrido alkyl metal complex. This process may occur via a five-center intermediate compound in which there is an agostic bond $M---H-C$ [structure (13.63)]. However, hydrogen migration is

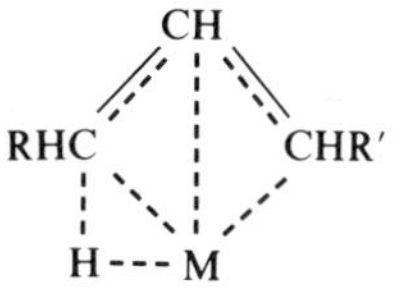

$$\text{(13.63)}$$

possible even without its coordination to the metal. Perhaps, isomerization of olefins catalyzed by iron carbonyls proceeds according to this type of mechanism.[80] It was proposed that the transfer occurs by the sigmatropic 1,3-suprafacial hydrogen shift. The hydrogen atom migrates above the plane of the olefin on the side opposite the metal [equation (13.64)].

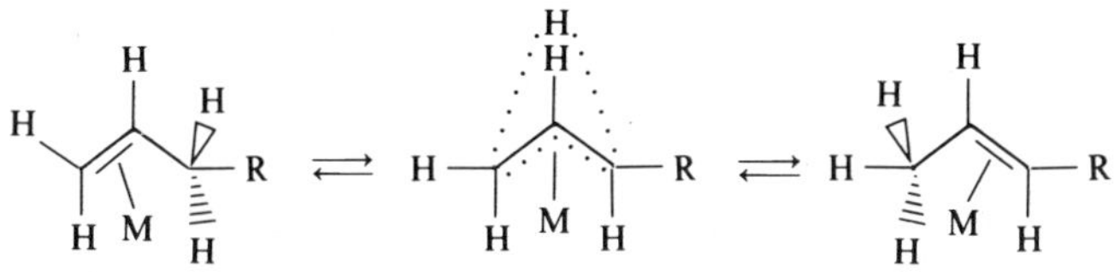

$$\text{(13.64)}$$

## b. *Cis–trans isomerization*

Initially, isomerization of 1-alkenes to 2-alkenes takes place with the formation of *cis*-isomers which, after some time, form an equilibrium mixture of the *cis* and *trans* compounds. Generally, *cis–trans* isomerization utilizing the same catalysts is slower than the migration of the double bond. A very rapid establishment of the *cis–trans* equilibrium was observed in the case of hydrido-type catalysts which were obtained *in situ* in the presence of strong acids. It is difficult to establish whether the influence of the coordination compound or that of the acid is more important.

Isomerization by means of hydrido complexes probably occurs through the formation of alkyl compounds in which the rotation about the C–C axis is possible, followed by reformation of the *trans*-alkene molecule as a result of $\beta$ hydride elimination [equation (13.65)].

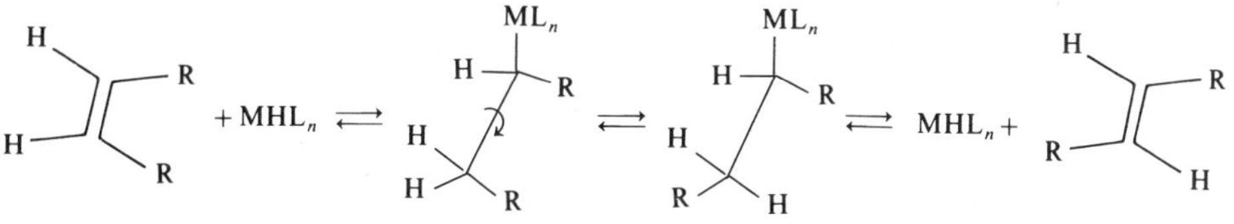

$$\text{(13.65)}$$

## 4. OLIGOMERIZATION AND POLYMERIZATION OF OLEFINS AND ACETYLENES

### a. Polymerization

Oligomerization and polymerization of olefins are usually carried out in the presence of Ziegler–Natta catalysts which were discovered by Ziegler in 1953.

Polymerization reaction may proceed according to the anionic, cationic, or radical mechanism:

$$R^* + nCH_2{=}CH_2 \longrightarrow R(CH_2CH_2)_n^* \xrightarrow{\;R^*\;} R(CH_2(CH_2)_nR \qquad (13.66)$$

Compounds, which at higher temperature or in the presence of light easily break down to afford radicals, for example, $Me_2C - N{=}N - CMe_2$ and ROOR, serve as initiators

$$\underset{CN}{\overset{|}{\phantom{x}}} \qquad \underset{CN}{\overset{|}{\phantom{x}}}$$

of this reaction.

Anionic polymerization occurs upon treatment with alkali metals, their amides, alcoholates, and organometallic compounds such as $M^+R^-$:

$$A^- + nCH_2{=}CH_2 \longrightarrow A(CH_2CH_2)_n^- \xrightarrow{\;M^+\;} A(CH_2CH_2)_{n-1}CH{=}CH_2 + MH \qquad (13.67)$$

Cationic polymerization is catalyzed by Lewis acids in the presence of cocatalysts such as water, alcohols, etc.:

$$MH^+ + nCH_2{=}CH_2 \xrightarrow{\;-M\;} H(CH_2CH_2)_n^+ \xrightarrow{\;M\;} H(CH_2CH_2)_{n-1}CH{=}CH_2 + MH^+ \qquad (13.68)$$

Ziegler–Natta-type catalysts are composed of complexes of the transition metals and organometallic compounds of the main group elements (usually those of groups 1, 2, and 13): $MX_n + M'R_mX'_r$ where M is the transition metal of groups 3–10,

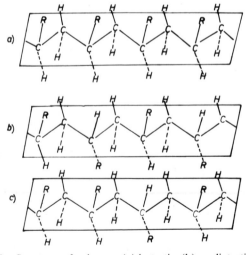

Figure 13.3. Structures of polymers: (a) isotactic, (b) syndiotactic, (c) atactic.

X = halogen, acac, OR, NCS, etc., R = aryl, alkyl, etc., and X' = halogen, OR, etc. Ziegler–Natta catalysts catalyze polymerization and oligomerization of unsaturated hydrocarbons under considerably milder conditions than other catalysts, ionic or radical. The structure of polymers obtained by employing Ziegler–Natta catalysts is more regular, and these polymers therefore have higher densities, higher melting points, and better mechanical properties. Isotactic, syndiotactic, and atactic polymers may be obtained (Figure 13.3).

The most commonly used transition metal compounds are those of titanium, vanadium, chromium, and zirconium. The transition metal–carbon bond is formed by the action of an organometallic compound of the main group elements, leading to the formation of the complex which has catalytic properties.

In the case of titanium, the catalytically active compound for homogeneous polymerization has structure (13.69).[8–10, 22, 81–86]

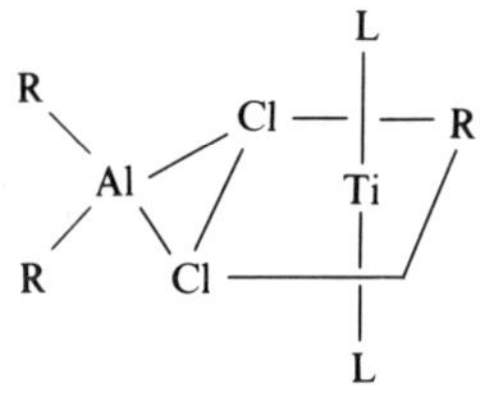

$$(13.69)$$

The free coordination sites may be occupied by olefin molecules. The catalytic increase of the chain of the polymer occurs by migration of the alkyl to the olefin [equation (13.70)]. It has been proposed by Green[92] that, during polymerization,

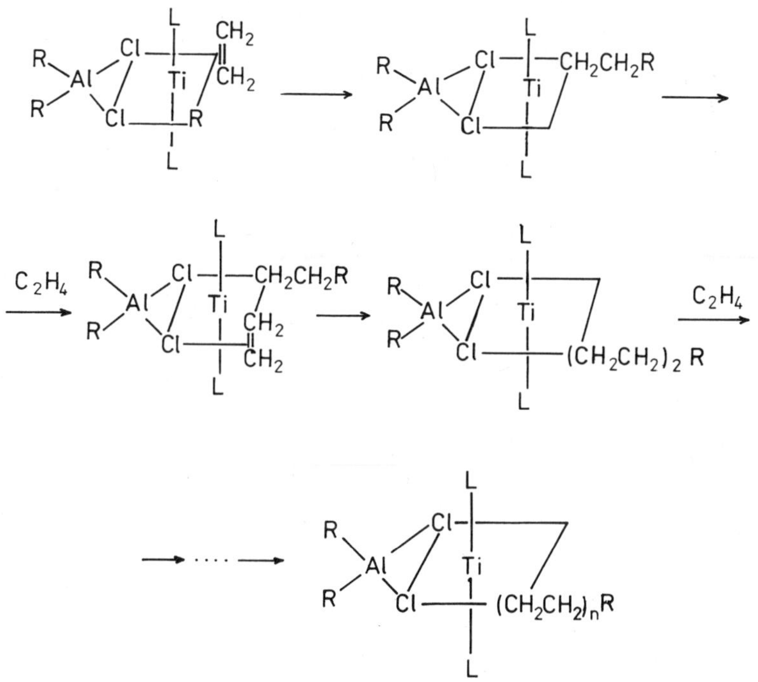

$$(13.70)$$

carbene complexes are formed as intermediates which catalyze this process.[86-88, 92] The carbene mechanism is given in equation (13.71).

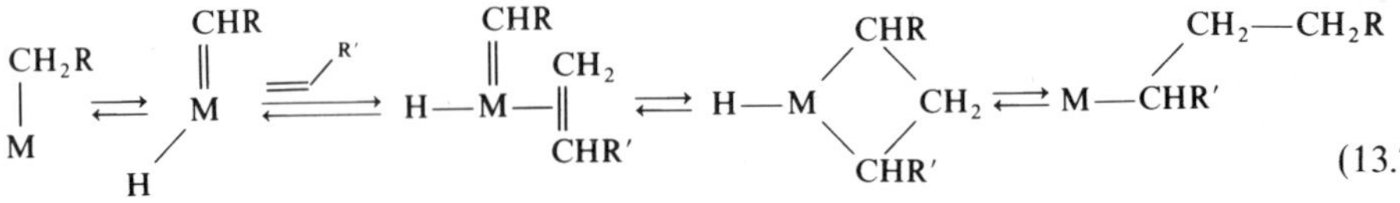

$$(13.71)$$

Carbene complexes also catalyze polymerization of cycloalkenes (see Section 8.8.b) and acetylenes[88, 90] [equation (13.72)].

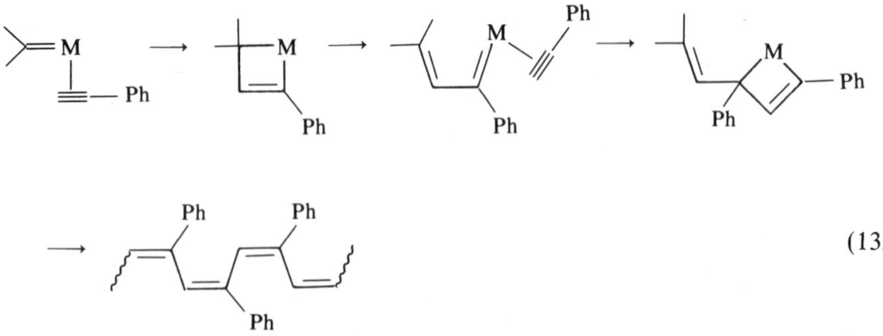

$$(13.72)$$

Both homogeneous (for example, $TiCl_4 + AlR_3$) and heterogeneous ($TiCl_3 + AlR_3$) catalysts as well as heterogenized homogeneous catalysts may be utilized. Table 13.3 gives the most commonly used Ziegler–Natta-type catalysts for polymerization reactions.

Ziegler–Natta catalysts are very sensitive to moisture, oxygen, and peroxides. As solvents for polymerization reactions, any compounds may be used provided they do not react with the catalysts and they do not cause their decomposition. For practical application, paraffins, cycloparaffins, and aromatic hydrocarbons are utilized. The choice of solvent depends on reaction conditions and the catalyst, on its availability in sufficient quantities, and on its sufficient purity. Usually, pressure from 0.1 to 3 MPa is utilized. Often, the concentration of monomers is low in the case of reactive olefins; however, their polymerization is sometimes carried out utilizing the excess of the olefin as a solvent (only small amounts of alkanes are present and are used to introduce the catalyst). The temperature of the process varies from 170 to 470 K depending on utilized catalysts, the olefin, and the desired product.

The preparation of syndiotactic polypropylene is carried out in the 195–230 K range, while the preparation of isotactic polypropylene takes place at 295–365 K if the catalyst is suspended and at 380–425 K in the presence of a homogeneous catalyst. These conditions are very mild compared to those for radical polymerization which is carried out at 295–470 K and 100–300 MPa. Ziegler–Natta catalysts are also utilized in polymerization of 1,3 dienes and cycloalkenes, for example, butadiene, isoprene, cyclobutene, cyclopentene, and dicyclopentadiene. Conjugated dienes may form the

*Table 13.3. Some Polymerization Catalysts*

| Transition metal compound | Main group metal compound | Monomer | Polymer | Reaction conditions |
|---|---|---|---|---|
| $TiCl_4$, $TiCl_3$, $VCl_3$ | $AlR_3$, $AlClR_2$ | Ethylene, 1-alkenes, styrene | Linear, regular, high molecular weight, isotactic | Mild conditions, 0.1–3 MPa, heterogeneous catalyst |
| $[TiCl_2cp_2]$, $[Ti(OR)_4]$, $VCl_4$, $[VO(OR)_3]$, $[Cr(acac)_3]$, $TiCl_4$ $TiMe_2cp_2$, $ZrMe_2cp_2$ $[VO(acac)_2]$ | $AlXR_2$, $AlX_2R$; (X = Cl, Br) $AlMe_2Cl + H_2O$ $AlMe_3 + H_2O$ $+Al(Me)O+_n$ | Ethylene; Styrene; propylene | Linear, molecular weight is smaller than for heterogenized catalysts Atactic, low molecular weight; atactic or syndiotactic | Homogeneous catalyst, mild conditions |
| $TiCl_4$, $TiCl_3$, $VCl_3$, $VCl_4$, $[VO(OR)_3]$, $[Cr(acac)_3]$, $[Fe(acac)_3]$, $[Co(acac)_3]$, $[Ti(OR)_4]$ | $AlXR_2$, $Al(OR)R_2$ | 1,3-Diene | Stereoregular | Homogeneous and heterogeneous catalysts, mild conditions |
| $[M(acac)_3]$ (M = V, Cr, Fe, Co), $[Ti(OR)_4]$, $TiCl_3$ | $Al(OR)R_2$, $AlClR_2$, $AlCl_2R$, $AlR_3$ | Vinyl monomers | Polymer structure depends on the character of the monomer and catalyst | Homogeneous catalysts |

following polymers: *trans*-1,4-polydienes, *cis*-1,4-polydienes, 1,2-polydienes, and 3,4-polydienes; see structures (13.73).

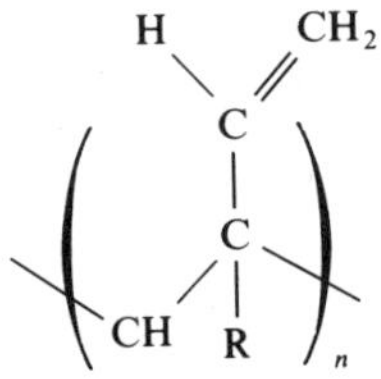

1,2-polybutadiene, R = H
1,2-polyisoprene, R = Me

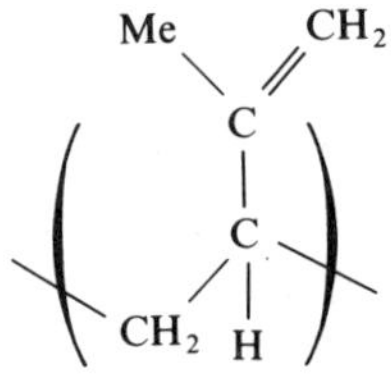

3,4-polyisoprene

(13.73)

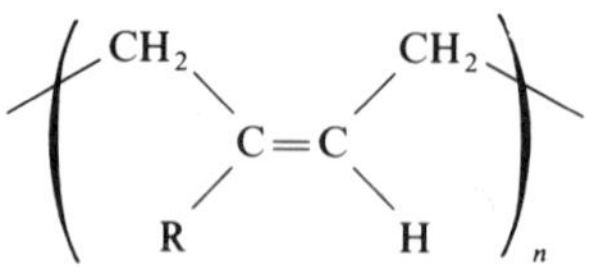

*cis*-1,4-polybutadiene, R = H
*cis*-1,4-polyisoprene, R = Me

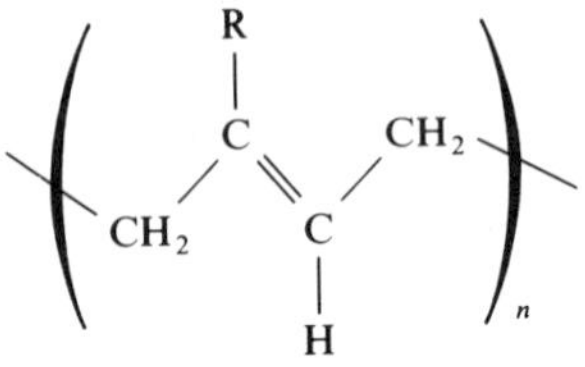

*trans*-1,4-polybutadiene, R = H
*trans*-1,4-polyisoprene, R = Me

Polybutadienes which occur in nature always have *cis*-1,4 and *trans*-1,4 structure while natural 1,2-polybutadiene has never been found. Natural rubber is also a 1,4-polymer, that is, *cis*-1,4-polyisoprene. In the presence of Ziegler–Natta catalysts, polymers of all structures have been obtained. The polymers *trans*-1,4-polybutadiene and *trans*-1,4-polyisoprene may be synthesized utilizing heterogeneous catalysts $VCl_3$ + $AlR_3$ and $VOCl_3 + AlR_3$, respectively. The polymer *cis*-1,4-polybutadiene is formed in the presence of the following catalysts: $TiCl_4 + AlR_3$, $Ni(acac)_2 + AlR_3$, $Co(dmgH)_3 + AlR_3$, and others. An amorphous polymer consisting of 3,4-polyisoprene units is formed during polymerization using catalysts such as $[Ti(OR)_4] + AlR_3$, $[Fe(dmgH)_2] + AlR_3$, $[Ni(dmgH)_2] + AlR_3$, $[Co(dmgH)_3] + AlR_3$, etc. *Trans*-1,4-polybutadiene may also be obtained with very high selectivity (above 99%) by means of a homogeneous catalyst, e.g., $RhCl_3$ in water, $RhCl_3$ in hydrated ethanol, and $IrCl_3$ in water. Isotactic 1,2-polybutadiene is formed with 97–99% selectivity in the presence of catalysts which are obtained from chromium(III) compounds: $[Cr(dipy)_3]^{3+} + AlR_3$ and $[Cr(PhCN)_6]^{3+} + AlR_3$. The polymer containing 76% of syndiotactic and 25% of atactic 1,2-polybutadiene is formed when the reaction is catalyzed by $MoO_2(OR)_2 + AlR_3$ and $MoO_2(acac)_2 + AlR_3$.

Many types of rubber which resist ozone were obtained by copolymerization of ethylene and propylene with asymmetric dienes such as 1,4-hexadiene, dicyclopentadiene, or ethylidenenorbornene. The addition to the reaction mixture of a certain amount of diene allows preparation of a very stable polymer, because the formation of a linear polymer requires breaking of one double bond while the other bond is utilized to net the copolymer which takes place during the polymerization process. Vanadium compounds such as $VOCl_3$ or $VCl_3$ with the addition of $Al_2Cl_3Et_3$ constitute catalysts for reactions leading to *three composite polymers*. In recent years, many unusually active homogeneous, heterogenized, and heterogeneous catalysts for olefin polymerization have been obtained.[35, 84–86, 89, 91] Exceptionally active are catalysts which are obtained by the addition of small amounts of water to catalytic systems of the type $Mcp_2X_2 + AlR_3$ ($Al:H_2O = 2:1–100:1$) where $M = Ti$, Zr, $X = Cl$, alkyl, etc.[84, 85] as well as $Mcp_2R_2$ containing alumoxanes, e.g. $\{Al(Me)O\}_n$, as the catalyst.[84, 85] Some of these systems allow preparation of $10^5$ kg of polyethylene per mol of the transition metal in one hour at normal pressure. This procedure enables preparation of polymers from which the catalyst is not separated because its content is of the order of several ppm. No less catalytically active are organometallic compounds of the lanthanides. The compound hydridobis(pentamethylcyclopentadienyl)neodymium(III) polymerizes ethylene with the rate of 80,000 cycles per min.[35] The complex $LuMe(pmcp)_2$ is also a very active catalyst for polymerization of ethylene.[91] Recently, heterogeneous and heterogenized catalysts have been investigated intensively and are prepared by deposition of transition metal compounds on metal compounds, for example, $MgX_2$ ($X = Cl$, Br, I), $CaI_2$, $MnCl_2$, $Mg(OR)_2$, and $MgR_2$ or on copolymers of dienes with 4-vinylpyridine, allyl alcohol, or methacrylic acid. Catalysts for polymerization and oligomerization which are obtained via electrochemical reduction of complexes of metals of groups 8–10 are interesting.[93]

Allyl complexes are also very broadly utilized as catalysts for polymerization of 1,3-dienes. Products of the reaction are polydienes 1,4-*cis* and 1,4-*trans* and polymers 1,2 [equation (13.73)].[86, 93–95] Very high stereospecificity may be achieved with application of allyl complexes as catalysts. This stereospecificity is influenced by solvents, ligands, and transition metal complexes which are added to the catalytic system. Complete structural change of the polymer occurs when some coordination sites are occupied by solvent or ligand molecules. Teysie[96, 97] found that $[Ni_2(\eta^3\text{-}C_3H_5)_2(CF_3COO)_2]$ in aliphatic hydrocarbons gives pure polymer 1,4-*cis*, while in the presence of strongly coordinating ligands, for example, $P(OPh)_3$, pure polymer 1,4-*trans* is obtained. If ligands possessing moderate complexing properties such as $CF_3COOH$, olefins, aromatic hydrocarbons, and chlorohydrocarbons are used, the polymer containing 50% of the isomer 1,4-*trans* and 50% of the isomer 1,4-*cis* is formed. The occurrence of *cis* and *trans* fragments is most commonly accidental; however, in the presence of chlorohydrocarbons, the polymer in which the *cis* and *trans* groupings occur alternatively is obtained. Such polydienes are called equibinary.

Acetylene polymers alternatively contain single and double bonds in the chain [equation (13.72)]. Therefore, they are characterized by the properties[98, 99] that include conductivity (semiconductivity), paramagnetism, energy transfer, reactivity, and the ability to form complexes. Acetylene polymers are often colored. Acetylene polymerizes in the presence of Ziegler–Natta catalysts. Substituted acetylenes form linear oligomers or cyclotrimers in many cases. Effective catalysts for polymerization of acetylene are molybdenum and tungsten compounds and those of mono- and disubstituted acetylenes, Nb(III) and Ta(III) organometallic compounds, as well as Nb(V) and Ta(V) halides.[90] Polymerization in the presence of molybdenum and tungsten compounds is catalyzed by carbene compounds of these elements[88, 90] [equation (13.72)]. In the case of Ziegler–Natta catalysts, polymerization probably occurs like for olefins as a result of migration of $\sigma$-carbyl ligand to the alkyne molecule.[88] Complexes of the type $[M(\text{mesitylene})(CO)_3]$ where $M = Cr$ or W, catalyze polymerization of phenylacetylene according to the mechanism of consecutive $2 + 2$ cycloaddition with the formation of cyclobutane derivative, Dewar's benzene, etc. [equation (13.74)].

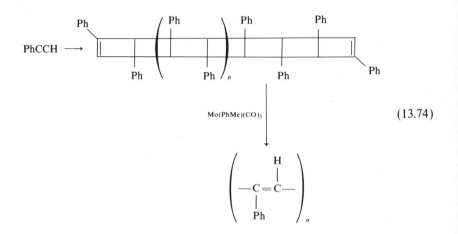

$$(13.74)$$

The ladder-like polymer having relatively small molecular weight in the presence of $[Mo(PhMe)(CO)_3]$ rapidly undergoes further polymerization and transformation into the final linear polymer.[100]

### b. *Oligomerization*

Dimerization, oligomerization, and cooligomerization reactions of olefins and dienes also constitute very important processes. Catalysts for oligomerization reaction of olefins should exhibit selectivity with respect to the degree of polymerization, that is, the degree of the increase of the chain should be small and relatively well determined so that the number of various other oligomer products is not too great. The control of the degree of polymerization is related to hydrogen transfer (migration) reactions. An increase in temperature leads to the lowering of the molecular weight of the polymer. This allows regulation of the composition of the obtained product. Resulting oligomers should be characterized by a small affinity to the catalyst. Often, this condition is not fulfilled. In such cases, the formation of the polymer is prevented by increasing the concentration of the substrate and by limiting the degree of conversion. Ziegler–Natta systems also catalyze oligomerization of olefins and dienes, cooligomerization of olefins, and cooligomerization of olefins with dienes. Most commonly utilized Ziegler–Natta catalysts are $Ti(OR)_4 + AlR_3$, $NiX_2 + Al_2Cl_xR_{6-x}$ $(x = 2, 3, 4)$, and $[NiX_2(PR_3)_2] + Al_2Cl_xR_{6-x}$. Additional catalysts are compounds of some noble metals such as $PdCl_2$, $[PdCl_2(PhCN)_2]$, $[Pd_2Cl_4(C_2H_4)_2]$, $RhCl_3 \cdot 3H_2O$, $RuCl_3$, and $IrCl_3 \cdot 4H_2O$.

The compound 1,4-hexadiene is an important product of codimerization of ethylene and butadiene. It is obtained in the presence of such catalysts as $FeX_3 + AlEt_3$ ($X = Cl$, acac), $CoCl_2 + AlEt_3$, $[NiCl_2(PR_3)_2] + AlClR_2$, $RhCl_3 \cdot 3H_2O$, $[Rh_2Cl_2(C_2H_4)_4]$, etc.

The mechanism of the formation of 1,4-hexadiene in the presence of $RhCl_3$ is as shown in scheme (13.75). Oligomerization of 1,3-butadiene which is catalyzed by com-

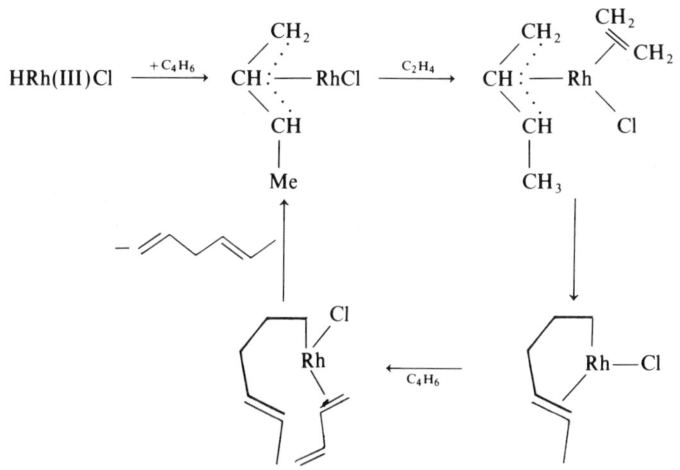

$$(13.75)$$

plexes of transition metals leads, depending upon conditions of the reactions, the solvent, and the ligands, to the formation of linear or cyclic oligomers.[22] Allyl derivatives are also formed as intermediate products which, in many cases, were isolated and investigated. For some of them, X-ray studies were conducted. During reduction of $CoCl_2$ by sodium tetrahydroborate, $NaBH_4$, in the presence of butadiene in ethyl alcohol as solvent, the complex $[Co(C_8H_{13})(C_4H_6)]$ possessing structure (13.76) is

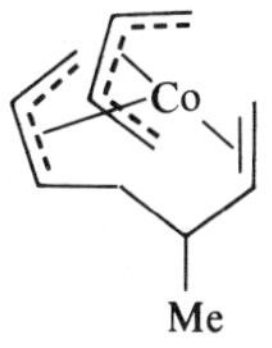

$$(13.76)$$

formed. The formation of the coordination compound which contains the branched allyl ligand strongly suggests that allyl is formed as a result of addition of butadiene to the hydrido cobalt complex [equation (13.77)]. An analogous mechanism may be operating

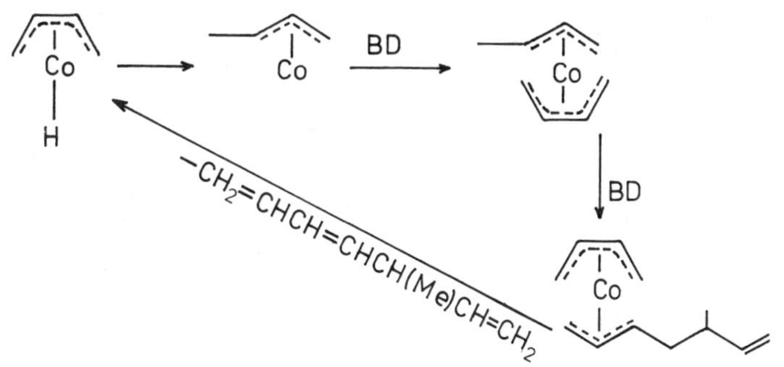

$$(13.77)$$

if the catalyst is a compound containing a $\sigma\,M-C$ bond, for example, alkyl or aryl compounds. In oligomerization processes which occur according to the above-mentioned mechanism (migration reaction, Figure 13.2) no change in the oxidation state of the metal takes place. However, the addition of two butadiene molecules and the formation of the diallyl complex is possible in the case of complexes containing metals in low oxidation states with labile ligands [equation (13.78)].

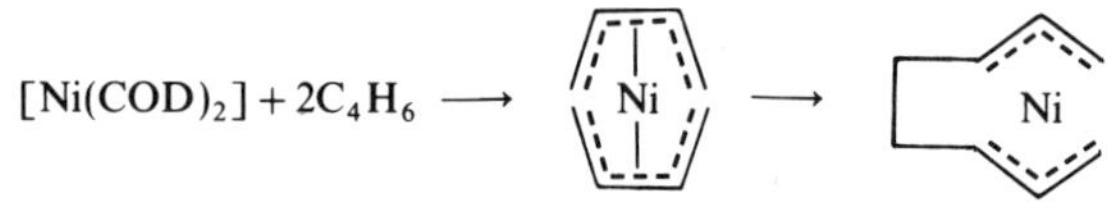

$$[Ni(COD)_2] + 2C_4H_6 \longrightarrow \qquad \longrightarrow \qquad \qquad (13.78)$$

At low temperature, intermediate products having formula (13.79) were identified.

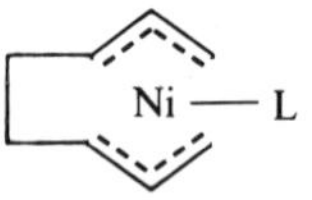

$$Ni-L \qquad (L = PR_3, P(OR)_3) \qquad\qquad (13.79)$$

The formation of an analogous palladium compound as an intermediate product is also assumed. These are oxidative cycloaddition reactions during which there is an increase in the formal oxidation state by 2 units (Table 1.4).

In the presence of allyl complexes oligomerization may occur according to both mechanisms. During the reaction with butadiene, first reductive elimination of the allyl group may take place followed by oxidative cycloaddition of the butadiene molecule; see equation (13.80).

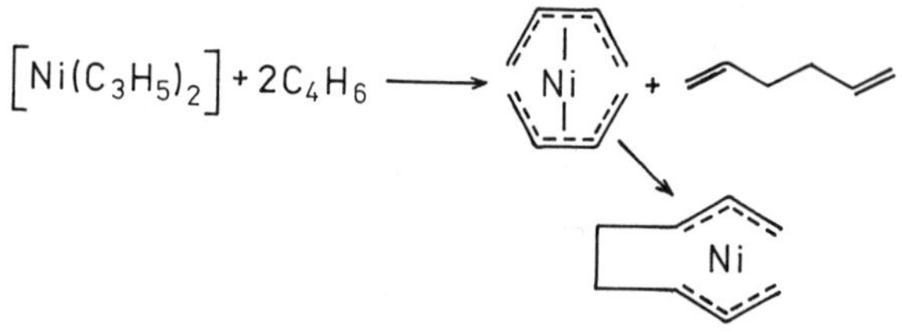

$$\left[Ni(C_3H_5)_2\right] + 2C_4H_6 \longrightarrow \qquad\qquad (13.80)$$

In addition to the described allyl complexes which are formed by dimerization of butadiene, allyl compounds consisting of trimers of butadiene may be obtained; see equation (13.81).

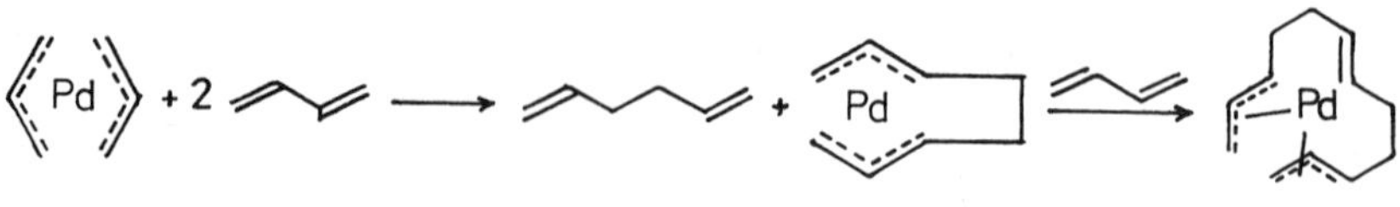

$$(13.81)$$

The ruthenium complex $[RuCl_2C_{12}H_{18}]$ [structure (13.82)] was obtained and its structure investigated.[101] The type of resulting products which come from complexes containing dimers as well as trimers depends on the metal, its oxidation state, properties

(13.82)

of ligands, the solvent, and other reagents [scheme (13.83)]. Ligands possessing strong electron-donor properties cause an increase in yields of DVCB and VCH.

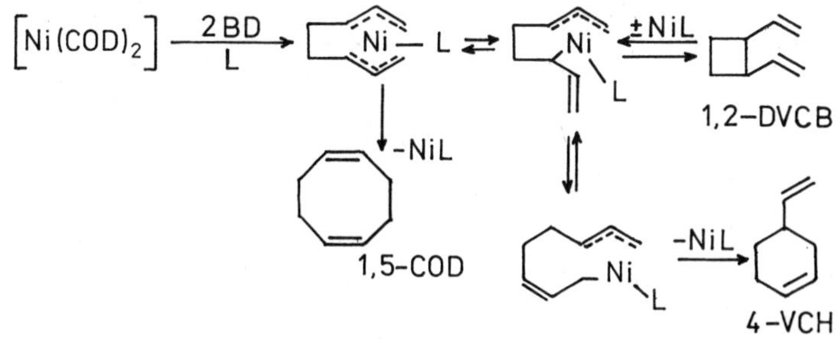

(13.83)

However, migration of the allyl group to the diene is also possible; no change of the oxidation state of the metal takes place, but the number of valence electrons in the complex changes.

If phosphines or phosphites are not present, Ni(0) compounds cause trimerization of butadiene [scheme (13.84)]. In addition to the isomer *trans,trans,trans-*

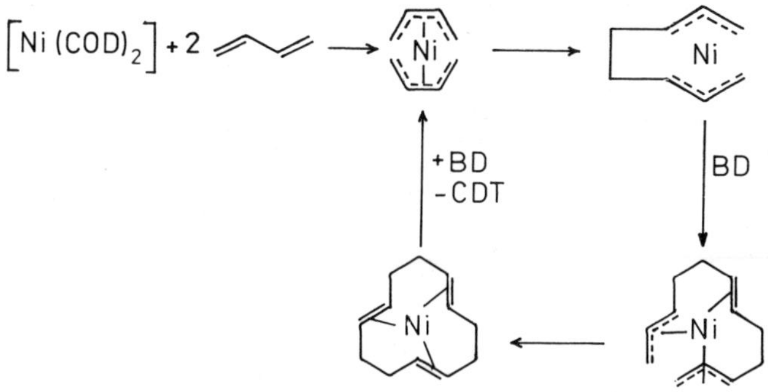

(13.84)

cyclododecatriene, also smaller amounts of isomers *trans,trans,cis* and *trans,cis,cis* are formed. However, the isomer *cis,cis,cis* is not obtained; see structures (13.85). Allyl

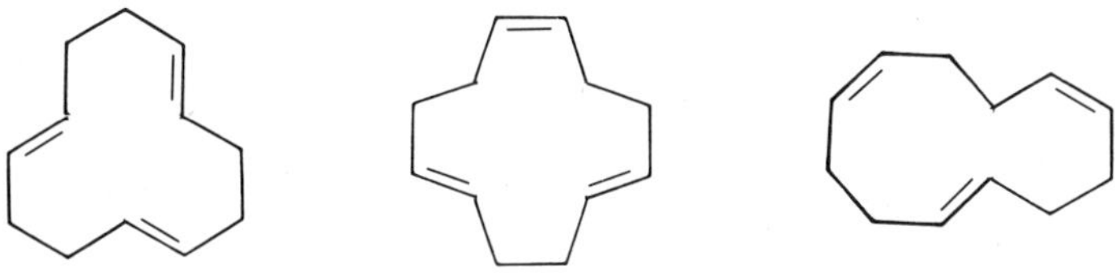

(13.85)

*t,t,t*-CDT          *t,t,c*-CDT          *t,c,c*-CDT

complexes also catalyze linear oligomerization and cooligomerization of butadiene (dienes) with olefins, acetylenes, and allene. During linear dimerization of butadiene to 1,3,6-octatriene there must be hydrogen transfer from $C_4$ to $C_8$. In the case of nickel compounds, such dimerization takes place in the presence of alcohols or secondary amines because these compounds are involved in hydrogen transfer [equation (13.86)].

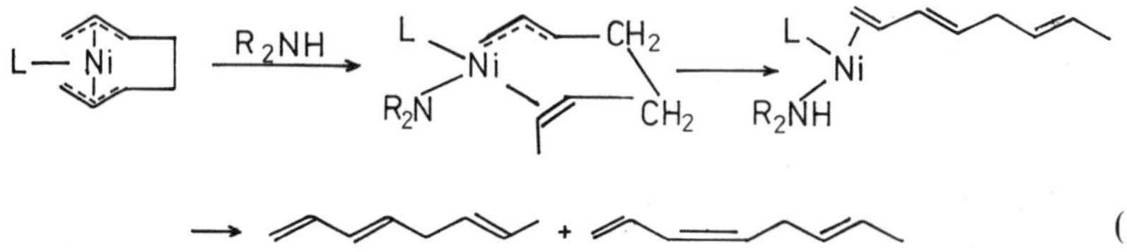

$$(13.86)$$

Very active and selective catalysts for oligomerization of dienes are formed during electrochemical reduction of complexes.[93] The rate of dimerization of 1,3-butadiene to 4-vinylcyclohexene in the presence of the catalyst which is obtained by electrochemical reduction of $[\{FeCl(NO)_2\}_2]$ is 20,000 moles of BD per mole of Fe per 1 h at 353 K.[93]

Cooligomerization (hetero-oligomerization) of olefins and acetylenes with butadiene in the presence of nickel complexes prevents the formation of butadiene trimers. During hetero-oligomerization chain and cyclic compounds may be formed; see equations (13.87) and (13.88). Dodeca-2,6,10-triene-1,12-diylnickel reacts with allene to give various products. Reaction (13.89) is an example.

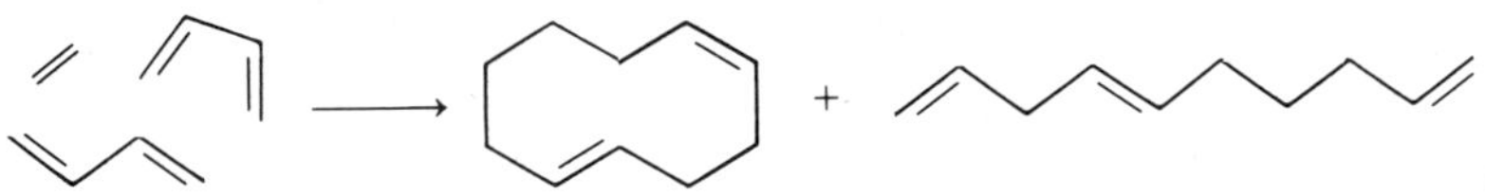

$$(13.87)$$

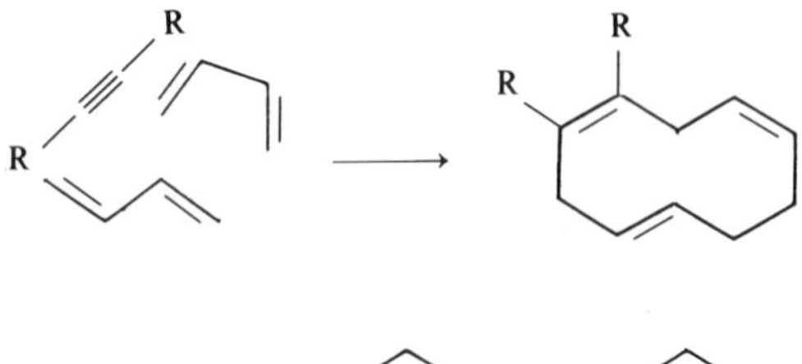

$$(13.88)$$

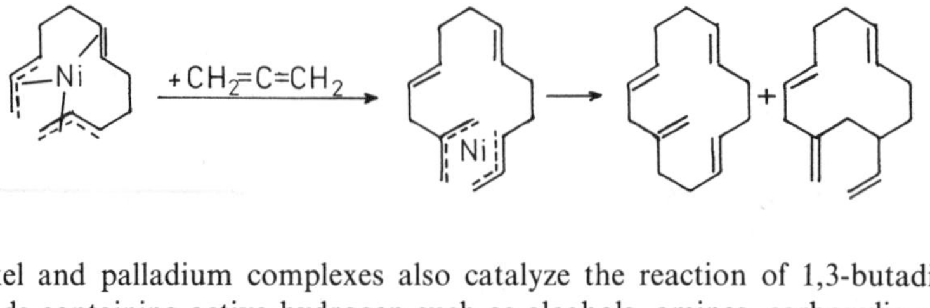

$$(13.89)$$

Nickel and palladium complexes also catalyze the reaction of 1,3-butadiene with compounds containing active hydrogen such as alcohols, amines, carboxylic acids, and active methylene and methyne compounds [equations (13.90)–(13.94)]. The mechanism

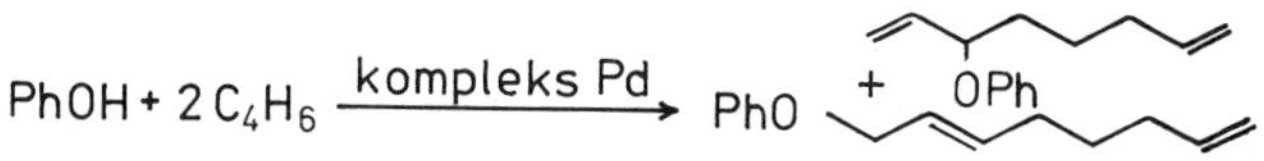

$$PhOH + 2\,C_4H_6 \xrightarrow{\text{kompleks Pd}} PhO\diagup\diagup + \diagup OPh\diagup \qquad (13.90)$$

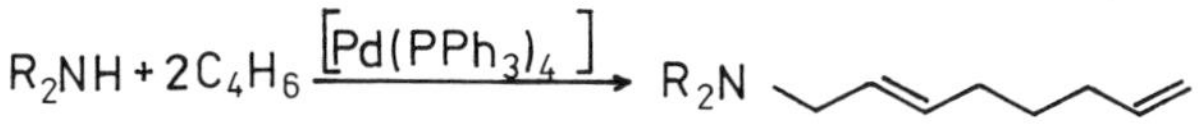

$$R_2NH + 2C_4H_6 \xrightarrow{\left[Pd(PPh_3)_4\right]} R_2N\diagup\diagup\diagup \qquad (13.91)$$

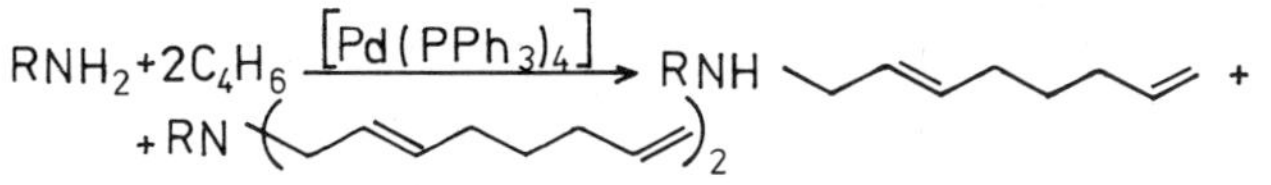

$$RNH_2 + 2C_4H_6 \xrightarrow{\left[Pd(PPh_3)_4\right]} RNH\diagup\diagup\diagup + \qquad (13.92)$$
$$+ RN\left(\diagup\diagup\diagup\right)_2$$

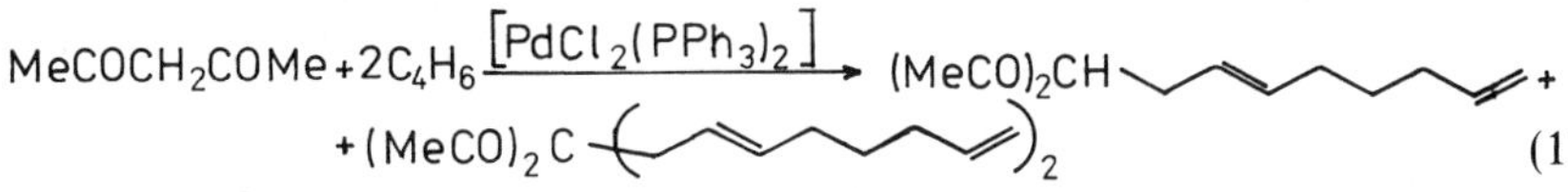

$$MeCOCH_2COMe + 2C_4H_6 \xrightarrow{\left[PdCl_2(PPh_3)_2\right]} (MeCO)_2CH\diagup\diagup\diagup + \qquad$$
$$+ (MeCO)_2C\left(\diagup\diagup\diagup\right)_2 \qquad (13.93)$$

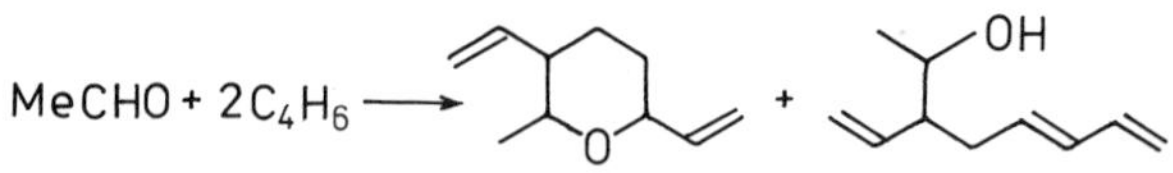

$$MeCHO + 2C_4H_6 \longrightarrow \qquad + \qquad OH \qquad (13.94)$$

of oligomerization reaction of acetylenes may be represented by equations (13.95) and (13.96) [see Section 6.6, 6.7, and reactions (6.182)–(6.196)]. Liner dimerization of phenylacetylene by the catalyst $[(RhCl(PPh_3)_3]$ may serve as an example [scheme (13.97)]. Metal complexes possessing low-valent $d^8$ and $d^{10}$ electronic structure or compounds of metals in higher oxidation states ($d^1$ and $d^2$ electronic structure) which begin a given transition series are catalysts for these reactions. In the first case, the catalytic activity is due to the antibonding character of the orbital c (Figure 6.21) which enables the reaction of thermal excitation [equation (6.193)]. However, in the second case, thermal excitation with the formation of a complex in which the carbon atom has a certain positive charge takes place because there is a strong $\sigma$-donor alkyne–metal interaction (a in Figure 6.21) and a weak $\pi$ back-bonding interaction (b in Figure 6.21); see equation (13.98). Linear oligomerization may proceed with the formation of the hydrido

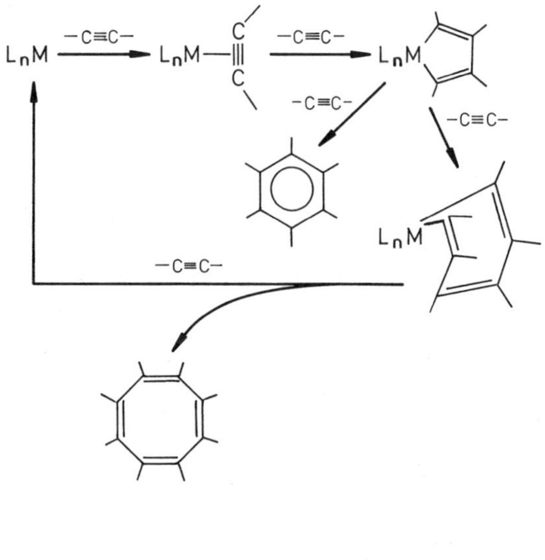

$$(13.95)$$

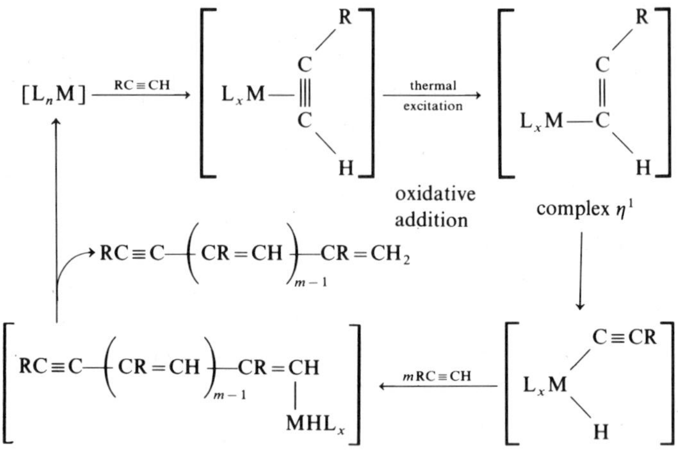

$$[L_nM] \xrightarrow{RC\equiv CH} \left[ L_xM-\underset{H}{\overset{R}{\underset{|||}{\overset{C}{\underset{C}{|}}}}} \right] \xrightarrow[\text{excitation}]{\text{thermal}} \left[ L_xM-\underset{H}{\overset{R}{\overset{C}{\underset{C}{\overset{||}{|}}}}} \right]$$

oxidative                complex $\eta^1$
addition

$$\rightarrow RC\equiv C-\left(CR=CH\right)_{m-1}-CR=CH_2 \tag{13.96}$$

$$\left[ RC\equiv C-\left(CR=CH\right)_{m-1}-\underset{MHL_x}{\overset{CR=CH}{\underset{|}{}}} \right] \xleftarrow{mRC\equiv CH} \left[ L_xM\underset{H}{\overset{C\equiv CR}{\diagup}} \right]$$

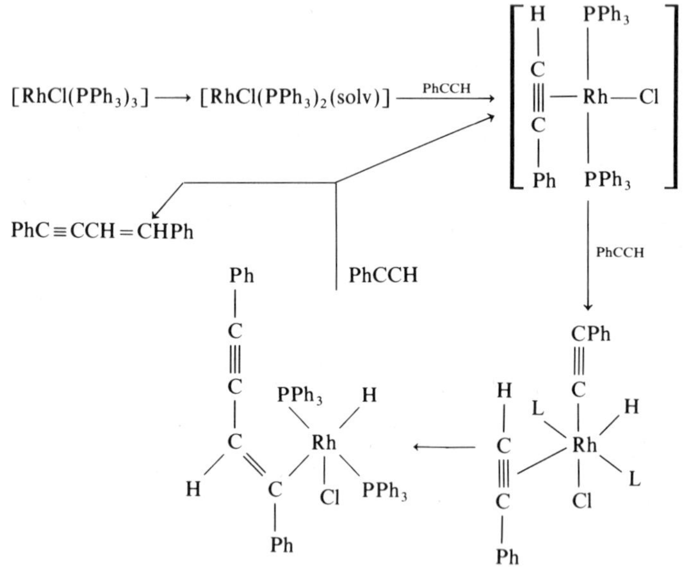

$$[RhCl(PPh_3)_3] \longrightarrow [RhCl(PPh_3)_2(solv)] \xrightarrow{PhCCH} \left[ \underset{Ph\quad PPh_3}{\overset{H\quad PPh_3}{\underset{|||}{\overset{|}{\underset{C}{\overset{C}{\underset{C}{|}}}}}}-Rh-Cl} \right]$$

$$PhC\equiv CCH=CHPh$$

Ph          PhCCH                                                              PhCCH
|                                                                                                                    (13.97) 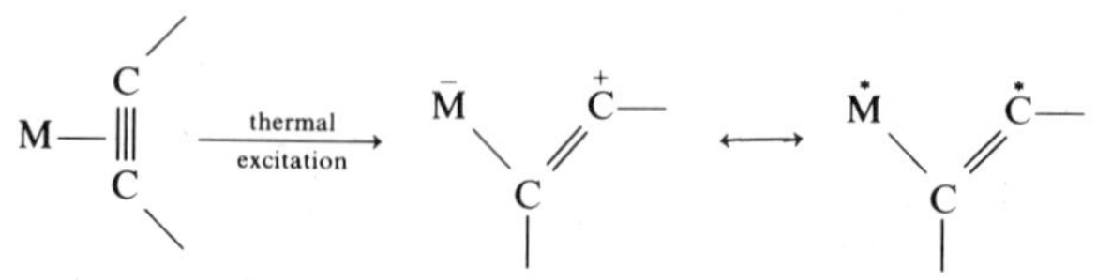
C
|||
C     PPh₃   H                                                      CPh
|     \   /                                             H   L    |||
C     Rh                          H   L     |   \ /   C
/ \   / \                   ←    C  \ / Rh      |   Rh   H
H    C   Cl  PPh₃              |||  |   \      C   |   \
|                              C   Cl    L   |||  Cl   L
Ph                             |                C
Ph                             |
Ph

$$M-\underset{C}{\overset{C}{\underset{\diagdown}{\overset{\diagup}{\underset{|||}{}}}}} \xrightarrow[\text{excitation}]{\text{thermal}} \quad \bar{M}\diagdown\overset{\overset{+}{C}-}{\underset{\underset{|}{C}}{}} \quad \longleftrightarrow \quad \dot{M}\diagdown\overset{\dot{C}-}{\underset{\underset{|}{C}}{}} \tag{13.98}$$

alkynyl complex as an intermediate. This explains the great catalytic activity of some 4*d* and 5*d* complexes of metals of groups 8–10 in linear oligomerization reactions. Such complexes show a great tendency for the formation of hydrido complexes. Catalysts for cyclooligomerization reactions include $[Ni(CO)_2(PPh_3)_2]$, $[Ni_3(acac)_6]$, $[Ni(CN)_2]$,

$[Ni(NCS)_2]$, $Hg[Co(CO)_4]_2$, $[Co_2(RCCR)(CO)_6]$, $[TiCl_4] + AlR_3$, $[CuI(PPh_3)_2]$, $[Pd(PR_3)_3]$, $[PdCl_2(PhCN)_2]$, and $[Pt(PR_3)_3]$. Linear oligomerization of acetylenes is catalyzed by the complexes $[Ni(CO)_2(PPh_3)_2]$, $[NiCl_2(PR_3)_2] + NaBH_4$, $[NiBr_2(PPh_3)_2]$, $[Ni_3(acac)_6] + py$, $CuCl + NH_4Cl + HCl$, $TiCl_4 + AlR_3$, etc. Also, these reactions are catalyzed by most metal carbonyls.

Alkynes also undergo catalytic cooligomerization with other compounds containing multiple bonds, e.g., CO, RCN, RNC, RSCN, RNCS, $NH_2CN$, as well as $S_8$.

Reactions of acetylenes with CO afford cyclic ketones or quinones; see equation (13.99). The formation of these compounds is best catalyzed by metal carbonyls of

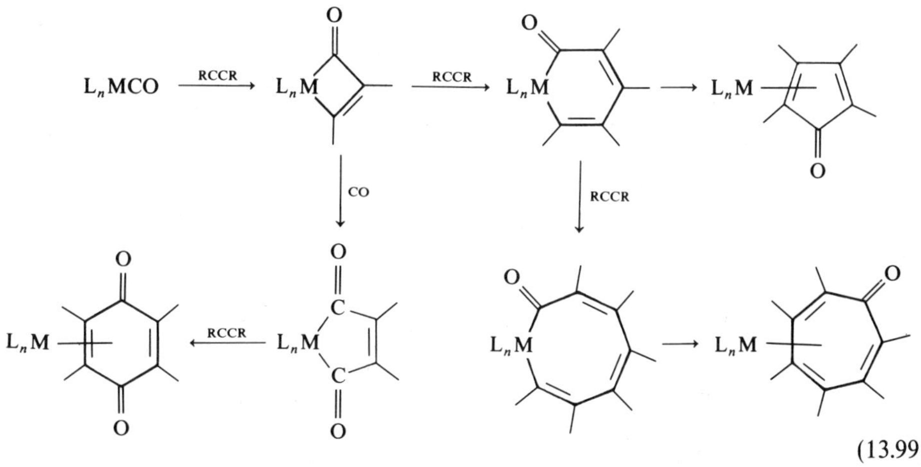

(13.99)

groups 8 and 9. Reactions of alkynes with isonitriles give compounds containing carbon–nitrogen double bonds. Low-valent nickel, palladium, and cobalt complexes catalyze reaction (13.100). The interaction of acetylenes with nitriles leads to the formation of derivatives of pyridine[102,103]; see equation (13.101). Reactions of acetylenes

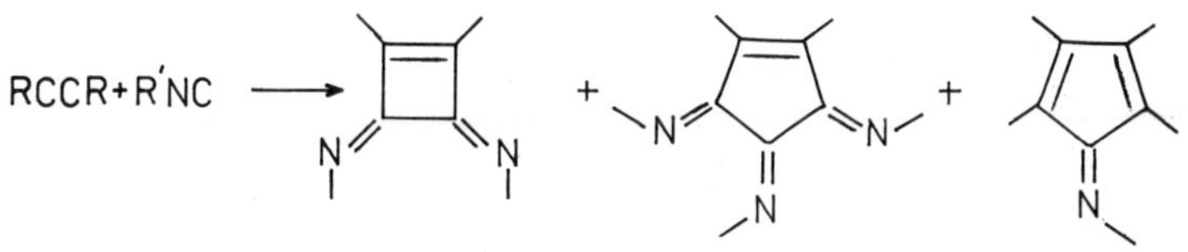

(13.100)

with $CS_2$ and RNCS also afford heterocyclic compounds [equations (13.101)–(13.107)]. Oligomerization of allene may afford interesting products. Thermal oligomerization has been known since 1913. Upon heating, cyclooligomers (13.108) are formed.

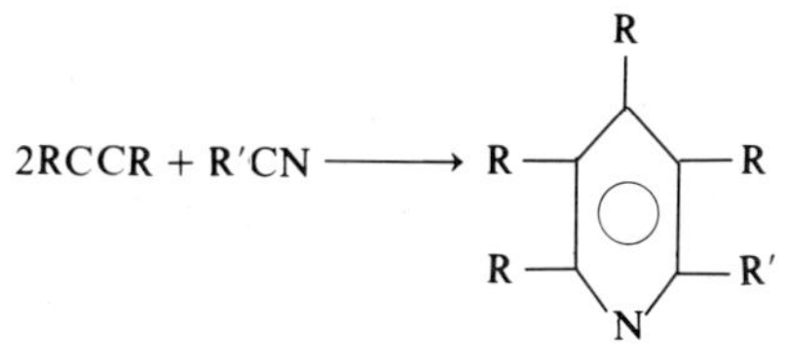

(13.101)

$$RCCR + R'NCS \longrightarrow \qquad (13.102)$$

$$2RCCR + CS_2 \longrightarrow \qquad (13.103)$$

$$2RCCH + H_2NCN \longrightarrow \qquad (13.104)$$

$$CocpL_2 + \tfrac{1}{4}S_8 + RCCR \longrightarrow cpCo \qquad (13.105)$$

$$2RCN + HCCH \longrightarrow \qquad (13.106)$$

$$CH_3CN \longrightarrow HC \equiv CNH_2$$

$$3MeCN \longrightarrow \qquad (13.107)$$

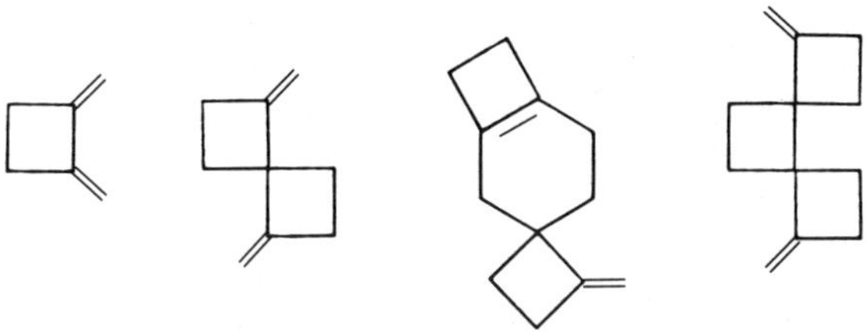

(13.108)

Oligomerization reactions catalyzed by complexes other than those described also give quite different cyclooligomerization and oligomerization products[104, 106]; see equations (13.109) and (13.110). The same products (the tetramer and trimers) are formed if $[Ni(CO)_x\{P(OPh)_3\}_{4-x}]$ $(x = 1-3)$ is utilized as a catalyst.

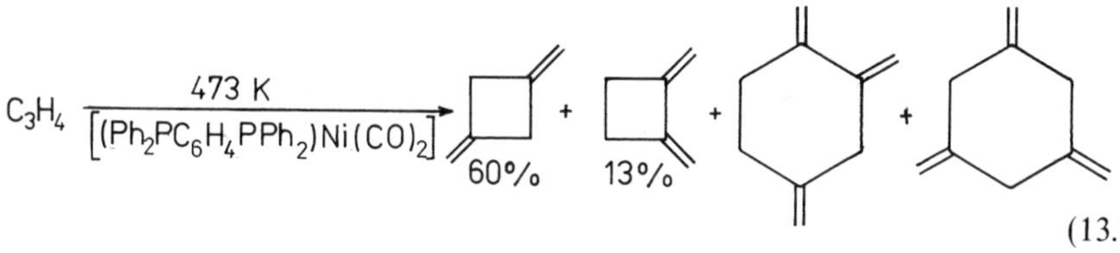

(13.109)

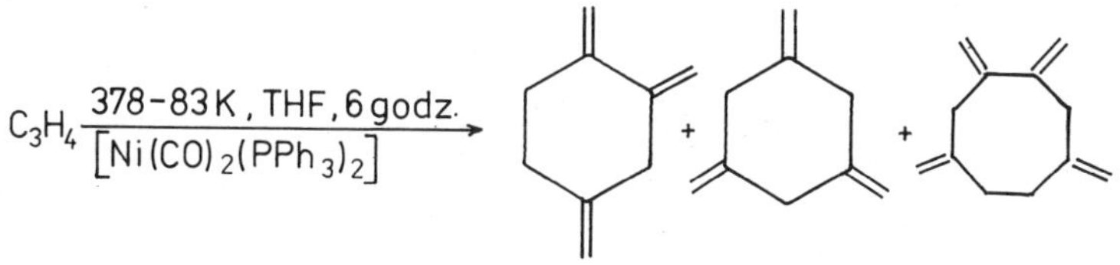

(13.110)

It is also possible to obtain pentamers and higher oligomers as well as polymers. The addition of Lewis bases (phosphines, etc.) to the catalyst or application of complexes containing such ligands, such as $[Ni\{P(OPh)_3\}_4]$, causes an increase in yield of the lower oligomers at the expense of the higher oligomers.

The oligomerization proceeds according to the mechanism of coordination and migration (insertion) of consecutive allene molecules which is proven by isolation of intermediate products; see, e.g., structure (13.111). Therefore, the mechanism of oligomerization may be represented as scheme (13.112).

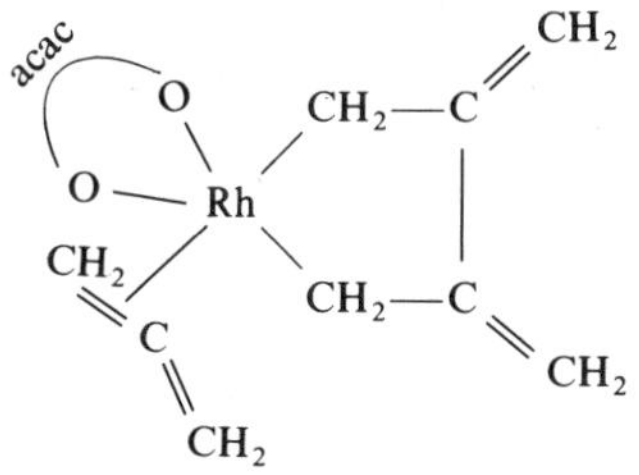

(13.111)

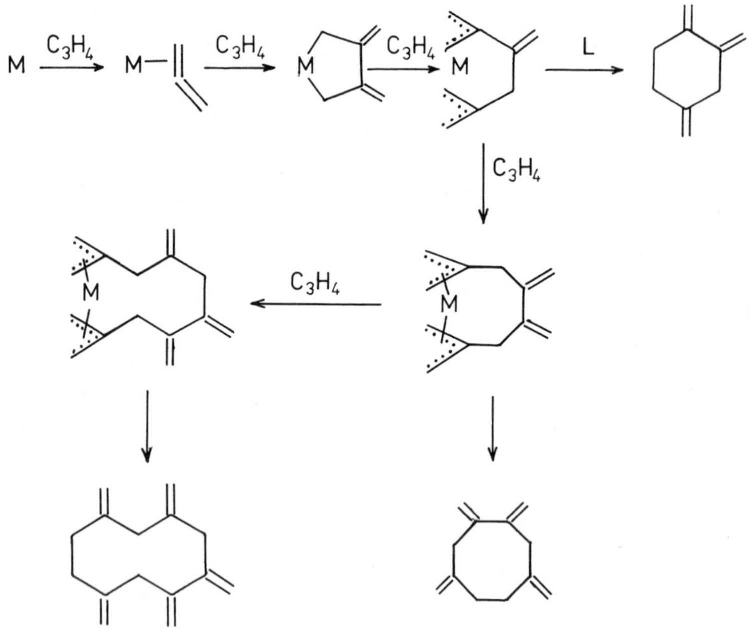

(13.112)

In the case of nickel catalysts, the rate of oligomerization increases in the series $Ni(O) + P(OPh)_3 < Ni(O) + PPh_3 < Ni(O)$. The pentamerization and tetramerization reactions are first order with respect to allene concentration while the trimerization reaction is almost of zero order.

## 5. CARBONYLATION REACTIONS

### a. Hydroformylation reactions

Addition reactions of carbon monoxide to various organic compounds are usually referred to as carbonylation reactions. The most important reaction of this type is hydroformylation of olefins:

$$RCH = CH_2 + CO + H_2 \xrightarrow{\text{cat.}} RCH_2CH_2CHO \tag{13.113}$$

This is an exothermic reaction. For various olefins the enthalpy of the reaction changes from $-117$ to $-147\,\text{kJ mol}^{-1}$.

The synthesis of aldehydes according to this method is utilized in industry. This reaction is catalyzed by metal carbonyls of groups 8–10 as well as by other compounds of these metals which, under the reaction conditions in the atmosphere of CO and $H_2$, always form metal carbonyls or hydrido carbonyls. The latter are the proper catalysts for the hydroformylation reaction. Compounds of other elements are not practically active as catalysts and therefore are not utilized in industry. The catalytic activity of elements of groups 8–10 changes in the following manner[8–10, 22, 107–113]:

$$
\begin{array}{ccc}
\text{Fe} \ll & \text{Co} \gg & \text{Ni} \\
\wedge & \wedge & \wedge \\
\text{Ru} \ll & \text{Rh} \gg & \text{Pd} \\
\vee & \vee\!\!\vee & \wedge \\
\text{Os} < & \text{Ir} & \text{Pt}
\end{array}
$$

The most active catalysts are rhodium compounds. The catalytic activity decreases in the series: $Rh \gg Co > Ir \sim Ru > Os > Pt > Pd > Fe > Ni$.

The presented series deals with the average catalytic activity of many compounds. Of course, one can find cobalt compounds which show lower activity than iridium compounds. Rhodium complexes are $10^2$–$10^4$ more active than $[Co_2(CO)_8]$ which is most commonly utilized in industry. Therefore, it is possible to obtain the same amounts of aldehydes with considerably smaller amounts of rhodium complexes and under milder conditions, because rhodium compounds catalyze the hydroformylation reaction at lower pressures and temperatures, even at room temperature and atmospheric pressure. The price of the rhodium catalyst is comparable to the cost of $[Co_2(CO)_8]$. Therefore,

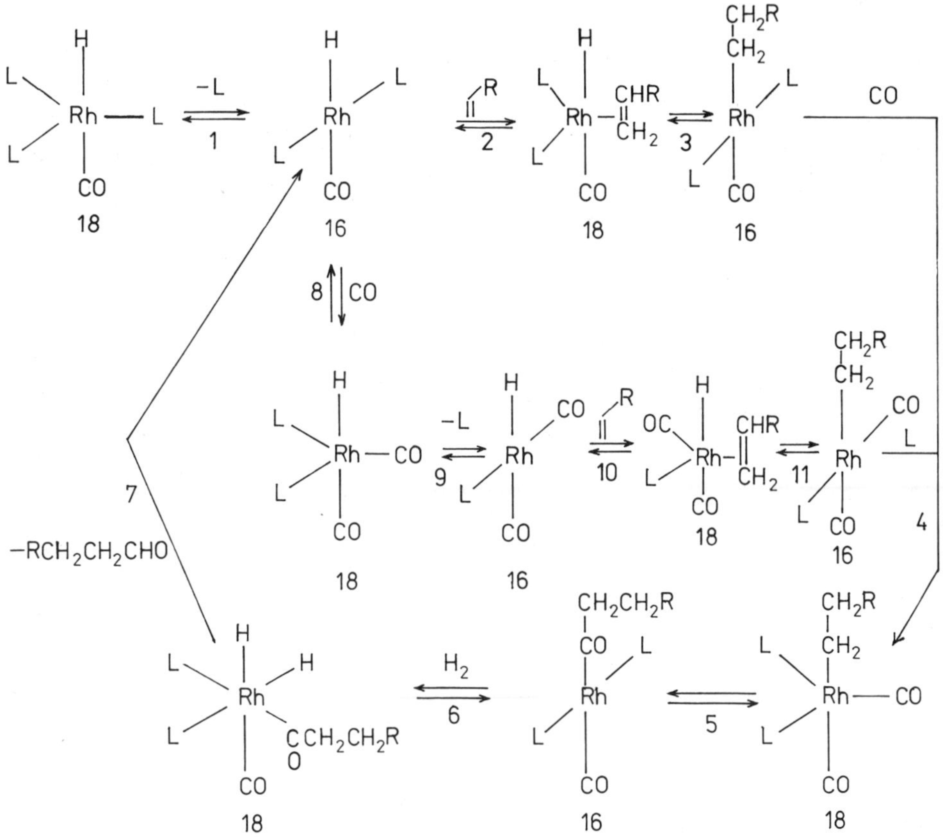

$$(13.114)$$

the investment expenses are smaller because of the milder reaction conditions. Union Carbide introduced $[RhH(CO)(PPh_3)_3]$ for catalytic preparation of butanol from propene in 1976. In the presence of this catalyst, considerably higher ratios of n-aldehyde to isoaldehyde are obtained, particularly at higher ratios of $PPh_3/Rh$, than for the cobalt catalyst $[Co_2(CO)_8]$. This process proceeds at *ca* 373 K and 1–2.5 MPa. Hydroformylation catalysts are stable. They are not reduced to the metal above certain CO pressure which is higher, the higher the temperature.

The oxo synthesis is generally carried out at temperatures up to 470 K and 20–30 MPa. The following solvents are utilized: saturated and aromatic hydrocarbons, ethers, alcohols, esters, ketones, lactones, lactams, nitriles, acid anhydrides, and water. The rate of the reaction is highest in primary and secondary alcohols probably because of the easier formation of the metal hydrido carbonyls in these solvents. However, utilization of alcohols is not always desired because in processes occurring at low temperatures, acetals are formed. This is beneficial if an unstable aldehyde is produced. In industry, the reaction is often carried out without solvents.

Terminal n-alkenes undergo hydroformylation reactions most rapidly followed by internal n-alkenes, while branched olefins react least rapidly. The greater number of alkyl groups that are bonded to the carbon atom which forms the double bond, the slower the reaction course. The mechanism of the reaction for the cobalt, rhodium, or manganese $[MnH(CO)_5]$ catalyst is basically the same. The proper catalysts are metal hydrido carbonyls.

In consecutive reactions of dissociation and association of the Lewis base as well as in reactions of migration and oxidative addition, an aldehyde molecule is formed. Many times, 18*e* and 16*e* complexes are formed. The numbers below the formulas in schemes (13.114) and (13.115) indicate the number of valence electrons. The dissociation of the Lewis base takes place in reactions 1 and 9, the association of the Lewis base in reactions 2, 4, 8, and 10, the migration of ligands in reactions 3, 11, 5, and 7, while

$$[Co_2(CO)_8] + H_2 \rightleftharpoons 2CoH(CO)_4$$

$$CoH(CO)_4 \xrightleftharpoons{-CO} CoH(CO)_3$$
$$\quad\; 18 \qquad\qquad\qquad 16$$

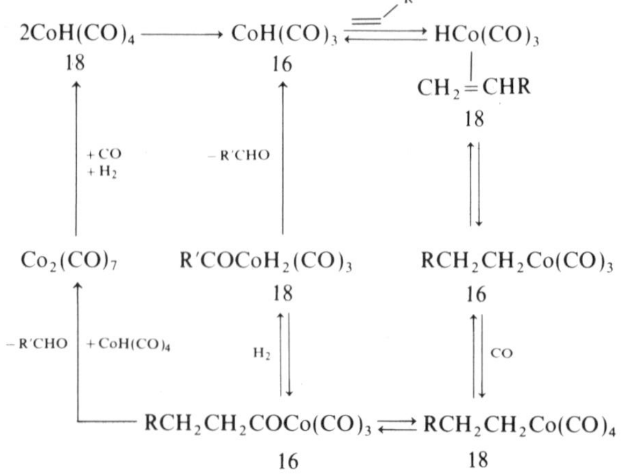

$$(13.115)$$

oxidative addition of the dihydrogen is represented in reaction 6. The hydroformylation reaction of olefins in the presence of $Co_2(CO)_8$ proceeds in a similar manner. The first stage of the reaction involves the formation of hydridotetracarbonylcobalt, which constitutes the direct source of the catalytically active compound $CoH(CO)_3$. Recent studies confirmed proposals that the formation of aldehyde occurs in the reaction of $[R'COCo(CO)_3]$ with $[CoH(CO)_4]$. Lately it was found that hydrogenation and hydroformylation of styrene via $[CoH(CO)_4]/CO$ proceeds with the formation in the first stage of the radical pair $[Ph\overset{*}{C}HMe\overset{*}{Co}(CO)_4]$.[111]

Studies of hydroformylation of (*E*)- and (*Z*)-3-methyl-2-pentene and 1,2-dimethyl-cyclohexene in the presence of $[RhH(CO)(PPh_3)_3]$ and $[Co_2(CO)_8]$ showed[114] that the addition of the hydrogen and the carbonyl group to the double bond occurs according to *cis* stereochemistry in the case of both catalytic systems. This is in agreement with the symmetry rules.

The rate of the formation of aldehyde for the cobalt and rhodium catalysts ($[Co_2(CO)_8]$ and $[Rh_4(CO)_{12}]$) is proportional to the concentration of olefin, the catalyst, and hydrogen, and inversely proportional to CO concentration:

$$\frac{d[\text{aldehyde}]}{dt} = k[\text{olefin}]^x [\text{cat}]^y \frac{p(H_2)}{p(CO)} \qquad (13.116)$$

cat = Co, $x = 1$, $y = 1$; cat = Rh, $x \gtrsim 0.1$; $y = 0.7$–$1.0$

The rate of the reaction changes depending upon the partial pressure of CO; it reaches a maximum at *ca* 3 MPa. The yield of the reaction decreases as the temperature increases. At low temperatures the rate of the hydroformylation is small and therefore higher temperatures are applied. Hydroformylation allows the preparation of various valuable products because the oxo synthesis may utilize different compounds containing a carbon–carbon double bond, for example, dienes, polyenes, and unsaturated aldehydes, ketones, nitriles, alcohols, esters, etc. For example, dienes may afford dialdehydes. A substantial amount of aldehydes is converted to alcohols which find considerable application in the preparation of detergents, plasticizers, and lubricants. The aldol condensation generally is not desired in hydroformylation processes. Nevertheless, via aldol condensation followed by hydrogenation, 2-ethylhexanol is obtained from n-butyraldehyde; see equation (13.117).

$$2CH_3CH_2CH_2CHO \longrightarrow C_3H_7CH=C-CHO + H_2O$$

$$\Bigg\downarrow {\scriptstyle H_2} \qquad \overset{|}{E}t$$

$$(13.117)$$

$$C_3H_7CH_2CHCH_2OH$$
$$\overset{|}{E}t$$

Condensation reactions often are carried out simultaneously with hydroformylation reactions in the presence of basic catalysts; the hydroformylation catalyst in this case also plays the role of the hydrogenation catalyst.

Other alcohols are obtained by reduction of aldehydes because the reaction usually proceeds via two steps,

$$RCH=CH_2 + H_2 + CO \longrightarrow RCH_2CH_2CHO \xrightarrow{H_2} RCH_2CH_2CH_2OH \qquad (13.118)$$

However, it is possible to obtain alcohols directly under different conditions in the presence of hydroformylation catalysts.

$$RCH=CH_2 + 2H_2 + CO \longrightarrow RCH_2CH_2CH_2OH \qquad (13.119)$$

This reaction is carried out at higher temperatures compared to the synthesis of aldehydes, at higher total pressure, and at a higher ratio of $H_2/CO$.

Serious difficulties are encountered in processes of separation of homogeneous catalysts from the products. Consequently, intensive studies of heterogenized catalysts, or catalysts which are soluble in water, have been conducted. Examples of the latter catalysts include complexes containing amino phosphines such as $Ph_2PCH_2CH_2\overset{+}{N}Me_3$,[115] and also sulfonated arylphosphines.[22, 107, 112] Heterogenized catalysts comprise complexes supported on organic polymers and inorganic materials such as carbon, aluminum oxide, silica, oxides of other metals, and zeolites.[116] Supported liquid-phase catalysts (SLPC) show very interesting properties. As supporting materials, organic polymers, $Al_2O_3$, $SiO_2$, etc., are used, while the liquid phase results from phosphines which constitute liquids at the temperature of the reaction.[117] The hydroformylation reaction proceeds at the boundary of the solid and liquid phases, in the liquid phase, and at the boundary of the liquid and gaseous phases. The ratio of n-aldehydes to isoaldehydes is very high, sometimes reaching values above 40.[117] Complexes containing asymmetric phosphine ligands that are analogous to those utilized for asymmetric hydrogenation are used for the preparation of optically active esters, aldehydes, or alcohols.[112, 113, 118, 119] The optical yields are higher in the case of chelating ligands possessing bulky groups connected to phosphorus atoms, for example, dbp–diop [see equations (13.45) and (13.46)].

### b. *Carbonylation reactions*[107–109, 112, 113]

Carbonylation of alcohols leads to the formation of aldehydes or alcohols containing one more methylene group compared to the starting alcohol, and therefore this reaction is called homologation reaction:

$$CH_3OH + CO + H_2 \longrightarrow CH_3CH_2OH \qquad (13.120)$$

The most reactive catalysts for these reactions are ruthenium compounds and ruthenium–cobalt systems[124] (Table 13.4).

The oxosynthesis may also be utilized for the preparation of ketones [equation (13.121)].

$$CH_2CH_2 + [HCo(CO)_4] \longrightarrow [CH_3CH_2Co(CO)_4] \xrightarrow{CO} [CH_3CH_2COCo(CO)_4]$$

$$\downarrow {\scriptstyle[EtCo(CO)_4]} \qquad (13.121)$$

$$Et_2CO + [Co_2(CO)_8]$$

*Table 13.4. Carbonylation Catalysts*

| Catalyst | Substrates | Products |
|---|---|---|
| $Co_2(CO)_8$ | Olefins $C_nH_{2n}$ | $C_nH_{2n+1}CHO$, $C_nH_{2n+1}CH_2OH$, $(C_nH_{2n+1})_2CO$ dialdehydes, dialcohols, aldehydes, alcohols |
| | Nonconjugated dienes | |
| | $C_nH_{2n+1}OH$ | $C_nH_{2n+1}CH_2OH$ |
| | $CR{\equiv}CH$, $H_2O$ | $CH_2{=}CRCOOH$ |
| | $CHR{=}CH_2$, $H_2O$ | $RCH_2CH_2COOH$ |
| | $CHR{=}CH_2$, $R'OH$ | $RCH_2CH_2COOR'$ |
| | $CHR{=}CH_2$, $R'SH$ | $RCH_2CH_2COSR'$ |
| | $CHR{=}CH_2$, $R'_2NH$ | $RCH_2CH_2CONR'_2$ |
| | MeOH | MeCOOH |
| | Oximes, Schiff bases, phenylhydrazones, semicarbazones | Phtalimidines |
| | Allylamines | Lactams |
| | Alcohols | Lactones |
| | Amides of unsaturated alkenes[121] | Imides |
| $[Co_3(\mu_3\text{-}CPh)(CO)_9]$ | Olefins | Aldehydes |
| $[RhH(CO)(PPh_3)_3]$ | Olefins | Aldehydes, alcohols |
| | MeOH | MeCOOH |
| | 1,3-dienes[a] | Monoaldehydes, dialdehydes |
| | Esters of $\alpha,\beta$-unsaturated acids | Esters of $\alpha$-formyl acids |
| $Rh_2O_3 + PR_3$ | Olefins | Aldehydes, alcohols |
| | 1,3-dienes | Aldehydes, dialdehydes |
| | | Alcohols, dialcohols |
| | $CH_2{=}CHCOOEt$ | $CH_3\underset{\underset{CHO}{\|}}{C}HCOOEt$ |
| | $CH_2{=}CMeCOOMe$ | $OHCCH_2CHMeCOOMe$ <br> $CH_3\underset{\underset{CHO}{\|}}{C}MeCOOMe$ <br> (α-methyl-γ-butyrolactone) 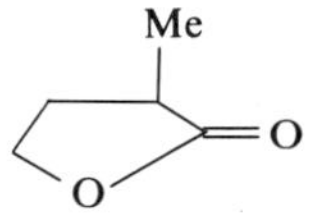 |
| $[RhH_2(O_2COH)(PPr^i_3)_2]$ | $n\text{-}BuCH{=}CH_2$, $(CH_2O)_n$[120] | $C_6H_{13}CHO$, $C_6H_{13}CH_2OH$, $C_6H_{13}COOMe$ |

[a] B. Fell, W. Boll, and J. Hagen, *Chem. Z.*, **99**, 485 (1975).

*(Table continued)*

*Table 13.4. (continued)*

| Catalyst | Substrates | Products |
|---|---|---|
| [Rh$_4$(CO)$_{12}$] | Olefins | Aldehydes, alcohols |
| [PdCl$_2$(PPh$_3$)$_2$], HCl | 1,5-COD | 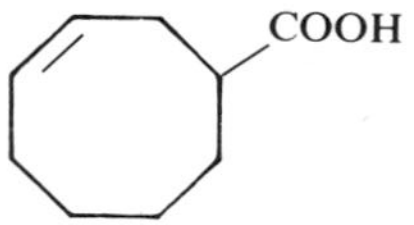 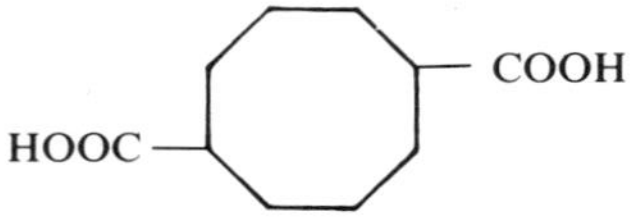 (cyclooctene ring with COOH); (cyclooctane ring with HOOC and COOH) |
| PdCl$_2$, HCl, O$_2$ | CH$_2$=CH—CH=CH$_2$, PhCH=CH$_2$, RC≡CH, RCH=CH$_2$, MeOH | CH$_3$CH=CHCH$_2$COOH, CH$_3$CH(Ph)COOH, CH$_2$=CRCOOH, RCH(OMe)CH$_2$COOMe, RCH(COOMe)CH$_2$COOMe |
| PdCl$_2$, CuCl$_2$ | Cyclopentene$^d$ | (cyclopentane with COOMe and OMe); (cyclopentane with two COOMe); (cyclopentane with COOMe); (cyclopentane with two COOMe) |
| [PtH(SnCl$_3$)(CO)(PPh$_3$)$_2$] | Olefins$^e$ | n-Aldehydes |
| [PtHCl(PPh$_3$)$_2$] + SnCl$_2$ | Olefins | n-Aldehydes |

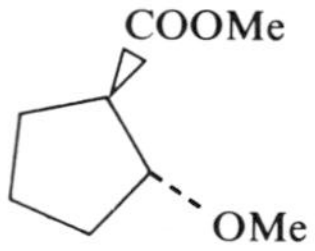

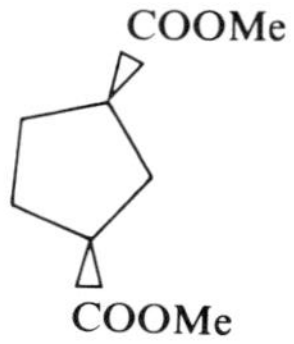

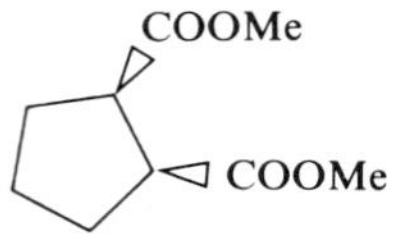

| Catalyst | Substrate | Product |
|---|---|---|
| $Rh_4(CO)_{12}$ | $CH_2=CHCOOEt$ | $CH_3CHCOOEt$ with $\mid$ CHO |
| $[Rh_2Cl_2(CO)_4] + R_2P(CH_2)_nPR_2$ <br> $[Rh_2(\mu\text{-}SBu^t)_2(CO)_2(PR'_3)_2]$ <br> $[Fe(CO)_5]$ | $\alpha,\beta$-Unsaturated esters[b] <br> Olefin[122] <br> $RCH=CH_2, H_2O$ <br> $RC\equiv CH, H_2O$ | $\alpha,\beta$-Formyl esters <br> Aldehydes <br> $RCH_2CH_2COOH$ <br> $CH_2=CRCOOH$ |
| $[Ni(CO)_4]$ | $RC\equiv CH, H_2O$ <br> $RC\equiv CH, R'OH$ <br> $RC\equiv CH, CH_2=CRCOOH$ <br> $CH\equiv CH, MeOH,$ <br> $CH_2=CHCH_2X^c$ <br> $CH\equiv CH, CH_2=CHCH_2X,$ <br> $MeOH^c$ <br> $X = halogen$ | $CH_2=CRCOOH$ <br> $CH_2=CRCOOR'$ <br> $(CH_2=CRCO)_2O$ <br> $CH_2=CHCH_2CH=CHCOOMe$ <br> (cyclopentenone with $-CH_2COOMe$) |
| $[RhI_2(CO)_2]^- + MeI$ <br> $PdI_2, HI, I_2$ <br> $[PtH(SnCl_3)(PPh_3)_2]$ <br> $[PtX_2(PPh_3)_2] + SnX_2$ <br> $(X = Cl, Br, I)$ <br> $[Ru_3(CO)_{12}]^g$ <br> $[Ru(OAc)(CO)_2]_n$ <br> $CuCl, CuBr^h$ <br> $PtCl_2 + Ph_2P(CH_2)_nPPh_2 + SnCl_2^i$ <br> $(n = 1\text{-}6, 10)$ | $MeOH$ <br> $C_2H_4$ <br> Olefins[e] <br> Olefins[f] <br><br> Piperidine <br> Morpholine <br> Pyrrolidine <br> Olefins | $MeCOOH$ <br> $EtCOOH$ <br> n-Aldehydes <br> Aldehydes <br><br> N-Formylamine <br><br> Aldehydes |

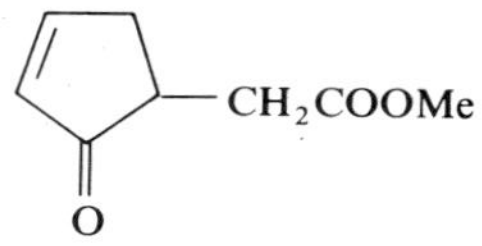

[b] M. Tanaka, T. Hayashi, and I. Ogata, *Bull. Chem. Soc. Jpn.*, **50**, 2351 (1977).

[c] G. P. Chiusoli, *Acc. Chem. Res.*, **6**, 422 (1973).

[d] D. E. James and J. K. Stille, *J. Am. Chem. Soc.*, **98**, 1810 (1976).

[e] C. Y. Hsu and M. Orchin, *J. Am. Chem. Soc.*, **97**, 3553 (1975).

[f] I. Schwager and J. F. Knifton, *J. Catal.*, **45**, 256 (1976).

[g] J. J. Byerley, G. L. Rempel, N. Takebe, and B. R. James, *Chem. Commun.*, 1482 (1971).

[h] B. K. Nefedov, N. S. Sergeeva, and Ya. T. Eidus, *Kinet. Katal.*, **15**, 1523 (1974).

[i] Y. Kawabata, T. Hayashi, and I. Ogata, *J. Chem. Soc., Chem. Commun.*, 462 (1979).

According to the given equation, the formation of diethylketone is favored by an excess of the olefin because the concentration of hydridocarbonyl needed for aldehyde formation is very small. At higher ratios of $C_2H_4/H_2$, it is possible to obtain a reaction mixture containing 85% of the ketone. As the chain length of the alkene increases, the yield of the ketone rapidly decreases.

Carbonylation reactions of acetylenes and olefins are used in the preparation of acids, esters, amides, and other compounds:

$$CH \equiv CH + CO + ROH \longrightarrow CH_2 = CHCOOR \qquad (13.122)$$

$$CH \equiv CH + CO + HNR_2 \longrightarrow CH_2 = CHCONR_2 \qquad (13.123)$$

$$C_2H_4 + CO + H_2O \longrightarrow EtCOOH \qquad (13.124)$$

$$CH_2 = CH_2 + CO + EtCOOH \longrightarrow (EtCO)_2O \qquad (13.125)$$

$$CH_2 = CH_2 + CO + HCl \longrightarrow EtCOCl \qquad (13.126)$$

$$\text{C=C} + MeOH + CO \longrightarrow MeO - C - C - COOMe \qquad (13.127)$$

$$HC \equiv CH + CO + RSH \longrightarrow CH_2 = CHCOSR \qquad (13.128)$$

Cobalt, nickel, iron, ruthenium, and rhodium carbonyls as well as palladium complexes are catalysts for hydrocarboxylation reactions and therefore reactions of olefins and acetylenes with CO and water, and also other carbonylation reactions. Analogously to hydroformylation reactions, better catalytic properties are shown by metal hydrido carbonyls having strong acidic properties. As in hydroformylation reactions, phosphine–carbonyl complexes of these metals are particularly active. Solvents for such reactions are alcohols, ketones, esters, pyridine, and acidic aqueous solutions. Stoichiometric carbonylation reaction by means of $[Ni(CO)_4]$ proceeds at atmospheric pressure at 308–353 K. In the presence of catalytic amounts of nickel carbonyl, this reaction is carried out at 390–490 K and 3 MPa. In the case of carbonylation which utilizes catalytic amounts of cobalt carbonyl, higher temperatures (up to 530 K) and higher pressures (3–90 MPa) are applied. Alkoxylcarbonylation reactions generally proceed under more drastic conditions than corresponding hydrocarboxylation reactions.

Many valuable organic compounds are also prepared by cyclocarbonylation reactions of organic compounds. Amides of unsaturated acids react with CO to give imides; for example, the amide of acrylic acid gives the imide of succinic acid (succinimide); see equation (13.129).

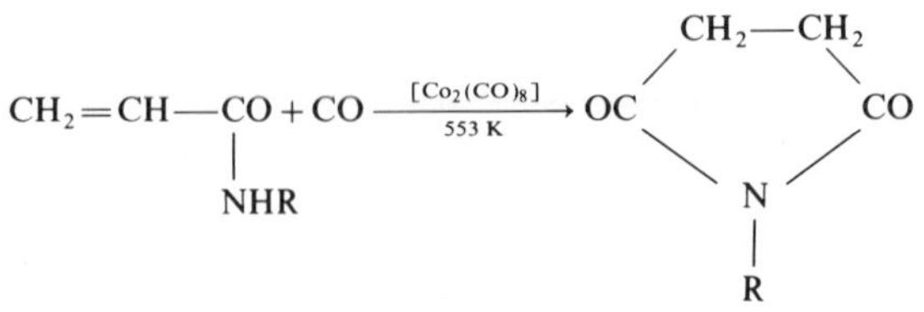

$$CH_2 = CH - CO + CO \xrightarrow[553\ K]{[Co_2(CO)_8]} OC \begin{matrix} CH_2 - CH_2 \\ \diagup \quad\quad \diagdown \\ \diagdown \quad\quad \diagup \\ N \end{matrix} CO \qquad (13.129)$$

$$| \atop NHR$$

$$| \atop R$$

Amides of cycloalkenyl acids react with CO to form bicyclic imides; see equation (13.130).

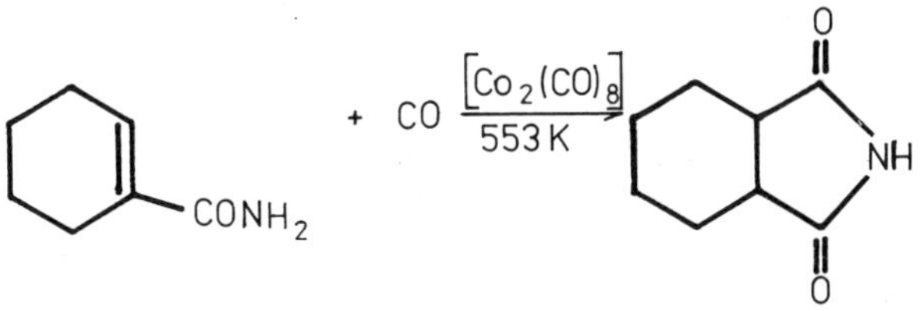

$$\text{[structure: cyclohexenyl-CONH}_2\text{]} + CO \xrightarrow[553\,K]{[Co_2(CO)_8]} \text{[bicyclic imide structure with NH]} \tag{13.130}$$

Unsaturated amines react with CO to afford cyclic five- and six-membered lactams. *N*-Alkylallyl amines give pyrrolidones:

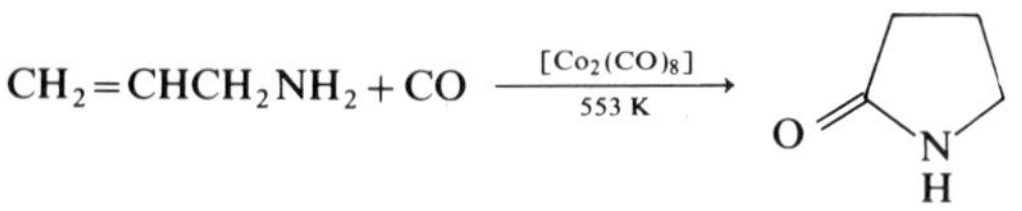

$$CH_2=CHCH_2NH_2 + CO \xrightarrow[553\ K]{[Co_2(CO)_8]} \text{[pyrrolidone, N-H]} \tag{13.131}$$

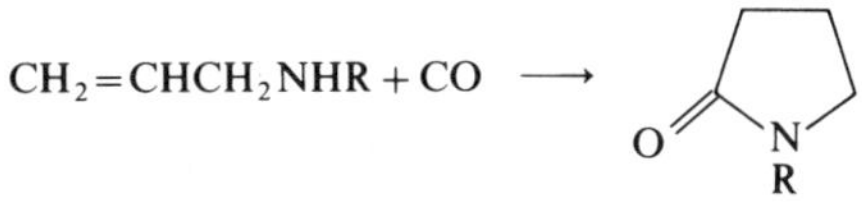

$$CH_2=CHCH_2NHR + CO \longrightarrow \text{[pyrrolidone, N-R]} \tag{13.132}$$

In addition to pyrrolidones, $\gamma$-alkylallylamines also form piperidones:

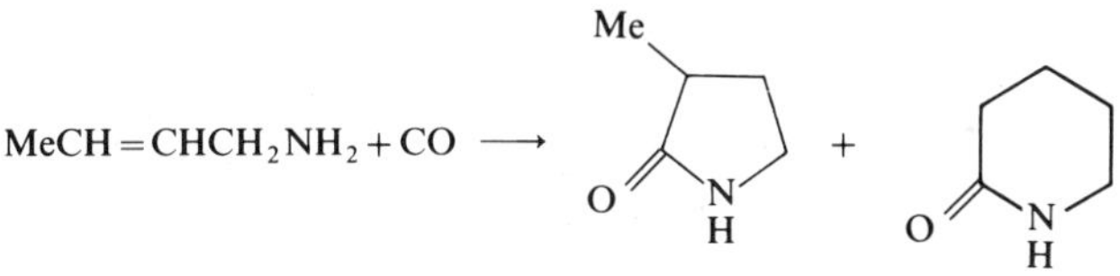

$$MeCH=CHCH_2NH_2 + CO \longrightarrow \text{[Me-pyrrolidone, N-H]} + \text{[piperidone, N-H]} \tag{13.133}$$

Unsaturated alcohols react with CO to give lactones:

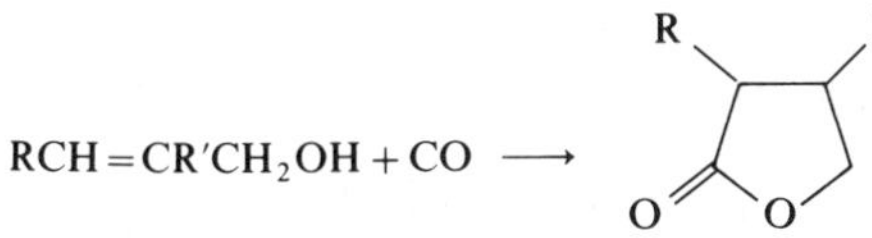

$$RCH=CR'CH_2OH + CO \longrightarrow \text{[lactone with R and R']} \tag{13.134}$$

Reactions of carbon monoxide and hydrogen with Schiff bases, aromatic nitriles, oximes, phenylhydrazones, semicarbazones, and azynes afford phthalimidines:

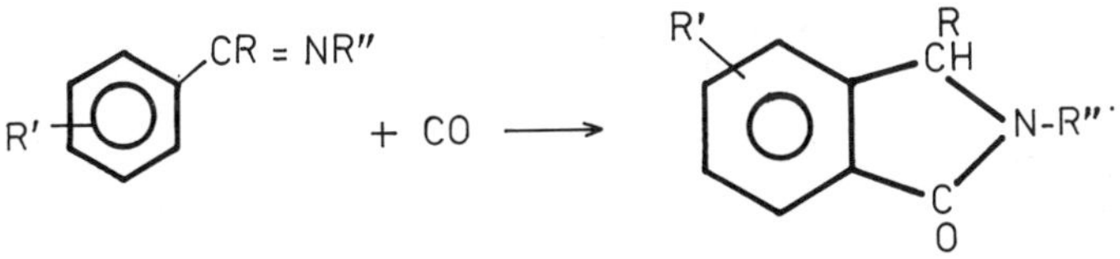

$$\text{[aryl CR=NR'', R']} + CO \longrightarrow \text{[phthalimidine structure with CHR, N-R'', C=O]} \tag{13.135}$$

$$(R'' = \text{alkyl, aryl}; R' = H, OH, OR, NR_2, Cl; R = H, Me, Ph)$$

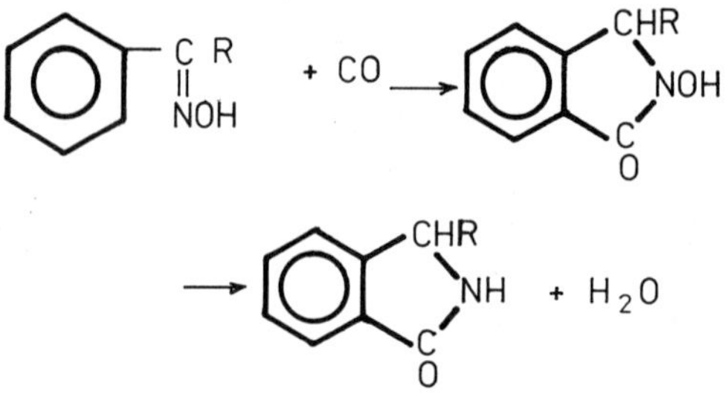

$$\tag{13.136}$$

In the case of phenylhydrazones, the reaction proceeds according to equation (13.137).

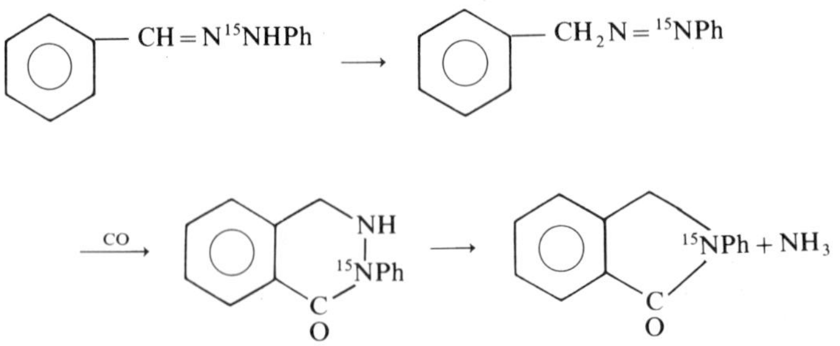

$$\tag{13.137}$$

The products of cyclocarbonylation find application in the pharmaceutical industry as well as in the chemical industry, for example, as solvents. The best catalysts for this reaction are cobalt carbonyls; however, carbonyls of rhodium and iron and palladium complexes have also been utilized. Sometimes nickel carbonyl may also be used, although it is less active in this type of reaction. Aromatic and aliphatic hydrocarbons as well as cyclic ethers are used as solvents. Reactions are carried out at 390–570 K and 10–30 MPa.

Carbonylation of methanol to acetic acid has considerable significance in industry.[123]

$$MeOH + CO \longrightarrow MeCOOH \tag{13.138}$$

At present, this process utilizes $Co_2(CO)_8$ as catalyst with $I_2$ as promoter, or $RhI_2(CO)_2^-$ as catalyst with MeI as promoter. Other rhodium carbonyls with the addition of iodine compounds may be utilized. Also, this reaction is catalyzed by nickel and iron carbonyls, although at considerably higher pressures and temperature. The process with the application of the cobalt catalyst is carried out at *ca* 460 K and 20 MPa, while in the presence of $RhI_2(CO)_2^-$, at 453 K and 3–4 MPa.

The mechanism of the reaction is as shown in equations (13.139) and (13.140). This reaction is also catalyzed by compounds of other metals of groups 8 and 9 such as ruthenium and iridium. Higher alcohols EtOH, $Pr^nOH$, $Pr^iOH$ also undergo carbonylation to give corresponding carboxylic acids.[125] However, the rate of the reaction is lower. It is assumed that in this case, the oxidative addition of alkyl iodide to the rhodium(I) complex proceeds according to a radical mechanism. Hydrocarboalkoxylation,[119] carbonylation of esters, reductive carbonylation of esters[126] and alcohols,[127] and carbonylation of formic aldehyde have also been intensively investigated.[128]

$$[RhI_2(CO)_2]^- + MeI \longrightarrow [Rh(Me)I_3(CO)_2]^- \xrightarrow{+CO} [Rh(COMe)I_3(CO)_2]^-$$

$$\longrightarrow [RhI_2(CO)_2]^- + MeCOI \qquad (13.139)$$

$$MeCOI + MeOH \longrightarrow MeCOOH + MeI \qquad (13.140)$$

$$RCH=CH_2 + CO + R'OH \longrightarrow RCH_2CH_2COOR'^{[119]} \qquad (13.141)$$

$$MeCOOMe + CO + 2H_2 \longrightarrow MeCOOEt + H_2O^{[126]} \qquad (13.142)$$

$$MeOH + CO + H_2 \xrightarrow[PPh_3]{CoI_2} MeCHO + H_2O^{[127]} \qquad (13.143)$$

$$H_2C=O + CO + 2H_2 + ROH \xrightarrow[14-24\ MPa]{470\ K} HOCH_2CH_2OR + H_2O^{[128]} \qquad (13.144)$$

Reaction (13.144) is catalyzed by $Cocp(CO)_2$ and $Co_2(CO)_8$, the latter being an effective catalyst in the presence of the following modifiers: $XPh_2$ ($X = O, S, Se$), $\eta^5\text{-}C_5R_5$ ligands ($R = H, Me$), and $GeR_3X$ ($R = Ph$, alkyl; $X = Cl, Br, H$). Hydroformylation of formaldehyde which leads to the formation of $HOCH_2CHO$ takes place under relatively mild conditions (383 K, $p(CO + H_2)(1:1) \approx 10\text{--}25$ MPa).[129] The main byproduct is methanol.

Many valuable products may be obtained by double carbonylation[130, 131] which transforms organic halides, amines, and CO into $\alpha$-ketoamides:

$$RX + 2HNR'_2 + 2CO \xrightarrow{PdL_n} RCOCONR'_2 + R'_2NH_2X \qquad (13.145)$$

The mechanism of this process is probably that shown in scheme (13.146).

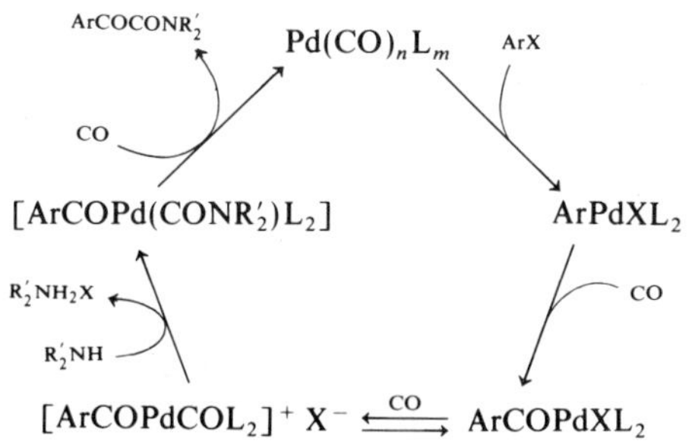

$$(13.146)$$

## c. *The stability of phosphine-containing catalysts for hydroformylation*

In spite of intensive studies of phosphine-containing catalysts for hydroformylation and their application in industry, publications concerning the stability of these catalysts under conditions of hydroformylation reactions have appeared only recently.[132–134] It was found that the phosphorus–carbon bond is relatively rapidly broken. The decomposition of triarylphosphines leads to the formation of alkyldiarylphosphines as well as aromatic hydrocarbons and their derivatives such as aldehydes and alcohols. Several possible compounds were proposed as intermediates during the scission of the $P-C$ bond which is catalyzed by transition metal complexes; see structures (13.147). The

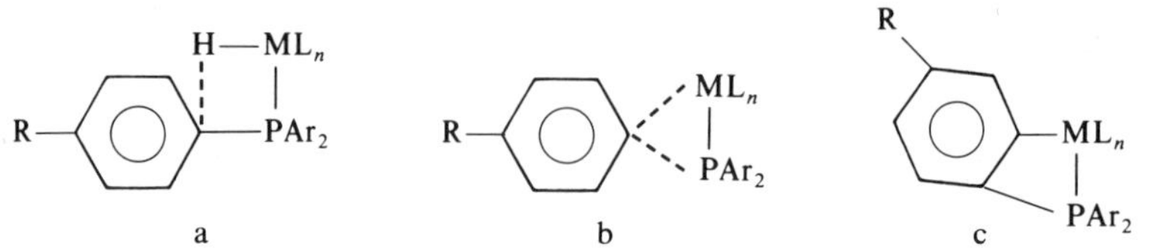

$$\text{(13.147)}$$

a          b          c

oxidative addition of the $P-C$ bond subsequently takes place. It may occur for both mono- and polynuclear complexes [equation (13.148)]. The activity of complexes in the

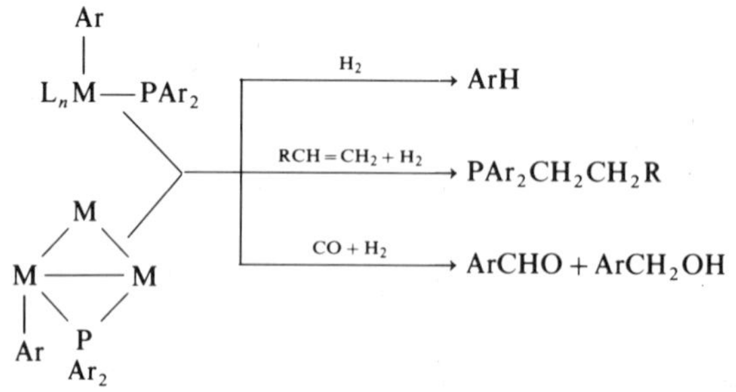

$$\text{(13.148)}$$

cleave reaction of the $P-C$ bond, depending upon the metal, varies as follows: $Rh > Ru > Co$, while the rate of the reaction for various phosphines of the type $PAr_3$ $(Ar = p\text{-}RC_6H_4)$ decreases in the series $R = CF_3 > Cl > H > Me > OMe$.

## 6. HYDROCYANATION

Hydrocyanation of olefins and acetylenes, i.e., the addition of hydrogen cyanide to olefins, is catalyzed by iron, cobalt, and nickel carbonyls, particularly in the presence of phosphines, as well as $[Ni\{R_2P(CH_2)_nPR_2\}_2]$, $[M\{P(OR)_3\}_4]$ ($M = Ni$, Pd), $[Ni(CN)_4]^{4-}$, $[Ni(C_2H_4)\{P(OC_6H_4Me\text{-}2)_3\}_2]$, CuCl, etc.[10,21,135]

$$\ce{C=C} + HCN \longrightarrow H-\ce{C-C}-CN \qquad \text{(13.149)}$$

Hydrogen cyanide may undergo oxidative addition to low-valent complexes or it may be coordinated as a Lewis base in the case of complexes of metals in higher oxidation states:

$$ML_n + HCN \longrightarrow L_nM\big\langle\begin{smallmatrix} H \\ CN \end{smallmatrix} \qquad \text{(13.150)}$$

$$ML_m + HCN \longrightarrow L_mMNCH \qquad \text{(13.151)}$$

Hydrocyanation proceeds via hydrido metal complexes as a result of the migration reaction of hydrogen to the olefin and reductive elimination of the nitrile[10, 21, 135, 136] [equation (13.152)].

$$[HM(CN)L_n] + CH_2 = CHR \longrightarrow \left[ \begin{array}{c} HM(CN)L_n \\ | \\ CH_2 = CHR \end{array} \right] \longrightarrow [RCH_2CH_2M(CN)L_n]$$

$$\longrightarrow ML_n + RCH_2CH_2CN \qquad (13.152)$$

Hydrocyanation of nonactivated olefins is catalyzed by metal carbonyls which include $[Co_2(CO)_8]$, $[Co_2(CO)_6(PPh_3)_2]$, $[Co_2(CO)_6\{P(OPh)_3\}_2]$, $[Mo(CO)_3\{P(OPh)_3\}_3] + TiCl_3$, and $[W(CO)_3\{P(OPh)_3\}_3] + BPh_3$. The following reactions may serve as examples:

$$C_2H_4 + HCN \longrightarrow C_2H_5CN \qquad (13.153)$$

$$RCH = CH_2 + HCN \longrightarrow RCH(CN)CH_3 \qquad (13.154)$$

$$CH_2 = CH(CH_2)_3CN \longrightarrow CH_3CH(CN)(CH_2)_3CN \qquad (13.155)$$

The addition of HCN to olefins in which the double bond is activated by substituents such as $-COOR$, aryl, $-NO_2$, $-CN$, and $-COR$ proceeds more easily than in the case of nonactivated olefins and is catalyzed by basic species, for example, $CN^-$. Metal carbonyls are also effective catalysts, for example, $[Co_2(CO)_8]$, $[Ni(CO)_4]$, and $[Ni(CO)_4] + PPh_3$.

$$CH_2 = CH - CH = CH_2 + HCN \longrightarrow CH_3CH = CHCH_2CN + CH_2CH(CH_2)_2CN \qquad (13.156)$$

$$CH_2 = C = CH_2 + HCN \longrightarrow CH_2 = CHCH_2CN + CH_2 = C(Me)CN + MeCH = CHCN \qquad (13.157)$$

Hydrocyanation of acetylene is catalyzed by an aqueous solution of copper(I) chloride and ammonium chloride; however, byproducts are also formed, viz., acetaldehyde and vinylacetylene, the latter arising by dimerization of acetylenes. Hydrocyanation, followed by reduction of alkynes leading to secondary nitriles, is catalyzed by $Co(CN)_5^{3-}$ in the atmosphere of $H_2$ or by $Ni(CN)_4^{2-}$ in the presence of $BH_4^-$ or cyanide.

## 7. HYDROSILATION

Hydrosilation denotes the process of addition of one or more $Si-H$ bonds to any other reagent.[10, 50, 137, 138] Most commonly utilized hydrosilating agents are $SiR_3H$, $SiHX_3$, and $SiHR_nX_{3-n}$ ($X =$ halogen, OR), and silated substrates are olefins, acetylenes, and ketones.

$$SiHR_3 + R'CH = CH_2 \longrightarrow Si(CH_2CH_2R')R_3$$

$$SiHR_3 + R_2'CO \longrightarrow R_2'CHOSiR_3 \qquad (13.158)$$

The mechanism of hydrosilation is similar to the previously discussed mechanisms of hydrogenation or hydroformylation reactions; see equations (13.159) and (13.160).

$$R_3SiH + Co_2(CO)_8 \longrightarrow CoH(CO)_4 + R_3SiCo(CO)_4 \tag{13.159}$$

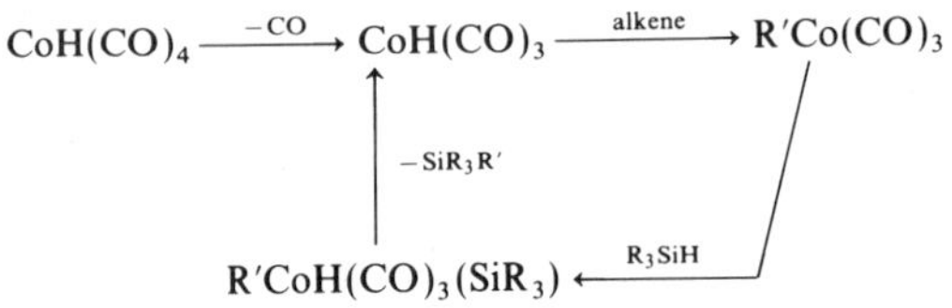

$$CoH(CO)_4 \xrightarrow{-CO} CoH(CO)_3 \xrightarrow{\text{alkene}} R'Co(CO)_3 \tag{13.160}$$

$$-SiR_3R' \uparrow \qquad \qquad \Bigg/$$

$$R'CoH(CO)_3(SiR_3) \xleftarrow{R_3SiH}$$

The synthesis of complexes of the type $[FeR(SiMe_3)(CO)_4]$ ($R = Me$, $CH_2Ph$) and the finding that they undergo reductive elimination to give $SiRMe_3$[139] serve as a strong argument for such a mechanism. Therefore, as in the case of hydroformylation, hydrocyanation, and hydrogenation, hydrido complexes of transition metals or complexes from which hydrides can be formed such as metal carbonyls, hydrido-carbonyls, phosphine complexes, etc., are catalysts for hydrosilation. Examples of such catalysts include $[Fe(CO)_5]$, $[Fe_2(CO)_9]$, $[Co_2(CO)_8]$, $[RhH(CO)(PPh_3)_3]$, $[RhCl(CO)(PPh_3)_2]$, and $[IrCl(CO)(PPh_2)_2]$. Tetracarbonylnickel catalyzes hydrosilation of olefins that are activated by electron-withdrawing substituents, for example, $CH_2=CHCN$, $CH_2=CHCOOR$, etc. Hydrosilation reactions usually take place at high pressures and higher temperatures. However, platinum group complexes (Pt, Rh, Ru, Ir, etc.) catalyze these reactions under mild conditions. Also, coordination compounds which do not contain carbonyl groups are utilized as catalysts; these include, especially, phosphine and halide complexes, for example, $H_2PtCl_6$, $PdCl_2$, $IrCl_3$, $H_2IrCl_6$, $RhCl_3$, $RuCl_3$, $[RuX_2(PR_3)_3]$, $[PtX_2(PR_3)_2]$, $[RhX(PR_3)_3]$, olefin complexes of these elements, etc.

Sometimes, during hydrosilation of olefins, alkenylsilanes are also formed in addition to alkylsilanes.[140]

$$SiEt_3H + Bu^nCH=CH_2 \longrightarrow SiEt_3(C_6H_{13}) + (E)\text{-}Et_3SiCH=CHC_4H_9 + (E)\text{-}Et_3SiCH_2CH=CHC_3H_7 \tag{13.161}$$

This reaction is catalyzed by rhodium complexes such as $[Rh_2(pmcp)_2X_4]$ ($X = Cl$, I), $[Rh(pmcp)H_2(SiEt_3)_2]$, $[Rh(pmcp)(C_2H_4)_2]$, $[Rh_2Cl_2(CO)_4]$, $[RhCl(PPh_3)_3]$, and $Rh(acac)_3$. The yield of alkenylsilanes increases as the ratio of $SiEt_3H$/olefin decreases.

## 8. METATHESIS

### a. Olefin metathesis

Olefin metathesis represents a reaction during which cleavage of two carbon–carbon double bonds occurs and the formation of two new $C=C$ bonds results.

Therefore, olefin metathesis leads to the exchange of alkylidene groups of two alkene molecules[8–10, 22, 87, 141–145] [equation (13.162)]. This process was first called dispro-portionation of olefins. However, the term metathesis (from Greek "metathesis"—transposition), which is used in chemistry to designate the exchange of atoms or groups

$$
\begin{array}{ccc}
RCH=CHR' & & RCH \quad CHR' \\
+ & \overset{cat.}{\rightleftarrows} & \| \quad + \quad \| \\
RCH=CHR' & & RCH \quad CHR'
\end{array}
\tag{13.162}
$$

between two molecules without their structural changes, is a more appropriate and unequivocal definition. Some authors describe this process by the term dismutation, which is a Latin equivalent for metathesis.

Olefin metathesis is a general reaction which takes place for a great number of olefins and is catalyzed by many complexes and metals. The most active homogeneous catalysts lead to the conversion of $10^4$ moles of olefin per mole of the complex (to the equilibrium mixture of the products) within several seconds at 298 K. The activation energy for very active homogeneous catalysts is quite low; it equals only 27–30 kJ mol$^{-1}$. The energy of activation for the remaining catalysts varies within the broader range of 30–90 kJ mol$^{-1}$. The metathesis reaction of propylene gives 2-butene and ethylene. In the case of heavier olefins, additional compounds are formed which prove that the metathesis reaction between the product obtained from the primary (at least two) substrates takes place. They also prove that isomerization of olefins occurs.

The most important properties of the metathesis reaction are as follows:

1. thermoneutrality,
2. low activation energy,
3. high reaction rate,
4. essentially first order of the reaction with respect to both the complex and the olefin.

Metathesis of acyclic olefins is thermoneutral, i.e., the enthalpy of the reaction is approximately equal to zero because double bonds are broken and reformed in the reaction. Therefore, alkene metathesis leads to an equilibrium which is determined by the entropy of the reaction. Metathesis of cyclic olefins proceeds differently; it leads to the polymerization of cyclic olefins owing to ring opening (however, such polymerization does not occur for cyclohexene); see equation (13.163). Although the number and type

$$
(CH_2)_n \overset{CH}{\underset{CH}{\|}} + \overset{CH}{\underset{CH}{\|}} (CH_2)_n \rightleftarrows (CH_2)_n \overset{CH=CH}{\underset{CH=CH}{}} (CH_2)_n
\tag{13.163}
$$

$$(n = 2, 3, 5, 6,...)$$

of bonds do not change, the enthalpy of the reaction is negative ($\Delta H < 0$) because the reaction products are characterized by smaller ring strains. For four- and five-membered rings, $\Delta H < -20$ kJ mol$^{-1}$, while for the remaining rings, $\Delta H$ assumes values in the $-20$ to $-12$ kJ mol$^{-1}$ range. The entropy changes are more complicated in metathesis reactions of cycloolefins. Because the number of molecules decreases, the entropy becomes lower, and at the same time it increases owing to the increase in the entropy of oscillation and bending of the polymer molecules. The latter factor is greater than the previous effect for larger rings, namely $> C_7$. Thus, the metathesis of cyclopentene occurs because the reaction is exothermic, although the entropy change for this process is negative. Cyclohexene does not undergo polymerization because $\Delta H = 0$ and $\Delta S < 0$. For cycloolefins containing more than seven carbon atoms in the ring, both the enthalpy and entropy changes favor polymerization caused by the dismutation. The metathesis of linear olefins always leads to thermodynamic equilibrium between the *cis* and *trans* isomers. Polymerization which is caused by ring opening of some cycloolefins leads to the formation of polymers characterized by a high degree of stereospecificity. In the presence of $MoCl_5 + AlEt_3$ (without the solvent) cyclopentene gives polypentamer which contains more than 99% of *cis* double bonds, while in the presence of $WCl_6 + CHOH(CH_2Cl)_2 + Al(i\text{-}Bu)_3$ it contains 80–85% of *trans* $C=C$ bonds.

Heterogeneous catalysts for olefin metathesis, like homogeneous catalysts, include molybdenum and tungsten compounds as well as some rhenium derivatives; compounds of other elements catalyze the dismutation of alkenes rather poorly. Tungsten homogeneous catalysts are more active than molybdenum ones, contrary to their heterogeneous catalysts. Generally, heterogeneous catalysts are more active under more drastic conditions (higher temperatures and pressures) compared to homogeneous catalysts.

The following metal compounds are used for the preparation of the catalysts: oxides, metal carbonyls, halides, alkyl and allyl complexes, as well as molybdenum, tungsten, and rhenium sulfides. Oxides of iridium, osmium, ruthenium, rhodium, niobium, tantalum, lanthanum, tellurium, and tin are effective promoters, although their catalytic activity is considerably lower. Oxides of aluminum, silicon, titanium, manganese, zirconium as well as silicates and phosphates of these elements are utilized as supports. Also, mixtures of oxides are used. The best supports are those of alumina oxide and silica.

Aluminum oxide itself catalyzes the metathesis reaction of propylene, although its activity is low. Catalysts that are obtained from oxides are active at temperatures *ca* 100 K higher than catalysts prepared from metal carbonyls. Compounds deposited on silica are catalytically active at temperatures which are higher by *ca* 200 K than compounds supported on aluminum oxide. However, because of the $SiO_2$ support these catalysts are more resistant to poisoning by polar compounds. Heterogeneous catalysts must be activated at 390–870 K before use. Catalysts obtained from metal carbonyls and alkyl compounds require lower temperatures of activation. Supported allyl complexes of Mo, W, and Re need no activation.

The ESCA spectra show that the active center in catalysts that are obtained from molybdenum and tungsten carbonyls do not contain coordinated CO groups and the oxidation state of the metal is higher than zero but lower than six. Transition metal atoms are bonded to the electron-accepting centers. Magnetic measurements and ESR spectra of catalysts containing molybdenum oxide with the addition of $TiO_2$ as the promoter show that the activity is proportional to the Mo(V) content and that the active center contains titanium–molybdenum complexes probably possessing oxygen bridges:

Homogeneous catalysts are mainly prepared from metal carbonyls, halides, halide carbonyl complexes, halide coordination compounds with phosphine, nitrosyl, arene, and pyridine ligands, as well as from carbonyl arene compounds of tungsten, molybdenum, rhenium, and tantalum. Catalytic activity of compounds of other metals of groups 4–10 has also been investigated; however, it is considerably lower than that for the above-mentioned complexes. Compounds of copper group metals and those of some lanthanides and actinides also exhibit low activity. Transition metal compounds catalyze the metathesis reaction in the presence of organometallic compounds of elements of groups 1–4, their hydrido compounds, or, in certain cases, their halides. The most commonly utilized compounds include $[Al_2Cl_xR_{6-x}]$ $(x = 1-6)$, LiR, MgXR, $Li[AlH_4]$, $Na[BH_4]$, and $SnR_4$. The following solvents are usually employed: chlorobenzene, carbon tetrachloride, chloroform, benzene, toluene, saturated hydrocarbons, ethyl ether, and acetonitrile. Often, the addition of a certain amount of alcohol, phenol, acetic acid, water, or other compounds containing the hydroxyl group as well as oxygen in the amount of up to several moles per mole of the transition metal complex causes a considerable increase in activity and stability of the catalytic system (Table 13.5).

## b. *Mechanism of the metathesis reaction*

It was first assumed that the olefin metathesis proceeds by concerted (synchronous) breaking of the original C=C bonds and the formation of new C=C bonds. This reaction is forbidden by symmetry rules; however, it is allowed in the presence of catalysts [equation (13.164)]. Another, pairwise (involving two alkene molecules), nonconcerted

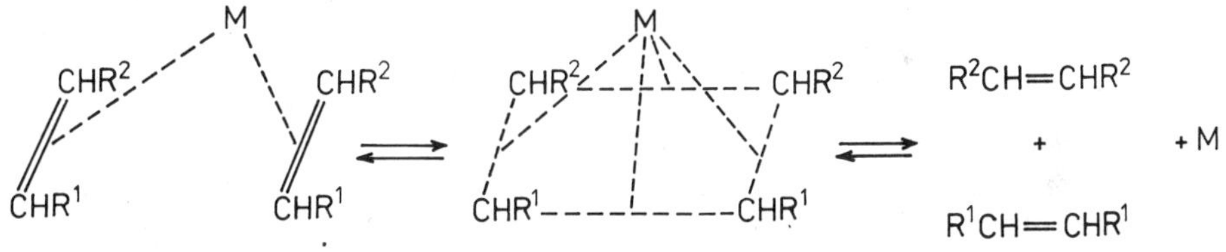

$$(13.164)$$

mechanism, in which the formation of the metalloacyclopentane ring was assumed, was also proposed [equation (13.165)]. At present, it is assumed that the metathesis reaction is catalyzed by carbene complexes and also that metallacyclobutane compounds are

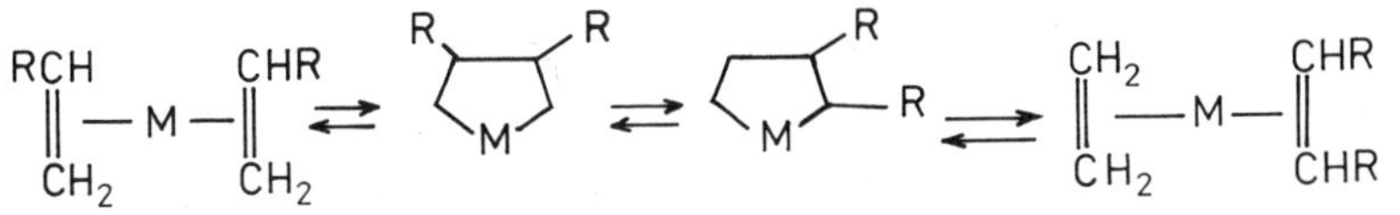

$$(13.165)$$

formed as intermediates. Thus, this process involves an odd number (3) of carbon atoms; see equations (13.166) and (13.167).

In order to explain reactions of olefins with carbene complexes, the fomation of metallacyclobutane compounds has been postulated [equations (5.127)–(5.131)]. The validity of the carbene mechanism is supported by metathesis reactions of 1,7-octadiene

Table 13.5. Catalysts for Metathesis Reactions[22,141-148]

| Catalyst (molar or percentage ratios) | Solvent | Substrates | Reaction rate | Pressure (MPa) | Temperature (K) |
|---|---|---|---|---|---|
| MoO₃ + CoO on Al₂O₃ (12–14%):(2.5–3.5%) | | HMeC=CH₂, R₂C=CR'₂ | | 3.2 | <473 |
| MoO₃ + Cr₂O₃ on Al₂O₃ | | | | 0.7 | |
| Mo(CO)₆ on Al₂O₃ | | HMeC=CH₂, R₂C=CR'₂ | | 3.5 | 300–410 |
| MoO₃ + TiO₂ on Al₂O₃ | | | | | Low |
| MoO₃ on SiO₂ | | R₂C=CH₂, R₂C=CR'₂, RC≡CR' | | 3.2 | 670–800 |
| WO₃ on SiO₂ | | R₂C=CH₂, R₂C=CR'₂, RC≡CR' | | | 670–800 |
| Re₂O₇ on Al₂O₃ (<20%) | | R₂C=CH₂, R₂C=CR'₂ | | 0.1 | 300–360 |
| [Re₂(CO)₁₀] on Al₂O₃ | | | | 0.1 | 370–420 |
| Re₂O₇ on Al₂O₃ + SnMe₄ᵃ | | CH₂=CH(CH₂)₂COOMe | | | 323 |
| [WMe₄{ON(Me)NO}₂] on Al₂O₃ᵇ | | MeCH=CH₂ | | | 300 |
| WCl₆ + EtOH + AlEtCl₂ (1:1:4) | PhH | R₂C=CR'₂ | Very high | 0.1 | 290–323 |
| WCl₆ + AlEt₃ + O₂ (2:1) | PhCl | | High | | |
| WCl₆ + Li(Bu-n) + O₂ (1:2) | PhH | R₂C=CR'₂ | Low | | 290–323 |
| WCl₆ + LiAlH₄ (1:1) | PhCl | | High | | |
| WCl₄py₂ + AlEtCl₂ + CO (1:8) | PhH + pentane | | High | 0.1 | |
| [W(CO)₆] + hv | CCl₄ | R₂C=CR'₂ | Low | | 353 |

| Catalyst system | Solvent | Substrate | Activity | | T (K) |
|---|---|---|---|---|---|
| trans-WX(CO)$_4$(CR) + TiCl$_4$ [c] | | Cyclic olefins | High | | 273 |
| [W(CO)$_5$(CPh$_2$)] [d] | | | | | |
| WCl$_6$ + SiEt$_3$H [k] | | $R_2C=CH_2$, $R_2C=CR_2$ | | | 290–323 |
| [W(OMe)$_6$ + AlEtCl$_2$] [e] | | $R_2C=CR_2'$ | | | |
| [WCl$_2$(CO)$_3$(PPh$_3$)$_2$] + AlEtCl$_2$ [f] | PhCl | $R_2C=CH_2$, $R_2C=CR_2'$ | | 0.1 | 273 |
| [MoCl$_2$(NO)$_2$(PPh$_3$)$_2$] [f] | | $R_2C=CH_2$ | | 0.1 | 323–343 |
| [Mo(CO)$_5$py] + AlEtCl$_2$ + [NR$_4$]Cl [g] | | | Very high | 0.1 | 298 |
| [MoCl$_3$(NO)] + AlEtCl$_2$ + CO [h] | | $R_2C=CR_2'$ | Low | 0.1 | 290–330 |
| MoCl$_5$ + AlEt$_3$ + O$_2$ (1:2) | PhCl | | High | 0.1 | 290–330 |
| ReCl$_5$ + AlEt$_3$ + O$_2$ (1:4) | PhCl | $R_2C=CH_2$, $R_2C=CR_2'$ | High | 0.1 | 363 |
| [ReCl(CO)$_5$] + AlEtCl$_2$ (1:8) | | | Low | | |
| ReCl$_5$ + Sn(Bu-$n$)$_4$ (1:2) | PhCl | | Low | | |
| [Mo(CO)$_6$] + $m$-C$_6$H$_4$(OH)$_2$ (1:6) [i] | Decaline | $RC \equiv CR'$ | Low | | 433 |

[a] E. Verkuijlen, F. Kapteijn, J. C. Mol, and C. Boelhouwer, *Chem. Commun.*, 198 (1977).
[b] W. Mowat, J. Smith, and D. A. Whan, *Chem. Commun.*, 34 (1974).
[c] E. O. Fischer and W. R. Wagner, *J. Organomet. Chem.*, **116**, C21 (1976).
[d] T. J. Katz, S. J. Lee, and N. Acton, *Tetrahedron Lett.*, 4247 (1976).
[e] M. T. Mocella, M. A. Busch, and E. L. Muetterties, *J. Am. Chem. Soc.*, **98**, 1283 (1976).
[f] L. Bencze and L. Marko, *J. Organomet. Chem.*, **28**, 271 (1971); L. Bencze, J. Pallos, and L. Marko, *J. Mol. Catal.*, **2**, 139 (1977).
[g] M. F. Farona and V. W. Motz, *Chem. Commun.*, 930 (1976); V. Motz and M. F. Farona, *Inorg. Chem.*, **16**, 2545 (1977).
[h] R. Taube, K. Seyferth, L. Bencze, and L. Marko, *J. Organomet. Chem.*, **111**, 215 (1976); R. Taube and K. Seyferth, *Z. Chem.*, **14**, 284 (1974).
[i] A. Mortreux and M. Blanchard, *Chem. Commun.*, 786 (1974).
[j] W. B. Hughes, *J. Am. Chem. Soc.*, **92**, 532 (1970); W. B. Hughes, *Adv. Chem. Ser.*, **112**, 192 (1974); T. J. Katz and J. McGinnis, *J. Am. Chem. Soc.*, **99**, 1903 (1977).
[k] N. S. Nametkin, V. M. Vdovin, E. D. Babich, V. N. Karelskii, and B. W. Kacharmin, *Dokl. Akad. Nauk SSSR*, **213**, 356 (1973).

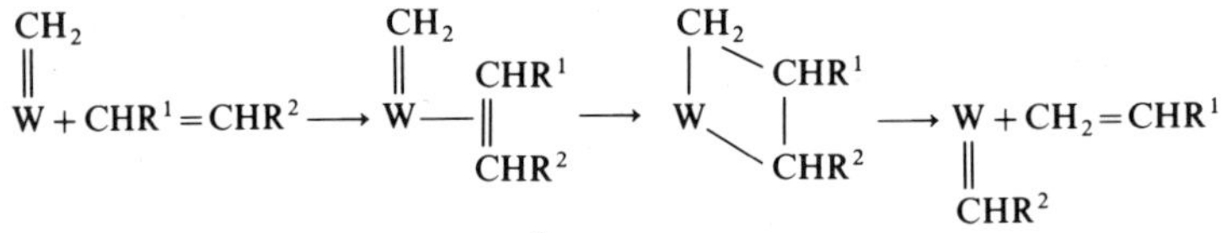

$$\begin{array}{c}
\text{CH}_2\\
\parallel\\
\text{W} + \text{CHR}^1 = \text{CHR}^2
\end{array}
\longrightarrow
\begin{array}{c}
\text{CH}_2\\
\parallel \quad \text{CHR}^1\\
\text{W} - \parallel\\
\text{CHR}^2
\end{array}
\longrightarrow
\begin{array}{c}
\text{CH}_2 \quad \text{CHR}^1\\
\text{W} \mid\\
\quad \text{CHR}^2
\end{array}
\longrightarrow
\text{W} + \begin{array}{c}\text{CH}_2 = \text{CHR}^1\\ \parallel\\ \text{CHR}^2\end{array}
\qquad (13.166)$$

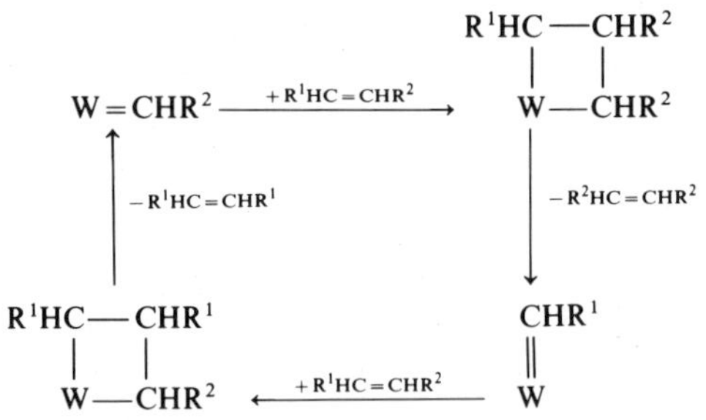

$$\begin{array}{ccc}
 & & \text{R}^1\text{HC} - \text{CHR}^2\\
 & & \mid \quad\quad \mid\\
\text{W} = \text{CHR}^2 & \xrightarrow{\;+\text{R}^1\text{HC}=\text{CHR}^2\;} & \text{W} - \text{CHR}^2\\
\uparrow & & \\
\big|\;-\text{R}^1\text{HC}=\text{CHR}^1 & & \big|\;-\text{R}^2\text{HC}=\text{CHR}^2 \qquad\qquad (13.167)\\
\big| & & \downarrow\\
\text{R}^1\text{HC} - \text{CHR}^1 & & \text{CHR}^1\\
\mid\quad\quad\mid & & \parallel\\
\text{W} - \text{CHR}^2 & \xleftarrow{\;+\text{R}^1\text{HC}=\text{CHR}^2\;} & \text{W}
\end{array}$$

[equation (13.168)] as well as those of a mixture of *trans*-2-butene, *trans*-4-octene, and cyclooctene [equation (13.169)]. In the case of 1,7-octadiene, it was experimentally found that the ratio of the ethylene $d_0:d_2:d_4$ is equal to $1:2:1$ and is in agreement with the ratio predicted by the carbene mechanism. For the "even" mechanism, this ratio should equal $1:1.6:1$.

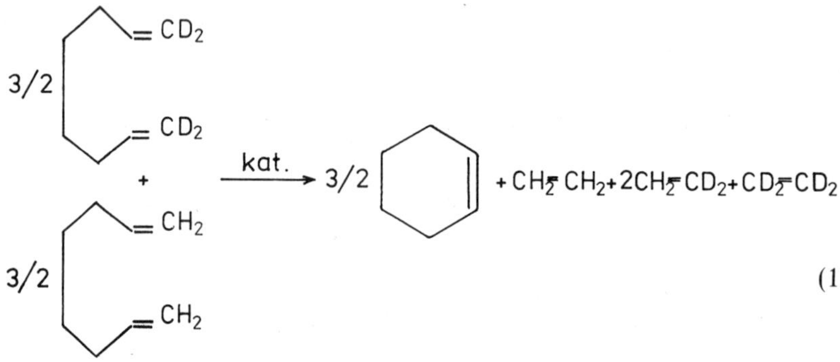

$$\begin{array}{c}
3/2 \;\{\; =\text{CD}_2,\; =\text{CD}_2 \;\}\\
+\\
3/2 \;\{\; =\text{CH}_2,\; =\text{CH}_2 \;\}
\end{array}
\xrightarrow{\text{kat.}}
3/2\; \bigcirc\!\!= \;+\, \text{CH}_2\!\!=\!\!\text{CH}_2 + 2\text{CH}_2\!\!=\!\!\text{CD}_2 + \text{CD}_2\!\!=\!\!\text{CD}_2$$

$$(13.168)$$

In the second case, after a short reaction period, a considerable amount of alkadiene $C_{14}$ was found; this is the product of cross metathesis resulting from the dismutation reaction between the products and substrates. Extrapolation of the reaction time to zero gives the ratios $C_{14}/C_{12} = 1.3$ and $C_{14}/C_{16} = 3.3$. These data are best explained by the carbene mechanism.

$$\bigcirc\!\!= \;+\; \underset{C_4}{\text{MeCH}=\text{CHMe}} + \underset{C_8}{\text{PrCH}=\text{CHPr}} \longrightarrow \underset{C_{12}}{\left(\begin{array}{c}\text{CHMe}\\ \text{CHMe}\end{array}\right)}$$

$$+\; \underset{C_{14}}{\left(\begin{array}{c}\text{CHMe}\\ \text{CHPr}\end{array}\right)} +\; \underset{C_{16}}{\left(\begin{array}{c}\text{CHPr}\\ \text{CHPr}\end{array}\right)} +\; \underset{C_6}{\text{MeCH}=\text{CHPr}}$$

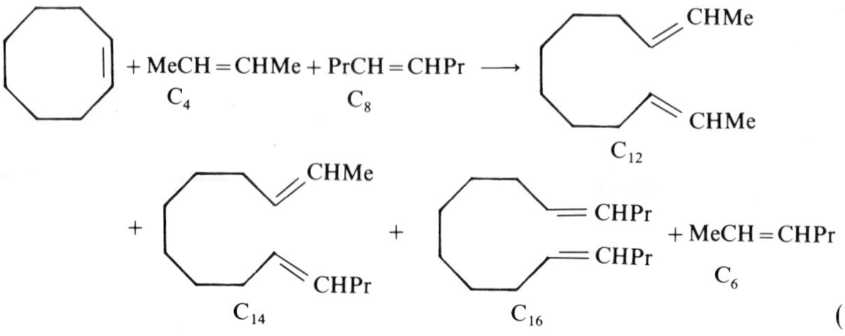

$$(13.169)$$

The mechanism of polymerization of cycloolefins may be represented by equation (13.170).

$$R_2C=M + \overset{C=C}{\underset{\smile}{\phantom{x}}} \longrightarrow \begin{array}{c} R_2C-M \\ | \quad | \\ C-C \\ \underset{\smile}{\phantom{x}} \end{array} \xrightarrow{C=C} \begin{array}{c} \overset{\frown}{C=CR_2} \\ C-M \\ | \quad | \\ C-C \\ \underset{\smile}{\phantom{x}} \end{array} \longrightarrow \begin{array}{c} \overset{C=CR_2}{\underset{|}{}} \\ C \quad M \\ || \quad || \\ C \quad C \\ \underset{\smile}{\phantom{x}} \end{array} \longrightarrow \begin{array}{c} \overset{\frown}{C=C} \\ C=C \\ \underset{\smile}{\phantom{x}} \end{array} + R_2C=M \text{ etc.}$$

$$(13.170)$$

*Ab initio* calculations by Rappe and Goddard show that the active catalysts for the metathesis reaction in the case of Mo, W, and Re compounds containing these metals in high oxidation states are oxo-alkylidene complexes.[149] The oxygen ligand stabilizes the intermediate metallacyclobutane compound. The addition of oxygen indeed increases the activity of many catalysts. According to Ivin,[149] this increased activity is due to the formation of a $\overline{\text{M-OCHRCHR}^1}$ ring whose scission leads to the formation of a carbene complex which initiates the metathesis reaction.

Owing to the nucleophilic character of the carbene ligands which do not contain heteroatoms [equations (5.112) and (5.113)] it is possible to form complexes containing a carbene bonded both to a transition metal as well as to a main group metal. If the carbene ligand is electrophilic, the formation of the $\overline{\text{M-C}_{\text{carb.}}\text{X-M}'}$ grouping is possible. Many authors believe that such complexes are formed during the metathesis reaction; see equations (13.171) and (13.172). Carbene complexes in a catalytic system may be

$$\begin{array}{cc} \text{M}-\text{CH}_2 & \text{M}-\text{CH}_2 \\ | \quad\quad | & | \quad\quad | \\ \text{X}-\text{AlX}_n & \text{X} \quad\quad \text{X} \\ & \quad \backslash \;\; / \\ & \quad \text{AlX}_m \end{array} \qquad (13.171)$$

$$\begin{array}{l} \text{M}-\text{CH}_2 \\ | \quad\quad | \;\; + \text{R}^1\text{CH}=\text{CHR}^2 \rightleftharpoons \text{X}\begin{array}{c} \diagup \text{M}-\text{CH}_2 \diagdown \\ \diagdown \quad\quad\quad \diagup \\ \text{M}'-\text{CHR}^2 \end{array} \text{CHR}^1 \rightleftharpoons \text{X}\begin{array}{c} \diagup \text{M} \diagdown \\ \diagdown \quad \diagup \\ \text{M}' \end{array} \text{CHR}^2 + \text{R}^1\text{CH}=\text{CH}_2 \\ \text{X}-\text{M}' \end{array}$$

$$(13.172)$$

formed as a result of eliminating the hydrogen atom in α-carbyl complexes. These are formed by the reaction of the transition metal compound with the organometallic derivative of a main group element or by migration of a M−H hydrogen atom (or the halogen atom) to the olefin molecule coordinated to this metal. In the case of carbonyl complexes, acyl ligand formation is possible from which carbene derivatives may be obtained by protonation of the oxygen [equation (13.173)].

$$\begin{array}{c} \text{L}_n\text{M}-\text{M}+\text{CH}_2=\text{CR}_2 \longrightarrow \text{L}_m\text{M}-\text{H} \longrightarrow \text{L}_n\text{M}-\text{CH}_2\text{CHR}_2 \\ \quad\quad\quad \diagdown \quad\quad\quad\quad\quad\quad\quad | \\ \quad\quad \text{CH}_2=\text{CR}_2 \quad\quad\quad\quad | \\ \quad\quad\quad\quad\quad\quad\quad\quad \text{H} \quad\quad | \quad\quad (13.173) \\ \quad\quad\quad\quad\quad\quad\quad\quad | \quad\quad \downarrow \\ \quad\quad\quad\quad\quad\quad \text{L}_x\text{M}=\text{CHCHR}_2 \end{array}$$

The first industrial metathesis of propylene was introduced by Philips Petroleum in 1966 (two years after the discovery of the metathesis reaction) and termed the "triolefin process" because ethylene and 2-butene are obtained from propylene. Ethylene may be used in other processes without additional purification, while butene is utilized in the preparation of butadiene. Olefin metathesis in conjunction with isomerization of olefins which possess an internal $C=C$ double bond leading to 1-alkenes allows the preparation of higher hydrocarbon alkenes. Isomerization of 2-butene to 1-butene followed by the dismutation reaction of the latter affords 3-hexene, etc. This process utilizes the heterogeneous catalysts $MoO_3 + Al_2O_3$, $CoO + MoO_3 + Al_2O_3$, $MoO_3 + Cr_2O_3$, $WO_3 + SiO_2$, and $Re_2O_7 + Al_2O_3$. Also, simultaneous metathesis of 2-butene and isobutene in the presence of $WO_3 + SiO_2$ leading to isopentenes has been utilized [equation (13.174)].

$$MeCH = CHMe$$
$$+ \quad \longrightarrow MeCH = CH_2 + MeCH = CMe_2 \qquad (13.174)$$
$$CH_2 = CMe_2$$

As a result of the relatively high cost of cycloolefins, large-scale metathesis leading to polymers is not utilized for these alkenes. Philips Petroleum also applies metathesis of $Me_2C = CHCMe_3$ with an excess of ethylene to synthesize $CH_2 = CHCMe_3$, which is employed in the perfume industry.

The synthesis (13.175) of styrene from toluene and ethylene seems to be a promising process.

$$PhMe \xrightarrow{\text{[O]}} PhCH = CHPh$$
$$\xrightarrow{CH_2 = CH_2} PhCH = CH_2 \qquad (13.175)$$

The process in which ethylene oligomerization is carried out first, followed by metathesis of the oligomerization products, has also been utilized. By this method, $C_{10}$–$C_{20}$ olefins are obtained and are further used for the synthesis of detergents. Because of relatively high prices of cyclic olefins, polymers from them are not obtained by metathesis. More useful is the case of cyclopentene whose price is comparable to that of butadiene. In addition, during polymerization, cyclopentene gives exclusively *trans*-polypentenamer, which may be utilized as a valuable elastomer for the preparation of synthetic rubber. The molecular weight of polyalkenamers may easily be controlled by the addition of lighter olefins. The reaction of the polymer with a stoichiometric quantity of lighter alkenes leads to complete depolymerization. This reaction (13.176) has been exploited for analytical and synthetic purposes.

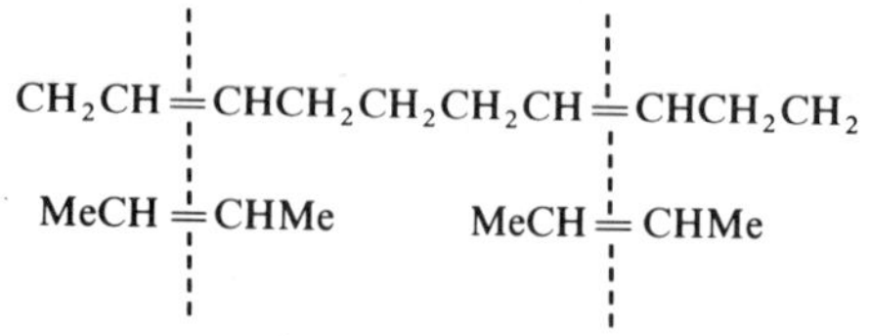

$$CH_2CH \doteq CHCH_2CH_2CH_2CH \doteq CHCH_2CH_2$$
$$MeCH \doteq CHMe \qquad MeCH \doteq CHMe \qquad (13.176)$$

## c. *Metathesis of functionalized olefins*

Metathesis of olefins containing functional groups leads to synthesis of many valuable compounds. However, such olefins undergo metathesis with difficulty. Therefore, examples of metathesis of functionalized olefins are scarce. This is due to the following reasons: (1) functional groups may react with the catalyst; in some cases, they may cause its decomposition; and (2) competitive complexation of functional groups rather than $C=C$ bonds may take place.

The metathesis reaction of alkyl esters of unsaturated carboxylic acids and of alkenyl acetates is catalyzed by the system $WCl_6/SnMe_4$. However, systems containing $SnR_4$ where $R = Et$, Bu, and Ph are not active. Catalysts which are obtained by reactions of $WCl_6$, $MoCl_5$, $Mo(OEt)_2Cl_3$, or $W(CO)_6$ with $Al_2Me_3Cl_3$ cause metathesis of esters of fatty acids as well as metathesis of nitriles, ketones, ethers, amides, and oxosilanes containing alkenyl groups. Unsaturated ammonium salts also undergo metathesis which is catalyzed by $[W(CO)_3(\text{mesitylene})] + AlEtCl_2 + O_2$ and $[MoCl_2(NO)_2(PPh_3)] + AlEtCl_2$. Esters of unsaturated carboxylic acids also undergo metathesis in the presence of the heterogeneous rhenium catalyst $Re_2O_7/SnMe_4$. The compounds $RN=C=NR$ and $PhN=NPh$ undergo metathesis as well [see reactions (5.161) and (5.162)].

## d. *Metathesis of alkynes*

Metathesis of alkynes proceeds according to a mechanism (13.177) analogous to that already described for olefins. The formation of catalytically active metallacyclobutadiene complexes was confirmed by both kinetic and isotopic studies of

$$
\begin{array}{cccc}
M\equiv CR & M=CR & M-CR & M\equiv CR' \\
| & | \quad | & \| \quad \| & | \\
R'C\equiv CR'' & R'C=CR'' & R'C-CR'' & RC\equiv CR''
\end{array} \tag{13.177}
$$

reactions of $[W(CBu')(NPr^i_2)_3]$ and $[W(CBu')(OR)_3]$ with alkynes as well as by X-ray studies.[150–153] The complexes $[W(CBu')(OBu')_3]$ and $[W(CR)(OBu')_3]$ $(R = Me, Et, \text{etc.})$ metathesize dialkylacetylenes. Metallacyclobutadiene complexes $[W(C_3Et_3)(OC_6H_3Pr^i_2\text{-}2,6)_3]$ and $[W(C_3Et_3)\{OCH(CF_3)_2\}_3]$ possessing a planar ring WCEtCEtCEt also catalyze metathesis of acetylenes.[150, 152] Additionally, metathesis of acetylene is effectively catalyzed by some molybdenum complexes in the presence of phenol, for instance, $[MoO_2(acac)_2] + AlEt_3 + PhOH$.[154]

## 9. *DISPROPORTIONATION REACTIONS*

Olefins,[155–157] including acrylonitrile,[158] may undergo disproportionation reaction, a process in which the oxidation state of carbon changes. Ethylene undergoes disproportionation to ethane and butadiene under relatively mild conditions in the presence of titanium cyclopentadienyl complexes such as $[cp_2Ti(\mu\text{-}\eta^1:\eta^5\text{-}C_5H_4)Ticp]$ and $[cpTi(\mu\text{-}C_5H_4)_2(\mu\text{-}H)_2Ticp]$ (see titanium cyclopentadienyl complexes). During passage of the mixture of ethylene and benzene over heated molybdenum, tungsten, and rhenium oxides, propene and 1,3-butadiene are formed, while cycloheptatriene undergoes disproportionation to methane, ethylene, propene, 1,3-butadiene, and benzene[156];

see equation (13.178). Platinum(II) complexes of the type $[PtCl_2(C_2H_4)L]$ $(L = PPh_3,\ Et_2NH)$, and ruthenium compounds such as $RuCl_3 \cdot 3H_2O$, $[Ru_3O(OOCR)_6(H_2O)_3]OOCR$ $(R = Me,\ Et)$, and $RuCl_2$ catalyze disproportionation of acrylnitrile [equation (13.179)]. Rhodium(III) and iridium(III) pentamethylcyclopentadienyl complexes catalyze disproportionation of acetaldehyde to ethanol and acetic acid.[159]

$$3C_2H_4 \xrightarrow[\text{0.8–7 MPa, 296–433 K}]{[\text{Ti}_2\text{cp}_3(\text{C}_5\text{H}_4)],\ \text{PhMe, or THF}} C_2H_6 + CH_2=CHCH=CH_2 \qquad (13.178)$$

$$3CH_2=CHCN \xrightarrow[\text{4 h}]{423\ K} MeCH_2CN + NCCH=CHCH=CHCN \qquad (13.179)$$

## 10. REDUCTION OF CARBON MONOXIDE; THE FISCHER–TROPSCH SYNTHESIS

Mixtures of carbon monoxide and hydrogen may undergo the following reactions in the presence of catalysts (Table 13.6):

1.  Methanol synthesis

$$CO + 2H_2 \longrightarrow CH_3OH \qquad (13.180)$$

2.  Methanation

$$CO + 3H_2 \longrightarrow CH_4 + H_2O \qquad (13.181)$$

3.  Fischer–Tropsch synthesis*

$$CO + H_2 \longrightarrow C_aH_b + C_cH_d(OH)_e + RCHO + RCOOH + RCOOR' \qquad (13.182)$$

Fischer–Tropsch reaction affords various products, viz., saturated hydrocarbons, alkenes, aromatic hydrocarbons, alcohols, aldehydes, ketones, acids, esters, and compounds which are formed by reactions between these products.[160–172]

The sum reactions of synthesis of hydrocarbons from carbon monoxide and hydrogen, which depend upon the catalyst and the conditions of the reaction, may be represented by various equations. These equations essentially may be comprised from the two reactions in equations (13.183) and (13.184):

$$CO + H_2O \longrightarrow CO_2 + H_2, \qquad \Delta H_{400\ K} = -39.8\ kJ \qquad (13.183)$$

$$CO + 2H_2 \longrightarrow (-CH_2-) + H_2O, \qquad \Delta H_{400\ K} = -165\ kJ \qquad (13.184)$$

---

* Reaction (3) was discovered by Fischer–Tropsch in 1923. The first plants which produced 200,000 tons of hydrocarbons were developed in Germany in 1936. The maximum production of hydrocarbons utilizing the Fischer–Tropsch method was achieved in Germany at the beginning of 1944 (ca 6,000,000 tons per year). After WWII, hydrocarbons were produced by this method in the United States (1948–1953) and in South Africa (from 1955). Owing to the rise in the price of oil after 1973, interest in the Fischer–Tropsch synthesis increased.

The following result from these two equations:

$$2CO + H_2 \longrightarrow (-CH_2-) + CO_2, \qquad \Delta H_{400\,K} = -204.8 \text{ kJ} \qquad (13.185)$$

$$3CO + H_2O \longrightarrow (-CH_2-) + 2CO_2, \qquad \Delta H_{400\,K} = -244.5 \text{ kJ} \qquad (13.186)$$

$$CO_2 + 3H_2 \longrightarrow (-CH_2-) + 2H_2O, \qquad \Delta H_{400\,K} = -125.2 \text{ kJ} \qquad (13.187)$$

General reactions of synthesis of hydrocarbons are as follows:

$$nCO + (2n+1)H_2 \longrightarrow C_nH_{2n+2} + nH_2O \qquad (13.188)$$

$$2nCO + (n+1)H_2 \longrightarrow C_nH_{2n+2} + nCO_2 \qquad (13.189)$$

$$3nCO + nH_2O \longrightarrow C_nH_{2n} + 2nCO_2 \qquad (13.190)$$

$$nCO_2 + (3n+1)H_2 \longrightarrow C_nH_{2n+2} + 2nH_2O \qquad (13.191)$$

$$nCO + 2nH_2 \longrightarrow C_nH_{2n} + nH_2O \qquad (13.192)$$

$$nCO_2 + 3nH_2 \longrightarrow C_nH_{2n} + 2nH_2O \qquad (13.193)$$

$$nCO + 2nH_2 \longrightarrow C_nH_{2n+1}OH + (n-1)H_2O \qquad (13.194)$$

$$(n+1)CO + (2n+1)H_2 \longrightarrow C_nH_{2n+1}CHO + nH_2O \qquad (13.194a)$$

The thermodynamic probability of the formation of products of many parallel and subsequent reactions is calculated by taking into consideration general and simultaneous equilibria among them. This is not possible for the Fischer–Tropsch reaction, because the number of reactions is theoretically unlimited. Therefore, for calculation purposes, it is assumed that the reactions are independent. The Fischer–Tropsch synthesis is strongly exothermic, and the removal of heat represents an important problem in the technology of this process. In order to facilitate a comparison of thermodynamic data for various reactions, such data are given with respect to one mole of carbon (Figure 13.4).

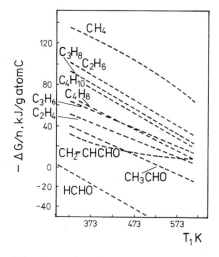

Figure 13.4. Dependence of the thermodynamic potential on temperature for the Fischer–Tropsch reaction.

The universal, macrokinetic equation of the Fischer–Tropsch reaction is not known. For each type of catalyst and reaction, separate equations are written which are only valid for a specific range of parameters. The activation energy is high ($80$–$105$ kJ mol$^{-1}$) and it governs the reaction rate; the reaction rate is not determined by the diffusion of reagents to the surface of the catalyst or by the reverse process of diffusion of products.

## a. *Mechanism of the Fischer–Tropsch reaction*

Elementary reactions occurring during reduction of carbon monoxide may be divided into five types:

1. Transfer of a hydrogen atom from the surface of the catalyst or the central atom in the complex to a carbon atom and the formation of a $C-H$ bond.

2. Transfer of a hydrogen atom to an oxygen atom and the formation of an $O-H$ bond.

3. Formation of a $C-C$ bond.

4. Cleavage of a $C-O$ bond.

5. Formation of a $C-O$ bond.

*Table 13.6. Catalysts for Fischer–Tropsch Reactions*[162–165]

| Catalyst (weight ratios) and Solvent | Reaction temperature (K) | Pressure (MPa) | Products (yield) (g m$^{-3}$) |
|---|---|---|---|
| $Co + ThO_2$ + kieselguhr[162–163] ($100 + 18 + 100 - 200$) | 463–470 | 0.7 | Hydrocarbons (97–124) |
| $Co + ZrO_2 + MgO$ + kieselguhr ($100 + 6 + 10 + 200$)[162,163] | 463–473 | 0.1 | Hydrocarbons (130–190) |
| $Fe + Cu + K_2CO_3$ ($100 + 0.3 + 0.6 - 1.2$)[162] | 503–543 | 0.3–1 | Liquid hydrocarbons (110–170) + gaseous hydrocarbons (10–45) |
| $Fe + Cu + ZnO + H_2O + SiO_2$[162] ($100 + 25 + 18 + 9 + 48$) | 543 | 2 | Hydrocarbons (145) |
| $[Ir_4(CO)_{12}]$, $NaAlCl_4$[186] | 453 | 0.15 | Hydrocarbons $C_1$–$C_4$ |
| $[Co(acac)_2] + AlEt_3$, alkylterphenyls[a] | 463 | 0.1 | Olefins $C_2$–$C_6$, methane |
| $[Co_2(CO)_8]$, dioxane[185] | 455 | 30 | MeOH, EtOH, PrOH, HCOOMe, HCOOEt |
| $[Fe_3(CO)_{12}] + Al_2O_3$[b] | 453–543 | 1 | $C_1$–$C_5$ hydrocarbons |
| $[Ru_3(CO)_{12}]$, $MeOC_2H_4OH$[184] | 423–463 | 18 | MeOH, $Me_2CO$, HCOOMe, $Me_2O$ |
| $[Co_2(CO)_8]$, $MeOC_2H_4OH$, diglyme[184] | <473 | <30 | EtOH (80%), HCOOEt, $Me_2CO$, $Et_2O$ |
| $Fe(NO_3)_3$ + graphite[c] | 673 | 0.1 | $C_1$–$C_5$ hydrocarbons |
| $Ir_4(CO)_{12}$[196], hydrocarbons (amines as promotors) | 523 | 200 | MeOH, HCOOMe, $HOCH_2CH_2OH$ |
| $Ru_3(CO)_{12} + Co_2(CO)_8$, $[PR_4]Br$[199] | 493 | 43.5 | Alcohols, esters |

[a] M. Blanchard, I. Vanhove, F. Petit, and A. Mortreux, *Chem. Commun.*, 908 (1980).
[b] D. Commereuc, Y. Chauvin, F. Hugues, J. M. Basset, and D. Olivier, *Chem. Commun.*, 154 (1980).
[c] E. Kikuchi, T. Ino, N. Ito, and Y. Morita, *J. Japan. Petroleum Institute*, **21**, 242 (1978).

Reactions (1) and (2) occur during methanol synthesis, while reactions (1), (2), and (4) are essential for the preparation of methane. Except for methane, hydrocarbons undergo dehydrogenation, isomerization, and aromatization under Fischer–Tropsch reaction conditions. These reactions are essentially influenced by the character of the support.

The breaking of the carbon–oxygen bond may occur both in the case of a heterogeneous catalytic reduction reaction[173–176] as well as in homogeneous solutions of cluster metal carbonyls.[167–172, 177, 178] Protonation of tetranuclear iron cluster $[HFe_4(CO)_{13}]^-$ (Figure 2.3) by means of trifluoromethylsulfonic acid, $HOSO_3CF_3$, causes reduction of the coordinated carbonyl group $\mu_4$-CO to methane. Protonation of the carbonyl group [equation (3.59), Figure 3.24] probably leads to the elimination of water and the formation as an intermediate of the reactive carbide which, after the addition of $H^+$, gives methane. Iron, which is oxidized to Fe(II) in this process, serves as a reducing agent.[117, 178] Therefore, the reduction of the CO molecule in the cluster $[Fe_4(CO)_{13}]^{2-}$ may be represented as in scheme (13.195). The $\pi$ olefin type coordination (side on) of the carbon monoxide ligand weakens the carbon–oxygen bond to a significant degree, causing it to break.[177, 178] The order of addition of protons to $[Fe_4(CO)_{13}]^{2-}$, first to the iron atom and subsequently to the carbon atom, was confirmed by Hückel and Fenske–Hall calculations.[179] Other carbides, such as $[Fe_6(CO)_{16}C]^{2-}$, do not undergo reduction to methane.

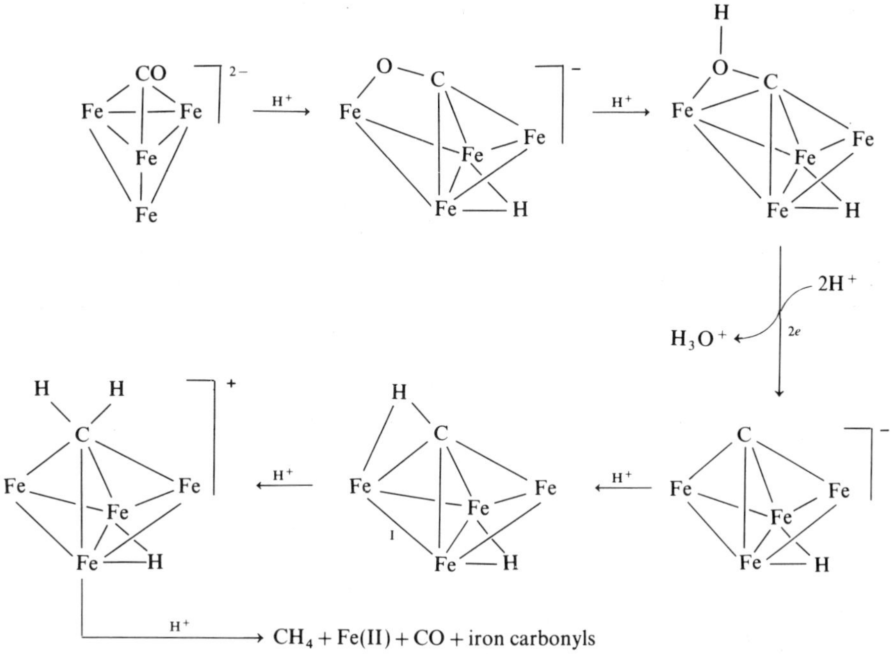

$$(13.195)$$

In the case of carbon monoxide reduction in the presence of metals such as iron, cobalt, nickel, and ruthenium, scission of the C—O bond occurs in the first stage. Oxygen is eliminated as water (in the case of electropositive metals, e.g., Fe) or carbon dioxide. The activation of the CO group may also occur owing to the interaction of its oxygen atom with the strong electropositive metal[186] [equations (2.173)–(2.175)] or through the formation of the $\pi$ olefin bond with the transition metal (Figure 2.3)[170, 177, 178, 187, 188]; see equation (13.196). Subsequently the carbon atoms may react

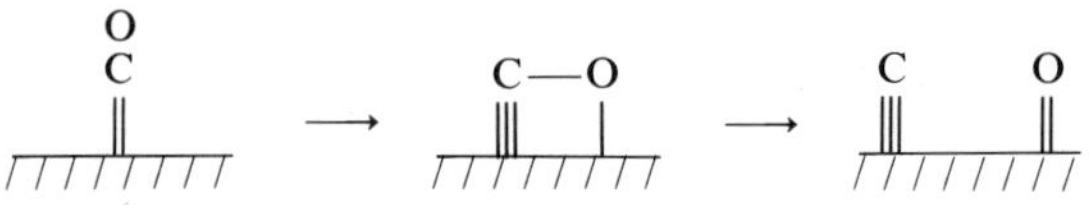

$$
\begin{array}{ccc}
\overset{\text{O}}{\underset{\text{C}}{\|}} & \overset{\text{C—O}}{\|} & \overset{\text{C}}{\|}\ \ \overset{\text{O}}{\|} \\
\overline{/////// } & \longrightarrow \quad \overline{///////} & \longrightarrow \quad \overline{///////}
\end{array}
\qquad (13.196)
$$

with absorbed hydrogen to form methylidyne, methylidene, and methyl compounds with metal atoms. This corresponds to carbyne, carbene, and $\sigma$-carbyl complexes in organometallic chemistry (Chapters 4 and 5); see structures (13.197). Alkyl groups may

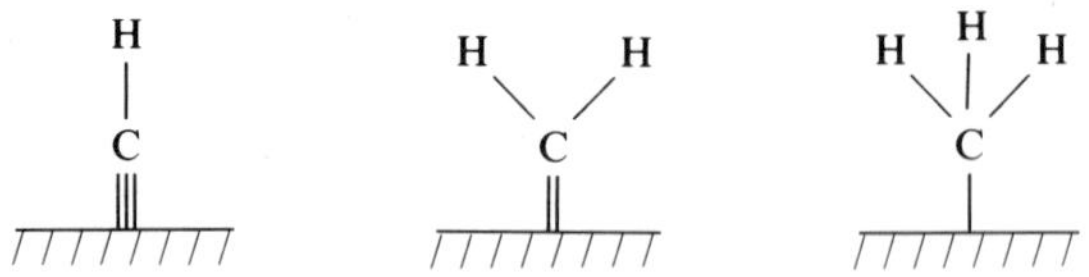

$$ (13.197) $$

migrate to carbene or carbyne to give alkyls and alkylidenes, respectively, which have higher molecular masses. Reactions leading to the formation of various unsaturated hydrocarbons are also possible; see equation (13.198).

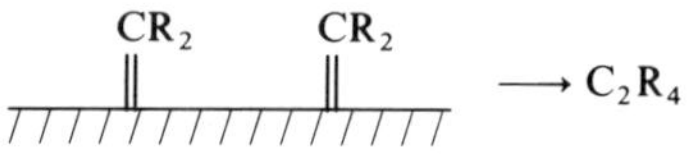

$$
\overset{CR_2}{\|}\quad \overset{CR_2}{\|} \quad \longrightarrow C_2R_4 \qquad (13.198)
$$

During CO reduction, formyl complexes of the type $L_nM-C\overset{H}{\underset{O}{}}$ are also formed as intermediates. At present, many formyl, carbene, hydroxymethyl, and $\pi$-acyl complexes and their derivatives are known. The formation of these complexes in the Fischer–Tropsch reaction is very probable. This was confirmed by the preparation of various model compounds[169, 170–172, 180–191]; see equations (13.199)–(13.201). Metal

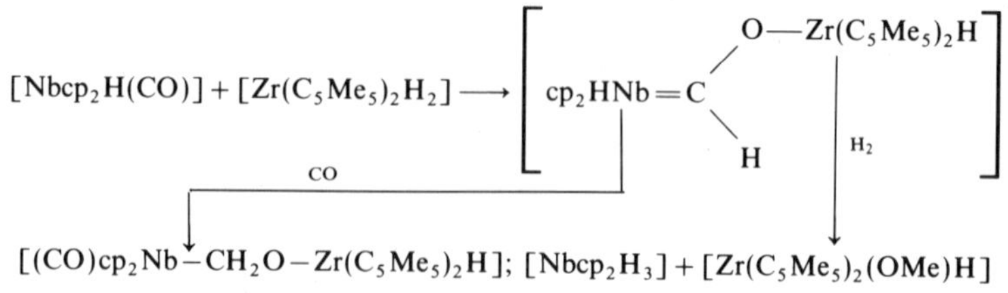

$$
[Nbcp_2H(CO)] + [Zr(C_5Me_5)_2H_2] \longrightarrow \left[ cp_2HNb=C \overset{O-Zr(C_5Me_5)_2H}{\underset{H}{}} \right] \qquad (13.199)
$$

$$
[(CO)cp_2Nb-CH_2O-Zr(C_5Me_5)_2H]; \ [Nbcp_2H_3] + [Zr(C_5Me_5)_2(OMe)H]
$$

$$
t\text{-}[IrX(CO)(dppe)_2]^{2+} + Li[B(Bu^s)_3H] \xrightarrow[CH_2Cl_2]{228\ K} t\text{-}[Ir(CHO)X(dppe)_2]^+ \qquad (13.200)
$$

$$
Et_3N \ \Big\updownarrow\ HBF_4Et_2O \quad {}^{(197)}
$$

$$
(X = H, Cl) \qquad\qquad t\text{-}[XIr\{=CH(OH)\}(dppe)_2]^{2+}
$$

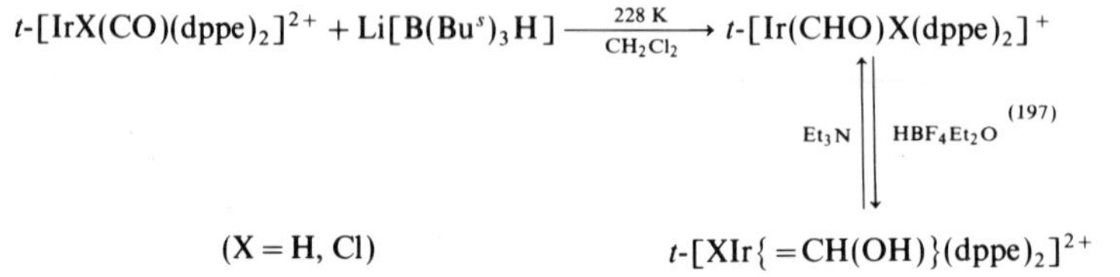

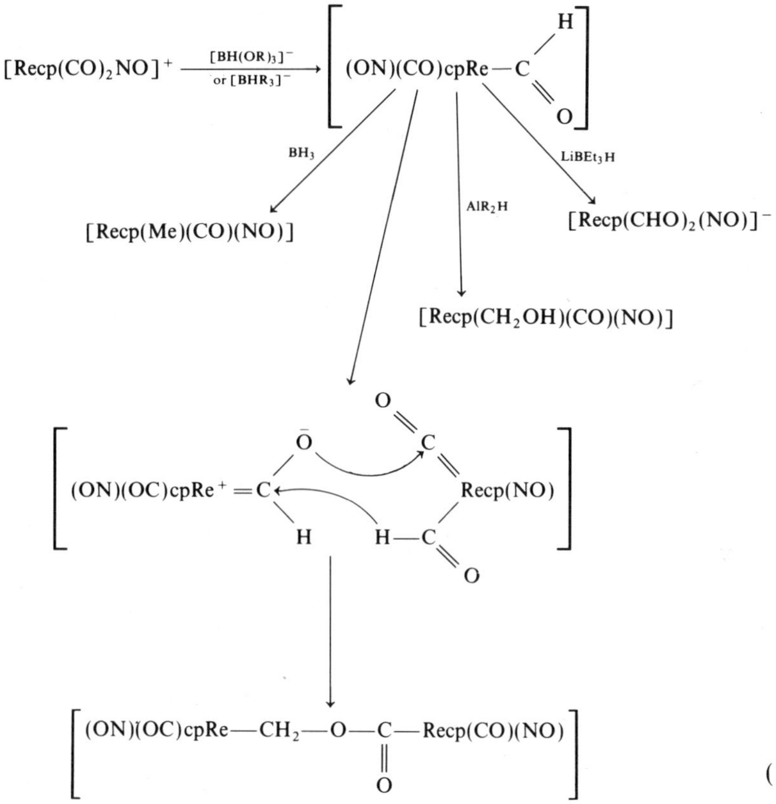

$$[Recp(CO)_2NO]^+ \xrightarrow[\text{or}\,[BHR_3]^-]{[BH(OR)_3]^-} \left[ (ON)(CO)cpRe\!-\!C\!\!\begin{array}{c}H\\ \diagup\\ \diagdown\\ O\end{array} \right]$$

$$\xrightarrow{BH_3} [Recp(Me)(CO)(NO)]$$

$$\xrightarrow{LiBEt_3H} [Recp(CHO)_2(NO)]^-$$

$$\xrightarrow{AlR_2H} [Recp(CH_2OH)(CO)(NO)]$$

$$\left[ (ON)(OC)cpRe^+ = C \overset{\bar{O}}{\underset{H}{\diagup}} \longleftarrow \overset{O}{\underset{H-C}{\diagdown}} C \overset{}{\underset{O}{}} Recp(NO) \right]$$

$$\left[ (ON)(OC)cpRe\!-\!CH_2\!-\!O\!-\!\underset{\underset{O}{\parallel}}{C}\!-\!Recp(CO)(NO) \right] \qquad (13.201)$$

acetyl and formyl complexes may exist in equilibrium involving alkyl and hydrido compounds [equations (13.202) and (13.203)]. Acetyl compounds are considerably

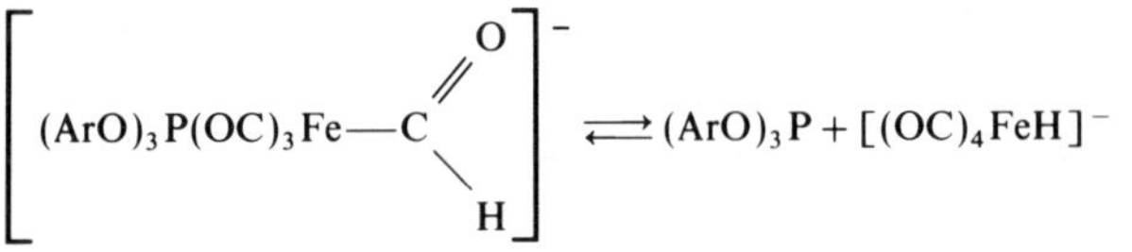

$$\left[ (ArO)_3P(OC)_3Fe\!-\!C\!\!\begin{array}{c}\diagup O\\ \diagdown H\end{array} \right]^- \rightleftharpoons (ArO)_3P + [(OC)_4FeH]^- \qquad (13.202)$$

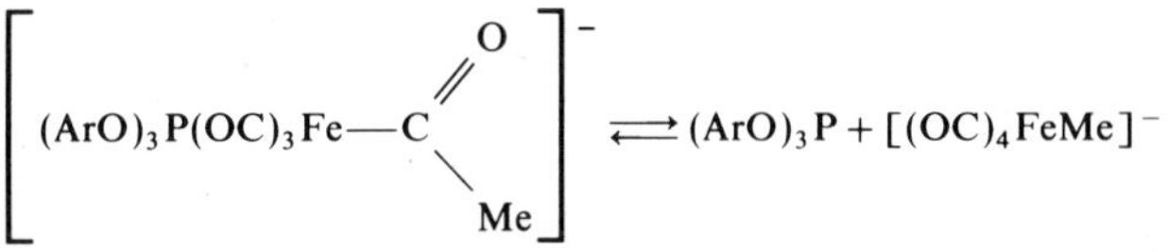

$$\left[ (ArO)_3P(OC)_3Fe\!-\!C\!\!\begin{array}{c}\diagup O\\ \diagdown Me\end{array} \right]^- \rightleftharpoons (ArO)_3P + [(OC)_4FeMe]^- \qquad (13.203)$$

more stable than formyl complexes because the $M-H$ bond energy (210–250 kJ mol$^{-1}$) is greater than that of $M-Me$ (125–170 kJ mol$^{-1}$). Thus, the equilibrium constant of reaction 13.202 ($>125$ M) is $10^5$ times greater than the equilibrium constant of reaction 13.203 ($<10^{-3}$ M). The decomposition of formyl complexes proceeds via coordinatively unsaturated compounds although, to a certain degree, there is also CO elimination from the formyl group [route *b*, equation (13.204)]. Formyl ligands may be reduced to hydroxymethyl compounds which, by treatment with acids, afford methyl

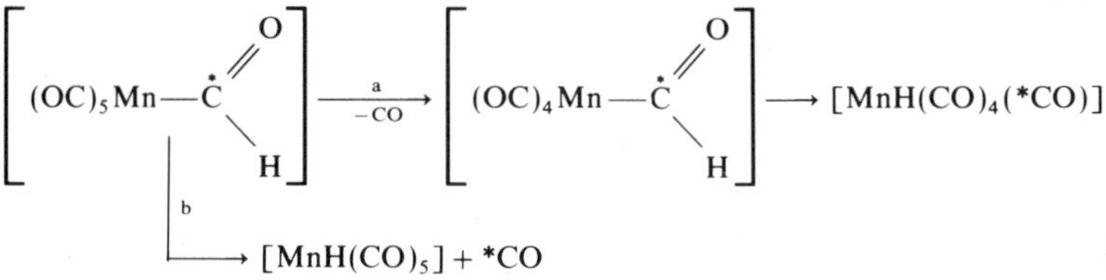

$$
\left[ (OC)_5 Mn-\overset{O}{\underset{H}{\overset{\parallel}{\dot{C}}}} \right] \xrightarrow[-CO]{a} \left[ (OC)_4 Mn-\overset{O}{\underset{H}{\overset{\parallel}{\dot{C}}}} \right] \longrightarrow [MnH(CO)_4(*CO)]
$$

$$
\xrightarrow{b} [MnH(CO)_5] + *CO \tag{13.204}
$$

complexes[198]; see equation (13.205). The formation of hydroxycarbyne complexes,[195] $L_n M \equiv C-OH$, is probably an alternative or a parallel process. Structural and thermodynamic data, as well as $v(M \equiv C)$ frequencies and $k(M \equiv C)$ force constants, show

$$
[Ir(PMe_3)_4][PF_6] \xrightarrow{(CH_2O)_x} c\text{-}[Ir(CHO)HL_4]^+
$$

$$
\xrightarrow[THF/H_2O]{NaBH_4} c\text{-}[Ir(CH_2OH)HL_4]^+ \xrightarrow[X^-]{HBF_4 \cdot Et_2O} c\text{-}[Ir(CH_3)XL_4]^+ \tag{13.205}
$$

that metal carbyne complexes such as $L_n M \equiv C-OH$ should not be less stable than metal formyl complexes, $L_n M - \overset{O}{\underset{H}{\overset{\parallel}{C}}}$. Calculations using the modified Hückel method (MEHT) showed that the 18*e* complex $(OC)_3 Co \equiv C-OH$ ($C_{3v}$) is more stable by *ca* 29 kJ mol$^{-1}$ than the 16*e* coordination compound $(OC)_3 Co - \overset{O}{\underset{H}{\overset{\parallel}{C}}}$ of preferred planar structure. Therefore, in the case of strongly acidic metal hydridocarbonyls, two reaction routes seem possible; see equation (13.206). Ligands that are formed by protonation of

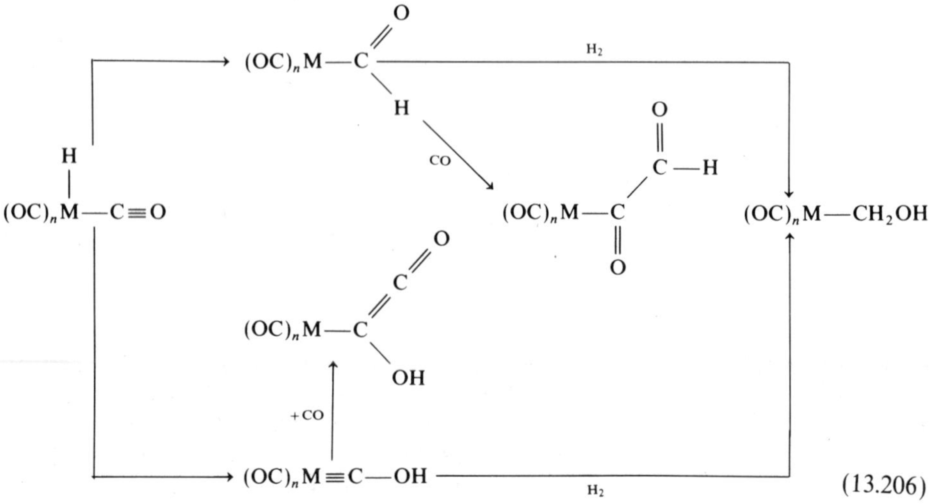

$$
\tag{13.206}
$$

metal carbonyl clusters by means of strong acids may be considered as bridging carbynes[117]; see structures (13.207).

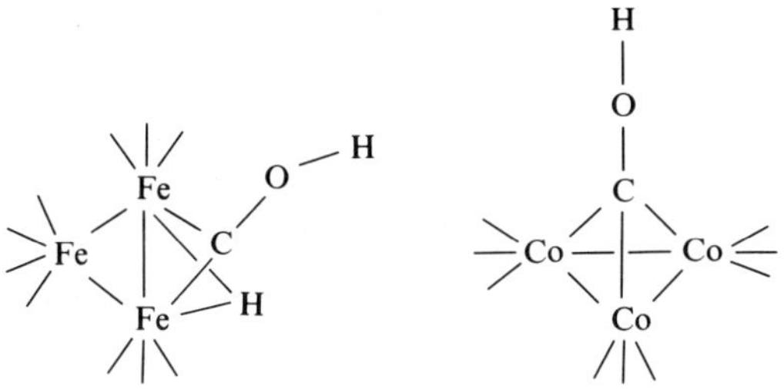

(13.207)

The formation of formaldehyde complexes by reduction of the formyl group cannot be ruled out. This is indicated by the reduction of HCHO to $CH_2OH$ in the complex.[191]

$$[Os(CO)_2(\eta^2\text{-}HCHO)(PPh_3)_2] + CF_3COOH \longrightarrow [Os(CH_2OH)(OOCCF_3)(CO)_2(PPh_3)_2] \quad (13.208)$$

In some cases during HCHO reduction, the same products are formed as in the reaction of carbon monoxide with hydrogen. This confirms the formation of formaldehyde from carbon monoxide in the intermediate stage.[171] The breaking of the C–O bond may occur as a result of elimination of the water molecule from coordinated hydroxycarbene and hydroxymethyl ligands[167–170, 172]; see equations (13.209) and (13.210). A radical mechanism for carbon monoxide reduction is proposed for reactions

$$L_nM\text{—}C(H){=}O \xrightarrow{H_2} L_nM{=}C(H)(OH) \xrightarrow{H_2} M{=}CH_2 + H_2O \quad (13.209)$$

$$L_nM{=}C(H)(OH) \xrightarrow{H_2} L_nM-CH_2OH \xrightarrow{H_2} M-CH_3 + H_2O \quad (13.210)$$

proceeding in the presence of some homogeneous catalysts. Probably, according to this mechanism, methanol, ethanol, and propanol, as well as their formic esters, are formed in the presence of $CoH(CO)_4$ and $MnH(CO)_5$ as catalysts.[185]

$$[HCo(CO)_4] + CO \xrightarrow[30\text{ MPa, }473\text{ K}]{\text{slow}} HCO^* + [{}^*Co(CO)_4] \quad (13.211)$$

$$HCO^* + [HCo(CO)_4] \xrightarrow{\text{fast}} \text{products} \quad (13.212)$$

$$2[Co(CO)_4] \longrightarrow [Co_2(CO)_8] \xrightarrow{H_2} 2[CoH(CO)_4] \quad (13.213)$$

The formation of C–C bonds during the Fischer–Tropsch synthesis occurs via migration of various groups bonded to the metal through the carbon atom. This is true for groups containing the oxygen atom as well as the $CH_x$ fragments, where $x = 0$–3. Therefore, it is possible to form, for example, ketenyl complexes by reaction of the carbyne ligands $\equiv CH$ with carbon monoxide [equation (5.143)], ketenes $(CR_2 = C = O)$ by reaction of carbenes with CO [equations (5.135) and (5.136)],[184] aldehydes and alcohols via the hydroformylation reaction, etc. (e.g., hydroformylation of formaldehyde)[107, 128, 190]; see equation (13.214). It is proposed that the formation of

$$[HCo(CO)_4] + HCHO \longrightarrow [HOCH_2Co(CO)_4] \xrightarrow{CO} \left[ \begin{array}{c} HOCH_2CCo(CO)_4 \\ \parallel \\ O \end{array} \right]$$

$$\xrightarrow[CO]{HCo(CO)_4} [HOCH_2CHO] + Co_2(CO)_8 \qquad (13.214)$$

methyl acetate during the decomposition of the $\mu$-methylene iron complex $Fe_2(\mu\text{-}CH_2)(CO)_8$ by the action of methanol proceeds via a $\mu\text{-}\eta^2$-ketene intermediate compound[192] [equation (13.215)]. A stable $\mu$-ketene osmium compound is formed by

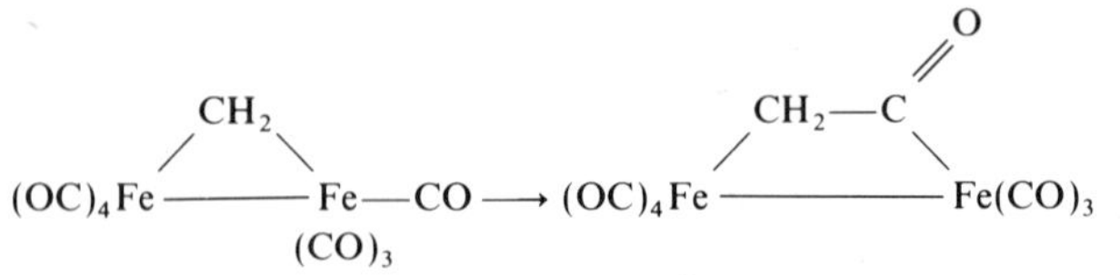

$$(13.215)$$

reaction of $[Os_3(\mu\text{-}CH_2)(CO)_{11}]$ with carbon monoxide [equations (3.74) and (3.75)]. The formation of C–C bonds also proceeds by CO migration to carbyne and carbene ligands, and leads to metal ketenylidene complexes [Section 3.10.b, Equations (3.71) and (3.72), Table 3.14]. The transient ketenylidene ligand probably forms during the preparation of the complex $[Fe_4(\mu_4\text{-CCOOMe})(CO)_{12}]$[193]; see equation (13.216).

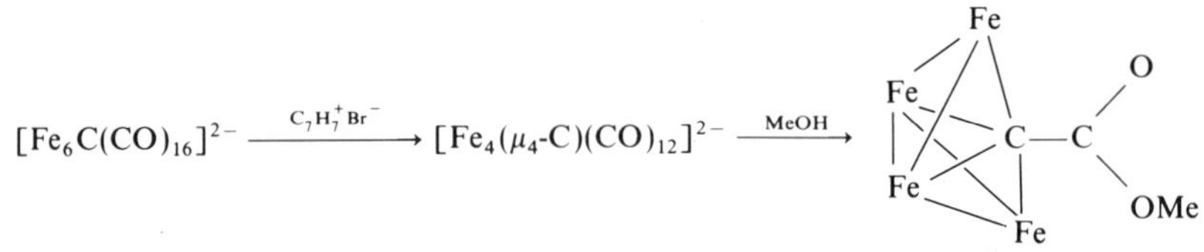

$$[Fe_6C(CO)_{16}]^{2-} \xrightarrow{C_7H_7^+Br^-} [Fe_4(\mu_4\text{-}C)(CO)_{12}]^{2-} \xrightarrow{MeOH}$$

$$(13.216)$$

Elongation of a hydrocarbon chain may also take place through reduction of a carbyne ligand to a carbyl group followed by CO insertion, C–O bond breaking, and coordination of the oxygen atom to the metal[194] [equations (13.217) and (13.218)]. Additionally, the formation of metalladiketones (a), metallaketoesters (b), and metallaethers

$$RC \xrightarrow{2H} RCH_2 \xrightarrow{CO} RCH_2C(=O) \xrightarrow{-O} RCH_2C \cdots \text{etc.} \qquad (13.217)$$

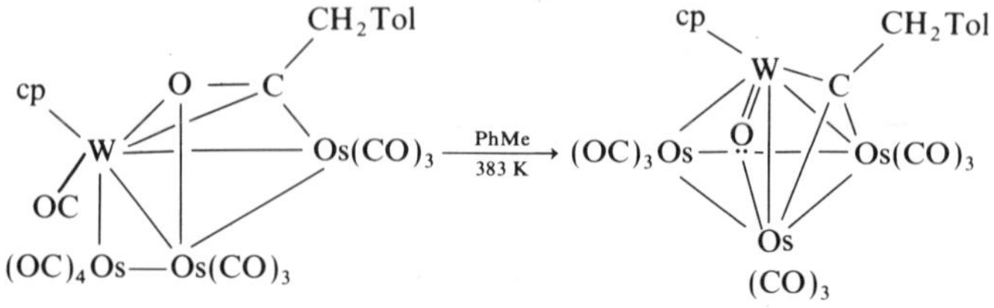

$$(13.218)$$

(c) [structures (13.219)] is assumed for reactions in diglyme and 2-methoxyethanol[168, 184] Therefore, the mechanism of the Fischer–Tropsch reaction is very complicated, and the possibility of a great variety of reactions accounts for the small selectivity of the process.

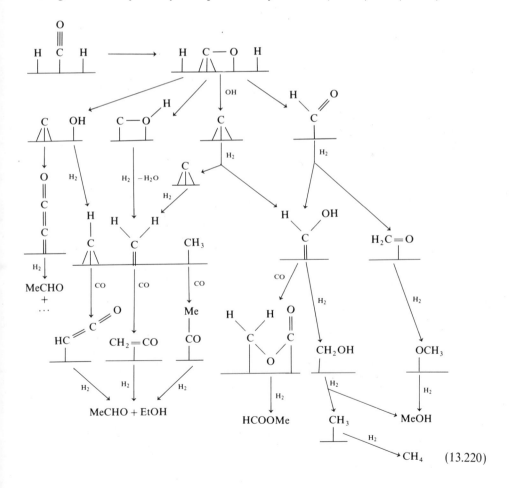

$$(13.219)$$

According to present data, the mechanism of carbon monoxide reduction on heterogeneous catalysts may be represented by schemes (13.220) and (13.221).

$$(13.220)$$

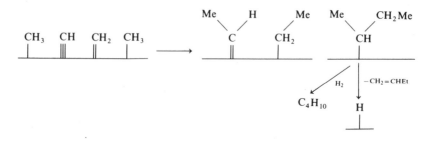

(13.221)

## 11. *REACTIONS OF CARBON DIOXIDE*

Carbon dioxide may be coordinated to the metal through the carbon atom as a Lewis acid, through the oxygen atom(s) as a Lewis base, and as a $\eta^2$-type $\pi$ olefin ligand (side-on) through the carbon and oxygen atoms[167, 200–205] [structures (13.222)]. Additionally, $CO_2$ may behave as a bridging ligand [structures (13.223)]. Bent

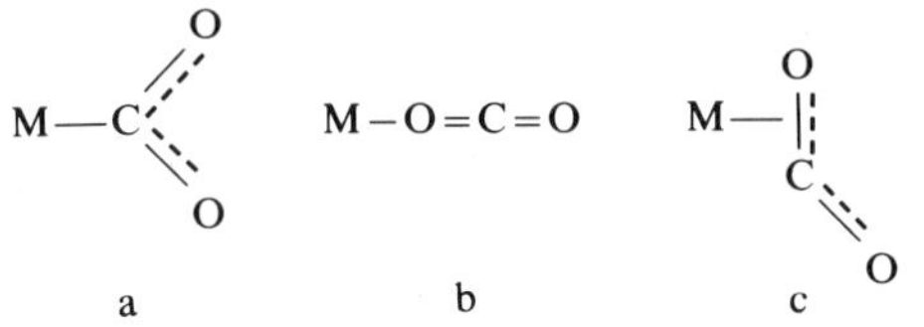

$$
\text{(13.222)}
$$

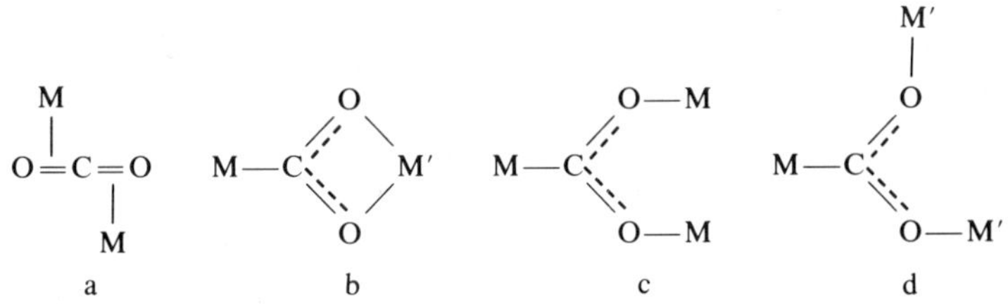

a      b      c

$$
\text{(13.223)}
$$

a      b      c      d

complexes of type (13.222a) are formed with the $ML_n$ fragments that possess electron-donating properties. The bending of the $CO_2$ molecule results in the lowering of the energy of the antibonding $\pi^*$ orbital. This enables more effective interaction of the $\pi^*$ orbital with the metal $d_\pi$ orbital. The complex $[RhCl(\eta^1\text{-}CO_2)(diars)_2]$[201] is an example of such a compound. Complexes of type (13.222c) are represented by the species $[Ni(\eta^2\text{-}CO_2)(PCy_3)_2]$, $[Nb(C_5H_4Me)_2(CH_2SiMe_3)(\eta^2\text{-}CO_2)]$,[206] and $[Mocp_2(CO_2)]$.[207] The $CO_2$ molecule forms a $\mu_3$ bridge (13.223c) in the cluster $[Os_6(CO)_{17}(\mu_3\text{-}CO_2)Os_3H(CO)_{10}]$ (Figure 3.24h), and the mode (13.223d) of bonding is shown by example (13.224).[208]

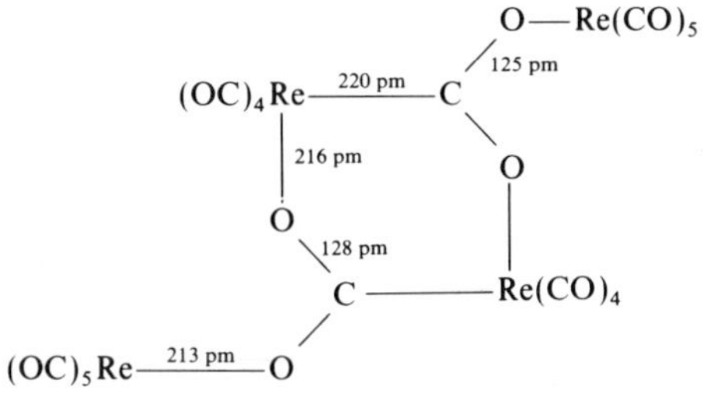

$$
\text{(13.224)}
$$

The cobalt(I) compound $K(THF)_{0.5}Co(Pr''\text{-salen})$ reacts with $CO_2$ to give $[Co(Pr''\text{-salen})(CO_2)K(THF)]_n$ which has structure (13.225).[209] Recently, the complex $[Mo(CO_2)_2(PMe_3)_4]$ containing two $CO_2$ molecules was obtained. Based on spectroscopic studies, a *trans* structure with two $\eta^2\text{-}CO_2$ ligands was proposed.[210]

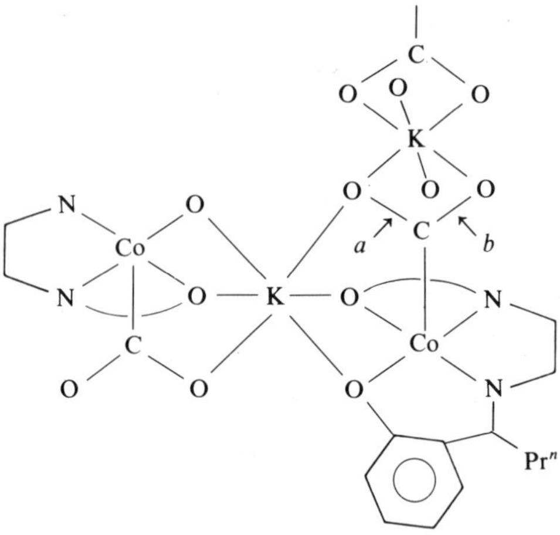

$a = 120$ pm
$b = 124$ pm
$\measuredangle OCO = 132°$

(13.225)

The iridium compound $[IrCl(C_8H_{14})(PMe_3)_3]$ reacts with carbon dioxide to afford the complex (13.226).[211] Coordinated $CO_2$ undergoes migratory insertion to the

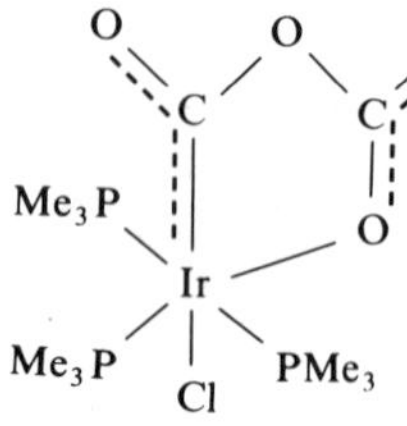

(13.226)

following bonds: M—H, M—C, M—O, M—N, etc. In insertion reactions with $CO_2$, either the C—X or O—X bond may be formed, depending upon the polarity of the M—X bond (X = H, C); see equations (13.227) and (13.228). Therefore, the reaction of

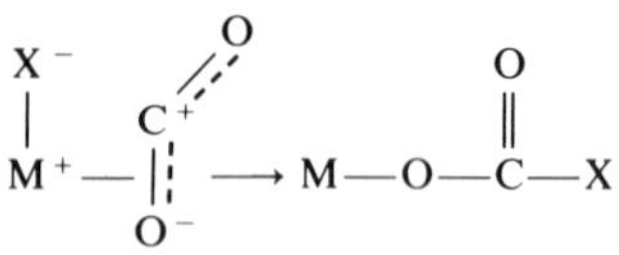

(13.227)

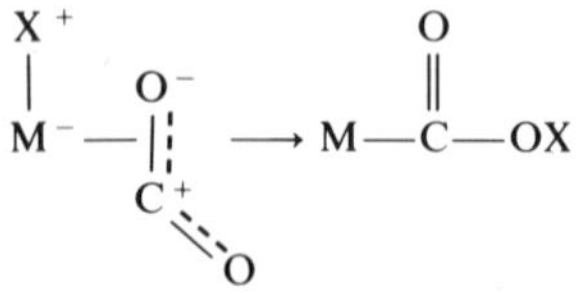

(13.228)

carbon dioxide with metal hydrido complexes leads to the formation of either metal formates or metallacarboxylic acids. The product of the reaction of $[CoH(N_2)(PPh_3)_3]$ with $CO_2$ gives methyl formate and methyl acetate upon treatment with methyl iodide;

thus, during this reaction, both kinds of complexes (13.229) and (13.230), are formed simultaneously.

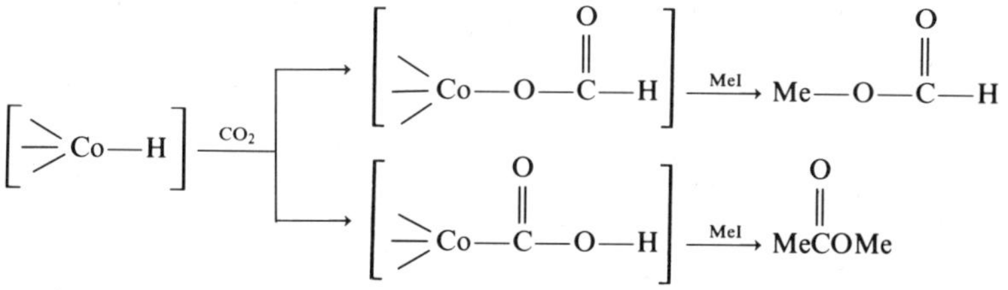

$$\left[\!\!\begin{array}{c}\diagdown\\\diagup\end{array}\!\!Co-H\right] \xrightarrow{CO_2} \left\{ \begin{array}{l} \longrightarrow \left[\!\!\begin{array}{c}\diagdown\\\diagup\end{array}\!\!Co-O-\overset{\displaystyle O}{\overset{\|}{C}}-H\right] \xrightarrow{MeI} Me-O-\overset{\displaystyle O}{\overset{\|}{C}}-H \qquad (13.229) \\[3em] \longrightarrow \left[\!\!\begin{array}{c}\diagdown\\\diagup\end{array}\!\!Co-\overset{\displaystyle O}{\overset{\|}{C}}-O-H\right] \xrightarrow{MeI} MeCOMe \qquad\qquad\quad (13.230) \end{array} \right.$$

The $\sigma$-carbyl complexes react similarly [see reactions (4.181)–(4.186)]. The insertion of $CO_2$ into the $M-O$ bond leads to the formation of metal carbonates. An analogous reaction involving the $M-N$ bond in metal amide complexes affords carbamates.

$$Fe(OR)_3 + CO_2 \longrightarrow Fe(OR)_2\left(\overset{\displaystyle O}{\overset{\|}{OCOR}}\right) \qquad\qquad (13.231)$$

$$Rh(OH)CO(PR_3)_3 + CO_2 \longrightarrow Rh\left(O-\overset{\displaystyle O}{\overset{\|}{C}}-OH\right)(CO)(PR_3)_3 \qquad (13.232)$$

$$W_2(OBu')_6 + CO_2 \longrightarrow W_2(OBu')_4\,(O_2COBu')_2 \qquad\qquad (13.233)$$

$$W_2(NMe_2)_6 + CO_2 \longrightarrow W_2(O_2CNMe_2)_6 \qquad\qquad (13.234)$$

$$W(NMe_2)_6 + CO_2 \longrightarrow W(NMe_2)_3\,(O_2CNMe_2)_3 \qquad\qquad (13.235)$$

The reaction of $CO_2$ with complexes containing $M-N$ bonds may proceed according to the mechanisms (13.236) and (13.237). Carbon dioxide reacts with 1,3-dienes,

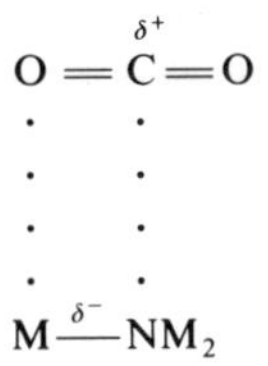

$$\begin{array}{c} O = \overset{\delta^+}{C} = O \\ \vdots\qquad\vdots \\ \vdots\qquad\vdots \\ \vdots\qquad\vdots \\ M \overset{\delta^-}{\longrightarrow} NM_2 \end{array} \qquad\qquad (13.236)$$

$$\begin{array}{c} CO_2 + HNR_2 \rightleftharpoons HO_2CNR_2 \\ \qquad\qquad\quad \downarrow{\scriptstyle MNR_2} \\ \qquad MO_2CNR_2 + HNR_2 \end{array} \qquad\qquad (13.237)$$

acetylenes, and activated $C_{sp^3}-H$ bonds to give many valuable compounds in the presence of some transition metal complexes. For example, carbon dioxide reacts with

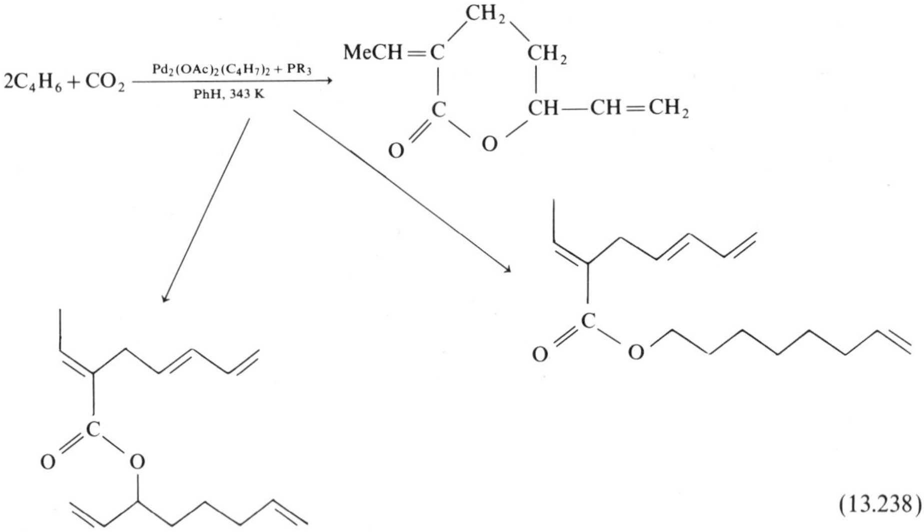

$$2C_4H_6 + CO_2 \xrightarrow[\text{PhH, 343 K}]{\text{Pd}_2(\text{OAc})_2(\text{C}_4\text{H}_7)_2 + \text{PR}_3}$$

(13.238)

butadiene in the presence of catalytic amounts of $[\text{Pd}(\text{PR}_3)_4]$ or $\text{Pd}_2(\text{OAc})_2 (\text{C}_4\text{H}_7)_2$ and $\text{PR}_3$ $(\text{C}_4\text{H}_7 = 2\text{-methylallyl})^{[212]}$; see scheme (13.238). The reaction of $CO_2$ and butadiene gives a five-membered lactone ring in the presence of $\text{Pd}(\text{dppe})_2^{[213]}$:

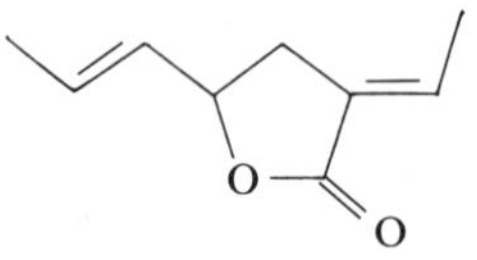

Carbon dioxide reacts with 1-hexyne to give 4,6-dibutyl-2-pyrone in addition to tributylbenzenes[214] [equation (13.239)]. A stoichiometric reaction of dienes with $CO_2$

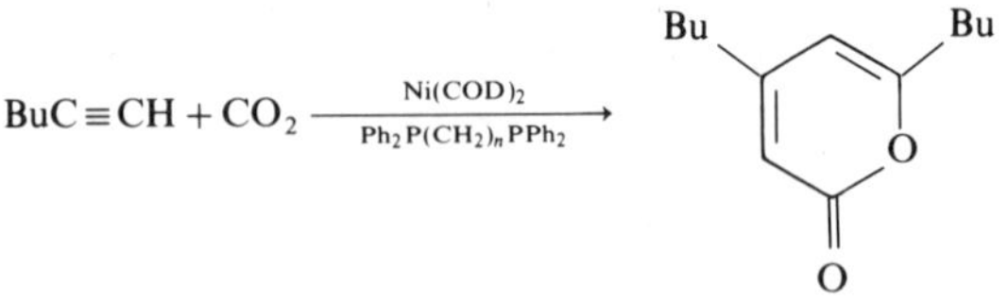

$$\text{BuC} \equiv \text{CH} + CO_2 \xrightarrow[\text{Ph}_2\text{P}(\text{CH}_2)_n\text{PPh}_2]{\text{Ni(COD)}_2}$$

(13.239)

in the presence of nickel compounds affords chelating nickel complexes (13.240).[215] Hydrolysis of these compounds furnishes corresponding carboxylic acids.

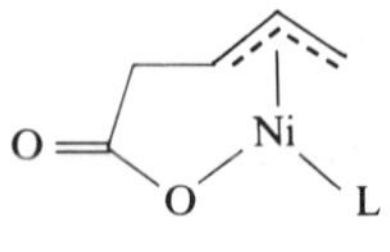

(13.240)

Electrolytic reduction of carbon dioxide is catalyzed by some transition metal complexes. Phthalocyanine and porphyrin complexes, as well as clusters such as

$[Fe_4S_4(SR)_4]^{2-}$, are good catalysts for reduction of $CO_2$. Recently it was found that other compounds also facilitate such reduction. Electrolytic reduction of $CO_2$ in acetonitrile in the presence of $[Rh(dppe)_2]Cl$ leads to the formation of $HCOO^-$.[216] The rhenium(I) compound $[ReCl(CO)_3(bipy)]$ catalyzes electrolytic reduction of $CO_2$ to carbon monoxide.[217] The electrolytic reduction of $CO_2$ in some cases probably proceeds via the radical anion $CO_2^{\cdot-}$; its formation explains various reduction products[218]; see scheme (13.241). Palladium complexes, for instance, $[Pd_2Cl_2(dppm)_2]$, $[Pd(dppm)_2]$, and $[PdCl_2(dppm)]$, slowly catalyze reduction of $CO_2$ to methane, ethyl formate, and traces of ethyl oxalate.[219]

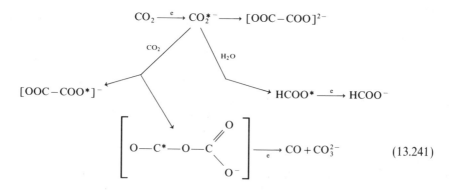

$$(13.241)$$

Strongly electropositive metals and their amalgams may reduce $CO_2$ to CO, hydrocarbons, carbides, carbon, and oxalates.[200–204] Similar products are formed if the reduction is carried out by means of these electropositive metals in the presence of transition metal compounds.[200–204, 220]

## 12. WATER–GAS SHIFT REACTION

The water–gas shift reaction (wgsr) may be represented by the equation:

$$CO_{(g)} + H_2O_{(g)} \longrightarrow CO_{2(g)} + H_{2(g)}$$

$$\Delta G^\circ_{298} = -28.53 \text{ kJ}, \qquad \Delta H^\circ_{298} = -41.17 \text{ kJ}, \qquad \Delta S^\circ_{298} = -42.26 \text{ J K}^{-1} \qquad (13.242)$$

$$CO_{(g)} + H_2O_{(c)} \longrightarrow CO_{2(g)} + H_{2(g)}$$

$$\Delta G^\circ_{298} = -19.92 \text{ kJ}, \qquad \Delta H^\circ_{298} = +2.86 \text{ kJ}, \qquad \Delta S^\circ_{298} = +76.57 \text{ J K}^{-1} \qquad (13.243)$$

The wgsr has great industrial potential. This reaction is thermodynamically favorable ($\Delta G^\circ < 0$) at low temperatures for water in both gaseous [equation (13.242)] and liquid [equation (13.243)] states. The wgsr is utilized in the nitrogen industry for the preparation of hydrogen. This reaction easily allows control of the $H_2:CO$ ratio of the water-gas. Thus, it allows preparation of reagents for the Fischer–Tropsch synthesis, the production of methanol, hydroformylation, etc. The wgsr is also used in the purification of exhaust gases, because it removes carbon monoxide and forms hydrogen, which reduces nitrogen and sulfur oxides. Studies involving utilization of wgsr in hydroformylation, hydroxymethylation, as well as hydrogenation of olefins and aldehydes,[232, 236, 237] where water serves as the hydrogen source, have also been conducted.

Industrial heterogeneous catalysts are active at relatively high temperatures ($>470$ K). This is unfavorable because, as the temperature increases, the yield of hydrogen decreases. The equilibrium constant of reaction (13.242) at 400, 500, and 600 K equals 1450, 130.6, and 26.9, respectively. Therefore, homogeneous and heterogeneous catalysts for carbon monoxide conversion that would be active at lower temperatures have been investigated intensively. Wgsr is catalyzed by metal carbonyl clusters,[221–227, 234] metal carbonyls,[221, 225–228, 234] as well as amine,[230, 231] phosphine,[226, 229, 232] and tris(2-pyridyl)phosphine[233] transition metal complexes.

The reaction may proceed in basic,[221–225, 227, 229, 230] neutral,[229, 231, 233] and acidic solutions.[228, 231, 234]

The experimental data show that, independent of the acidity of the solution,

$$\sigma\text{-carbyl complexes of the type } L_n M - C \overset{\displaystyle O}{\underset{\displaystyle OH}{\big\|}} \quad \text{(metallacarboxylic acids) are formed as}$$

intermediates during carbon monoxide conversion, owing to nucleophilic attack of $OH^-$ or water on the carbon atom of the CO molecules. The complex

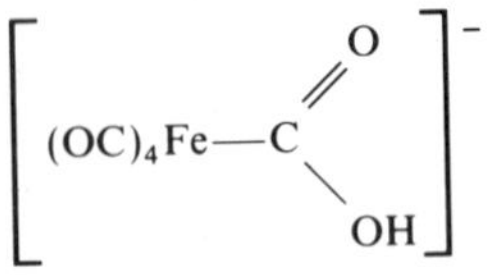

$$\left[ (OC)_4 Fe - C \overset{\displaystyle O}{\underset{\displaystyle OH}{\big\|}} \right]^-$$

is formed by reaction of $Fe(CO)_5$ with cluster ions of the type $[OH(H_2O)_n]^-$ ($n = 1\text{–}4$) in the gaseous phase. The ion $OH^-$ reacts with $Fe(CO)_5$[235] to give the complex $[Fe(CO)_3OH]^-$.

The reactivity of the complex $[(OC)_4FeCOOH]^-$ in the gaseous phase confirms the earlier postulated mechanism of its decomposition. According to this mechanism, the first dissociation of a proton occurs [scheme (13.244)]. Therefore, the presence of bases, which promote dissociation of the proton, should facilitate the conversion of carbon monoxide.

$$[(OC)_4 Fe(COOH)]^- \longrightarrow [Fe(CO)_4(CO_2)]^{2-} + H^+$$

$$\downarrow \qquad\qquad\qquad (13.244)$$

$$[FeH(CO)_4]^- \xleftarrow{\ H^+\ } [Fe(CO)_4]^{2-} + CO_2$$

The mechanism of the wgsr in basic medium may be represented by scheme (13.245). The considerable activity of such catalytic systems as $Rh_2(OAc)_4 + L$ and $RhCl_2(C_4H_7) + LRhCl_3 + L$ [$L = P(2\text{-}C_5H_4N)_3$], as well as of the nickel complexes $[NiX\{2,6\text{-}C_5H_3N(CH_2PPh_2)_2\}]$ ($X = Cl$, Br), may be explained by the binding of protons to the nitrogen atoms of the ligands tris(2-pyridyl)phosphine[233] or 2,6-bis(diphenylphosphinemethyl)pyridine.[229]

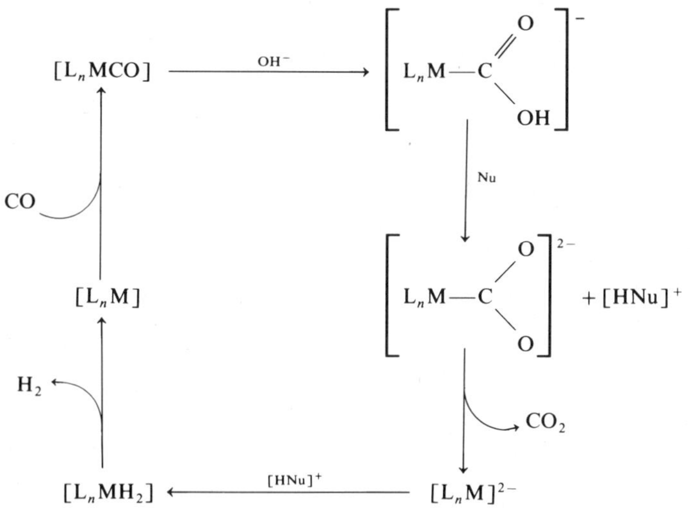

The following compounds are effective catalysts in the wgsr in the 320–470 K temperature range: $[Fe(CO)_5]$,[225] $[M(CO)_6]$ (M = Cr, Mo, W),[225] and $[Re_2(CO)_{10}]$,[221] as well as cluster carbonyls of iron, ruthenium, osmium, rhodium, and iridium.[221, 227, 234] Catalysts that are obtained in basic aqueous–methanol solution of metal carbonyls of the type $M(CO)_6$ (M = Cr, Mo, W) and $M_3(CO)_{12}$ (M = Ru, Os) are active even in the presence of sodium sulfide.[225] Most commonly utilized solvents have been mixtures of water with alcohols, methoxyethanol, pyridine, etc. (mainly in the case of neutral solutions and those containing bases, such as KOH, $K_2CO_3$). Catalysts that are active in acidic media include $Rh_2Cl_2(CO)_4 + HCl + NaI$ (solvent $HOAc + HCl + H_2O$),[228] $[Rh(bipy)_2]Cl$,[231] and $K_2PtCl_4 + SnCl_4$.[228]

## 13. AMINATION AND AMOXIDATION REACTIONS

Reactions of ammonia and amines with C=C double and C≡C triple bonds are interesting because they allow the preparation of many valuable compounds [see Section 6.A.11.c and reactions (13.91) and (13.92)]. For example, the addition of ammonia to unsaturated acids would allow the preparation of amino acids. Application of catalysts with asymmetric ligands for this reaction could lead to synthesis of optically active amino acids, which already exist in nature.

The equilibrium of the reaction of olefins with ammonia

$$RCH=CH_2 + NH_3 \longrightarrow RCH_2CH_2NH_2 \qquad (13.246)$$

at low temperatures is shifted decisively to the right. However, amination proceeds at high temperatures and high pressures, and therefore under conditions in which the equilibrium constant is smaller.

Propylene reacts with ammonia in the presence of potassium amide and $[Pt(C_2H_4)(PPh_3)_2]$ to give isopropylamine.[238] The formation of 2-aminopropane occurs probably via nucleophilic attack of the amide group $NH_2^-$ on the coordinated propene molecule followed by ammonolysis of the $Pt-CMe_2$ bond. Nucleophilic attack

$$| \\ NH_2$$

of the amine or ammonia molecule on olefins also takes place in the case of reactions catalyzed by palladium complexes[239-241]; see equation (13.247). Reactions of alkylation

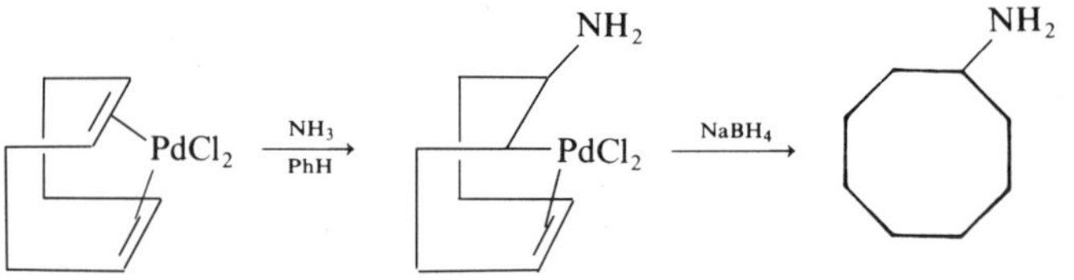

(13.247)

of primary and secondary amines are mainly catalyzed by nickel, palladium, platinum, and rhodium complexes [equation (13.248)].

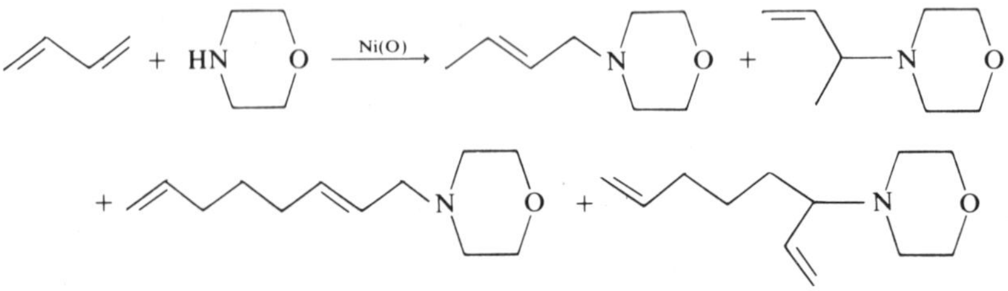

(13.248)

The amoxidation of propylene is an important industrial process of acrylonitrile synthesis. One of the methods of acrylonitrile synthesis involves the reaction of propylene with ammonia and oxygen in the presence of a heterogeneous catalyst (bismuth molybdate containing compounds of Ni, Co, Fe, Sn, etc.)[242, 243]; see equation (13.249). The amoxidation process is also carried out in the case of other compounds, for instance, 3-methylpyridine and p-xylene, which allows the preparation of nicotinic and terephthalic acids, respectively, after hydrolysis of the nitriles.

$$CH_2 = CHCH_3 + NH_3 + \tfrac{3}{2}O_2 \xrightarrow[670-770 \text{ K}, 0.2 \text{ MPa}]{Bi_2O_3/MoO_3} CH_2 = CHCN + 3H_2O \qquad (13.249)$$

Recent studies involving syntheses of homogeneous catalysts for amoxidation have begun.[244, 245] Chromium and molybdenum imide complexes $[M(NBu')_2 (OSiMe_3)_2]$ have been utilized as model compounds. The reaction of such complexes with toluene and benzoyl peroxide affords $PhCH = NBu'$.[245]

## 14. HETEROGENIZED HOMOGENEOUS CATALYSTS

Investigations involving properties of catalysts obtained by deposition of coordination and organometallic compounds on supports have been rapidly developed.

Both inorganic compounds and organic polymers may be used as such supports. The following substances have been most commonly utilized:

| Inorganic | Organic |
|---|---|
| Silica gels | Polystyrene |
| Zeolites | Vinyl polymers |
| Aluminum oxide | Polybutadiene |
| Silica and other metal oxides | Polyaminoacids |
| Glass | Acrylic polymers |
| Ceramics | Urethanes, cellulose |

Heterogenized homogeneous catalysts possess properties of both homogeneous and heterogeneous catalysts.[246, 252] Like homogeneous catalysts, they catalyze reactions under mild conditions at low temperatures and pressures, and are characterized by high selectivity and stereospecificity, and, like heterogeneous catalysts, they may be easily separated from reagents and solvents. However, in general, heterogenized catalysts are less active than homogeneous catalysts because not all molecules of the complex may participate in the reaction owing to steric factors, for instance, some molecules of the catalyst may be located in the pores and thus would be inaccessible for reagents. Heterogenized catalysts most commonly attain 80% of the activity of the same catalysts in the homogeneous phase.

However, in some cases, an opposite trend is observed because the molecule which is bonded to the support as a ligand has different properties. Stated simply, a complex containing $PR_3$ may have different properties than the same complex possessing a phosphine which is attached to the support:

$$ML_nPR_3 \qquad \overset{R}{\underset{R}{|}}-CH_2-\overset{R}{\underset{R}{\overset{|}{\underset{|}{P}}}}-ML_n \qquad (13.250)$$

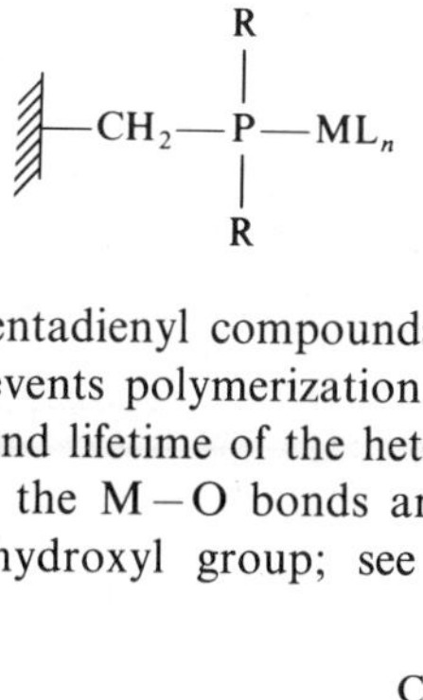

In many cases, such as for cyclopentadienyl compounds of titanium, the attachment of the complex to the support prevents polymerization and other side reactions. This additionally increases the activity and lifetime of the heterogenized catalyst.

Very often, for inorganic supports, the $M-O$ bonds are formed by reactions of organometallic compounds with the hydroxyl group; see equations (13.251) and (13.252).

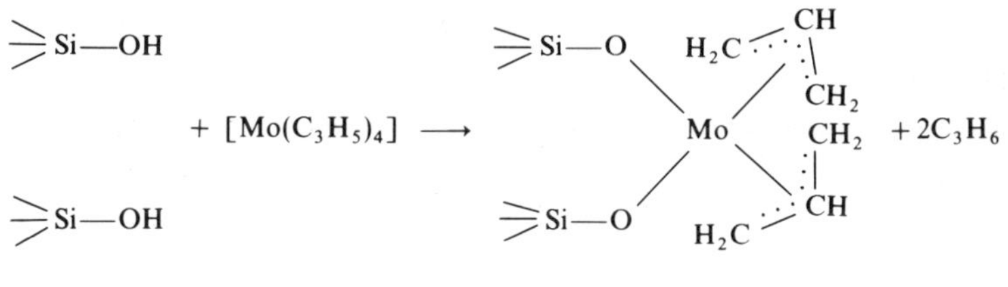

$$(13.251)$$

$$\geqslant Si-OH + [RCo(CO)_4] \longrightarrow \geqslant Si-O-Co(CO)_4 + RH \qquad (13.252)$$

Other inorganic supports containing various organic functional groups may be obtained [equation (13.253)].

$$(EtO)_3SiCH=CH_2 + PPh_2H \longrightarrow (EtO)_3SiCH_2CH_2PPh_2$$

$$\geqslant SiOH + (EtO)_3SiCH_2CH_2PPh_2 \longrightarrow \geqslant SiO-Si(OEt)_2CH_2CH_2PPh_2 \qquad (13.253)$$

$$\xrightarrow{[RhCl(CO)(PPh_3)_2]} \geqslant Si-O-Si(OEt)_2CH_2CH_2PPh_2RhCl(CO)(PPh_3)$$

Utilizing various reagents, functional groups that contain P, N, S, and O atoms and play the role of ligands may be introduced to organic supports; see equations (13.254)–(13.257). Aromatic rings may also play the role of ligands [equation (13.258)].

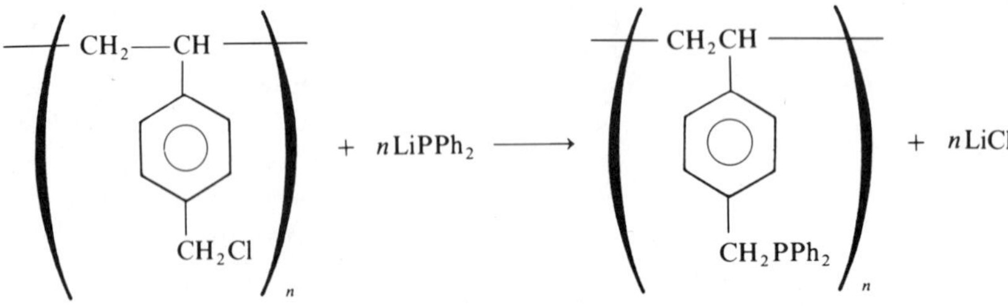

$$(13.254)$$

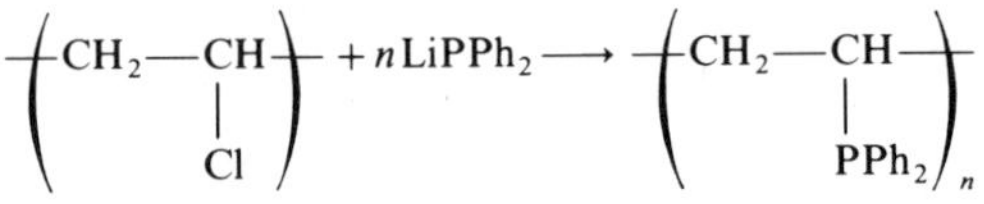

$$(13.255)$$

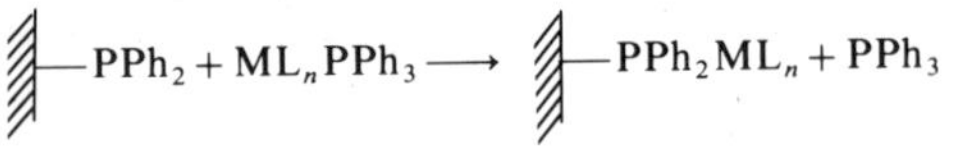

$$(13.256)$$

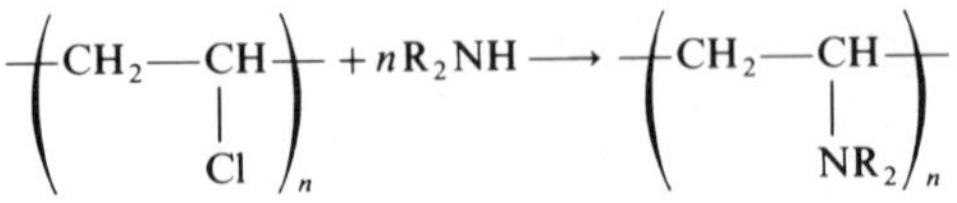

$$(13.257)$$

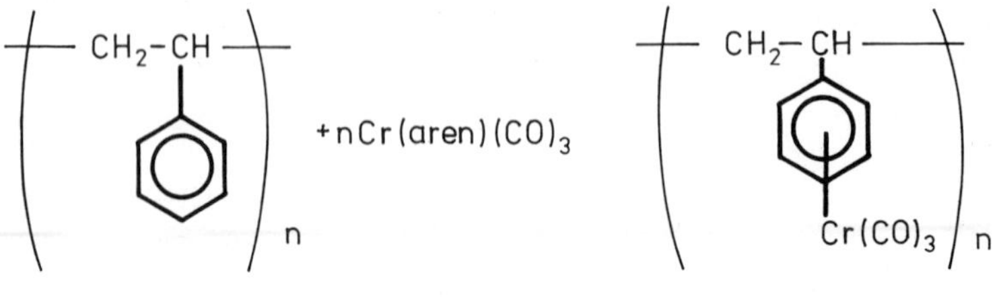

$$(13.258)$$

Polymerization of vinyl monomers containing functional groups or coordinated metal atoms is also commonly used; see equations (13.259) and (13.260).

$$p\text{-}Ph_2PC_6H_4CH=CH_2 + PhCH=CH_2 + CH_2=CHR_6H_4CH=CH_2 \longrightarrow \text{\textasciitilde\textasciitilde\textasciitilde} CH-CH_2\text{\textasciitilde\textasciitilde} $$
$$\underset{\displaystyle C_6H_4PPh_2}{|}$$

$$(13.259)$$

$$[L_nMC_5H_4CH=CH_2] \longrightarrow \left( \underset{\displaystyle C_5H_4ML_n}{\overset{\displaystyle CH-CH_2}{|}} \right)_m \qquad (13.260)$$

$$M = Fe,\ Mn,\ Cr,... \qquad L_n = cp,\ (CO)_3,\ X(CO)_3,...$$

In polymerization, oligomerization, hydroformylation, hydrogenation, and hydrosilation, many metal carbonyl heterogenized catalysts, such as the following, have been utilized: $[LNi(CO)_2PPh_3]$, $[LRhH(CO)(PPh_3)_2]$, $[LRuCl_2(CO)_2(PPh_3)]$, $[LRhCl(CO)PPh_3]$, $[LRh(acac)(CO)]$, $[LIrCl(CO)PPh_3]$, $[L_2IrCl(CO)]$, $[L_2Ni(CO)_2]$, $[LRh_4(CO)_{11}]$, $[LRh_6(CO)_{15}]$, $[L_2Co_2(CO)_6]$, and $[LCo(CO)_2NO]$, where L denotes ⨋–$PPh_2$.

Most of these catalysts show activity similar to their homogeneous congeners, but the time of their operation is longer. Analogous catalysts, supported by substances containing other functional groups such as amines, amides, and nitriles, etc., are also used in the aforementioned reactions. Also, other organometallic and inorganic complexes may be attached to supports; they include $RhCl_3 \cdot 3H_2O$, $[RhCl(PPh_3)_3]$, $PtCl_2$, $H_2[PtCl_6]$, $K_2[PdCl_4]$, $NiCl_2 + NaBH_4$, $RuCl_3 \cdot 3H_2O$, $[M(RCOO)_2]$ (M = Ni, Cu, Pd), $[PdCl_2(PPh_3)_2]$, $Rh_2Cl_2(COD)_2$, $[Rh_2Cl_2(C_2H_4)_4]$, $[Pd_2Cl_2(C_3H_5)_2]$, and $[Rh(acac)(NBD)]$. The catalytic activity of rhodium(I) complexes (supported on phosphine containing polystyrene) in the hydrogenation reaction decreases in the series $[RhH(CO)(PPh_3)_3] > RhCl_3 > RhCl_3 + PPh_3 > RhCl_3 + PHPh_2 > RhCl_3 + C_2H_4 > [RhCl(PPh_3)_3] > [RhCl(PHPh_2)_3]$.

Catalysts for hydrogenation and hydroformylation reactions also catalyze isomerization of olefins. In the case of rhodium(I) catalysts for hydroformylation, the rate of isomerization decreases, depending upon the ligands, as follows: $CO > NR_3 > S > PR_3$.

Hydroformylation catalysts, which are obtained by deposition of $[Co_2(CO)_8]$ or $[CoH(CO)_4]$ on polyvinylpyridine, are actually homogeneous because, during reaction, the cobalt carbonyl reversibly enters the solution in the presence of an excess of carbon monoxide. The lowering of CO pressure causes attachment of cobalt complexes to the polymers. This allows simple separation of the catalyst from products by filtration. Cobalt and rhodium complexes containing phosphine groups are bonded more strongly to the polymer, and therefore they are not easily leached out during the reaction. The cobalt catalyst has analogous properties to the homogeneous system containing $[Co_2(CO)_6(PR_3)_2]$.

$$\left( \text{⨋}-PPh_2 \right)_2 RhH(CO)PPh_3 + \text{⨋}-PPh_2 \rightleftharpoons \left( \text{⨋}-PPh_2 \right)_3 RhH(CO) + PPh_3$$

In the case of rhodium complexes, their content in the polymer may even increase during the hydroformylation owing to the reaction of phosphine elimination. Solvents possessing strong coordinating properties considerably accelerate leaching out of the catalyst from the support.

Heterogeneous catalysts, which are obtained by deposition of cluster-type metal complexes on inorganic substances such as aluminum, titanium, magnesium, zinc, silicon, tin, and tungsten oxides[246–248, 253–255] or on polymers,[266, 246–248] have also been utilized. Most commonly applied clusters include $[Pt_3(CO)_6]_n^{2-}$, $[M_4(CO)_{12}]$, $[M_6(CO)_{16}]$ (M = Co, Rh, Ir), $[M_3(CO)_{12}]$ (M = Fe, Ru, Os), amine-carbonyl, phosphine carbonyl and hydrido carbonyl clusters of elements of groups 8 and 9.

Heterogenized clusters are used as catalysts for many reactions, for instance, the Fischer–Tropsch synthesis, the water–gas shift reaction, hydroformylation, hydrogenation, polymerization, oligomerization, isomerization, etc. In the case of the Fischer–Tropsch synthesis, catalysts that are prepared from clusters often show considerably higher selectivity than other catalysts.

## REFERENCES

1. C. A. Tolman, *Chem. Soc. Rev.*, **1**, 337 (1972).
2. R. G. Pearson, *Chem. Br.*, **12**, 160 (1976); R. G. Pearson, *Top. Curr. Chem.*, **41**, 75 (1973); R. G. Pearson, *Symmetry Rules for Chemical Reactions*, Wiley, New York (1976).
3. D. M. P. Mingos, in: *Comprehensive Organometallic Chemistry* (G. Wilkinson, F. G. A. Stone, and E. W. Abel, eds.), Vol. 3, Chapter 19, Pergamon Press, Oxford (1982).
4. G. K. Anderson and R. J. Cross, *Acc. Chem. Res.*, **17**, 67 (1984).
5. A. Wojcicki, *Adv. Organomet. Chem.*, **11**, 87 (1973).
6. D. C. Sonnenberger, E. A. Nintz, and T J. Marks, *J. Am. Chem. Soc.*, **106**, 3484 (1984).
7. N. M. Doherty and J. E. Bercaw, *J. Am. Chem. Soc.*, **107**, 2670 (1985).
8. C. Masters, *Homogeneous Transition-metal Catalysis*, Chapman and Hall, London (1981).
9. A. Nakamura and M. Tsutsui, *Principles and Applications of Homogeneous Catalysis*, Wiley, New York (1980).
10. G. W. Parshall, *Homogeneous Catalysis*, Wiley, New York (1980); G. W. Parshall, *J. Mol. Catal.*, **4**, 243 (1978).
11. F. D. Mango, *Coord. Chem. Rev.*, **15**, 109 (1975).
12. J. U. Mondal and D. M. Blake, *Coord. Chem. Rev.*, **47**, 205 (1982).
13. D. Milstein, *Acc. Chem. Res.*, **17**, 221 (1984).
14. S. Obara, K. Kitaura, and K. Morokuma, *J. Am. Chem. Soc.*, **106**, 7482 (1984).
15. J. J. Low and W. A. Goddard III, *J. Am. Chem. Soc.*, **106**, 8321 (1984).
16. J. W. Suggs and C.-H. Jun, *J. Am. Chem. Soc.*, **106**, 3055 (1984).
17. D. H. Gerlach, A. R. Kane, A. W. Parshall, J. P. Jesson, and E. L. Muetterties, *J. Am. Chem. Soc.*, **93**, 3543 (1971).
18. M. Chanon and M. L. Tobe, *Angew. Chem., Int. Ed. Engl.*, **21**, 1 (1982); M. Chanon, *Bull. Soc. Chim. Fr.*, 197 (1982); M. Juliard and M. Chanon, *Chem. Rev.*, **83**, 425 (1983).
19. K. Hata, *Urushibara Catalysts*, University of Tokyo Press (1971).
20. V. I. Labunskaya, A. D. Shebaldova, and M. L. Khidekel, *Usp. Khim.*, **43**, 3 (1974).
21. B. R. James, *Homogeneous Hydrogenation*, Wiley, New York (1973); B. R. James, in: *Comprehensive Organometallic Chemistry* (G. Wilkinson, F. G. A. Stone, and E. W. Abel, eds.), Vol. 8, Chapter 51, Pergamon Press, Oxford (1982); B. R. James, *Adv. Organomet. Chem.*, **17**, 319 (1979).
22. G. Henrici-Olive and S. Olive, *Coordination and Catalysis*, Verlag Chemie, Weinheim (1977).
23. R. S. Coffey, in: *Aspects of Homogeneous Catalysis* (R. Ugo, ed.), Vol. 1, Carlo Manfredi, Milano (1970).
24. M. E. Volpin and I. S. Kolomnikov, *Usp. Khim.*, **38**, 561 (1969).
25. J. A. Osborn, F. H. Jardine, J. F. Young, and G. Wilkinson, *J. Chem. Soc. (A)*, 1711 (1966).

26. G. Wilkinson, *J. Chem. Soc. (A)*, 1711 (1966); J. Halpern, in: *Organotransition Metal Chemistry* (Y. Ishii and M. Tsutsui, eds.), Plenum, New York (1975).
27. C. Rousseau, M. Evrard, and F. Petit, *J. Mol. Catal.*, **3**, 309 (1977/1978); C. Rousseau, M. Evrard, and F. Petit, *J. Mol. Catal.*, **5**, 163 (1979).
28. M. H. J. M. de Croon, H. J. A. M. Kuipers, P. F. M. T. Nisselrooij, and J. W. E. Coenen, *Proceedings Symposium on Rhodium in Homogeneous Catalysis*, p. 44, Veszprem, Hungary (1978).
29. F. Pruchnik, M. Zuber, and S. Krzysztofik, *Rocz. Chem.*, **51**, 1177 (1977); M. Zuber, W. A. Szuberla, and F. Pruchnik, *J. Mol. Catal.*, **38**, 309 (1986).
30. M. Zuber and F. Pruchnik, *React. Kinet. Catal. Lett.*, **4**, 281 (1976).
31. J. C. Poulin, T. P. Dang, and H. B. Kagan, *J. Organomet. Chem.*, **84**, 87 (1975).
32. M. Zuber, B. Banas, and F. Pruchnik, *J. Mol. Catal.*, **10**, 143 (1981).
33. F. Pruchnik, *Inog. Nucl. Chem. Lett.*, **9**, 1229 (1973); F. Pruchnik, *Inorg. Nucl. Chem. Lett.*, **10**, 661 (1974); F. Pruchnik, *Bull. Acad. Polon. Sci., Ser. Chim.*, **21**, 793 (1973); H. Pasternak, T. Glowiak, and F. Pruchnik, *Inorg. Chim. Acta*, **19**, 11 (1976).
34. A. Dedieu, *Inorg. Chem.*, **19**, 375 (1980).
35. H. Mauermann, P. N. Swepston, and T. J. Marks, *Organometallics*, **4**, 200 (1985).
36. G. Folcher, J. F. Le Marechal, and H. Marquet-Ellis, *J. Chem. Soc., Chem. Commun.*, 323 (1982).
37. E. A. Karachanow, A. G. Dedow, and A. C. Loktew, *Usp. Khim.*, **54**, 289 (1985).
38. F. Pruchnik and A. Litewka, *Wiad. Chem.*, **36**, 525 (1982).
39. J. R. Bleeke and E. L. Muetterties, *J. Am. Chem. Soc.*, **103**, 556 (1981).
40. S. L. Grundy, A. J. Smith, H. Adams, and P. M. Maitlis, *J. Chem. Soc., Dalton Trans.*, 1747 (1984).
41. E. L. Muetterties and J. R. Bleeke, *Acc. Chem. Res.*, **12**, 324 (1979).
42. M. Bennett, *Chimetech.*, **10**, 444 (1980).
43. J. Wrister, *J. Chem. Soc., Chem. Commun.*, 575 (1977).
44. F. Pruchnik, M. Zuber, K. Wajda-Hermanowicz, and G. Kluczewska-Patrzalek, *Polish J. Chem.*, in press, Patent 132873.
45. P. Pino and G. Consiglio, in: *Fundamental Research in Homogeneous Catalysis* (M. Tsutsui, ed.), Plenum, New York (1977).
46. B. R. James, R. S. McMillan, R. H. Morris, and D. K. W. Wang, in: *Advances in Chemistry, Series No. 167, Transition Metal Hydrides* (R. Bau, ed.), Americain Chemical Society (1978).
47. M. Kumada and T. Hayashi, *Proceedings Symposium on Rhodium in Homogeneous Catalysis*, p. 157, Veszprem, Hungary (1978).
48. R. Selke, G. Pracejus, and H. Pracejus, *Proceedings Symposium on Rhodium in Homogeneous Catalysis*, p. 143, Veszprem, Hungary (1978); H. Pracejus, in: *Metal Complex Catalysis* (J. J. Ziółkowski, ed.), pp. 223–260, University of Wrocław, Wrocław (1980); H. Pracejus, *Wissen. Z. Ernst-Moritz-Arndt-Univ. Greifswald*, **31**, 5 (1982).
49. W. R. Cullen, Y. Sugi, and E. S. Yeh, *Proceedings Symposium on Rhodium in Homogeneous Catalysis*, p. 149, Veszprem, Hungary (1978).
50. H. B. Kagan, in: *Comprehensive Organometallic Chemistry* (G. Wilkinson, F. G. A. Stone, and E. W. Abel, eds.), Vol. 8, Chapter 53, Pergamon Press, Oxford (1982). H. B. Kagan. *Asymmetric Synthesis*, **5**, 1 (1985).
51. K. Drauz, A. Kleeman, and J. Martens, *Angew. Chem., Int. Ed. Engl.*, **21**, 584 (1982).
52. W. Bergstein, A. Kleeman, and J. Martens, *Synthesis*, 76 (1981).
53. M. D. Fryzuk and B. Bosnich, *J. Am. Chem. Soc.*, **99**, 6262 (1977).
54. Y. Amurani, D. Lafont, and D. Sinou, *J. Mol. Catal.*, **32**, 333 (1985).
55. M. Kumada, T. Hayashi, and K. Tamao, in: *Fundamental Research in Organometallic Chemistry* (M. Tsutsui, Y. Ishii, and H. Yaozeng), p. 175, Van Nostrand Reinhold, New York and Sciences Press, Beijing (1982).
56. K. Saito, S. Saigo, K. Kotera, and T. Date, *Chem. Pharm. Bull.*, **33**, 1342 (1985).
57. T. D. Appleton, W. R. Cullen, S. V. Evans, T.-J. Kim, and J. Trotter, *J. Organomet. Chem.*, **279**, 5 (1985).
58. T. Yamagishi, M. Yatagai, H. Hatakayama, and M. Hida, *Bull. Chem. Soc. Jpn.*, **57**, 1897 (1984).
59. I. S. Kołomnikov, V. P. Kukolev, and M. E. Volpin, *Usp. Khim.*, **43**, 904 (1974); G. Brieger and T. J. Nestrick, *Chem. Rev.*, **74**, 567 (1974).
60. A. S. C. Chan, J. J. Pluth, and J. Halpern, *J. Am. Chem. Soc.*, **102**, 5952 (1980).

738

Chapter 13

61. W. S. Knowles, B. D. Vineyard, M. J. Sabacky, and B. R. Stults, in: *Fundamental Research in Homogeneous Catalysis* (Y. Ishii and M. Tsutsui, eds.), Vol. 3, p. 537, Plenum, New York (1979).
62. R. A. W. Johnstone, A. W. Wilby, and I. D. Entwistle, *Chem. Rev.*, **85**, 129 (1985).
63. B. Heil, L. Marko, and S. Toros, *Homogeneous Catalysis with Metal Phosphine Complexes*, p. 317, Plenum, New York (1983).
64. H. Fujitsu, S. Shirahama, E. Matsurama, K. Takeshita, and I. Mochida, *J. Org. Chem.*, **46**, 2287 (1981).
65. W. Strohmeier and L. Weihelt, *J. Organomet. Chem.*, **145**, 189 (1978).
66. W. Strohmeier and H. Steigerwald, *J. Organomet. Chem.*, **129**, C43 (1977).
67. R. A. Sanchez-Delgado, A. Andriollo, O. L. de Ochoa, T. Suarez, and N. Valencia, *J. Organomet. Chem.*, **209**, 77 (1981).
68. G. Zassinovich, A. Camus, and G. Mestroni, *Inorg. Nucl. Chem. Lett.*, **12**, 865 (1976); G. Zassinovich, G. Mestroni, and A. Camus, *J. Mol. Catal.*, **2**, 63 (1977); G. Mestroni, G. Zassinovich, and A. Camus, *J. Organomet. Chem.*, **140**, 63 (1977); G. Mestroni, R. Spogliarich, A. Camus, F. Martinelli, and G. Zassinovich, *J. Organomet. Chem.*, **157**, 345 (1978).
69. H. Paternak, E. Lancman, and F. Pruchnik, *J. Mol. Catal.*, **29**, 13 (1985).
70. A. Camus, G. Mestroni, and H. Zassinovich, *J. Mol. Catal.*, **6**, 231 (1979).
71. M. Visintin, R. Spogliarich, J. Kaspar, and M. Graziani, *J. Mol. Catal.*, **32**, 349 (1985).
72. Z. Brouckova, M. Czakova, and M. Capka, *J. Mol. Catal.*, **30**, 241 (1985).
73. S. Toros, L. Kollar, B. Heil, and L. Marko, *J. Organomet. Chem.*, **232**, C17 (1982); P. Kvintovics, J. Bakos, and B. Heil, *J. Mol. Catal.*, **32**, 111 (1985).
74. A. Baranyai, F. Ungvary, and L. Marko, *J. Mol. Catal.*, **32**, 343 (1985).
75. K. Jothimony, S. Vancheesan, and J. C. Curiacose, *J. Mol. Catal.*, **32**, 11 (1985).
76. Yu. M. Zhorov, G. M. Panchenkov, and G. S. Volokhova, *Isomerizacja Olefinov*, Izd. Chimia, Moscow (1977).
77. C. Tolman, in: *Transition Metal Hydrides* (E. L. Muetterties, ed.), Marcel Dekker, New York (1971).
78. H. Boennemann, *Angew. Chem., Int. Ed. Engl.*, **9**, 736 (1970).
79. N. R. Davies, *Rev. Pure Appl. Chem.*, **17**, 83 (1967).
80. M. Green and R. P. Hughes, *J. Chem., Soc., Dalton Trans.*, 1907 (1976).
81. J. Boor, Jr., *Ziegler–Natta Catalysts and Polymerizations*, Academic Press, New York (1979).
82. W. W. Mazurek, *Polimeryzacja pod Deistvyem Soedinenii Perekhodnykh Metallov*, Izd. Nauka, Leningrad (1974).
83. M. M. Taqui Khan and A. E. Martell, *Homogeneous Catalysis by Metal Complexes*, Vols. 1 and 2, Academic Press, New York (1974).
84. H. Sinn and W. Kaminsky, *Adv. Organomet. Chem.*, **18**, 99 (1980); W. Kaminsky and H. Lueker, *Makromol. Chem. Rapid. Commun.*, **5**, 225 (1984); J. Deutschke, W. Kaminsky, and H. Lueker, in: *Polymer Reaction Engineering* (K. H. Reichert and W. Geiseler, eds.), p. 207, Hanser Publishers, Munich (1983).
85. I. Pasquon and U. Giannini, *Catalysis—Science and Technology*, **6**, 65 (1985).
86. B. A. Dolgoplosk and E. I. Tinyakova, *Metallorganicheskii Kataliz w Processakh Polimerizacii*, Izd. Nauka, Moscow (1982); B. A. Dolgoplosk and E. I. Tinyakova, *Usp. Khim.*, **53**, 40 (1984).
87. K. Weiss, in: *Transition Metal Carbene Complexes* (K. H. Doetz, ed.), Verlag Chemie, Weinheim (1983).
88. T. J. Katz, S. M. Hacker, R. D. Kendrick, and C. S. Yannoni, *J. Am. Chem. Soc.*, **107**, 2182 (1985); T. J. Ho and T. J. Kotz, *J. Mol. Catal.*, **28**, 359 (1985).
89. W. Kaminsky, K. Kuelper, H. H. Brintzinger, and F. R. W. P. Wild, *Angew. Chem., Int. Ed. Engl.*, **24**, 507 (1985).
90. T. Masuda and T. Higashimura, *Acc. Chem. Res.*, **17**, 51 (1984); T. Masuda, E. Isobe, and T. Higashimura, *Macromolecules*, **18**, 841 (1985).
91. P. L. Watson and G. W. Parshall, *Macromolecules*, **18**, 51 (1985).
92. K. J. Ivin, I. I. Rooney, C. D. Stewart, M. L. H. Green, and R. Mahtab, *J. Chem. Soc., Chem. Commun.*, 604 (1978); M. L. H. Green, *Pure Appl. Chem.*, **50**, 27 (1978).
93. A. Mortreux and F. Petit, *Symposium on Industrial Applications of Homogeneous Catalysis and Related Topics*, Villeneuve, p. 476 (September 23–26, 1985); E. Leroy, J. Hennion, J. Nicole, and

F. Petit, *Tetrahedron Lett.*, **27**, 2403 (1978); A. Mortreux, J. C. Bavay, and F. Petit, *Nouv. J. Chim.*, 671 (1980); M. Gilet, A. Mortreux, J. C. Folest, and F. Petit, *J. Am. Chem. Soc.*, **105**, 3876 (1983); M. Petit, A. Mortreux, and F. Petit, *J. Chem. Soc., Chem. Commun.*, 341 (1984).

94. B. A. Dolgoplosk, K. L. Makowieckij, E. I. Tinyakova, and O. K. Sharaev, *Polimerizacja Dienov pod Vliyaniem Allilnykh Komplekov*, Izd. Nauka, Moscow (1968).

95. D. H. Richards, *Chem. Soc. Rev.*, **6**, 235 (1976).

96. J. M. Thomassin, E. Walckiers, R. Warin, and P. Teyssie, *J. Polym. Sci.*, **13**, 1147 (1975).

97. Ph. Teyssie, M. Julemont, J. M. Thomassin, E. Walckiers, and R. Warin, in: *The Specific Polymerization of Diolefins by η³-allylic Coordination Complexes* (J. C. W. Chien, ed.), *Coordination Polymerization: A Memorial to K. Ziegler*, Academic Press, New York (1975).

98. C. I. Simionescu and V. Percec, *Prog. Polym. Sci.*, **8**, 133 (1982).

99. G. Wegner, *Angew. Chem., Int. Ed. Engl.*, **20**, 361 (1981).

100. M. F. Farona, P. A. Lofgren, and P. S. Woon, *J. Chem. Soc., Chem. Commun.*, 246 (1977).

101. J. E. Lydon, J. K. Nicholson, B. L. Shaw, and M. R. Truter, *Proc. Chem. Soc. London*, 421 (1964).

102. H. Boennemann, R. Brinkmann, and H. Schenkluhn, *Synthesis*, 575 (1974); H. Boennemann and R. Brinkmann, *Synthesis*, 600 (1975); H. Boennemann and H. Schenkluhn, Ger. Patent 2416295 (1975); H. Boennemann, VIII Polish–GDR Colloguy on Organometallic Chemistry, Mausz (1977).

103. H. Boennemann, *Angew. Chem., Int. Ed. Engl.*, **24**, 248 (1985).

104. S. Ostuka and A. Nakamura, *Adv. Organomet. Chem.*, **14**, 245 (1976).

105. F. L. Bowden and R. Giles, *Coord. Chem. Rev.*, **20**, 81 (1976).

106. R. Baker, *Chem. Rev.*, **73**, 487 (1973).

107. J. Falbe, *Sintezy na Osnowie Okisi Ugleroda*, Chimia, Leningradskoje otdielenie, 1971; J. Falbe, in: *New Syntheses with Carbon Monoxide* (J. Falbe, ed.), Springer-Verlag, Berlin (1980).

108. F. Calderazzo, *Angew. Chem., Int. Ed. Engl.*, **16**, 299 (1977).

109. B. K. Nefedov, *Sintiezy Organiczeskich Sojedinienij na Osnowie Okisi Ugleroda*, Izd. Nauka, Moscow (1978).

110. E. Kwaskowska-Chec and J. Ziolkowski, *Chemia Stosowana*, **24**, 433 (1980).

111. L. Marko, in: *Aspects of Homogeneous Catalysis* (R. Ugo, ed.), Vol. 2, p. 3, Reidel, Dordrecht (1974); T. M. Bockman, J. F. Garst, R. B. King, L. Marko, and F. Ungvary, *J. Organomet. Chem.*, **279**, 165 (1985).

112. I. Tkatchenko, in: *Comprehensive Organometallic Chemistry* (G. Wilkinson, F. G. A. Stone, and E. W. Abel, eds.), Vol. 8, Chapter 503, Pergamon Press, Oxford (1982).

113. G. Consiglio and P. Pino, *Top. Curr. Chem.*, 105, 77 (1982).

114. A. Stefani, G. Consiglio, C. Botteghi, and A. Pino, *J. Am. Chem. Soc.*, **99**, 1058 (1977).

115. R. T. Smith, R. K. Ungar, L. J. Sanderson, and M. C. Baird, *Organometallics*, **2**, 1138 (1983).

116. M. E. Davis, P. M. Butter, J. A. Rossin, and B. E. Hanson, *J. Mol. Catal.*, **31**, 385 (1985).

117. H. L. Pelt, J. J. J. J. Brockhus, R. P. J. Verburg, and J. J. F. Scholten, *J. Mol. Catal.*, **31**, 107 (1985); H. L. Pelt, N. A. DeMunck, R. P. J. Verburg, J. J. J. J. Brockhus, and J. J. F. Scholten, *J. Mol. Catal.*, **31**, 371 (1985); H. L. Pelt, R. P. Verburg, and J. J. F. Scholten, *J. Mol. Catal.*, **32**, 77 (1985).

118. T. Hayashi, M. Tanaka, I. Ogata, T. Kodama, T. Takahashi, Y. Uchida, and T. Uchida, *Bull. Chem. Soc. Jpn.*, **56**, 1780 (1983).

119. T. Hayashi, M. Tanaka, and I. Ogata, *J. Mol. Catal.*, **26**, 17 (1984).

120. T. O. Okano, T. Kobayashi, H. Konishi, and J. Kiji, *Tetrahedron Lett.*, **23**, 4967 (1982).

121. H. P. Withers, Jr. and D. Seyferth, *Inorg. Chem.*, **22**, 2931 (1983).

122. P. Escaffre, A. Dediu, J. M. Frances, P. Kalck, and A. Thorez, 12 ICOMC, Vienna, *Abstracts* p. 458 (September 8–13, 1985); P. Kalck, A. Thorez, M. T. Pinnillos, and L. A. Oro, *J. Mol. Catal.*, **31**, 311 (1985); P. Kalck, J. M. Frances, P.-M. Pfister, T. G. Southern, and A. Thorez, *J. Chem. Soc., Chem. Commun.*, 510 (1983).

123. D. Forster and T. C. Singleton, *J. Mol. Catal.*, **17**, 299 (1982).

124. G. Braca, G. Sbrana, G. Valentini, and C. Barberini, *C₁ Mol. Chem.*, **1**, 9 (1984); G. Braca, B. Dattilo, G. Guainai, G. Sbrana, and G. Valentini, *Proceedings of the 9th Iberoamerican Symposium on Catalysis*, Vol. 1, p. 416, Lisbon (July 16–21, 1984).

125. T. W. Dekleva and D. Forster, *J. Am. Chem. Soc.*, **107**, 3565 (1985); T. W. Dekleva and D. Forster, *J. Am. Chem. Soc.*, **107**, 3568 (1985).

126. A. M. R. Galletti, G. Braca, G. Sbrana, and F. Marchetti, *J. Mol. Catal.*, **32**, 291 (1985).

127. R. W. Wegman and D. C. Busby, *J. Mol. Catal.*, **32**, 125 (1985).
128. J. F. Knifton, *J. Mol. Catal.*, **30**, 281 (1985).
129. A. S. C. Chan, W. E. Carroll, and D. E. Willis, *J. Mol. Catal.*, **19**, 377 (1983).
130. F. Ozawa, T. Sugimoto, Y. Yuasa, M. Santra, T. Yamamoto, and A. Yamamoto, *Organometallics*, **3**, 683 (1984); F. Ozawa, T. Sugimoto, T. Yamamoto, and A. Yamamoto, *Organometallics*, **3**, 692 (1984); F. Ozawa, H. Soyama, H. Yanagihara, I. Aoyama, H. Takino, K. Izawa, T. Yamamoto, and A. Yamamoto, *J. Am. Chem. Soc.*, **107**, 3235 (1985).
131. J.-T. Chen and A. Sen, *J. Am. Chem. Soc.*, **106**, 1506 (1984).
132. P. E. Garrou, *Chem. Rev.*, **85**, 171 (1985).
133. R. A. Dubois, P. E. Garrou, K. D. Lavin, and H. R. Allcock, *Organometallics*, **3**, 649 (1984).
134. A. G. Abatjoglou and D. R. Bryant, *Organometallics*, **3**, 932 (1984); A. G. Abatjoglou, E. Billig, and D. R. Bryant, *Organometallics*, **3**, 923 (1984).
135. E. S. Brown, in: *Aspects of Homogeneous Catalysis* (R. Ugo, ed.), Vol. 2, p. 57, Reidel, Dordrecht (1974); E. S. Brown, in: *Organic Synthesis by Metal Carbonyls* (I. Wender and P. Pino, eds.), Vol. II, p. 655, Wiley, New York (1977).
136. R. J. McKinney and D. C. Roe, *J. Am. Chem. Soc.*, **107**, 261 (1985).
137. J. L. Speier, *Adv. Organomet. Chem.*, **17**, 407 (1979).
138. B. Marciniec, *Hydrosililowanie*, PWN, Warsaw (1981).
139. A. J. Blakeney and J. A. Gladysz, *Inorg. Chim. Acta*, **53**, L25 (1981).
140. A. Millan, M.-J. Fernandez, P. Bentz, and P. M. Maitlis, *J. Mol. Catal.*, **26**, 89 (1984).
141. B. A. Dolgoplosk and Yu. W. Korshak, *Usp. Khim.*, **53**, 65 (1984).
142. R. H. Grubbs, in: *Comprehensive Organometallic Chemistry* (G. Wilkinson, F. G. A. Stone, and E. W. Abel, eds.), Vol. 8, p. 499, Pergamon Press, Oxford (1982).
143. K. H. Doetz, *Angew. Chem., Int. Ed. Engl.*, **23**, 587 (1984).
144. N. Calderon, J. P. Lawrence, and E. A. Ofstead, *Adv. Organomet. Chem.*, **17**, 449 (1979); N. Calderon, E. A. Ofstead, and W. A. Judy, *Angew. Chem., Int. Ed. Engl.*, **15**, 401 (1976).
145. W. Sz. Feldblum, *Dimerizacija i Disproporcjonirowanie Olefinov*, Chimia, Moscow (1978).
146. R. J. Haines and G. J. Leigh, *Chem. Soc. Rev.*, **4**, 155 (1975).
147. T. J. Katz, *Adv. Organomet. Chem.*, **16**, 283 (1977).
148. A. Korda, Thesis, Warsaw Technical University (1979).
149. A. K. Rappe and W. A. Goddard, III, *J. Am. Chem. Soc.*, **104**, 448 (1982); K. J. Ivin, B. S. R. Reddy, and J. J. Rooney, *J. Chem. Soc., Chem. Commun.*, 1062 (1981).
150. M. R. Churchill, J. W. Ziller, J. H. Freudenberger, and R. R. Schrock, *Organometallics*, **3**, 1554 (1984).
151. J. H. Wengrovius, J. Sancho, and R. R. Schrock, *J. Am. Chem. Soc.*, **103**, 3932 (1981); J. Sancho and R. R. Schrock, *J. Mol. Catal.*, **15**, 75 (1982); R. R. Schrock, D. N. Clark, J. Sancho, J. H. Wengrovius, S. M. Rocklage, and S. F. Pedersen, *Organometallics*, **1**, 1645 (1982); R. R. Schrock, M. L. Listemann, and L. G. Sturgeoff, *J. Am. Chem. Soc.*, **104**, 4291 (1982).
152. J. H. Freudenberger, R. R. Schrock, M. R. Churchill, A. L. Rheingold, and J. W. Ziller, *Organometallics*, **3**, 1563 (1984).
153. G. J. Leigh, M. T. Rahman, and D. R. Walton, *J. Chem. Soc., Chem. Commun.*, 541 (1982).
154. A. Bencheick, M. Petit, A. Mortreux, and F. Petit, *J. Mol. Catal.*, **15**, 93 (1982)).
155. G. P. Pez, *Chem. Commun.*, 560 (1977).
156. J. I. C. Archibald, J. J. Roney, and A. Stewart, *Chem. Commun.*, 547 (1975).
157. M. G. Clerici, S. DiGioacchino, F. Maspero, E. Perroti, and A. Zanobi, *J. Organomet. Chem.*, **84**, 379 (1975).
158. D. J. Milner and R. Whelan, *J. Organomet. Chem.*, **152**, 193 (1978).
159. J. Cook, J. E. Hamlin, A. Nutton, and P. M. Maitlis, *J. Chem. Soc., Dalton Trans.*, 2342 (1981).
160. E. L. Muetterties and J. Stein, *Chem. Rev.*, **79**, 479 (1979).
161. C. Masters, *Adv. Organomet. Chem.*, **17**, 61 (1979).
162. K. D. Froning, G. Koelbel, M. Ralek, F. Schnur, and H. Schulz, *Chemierohstoffe aus Kohle* (J. Falbe, ed.), Georg. Thiem, Stuttgart (1977).
163. S. M. Loktev, in K. D. Froning, G. Koelbel, M. Ralek, W. Rottig, F. Schnur, and H. Schulz, *Khimicheskie Veshchestva iz Ugla*, p. 568, Izd. Chimia, Moscow (1980).

164. H. Schulz and A. Zein El Deen, *Fuel Processing Technol.*, **1**, 31 (1977).
165. B. K. Nefedov, *Sintiezy Organicheskikh Soedinenii na Osnowie Okisi Ugleroda*, Izd. Nauka, Moscow (1978).
166. G. Henrici-Olive and S. Olive, *Angew. Chem., Int. Ed. Engl.*, **15**, 136 (1976).
167. A. L. Lapidus and M. M. Sawieliew, *Usp. Khim.*, **53**, 925 (1984).
168. J. R. Blackborow, R. J. Daroda, and G. Wilkinson, *Coord. Chem. Rev.*, **43**, 17 (1982).
169. W. A. Herrmann, *Angew. Chem., Int. Ed. Engl.*, **21**, 117 (1982).
170. G. Henrici-Olive and S. Olive, *Catalyzed Hydrogenation of Carbon Monoxide*, Springer-Verlag, Berlin (1984).
171. B. D. Dombek, *Adv. Catal.*, **32**, 325 (1983).
172. J. A. Gladysz, *Adv. Organomet. Chem.*, **20**, 1 (1982).
173. P. Biloen, J. N. Helle, and W. M. H. Sachtler, *J. Catal.*, **58**, 95 (1979).
174. G. Low and A. T. Bell, *J. Catal.*, **57**, 397 (1979).
175. J. A. Rabo, A. P. Risch, and M. L. Poutsma, *J. Catal.*, **53**, 295 (1978).
176. J. G. Ekerdt and A. T. Bell, *J. Catal.*, **58**, 170 (1979).
177. K. Whitmire and D. F. Shriver, *J. Am. Chem. Soc.*, **103**, 6754 (1981); K. Whitmire and D. F. Shriver, *J. Am. Chem. Soc.*, **102**, 1456 (1980); C. P. Horwitz and D. F. Shriver, *J. Am. Chem. Soc.*, **107**, 8147 (1985); E. M. Holt, K. H. Whitmire, and D. F. Shriver, *J. Organomet. Chem.*, **213**, 125 (1981).
178. C. P. Horwitz and D. F. Shriver, *Adv. Organomet. Chem.*, **23**, 219 (1984).
179. S. D. Wijeyesekera, R. Hoffmann, and C. N. Wilker, *Organometallics*, **3**, 962 (1984); S. Harris and J. S. Bradley, *Organometallics*, **3**, 1086 (1984).
180. C. P. Casey, M. A. Andrews, D. R. McAlister, and J. E. Rinz, *J. Am. Chem. Soc.*, **102**, 1927 (1980); C. P. Casey and S. M. Naumann, *J. Am. Chem. Soc.*, **100**, 2544 (1978).
181. R. A. Fiato, J. L. Vidal, and R. L. Pruett, *J. Organomet. Chem.*, **172**, C4 (1979); J. N. Cawse, R. A. Fiato, and R. L. Pruett, *J. Organomet. Chem.*, **172**, 405 (1979); R. C. Schoening, J. L. Vidal, and R. A. Fiato, *Fundamental Research in Metal Complex Catalysis*, Workshop, Wroclaw (5–8. IX, 1980).
182. K. G. Caulton, *Metal Complex Catalysis*, Proceedings of the First International Summer School of Metal Complex Catalysis (J. J. Ziolkowski, ed.), p. 43, University of Wrocław, Wrocław (1980); J. A. Marsella, K. G. Moloy, and K. G. Caulton, *J. Organomet. Chem.*, **201**, 389 (1980).
183. W. K. Wong, W. Tam, C. E. Strouse, and J. A. Gladysz, *Chem. Commun.*, 530 (1979); W. K. Wong, W. Tam, C. E. Strouse, and J. A. Gladysz, *J. Am. Chem. Soc.*, **101**, 5440 (1979); J. A. Gladysz and W. Tam, *J. Am. Chem. Soc.*, **100**, 2545 (1978).
184. R. J. Daroda, J. R. Blackborow, and G. Wilkinson, *Chem. Commun.*, 1098 (1980); R. J. Daroda, J. R. Blackborow, and G. Wilkinson, *Chem. Commun.*, 1101 (1980).
185. J. W. Rathke and H. M. Feder, *J. Am. Chem. Soc.*, **100**, 3623 (1978).
186. G. C. Demitras and E. L. Muetterties, *J. Am. Chem. Soc.*, **99**, 2796 (1977).
187. L. N. Lewis and K. G. Caulton, *Inorg. Chem.*, **19**, 3201 (1980).
188. W. Herrmann, M. Ziegler, K. Weidenhammer, and K. Biersack, *Angew. Chem., Int. Ed. Engl.*, **18**, 960 (1979).
189. J. M. Manriquez, D. R. McAlister, R. D. Sanner, and J. E. Bercaw, *J. Am. Chem. Soc.*, **100**, 2716 (1978); P. T. Wolczanski and J. E. Bercaw, *Acc. Chem. Res.*, **13**, 121 (1980); P. T. Wolczanski, R. S. Threlkel, and J. E. Bercaw, *J. Am. Chem. Soc.*, **101**, 218 (1979).
190. J. A. Toth and M. Orchin, *J. Organomet. Chem.*, **172**, C27 (1979).
191. K. L. Brown, G. R. Clark, C. E. Headford, K. Marsden, and W. R. Roper, *J. Am. Chem. Soc.*, **101**, 503 (1979).
192. W. Keim, M. Roper, and H. Strutz, *J. Organomet. Chem.*, **219**, C5 (1981).
193. J. S. Bradley, G. B. Ansell, and E. W. Hill, *J. Am. Chem. Soc.*, **101**, 7417 (1979).
194. J. R. Shapley and J. T. Park, *J. Am. Chem. Soc.*, **106**, 1144 (1984).
195. K. M. Nicholas, *Organometallics*, **1**, 1713 (1982).
196. W. Keim, M. Anstock, M. Roper, and J. Schlupp, $C_1$ *Mol. Chem.*, **1**, 21 (1984).
197. M. A. Lilga and J. A. Hers, *Organometallics*, **4**, 590 (1985).
198. D. L. Thorn and T. H. Tulip, *Organometallics*, **1**, 1580 (1982).

199. J. F. Knifton, R. A. Grisby, Jr., and J. J. Lin, *Organometallics*, **3**, 62 (1984).
200. D. A. Palmer and R. van Eldik, *Chem. Soc. Rev.*, **83**, 651 (1983).
201. D. J. Darensbourg and R. A. Kudaroski, *Adv. Organomet. Chem.*, **22**, 129 (1983).
202. J. A. Ibers, *Chem. Soc. Rev.*, **11**, 57 (1982).
203. M. E. Volpin and I. S. Kolomnikov, *Organomet. React.*, **5**, 313 (1975).
204. R. P. A. Sneeden, in: *Comprehensive Organometallic Chemistry* (G. Wilkinson, F. G. A. Stone, and E. W. Abel, eds.), Vol. 8, p. 225, Pergamon Press, Oxford (1982).
205. C. Mealli, R. Hoffmann, and A. Stockis, *Inorg. Chem.*, **23**, 56 (1984).
206. G. S. Bristow, P. B. Hitchcock, and M. F. Lappert, *J. Chem. Soc., Chem. Commun.*, 1145 (1981).
207. S. Gambarotta, C. Floriani, A. ChiesiVilla, and C. Guastini, *J. Am. Chem. Soc.*, **107**, 2985 (1985).
208. W. Beck, K. Raab, U. Nagel, and M. Steimann, *Angew. Chem., Int. Ed. Engl.*, **21**, 526 (1982).
209. S. Gambarotta, F. Arena, C. Floriani, and P. F. Zanazzi, *J. Am. Chem. Soc.*, **104**, 5082 (1982).
210. R. Alvarez, E. Carmona, and M. L. Poveda, *J. Am. Chem. Soc.*, **106**, 2731 (1984).
211. T. Herskovitz and L. J. Guggenberger, *J. Am. Chem. Soc.*, **98**, 1615 (1976).
212. A. Musco, C. Perego, and V. Tartiarai, *Inorg. Chim. Acta*, **28**, L147 (1978).
213. Y. Sasaki, Y. Inoue, and H. Hashimoto, *J. Chem. Soc., Chem. Commun.*, 605 (1976).
214. Y. Inouo, Y. Itoh, and H. Hashimoto, *Chem. Lett.*, 855 (1977).
215. D. Walther, E. Dinjus, J. Sieler, L. Andersen, and O. Lindqvist, *J. Organomet. Chem.*, **276**, 99 (1984); D. Walther, E. Dinjus, H. Goerls, J. Sieler, O. Lindqvist, and L. Andersen, *J. Organomet. Chem.*, **286**, 103 (1985); D. Walther and E. Dinjus, *Z. Chem.*, **24**, 63 (1984); D. Walther, E. Dinjus, and V. Herzog, *Z. Chem.*, **24**, 260 (1984).
216. S. Slater and J. H. Wagenknecht, *J. Am. Chem. Soc.*, **106**, 5367 (1984).
217. J. Hawecker, J.-M. Lehn, and R. Ziessel, *J. Chem. Soc., Chem. Commun.*, 328 (1984).
218. J. C. Gressin, D. Michelet, L. Najdo, and J. M. Saveant, *Nouv. J. Chem.*, **3**, 545 (1979).
219. B. Denise and R. P. A. Sneeden, *J. Organomet. Chem.*, **221**, 111 (1981).
220. P. Sobota and B. Jezowska-Trzebiatowska, *Coord. Chem. Rev.*, **26**, 71 (1978); P. Sobota, *Wiad. Chem.*, **31**, 101 (1977).
221. P. C. Ford, *Acc. Chem. Res.*, **14**, 31 (1981); C. Ungermann, V. Landis, S. A. Moya, H. Cohen, H. Walker, R. G. Pearson, R. G. Rinker, and P. C. Ford, *J. Am. Chem. Soc.*, **101**, 5922 (1979).
222. D. C. Gross and P. C. Ford, *J. Am. Chem. Soc.*, **107**, 585 (1985).
223. D. M. Vandenberg, T. M. Suzuki, and P. C. Ford, *J. Organomet. Chem.*, **272**, 309 (1984).
224. J. Bricker, C. C. Nagel, A. A. Bhattacharyya, and C. G. Shore, *J. Am. Chem. Soc.*, **107**, 377 (1985).
225. D. A. King, R. B. King, and D. B. Yang, *J. Chem. Soc., Chem. Commun.*, 529 (1980); D. A. King, R. B. King, and D. B. Yang, *J. Am. Chem. Soc.*, **102**, 1028 (1980); R. B. King, C. C. Frazier, R. M. Hanes, and A. D. King, Jr., *J. Am. Chem. Soc.*, **100**, 2925 (1978).
226. Y. Doi, A. Yokota, H. Miyake, and K. Soga, *Inorg. Chim. Acta*, **90**, L7 (1984); Y. Doi, A. Yokota, H. Miyake, and K. Soga, *J. Chem. Soc., Chem. Commun.*, 394 (1984).
227. J. C. Bricker, N. Bhattacharyya, and S. G. Shore, *Organometallics*, **3**, 201 (1984).
228. C.-H. Cheng, D. E. Hendriksen, and R. Eisenberg, *J. Am. Chem. Soc.*, **99**, 2791 (1977); E. C. Baker, D. E. Hendriksen, and R. Eisenberg, *J. Am. Chem. Soc.*, **102**, 1020 (1980); C.-H. Cheng and R. Eisenberg, *J. Am. Chem. Soc.*, **100**, 5968 (1978).
229. P. Giannoccaro, G. Vasapollo, and A. Sacco, *J. Chem. Soc., Chem. Commun.*, 1136 (1980).
230. W. R. Tikkanen, E. Binamira-Soriaga, W. C. Kaska, and P. C. Ford, *Inorg. Chem.*, **22**, 1147 (1983).
231. D. Mahajan, C. Creutz, and N. Sutin, *Inorg. Chem.*, **24**, 2063 (1985).
232. T. Okano, T. Kobayashi, H. Konishi, and J. Kiji, *Bull. Chem. Soc. Jpn.*, **54**, 3799 (1981).
233. F. Pruchnik, G. Kluczewska-Patrzalek, 3rd International Symposium on Homogeneous Catalysis August 30–September 3, 1982, Milano, *Abstracts*, p. 179; submitted for publication.
234. A. Basinska and F. Domka, *Wiad. Chem.*, **40**, 351 (1986).
235. K. R. Lane, R. E. Lee, L. Sallans, and R. R. Squires, *J. Am. Chem. Soc.*, **106**, 5767 (1984).
236. R. M. Laine, *J. Am. Chem. Soc.*, **100**, 6451 (1978).
237. K. Kaneda, M. Yasumura, T. Imanake, and S. Teranishi, *J. Chem. Soc., Chem. Commun.*, 935 (1982).
238. I. E. Ruyter, XXII ICCC, *Abstracts*, Vol. 1, p. 170 (August 23–27, 1982).
239. B. M. Trost, T. R. Verhoeven, in: *Comprehensive Organometallic Chemistry* (G. Wilkinson, F. G. A. Stone, and E. W. Abel, eds.), Vol. 8, p. 799, Pergamon Press, Oxford (1982).

240. B. Akermark and K. Zetterberg, *J. Am. Chem. Soc.*, **106**, 5560 (1984).
241. L. S. Hegedus, B. Akermark, K. Zetterberg, and L. F. Olsson, *J. Am. Chem. Soc.*, **106**, 7122 (1984).
242. J. Haber and A. Bielanski, *Catal. Rev. Sci. Eng.*, **19**, 1 (1979).
243. P. N. Rylander, *Catalysis—Science and Technology*, Vol. 4, p. 1, Akademie-Verlag, Berlin (1983).
244. D. M.-T. Chan, W. C. Fultz, W. A. Nugent, D. C. Roe, and T. H. Tulip, *J. Am. Soc.*, **107**, 251 (1985).
245. D. M.-T. Chan and W. A. Nugent, *Inorg. Chem.*, **24**, 1422 (1985).
246. C. U. Pittman, Jr., in: *Comprehensive Organometallic Chemistry* (G. Wilkinson, F. G. A. Stone, and E. W. Abel, eds.), Vol. 8, p. 553, Pergamon Press, Oxford (1982).
247. N. L. Holy, in: *Homogeneous Catalysis with Metal Phosphine Complexes* (L. H. Pignolet, ed.), p. 443, Plenum, New York (1983).
248. Yu. I. Yermakov, *J. Mol. Catal.*, **21**, 35 (1983); Yu. I. Yermakov, B. N. Kuznetsov, and V. A. Zakharov, *Catalysis by Supported Complexes*, Elsevier, Amsterdam (1981).
249. Z. M. Michalska, *Wiad. Chem.*, **39**, 545 (1985); Z. M. Michalska and D. E. Webster, *Platinum Met. Rev.*, **18**, 65 (1975); Z. M. Michalska and D. E. Webster, *Wiad. Chem.*, **29**, 747 (1975).
250. F. R. Hartley and P. N. Vezey, *Adv. Organomet. Chem.*, **15**, 189 (1977).
251. B. Delmon and G. Jannes, *Catalysis, Heterogeneous and Homogeneous*, Elsevier, Amsterdam (1975).
252. A. Ja. Juffa and G. W. Lisiczkin, *Usp. Khim.*, **47**, 1414 (1978).
253. R. Whyman, in: *Transition Metal Clusters* (B. F. G. Johnson, ed.), p. 545, Wiley, Chichester (1980).
254. J. M. Basset, *Industrial Applications of Homogeneous Catalysis and Related Topics*, Villeneuve d'Ascq E.N.S.C.L.U.S.T.L., p. 494, (September 23–26, 1985); J. M. Basset, B. Besson, A. Choplin, and A. Theolier, *Phil. Trans. R. Soc. London, Ser. A*, **308**, 115 (1982); J. M. Basset, B. Besson, A. Choplin, and A. Theolier, *Mechanism Kataliza*, Vol. 2, p. 154, Izd. Nauka, Sibirskoe Otdelenie, Novosibirsk (1984).
255. I. I. Moiseev, *Mechanism Kataliza*, Vol. 1, p. 72, Izd. Nauka, Sibirskoe Otdelenie, Novosibirsk (1984).

# Abbreviations

| | |
|---|---|
| A | Acetone |
| Ac | Acetyl, $CH_3CO$ |
| acac | Acetylacetonate anion |
| adme | 1-Adamantylmethyl |
| AN | Acrylonitrile |
| Ar | Aryl |
| BD | 1, 3-butadiene |
| bipy | 2, 2'-bipyridine |
| BM | Bohr magneton |
| Bu | Butyl |
| CBD | Cyclobutadiene |
| CDO | Chelating diolefin, e.g.: COD, NBD, etc. |
| CDT | Cyclododecatriene |
| CH | Saturated hydrocarbons |

chp      6-Chloro-2-hydroxypyridine anion,

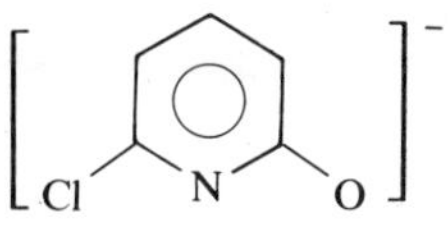

| | |
|---|---|
| CHT | Cycloheptatrienyl |

CMO      Chelating monoolefin, e.g.:

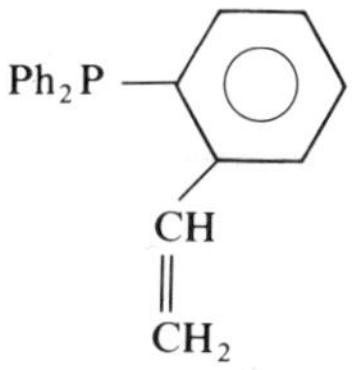

| | |
|---|---|
| COD | 1, 5-Cyclooctadiene |
| COT | Cyclooctatetraene |
| cp | Cyclopentadienyl |

| | |
|---|---|
| CVE | Cluster valence electrons |
| CVMO | Cluster valence molecular orbitals |
| Cy | Cyclohexyl |
| DAB | $RN = CH - CH = NR$, 1, 4-diazo-1, 3-butadiene |
| dec | Decomposition |
| depe | 1, 2-bis(diethylphosphino)ethane |
| depm | 1, 2-bis(diethylphosphino)methane |
| DHMB | Dewar hexamethylbenzene |
| diglyme | $MeOCH_2CH_2OCH_2CH_2OMe$ |
| diphos / dppe | 1, 2-bis(diphenylphosphino)ethane |
| DMA | Dimethylacetamide |
| DMF | Dimethylformamide |

dmp      2, 6-Dimethoxyphenyl anion,

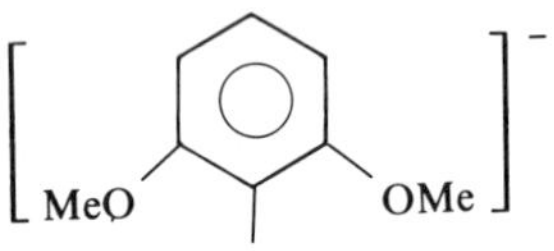

dmhp      2, 4-Dimethyl-6-hydroxypyrimidine anion,

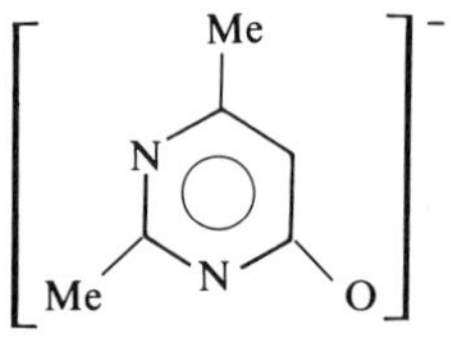

| | |
|---|---|
| $dmgH_2$ | Dimethylglyoxime |
| dmpe | 1, 2-bis(dimethylphosphino)ethane |
| dmpm | 1, 2-bis(dimethylphosphino)methane |
| dppm | bis(diphenylphosphino)methane |
| DUR | Duroquinone |
| ee | Enantiomeric excess |
| en | Ethylenediamine |
| Et | Ethyl |
| FN | Fumaronitrile (*trans*-1, 2-dicyanoethylene) |
| HMPA | Hexamethylphosphoric triamide |
| HOMO | Highest occupied molecular orbital |
| L | 2e ligand, e.g., $PR_3$, $NR_3$, and so on |
| LUMO | Lowest unoccupied molecular orbital |
| m | Medium, moderate |
| MA | Maleic anhydride |
| Me | Methyl |

| | | |
|---|---|---|
| map | 6-Methyl-2-aminopyridine anion, 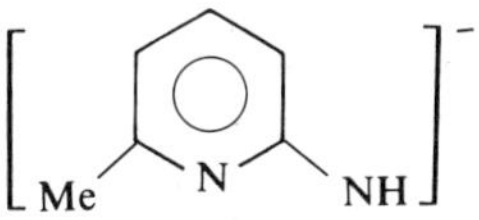 | |

mcp      Methylcyclopentadienyl, $C_5H_4Me$

mes      Mestityl, 2, 4, 6-$C_6H_2Me_3$

| | | |
|---|---|---|
| mhp | 6-Methyl-2-hydroxypyridine anion, 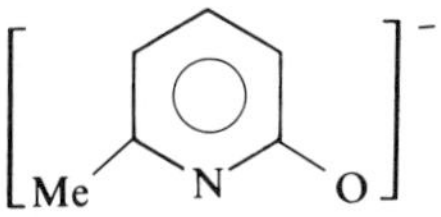 | |

MN      Maleonitrile (cis-1, 2-dicyanoethylene)

NBD      Norbornadiene

np      Neopentyl

OEP      2, 3, 7, 8, 12, 13, 17, 18-Octaethylporphyrinato dianion

Ph      Phenyl

phen      1, 10-Penantroline

pmcp      Pentamethylcyclopentadienyl

PMDT      Pentamethyldiethylenetriamine

pol      Polarizable (band)

PPN      $[N(PPh_3)_2]^+$

Pr      Propyl

py      Pyridine

s      Strong

salen

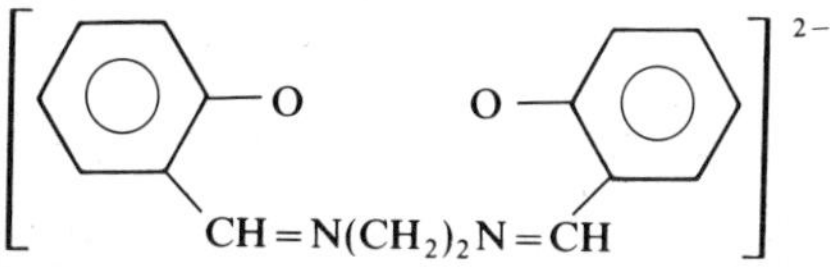

saloph

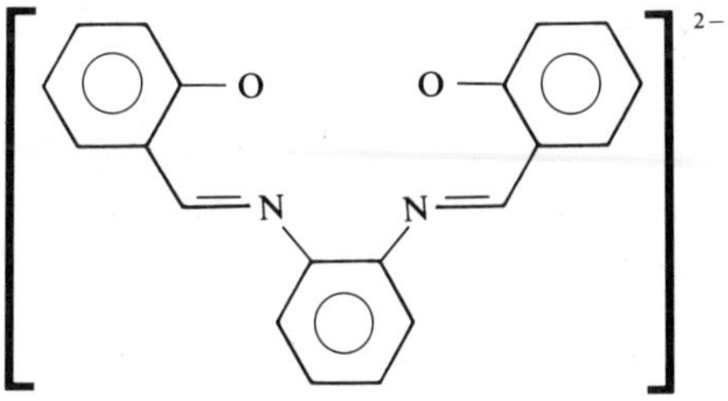

tcbiH$_2$      Tetracyanobiimidazole

TCNE      Tetracyanoethylene

TDPME      tris(diphenylphosphinomethyl)ethane

TEED      Tetraethylethylenediamine

THF      Tetrahydrofuran

TMED      Tetramethylethylenediamine

tmm      Trimethylenemethane

| | |
|---|---|
| tmp | 2, 4, 6-Trimethoxyphenyl |
| *p*-tol | *p*-Tolyl, p-C$_6$H$_4$Me |
| tripod | MeC(CH$_2$PPh$_2$)$_3$ |
| TTP | 5,10,15,20-Tetratolylporphyrinato dianion |
| v | Very |
| 4-VCH | 4-Vinylcyclohexene |
| w | Weak |
| XPE | X-ray photoelectron spectroscopy (ESCA) |

# Index

Acetylene complexes
 bonding, 389, 390
 enthalpy of decomposition, 396
 ESCA, 396
 IR, 394–396
 NMR, 396
 preparation, 396–398
  alkyne addition, 396, 397
  ligand substitution, 397, 398
 properties, 398–404
 reactions, 404–409
  acylation, 406
  decomposition, 404
  metallacyclization, 409–412
  migration, 405, 407–409
 structure, 389–391, 399–401
Actinide complexes
 allyl, 460, 461
 arene, 586
 cyclooctatetrane, 606–608
 cyclopentadienyl, 524, 551–554
 σ-hydrocarbyl, 227, 228
 isocyanide, 624
Activation of the C−H bond in alkanes,
  262–267
 by electron-deficient complexes, 265–267
 mechanisms of, 262–264
 by metal atoms, 267, 268
 solar energy storage, 269
 by superacids, 262
A frames, 91, 95
Agostic complexes, 264, 265
Allene complexes
 bonding, 415, 416
 dynamic properties, 418, 419

Allene complexes (cont.)
 IR, 417
 NMR, 418, 419
 reactions, 419–421
  allene exchange, 419
  migratory insertion, 420
  oligomerization, 421, 684, 687, 689, 690
  polymerization, 421
  protonation, 420
 structure, 415, 416
Allyl complexes
 bonding, 427–431
 dynamic properties, 434–439
 electronic spectra, 439, 440
 group 3 metals, 460, 461
 group 4 metals, 450, 451
 group 5 metals, 451
 group 6 metals, 451–453
 group 7 metals, 453–454
 group 8 metals, 454–456
 group 9 metals, 457–459
 group 10 metals, 458–460
 IR, 431, 432
 NMR, 432–434
 photoelectron spectra, 441, 443
 preparation, 444–449
  from allenes, 448
  from 1,3-dienes, 446, 447
  Grignard reagents, 444
  lithium allyls, 444
  from olefins, 445, 446
  oxidative addition, 445
 reactions, 463–468
  with electrophiles, 466
  formation of σ-allyl complexes, 466

Allyl complexes *(cont.)*
  reactions *(cont.)*
    migration, 467, 468
    with nucleophiles, 464
    reduction, 466
    reductive elimination, 467
    thermal decomposition, 463, 464
  structure, 440–443
Amination reactions, 731, 732
Amoxidation reactions, 732
Arene complexes
  application, 599, 600
  bonding, 575–577
  electronic spectra, 578
  ESR, 582, 583
  IR, 579
  magnetic properties, 578, 579
  mass spectra, 585
  NMR, 579–581
  photoelectron spectra, 580, 582
  preparation, 590–594
  properties, 586, 587
  reactions, 594–599
    displacement of ligands, 598, 599
    electrophilic substitution, 596
    nucleophilic substitution, 596–598
    redox, 594
  structure, 576, 583, 584
  thermochemistry, 584, 585
  triple layer, 558, 589
Asymmetric ligands, 662–666

Butadiene complexes
  bonding, 471–476
  IR, 485
  NMR, 480–484
  photoelectron spectra, 485
  preparation, 485–489
  properties, 486, 504, 505
  reactions, 498, 499
  structure, 476–481

Carbene complexes
  applications, 322–324
  bonding, 277–280
  electronic spectra, 290, 291
  ESR, 291
  IR, 285
  magnetic properties, 291
  NMR, 285–287
  photoelectron spectra, 287, 290
  preparation, 292–299
  properties of, 302–305, 307, 308
  reactions, 308–317
    with electrophiles, 314
    migration, 317

Carbene complexes *(cont.)*
  reactions *(cont.)*
    nucleophilic substitution, 308–310
    oxidation, 313, 314
    rearrangement, 311, 312
    reduction, 312, 313
    release of carbene, 315–317
    substitution, 314, 315
    substitution at $\alpha$-C, 312
Carbon dioxide
  complexes, 725, 726
  reduction of, 729
Carbon dioxide reactions, 725–729
  with acetylenes, 728
  with alkoxides, 727
  with amides, 727
  with dienes, 728
  with hydride complexes, 726, 727
Carbonyl hydrides, 96–100
  acidic properties of, 100, 107
  bonding, 96, 97
  identification of, 97, 100, 104
  preparation of, 101–107
  properties of, 101–103, 108–110
  reactions of, 108–110
  structure of, 97–99, 107
Carbonyl ligand, modes of bonding, 24–26
Carbonylation, 690–702
Carbonylation reactions, 694–701
  alcohol, 694–698, 700, 701
  unsaturated amides, 698, 699
Carbonyls
  A frames, 91, 95
  applications, 119–121
    inorganic syntheses, 119–121
  bonding, 24–32
  chain complexes, 121–122
  clusters, 93, 94
  column structures, 121–122
  dynamic properties, 48, 49
  electronic spectra, 41, 42, 45
  ESCA, 39, 42, 49, 52, 53
  exchange of CO with, 55, 56
  group 3 metal, 60, 61, 63
  group 4 metal, 63, 64
  group 5 metal, 64–69
  group 6 metal, 69–75
  group 7 metal, 75–80
  group 8 metal, 79–85
    clusters of, 83, 84
    polynuclear, 80–84
    structure of, 81, 82, 84
  group 9 metal, 84–92
    clusters of, 89
    polynuclear, 84–89
    structure of, 84, 88, 90, 91

Carbonyls *(cont.)*
  group 10 metal, 92–95
  group 11 metal, 95, 96
  ionic metal, 62
  IR, 32–41
  mass spectra, 53–55
  M–CO bonding, 28
  migration–insertion reactions, 58–60
  MO in polynuclear, 31
  MO scheme for $M(CO)_6$, 28–29
  MO scheme for $M(CO)_4$, 29–30
  NMR, 45–48
  nucleophile attack on CO, 57, 60
  oxidation–reduction reactions, 56–58
  oxygen-to-metal bonds, 110–112
  photoelectron spectroscopy, 49–52
  polynuclear, 31
  preparation, 23, 24
  preparation of metals from, 119–121
  preparation of oxides from, 119–121
  properties, 55
  samarium(II),
    pentametylcyclopentadienyl complex,
    reaction with CO, 63
  stable metal, 60
  structure of, 26, 27
  substitution reactions, 58
  thermodynamic data for, 54
  unstable metal, 61
Carbyne complexes
  applications, 332
  bonding,, 277
  IR, 285
  NMR, 285–287
  preparation, 300, 301, 306
  properties of, 307, 308
  reactions, 317–320
  structure, 281–286
Chromium complexes
  acetylene, 399, 402
  allyl, 438, 440, 451–453
  arene, 586
  butadiene, 486
  carbene, 281–292, 296, 298, 300–302
  carbonyls, 69, 70, 72, 73
  carbyne, 307
  clusters, 70, 72, 73, 132, 170, 171
  cyclobutadiene, 483, 494
  cycloheptatrienyl, 606
  hydridocarbonyls, 70, 98, 101
  hydrocarbyl, 232–234
  isocyanide, 625, 626
  olefin, 353–356
  selenocarbonyls, 113–115
  thiocarbonyls, 113–116
  trimethylenemethane, 483, 495

Clusters
  acide–base reactions, 187–189
  *arachno,* 133–135
  boron, 133–135
  *closo,* 133–135
  *closo-nido-arachno* transformation, 137
  CVMO, 133–135
  dinuclear, 129–132
  dynamic properties, 48, 82, 183–185
  "electron poor," 129, 160–163
  "electron rich," 129
  interstitial, 157–160
  IR, 165, 166
  isolobal analogy, 141–147
  main group elements, 137
  metal–metal bonding, 130–132
  metal–metal energy in, 54, 166
  migration of CO, 183–187
  migration of ligands, 183–187
  models of catalysts, 155–157
  molecular orbital theory, 150–155
  *nido,* 133–135
  NMR, 163–165
  oxidation–reduction reactions, 137, 180, 181
  oxidative addition, 181, 190–193
  reaction between, 149–151
  reactions of, 179–193
  Sidgwick's rule, 136
  skeletal electron pairs theory, 133–140
  synthesis of, 166–169
  theory of skeletal electron pairs, 133–140
  topological calculations of CVMO, 147–149
  transition metal, 138, 139
  Wade's rule, 136
Cobalt complexes
  acetylene, 400–402
  allyl, 457, 458
  arene, 586, 587
  butadiene, 478, 482, 486
  carbene, 293, 303, 304
  carbonyls, 84–90
  clusters, 54, 85–89, 158, 164, 174, 175
  cyclobutadiene, 478, 479, 482, 494
  cyclopentadienyl, 548–551
  hydridocarbonyls, 97, 100, 102, 104, 109, 110
  hydrocarbyl, 241–244
  isocyanide, 627, 628
  olefin, 361–368
  pentadienyl, 555
  thiocarbonyls, 113
Cooligomerization, 680, 684–688
  acetylene and heteroacetylene, 687, 688
  allene, 684, 687, 689, 690
  diene and acetylene, 684
  diene and allene, 684
  diene and olefin, 680, 684

Copper complexes
    acetylene, 396, 397, 403, 404
    arene, 586
    carbonyls, 95, 96
    clusters, 177–179
    cyclopentadienyl, 551
    hydrocarbyl, 249–253
    isocyanide, 629
    olefin, 277–279
Cotton–Kraihanzel method, 37, 39
Cyclobutadiene complexes
    bonding, 471–476
    IR, 485
    NMR, 480–484
    photoelectron spectra, 485
    preparation, 489–493
    properties, 494, 495, 504, 505
    reactions, 500–504
    structure, 476–481
Cycloheptatrienyl complexes
    bonding, 603, 604
    magnetic properties, 604, 605
    preparation, 609, 610
    properties, 607, 608
    reactions, 611–616
    spectroscopy, 604–606
    structure, 603–605
Cyclooctatetraene complexes
    bonding, 603–605
    magnetic properties, 604, 605
    preparation, 610
    properties, 607
    reactions, 613–616
    spectroscopy, 604–605
    structure, 603–605
Cyclopentadienyl complexes
    actinide, 551, 553, 554
    anticancer properties, 569
    application, 565–569
    bonding, 509–516, 551
    cluster, 546, 547, 549, 551, 552, 554
    dynamic properties, 522
    electronic spectra, 525, 526
    ESR, 525
    group 3 metals, 551–554
    group 4 metals, 538–541
    group 5 metals, 541, 542
    group 6 metals, 543–545
    group 7 metals, 545, 546
    group 8 metals, 545–548
    group 9 metals, 548–550
    group 10 metals, 550–552
    group 11 metals, 551
    IR, 522–524
    lanthanide, 551–553
    magnetic properties, 524

Cyclopentadienyl complexes *(cont.)*
    mass spectra, 529
    multidecker, 515, 516, 529, 534–536, 556
    NMR, 520–522
    photoelectron spectra, 527
    preparation, 530–537
    reactions, 557–565
        activation of C−H bonds, 565
        aromatic substitution, 563
        metallation, 561–563
        oxidation, 557, 558
        photochemical, 565
        protonation, 559
        reduction, 558
        rupture of M-cp bond, 559, 560
    structure, 510, 512–516, 518–520, 551
    thermochemistry, 528
Cyclopropenyl complexes, 461, 462

Dienyl complexes
    bonding, 509, 510, 516, 517
    dynamic properties, 522
    NMR, 522
    preparation, 537, 538
    properties, 554, 555
    structure, 510, 517, 518, 520, 554, 555
Dinuclear clusters
    bond order, 130–132
    eclipsed structure, 132
    electronic configuration, 130–132
    staggered structure, 132
Disproportionation reactions, 713, 714
Double carbonylation, 701

EAN rule, 2, 30
Electroneutrality rule, 11
Elementary reactions, 15, 16, 652–655

Fischer–Tropsch synthesis, 714–724
    catalysts, 716
    mechanism, 717–724
    thermochemistry, 714–716

Gold complexes
    acetylene, 397, 403, 404
    carbene, 305
    carbonyls, 95, 96
    clusters, 177–179
    cyclopentadienyl, 551
    hydrocarbyl, 250–253
    isocynide, 629
    olefin, 377–379

Hafnium complexes
    allyl, 450, 451
    arene, 586
    butadiene, 486

Hafnium complexes *(cont.)*
  carbonyls, 63, 64
  cyclooctatetraene, 608
  cyclopentadienyl, 538–541
  hydrocarbyl, 229–232
Heteroacetylene complexes, 412–415
  isocyanide, 414
  nitrile, 412, 413
  phosphaalkyne, 413, 414
Heteroallyl complexes, 462, 463
Heteroarene complexes, 589
Heterogenized homogeneous catalysts, 732–736
  catalysts, 735, 736
  supports for, 733, 734
Homogeneous catalysis, basic principles, 647–652
Hydrocarboxylation, 698
Hydrocarbyl complexes
  actinide, 227, 228
  bond dissociation enthalpies, 201, 202
  bond energies, 201, 202
  bonding in, 199–201
  electronic spectra, 218, 219
    α-elimination, 204–207
    β-elimination, 203–207
  enthalpies for insertion reaction, 203
  ESR, 215
  group 4 metals, 229–232
  group 5 metals, 231, 232
  group 6 metals, 232–234
  group 7 metals, 234–236, 238
  group 8 metals, 236–241
  group 9 metals, 241–244
  group 10 metals, 244–249
  group 11 metals, 249–253
  IR, 212–213
  lanthanide, 227–229
  magnetic properties, 217, 218
  mechanisms of decomposition, 203–208
  NMR, 213
  photoelectron spectra, 215–217
  preparation, 219–226
  reactions of, 253–262
    breaking of the M−C bond, 253–255
    migration, 255–260
    oxidation, 258
    oxidative addition, 260, 261
    σ−π rearrangement, 261, 262
  reactions with dinitrogen, 259, 267, 568
  reductive elimination, 204, 205
  stability of, 201–208
  structure of, 208–211
Hydrocyanation, 702, 703
Hydrogenation, 650, 651, 655–669
  arene, 650, 651, 662
  asymmetric, 662–667
  catalysts for, 650, 651, 657–662

Hydrogenation *(cont.)*
  of the C=O, C=N, C≡N bonds, 667–669
  olefin, 650, 651, 655–662
Hydroformylation, 690–697, 701, 702
  catalyst decomposition, 701, 702
  catalysts for, 690, 691, 695–697
  mechanism, 691, 692
Hydrosilation, 703, 704

Iridium complexes
  acetylene, 396, 398, 401, 403
  allyl, 433, 440, 442, 457–459
  arene, 586
  butadiene, 486
  carbene, 281, 304
  carbonyls, 84–92
  clusters, 54, 85–89, 174–175
  cyclopentadienyl, 548, 549
  hydridocarbonyls, 99, 100, 103–107, 109
  hydrocarbyl, 242–244
  isocyanide, 628
  olefin, 362–368
  thiocarbonyls, 113, 116, 117
  trimethylenemethane, 483, 495
Iron complexes
  acetylene, 397, 400, 401, 402
  allyl, 438, 440, 454–456
  arene, 586, 587
  butadiene, 477, 478, 482, 484, 486
  carbene, 290, 291, 296, 298, 303
  carbonyls, 79–85
  carbyne, 307
  clusters, 54, 80–84, 158, 164, 172–174
  cyclobutadiene, 477–480, 482–484, 494
  cyclopentadienyl, 545–548
  hydridocarbonyls, 81, 83, 84, 97, 98, 102, 106, 108, 109
  hydrocarbyl, 236, 238–241
  isocyanide, 625, 627
  olefin, 358–361
  pentadienyl, 518, 520, 555
  thiocarbonyls, 113, 117, 118
  trimethylenemethane, 477, 483, 495
Isocyanide complexes
  application, 640, 641
  bonding, 617, 618
  electronic spectra, 620, 621
  ESR, 624
  group 3 metals, 624
  group 4 metals, 625
  group 5 metals, 625, 626
  group 6 metals, 625, 626
  group 7 metals, 625–627
  group 8 metals, 625, 627
  group 9 metals, 627, 628
  group 10 metals, 625, 628, 629

Isocyanide complexes *(cont.)*
   group 11 metals, 629
   IR, 622, 623
   magnetic properties, 624
   NMR, 621, 622
   photoelectron spectra, 624
   preparation, 629–632
   properties, 624–629
   reactions, 632–641
      carbyne formation, 635, 636
      cleavage of the M−CNR bond, 632, 633
      dimerization, 639, 640
      insertion, 636–639
      nucleophilic addition, 634, 635
      oxidative addition, 634
      redox, 633, 634
      structure, 617–620
Isomerization of olefins, 670–673
   *cis-trans,* 673
   mechanisms of, 670, 671, 673

Lathanide complexes
   allyl, 460, 461
   butadiene, 486
   cyclooctatetraene, 606–608
   cyclopentadienyl, 528, 552, 553
   σ-hydrocarbyl, 227–229
   isocyanide, 624
Ligands
   names of, 17
   properties of, 2–5

Main group elements, 7
   character of M-L bonds, 7–11
   organometallic compounds of, 8–10
Manganese complexes
   acetylene, 401, 402
   allyl, 433, 453
   arene, 586, 587
   butadiene, 478, 482, 486
   carbene, 284, 286, 290, 292, 296, 298, 303
   carbonyls, 75–80
   carbyne, 307
   clusters, 54, 76–79, 171, 172
   cyclopentadienyl, 545, 546
   cycloheptatrienyl, 607
   hydridocarbonyls, 79, 98–101, 104, 109, 110
   hydrocarbyl, 234–236
   isocyanide, 625, 627
   olefin, 356–358
   pentadienyl, 520, 555
   selenocarbonyls, 113–115
   thiocarbonyls, 113–115, 118
Metals
   organometallic compounds from, 23, 488, 489,
      533, 534, 594

Metals *(cont.)*
   preparation from coordination compounds,
      119–121
   preparation from organometallic compounds,
      119–121
Metathesis, 704–713
   alkynes, 713
   functionalized olefins, 713
   olefin, 704–713
      application of, 711, 712
      catalysts for, 706–709
   mechanism of, 707, 710, 711
   thermochemistry of, 705–706
μ-Methylene complexes
   reactions, 320–322
   synthesis of, 299
Migration reactions, 16, 58–60, 222–224,
      255–260, 420, 467, 468, 652, 653
Molybdenum complexes
   acetylene, 399–402
   allyl, 433, 443, 451–453
   arene, 586
   butadiene, 477, 478, 486
   carbene, 286, 289, 292, 298, 302
   carbonyls, 70–75
   carbyne, 283, 286, 307
   clusters, 72–74, 132, 133, 160–163, 170, 171
   cyclobutadiene, 479, 483, 484, 494
   cyclopentadienl, 543–545
   cycloheptatrienyl, 606–608, 610
   dihydrogen, 96
   hydridocarbonyls, 73, 101, 105, 109
   hydrocarbyl, 232–234
   isocyanide, 625, 626
   olefin, 353–356
   pentadienyl, 555
   thiocarbonyls, 113, 116
   trimethylenemethane, 483, 495
Multidecker complexes, 515, 516, 529, 534–536,
      556, 588, 589
   μ-cyclopentadienyl, 515, 516. 529, 534
   μ-arene, 516, 588, 589
   μ-*cyclo*-E$_3$ (E = P, As), 463, 556
   μ-*cyclo*-E$_5$ (E = P, As), 516, 566
   μ-*cyclo*-P$_6$, 516, 556, 589
   μ-2,3-dihydro-1,3-diborolyl, 535, 536, 556

Nickel complexes
   acetylene, 402–403
   allyl, 438–440, 443, 458–460
   arene, 586, 587
   carbene, 290, 293, 304
   carbonyls, 92–94
   clusters, 93, 158, 164, 175–177
   cyclobutadiene, 477, 479, 494
   cyclopentadienyl, 550–552

Nickel complexes *(cont.)*
  hydridocarbonyls, 93, 97
  hydrocarbyl, 244–249
  isocyanide, 625, 628
  olefin, 368–374
  pentadienyl, 555
  thiocarbonyl, 113
Niobium complexes
  acetylene, 391, 392, 399, 402
  allyl, 451
  arene, 586
  butadiene, 486
  carbene, 288, 289, 302
  carbonyls, 64–69
  carbyne, 288
  clusters, 160–162, 170, 324
  cyclobutadiene, 479, 494
  cycloheptatrienyl, 607
  cyclopentadienyl, 541, 542
  hydridocarbonyls, 68, 101
  hydrocarbyl, 231, 232
  isocyanide, 626
  olefin, 353
  pentadienyl, 555
Nomenclature of organometallic compounds, 15,
    17–20

Olefin complexes
  bonding, 327–331
  $^{13}$C NMR, 337–339
  electronic spectra, 341, 342
  group 4 metals, 352
  group 5 metals, 352, 353
  group 6 metals, 353–356
  group 7 metals, 356–358
  group 8 metals, 358–361
  group 9 metals, 361–368
  group 10 metals, 368–377
  group 11 metals, 377–379
  heteroolefin complexes, 379–381
  $^{1}$H NMR, 335–337
  IR, 333–335
  reactions, 381–388
    amination, 386
    with electrophiles, 384, 385
    isotope H/D exchange, 387, 388
    with nucleophiles, 381–384
    olefin exchange, 386
    olefin substitution, 344–346, 350, 386
    olefin transformation, 386, 387
    photochemical, 388
  photoelectron spectra, 339–341
  stability, 343–346
  structure, 328–333
  synthesis, 347–351
    addition of olefins, 347

Olefin complexes *(cont.)*
  synthesis *(cont.)*
    dehydration of alcohols, 351
    electrochemical, 351
    exchange, 350
    reduction, 349, 350
    substitution, 347–349
Oligomerization, 680–690
  acetylenes, 685–687
  allenes, 421, 684, 687, 689, 690
  dienes, 681–684
  olefin, 680, 681
Organometallic compounds, definition of, 1
Osmium complexes
  allyl, 454–456
  arene, 586
  carbene, 303
  carbonyls, 80–84
  clusters, 54, 80–84, 164, 172–174
  cyclopentadienyl, 545–547
  hydridocarbonyls, 84, 97, 98, 102, 104, 110,
    113
  hydrocarbyl, 237–241
  isocyanide, 625, 627
  olefin, 358–361
  thiocarbonyls, 113, 117
  trimethylenemethane, 495
Oxidative addition, 15, 16, 220–222, 654, 655
Oxidative coupling, 15, 16, 652–655

Palladium complexes
  acetylene, 391, 403, 404
  allyl, 431–433, 438–440, 442, 443, 454–456
  arene, 576, 586
  carbene, 293, 304
  carbonyls, 92–95
  clusters, 93–94, 175–177
  cyclobutadiene, 494
  cyclopentadienyl, 550–552
  hydrocarbyl, 244–248
  isocyanide, 625, 629
  olefin, 368–374
Pauling's electroneutrality rule 11
Platinum complexes
  acetylene, 390–396, 403, 404
  allyl, 438, 440, 442, 459, 460
  carbene, 293, 297, 304, 305
  carbonyls, 92–95
  clusters, 93, 94, 133, 175–177
  cyclobutadiene, 494, 495
  cyclopentadienyl, 550–552
  hydridocarbonyls, 94, 103, 107
  hydrocarbyl, 244–249
  isocyanide, 625, 629
  olefin, 368–377
  thiocarbonyl, 113, 117

Polymerization of acetylenes, 676, 679, 680
Polymerization of dienes, 674–679
Polymerization of olefins, 674–679
  anionic, 674
  carbene mechanism, 676
  catalysts for, 675–679
  cationic, 674
  mechanisms of, 674–676
  migration mechanism, 675
  radical, 674
Pyrrole complexes, 555, 556

Reduction of CO, 714–724
  carbonyl clusters for, 717, 721, 722
  carbonyl complexes for, 718, 722
  mechanism, 717–723
Rhenium complexes
  acetylene, 401
  allyl, 433, 440, 443, 453, 454
  arene, 586, 587
  carbene, 294, 303
  carbonyls, 76–79
  clusters, 54, 76–79, 132, 133, 160–163, 171–172
  cyclopentadienyl, 545, 546
  hydridocarbonyls, 78, 79, 98, 99, 101, 102,
    104, 105, 109, 110
  hydrocarbyl, 234–236, 238
  isocyanide, 625–627
  olefin, 356–358
  selenocarbonyls, 113–115
  thiocarbonyls, 114, 115
Rhodium complexes
  acetylene, 391, 393, 396, 400–402
  allyl, 433, 438–440, 442, 457, 458
  arene, 586, 587
  butadiene, 486
  carbene, 293, 304
  carbonyls, 84–92
  clusters, 54, 85–89, 133, 158, 164, 174–175
  cyclobutadiene, 479, 483, 494
  cyclopentadienyl, 548–551
  hydridocarbonyls, 89, 97, 99, 102, 103, 105,
    106, 107
  hydrocarbyl, 242–244
  isocyanide, 625, 627, 628
  olefin, 362–368
  thiocarbonyl complex, 113, 116
Ruthenium complexes
  acetylene, 401, 402
  allyl, 440, 442, 454–456
  arene, 586, 587
  butadiene, 484, 486
  carbene, 281, 303
  carbonyls, 80–84
  clusters, 54, 80–84, 133, 158, 164, 172–174
  cyclobutadiene, 483, 494

Ruthenium complexes *(cont.)*
  cyclopentadienyl, 545–547
  hydridocarbonyls, 84, 97, 102, 106–108, 110
  hydrocarbyl, 237–240
  isocyanide, 625, 627
  olefin, 358–361
  thiocarbonyls, 113
  selenocarbonyl complex, 113–115

Scandium complexes
  allyl, 460, 461
  cyclooctatetraene, 608
  cyclopontadienyl, 551–553
  $\sigma$-hydrocarbyl, 227
Selenocarbonyl complexes, 112–115
  IR, 113–115
  preparation of, 115
  structur of, 114, 115
Sidgwick rule, 2, 30, 136
Silver complexes
  acetylene, 397, 404
  carbonyls, 95
  clusters, 177–179
  hydrocarbyl, 250–253
  isocyanide, 629
  olefin, 377–379

Tantalum complexes
  acetylene, 399, 402
  allyl, 451
  arene, 586
  butadiene, 478, 486
  carbene, 281, 282, 288, 289, 302
  carbonyls, 64–69
  carbyne, 288, 307
  clusters, 129, 160–163, 170
  cyclobutadiene, 494
  cyclopentadienyl, 541, 542
  hydorcarbyl, 231–232
  isocyanide, 626
  olefin, 353
  trimethylenemethane, 495
Technetium complexes
  arene, 586, 587
  carbonyls, 76–79
  clusters, 76, 78, 79, 105, 132, 133, 171
  cyclopentadienyl, 545, 546
  hydridocarbonyls, 78, 79, 105
  isocyanide, 625–627
  olefin, 356, 357
Thiocarbonyl complexes, 112–118
  bonding, 112, 114, 115
  preparation, 115, 116
  reactions, 116–118
  spectroscopy, 113, 114
  structure, 112, 115

Timney method, 39–41
Titanium complexes
  acetylene, 391, 394, 398–400
  allyl, 450, 451
  arene, 586
  butadiene, 486
  carbonyls, 63, 64
  clusters, 160, 169, 170
  cyclobutadiene, 494
  cycloheptatrienyl, 607, 608
  cyclooctatetraene, 606–608
  cyclopentadienyl, 538–541
  hydrocarbyl, 229–232
  isocyanide, 625
  olefin, 351, 352
  pentadienyl, 555
Tolman's rule, 15, 16, 652
Transition metals
  character of M−L bonds, 11–15
  energy of the valence orbitals of, 11–12
Transition metals complexes
  with $\sigma$-acceptor ligands, 13–15
  with $\sigma$-donor-$\pi$-acceptor ligands, 13–15
  with $\sigma$-donor ligands, 13–15
Trimethylenemethane complexes
  bonding, 471–476
  IR, 485
  NMR, 480–484
  photoelectron spectra, 485
  preparation, 496, 497
  properties, 495, 504, 505
  reactions, 504
  structure, 476–481
Tungsten complexes
  acetylene, 393–395, 397, 399–402
  allyl, 451–453
  arene, 586
  butadiene, 486
  carbene, 282–284, 288, 302, 303
  carbonyls, 71–75
  carbyne, 288, 289, 307
  clusters, 72–74, 132, 133, 160, 161, 170, 171
  cyclobutadiene, 483, 494
  cycloheptatrienyl, 607, 608
  cyclopentadienyl, 543–545
  dihydrogen, 96, 97
  hydridocarbonyls, 73, 101, 105
  hydrocarbyl, 232–234

Tungsten complexes *(cont.)*
  isocyanide, 625, 626
  olefin, 353–356
  thiocarbonyls, 113, 114, 116, 118

Valence electrons
  count, 2, 6
  number of, 2
Vanadium complexes
  acetylene, 392, 394, 397, 399–402
  allyl, 451
  arene, 586
  butadiene, 482, 486
  carbonyls, 64–69
  clusters, 170
  cyclobutadiene, 479, 494
  cycloheptatrienyl, 606, 608
  cyclooctatetraene, 607, 608
  cyclopentadienyl, 541–542
  hydridocarbonyls, 65, 100, 101, 109
  hydrocarbyl, 231, 232
  isocyanide, 625, 626
  olefin, 352, 353
  pentadienyl, 518, 555
Vitaminum $B_{12}$, 241, 242

Wade's rule, 136
Water gas shift reaction, 729
Wilkinson catalyst, 650, 657–659

Yttrium complexes
  cyclooctatetraene, 607, 608
  cyclopentadienyl, 528, 552, 553
  $\sigma$-hydrocarbyl, 227
  isocyanide, 624

Ziegler–Natta catalysts, 675–679
Zirconium complexes
  allyl, 433, 450, 451
  arene, 586
  butadiene, 477, 481
  carbonyls, 63, 64
  clusters, 160, 169, 170
  cycloheptatriene, 607, 608
  cyclooctatetraene, 607–608
  cyclopentadienyl, 538–541
  hydrocarbyl, 229–232
  isocyanide, 625
  pentadienyl, 555

1DR